TABLE II (cont.)
Areas under the
standard normal curve

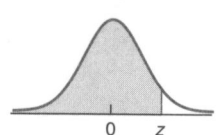

z	|	0.00	0.01	0.02	0.03	0.04	0.05	0.06	0.07	0.08	0.09
						Second decimal place in z					
0.0		0.5000	0.5040	0.5080	0.5120	0.5160	0.5199	0.5239	0.5279	0.5319	0.5359
0.1		0.5398	0.5438	0.5478	0.5517	0.5557	0.5596	0.5636	0.5675	0.5714	0.5753
0.2		0.5793	0.5832	0.5871	0.5910	0.5948	0.5987	0.6026	0.6064	0.6103	0.6141
0.3		0.6179	0.6217	0.6255	0.6293	0.6331	0.6368	0.6406	0.6443	0.6480	0.6517
0.4		0.6554	0.6591	0.6628	0.6664	0.6700	0.6736	0.6772	0.6808	0.6844	0.6879
0.5		0.6915	0.6950	0.6985	0.7019	0.7054	0.7088	0.7123	0.7157	0.7190	0.7224
0.6		0.7257	0.7291	0.7324	0.7357	0.7389	0.7422	0.7454	0.7486	0.7517	0.7549
0.7		0.7580	0.7611	0.7642	0.7673	0.7704	0.7734	0.7764	0.7794	0.7823	0.7852
0.8		0.7881	0.7910	0.7939	0.7967	0.7995	0.8023	0.8051	0.8078	0.8106	0.8133
0.9		0.8159	0.8186	0.8212	0.8238	0.8264	0.8289	0.8315	0.8340	0.8365	0.8389
1.0		0.8413	0.8438	0.8461	0.8485	0.8508	0.8531	0.8554	0.8577	0.8599	0.8621
1.1		0.8643	0.8665	0.8686	0.8708	0.8729	0.8749	0.8770	0.8790	0.8810	0.8830
1.2		0.8849	0.8869	0.8888	0.8907	0.8925	0.8944	0.8962	0.8980	0.8997	0.9015
1.3		0.9032	0.9049	0.9066	0.9082	0.9099	0.9115	0.9131	0.9147	0.9162	0.9177
1.4		0.9192	0.9207	0.9222	0.9236	0.9251	0.9265	0.9279	0.9292	0.9306	0.9319
1.5		0.9332	0.9345	0.9357	0.9370	0.9382	0.9394	0.9406	0.9418	0.9429	0.9441
1.6		0.9452	0.9463	0.9474	0.9484	0.9495	0.9505	0.9515	0.9525	0.9535	0.9545
1.7		0.9554	0.9564	0.9573	0.9582	0.9591	0.9599	0.9608	0.9616	0.9625	0.9633
1.8		0.9641	0.9649	0.9656	0.9664	0.9671	0.9678	0.9686	0.9693	0.9699	0.9706
1.9		0.9713	0.9719	0.9726	0.9732	0.9738	0.9744	0.9750	0.9756	0.9761	0.9767
2.0		0.9772	0.9778	0.9783	0.9788	0.9793	0.9798	0.9803	0.9808	0.9812	0.9817
2.1		0.9821	0.9826	0.9830	0.9834	0.9838	0.9842	0.9846	0.9850	0.9854	0.9857
2.2		0.9861	0.9864	0.9868	0.9871	0.9875	0.9878	0.9881	0.9884	0.9887	0.9890
2.3		0.9893	0.9896	0.9898	0.9901	0.9904	0.9906	0.9909	0.9911	0.9913	0.9916
2.4		0.9918	0.9920	0.9922	0.9925	0.9927	0.9929	0.9931	0.9932	0.9934	0.9936
2.5		0.9938	0.9940	0.9941	0.9943	0.9945	0.9946	0.9948	0.9949	0.9951	0.9952
2.6		0.9953	0.9955	0.9956	0.9957	0.9959	0.9960	0.9961	0.9962	0.9963	0.9964
2.7		0.9965	0.9966	0.9967	0.9968	0.9969	0.9970	0.9971	0.9972	0.9973	0.9974
2.8		0.9974	0.9975	0.9976	0.9977	0.9977	0.9978	0.9979	0.9979	0.9980	0.9981
2.9		0.9981	0.9982	0.9982	0.9983	0.9984	0.9984	0.9985	0.9985	0.9986	0.9986
3.0		0.9987	0.9987	0.9987	0.9988	0.9988	0.9989	0.9989	0.9989	0.9990	0.9990
3.1		0.9990	0.9991	0.9991	0.9991	0.9992	0.9992	0.9992	0.9992	0.9993	0.9993
3.2		0.9993	0.9993	0.9994	0.9994	0.9994	0.9994	0.9994	0.9995	0.9995	0.9995
3.3		0.9995	0.9995	0.9995	0.9996	0.9996	0.9996	0.9996	0.9996	0.9996	0.9997
3.4		0.9997	0.9997	0.9997	0.9997	0.9997	0.9997	0.9997	0.9997	0.9997	0.9998
3.5		0.9998	0.9998	0.9998	0.9998	0.9998	0.9998	0.9998	0.9998	0.9998	0.9998
3.6		0.9998	0.9998	0.9999	0.9999	0.9999	0.9999	0.9999	0.9999	0.9999	0.9999
3.7		0.9999	0.9999	0.9999	0.9999	0.9999	0.9999	0.9999	0.9999	0.9999	0.9999
3.8		0.9999	0.9999	0.9999	0.9999	0.9999	0.9999	0.9999	0.9999	0.9999	0.9999
3.9		1.0000[†]									

[†] For $z \geq 3.90$, the areas are 1.0000 to four decimal places.

Case Studies

. . . . .

Biographical Sketches

. . . . .

Introductory Statistics

. . . . .

Introductory Statistics

• • • • •

Fourth Edition

Neil A. Weiss
Arizona State University

Biographies by Carol A. Weiss

Addison-Wesley Publishing Company, Inc.

Reading, Massachusetts   Menlo Park, California   New York
Don Mills, Ontario   Wokingham, England   Amsterdam   Bonn
Sydney   Singapore   Tokyo   Madrid   San Juan   Milan   Paris

Executive Editor: Michael Payne
Sponsoring Editor: Julia Berrisford
Editorial Assistants: Maureen Lawson and Marcia Cole
Senior Production Supervisors: Mona Zeftel and Loren Hilgenhurst Stevens
Production Coordinator: Barbara Ames
Technical Art Consultant: Susan London-Payne
Text Designer: Rebecca Lemna
Cover Designer: Marshall Henrichs
Cover Design Director: Peter M. Blaiwas
Copy Editor: Sally Stickney
Compositor: Carol and Neil Weiss
Illustrator: Scientific Illustrators
Art Supervisor: Joseph Vetere
Proofreader: Phyllis Coyne
Manufacturing Manager: Roy Logan

Photographs:

Florence Nightingale, page 2, courtesy of The Bettmann Archive.
Adolphe Quetelet, page 42, Copyright Bibliothèque royale Albert Ier,
 Cabinet des Estampes, Bruxelles, J. Odevaére.
John Tukey, page 100, courtesy of John W. Tukey.
Andrei Kolmogorov, page 180, courtesy of The Bettmann Archive.
James Bernoulli, page 264, courtesy of The Bettmann Archive.
Carl Friedrich Gauss, page 334, courtesy of The Bettmann Archive.
Pierre-Simon Laplace, page 402, courtesy of The Bettmann Archive.
William Gosset, page 438, courtesy of The Granger Collection, New York.
Jerzy Neyman, page 480, courtesy of the Department of Statistics, University of
 California, Berkeley.
Gertrude Cox, page 572, courtesy of Research Triangle Institute.
Abraham de Moivre, page 652, courtesy of The Granger Collection, New York.
Karl Pearson, page 686, courtesy of Brown Brothers.
Adrien Legendre, page 734, courtesy of The Bettmann Archive.
Sir Francis Galton, page 792, courtesy of Stock Montage.
Sir Ronald Fisher, page 864, courtesy of The Bettmann Archive.

Library of Congress Cataloging-in-Publication Data

Weiss, N. A. (Neil A.)
 Introductory Statistics / Neil A. Weiss.—4th ed.
  p.   cm.
 Includes index.
 ISBN 0–201–53270–0
 1. Statistics.   I. Title.
QA276.12.W45   1995
519.5—dc20   94-16050
                                    CIP

1 2 3 4 5 6 7 8 9 10 DOW 98 97 96 95 94

Preface

· · · · ·

Statistics has become an indispensable tool in business, government, and virtually every academic discipline. Some familiarity with statistics is essential for all of us if we are to comprehend the world around us.

The purpose of this book is to provide a clear understanding of basic statistical concepts and techniques and to present well-organized procedures for applying them. Introductory high-school algebra is a sufficient prerequisite.

The text is designed to be flexible. It can be used in a one-quarter, one-semester, two-quarter, two-semester, or three-quarter course. The amount of time devoted to the book can be varied by both choice of topics and depth of coverage.

Technological advances and ever increasing calls for new approaches to presenting statistics have made this an exciting time to learn, practice, and teach statistics. In writing the fourth edition of *Introductory Statistics,* we have incorporated many of the techniques and attitudes that reflect recent developments in statistics.

Features

The text contains the following features that will provide valuable assistance for the reader in learning introductory statistics.

Emphasis on application.   We have concentrated on the application of statistical techniques to the analysis of data. Although statistical theory has been kept to a minimum, we have provided a thorough explanation of the rationale for using each statistical procedure.

Data analysis and exploration.   We agree wholeheartedly with the trend of including more exploratory and confirmatory data analysis in statistics courses and have incorporated an extensive amount into the text and exercises. We also recognize, however, that not all readers will have access to computers and therefore have presented data analysis in a way that does not require using a computer even though access to one is recommended.

Detailed and careful explanations.   We have included every step of explanation we think a typical reader might need. Our guiding principle is to avoid "cognitive jumps" and thereby make the learning process smooth and enjoyable. We believe detailed and careful explanations result in better understanding.

Data sets.   In most examples and exercises, we have presented raw data in addition to summary statistics. This gives a more realistic view of statistics and provides an opportunity for the problems to be solved by computer, if so desired.

Procedure boxes.   To help the reader learn to apply statistical procedures, we have developed easy-to-follow, step-by-step methods for carrying out those procedures. For ease in locating, each procedure is displayed with a color background. A unique feature of this book is that when a procedure is illustrated by an example, each step in the procedure is presented again within the example. This serves a twofold purpose: It shows how the procedure is applied and helps the reader master the steps in the procedure.

Procedure index.   Given the numerous statistical procedures, it is sometimes difficult to find a specific one, especially when the book is being used for reference purposes. Consequently, we have included a *procedure index* on the back inside cover of the book. This provides a quick and easy way to find the required procedure for performing any particular statistical analysis.

Computer simulations.   We have incorporated many computer simulations in both the text and the exercises. These serve as pedagogical aids for understanding complex concepts such as sampling distributions. Readers can benefit from this material even if they do not have access to a computer.

Computer usage.   Today, virtually all professional applications of statistics are done by computer. It is therefore important that every student of statistics have at least some familiarity with statistical software. Although we have chosen Minitab[†] to illustrate the use of statistical software, the text has been written so that the instructor is free to select other statistical software packages. The Minitab discussion in this book is self-contained.[‡] All computer material is *optional,* but recommended.

The computer sections are integrated as optional subsections occurring immediately following the particular statistical concept under consideration. In each subsection we explain how Minitab can be used to solve problems that were solved by hand earlier in the section. Each solution consists of introducing the required commands, displaying the computer output, and interpreting the results.

Additionally, computer exercises (clearly marked as such) are incorporated into the exercise sets. Three types of computer exercises have been included. The first type asks the reader to interpret computer printouts; no knowledge of or access to statistical software is necessary for these exercises. The second type asks the reader to use Minitab or some other statistical software to solve exercises that were presented previously for hand solution; all Minitab commands required for these computer exercises will have already been discussed in the text. The third type of computer exercise asks the reader to use statistical software to perform a computer simulation; these exercises are designed to provide concrete illustrations of some of the more complex concepts (e.g., sampling distributions) and to show the reader how the computer can be used to reveal statistical facts.

[†] Minitab is a registered trademark of Minitab Inc., 3081 Enterprise Drive, State College, PA 16801. Telephone: 814-238-3280. Fax: 814-238-4383. We would like to thank Minitab Inc., for their assistance.

[‡] Additional details and further topics, such as writing and using macros, are provided in the *Minitab Supplement* to this book.

For maximum flexibility we have allowed for three options in computer coverage: (1) no coverage, (2) coverage omitting the Minitab-command discussions but including the interpretation of computer printouts, and (3) coverage of Minitab commands (or those for some other statistical software) and the interpretation of computer printouts.

Choice of session- or menu-command interface. Traditionally, Minitab has been applied by typing (session) commands at the MTB > prompt. But in response to the strong emergence of graphical user interface (GUI), Minitab has recently introduced a menu interface as well, so that commands can also be executed by choosing them from menus and completing dialogue boxes. For each application of Minitab, we explain separately both the required session commands and the required menu commands. It is not necessary to cover both session commands and menu commands (although experienced users of Minitab often intersperse them).

Biographical sketches. Each chapter begins with a brief biography of a famous statistician. Besides being of general interest, these biographies help the reader obtain a perspective on how the science of statistics developed.

Chapter introductions and chapter outlines. Also included at the beginning of each chapter is a description of the chapter and an explanation of how the chapter relates to the text as a whole. As a further aid, a chapter outline, which follows the chapter introduction, lists the sections in the chapter.

Case studies. We have presented case studies ranging from the classic to the current. At the beginning of each chapter, a case study is described briefly; at the end of the chapter, when the reader has studied the required concepts, the case study is considered in detail. Exercises are provided for all case studies.

Definitions, formulas, key facts. As an aid to learning and for reference, we have prominently displayed all definitions, formulas, and key facts. These items are enclosed by a color rectangle to make them easy to locate.

Real examples. Because we believe that the majority of students learn by example, every concept discussed in the book is illustrated by at least one detailed example. The examples are, for the most part, based on real-life situations and have been chosen for their interest as well as for their illustrative value.

Extensive and diverse exercise sets. We have constructed exercise sets that are both extensive and diverse. Most of the exercises are based on information found in newspapers, magazines, statistical abstracts, and journal articles; sources are explicitly cited. The exercises are designed not only to help the reader learn the material but also to show that statistics is a lively and relevant discipline.

Since students in introductory statistics courses often have different mathematical backgrounds, we have included three levels of exercises: basic, intermediate, and advanced. Every exercise set contains several basic exercises. The *basic exercises* provide applications of material presented in the text, and every reader should master these. We have organized the basic exercises so that each concept is covered by at least two problems; for each odd-numbered basic exercise that involves a particular concept, there is also an even-numbered basic exercise that involves that same concept. A single color bullet (·) preceding an exercise number identifies the exercise as a basic one. The answers to the odd-numbered basic exercises are presented in Appendix C; the answers to the even-numbered basic exercises are in the *Instructor's Solutions Manual.*

Most exercise sets also include intermediate and advanced problems. The *intermediate exercises* contain supplementary material that is not necessarily covered in the text but that may interest some of the more highly motivated students. Two color bullets (· ·) preceding an exercise number identify the exercise as an intermediate one. The *advanced exercises* cover abstract concepts, theory, and algebraic derivations. Those exercises are intended for students with special mathematical background and aptitude. Three color bullets (· · ·) preceding an exercise number identify the exercise as an advanced one. The solutions to all intermediate and advanced exercises are in the *Instructor's Solutions Manual.*

Chapter reviews.   Frequently, students in introductory statistics courses feel a certain amount of anxiety and confusion about how to study and review. To help the student, a chapter-review section appears at the end of every chapter. The chapter reviews include (1) a list of key terms with page references, (2) formulas with page references, (3) chapter objectives, and (4) a review test. These pedagogical aids provide the student with an organized method for reviewing and studying. The answers to the review tests are given in Appendix C.

Database exercises.   Appendix B contains a printout of a database obtained by randomly selecting 500 Arizona State University sophomores. Seven variables are considered for each student: sex, high-school GPA, SAT math score, cumulative GPA, SAT verbal score, age, and total hours. At the end of each chapter review, a section entitled "Using the Focus Database" presents several exercises about the database. These exercises are optional and are to be done by computer. We assume that the student knows or will be taught any additional commands required to carry out these exercises.

Formula/table card.   A detachable formula/table card is provided with the book. This card contains all of the formulas and many of the tables that appear in *Introductory Statistics.* The formula/table card is useful for quick-reference purposes; many instructors also find it convenient for use with examinations.

Organization

The text offers a great deal of flexibility in choosing material to cover. Chapter 1 introduces the nature of statistics, sampling designs, observational studies, and designed experiments; an optional introduction to Minitab is presented as well. Chapters 2 and 3 provide the fundamentals of descriptive statistics.

Chapters 4–6 present probability, discrete random variables, and the normal distribution. Chapter 7 introduces the concept of sampling distributions and provides a detailed discussion of the sampling distribution of the mean. Following that, Chapters 8 and 9 examine confidence intervals and hypothesis tests for one population mean. We consider Chapters 1–9 the core of an introductory statistics course.

Chapter 10 presents inferences for two population means, Chapter 11 inferences for population proportions, and Chapter 12 chi-square procedures (goodness-of-fit test, independence test, and inferences for a population standard deviation).

We have divided the traditional material on regression and correlation into two chapters. Chapter 13 examines descriptive methods in regression and correlation and can be covered at any time after Chapter 3. Chapter 14 presents inferential methods in regression and correlation and can be covered once Chapters 9 and 13 have been completed.

Chapter 15 examines one-way ANOVA and two-way ANOVA with interaction. Also included are optional sections on multiple comparisons and the Kruskal–Wallis test.

The following flowchart summarizes the preceding discussion and shows the interdependence among chapters. In the flowchart the prerequisites for a given chapter consist of all chapters having a path leading to that chapter.

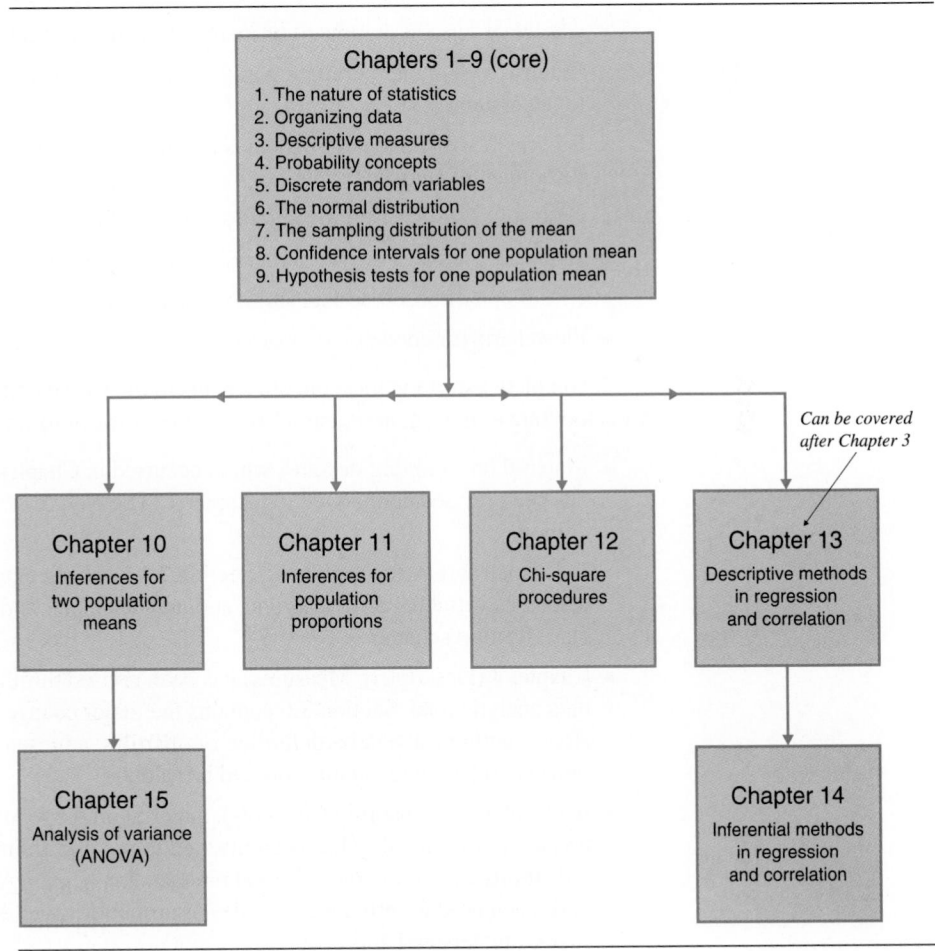

Changes in the fourth edition

We have made significant changes in the fourth edition of *Introductory Statistics*. All chapters have been rewritten for the purpose of updating and expanding the material, fine-tuning the organization, and adding new sections where appropriate. The entire manuscript has been edited to provide a smoother and more succinct presentation.

Among the global enhancements to the fourth edition of *Introductory Statistics* are:

- An increase in the number of exercises from approximately 1500 in the third edition to almost 1700 (not including parts) in the fourth edition.
- Extensive use of graphics to illustrate and explain concepts.
- Expanded coverage of data analysis, both exploratory and confirmatory.
- Increased interpretation of concepts in text and in exercises.
- Optional integration of nonparametric methods with the corresponding parametric methods.
- Fully expanded treatment of *P*-values with all hypothesis tests.
- Choice of session-command or menu-command Minitab interface.
- Extensive computer simulation and resampling to make complex concepts easier to understand.
- Biographies of famous statisticians to help students gain a perspective of the development of statistics.
- Case studies—classical to contemporary—with exercises.
- Database exercises to give students practice with very large data sets.
- Discussion of observational studies and designed experiments.
- Flowcharts for choosing the correct statistical procedure.

A complete list of the local enhancements to the fourth edition of *Introductory Statistics* is too long to present here; but a list of some of the most important ones follows.

- Material on sampling designs, which occurred in Chapter 16 of the third edition, has been revised and moved to Chapter 1 (The Nature of Statistics) in the fourth edition.
- In Chapter 2 (Organizing Data), Section 2.1 has been extensively revised: it now discusses variables as well as data and uses a simpler and more modern data-classification scheme.
- Chapter 3 (Descriptive Measures) has been revised throughout to provide a more data analytic tone. Section 3.6 contains the major changes: a thorough discussion of outliers, a general definition of quartiles, a presentation of the interquartile range, and the inclusion of modified boxplots.
- In Chapter 4 (Probability Concepts), material on the relative-frequency interpretation of probability has been inserted to provide an intuitive description of probability before the more formal presentation begins. Also, Venn diagrams have been used to introduce the rules of probability, yielding an informal and easy-to-understand approach.

- Probability histograms have been included throughout Chapter 5 (Discrete Random Variables) to provide visual displays of discrete probability distributions. A new and optional Section 5.6 covers the Poisson distribution.

- In Chapter 6 (The Normal Distribution), we have gone to a cumulative normal table, areas between $-\infty$ and z, because it is simpler to learn and more efficient to use than the normal table giving areas between 0 and z; the cumulative table is also consistent with Minitab's CDF and INVCDF commands, which are discussed in a new and optional subsection of Section 6.1. A new Section 6.5 presents normal probability plots.

- To give the reader an intuitive understanding of the sampling distribution of the mean, computer simulations in both the normal and nonnormal cases have been incorporated into Chapter 7 (The Sampling Distribution of the Mean).

- In Chapter 8 (Confidence Intervals for One Population Mean), explanations have been included on how to deal with outliers when obtaining confidence intervals. In Section 8.4 a computer simulation has been used to compare the distributions of the z-statistic (standard normal) and the t-statistic (Student's t).

- In Chapter 9 (Hypothesis Tests for One Population Mean), data analytic techniques have been employed to help decide on the correct procedure to apply. An explanation of the difference between statistical and practical significance has been included in Section 9.3. An optional Section 9.4 discusses Type II error probabilities and power; an optional Section 9.7 covers the Wilcoxon one-sample signed-rank test; and a new and optional Section 9.8 summarizes the hypothesis-testing procedures and presents a flowchart for deciding which procedure to use.

- In Chapter 10 (Inferences for Two Population Means), an optional Section 10.4 provides a discussion of the Mann–Whitney test; a new and optional subsection of Section 10.5 presents the Wilcoxon paired-sample signed-rank test; and an optional Section 10.6 summarizes the hypothesis-testing procedures and presents a flowchart for deciding which procedure to use.

- Chapter 11 (Inferences for Population Proportions) is new; it is comprised roughly of the proportion material found in Chapters 8, 9, and 10 of the third edition. New topics include margin of error and sample-size determination for one and two proportions.

- In Chapter 12 (Chi-Square Procedures), segmented bar graphs have been included in Section 12.3 to provide a visual interpretation of statistical independence for two characteristics of a population. Section 12.4 contains a simulation of a chi-square distribution, obtained by resampling a normal distribution and calculating $(n - 1)s^2/\sigma^2$ for each sample.

- Chapter 13 (Descriptive Methods in Regression and Correlation) now contains a discussion of influential observations and outliers. Also, the terms *predictor variable* and *response variable* have been introduced explicitly in Section 13.2.

- Chapter 14 (Inferential Methods in Regression and Correlation) now includes a discussion of analysis of residuals for checking the assumptions for regression inferences. A new and optional Section 14.6 presents a test for normality based on the correlation between the sample data and its normal scores; the test is essentially equivalent to the powerful Shapiro–Wilk test.

- In Chapter 15 (Analysis of Variance), the material on one-way ANOVA has been revised and split into two sections, one presenting the logic and one the procedure. A new and optional Section 15.4 covers multiple comparisons; a new and optional Section 15.5 examines the Kruskal–Wallis test; and two new sections on two-way ANOVA with interaction present the logic (Section 15.6) and the procedure (Section 15.7).

Supplements and other support

The following supplements have been prepared specifically to accompany the fourth edition of *Introductory Statistics*.

Minitab Supplement.   This supplement, written by Professor Peter W. Zehna, provides in-depth coverage of Minitab, augmenting that given in the book. It is designed to be used in conjunction with *Introductory Statistics* and is keyed to the book. No prerequisite knowledge of computers or statistical software is presumed.

Instructor's Solutions Manual.   This manual, prepared by Professor Chris Franklin, contains detailed, worked-out solutions to all exercises in *Introductory Statistics*.

Student's Solutions Manual.   This manual, also by Professor Chris Franklin, presents detailed, worked-out solutions to every fourth exercise in *Introductory Statistics*.

OmniTest II Computerized Testing.   This unique testing software offers a virtually endless supply of quizzes, tests, final examinations, and additional instructional exercises. Features include multiple-exam versions, customized question editing, on-screen preview and edit functions, and pull-down menus. *OmniTest II* is available for use with IBM PCs or compatibles.

Printed Test Bank.   This supplement provides several printed examinations for each chapter of *Introductory Statistics*.

DataDisk.   This floppy disk contains files for the Focus database and for the data sets appearing in the case studies and odd-numbered basic exercises in *Introductory Statistics*. DataDisk makes it possible to store and use these data sets without having to enter them manually.

Transparency Masters.   Many of the text's figures, tables, and procedure boxes have been reproduced on transparency masters for classroom use.

Additional support is available with the following items.

STAT101.   This inexpensive statistical software package is based on Release 6 of Minitab and is distributed by Addison-Wesley. Working with a spreadsheet-like Data Editor that allows up to 2000 data points, STAT101 provides an excellent foundation for progression to other, more powerful Minitab software products.

The Student Editions of Minitab. These statistical software packages, also distributed by Addison-Wesley, are available in three versions: for Windows, Release 8 for DOS, and Release 8 for the Macintosh. The student editions of Minitab allow the user access to 3500 data points and to a full range of Minitab functionalities and graphic capabilities, as well as a wide range of data sets drawn from business, the social sciences, and the physical sciences. Also accompanying the software is a user's manual that includes case studies and hands-on tutorials for using the software.

Acknowledgments

It is our pleasure to thank the following reviewers, whose comments and suggestions most certainly improved the fourth edition of *Introductory Statistics:*

Jasper Adams
Stephen F. Austin State University

Larry Ammann
University of Texas, Dallas

Jerald T. Ball
Las Positas College

Mary Sue Beersman
Northeast Missouri State University

Gary B. Beus
Brigham Young University

Jerry Bloomberg
Essex Community College

Patricia Buchanan
Penn State University

Curtis Church
Middle Tennessee State University

Gabie Church
Louisiana State University

Patti Collings
Brigham Young University

Rickie J. Domangue
James Madison University

Eugene Enneking
Portland State University

Ruby Evans
Sante Fe Community College

Dale O. Everson
University of Idaho

Charles M. Farmer
James Madison University

Chris Franklin
University of Georgia

Jeff Frost
Johnson County Community College

Joe Fred Gonzalez, Jr.
University of Maryland

Shu-Ping Hodgson
Central Michigan University

Mark E. Johnson
University of Central Florida

Debra Landre
San Joaquin Delta College

Benny Lo
Ohlone College

David R. Lund
University of Wisconsin, Eau Claire

Rhonda Magel
North Dakota State University

Linda Malone
University of Central Florida

Bernard J. Morzuch
University of Massachusetts, Amherst

Tom Ribley
Valencia Community College

Larry Ringer
Texas A&M University

(continued)

Gaspard Rizzuto
University of Southwestern Louisiana

Duane Steffey
San Diego State University

C. Bradley Russell
Clemson University

Larry Stephens
University of Nebraska, Omaha

Leroy Sathre
Valencia Community College

Ram C. Tiwari
University of North Carolina

Robert L. Schaefer
Miami University

Joseph J. Walker
Georgia State University

Franklin Sheehan
San Francisco State University

Lyndon C. Weberg
University of Wisconsin, River Falls

Donald Sisson
Utah State University

Calvin L. Williams
Clemson University

We also wish to thank Professor Michael Driscoll for helping us select the statisticians for the biographical sketches; and Professors Sharon Lohr and Dennis Young, with whom we had several illuminating discussions. Our special thanks go to Professor Ronald Jacobowitz for his many helpful comments and suggestions.

To Professor Larry Griffey, we express our appreciation for his formula/table card. We are grateful to Professor Peter Zehna for preparing the *Minitab Supplement* and for many interesting and informative conversations. Our thanks go as well to Professor Chris Franklin and her two assistants, Kristie Ball and Amy Deitrick, for writing the *Instructor's Solutions Manual* and *Student's Solutions Manual*. In addition, we thank Professor Bernard Morzuch for overseeing the development of the OmniTest computerized-testing software. Professors Joe Fred Gonzalez, Jr., and Larry Stephens did an outstanding job verifying the solutions to the examples and the answers to the exercises; our sincere thanks to them.

We are grateful to Dr. William Feldman and Mr. Frank Crosswhite for providing the data from their study on the Golden Torch Cactus; to Professor Thomas A. Ryan, Jr., for his correspondence concerning the correlation test for normality; and to Dr. George McManus and Mr. Gregory Weiss for supplying the data from their study of zooplankton nutrition in the Gulf of Mexico.

We also extend our appreciation to Professor John Tukey for taking the time to provide us with autobiographical information and a photo of himself; to Ms. Alison Stern-Dunyak of the American Statistical Association for supplying biographical information on Gertrude Cox; to Ms. Maureen Quinn of The Gallup Organization for providing the information used in the case study of Chapter 12; to Ms. Cathy Akritas (Minitab Technical Services) and Ms. Beth Solt (Minitab Author's Assistance Program); and to Ms. Jennifer Wang for obtaining and updating much of the data used in the examples and exercises in the book. Our appreciation also goes to Mr. Howard Blaut and Mr. Rick Hanna for providing data on real estate; to Mr. Jeffrey Jirele for supplying data on automobile insurance; and to Ms. Mary Neary for furnishing the Focus database.

Our thanks to our text designer, Ms. Rebecca Lemna; to our cover designers, Mr. Marshall Henrichs and Mr. Peter Blaiwas; to our technical art consultant, Ms. Susan London-Payne; and to our art supervisor, Joe Vetere. Thanks also to George Morris and the others at Scientific Illustrators. We are grateful as well to our copy editor, Ms. Sally Stickney; and to our proofreader, Ms. Phyllis Coyne.

Thanks to our production supervisors Ms. Mona Zeftel and Ms. Loren Hilgenhurst Stevens; and to our production coordinator Ms. Barbara Ames. To Mr. Michael Payne and Ms. Julia Berrisford, our editors, we extend our heartfelt thanks; and as well to our editorial and project assistants, Ms. Maureen Lawson and Ms. Marcia Cole.

Finally, we would like to thank Ms. Carol Weiss. Apart from writing the text, she was involved in every aspect of development and production. Moreover, Carol researched and wrote the biographies and took on the task of typesetter using the TEX typesetting system.

Tempe, Arizona   *N.A.W.*

Contents

. . . . .

Introductory Statistics

Part I

Introduction

Chapter 1

The Nature of Statistics

Florence Nightingale

Florence Nightingale (1820–1910), the founder of modern nursing, was born in Florence, Italy, into a wealthy English family. In 1849, over the objections of her parents, she entered the Institution of Protestant Deaconesses at Kaiserswerth, Germany, which "...trained country girls of good character to nurse the sick."

The Crimean War began in March 1854 when England and France declared war on Russia. Nightingale, after serving as superintendent of the Institution for the Care of Sick Gentlewomen in London, was appointed by the English secretary of state at war, Sidney Herbert, to be in charge of 38 nurses who were to be stationed at military hospitals in Turkey.

Nightingale found the conditions in the hospitals appalling—overcrowded, filthy, and without sufficient facilities. In addition to the administrative duties she undertook to alleviate those conditions, she spent many hours tending patients; after 8:00 P.M. she allowed none of her nurses in the wards, but made rounds herself every night, a deed that earned her the epithet Lady of the Lamp.

Nightingale was an ardent believer in the power of statistics and used statistics extensively to gain an understanding of social and health issues. She also lobbied to introduce statistics into the curriculum at Oxford and invented the coxcomb chart, a type of pie chart.

In May 1857, as a result of Nightingale's interviews with officials ranging from the secretary of state to Queen Victoria herself, the Royal Commission on the Health of the Army was established. Under the auspices of the commission, the Army Medical School was founded. In 1860 Nightingale used a fund set up by the public to honor her work in the Crimea to create the Nightingale School for Nurses at St. Thomas's Hospital. During that same year, at the International Statistical Congress in London, she authored one of the three papers discussed in the Sanitary Section and also met Adolphe Quetelet (see Chapter 2 biography) who had greatly influenced her work.

Nightingale was elected an Honorary Member of the American Statistical Association in 1874. In 1907 she was presented the Order of Merit, an order for meritorious service established by King Edward VII; she was the first woman to receive that award. Nightingale died in 1910 and was buried in the family plot in East Mellow, Hampshire, England.

The Nature of Statistics

. . . . .

General Objective

What does the word *statistics* bring to mind? Most people immediately think of numerical facts or data, such as unemployment figures, farm prices, or the number of marriages and divorces. *Webster's New World Dictionary* gives two definitions of the word *statistics:*

> 1. facts or data of a numerical kind, assembled, classified, and tabulated so as to present significant information about a given subject. 2. [construed as sing.], the science of assembling, classifying, and tabulating such facts or data.

But statistics encompasses much more than these definitions convey. Not only do statisticians assemble, classify, and tabulate data, but they also analyze data in order to make generalizations and decisions. For example, a political analyst can use data from a portion of the voting population to predict the political preferences of the entire voting population. And a city council can decide where to build a new airport runway based on environmental impact statements and demographic reports that include a variety of statistical data. In this chapter we introduce some basic terminology so that the various meanings of the word *statistics* will become clear to you.

Chapter Outline

1.1 Two Kinds of Statistics

1.2 Classifying Statistical Studies

1.3 The Development of Statistics

1.4 Using the Computer (Optional)

1.5 Is a Study Necessary?

1.6 Simple Random Sampling

1.7 Other Sampling Procedures

1.8 Observational Studies and Designed Experiments

Case Study *Are Americans Lazy?*

Do Americans work hard, or are they lazy? Near the beginning of 1992, two high-ranking Japanese officials made negative remarks about the work ethic of Americans. In response to those comments, the Associated Press commissioned a poll to gauge the feelings of Americans on that issue. We will examine the results of that poll at the end of this chapter.

1.1   TWO KINDS OF STATISTICS

. . . . .

You probably already know something about statistics. If you read newspapers, watch the news on television, or follow sports, then you see and hear the word *statistics* frequently. In this section we will use familiar examples such as baseball statistics and voter polls to introduce the two major types of statistics: *descriptive statistics* and *inferential statistics.*

Each spring in the late 1940s, the major-league baseball season was officially opened when President Harry S Truman threw out the "first ball" of the season at the opening game of the Washington Senators. Both President Truman and the Washington Senators had reason to be interested in statistics. Consider, for instance, the year 1948.

Example 1.1   Descriptive Statistics

In 1948 the Washington Senators played 153 games, winning 56 and losing 97. They finished seventh in the American League and were led in hitting by Bud Stewart, whose batting average was .279. These and many other statistics were compiled by baseball statisticians who took the complete records for each game of the season and organized that large mass of information effectively and efficiently.

Although baseball fans take baseball statistics for granted, a great deal of time and effort is required to gather and organize them. Moreover, without such statistics, baseball would be much harder to understand. For instance, picture yourself trying to select the best hitter in the American League with only the official score sheets for each game. (More than 600 games were played in 1948; the best hitter was Ted Williams, who led the league with a batting average of .369.)   ■

The work of baseball statisticians provides an excellent illustration of descriptive statistics. A formal definition of the term *descriptive statistics* is presented in Definition 1.1.

DEFINITION 1.1   **DESCRIPTIVE STATISTICS**

> *Descriptive statistics* consists of methods for organizing and summarizing information.

Descriptive statistics includes the construction of graphs, charts, and tables, and the calculation of various descriptive measures such as averages, measures of dispersion, and percentiles. We will discuss descriptive statistics in detail in Chapters 2 and 3.

As we said, descriptive statistics is one of the two major types of statistics. The other major type, inferential statistics, is illustrated in Example 1.2.

Example 1.2   Inferential Statistics

In the fall of 1948, President Truman was also concerned about statistics. The Gallup Poll taken just prior to the election predicted that he would win only 44.5% of the vote and be defeated by the Republican nominee, Thomas E. Dewey. But this time the statisticians had predicted incorrectly. Truman won more than 49% of the vote and with it the presidency. The Gallup Organization modified some of its procedures and has correctly predicted the winner ever since. ∎

Political polling provides an example of inferential statistics. It would be expensive and unrealistic to interview all Americans on their voting preferences. Statisticians who wish to gauge the sentiment of the entire *population* of U.S. voters can afford to interview only a carefully chosen group of a few thousand voters. This group is referred to as a *sample* of the population. Statisticians analyze the information obtained from a sample of the voting population to make inferences (draw conclusions) about the preferences of the entire voting population. Inferential statistics provides methods for making such inferences.

The terminology introduced in the context of political polling is used in general in statistics. Specifically, we have the following definitions.

DEFINITION 1.2

POPULATION AND SAMPLE

> ***Population:*** The collection of all individuals, items, or data under consideration in a statistical study.
>
> ***Sample:*** That part of the population from which information is collected.

In Fig. 1.1 we have depicted the relationship between a population and a sample from the population.

FIGURE 1.1

Relationship between population and sample

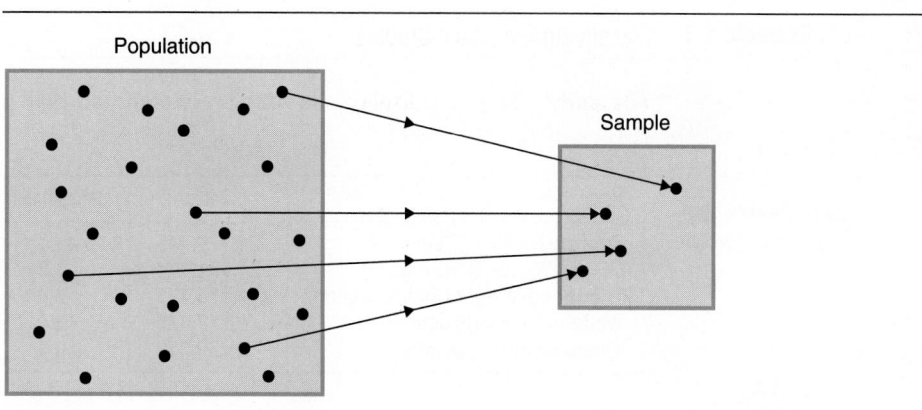

With Definition 1.2 in mind, we now present the definition of inferential statistics.

DEFINITION 1.3	**INFERENTIAL STATISTICS**

> *Inferential statistics* consists of methods for drawing and measuring the reliability of conclusions about a population based on information obtained from a sample of the population.

Descriptive statistics and inferential statistics are quite interrelated. It is almost always necessary to invoke techniques of descriptive statistics to organize and summarize the information obtained from a sample before carrying out an inferential analysis. Furthermore, the preliminary descriptive analysis of a sample often reveals features that lead to the choice of (or to a reconsideration of the choice of) the appropriate inferential method.

EXERCISES 1.1

1.1   Define the following terms.
a. population   b. sample

1.2   What are the two major types of statistics? Describe them in detail.

1.2   CLASSIFYING STATISTICAL STUDIES

In subsequent chapters we will thoroughly examine both descriptive and inferential statistics. At this point, however, you need only to be able to classify statistical studies as either descriptive or inferential. Examples 1.3 and 1.4 will give you some practice in making this distinction. In each example we have presented the result of a statistical study and have classified the study as either descriptive or inferential. Try to classify each study yourself before reading our explanation.

Example 1.3   Classifying Statistical Studies

The study   Table 1.1 displays the voting results for the 1948 presidential election.

TABLE 1.1
Final results of the 1948
presidential election

Ticket	Votes	Percentage
Truman-Barkley (Democratic)	24,179,345	49.7
Dewey-Warren (Republican)	21,991,291	45.2
Thurmond-Wright (States Rights)	1,176,125	2.4
Wallace-Taylor (Progressive)	1,157,326	2.4
Thomas-Smith (Socialist)	139,572	0.3

Classification   This study is descriptive. It is a summary of the votes cast by U.S. voters in the 1948 presidential election. No inferences were made.   ■

Example 1.4 Classifying Statistical Studies

The study For the 101 years preceding 1977, baseballs used by the major leagues were purchased from the Spalding Company. In 1977 that company stopped manufacturing major-league baseballs, and the major leagues arranged to buy their baseballs from the Rawlings Company.

Early in the 1977 season, pitchers began to complain that the Rawlings ball was "livelier" than the Spalding ball. They claimed it was harder, bounced farther and faster, and gave hitters an unfair advantage. There was some evidence for this. In the first 616 games of 1977, 1033 home runs were hit, compared to only 762 home runs in the first 616 games of 1976.

Sports Illustrated magazine sponsored a careful study of the liveliness question, and the results appeared in the June 13, 1977, issue. In this study an independent testing company randomly selected 85 baseballs from the current (1977) supplies of various major-league teams. The bounce, weight, and hardness of the baseballs chosen were carefully measured. Those measurements were then compared with measurements obtained from similar tests on baseballs used in the years 1952, 1953, 1961, 1963, 1970, and 1973. The conclusion, presented on page 24 of the *Sports Illustrated* article, was that "... the 1977 Rawlings ball is livelier than the 1976 Spalding, but not as lively as it could be under big league rules, or as the ball has been in the past."

Classification This is an inferential study. The independent testing company used a sample of 85 baseballs from the 1977 supplies of major-league teams to make an inference about the population of all such baseballs. (It has been estimated that approximately 360,000 baseballs were used by the major leagues in 1977.) ∎

The *Sports Illustrated* study also provides an excellent illustration of a situation in which it is not feasible to obtain data for the entire population. Indeed, after the bounce and hardness tests, all of the baseballs sampled were taken to a butcher in Plainfield, New Jersey, to be sliced in half so that researchers could look inside them. Clearly, it would not have been practical to test every baseball in this way.

In closing, we emphasize that it is possible to perform a descriptive study on a sample as well as on a population. Only when an inference is made about the population based on information obtained from the sample does the study become inferential.

EXERCISES 1.2

In Exercises 1.3–1.10, classify each of the studies as either descriptive or inferential.

· **1.3** The A. C. Nielsen Company collects and publishes information on the television-viewing habits of Americans. Data from a sample of Americans yielded the estimates of average TV viewing time per week for all Americans shown in the table in the first column at the top of the next page. The times are in hours and minutes. [SOURCE: Nielsen Media Research, *Nielsen Report on Television.*]

Group (by age)		Time
Average all persons		*30:20*
Women	Total 18+	34:56
	18–24	26:05
	55+	43:58
Men	Total 18+	29:21
	18–24	20:20
	55+	39:14
Teens	Female	20:33
	Male	22:38
Children	2–5	28:06
	6–11	23:31

Type of drug	Percentage of young adults			
	Ever used		*Current user*	
	1974	1991	1974	1991
Marijuana	52.7	50.5	25.2	13.0
Inhalants	9.2	10.9	(NA)	1.5
Hallucinogens	16.6	13.2	2.5	1.2
Cocaine	12.7	17.9	3.1	2.0
Heroin	4.5	0.8	(NA)	0.1
Analgesics	(NA)	10.2	(NA)	1.5
Stimulants[1]	17.0	9.4	3.7	0.8
Sedatives[1]	15.0	4.3	1.6	0.6
Tranquilizers[1]	10.0	7.5	1.2	0.6
Alcohol	81.6	90.2	69.3	63.6
Cigarettes	68.8	71.2	48.8	32.2

NA = Not available.   1 = Prescription drugs.

1.4   In 1936 the voters of North Carolina cast their presidential votes as follows.

Candidate and party	Number of votes
Roosevelt (Democratic)	616,414
Landon (Republican)	223,283
Thomas (Socialist)	21
Browder (Communist)	11
Lemke (Union)	2

1.5   The U.S. National Center for Health Statistics published the following rate estimates in *Vital Statistics of the United States* for the leading causes of death in 1990. The estimates are based on a 10% sampling of all 1990 U.S. death certificates. Rates are per 100,000 population.

Cause	Rate
Major cardiovascular diseases	366.9
Malignancies (cancers)	201.7
Accidents	37.3
Chronic obstructive pulmonary diseases	35.5
Influenza and pneumonia	31.3

1.6   The U.S. National Institute on Drug Abuse collects and publishes data on drug use, by type of drug, in *National Household Survey on Drug Abuse*. We have displayed information for the years 1974 and 1991 in the table at the top of the next column. The percentages shown are estimates obtained from national samples.

1.7   The following table displays the 1990 attendance figures for selected spectator sports. Data are in thousands.

Sport	Attendance (thousands)
Baseball, major leagues	55,509
Basketball	
College	37,558
Professional	18,586
Football	
College	36,627
Professional	17,666
Horse racing	63,803
Greyhound racing	28,660

1.8   Newspapers publish weather data for cities all over the world. Below are high and low temperatures and sky-condition readings for some selected cities on March 18, 1992.

City	High	Low	Sky condition
Athens	63	45	Rain
Bangkok	97	79	Rain
Buenos Aires	82	66	Clear
Cairo	72	54	Clear
Helsinki	30	21	Sunny
Johannesburg	82	55	Clear
Moscow	33	27	Sunny
Paris	52	46	Cloudy
Rome	61	41	Clear
Santiago	82	52	Clear
Sydney	70	64	Clear
Tokyo	57	36	Rain

1.9 The New York Stock Exchange keeps records of the selling prices for seats on the exchange. The following table provides the high and low prices for selected years in this century.

Year	High price ($)	Low price ($)
1900	47,500	37,500
1920	115,000	85,000
1929	625,000	550,000
1935	140,000	65,000
1940	60,000	33,000
1960	162,000	135,000
1970	320,000	130,000
1987	1,150,000	575,000
1991	440,000	345,000

1.10 A 1984 study by the Gallup Organization concluded that an estimated 83% of U.S. households are involved in at least one form of gardening. [SOURCE: National Gardening Association, *National Gardening Survey*.]

1.11 In each part below, decide whether the specified study would be descriptive or inferential. Provide a reason for each of your answers.
a. A tire manufacturer wants to estimate the average life of a new type of steel-belted radial.
b. A sports writer plans to list the winning times for all swimming events in the 2000 Olympics.
c. A politician obtains the exact number of votes that were cast for her opponent in 1992.
d. A medical researcher tests an anticancer drug that may have harmful side effects.

1.12 The chairperson of a mathematics department at a large state university wants to estimate the average final exam score for the 2476 students in basic algebra. She randomly selects 50 exams from the 2476 and finds that the average score on the 50 exams is 78.3%. From this she estimates that the average score for all 2476 students is roughly 78.3%.
a. What kind of study has the chairperson done?
b. What kind of study would she have done had she averaged all 2476 exam scores?

1.13 A 1988 *Newsweek* poll of a sample of Americans revealed that 84% of those surveyed would choose organically grown produce over produce grown using chemical fertilizers, pesticides, and herbicides. Is this an inferential statement, or is it simply descriptive? What if, based on the same information, the statement had been that 84% of Americans would choose organically grown produce over produce grown using chemical fertilizers, pesticides, and herbicides?

1.14 A headline in the December 24, 1991, issue of the *The Arizona Republic* read "Most wait to unwrap gifts on Christmas."
a. Do you think this statement is inferential or descriptive? Can you be sure?
b. Actually, *The Arizona Republic* polled 758 Arizona families who celebrate Christmas and found that 52% of them wait until Christmas Day to open their gifts. This resulted in the statement in the headline. How would you rephrase the statement to make it clear that it is a descriptive statement? By the way, we will discover in Chapter 11 that we cannot infer from the data that a majority of all Arizona families who celebrate Christmas wait until Christmas Day to unwrap their gifts.

1.3 THE DEVELOPMENT OF STATISTICS

According to the *Dictionary of Scientific Biography,* "The word '*Statistik*,' first printed in 1672, meant *Staatswissenschaft,* or, rather, a science concerning the states. It was cultivated at the German universities, where it consisted of more or less systematically collecting 'state curiosities' rather than quantitative material."

As we know, the modern science of statistics is much broader than just collecting "state curiosities" and includes both descriptive statistics and inferential statistics. Historically, descriptive statistics appeared first. Censuses were taken as long ago as Roman times. Over the years, records of such things as births, deaths, marriages, and taxes have led naturally to the development of descriptive statistics.

Inferential statistics is a newer arrival. Major developments began to occur with the research of Karl Pearson (1857–1936) and Ronald Fisher (1890–1962), who published their results in the early years of the twentieth century. Since the work of Pearson and Fisher, inferential statistics has evolved rapidly and is now applied in many fields. In fact, an understanding of the basic concepts of inferential statistics has become mandatory for virtually every professional.

Familiarity with statistics will also help you make more sense of many things you read in newspapers and magazines. For instance, in the description of the *Sports Illustrated* baseball test (Example 1.4), it may have struck you as unreasonable that a sample of only 85 baseballs could be used to draw a conclusion about a population of roughly 360,000 baseballs. By the time you have completed Chapter 9, you will understand why such inferences are not unreasonable.

1.4   USING THE COMPUTER (OPTIONAL)

Computers are ideal for performing statistical calculations and analyses. A person rarely needs to write his or her own computer programs, since they already exist for almost all aspects of statistics. The most commonly used programs for statistical work are obtained from **statistical software packages,** collections of statistical computer programs written by some organization or individual. Statistical software packages are currently available for use on mainframes, minicomputers, and microcomputers.

Many high-quality statistical software packages are on the market. We have chosen Minitab[†] to illustrate the basic ideas of statistical software. At the end of most sections, we will briefly discuss how Minitab can be used to solve problems that were solved by hand earlier in that section. Each solution first introduces the appropriate commands and then displays and interprets the computer output. For best results you should be at a computer and perform the steps taken in the example under consideration.

To use Minitab you must first gain access to it on the computer. Because the procedure for accessing Minitab depends on the type of computer system being used, the precise method should be obtained from your instructor or an appropriate member of the staff at the computer center. Once Minitab has been accessed, the prompt MTB > will be displayed on the screen.

Traditionally, Minitab has been applied by typing commands at the MTB > prompt, so-called **session commands.** Session-command interface is available on all platforms. Minitab has recently introduced a menu interface as well, so that commands can also be executed by choosing them from menus and completing dialogue boxes. Commands chosen from menus are called **menu commands.** Menu-command interface is not available on all platforms.

For each application of Minitab, we will explain separately the required session commands and the required menu commands.[‡] Although experienced users of Minitab often intersperse session and menu commands, the intent is that you will study either session

---

[†] Minitab is a registered trademark of Minitab Inc.

[‡] The menu commands presented in this book are for Release 8 of the PC version of Minitab. If you are using a different version of Minitab, you may encounter some discrepancies.

commands or menu commands, but not both. A reader covering session commands can ignore all references to menu commands; a reader covering menu commands can ignore all references to session commands.

Several typographical conventions used for both session and menu commands are presented in Table 1.2.

Item	Typographical convention
Key	A key is denoted by enclosing the name of the key in a box; e.g., [Enter] denotes the Enter key.
Multiple keys	Pressing a key while holding down another key is denoted by placing a plus sign between the two keys; e.g., [Alt]+[D] represents pressing the D key while holding down the Alt key.
Session command	Session commands are set in all uppercase letters; e.g., ZTEST.
Menu command	Menu commands are set with only the first letter capitalized; e.g., Exit.
Underlined text	Underlined text is to be typed by the person using the computer; e.g., BOXPLOT 'TIMES'.
Menu instructions	A sequence of menu instructions is set in bold-face type with the entries separated by pointers; e.g., **Stat ▶ Tables ▶ Chisquare Test...** means first pull down the Stat menu, then open the Tables submenu, and then choose Chisquare Test.

Storing Data Sets

To prepare you for the Minitab sections in this book, we will now explain how to store one or more data sets. To begin, you need to understand that a data set is stored in a **column.** A column is designated by a "C" followed by a number; thus C1 stands for Column 1, C2 for Column 2, and so forth.

Two commands can be used to store data. One is the **Set** command, which is employed to store one data set at a time. Example 1.5 explains how to use the Set command.

Example 1.5   The Set Command

Table 1.3 at the top of the following page displays the prices, in hundreds of dollars, for a sample of 11 Nissan Zs. The data were obtained from the *Asian Import* edition of the *Auto Trader* magazine. Use the Set command to store these data in Column 4.

TABLE 1.3
Prices ($100s) for 11 Nissan Zs

85	82	66	70
103	89	95	48
70	98	169	

SOLUTION   To store the price data from Table 1.3 into C4 using session commands, we proceed in the following manner.

Session commands: First we type the command **SET** followed by the column in which the data are to be stored; that is, we type SET C4. Next we press the [Enter] key to signify that, for the moment, we are done typing. (On some computers the [Enter] key is replaced by a [Return] key, on others by an [Endline] key, and on still others by a different key.)

After the [Enter] key is pressed, the prompt DATA> is displayed on the screen to indicate that the data may be entered. Now we type the numbers in the data set, separating the numbers by spaces; any number of spaces will do as long as the numbers are separated. We do not need to enter all the data on a single line; for instance, we can enter the data as they appear in Table 1.3. Remember, though, that the [Enter] key must be pressed after each line!

When we have finished entering the data, we use the **END** command; that is, we type END and press [Enter]. The entire procedure is depicted in Printout 1.1.

PRINTOUT 1.1
Storing the price data
from Table 1.3 in C4
using the SET command

```
MTB > SET C4
DATA> 85 82 66 70
DATA> 103 89 95 48
DATA> 70 98 169
DATA> END
```

Alternatively, we can store the price data from Table 1.3 into C4 using menu commands as follows.[†]

Menu commands: We choose **Edit ▸ Set Patterned Data...**, specify C4 in the **Put data into column** text box, select **Arbitrary list of constants**, type the price data in the box (not necessarily all on one line), and then select **OK**.

The price data from Table 1.3 are now stored in column C4. Those data will be available for analysis until we store another data set in C4 or sign off the computer. Finally, we emphasize that although the price data may have been entered using more than one line, all 11 numbers are stored in C4 in a single column.   ■

Note: If an error is made while entering a data set, the error can be corrected in several ways. One way is to start over again from the beginning; the old data will be erased and the

---

[†] This method works only with relatively small data sets. For larger data sets we can use either the Read command or the Data Editor, both to be discussed shortly.

new data inserted in their place. More efficient ways of correcting an error are discussed at the end of Example 1.7 and in the *Minitab Supplement.*

The other Minitab command that can be used to store data is the **Read** command, which is employed to store two or more data sets simultaneously, provided the data sets are all the same size. (Actually, the Read command can also be used to store a single data set.) Example 1.6 explains how to use the Read command.

Example 1.6   The Read Command

Current Population Reports, a publication of the U.S. Bureau of the Census, provides information on the ages of married people. The ages of 10 married couples are displayed in Table 1.4. Simultaneously store the ages of the husbands and wives in Column 1 and Column 2, respectively.

TABLE 1.4
Ages of 10 married couples

Couple	Husband	Wife
1	54	53
2	21	22
3	32	33
4	78	74
5	70	64
6	33	35
7	68	67
8	32	28
9	54	41
10	52	44

SOLUTION   To simultaneously store the husbands' ages and the wives' ages into C1 and C2, we proceed in the following manner.

Session commands: We first type the command **READ** followed by the columns in which the data are to be stored; that is, we type READ C1 C2 and press [Enter]. Then we enter the data row by row; that is, we type 54 53 and press [Enter], type 21 22 and press [Enter], and so on. When all 10 rows of data have been entered, we type END and press [Enter]. Printout 1.2 on the next page depicts the entire procedure.

(or)

Menu commands: We choose **File ▸ Import ASCII Data...**, specify C1 and C2 in the **Put data in column(s)** text box, select **Read data from keyboard instead of file**, and then select **OK**. The prompt DATA> is then displayed on the screen to indicate that the data may be entered. We enter the data row by row; that is, we type 54 53 and press the [Enter] key, type 21 22 and press the [Enter] key, and so on. When all 10 rows of data have been entered, we type END and press the [Enter] key. Printout 1.2 depicts the entire procedure. [*Note:* The first line of Printout 1.2 displays the Read command. This line results automatically from the menu commands but may appear on your screen in a slightly different form than shown in the printout (e.g., on your screen it may appear as Read C1 C2. instead of READ C1 C2).

Although, in general, the session commands generated by the menu commands may look somewhat different on your screen than displayed in a printout, such differences are minor and should cause no confusion.]

```
MTB > READ C1 C2
DATA> 54 53
DATA> 21 22
DATA> 32 33
DATA> 78 74
DATA> 70 64
DATA> 33 35
DATA> 68 67
DATA> 32 28
DATA> 54 41
DATA> 52 44
DATA> END
```

The husbands' ages from the second column of Table 1.4 are now stored in C1, and the wives' ages from the third column of Table 1.4 are now stored in C2. ■

Naming a Column

Naming a column is often useful because it allows us to refer to the column by its name instead of trying to remember its number. A column name can consist of between one and eight characters. Any character except an apostrophe (') or a number sign (#) can be used to form a name, but a name cannot begin or end with a blank.

Minitab doesn't have a menu command for naming a column; naming must be done either using the session command **NAME** or in the Data Editor. We will now explain how to name a column using the NAME command; naming a column in the Data Editor will be discussed shortly.

To illustrate, recall that we stored price data for 11 Nissan Zs in C4. To name that column PRICE, we type the command NAME followed first by the column to be named and then by the name; that is, we type NAME C4 'PRICE'. Notice that the name is enclosed by apostrophes. Apostrophes must be used to enclose the name; otherwise the command will not work and an error message will result. The apostrophes must also be included when referencing the column using session commands.

To change the name of a column, we simply use the NAME command again. For example, to change the name of C4 from PRICE to COST, we would type NAME C4 'COST'.

We can name more than one column at a time. For instance, consider again the ages of husbands and wives displayed in Table 1.4. We stored the husbands' ages in C1 and the wives' ages in C2. To name C1 HUSBAND and C2 WIFE, we type NAME C1 'HUSBAND' C2 'WIFE'.

Using the Data Editor

One of the simplest ways, although not necessarily the most efficient, to store and name a data set is to use the Data Editor.[†] Before explaining how to use the Data Editor, we recommend that you restart Minitab. To do that, employ Minitab's **Restart** command: if you are using session commands, type RESTART and press Enter; if you are using menu commands, choose **File ▶ Restart Minitab** and then select **OK**.

When Minitab is accessed, we are in the **Session window,** where session commands and their corresponding output are displayed. To use the Data Editor, we must switch to the **Data screen,** where the columns of the worksheet are displayed. This can be accomplished by pressing Alt+D. (To return to the Session window, we press Alt+M.) A portion of the Data screen, with all columns empty, is depicted in Fig. 1.2.

FIGURE 1.2
Minitab's Data screen

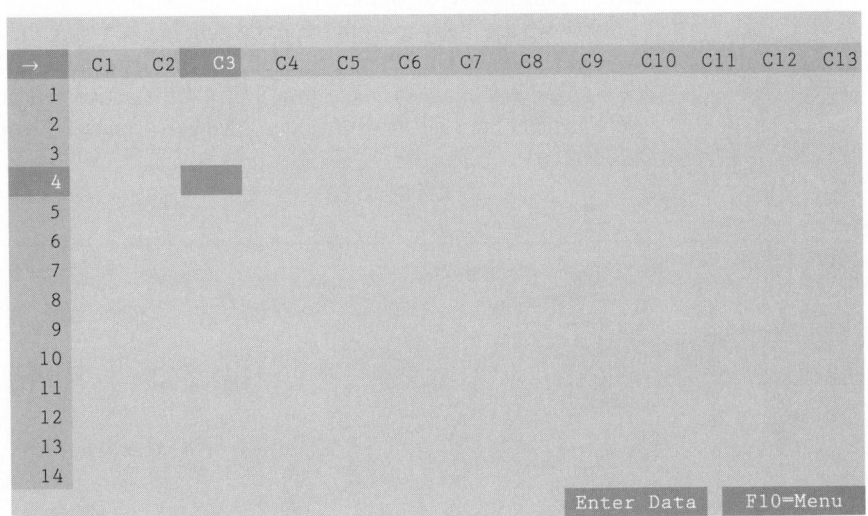

Notice that the columns are displayed along the top of the Data screen. The numbers on the left represent positions within a column and are referred to as **rows.** A **cell** in the Data screen is specified by giving its column number and row number. The **active cell** is the one that is highlighted at any particular time. In Fig. 1.2 the active cell is the fourth row of Column 3 (C3). To enter a new data value into the active cell, we type the value; the new value overwrites the previous contents, if any. An easy way to change the location of the active cell is to use the arrow keys.

[†] The Data Editor is not available on all platforms; in particular, not on mainframes.

Example 1.7   Using the Data Editor to Store and Name a Data Set

Table 1.3 listed the prices, in hundreds of dollars, for a sample of 11 Nissan Zs. For ease of reference, we repeat that table here as Table 1.5. Use the Data Editor to store these price data in a column named PRICE.

TABLE 1.5

Prices ($100s) for 11 Nissan Zs

85	82	66	70
103	89	95	48
70	98	169	

SOLUTION   Suppose we decide to store the price data in C4. To name C4 PRICE, we first use the arrow keys to move the cursor to the cell directly above "C4" (thus making that cell the active cell), and then we type PRICE.

Now we are ready to store the price data in C4. Using the down-arrow key, we make row 1 of C4 the active cell and type 85, the first data value in Table 1.5. Next we use the down-arrow key again to make row 2 of C4 the active cell and type 103. We continue in this manner until all 11 data values have been entered. Figure 1.3 depicts the resulting Data screen.

FIGURE 1.3

Data screen with price data for Nissan Zs stored in C4, and C4 named PRICE

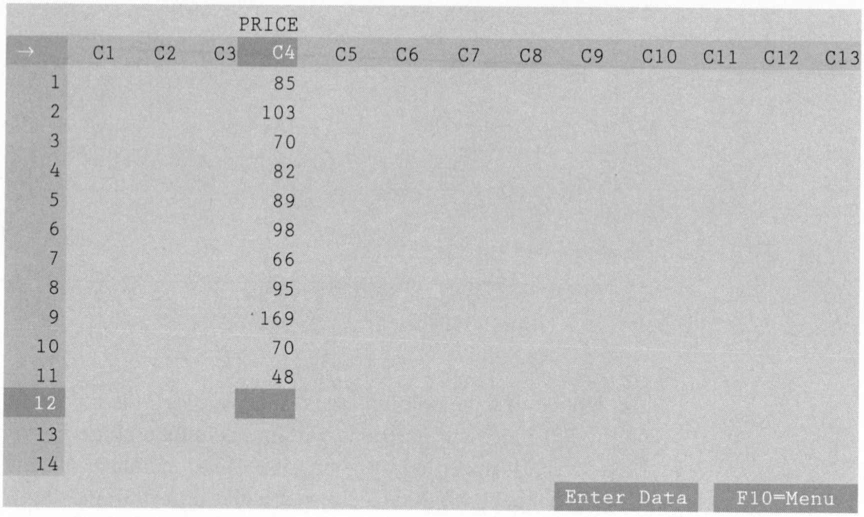

The price data from Table 1.5 are now stored in a column named PRICE and are available for analysis. *Note:* If a data value was incorrectly entered, we can rectify the error by making its cell active and typing the correct data value.   ■

Exiting Minitab

When we are finished with Minitab, we must exit the software. This is accomplished in the following manner.

Session commands: We use the **STOP** command; that is, we type STOP and press Enter.

(or)

Menu commands: We choose **File ▶ Exit** and then select **OK**.

It may also be necessary to sign off the computer. This will involve a special local procedure that you can obtain from your instructor or an appropriate member of the staff at the computer center.

EXERCISES 1.4

1.15 (Computer exercise) The U.S. Energy Information Administration collects data on residential energy consumption and expenditures. Results are published in *Residential Energy Consumption Survey: Consumption and Expenditures.* The following table shows one year's energy consumption for a sample of 50 households in the South. Data are in millions of BTU.

130	55	45	64	155	66	60	80	102	62
58	101	75	111	151	139	81	55	66	90
97	77	51	67	125	50	136	55	83	91
54	86	100	78	93	113	111	104	96	113
96	87	129	109	69	94	99	97	83	97

Use Minitab or some other statistical software to store the data on energy consumption in a column named ENERGY.

1.16 (Computer exercise) The Bureau of Economic Analysis publishes data on the length of stay in Europe and the Mediterranean by U.S. travelers in *Survey of Current Business.* A sample of 36 U.S. residents who traveled to Europe and the Mediterranean one year yielded the following data on length of stay, in days.

41	3	32	6	15	48	1	18	12
5	1	44	20	14	56	17	64	12
3	16	13	2	21	10	8	21	3
21	31	11	1	27	12	5	10	10

Use Minitab or some other statistical software to store the data on length of stay in a column named STAYS.

1.17 (Computer exercise) The U.S. National Center for Health Statistics publishes data on heights and weights by age and sex in *Vital and Health Statistics.* A sample of 11 males age 18–24 years yielded the following data on height (in inches) and weight (in pounds).

Height	Weight	Height	Weight
65	175	67	153
67	133	70	163
71	185	71	159
71	163	69	151
66	126	69	155
75	198		

Use Minitab or some other statistical software to simultaneously store the height and weight data in columns named HEIGHT and WEIGHT, respectively.

1.18 (Computer exercise) Hanna Properties specializes in custom-home resales in the Equestrian Estates, a subdivision in Phoenix, Arizona. A sample of nine custom homes currently listed for sale provided the following information on size and price. The size data are in hundreds of square feet; the price data are in thousands of dollars.

Size	Price	Size	Price
26	235	34	345
27	249	30	415
33	267	40	475
29	269	22	195
29	295		

Use Minitab or some other statistical software to simultaneously store the size and price data in columns named SIZE and PRICE, respectively.

1.5 IS A STUDY NECESSARY?

Throughout this book we will see examples of organizations or people conducting studies: a consumer group wants information about the gas mileage of a particular make of car, so it performs mileage tests on a sample of such cars and statistically analyzes the resulting data; or a teacher wants to know about the comparative merits of two teaching methods, so he tests those methods on two randomly selected samples of students.

This reflects a healthy attitude—to obtain information about a subject of interest, plan and conduct a study. However, the possibility always exists that a study being considered has already been done. Repeating it would be a waste of time, energy, and money. Therefore before a study is planned and conducted, a literature search should be made. This does not require going through all the books in the library. Many information-collection agencies specialize in finding studies on specific topics in specific areas.

For instance, the Educational Resources Information Center assembles educational studies; publications entitled *Psychological Abstracts* gather the results of studies in psychology; the National Library of Medicine compiles lists of medical studies and makes them accessible to research centers and universities. A considerable amount of information can also be found in publications by government agencies such as the Bureau of the Census and the Environmental Protection Agency. Data are published on income, age, energy consumption, and hundreds of other variables. Many of the examples and exercises in this book are based on information obtained from the Census Bureau's *Statistical Abstract of the United States* publications.

It is not our purpose to explain how to search through journal articles, abstracts, or census data. The important point here is this: *It is often possible to avoid the effort and expense of a study if someone else has already done that study and published the results.*

1.6 SIMPLE RANDOM SAMPLING

If it has been determined that the information required is not already available from a previous study, a new study can be planned to obtain the information. Broadly speaking, there are two methods for acquiring information: select a sample or take a **census** (collect data for the entire population). Sampling is far more common because it is less costly and can be done more quickly than a census; it is often the only practical way to gather information.

If sampling has been deemed appropriate, it must then be decided how to select a sample, that is, the method to use to obtain a sample from the population. In making that decision, it should be kept in mind that the sample will be employed to draw conclusions about the entire population. Consequently, it is crucial that the sample be a **representative sample**—it should reflect as closely as possible the relevant characteristics of the population under consideration.

For instance, it would not make much sense to use the average weight of a sample of football players to make an inference about the average weight of all adult males. Nor would it be reasonable to try to estimate the median income of California residents by sampling the incomes of residents of Beverly Hills.

To see what can happen when a sample is not representative, let's consider the presidential election of 1936. Before the election the *Literary Digest* magazine conducted an opinion poll of the voting population. Its survey team asked a sample of the voting population whether they would vote for Franklin D. Roosevelt, the Democratic candidate, or for Alfred Landon, the Republican candidate. Based on the results of the survey, the magazine predicted an easy win for Landon. The actual election results, of course, were that Roosevelt won by the greatest landslide in the history of presidential elections!

What happened? Here are two reasons given for why the poll failed: (1) The sample was obtained from among people who owned a car or had a telephone. In 1936 that group included only the more well-to-do people, and historically such people tend to vote Republican. (2) The response rate was low (less than 25% of those polled responded) and there was a nonresponse bias (a disproportionate number of those responding to the poll were Landon supporters). Whatever the reason for the poll's failure, the sample obtained by the *Literary Digest* was obviously not representative.

Most modern sampling procedures employ **probability sampling,** where a random device, such as tossing a coin or consulting a random-number table, is used to decide which members of the population will constitute the sample instead of leaving such decisions to human judgment. It is still possible to obtain a nonrepresentative sample when probability sampling is used. However, probability sampling eliminates unintentional selection bias and permits the researcher to control the chance of obtaining a nonrepresentative sample. Furthermore, the use of probability sampling guarantees that the techniques of inferential statistics can be applied. In this section and the next, we will examine the most important probability-sampling methods.

Simple Random Sampling

The statistical-inference techniques considered in this book are intended for use with only one particular sampling procedure: **simple random sampling,** or more briefly, **random sampling.** Simple random sampling is the simplest type of probability sampling and is also the basis for the more complex types of probability sampling.

DEFINITION 1.4 **SIMPLE RANDOM SAMPLING AND SIMPLE RANDOM SAMPLES**

> A *simple-random-sampling* procedure is a sampling procedure for which each possible sample is equally likely to be the one selected. A sample obtained by simple random sampling is called a *simple random sample.*

Actually, there are two types of simple random sampling. One is **simple random sampling with replacement,** where a member of the population can be selected more than once; the other is **simple random sampling without replacement,** where a member of the population can be selected at most once. *Unless otherwise specified we will assume that simple random sampling is done without replacement.*

In Example 1.8 we have chosen a very small population—the annual salaries of the five top Oklahoma state officials—to illustrate simple random sampling. In practice we would not sample from such a small population but would take a census. We are using a small population here to make it easier to understand the concept of simple random sampling.

Example 1.8   Illustrates Definition 1.4

According to *The World Almanac,* the annual salaries for the five top Oklahoma state officials are as shown in the second column of Table 1.6. Values are in thousands of dollars, rounded to the nearest thousand.

TABLE 1.6

Five top Oklahoma state officials and their annual salaries

Official	Salary ($thousands)
Governor (G)	70
Lieutenant governor (L)	40
Secretary of state (S)	37
Attorney general (A)	55
Treasurer (T)	50

a. List the possible samples (without replacement) of two salaries that can be obtained from the population of five salaries.

b. Describe a method for obtaining a simple random sample of two salaries from the population of five salaries.

c. For the sampling method described in part (b), what are the chances that any particular sample of two salaries will be the one selected?

d. Repeat parts (a)–(c) for samples of size four.

SOLUTION   For convenience we will use the letters placed parenthetically after each official in Table 1.6 to represent the officials.

a. There are 10 possible samples of two salaries from the population of five salaries.[†] They are listed in the second column of Table 1.7.

TABLE 1.7

Possible samples of two salaries from the population of five salaries

Officials selected	Sample obtained
G, L	70, 40
G, S	70, 37
G, A	70, 55
G, T	70, 50
L, S	40, 37
L, A	40, 55
L, T	40, 50
S, A	37, 55
S, T	37, 50
A, T	55, 50

[†] We will learn how to determine the number of possible samples in Section 4.8.

b. Here is one method we could use to obtain a simple random sample of size two: First write each of the letters corresponding to the five officials, G, L, S, A, and T, on separate pieces of paper. Next place the five slips of paper into a box and shake the box. Then, while blindfolded, pick two of the slips of paper.

c. The sampling procedure described in part (b) ensures that we are taking a simple random sample. Consequently, each of the possible samples of two salaries is equally likely to be the one selected. Since there are 10 possible samples, the chances are $\frac{1}{10}$ (1 in 10) that any particular sample will be the one selected.

d. For samples of size four, there are five possibilities, as indicated in Table 1.8.

TABLE 1.8

Possible samples of four salaries from the population of five salaries

Officials selected	Sample obtained
G, L, S, A	70, 40, 37, 55
G, L, S, T	70, 40, 37, 50
G, L, A, T	70, 40, 55, 50
G, S, A, T	70, 37, 55, 50
L, S, A, T	40, 37, 55, 50

†

In this case a simple-random-sampling procedure, such as picking four slips of paper out of a box, gives each of the five possible samples in the second column of Table 1.8 a 1 in 5 chance of being the one selected. ∎

Obtaining a simple random sample by picking slips of paper out of a box is usually not practical, especially when the population being sampled is large. But there are several practical procedures for getting simple random samples. One common method uses a **table of random numbers,** a table of randomly chosen digits. Example 1.9 explains how a table of random numbers can be employed to obtain a simple random sample.

Example 1.9   Using Random-Number Tables

Student questionnaires, known as "teacher evaluations," gained widespread use about 30 years ago. Generally, student evaluations of teachers are not done at final exam time. It is more common for professors to hand out evaluation forms a week or so before the final.

That practice, however, poses several problems. On some days less than 60% of the students registered for a class actually attend. Moreover, because many of those who are present have other classes to prepare for, they often fill out their teacher-evaluation forms in a hurry so that they can leave class early. It may well be better, therefore, to select a sample of students from the class and interview them individually. This is the kind of situation in which a simple random sample should be obtained.

---

† This logo, a registered trademark of Minitab Inc., is used for cross referencing between the book and the *Minitab Supplement.* If you are not using that supplement, just ignore the logo.

During one semester, Professor Hassett wanted to sample the attitudes of the students taking college algebra at his school. He decided to interview 15 of the 728 students enrolled in the course. Since Professor Hassett had a registration list on which the 728 students were numbered 1–728, he could obtain a simple random sample of 15 students by randomly selecting 15 numbers between 1 and 728. To do this he used a table of random numbers. The random-number table employed by Professor Hassett is presented as Table IX in Appendix A. For ease of reference, we repeat it here as Table 1.9.

TABLE 1.9
Random numbers

Line number	Column number									
	00–09		10–19		20–29		30–39		40–49	
00	15544	80712	97742	21500	97081	42451	50623	56071	28882	28739
01	01011	21285	04729	39986	73150	31548	30168	76189	56996	19210
02	47435	53308	40718	29050	74858	64517	93573	51058	68501	42723
03	91312	75137	86274	59834	69844	19853	06917	17413	44474	86530
04	12775	08768	80791	16298	22934	09630	98862	39746	64623	32768
05	31466	43761	94872	92230	52367	13205	38634	55882	77518	36252
06	09300	43847	40881	51243	97810	18903	53914	31688	06220	40422
07	73582	13810	57784	72454	68997	72229	30340	08844	53924	89630
08	11092	81392	58189	22697	41063	09451	09789	00637	06450	85990
09	93322	98567	00116	35605	66790	52965	62877	21740	56476	49296
10	80134	12484	67089	08674	70753	90959	45842	59844	45214	36505
11	97888	31797	95037	84400	76041	96668	75920	68482	56855	97417
12	92612	27082	59459	69380	98654	20407	88151	56263	27126	63797
13	72744	45586	43279	44218	83638	05422	00995	70217	78925	39097
14	96256	70653	45285	26293	78305	80252	03625	40159	68760	84716
15	07851	47452	66742	83331	54701	06573	98169	37499	67756	68301
16	25594	41552	96475	56151	02089	33748	65289	89956	89559	33687
17	65358	15155	59374	80940	03411	94656	69440	47156	77115	99463
18	09402	31008	53424	21928	02198	61201	02457	87214	59750	51330
19	97424	90765	01634	37328	41243	33564	17884	94747	93650	77668

To select 15 random numbers between 1 and 728, we first pick a random starting point, say, by closing our eyes and putting our finger down on Table 1.9. Then beginning with the three digits under our finger, we go down the table and record the numbers as we go. Since we want numbers between 1 and 728 only, we discard the number 000 and numbers between 729 and 999. To avoid repetition we also eliminate numbers that have occurred previously. If not enough numbers have been found by the time we reach the bottom of the table, we move over to the next column of three-digit numbers and go up.

Using this procedure, Professor Hassett obtained 069, circled in Table 1.9, as a starting point. Reading down from 069 to the bottom of Table 1.9 and then up the next column of three-digit numbers, he found the 15 random numbers displayed in Fig. 1.4 and in Table 1.10.

FIGURE 1.4

Procedure used by
Professor Hassett to obtain
15 random numbers between 1
and 728 from Table 1.9

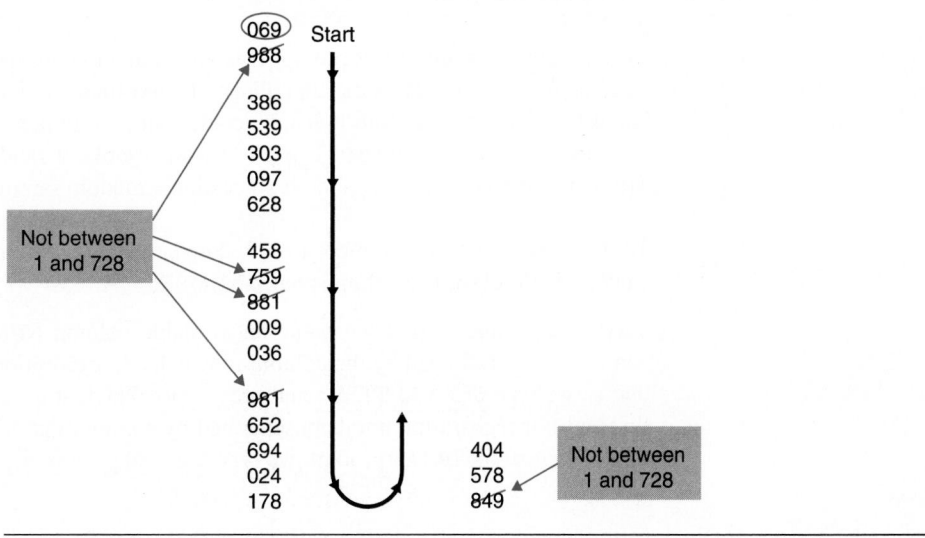

TABLE 1.10

Registration numbers
of students interviewed

69	303	458	652	178
386	97	9	694	578
539	628	36	24	404

Thus Professor Hassett interviewed the 15 students whose numbers on the registration list are the ones shown in Table 1.10. ■

Many calculators and most computers have a **random-number generator,** which makes it possible to automatically obtain a list of random numbers within any specified range. Random-number tables, such as Table 1.9, were used much more frequently before the availability of inexpensive machines with random-number generators. If you have a calculator or computer with a random-number generator, you may find it easier to use than a table of random numbers. Using random-number generators is discussed in the intermediate exercises at the end of this section.

Using the Computer (Optional)

Minitab can be employed to obtain a simple random sample from a population. The appropriate command for doing this is called **Sample.** Example 1.10 shows how the Sample command is applied.

Example 1.10   The Sample Command

Recall that in Example 1.9 Professor Hassett wanted to interview a simple random sample of 15 students from a class of 728 college-algebra students. Since the students were numbered 1–728 on the registration list, he could (and did) obtain a simple random sample by selecting 15 numbers between 1 and 728 from a table of random numbers. Explain how Minitab could have been used to obtain a simple random sample.

SOLUTION   To begin, we store the numbers 1–728, corresponding to the registration numbers of the students in the class, in a column named NUMBERS. An easy way to do that is the following.

Session commands: First we name an available column NUMBERS. Then we type the command SET followed by the column in which the registration numbers are to be stored; that is, we type SET 'NUMBERS' and press Enter. Next, at the DATA> prompt, we type the first and last registration numbers separated by a colon; that is, we type 1:728 and press Enter. Then, at the DATA> prompt, we type END and press Enter.

(or)

Menu commands: We choose **Edit ▸ Set Patterned Data...**, specify NUMBERS in the **Put data into column** text box, select **Patterned sequence**, type 1 in the **Start at** text box, type 728 in the **End at** text box, and then select **OK**.

Now we can obtain the required simple random sample by proceeding as follows.

Session commands: We first name an available column SRS, standing for "simple random sample." To obtain the simple random sample, we type the command **SAMPLE** followed by the sample size, the storage location of the population data, and the storage location for the sample data; that is, we type SAMPLE 15 from 'NUMBERS' put into 'SRS' and press Enter. Notice that the column names have apostrophes around them; the apostrophes must be included or the command will not work. The simple random sample of 15 registration numbers is now stored in SRS. We can observe that sample by typing PRINT 'SRS'. These commands and the resulting output are displayed in Printout 1.3.

(or)

Menu commands: To obtain the simple random sample, we choose **Calc ▸ Random Data ▸ Sample From Columns...**, type 15 for the number of rows (registration numbers) to be sampled, specify NUMBERS as the column to be sampled from, specify SRS in the **Put samples in** text box, and then select **OK**. To print the sample data, we choose **Edit ▸ Display Data...**, specify SRS in the text box, and then select **OK**. Printout 1.3 shows the resulting output.

PRINTOUT 1.3

Simple random sample obtained using Minitab's Sample command

```
MTB > SAMPLE 15 from 'NUMBERS' put into 'SRS'
MTB > PRINT 'SRS'

SRS
    147    509     18    252    630    423    678    376    577    558     35
    361    447    211    478
```

Note: In the first line of Printout 1.3, some of the words are in uppercase letters and some are in lowercase letters. Only the words in uppercase letters need be typed (by session-command users); the words in lowercase letters are included for purposes of clarity. Menu-command users will see something like `Sample 15 'NUMBERS' 'SRS'.` on the screen.

Printout 1.3 provides a list of numbers that Professor Hassett could have used as the registration numbers of the students to be interviewed. ■

EXERCISES 1.6

· **1.19** Explain why the following sample is not representative: A sample of 30 dentists from Seattle is taken in order to estimate the median income of all Seattle residents.

· **1.20** The political opinions of 150 voters in the retirement community of Sun City, Arizona, is used as a sample of the political opinions of all Arizona voters. Is the sample representative? Explain your answer.

· **1.21** The annual salaries of the five top Oklahoma state officials are displayed in Table 1.6 on page 20. Use that table to solve the following problems.
a. List the 10 possible samples (without replacement) of size three that can be obtained from the population of five salaries. Construct a table similar to Table 1.7 on page 20.
b. If a simple-random-sampling procedure is used to obtain a sample of three salaries, what are the chances that it is the first sample on your list in part (a)? the second sample? the tenth sample?

· **1.22** According to *Congressional Quarterly Weekly Report,* the current members of the U.S. House of Representatives from South Carolina are Arthur Ravenel, Jr. (AR), Floyd Spence (FS), Butler Derrick (BD), Bob Inglis (BI), John Spratt, Jr. (JS), and James Clyburn (JC).
a. List the 15 possible samples (without replacement) of two representatives that can be selected from the six. For brevity use the initials provided.
b. Describe a procedure for taking a simple random sample of two representatives from the six.
c. If a simple-random-sampling procedure is used to obtain two representatives, what are the chances of selecting AR and FS? BD and JC?

· **1.23** Refer to Exercise 1.22.
a. List the 15 possible samples (without replacement) of four representatives that can be selected from the six.
b. Describe a procedure for taking a simple random sample of four representatives from the six.

c. If a simple-random-sampling procedure is used to obtain four representatives, what are the chances of selecting AR, BI, JS, and JC? FS, BD, BI, and JC?

· **1.24** Refer to Exercise 1.22.
a. List the 20 possible samples (without replacement) of three representatives that can be selected from the six.
b. Describe a procedure for taking a simple random sample of three representatives from the six.
c. If a simple-random-sampling procedure is used to obtain three representatives, what are the chances of selecting AR, FS, and BD? BD, JS, and JC?

In each of Exercises 1.25–1.28, use Table IX in Appendix A to obtain the required list of random numbers.

· **1.25** The owner of a business that employs 685 people wants to select 25 of them at random for extensive interviewing. Construct a list of 25 random numbers between 1 and 685 that can be used in obtaining the required simple random sample.

· **1.26** A university committee on parking has been formed to gauge the sentiment of the people using the university's parking facilities. Each person that uses the facilities has a parking sticker with a number. This year the numbers range from 1 to 8493. Make a list of 30 numbers that can be employed to obtain a simple random sample of 30 people who use the parking facilities.

· **1.27** Each year *Fortune Magazine* publishes an article entitled "The International 500" that provides a ranking by sales of the top 500 firms outside the United States. Suppose you want to examine various characteristics of successful firms. Further suppose that for your study you decide to take a simple random sample of 10 firms from *Fortune Magazine*'s list of "The International 500." Determine 10 numbers you can use to obtain your sample.

1.28   In the game of keno, there are 80 balls, numbered 1–80, and 20 of the 80 balls are selected at random. Simulate one game of keno by obtaining 20 random numbers between 1 and 80.

1.29 (Computer exercise)   Refer to Exercise 1.27. Use Minitab or some other statistical software to determine 10 random numbers that can be used to obtain a simple random sample of 10 firms from *Fortune Magazine*'s list of "The International 500."

1.30 (Computer exercise)   Refer to Exercise 1.28. Use Minitab or some other statistical software to simulate one game of keno, that is, to obtain a simple random sample of 20 numbers between 1 and 80.

Random-number generators. As we mentioned earlier, a random-number generator makes it possible to automatically obtain a list of random numbers within any specified range.

Usually a random-number generator returns a number, r, between 0 and 1. To obtain random integers in an arbitrary range, A to B, use the conversion $A + (B - A + 1)r$ and round down to the nearest integer. For example, to obtain random integers between 1 and 728, use the conversion $1 + (728 - 1 + 1)r = 1 + 728r$ and round down to the nearest integer.

1.31   Refer to Exercise 1.25.
a. Explain how a random-number generator can be used to obtain 25 random numbers between 1 and 685, and thereby identify 25 employees to be interviewed.
b. Implement part (a).

1.32   Refer to Exercise 1.26.
a. Explain how a random-number generator can be used to obtain 30 random numbers between 1 and 8493, and thereby identify 30 people who use the parking facilities.
b. Implement part (a).

1.7   OTHER SAMPLING PROCEDURES

Simple random sampling is the most natural and easiest to understand probability-sampling method—it corresponds to our intuitive notion of random selection by lot. However, simple random sampling does have drawbacks. For instance, it may fail to provide sufficient coverage when information about subpopulations is required and may be impractical when the members of the population are widely scattered geographically. In this section we will examine some other commonly used sampling procedures that are often more appropriate than simple random sampling.

Systematic Random Sampling

One method that takes less effort to implement than simple random sampling is **systematic random sampling.** Consider Example 1.11.

Example 1.11   Systematic Random Sampling

Let's return to the situation in Example 1.9, where Professor Hassett wanted to obtain a sample of 15 of the 728 students enrolled in college algebra at his school. Use systematic random sampling to obtain the sample.

SOLUTION   To begin, we divide the population size by the sample size and round the answer down to the nearest whole number: $\frac{728}{15} = 48$ (rounded down). Next we select a number at random between 1 and 48 using, say, a table of random numbers. We did this and obtained the number 22. Then we list every 48th number, starting at 22, until we have 15 numbers. This yields the 15 numbers displayed in Table 1.11.

TABLE 1.11

Numbers obtained by
systematic random sampling

22	166	310	454	598
70	214	358	502	646
118	262	406	550	694

Had Professor Hassett used systematic random sampling to obtain his sample of students and had he obtained the number 22 as his starting point, he would have interviewed the 15 students whose numbers on the registration list are the ones shown in Table 1.11.   ■

As illustrated in Example 1.11, we implement systematic random sampling using three steps: (1) divide the population size by the sample size and round the result down to the nearest whole number, m; (2) use a random number table (or a similar device) to obtain a number, k, between 1 and m; (3) select for the sample those members of the population numbered $k, k + m, k + 2m, \ldots$.

Systematic random sampling is not only easier to execute than simple random sampling, but it also provides results comparable to simple random sampling unless there is some kind of cyclical pattern in the listing of the members of the population (e.g., male, female, male, female, ...). This phenomenon is relatively rare, though.

Cluster Sampling

Another alternative to simple random sampling is **cluster sampling.** This method is particularly useful when the members of the population are widely scattered geographically.

Example 1.12   Cluster Sampling

At one time the city council of Tempe, Arizona, was being pressured by citizens' groups to install bike paths in the city. The members of the council wanted to be sure they had the support of a majority of the taxpayers, so they decided to poll the homeowners in the city.

Their first attempt at surveying public opinion was a questionnaire mailed out with the city's 18,000 homeowner water bills. Unfortunately, this method did not work very well. Only 19.4% of the questionnaires were returned, and a large number of those had comments written on them indicating that they came from avid bicyclists or people strongly resenting bicyclists. The questionnaire generally had not been returned by the average voter, and the city council realized that.

The city had an employee in the planning department with sample-survey experience. The council called her in and asked her to do a survey. She was given two assistants to help interview a representative sample of voters and was instructed to report back in 10 days.

The planner thought about taking a simple random sample of 300 voters—100 interviews for herself and for each of her two assistants. However, using a simple random sample created some time problems. The city was so spread out that an interviewer with a list of 100 voters randomly scattered around the city would have to drive an average of 18 minutes from one interview to the next. This would require approximately 30 hours of driving time for each interviewer and could delay completion of the report. Obviously, simple random sampling would not do.

To save time the planner decided to use cluster sampling. The residential portion of the city was divided into 947 blocks, each containing approximately 20 houses, as seen in Fig. 1.5.

FIGURE 1.5
A typical block of homes

The planner numbered the blocks (clusters) on the city map from 1 to 947 and then used a table of random numbers to obtain a simple random sample of 15 of the 947 blocks. Each of the three interviewers was then assigned 5 of the 15 blocks obtained. This method gave each interviewer roughly 100 homes to visit but saved a great deal of travel time; an interviewer could work on a block for nearly a full day without having to drive to another neighborhood. The report was finished on time.   ∎

In the simplest case, as illustrated by Example 1.12, cluster sampling is implemented using three steps: (1) divide the population into groups (clusters); (2) obtain a simple random sample of the clusters; (3) use all of the members of the clusters obtained in step 2 as the sample.

Although cluster sampling can save time and money, it does have disadvantages. Ideally, each cluster should mirror the entire population. However, that is often not the case—members of a cluster are frequently more homogeneous than the members of the population as a whole. This situation can cause problems.

For instance, let's look at a simplified small town, as depicted in Fig. 1.6. The town council is thinking about building a town swimming pool. A planner for the town needs to sample voter sentiment on using public funds to build the pool. Many upper-income and middle-income homeowners will probably say "No" because they own pools or can use a neighbor's. Many low-income voters will probably say "Yes" because they generally do not have access to pools.

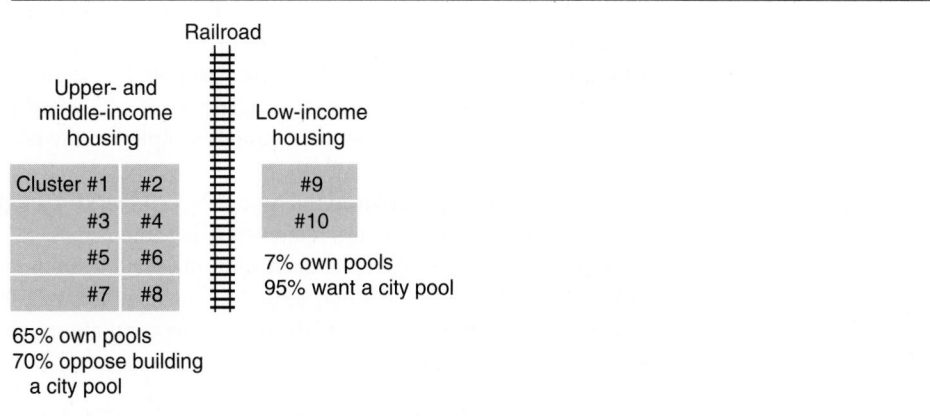

FIGURE 1.6

Clusters for a small town

If the planner uses cluster sampling and interviews the voters of, say, three randomly selected clusters, then there is a good chance that no low-income voters will be interviewed (because 46.7% of all possible three-cluster samples will contain neither of the low-income clusters, #9 and #10). And if no low-income voters are interviewed, the results of the survey will be misleading. Suppose, for instance, the planner obtained clusters #3, #5, and #8. Then his survey would show that only about 30% of the voters want a pool. But that is not true. More than 40% of the voters want a pool. The clusters that most strongly support the town swimming pool would not have been included in the survey.

In this hypothetical example, the town is so small that common sense indicates that a cluster sample may not be representative. However, in situations where there are hundreds of clusters, such problems may be more difficult to detect.

Stratified Sampling

Another sampling method, known as **stratified sampling,** is often more reliable than cluster sampling. In stratified sampling the population is first divided into subpopulations, called **strata,** and then sampling is done from each stratum. Ideally, the members of each stratum should be homogeneous relative to the characteristic under consideration.

Example 1.13   Stratified Sampling

Consider again the town-swimming-pool illustration. In stratified sampling, voters could be divided into three strata: upper-income, middle-income, and low-income. A simple random sample could then be taken from each of the three strata.

This stratified-sampling procedure would ensure that no income group is missed. It would also improve the precision of the statistical estimates (since the voters within each income group tend to be somewhat homogeneous) and would make it possible to estimate the separate opinions of each of the three strata.   ■

In stratified sampling the strata are often sampled in proportion to their size; this is called **proportional allocation.** For instance, suppose the strata consisting of the three income groups (upper, middle, and low) in Example 1.13 comprise, respectively, 10%, 70%, and 20% of the town. Then for a sample size of 50, say, the number of upper-income, middle-income, and low-income individuals sampled would be, respectively, 5 (10% of 50), 35 (70% of 50), and 10 (20% of 50).

The simplest type of stratified sampling, called **stratified random sampling with proportional allocation,** is implemented using three steps: (1) divide the population into subpopulations (strata); (2) from each stratum obtain a simple random sample of size proportional to the size of the stratum; that is, the sample size for a stratum equals the total sample size times the stratum size divided by the population size; (3) use all of the members obtained in step 2 as the sample.

Multistage Sampling

Most large-scale surveys combine simple random sampling, systematic random sampling, cluster sampling, and stratified sampling in ways that can be quite complex. Such **multistage sampling** is used quite frequently by pollsters and government agencies.

For instance, the U.S. National Center for Health Statistics conducts surveys of the civilian noninstitutional U.S. population to obtain information on illnesses, injuries, and other health issues. Data collection is by a multistage probability sample of approximately 42,000 households. Information obtained from the surveys is published in *National Health Interview Survey.*

EXERCISES 1.7

1.33 In Exercise 1.25 on page 25, we used simple random sampling to obtain 25 numbers between 1 and 685, and thereby identified 25 employees to be interviewed.
a. Employ systematic random sampling to accomplish that same task.
b. Which method is easier: simple random sampling or systematic random sampling?
c. Does it seem reasonable to use systematic random sampling to obtain a representative sample? Why or why not?

1.34 In Exercise 1.26 on page 25 we used simple random sampling to obtain 30 numbers between 1 and 8493, and thereby identified 30 people who use a university's parking facilities.
a. Employ systematic random sampling to accomplish the same thing.
b. Which method is easier: simple random sampling or systematic random sampling?
c. Does it seem reasonable to use systematic random sampling to obtain a representative sample? Why or why not?

1.35 Refer to Exercise 1.27 on page 25. Would it be reasonable to use systematic random sampling? Why?

1.36 Refer to Exercise 1.28 on page 26. Would it be reasonable to use systematic random sampling? Why?

1.37 Students in the dormitories of a university in the state of New York live in clusters of four double rooms, called *suites.* There are 48 suites, with eight students per suite.
a. Describe a cluster-sampling procedure for obtaining a sample of 24 dormitory residents.
b. Students typically choose friends from their classes as suitemates. With that in mind, do you think cluster sampling is a good procedure for obtaining a representative sample of dormitory residents? Explain your answer.
c. The university housing office has separate lists of dormitory residents by class level. Using those lists, the following table was obtained showing the number of dormitory residents at each class level.

Class level	Number of dorm residents
Freshman	128
Sophomore	112
Junior	96
Senior	48

Employ the table to design a procedure for obtaining a stratified sample of 24 dormitory residents. Use stratified random sampling with proportional allocation.

1.38 In simple random sampling, all samples of a given size are equally likely. Is that true in systematic random sampling? Explain your answer.

1.8 OBSERVATIONAL STUDIES AND DESIGNED EXPERIMENTS

Often the purpose of a statistical study is to investigate whether a relationship exists between two variables or characteristics, such as smoking and lung cancer, height and weight, or educational attainment and annual income. For such studies it is important to distinguish between two types of procedures: observational studies and designed experiments.

In an **observational study,** researchers simply observe; in a **designed experiment,** researchers design and control. Observational studies can reveal only *association,* whereas designed experiments can establish *causation* (cause-and-effect). Examples 1.14 and 1.15 illustrate the difference between observational studies and designed experiments.

Example 1.14   An Observational Study

Approximately 450,000 vasectomies are performed each year in the United States. In this surgical procedure for contraception, the tube carrying sperm from the testicles is cut. Several studies have been conducted to analyze the relationship between vasectomies and prostate cancer. One such study appeared in a February 1993 issue of *The Journal of the American Medical Association.* Dr. Edward Giovannucci, leader of the study and epidemiologist at Harvard-affiliated Brigham and Women's Hospital, said that "...we found 113 cases of prostate cancer among 22,000 men who had a vasectomy. This compares to a rate of 70 cases per 22,000 among men who didn't have a vasectomy."

Dr. Giovannucci's study shows about a 60% elevated risk of prostate cancer for men who have had a vasectomy, thereby revealing an association between vasectomy and prostate cancer. But does it establish causation: that having a vasectomy causes an increased risk of prostate cancer?

The answer is no, because the study is *observational.* Dr. Giovannucci simply observed two groups of men, one having vasectomies and the other not. Thus, although an association was established between vasectomy and prostate cancer, the association might be due to other factors (e.g., temperament) that make some men more likely to have vasectomies and also put them at greater risk of prostate cancer. In the words of Dr. Stuart Howards, a urology professor at the University of Virginia Medical School who did not participate in the study, "...[these results] have to be considered seriously but do not prove that vasectomy causes prostate cancer."   ■

Example 1.15   A Designed Experiment

For several years, evidence has been mounting that folic acid reduces major birth defects. In December 1992 *The Arizona Republic* reported on a Hungarian study that provides the strongest evidence yet. The results of the study, directed by Dr. Andrew E. Czeizel and Dr. Istvan Dudas of the National Institute of Hygiene in Budapest, were published in the *New England Journal of Medicine.*

For the study, the doctors enrolled 4753 women prior to conception. The women were divided randomly into two groups. One group took daily multivitamins containing 0.8 mg of folic acid, whereas the other group received only trace elements. A drastic reduction in the rate of major birth defects occurred among the women who took folic acid: 13 per 1000 as compared to 23 per 1000 for those women who did not take folic acid.

In contrast to the observational study considered in Example 1.14, this is a *designed experiment* and does help establish causation. The researchers did not simply observe two groups of women, but instead randomly assigned one group to take daily doses of folic acid and the other group to take only trace elements.   ∎

The study in Example 1.15 illustrates three basic principles of experimental design: control, randomization, and replication.

- *Control:* The doctors compared the rate of major birth defects for the women who took folic acid, called the *treatment group,* to that for the women who did not take folic acid, called the *control group.* This controlled for the effects of other factors, such as the *placebo effect,* where subjects respond to the idea of a treatment rather than to the treatment itself.

- *Randomization:* The women were divided randomly into two groups to avoid unintentional selection bias in constituting the groups and thereby help eliminate the problem of potential confounding factors such as life-style and emotional state.

- *Replication:* A large number of women were recruited for the study to make it likely that randomization would create treatment and control groups that were similar and to increase the chances of detecting an effect due to the folic acid if such an effect exists.

By using these three principles of experimental design, the doctors could conclude that a difference in the rates of major birth defects between the two groups not reasonably attributable to chance is likely caused by the folic acid.

In the folic-acid study, one group of women took folic acid and the other group took only trace elements. Both the folic acid and trace elements are referred to as treatments in the context of experimental design. Generally, each specific experimental condition is called a **treatment,** of which there may be several.

We summarize the above discussion in Key Fact 1.1.

KEY FACT 1.1	PRINCIPLES OF EXPERIMENTAL DESIGN

The following principles of experimental design enable a researcher to conclude that differences in the results of an experiment not reasonably attributable to chance are likely caused by the treatments.

- *Control:* Some method should be used to control for effects due to factors other than the ones of primary interest.

- *Randomization:* Subjects should be randomly divided into groups to avoid unintentional selection bias in constituting the groups, that is, to make the groups as similar as possible.

- *Replication:* A sufficient number of subjects should be used to ensure that randomization creates groups that resemble each other closely and to increase the chances of detecting differences among the treatments when such differences actually exist.

In this section we have introduced some of the basic terminology and principles of experimental design. However, we have just scratched the surface of this vast and important topic to which entire courses and books are devoted.

EXERCISES 1.8

1.39 Define the following terms:
a. observational study   b. designed experiment.

1.40 State and explain the significance of the three basic principles of experimental design.

1.41 Explain the important difference between the conclusion that can be drawn in an observational study and that which can be drawn in a designed experiment.

1.42 How could the study in Example 1.14 be modified to make it a designed experiment? Is this designed experiment feasible?

1.43 In the folic-acid study, 4753 women were enrolled by the doctors. Explain how a table of random numbers or a random-number generator could be used to divide the women randomly into two groups, one of size 2376 and the other of size 2377.

1.44 (Computer exercise) Use Minitab or some other statistical software to carry out the process described in Exercise 1.43.

In Exercises 1.45–1.48, identify each study as an observational study or a designed experiment. Justify your answers.

1.45 In the 1940s and early 1950s, there was great public concern over epidemics of polio. In an attempt to alleviate this serious problem, Jonas Salk of the University of Pittsburgh developed a vaccine for polio. Various preliminary experiments indicated that the vaccine was safe and potentially effective. Nonetheless, it was deemed necessary to conduct a large-scale study to determine whether the vaccine would truly work. A test involving nearly 2 million grade-school children was devised. All of the children were inoculated, but only half received the Salk vaccine; the other half were given a placebo, in this case an injection of salt dissolved in water. Neither the children nor the doctors performing the diagnoses knew which children belonged to which group. Instead an evaluation center kept records of who received the Salk vaccine and who did not. The center found that the incidence of polio was far less among the children inoculated with the Salk vaccine. From that information it was concluded that the Salk vaccine would be effective in preventing polio for all U.S. schoolchildren and, consequently, was then made available for general use.

1.46   According to a study published in the February 1993 issue of the *Journal of the American Public Health Association*, left-handed people do not die at an earlier age than right-handed people, contrary to the conclusion of a highly publicized report done two years prior. The investigation involved a 6-year study of 3800 people in East Boston older than age 65. Researchers at Harvard University and the National Institute of Aging found that the "lefties" and "righties" died at exactly the same rate. "There was no difference, period," said Dr. Jack Guralnik, an epidemiologist at the institute and one of the co-authors of the report.

1.47   Catherine E. Ross, a sociologist at the University of Illinois, presented the results of her research on mental depression in families at the 1993 annual meeting of the American Association for the Advancement of Science. Her results were based on a 1990 study of mental depression in 1000 families and showed in part that, on the average, the most depressed women are those that remain at home with their children, whereas the least depressed are those that have no children and a job. Ross emphasized that her conclusions concern averages, meaning, for example, that some mothers are quite happy staying home with their children. But she added that the data show that ". . . any plea to return to the 'traditional' family of the 1950s is a plea to return wives and mothers to a psychologically disadvantaged position in which husbands have much better mental health than wives."

1.48   For environmental purposes, efforts are being made to use industrial wastes on agricultural soils, since many such wastes contain nutrients that enhance crop growth. In a 1984 issue of *Environmental Pollution* (*Series A*), Mohammad Ajmal and Ahsan Ullah Khan reported their findings on experiments with brewery wastes used for agricultural purposes. The researchers studied the physico-chemical properties of effluent from Mohan Meakin Breweries Ltd, Ghazibad, UP, India, and ". . . its effects on the physico-chemical characteristics of agricultural soil, seed germination pattern, and the growth of two common crop plants." They assessed the impact using different concentrations of the effluent: 25%, 50%, 75%, and 100%. Various chemical properties of the treated soil were measured, in particular, available nutrients.

1.49   In Exercise 1.45 we discussed the Salk-vaccine experiment. As we mentioned, neither the children nor the doctors performing the diagnoses knew which children had been given the Salk vaccine and which had been given the placebo. This technique is called **double-blinding.** Explain the advantages of using double-blinding here.

Chapter Review

Key Terms

active cell,* *15*	probability sampling, *19*
cell,* *15*	proportional allocation, *30*
census, *18*	random-number generator, *23*
cluster sampling, *27*	random sampling, *19*
column,* *11*	randomization, *33*
control, *33*	Read,* *13*
Data screen,* *15*	READ,* *13*
descriptive statistics, *4*	replication, *33*
designed experiment, *31*	representative sample, *18*
END,* *12*	Restart,* *15*
inferential statistics, *6*	rows,* *15*
menu commands,* *10*	sample, *5*
multistage sampling, *30*	Sample,* *23*
NAME,* *14*	SAMPLE,* *24*
observational study, *31*	session commands,* *10*
population, *5*	Session window,* *15*

---

* An asterisk indicates material that is optionally covered.

You Should Be Able To

1. classify statistical studies as either descriptive or inferential.
2. identify the population and the sample in an inferential study.
3. explain what is meant by a representative sample.
4. describe simple random sampling, systematic random sampling, cluster sampling, and stratified sampling.
5. use a table of random numbers to obtain a simple random sample.
6. explain the difference between an observational study and a designed experiment.
7. classify a study concerning the relationship between two variables or characteristics as either an observational study or a designed experiment.
8. state the three basic principles of experimental design.
9. store and name a data set in Minitab.*
10. use the Minitab commands covered in this chapter.*

REVIEW TEST

1. Give an example of
   a. a descriptive study.   b. an inferential study.

2. Almost any inferential study involves aspects of descriptive statistics. Explain why this is so.

In Problems 3–7, classify each of the studies as either descriptive or inferential.

3. A newspaper displayed the following college football scores on the front page of the Sunday paper.

COLLEGE FOOTBALL			
Michigan St.	12	Arizona	12
ASU	3	Washington St.	7
BYU	31	Rutgers	28
Washington	3	Florida	28
Michigan	20	UCLA	26
Notre Dame	12	Tennessee	26

4. A National Institute of Mental Health survey concluded that "about 20% of adult Americans suffer from at least one psychiatric disorder." This and other estimates were obtained from results of interviews with thousands of Americans in St. Louis, Baltimore, and New Haven.

5. On February 14, 1985, the Census Bureau released a survey stating that "fewer Americans have health insurance coverage than previously thought." The survey, which was based on a multistage probability sample of 20,000 households, concluded that approximately 85% of the population is covered by health insurance—a far cry from the 97.3% figure found in a 1978 survey by the Department of Health and Human Services. By the way, the 1991 estimate given by the Census Bureau is that roughly 86.5% of the U.S. population is covered by health insurance.

6. A newspaper reporter doing library research for an article on civil aviation obtained the following data on fatalities. [SOURCE: U.S. Federal Aviation Administration.]

Year	1982	1983	1985	1986	1989
Fatalities	1480	1135	1267	1029	1123

7. The U.S. Bureau of Justice Statistics (BJS) conducts monthly surveys of approximately 60,000 U.S. house-

holds. These monthly surveys are then used to determine annual estimates of criminal victimization, which are subsequently published in *Criminal Victimization in the United States*. For example, the following table shows the victimization-rate estimates reported by the BJS for crimes against households between 1986 and 1990. Rates are per 1000 households.

Year	Burglary	Larceny	Motor vehicle theft
1986	61.5	93.5	15.0
1987	62.1	95.7	16.0
1988	61.9	90.2	17.5
1989	56.4	94.4	19.2
1990	53.8	86.7	20.5

*8. **(Computer problem)** A research physician conducted a study on the ages of people with diabetes. The following data were obtained for the ages of a sample of 35 diabetics.

48	41	57	83	41	55	59
61	38	48	79	75	77	7
54	23	47	56	79	68	61
64	45	53	82	68	38	70
10	60	83	76	21	65	47

Use Minitab or some other statistical software to store the data set in a column named AGES.

*9. **(Computer problem)** A *Consumer Reports* article revealed the following data on horsepower and gas mileage for a sample of 12 automobiles.

Horsepower	Mileage (mpg)
155	16.9
68	30.0
95	27.5
97	27.4
125	17.0
115	21.6
110	18.6
120	18.1
68	34.1
80	27.4
70	34.2
78	30.5

Use Minitab or some other statistical software to simultaneously store the horsepower and mileage data in columns named HP and MPG, respectively.

10. An economist wants to estimate the average income of parents of college students. To accomplish that, he surveys a sample of 250 students at Yale. Is this a representative sample? Why or why not?

11. Which of the following sampling procedures employ probability sampling?
    a. A college student is hired to interview a sample of voters in her town. She stays on campus and interviews 100 students in the cafeteria.
    b. A pollster wants to interview a sample of 20 gas-station managers in Baltimore. He posts a list of all such managers on his wall, closes his eyes, and tosses a dart at the list 20 times. He interviews the people whose names the dart hits.

12. The mayor of a small town having 7246 registered voters wants to send a detailed questionnaire to a simple random sample of 50 voters.
    a. Explain how Table IX in Appendix A can be employed to obtain the sample.
    b. Starting at the four-digit number in line number 14 and column numbers 16–19 of Table IX, read down the column, up the next, and so on to find 50 numbers that can be used to identify the voters who will receive the questionnaire.

*13. **(Computer problem)** Refer to Problem 12. Use Minitab or some other statistical software to obtain a simple random sample of 50 numbers from the numbers 1–7246, thus identifying 50 registered voters to receive the questionnaire. *(Note: Some statistical software will not have sufficient storage to carry out this exercise.)*

14. Refer to Problem 12.
    a. Use systematic random sampling to obtain 50 numbers from the numbers 1–7246, thereby identifying 50 registered voters to receive the questionnaire.
    b. In this case do you think systematic random sampling is an appropriate alternative to simple random sampling? Explain your answer.

15. The faculty at a university consists of 820 members. A new president has just been appointed. The president wants to get an idea of what the faculty considers the most important issues currently facing the school. She does not have time to interview all the faculty members and so decides to stratify the faculty by rank and use stratified random sampling with proportional allocation to obtain a sample of 40 faculty members. There are 205 full professors, 328 associate professors, 246 assistant professors, and 41 instructors.

a. How many of each rank should be selected for the interviewing?

b. Use Table IX in Appendix A to obtain the required sample. Explain your procedure in detail.

16. An article in the March 27, 1993, issue of *The Arizona Republic* reported on a study conducted by Greg Duncan of the University of Michigan. According to the report, "Persistent poverty during the first 5 years of life leaves children with IQs 9.1 points lower at age 5 than children who suffer no poverty during that period . . . ."

a. Identify the two variables under consideration.

b. Is this an observational study or is it a designed experiment?

Using the Focus Database

In Appendix B we have printed a database obtained by randomly selecting 500 Arizona State University sophomores. The report was created by the *Focus Student Database* on December 8, 1991. Seven variables are considered for each student: sex, high-school GPA, SAT math score, cumulative GPA, SAT verbal score, age, and total hours completed. For reference purposes we will call the database the **Focus database.**

We will employ this sample data in the "Using the Focus Database" exercises that appear at the end of each chapter. Large data sets such as these are almost always processed by computer, and that is how you should handle the "Using the Focus Database" exercises.

The Focus database is available in the file FOCUS.DAT residing in the FOCUS directory on *DataDisk,* a floppy disk containing hundreds of data sets found in the book. We will explain how to store the Focus database in Columns C1–C7 of Minitab installed on a DOS microcomputer (for other configurations consult your software documentation or your instructor). To begin, we assume Minitab has been accessed and *DataDisk* is in floppy drive A. Then we proceed as follows.

Session commands: We type the command READ followed by the file to be read and the storage locations to be used; that is, we type `READ 'A:\FOCUS\FOCUS.DAT' into C1-C7` and press `Enter`.

(or)

Menu commands: We choose **File ▶ Import ASCII Data...**, type `C1-C7` in the **Put data in column(s)** text box, select **Replace existing data, if any**, and then select **OK**. Next we type `A:\FOCUS\FOCUS.DAT` in the **File name** text box and again select **OK**.

The seven data sets in FOCUS.DAT are now stored in columns C1–C7. Notice that for the sex data, which is in C1, we have employed the coding 1 for female and 2 for male.

For ease of reference, it is useful to name the columns containing the data sets in the Focus database. The naming can be accomplished as explained on page 14. Using that or any other method, you should name columns C1–C7 as suggested in Table 1.12.

TABLE 1.12
Names for data sets in the Focus database

Column	Name
C1	SEX
C2	HS GPA
C3	SAT MATH
C4	CUM GPA
C5	SAT VERB
C6	AGE
C7	HOURS

Now that we have the data sets in the Focus database stored and named in Minitab, we will use Minitab's **Save** command to save the worksheet, which includes both the data sets and their names, to a file named FOCUS.MTW.[†] Then any time we want to consider the Focus database, we need only access Minitab and apply the **Retrieve** command. Here, first, are the details for using the Save command. We remove *DataDisk* from drive A and then proceed in the following manner.

Session commands: We type the command **SAVE** followed by the name of the file in which we want the worksheet saved; that is, we type `SAVE 'FOCUS.MTW'` and press Enter.

(or)

Menu commands: We choose **File ▶ Save Worksheet As...**, select **Minitab worksheet** from the choices presented in the **Save As** list box, select **Select File**, type `FOCUS.MTW` in the **File name** text box, and then select **OK**.

The data sets in the Focus database and their names are now stored in the file FOCUS.MTW. That information can be recovered at any time by employing the Retrieve command as follows.

Session commands: We type the command **RETRIEVE** followed by the name of the worksheet file; that is, we type `RETRIEVE 'FOCUS.MTW'` and press Enter.

(or)

Menu commands: We choose **File ▶ Open Worksheet...**, select **Minitab worksheet** from the choices presented in the **Open** list box, select **Select File**, specify FOCUS.MTW in the **File name** text box, and then select **OK**.

Case Study   Are Americans Lazy?

In early 1992, Yoshio Sakurauchi, speaker of the Japanese Parliament lower house, called American workers lazy. Two weeks later, Prime Minister Kiichi Miyazawa commented to his Parliament that Americans are losing their work ethic. Japan subsequently apologized for the prime minister's remark.

But those controversial statements prompted the Associated Press to commission ICR Survey Research Group of Media, Pennsylvania, to conduct a nationwide poll to ask U.S. adults their views on these and related issues. The poll was taken February 12–16 and involved 1009 adults nationwide. Figure 1.7 depicts a pie chart that was used to summarize the results of the survey.

As we see from Fig. 1.7, a majority of Americans disagree with the views held by the two Japanese officials. In fact, 72% of those surveyed felt that most American workers are hard-working, whereas only 17% thought that most American workers are lazy.

a. Identify the population of interest in the Associated Press poll.
b. Identify the sample.
c. The results of the Associated Press poll appeared in *The Arizona Republic* on February 26, 1992. In the newspaper article, the following statement was made: "... 72 percent of Americans in the poll said they would describe most American workers as hard-working...." Is this an inferential statement, or is it simply descriptive? What if the statement had been "... 72 percent of Americans would describe most American workers as hard-working...."?

[†] In what follows, we will assume you have access to a hard disk; if not, consult your instructor.

FIGURE 1.7

Pie chart summarizing
Associated Press poll

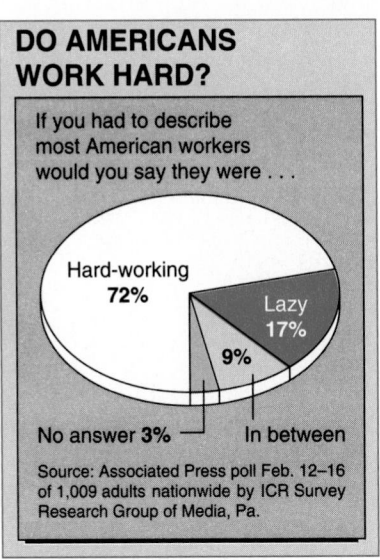

Figure 1.7 Pie chart summarizing Associated Press poll

DO AMERICANS WORK HARD?

If you had to describe most American workers would you say they were . . .

Hard-working 72%

Lazy 17%

9%

No answer 3% In between

Source: Associated Press poll Feb. 12–16 of 1,009 adults nationwide by ICR Survey Research Group of Media, Pa.

Descriptive Statistics

Adolphe Quetelet

Lambert Adolphe Jacques Quetelet was born in Ghent, Belgium, on February 22, 1796. He attended school locally and in 1819 received the first doctorate of science degree granted at the newly established University of Ghent. In that same year, he obtained a position as a professor of mathematics at the Brussels Athenaeum.

Quetelet was elected to the Belgian Royal Academy in 1820 and served as its secretary from 1834 until his death in 1874. He was founder and director of the Royal Observatory in Brussels, founder and a major contributor to the journal *Correspondance mathématique et physique,* and, according to Stephen M. Stigler in *The History of Statistics,* "... active in the founding of more statistical organizations than any other individual in the nineteenth century." Among the organizations he established was the International Statistical Congress, initiated in 1853.

When Quetelet conceived the idea of an observatory in Brussels, he knew nothing of practical astronomy; in 1823, after obtaining government support for the observatory, he went to the Observatory of Paris to learn about astronomy and how to run an observatory. It was during the 3 months he spent there that he became acquainted with the two eminent scientists Charles Fourier and Pierre Laplace and first heard of probability and its applications. This incidental knowledge was the beginning of his work as a statistician.

The observatory was not completed until 1833. During the intervening years, Quetelet scrutinized masses of census data and other recorded data by constructing tables and drawing diagrams.

In 1835 Quetelet wrote a two-volume set entitled *A Treatise on Man and the Development of His Faculties,* the publication in which he introduced his concept of the "average man" and that firmly established his international reputation as a statistician and sociologist. A review in the *Athenaeum* stated, "We consider the appearance of these volumes as forming an epoch in the literary history of civilization."

In 1855 Quetelet suffered a stroke that limited his work but not his popularity. He died on February 17, 1874. His funeral was attended by royalty and famous scientists from around the world. A monument to his memory was erected in Brussels in 1880.

Chapter 2

Organizing Data

. . . . .

General Objective As we discovered in Chapter 1, descriptive statistics consists of methods for organizing and summarizing information clearly and effectively. In this chapter we will begin our study of descriptive statistics. Specifically, we will learn how to classify data by type, organize data into tables, and summarize data with graphical displays. And because graphical displays can often be misleading, we will examine ways of analyzing and interpreting them carefully.

Chapter Outline

Case Study *Infant Mortality in Developed Nations*

How do the developed nations compare with regard to infant mortality? Are infant mortality rates much the same from country to country, or do they vary substantially? Where does the United States rank? A report by the National Commission to Prevent Infant Mortality, which we will examine in the case study at the end of this chapter, answers these and other questions about infant mortality.

2.1   VARIABLES AND DATA

A characteristic that varies from one person or thing to another is called a **variable.** Examples of variables for humans are height, weight, number of siblings, sex, marital status, and eye color. The first three of these variables yield numerical information and are examples of **quantitative variables;** the last three yield nonnumerical information and are examples of **qualitative variables.**

Quantitative variables can be classified as either discrete or continuous. A **discrete variable** is one whose possible values form a finite (or countably infinite[†]) set of numbers, usually some collection of whole numbers. The number of siblings a person has is an example of a discrete variable. A **continuous variable** is a variable whose possible values form some interval of numbers. The height of a person is an example of a continuous variable.

The preceding discussion is summarized in Definition 2.1 and in the diagram that follows the definition.

DEFINITION 2.1   **VARIABLES**

> *Variable:* A characteristic that varies from one person or thing to another.
>
> *Qualitative variable:* A nonnumerically valued variable.
>
> *Quantitative variable:* A numerically valued variable.
>
> *Discrete variable:* A quantitative variable whose possible values form a finite (or countably infinite) set of numbers.
>
> *Continuous variable:* A quantitative variable whose possible values form some interval of numbers.

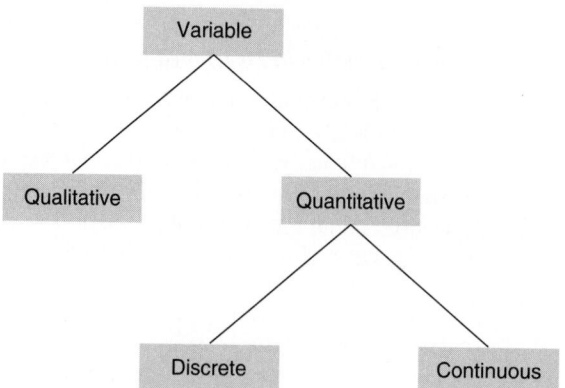

Observing the values of a variable for one or more people or things yields **data.** Thus the information collected, organized, and analyzed by statisticians is data. The terms *qualitative, quantitative, discrete,* and *continuous* are used to describe data as well as variables:

[†]   A countably infinite set is one that can be put in one-to-one correspondence with the positive integers.

qualitative data are data obtained by observing values of a qualitative variable; quantitative data are data obtained by observing values of a quantitative variable; and so forth.

DEFINITION 2.2 **DATA**

> *Data:* Information obtained by observing values of a variable.
> *Qualitative data:* Data obtained by observing values of a qualitative variable.
> *Quantitative data:* Data obtained by observing values of a quantitative variable.
> *Discrete data:* Data obtained by observing values of a discrete variable.
> *Continuous data:* Data obtained by observing values of a continuous variable.

Example 2.1 Variables and Data

At noon on April 20, 1992, more than 9600 men and women set out to run from Hopkinton Center to the John Hancock Building in Boston. Their run, covering 26 miles and 385 yards, would be watched by thousands of people lining the streets leading into Boston and by millions more on television. It was the 96th running of the Boston Marathon.

A great deal of information was accumulated and recorded that afternoon by the Boston Athletic Association. The men's competition was won by Ibraham Hussein of Kenya with a time of 2 hours, 8 minutes, and 14 seconds. The winner of the women's competition was Olga Markova of Russia; her time was 2 hours, 23 minutes, and 43 seconds. There were 6562 men and 1561 women who finished the marathon before the official cutoff time of 5 hours.

The Boston Marathon provides examples of different types of variables (and data). The simplest type is illustrated by the classification of each entrant as either male or female. "Sex" is a qualitative variable because its possible values (male or female) are nonnumerical. Thus, for instance, the information that Olga Markova is a female is qualitative data—data obtained by observing the value of the variable "sex" for Olga Markova.

Most racing fans are interested in the places of the finishers. "Place of finish" is a quantitative variable, which is also a discrete variable because there are only a finite number of possible finishing places. The information that among the men, Ibraham Hussein and Joaquim Pinheiro finished first and second, respectively, and Noel Robinson and Bob Ellis finished 500th and 501st, respectively, is discrete, quantitative data—data obtained by observing the values of the variable "place of finish" for the four runners.

More can be learned about what happened in a race by looking at the times of the finishers. For instance, Ibraham Hussein finished 2 minutes and 25 seconds ahead of Joaquim Pinheiro, but Noel Robinson beat Bob Ellis by only 1 second. Differences between times indicate exactly how far apart two runners finished, whereas differences in places do not. "Finishing time" is a quantitative variable, which is also a continuous variable because the finishing time of a runner can conceptually be any positive real number. The information that Ibraham Hussein ran his race in 2:08:14 and Olga Markova ran hers in 2:23:43 is continuous, quantitative data—data obtained by observing the values of the variable "finishing time" for Ibraham Hussein and Olga Markova. ∎

Example 2.2   Variables and Data

Humans are classified as having one of four blood types: A, B, AB, or O. What kind of data do you receive when you are told your blood type?

SOLUTION   Your blood type is qualitative data, data obtained by observing your value of the variable "blood type." ∎

Example 2.3   Variables and Data

The U.S. Bureau of the Census collects data on household size and publishes the information in *Current Population Reports*. What kind of data is the number of people in your household?

SOLUTION   The number of people in your household is discrete, quantitative data—data obtained by observing the value of the variable "household size" for your household. ∎

Example 2.4   Variables and Data

The *Information Please Almanac* lists the world's highest waterfalls. The list shows that Angel Falls in Venezuela is 3281 feet high, more than twice as high as Ribbon Falls in Yosemite, California, which is 1612 feet high. What kind of data are these heights?

SOLUTION   The waterfall heights are continuous, quantitative data—data obtained by observing the values of the variable "height" for the two waterfalls. ∎

Classification and the Choice of a Statistical Method

Some of the descriptive and inferential procedures that we will study are valid for only certain types of data (and variables); that is one reason why it is important to be able to correctly classify data. Statisticians use other classifications besides the ones presented here. But the types we have discussed are sufficient for the majority of applications.

Data classification is sometimes difficult; statisticians themselves occasionally disagree over data type. For example, some classify data involving amounts of money as discrete data, whereas others classify such data as continuous data. In most cases, however, the appropriate classification of data is fairly clear and may aid in the choice of the correct statistical method.

Variables and Populations

Suppose we are studying heights of Americans. We can think of the population under consideration as all Americans; or we can think of it as the heights of all Americans, that is, as all the possible values of the variable "height" for Americans. Selecting a sample from this latter population is equivalent to observing the values of the variable "height" for a sample from the former population. Although this distinction is subtle, it is important to keep in mind because sometimes we think of the population as the people and sometimes as the heights of the people. These comments apply in general, not just to heights and Americans.

EXERCISES 2.1

2.1 Give a reason why the classification of data is important.

For each part of Exercises 2.2–2.9, classify the data as either qualitative or quantitative; if quantitative, further classify it as discrete or continuous. Also identify the variable under consideration.

2.2 According to Sidney S. Culbert of the University of Washington, the principal languages of the world in 1993 were as follows.

Rank	Language	Speakers (millions)
1	Mandarin	930
2	English	463
3	Hindi	400
4	Spanish	371
5	Russian	291
6	Arabic	214
7	Bengali	192

a. What type of data is given in the first column of the table?
b. What type of data is provided by the information that Sally Ride speaks English?
c. What type of data is provided by the information in the third column of the table?

2.3 As reported by the U.S. Department of Agriculture in *Agricultural Statistics,* tobacco production in the United States for the years 1986–1991 is as displayed in the following table. What type of data is provided by the second column of the table?

Year	Pounds (millions)
1986	1164
1987	1226
1988	1370
1989	1367
1990	1625
1991	1638

2.4 On May 4, 1961, Commander Malcolm Ross, U.S. Naval Reserves, ascended 113,739.9 ft in a free balloon. What kind of data is the height given here?

2.5 According to *Employment and Earnings,* published by the U.S. Bureau of Labor Statistics, figures on industrial employment in the United States for 1991 are as displayed in the following table. What kind of data is given by the employee numbers?

Industry	Employees (millions)
Agriculture	3.2
Mining	0.7
Construction	7.1
Manufacturing	20.4
Trade	24.1
Services	39.7

2.6 What type of data is provided by the numbers in the following table?

DO YOUR WORRIES MATCH THOSE OF THE EXPERTS?

Experts and laypeople were asked to rank the risk of dying in any year from various activities and technologies. The experts' ranking closely matches known fatality statistics.

Public		Experts
1	Nuclear power	20
2	Motor vehicles	1
3	Handguns	4
4	Smoking	2
5	Motorcycles	6
6	Alcoholic beverages	3
7	General (private) aviation	12
8	Police work	17
9	Pesticides	8
10	Surgery	5
11	Fire fighting	18
12	Large construction	13
13	Hunting	23
14	Spray cans	26
15	Mountain climbing	29
16	Bicycles	15
17	Commercial aviation	16
18	Electric power (nonnuclear)	9
19	Swimming	10
20	Contraceptives	11
21	Skiing	30
22	X rays	7
23	High school and college football	27
24	Railroads	19
25	Food preservatives	14
26	Food coloring	21
27	Power mowers	28
28	Prescription antibiotics	24
29	Home appliances	22
30	Vaccinations	25

Note: The table above was reprinted by permission from the October issue of *Science 85.* Copyright © 1985 by the American Association for the Advancement of Science.

2.7 The following table of continental statistics provides approximate land areas and 1992 population estimates. [SOURCE: Population Reference Bureau, Inc.]

Continent	Land area (1000 sq mi)	Population (millions)
Asia	10,644	3207
Africa	11,707	654
North America	9,360	436
South America	6,883	300
Europe	1,905	511
Oceania	3,284	28

a. What type of data is provided by the land area figures displayed in the second column of the table?

b. What type of data is contained in the statement, "Africa is largest in land area and second largest in population"?

c. What type of data is provided by the population figures shown in the third column of the table?

d. What type of data do we obtain from the fact that Marie Curie was born in Europe?

2.8 The following table displays the rank by population and the population in thousands of the 10 largest cities in the United States in 1980 and 1990. [SOURCE: U.S. Bureau of the Census, *Census of Population.*]

City	1980		1990	
	Rank	Population (thous.)	Rank	Population (thous.)
New York	1	7,072	1	7,323
Los Angeles	3	2,967	2	3,485
Chicago	2	3,005	3	2,784
Houston	5	1,595	4	1,631
Philadelphia	4	1,688	5	1,586
San Diego	8	876	6	1,111
Detroit	6	1,203	7	1,028
Dallas	7	904	8	1,007
Phoenix	9	790	9	983
San Antonio	11	786	10	936

a. What type of data is provided by the statement, "In 1980 Houston was the fifth largest city in the United States"?

b. What type of data is given in the 1990 "Population" column of the table?

c. What type of data is provided by the information that Theodore Roosevelt was born in New York?

·   **2.9** The *American Banker* reports that the five largest U.S. commercial banks, by deposits, as of December 31, 1990, were as follows.

Bank	Rank	Deposits ($millions)
Citibank NA, New York	1	112,586
Bank of America NT&SA, San Francisco	2	77,027
Chase Manhattan Bank NA, New York	3	59,862
Security Pacific National Bank, Los Angeles	4	45,376
Wells Fargo Bank NA, San Francisco	5	42,716

a. What kind of data is displayed in the second column of the table?

b. What kind of data is given in the third column of the table?

·   **2.10** What kinds of data would be collected in each of the following situations?

a. A quality-control engineer measures the lifetimes of electric light bulbs.

b. A businessperson wants to know the number of families with preteen children in Pueblo, Colorado.

c. A manufacturer of sporting goods classifies each major-league baseball player as either right-handed or left-handed and counts the number of players in each category.

d. A sociologist needs to estimate the average annual income of the residents of Ossining, New York.

e. A pollster plans to classify each individual in a sample of voters as Democrat or Republican and count the total number in each group.

f. An administrator at a community college needs to know how many men and women participated in varsity sports during the spring semester and how much money was spent on men's sports and on women's sports.

·   **2.11** Of the variables we have studied so far, which type yields nonnumerical data?

Ordinal data. Another important type of data is **ordinal data,** data about order or rank given on a scale such as 1, 2, 3, . . . or A, B, C, ...

· ·   **2.12** In each of Exercises 2.2–2.9, identify ordinal data, if any.

· ·   **2.13** Following are several variables. Which, if any, yield ordinal data? Explain your answer.

a. height   b. weight

c. age   d. sex

e. number of siblings   f. religion

g. place of birth   h. high-school class rank

2.2 GROUPING DATA

When we studied data types in Section 2.1, we used examples that did not present large quantities of data. But the amount of data collected in real-world situations can sometimes be overwhelming. For example, a list of U.S. colleges and universities with information on enrollment, number of teachers, highest degree offered, and governing official can be found in *The World Almanac*. These data occupy 28 pages of small type!

By suitably organizing data, we can often make a large and complicated set of data more compact and easier to understand. In this section we will discuss **grouping,** which involves, as the term implies, putting data into groups rather than treating each piece of data individually. Grouping is one of the most common methods for organizing data.

Example 2.5   Grouping Data

Table 2.1 displays the number of days to maturity for 40 short-term investments. The data are from *Barron's National Business and Financial Weekly*.

TABLE 2.1

Days to maturity for
40 short-term investments

70	64	99	55	64	89	87	65
62	38	67	70	60	69	78	39
75	56	71	51	99	68	95	86
57	53	47	50	55	81	80	98
51	36	63	66	85	79	83	70

It is difficult to get a clear picture of the data in Table 2.1. By grouping the data into categories, or **classes,** we can make it much simpler to comprehend. The first step is to decide on the classes. One convenient way to group these data is by 10s. Since the smallest piece of data is 36 and the largest is 99, grouping by 10s results in the classes 30–39, 40–49, and so on up to 90–99. These classes are given in the first column of Table 2.2.

Days to maturity	Tally	Number of investments
30–39	|||	3
40–49			1
50–59	ⅢⅡ |||	8
60–69	ⅢⅡ ⅢⅡ	10
70–79	ⅢⅡ ||	7
80–89	ⅢⅡ ||	7
90–99	||||	4
		40

The second (and final) step in grouping the data is to determine how many investments are in each class. We do this by going through the data in Table 2.1 and placing a tally mark in the appropriate line of Table 2.2 for each investment. For instance, the first investment in Table 2.1 has a 70-day maturity period. This calls for a tally mark on the line for the class 70–79. The results of the tallying procedure are shown in the second column of Table 2.2. Now we count the tallies for each class and record the totals in the third column of Table 2.2.

By simply glancing at Table 2.2, we can obtain various pieces of useful information. For instance, we see that more investments are in the 60–69 days-to-maturity range than in any other. Comparing Tables 2.1 and 2.2, we see that grouping the data makes it much easier to read and understand.   ■

In Example 2.5 we used a commonsense approach to grouping data into classes. Some of that common sense can be used as guidelines for grouping. Three of the most important guidelines follow.

1. *The number of classes should be small enough to provide an effective summary but large enough to display the relevant characteristics of the data.*

In Example 2.5, seven classes are used. Usually, the number of classes should be between 5 and 20; but that is only a rule of thumb.

2. Each piece of data must belong to one, and only one, class.

Careless planning in Example 2.5 could have led to classes like 30–40, 40–50, 50–60, and so on. Then, for instance, to which class would the investment with a 50-day maturity period belong? The classes in Table 2.2 do not cause such confusion; they cover all maturity periods and do not overlap.

3. Whenever feasible, all classes should have the same width.

The classes in Table 2.2 all have a width of 10 days. Among other things, choosing classes of equal width facilitates the graphical display of the data.

The list could go on, but for our purposes these three guidelines provide a solid basis for grouping data. And we should always keep in mind that the reason for grouping is to organize the data into a sensible number of classes in order to make the data more accessible and understandable.

Frequency and Relative-Frequency Distributions

The number of pieces of data that fall into a particular class is called the **frequency** of that class. For example, as we see from Table 2.2, the frequency of the class 50–59 is eight, since eight investments are in the 50–59 days-to-maturity range. A table listing all classes and their frequencies is called a **frequency distribution.** The first and third columns of Table 2.2 constitute a frequency distribution for the days-to-maturity data.

In addition to the frequency of a class, we are often interested in the **percentage** of a class. We find the percentage by dividing the frequency of the class by the total number of pieces of data and multiplying the result by 100. Referring again to Table 2.2, we see that the percentage of investments in the class 50–59 is

$$\frac{8}{40} = 0.20 \text{ or } 20\%.$$

Thus 20% of the investments have a maturity period between 50 and 59 days, inclusive.

The percentage of a class, expressed as a decimal, is called the **relative frequency** of the class. For the class 50–59, the relative frequency is 0.20. A table listing all classes and their relative frequencies is called a **relative-frequency distribution.** Table 2.3 displays a relative-frequency distribution for the days-to-maturity data. Notice that the relative frequencies sum to 1 (100%).

TABLE 2.3

Relative-frequency distribution for the days-to-maturity data in Table 2.1

Days to maturity	Relative frequency		
30–39	0.075	←	3/40
40–49	0.025	←	1/40
50–59	0.200	←	8/40
60–69	0.250	←	10/40
70–79	0.175	←	7/40
80–89	0.175	←	7/40
90–99	0.100	←	4/40
	1.000		

When comparing two data sets, relative-frequency distributions are better than frequency distributions. This is because relative frequencies are always between 0 and 1 and hence provide a standard for comparison. Two data sets having identical frequency distributions will, of course, have identical relative-frequency distributions, but two data sets having identical relative-frequency distributions will have identical frequency distributions only if both data sets have the same number of pieces of data.

Grouping Terminology

To become adept at grouping, you must first become familiar with and understand the various terms associated with it. We have already discussed several of these terms. To introduce some additional ones, let's return to the days-to-maturity data. Consider, for example, the class 50–59. The smallest maturity period that can go in this class is 50. This value is called the **lower class limit** of the class. The largest maturity period that can go in this class is 59. This value is called the **upper class limit** of the class.

The midpoint of the class 50–59 is $(50 + 59)/2 = 54.5$, and this is called the **class mark** of the class. Class marks provide single numbers for representing classes and are sometimes used in graphical displays and in computing descriptive measures.

The width of the class 50–59, obtained by subtracting its lower class limit from the lower class limit of the next higher class, is $60 - 50 = 10$. This is called the **class width** of the class.

Definition 2.3 summarizes the terminology of grouping.

DEFINITION 2.3   **TERMS USED IN GROUPING**

> *Classes:* Categories for grouping data.
>
> *Frequency:* The number of pieces of data in a class.
>
> *Frequency distribution:* A listing of classes and their frequencies.
>
> *Relative frequency:* The ratio of the frequency of a class to the total number of pieces of data.
>
> *Relative-frequency distribution:* A listing of classes and their relative frequencies.
>
> *Lower class limit:* The smallest value that can go into a class.
>
> *Upper class limit:* The largest value that can go into a class.
>
> *Class mark:* The midpoint of a class.
>
> *Class width:* The difference between the lower class limit of the given class and the lower class limit of the next higher class.

A table giving the classes, frequencies, relative frequencies, and class marks for a data set is called a **grouped-data table.** A grouped-data table for the days-to-maturity data is presented in Table 2.4.

TABLE 2.4
Grouped-data table for
the days-to-maturity data

Days to maturity	Frequency (no. of investments)	Relative frequency	Class mark
30–39	3	0.075	34.5
40–49	1	0.025	44.5
50–59	8	0.200	54.5
60–69	10	0.250	64.5
70–79	7	0.175	74.5
80–89	7	0.175	84.5
90–99	4	0.100	94.5
	40	1.000	

TABLE 2.4
Grouped-data table for
the days-to-maturity data

Example 2.6   Grouped-Data Tables

A pediatrician who tested the cholesterol levels of several young patients was alarmed to find that many had levels over 200 mg per 100 mL. The readings of 20 patients with high cholesterol levels are presented in Table 2.5. Construct a grouped-data table. Use a class width of five and start at 195.

TABLE 2.5
Cholesterol levels for
20 high-level patients

210	209	212	208
217	207	210	203
208	210	210	199
215	221	213	218
202	218	200	214

SOLUTION   Because we are to use a class width of five and start at 195, the first class will be 195–199. From Table 2.5 we see that the highest cholesterol level is 221. Thus we choose the classes displayed in the first column of Table 2.6.

TABLE 2.6
Classes and frequencies for the
cholesterol-level data in Table 2.5

Cholesterol level	Tally	Frequency
195–199	I	1
200–204	III	3
205–209	IIII	4
210–214	LHI II	7
215–219	IIII	4
220–224	I	1
		20

The results of tallying the data in Table 2.5 are shown in the second and third columns of Table 2.6. From Table 2.6 we obtain the grouped-data table given in Table 2.7.

TABLE 2.7

Grouped-data table for
the cholesterol-level data

Cholesterol level	Frequency	Relative frequency	Class mark
195–199	1	0.05	197
200–204	3	0.15	202
205–209	4	0.20	207
210–214	7	0.35	212
215–219	4	0.20	217
220–224	1	0.05	222
	20	1.00	

To illustrate some typical computations for the third and fourth columns of Table 2.7, let's consider the class 210–214. We have

$$\text{relative frequency} = \frac{7}{20} = 0.35$$

and

$$\text{class mark} = \frac{210 + 214}{2} = 212.$$

■

Single-Value Grouping

Up to this point, each class we have used for grouping data represents several possible numerical values. For instance, in Table 2.7 each cholesterol-level class represents five possible cholesterol levels. The first class, 195–199, is for levels of 195, 196, 197, 198, or 199; the second class, 200–204, is for levels of 200, 201, 202, 203, or 204; and so on.

In some cases, however, it is more appropriate to use classes that each represent a single possible numerical value. This is often true for discrete data. Consider, for instance, Example 2.7.

Example 2.7   Single-Value Grouping

A city planner is collecting data on the number of school-age children per family in a small town. Thirty families are selected at random. Table 2.8 displays the number of school-age children in each of the 30 families chosen.

TABLE 2.8

Number of school-age children
in each of 30 families

0	3	0	0	3	0
2	2	0	1	2	1
0	0	1	2	4	0
4	2	1	0	1	0
0	2	0	1	3	2

a. Group these data using classes that each represent a single numerical value.

b. Identify the class limits and the class marks.

c. Construct a grouped-data table.

SOLUTION   a. We first note that since each class is to represent a single numerical value, the classes must be 0, 1, 2, 3, and 4. These are displayed in the first column of Table 2.9.

TABLE 2.9

Frequency and relative-frequency distributions for number of school-age children

Number of school-age children	Frequency	Relative frequency
0	12	0.400
1	6	0.200
2	7	0.233
3	3	0.100
4	2	0.067
	30	1.000

Tallying the data in Table 2.8, we obtain the frequencies in the second column of Table 2.9. Dividing each frequency by the total number of pieces of data, 30, yields the relative frequencies shown in the third column of Table 2.9. The table indicates, for example, that 7 of the 30 families, or 23.3%, have two school-age children.

b. For this part we are to identify the class limits and class marks. Consider, for instance, the class "3" (i.e., three school-age children). We have

lower class limit = 3 (the smallest value that can go into the class),

upper class limit = 3 (the largest value that can go into the class),

and

$$\text{class mark} = \frac{3+3}{2} = 3 \text{ (the midpoint of the class).}$$

Thus for the class "3," the lower class limit, the upper class limit, and the class mark are all equal to 3. A similar statement is true for the other classes.

c. Here we are to construct a grouped-data table. This requires us to append a class-mark column to Table 2.9. However, by part (b), the class mark for each class is the same as the class itself. Thus it is unnecessary to include a class-mark column in Table 2.9, since such a column would be identical to the first column. In other words, Table 2.9 can serve as a grouped-data table. ■

We now summarize the important points made about single-value grouping in Example 2.7: When a class for grouping data represents a single numerical value, then that value is also the lower class limit, the upper class limit, and the class mark for the class. If every class for grouping a data set is based on a single value, then the class-mark column may be omitted from the grouped-data table because the first column gives the class marks as well as the classes.

Another Way of Grouping Data

It is often convenient to group data using classes that consist of values from one number up to, but not including, another number. This is particularly true when dealing with continuous data in which one or more decimal places have been retained. Example 2.8 shows such a case.

Example 2.8　Another Way to Group Data

The U.S. National Center for Health Statistics publishes data on weights and heights by age and sex in *Vital and Health Statistics.* The weights in Table 2.10, given to the nearest tenth of a pound, were obtained from a sample of 18–24-year-old males. Group the weights using the classes "120–under 140," "140–under 160," and so on.

TABLE 2.10
Weights of 37 males, age 18–24 years

129.2	185.3	218.1	182.5	142.8	155.2	170.0	151.3
187.5	145.6	167.3	161.0	178.7	165.0	172.5	191.1
150.7	187.0	173.7	178.2	161.7	170.1	165.8	
214.6	136.7	278.8	175.6	188.7	132.1	158.5	
146.4	209.1	175.4	182.0	173.6	149.9	158.6	

SOLUTION　The class "120–under 140" is for weights of at least 120 pounds, but less than 140 pounds; the class "140–under 160" is for weights of at least 140 pounds, but less than 160 pounds; and so forth. These classes are displayed in the first column of Table 2.11.

TABLE 2.11
Grouped-data table for the weights of 37 males, age 18–24 years

Weight (lb)	Frequency	Relative frequency	Class mark
120–under 140	3	0.081	130
140–under 160	9	0.243	150
160–under 180	14	0.378	170
180–under 200	7	0.189	190
200–under 220	3	0.081	210
220–under 240	0	0.000	230
240–under 260	0	0.000	250
260–under 280	1	0.027	270
	37	0.999	

Applying the tallying procedure to the data in Table 2.10, we obtain the frequencies in the second column of Table 2.11. The relative frequencies, given in the third column of Table 2.11, are then found in the usual manner.

Finally, let's discuss the class marks. For a class of the form "a–under b," we define the class mark to be $(a + b)/2$. For instance, the class mark for the class "120–under 140" is $(120 + 140)/2 = 130$. Similar calculations yield the remaining class marks shown in the fourth column of Table 2.11. ∎

As we know, the relative frequencies must always sum to 1 (100%). However, the sum of the relative frequencies in the third column of Table 2.11 is given as 0.999. The reason for this discrepancy is that each relative frequency is rounded to three decimal places; and when those rounded relative frequencies are added, the resulting sum differs from 1 by a little. This phenomenon is usually referred to as *rounding error* or *roundoff error*.

Frequency and Relative-Frequency Distributions for Qualitative Data

Although the concepts of class limits and class marks are applicable to quantitative data, they are not appropriate for qualitative data. For instance, with data that categorize people as male or female, the classes are "male" and "female." For qualitative-data classes such as these, it makes no sense to look for class limits or class marks. We can, of course, still compute frequencies and relative frequencies for qualitative data. Example 2.9 provides an illustration.

Example 2.9   Frequency and Relative-Frequency Distributions for Qualitative Data

Professor Weiss asked his introductory statistics students to state their political party affiliations as Democratic (D), Republican (R), or Other (O). The responses are given in Table 2.12. Determine the frequency and relative-frequency distributions for these data.

TABLE 2.12
Political party affiliations of the students in introductory statistics

D	R	O	R	R	R	R	R
D	O	R	D	O	O	R	D
D	R	O	D	R	R	O	R
D	O	D	D	D	R	O	D
O	R	D	R	R	R	R	D

SOLUTION   The classes for grouping the data are "Democratic," "Republican," and "Other." Tallying the data in Table 2.12, we obtain the frequency distribution displayed in the first two columns of Table 2.13.

TABLE 2.13
Frequency and relative-frequency
distributions for political
party affiliations

Party	Frequency	Relative frequency
Democratic	13	0.325
Republican	18	0.450
Other	9	0.225
	40	1.000

Dividing each frequency in the second column of Table 2.13 by the total number of students (40), we get the relative frequencies in the third column. The first and third columns of Table 2.13 constitute the relative-frequency distribution for the data.   ∎

Statistical software packages, such as Minitab, contain programs for automatically obtaining frequency and relative-frequency distributions. Interested readers should consult the *Minitab Supplement.*

EXERCISES 2.2

2.14   Identify an important reason for grouping data.

2.15   Do class limits and class marks make sense for qualitative-data classes? Explain your answer.

2.16   State the three most important guidelines in choosing the classes for grouping a data set.

2.17   Explain the difference between each of the following pairs of terms.
a. frequency and relative frequency
b. percentage and relative frequency

2.18   A research physician conducted a study on the ages of people with diabetes. The following data were obtained for the ages of a sample of 35 diabetics.

48	41	57	83	41	55	59
61	38	48	79	75	77	7
54	23	47	56	79	68	61
64	45	53	82	68	38	70
10	60	83	76	21	65	47

Construct a grouped-data table for these ages. Use classes of equal width, beginning with the class 0–9.

2.19   A soft-drink bottler sells "one-liter" bottles of soda. A consumer group is concerned that the bottler may be shortchanging customers. Thirty bottles of soda are randomly selected. The contents, in milliliters, of the bottles obtained are shown below.

1025	977	1018	975	977
990	959	957	1031	964
986	914	1010	988	1028
989	1001	984	974	1017
1060	1030	991	999	997
996	1014	946	995	987

Construct a grouped-data table for these soft-drink data. Use classes of equal width, starting with the class 910–929.

2.20   The Bureau of Economic Analysis gathers information on the length of stay in Europe and the Mediterranean by U.S. travelers. Data are published in *Survey of Current Business.* A sample of 36 U.S. residents who traveled to Europe and the Mediterranean one year yielded the following data, in days, on length of stay.

41	16	6	21	1	21
5	31	20	27	17	10
3	32	2	48	8	12
21	44	1	56	5	12
3	13	15	10	18	3
1	11	14	12	64	10

Using classes of equal width and beginning with the class 1–7, construct a grouped-data table for these data on length of stay.

2.21 The U.S. Energy Information Administration collects data on residential energy consumption and expenditures. Results are published in the document *Residential Energy Consumption Survey: Consumption and Expenditures*. The following table gives one year's energy consumptions for a sample of 50 households in the South. Data are in millions of BTUs.

130	55	45	64	155	66	60	80	102	62
58	101	75	111	151	139	81	55	66	90
97	77	51	67	125	50	136	55	83	91
54	86	100	78	93	113	111	104	96	113
96	87	129	109	69	94	99	97	83	97

Using classes of equal width and starting with the class 40–49, construct a grouped-data table for these data on energy consumption.

2.22 The U.S. Bureau of the Census conducts nationwide surveys on characteristics of U.S. households. Following are data on the number of people per household for a sample of 40 households.

2	5	2	1	1	2	3	4
1	4	4	2	1	4	3	3
7	1	2	2	3	4	2	2
6	5	2	5	1	3	2	5
2	1	3	3	2	2	3	3

Construct a grouped-data table for these household sizes using classes based on a single value.

2.23 A car salesperson keeps track of the number of cars she sells per week. The number of cars she sold per week last year are as follows.

1	0	3	3	1	0	2	1	4	0	4	1	2
3	6	4	3	0	2	2	1	1	2	2	2	3
5	1	0	2	5	3	1	3	1	1	1	1	2
2	3	0	4	4	1	0	1	1	3	2	5	2

Construct a grouped-data table for the number of sales per week. Use classes based on a single value.

2.24 A fifth-grade class consisting of 25 students was given an eight-question quiz on fractions. The number of incorrect answers for each student follows.

3	4	2	3	1
1	1	6	1	2
0	5	1	2	1
0	3	4	0	3
2	0	3	1	1

Construct an appropriate grouped-data table.

2.25 Cudahey Masonry employs 80 bricklayers. The number of days each employee misses is recorded. Absentee records for the past year are as follows.

2	3	6	2	6	5	5	2	4	7	5	3	6	4	4	4
2	2	4	5	4	0	2	1	6	3	5	3	6	6	4	7
5	2	5	0	5	6	5	2	4	2	6	2	4	3	5	4
2	4	4	3	3	4	0	5	6	3	5	5	2	4	4	2
0	7	5	5	7	6	1	5	3	3	4	7	7	2	5	5

Construct an appropriate grouped-data table for these data.

2.26 The Food and Nutrition Board of the National Academy of Sciences states that the recommended daily allowance of iron is 18 mg for adult females under the age of 51. The amounts of iron intake, in milligrams, during a 24-hour period for a sample of 45 such females follows.

15.0	18.1	14.4	14.6	10.9	18.1	18.2	18.3	15.0
16.0	12.6	16.6	20.7	19.8	11.6	12.8	15.6	11.0
15.3	9.4	19.5	18.3	14.5	16.6	11.5	16.4	12.5
14.6	11.9	12.5	18.6	13.1	12.1	10.7	17.3	12.4
17.0	6.3	16.8	12.5	16.3	14.7	12.7	16.3	11.5

Construct a grouped-data table for these iron intakes. Use classes of equal width, beginning with the class 6–under 8.

2.27 *The Northwestern Endicott-Lindquist Report* provides data on starting salaries for college graduates. A sample of 35 liberal-arts graduates yielded the following starting annual salaries. Data are in thousands of dollars, rounded to the nearest hundred dollars.

20.0	16.8	21.3	20.6	21.0	18.7	23.8
18.3	17.7	18.0	19.1	21.1	19.6	19.0
18.7	20.8	20.4	17.1	19.5	19.9	19.2
19.1	17.2	18.3	22.7	20.0	19.2	20.9
19.1	20.8	20.5	21.4	16.3	16.3	20.5

Using classes of equal width and beginning with the class 16–under 17, construct a grouped-data table for these starting annual salaries.

The following Associated Press article about the New York Stock Exchange appeared in a major metropolitan newspaper. We will use the article in Exercises 2.28 and 2.29.

DOLLAR LEADERS

New York (AP)—The following is a list of the most active NYSE stocks based on the dollar volume. The total is based on the median price of the stock traded multiplied by the shares traded.

	Tot. ($1000)	Sales (hds.)	Last
Reyl pfc	$1,009,720	80536	127
IBM	$705,516	55011	127 ½
RchVck	$523,317	10755	48 ¾
SCM	$358,257	49245	72 ⅝
Revlon	$271,385	63113	43
WstgE	$261,517	68148	38 ⅜
CocaCl	$259,958	37607	68 ⅜
GnFds	$251,132	29545	83 ¼
Digital	$218,483	20636	105 ⅜
RckCtr n.........	$217,851	09611	19 ⅞
CessAir..........	$206,341	73366	29 ½
GMot	$193,443	28500	68 ¼
PhilMr	$173,890	22258	75 ⅝
AmExp	$158,911	37949	42
McDnld	$148,393	22273	65
MCA	$143,096	20190	68
GenEl	$142,523	23509	60
Exxon	$140,513	27087	51 ⅛
AtlRich	$138,506	23181	59
MartM s	$137,368	38832	33 ⅝
Morgn s	$126,692	26672	46 ⅝
Burrgh	$126,061	19469	63 ⅛
BeafCo..........	$125,221	36695	34 ⅜
CBS	$120,761	10234	117 ⅛

2.28 The column headed "Tot." in the article shows the total dollar volumes in thousands of dollars for the most active New York Stock Exchange (NYSE) stocks.

a. Group these totals using the classes "100–under 150," "150–under 200," ..., "350–under 400," and "400 and over," where the values are in millions of dollars (e.g., the class "100–under 150" is for volumes of at least $100 million but less than $150 million).

b. Why is there no class mark for the last class?

2.29 The "Sales" column of the Associated Press article displays the numbers of shares traded for the most active New York Stock Exchange (NYSE) stocks. The sales are given in hundreds. Group these sales data into a grouped-data table using classes of equal width. Begin with the class "0–under 1," where the values are in millions of sales.

2.30 The following table depicts the all-time top TV programs by number of viewers as of January 1992. [SOURCE: A. C. Nielsen.]

Program	Date	Network	Households
M*A*S*H Special	2/28/83	CBS	50,150,000
Super Bowl XX	1/26/86	NBC	41,490,000
Dallas	11/21/80	CBS	41,470,000
Super Bowl XVII	1/30/83	NBC	40,480,000
Super Bowl XXI	1/25/87	CBS	40,030,000
Super Bowl XVI	1/24/82	CBS	40,020,000
Super Bowl XIX	1/20/85	ABC	39,390,000
Super Bowl XXIII	1/22/89	NBC	39,320,000
Super Bowl XVIII	1/22/84	CBS	38,800,000
The Day After	11/20/83	ABC	38,550,000
Roots Pt. VIII	1/30/77	ABC	36,380,000
Super Bowl XIV	1/20/80	CBS	35,330,000
Super Bowl XIII	1/21/79	NBC	35,090,000
Super Bowl XV	1/25/81	NBC	34,540,000
Super Bowl XII	1/15/78	CBS	34,410,000
Gone with the Wind Pt. 1	11/7/76	NBC	33,960,000
Gone with the Wind Pt. 2	11/8/76	NBC	33,750,000
Roots Pt. VI	1/28/77	ABC	32,680,000
Roots Pt. V	1/27/77	ABC	32,540,000
Roots Pt. III	1/25/77	ABC	31,900,000

Construct a frequency and relative-frequency distribution for the network data. *(Hint:* The classes are "ABC," "CBS," and "NBC.")

2.31 According to *The World Almanac, 1993,* the National Collegiate Athletic Association wrestling champions for the years 1963–1992 are as follows.

Year	Champion	Year	Champion
1963	Oklahoma	1978	Iowa
1964	Oklahoma State	1979	Iowa
1965	Iowa State	1980	Iowa
1966	Oklahoma State	1981	Iowa
1967	Michigan State	1982	Iowa
1968	Oklahoma State	1983	Iowa
1969	Iowa State	1984	Iowa
1970	Iowa State	1985	Iowa
1971	Oklahoma State	1986	Iowa
1972	Iowa State	1987	Iowa State
1973	Iowa State	1988	Arizona State
1974	Oklahoma	1989	Oklahoma State
1975	Iowa	1990	Oklahoma State
1976	Iowa	1991	Iowa
1977	Iowa State	1992	Iowa

Construct a frequency and relative-frequency distribution for the champions.

2.32 The exam scores for the students in an introductory statistics class are as follows.

88	82	89	70	85
63	100	86	67	39
90	96	76	34	81
64	75	84	89	96

a. Group these exam scores using the classes 30–39, 40–49, 50–59, 60–69, 70–79, 80–89, and 90–100.
b. What are the widths of the classes?
c. If you wanted all the classes to have the same width, what classes would you use?

Contingency tables. The methods we have examined in this section apply to grouping data obtained from observing values of one variable. Such data are called **univariate.** For instance, in Example 2.8 we considered data obtained from observing the values of the variable "weight" for a sample of 18–24-year-old males. Consequently, those data are univariate. We could have considered not only the weights of the males, but also their heights. Then we would have data on two variables, "height" and "weight." Data obtained from observing values of two variables are called **bivariate.** Bivariate data can be grouped using tables called **contingency tables.** Exercises 2.33 and 2.34 deal with the grouping of bivariate data using contingency tables.

2.33 The following bivariate data on age (in years) and sex were obtained from the students in a freshman calculus course. The data show, for example, that the first student on the list is 21 years old and is a male.

Age	Sex	Age	Sex	Age	Sex	Age	Sex	Age	Sex
21	M	29	F	22	M	23	F	21	F
20	M	20	M	23	M	44	M	28	F
42	F	18	F	19	F	19	M	21	F
21	M	21	M	21	M	21	F	21	F
19	F	26	M	21	F	19	M	24	F
21	F	24	F	21	F	25	M	24	F
19	F	19	M	20	F	21	M	24	F
19	M	25	M	20	F	19	M	23	M
23	M	19	F	20	F	18	F	20	F
20	F	23	M	22	F	18	F	19	M

We will discuss the grouping of these data into the following contingency table.

	Age (yrs)			
	Under 21	**21–25**	**Over 25**	**Total**
Male		|		
Female				
Total				

(Sex labels the rows Male and Female)

a. To tally the data for the first student, place a tally mark in the box labeled by the "21–25" column and the "Male" row, as indicated. Tally the data for the other 49 students.
b. Construct a table like the one in part (a) but with frequencies replacing the tally marks. Add the frequencies in each row and column of your table and record the sums in the proper "Total" boxes.
c. What do the row and column totals represent?
d. Add the row totals and add the column totals. Why are those two sums equal, and what does their common value represent?
e. Construct a table giving the relative frequencies for the data. *(Hint:* Divide each frequency obtained in part (b) by the grand total of 50 students.*)*
f. Interpret the entries in your table from part (e) in terms of percentages.

2.34 The heights (in inches) and weights (in pounds) of the students in Exercise 2.33 are as follows.

Height	Weight	Height	Weight	Height	Weight
68	140	67	155	74	215
67	140	67	130	67	129
72	145	68	160	72	275
69	145	64	127	68	135
66	115	74	170	60	95
72	185	73	180	75	175
64	130	63	142	61	120
65	145	69	170	73	180
62	127	62	103	64	125
69	135	68	160	66	130
66	110	75	185	63	105
69	178	64	122	69	155
63	130	64	130	70	170
72	185	70	215	64	105
67	120	63	105	65	132
68	135	76	200	65	115
64	130	71	169		

Repeat parts (a)–(f) of Exercise 2.33 for the bivariate data on heights and weights. Use a contingency table with classes for weight of equal width 40 starting with 90–129 and classes for height of equal width 6 starting with 60–65.

2.3   GRAPHS AND CHARTS

Besides grouping, another method for organizing and summarizing data is to draw a picture of some kind. The old saying "a picture is worth a thousand words" has particular relevance in statistics—a graph or chart of a data set often provides the simplest and most efficient display. In this section we will examine various techniques for organizing and summarizing data using graphs and charts. We begin by discussing *histograms.*

Example 2.10   Histograms

Table 2.4 showed a grouped-data table for the number of days to maturity for 40 short-term investments. The first three columns of that table are repeated in Table 2.14. Obtain graphical displays for these grouped data.

TABLE 2.14
Frequency and relative-frequency distributions for the days-to-maturity data

Days to maturity	Frequency (no. of investments)	Relative frequency
30–39	3	0.075
40–49	1	0.025
50–59	8	0.200
60–69	10	0.250
70–79	7	0.175
80–89	7	0.175
90–99	4	0.100
	40	1.000

SOLUTION   One way to display these grouped data pictorially is to construct a graph with the classes depicted on the horizontal axis and the frequencies depicted on the vertical axis. This can be done using a **frequency histogram.** A frequency histogram for the days-to-maturity data is shown in Fig. 2.1(a).

The height of each bar is equal to the frequency of the class it represents. Notice that each bar extends from the lower class limit of a class to the lower class limit of the next higher class.[†]

Some other important observations about Fig. 2.1(a) are that each axis of the frequency histogram has a label, and the frequency histogram as a whole has a title. All graphical displays should possess these elements.

A frequency histogram displays the frequencies of the classes. To display the relative frequencies (or percentages), we can use a **relative-frequency histogram,** which is similar to a frequency histogram. The only difference is that the height of each bar in a relative-frequency histogram is equal to the relative frequency of the class instead of

---

[†] This is only one of several methods for depicting the classes. Additional methods are discussed in Exercises 2.64 and 2.65.

FIGURE 2.1
Days-to-maturity:
(a) frequency histogram
(b) relative-frequency histogram

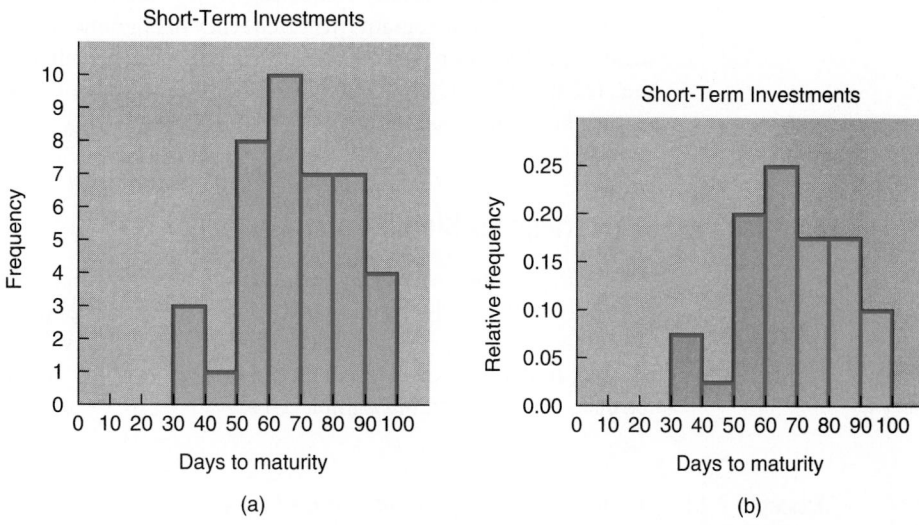

the frequency of the class. A relative-frequency histogram for the days-to-maturity data is shown in Fig. 2.1(b).

Notice that the shapes of the relative-frequency histogram in Fig. 2.1(b) and the frequency histogram in Fig. 2.1(a) are identical. This is because the frequencies and relative frequencies are in the same proportion.

Finally, we present a word of caution. We have used the lower class limits, 30, 40, 50, ..., to label the horizontal axes of the histograms in Figs. 2.1(a) and (b), since it is natural to do so. However, you must be careful when reading histograms. For instance, the bar for 80 to 90 represents only maturity periods in the 80s. A maturity period of 90 days is included in the next bar to the right. ∎

With Example 2.10 in mind, we now formally present the definitions of frequency histogram and relative-frequency histogram.

| DEFINITION 2.4 | **FREQUENCY AND RELATIVE-FREQUENCY HISTOGRAMS** |

Frequency histogram: A graph that displays the classes on the horizontal axis and the frequencies of the classes on the vertical axis. The frequency of each class is represented by a vertical bar whose height is equal to the frequency of the class.

Relative-frequency histogram: A graph that displays the classes on the horizontal axis and the relative frequencies of the classes on the vertical axis. The relative frequency of each class is represented by a vertical bar whose height is equal to the relative frequency of the class.

For purposes of visually comparing the distributions of two data sets, it is better to use relative-frequency histograms than frequency histograms. This is because the same vertical scale is used for all relative-frequency histograms, a minimum of 0 and a maximum of 1 (i.e., 0% to 100%). On the other hand, the vertical scale of a frequency histogram depends on the number of pieces of data.

Histograms for Single-Value Grouping

For the days-to-maturity data, each class represents 10 possible days to maturity, and in Figs. 2.1(a) and (b) the histogram bar for each class extends over those 10 possible days. When data are grouped using classes based on a single value, we proceed somewhat differently. In that case each bar is placed directly over the only possible numerical value in the class, as illustrated in Example 2.11.

Example 2.11   Histograms for Single-Value Grouped Data

In Example 2.7 we considered data on the number of school-age children in each of 30 families. We grouped those data using classes based on a single value. The frequency and relative-frequency distributions were given in Table 2.9 and are repeated in Table 2.15. Construct a frequency histogram and a relative-frequency histogram for these grouped data.

TABLE 2.15
Frequency and relative-frequency distributions for number of school-age children

Number of school-age children	Frequency	Relative frequency
0	12	0.400
1	6	0.200
2	7	0.233
3	3	0.100
4	2	0.067
	30	1.000

SOLUTION   As we just mentioned, for single-value grouping, we place the middle of each histogram bar directly over the single numerical value represented by the class. Hence the frequency and relative-frequency histograms for the grouped data in Table 2.15 are as pictured in Figs. 2.2(a) and (b).

Notice the symbol // on the horizontal axes in Figs. 2.2(a) and (b). This symbol indicates that the zero point on that axis is not in its usual position at the intersection of the horizontal and vertical axes. Whenever any such modification is made, whether on the horizontal or vertical axis, the symbol // or some similar symbol should be used to indicate that fact.   ■

FIGURE 2.2
School-age children:
(a) frequency histogram
(b) relative-frequency histogram

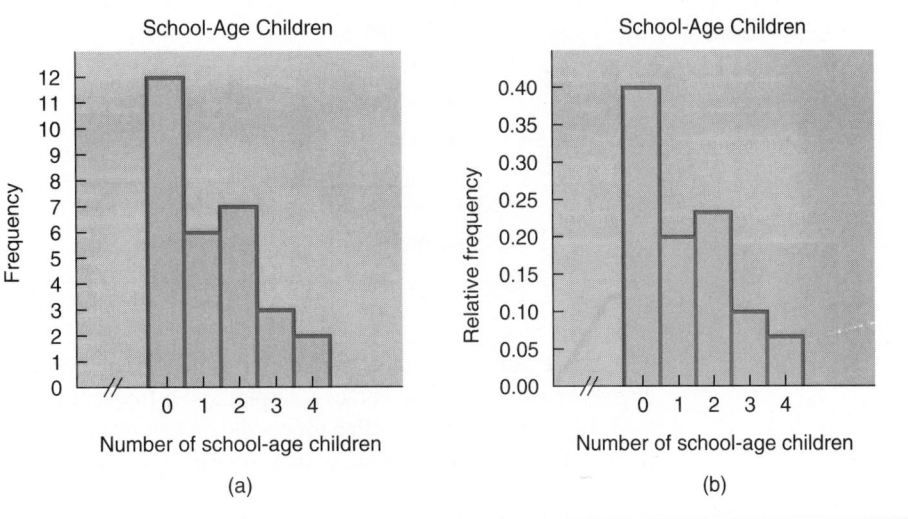

FIGURE 2.2
School-age children:
(a) frequency histogram
(b) relative-frequency histogram

Dotplots

Another type of graphical display for quantitative data is the **dotplot.** Dotplots are particularly useful for showing the relative positions of the data in a data set or for comparing two or more data sets. We introduce dotplots in Example 2.12.

Example 2.12   Dotplots

A farmer is interested in a newly developed fertilizer that supposedly will increase his yield of oats. He uses the fertilizer on a sample of 15 one-acre plots. The yields, in bushels, are depicted in Table 2.16. Construct a dotplot for the data.

TABLE 2.16
Oat yields

67	65	55	57	58
61	61	61	64	62
62	60	62	60	67

SOLUTION

To construct a dotplot for the data in Table 2.16, we begin by drawing a horizontal axis that displays the possible oat yields. Then we record each yield by placing a dot over the appropriate value on the horizontal axis. For instance, the first yield is 67 bushels. This calls for a dot over the "67" on the horizontal axis. The dotplot for the data in Table 2.16 is pictured in Fig. 2.3.   ■

FIGURE 2.3
Dotplot for oat yields

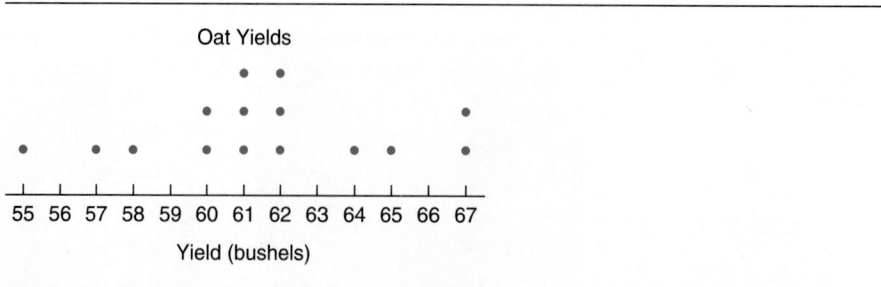

As we can see, dotplots are similar to histograms. In fact, when data are grouped in classes based on a single value, a dotplot and a frequency histogram are essentially identical. However, for single-value grouped data that involve decimals, dotplots are generally preferable to histograms because they are easier to construct and use.

Graphical Displays for Qualitative Data

Histograms and dotplots are designed for use with quantitative data. Qualitative data are portrayed using different techniques. Two common methods for displaying qualitative data graphically are *pie charts* and *bar graphs.*

Example 2.13   Pie Charts and Bar Graphs

In Example 2.9 we obtained frequency and relative-frequency distributions for the political party affiliations of the students in Professor Weiss's introductory statistics class. We repeat those distributions in Table 2.17.

TABLE 2.17
Frequency and relative-frequency distributions for political party affiliations

Party	Frequency	Relative frequency
Democratic	13	0.325
Republican	18	0.450
Other	9	0.225
	40	1.000

Display the relative-frequency distribution of these qualitative data using

a. a pie chart.   b. a bar graph.

SOLUTION   a. A **pie chart** is a disk divided into pie-shaped pieces proportional to the relative frequencies. In this case we need to divide a disk into three pie-shaped pieces comprising 32.5%, 45.0%, and 22.5% of the disk. We can do this by using a protractor and the fact that there are 360° in a circle. Thus, for instance, the first piece of the disk is obtained by marking off 117° (32.5% of 360°). The pie chart for the relative-frequency distribution in Table 2.17 is shown in Fig. 2.4(a).

FIGURE 2.4

Political party affiliations:
(a) pie chart
(b) bar graph

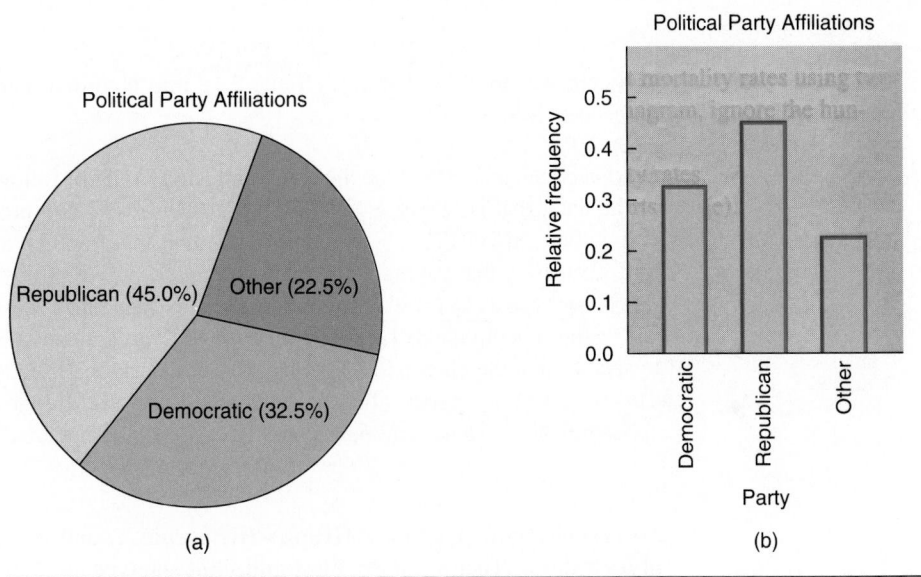

(a)   (b)

b. A **bar graph** is like a histogram except that its bars do not abut each other. The bar graph for the relative-frequency distribution in Table 2.17 is pictured in Fig. 2.4(b).   ∎

Histograms, dotplots, pie charts, and bar graphs are only a few of the countless ways that data can be portrayed pictorially. We will consider some additional graphical displays in Exercises 2.66–2.69.

Using the Computer (Optional)

Minitab has a command called **Histogram** that can be applied to simultaneously group a data set and obtain a frequency histogram for the data. To illustrate we return once again to the days-to-maturity data from Table 2.1.

Example 2.14   The Histogram Command

The days to maturity for 40 short-term investments are repeated in Table 2.18 at the top of the next page. Use Minitab to construct a frequency distribution and a frequency histogram for these data based on the classes 30–39, 40–49, . . . , 90–99.

TABLE 2.18
Days to maturity for
40 short-term investments

70	64	99	55	64	89	87	65
62	38	67	70	60	69	78	39
75	56	71	51	99	68	95	86
57	53	47	50	55	81	80	98
51	36	63	66	85	79	83	70

SOLUTION   To begin we store the data from Table 2.18 into a column named MATURITY. Then we proceed as follows.

Session commands: We type the command **HISTOGRAM** followed by the storage location of the data; that is, we type `HISTOGRAM 'MATURITY';` and press Enter. Notice the semicolon after `'MATURITY'`. This tells Minitab that there will be a *subcommand* on the next line. To specify that the first class mark is to be 34.5, we employ the **START** subcommand; that is, we type `START=34.5;` and press Enter. Notice that a semicolon again appears at the end of the subcommand. This is because we will use another subcommand, **INCREMENT**, to specify that the class width is to be 10. Thus we type `INCREMENT=10.` and press Enter. This time we put a period after the subcommand to tell Minitab that there will be no more subcommands.[†] These commands and the resulting output are displayed in Printout 2.1.

(or)

Menu commands: We choose **Graph ▶ Histogram...** and specify MATURITY in the **Variables** text box. Then we select **First midpoint** and type `34.5` to indicate that the first class mark is to be 34.5. Next we select **Interval width** and type `10` to indicate that the class width is to be 10. Then, if high resolution is enabled, we select **High resolution** and press Spacebar to disable it. Finally, we select **OK**.[‡] Printout 2.1 shows the output that results.

The column headed `Midpoint` in Printout 2.1 gives the class marks corresponding to the classes 30–39, 40–49, . . . , 90–99; the column headed `Count` displays the frequencies of those classes. Finally, by turning the output 90° counterclockwise, we obtain, from the asterisks, a frequency histogram. Compare Minitab's frequency histogram to the one we drew by hand in Fig. 2.1(a) on page 63.   ■

Earlier in this section we discussed dotplots. The Minitab command for constructing dotplots is called **DotPlot**. We will apply that command to obtain a dotplot of the data on oat yields considered in Example 2.12.

[†] The two subcommands START and INCREMENT are optional. We have used them to obtain a frequency distribution and a frequency histogram that are based on the classes 30–39, 40–49, . . . , 90–99. If the subcommands are not used, Minitab will choose its own classes.

[‡] Specifying the first midpoint and interval width is optional. We have specified them in order to obtain a frequency distribution and a frequency histogram that are based on the classes 30–39, 40–49, . . . , 90–99. If the first midpoint and interval width are not specified, Minitab will choose its own classes.

PRINTOUT 2.1

Minitab output for the
Histogram command

```
MTB > HISTOGRAM 'MATURITY';
SUBC> START=34.5;
SUBC> INCREMENT=10.

Histogram of MATURITY   N = 40

Midpoint   Count
    34.5       3   ***
    44.5       1   *
    54.5       8   ********
    64.5      10   **********
    74.5       7   *******
    84.5       7   *******
    94.5       4   ****
```

Example 2.15   The DotPlot Command

A farmer is interested in a new fertilizer that supposedly will increase his yield of oats. He uses the fertilizer on a sample of 15 one-acre plots. The yields, in bushels, are depicted in Table 2.19. Use Minitab to construct a dotplot of the data.

TABLE 2.19
Oat yields

67	65	55	57	58
61	61	61	64	62
62	60	62	60	67

SOLUTION   To begin we store the data from Table 2.19 into a column named OATS. Then we proceed in the following manner.

Session commands: We type the command **DOTPLOT** followed by the storage location of the data; that is, we type DOTPLOT 'OATS' and press Enter. This command and the resulting output are displayed in Printout 2.2.

(or)

Menu commands: We choose **Graph ▸ Dotplot...**, specify OATS in the **Variables** text box, and then select **OK**. Printout 2.2 shows the output that results.

PRINTOUT 2.2

Minitab output for the
DotPlot command

```
MTB > DOTPLOT 'OATS'

                                      .   .
           .        .   .      :   :  :       .   .       :
        ---+---------+---------+---------+---------+---------+---OATS
          55.0      57.5      60.0      62.5      65.0      67.5
```

Notice that Minitab uses 10 dashes (−) from one tick mark (+) to the next (counting the − part of a + as a last dash). Since in this case there are 2.5 units between tick marks, each dash represents 0.25 (= 2.5/10) units. Compare the dotplot in Printout 2.2 to the one we drew by hand in Fig. 2.3 on page 66.   ■

EXERCISES 2.3

2.35 Explain the difference between a frequency histogram and a relative-frequency histogram.

In Exercises 2.36–2.45, we have presented the frequency and relative-frequency distributions obtained in Exercises 2.18–2.27, respectively. For each exercise,
a. *construct a frequency histogram.*
b. *construct a relative-frequency histogram.*

2.36 The ages of a sample of 35 diabetics obtained by a research physician:

Age (yrs)	Frequency	Relative frequency
0–9	1	0.029
10–19	1	0.029
20–29	2	0.057
30–39	2	0.057
40–49	7	0.200
50–59	6	0.171
60–69	7	0.200
70–79	6	0.171
80–89	3	0.086

2.37 The contents, in milliliters, of a sample of 30 "one-liter" bottles of soda obtained by a consumer group from a soft-drink bottler:

Content (mL)	Frequency	Relative frequency
910–929	1	0.033
930–949	1	0.033
950–969	3	0.100
970–989	9	0.300
990–1009	7	0.233
1010–1029	6	0.200
1030–1049	2	0.067
1050–1069	1	0.033

2.38 The lengths of stay in Europe and the Mediterranean obtained from a sample of 36 U.S. residents who traveled there one year:

Length of stay (days)	Frequency	Relative frequency
1–7	10	0.278
8–14	10	0.278
15–21	8	0.222
22–28	1	0.028
29–35	2	0.056
36–42	1	0.028
43–49	2	0.056
50–56	1	0.028
57–63	0	0.000
64–70	1	0.028

2.39 One year's energy consumptions for a sample of 50 households in the South:

Energy consumption (millions of BTU)	Frequency	Relative frequency
40–49	1	0.02
50–59	7	0.14
60–69	7	0.14
70–79	3	0.06
80–89	6	0.12
90–99	10	0.20
100–109	5	0.10
110–119	4	0.08
120–129	2	0.04
130–139	3	0.06
140–149	0	0.00
150–159	2	0.04

2.40 The number of people per household for a sample of 40 U.S. households:

Number of people	Frequency	Relative frequency
1	7	0.175
2	13	0.325
3	9	0.225
4	5	0.125
5	4	0.100
6	1	0.025
7	1	0.025

2.41 The number of cars sold per week one year by a car salesperson:

Number of cars sold	Frequency	Relative frequency
0	7	0.135
1	15	0.288
2	12	0.231
3	9	0.173
4	5	0.096
5	3	0.058
6	1	0.019

2.42 The number of questions answered incorrectly on an eight-question fraction quiz by each of the 25 students in a fifth-grade class:

Number incorrect	Frequency	Relative frequency
0	4	0.16
1	8	0.32
2	4	0.16
3	5	0.20
4	2	0.08
5	1	0.04
6	1	0.04

2.43 The number of days missed last year by each of the 80 bricklayers employed at Cudahey Masonry:

Number of days missed	Frequency	Relative frequency
0	4	0.050
1	2	0.025
2	14	0.175
3	10	0.125
4	16	0.200
5	18	0.225
6	10	0.125
7	6	0.075

2.44 The 24-hour iron intakes, in milligrams, for a sample of 45 women under the age of 51:

Iron intake (mg)	Frequency	Relative frequency
6–under 8	1	0.022
8–under 10	1	0.022
10–under 12	7	0.156
12–under 14	9	0.200
14–under 16	9	0.200
16–under 18	9	0.200
18–under 20	8	0.178
20–under 22	1	0.022

2.45 The starting annual salaries for a sample of 35 liberal-arts graduates:

Starting salary ($thousands)	Frequency	Relative frequency
16–under 17	3	0.086
17–under 18	3	0.086
18–under 19	5	0.143
19–under 20	9	0.257
20–under 21	9	0.257
21–under 22	4	0.114
22–under 23	1	0.029
23–under 24	1	0.029

In Exercises 2.46–2.49, construct a dotplot for each of the data sets presented.

2.46 The exam scores for the students in an introductory statistics class are as follows.

88	82	89	70	85
63	100	86	67	39
90	96	76	34	81
64	75	84	89	96

2.47 A paint manufacturer claims that the average drying time for his new latex paint is 2 hours. To test this claim, the drying times, in minutes, are obtained for a sample of 20 cans of paint.

123	109	115	121	130
127	106	120	116	136
131	128	139	110	133
122	133	119	135	109

2.48 A consumer advocacy group thinks that Wheat Flakes brand cereal contains, on the average, less than the advertised weight of 15 oz per box. A sample of 40 boxes of Wheat Flakes provided the weights, in ounces, displayed in the following table.

15.8	15.1	15.2	15.4	14.8	15.6	15.7	14.5
14.8	15.4	15.3	15.5	15.2	14.6	15.4	15.4
15.5	14.7	14.7	15.1	14.7	15.3	15.3	15.5
14.0	14.2	14.6	15.0	15.1	14.9	14.9	15.8
15.0	14.4	15.4	14.3	15.4	15.9	15.2	15.6

2.49 The Motor Vehicle Manufacturers Association of the United States publishes information on the ages of cars and trucks currently in use in *Motor Vehicle Facts and Figures.* A sample of 37 trucks provided the ages, in years, displayed in the following table.

8	12	14	16	15	5	11	13
4	12	12	15	12	3	10	9
11	3	18	4	9	11	17	
7	4	12	12	8	9	10	
9	9	1	7	6	9	7	

In Exercises 2.50 and 2.51, we have displayed the frequency and relative-frequency distributions obtained in Exercises 2.30 and 2.31, respectively. For each exercise,
a. draw a pie chart for the relative frequencies.
b. construct a bar graph for the relative frequencies.

2.50 The network data for the all-time top TV programs by number of viewers as of January 1992:

Network	Frequency	Relative frequency
ABC	6	0.30
CBS	7	0.35
NBC	7	0.35

2.51 The winners of the NCAA wrestling championships for the years 1963–1992:

Champion	Frequency	Relative frequency
Oklahoma	2	0.067
Oklahoma State	6	0.200
Iowa State	7	0.233
Michigan State	1	0.033
Iowa	13	0.433
Arizona State	1	0.033

2.52 The IRS groups income tax returns by adjusted gross income. Following is a relative-frequency histogram for one year's federal individual income tax returns showing an adjusted gross income less than $50,000. [SOURCE: U.S. Internal Revenue Service, *Statistics of Income, Individual Income Tax Returns.*]

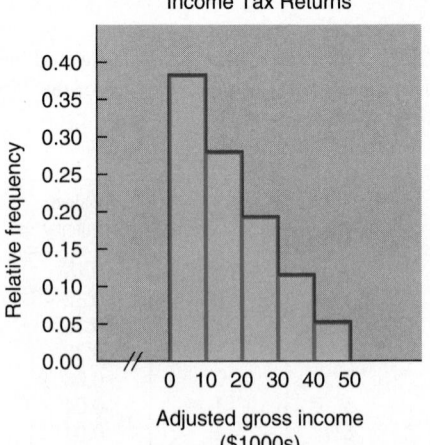

a. Approximately what percentage of the individual income tax returns had an adjusted gross income between $10,000 and $19,999, inclusive?
b. Approximately what percentage had an adjusted gross income less than $30,000?
c. Given that 89,928,000 individual income tax returns had an adjusted gross income of less than $50,000, roughly how many had an adjusted gross income between $30,000 and $49,999, inclusive?

· **2.53** Below is a relative-frequency histogram for the cholesterol-level data discussed in Example 2.6.

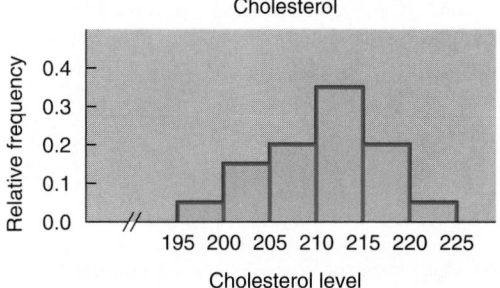

Answer the following questions using only the graph.
a. What percentage of the patients have cholesterol levels between 205 and 209, inclusive?
b. What percentage have levels of 215 or higher?
c. Given that the number of patients is 20, how many have levels between 210 and 214, inclusive?

· **2.54 (Computer exercise)** A study was conducted by a research physician on the ages of people with diabetes. The following data were obtained for the ages of a sample of 35 diabetics.

48	41	57	83	41	55	59
61	38	48	79	75	77	7
54	23	47	56	79	68	61
64	45	53	82	68	38	70
10	60	83	76	21	65	47

Use Minitab or some other statistical software to obtain a frequency distribution and a frequency histogram for these data using classes of equal width beginning with the class 0–9.

· **2.55 (Computer exercise)** A soft-drink bottler sells "one-liter" bottles of soda. A consumer group is concerned that the bottler may be shortchanging the customers. Thirty bottles of soda are randomly selected. The contents of the bottles chosen, in milliliters, are shown below.

1025	977	1018	975	977
990	959	957	1031	964
986	914	1010	988	1028
989	1001	984	974	1017
1060	1030	991	999	997
996	1014	946	995	987

Use Minitab or some other statistical software to obtain a frequency distribution and a frequency histogram for these data using classes of equal width starting with the class 910–929.

· **2.56 (Computer exercise)** The U.S. Bureau of the Census conducts nationwide surveys to obtain data on characteristics of U.S. households. Following are data on the number of people per household for a sample of 40 households.

2	5	2	1	1	2	3	4
1	4	4	2	1	4	3	3
7	1	2	2	3	4	2	2
6	5	2	5	1	3	2	5
2	1	3	3	2	2	3	3

Use Minitab or some other statistical software to obtain a frequency distribution and a frequency histogram for these data using classes based on a single value. *(Note: You may have to specify an increment in order to get classes based on a single value.)*

· **2.57 (Computer exercise)** A new-car salesperson keeps track of the number of cars she sells per week. The number of cars she sold per week one year are as follows.

1	0	3	3	1	0	2	1	4	0	4	1	2
3	6	4	3	0	2	2	1	1	2	2	2	3
5	1	0	2	5	3	1	3	1	1	1	1	2
2	3	0	4	4	1	0	1	1	3	2	5	2

Use Minitab or some other statistical software to obtain a frequency distribution and a frequency histogram for these data using classes based on a single value. *(Note: You may have to specify an increment in order to get classes based on a single value.)*

· **2.58 (Computer exercise)** The Bureau of Economic Analysis gathers information on the length of stay in Europe and the Mediterranean by U.S. travelers. Data are published in *Survey of Current Business*. We used Minitab's Histogram command to obtain a frequency distribution and a frequency histogram for the lengths of stay, in days, of a sample of

U.S. residents who traveled to Europe and the Mediterranean one year. The Minitab output follows.

```
Histogram of STAYS   N = 36

Midpoint  Count
    4.00     10  **********
   11.00     10  **********
   18.00      8  ********
   25.00      1  *
   32.00      2  **
   39.00      1  *
   46.00      2  **
   53.00      1  *
   60.00      0
   67.00      1  *
```

a. How many U.S. residents were sampled?
b. Identify the class mark for the sixth class.
c. What common class width was used to construct the frequency distribution?
d. How many stayed between 15 and 21 days?

2.59 (Computer exercise) The U.S. Energy Information Administration collects data on residential energy consumption and expenditures. Results are published in *Residential Energy Consumption Survey: Consumption and Expenditures*. Data, in millions of BTU, on one year's energy consumptions were collected for a sample of households in the South. We employed Minitab's Histogram command to obtain a frequency distribution and a frequency histogram of the data. Here is the computer output.

```
Histogram of ENERGY   N = 50

Midpoint  Count
   44.5      1  *
   54.5      7  *******
   64.5      7  *******
   74.5      3  ***
   84.5      6  ******
   94.5     10  **********
  104.5      5  *****
  114.5      4  ****
  124.5      2  **
  134.5      3  ***
  144.5      0
  154.5      2  **
```

a. How many households were sampled?
b. Identify the class mark for the sixth class.

c. What common class width was used to construct the frequency distribution?
d. How many of the households sampled had an energy consumption of between 100 million and 109 million BTU, inclusive?

2.60 (Computer exercise) Use Minitab or some other statistical software to obtain a dotplot for the data from Exercise 2.46.

2.61 (Computer exercise) Use Minitab or some other statistical software to obtain a dotplot for the data from Exercise 2.47.

2.62 (Computer exercise) We used Minitab to obtain the following dotplot for the weights of the sampled boxes of Wheat Flakes from Exercise 2.48.

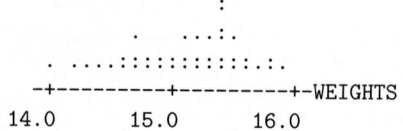

a. How many boxes of Wheat Flakes were sampled?
b. How many of the boxes weighed between 15 and 15.5 oz, inclusive?

2.63 (Computer exercise) The dotplot below was obtained for the ages of the trucks from Exercise 2.49 by employing Minitab's DotPlot command.

```
            :       :
         .     .  :  . :
  .   : : . . : : : : :  . . : . . .
+---------+---------+---------+------AGES
0.0       5.0      10.0      15.0
```

a. How many trucks were sampled?
b. How many of the trucks were between 5 and 8 years old, inclusive?

Class boundaries. Another method of depicting the classes on the horizontal axis of a histogram is to use class boundaries. The **lower class boundary** of a class is the number halfway between the lower class limit of the class and the upper class limit of the next lower class; the **upper class boundary** of a class is the number halfway between the upper class limit of the class and the lower class limit of the next higher class. For instance, consider the class 50–59 of the days-to-maturity data (see Table 2.14 on page 62). We have

$$\text{lower class boundary} = \frac{49 + 50}{2} = 49.5$$

and

$$\text{upper class boundary} = \frac{59 + 60}{2} = 59.5.$$

· · **2.64** Refer to Exercise 2.36.

a. Add a lower-class-boundary column and an upper-class-boundary column to the table given in Exercise 2.36. Then fill in those columns by computing the lower and upper class boundaries for each class.

b. Construct a frequency histogram for the data using class boundaries instead of class limits on the horizontal axis.

c. What are the advantages and disadvantages of using class boundaries instead of class limits?

· · **2.65** Refer to Exercise 2.37.

a. Add a lower-class-boundary column and an upper-class-boundary column to the table given in Exercise 2.37. Then fill in those columns by computing the lower and upper class boundaries for each class.

b. Construct a frequency histogram for the data using class boundaries instead of class limits on the horizontal axis.

c. What are the advantages and disadvantages of using class boundaries instead of class limits?

Relative-frequency polygons. Another graphical display in common use is the relative-frequency polygon. In a **relative-frequency polygon,** a point is plotted above each class mark at a height equal to the relative frequency of the class. Then the points are joined with connecting lines. For instance, the relative-frequency polygon for the days-to-maturity data in Table 2.4 on page 53 is shown below.

Short-Term Investments

· · **2.66** Construct a relative-frequency polygon for the length-of-stay data in Exercise 2.38.

· · **2.67** Construct a relative-frequency polygon for the energy-consumption data in Exercise 2.39.

Ogives. Cumulative information can be portrayed using a graph called an ogive (ō′jīv). To construct an ogive, we first make a table displaying cumulative frequencies and cumulative relative frequencies. Table 2.20 provides such a table for the days-to-maturity data shown in Table 2.1 on page 50.

TABLE 2.20   Cumulative information for days-to-maturity data

Less than	Cumulative frequency	Cumulative relative frequency
30	0	0.000
40	3	0.075
50	4	0.100
60	12	0.300
70	22	0.550
80	29	0.725
90	36	0.900
100	40	1.000

The first column of Table 2.20 gives the class limits, and the second column gives the cumulative frequencies. A **cumulative frequency** is obtained by summing the frequencies of all classes representing values less than the specified class limit. For instance, by referring to Table 2.14 on page 62, we can find the cumulative frequency of investments with a maturity period of less than 50 days. We see that the

$$\text{cumulative frequency} = 3 + 1 = 4.$$

This means that four of the investments have a maturity period of less than 50 days.

The third column of Table 2.20 gives the cumulative relative frequencies. A **cumulative relative frequency** is found by dividing the corresponding cumulative frequency by the total number of pieces of data. For instance, the cumulative relative frequency of investments with a maturity period of less than 50 days is obtained as follows:

$$\text{cumulative relative frequency} = \frac{4}{40} = 0.100.$$

This means that 10% of the investments have a maturity period of less than 50 days.

Using Table 2.20, we can now construct an ogive for the days-to-maturity data. In an **ogive** a point is plotted above each class limit at a height equal to the cumulative relative frequency. Then the points are joined with connecting lines. Consequently, the ogive for the days-to-maturity data is as displayed in the figure at the right.

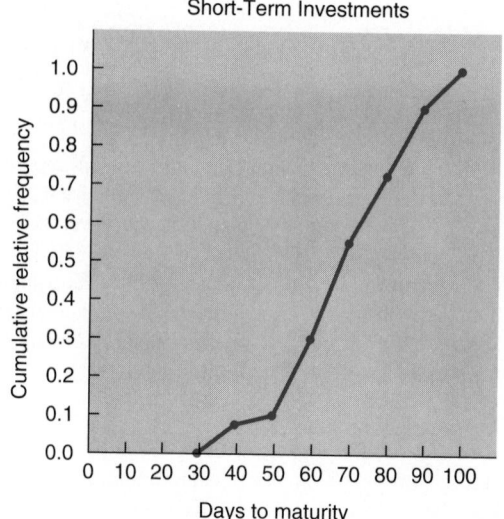

Short-Term Investments

· · **2.68** Refer to Exercise 2.38.
a. Construct a table similar to Table 2.20 for the length-of-stay data. Interpret your results.
b. Draw an ogive for the data.

· · **2.69** Refer to Exercise 2.39.
a. Construct a table similar to Table 2.20 for the energy-consumption data. Interpret your results.
b. Draw an ogive for the data.

2.4 STEM-AND-LEAF DIAGRAMS

New ways of displaying data are constantly being invented. One method, developed in the late 1960s by Professor John Tukey of Princeton University, is called a *stem-and-leaf diagram,* or *stemplot.* This ingenious diagram is often easier to construct than either a frequency distribution or a histogram and generally displays more information. To illustrate stem-and-leaf diagrams, we return once more to the days-to-maturity data first presented in Table 2.1.

Example 2.16   Stem-and-Leaf Diagrams

The data on the number of days to maturity for 40 short-term investments are repeated in Table 2.21.

TABLE 2.21

Days to maturity for
40 short-term investments

70	64	99	55	64	89	87	65
62	38	67	70	60	69	78	39
75	56	71	51	99	68	95	86
57	53	47	50	55	81	80	98
51	36	63	66	85	79	83	70

In Table 2.2 on page 50, we grouped these data using the classes 30–39, 40–49, …, 90–99, and in Fig. 2.1(a) on page 63, we portrayed the data graphically with a frequency histogram. Now we will construct a stem-and-leaf diagram for the data. With a stem-and-leaf diagram, we can simultaneously group the data and obtain a graphical display similar to a histogram.

First we select the leading digits from the data in Table 2.21. This yields the numbers 3, 4, ..., 9. Next we list those leading digits in a column, as shown by the colored numbers in Fig. 2.5(a).

Then we go through the data in Table 2.21 and write the final digit of each number to the right of the appropriate leading digit: The first investment has a maturity period of 70 days. This calls for a "0" to the right of the "7" in colored type in Fig. 2.5(a). Reading down the first column of Table 2.21, the second investment has a maturity period of 62 days, so this calls for a "2" to the right of the "6" in colored type. Continuing in this manner, we obtain the diagram displayed in Fig. 2.5(a). As indicated in the figure, the leading digits are called **stems,** the final digits **leaves.** The entire diagram is called a **stem-and-leaf diagram.**

FIGURE 2.5

Days-to-maturity:
(a) stem-and-leaf diagram
(b) shaded stem-and-leaf diagram
(c) ordered stem-and-leaf diagram

Stems	Leaves		Leaves		Leaves
3	8 6 9	3	8 6 9	3	6 8 9
4	7	4	7	4	7
5	7 1 6 3 5 1 0 5	5	7 1 6 3 5 1 0 5	5	0 1 1 3 5 5 6 7
6	2 4 7 3 6 4 0 9 8 5	6	2 4 7 3 6 4 0 9 8 5	6	0 2 3 4 4 5 6 7 8 9
7	0 5 1 0 9 8 0	7	0 5 1 0 9 8 0	7	0 0 0 1 5 8 9
8	5 9 1 7 0 3 6	8	5 9 1 7 0 3 6	8	0 1 3 5 6 7 9
9	9 9 5 8	9	9 9 5 8	9	5 8 9 9
	(a)		(b)		(c)

The stem-and-leaf diagram for the days-to-maturity data is similar to the frequency histogram for those data. This is because the length of the row of leaves for a class equals the frequency of the class. [Turn the stem-and-leaf diagram 90° counterclockwise and compare it to the frequency histogram in Fig. 2.1(a) on page 63.]

By shading each row of leaves, as in Fig. 2.5(b), we get a diagram that looks even more like the frequency histogram of the data. The diagram in Fig. 2.5(b) is called a **shaded stem-and-leaf diagram.** Because the numbers in a shaded stem-and-leaf diagram are still visible under the shading, a shaded stem-and-leaf diagram exhibits the original, or raw, data in addition to providing a graphical display of a frequency distribution. On the other hand, although a frequency histogram provides a graphical display of a frequency distribution, it is generally not possible to recover the raw data from a frequency histogram. Why is this so?

Another form of stem-and-leaf diagram is called an **ordered stem-and-leaf diagram.** For this type of stem-and-leaf diagram, the leaves in each row are ordered from smallest to largest. This makes it easier to comprehend the data and also facilitates the computation of descriptive measures such as the median (to be discussed in Chapter 3). The ordered stem-and-leaf diagram for the days-to-maturity data is presented in Fig. 2.5(c). ∎

Example 2.17   Stem-and-Leaf Diagrams

The cholesterol-level data from Example 2.6 are repeated in Table 2.22. The data are in milligrams per 100 milliliters. Construct a stem-and-leaf diagram for these cholesterol-level data.

TABLE 2.22
Cholesterol levels for 20 high-level patients

210	209	212	208
217	207	210	203
208	210	210	199
215	221	213	218
202	218	200	214

SOLUTION   Because these data are three-digit numbers, we use the first two digits as the stems and the third digit as the leaves. A stem-and-leaf diagram for the cholesterol levels is displayed in Fig. 2.6(a).

FIGURE 2.6

Stem-and-leaf diagram for cholesterol levels: (a) using one line per stem (b) using two lines per stem

```
                                        19 |
                                        19 | 9
                                        20 | 2 0 3
                                        20 | 8 9 7 8
    19 | 9                              21 | 0 0 2 0 0 3 4
    20 | 8 2 9 7 0 8 3                  21 | 7 5 8 8
    21 | 0 7 5 0 8 2 0 0 3 8 4          22 | 1
    22 | 1                              22 |
            (a)                                 (b)
```

The stem-and-leaf diagram in Fig. 2.6(a) is only moderately helpful because there are so few stems. We can construct a better stem-and-leaf diagram by using two lines for each stem, with the first line for the leaf digits 0–4 and the second line for the leaf digits 5–9. This stem-and-leaf diagram is shown in Fig. 2.6(b).   ■

As we have seen, stem-and-leaf diagrams have several advantages over the more classical techniques for grouping and graphing. However, they do have some drawbacks. For instance, they are generally not useful with large data sets and can be awkward with data containing decimals. Histograms are usually preferable to stem-and-leaf diagrams in these latter two cases.

Using the Computer (Optional)

Minitab has a command, called **Stem-and-Leaf**, that can be used to obtain a stem-and-leaf diagram for a set of data. In Example 2.18 we use the cholesterol-level data from Example 2.17 to illustrate the application of the Stem-and-Leaf command.

Example 2.18 The Stem-and-Leaf Command

Consider again the data on cholesterol levels given in Table 2.22. Apply Minitab to obtain a stem-and-leaf diagram for those data with two lines per stem.

SOLUTION We begin by storing the cholesterol-level data in a column named CHOLLEVS. Then we proceed as follows.

Session commands: We type the command **STEM-AND-LEAF** followed by the storage location of the data; that is, we type STEM-AND-LEAF 'CHOLLEVS'; and press Enter. Since in this case we are to use two lines per stem, we also employ the (optional) **INCREMENT** subcommand. Specifically, we type INCREMENT=5. and press Enter. This instructs Minitab to use a distance of 5 between the smallest possible number on one line and the smallest possible number on the next line. Printout 2.3 shows these commands and the output that results.

(or)

Menu commands: We choose **Graph ▶ Stem-and-Leaf...** and specify CHOLLEVS in the **Variables** text box. Then, in order to obtain a stem-and-leaf diagram with two lines per stem, we select **Increment** and type 5. This tells Minitab that the distance between the smallest possible number on one line and the smallest possible number on the next line should be 5. Finally, we select **OK**. The output resulting is depicted in Printout 2.3.

PRINTOUT 2.3
Minitab output for the
Stem-and-Leaf command

```
MTB > STEM-AND-LEAF 'CHOLLEVS';
SUBC> INCREMENT=5.

Stem-and-leaf of CHOLLEVS   N  = 20
Leaf Unit = 1.0

     1    19 9
     4    20 023
     8    20 7889
    (7)   21 0000234
     5    21 5788
     1    22 1
```

The third line in Printout 2.3 describes what follows (Stem-and-leaf of CHOLLEVS) and also displays the number of pieces of data (N = 20). The fourth line (Leaf Unit = 1.0) indicates where the decimal point goes, in this case directly after each leaf digit. The second and third columns of numbers in Printout 2.3 give the stems and leaves, respectively. Since the leaves in each row are ordered, we see that the output in Printout 2.3 is actually an ordered stem-and-leaf diagram.

The numbers in the first column of the stem-and-leaf diagram, called **depths,** are used to display cumulative frequencies. Starting from the top, the depths indicate the number of leaves (pieces of data) that lie in a given row or before. For instance, the 8 in the third row shows that there are eight leaves (pieces of data) in the first three rows.

When the row containing the middle data value(s) is reached, the cumulative frequency is replaced by the number of leaves in that row. Thus in this case we see that the middle data values lie somewhere between 210 and 214, inclusive, and that seven pieces of data are in that range. The depths following the row that contains the middle data value(s) indicate the number of leaves (pieces of data) that lie in a given row or after. For instance, the 5 in the fifth row shows that there are five leaves (pieces of data) in the last two rows.

Compare the stem-and-leaf diagram in Printout 2.3 to the one obtained by hand in Fig. 2.6(b) on page 78.   ■

EXERCISES 2.4

In each of Exercises 2.70–2.73,
a. construct a stem-and-leaf diagram.
b. construct an ordered stem-and-leaf diagram.

2.70 A research physician conducted a study on the ages of people with diabetes. The following data were obtained for the ages of a sample of 35 diabetics.

48	41	57	83	41	55	59
61	38	48	79	75	77	7
54	23	47	56	79	68	61
64	45	53	82	68	38	70
10	60	83	76	21	65	47

2.71 A soft-drink bottler sells "one-liter" bottles of soda. A consumer group is concerned that the bottler may be shortchanging customers. Thirty bottles of soda are randomly selected. The contents, in milliliters, of the bottles chosen are shown below. (*Hint:* Use the stems 91, 92, ..., 106.)

1025	977	1018	975	977
990	959	957	1031	964
986	914	1010	988	1028
989	1001	984	974	1017
1060	1030	991	999	997
996	1014	946	995	987

2.72 The Bureau of Economic Analysis gathers information on the length of stay in Europe and the Mediter-

ranean by U.S. travelers. Data are published in *Survey of Current Business.* A sample of 36 U.S. residents who traveled to Europe and the Mediterranean one year yielded the following data, in days, on length of stay.

41	16	6	21	1	21
5	31	20	27	17	10
3	32	2	48	8	12
21	44	1	56	5	12
3	13	15	10	18	3
1	11	14	12	64	10

2.73 The U.S. Energy Information Administration collects data on residential energy consumption and expenditures. Results are published in *Residential Energy Consumption Survey: Consumption and Expenditures.* The following table provides the data on one year's energy consumptions for a sample of 50 households in the South. Data are given in millions of BTU.

130	55	45	64	155	66	60	80	102	62
58	101	75	111	151	139	81	55	66	90
97	77	51	67	125	50	136	55	83	91
54	86	100	78	93	113	111	104	96	113
96	87	129	109	69	94	99	97	83	97

2.74 As reported by the U.S. Bureau of the Census in *Current Population Reports,* the percentage of the adult population completing high school in each state is as follows.

State	Percentage	State	Percentage	State	Percentage
AL	57	LA	58	OH	67
AK	83	ME	69	OK	66
AZ	72	MD	67	OR	76
AR	56	MA	72	PA	65
CA	74	MI	68	RI	61
CO	79	MN	73	SC	54
CT	70	MS	55	SD	68
DE	69	MO	64	TN	56
FL	67	MT	74	TX	63
GA	56	NE	73	UT	80
HI	74	NV	76	VT	71
ID	74	NH	72	VA	62
IL	67	NJ	67	WA	78
IN	66	NM	69	WV	56
IA	72	NY	66	WI	70
KS	73	NC	55	WY	78
KY	53	ND	66		

Construct a stem-and-leaf diagram for the percentages
a. using one line per stem.
b. using two lines per stem.

2.75 The U.S. Federal Bureau of Investigation published the following annual crime rates in *Crime in the United States.* Rates are per 1000 population.

State	Rate	State	Rate	State	Rate
AL	43	LA	61	OH	44
AK	62	ME	35	OK	60
AZ	73	MD	56	OR	71
AR	39	MA	47	PA	31
CA	68	MI	65	RI	49
CO	70	MN	44	SC	51
CT	48	MS	33	SD	27
DE	48	MO	47	TN	45
FL	82	MT	45	TX	74
GA	55	NE	39	UT	55
HI	57	NV	63	VT	40
ID	42	NH	33	VA	39
IL	55	NJ	52	WA	69
IN	39	NM	66	WV	23
IA	42	NY	58	WI	41
KS	48	NC	43	WY	44
KY	31	ND	26		

Construct a stem-and-leaf diagram for the crime rates
a. using one line per stem.
b. using two lines per stem.

2.76 The U.S. National Oceanic and Atmospheric Administration publishes temperature data in *Climatography of the United States.* According to that document, the annual average maximum temperatures for selected cities in the United States are as follows.

City	Annual average max. temp.	City	Annual average max. temp.
Mobile, AL	77	Reno, NV	67
Juneau, AK	47	Concord, NH	57
Phoenix, AZ	85	Atlantic City, NJ	63
Little Rock, AR	73	Albuquerque, NM	70
Los Angeles, CA	70	Albany, NY	58
Sacramento, CA	73	Buffalo, NY	56
San Francisco, CA	65	New York, NY	62
Denver, CO	64	Charlotte, NC	71
Hartford, CT	60	Raleigh, NC	70
Wilmington, DE	64	Bismarck, ND	54
Washington, DC	67	Cincinnati, OH	64
Jacksonville, FL	79	Cleveland, OH	59
Miami, FL	83	Columbus, OH	62
Atlanta, GA	71	Oklahoma City, OK	71
Honolulu, HI	84	Portland, OR	62
Boise, ID	63	Philadelphia, PA	63
Chicago, IL	59	Pittsburgh, PA	60
Peoria, IL	60	Providence, RI	59
Indianapolis, IN	62	Columbia, SC	75
Des Moines, IA	59	Sioux Falls, SD	57
Wichita, KS	68	Memphis, TN	72
Louisville, KY	66	Nashville, TN	70
New Orleans, LA	78	Dallas-Ft. Worth, TX	77
Portland, ME	55	El Paso, TX	78
Baltimore, MD	65	Houston, TX	79
Boston, MA	59	Salt Lake City, UT	64
Detroit, MI	58	Burlington, VT	54
Sault Ste. Marie, MI	49	Norfolk, VA	68
Duluth, MN	48	Richmond, VA	69
Mnpls-St. Paul, MN	54	Seattle-Tacoma, WA	59
Jackson, MS	76	Spokane, WA	57
Kansas City, MO	64	Charleston, WV	66
St. Louis, MO	66	Milwaukee, WI	55
Great Falls, MT	56	Cheyenne, WY	58
Omaha, NE	62	San Juan, PR	86

Construct a stem-and-leaf diagram for these data
a. using two lines per stem.
b. using five lines per stem.

2.77 The U.S. National Oceanic and Atmospheric Administration publishes temperature data in *Climatography of the United States*. According to that document, the annual average minimum temperatures for selected cities in the United States are as follows.

City	Annual average min. temp.	City	Annual average min. temp.
Mobile, AL	58	Reno, NV	32
Juneau, AK	33	Concord, NH	33
Phoenix, AZ	57	Atlantic City, NJ	43
Little Rock, AR	51	Albuquerque, NM	42
Los Angeles, CA	55	Albany, NY	37
Sacramento, CA	48	Buffalo, NY	39
San Francisco, CA	48	New York, NY	47
Denver, CO	36	Charlotte, NC	49
Hartford, CT	40	Raleigh, NC	48
Wilmington, DE	45	Bismarck, ND	29
Washington, DC	49	Cincinnati, OH	45
Jacksonville, FL	57	Cleveland, OH	41
Miami, FL	69	Columbus, OH	42
Atlanta, GA	51	Oklahoma City, OK	49
Honolulu, HI	70	Portland, OR	44
Boise, ID	39	Philadelphia, PA	45
Chicago, IL	40	Pittsburgh, PA	41
Peoria, IL	41	Providence, RI	41
Indianapolis, IN	42	Columbia, SC	51
Des Moines, IA	40	Sioux Falls, SD	34
Wichita, KS	45	Memphis, TN	52
Louisville, KY	46	Nashville, TN	49
New Orleans, LA	59	Dallas-Ft. Worth, TX	55
Portland, ME	35	El Paso, TX	49
Baltimore, MD	45	Houston, TX	57
Boston, MA	44	Salt Lake City, UT	39
Detroit, MI	39	Burlington, VT	35
Sault Ste. Marie, MI	31	Norfolk, VA	51
Duluth, MN	29	Richmond, VA	47
Mnpls-St. Paul, MN	35	Seattle-Tacoma, WA	44
Jackson, MS	53	Spokane, WA	37
Kansas City, MO	44	Charleston, WV	44
St. Louis, MO	45	Milwaukee, WI	38
Great Falls, MT	33	Cheyenne, WY	33
Omaha, NE	40	San Juan, PR	73

Construct a stem-and-leaf diagram for these data
a. using two lines per stem.
b. using five lines per stem.

2.78 (**Computer exercise**) Use Minitab or some other statistical software to construct a stem-and-leaf diagram for the data in Exercise 2.70 with
a. one line per stem.
b. two lines per stem.
c. five lines per stem.

2.79 (**Computer exercise**) Use Minitab or some other statistical software to construct a stem-and-leaf diagram for the data in Exercise 2.71 with
a. one line per stem.
b. two lines per stem.

2.80 (**Computer exercise**) A hardware manufacturer produces 10-mm-diameter bolts. The manufacturer knows that the diameters of the bolts produced vary somewhat from 10 mm and also from each other. To get an idea of the variation, the manufacturer takes a sample of bolts and obtains the following stem-and-leaf diagram for the diameters.

```
Stem-and-leaf of BOLTDIAM  N  = 20
Leaf Unit = 0.010

    1     98  9
    1     99
    1     99
    3     99  45
    5     99  67
    9     99  8899
   (2)   100  01
    9    100  2333
    5    100  55
    3    100
    3    100  88
    1    101  0
```

a. How many bolts were sampled?
b. What is the smallest diameter, in millimeters, of the bolts sampled? (*Hint:* Note that the leaf unit is 0.010.)
c. How many lines per stem are used in this stem-and-leaf diagram?
d. Use the depths to determine the number of bolts sampled having diameters of 10.04 mm or greater.
e. Use the depths to determine the number of bolts sampled having diameters of at most 9.97 mm.
f. List the diameters that are less than 10 mm.

2.81 (**Computer exercise**) The U.S. Bureau of the Census reports the percentage of the adult population completing high school in each state in *Current Population Reports*.

We applied Minitab's Stem-and-Leaf command to obtain the stem-and-leaf diagram for those data shown at the right.

a. How many pieces of data are there?
b. How many lines per stem are used?
c. Use the depths to determine how many of the percentages are at least 75%, that is, 75% or greater.
d. Use the depths to determine how many of the percentages are at most 64%, that is, 64% or less.
e. Identify the largest percentage.
f. List the percentages that are in the 50s.

```
Stem-and-leaf of HSCOMP    N  = 50
Leaf Unit = 1.0

    2     5 34
   10     5 55666678
   14     6 1234
  (15)    6 566667777788999
   21     7 00122223334444
    7     7 66889
    2     8 03
```

2.5 DISTRIBUTION SHAPES; SYMMETRY AND SKEWNESS

An important aspect of the distribution of a data set is its shape. As we will see in later chapters, the shape of a distribution often plays a role in determining the appropriate method of statistical analysis.

In Sections 2.3 and 2.4, we learned several techniques for displaying distributions graphically, among them histograms, dotplots, and stem-and-leaf diagrams. In discussing distribution shapes, it is better to use smooth curves that approximate the overall shape.

For instance, Fig. 2.7 displays a relative-frequency histogram for the heights of the 3264 female students attending a midwestern college. Also included in Fig. 2.7 is a smooth curve that approximates the overall shape of the distribution. Both the histogram and the smooth curve show that this distribution of heights is bell-shaped (or mound-shaped), but the smooth curve makes it a little easier to see the shape.

FIGURE 2.7

Relative-frequency histogram and approximating smooth curve for the distribution of heights

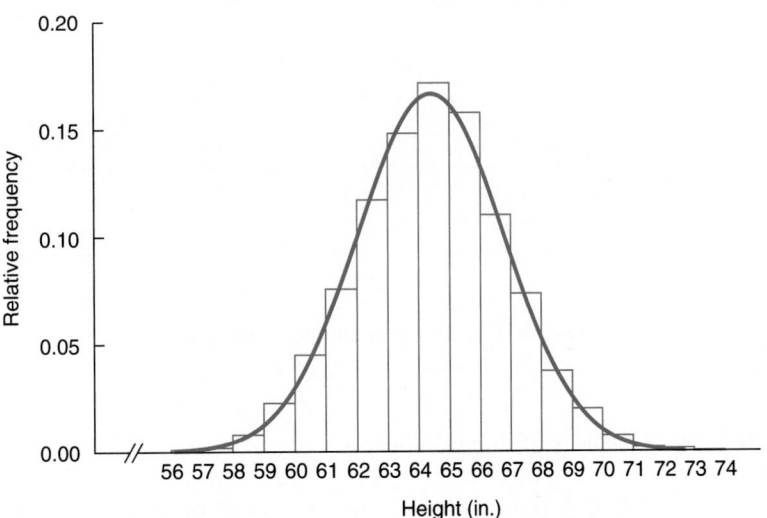

Another advantage of using smooth curves to describe distribution shapes is that we need not worry about minor differences in shape but can instead concentrate on overall patterns. This, in turn, allows us to designate relatively few shapes in order to classify most distributions.

Distribution Shapes

Figure 2.8 displays some common distribution shapes: **bell-shaped, triangular, uniform, reverse J-shaped, J-shaped, right skewed, left skewed, bimodal,** and **multimodal.** These shapes are idealized forms; in practice, distributions rarely have these exact shapes. Consequently, in identifying the shape of a distribution, we do not require exact conformance, especially when considering small data sets. So, for example, we describe the distribution of heights displayed in Fig. 2.7 as bell-shaped even though the histogram does not form a perfect bell.

FIGURE 2.8
Common distribution shapes

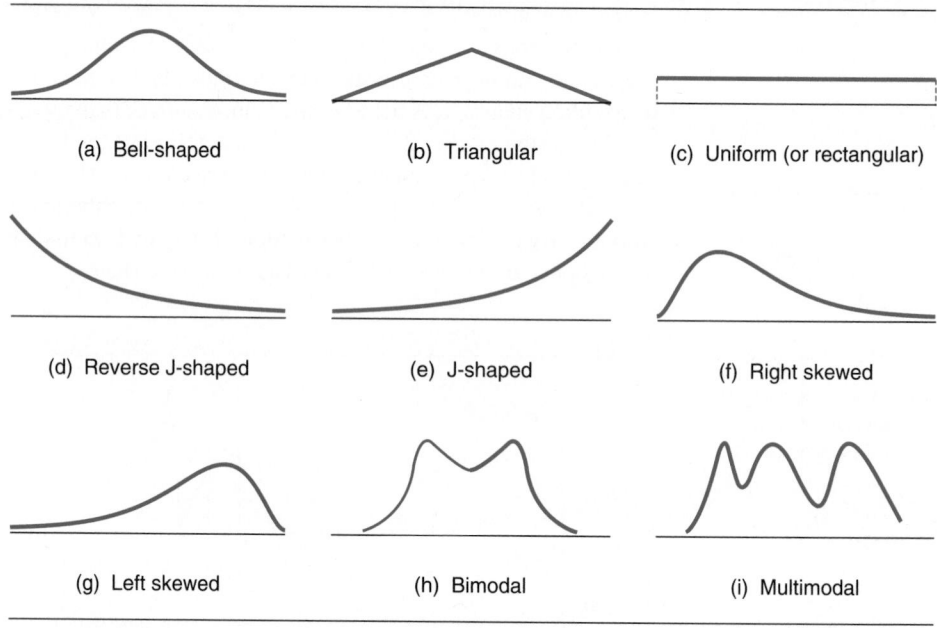

Example 2.19 provides another illustration of identifying the shape of a distribution.

Example 2.19   Identifying Distribution Shapes

The U.S. Bureau of the Census collects data on household size and publishes the results in *Current Population Reports.* Based on information found in that document, we constructed the relative-frequency histogram for household size in the United States shown in Fig. 2.9(a). Describe the distribution shape for sizes of U.S. households.

FIGURE 2.9
Relative-frequency histogram
for household size

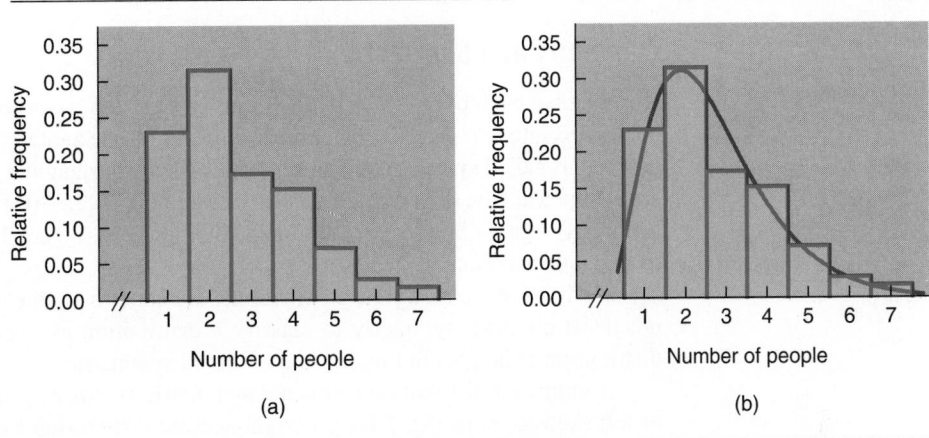

(a)   (b)

SOLUTION   First we draw a smooth curve through the histogram in Fig. 2.9(a), as seen in Fig. 2.9(b). Then, by referring to Fig. 2.8, we can conclude that the distribution of household sizes is right skewed.   ∎

There are other distribution shapes besides those represented in Fig. 2.8. However, the types in Fig. 2.8 do comprise the most commonly encountered distribution shapes and will suffice for our purposes.

Modality

In considering the shape of a distribution, it is helpful to observe the number of peaks (highest points). A distribution is said to be **unimodal** if it has one peak; **bimodal** if it has two peaks; and **multimodal** if it has three or more peaks.

The distribution of heights in Fig. 2.7 on page 83 is unimodal. More generally, we see from Fig. 2.8 that bell-shaped, triangular, reverse-J-shaped, J-shaped, right-skewed, and left-skewed distributions are unimodal. Representations of bimodal and multimodal distributions are displayed in Figs. 2.8(h) and (i), respectively.[†]

[†] A uniform distribution has either no peaks or infinitely many peaks, depending on how one looks at it. In any case, we do not classify a uniform distribution according to modality.

Technically, a distribution is bimodal or multimodal only if the peaks are the same height. However, in practice, distributions with pronounced but not necessarily equal-height peaks are often referred to as bimodal or multimodal.

Symmetry and Skewness

Observe that each of the three distributions in Figs. 2.8(a)–(c) has the property that it can be divided into two pieces that are mirror images of one another. A distribution having that property is called **symmetric.** Therefore, bell-shaped, triangular, and uniform distributions are symmetric. The bimodal distribution pictured in Fig. 2.8(h) is also symmetric, but that is not always true for bimodal or multimodal distributions. Figure 2.8(i) shows an asymmetric trimodal distribution.

Again, when classifying distributions, we must be somewhat flexible. Thus we do not insist on exact symmetry to classify a distribution as symmetric. For example, the distribution of heights in Fig. 2.7 is considered symmetric.

A unimodal distribution that is not symmetric is either right skewed, as in Fig. 2.8(f), or left skewed, as in Fig. 2.8(g). A right-skewed distribution rises to its peak rapidly and comes back toward the horizontal axis more slowly; or, equivalently, its "right tail" is longer than its "left tail." On the other hand, a left-skewed distribution rises to its peak slowly and comes back toward the horizontal axis more rapidly; or, equivalently, its "left tail" is longer than its "right tail."

Notice that a reverse-J-shaped distribution, as in Fig. 2.8(d), is also right skewed; and that a J-shaped distribution, as in Fig. 2.8(e), is also left skewed.

Population and Sample Distributions

When a (simple) random sample is taken from a population, we expect the relative-frequency distribution of the sample to be similar, although not identical, to that of the population. Example 2.20 provides an illustration.

Example 2.20   Population and Sample Distributions Compared

Consider again the distribution of the number of people per U.S. household. A relative-frequency histogram for that distribution is displayed in Fig. 2.9(a) on page 85. We obtained six random samples of size 100 from the population of U.S. households. Figure 2.10 shows the relative-frequency histograms for household size for all six samples. Compare the distributions of the samples to each other and to the distribution of the population.

SOLUTION   As we see from Fig. 2.10, the distributions of the six samples, although similar, do have definite differences. This is not surprising since we would expect variation from one sample to another. Nonetheless, the overall shapes of the six sample distributions are roughly the same and are also similar in shape to the population distribution—all are right skewed.   ∎

FIGURE 2.10

Relative-frequency histograms for household size for six random samples of size 100 from the population of U.S. households

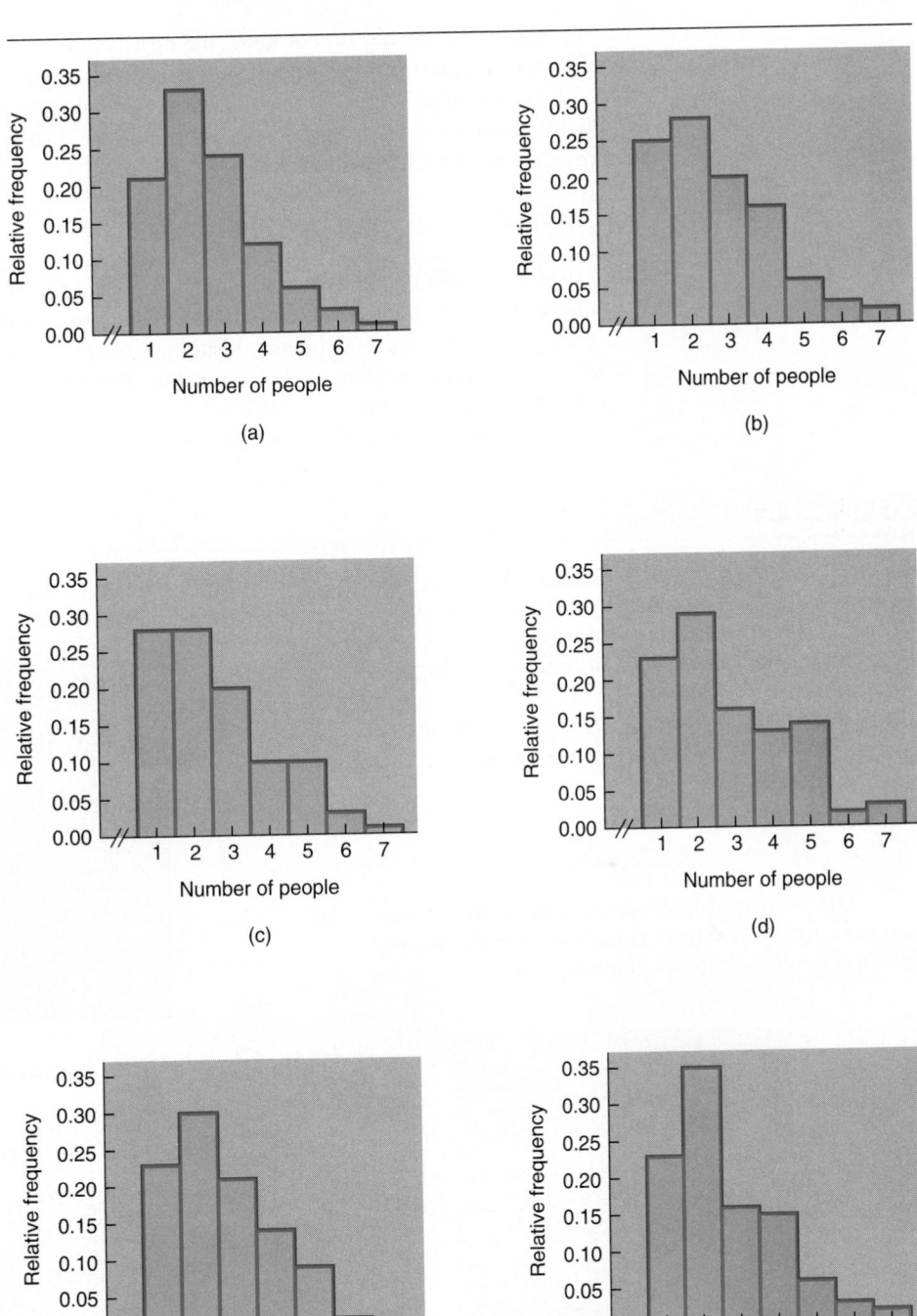

In practice, we usually do not know the distribution of the population. Under those circumstances we can use the distribution of a random sample from the population to get a rough idea of the distribution of the population. Generally, the larger the sample the more closely the distribution of the sample will resemble the distribution of the population. In summary, we have the following key fact.

KEY FACT 2.1   **POPULATION AND SAMPLE DISTRIBUTIONS**

> If a random sample is taken from a population, then a relative-frequency distribution of the sample will approximate the relative-frequency distribution of the population, or equivalently, a relative-frequency histogram of the sample will approximate the relative-frequency histogram of the population. The larger the sample, the better the approximation tends to be.

EXERCISES 2.5

In each of Exercises 2.82–2.91, we have provided a graphical display of a data set considered previously in the exercises of Section 2.3 or Section 2.4. For each exercise,

a. identify the overall shape of the distribution by referring to Fig. 2.8 on page 84.

b. state whether the distribution is (approximately) symmetric, right skewed, or left skewed.

2.82   A frequency histogram for the number of questions answered incorrectly on an eight-question fraction quiz by each of the 25 students in a fifth-grade class:

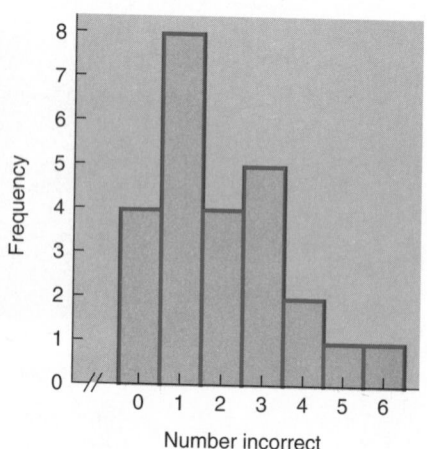

2.83   A frequency histogram for the number of cars sold per week one year by a car salesperson:

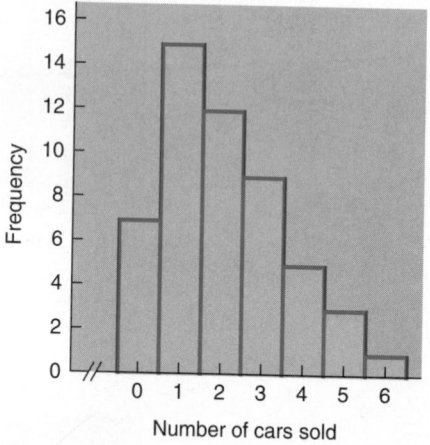

2.84   A relative-frequency histogram for the ages of a sample of 35 diabetics obtained by a research physician:

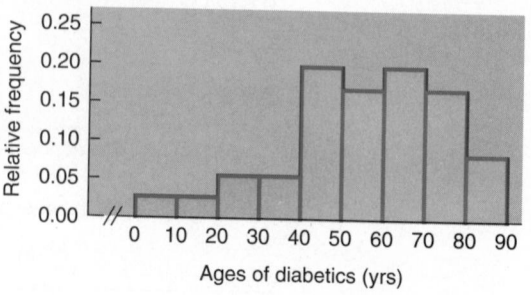

2.85   A relative-frequency histogram for the starting salaries of a sample of 35 liberal-arts graduates:

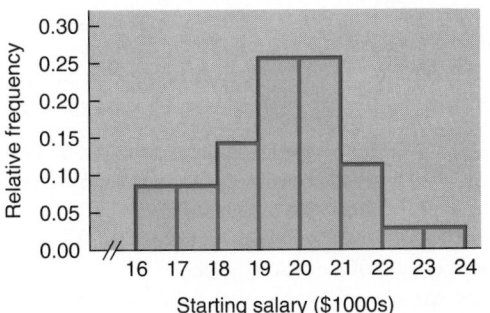

2.86   A relative-frequency histogram for one year's federal individual income tax returns showing an adjusted gross income less than $50,000:

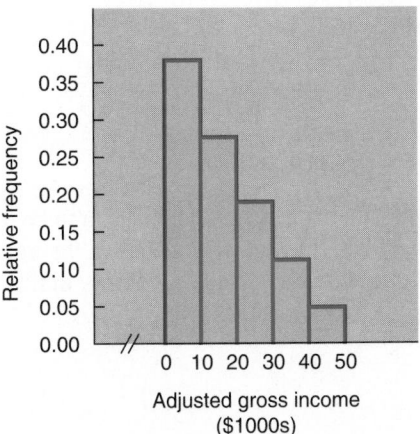

2.87   A relative-frequency histogram for the cholesterol levels of 20 high-level patients:

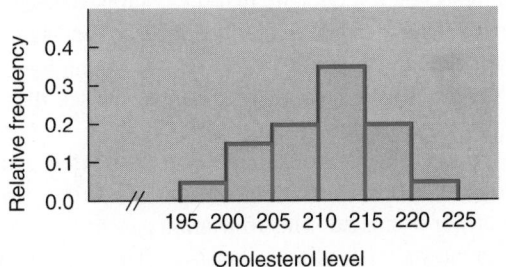

2.88   A stem-and-leaf diagram for the lengths of stay in Europe and the Mediterranean obtained from a sample of 36 U.S. residents who traveled there one year:

```
0 | 5 3 3 1 6 2 1 1 8 5 3
1 | 6 3 1 5 4 0 2 7 8 0 2 2 0
2 | 1 0 1 7 1
3 | 1 2
4 | 1 4 8
5 | 6
6 | 4
```

2.89   A stem-and-leaf diagram for the contents, in milliliters, of a sample of 30 "one-liter" bottles obtained by a consumer group from a soft-drink bottler:

```
 91 | 4
 92 |
 93 |
 94 | 6
 95 | 9 7
 96 | 4
 97 | 7 5 4 7
 98 | 6 9 4 8 7
 99 | 0 6 1 9 5 7
100 | 1
101 | 4 8 0 7
102 | 5 8
103 | 0 1
104 |
105 |
106 | 0
```

2.90   A stem-and-leaf diagram for the percentage of adults in each state that has completed high school:

```
5 | 3 4
5 | 7 6 6 8 5 5 6 6
6 | 4 1 3 2
6 | 9 7 7 6 9 7 8 7 9 6 6 7 6 5 8
7 | 2 4 0 4 4 2 3 2 3 4 3 2 1 0
7 | 9 6 6 8 8
8 | 3 0
8 |
```

2.91   A stem-and-leaf diagram for the annual crime rates of the states in the United States:

```
2 | 3
2 | 6 7
3 | 1 3 3 1
3 | 9 9 5 9 9
4 | 3 2 2 4 3 4 0 1 4
4 | 8 8 8 7 7 5 9 5
5 | 2 1
5 | 5 7 5 6 8 5
6 | 2 1 3 0
6 | 8 5 6 9
7 | 3 0 1 4
7 |
8 | 2
8 |
```

2.92   Use a table of random numbers, a random-number generator, or a computer to obtain 60 random integers between 0 and 9.

a. Without graphing the distribution of the 60 numbers you obtained, guess its shape and explain your reasoning.

b. Construct a relative-frequency histogram for the 60 numbers you obtained. Is its shape roughly what you expected?

2.93   Refer to the height data given in Exercise 2.34 on page 61. Those data consist of the heights of both the males and females in a freshman calculus course.

a. What shape would you expect for the distribution of the male heights? of the female heights?

b. What shape would you expect for the distribution of the male and female heights combined?

c. Using equal-width classes starting with the class 60–61, construct a relative-frequency histogram for the heights of males and females combined. Does this confirm your answer to part (b)?

2.6   MISLEADING GRAPHS

Graphs and charts are frequently constructed in a manner that causes them to be misleading. Sometimes this is intentional and sometimes it is not. Regardless of the intent, it is important to read and interpret graphs and charts with a great deal of care. In this section we will examine some misleading graphs and charts.

Example 2.21   Truncated Graphs

Figure 2.11(a) shows a bar graph from an article in a major metropolitan newspaper. The graph displays the unemployment rates in the United States from September of one year to March of the next year.

A quick look at Fig. 2.11(a) might lead you to conclude that the unemployment rate dropped by roughly one-fourth between January and March, since the bar for March is about one-fourth smaller than the bar for January. In reality, however, the unemployment rate dropped less than one-thirteenth, from 5.4% to 5.0%. Consequently, we see that we must analyze the graph more carefully to figure out what it truly represents.

The unemployment-rate graph in Fig. 2.11(a) is an example of a **truncated graph** because the vertical axis, which should start at 0%, starts at 4% instead. Thus the part of the graph from 0% to 4% has been cut off, or truncated. This truncation causes the bars to be out of correct proportion and hence creates a misleading impression. The graph would

FIGURE 2.11
Unemployment rates:
(a) truncated graph,
(b) nontruncated graph

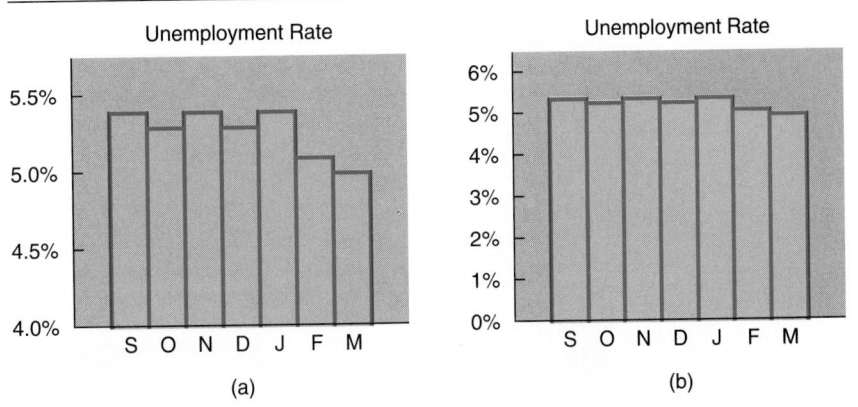

be even more deceptive if it started at 4.5%. To see this, slide a piece of paper over the bottom of Fig. 2.11(a) so that the bars begin at 4.5%. By how much does it now appear that the unemployment rate dropped between January and March?

As we have observed, the truncated graph in Fig. 2.11(a) is potentially misleading. However, it is probably safe to say that the truncation was done to present a picture of the "ups" and "downs" in the unemployment-rate pattern rather than to intentionally mislead the reader.

A nontruncated version of Fig. 2.11(a) is shown in Fig. 2.11(b). Although Fig. 2.11(b) provides a correct graphical display of the unemployment-rate data, the "ups" and "downs" are not as easy to spot as they are in the truncated graph in Fig. 2.11(a). ■

Truncated graphs have long been a target of statisticians. Many statistics books warn against their use. Nonetheless, as we saw in Example 2.21, truncated graphs are still used today, even in reputable publications.

On the other hand, Example 2.21 also suggests that it may be desirable to cut off part of the vertical axis of a graph in order to more easily convey relevant information, such as the "ups" and "downs" of the monthly unemployment rates. In these cases a truncated graph should not be used. Instead, a special symbol, such as //, should be employed to signify that the vertical axis has been modified.

The two graphs shown in Fig. 2.12 at the top of the following page provide an excellent illustration. Both graphs portray the number of new single-family homes sold per month over several months. The first graph, Fig. 2.12(a), is a truncated graph, truncated most likely in an attempt to present a clear visual display of the variation in sales. The second graph, Fig. 2.12(b), accomplishes the same result but is less subject to misinterpretation. The reader is aptly warned by the slashes that part of the vertical axis between 0 and 500 has been removed.

FIGURE 2.12
New single-family home sales

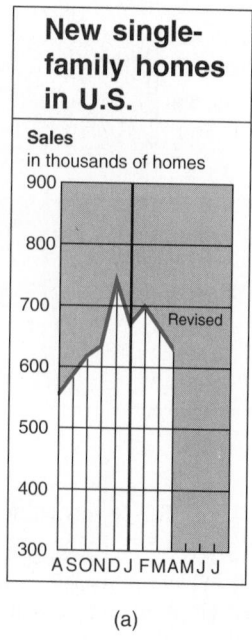

(a)

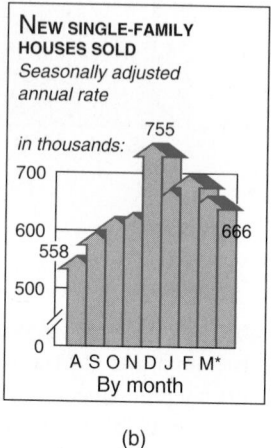

(b)

SOURCES: Figure 2.12(a) reprinted by permission of Tribune Media Services. Figure 2.12(b) data from U.S. Department of Commerce and U.S. Department of Housing and Urban Development.

Improper Scaling

Misleading graphs and charts can also result from **improper scaling.** Example 2.22 shows how this can happen.

Example 2.22   Improper Scaling

A developer is preparing a brochure to attract investors for a new shopping center that is to be built in an area of Denver, Colorado. The area is growing rapidly; this year twice as many homes will be built there as last year. To illustrate that fact, the developer draws a *pictogram*, as in Fig. 2.13.

FIGURE 2.13
Pictogram for home building

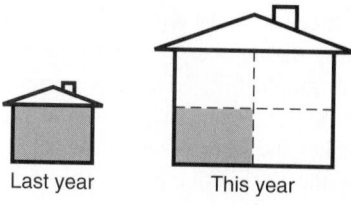

The house on the left represents the number of homes built last year. Since the number of homes that will be built this year is double the number built last year, the developer makes the house on the right twice as tall and twice as wide as the house on the left. However, this scaling is improper because it gives the visual impression that four times as many homes will be built this year as last. So the developer's brochure may mislead the unwary investor. ■

Graphs and charts can be misleading in countless ways besides the two ways we have discussed. Many more examples of misleading graphs can be found in the entertaining and classic book *How to Lie with Statistics* by Darrell Huff (New York: Norton, 1955). The main purpose of this section has been to show you that graphs and charts should be constructed and read carefully.

EXERCISES 2.6

2.94 Find two examples of graphs in a current newspaper or magazine that might be misleading. Explain why you think the graphs are potentially misleading.

2.95 Each year the director of the reading program in a school district administers a standard test of reading skills. Then the director compares the average score for his district with the national average. Figure 2.14 was presented to the school board in 1993.

FIGURE 2.14   Average reading scores

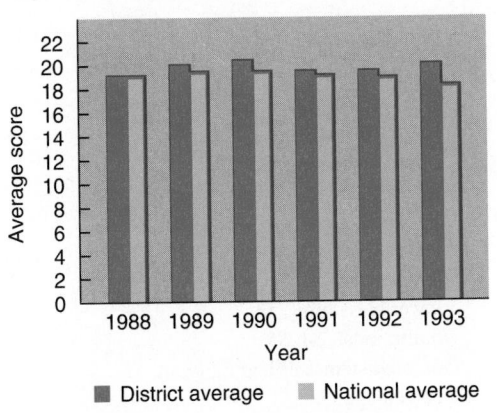

a. Obtain a truncated version of Fig. 2.14 by sliding a piece of paper over the bottom of the graph so that the bars start at 16.

b. Repeat part (a) but have the bars start at 18.
c. What misleading impression about the 1993 scores is given by the truncated graphs you observed in parts (a) and (b)?

2.96 The following bar graph was taken from a newspaper article entitled "Immigrants add seasoning to America's melting pot." [Used with permission from American Demographics, © 1985, Ithaca, NY.]

Race and Ethnicity in America
(in millions)

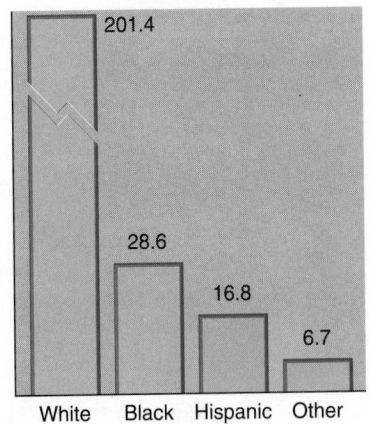

Data from Census Bureau July 1984 Estimates

a. Explain why a break is shown in the first bar.
b. Why was the graph constructed with a broken bar?
c. Do you think this graph is potentially misleading?

2.97   The following bar graph, taken from *The Arizona Republic,* provides data on the M2 money supply over several months. M2 consists of cash in circulation, deposits in checking accounts, nonbank traveler's checks, accounts such as savings deposits, and money-market mutual funds.

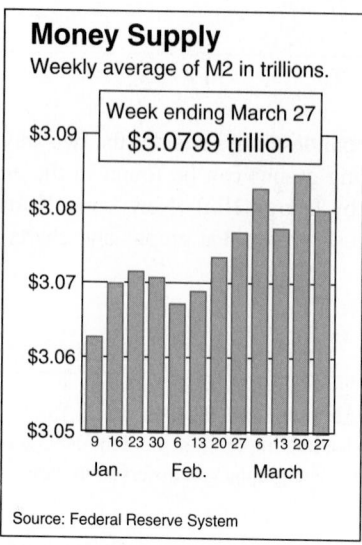

a. What is wrong with the bar graph?

b. Construct a version of the bar graph with a nontruncated and unmodified vertical axis.

c. Construct a version of the bar graph in which the vertical axis is modified in an acceptable manner.

2.98   Refer to Example 2.22 on page 92. Indicate a way in which the developer can accurately illustrate that twice as many homes will be built in the area this year as last.

2.99   A manufacturer of golf balls has determined that a newly developed process results in a ball that lasts roughly twice as long as a ball produced using the current process. To illustrate this graphically, she designs a brochure showing a "new" ball with twice the radius of the "old" ball.

Old ball   New ball

a. What is wrong with this? *(Hint:* The volume of a sphere is proportional to the cube of its radius.*)*

b. How can the manufacturer accurately illustrate the fact that the "new" ball lasts twice as long as the "old" ball?

| **Chapter** | **Review** |

relative frequency, *52*   stem-and-leaf diagram, *77*
relative-frequency distribution, *52*   stems, *77*
relative-frequency histogram, *63*   symmetric, *86*
reverse J-shaped, *84*   triangular, *84*
right skewed, *84*   truncated graph, *90*
shaded stem-and-leaf diagram, *77*   uniform, *84*
START,* *68*   unimodal, *85*
Stem-and-Leaf,* *79*   upper class limit, *52*
STEM-AND-LEAF,* *79*   variable, *44*

You Should Be
Able To

1. classify variables and data as either qualitative or quantitative.
2. distinguish between discrete and continuous variables and data.
3. group data into a frequency distribution and a relative-frequency distribution.
4. construct a grouped-data table.
5. draw a frequency histogram and a relative-frequency histogram.
6. construct a dotplot.
7. draw a pie chart and a bar graph.
8. construct stem-and-leaf, shaded stem-and-leaf, and ordered stem-and-leaf diagrams.
9. identify the shape and modality of the distribution of a data set.
10. specify whether a distribution is symmetric, right skewed, or left skewed.
11. identify and correct misleading graphs.
12. use the Minitab commands covered in this chapter.*
13. interpret the output obtained from the application of the Minitab commands discussed in this chapter.*

REVIEW TEST

1. The world's five largest hydroelectric plants, based on ultimate capacity, are as shown in the following table. Capacities are in megawatts. [SOURCE: T. W. Mermel, *Intl. Waterpower & Dam Construction Handbook.*]

Rank	Name	Country	Capacity
1	Turukhansk	Russia	20,000
2	Itaipu	Brazil/Para.	13,320
3	Grand Coulee	U.S.A.	10,830
4	Guri	Venezuela	10,300
5	Tucurui	Brazil	7,260

a. What type of data is given in the first column of the table?
b. What type of data is given in the fourth column?
c. What type of data is given in the third column?

2. The ages at inauguration for the first 42 presidents of the United States are shown in the table at the right.
   a. Construct a grouped-data table for these inauguration ages using equal-width classes and beginning with the class 40–44.

President	Age at inaug.	President	Age at inaug.
G. Washington	57	G. Cleveland	47
J. Adams	61	B. Harrison	55
T. Jefferson	57	G. Cleveland	55
J. Madison	57	W. McKinley	54
J. Monroe	58	T. Roosevelt	42
J. Q. Adams	57	W. Taft	51
A. Jackson	61	W. Wilson	56
M. Van Buren	54	W. Harding	55
W. Harrison	68	C. Coolidge	51
J. Tyler	51	H. Hoover	54
J. Polk	49	F. Roosevelt	51
Z. Taylor	64	H. Truman	60
M. Fillmore	50	D. Eisenhower	62
F. Pierce	48	J. Kennedy	43
J. Buchanan	65	L. Johnson	55
A. Lincoln	52	R. Nixon	56
A. Johnson	56	G. Ford	61
U. Grant	46	J. Carter	52
R. Hayes	54	R. Reagan	69
J. Garfield	49	G. Bush	64
C. Arthur	50	W. Clinton	46

b. Draw a frequency histogram for the inauguration ages based on your grouping in part (a).

3. Refer to Problem 2. Construct a dotplot for the ages at inauguration of the first 42 U.S. presidents.

4. Refer to Problem 2. Construct an ordered stem-and-leaf diagram for the inauguration ages of the first 42 presidents of the United States using
   a. one line per stem.
   b. two lines per stem.
   c. Which of the two stem-and-leaf diagrams you just constructed corresponds to the frequency distribution of Problem 2(a)?

*5. (Computer problem) Refer to the age data in Problem 2. Use Minitab or some other statistical software to obtain a
   a. frequency distribution and histogram of the data similar to the ones found in Problem 2.
   b. dotplot of the data.
   c. stem-and-leaf diagram similar to the one constructed in Problem 4(b).

*6. (Computer problem) A city planner working on bikeways needs information about local bicycle commuters. She designs a questionnaire. One of the questions asks how many minutes it takes the rider to pedal from home to his or her destination. The times, rounded to the nearest minute, for a sample of local bicycle commuters are stored in Minitab, and the following output is obtained by applying the Histogram command.

```
Histogram of TIMES   N = 22


Midpoint   Count
   12.00      1   *
   17.00      3   ***
   22.00      6   ******
   27.00      7   *******
   32.00      3   ***
   37.00      1   *
   42.00      0
   47.00      1   *
```

a. How many times are in the sample?
b. Identify the class mark for the fourth class.
c. How many times are between 30 and 34 minutes, inclusive?
d. What column name was chosen for the data?

*7. (Computer problem) Refer to Problem 6. The Minitab output below provides a stem-and-leaf diagram for the sample of bicycle-commuter times.

```
Stem-and-leaf of TIMES     N = 22
Leaf Unit = 1.0


    1      1 2
    4      1 569
   10      2 122334
   (7)     2 6678999
    5      3 011
    2      3 7
    1      4
    1      4 8
```

a. How many lines per stem are used in this stem-and-leaf diagram?
b. Use the depths to determine how many of the times are 35 minutes or more.
c. Use the depths to determine how many of the times are less than 20 minutes.
d. What is the longest time in the sample?
e. List the times that are in the 30s.

8. The Prescott National Bank has six tellers available to serve customers. The data in the following table provide the number of busy tellers observed at 25 spot checks.

6	5	4	1	5
6	1	5	5	5
3	5	2	4	3
4	5	0	6	4
3	4	2	3	6

a. Construct a grouped-data table for these data using single-value grouping.
b. Draw a relative-frequency histogram for the data based on the grouping in part (a).

9. The National Safety Council reports that in 1988 the number of accidental deaths by type in the United States were as follows.

Type	Frequency
Motor vehicle	49,000
Work	10,600
Home	22,500
Public	18,000

a. Draw a pie chart of the data that displays the percentages of accidental deaths in the four categories.

b. Draw a bar graph of the data that displays the relative frequencies of accidental deaths in the four categories.

10. According to *The World Almanac, 1993,* the highs for the Dow Jones Industrial Averages for the years 1956–1991 are as follows.

Year	High	Year	High
1956	521.05	1974	891.66
1957	520.77	1975	881.81
1958	583.65	1976	1014.79
1959	679.36	1977	999.75
1960	685.47	1978	907.74
1961	734.91	1979	897.61
1962	726.01	1980	1000.17
1963	767.21	1981	1024.05
1964	891.71	1982	1071.55
1965	969.26	1983	1287.20
1966	995.15	1984	1286.64
1967	943.08	1985	1553.10
1968	985.21	1986	1955.57
1969	968.85	1987	2722.42
1970	842.00	1988	2183.50
1971	950.82	1989	2791.41
1972	1036.27	1990	2999.75
1973	1051.70	1991	3168.83

a. Construct a grouped-data table for the highs using classes of equal width. Take "400–under 600" as the first class.

b. Draw a relative-frequency histogram for the highs based on your result from part (a).

11. Identify the shape of the distribution for each of the following data sets.

a. The inauguration ages of the first 42 presidents of the United States (from Problem 2).

b. The number of tellers busy with customers at Prescott National Bank during 25 spot checks (from Problem 8).

12. Draw a smooth curve that represents a symmetric trimodal distribution.

13. The following graph is from a newspaper article entitled "Hand that rocked cradle turns to work as women reshape U.S. labor force." [Used with permission from American Demographics, © 1985.]

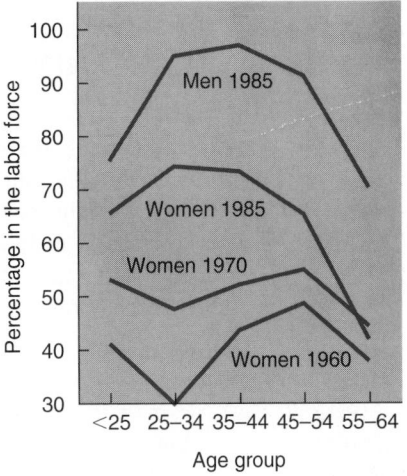

Working Men and Women by Age, 1960–1985

a. Cover up the numbers on the vertical axis of the graph with a piece of paper.

b. Look at the 1970 and 1985 graphs for women. Focus on the 25–34-year-old age group. What impression does the graph convey regarding the ratio of the percentages of women in the labor force for 1985 and 1970?

c. Now remove the piece of paper from the graph. Using the vertical scale, find the actual ratio of the percentages of 25–34-year-old women in the labor force for 1985 and 1970.

d. Why is the graph potentially misleading?

e. What can be done to make the graph less potentially misleading?

Using the Focus Database

Appendix B contains a printout of a database obtained by randomly selecting 500 Arizona State University sophomores. Seven variables are considered for each student: sex, high-school GPA, SAT math score, cumulative GPA, SAT verbal score, age, and total hours. Use Minitab or some other statistical software to solve the following problems.

a. Obtain a histogram of the ages of the sophomores in the sample.

b. Obtain individual histograms of the ages of the female sophomores and the male sophomores in the sample. Compare the two histograms and discuss the differences you observe.

c. Construct a stem-and-leaf diagram for the cumulative GPAs of the sophomores in the sample. Use five lines per stem.

d. Construct individual stem-and-leaf diagrams for the cumulative GPAs of the female sophomores and the male sophomores in the sample. Use five lines per stem for both diagrams. Compare the two diagrams and discuss any differences you observe.

e. Obtain dotplots for both the SAT math and SAT verbal scores of the sophomores in the sample. Compare the two dotplots.

f. Identify the shapes of the distributions in parts (a)–(e). Which distributions are symmetric?

Case Study   **Infant Mortality in Developed Nations**

Infant mortality is concerned with infant deaths during the first year of life. Generally, the infant mortality rate provides the number of such deaths per 1000 live births. In 1987, Congress established the National Commission to Prevent Infant Mortality, whose charge is to create a national strategy for reducing the nation's infant mortality rate. A study conducted by the commission, released in March 1992, ranked the developed nations by their infant mortality rates. Table 2.23 presents the results.

TABLE 2.23

Infant mortality rates of developed nations, 1989

Rank	Country	Infant mortality	Rank	Country	Infant mortality
1	Japan	4.59	13	Norway	7.72
2	Sweden	5.77	14	Australia	7.99
3	Finland	6.03	15	Spain	8.07[†]
4	Singapore	6.61	16	Austria	8.31
5	Netherlands	6.78	17	United Kingdom	8.42
6	Canada	7.20[†]	18	Denmark	8.45
7	Switzerland	7.34	19	Belgium	8.64
8	France	7.36	20	Italy	8.80
9	Hong Kong	7.43	21	Greece	9.77
10	East Germany	7.44	22	United States	9.80
11	Ireland	7.55	23	Israel	9.94
12	West Germany	7.56	24	New Zealand	10.19

[†] Rate is for 1988.

Note: Singapore and Hong Kong, although not defined as "developed" by the United Nations, are included since they have infant mortality rates lower than that of the United States.

As we see from Table 2.23, the infant mortality rates for developed countries vary considerably, from a low of 4.59 (Japan) to a high of 10.19 (New Zealand). What is surprising is the poor showing of the United States, which is purported to have the most sophisticated medical technology and most highly trained medical personnel in the world. In fact, the commission's report states that "... our nation's mothers and children are in trouble. The signs warning of this have been clearly visible for some time now, but we have not taken them seriously enough."

a. What type of data is displayed in the second column of Table 2.23?
b. What type of data is provided by the statement that the United States ranks 22nd in infant mortality rate among the developed nations?
c. Construct a grouped-data table for the infant mortality rates using classes of equal width and starting with the class "4–under 5."
d. Construct a frequency histogram for the infant mortality rates based on your grouping in part (c).
e. Construct an ordered stem-and-leaf diagram for the infant mortality rates using two lines per stem. *(Note:* In constructing the stem-and-leaf diagram, ignore the hundredths digit in the numbers.*)*
f. Identify the shape of the distribution of the infant mortality rates.
g. Use Minitab or some other statistical software to solve parts (c)–(e).

John Tukey

John Wilder Tukey was born on June 16, 1915, in New Bedford, Massachusetts. After earning bachelor's and master's degrees in chemistry from Brown University in 1936 and 1937, respectively, he enrolled in the mathematics program at Princeton University, where he received a master's degree in 1938 and a doctorate in 1939.

After graduating, Tukey was appointed Henry B. Fine Instructor in Mathematics at Princeton; 10 years later he was advanced to a full professorship. In 1965, Princeton established a department of statistics, and Tukey was named its first chairperson. In addition to his position at Princeton, he was a member of the Technical Staff at AT&T Bell Laboratories from 1945 until his retirement in 1985 as Associate Executive Director–Research in the Information Sciences Division.

Tukey is among those who have led the way in the field of exploratory data analysis, which provides techniques, such as stem-and-leaf diagrams, for effectively investigating data. He has made fundamental contributions to the areas of robust estimation and time series analysis. Tukey has written numerous books and more than 350 technical papers on mathematics, statistics, and other scientific subjects. He also coined the word *bit,* a contraction of *binary digit* (a unit of information, often as processed by a computer).

Tukey's participation in educational, public, and government service is most impressive. He was appointed to serve on the President's Science Advisory Committee by President Eisenhower; was chairperson of the committee that prepared "Restoring the Quality of our Environment" in 1965; helped develop the National Assessment of Educational Progress; and was a member of the Special Advisory Panel on 1990 Census of the U.S. Department of Commerce, Bureau of the Census—to name only a few of his involvements.

Among many honors, Tukey has received the National Medal of Science, the IEEE Medal of Honor, Princeton University's James Madison Medal, and Foreign Member, The Royal Society (London). He was the first recipient of the Samuel S. Wilks Award of the American Statistical Association. Tukey remains on the faculty at Princeton as Donner Professor of Science, Emeritus; Professor of Statistics, Emeritus; and Senior Research Statistician.

Chapter 3

Descriptive Measures

. . . . .

General Objective   In Chapter 2 we began our study of descriptive statistics. We learned how to organize data into tables and summarize data with graphical displays. Another method of summarizing data is to compute numbers, such as an average, that describe the data set. Numbers that are used to describe data sets are called **descriptive measures.** In this chapter we will continue our study of descriptive statistics by examining some of the most important descriptive measures.

Chapter Outline   3.1   Measures of Central Tendency

3.2   Summation Notation; the Sample Mean

3.3   Measures of Dispersion; the Sample Standard Deviation

3.4   Interpretation of the Standard Deviation; z-Scores

3.5   Computing $\overline{x}$ and s for Grouped Data

3.6   The Five-Number Summary; Boxplots

3.7   Descriptive Measures for Populations; Use of Samples

Case Study   *Per Capita Income by State*

Per capita income is defined as the annual total personal income of the residents of a region divided by the resident population of that region as of July 1 of that year. How does per capita income vary from state to state? Where does your state rank? A report by the Commerce Department and The Associated Press provided a state-by-state breakdown of per capita income. We will analyze those data in the case study at the end of this chapter.

3.1 MEASURES OF CENTRAL TENDENCY

Descriptive measures that indicate where the center or most typical value of a data set lies are called **measures of central tendency,** often more simply referred to as averages. In this section we will discuss the three most important measures of central tendency: the *mean,* the *median,* and the *mode.* The mean and median apply only to quantitative data, whereas the mode can be used with either quantitative or qualitative data.

The Mean

The most commonly used measure of central tendency is the mean. When people speak of taking an average, it is the **mean** that they are most often referring to. The definition of the mean follows.

DEFINITION 3.1 **MEAN OF A DATA SET**

> The *mean* of a data set is defined as the sum of the data divided by the number of pieces of data:
>
> $$\text{Mean} = \frac{\text{Sum of the data}}{\text{Number of pieces of data}}.$$

Example 3.1   Illustrates Definition 3.1

A mathematician spent one summer working for a small mathematical consulting firm. The firm employed a few senior consultants, who made between $700 and $950 per week; a few junior consultants, who made between $300 and $350 per week; and several clerical workers, who made $200 per week.

Because the first half of the summer was busier than the second half, more employees were required during the first half. Tables 3.1 and 3.2 display typical lists of weekly earnings for the two halves of the summer. Find the mean of each of the two data sets.

TABLE 3.1
Data Set I

$200	200	200	840	200	200	300
200	300	350	700	350	950	

TABLE 3.2
Data Set II

$200	200	840	350	300
300	200	200	950	200

SOLUTION   According to Definition 3.1, the mean of a data set is obtained by summing all the data and then dividing that sum by the total number of pieces of data. As we see from Table 3.1, Data Set I has 13 pieces of data. The sum of the 13 pieces of data in that data set is $4990.

Consequently,

$$\text{Mean of Data Set I} = \frac{\$4990}{13} = \$383.85 \text{ (rounded to the nearest cent).}$$

Similarly,

$$\text{Mean of Data Set II} = \frac{\$3740}{10} = \$374.00.$$

 Thus the mean salary of the 13 employees in Data Set I is $383.85 and that of the 10 employees in Data Set II is $374.00. ∎

The Median

Another frequently used measure of central tendency is the median. Essentially, the **median** of a data set is the number that divides the bottom 50% of the data from the top 50%. To obtain the median of a data set, we arrange the data in increasing order and then determine the middle value in the ordered list. A more precise definition of the median is found in Definition 3.2.

DEFINITION 3.2

MEDIAN OF A DATA SET

Arrange the data in increasing order.

- If the number of pieces of data is odd, then the *median* is the data value exactly in the middle of the ordered list.
- If the number of pieces of data is even, then the *median* is the mean of the two middle data values in the ordered list.

In both cases, if we let n denote the number of pieces of data, then the median is at position $(n + 1)/2$ in the ordered list.

Example 3.2 Illustrates Definition 3.2

Consider again the two sets of salary data shown in Tables 3.1 and 3.2. Determine the median of each of the two data sets.

SOLUTION To find the median of Data Set I, we apply Definition 3.2. First we arrange the data in increasing order:

200 200 200 200 200 200 **300** 300 350 350 700 840 950

The number of pieces of data in Data Set I is 13, which is an odd number. Since $n = 13$, $(n + 1)/2 = (13 + 1)/2 = 7$. Consequently, the median is the seventh data value in the ordered list, which is 300 (shown in boldface type). The median salary of the 13 employees in Data Set I is $300.

To find the median of Data Set II, we again apply Definition 3.2. First we arrange the data in increasing order:

200   200   200   200   **200   300**   300   350   840   950

The number of pieces of data in Data Set II is 10, which is an even number. Since $n = 10$, $(n + 1)/2 = (10 + 1)/2 = 5.5$. Consequently, the median is halfway between the fifth and sixth data values (shown in boldface type) in the ordered list. In other words, the median salary of the 10 employees in Data Set II is $(200 + 300)/2 = \$250$.   ∎

As we have just seen, to determine the median of a data set we must first arrange the data in increasing order. It is often helpful to construct a stem-and-leaf diagram as a preliminary step to ordering the data.

The Mode

The final measure of central tendency we will discuss is the mode. Basically, the **mode** is the value that occurs most frequently in a data set. A more exact definition of the mode is provided in Definition 3.3.

DEFINITION 3.3   **MODE OF A DATA SET**

> If no value in a data set occurs more than once, then we say that the data set has no mode. Otherwise, any value that occurs with maximal frequency is called a *mode* of the data set. In other words, a mode of a data set is any value whose frequency of occurrence is greater than 1 and is as large or larger than any other value's frequency of occurrence.

To obtain the mode(s) of a data set, we first construct a frequency distribution for the data using classes based on a single value. The mode(s) can then be determined easily from the frequency distribution, as explained in Example 3.3.

Example 3.3   Illustrates Definition 3.3

Determine the mode(s) of each of the two sets of salary data given in Tables 3.1 and 3.2 on page 102.

SOLUTION   First we consider the salary data in Data Set I. Referring to Table 3.1, we obtain the frequency distribution of the data using classes based on a single value, as shown in Table 3.3.

TABLE 3.3
Frequency distribution for Data Set I using single-value grouping

Salary	200	300	350	700	840	950
Frequency	6	2	2	1	1	1

We see from Table 3.3 that the most frequently occurring value in Data Set I is 200 (which occurs six times). So the mode of the 13 salaries in Data Set I is $200.

Next we determine the mode for Data Set II. Referring to Table 3.2, we construct the frequency distribution using classes based on a single value, as shown in Table 3.4.

TABLE 3.4
Frequency distribution for Data Set II using single-value grouping

Salary	200	300	350	840	950
Frequency	5	2	1	1	1

Table 3.4 shows that 200 is the value that occurs most often in Data Set II. Therefore the mode of the 10 salaries in Data Set II is $200. ■

As we noted, a data set can have more than one mode if there is a tie for the most frequently occurring value. For instance, suppose two of the clerical workers in Data Set I, who make $200 per week, were promoted to $300-per-week jobs. Then both the value 200 and the value 300 would occur with maximal frequency (four times each). This new data set would have two modes, $200 and $300.

Comparison of the Mean, Median, and Mode

The mean, median, and mode of a data set are frequently different. Table 3.5 summarizes the definitions of these three measures of central tendency and gives their values for Data Set I and Data Set II, which we computed in Examples 3.1–3.3.

TABLE 3.5
Means, medians, and modes of salaries in Data Set I and Data Set II

Measure of central tendency	Definition	Data Set I	Data Set II
Mean	$\dfrac{\text{Sum of the data}}{\text{Number of pieces of data}}$	$383.85	$374.00
Median	Middle value in ordered list	$300.00	$250.00
Mode	Most frequent value	$200.00	$200.00

In both Data Sets I and II, the mean is larger than the median. This is because the mean is strongly affected by the few large salaries in each data set. Generally speaking, the mean is sensitive to extreme (very large or very small) data values, whereas the median is not. Consequently, *when the choice for the measure of central tendency is between the mean and the median, the median is usually preferred for data sets that have extreme values.*

Figure 3.1 shows the relative positions of the mean and median for right-skewed, symmetric, and left-skewed distributions. As we see from the figure, the mean is pulled in the direction of skewness, that is, in the direction of the extreme values. For a right-skewed distribution, the mean is greater than the median; for a symmetric distribution, the mean and median are equal; and for a left-skewed distribution, the mean is less than the median.

FIGURE 3.1

Relative positions of the mean and median for (a) right-skewed, (b) symmetric, and (c) left-skewed distributions

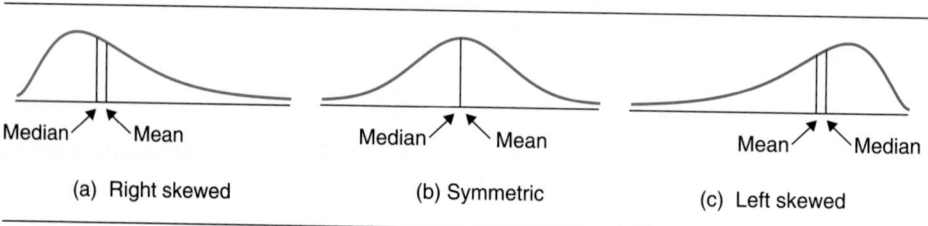

A descriptive measure is called **resistant** if it is not sensitive to the influence of a few extreme data values. Thus the median is a resistant measure of central tendency, whereas the mean is not. The resistance of the mean can be improved by using **trimmed means,** where a specified percentage of the smallest and largest data values are removed before computing the mean. Exercise 3.20 discusses trimmed means in more detail.

The mode for each of Data Sets I and II differs from both the mean and the median. Whereas the mean and the median are aimed at finding the center of a data set, the mode is not—the most frequently occurring data value may not be near the center.

It should now be clear that the mean, median, and mode generally provide different information. There is no simple rule for deciding which measure of central tendency should be used in any given situation. Skill in making such decisions is attained through practice. However, even experts may disagree on the most suitable measure of central tendency for a particular data set. Example 3.4 discusses three data sets and suggests the most appropriate measure of central tendency for each.

Example 3.4   Selecting an Appropriate Measure of Central Tendency

a. A student takes four exams in a biology class. His grades are 88, 75, 95, and 100. If asked for his average, which measure of central tendency is the student likely to report?

b. The National Association of REALTORS® publishes data on resale prices of U.S. homes. Which measure of central tendency is most appropriate for such resale prices?

c. In the 1992 Boston Marathon, there were two categories of official finishers: male and female. The following table provides a frequency distribution for those data. Which measure of central tendency should be used here?

Sex	Frequency
Male	6562
Female	1561

SOLUTION

a. Chances are that the student would report the mean of his four exam scores, which is 89.5. The mean is probably the most suitable measure of central tendency for the student to use since it takes into account the numerical value of each score and therefore indicates total overall performance.

b. The most appropriate measure of central tendency for resale home prices is the median, because it is aimed at finding the center of the data on resale home prices and because it is not strongly affected by the relatively few homes with extremely high or low resale prices. Thus the median provides a better indication of the "typical" resale price than either the mean or the mode does.

c. The only suitable measure of central tendency for these data is the mode, which is "male." Each piece of data in this data set is either "male" or "female." There is no way to compute a mean or median for such data. The mode is the only measure of central tendency that can be used for qualitative data. ■

Many averages appearing in newspapers or reported by government agencies are medians; for example, household income and years of school completed. Furthermore, in an attempt to provide a clearer picture of the data, some studies report both the mean and the median. For instance, the National Center for Health Statistics does this for average daily intake of nutrients in the publication *Vital and Health Statistics*.

Population and Sample Averages

We have defined the mean, the median, and the mode of a data set. Whether those measures of central tendency are considered population averages or sample averages depends on various things, such as how they are to be used. For example, suppose the data set consists of the ages of the students in Professor Weiss's introductory statistics class, as shown in Table 3.6.

TABLE 3.6
Ages of the students in Professor Weiss's introductory statistics class

19	24	24	24	23	20	22	21
18	20	19	19	21	19	19	23
36	22	20	35	22	23	19	26
22	17	19	20	20	21	19	21
20	20	21	19	24	21	22	21

We computed the mean, median, and mode of this data set and obtained the following values: mean = 21.6 years, median = 21.0 years, and mode = 19.0 years.

The age data in Table 3.6 may constitute the entire population, or it may be only a sample from a population. For a researcher interested only in the ages of the students in this particular introductory statistics class, the data set is a population; and its mean, median, and mode are a *population mean, population median,* and *population mode.* For another researcher, interested in the ages of students in all introductory statistics classes, the data set is only a sample; and its mean, median, and mode are a *sample mean, sample median,* and *sample mode.* Figure 3.2 shows the two ways in which the mean of a data set may be interpreted.

FIGURE 3.2

Possible interpretations
for the mean of a data set

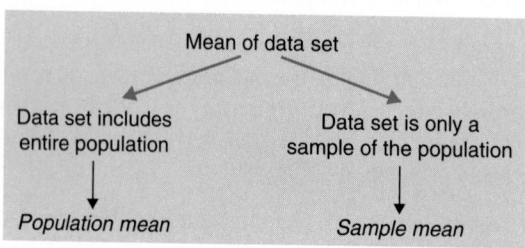

Recall that inferential statistics consists of methods for making inferences about a population based on information obtained from a sample of the population. Thus, in inferential statistics, the data analyzed are sample data. Because most of this text is devoted to inferential statistics, we will concentrate on the calculations required for determining descriptive measures of sample data.

However, keep in mind that descriptive measures of sample data are rarely an end in themselves. Rather, they are usually a means of drawing conclusions about the population from which the sample was taken. The relationships between a population and a sample will be discussed further in Section 3.7.

Using the Computer (Optional)

Minitab can be applied in several ways to obtain measures of central tendency for a data set. Here we will illustrate how the **Mean** and **Median** commands can be used to determine a mean and median.

Example 3.5   The Mean and Median Commands

The salary data in Data Set I, displayed in Table 3.1, are repeated here in Table 3.7. Use Minitab to find the mean and median of these salaries.

TABLE 3.7 Data Set I	$200	200	200	840	200	200	300
	200	300	350	700	350	950	

SOLUTION First we store the salary data from Table 3.7 in a column named DATASETI. Then we proceed in the following manner.

Session commands: To determine the mean of Data Set I, we type the command **MEAN** followed by the storage location of the data; that is, we type MEAN 'DATASETI' and press Enter. Similarly, to determine the median of Data Set I, we type the command **MEDIAN** followed by the storage location of the data; that is, we type MEDIAN 'DATASETI' and press Enter. Printout 3.1 shows these commands and the resulting output.

(or)

Menu commands: To get the mean of Data Set I, we choose **Calc ▸ Functions and Statistics ▸ Column Statistics...**, select **Mean** from the **Statistic** list, specify DATASETI as the **Input variable**, and then select **OK**. Using the same steps except selecting **Median** instead of **Mean**, we obtain the median of Data Set I.[†] The resulting output is displayed in Printout 3.1.

PRINTOUT 3.1
Minitab output for the Mean and Median commands

```
MTB > MEAN 'DATASETI'
    MEAN    =      383.85
MTB > MEDIAN 'DATASETI'
    MEDIAN =      300.00
```

From Printout 3.1 we see that for Data Set I, the mean salary is $383.85 and the median salary is $300.00. ■

Minitab does not have a Mode command. Nonetheless, we can still use Minitab to obtain the mode of a data set. The details for doing this are discussed in Exercise 3.19.

EXERCISES 3.1

· **3.1** Explain in detail the purpose of a measure of central tendency.

· **3.2** Name and describe the three most important measures of central tendency.

[†] Once we have obtained the mean, it is easier to obtain the median by editing the last command dialogue. Consult your Minitab documentation for details.

Determine the mean, median, and mode(s) for each of the data sets in Exercises 3.3–3.10. For the mean and the median, round each answer to one more decimal place than the original data.

3.3   The U.S. National Science Foundation, Division of Science Resources Studies, collects data on the ages of recipients of science and engineering doctoral degrees. Results are published in *Survey of Earned Doctorates*. A sample of one year's recipients yields the following ages.

37	28	36	33
37	43	41	28
24	44	27	24

3.4   The American Hospital Association publishes figures on the costs to community hospitals per patient per day in *Hospital Statistics*. A sample of 10 such costs in New York yields the data below.

$602	539	569	916	768
335	776	887	806	422

3.5   The average retail price for oranges in 1983 was 38.5 cents per pound, as reported by the U.S. Department of Agriculture in *Food Cost Review*. Recently, a random sample of 15 markets gave the following prices for oranges in cents per pound.

43.0	40.0	42.6	40.2	37.5
44.1	45.2	41.8	35.6	34.6
37.9	44.2	44.5	38.2	42.4

3.6   A biologist is studying the gestation period (duration of pregnancy) of domestic dogs. Fifteen dogs are observed during pregnancy and are found to have the following gestation periods, in days.

62.0	61.4	59.8	62.2	60.3
60.4	59.4	60.2	60.4	60.8
61.8	59.2	61.1	60.4	60.9

3.7   A liquid-soap manufacturer produces a bottle with an advertised content of 310 mL. A sample of 16 bottles yields the following contents.

297	318	306	300
311	303	291	298
322	307	312	300
315	296	309	311

3.8   A city planner working on bikeways needs information about local bicycle commuters. She designs a questionnaire. One of the questions asks how long it takes the rider to pedal from home to his or her destination. A sample of 22 local bicycle commuters yields the times, in minutes, displayed in the following table.

22	19	24	31	12	48
29	29	21	15	22	28
27	23	37	31	23	
30	26	16	26	29	

3.9   The National Center for Education Statistics surveys college and university libraries to obtain information on the number of volumes held. The number of volumes, in thousands, for a sample of seven public colleges and universities are as follows.

79	516	24	265
41	15	411	

3.10   As reported by the College Entrance Examination Board in *National College-Bound Senior*, the average verbal score on the Scholastic Aptitude Test in 1991 was 422 points out of a possible 800. A sample of 25 verbal scores for last year follows.

346	496	352	378	315
491	360	385	500	558
381	303	434	562	496
420	485	446	479	422
494	289	436	516	615

3.11   The U.S. Energy Information Administration conducts annual surveys to estimate the average number of liveable square feet for housing units. Results are published in *Residential Energy Consumption Survey: Housing Characteristics*. In 1984 it was reported that the mean was 1440 sq ft and the median was 1225 sq ft. Which measure of central tendency do you think is more appropriate? Why?

3.12   The U.S. Bureau of the Census publishes figures on the average annual income of all U.S. households in *Current Population Reports*. In 1990 the mean household income was $37,403 and the median household income was $29,943. Which measure of central tendency do you think is more appropriate? Explain your answer.

3.13 According to *The World Almanac, 1993,* the National Collegiate Athletic Association wrestling champions for the years 1963–1992 are as follows.

Year	Champion	Year	Champion
1963	Oklahoma	1978	Iowa
1964	Oklahoma State	1979	Iowa
1965	Iowa State	1980	Iowa
1966	Oklahoma State	1981	Iowa
1967	Michigan State	1982	Iowa
1968	Oklahoma State	1983	Iowa
1969	Iowa State	1984	Iowa
1970	Iowa State	1985	Iowa
1971	Oklahoma State	1986	Iowa
1972	Iowa State	1987	Iowa State
1973	Iowa State	1988	Arizona State
1974	Oklahoma	1989	Oklahoma State
1975	Iowa	1990	Oklahoma State
1976	Iowa	1991	Iowa
1977	Iowa State	1992	Iowa

a. Determine the mode for the data.

b. Would it be appropriate to use either the mean or the median here? Explain your answer.

3.14 The following table depicts the all-time top TV programs by number of viewers as of January 1992. [SOURCE: A. C. Nielsen.]

Program	Date	Network	Households
M*A*S*H Special	2/28/83	CBS	50,150,000
Super Bowl XX	1/26/86	NBC	41,490,000
Dallas	11/21/80	CBS	41,470,000
Super Bowl XVII	1/30/83	NBC	40,480,000
Super Bowl XXI	1/25/87	CBS	40,030,000
Super Bowl XVI	1/24/82	CBS	40,020,000
Super Bowl XIX	1/20/85	ABC	39,390,000
Super Bowl XXIII	1/22/89	NBC	39,320,000
Super Bowl XVIII	1/22/84	CBS	38,800,000
The Day After	11/20/83	ABC	38,550,000
Roots Pt. VIII	1/30/77	ABC	36,380,000
Super Bowl XIV	1/20/80	CBS	35,330,000
Super Bowl XIII	1/21/79	NBC	35,090,000
Super Bowl XV	1/25/81	NBC	34,540,000
Super Bowl XII	1/15/78	CBS	34,410,000
Gone with the Wind Pt. 1	11/7/76	NBC	33,960,000
Gone with the Wind Pt. 2	11/8/76	NBC	33,750,000
Roots Pt. VI	1/28/77	ABC	32,680,000
Roots Pt. V	1/27/77	ABC	32,540,000
Roots Pt. III	1/25/77	ABC	31,900,000

a. Determine the mode for the network data.

b. Would it be appropriate to use either the mean or the median here? Explain your answer.

Ordinal data are data about order or rank. Most statisticians recommend using the median for indicating the center of an ordinal data set, but some researchers also use the mean. In Exercises 3.15 and 3.16, we have presented ordinal data sets. For each exercise,

a. compute the mean of the data.

b. compute the median of the data.

c. decide which measure of central tendency is best.

3.15 A distance runner entered seven marathons. His finishing places in the first six races were 4, 5, 3, 2, 7, and 4. In the seventh race, he decided to go all out to win and ran in first place for 20 miles. This tired him out so badly that he ended up walking parts of the last 6 miles. He did finish, but in 72nd place. The runner's places provide the following ordinal data set: 4, 5, 3, 2, 7, 4, 72.

3.16 Twenty-one algebra students were asked to rate the change in "test anxiety" produced by their algebra course. Negative ratings meant that they worried more about tests at the end of the course than at the beginning, whereas positive ratings meant they worried less at the end of the course than at the beginning. Their ratings were as follows.

0	0	0	−1	−1	1	0
−2	1	0	−2	0	−2	2
−2	1	0	−1	−3	−3	−1

3.17 (Computer exercise) Use Minitab or some other statistical software to determine the mean and the median of the age data in Exercise 3.3.

3.18 (Computer exercise) Use Minitab or some other statistical software to determine the mean and the median of the cost data in Exercise 3.4.

3.19 (Computer exercise) One way that Minitab can be employed to determine the mode of a data set is the following: (1) store the data in a column; (2) apply the Histogram command using a suitable increment to obtain a frequency distribution in which each class is based on a single value; and (3) scan the asterisks in the output to find the mode(s).

a. The following printout was obtained by applying steps (1) and (2) to the test-anxiety data given in Exercise 3.16.

```
Histogram of ANXIETY    N = 21

Midpoint   Count
      -3      2   **
      -2      4   ****
      -1      4   ****
       0      7   *******
       1      3   ***
       2      1   *
```

Apply step (3) to obtain the mode of this data set.

b. Use Minitab or some other statistical software to obtain the mode of the data on bicycle-commuter times from Exercise 3.8.

· ·   **3.20**   Some data sets contain *outliers*, data values that fall well outside the overall pattern of the data. Suppose, for instance, you are interested in the ability of high-school algebra students to compute square roots. You decide to give a square-root exam to 10 of these students. Unfortunately, one of the students had a fight with his girlfriend and cannot concentrate—he gets a 0. The 10 scores are displayed in order in the following table.

0	58	61	63	67	69	70	71	78	80

The score of 0 is an outlier. (Outliers are discussed in more detail in Section 3.6.)

Statisticians have a systematic method for avoiding extreme values and outliers when they calculate means. They compute *trimmed means*, in which high and low values are deleted or "trimmed off" before the mean is calculated. For instance, to compute the 10% trimmed mean of the test-score data, we first delete both the top 10% and the bottom 10% of the ordered data, that is, 0 and 80. Then we calculate the mean of the remaining data. Thus the 10% trimmed mean of the test-score data is

$$\frac{58 + 61 + 63 + 67 + 69 + 70 + 71 + 78}{8} = 67.1.$$

The following table displays a set of algebra final-exam scores for a 40-question test.

2	15	16	16	19	21	21	25	26	27
4	15	16	17	20	21	24	25	27	28

a. Do any of the scores look like outliers?
b. Compute the usual mean of the data.
c. Compute the 5% trimmed mean of the data.
d. Compute the 10% trimmed mean of the data.

e. Compare the three means you obtained in parts (b)–(d). Which of the three means do you think provides the best measure of central tendency for the data?

· ·   **3.21**   Another measure of central tendency is the mid-range. The **midrange** of a data set is defined as the mean of the minimum (smallest) and maximum (largest) data values in the data set:

$$\text{Midrange} = \frac{\text{Min} + \text{Max}}{2}.$$

For instance, the midrange of the four exam scores, 88, 75, 95, and 100, in Example 3.4(a) on page 106 is

$$\text{Midrange} = \frac{75 + 100}{2} = 87.5.$$

a. Compute the midrange of the ages in Exercise 3.3.
b. Compute the midrange of the costs in Exercise 3.4.
c. Compute the midrange of the gestation periods in Exercise 3.6.
d. Identify some of the advantages and disadvantages of the midrange as a measure of central tendency.

The modal class. Suppose we are given a data set that is already grouped into a frequency distribution and we do not have access to the raw data. Then, in general, it is not possible to determine the mode(s). In such cases we can find the modal class instead. The **modal class** is defined as the class(es) having the largest frequency. Note that the mode of a data set may or may not be contained in the modal class. We will apply the concept of the modal class in Exercises 3.22–3.25.

· ·   **3.22**   A frequency distribution for the number of days to maturity for 40 short-term investments is displayed in the following table.

Days to maturity	Frequency
30–39	3
40–49	1
50–59	8
60–69	10
70–79	7
80–89	7
90–99	4

a. Determine the modal class.
b. Now refer to the raw data presented in Table 2.1 on page 50. Obtain the mode of the data.
c. Is the mode contained in the modal class?

. .   **3.23**   The following table gives a frequency distribution for the cholesterol levels of 20 patients with high levels.

Cholesterol level	Frequency
195–199	1
200–204	3
205–209	4
210–214	7
215–219	4
220–224	1

a. Determine the modal class.
b. Now refer to the raw data presented in Table 2.5 on page 53. Find the mode of the data.
c. Is the mode contained in the modal class?

. .   **3.24**   There is a case in which it is always possible to obtain the mode of a data set from a frequency distribution. When is that?

. .   **3.25**   True or false: Suppose all the classes in a frequency distribution are based on a single value. Then the modal class and the mode are identical.

3.2   SUMMATION NOTATION; THE SAMPLE MEAN

In Definition 3.1 we defined the mean of a data set with an equation written in words:

$$\text{Mean} = \frac{\text{Sum of the data}}{\text{Number of pieces of data}}.$$

Such equations can be written more concisely using mathematical notation. To begin, we introduce the mathematical notation for "sum of the data."

Example 3.6   Summation Notation

The exam scores for the student in Example 3.4(a) are 88, 75, 95, and 100. In mathematical notation, the symbol x_i denotes the ith value in the data set. For the exam scores, we have

$$x_1 = \text{score on Exam 1} =\ 88,$$
$$x_2 = \text{score on Exam 2} =\ 75,$$
$$x_3 = \text{score on Exam 3} =\ 95,$$
$$x_4 = \text{score on Exam 4} = 100.$$

More simply, we can just write $x_1 = 88$, $x_2 = 75$, $x_3 = 95$, and $x_4 = 100$. The numbers 1, 2, 3, and 4 written below the xs are called **subscripts.** Using this notation, the sum of the exam-score data can be expressed symbolically as

$$x_1 + x_2 + x_3 + x_4.$$

Summation notation provides a shorthand description for this last sum. The notation uses the uppercase Greek letter Σ (sigma). That letter, which corresponds to the English letter S, stands for the phrase "the sum of." Thus in place of the lengthy expression $x_1 + x_2 + x_3 + x_4$, we can use Σx, read as "summation x" or "the sum of the x values":

$$\Sigma x$$

The sum of   the x values

For the exam-score data,

$$\Sigma x = x_1 + x_2 + x_3 + x_4 = 88 + 75 + 95 + 100 = 358.$$

In words, the sum of the four exam scores is 358. ∎

For clarity it is sometimes useful to incorporate subscripts into the summation notation. This can be done by writing Σx_i instead of Σx. The subscript i is a generic subscript. To be even more precise, we can use *indices* and write $\sum_{i=1}^{n} x_i$, where n denotes the number of pieces of data.

Notation for the Sample Mean

To save us the trouble of writing the phrase "sample mean," we use the symbol $\bar{x}$, read as "x bar." If we also use the letter n to denote the sample size (number of pieces of data), then the definition of the sample mean can be expressed in the following concise form:

Sum of the data

$$\text{Sample mean} \longrightarrow \bar{x} = \frac{\Sigma x}{n}$$

Number of pieces of data

We summarize this discussion in Definition 3.4.

DEFINITION 3.4   **SAMPLE MEAN**

The *sample mean, $\bar{x}$,* of a sample of size n is given by

$$\bar{x} = \frac{\Sigma x}{n}.$$

Example 3.7   Illustrates Definition 3.4

Each year, automobile manufacturers perform mileage tests on their new car models and submit the results of their analyses to the Environmental Protection Agency (EPA). The EPA then tests the vehicles to determine whether the manufacturers' results are correct. In 1992 one company reported that a particular model, equipped with a four-speed manual transmission, averaged 29 miles per gallon (mpg) on the highway. Let's suppose the EPA tested 15 of the cars and obtained the gas mileages shown in Table 3.8. Determine the sample mean of these gas mileages.

TABLE 3.8
Gas mileages

27.3	31.2	29.4	31.6	28.6
30.9	29.7	28.5	27.8	27.3
25.9	28.8	28.9	27.8	27.6

SOLUTION   Summing the gas mileages in Table 3.8, we obtain $\Sigma x = 431.3$. Since the number of pieces of data is $n = 15$, the sample mean gas mileage equals

$$\bar{x} = \frac{\Sigma x}{n} = \frac{431.3}{15} = 28.75 \text{ mpg.}$$

∎

Other Important Sums

We must often find sums other than the sum of the data, Σx. One such sum is the sum of the squares of the data, Σx^2. In Section 3.3 we will need to obtain Σx, Σx^2, and various other sums. So that we can concentrate on the concepts to be presented there instead of the computations, we will discuss computing those sums now.

Example 3.8   Some Additional Important Sums

The exam-score data from Example 3.6 are repeated in the first column of Table 3.9. The remaining columns of the table contain some related quantities whose significance will become apparent in Section 3.3.

TABLE 3.9
Exam-score data and
related quantities

x	x^2	$x - \bar{x}$	$(x - \bar{x})^2$
88	7,744	−1.5	2.25
75	5,625	−14.5	210.25
95	9,025	5.5	30.25
100	10,000	10.5	110.25
358	32,394	0	353.00

In Example 3.6 we determined that the sum of the exam-score data is 358. This fact is recorded at the bottom of the first column of Table 3.9. The second column of Table 3.9 displays the squares, x^2, of the exam scores. The sum of those squares is 32,394; that is, $\Sigma x^2 = 32,394$.

To obtain the third column of Table 3.9, we must first compute the mean, $\bar{x}$, of the four exam scores. Since $n = 4$ and $\Sigma x = 358$,

$$\bar{x} = \frac{\Sigma x}{n} = \frac{358}{4} = 89.5.$$

Subtracting 89.5 from each of the four exam scores in the first column of Table 3.9, we get the $x - \bar{x}$ values shown in the third column. The sum of those values is 0; that is, $\Sigma(x - \bar{x}) = 0$. Finally, the fourth column of Table 3.9 gives the squares, $(x - \bar{x})^2$, of the $x - \bar{x}$ values. The sum of those squares is 353; that is, $\Sigma(x - \bar{x})^2 = 353$. ∎

EXERCISES 3.2

3.26 Let $x_1 = 1$, $x_2 = 7$, $x_3 = 4$, $x_4 = 5$, and $x_5 = 10$.
a. Compute Σx.   b. Find n.   c. Determine $\bar{x}$.

3.27 Let $x_1 = 12$, $x_2 = 8$, $x_3 = 9$, and $x_4 = 17$.
a. Compute Σx.   b. Find n.   c. Determine $\bar{x}$.

For each data set in Exercises 3.28–3.31,
a. compute Σx.
b. find n.
c. determine the sample mean. Round your answer to one more decimal place than the original data.

3.28 Atlas Fishing Line manufactures a 10-lb test line. A sample of 12 spools is subjected to tensile-strength tests. The test results follow.

9.8	10.2	9.8	9.4
9.7	9.7	10.1	10.1
9.8	9.6	9.1	9.7

3.29 A tire manufacturer needs to obtain information about the life of a new steel-belted radial he is going to sell. The results of tests on a sample of these tires are given below. Data are in miles.

43,725	37,732	44,473	37,396
40,652	41,868	43,097	42,200

3.30 In 1989 the average annual gasoline and motor-oil expenditure per U.S. household was $985, as reported by the Bureau of Labor Statistics in *Consumer Expenditure Survey.* That same year a sample of 16 households within metropolitan areas yielded the following gasoline and motor-oil expenditures, in dollars.

1101	1170	1754	1262
126	1070	1489	1248
878	1427	1174	615
271	1303	1421	349

3.31 According to the Salt River Project (SRP), a supplier of electricity to the greater Phoenix area, the mean annual electric bill in 1984 was $852.31. An economist wants to estimate the mean for last year. He takes a sample of SRP customers and obtains the following amounts for last year's electric bills.

$1875	1478	2206	1740	1830	1516
1738	1486	1941	1608	1794	1828
1264	1999	1798	1794	1598	1568

In Exercises 3.32–3.35,
a. compute $\bar{x}$.
b. compute Σx^2, $\Sigma(x - \bar{x})$, and $\Sigma(x - \bar{x})^2$ by constructing a table similar to Table 3.9 on page 115.

3.32 A sample of five families have the following number of children: 2, 3, 4, 4, 3.

3.33 The amount of money, in dollars, a salesperson earned on six randomly selected days yielded the following data set: 75, 98, 130, 63, 112, 107.

3.34 As reported by the R. R. Bowker Company of New York in *Library Journal,* the mean annual subscription rate to law periodicals was $39.82 in 1987. A sample of this year's law periodicals yields the following subscription rates.

$30	46	44	47
42	38	62	55
52	48	43	54

3.35 A team of medical researchers has developed an exercise program to help reduce hypertension. To ascertain whether the program is effective, the team selects a sample of 10 hypertensive individuals and places them on the exercise program for 1 month. The following table displays the diastolic blood pressures of the 10 hypertensive individuals before they began the exercise program.

106	118	118	99	109
94	109	95	97	106

In Exercises 3.36–3.40, assume we are considering two data sets, each having n data values. Denote the data values in one data set by $x_1, x_2, \ldots, x_n$ and the data values in the other data set by $y_1, y_2, \ldots, y_n$.

· · **3.36** Explain the difference between the quantities $(\Sigma x)^2$ and Σx^2. Construct an example to show that, in general, those two quantities are unequal.

· · **3.37** Explain the difference between the quantities Σxy and $\Sigma x \Sigma y$. Provide an example to show that, in general, those two quantities are unequal.

· · · **3.38** Prove the identity $\Sigma(x + y) = \Sigma x + \Sigma y$.

· · · **3.39** If c is a constant, prove that $\Sigma cx = c\Sigma x$.

· · · **3.40** If c is a constant, prove that $\Sigma c = nc$.

3.3 MEASURES OF DISPERSION; THE SAMPLE STANDARD DEVIATION

· · · · ·

Up to this point, the only descriptive measures we have discussed are measures of central tendency; namely, the mean, the median, and the mode. Those descriptive measures indicate where the center or most typical value of a data set lies.

However, two data sets can have the same mean, the same median, or the same mode and yet still be quite different in other respects. For example, consider the heights of the five starting players on each of two men's college basketball teams, as shown in Fig. 3.3.

FIGURE 3.3

Heights of the five starting players on each of two men's college basketball teams

	Team I					Team II				
Feet and inches	6'	6'1"	6'4"	6'4"	6'6"	5'7"	6'	6'4"	6'4"	7'
Inches	72	73	76	76	78	67	72	76	76	84

The two teams have the same mean heights, 75 inches (6′ 3″); the same median heights, 76 inches (6′ 4″); and the same modes, 76 inches (6′ 4″). Nonetheless, it is clear that the two data sets differ. In particular, the heights of the players on Team II vary much more than those on Team I. To describe that difference quantitatively, we use a **measure of dispersion**—a descriptive measure that indicates the amount of variation or spread in a data set.

Just as there are several different measures of central tendency, there are also several different measures of dispersion. In this section we will examine two of the most frequently used measures of dispersion.

The Range

The first measure of dispersion we will discuss is the *range,* since it is the simplest to understand and compute. In Fig. 3.3 the contrast between the two teams becomes clear if we place the shortest player on each team next to the tallest, as shown in Fig. 3.4.

FIGURE 3.4

Heights of the shortest and tallest starting players on each of two men's college basketball teams

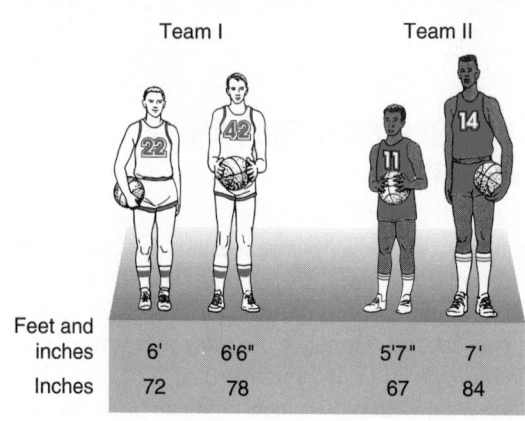

	Team I		Team II	
Feet and inches	6'	6'6"	5'7"	7'
Inches	72	78	67	84

The **range** of a data set is obtained by computing the difference between the maximum (largest) and minimum (smallest) data values in the data set. Hence, as we see from Fig. 3.4,

$$\text{Team I: Range} = 78 - 72 = 6 \text{ inches},$$
$$\text{Team II: Range} = 84 - 67 = 17 \text{ inches}.$$

In general, we use the following definition of the range.

DEFINITION 3.5   **RANGE OF A DATA SET**

> The *range* of a data set is defined as the difference between the maximum and minimum data values in the data set:
>
> $$\text{Range} = \text{Max} - \text{Min}.$$

The range of a data set is quite easy to compute. However, in using the range, a great deal of information is ignored—only the largest and smallest data values are considered; the remaining data are disregarded.

For that reason, two other measures of dispersion, the *standard deviation* and the *interquartile range,* are generally favored over the range. The standard deviation is the preferred measure of dispersion when the mean is used as the measure of central tendency; the interquartile range is preferred when the median is used as the measure of central tendency. We will study the standard deviation in this section and the interquartile range in Section 3.6.

The Sample Standard Deviation

In contrast to the range, the standard deviation takes into account all the data. The calculations required to determine the standard deviation are more involved than those needed to find the range. However, this problem is not serious since computers and sophisticated calculators can do the necessary computations.

Roughly speaking, the **standard deviation** measures the variation in a data set by indicating how far, on the average, the data values are from the mean. For a data set with a large amount of variation, the data values will, on the average, be far from the mean; hence the standard deviation will be large. For a data set with a small amount of variation, the data values will, on the average, be close to the mean; consequently, the standard deviation will be small.

To compute the standard deviation of a data set, we need to know whether the data set constitutes an entire population or whether it is only a sample from a population. This information is necessary because the formula used to obtain the standard deviation of a sample is slightly different from the one used to obtain the standard deviation of a population. In this section we will concentrate on the sample standard deviation. The population standard deviation will be discussed in Section 3.7.

The first step in computing the sample standard deviation is to find how far each data value is from the mean, the so-called **deviations from the mean.** We show how to calculate the deviations from the mean in Example 3.9.

Example 3.9 Deviations from the Mean

The heights, in inches, of the five starting players on Team I are 72, 73, 76, 76, and 78, as we see from Fig. 3.3 on page 117. Find the deviations from the mean.

SOLUTION The mean height of the starting players on Team I is

$$\bar{x} = \frac{\Sigma x}{n} = \frac{72 + 73 + 76 + 76 + 78}{5} = \frac{375}{5} = 75 \text{ inches.}$$

To obtain the deviation from the mean for a particular data value, we subtract the mean from that data value; that is, we compute $x - \bar{x}$. For instance, the deviation from the mean for the height of 72 inches is $x - \bar{x} = 72 - 75 = -3$. The deviations from the mean for all five data values are given in the second column of Table 3.10 and are displayed graphically in Fig. 3.5. ∎

TABLE 3.10
Deviations from the mean

Height x	Deviation from the mean $x - \bar{x}$
72	−3
73	−2
76	1
76	1
78	3

FIGURE 3.5

Graphical display of the
deviations from the mean
(dots represent data values)

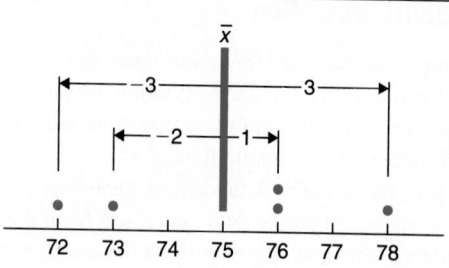

The second step in computing the sample standard deviation is to obtain a measure of the total deviation from the mean for all the data values. Although the quantities $x - \bar{x}$ represent deviations from the mean, adding them to get a total deviation from the mean is of no value because their sum, $\Sigma(x - \bar{x})$, always equals zero. Summing the data in the second column of Table 3.10 shows this to be true for the height data of Team I. In Exercise 3.60 you are asked to verify that, in general, $\Sigma(x - \bar{x}) = 0$.

In computing the sample standard deviation, the deviations from the mean, $x - \bar{x}$, are squared to obtain quantities that do not sum to zero. The sum of the squared deviations from the mean, $\Sigma(x - \bar{x})^2$, is called the **sum of squared deviations** and provides a measure of the total deviation from the mean for all the data.

Example 3.10   The Sum of Squared Deviations

Compute the sum of squared deviations for the heights of the starting players on Team I.

SOLUTION   In Table 3.11 we have appended a column for $(x - \bar{x})^2$ to Table 3.10.

TABLE 3.11

Table for computing the
sum of squared deviations
for the heights of Team I

Height x	Deviation from mean $x - \bar{x}$	Squared deviation $(x - \bar{x})^2$
72	-3	9
73	-2	4
76	1	1
76	1	1
78	3	9
		24

From the third column of Table 3.11, we find that

$$\text{Sum of squared deviations} = \Sigma(x - \bar{x})^2 = 24 \text{ inches}^2.$$

The third step in computing the sample standard deviation is to take an average of the squared deviations. This is accomplished by dividing the sum of squared deviations by $n - 1$, one less than the sample size, n. The resulting quantity is called the **sample variance** and is denoted by s^2. In symbols,

$$s^2 = \frac{\Sigma(x - \overline{x})^2}{n - 1}.$$

Example 3.11   The Sample Variance

Compute the sample variance of the heights of the starting players on Team I.

SOLUTION   From Example 3.10 the sum of squared deviations is $\Sigma(x - \overline{x})^2 = 24$ inches^2. Also, $n = 5$ since there are five pieces of data. Thus the sample variance equals

$$s^2 = \frac{\Sigma(x - \overline{x})^2}{n - 1} = \frac{24}{5 - 1} = 6 \text{ inches}^2.$$

∎

Note: If we divided by n instead of by $n - 1$, then the sample variance would be the mean of the squared deviations. Although dividing by n seems more natural, we divide by $n - 1$ for the following reason: One of the main uses of the sample variance is to estimate the population variance (to be defined in Section 3.7). Division by n tends to underestimate the population variance, whereas division by $n - 1$ does not.

It is important to realize that the sample variance is in units that are the square of the original units. This results from squaring the deviations from the mean. For instance, as we know from Example 3.11, the sample variance of the heights of the players on Team I is 6 inches^2. Since it is desirable to have descriptive measures in the original units, the final step in computing the sample standard deviation is to take the square root of the sample variance. In other words, the **sample standard deviation, s,** is

$$s = \sqrt{\frac{\Sigma(x - \overline{x})^2}{n - 1}}.$$

Example 3.12   The Sample Standard Deviation

Compute the sample standard deviation of the heights of the starting players on Team I.

SOLUTION   From Example 3.11 the sample variance is 6 inches^2. Thus the sample standard deviation equals

$$s = \sqrt{\frac{\Sigma(x - \overline{x})^2}{n - 1}} = \sqrt{6} = 2.4 \text{ inches,}$$

rounded to the nearest tenth of an inch.

∎

Definition 3.6 summarizes our discussion of the sample standard deviation.

DEFINITION 3.6   **SAMPLE STANDARD DEVIATION**

> The *sample standard deviation, s,* of a sample of size n is defined by
>
> $$s = \sqrt{\frac{\Sigma (x - \overline{x})^2}{n - 1}}.$$

The steps required to obtain a sample standard deviation were illustrated in Examples 3.9–3.12. The computations were performed in four separate examples to explain the interpretation of the sample standard deviation as well as the calculations involved. Now that we have done that, we can present a simple procedure for computing a sample standard deviation:

1. Calculate the sample mean, $\overline{x}$.

2. Construct a table to determine the sum of squared deviations, $\Sigma (x - \overline{x})^2$.

3. Apply Definition 3.6 to obtain the sample standard deviation, s.

Example 3.13   Illustrates Definition 3.6

The heights, in inches, of the five starting players on Team II are 67, 72, 76, 76, and 84. Compute the sample standard deviation of these heights.

SOLUTION   We apply the three-step procedure just described. First we calculate the sample mean, $\overline{x}$:

$$\overline{x} = \frac{\Sigma x}{n} = \frac{67 + 72 + 76 + 76 + 84}{5} = \frac{375}{5} = 75 \text{ inches.}$$

Next we construct a table, with columns for x, $x - \overline{x}$, and $(x - \overline{x})^2$, in order to determine the sum of squared deviations, $\Sigma (x - \overline{x})^2$. This is done in Table 3.12.

TABLE 3.12
Table for computing the sum of squared deviations for the heights of Team II

x	$x - \overline{x}$	$(x - \overline{x})^2$
67	−8	64
72	−3	9
76	1	1
76	1	1
84	9	81
		156

From the third column of Table 3.12, we see that the sum of squared deviations is

$$\Sigma (x - \overline{x})^2 = 156 \text{ inches}^2.$$

Finally, we apply Definition 3.6. We have $n = 5$ and $\Sigma(x - \bar{x})^2 = 156$. Consequently, the sample standard deviation of the heights for Team II equals

$$s = \sqrt{\frac{\Sigma(x - \bar{x})^2}{n - 1}} = \sqrt{\frac{156}{5 - 1}} = \sqrt{39} = 6.2 \text{ inches},$$

rounded to the nearest tenth of an inch. ∎

In Example 3.12 we found that the sample standard deviation of the heights of the players on Team I is 2.4 inches; and in Example 3.13 we found that the sample standard deviation of the heights of the players on Team II is 6.2 inches. Hence we see that Team II, which has more variation in height than Team I, also has a larger standard deviation. That is the way a measure of dispersion is supposed to work.

KEY FACT 3.1 **VARIATION AND THE STANDARD DEVIATION**

> The more variation there is in a data set, the larger its standard deviation.

The standard deviation satisfies the basic criterion for a measure of dispersion—it measures variation. It is the most commonly used measure of dispersion. Nonetheless, the standard deviation does have its drawbacks, such as not being resistant; its value can be strongly affected by a few extreme data values.

A Shortcut Formula for s

We are going to present an alternative formula for computing a sample standard deviation, s. Thus it will be convenient to have a name for the original formula,

$$s = \sqrt{\frac{\Sigma(x - \bar{x})^2}{n - 1}}.$$

Since this is the formula used to define the sample standard deviation, we will call it the *defining formula* for s.

The alternative formula for computing the sample standard deviation is given in Formula 3.1. We will call that formula the *shortcut formula* for s.

FORMULA 3.1 **SHORTCUT FORMULA FOR THE SAMPLE STANDARD DEVIATION**

> The sample standard deviation, s, of a sample of size n can be computed from the formula
>
> $$s = \sqrt{\frac{\Sigma x^2 - (\Sigma x)^2/n}{n - 1}}.$$

The shortcut formula for s is equivalent to the defining formula; that is, both formulas give the same result (differences due to roundoff error are possible). However, the shortcut formula is usually faster and easier for doing calculations and also reduces the chance for roundoff error.

Before illustrating the shortcut formula for s, we should comment on the similar-looking expressions, Σx^2 and $(\Sigma x)^2$, that occur in the formula. The expression Σx^2 represents the sum of the squares of the data; it is obtained by first squaring each data value and then summing those squared values. The expression $(\Sigma x)^2$ represents the square of the sum of the data; it is obtained by first summing the data values and then squaring that sum.

We should also emphasize that in the numerator of the shortcut formula, the division of $(\Sigma x)^2$ by n should be performed before the subtraction from Σx^2. In other words, first compute $(\Sigma x)^2/n$ and then subtract the result from Σx^2.

Example 3.14   Illustrates Formula 3.1

In Example 3.13 we computed the sample standard deviation of the heights for the five starting players on Team II using the defining formula for s. Compute that sample standard deviation using the shortcut formula for s.

SOLUTION   To apply the shortcut formula, we need the sums Σx and Σx^2. These are determined in Table 3.13.

TABLE 3.13
Table for computation of s using the shortcut formula

x	x^2
67	4,489
72	5,184
76	5,776
76	5,776
84	7,056
375	28,281

We have $n = 5$, and from the bottom row of Table 3.13 we see that $\Sigma x = 375$ and $\Sigma x^2 = 28,281$. Thus by the shortcut formula,

$$s = \sqrt{\frac{\Sigma x^2 - (\Sigma x)^2/n}{n-1}} = \sqrt{\frac{28,281 - (375)^2/5}{5-1}}$$

$$= \sqrt{\frac{28,281 - 28,125}{4}} = \sqrt{\frac{156}{4}} = \sqrt{39} = 6.2 \text{ inches,}$$

rounded to the nearest tenth of an inch.

■

We have now computed the sample standard deviation of the heights of the players on Team II in two ways: In Example 3.13 we used the defining formula for s, and in Example 3.14 we used the shortcut formula for s. As we see, both formulas give the same value for the sample standard deviation (6.2 inches). For these height data, either formula is relatively easy to apply. However, for most data sets, and especially for those in which the mean is not a whole number, the shortcut formula is preferable.

When computing a sample standard deviation it is important not to do any rounding until the computation is completed; otherwise, significant roundoff error can result. This same advice applies to all calculations.

Using the Computer (Optional)

In this section we discussed two measures of dispersion: the range and the sample standard deviation. Minitab can be used to obtain those two measures of dispersion.

First we consider the range. Recall that the range of a data set is the difference between the maximum and minimum data values in the data set. Minitab does not have a Range command, but it is easy to determine the range using two other Minitab commands—the Maximum and Minimum commands.[†] The **Maximum** command yields the largest data value (maximum) in a data set, and the **Minimum** command yields the smallest data value (minimum) in a data set. Consequently, to obtain the range of a data set, we need only compute the difference between the two numbers resulting from the application of the Maximum and Minimum commands. Example 3.15 shows how this can be done.

Example 3.15 How to Find the Range Using Minitab

The heights, in inches, of the five starting players on Team II are 67, 72, 76, 76, and 84. Use Minitab to obtain the range of these height data.

SOLUTION First we store the height data in a column named TEAM II, and then we proceed in the following manner.

Session commands: To obtain the maximum height, we type the command **MAX** followed by the storage location of the data; that is, we type MAX 'TEAM II' and press Enter. Similarly, to obtain the minimum height, we type the command **MIN** followed by the storage location of the data; that is, we type MIN 'TEAM II' and press Enter. Printout 3.2 on the next page shows these commands and the resulting output.

(or)

Menu commands: To find the maximum height, we choose **Calc ▶ Functions and Statistics ▶ Column Statistics...**, select **Maximum** from the **Statistic** list, specify TEAM II as

---

[†] Some of the newest versions of Minitab (e.g., Minitab for Windows, Release 9) now include a Range command.

the **Input variable**, and then select **OK**. Using the same steps, except selecting **Minimum** instead of **Maximum**, we obtain the minimum height.[†] The output that results is displayed in Printout 3.2.

```
MTB > MAX 'TEAM II'
    MAXIMUM =      84.000
MTB > MIN 'TEAM II'
    MINIMUM =      67.000
```

The output in Printout 3.2 reveals that the heights of the tallest and shortest players on Team II are 84 inches and 67 inches, respectively. Subtracting the height of the shortest player from the height of the tallest player, we obtain the range: Range $= 84 - 67 = 17$ inches.

For the height data considered in Example 3.15, it is actually simpler to determine the range manually than by using Minitab. This is because the data set is so small. But for large, unordered data sets, it is much easier to use Minitab.

Next we discuss how Minitab can be used to obtain the sample standard deviation of a data set using the **StDev** command. Example 3.16 provides an illustration.

Example 3.16   The StDev Command

The heights, in inches, of the five starting players on Team II are 67, 72, 76, 76, and 84. Use Minitab to determine the sample standard deviation of these heights.

SOLUTION   In Example 3.15 we stored the heights of the players on Team II in a column named TEAM II. To obtain the sample standard deviation of those heights, we do the following.

Session commands: We type the command **STDEV** followed by the storage location of the data; that is, we type <u>STDEV 'TEAM II'</u> and press [Enter]. Printout 3.3 shows this command and the output that results.

(or)

Menu commands: We choose **Calc ▸ Functions and Statistics ▸ Column Statistics...**, select **Standard deviation** from the **Statistic** list, specify TEAM II as the **Input variable**, and then select **OK**. Printout 3.3 displays the resulting output.

[†] Once we have obtained the maximum, it is easier to obtain the minimum by editing the last command dialogue. See your Minitab documentation for details.

PRINTOUT 3.3

Minitab output for
the StDev command

```
MTB > STDEV 'TEAM II'
   ST.DEV. =      6.2450
```

Thus the sample standard deviation of the heights of the five starting players on Team II is 6.2450 inches (rounded to four decimal places). ■

We should point out that Minitab has a command called Describe that provides, among other things, the mean, 5% trimmed mean, standard deviation, max, and min. Refer to the *Minitab Supplement* or your Minitab documentation for details.

EXERCISES 3.3

· **3.41** Explain in detail the purpose of a measure of dispersion.

· **3.42** Why is the standard deviation preferable to the range as a measure of dispersion?

In Exercises 3.43–3.50, we have repeated the data from Exercises 3.3–3.10. For each exercise,
a. *determine the range.*
b. *compute the sample standard deviation, s, using the defining formula.*
c. *compute the sample standard deviation, s, using the short-cut formula.*
d. *state which formula you found easier to use in computing s.*
Note: In parts (b) and (c), round your final answers to one more decimal place than the original data.

· **3.43** The National Science Foundation, Division of Science Resources Studies, collects data on the ages of recipients of science and engineering doctoral degrees. Results are published in *Survey of Earned Doctorates*. A sample of one year's recipients yields the following ages.

37	28	36	33
37	43	41	28
24	44	27	24

· **3.44** The American Hospital Association reports data on the costs to community hospitals per patient per day

in *Hospital Statistics*. A sample of 10 such costs in New York yields the data below.

$602	539	569	916	768
335	776	887	806	422

· **3.45** The average retail price for oranges in 1983 was 38.5 cents per pound, as reported by the U.S. Department of Agriculture in *Food Cost Review*. Recently, a random sample of 15 markets yielded the following prices for oranges in cents per pound.

43.0	40.0	42.6	40.2	37.5
44.1	45.2	41.8	35.6	34.6
37.9	44.2	44.5	38.2	42.4

· **3.46** A biologist is studying the gestation period of domestic dogs. Fifteen dogs are observed during pregnancy and are found to have the gestation periods, in days, given in the following table.

62.0	61.4	59.8	62.2	60.3
60.4	59.4	60.2	60.4	60.8
61.8	59.2	61.1	60.4	60.9

· **3.47** A manufacturer of liquid soap produces a bottle with an advertised content of 310 mL. A sample of 16 bottles yields the following contents.

297	318	306	300
311	303	291	298
322	307	312	300
315	296	309	311

3.48 A city planner working on bikeways needs information about local bicycle commuters. She designs a questionnaire. One of the questions asks how many minutes it takes the rider to pedal from home to his or her destination. A sample of 22 local bicycle commuters yields the following times.

22	19	24	31	12	48
29	29	21	15	22	28
27	23	37	31	23	
30	26	16	26	29	

3.49 The National Center for Education Statistics surveys college and university libraries to obtain information on the number of volumes held. The number of volumes, in thousands, for a sample of seven public colleges and universities are as follows.

79	516	24	265
41	15	411	

3.50 As reported by the College Entrance Examination Board in *National College-Bound Senior*, the average verbal score on the Scholastic Aptitude Test in 1991 was 422 points out of a possible 800. A sample of 25 verbal scores for last year follows.

346	496	352	378	315
491	360	385	500	558
381	303	434	562	496
420	485	446	479	422
494	289	436	516	615

3.51 TSI, an independent testing agency, tested the lifetimes of two brands of light bulbs. Light-bulb life is defined as the number of hours a bulb will operate continuously before it burns out. The results of tests on seven bulbs of each brand are as follows. Data are in hundreds of hours.

Brand *A*	10.5	9.1	10.0	10.3	9.4	9.6	9.7
Brand *B*	11.3	7.0	9.7	9.6	10.5	11.8	8.7

a. Compute $\bar{x}$ for each data set.
b. Determine the median of each data set.
c. Although the two data sets have the same means and medians, they are quite different in another respect. How are they different?
d. Which data set appears to have less variation?
e. Compute s for each data set.
f. Are your answers in parts (d) and (e) consistent?

3.52 Consider the following four data sets.

Data Set I		Data Set II		Data Set III		Data Set IV	
1	5	1	9	5	5	2	4
1	8	1	9	5	5	4	4
2	8	1	9	5	5	4	4
2	9	1	9	5	5	4	10
5	9	1	9	5	5	4	10

a. Compute $\bar{x}$ for each data set.
b. Although the four data sets have the same means, they are quite different in another respect. How are they different?
c. Which data set appears to have the least variation? the greatest variation?
d. Compute the range of each data set.
e. Compute s for each data set.
f. From your answers to parts (d) and (e), which measure of dispersion better distinguishes the variation in the four data sets—the range or the standard deviation? Why?
g. Are your answers from parts (c) and (e) consistent?

3.53 Below are 10 IQ scores.

110	122	132	107	101
97	115	91	125	142

Time each of the following calculations.
a. Find the sample standard deviation of the 10 IQs using the defining formula for s.
b. Find the sample standard deviation of the 10 IQs using the shortcut formula for s.
c. Did the shortcut formula really save time?

3.54 (Computer exercise) Use Minitab or some other statistical software to determine the range of the cost data in Exercise 3.44.

3.55 (Computer exercise) Use Minitab or some other statistical software to determine the range of the price data in Exercise 3.45.

3.56 (Computer exercise) Use Minitab or some other statistical software to determine the sample standard deviation of the cost data in Exercise 3.44.

3.57 (Computer exercise) Use Minitab or some other statistical software to determine the sample standard deviation of the price data in Exercise 3.45.

3.58 Another measure of dispersion is the **mean absolute deviation (MAD).** The MAD of a data set is defined as the mean of the absolute values of the deviations from the mean; that is, the MAD is defined by

$$MAD = \frac{\Sigma|x - \bar{x}|}{n}.$$

a. Compute the MAD of the ages in Exercise 3.43.
b. Compute the MAD of the costs in Exercise 3.44.

· ·   **3.59**   In Exercise 3.20 on page 112 we discussed *outliers*, data values that fall well outside the overall pattern of the data. The following table contains two data sets. Data Set II is obtained by removing the outliers from Data Set I.

Data Set I					Data Set II			
0	12	14	15	23	10	14	15	17
0	14	15	16	24	12	14	15	
10	14	15	17		14	15	16	

a. Compute the sample standard deviation for both data sets.

b. Compute the range for both data sets.
c. What effect do outliers have on the variation in a data set? Explain your answer.

· · ·   **3.60**   On page 120 we pointed out that the sum of the deviations from the mean is always equal to zero; that is, $\Sigma(x - \bar{x}) = 0$ for any data set. Prove that this is true. *(Hint: Use Exercises 3.38–3.40 on page 117.)*

· · ·   **3.61**   This exercise shows that the shortcut formula for s is equivalent to the defining formula.
a. Verify that

$$\Sigma(x - \bar{x})^2 = \Sigma x^2 - (\Sigma x)^2/n.$$

(Hint: Expand the square on the left and then apply Exercises 3.38–3.40 on page 117.)
b. Deduce from part (a) that the defining formula and the shortcut formula for s are equivalent.

3.4   INTERPRETATION OF THE STANDARD DEVIATION; z-SCORES

Until now we have been concentrating more on the calculation of the standard deviation than on its meaning. In this section we will learn how to interpret the standard deviation. As we know, the standard deviation is a measure of dispersion; that is, it is a descriptive measure that indicates the amount of variation in a data set. Thus the more variation there is in a data set, the larger its standard deviation.

Table 3.14 displays two data sets, each of which has 10 data values. A brief inspection of the table reveals that Data Set II has more variation than Data Set I.

TABLE 3.14
Data sets having different variation

Data Set I	51	44	41	58	48	47	53	47	45	66
Data Set II	37	61	49	20	70	53	48	48	50	64

Using the formulas

$$\bar{x} = \frac{\Sigma x}{n} \quad \text{and} \quad s = \sqrt{\frac{\Sigma x^2 - (\Sigma x)^2/n}{n - 1}},$$

we computed the sample mean and sample standard deviation of each data set. The results are depicted in Table 3.15. As expected, the standard deviation of Data Set II is larger than that of Data Set I.

TABLE 3.15

Means and standard deviations
of the data sets in Table 3.14

Data Set I	Data Set II
$\bar{x} = 50.0$	$\bar{x} = 50.0$
$s = 7.4$	$s = 14.2$

To enable us to visually compare the variations in the two data sets, we have drawn the graphs in Figs. 3.6 and 3.7. On each graph we have marked the data values with dots. In addition, we have located the sample mean, $\bar{x} = 50$, and have measured off intervals equal in length to the standard deviation: 7.4 for Data Set I and 14.2 for Data Set II.

FIGURE 3.6   Data Set I; $\bar{x} = 50$, $s = 7.4$

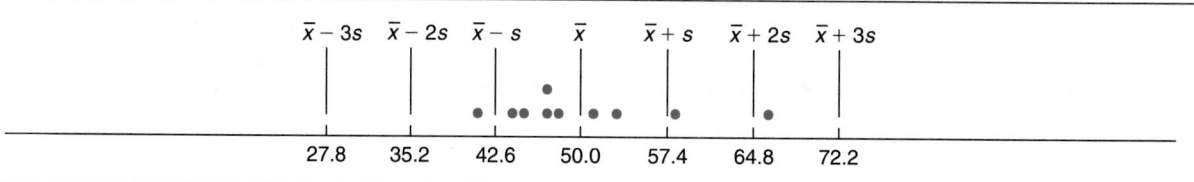

FIGURE 3.7   Data Set II; $\bar{x} = 50$, $s = 14.2$

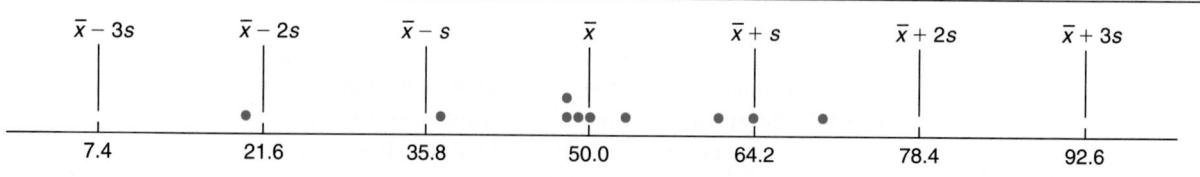

Let's examine Fig. 3.6. Notice that the horizontal position labeled $\bar{x} + 2s$ represents the number that is two standard deviations to the right of the mean, which in this case is $\bar{x} + 2s = 50.0 + 2 \cdot 7.4 = 50.0 + 14.8 = 64.8$.[†] Likewise, the horizontal position labeled $\bar{x} - 3s$ represents the number that is three standard deviations to the left of the mean, which in this case is $\bar{x} - 3s = 50.0 - 3 \cdot 7.4 = 50.0 - 22.2 = 27.8$. Figure 3.7 is interpreted in a similar manner.

The graphs in Figs. 3.6 and 3.7 vividly illustrate that Data Set II has more variation than Data Set I. Furthermore, they show that for each data set, all the data values lie within a few standard deviations to either side of the mean. This is no accident: *Almost all of the data in any data set will lie within three standard deviations to either side of the mean.*

A data set with a great deal of variation will have a large standard deviation and, consequently, three standard deviations to either side of its mean will be extensive, as in

---

[†] Recall that for an expression of the form $a + b \cdot c$, the multiplication should be done before the addition. Thus $50.0 + 2 \cdot 7.4 = 50.0 + 14.8 = 64.8$. Similarly, for an expression of the form $a - b \cdot c$, the multiplication should be done before the subtraction.

Fig. 3.7. A data set with little variation will have a small standard deviation and, hence, three standard deviations to either side of its mean will be narrow, as in Fig. 3.6.

Chebychev's Rule

As we said, almost all of the data in any data set will lie within three standard deviations to either side of the mean. The Russian mathematician Pafnuty Lvovich Chebychev (1821–1894) developed a rule that generalizes this statement and also makes it more precise.

KEY FACT 3.2 **CHEBYCHEV'S RULE**

For any data set:

Property 1: At least 75% of the data lie within *two* standard deviations to either side of the mean, that is, between $\bar{x} - 2s$ and $\bar{x} + 2s$.

Property 2: At least 89% of the data lie within *three* standard deviations to either side of the mean, that is, between $\bar{x} - 3s$ and $\bar{x} + 3s$.

Property 3: In general, for any number $k > 1$, at least $1 - 1/k^2$ of the data lie within k standard deviations to either side of the mean, that is, between $\bar{x} - k \cdot s$ and $\bar{x} + k \cdot s$.

Figure 3.8 provides a graphical display of the first two properties in Chebychev's rule.

FIGURE 3.8

Graphical illustration of Properties 1 and 2 of Chebychev's rule

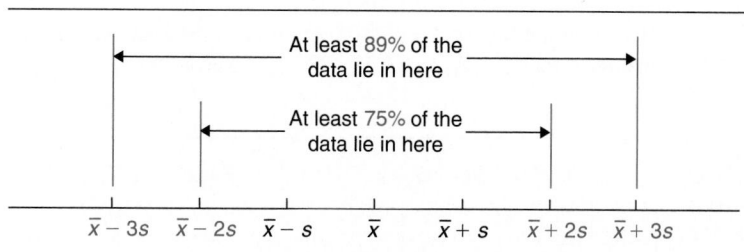

The first two properties of Chebychev's rule are special cases of the third property. For instance, let's apply Property 3 with $k = 2$. Then

$$1 - \frac{1}{k^2} = 1 - \frac{1}{2^2} = 1 - \frac{1}{4} = 0.75,$$

or 75%. Thus by Property 3, for any data set, at least 75% of the data lie within two standard deviations to either side of the mean. This statement is the one given in Property 1. Similarly, Property 2 can be obtained from Property 3 by setting $k = 3$.

Property 3 of Chebychev's rule can be applied using any value of k that is greater than 1. For instance, assume $k = 2.5$. Then

$$1 - \frac{1}{k^2} = 1 - \frac{1}{2.5^2} = 1 - \frac{1}{6.25} = 0.84,$$

or 84%. Thus by Property 3, for any data set, at least 84% of the data lie within 2.5 standard deviations to either side of the mean.

Now consider again the data set portrayed in Fig. 3.6 on page 130. Chebychev's rule says that at least 75% of the data lie within two standard deviations of the mean. However, as we see from Fig. 3.6, 90% of the data actually lie within two standard deviations of the mean. Similarly, Chebychev's rule says that at least 89% of the data lie within three standard deviations of the mean, whereas, in fact, 100% of the data actually lie within three standard deviations of the mean.

So we see that Chebychev's rule does not necessarily provide precise estimates for the percentage of data that lies within a specified number of standard deviations to either side of the mean. Its power is its generality—Chebychev's rule holds for any data set.

Chebychev's rule also permits us to make pertinent statements about a data set when we know only its mean and standard deviation; and frequently that is all we do know. Example 3.17 illustrates this point.

Example 3.17   Illustrates Key Fact 3.2

The R. R. Bowker Company of New York collects data on annual subscription rates to periodicals. Results are published in *Library Journal.* A sample of 40 sociology periodicals taken this year has a mean subscription rate of $63.75 and a standard deviation of $10.42. Using this information only, apply Properties 1 and 2 of Chebychev's rule to make some observations about the subscription rates of the sample of 40 sociology periodicals.

SOLUTION   Based on the fact that $\bar{x} = \$63.75$ and $s = \$10.42$, we can construct Fig. 3.9. Notice, for instance, that $\bar{x} - 2s = 63.75 - 2 \cdot 10.42 = 42.91$.

FIGURE 3.9

Graph showing the mean and one, two, and three standard deviations to either side of the mean for the 40 subscription rates

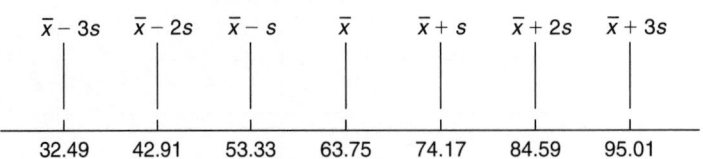

By Property 1 of Chebychev's rule, at least 75% of the 40 sociology periodicals sampled have subscription rates within two standard deviations to either side of the mean. Since 75% of 40 is 30, we can conclude in view of Fig. 3.9 that at least 30 of the 40 sociology periodicals have subscription rates between $42.91 and $84.59.

Applying Property 2 of Chebychev's rule, we can say that at least 89% of the 40 sociology periodicals sampled have subscription rates within three standard deviations to either side of the mean. Now, 89% of 40 is 35.6. Therefore, in view of Fig. 3.9, we can conclude that at least 36 of the 40 sociology periodicals have subscription rates between $32.49 and $95.01.   ■

One final remark about Chebychev's rule: Many data sets that occur in practice have bell-shaped distributions. Because Chebychev's rule applies to all data sets, it will apply to those having bell-shaped distributions. But we can provide more precise estimates for such data sets than those given by Chebychev's rule. These more precise estimates are stated in the *empirical rule,* which we will examine when we study normal distributions in Chapter 6.

z-Scores

A frequently used quantity in statistical analysis is the z-score. The **z-score, or standard score,** for a data value is the number of standard deviations that the data value is away from the mean of the data set. Example 3.18 introduces z-scores.

Example 3.18 z-Scores

z-Scores

A researcher in nutrition is studying daily protein intake. She takes a sample of 500 daily protein intakes. The mean intake turns out to be 77 g (grams) with a standard deviation of 8 g. Determine the z-score for an intake of 93 g.

SOLUTION   The z-score for the intake of 93 g is the number of standard deviations that intake is away from the mean. We know that the mean intake is 77 g. Consequently, the intake of 93 g is $93 - 77 = 16$ g away from the mean; and since the standard deviation is 8 g, 16 g away from the mean is

$$\frac{16}{8} = 2 \text{ standard deviations}$$

away from the mean. Thus the z-score for the intake of 93 g is $z = 2$ (see Fig. 3.10).

FIGURE 3.10

z-score for the protein intake of 93 g

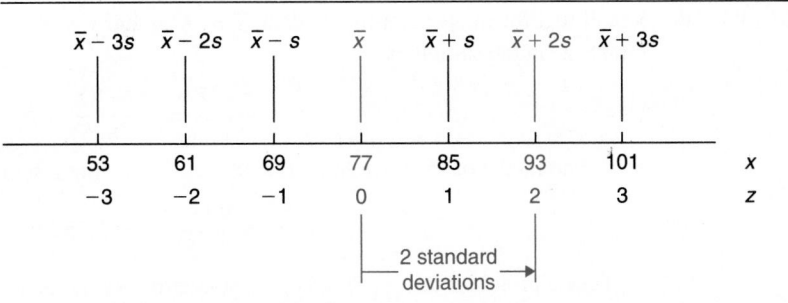

Notice in Fig. 3.10 that the numbers in the row labeled x represent protein intakes in grams and the numbers in the row labeled z represent z-scores (i.e., number of standard deviations away from the mean). ∎

By studying the solution in Example 3.18, we can obtain a general formula for computing z-scores. First we subtracted the mean of 77 g from the specified data value of 93 g in order to determine how far that data value is away from the mean. Then we divided that difference by the standard deviation of 8 g to find how many standard deviations the specified data value of 93 g is away from the mean. In summary, we computed the z-score for the intake of 93 g as follows:

$$z = \frac{93 - \bar{x}}{s} = \frac{93 - 77}{8} = \frac{16}{8} = 2.$$

More generally, if we use x to denote a data value, then we have the following definition.

DEFINITION 3.7   **SAMPLE z-SCORE**

> The *sample z-score* for a data value, x, is defined as the number of standard deviations that x is away from the mean. The sample z-score is computed by using the formula
>
> $$z = \frac{x - \bar{x}}{s},$$
>
> where $\bar{x}$ and s are, respectively, the mean and standard deviation of the sample data. A negative z-score indicates that a data value is smaller than the mean, whereas a positive z-score indicates that a data value is larger than the mean.

Note: When computing a z-score, round the answer to two decimal places.

Example 3.19   Illustrates Definition 3.7

Consider again the protein-intake data from Example 3.18. Determine the z-score for an intake of

a. 99 g.   b. 65 g.

SOLUTION   Recall that for the protein-intake data, $\bar{x} = 77$ g and $s = 8$ g. So the z-score for a protein intake, x, in the sample is

$$z = \frac{x - \bar{x}}{s} = \frac{x - 77}{8}.$$

a. For an intake of 99 g, we have $x = 99$, and hence the z-score is

$$z = \frac{99 - 77}{8} = 2.75.$$

Thus a protein intake of 99 g is 2.75 standard deviations above the mean.

b. For an intake of 65 g, the z-score is

$$z = \frac{65 - 77}{8} = -1.50.$$

So a protein intake of 65 g is 1.50 standard deviations below the mean.   ■

The *z*-Score as a Measure of Relative Standing

We often want to describe the position of a particular data value in a data set relative to the other data values. Descriptive measures that indicate the relative position of a data value are called **measures of relative standing.**

The *z*-score can be used as a measure of relative standing. If a data value has a large positive *z*-score, then it is larger than most of the other data values in the data set; if a data value has a large negative *z*-score, then it is smaller than most of the other data values in the data set; and if a data value has a *z*-score near 0, then it is located near the mean of the data set.

We can make the statements in the preceding paragraph more precise by applying Chebychev's rule to the *z*-score of a data value, as is done in Example 3.20.

Example 3.20   The *z*-Score as a Measure of Relative Standing

Consider once more the data on protein intakes. Use the *z*-score and Chebychev's rule to estimate the relative standing of a protein intake of 51 g.

SOLUTION   For the protein-intake data, $\bar{x} = 77$ g and $s = 8$ g. Thus the *z*-score for an intake of 51 g is $z = (51 - 77)/8 = -3.25$. This shows that a protein intake of 51 g is 3.25 standard deviations below the mean.

By Property 2 of Chebychev's rule, at least 89% of the protein intakes in the sample are within three standard deviations to either side of the mean. Consequently, a protein intake of 51 g, which is 3.25 standard deviations below the mean, is smaller than at least 89% of all the protein intakes in the sample, as can be seen in Fig. 3.11(a).

FIGURE 3.11

Estimate of the relative standing of 51 g using (a) Property 2 of Chebychev's rule, (b) Property 3 of Chebychev's rule

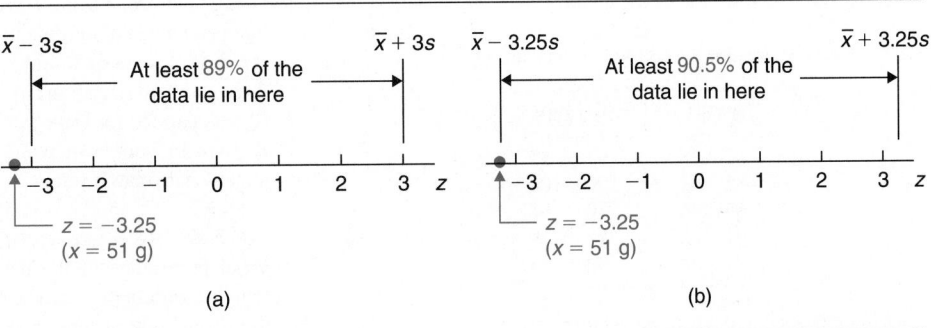

(a)   (b)

We can use Property 3 of Chebychev's rule to get an even better idea of the relative standing of a protein intake of 51 g. Taking $k = 3.25$ (the absolute value of the *z*-score for 51 g), we find that

$$1 - \frac{1}{k^2} = 1 - \frac{1}{3.25^2} = 1 - \frac{1}{10.5625} = 0.905,$$

or 90.5%. So by Property 3 of Chebychev's rule, at least 90.5% of the protein intakes in the sample are within 3.25 standard deviations to either side of the mean. Therefore a protein intake of 51 g, which is 3.25 standard deviations below the mean, is smaller than at least 90.5% of all the protein intakes in the sample, as Fig. 3.11(b) shows.   ∎

We can also use z-scores to compare the relative standings of two data values from different data sets. For example, we could use z-scores to compare the exam scores of two students in different sections of a beginning English course.

EXERCISES 3.4

3.62   Consider the following data sets.

Data Set 1		Data Set 2	
30	20	14	9
16	24	56	32
22	19	13	8
23	13	26	3
18	9	9	16
18	28	31	23

a. Which data set appears to have more variation?
b. Compute $\bar{x}$ and s for each data set.
c. Draw graphs similar to Fig. 3.6 on page 130.
d. Interpret your graphs from part (c).
e. Do most of the data in each data set lie within three standard deviations to either side of the mean?

3.63   Consider the following data sets.

Data Set 3		Data Set 4	
82	78	97	59
85	94	100	100
65	84	30	95
91	86	87	79
81	84	90	93

a. Which data set appears to have more variation?
b. Compute $\bar{x}$ and s for each data set.
c. Draw graphs similar to Fig. 3.6 on page 130.
d. Interpret your graphs from part (c).
e. Do most of the data in each data set lie within three standard deviations to either side of the mean?

3.64   What does Chebychev's rule say about the percentage of data in a data set that lies within

a. 4 standard deviations to either side of the mean?
b. 2.5 standard deviations to either side of the mean?
c. 3.75 standard deviations to either side of the mean?

3.65   What does Chebychev's rule say about the percentage of data in a data set that lies within
a. 1.25 standard deviations to either side of the mean?
b. 3.5 standard deviations to either side of the mean?
c. 5 standard deviations to either side of the mean?

3.66   Refer to Exercise 3.62.
a. What percentage of the data does Chebychev's rule say will lie within two standard deviations to either side of the mean? within three standard deviations to either side of the mean?
b. Using your graph from Exercise 3.62(c), determine the actual percentage of the data in Data Set 1 that lies within two standard deviations to either side of the mean and within three standard deviations to either side of the mean.
c. Repeat part (b) for Data Set 2.
d. Explain in your own words what parts (a)–(c) illustrate about Chebychev's rule.

3.67   Refer to Exercise 3.63.
a. What percentage of the data does Chebychev's rule say will lie within two standard deviations to either side of the mean? within three standard deviations to either side of the mean?
b. Using your graph from Exercise 3.63(c), determine the actual percentage of the data in Data Set 3 that lies within two standard deviations to either side of the mean and within three standard deviations to either side of the mean.
c. Repeat part (b) for Data Set 4.
d. Explain in your own words what parts (a)–(c) illustrate about Chebychev's rule.

3.68 We stated on page 130 that almost all of the data in any data set will lie within three standard deviations to either side of the mean. But in most examples and exercises so far, all of the data have been within three standard deviations to either side of the mean. Verify that this is not the case for the following data set.

100	28	69	85	85	98
97	100	87	97	96	94
89	97	93	92	95	94
79	83	97	90	96	57
64	74	80	87	89	58

(Note: For this data set, the sum of the data is 2550 and the sum of the squares of the data is 224,272.)

In each of Exercises 3.69–3.74, we have given the sample mean, sample standard deviation, and sample size for a data set. For each exercise,

a. *construct a graph similar to Fig. 3.9 on page 132.*
b. *apply Property 1 of Chebychev's rule to make some observations about the data.*
c. *apply Property 2 of Chebychev's rule to make some observations about the data.*

3.69 The Philadelphia CPA firm Laventhol and Horwath has conducted annual surveys to determine characteristics of full-service and economy lodging establishments. The room rates for a double at 300 lodging establishments have a mean of $40.97 and a standard deviation of $8.66.

3.70 The U.S. National Center for Health Statistics collects data on cigarette smokers by sex and age and publishes the results in *Vital and Health Statistics.* A sample of 2760 females who presently smoke yields a mean age of 38.7 years and a standard deviation of 12.6 years.

3.71 The Gallup Organization conducts surveys on home gardening for the National Association for Gardening. Data appear in *National Gardening Survey.* A random sample of 250 households with vegetable gardens yields a mean garden size of 643 sq ft and a standard deviation of 247 sq ft.

3.72 As reported by the Health Insurance Association of America in *Survey of Hospital Semi-Private Room Charges,* the average daily charge for a semi-private room in U.S. hospitals was $253 in 1988. In that same year, a sample of 30 Massachusetts hospitals yielded a mean semi-private room charge of $260.68 with a standard deviation of $12.77.

3.73 According to *Annual Report of the Commissioner and Chief Counsel of the Internal Revenue Service,* the average income tax per taxable return was $4471 in 1986. A sample of 30 taxable returns from last year has a mean income tax of $4239.8 and a standard deviation of $2215.6.

3.74 *The World Almanac, 1985,* reports that in 1980 the average travel time to work for residents of South Dakota was 13 minutes. For this year, a sample of 35 travel times for South Dakota residents gave a mean of 14.0 minutes and a standard deviation of 13.3 minutes.

3.75 Refer to Exercise 3.69.
a. Find the sample z-scores for room rates in the sample of $67, $20, and $37. Interpret your results.
b. Use your results from part (a) to estimate the relative standing in the sample of each of the room rates in part (a).

3.76 Refer to Exercise 3.70.
a. Determine the sample z-scores for female smokers in the sample whose ages are 58, 29, and 73 years. Interpret your results in words.
b. Use your results from part (a) to estimate the relative standing in the sample of each of the ages in part (a).

3.77 Refer to Exercise 3.71.
a. Determine the sample z-scores for vegetable gardens in the sample whose sizes are 1631, 211, and 618 sq ft. Interpret your results in words.
b. Use your results from part (a) to estimate the relative standing in the sample of each of the sizes in part (a).

3.78 Refer to Exercise 3.72.
a. Find the sample z-scores for semi-private rooms in the sample whose daily charges are $311, $343, and $262. Interpret your results in words.
b. Use your results from part (a) to estimate the relative standing in the sample of each of the daily charges in part (a).

3.79 Suppose you are thinking of buying a resale home in a large tract. The owner is asking $95,500. Your realtor obtains the sale prices of 25 comparable homes in the area that have sold recently. The sample mean of the 25 sale prices is $110,258 with a sample standard deviation of $5237. Does it appear that the home you are contemplating buying is a bargain? Explain your answer using the z-score and Chebychev's rule.

3.80 Suppose you take an exam with 400 possible points. The instructor tells you that the mean score is 280 and the standard deviation is 20. He also tells you that you got 350. Did you do well on the exam? Explain your answer using the z-score and Chebychev's rule.

· **3.81** Each year, thousands of high-school students bound for college take the Scholastic Aptitude Test (SAT). The test measures the verbal and mathematical abilities of prospective college students. Student scores are reported on a scale that ranges from a low of 200 to a high of 800. In one high-school graduating class, the mean mathematics score on the SAT was 493 and the standard deviation was 105; the mean verbal score was 420 and the standard deviation was 98. A student in the graduating class scored 703 on the math and 665 on the verbal. Relative to the other students in the graduating class, on which test did the student do better? Why?

· **3.82** It is often useful to transform a data set so that the transformed data set has mean 0 and standard deviation 1. This is called *standardizing* the data set, and it can be accomplished by replacing each data value by its *z*-score. As a simple example, consider the data set consisting of the three numbers 5, 10, and 15.
a. Compute the mean and standard deviation.
b. Standardize the data set by replacing each data value by its *z*-score.
c. Verify that the standardized data set has mean 0 and standard deviation 1.

· **3.83** Refer to Exercise 3.82. Once we have standardized a data set, it is a simple matter to transform the standardized data set to a data set having any specified mean, *a*, and standard deviation, *b*. This is accomplished by replacing each *z*-score, *z*, in the standardized data set by $a + bz$. For example, consider again the simple data set consisting of the three numbers 5, 10, and 15. As we discovered in Exercise 3.82, the corresponding standardized data set is -1, 0, and 1.
a. Transform the standardized data set to one having mean 4 and standard deviation 3.
b. Verify that the data set you obtained in part (a) does indeed have mean 4 and standard deviation 3.

· **3.84 (Computer exercise)** Minitab has a command called **Center** that can be used to standardize a data set (as discussed in Exercise 3.82). To apply the Center command,

we first store the data set in a column named, say, DATA. Then we proceed in the following manner.

Session commands: Using the NAME command, we name an available column ZSCORES. Then we type the command **CENTER** followed by the storage location of the data to be standardized and the storage location for the standardized data; that is, we type CENTER 'DATA' store in 'ZSCORES' and press ⌐Enter⌐.

(or)

Menu commands: We choose **Calc ▶ Standardize...**, specify DATA in the **Input column(s)** text box, specify ZSCORES in the **Put results in** text box, select **Subtract mean and divide by SD**, and then select **OK**.

Use Minitab or some other statistical software to
a. standardize the data set 5, 10, and 15.
b. obtain the standardized data set.
c. verify that the standardized data set has mean 0 and standard deviation 1.

· · **3.85** Given a data set, how many standard deviations to either side of the mean must we go to be assured that
a. at least 95% of the data lie within?
b. at least 99% of the data lie within?

· · · **3.86** Consider a data set with *n* data values.
a. Show that if all the data values are equal, then the standard deviation is 0.
b. Show that if the standard deviation of the data set is 0, then all the data values must be equal.

· · · **3.87** Suppose a data set consists of $2m^2 - 1$ zeros, one $-m$, and one m, where *m* is a positive integer.
a. Compute $\bar{x}$ and *s* for this data set.
b. How many standard deviations from the mean is the data value *m*?
c. Assuming $m \geq 4$, what percentage of the data lies within three standard deviations to either side of the mean? What does this percentage approach as *m* increases?

3.5 COMPUTING $\bar{x}$ AND *s* FOR GROUPED DATA

· · · · ·

The formulas we have developed to compute a sample mean, $\bar{x}$, and a sample standard deviation, *s*, apply only to raw (ungrouped) data. Often, however, the data we work with are grouped into a frequency distribution. In this section we will present methods for computing $\bar{x}$ and *s* when the data are in grouped form.

Computing $\bar{x}$ for Grouped Data

In Example 3.1 we considered two sets of salary data representing weekly earnings during two halves of a summer for the employees of a small mathematical consulting firm. The salary data for Data Set II are repeated in Table 3.16.

TABLE 3.16
Data Set II

$200	200	840	350	300
300	200	200	950	200

A frequency distribution for Data Set II using classes based on a single value is presented in Table 3.17.

TABLE 3.17
Frequency distribution for Data Set II using single-value grouping

Salary ($)	Frequency
200	5
300	2
350	1
840	1
950	1
	10

We computed the mean of Data Set II in Example 3.1. In doing so we had to determine the sum of the salaries displayed in Table 3.16. That sum can be found more quickly using the frequency distribution given in Table 3.17. To see how, we first express the sum of the salaries in Data Set II as follows:

$$\text{Sum of salaries} = 200 + 200 + 840 + 350 + 300 + 300 + 200 + 200 + 950 + 200$$

$$= \overbrace{200 + 200 + 200 + 200 + 200}^{5} + \overbrace{300 + 300}^{2} + \overbrace{350}^{1} + \overbrace{840}^{1} + \overbrace{950}^{1}.$$

In the second sum, we combined like salaries and placed the frequency of each of the different salaries above each group. Consequently, the sum of the salaries can be rewritten as shown in the following diagram:

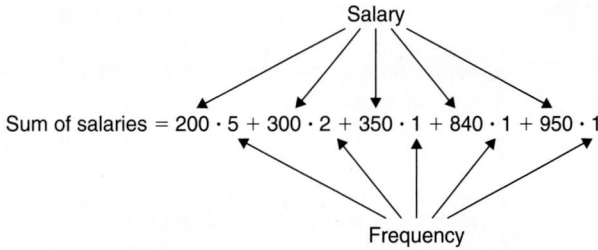

In other words, we can determine the sum of the salaries by first multiplying each of the different salaries by its corresponding frequency and then adding the results. This can be done most efficiently by appending a salary-times-frequency column to the frequency distribution in Table 3.17, as shown in Table 3.18.

TABLE 3.18

Frequency distribution for Data Set II with appended salary-times-frequency column

Salary x	Frequency f	Salary · Frequency xf
200	5	1000
300	2	600
350	1	350
840	1	840
950	1	950
	10	3740

⟵ *Sum of salaries*

We have also introduced some mathematical notation in Table 3.18: x represents salary-class value, not each individual salary as before; f represents class frequency; and xf represents salary-class value times class frequency. Note that for each class, xf equals the sum of the salaries in that class.

We will now explain how Table 3.18 can be used to compute the mean of the salary data. The sum of all the salaries is Σxf, the sum of the third column in Table 3.18. The number of salaries is Σf, the sum of the second column in Table 3.18. Consequently, the mean of the salaries in Data Set II is

$$\bar{x} = \frac{\text{Sum of salaries}}{\text{Number of salaries}} = \frac{\Sigma xf}{\Sigma f} = \frac{3740}{10} = \$374.$$

As we have just seen, when data are in grouped form, the sample size is equal to the sum of the class frequencies; that is, $n = \Sigma f$. To minimize the amount of notation, we generally use n instead of Σf to denote the sample size. But remember that n is the number of pieces of data (sample size), not the number of classes. For the grouped data in Table 3.18, $n = 10$ (the number of salaries), not 5 (the number of classes).

With the preceding discussion in mind, we now present a formula for computing the mean of a data set when the data are grouped in a frequency distribution.

FORMULA 3.2

COMPUTING THE SAMPLE MEAN FOR GROUPED DATA

The sample mean, $\bar{x}$, of a data set that is grouped in a frequency distribution can be computed using the formula

$$\bar{x} = \frac{\Sigma xf}{n},$$

where f denotes class frequency and n ($= \Sigma f$) denotes the sample size.

Example 3.21   Illustrates Formula 3.2

Table 3.19 gives a frequency distribution using classes based on a single value for the salary data in Data Set I (in Table 3.1 on page 102). Find the sample mean of these salaries.

TABLE 3.19

Frequency distribution for Data Set I using single-value grouping

Salary ($) x	Frequency f
200	6
300	2
350	2
700	1
840	1
950	1

SOLUTION   Since the data are in grouped form, we apply Formula 3.2. Appending an xf column to Table 3.19, we obtain Table 3.20.

TABLE 3.20

Table for calculating $\bar{x}$

x	f	xf
200	6	1200
300	2	600
350	2	700
700	1	700
840	1	840
950	1	950
	13	4990

Using the final row of Table 3.20, we can compute $\bar{x}$:

$$\bar{x} = \frac{\Sigma xf}{n} = \frac{4990}{13} = \$383.85.$$

Thus the mean of the salaries in Data Set I is $383.85. ■

Computing s for Grouped Data

We have just seen how to compute the sample mean of a data set when the data are grouped into a frequency distribution. Using similar reasoning we can obtain a formula for computing the sample standard deviation of a data set when the data are grouped into a frequency distribution.

FORMULA 3.3   **COMPUTING THE SAMPLE STANDARD DEVIATION
FOR GROUPED DATA**

> The sample standard deviation, s, of a data set that is grouped in a frequency distribution can be computed using the formula
>
> $$s = \sqrt{\frac{\Sigma(x - \bar{x})^2 f}{n - 1}},$$
>
> or its shortcut version,
>
> $$s = \sqrt{\frac{\Sigma x^2 f - (\Sigma x f)^2 / n}{n - 1}},$$
>
> where f denotes class frequency and n $(= \Sigma f)$ denotes the sample size.

Example 3.22   Illustrates Formula 3.3

A frequency distribution for the salary data in Data Set I is displayed in Table 3.19 and is repeated in the first two columns of Table 3.21. Compute the sample standard deviation of these salaries.

SOLUTION   We will apply the shortcut formula presented in Formula 3.3. As we see from that formula, we will need a table with columns for x, f, xf, x^2, and $x^2 f$. We will also need to sum the second, third, and fifth columns of that table to obtain n, $\Sigma x f$, and $\Sigma x^2 f$, as is done in Table 3.21.

TABLE 3.21
Table for calculating s
using the shortcut formula

x	f	xf	x^2	$x^2 f$
200	6	1200	40,000	240,000
300	2	600	90,000	180,000
350	2	700	122,500	245,000
700	1	700	490,000	490,000
840	1	840	705,600	705,600
950	1	950	902,500	902,500
	13	4990		2,763,100

From the final row of Table 3.21, we find that

$$s = \sqrt{\frac{\Sigma x^2 f - (\Sigma x f)^2 / n}{n - 1}} = \sqrt{\frac{2,763,100 - (4990)^2 / 13}{12}} = 265.79.$$

Thus the sample standard deviation of the salaries in Data Set I is $265.79. ∎

Computing $\bar{x}$ and s When the Classes Are Not Based on a Single Value

Up to this point we have considered the computation of $\bar{x}$ and s for grouped data in which each class is based on a single value. The appropriate formulas are Formulas 3.2 and 3.3. If each class represents a number of different values, such as the grouped data in Table 2.2 on page 50, then those formulas cannot be used to compute the exact values of $\bar{x}$ and s. However, they can be used to compute the approximate values of $\bar{x}$ and s. This is accomplished by substituting class marks for the x-values in Formulas 3.2 and 3.3. (See Exercises 3.94–3.99.)

EXERCISES 3.5

In each of Exercises 3.88–3.91, you will be given a frequency distribution for a data set. For each exercise,
a. compute the sample mean using Formula 3.2 on page 140.
b. compute the sample standard deviation using one of the formulas in Formula 3.3 on page 142.
Round your answers to one more decimal place than the original data.

· **3.88** An English professor wants to estimate the average study time per week for students in basic English courses at her school. She selects a sample of 25 students and records their weekly study times. The times, to the nearest hour, are displayed in the following table.

Study time	2	3	4	5	6	7	8	9	11
Frequency	2	1	3	1	5	5	5	2	1

· **3.89** The U.S. Bureau of the Census collects data on family size and publishes the results in *Current Population Reports*. A sample of 500 U.S. families yields the sizes indicated here.

Size of family	2	3	4	5	6	7	8	9
Frequency	198	118	101	59	12	3	8	1

· **3.90** The heights, to the nearest inch, of the basketball players on the 1992–1993 Boston Celtics roster are presented in the following frequency distribution. [SOURCE: *The Sporting News Official NBA Guide.*]

Height	72	73	77	79	81	82	84	86
Frequency	1	2	2	2	1	2	2	1

· **3.91** The ages of the students in Professor Weiss's introductory statistics class are displayed in Table 3.6 on page 107. A frequency distribution for those ages is provided by the following table.

Age	Frequency	Age	Frequency
17	1	23	3
18	1	24	4
19	9	26	1
20	7	35	1
21	7	36	1
22	5		

In Exercises 3.92 and 3.93, you will be given raw (ungrouped) data. For each data set,
a. compute $\bar{x}$ and s by applying Definition 3.4 on page 114 and Formula 3.1 on page 123.
b. group the data into a frequency distribution in which each class is based on a single value.
c. compute $\bar{x}$ and s using your frequency distribution from part (b) and Formulas 3.2 and 3.3.
d. compare your answers from parts (a) and (c).

· **3.92** The Prescott National Bank has six tellers available to serve customers. The data below give the number of busy tellers observed during 25 spot checks.

6	5	4	1	5
6	1	5	5	5
3	5	2	4	3
4	5	0	6	4
3	4	2	3	6

3.93 A wheat farmer tests a new fertilizer on a sample of 30 one-acre plots. The yields of wheat, in bushels, are as follows.

39	41	38	39	38
41	40	38	37	39
39	41	43	38	40
39	39	43	39	37
38	37	40	43	40
42	40	39	36	43

Use class marks in Formulas 3.2 and 3.3 to compute the approximate values of $\bar{x}$ and s for the grouped data in Exercises 3.94–3.97.

3.94 The National Center for Education Statistics publishes cost estimates for various aspects of higher education in *Digest of Education Statistics.* One such cost is the annual tuition and fees in private, four-year universities. The annual tuitions for a sample of 35 private, four-year universities yield the following frequency distribution.

Annual tuition	Frequency
$2000–$2999	2
$3000–$3999	2
$4000–$4999	6
$5000–$5999	7
$6000–$6999	7
$7000–$7999	6
$8000–$8999	4
$9000–$9999	1

3.95 A study by Lewis M. Terman, published in his book *The Intelligence of School Children* (Boston: Houghton Mifflin, 1946), gave the following frequency distribution for the IQs of 112 children attending kindergarten in San Jose and San Mateo, California.

IQ	Frequency
60–69	1
70–79	5
80–89	13
90–99	22
100–109	28
110–119	23
120–129	14
130–139	3
140–149	2
150–159	1

3.96 According to the Food and Nutrition Board of the National Academy of Sciences, the recommended daily allowance (RDA) of calcium for adults is 800 mg (milligrams). A nutritionist thinks that people with incomes below the poverty level average less than the RDA of 800 mg. To test her suspicion, the daily intakes of calcium are determined for a sample of 50 people with incomes below the poverty level. The results she obtained are displayed in the following frequency distribution.

Intake (mg)	Frequency
Under 200	1
200–under 400	1
400–under 600	12
600–under 800	16
800–under 1000	12
1000–under 1200	7
1200–under 1400	1

3.97 *Runner's World* magazine conducts studies on the finishing times of runners in the New York City 10-km run. A sample of 40 finishing times yields the following frequency distribution.

Finishing time (minutes)	Frequency
30–under 40	1
40–under 50	2
50–under 60	14
60–under 70	13
70–under 80	8
80–under 90	2

3.98 The Bureau of Economic Analysis gathers information on the length of stay in Europe and the Mediterranean by U.S. travelers. Data are published in *Survey of Current Business.* A sample of 36 U.S. residents who traveled to Europe and the Mediterranean one year yielded the following data, in days, on length of stay.

Raw Data					
41	16	6	21	10	21
5	31	20	27	17	10
3	32	2	48	8	12
21	44	1	56	5	12
3	13	15	10	18	3
1	11	14	12	64	10

(Note: $\Sigma x = 643$, $\Sigma x^2 = 20,185$)

Grouped Data

Length of stay (days)	Frequency
1–7	10
8–14	10
15–21	8
22–28	1
29–35	2
36–42	1
43–49	2
50–56	1
57–63	0
64–70	1

a. Compute the exact values of $\bar{x}$ and s by applying Definition 3.4 on page 114 and Formula 3.1 on page 123 to the raw data.
b. Compute the approximate values of $\bar{x}$ and s using the frequency distribution and Formulas 3.2 and 3.3.
c. Compare your answers from parts (a) and (b).

· **3.99** The U.S. Energy Information Administration collects data on residential energy consumption and expenditures. Results are published in *Residential Energy Consumption Survey: Consumption and Expenditures*. The tables below give one year's energy consumptions for a sample of 50 households in the South. Data are in millions of BTUs.

Raw Data

130	55	45	64	155	66	60	80	102	62
58	101	75	111	151	139	81	55	66	90
97	77	51	67	125	50	136	55	83	91
54	86	100	78	93	113	111	104	96	113
96	87	129	109	69	94	99	97	83	97

(Note: $\Sigma x = 4486$, $\Sigma x^2 = 438{,}942$)

Grouped Data

Energy consumption (million BTUs)	Frequency
40–49 44.5	1
50–59 54.5	7
60–69	7
70–79	3
80–89	6
90–99	10
100–109	5
110–119	4
120–129	2
130–139	3
140–149	0
150–159	2

a. Compute the exact values of $\bar{x}$ and s by applying Definition 3.4 on page 114 and Formula 3.1 on page 123 to the raw data.
b. Compute the approximate values of $\bar{x}$ and s using the frequency distribution and Formulas 3.2 and 3.3.
c. Compare your answers from parts (a) and (b).

· **3.100** If data are grouped into a frequency distribution using classes based on a single value, is it possible to compute $\bar{x}$ and s without using the grouped-data formulas?

· **3.101** If data are grouped into a frequency distribution using classes that are not based on a single value, why do the grouped-data formulas of this section generally yield only the approximate values of $\bar{x}$ and s?

Computing $\bar{x}$ for data grouped into a relative-frequency distribution. Formula 3.2, used for computing the mean of grouped data, is $\bar{x} = (\Sigma xf)/n$. It can be rewritten as

$$(1) \qquad \bar{x} = \Sigma x \cdot \frac{f}{n}.$$

Since f/n represents the relative frequency of a class, we see that the sample mean of data that are grouped into a relative-frequency distribution can be obtained as follows: For each class, multiply the class mark by the relative frequency of the class and then sum all of the resulting products. We will apply Formula (1) in Exercises 3.102 and 3.103.

· · **3.102** Refer to Exercise 3.88.
a. Construct a relative-frequency distribution.
b. Use Formula (1) to compute $\bar{x}$, and compare your answer to the one obtained in Exercise 3.88.

· · **3.103** Refer to Exercise 3.89.
a. Construct a relative-frequency distribution.
b. Use Formula (1) to compute $\bar{x}$, and compare your answer to the one obtained in Exercise 3.89.

· · · **3.104** The table below displays a set of 20 scores on a 100-point, true-false psychology quiz in which each question was worth 10 points.

70	70	70	70	70	70	70	70	80	80
80	80	80	80	80	80	80	80	90	90

a. Compute the sample mean, $\bar{x}$, of the exam scores.
b. Construct a frequency distribution for the data using the classes 70–79, 80–89, and 90–99.
c. Use your frequency distribution from part (b) to compute the approximate value of $\bar{x}$.
d. Compare your answers from parts (a) and (c). Why is the approximation in part (c) so poor?

3.6 THE FIVE-NUMBER SUMMARY; BOXPLOTS

In Sections 3.3–3.5, we concentrated on the mean, standard deviation, and related descriptive measures such as z-scores. Now we will examine several descriptive measures based on percentiles. Unlike the mean and standard deviation, descriptive measures based on percentiles are resistant; that is, they are not sensitive to the influence of a few extreme data values. For this reason, descriptive measures based on percentiles are often preferred over those based on the mean and standard deviation.

Percentiles, Deciles, and Quartiles

As we learned in Section 3.1, the median of a data set divides the data into two equal parts—the bottom 50% and the top 50%. The **percentiles** of a data set divide it into hundredths, or 100 equal parts. A data set has 99 percentiles, denoted by P_1, P_2, ..., P_{99}. Roughly speaking, the first percentile, P_1, is the number that divides the bottom 1% of the data from the top 99%; the second percentile, P_2, is the number that divides the bottom 2% of the data from the top 98%; and so forth. Note that the median is the 50th percentile.

Deciles are also useful. The **deciles** of a data set divide it into tenths, or 10 equal parts. A data set has nine deciles, which we denote by D_1, D_2, ..., D_9. Basically, the first decile, D_1, is the number that divides the bottom 10% of the data from the top 90%; the second decile, D_2, is the number that divides the bottom 20% of the data from the top 80%; and so on. Note that the first decile is the 10th percentile, the second decile is the 20th percentile, and so forth.

The most commonly used percentiles are quartiles. The **quartiles** of a data set divide it into quarters, or four equal parts. A data set has three quartiles, which we denote by Q_1, Q_2, and Q_3. Roughly speaking, the first quartile, Q_1, is the number that divides the bottom 25% of the data from the top 75%; the second quartile, Q_2, is the median, which, as we know, is the number that divides the bottom 50% of the data from the top 50%; and the third quartile, Q_3, is the number that divides the bottom 75% of the data from the top 25%. Note that the first and third quartiles are the 25th and 75th percentiles, respectively. Figure 3.12 depicts the quartiles for uniform, bell-shaped, right-skewed, and left-skewed distributions.

At this point our intuitive definitions of percentiles and deciles will suffice. However, quartiles need to be defined more precisely, which is done in Definition 3.8.

DEFINITION 3.8   **QUARTILES**

> Arrange the data in increasing order and determine the median.
>
> - The *first quartile* is the median of the data lying at or below the median of the entire data set.
> - The *second quartile* is the median of the entire data set.
> - The *third quartile* is the median of the data lying at or above the median of the entire data set.

FIGURE 3.12

Quartiles for (a) uniform,
(b) bell-shaped, (c) right-skewed,
and (d) left-skewed distributions

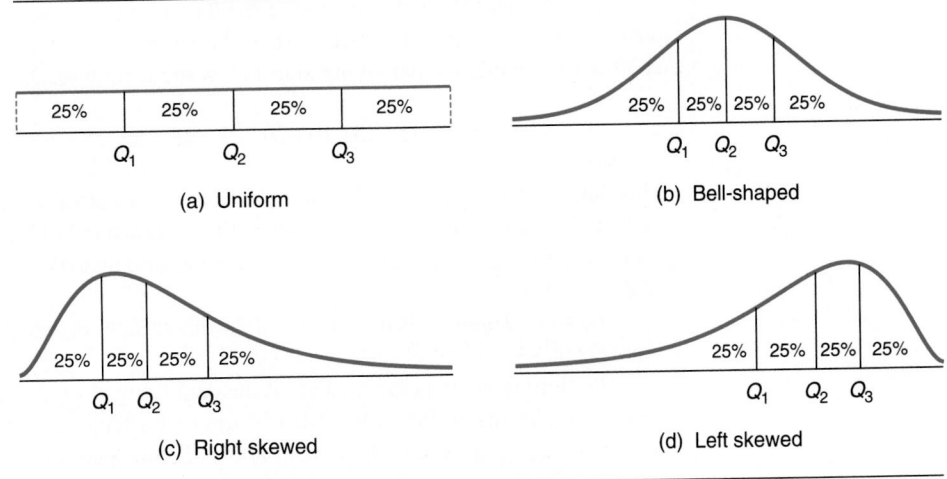

(a) Uniform

(b) Bell-shaped

(c) Right skewed

(d) Left skewed

Note: Not all statisticians define quartiles in exactly the same way. We have defined them so that the first and third quartiles are identical, respectively, to the lower and upper hinges (to be discussed shortly). Other definitions of quartiles may lead to slightly different values.

Example 3.23 Illustrates Definition 3.8

The A. C. Nielsen Company publishes data on the TV viewing habits of Americans by various characteristics in *Nielsen Report on Television*. A sample of 20 people yields the weekly viewing times, in hours, displayed in Table 3.22. Determine and interpret the quartiles for these data.

TABLE 3.22
Weekly TV viewing times

25	41	27	32	43
66	35	31	15	5
34	26	32	38	16
30	38	30	20	21

SOLUTION To find the quartiles, we apply Definition 3.8. First we arrange the data in increasing order. The ordered list of the entire data set is

5 15 16 20 21 25 26 27 30 **30 31** 32 32 34 35 38 38 41 43 66

Next we determine the median of the entire data set. The number of pieces of data is 20, and so the position of the median is at $(20 + 1)/2 = 10.5$, halfway between the tenth and eleventh data values (shown in boldface type) in the ordered list. Thus the median of the entire data set is $(30 + 31)/2 = 30.5$.

The first quartile is the median of the data lying at or below the median of the entire data set. Referring to the ordered list of the entire data set and recalling that the median is 30.5, we see that the data lying at or below the median of the entire data set is

$$5 \quad 15 \quad 16 \quad 20 \quad \mathbf{21} \quad \mathbf{25} \quad 26 \quad 27 \quad 30 \quad 30$$

This data set has 10 pieces of data. Its median is therefore at position $(10 + 1)/2 = 5.5$, halfway between the fifth and sixth data values (shown in boldface type) in the ordered list. Thus the median of this data set, and hence the first quartile, is $(21 + 25)/2 = 23.0$; that is, $Q_1 = 23.0$.

The second quartile is the median of the entire data set, which, as we have seen, is 30.5. Thus $Q_2 = 30.5$.

The third quartile is the median of the data lying at or above the median of the entire data set. Referring to the ordered list of the entire data set and recalling that its median is 30.5, we see that the data lying at or above the median of the entire data set is

$$31 \quad 32 \quad 32 \quad 34 \quad \mathbf{35} \quad \mathbf{38} \quad 38 \quad 41 \quad 43 \quad 66$$

This data set has 10 pieces of data. Its median is therefore at position $(10 + 1)/2 = 5.5$, halfway between the fifth and sixth data values (shown in boldface type) in the ordered list. Thus the median of this data set, and hence the third quartile, is $(35 + 38)/2 = 36.5$; that is, $Q_3 = 36.5$.

Interpreting our results, we conclude that 25% of the viewing times are less than 23.0 hours, 25% are between 23.0 and 30.5 hours, 25% are between 30.5 and 36.5 hours, and 25% are greater than 36.5 hours.   ■

In Example 3.23 the number of pieces of data is 20, an even number. To illustrate the determination of quartiles for a data set with an odd number of pieces of data, let's consider the TV-viewing-time data again, but this time without the largest data value, 66. Then the ordered data are

$$5 \quad 15 \quad 16 \quad 20 \quad 21 \quad 25 \quad 26 \quad 27 \quad 30 \quad \mathbf{30} \quad 31 \quad 32 \quad 32 \quad 34 \quad 35 \quad 38 \quad 38 \quad 41 \quad 43$$

The median of the entire data set (also the second quartile) is 30, shown in boldface type. The first quartile is the median of the data lying at or below the median of the entire data set, that is, the median of the 10 data values from 5 to the boldfaced 30, which is $(21 + 25)/2 = 23.0$. The third quartile is the median of the data lying at or above the median of the entire data set, that is, the median of the 10 data values from the boldfaced 30 to 43, which is $(34 + 35)/2 = 34.5$.

The Interquartile Range

The **interquartile range, IQR,** is the preferred measure of dispersion when the median is used as the measure of central tendency. Like the median, the IQR is a resistant measure.

| DEFINITION 3.9 | **INTERQUARTILE RANGE** |

> The *interquartile range,* denoted by **IQR,** is defined as the difference between the first and third quartiles; that is,
>
> $$IQR = Q_3 - Q_1.$$
>
> Thus, roughly speaking, the IQR gives the range of the middle 50% of the data.

Example 3.24 Illustrates Definition 3.9

Obtain the IQR for the TV-viewing-time data displayed in Table 3.22 on page 147.

SOLUTION As we discovered in Example 3.23, the first and third quartiles are 23.0 and 36.5, respectively. Therefore the interquartile range is

$$IQR = Q_3 - Q_1 = 36.5 - 23.0 = 13.5.$$ ∎

The Five-Number Summary

From the three quartiles, we can obtain a measure of central tendency (the median, Q_2) and measures of variation of the two middle quarters of the data, $Q_2 - Q_1$ for the second quarter and $Q_3 - Q_2$ for the third quarter. But the three quartiles don't tell us anything about the first and fourth quarters, that is, the tails of the distribution. To gain that information, we need only include the minimum and maximum values also. Then the variation of the first quarter (left tail) can be measured as the difference between the minimum and the first quartile, $Q_1 - \text{Min}$, and the variation of the fourth quarter (right tail) can be measured as the difference between the third quartile and the maximum, $\text{Max} - Q_3$.

For the TV-viewing-time data in Table 3.22, we have $\text{Min} = 5$, $Q_1 = 23$, $Q_2 = 30.5$, $Q_3 = 36.5$, and $\text{Max} = 66$. Consequently, the variations of the four quarters are, respectively, 18, 7.5, 6, and 29.5. There is less variation in the middle two quarters than in the first and fourth quarters, and the fourth quarter has the greatest variation of all.

Thus we see that the minimum, maximum, and quartiles together provide, among other things, information on central tendency and dispersion. Written in increasing order, they comprise what is called the **five-number summary** of a data set.

| DEFINITION 3.10 | **FIVE-NUMBER SUMMARY** |

> The *five-number summary* of a data set consists of the minimum, maximum, and quartiles written in increasing order: Min, Q_1, Q_2, Q_3, Max.

So, for example, the five-number summary of the data on TV-viewing times is 5, 23, 30.5, 36.5, and 66.

Outliers

In data analysis it is important to identify **outliers,** data values that fall well outside the overall pattern of the data. An outlier requires special attention: It may be the result of a measurement or recording error, a member from a different population than the rest of the sample, or simply an unusual extreme value. Note that an extreme value need not be an outlier; it may instead be an indication of skewness.

As an example of an outlier, consider individual wealth (in dollars). For this data set, the wealth of H. Ross Perot is an outlier, in this case an unusual extreme value.

When an outlier is found, its cause should be determined if possible. If it is discovered that an outlier is due to a measurement or recording error or that for some other reason it clearly does not belong in the data set, then the outlier can be removed without further ado. However, if no explanation for the outlier is apparent, then the decision whether to retain it in the data set can often be difficult and calls for a judgment by the researcher.

We can use quartiles and the IQR to identify potential outliers. To accomplish that we need to define the inner and outer fences. The number that lies 1.5 IQRs below the first quartile and the number that lies 1.5 IQRs above the third quartile together comprise the **inner fences.** The number that lies 3 IQRs below the first quartile and the number that lies 3 IQRs above the third quartile together comprise the **outer fences.** More formally, we have the following definitions.

DEFINITION 3.11   **INNER AND OUTER FENCES; POSSIBLE AND PROBABLE OUTLIERS**

The *inner fences* and *outer fences* are defined as follows:

Inner fences:
$$Q_1 - 1.5 \cdot \text{IQR}$$
$$Q_3 + 1.5 \cdot \text{IQR}$$

Outer fences:
$$Q_1 - 3 \cdot \text{IQR}$$
$$Q_3 + 3 \cdot \text{IQR}$$

Data values that lie between the inner and outer fences are considered *possible outliers;* those that lie outside the outer fences are considered *probable outliers.*

For the TV-viewing-time data in Table 3.22, we have $Q_1 = 23.0$, $Q_3 = 36.5$, and $\text{IQR} = 13.5$. Therefore the inner fences are

Inner fences:
$$Q_1 - 1.5 \cdot \text{IQR} = 23.0 - 1.5 \cdot 13.5 = 2.75$$
$$Q_3 + 1.5 \cdot \text{IQR} = 36.5 + 1.5 \cdot 13.5 = 56.75$$

and the outer fences are

Outer fences:
$$Q_1 - 3 \cdot \text{IQR} = 23.0 - 3 \cdot 13.5 = -17.5$$
$$Q_3 + 3 \cdot \text{IQR} = 36.5 + 3 \cdot 13.5 = 77.0.$$

These fences are portrayed graphically in Fig. 3.13.

FIGURE 3.13
Inner and outer fences
for TV-viewing times

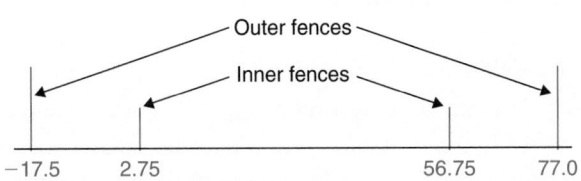

Referring now to Table 3.22, or more simply to the ordered list of the data on page 147, we see that there is one data value, 66, that lies outside the inner fences. Since that data value does not lie outside the outer fences, we classify it as a possible outlier (not as a probable outlier).

Boxplots

A **boxplot,** also called a **box-and-whisker diagram,** is based on the five-number summary and can be used to provide a graphical display of the center and variation of a data set. These diagrams, like stem-and-leaf diagrams, were invented by John Tukey.

Actually, two types of boxplots are in common use. One is simply called a boxplot; the other is called a **modified boxplot.** The main difference between the two types of boxplots is that possible and probable outliers are plotted individually in a modified boxplot, but not in a boxplot. Thus when outliers are of concern, modified boxplots are the preferred type of boxplot. Procedures 3.1 and 3.2 provide, respectively, step-by-step methods for constructing boxplots and modified boxplots.

PROCEDURE 3.1   **TO CONSTRUCT A BOXPLOT**

Step 1   Determine the quartiles for the data.

Step 2   Determine the minimum and maximum of the data.

Step 3   Draw a horizontal axis on which the values obtained in Steps 1 and 2 can be located. Above this axis, mark the quartiles and the minimum and maximum with vertical lines.

Step 4   Connect the quartiles to each other to make a box, and then connect the box to the minimum and maximum with lines.

To construct a modified boxplot, we need to introduce the concept of adjacent values. The **adjacent values** of a data set are the most extreme values still lying within the inner fences, that is, the most extreme values that are not possible or probable outliers. Note that if a data set has no possible or probable outliers, then the adjacent values are just the minimum and maximum values of the data set.

PROCEDURE 3.2	TO CONSTRUCT A MODIFIED BOXPLOT

Step 1   Determine the quartiles for the data.

Step 2   Determine the possible and probable outliers and the adjacent values.

Step 3   Draw a horizontal axis on which the values obtained in Steps 1 and 2 can be located. Above this axis, mark the quartiles and the adjacent values with vertical lines.

Step 4   Connect the quartiles to each other to make a box, and then connect the box to the adjacent values with lines.

Step 5   Plot each possible outlier with an asterisk and each probable outlier with a hollow dot.

If a data set has no possible or probable outliers, then its boxplot and modified boxplot are identical. (Why is this so?) Hence, in such cases, there is no need to construct both a boxplot and a modified boxplot.

In both types of boxplots, the two lines emanating from the box are called **whiskers.** Also, in the context of boxplots, the first quartile is called the **lower hinge** and the third quartile is called the **upper hinge.** (See the note on page 147.)

Example 3.25   Illustrates Procedures 3.1 and 3.2

SOLUTION   The weekly TV-viewing times for a sample of 20 people are displayed in Table 3.22 on page 147. Construct a boxplot and, if appropriate, a modified boxplot.

To obtain a boxplot for the TV-viewing times, we apply the step-by-step method presented in Procedure 3.1 on page 151.

Step 1   *Determine the quartiles for the data.*

The quartiles for the TV-viewing times were obtained in Example 3.23: $Q_1 = 23.0$, $Q_2 = 30.5$, and $Q_3 = 36.5$.

Step 2   *Determine the minimum and maximum of the data.*

From Table 3.22, or more simply by referring to the ordered list on page 147, we see that Min = 5 and Max = 66.

Step 3   *Draw a horizontal axis on which the values obtained in Steps 1 and 2 can be located. Above this axis, mark the quartiles and the minimum and maximum with vertical lines.*

This is done in Fig. 3.14(a).

Step 4   *Connect the quartiles to each other to make a box, and then connect the box to the minimum and maximum with lines.*

This is also done in Fig. 3.14(a).

FIGURE 3.14
(a) Boxplot for TV-viewing times
(b) Modified boxplot
for TV-viewing times

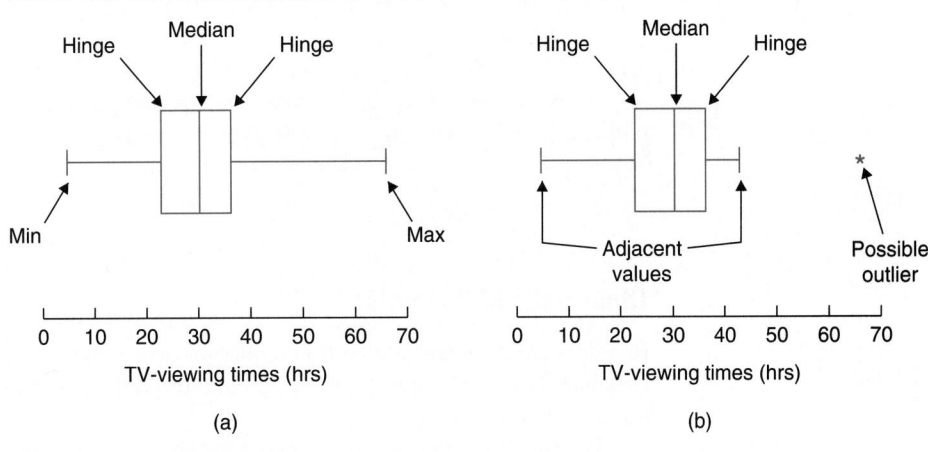

(a)   (b)

Figure 3.14(a) is the boxplot for the data on TV-viewing times. Observe that the two boxes in the boxplot indicate the spread of the second and third quarters of the data and that since the ends of the combined box are at the quartiles, the width of the combined box equals the interquartile range, IQR. Also notice that the two whiskers indicate the spread of the first and fourth quarters. Thus we see easily from the boxplot that there is less variation in the middle two quarters of the TV-viewing times than in the first and fourth quarters, and that the fourth quarter has the greatest variation of all.

We discovered on page 151 that the TV-viewing times contain a possible outlier. Therefore it is appropriate to construct a modified boxplot since it will be different from the (ordinary) boxplot we just obtained. We apply Procedure 3.2.

Step 1 *Determine the quartiles for the data.*

The quartiles for the TV-viewing times were obtained in Example 3.23: $Q_1 = 23.0$, $Q_2 = 30.5$, and $Q_3 = 36.5$.

Step 2 *Determine the possible and probable outliers and the adjacent values.*

We found earlier that the TV-viewing-times data set has one possible outlier, 66, and no probable outliers. Referring now to the ordered list of the data on page 147, we see that the adjacent values are 5 and 43.

Step 3 *Draw a horizontal axis on which the values obtained in Steps 1 and 2 can be located. Above this axis, mark the quartiles and the adjacent values with vertical lines.*

This is done in Fig. 3.14(b).

Step 4 *Connect the quartiles to each other to make a box, and then connect the box to the adjacent values with lines.*

This is also done in Fig. 3.14(b).

Step 5 *Plot each possible outlier with an asterisk and each probable outlier with a hol-*
low dot.

As we noted in Step 2, the data contains one possible outlier (66) and no probable
outliers. The possible outlier of 66 is plotted with an asterisk in Fig. 3.14(b). ■

Other Uses of Boxplots

Boxplots are especially suited for comparing two or more data sets. In doing so, it is important to plot all of the boxplots using the same scale. Exercises 3.121 and 3.122 examine
this use of boxplots.

We can also use a boxplot to identify the approximate shape of the distribution of a data
set. Figure 3.15 displays uniform, bell-shaped, right-skewed, and left-skewed distributions
and their corresponding boxplots. Study Fig. 3.15 carefully, noticing especially how box
width and whisker length relate to skewness and symmetry.

FIGURE 3.15

Distribution shapes and
boxplots for (a) uniform,
(b) bell-shaped, (c) right-skewed,
and (d) left-skewed distributions

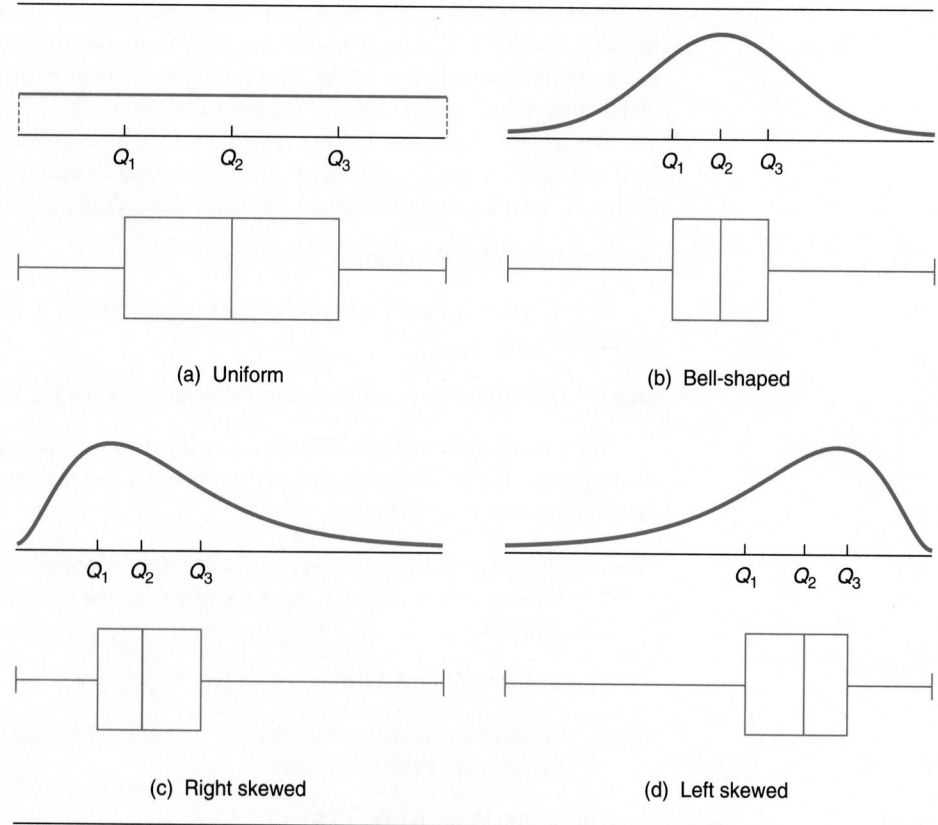

Employing boxplots to identify the shape of a distribution is most useful with large data sets. For small data sets, boxplots can be unreliable in identifying distribution shape; it is generally better to use a histogram or a stem-and-leaf diagram to ascertain distribution shape for a small data set.

Using the Computer (Optional)

Minitab has a command called **BoxPlot** that can be used to obtain a boxplot, actually a modified boxplot. We will illustrate the use of that command by applying it to the TV-viewing-times data considered in Examples 3.23–3.25.

Example 3.26   The BoxPlot Command

Use Minitab to obtain a (modified) boxplot for the TV-viewing times displayed in Table 3.22 on page 147.

SOLUTION   First we store the TV-viewing times in a column named TIMES. Then we proceed in the following way.

Session commands: We type the command **BOXPLOT** followed by the storage location of the sample data; that is, we type BOXPLOT 'TIMES' and press Enter. Printout 3.4 displays this command and the output that results.

(or)

Menu commands: We choose **Graph ▶ Boxplot...** and specify TIMES in the **Variable** text box. If high resolution is enabled, we then select **High resolution** and press Spacebar to disable it.[†] Finally, we select **OK**. The resulting output is shown in Printout 3.4.

PRINTOUT 3.4
Minitab output for the BoxPlot command

```
MTB > BOXPLOT 'TIMES'


                               -----------
               ---------------I    +    I------                        *
                               -----------
       --------+---------+---------+---------+---------+--------+--------TIMES
               12        24        36        48        60
```

Notice that in this boxplot there are 12 units between tick marks (+) on the horizontal axis. Also notice that there are 10 dashes (–) from one tick mark to the next (counting the – part of a + as a last dash). Hence each dash represents 1.2 units.

[†] We are disabling high resolution so that all readers will obtain the same output. If you would like high-resolution output, omit the disabling step.

Minitab's boxplot uses Is to mark the hinges (first and third quartiles) and a + to mark the median. Also, the data value 66 is plotted with an asterisk, indicating that it is a possible outlier. Minitab uses an 0 to plot a probable outlier, of which there are none in this data set. Compare the boxplot in Printout 3.4 to the one we obtained by hand in Fig. 3.14(b) on page 153. ∎

We can use a boxplot generated by Minitab's BoxPlot command to estimate the five-number summary: the minimum, maximum, and three quartiles. For instance, consider the boxplot for the TV-viewing times, shown in Printout 3.4. Recalling that each dash (-) in that diagram represents 1.2 units, we find that the minimum is approximately 4.8, the first quartile is approximately 22.8, the median is approximately 30.0, the third quartile is approximately 36.0, and the maximum is approximately 66.0.

But we don't have to be content with estimating those values—Minitab has a command called **LVals** that among other things will determine exactly the five-number summary.[†] In Example 3.27 we will apply the LVals command to the TV-viewing-times data.

Example 3.27   The LVals Command

Use Minitab's LVals command to obtain the five-number summary for the TV-viewing times displayed in Table 3.22 on page 147.

SOLUTION   In Example 3.26 we stored the TV-viewing times in a column named TIMES. To apply LVals to that data, we proceed as follows.

Session commands: We type the command **LVALS** followed by the storage location of the data; that is, we type LVALS 'TIMES' and press [Enter]. This command and the resulting output is displayed in Printout 3.5.

(or)

Menu commands: We choose **Stat ▸ EDA ▸ Letter Values...**, specify TIMES in the **Variable** text box, and then select **OK**. Printout 3.5 shows the output obtained.

PRINTOUT 3.5
Minitab output for
the LVals command

```
MTB > LVALS 'TIMES'
        DEPTH      LOWER        UPPER        MID       SPREAD
  N=    20
  M     10.5              30.500           30.500
  H     5.5      23.000      36.500        29.750     13.500
  E     3.0      16.000      41.000        28.500     25.000
  D     2.0      15.000      43.000        29.000     28.000
        1         5.000      66.000        35.500     61.000
```

[†] The LVals command is not available in all versions of Minitab.

The row of the output that reads N= 20 indicates that there are 20 pieces of data (TV-viewing times). The row beneath that begins with the letter M, which stands for "median." To the right of the letter M, under the column labeled DEPTH, is the number 10.5. This signifies that the median of the data set lies halfway between the tenth and eleventh data values when the data are arranged in increasing order. In other words, the median is equal to the mean of the tenth and eleventh data values in an ordered list. The numerical value of the median is the number to the right of the depth of 10.5, namely, 30.5.

In the next row, we find the information about the hinges (first and third quartiles). As you might suspect, the letter H stands for "hinges." The depth of 5.5, which appears next, indicates that the first quartile lies halfway between the fifth and sixth data values in an ordered list, counting in from the left; and that the third quartile lies halfway between the fifth and sixth data values in an ordered list, counting in from the right. Following the depth of 5.5 are the first and third quartiles. The first quartile equals 23.0 and the third quartile equals 36.5.

To obtain the minimum and maximum values in the data set, we refer to the last row of Printout 3.5. The minimum is the entry in that row under the column labeled LOWER, and the maximum is the entry in that row under the column labeled UPPER. Thus the minimum and maximum are 5 and 66, respectively.   ■

As we observe from Printout 3.5, the output of the LVals command contains a large amount of additional information. For a detailed discussion of that information, consult your Minitab documentation.

EXERCISES 3.6

In each of Exercises 3.105–3.112, determine the quartiles.

3.105 The exam scores for the students in an introductory statistics course are as follows.

88	67	64	76	86
85	82	39	75	34
90	63	89	90	84
81	96	100	70	96

3.106 The table below gives the hourly temperature readings (in degrees Fahrenheit) for Colorado Springs, Colorado, on Tuesday, August 22, 1978.

69	63	61	74	88	87	74	68
66	61	64	79	87	85	71	65
65	60	70	84	85	73	70	62

3.107 The U.S. Federal Highway Administration conducts studies on motor vehicle travel by type of vehicle.

Results are published annually in *Highway Statistics*. A sample of 15 cars yields the following data on number of miles driven, in thousands, for last year.

10.2	10.3	8.9	12.7	8.3
9.2	13.7	7.7	3.3	10.6
11.8	6.6	8.6	5.7	12.0

3.108 A water-heater manufacturer guarantees the electric heating element for 5 years. The lifetimes, in months, for a sample of 19 such elements follow.

49.3	79.3	86.4	68.4	62.6
64.1	53.2	30.0	66.6	66.9
65.1	67.8	95.9	40.1	79.3
92.1	92.0	93.0	48.9	

3.109 The U.S. Energy Information Administration reports figures on residential energy consumption and expenditures in *Residential Energy Consumption Survey: Con-*

sumption and Expenditures. A sample of 18 households using electricity as their primary energy source yields the following data on one year's energy expenditures.

$1376	1452	1235	1480	1185	1327
1059	1400	1227	1102	1168	1070
949	1351	1259	1179	1393	1456

· **3.110**   In January 1984 the U.S. Department of Agriculture reported in *Family Economic Review* that a typical American family of four with an intermediate budget would spend about $92 per week for food. A consumer researcher in Kansas suspected that the median weekly cost was less in her state. She took a sample of 10 Kansas families of four, each with an intermediate budget, and obtained the following weekly food costs.

$78	104	84	70	96
73	87	85	76	94

· **3.111**   The U.S. National Center for Health Statistics compiles data on the length of stay by patients in short-term hospitals and publishes its findings in *Vital and Health Statistics.* A random sample of 21 patients yielded the following data on length of stay, in days.

4	4	12	18	9	6	12
3	6	15	7	3	55	1
10	13	5	7	1	23	9

· **3.112**   IQs measured on the Stanford Revision of the Binet-Simon Intelligence Scale have a mean of 100 points and a standard deviation of 16 points. Suppose that 25 randomly selected people are given that IQ test and that the results are as follows.

91	96	106	116	97
102	96	124	115	121
95	111	105	101	86
88	129	112	82	98
104	118	127	66	102

In each of Exercises 3.113–3.116,
a. *determine the interquartile range.*
b. *obtain the five-number summary.*
c. *identify possible and probable outliers, if any.*
d. *construct and interpret a boxplot and, if appropriate, a modified boxplot.*

· **3.113**   The exam scores for the 20 students in an introductory statistics class, given in Exercise 3.105.

· **3.114**   The hourly temperature data for Colorado Springs, Colorado, on Tuesday, August 22, 1978, given in Exercise 3.106.

· **3.115**   The lengths of stay in short-term hospitals by 21 randomly selected patients, given in Exercise 3.111.

· **3.116**   The IQs on the Stanford Revision of the Binet-Simon Intelligence Scale for a sample of 25 people, given in Exercise 3.112.

· **3.117 (Computer exercise)**   Refer to the exam-score data in Exercise 3.105. Use Minitab or some other statistical software to
a. obtain a boxplot for the data.
b. determine the five-number summary of the data.

· **3.118 (Computer exercise)**   Refer to the data on hourly temperatures in Exercise 3.106. Use Minitab or some other statistical software to
a. obtain a boxplot for the data.
b. determine the five-number summary of the data.

· **3.119 (Computer exercise)**   The U.S. Department of Agriculture collects data on annual per capita beef consumption and publishes the results in *Food Consumption, Prices, and Expenditures.* Each person in a sample is asked to estimate his or her beef consumption, in pounds, for last year.
a. Printout 3.6 depicts a boxplot, generated by Minitab's BoxPlot command, for the beef-consumption data. Use the printout to discuss the variation in the data set and to determine approximately the five-number summary.
b. Are there any possible or probable outliers in the data? If so, identify them approximately and provide an explanation for their possible cause.
c. In Printout 3.7 we have displayed the Minitab output obtained by applying the LVals command to the beef-consumption data. Use the output to determine exactly the five-number summary of the data. Compare your answers to the approximate values you obtained in part (a).

· **3.120 (Computer exercise)**   A prospective investor in a limited partnership wants information on the monthly rental charges for three-bedroom homes in the area. She obtains the monthly rental charges for a sample of three-bedroom homes.
a. Printout 3.8 provides the output obtained by applying Minitab's BoxPlot command to the sample of monthly rental charges. Use the printout to discuss the variation in the data set and to find approximately the five-number summary.

PRINTOUT 3.6   Minitab output for Exercise 3.119(a)

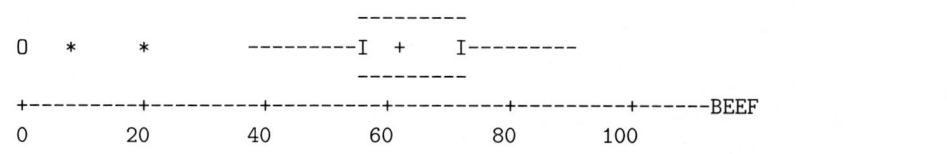

PRINTOUT 3.7   Minitab output for Exercise 3.119(c)

	DEPTH	LOWER	UPPER	MID	SPREAD
N=	40				
M	20.5	62.000		62.000	
H	10.5	55.000	72.500	63.750	17.500
E	5.5	42.000	76.500	59.250	34.500
D	3.0	8.000	78.000	43.000	70.000
C	2.0	0.000	79.000	39.500	79.000
	1	0.000	89.000	44.500	89.000

PRINTOUT 3.8   Minitab output for Exercise 3.120(a)

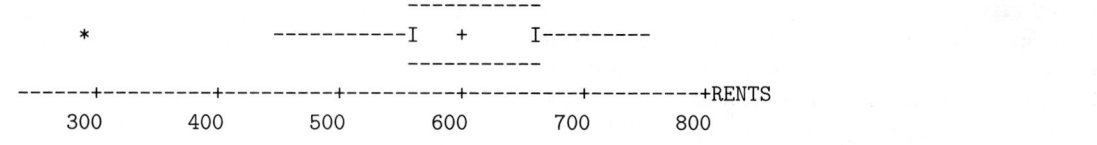

PRINTOUT 3.9   Minitab output for Exercise 3.120(c)

	DEPTH	LOWER	UPPER	MID	SPREAD
N=	32				
M	16.5	603.500		603.500	
H	8.5	562.500	662.000	612.250	99.500
E	4.5	505.000	675.500	590.250	170.500
D	2.5	461.000	727.500	594.250	266.500
C	1.5	369.500	737.000	553.250	367.500
	1	289.000	745.000	517.000	456.000

b. Are there any possible or probable outliers in the data? If so, identify them approximately and provide an explanation for their possible cause.

c. Printout 3.9 shows the Minitab output obtained by applying the LVals command to the monthly rental-charge data. Use the output to determine exactly the five-number summary of the data. Compare your answers to the approximate values determined in part (a).

·   **3.121**   Surveys are conducted by the Northwestern University Placement Center, Evanston, Illinois, on starting salaries for college graduates. Results of the surveys can be found in *The Northwestern Lindquist-Endicott Report.* The following diagram shows boxplots for the starting annual salaries obtained from samples of 32 accounting graduates (top boxplot) and 35 liberal-arts graduates (bottom boxplot). The data are in thousands of dollars.

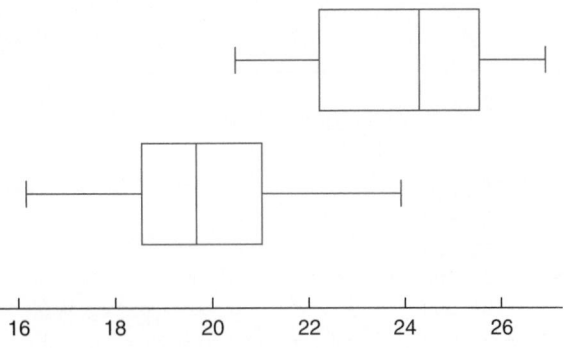

Use the boxplots to compare the starting salaries of the accounting and liberal-arts graduates sampled.

·   **3.122**   Researchers in obesity wanted to compare the effectiveness of dieting with exercise against dieting without exercise. Seventy-three patients were randomly divided into two groups. Group 1, composed of 37 patients, was put on a program of dieting with exercise. Group 2, composed of 36 patients, dieted only. The results for weight loss, in pounds, after 2 months are summarized in the following boxplots. The top boxplot is for Group 1 and the bottom boxplot is for Group 2.

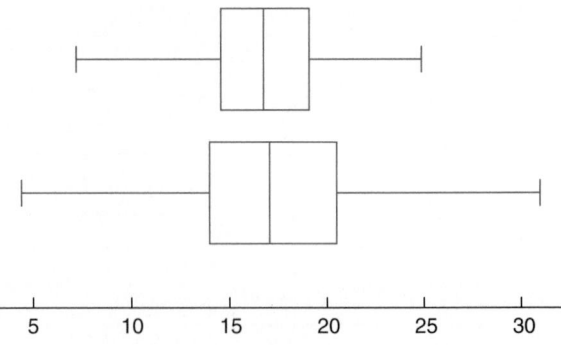

Employ the boxplots to compare the weight losses for the two groups.

· ·   **3.123**   Each boxplot below was obtained from a very large data set. Use the boxplots to identify the approximate shape of the distribution of each data set.

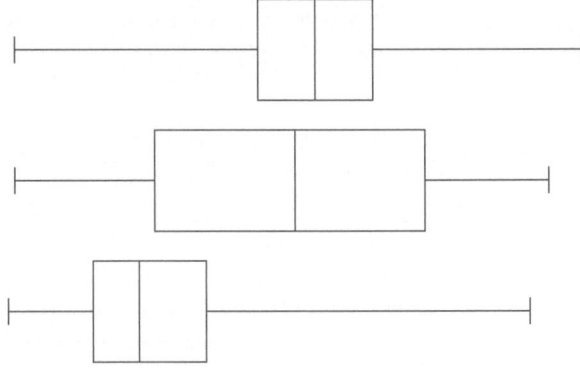

· ·   **3.124**   What can you say about the boxplot of a symmetric distribution?

3.7   DESCRIPTIVE MEASURES FOR POPULATIONS; USE OF SAMPLES

·   ·   ·   ·   ·

In this section we will discuss several descriptive measures of populations. Although in reality we often don't have access to the data for an entire population, it is nonetheless helpful to see the notation and formulas used for descriptive measures of populations.

The Population Mean

Recall that for a sample of size *n*, the sample mean, $\overline{x}$, is defined by

$$\overline{x} = \frac{\Sigma x}{n}.$$

The mean of a finite population is computed in the same way: first sum the data and then divide by the total number of pieces of data. However, to distinguish a population mean from a sample mean, we use the Greek letter μ (pronounced "mew") to denote the mean of a population. We also use the uppercase English letter N to represent the size of a population. Table 3.23 summarizes the notation employed for both a sample and a population.

TABLE 3.23

Notation used for a sample and for a population

	Size	Mean
Sample	n	$\bar{x}$
Population	N	μ

Using the notation displayed in the second row of Table 3.23, we now define the mean of a finite population.

DEFINITION 3.12

POPULATION MEAN

The *population mean, μ,* of a finite population of size N is defined by

$$\mu = \frac{\Sigma x}{N}.$$

In inferential studies the size of the population is typically quite large, and the mean of the population is usually difficult or impossible to determine exactly. However, for the sake of illustration, in Example 3.28 we will consider a small population for which the mean, μ, can easily be computed.

Example 3.28   Illustrates Definition 3.12

Table 3.24 presents some data for the starting offense of the 1992–1993 Dallas Cowboys football team. Compute the population mean weight of these 11 players.

TABLE 3.24

Dallas Cowboys offense, 1992–1993

Name	Position	Height	Weight (lb)
Troy Aikman	QB	6'4"	222
Mark Stepnoski	C	6'2"	269
Erik Williams	RT	6'6"	321
John Gesek	RG	6'5"	282
Nate Newton	LG	6'3"	303
Mark Tuinei	LT	6'5"	298
Jay Novacek	TE	6'4"	231
Michael Irvin	WR	6'2"	199
Alvin Harper	WR	6'3"	207
Emmitt Smith	RB	5'9"	209
Daryl Johnston	FB	6'2"	238

SOLUTION   The sum of the weights in Table 3.24 is 2779 lb. Since there are 11 players, $N = 11$. Thus the population mean weight of the players is

$$\mu = \frac{\Sigma x}{N} = \frac{2779}{11} = 252.6 \text{ lb,}$$

to the nearest tenth of a pound.   ∎

If we were interested in the mean weight of all starting-offense players in professional football, then the 11 weights of the starting-offense players on the Cowboys would be considered a sample instead of a population. The mean weight of those 11 players would then be a sample mean instead of a population mean, and we would write $\overline{x} = 252.6$ lb instead of $\mu = 252.6$ lb.

Using a Sample Mean to Estimate a Population Mean

Although we usually deal with sample data in inferential studies, the objective is to describe the entire population. The reason for resorting to a sample is that it is generally more practical. We illustrate this point in Example 3.29.

Example 3.29   A Use of the Sample Mean

The U.S. Bureau of the Census reports the mean income of U.S. households in its annual publication *Current Population Reports*. To obtain the complete population data—the incomes of all U.S. households—would be extremely expensive and time-consuming. It is also unnecessary, because accurate estimates for the mean income of all U.S. households can be obtained from the mean income of a sample of such households. In reality the Census Bureau samples 72,000 households out of a total of more than 83 million U.S. households.

The population of interest here is composed of the incomes of all U.S. households, and the mean of those incomes is the population mean, μ. The sample consists of the incomes of the 72,000 households sampled by the Census Bureau, and the mean of those incomes is a sample mean, $\overline{x}$. Figure 3.16, shown at the top of the following page, summarizes this discussion graphically.

Once the sample has been taken, the Census Bureau can compute the sample mean income, $\overline{x}$, of the 72,000 households obtained. Using the value of $\overline{x}$, the Census Bureau can then estimate the population mean income, μ, of all U.S. households. We will study these kinds of inferences in Chapter 8.   ∎

FIGURE 3.16
Population and sample for
incomes of U.S. households

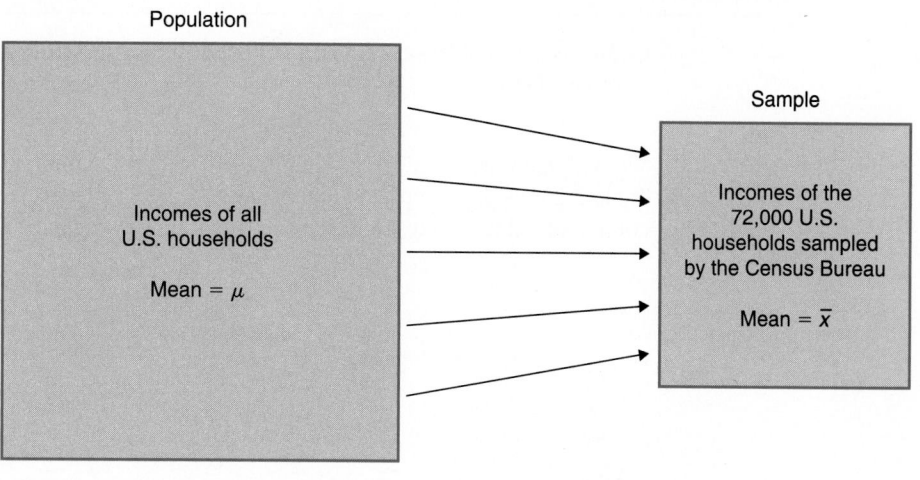

The Population Standard Deviation

Recall that the standard deviation of a sample is denoted by the letter s and is defined by

$$s = \sqrt{\frac{\Sigma(x - \bar{x})^2}{n - 1}}.$$

The standard deviation of a finite population is computed in a similar but slightly different way. It is denoted by the Greek letter σ (pronounced "sigma") and is defined as follows.

DEFINITION 3.13   **POPULATION STANDARD DEVIATION**

> The *population standard deviation, σ,* of a finite population of size N is defined by
>
> $$\sigma = \sqrt{\frac{\Sigma(x - \mu)^2}{N}}$$
>
> and can also be computed using the shortcut formula
>
> $$\sigma = \sqrt{\frac{\Sigma x^2}{N} - \mu^2}.$$

Just as s^2 is called the sample variance, σ^2 is called the **population variance.**

Notice that in the defining formula for the population standard deviation, σ, we divide by N, the size of the population. On the other hand, in the defining formula for the sample standard deviation, s, we divide by $n - 1$, one less than the size of the sample. This is because in inferential studies the main use of s is as an estimate of σ. Dividing by $n - 1$ instead of by n in the formula for s results in a better estimate of σ, on the average.

Example 3.30   Illustrates Definition 3.13

Compute the population standard deviation, σ, of the weights of the 11 players on the starting offense of the 1992–1993 Dallas Cowboys football team.

SOLUTION   We will apply the shortcut formula for σ given in Definition 3.13. In the first column of Table 3.25, we have repeated the weights of the 11 players from Table 3.24. The second column of Table 3.25 displays the squares of those weights.

TABLE 3.25
Table for computing σ
using the shortcut formula

x	x^2
222	49,284
269	72,361
321	103,041
282	79,524
303	91,809
298	88,804
231	53,361
199	39,601
207	42,849
209	43,681
238	56,644
2779	720,959

Recalling that $\mu = \Sigma x/N$, we see from the last row of Table 3.25 that

$$\sigma = \sqrt{\frac{\Sigma x^2}{N} - \mu^2} = \sqrt{\frac{720{,}959}{11} - \left(\frac{2779}{11}\right)^2} = 41.4.$$

The population standard deviation of the weights of the 11 players is 41.4 lb. ∎

In Example 3.28 we found that $\mu = 252.6$ lb for the weights of the 11 players on the starting offense of the Dallas Cowboys team. But when we computed σ in Example 3.30, we didn't use that value for μ because it is a rounded value and, as we said earlier, no rounding should be done until the computation is complete.

Using a Sample Standard Deviation to Estimate a Population Standard Deviation

In Example 3.31, we will illustrate how a sample standard deviation can be used to estimate a population standard deviation.

Example 3.31   A Use of the Sample Standard Deviation

A hardware manufacturer produces 10-mm bolts. The manufacturer knows that the diameters of the bolts produced vary somewhat from 10 mm and also from each other. But even if he is willing to accept some variation in bolt diameters, he cannot tolerate too much variation—if the variation is too large, too many of the bolts will be unusable.

Therefore the manufacturer must ensure that the standard deviation, σ, of the bolt diameters is not unduly large. How can the manufacturer obtain the value of σ? Since in this case it is not possible to determine σ exactly (do you know why?), inferential statistics must be employed to estimate it.

Suppose the manufacturer decides to estimate σ by taking a sample of 20 bolts. The population of interest here consists of the diameters of all bolts that have been or ever will be produced by the manufacturer, and the standard deviation of those diameters is the population standard deviation, σ. The sample consists of the diameters of the 20 bolts obtained by the manufacturer, and the standard deviation of those diameters is a sample standard deviation, s (see Fig. 3.17).

FIGURE 3.17
Population and sample
for bolt diameters

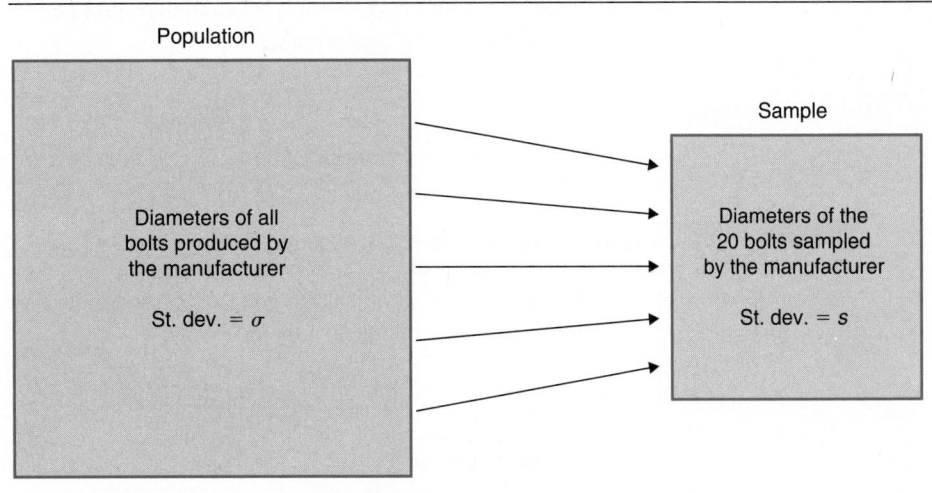

After the sample has been taken, the manufacturer can compute the sample standard deviation, s, of the diameters of the 20 bolts obtained. Using the value of s, he can then estimate the population standard deviation, σ, of the diameters of all bolts being produced. Such inferences will be examined in Chapter 12.   ■

Parameter and Statistic

Some specific statistical terminology helps us distinguish between descriptive measures for populations and descriptive measures for samples.

DEFINITION 3.14

PARAMETER AND STATISTIC

> *Parameter:*   A descriptive measure for a population.
>
> *Statistic:*   A descriptive measure for a sample.

Thus, for example, μ and σ are parameters, whereas $\bar{x}$ and s are statistics.

Formulas for Grouped Population Data

There are formulas for computing the mean and standard deviation of a population when the data are grouped in a frequency distribution. These formulas are derived in the same way as the grouped-data formulas for the sample mean and sample standard deviation, as discussed in Section 3.5.

FORMULA 3.4

COMPUTING μ FOR GROUPED DATA

> The mean of a finite population whose data are grouped in a frequency distribution can be computed using the formula
>
> $$\mu = \frac{\Sigma xf}{N},$$
>
> where f denotes class frequency and $N\ (= \Sigma f)$ denotes the population size.

FORMULA 3.5

COMPUTING σ FOR GROUPED DATA

> The standard deviation of a finite population whose data are grouped in a frequency distribution can be computed using the formula
>
> $$\sigma = \sqrt{\frac{\Sigma (x - \mu)^2 f}{N}},$$
>
> or its shortcut version
>
> $$\sigma = \sqrt{\frac{\Sigma x^2 f}{N} - \mu^2},$$
>
> where f denotes class frequency and $N\ (= \Sigma f)$ denotes the population size.

Example 3.32   Illustrates Formulas 3.4 and 3.5

A real-estate development consists of 385 homes. The first two columns of Table 3.26 give a frequency distribution for the number of bedrooms in each of the 385 homes. The table shows, for instance, that 38 of the homes have two bedrooms. Regarding this development as a population of interest, compute the mean and standard deviation of the number of bedrooms per home.

TABLE 3.26

Table for computing μ
and σ for the number
of bedrooms per home

No. bedrooms x	Frequency f	xf	x^2	x^2f
2	38	76	4	152
3	177	531	9	1593
4	128	512	16	2048
5	42	210	25	1050
	385	1329		4843

SOLUTION   Because the population data are in grouped form, we apply Formulas 3.4 and 3.5 (we will use the shortcut formula to compute σ). As the formulas indicate, we will need a table with columns for x, f, xf, x^2, and x^2f. These columns are all included in Table 3.26.

From the second and third columns of Table 3.26, we see that

$$\mu = \frac{\Sigma xf}{N} = \frac{1329}{385} = 3.5.$$

Now using the fifth column also, we find that

$$\sigma = \sqrt{\frac{\Sigma x^2 f}{N} - \mu^2} = \sqrt{\frac{4843}{385} - \left(\frac{1329}{385}\right)^2} = 0.8.$$

 Thus the (population) mean number of bedrooms per home in the development is 3.5 and the standard deviation is 0.8.   ■

The frequency distribution in the first two columns of Table 3.26 uses classes based on a single value. When population data are grouped into a frequency distribution in which each class represents several values, we substitute class marks for the x values in Formulas 3.4 and 3.5. However, keep in mind that the values for μ and σ so obtained generally only approximately equal the actual population mean and standard deviation.

Interpretation of the Population Standard Deviation; z-Scores

The interpretation of the standard deviation of a population is essentially identical to that of the standard deviation of a sample. In particular, the more variation a population has, the larger its standard deviation. Chebychev's rule also holds for populations, as it does for any data set: In Key Fact 3.2 on page 131, simply replace $\bar{x}$ by μ, and s by σ.

We will now consider population z-scores. Recall from Section 3.4 that the z-score for a data value is the number of standard deviations that the data value is away from the mean of the data set. The sample z-score is obtained by first subtracting the sample mean, $\bar{x}$, from the data value, x, and then dividing the result by the sample standard deviation, s:

$$z = \frac{x - \bar{x}}{s}.$$

The population z-score is defined and obtained in the same way except μ is used in place of $\overline{x}$, and σ in place of s.

DEFINITION 3.15   **POPULATION z-SCORE**

> The *population z-score* for a value, x, is defined as the number of standard deviations that x is away from the mean. The population z-score is computed by using the formula
>
> $$z = \frac{x - \mu}{\sigma},$$
>
> where μ and σ are, respectively, the mean and standard deviation of the population. A negative z-score indicates that the population value is smaller than the mean, whereas a positive z-score indicates that the population value is larger than the mean.

The population z-score can be used as a measure of relative standing. If a population value has a large positive z-score, then it is larger than most of the other members of the population; if a population value has a large negative z-score, then it is smaller than most of the other members of the population; and if a population value has a z-score near 0, then it is located near the mean of the population.

We can make the statements in the preceding paragraph more precise by applying Chebychev's rule to the z-score of a population value. This is done in the same way as explained in Example 3.20 on page 135.

The Population Mean and Population Standard Deviation in Inferential Studies

We emphasize that in inferential studies the population mean, μ, and population standard deviation, σ, are rarely known. In fact, two major topics in inferential statistics are

1. using the mean, $\overline{x}$, of a sample to make inferences about the population mean, μ, and

2. using the standard deviation, s, of a sample to make inferences about the population standard deviation, σ.

Topic 1 will be examined in Chapters 8 and 9, and Topic 2 in Chapter 12.

Other Descriptive Measures for Populations

Up to this point, we have concentrated our discussion of descriptive measures for populations on the mean and standard deviation. This is because many of the classical inference procedures for center and spread concern those two parameters.

However, modern statistical analyses also rely heavily on descriptive measures based on percentiles, such as the median. Percentiles, deciles, quartiles, the IQR, and other descriptive measures based on percentiles are defined in the same way for (finite) populations as they are for samples. For simplicity we will use the same notation for percentiles, deciles, and quartiles whether we are considering a sample or a population. But we will employ different notations for a sample median and a population median—we'll use M to denote a sample median and η (eta) to denote a population median.

EXERCISES 3.7

3.125 Although in practice we generally deal with sample data in inferential studies, what is the ultimate objective of such studies?

3.126 In Section 3.3 we analyzed the heights of the starting five players on each of two men's college basketball teams. The heights, in inches, of the players on Team I are 72, 73, 76, 76, and 78. Regarding the heights as a sample of the heights of all male starting college basketball players,
a. compute the sample mean height, $\bar{x}$.
b. compute the sample standard deviation, s.

Regarding the heights now as a population,
c. compute the population mean height, μ.
d. compute the population standard deviation, σ.

Comparing your answers from parts (a) and (c) and from parts (b) and (d),
e. why are the values for the sample mean, $\bar{x}$, and the population mean, μ, equal?
f. why are the values for the sample standard deviation, s, and the population standard deviation, σ, different?

3.127 In Section 3.3 we analyzed the heights of the starting five players on each of two men's college basketball teams. The heights, in inches, of the players on Team II are 67, 72, 76, 76, and 84. Regarding the heights as a sample of the heights of all male starting college basketball players,
a. compute the sample mean height, $\bar{x}$.
b. compute the sample standard deviation, s.

Regarding the heights now as a population,
c. compute the population mean height, μ.
d. compute the population standard deviation, σ.

Comparing your answers from parts (a) and (c) and from parts (b) and (d),
e. why are the values for the sample mean, $\bar{x}$, and the population mean, μ, equal?
f. why are the values for the sample standard deviation, s, and the population standard deviation, σ, different?

In each of Exercises 3.128–3.131, consider the given data set a population of interest. For each exercise,
a. compute the population mean, μ.
b. compute the population standard deviation, σ.

3.128 One year's monthly electric bills for a family living in the Southeast are as follows.

$87	54	125	89
92	68	142	63
65	95	140	84

3.129 One year's monthly telephone bills for the family in Exercise 3.128 are presented below.

$81	35	75	48
55	46	61	41
24	25	33	32

3.130 As reported by the U.S. Department of Agriculture in *Crop Production,* the acreages (in thousands) of tobacco harvested in 1990 by each of the tobacco-producing states are as shown in the following table.

State	SC	TN	NC	GA	VA	KY
Acreage	51	49	286	43	53	194

3.131 The U.S. Agency for International Development compiles information on U.S. foreign aid commitments for economic assistance. Data are published in *U.S. Overseas Loans and Grants and Assistance from International Organizations.* In 1990 the commitments to selected Latin American countries were as follows.

Country	Aid ($millions)	Country	Aid ($millions)
Bolivia	58	Haiti	42
Costa Rica	78	Honduras	167
Domin. Rep.	18	Jamaica	28
Ecuador	17	Panama	396
El Salvador	200	Peru	22
Guatemala	86		

3.132 In Example 3.28 we found that the population mean weight of the starting offensive players on the 1992–1993 Dallas Cowboys football team is 252.6 lb. In this context is the number 252.6 a parameter or a statistic? Why?

3.133   In Example 3.30 we determined that the population standard deviation of the weights of the starting offensive players on the 1992–1993 Dallas Cowboys football team is 41.4 lb. In this context is the number 41.4 a parameter or a statistic? Explain your answer.

3.134   The ages of the students in Professor Weiss's introductory statistics class can be found in Table 3.6 on page 107. A frequency distribution for those ages follows.

Age	17	18	19	20	21	22	23	24	26	35	36
Freq.	1	1	9	7	7	5	3	4	1	1	1

Regarding the ages of these students as a population,
a. compute the population mean age, μ.
b. compute the population standard deviation, σ.

3.135   An eight-question quiz on fractions is given to a fifth-grade class. A frequency distribution for the number of incorrect answers follows.

Number incorrect	0	1	2	3	4	5	6
Frequency	4	8	4	5	2	1	1

a. Compute μ.   b. Compute σ.

3.136   The U.S. Bureau of the Census collects and publishes annual figures for state expenditures on highways. The following is a frequency distribution of the 1990 highway expenditures by the 50 states.

Expenditures ($millions)	Number of states
0–under 400	14
400–under 800	17
800–under 1200	6
1200–under 1600	5
1600–under 2000	3
2000–under 2400	3
2400–under 2800	1
2800–under 3200	0
3200–under 3600	1

a. Compute the approximate value of the mean 1990 highway expenditure, μ, per state.
b. Compute the approximate value of the standard deviation, σ, of these highway expenditures.
c. Why are the values for μ and σ that you obtained in parts (a) and (b) only approximate?

3.137   The following table provides the 1991 age distribution for Americans between 3 and 34 years old who are enrolled in school. [SOURCE: U.S. Bureau of the Census, *Current Population Reports.*]

Age (yrs)	Frequency (thousands)
3–4	3,068
5–6	7,178
7–13	25,445
14–15	6,634
16–17	6,155
18–19	3,969
20–21	3,041
22–24	2,365
25–29	2,045
30–34	1,377

a. Determine the approximate value of the mean age, μ, for Americans between 3 and 34 years old who are enrolled in school.
b. Compute the approximate value of the standard deviation, σ, of the ages.
c. Why are the values for μ and σ that you obtained in parts (a) and (b) only approximate?

3.138   As reported in *News,* by the Department of Agriculture, the mean weekly food cost is $95.40 for a couple with two children 6–11 years old. Assume a standard deviation of $17.20.
a. Fill in the blanks: At least 75% of such couples have weekly food costs between $_____ and $_____.
b. Fill in the blanks: At least 89% of such couples have weekly food costs between $_____ and $_____.
c. Obtain and interpret the population z-scores for weekly food costs of $149.59, $96.60, and $56.70.
d. Use your results from part (c) to estimate the relative standing of each of the weekly food costs in part (c).

3.139   According to a 1973 study by R. R. Paul, the mean gestation period of the Morgan mare is 339.6 days, and the standard deviation is 13.3 days.
a. Fill in the blanks: At least 75% of such gestation periods are between _____ and _____ days.
b. Fill in the blanks: At least 89% of such gestation periods are between _____ and _____ days.
c. Find and interpret the population z-scores for gestation periods of 293 days, 368 days, and 338 days.
d. Use your results from part (c) to estimate the relative standing of each of the gestation periods in part (c).

3.140   A company produces cans of stewed tomatoes with an advertised weight of 14 oz. The standard deviation of the weights is known to be 0.4 oz. A quality-control engineer selects a can of stewed tomatoes at random and finds its net weight to be 17.28 oz.

a. Estimate the relative standing of that can of stewed tomatoes, assuming the true mean weight is 14 oz. Use the z-score and Chebychev's rule.

b. Does the quality-control engineer have reason to suspect that the true mean weight of all cans of stewed tomatoes being produced is not 14 oz? Explain your answer.

· **3.141** Suppose you buy a new car whose advertised mileage is 25 miles per gallon (mpg). After driving the car for several months, you find that its mileage is 21.4 mpg. You telephone the manufacturer and learn that the standard deviation of gas mileages for all cars of the model you bought is 1.15 mpg.

a. Find and interpret the z-score for the gas mileage of your car.

b. Does it appear that your car is getting unusually low gas mileage? Explain your answer. *(Hint: Use part (a) and Chebychev's rule.)*

Computing μ for data grouped into a relative-frequency distribution. Formula 3.4, a formula for computing the mean of a finite population whose values are grouped in a frequency distribution, is $\mu = (\Sigma xf)/N$. Using algebra we can rewrite that formula as

$$(2) \qquad \mu = \Sigma x \cdot \frac{f}{N}.$$

Since f/N represents the relative frequency of a class, we see that the mean of a population whose values are grouped in a relative-frequency distribution can be obtained as follows: Multiply the class mark of each class by its relative frequency and then sum all of the resulting products. We will apply Formula (2) in Exercises 3.142 and 3.143.

· · **3.142** Refer to Exercise 3.134.

a. Construct a relative-frequency distribution for the population of ages.

b. Use Formula (2) to compute μ and compare your answer to the one obtained in Exercise 3.134(a).

· · **3.143** Refer to Exercise 3.135.

a. Construct a relative-frequency distribution for the number of incorrect answers.

b. Use Formula (2) to compute μ and compare your answer to the one obtained in Exercise 3.135(a).

· · **3.144** Derive two formulas that can be used to compute the standard deviation, σ, of a finite population whose values are grouped in a relative-frequency distribution.

· · **3.145** Use one of the formulas you derived in Exercise 3.144 to compute σ from the relative-frequency distribution you constructed in Exercise 3.143(a). Compare your answer to the one you obtained in Exercise 3.135(b).

· · **3.146** Consider the following three data sets.

Data Set 1	Data Set 2	Data Set 3
2 4	7 5 5 3	4 7 8 9 7
7 3	9 8 6	4 5 3 4 5

a. Assuming these data sets are each a sample of a population, compute their standard deviations. (Round your final answers to two decimal places.)

b. Assuming these data sets are each a population, compute their standard deviations. (Round your final answers to two decimal places.)

c. Using your results from parts (a) and (b), make an educated guess about the answer to the following question: Suppose both s and σ are computed for the same data set. Will they tend to be closer together if the data set is large or if it is small?

· · · **3.147** Consider a data set with m data values. If the data set is a sample from a population, then we compute the sample standard deviation, s. If the data set is a population, then we compute the population standard deviation, σ.

a. Derive a formula that gives the precise relationship between the values of s and σ calculated from the same data set.

b. Refer to the three data sets displayed in Exercise 3.146. Verify that your formula in part (a) works for each of the three data sets.

c. Suppose a data set consists of 15 data values. You compute the population standard deviation of the data and obtain $\sigma = 38.6$. Then you realize that the data set is actually a sample of a population and that you should have computed the sample standard deviation, s, instead. Use your result from part (a) to obtain s.

· · · **3.148 (Computer exercise)** Many statistical software packages, including Minitab, have a command for obtaining a sample standard deviation but do not have one for obtaining a population standard deviation. Employing your statistical software, how would you determine a population standard deviation using the command that provides the sample standard deviation? *(Hint: Refer to Exercise 3.147(a).)*

Chapter Review

Key Terms

adjacent values, *151*
box-and-whisker diagram, *151*
boxplot, *151*
BoxPlot,* *155*
BOXPLOT,* *155*
Chebychev's rule, *131*
deciles, *146*
descriptive measures, *101*
deviations from the mean, *119*
five-number summary, *149*
inner fences, *150*
interquartile range (IQR), *149*
lower hinge, *152*
LVals,* *156*
LVALS,* *156*
MAX,* *125*
Maximum,* *125*
mean, *102*
Mean,* *108*
MEAN,* *109*
measures of central tendency, *102*
measures of dispersion, *117*
measures of relative standing, *135*
median, *103*
Median,* *108*
MEDIAN,* *109*
MIN,* *125*
Minimum,* *125*
mode, *104*
modified boxplot, *151*
outer fences, *150*
outliers, *150*
parameter, *166*
percentiles, *146*
population mean (μ), *161*
population standard deviation (σ), *163*
population variance (σ^2), *163*
population z-score, *168*
possible outliers, *150*
probable outliers, *150*
quartiles, *146*
range, *118*
resistant, *106*
sample mean ($\bar{x}$), *114*
sample standard deviation (s), *122*
sample variance (s^2), *121*
sample z-score, *134*
standard deviation, *119*
standard score, *133*
statistic, *166*
StDev,* *126*
STDEV,* *126*
subscripts, *113*
sum of squared deviations, *120*
summation notation, *113*
trimmed means, *106*
upper hinge, *152*
whiskers, *152*
z-score, *133*

Formulas

In the formulas below,

x = data value
$\bar{x}$ = sample mean
n = sample size
s = sample standard deviation
z = z-score
f = class frequency
IQR = interquartile range
Q_1 = first quartile
Q_3 = third quartile
μ = population mean
N = population size
σ = population standard deviation

Sample mean, *114*

$$\bar{x} = \frac{\Sigma x}{n}$$

Range, *118*

$$\text{Range} = \text{Max} - \text{Min}$$

Sample standard deviation, *122, 123*

$$s = \sqrt{\frac{\Sigma(x - \overline{x})^2}{n - 1}} \qquad \text{or} \qquad s = \sqrt{\frac{\Sigma x^2 - (\Sigma x)^2/n}{n - 1}}$$

Sample *z*-score, *134*

$$z = \frac{x - \overline{x}}{s}$$

Sample mean (grouped form), *140*

$$\overline{x} = \frac{\Sigma xf}{n}$$

(x = class mark)

Sample standard deviation (grouped form), *142*

$$s = \sqrt{\frac{\Sigma(x - \overline{x})^2 f}{n - 1}} \qquad \text{or} \qquad s = \sqrt{\frac{\Sigma x^2 f - (\Sigma xf)^2/n}{n - 1}}$$

(x = class mark)

Interquartile range, *149*

$$\text{IQR} = Q_3 - Q_1$$

Inner and outer fences, *150*

$$\text{Inner fences:} \quad \begin{array}{c} Q_1 - 1.5 \cdot \text{IQR} \\ Q_3 + 1.5 \cdot \text{IQR} \end{array} \qquad \text{Outer fences:} \quad \begin{array}{c} Q_1 - 3 \cdot \text{IQR} \\ Q_3 + 3 \cdot \text{IQR} \end{array}$$

Population mean, *161*

$$\mu = \frac{\Sigma x}{N}$$

Population standard deviation, *163*

$$\sigma = \sqrt{\frac{\Sigma(x - \mu)^2}{N}} \qquad \text{or} \qquad \sigma = \sqrt{\frac{\Sigma x^2}{N} - \mu^2}$$

Population mean (grouped form), *166*

$$\mu = \frac{\Sigma xf}{N}$$

(x = class mark)

Population standard deviation (grouped form), *166*

$$\sigma = \sqrt{\frac{\Sigma(x - \mu)^2 f}{N}} \qquad \text{or} \qquad \sigma = \sqrt{\frac{\Sigma x^2 f}{N} - \mu^2}$$

(x = class mark)

Population *z*-score, *168*

$$z = \frac{x - \mu}{\sigma}$$

You Should Be Able To

1. use and understand each of the preceding formulas.
2. explain the purpose of a measure of central tendency.
3. obtain and interpret the mean, the median, and the mode(s) of a data set.
4. choose an appropriate measure of central tendency for a data set.
5. use and understand summation notation.
6. explain the purpose of a measure of dispersion.
7. compute and interpret the range and the sample standard deviation of a data set.
8. state and apply Chebychev's rule.
9. compute and interpret sample z-scores.
10. define percentiles, deciles, and quartiles.
11. obtain and interpret the quartiles, IQR, and five-number summary of a data set.
12. obtain the inner and outer fences of a data set and classify outliers as possible or probable.
13. construct and interpret a boxplot and a modified boxplot.
14. use a boxplot to identify the shape of the distribution of a data set.
15. compute the population mean and standard deviation of a finite population.
16. distinguish between a parameter and a statistic.
17. compute and interpret population z-scores.
18. use the Minitab commands covered in this chapter.*
19. interpret the output obtained from the application of the Minitab commands discussed in this chapter.*

REVIEW TEST

1. Euromonitor Publications Limited, London, compiles data on per capita food consumption of major food commodities in various countries. Data are published in *European Marketing Data and Statistics*. Samples of 10 Germans and 15 Russians yield the following fish consumptions, in kilograms, for last year.

Germans		Russians		
10	12	16	21	12
17	12	11	5	23
14	11	19	19	22
13	8	16	23	12
9	8	18	7	17

   a. Determine the mean of each data set.
   b. Determine the median of each data set.
   c. Determine the mode(s) of each data set.

2. The National Center for Health Statistics publishes information on the duration of marriages in *Vital Statistics of the United States*. Which measure of central tendency is more appropriate for data on the duration of marriages: the mean or the median? Explain your answer.

3. Death certificates provide data on the causes of death. Which measure of central tendency is appropriate here?

4. Telephone companies conduct surveys to obtain information on the durations of telephone conversations. A sample of 12 phone calls yields the following durations to the nearest minute.

4	2	1	2	2	8
6	3	1	3	1	15

   a. Compute the sample mean duration, $\bar{x}$, of the 12 calls.
   b. Compute the range of the durations.
   c. Compute the sample standard deviation, s.

5. Refer to Problem 4.
   a. Draw a graph similar to Fig. 3.6 on page 130.
   b. By Chebychev's rule, at least _____% of the data lie within two standard deviations to either side of the mean.
   c. What percentage of the telephone data lie within two standard deviations to either side of the mean?
   d. Parts (b) and (c) show that Chebychev's rule does not necessarily provide precise estimates. What then makes it so valuable?

6. Dr. Thomas Stanley of Georgia State University has collected information on millionaires, including their ages, since 1973. A sample of 36 millionaires has a mean age of 58.5 years and a standard deviation of 13.4 years.

a. Complete the graph below.

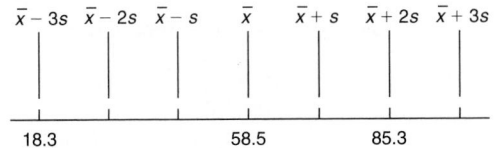

18.3   58.5   85.3

b. By Chebychev's rule, at least 27 of the 36 millionaires are between _____ and _____ years old.

c. By Chebychev's rule, at least _____% of the 36 millionaires are between 18.3 and 98.7 years old.

7. Refer to Problem 6.

a. Determine and interpret the sample z-scores for millionaires in the sample whose ages are 59, 31, and 79 years.

b. Use your results from part (a) to estimate the relative standing in the sample of each of the ages in part (a).

8. Referring to Problem 6, the actual ages of the 36 millionaires sampled are as follows.

31	64	39	66	68	48	69	71	52
68	45	60	54	66	79	38	48	77
53	52	79	75	67	71	42	39	57
47	74	59	64	42	55	61	79	48

a. Determine the quartiles for the data.

b. Obtain the interquartile range.

c. Find the five-number summary.

d. Calculate the inner and outer fences.

e. Identify and classify potential outliers, if any.

f. Construct and interpret a boxplot and, if appropriate, a modified boxplot.

*9. **(Computer problem)** Use Minitab or some other statistical software to determine the following statistics for the age data in Problem 8.

a. mean   b. median   c. range

d. sample standard deviation

*10. **(Computer problem)** Refer to the data in Problem 8. Use Minitab or some other statistical software to obtain

a. a (modified) boxplot for the data.

b. the five-number summary of the data.

*11. **(Computer problem)** Refer to Problem 8.

a. Printout 3.10 shows a boxplot for the age data generated by Minitab's BoxPlot command. Use the boxplot to discuss the variation in the data set and to determine approximately the five-number summary.

b. Are there possible or probable outliers in the data? If so, identify them approximately and provide a possible explanation for their cause.

c. In Printout 3.11 we have given the Minitab output obtained by applying the LVals command to the age data. Use the output to determine exactly the five-number summary of the data. Compare your answers to the approximate values you obtained in part (a).

PRINTOUT 3.10   Minitab output for Problem 11(a)

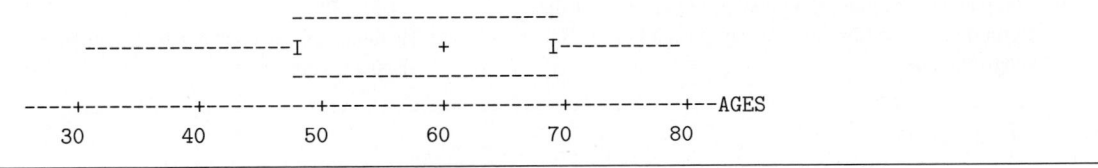

PRINTOUT 3.11   Minitab output for Problem 11(c)

	DEPTH	LOWER	UPPER	MID	SPREAD
N=	36				
M	18.5		59.500	59.500	
H	9.5	48.000	68.500	58.250	20.500
E	5.0	42.000	75.000	58.500	33.000
D	3.0	39.000	79.000	59.000	40.000
C	2.0	38.000	79.000	58.500	41.000
	1	31.000	79.000	55.000	48.000

12. An exercise physiologist measured the heart rates of 30 people who had taken part in a long-distance running program. The following table provides a frequency distribution for those rates.

Rate	67	72	68	73	69	74	70	75	71	77
Freq.	2	3	2	4	2	4	3	3	6	1

a. Compute the mean heart rate, $\bar{x}$.
b. Compute the sample standard deviation, s.

13. According to *Peterson's Guides,* the 1991–1992 enrollment figures for the University of California campuses were as follows.

Campus	Enrollment (thousands)
Berkeley	30.3
Davis	23.3
Irvine	16.9
Los Angeles	36.3
Riverside	8.8
San Diego	17.9
San Francisco	3.6
Santa Barbara	18.5
Santa Cruz	10.1

a. Compute the population mean enrollment, μ, of the UC campuses.
b. Compute σ.
c. Compute and interpret the population z-scores for the enrollment at the Los Angeles campus and at the San Diego campus.

14. The following table provides a frequency distribution for the 1992 annual salaries of the state governors in the United States. [SOURCE: *The World Almanac, 1993.*]

Salary ($thousands)	Frequency
30–under 40	1
40–under 50	0
50–under 60	1
60–under 70	4
70–under 80	17
80–under 90	7
90–under 100	11
100–under 110	5
110–under 120	0
120–under 130	3
130–under 140	1

a. Find the approximate value of μ for this population of salaries.
b. Find the approximate value of σ.
c. Why are the values you obtained for μ and σ in parts (a) and (b) only approximate?

15. The Energy Information Administration reports figures on retail gasoline prices in *Monthly Energy Review.* Data are obtained by sampling 10,000 gasoline service stations from a total of more than 185,000. For the 10,000 stations sampled, suppose the mean price per gallon for unleaded regular gasoline is $1.13.

a. Is the mean price given here a sample mean or a population mean? Why?
b. What letter would you use to designate the mean of $1.13?
c. Is the mean price given here a statistic or a parameter? Explain your answer.

Using the Focus Database

Appendix B contains a printout of a database obtained by randomly selecting 500 Arizona State University sophomores. Seven variables are considered for each student: sex, high-school GPA, SAT math score, cumulative GPA, SAT verbal score, age, and total hours. Use Minitab or some other statistical software to solve the following problems.

a. Determine the means and medians of the SAT math scores and SAT verbal scores. Use these measures of central tendency to compare the two sets of scores.
b. Find the ranges, sample standard deviations, and interquartile ranges of the SAT math scores and SAT verbal scores. Use these measures of dispersion to compare the two sets of scores.
c. Construct boxplots for the SAT math scores and SAT verbal scores. Use the boxplots to compare the two sets of scores. Be sure to consider variation, distribution shape, and outliers.

Case Study   **Per Capita Income by State**
· · · ·

As we mentioned in the introduction to this case study at the beginning of the chapter, *per capita income* is defined as the annual total personal income of the residents of a region divided by the resident population of that region as of July 1 of that year. Table 3.27 displays the per capita income for each state in the United States and for the District of Columbia. The data were printed in *The Arizona Republic* on September 7, 1991. The sources for the data were the Commerce Department and The Associated Press.

TABLE 3.27
Per capita income by state, 1990

1. Conn.	$25,484	18. Minn.	$18,731	35. N.C.	$16,293
2. N.J.	$24,936	19. Pa.	$18,686	36. Ariz.	$16,012
3. D.C.	$23,243	20. Fla.	$18,530	37. Tenn.	$15,866
4. Mass.	$22,569	21. Mich.	$18,360	38. S.D.	$15,797
5. N.Y.	$22,086	22. Kan.	$18,162	39. Okla.	$15,457
6. Md.	$21,789	23. Ohio	$17,564	40. Mont.	$15,270
7. Alaska	$21,688	24. Wis.	$17,560	41. Idaho	$15,249
8. N.H.	$20,827	25. Neb.	$17,549	42. N.D.	$15,215
9. Calif.	$20,677	26. Vt.	$17,511	43. S.C.	$15,151
10. Ill.	$20,419	27. Mo.	$17,472	44. Ala.	$15,021
11. Hawaii	$20,356	28. Iowa	$17,218	45. Ky.	$15,001
12. Del.	$20,022	29. Ore.	$17,196	46. La.	$14,542
13. Va.	$19,671	30. Maine	$17,175	47. N.M.	$14,265
14. Nev.	$19,035	31. Ga.	$17,049	48. Ark.	$14,188
15. Colo.	$18,890	32. Ind.	$16,890	49. Utah	$13,993
16. R.I.	$18,802	33. Texas	$16,716	50. W.Va.	$13,755
17. Wash.	$18,775	34. Wyo.	$16,314	51. Miss.	$12,823

The newspaper article noted that the Southwest and Rocky Mountain regions posted the fastest increases in per capita income. It also noted that people in South Dakota experienced the fastest per capita income growth, 9.8%, whereas New Hampshire residents had the slowest advance, 2.0%.

a. Determine the mean and median of the per capita incomes. Explain any difference between these two measures of central tendency.
b. Obtain the range and sample standard deviation of the per capita incomes.
c. Find and interpret the z-score for your state's per capita income.
d. Determine and interpret the quartiles of the per capita incomes.
e. Find and interpret the five-number summary of the data.
f. Find the inner and outer fences. Use them to identify possible or probable outliers.
g. Construct a boxplot for the per capita incomes and interpret your result in terms of the variation in the data.
h. Use Minitab or some other statistical software to solve parts (a)–(g).

Part III

Probability, Random Variables, and Sampling Distributions

Andrei Kolmogorov

Andrei Nikolaevich Kolmogorov was born on April 25, 1903, in Tambov, Russia. At the age of 17, Kolmogorov entered Moscow State University and graduated from there in 1925. His contributions to the world of mathematics, many of which appear in his numerous articles and books, encompass a formidable range of subjects.

Kolmogorov revolutionized probability theory by introducing the modern axiomatic approach to probability and by proving many of the fundamental theorems that are a consequence of that approach. He also developed two systems of partial differential equations, which bear his name. Those systems extended the development of probability theory and allowed its broader application to the fields of physics, chemistry, biology, and civil engineering.

In 1938 Kolmogorov published an extensive article entitled "Mathematics," which appeared in the first edition of the *Bolshaya Sovyetskaya Entsiklopediya* (Great Soviet Encyclopedia). This article discussed the development of mathematics from ancient to modern times and interpreted it in terms of dialectical materialism, the philosophy originated by Karl Marx and Friedrich Engels.

In addition to his work in higher mathematics, Kolmogorov was interested in the mathematical education of schoolchildren. He was chairman of the Commission for Mathematical Education under the Presidium of the Academy of Sciences of the U.S.S.R. During his tenure as chairman, he was instrumental in the development of a new mathematics training program that was introduced into Soviet schools.

Kolmogorov became a member of the faculty at Moscow State University in 1925, at the age of 22. In 1931 he was promoted to professor; in 1933 he was appointed a director of the Institute of Mathematics of the university; and in 1937 he became Head of the University. Kolmogorov remained on the faculty at Moscow State University until his death in 1987.

Chapter 4

Probability Concepts

. . . . .

General Objective   Up to this point, we have been concentrating on descriptive statistics, methods for organizing and summarizing data. However, one of the main purposes of this text is to present the fundamentals of inferential statistics, methods of drawing conclusions about a population based on information obtained from a sample of the population.

Because inferential statistics involves using information obtained from part of a population (a sample) to draw conclusions about the entire population, we can never be certain that our conclusions are correct—uncertainty is inherent in inferential statistics. So before we can understand, develop, and apply the methods of inferential statistics, we need to become familiar with uncertainty.

The science of uncertainty is called **probability theory.** Probability theory enables us to evaluate and control the likelihood that a statistical inference is correct. This chapter begins our study of probability.

Chapter Outline

4.1   Introduction; Classical Probability

4.2   Events

4.3   Some Rules of Probability

4.4   Contingency Tables; Joint and Marginal Probabilities

4.5   Conditional Probability

4.6   The Multiplication Rule; Independence

4.7   Bayes's Rule (Optional)

4.8   Counting Rules (Optional)

Case Study   *Lotto—The Arizona State Lottery*

What are the chances of winning the jackpot in a lottery? What are the chances of winning any prize whatsoever? In the case study at the end of this chapter, we will answer these and other questions for the Arizona state lottery, called *Lotto*.

4.1 INTRODUCTION; CLASSICAL PROBABILITY

Although most applications of probability theory to statistical inference involve large populations, the fundamental concepts of probability are most easily illustrated and explained using relatively small populations and games of chance. So keep in mind that many of the examples in this chapter are designed expressly to present the principles of probability in a lucid manner. Building on the fundamentals described in this chapter, we will continue our discussion of probability in Chapters 5, 6, and 7 and will gradually learn how to apply these principles to solve typical problems arising in inferential statistics.

The Meaning of Probability

An *event* is some specified result that may or may not occur when an experiment is performed. For example, in the experiment of tossing a coin once, the coin landing with heads facing up is an event, since it may or may not occur.

The *probability of an event* is a measure of the likelihood of its occurrence. A probability near 0 indicates that the event is very unlikely to occur, whereas a probability near 1 (100%) suggests that the event is quite likely to occur. To gain further insight into the meaning of probability, we will present the **relative-frequency interpretation of probability,** which construes the probability of an event to be the relative frequency of its occurrence in a large number of repetitions of the experiment.

As an illustration, consider the experiment of tossing a balanced coin once. Because the coin is balanced, we reason that there is a 50-50 chance the coin will land with heads facing up. Consequently, we attribute a probability of 0.5 to that event. The relative-frequency interpretation is that in a large number of tosses, the coin will land with heads facing up about half the time.

We used a computer to perform two simulations of tossing a balanced coin 100 times. The results are displayed in Fig. 4.1. Each graph shows the number of tosses of the coin versus the relative frequency (proportion) of heads. Both graphs seem to corroborate the relative-frequency interpretation.

FIGURE 4.1
Two computer simulations of tossing a balanced coin 100 times

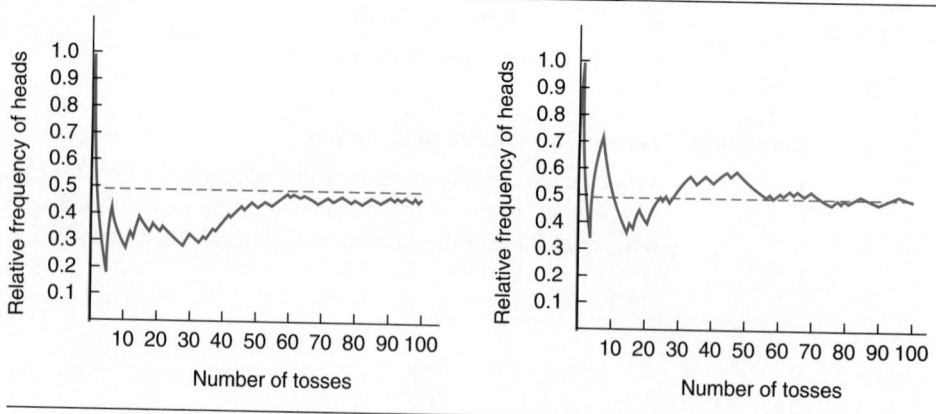

Although the relative-frequency interpretation is helpful for understanding the meaning of probability, it cannot be used as a definition of probability. One common way probabilities *are* defined is by specifying a **probability model,** a mathematical description of the experiment based on certain primary aspects and assumptions.

For instance, in the coin-tossing example, we postulated that the coin is balanced. Under that assumption we specify a probability of 0.5 to the coin coming up heads and a probability of 0.5 to the coin coming up tails. This model is a special case of a more general probability model, which we will examine next.

Classical Probability: The Equal-Likelihood Model

We will first discuss **classical probability,** which utilizes the **equal-likelihood model.** That model applies when the possible outcomes of an experiment are equally likely to occur.

Example 4.1 Classical Probability
_____

The ages of the students in Professor Weiss's introductory statistics class are displayed in Table 3.6 on page 107. A grouped-data table for those ages is presented in Table 4.1. Suppose one of the students is selected **at random,** meaning that each student is equally likely to be the one selected. Find the probability that the student selected is 20 years old.

TABLE 4.1
Grouped-data table for the ages of the students in Professor Weiss's introductory statistics class

Age (yrs) x	Frequency f	Relative frequency
17	1	0.025
18	1	0.025
19	9	0.225
20	7	0.175
21	7	0.175
22	5	0.125
23	3	0.075
24	4	0.100
26	1	0.025
35	1	0.025
36	1	0.025
	40	1.000

SOLUTION As we see from the second column of Table 4.1, 7 of the 40 students in the class are 20 years old. Thus the chances are 7 out of 40 for selecting a student who is 20 years old. The probability is therefore

$$\frac{\text{Number of 20-year-olds}}{\text{Total number of students}} = \frac{7}{40}.$$

Note that the probability, $\frac{7}{40}$, of randomly selecting a student age 20 is exactly the same as the relative frequency, 0.175, of the students age 20. ∎

With Example 4.1 in mind, we provide the following definition.

DEFINITION 4.1

CLASSICAL PROBABILITY: THE EQUAL-LIKELIHOOD MODEL

Suppose N equally likely outcomes for an experiment are possible. Then the probability that a specified event occurs equals the number of ways, f, that the event can occur, divided by the total number, N, of possible outcomes. In other words,

Number of ways event can occur

$$\text{Probability of an event} = \frac{f}{N}.$$

Total number of possible outcomes

For convenience we usually refer to Definition 4.1 as the *f/N* **rule.**

Let's apply the f/N rule to Example 4.1. The experiment consists of selecting one student at random from Professor Weiss's introductory statistics class. $N = 40$ since there are 40 possible outcomes, the 40 students in the class. Those outcomes are equally likely because, by assumption, the selection is done at random. The event that the student selected is 20 years old can occur in seven ways since seven of the students in the class are 20 years old; hence $f = 7$ for that event. Therefore, by the f/N rule, the probability that the student selected is 20 years old equals

$$\frac{f}{N} = \frac{7}{40} = 0.175,$$

as we noted in Example 4.1.

Example 4.2   Illustrates Definition 4.1

Refer again to Example 4.1. Find the probability that a student selected at random is younger than 21.

SOLUTION   From Table 4.1 we see that 18 (1 + 1 + 9 + 7) students in the class are younger than 21. So $f = 18$ and the probability equals

$$\frac{f}{N} = \frac{18}{40} = 0.450.$$

In terms of percentages, this means that 45.0% of the students in the class are under 21 years of age. ∎

Example 4.3 Illustrates Definition 4.1

A frequency distribution for the incomes of the families in a small city is displayed in Table 4.2. Determine the probability that a randomly selected family makes between $15,000 and $49,999, inclusive (i.e., including $15,000 and $49,999).

TABLE 4.2
Frequency distribution
for family incomes

Income	Frequency
Under $5,000	285
$5,000–$9,999	474
$10,000–$14,999	584
$15,000–$24,999	1213
$25,000–$34,999	1135
$35,000–$49,999	1310
$50,000–$74,999	980
$75,000 & over	506
	6487

SOLUTION The second column of Table 4.2 shows that 6487 families live in the city; so $N = 6487$. The event in question is that the family selected makes between $15,000 and $49,999. We see from Table 4.2 that the number of such families is $1213 + 1135 + 1310 = 3658$; so $f = 3658$. Therefore, by the f/N rule, the probability that the family selected makes between $15,000 and $49,999 equals

$$\frac{f}{N} = \frac{3658}{6487} = 0.564,$$

to three decimal places. In other words, approximately 56.4% of the families in the city make between $15,000 and $49,999. ∎

Example 4.4 Illustrates Definition 4.1

When a pair of balanced dice is rolled, 36 equally likely outcomes are possible, as depicted in Fig. 4.2 at the top of the next page. Find the probability that

a. the sum of the dice is 11.

b. doubles are rolled; that is, both dice come up the same number.

SOLUTION For this experiment, $N = 36$.

a. The sum of the dice can be 11 in two ways, as is apparent from Fig. 4.2. Thus the probability that the sum of the dice is 11 equals

$$\frac{f}{N} = \frac{2}{36} = 0.056.$$

FIGURE 4.2
Possible outcomes for
rolling a pair of dice

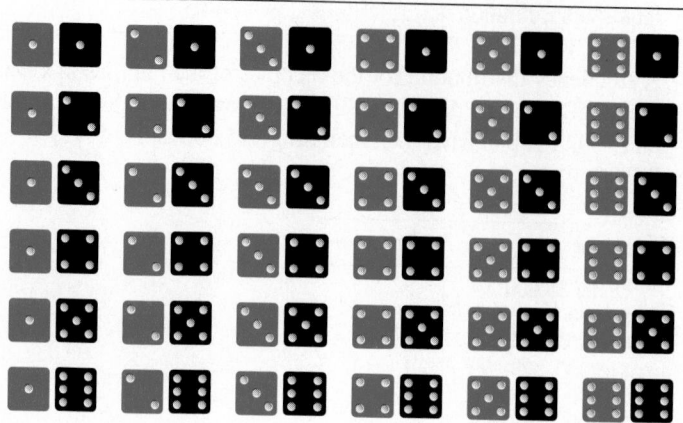

b. Doubles can be rolled in six ways; hence the probability of rolling doubles equals

$$\frac{f}{N} = \frac{6}{36} = 0.167.$$

■

Probabilities and Percentages

We noted in Example 4.1 that the probability of randomly selecting a student age 20 equals the relative frequency of the students age 20. More generally, we have the following fact.

KEY FACT 4.1

PROBABILITIES AND PERCENTAGES

> Suppose an experiment consists of selecting one member at random from a finite population. Then the probability that a specified event occurs is equal to the relative frequency (percentage) of members of the population that satisfy the conditions prescribed by the event.

Therefore, for example, the fact that 12.4% of the U.S. population is African-American also means that the probability is 0.124 that a randomly selected U.S. resident will be an African-American.

Basic Properties of Probabilities

Probabilities have some simple but basic properties, as listed in Key Fact 4.2.

KEY FACT 4.2   **BASIC PROPERTIES OF PROBABILITIES**

> *Property 1:* The probability of an event is always between 0 and 1, inclusive.
>
> *Property 2:* The probability of an event that cannot occur is 0. (An event that cannot occur is called an **impossible event.**)
>
> *Property 3:* The probability of an event that must occur is 1. (An event that must occur is called a **certain event.**)

Property 1 indicates that numbers such as 5 or −0.23 could not possibly be probabilities. Thus if you calculate a probability and get an answer like 5 or −0.23, then you made an error. Example 4.5 illustrates Properties 2 and 3.

Example 4.5   Illustrates Properties 2 and 3 of Key Fact 4.2

Refer once more to Example 4.1. Find the probability that a student selected at random from this particular class is

a. 25 years old.   b. younger than 50.

SOLUTION   a. From Table 4.1 on page 183, we see that *none* of the students in the class are 25 years old. Thus the probability that a student selected at random is 25 years old equals

$$\frac{f}{N} = \frac{0}{40} = 0.$$

This illustrates Property 2 of Key Fact 4.2.

b. Again from Table 4.1, we observe that *all* of the students in the class are younger than 50. So the probability that a student selected at random is younger than 50 equals

$$\frac{f}{N} = \frac{40}{40} = 1.$$

This illustrates Property 3 of Key Fact 4.2.   ■

In all examples considered thus far, either we have been given both the number of ways, f, that an event can occur and the total number, N, of possible outcomes or we have been able to obtain those numbers by a simple computation or direct listing. Often, however, neither f nor N will be given and a direct listing will be impractical because the number of possible outcomes is so large. In those situations we must employ alternative techniques, such as the ones we will describe in Section 4.8.

Finally, the procedure we have learned for computing probabilities—the f/N rule—applies only to experiments whose possible outcomes are equally likely to occur, that is, where the equal-likelihood model is appropriate. When that is not the case, we must use other methods to determine probabilities. Some of those methods will be examined later in this chapter and in Chapter 5.

EXERCISES 4.1

In the following exercises, express each probability as a decimal rounded to three places.

• **4.1** The absentee records over the past year for the employees of Cudahey Masonry are presented below in grouped form.

Days missed	0	1	2	3	4	5	6	7
No. of employees	4	2	14	10	16	18	10	6

If an employee is selected at random, find the probability that, over the past year, the employee missed
a. 3 days of work.
b. at most 2 days of work, that is, 2 or fewer.
c. between 1 and 5 days of work, inclusive.
d. 8 days of work.
e. at most 8 days of work.

• **4.2** A frequency distribution for the number of cars owned by each of the 6487 families in Example 4.3 is shown in the following table.

Cars owned	0	1	2	3	4	5
No. of families	27	1422	2865	1796	324	53

Suppose one of these families is selected at random. Find the probability that the family obtained owns
a. two cars.   b. more than three cars.
c. between one and three cars, inclusive.
d. seven cars.
e. at most seven cars, that is, seven or fewer.

• **4.3** As reported by the Bureau of Labor Statistics in *Employment and Earnings,* the age distribution of employed persons 16 years old and over is as shown in the following table. The frequencies are in thousands.

Age (yrs) x	Number of persons f
16–19	5,899
20–24	11,748
25–34	28,429
35–44	23,597
45–54	15,216
55–64	10,163
65 & over	2,737
	97,789

If an employed person is selected at random, find the probability that the person obtained is
a. between 25 and 34 years old, inclusive.
b. at least 45 years old, that is, 45 years old or older.
c. between 20 and 44 years old, inclusive.
d. under 20 or over 54.

• **4.4** According to *U.S. Scientists and Engineers,* published by the National Science Foundation, the distribution of U.S. scientists by field is as follows. Frequencies are in thousands.

Type	Frequency
Life scientists	452
Physical scientists	313
Computer scientists	576
Social scientists	282
Psychologists	274
Mathematical scientists	140
Environmental scientists	123
	2160

Find the probability that a randomly selected scientist is
a. a psychologist.   b. a life or social scientist.
c. not a computer scientist.

• **4.5** Suppose two balanced dice are rolled. Referring to Fig. 4.2 on page 186, determine the probability that the sum of the dice is
a. 6.   b. even.   c. 7 or 11.   d. 2, 3, or 12.

• **4.6** A balanced dime is tossed three times. The possible outcomes can be represented as follows.

HHH	HTH	THH	TTH
HHT	HTT	THT	TTT

Here, for example, HHT means that the first two tosses come up heads and the third tails. Obtain the probability that
a. exactly two of the three tosses are heads.
b. the last two tosses come up tails.
c. all three tosses come up the same.
d. the second toss is heads.

• **4.7** The U.S. Senate consists of 100 senators, two from each state. If a senator is selected at random to serve as the chair of a subcommittee, what is the probability that the senator is from Georgia?

· **4.8** According to the Census Bureau publication *1990 Census of Housing, General Housing Characteristics,* housing in the United States is occupied as follows. Frequencies are in thousands.

Owner-occupied	Renter-occupied	Vacant year-round
59,025	32,923	10,316

Find the probability that a housing unit selected at random is
a. owner-occupied.　b. not renter-occupied.
c. vacant year-round.

· **4.9** The Internal Revenue Service reports in *Statistics of Income* that the number of U.S. businesses in various categories are as follows. Frequencies are in thousands.

Proprietorships	Partnerships	Corporations
13,679	1,654	3,563

Find the probability that a business selected at random is
a. a proprietorship.　b. not a partnership.
c. a corporation.

· **4.10** Which of the following numbers could not possibly be probabilities? Justify your answer.
a. 0.462　b. −0.201　c. 1
d. $\frac{5}{6}$　e. 3.5　f. 0

· **4.11** Which of the following numbers could not possibly be probabilities? Justify your answer.
a. 0.732　b. 4.75　c. 0.0
d. $-\frac{3}{4}$　e. 1　f. $\frac{4}{5}$

· **4.12** Refer to Exercise 4.2. Which, if any, of the events in parts (a)–(e) are certain? impossible?

· **4.13** Refer to Exercise 4.1. Which, if any, of the events in parts (a)–(e) are certain? impossible?

· · **4.14** Explain what is wrong with the following argument: When two balanced dice are rolled, the sum of the dice can be 2, 3, 4, 5, 6, 7, 8, 9, 10, 11, or 12. This gives 11 possibilities. Therefore the probability that the sum is 12 equals $\frac{1}{11}$.

· · **4.15** Explain what is wrong with the following argument: When a balanced coin is tossed twice, the total number of heads obtained can be 0, 1, or 2. This gives three possibilities. So the probability of getting two heads is $\frac{1}{3}$.

· · **4.16** A study conducted by the Census Bureau on the methods that Americans use to get to work revealed the following data, in hundreds of thousands of workers.

Type of worker

Method		Urban	Rural	Total
	Automobile	450	150	600
	Public trans.	65	5	70
	Total	515	155	670

If a worker is selected at random, determine the probability that he or she
a. is an urban worker.
b. drives an automobile to work.
c. is an urban worker who drives an automobile to work.
d. is an urban worker or drives an automobile to work.
e. is a rural worker using public transportation to get to work.

· · **4.17** The table that follows provides a frequency distribution for institutions of higher education in the United States by region and type. [SOURCE: U.S. National Center for Education Statistics, *Digest of Education Statistics.*]

Type

Region		Public	Private	Total
	Northeast	266	555	821
	Midwest	359	504	863
	South	533	502	1035
	West	313	242	555
	Total	1471	1803	3274

If an institution of higher education is selected at random, what is the probability that it is
a. a public school?
b. in the Midwest?
c. a public school and in the Midwest?
d. a public school or in the Midwest?
e. a private school in the Northeast?

· · · **4.18** Consider an experiment having *N* equally likely possible outcomes. Use Definition 4.1 (page 184) to verify that Properties 1–3 of Key Fact 4.2 (page 187) hold for such an experiment.

4.2   EVENTS

Before continuing our study of probability, we need to discuss events in greater detail. In Section 4.1 we used the word *event* intuitively. To be more precise, in probability an **event** consists of a collection of outcomes.

Example 4.6   Events

A deck of playing cards consists of 52 cards, as seen in Fig. 4.3. When we perform the experiment of randomly selecting one card from the deck, exactly one of these 52 cards will be obtained. The collection of all 52 cards—the possible outcomes—is called the **sample space** for this experiment.

FIGURE 4.3
A deck of playing cards

Many different events can be associated with this card-selection experiment. Let's consider the following four:

1. The event that the card selected is the king of hearts.

2. The event that the card selected is a king.

3. The event that the card selected is a heart.

4. The event that the card selected is a face card.

List the outcomes comprising each of these four events.

SOLUTION   The first event, the event that the card selected is the king of hearts, consists of the single outcome, "king of hearts," and is pictured in Fig. 4.4.

FIGURE 4.4
The event the king
of hearts is selected

The second event, the event that the card selected is a king, consists of the four outcomes, "king of spades," "king of hearts," "king of clubs," and "king of diamonds." This event is depicted in Fig. 4.5.

FIGURE 4.5
The event a king is selected

Thirteen outcomes comprise the third event, the event that the card selected is a heart; namely, the outcomes, "ace of hearts," "two of hearts," ..., "king of hearts." This event is shown in Fig. 4.6.

FIGURE 4.6
The event a heart is selected

Finally, the fourth event, the event that the card selected is a face card, consists of 12 outcomes; namely, the 12 face cards shown in Fig. 4.7.

FIGURE 4.7
The event a face card is selected

When the experiment of selecting a card from the deck is performed, an event *occurs* if it includes the card selected. For instance, if the card selected turns out to be the king of spades, then the second and fourth events (Figs. 4.5 and 4.7) occur, whereas the first and third events (Figs. 4.4 and 4.6) do not. ∎

Definition 4.2 summarizes the terminology discussed so far in this section.

DEFINITION 4.2   **SAMPLE SPACE AND EVENTS**

> *Sample space:* The collection of all possible outcomes for an experiment.
>
> *Event:* Any collection of outcomes for the experiment; in other words, any subset of the sample space.

Notation and Graphical Displays for Events

It is convenient and less cumbersome to employ letters such as *A, B, C, D,* ... to represent events. For instance, in the card-selection experiment of Example 4.6, we might let

A = event the card selected is the king of hearts,

B = event the card selected is a king,

C = event the card selected is a heart,

D = event the card selected is a face card.

Graphical displays of events are useful for explaining and understanding probability. **Venn diagrams,** named after English logician John Venn (1834–1923), are one of the best ways to visually portray events and relationships among events. The sample space is depicted as a rectangle, and the various events are drawn as disks (or other geometric shapes) inside the rectangle. In the simplest case, only one event is displayed, as in Fig. 4.8. The colored portion of Fig. 4.8 represents event E.

FIGURE 4.8

Venn diagram for event E

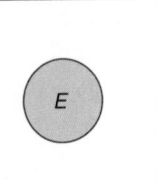

Relationships Among Events

To each event, E, there corresponds another event defined by the condition that "E does not occur." That event is called the **complement** of E and is denoted by **(not E).** Event (not E) consists of all outcomes not in E. A Venn diagram, as in Fig. 4.9(a), makes this idea clearer.

With any two events, say, A and B, we can associate two new events. One new event is defined by the condition that "both event A and event B occur" and is denoted by **(A & B).** Event (A & B) consists of all outcomes common to both event A and event B, as illustrated in Fig. 4.9(b).

The other new event associated with two events, A and B, is defined by the condition that "either event A or event B or both occur" or, equivalently, that "at least one of events A

FIGURE 4.9

Venn diagrams for
(a) event (not *E*),
(b) event (*A* & *B*),
(c) event (*A* or *B*)

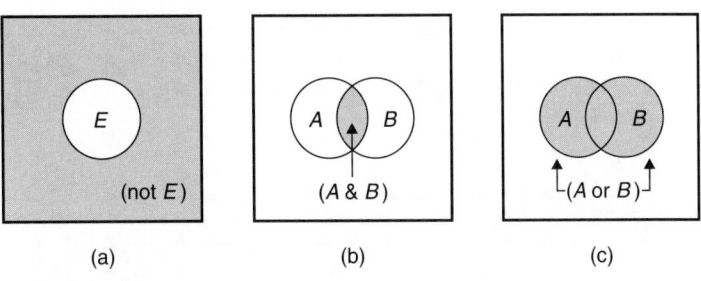

(a)   (b)   (c)

and *B* occurs." That event is denoted by **(*A* or *B*)** and consists of all outcomes in either event *A* or event *B* or both, as Fig. 4.9(c) shows.

Definition 4.3 summarizes the terms describing relationships among events.

DEFINITION 4.3

RELATIONSHIPS AMONG EVENTS

Let *E*, *A*, and *B* be events. Then

(not *E*) is the event that "*E* does not occur."

(*A* & *B*) is the event that "both *A* and *B* occur."

(*A* or *B*) is the event that "either *A* or *B* or both occur."

Because the event that "both *A* and *B* occur" is the same as the event that "both *B* and *A* occur," event (*A* & *B*) is the same as event (*B* & *A*). Similarly, event (*A* or *B*) is the same as event (*B* or *A*).

Example 4.7   Illustrates Definition 4.3

For the experiment of randomly selecting one card from a deck of 52, let

A = event the card selected is the king of hearts,

B = event the card selected is a king,

C = event the card selected is a heart,

D = event the card selected is a face card.

The outcomes constituting each of those four events are shown in Figs. 4.4–4.7, respectively, on page 191. Determine the following events.

a. (not *D*)   b. (*B* & *C*)   c. (*B* or *C*)   d. (*C* & *D*)

SOLUTION   a. (not *D*) is the event that "*D* does not occur"—the event that a face card is not selected. Event (not *D*) consists of the 40 cards in the deck that are not face cards, as depicted in Fig. 4.10.

FIGURE 4.10
Event (not *D*)

b. (*B* & *C*) is the event that "both *B* and *C* occur"—the event that the card selected is both a king and a heart. This can happen only if the card selected is the king of hearts. Consequently, (*B* & *C*) is the event that the card selected is the king of hearts and consists of the single outcome shown in Fig. 4.11. *Note:* Because event (*B* & *C*) is the same as event *A*, we can write *A* = (*B* & *C*).

FIGURE 4.11
Event (*B* & *C*)

c. (*B* or *C*) is the event that "either *B* or *C* or both occur"—the event that the card selected is either a king or a heart or both. Event (*B* or *C*) consists of 16 outcomes; namely, the 4 kings and the 12 non-king hearts, as illustrated in Fig. 4.12. *Note:* Event (*B* or *C*) can occur in 16, not 17, ways since the outcome "king of hearts" is common to both event *B* and event *C*.

FIGURE 4.12
Event (*B* or *C*)

d. $(C \& D)$ is the event that "both C and D occur"—the event that the card selected is both a heart and a face card. For that event to occur, the card selected must be either the jack, queen, or king of hearts. Thus event $(C \& D)$ consists of the three outcomes displayed in Fig. 4.13.

FIGURE 4.13
Event $(C \& D)$

These three outcomes are the ones common to events C and D. ∎

In Example 4.7, we described four events by listing their outcomes (see Figs. 4.10–4.13). Sometimes it is preferable to describe events verbally. Consider Example 4.8.

Example 4.8 Illustrates Definition 4.3

A frequency distribution for the ages of the 40 students in Professor Weiss's introductory statistics class is presented in Table 4.3.

TABLE 4.3
Frequency distribution
for students' ages

Age (yrs)	17	18	19	20	21	22	23	24	26	35	36
Frequency	1	1	9	7	7	5	3	4	1	1	1

Suppose a student is selected at random, and let

A = event the student selected is under 21,

B = event the student selected is over 30,

C = event the student selected is in his or her 20s,

D = event the student selected is over 18.

Determine the following events.

a. (not D)   b. $(A \& D)$ c. $(A$ or $D)$ d. $(B$ or $C)$

SOLUTION   a. (not D) is the event that "D does not occur"—the event that the student selected is not over 18, that is, is 18 or under. As we see from Table 4.3, (not D) is composed of the two students in the class who are 18 or under.

b. $(A \& D)$ is the event that "both A and D occur"—the event that the student selected is both under 21 and over 18, that is, is either 19 or 20. Event $(A \& D)$ is composed of the 16 students in the class who are 19 or 20.

c. (*A* or *D*) is the event that "either *A* or *D* or both occur"—the event that the student selected is either under 21 or over 18 or both. But every student in the class is either under 21 or over 18. Consequently, event (*A* or *D*) is composed of all 40 students in the class and hence is certain to occur.

d. (*B* or *C*) is the event that "either *B* or *C* or both occur"—the event that the student selected is either over 30 or in his or her 20s. Table 4.3 shows that event (*B* or *C*) is composed of the 29 students in the class who are 20 or over. ∎

Mutually Exclusive Events

Two events are **mutually exclusive** (or **disjoint**) if at most one of them can occur when the experiment is performed. Equivalently, we have the following definition.

DEFINITION 4.4   **TWO MUTUALLY EXCLUSIVE EVENTS**

> Two events are said to be *mutually exclusive* if they cannot both occur when the experiment is performed, that is, if they have no outcomes in common.

The Venn diagrams in Fig. 4.14 portray the difference between events that are mutually exclusive (a) and events that are not mutually exclusive (b).

FIGURE 4.14
(a) Mutually exclusive events
(b) Non–mutually exclusive events

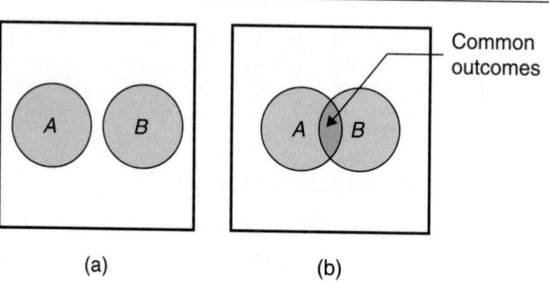

(a)   (b)

Example 4.9   Illustrates Definition 4.4

For the experiment of randomly selecting one card from a deck of 52, let

$$C = \text{event the card selected is a heart,}$$
$$D = \text{event the card selected is a face card,}$$
$$E = \text{event the card selected is an ace.}$$

Determine which of the following pairs of events are mutually exclusive.

a. *C*, *D*   b. *C*, *E*   c. *D*, *E*

SOLUTION a. Event C and event D are *not* mutually exclusive because they have the common out-comes "king of hearts," "queen of hearts," and "jack of hearts." Both events occur if the card selected is the king, queen, or jack of hearts.

b. Event C and event E are *not* mutually exclusive; they have the common outcome "ace of hearts." Both events occur if the card selected is the ace of hearts.

c. Event D and event E *are* mutually exclusive since they have no common outcomes. They cannot both occur when the experiment is performed because it is impossible to select a card that is both a face card and an ace. ■

The concept of mutually exclusive events can be extended to more than two events. For instance, three events, A, B, and C, are said to be mutually exclusive if at most one of them can occur when the experiment is performed. This requires that no two of them have outcomes in common. Figure 4.15 portrays three mutually exclusive events.

FIGURE 4.15
Three mutually exclusive events

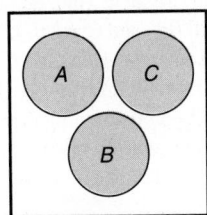

More generally, we have the following definition.

DEFINITION 4.5

MUTUALLY EXCLUSIVE EVENTS

Two or more events are said to be ***mutually exclusive*** if no two of them can occur when the experiment is performed, that is, if no two of them have outcomes in common.

Example 4.10 Illustrates Definition 4.5

In the card-selection experiment, let

D = event the card selected is a face card,

E = event the card selected is an ace,

F = event the card selected is an 8,

G = event the card selected is a 10 or a jack.

a. Are the three events D, E, and F mutually exclusive?

b. Are the four events D, E, F, and G mutually exclusive?

SOLUTION   a. Events *D*, *E*, and *F are* mutually exclusive because no two of them can occur simultaneously.

b. Events *D*, *E*, *F*, and *G* are *not* mutually exclusive since event *D* and event *G* both occur if the card selected is a jack. ∎

EXERCISES 4.2

4.19 When one die is rolled, the following six outcomes are possible:

List the outcomes comprising each of the events below.

   A = event the die comes up even
   B = event the die comes up 4 or more
   C = event the die comes up at most 2
   D = event the die comes up 3

4.20 In a horse race, the odds against winning are:

Horse	#1	#2	#3	#4	#5	#6	#7	#8
Odds	8	15	2	3	30	5	10	5

List the outcomes comprising each of the events below.

A = event that one of the top two favorites wins (the top two favorites are the two horses with the lowest odds against winning)

B = event that the winning horse's number is above 5

C = event that the winning horse's number is at most 3, that is, 3 or less

D = event that one of the two long shots wins (the two long shots are the two horses with the highest odds against winning)

4.21 When a dime is tossed four times, 16 outcomes are possible:

HHHH	HTHH	THHH	TTHH
HHHT	HTHT	THHT	TTHT
HHTH	HTTH	THTH	TTTH
HHTT	HTTT	THTT	TTTT

Here, for example, HTTH represents the outcome that the first toss is heads, the next two tosses are tails, and the fourth toss is heads. List the outcomes that constitute each of the following four events.

   A = event exactly two heads are tossed
   B = event the first two tosses are tails
   C = event the first toss is heads
   D = event all four tosses come up the same

4.22 A committee consists of five executives, three women and two men. Their names are Maria (M), John (J), Susan (S), Bill (B), and Carol (C). The committee needs to select a chairperson and a secretary. It decides to make the selection randomly by drawing straws. The person getting the longest straw will be appointed chairperson and the one getting the shortest straw will be appointed secretary. We can represent the possible outcomes in the following manner.

MS	SM	CM	JM	BM
MC	SC	CS	JS	BS
MJ	SJ	CJ	JC	BC
MB	SB	CB	JB	BJ

Here, for example, MS represents the outcome that Maria is appointed chairperson and Susan is appointed secretary. List the outcomes comprising each of the following four events.

   A = event a male is appointed chairperson
   B = event Carol is appointed chairperson
   C = event Bill is appointed secretary
   D = event only females are appointed

4.23 Refer to Exercise 4.19. For each of the following events, list the outcomes that constitute the event and describe the event in words.
a. (not *A*)   b. (*A* & *B*)   c. (*B* or *C*)

4.24 Refer to Exercise 4.20. For each of the following events, list the outcomes that constitute the event and describe the event in words.
a. (not *C*)   b. (*C* & *D*)   c. (*A* or *C*)

4.25 Refer to Exercise 4.21. For each of the following events, list the outcomes that comprise the event and describe the event in words.
a. (not *B*)   b. (*A* & *B*)   c. (*C* or *D*)

4.26 Refer to Exercise 4.22. For each of the following events, list the outcomes that comprise the event and describe the event in words.
a. (not *A*)   b. (*B* & *D*)   c. (*B* or *C*)

4.27 The absentee records over the past year for the employees of Cudahey Masonry are as shown in the following frequency distribution.

Days missed	0	1	2	3	4	5	6	7
No. of employees	4	2	14	10	16	18	10	6

For an employee selected at random, let

A = event that the employee missed at most 3 days,
B = event that the employee missed at least 1 day,
C = event that the employee missed between 4 and 6 days, inclusive,
D = event that the employee missed more than 6 days.

Describe each of the following events in words and determine the number of outcomes (employees) that comprise each event.
a. (not *A*)   b. (*A* & *B*)   c. (*C* or *D*)

4.28 A frequency distribution for the number of cars owned by each of the families in a small city is shown in the following table.

Cars owned	0	1	2	3	4	5
No. of families	27	1422	2865	1796	324	53

For a family selected at random, let

A = event the family owns at most three cars,
B = event the family owns at least one car,
C = event the family owns between two and four cars, inclusive,
D = event the family owns at least three cars.

Describe each of the following events in words and determine the number of outcomes (families) that comprise each event.
a. (not *B*)   b. (*C* & *D*)   c. (*A* or *D*)

4.29 As reported by the Bureau of Labor Statistics in *Employment and Earnings,* the age distribution of employed persons 16 years old and over is as shown in the following table. The frequencies are in thousands.

Age (yrs) *x*	Number of persons *f*
16–19	5,899
20–24	11,748
25–34	28,429
35–44	23,597
45–54	15,216
55–64	10,163
65 & over	2,737

An employed person is selected at random. Let

A = event the person is under 20,
B = event the person is between 20 and 54, inclusive,
C = event the person is under 45,
D = event the person is at least 55.

Describe each of the following events in words and determine the number of outcomes (persons) that comprise each event.
a. (not *C*)   b. (not *B*)   c. (*B* & *C*)   d. (*A* or *D*)

4.30 Each part of this exercise lists events from Exercise 4.20. Indicate whether or not the events are mutually exclusive.
a. *A* and *B*   b. *B* and *C*   c. *A*, *B*, and *C*
d. *A*, *B*, and *D*   e. *A*, *B*, *C*, and *D*

4.31 Refer to Exercise 4.19.
a. Are events *A* and *B* mutually exclusive?
b. Are events *B* and *C* mutually exclusive?
c. Are events *A*, *C*, and *D* mutually exclusive?
d. Are there three mutually exclusive events among *A*, *B*, *C*, and *D*? How about four?

4.32 For the following groups of events from Exercise 4.28, determine which are mutually exclusive.
a. *A* and *B*   b. (not *B*) and *C*
c. *A* and *D*   d. (not *B*), *C*, and *D*

4.33 For the following groups of events from Exercise 4.29, determine which are mutually exclusive.
a. *C* and *D*   b. *B* and *C*   c. *A*, *B*, and *D*
d. *A*, *B*, and *C*   e. *A*, *B*, *C*, and *D*

4.34 Draw a Venn diagram portraying four mutually exclusive events.

4.35 Construct a Venn diagram that portrays four events, *A*, *B*, *C*, and *D*, having the following properties: Events *A*, *B*, and *C* are mutually exclusive; events *A*, *B*, and *D* are mutually exclusive; no other three of the four events are mutually exclusive.

4.36 Suppose *A*, *B*, and *C* are three events with the property that they cannot all occur simultaneously. Does this necessarily imply that *A*, *B*, and *C* are mutually exclusive? Provide a detailed explanation for your answer and illustrate it with a Venn diagram.

4.3   SOME RULES OF PROBABILITY

In this section we will discuss several rules of probability. Before beginning, however, we need to introduce an additional notation used in probability.

Example 4.11   An Additional Probability Notation

When a balanced die is rolled once, six equally likely outcomes are possible, as shown in Fig. 4.16.

FIGURE 4.16
Sample space for
rolling a die once

Consider, for instance, the event that the die comes up even. This event can occur in three ways; namely, if 2, 4, or 6 is rolled. Since $f/N = 3/6 = 0.5$, we see that *the probability is 0.5 that the die comes up even.*

Employing probability notation enables us to express the italicized phrase much more concisely. Let *A* denote the event that the die comes up even. We use the notation $P(A)$ to represent the probability that event *A* occurs. Thus the italicized statement can be written simply as $P(A) = 0.5$, read "the probability of *A* is 0.5." Keep in mind that *A* refers to the event that the die comes up even, whereas $P(A)$ refers to the probability of that event occurring. ■

DEFINITION 4.6   PROBABILITY NOTATION

> If *E* is an event, then $P(E)$ stands for the probability that event *E* occurs. It is read "the probability of *E*."

The Special Addition Rule

The first rule of probability we will study is the **special addition rule,** which states that if two events are mutually exclusive, then the probability that at least one of them occurs equals the sum of their probabilities. We can use a Venn diagram to see the validity of the special addition rule. Figure 4.17 shows two mutually exclusive events, A and B.

FIGURE 4.17
Two mutually exclusive events

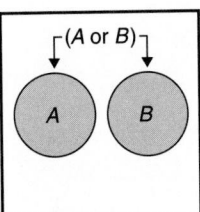

If we think of the colored regions in Fig. 4.17 as probabilities, then the colored disk on the left is $P(A)$, the colored disk on the right is $P(B)$, and the total colored region is $P(A \text{ or } B)$. Because event A and event B are mutually exclusive, the total colored region equals the sum of the two colored disks; that is, $P(A \text{ or } B) = P(A) + P(B)$. Thus we have the following fact.

FORMULA 4.1

THE SPECIAL ADDITION RULE

If event A and event B are mutually exclusive, then
$$P(A \text{ or } B) = P(A) + P(B).$$
More generally, if events A, B, C, ... are mutually exclusive, then
$$P(A \text{ or } B \text{ or } C \text{ or } \cdots) = P(A) + P(B) + P(C) + \cdots.$$

Example 4.12   Illustrates Formula 4.1

The U.S. Bureau of the Census compiles information about U.S. farms and publishes the results in *Census of Agriculture*. According to that publication, a relative-frequency distribution for the size of farms in the United States is as presented in the first two columns of Table 4.4, located at the top of the next page. The table shows, for instance, that 17.3% (0.173) of U.S. farms have between 100 and 179 acres, inclusive.

In the third column of Table 4.4, we have introduced events that correspond to each size class. For example, if a farm is selected at random, then D denotes the event that the farm obtained has between 100 and 179 acres. The probabilities of the events in the third column of Table 4.4 equal the relative frequencies displayed in the second column. Thus the probability that a randomly selected farm has between 100 and 179 acres is $P(D) = 0.173$. Use Table 4.4 and the special addition rule to determine the probability that a randomly selected farm has between 100 and 499 acres, inclusive.

TABLE 4.4
Size of farms in the United States

Size (acres)	Relative frequency	Event
Under 10	0.087	A
10–49	0.192	B
50–99	0.156	C
100–179	0.173	D
180–259	0.098	E
260–499	0.143	F
500–999	0.085	G
1000–1999	0.040	H
2000 & over	0.026	I

SOLUTION   As we see from Table 4.4, the event that the farm obtained has between 100 and 499 acres can be expressed as $(D$ or E or $F)$. Events D, E, and F are mutually exclusive and so by the special addition rule,

$$P(D \text{ or } E \text{ or } F) = P(D) + P(E) + P(F)$$
$$= 0.173 + 0.098 + 0.143 = 0.414.$$

In other words, 41.4% of U.S. farms have between 100 and 499 acres, inclusive.   ■

The Complementation Rule

The second rule of probability we will study is the **complementation rule,** which states that the probability that an event occurs equals 1 minus the probability that it does not occur. We can use a Venn diagram to see the validity of the complementation rule. Figure 4.18 shows an event, E, and its complement, (not E).

FIGURE 4.18
An event and its complement

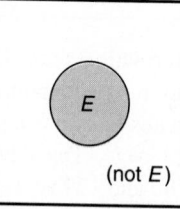

If we think of the regions in Fig. 4.18 as probabilities, then the entire region enclosed by the rectangle is the probability of the sample space, which is 1. Furthermore, the colored region is $P(E)$ and the uncolored region is $P(\text{not } E)$. So we see that $P(E) + P(\text{not } E) = 1$ or, equivalently, $P(E) = 1 - P(\text{not } E)$.

| FORMULA 4.2 | **THE COMPLEMENTATION RULE** |

For any event, E,

$$P(E) = 1 - P(\text{not } E).$$

In words, the probability that an event, E, occurs equals 1 minus the probability that its complement, (not E), occurs.

The complementation rule is significant because for an event, E, it is sometimes easier to compute the probability that it does not occur, $P(\text{not } E)$, than the probability that it does occur, $P(E)$. In such cases we can, in view of the complementation rule, obtain $P(E)$ by first computing $P(\text{not } E)$ and then subtracting the result from 1.

Example 4.13 Illustrates Formula 4.2

The first two columns of Table 4.4 on page 202 provide a relative-frequency distribution for the size of U.S. farms. Suppose a farm is selected at random. Determine the probability that the farm obtained has

a. less than 2000 acres.   b. at least 50 acres.

SOLUTION   a. Let

$$J = \text{event the farm obtained has less than 2000 acres.}$$

To find $P(J)$ we will apply the complementation rule, since it is easier to compute $P(\text{not } J)$. Note that (not J) is the event that the farm obtained has 2000 or more acres, which is event I in Table 4.4. So $P(\text{not } J) = P(I) = 0.026$. Applying the complementation rule, we find that

$$P(J) = 1 - P(\text{not } J) = 1 - 0.026 = 0.974.$$

The probability is 0.974 that a randomly selected farm has less than 2000 acres.

b. Let

$$K = \text{event the farm obtained has at least 50 acres.}$$

We will apply the complementation rule to find $P(K)$. Now, (not K) is the event that the farm obtained has less than 50 acres. From Table 4.4 we see that event (not K) is the same as event (A or B). Since event A and event B are mutually exclusive, the special addition rule implies that

$$P(\text{not } K) = P(A \text{ or } B) = P(A) + P(B) = 0.087 + 0.192 = 0.279.$$

Using this fact and the complementation rule, we conclude that

$$P(K) = 1 - P(\text{not } K) = 1 - 0.279 = 0.721.$$

The probability is 0.721 that a randomly selected farm has at least 50 acres. ∎

The General Addition Rule

The special addition rule, $P(A \text{ or } B) = P(A) + P(B)$, is a formula for obtaining the probability of event $(A \text{ or } B)$ from the probabilities of event A and event B. That rule is valid provided event A and event B are mutually exclusive. For events that are not mutually exclusive, we must use a different rule—the *general addition rule*. To introduce the general addition rule, we will employ the Venn diagram shown in Fig. 4.19.

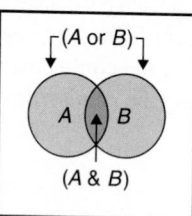

If we think of the colored regions in Fig. 4.19 as probabilities, then the colored disk on the left is $P(A)$, the colored disk on the right is $P(B)$, and the total colored region is $P(A \text{ or } B)$. To obtain the total colored region, $P(A \text{ or } B)$, we first sum the two colored disks, $P(A)$ and $P(B)$. In doing this, the common colored region, $P(A \& B)$, is counted twice. To account for that fact, we must subtract $P(A \& B)$ from the sum. So we see that $P(A \text{ or } B) = P(A) + P(B) - P(A \& B)$. This formula is the **general addition rule.**

FORMULA 4.3   **THE GENERAL ADDITION RULE**

> If A and B are any two events, then
>
> $$P(A \text{ or } B) = P(A) + P(B) - P(A \& B).$$
>
> In words, the probability that either event A or event B occurs equals the probability that event A occurs plus the probability that event B occurs minus the probability that both occur.

Example 4.14   Illustrates Formula 4.3

Consider again the experiment of selecting one card at random from a deck of 52 playing cards. Find the probability that the card selected is either a spade or a face card

a. without using the general addition rule.

b. using the general addition rule.

SOLUTION   a. Let

E = event the card selected is either a spade or a face card.

Event E consists of 22 cards; namely, the 13 spades plus the other 9 face cards that are not spades, as shown in Fig. 4.20.

FIGURE 4.20
Event E

Consequently, by the f/N rule,

$$P(E) = \frac{f}{N} = \frac{22}{52} = 0.423.$$

b. To determine $P(E)$ using the general addition rule, we first note that we can write $E = (C \text{ or } D)$, where

$$C = \text{event the card selected is a spade,}$$

$$D = \text{event the card selected is a face card.}$$

Event C consists of the 13 spades and event D consists of the 12 face cards. Also, event $(C \& D)$ consists of the three spades that are face cards—the jack, queen, and king of spades. Applying the general addition rule, we get

$$P(E) = P(C \text{ or } D) = P(C) + P(D) - P(C \& D)$$

$$= \frac{13}{52} + \frac{12}{52} - \frac{3}{52} = 0.250 + 0.231 - 0.058 = 0.423.$$

This agrees with the result obtained in part (a). ∎

In Example 4.14, we computed the probability of selecting either a spade or a face card in two ways: first without using the general addition rule and then using it. There, computing the probability was simpler without using the general addition rule. Frequently, however, the general addition rule is the easier or even the only way to compute a probability. This point is illustrated in Example 4.15.

Example 4.15   Illustrates Formula 4.3

Data on people arrested are published by the U.S. Federal Bureau of Investigation in *Crime in the United States*. In 1991, 81.3% of people arrested were male, 16.3% were under 18, and 12.6% were males under 18. Suppose a person arrested in 1991 is selected at random. Find the probability that the person obtained is either male or under 18.

SOLUTION   Let

$$M = \text{event the person obtained is male,}$$
$$E = \text{event the person obtained is under 18.}$$

The event that the person obtained is either male or under 18 can be expressed as $(M \text{ or } E)$. We want to determine $P(M \text{ or } E)$. From the percentage data given, we know that $P(M) = 0.813$, $P(E) = 0.163$, and $P(M \& E) = 0.126$. Employing the general addition rule, we conclude that

$$P(M \text{ or } E) = P(M) + P(E) - P(M \& E)$$
$$= 0.813 + 0.163 - 0.126 = 0.850.$$

In terms of percentages, this means that 85.0% of those arrested in 1991 were either male or under 18.   ■

The general addition rule is consistent with the special addition rule—if two events are mutually exclusive, then both rules yield the same result. To see this, suppose event A and event B are mutually exclusive. Then event $(A \& B)$ is impossible and hence has a probability of 0. Applying the general addition rule, we get

$$P(A \text{ or } B) = P(A) + P(B) - P(A \& B) = P(A) + P(B) - 0 = P(A) + P(B),$$

which is the same result given by the special addition rule. Thus, when considering two events, if you are not sure whether to use the special addition rule or the general addition rule, it is always safe to use the general addition rule.

EXERCISES 4.3

· **4.37**   A bowl contains 10 marbles of which 3 are red, 2 are white, and 5 are blue. Suppose a marble is selected at random from the bowl, and let E denote the event that the marble obtained is white. Determine the probability that the marble obtained is white. Use probability notation to express your answer.

· **4.38**   Suppose you hold 20 out of a total of 500 tickets sold for a lottery. The grand-prize winner is determined by the random selection of 1 of the 500 tickets. Let G be the event that you win the grand prize. Find the probability that you win the grand prize. Express your answer in probability notation.

· **4.39**   An age distribution for U.S. senators in the 102nd Congress is provided by the following table. [SOURCE: U.S. Congress, Joint Committee on Printing, *Congressional Directory.*]

Age (yrs)	Number of senators
Under 40	0
40–49	23
50–59	46
60–69	24
70–79	5
80 and over	2
	100

Suppose a senator is selected at random. Let

$$A = \text{event the senator is under 40,}$$
$$B = \text{event the senator is in his or her 40s,}$$
$$C = \text{event the senator is in his or her 50s,}$$
$$S = \text{event the senator is under 60.}$$

a. Use the table and the f/N rule to find $P(S)$.
b. Express event S in terms of events A, B, and C.

c. Determine $P(A)$, $P(B)$, and $P(C)$.

d. Compute $P(S)$ using the special addition rule and your results from parts (b) and (c). Compare your answer with the one you found in part (a).

4.40 As reported by Dun & Bradstreet in *Business Failure Record,* the number of commercial failures for the year 1990 by type of industry are as follows.

Industry	Failures
Mining	381
Wholesale trade	4,376
Retail trade	12,826
Construction	8,072
Transportation, public utilities	2,610
	28,265

Suppose a failed business is selected at random. Let

A = event it was in wholesale trade,
B = event it was in retail trade,
T = event it was in either wholesale trade or retail trade.

a. Use the table and the f/N rule to find $P(T)$.

b. Express event T in terms of events A and B.

c. Determine $P(A)$ and $P(B)$.

d. Compute $P(T)$ using the special addition rule and your results from parts (b) and (c). Compare your answer with the one you found in part (a).

4.41 According to *Statistics of Income,* compiled by the Internal Revenue Service, a relative-frequency distribution for U.S. businesses by receipts received from sales and services is as follows.

Receipts	Relative frequency	Event
Under $25,000	0.593	A
$25,000–$49,999	0.107	B
$50,000–$99,999	0.092	C
$100,000–$499,999	0.144	D
$500,000–$999,999	0.031	E
$1,000,000 or more	0.033	F

The table shows, for example, that 59.3% of U.S. businesses had receipts of under $25,000. Suppose a business is selected

at random. Let

A = event the business obtained had receipts under $25,000,
B = event the business obtained had receipts between $25,000 and $49,999,

and so on (see the third column of the table). Find the probability that the business obtained had receipts of

a. under $100,000.

b. at least $500,000.

c. between $25,000 and $499,999, inclusive.

d. Interpret each of your results in parts (a)–(c) in terms of percentages.

4.42 The following is a percentage distribution for the number of years of school completed by U.S. adults, 25 years old and over. [SOURCE: U.S. Bureau of the Census, *Current Population Reports.*]

Years completed	Percentage	Event
0–4	2.4	A
5–7	3.8	B
8	4.4	C
9–11	11.0	D
12	38.6	E
13–15	18.4	F
16 or more	21.4	G

The table shows, for instance, that 38.6% of adults 25 years old and over have completed exactly 12 years of school. Suppose an adult 25 years old or over is randomly selected from the population. Let

A = event the person obtained has completed between 0 and 4 years of school,
B = event the person obtained has completed between 5 and 7 years of school,

and so forth (see the third column of the table). Determine the probability that the person obtained

a. has at most an elementary-school education, that is, has completed at most 8 years of school.

b. has at most a high-school education, that is, has completed at most 12 years of school.

c. has completed at least 1 year of college, that is, at least 13 years of school.

d. Interpret each of your results in parts (a)–(c) in terms of percentages.

In Exercises 4.43–4.46, find the designated probabilities using the complementation rule.

·   **4.43**   Refer to Exercise 4.39. Find the probability that a randomly selected senator in the 102nd Congress is
a. at least 40 years old.
b. under 70 years old.

·   **4.44**   Refer to Exercise 4.40. Find the probability that a failed business selected at random was not in construction.

·   **4.45**   Refer to Exercise 4.41. Find the probability that a randomly selected business had receipts of
a. under $1,000,000.
b. at least $50,000.

·   **4.46**   Refer to Exercise 4.42. Determine the probability that a randomly selected adult 25 years old or over has completed
a. less than 4 years of college.
b. at least 5 years of school.

·   **4.47**   In the game of craps, a player rolls two balanced dice. There are 36 equally likely outcomes possible, as shown in Fig. 4.2 on page 186. Let

A = event the sum of the dice is 7,
B = event the sum of the dice is 11,
C = event the sum of the dice is 2,
D = event the sum of the dice is 3,
E = event the sum of the dice is 12,
F = event the sum of the dice is 8,
G = event doubles are rolled.

a. Compute the probability of each of the seven events.
b. The player wins on the first roll if the sum of the dice is 7 or 11. Find the probability of that event using the special addition rule and your results from part (a).
c. The player loses on the first roll if the sum of the dice is 2, 3, or 12. Determine the probability of that event using the special addition rule and your results from part (a).
d. Compute the probability that either the sum of the dice is 8 or doubles are rolled
   i. without using the general addition rule.
   ii. using the general addition rule.

·   **4.48**   As reported by the U.S. Bureau of Justice Statistics in *Profile of Jail Inmates,* approximately 56.5% of jail inmates are white, 94.0% are male, and 53.5% are white males. Suppose a jail inmate is selected at random. Let

W = event the inmate obtained is white,
M = event the inmate obtained is male.

a. Find $P(W)$, $P(M)$, and $P(W \& M)$.
b. Determine $P(W$ or $M)$ and interpret your result in terms of percentages.
c. Obtain the probability that a randomly selected inmate is female.

·   **4.49**   According to *Current Population Reports,* published by the U.S. Bureau of the Census, 52.6% of U.S. adults are female, 7.0% are divorced, and 4.2% are divorced females. For a U.S. adult selected at random, let

F = event the person is female,
D = event the person is divorced.

a. Obtain $P(F)$, $P(D)$, and $P(F \& D)$.
b. Determine $P(F$ or $D)$ and interpret your result in terms of percentages.
c. Find the probability that a randomly selected adult is male.

·   **4.50**   Suppose A and B are events and $P(A) = \frac{1}{4}$, $P(B) = \frac{1}{3}$, and $P(A$ or $B) = \frac{1}{2}$.
a. Are A and B mutually exclusive?
b. Determine $P(A \& B)$.

·   **4.51**   Suppose $P(A) = \frac{1}{3}$, $P(A$ or $B) = \frac{1}{2}$, and $P(A \& B) = \frac{1}{10}$. Find $P(B)$.

· ·   **4.52**   *Bottom Line/Personal* newsletter interviewed Gerald Kushel, Ed.D., on the secrets of successful people. To study success, Kushel questioned 1200 people, among whom were lawyers, artists, teachers, and students. He found that 15% enjoy neither their jobs nor their personal lives, 80% enjoy their jobs but not their personal lives, and 4% enjoy both their jobs and their personal lives. Determine the percentage of the 1200 people interviewed who
a. enjoy either their jobs or their personal lives. $\setminus 5$
b. enjoy their personal lives but not their jobs.

· · ·   **4.53**   The general addition rule for three events is

$$P(A \text{ or } B \text{ or } C) = P(A) + P(B) + P(C)$$
$$- P(A \& B) - P(A \& C)$$
$$- P(B \& C) + P(A \& B \& C).$$

Prove the above rule. *Hint:* Let $D = (B$ or $C)$ and apply the general addition rule to $(A$ or $D)$. Also note that $(A \& D) = ((A \& B) \text{ or } (A \& C))$.

· · ·   **4.54**   When a balanced dime is tossed three times, eight equally likely outcomes are possible:

HHH	HTH	THH	TTH
HHT	HTT	THT	TTT

Let

A = event the first toss is heads,
B = event the second toss is tails,
C = event the third toss is heads.

a. List the outcomes comprising each of those events.
b. Find $P(A)$, $P(B)$, and $P(C)$.
c. Describe each of the following events in words and list

the outcomes comprising each one: $(A \& B)$, $(A \& C)$, $(B \& C)$, $(A \& B \& C)$.
d. Find the probability of each event in part (c).
e. Describe event $(A$ or B or $C)$ in words and list the outcomes that comprise it.
f. Find $P(A$ or B or $C)$ using your result from part (e) and the f/N rule.
g. Find $P(A$ or B or $C)$ using your results from parts (b) and (d) and the general addition rule for three events (Exercise 4.53).

4.4 CONTINGENCY TABLES; JOINT AND MARGINAL PROBABILITIES

We often need to analyze data obtained by cross classifying the members of a population or sample according to two characteristics (variables). Some examples of cross classifications are annual income versus educational level, automobile-accident frequency versus driver's age, and political affiliation versus religion. Frequencies for cross-classified data are most easily displayed using contingency tables, which we discuss in Example 4.16.

Example 4.16   Contingency Tables

Table 4.5 displays data adapted from the *Arizona State University Statistical Summary*. The table provides a frequency distribution obtained by cross classifying the faculty according to two characteristics: age and rank.

TABLE 4.5
Contingency table for cross classification of ASU faculty by age and rank

		Rank				
		Full professor R_1	Associate professor R_2	Assistant professor R_3	Instructor R_4	Total
Age	Under 30 A_1	2	3	57	6	68
	30–39 A_2	52	170	163	17	402
	40–49 A_3	156	125	61	6	348
	50–59 A_4	145	68	36	4	253
	60 & over A_5	75	15	3	0	93
	Total	430	381	320	33	1164

The number 2 in the upper left-hand corner of Table 4.5 indicates that two faculty members are full professors under the age of 30. The number 170, diagonally below and to the right of the 2, shows that 170 faculty members are associate professors in their 30s.

The row total in the first row of the table indicates that 68 $(2+3+57+6)$ of the faculty members are under 30. Similarly, the column total in the third column shows that 320 of the faculty members are assistant professors. The number 1164 in the lower right-hand corner of the table gives the total number of faculty. That total can be found by summing either the row totals or the column totals. It can also be found by summing all the numbers inside the box formed by the heavy lines.

Table 4.5 is an example of a **contingency table,** a frequency distribution for cross-classified data. A contingency table includes all the different possibilities in the cross classification—it accounts for all contingencies. The small boxes inside the heavy lines of a contingency table, called **cells,** give the frequencies for the various contingencies. Table 4.5 has 20 cells.

■

Joint and Marginal Probabilities

We will now use the age and rank data from Table 4.5 to introduce the concepts of joint probabilities and marginal probabilities.

Example 4.17   Joint and Marginal Probabilities

Suppose an Arizona State University faculty member is selected at random. Notice that the rows and columns of Table 4.5 are labeled with letters. The first row, labeled A_1, represents the event that the faculty member selected is under 30:

$$A_1 = \text{event the faculty member selected is under 30.}$$

Similarly,

$$R_2 = \text{event the faculty member selected is an associate professor,}$$

and so forth. The events A_1, A_2, A_3, A_4, and A_5 are mutually exclusive, as are the events R_1, R_2, R_3, and R_4. (Why is this so?)

In addition to considering events A_1 through A_5 and R_1 through R_4 separately, we can also consider them jointly. For example, the event that the faculty member selected is under 30 (event A_1) *and* is also an associate professor (event R_2) can be expressed as $(A_1 \ \& \ R_2)$:

$$(A_1 \ \& \ R_2) = \text{event the faculty member selected is}$$
$$\text{an associate professor under 30.}$$

Event $(A_1 \ \& \ R_2)$ is represented by the cell in the first row and second column of Table 4.5. Here there are 20 different joint events, one for each cell of the contingency table.

It is sometimes useful to think of a contingency table as a Venn diagram. The Venn diagram corresponding to the contingency table in Table 4.5 is as shown in Fig. 4.21. That Venn diagram makes it clear that the 20 joint events, $(A_1 \& R_1)$, $(A_1 \& R_2)$, ..., $(A_5 \& R_4)$, are mutually exclusive.

FIGURE 4.21

Venn diagram corresponding to Table 4.5

	R_1	R_2	R_3	R_4
A_1	$(A_1 \& R_1)$	$(A_1 \& R_2)$	$(A_1 \& R_3)$	$(A_1 \& R_4)$
A_2	$(A_2 \& R_1)$	$(A_2 \& R_2)$	$(A_2 \& R_3)$	$(A_2 \& R_4)$
A_3	$(A_3 \& R_1)$	$(A_3 \& R_2)$	$(A_3 \& R_3)$	$(A_3 \& R_4)$
A_4	$(A_4 \& R_1)$	$(A_4 \& R_2)$	$(A_4 \& R_3)$	$(A_4 \& R_4)$
A_5	$(A_5 \& R_1)$	$(A_5 \& R_2)$	$(A_5 \& R_3)$	$(A_5 \& R_4)$

Let's now move on to an examination of probabilities. Because the total number of faculty members is 1164, $N = 1164$. To determine, for instance, the probability that the faculty member selected is an associate professor (event R_2), we first note from Table 4.5 that $f = 381$ and then apply the f/N rule:

$$P(R_2) = \frac{f}{N} = \frac{381}{1164} = 0.327.$$

Similarly, the probability that the faculty member selected is under 30 equals

$$P(A_1) = \frac{f}{N} = \frac{68}{1164} = 0.058.$$

We can also find probabilities for joint events, so-called **joint probabilities.** For instance, the probability that the faculty member selected is an associate professor under 30 equals

$$P(A_1 \& R_2) = \frac{f}{N} = \frac{3}{1164} = 0.003.$$

In Table 4.6, shown at the top of the next page, we have replaced the joint frequency distribution in Table 4.5 with a **joint probability distribution.** The probabilities in Table 4.6 are determined in the same way as the three probabilities we just computed. Notice that the joint probabilities are displayed in the cells of Table 4.6. Also observe that the row and column labels "Total" in Table 4.5 have been changed in Table 4.6 to $P(R_j)$ and $P(A_i)$, respectively. This is because the last row of Table 4.6 gives the probabilities of events R_1 through R_4 and the last column gives the probabilities of events A_1 through A_5. Those probabilities are often called **marginal probabilities** because they are in the margin of the joint probability distribution.

TABLE 4.6
Joint probability distribution
corresponding to Table 4.5

Rank

Age		Full professor R_1	Associate professor R_2	Assistant professor R_3	Instructor R_4	$P(A_i)$
Under 30 A_1		0.002	0.003	0.049	0.005	0.058
30–39 A_2		0.045	0.146	0.140	0.015	0.345
40–49 A_3		0.134	0.107	0.052	0.005	0.299
50–59 A_4		0.125	0.058	0.031	0.003	0.217
60 & over A_5		0.064	0.013	0.003	0.000	0.080
$P(R_j)$		0.369	0.327	0.275	0.028	1.000

The sum of the joint probabilities in a row or column of a joint probability distribution equals the marginal probability in that row or column (any observed discrepancy is due to roundoff error). For example, consider the A_4 row of Table 4.6. The sum of the joint probabilities in that row is $0.125 + 0.058 + 0.031 + 0.003 = 0.217$, which is precisely the marginal probability at the end of the A_4 row. ∎

EXERCISES 4.4

4.55 The contingency table at the right cross classifies institutions of higher education in the United States by region and type. [SOURCE: U.S. National Center for Education Statistics, *Digest of Education Statistics*.]
a. How many cells does this contingency table have?
b. What is the total number of institutions of higher education in the United States?
c. How many institutions are in the Midwest?
d. How many are public?
e. How many are private schools in the South?

4.56 As reported by the Motor Vehicle Manufacturers Association of the United States in *Motor Vehicle Facts and Figures*, the number of cars and trucks in use by age are as shown in the contingency table at the top of the first column on the next page. Frequencies are in millions.

Type

Region		Public T_1	Private T_2	Total
Northeast R_1		266	555	821
Midwest R_2		359	504	863
South R_3		533	502	1035
West R_4		313	242	555
Total		1471	1803	3274

Type

	Car V_1	Truck V_2	Total
Under 3 A_1	21.5	3.2	24.7
3–5 A_2	29.9	2.5	32.4
6–8 A_3	22.2	2.0	24.2
9–11 A_4	17.9	1.8	19.7
12 & over A_5	15.4	3.6	19.0
Total	106.9	13.1	120.0

Age (yrs)

a. How many cells does this contingency table have?
b. What is the total number of cars and trucks in use?
c. How many vehicles are trucks?
d. How many vehicles are between 3 and 5 years old?
e. How many vehicles are trucks between 9 and 11 years old?

4.57 The following contingency table cross classifies employment status and educational level for the civilian labor force. Frequencies are in thousands. [SOURCE: U.S. Bureau of Labor Statistics.]

Employment status

	Employed E_1	Unemployed E_2	Total
Less than 8 S_1	3,535.5	612.2	4,147.7
8 S_2	3,240.9	517.1	3,758.0
9–11 S_3	12,767.0	2,807.4	15,574.4
12 S_4	40,068.9	4,601.5	44,670.4
13–15 S_5	18,266.7	1,340.4	19,607.1
16 or more S_6	20,230.8	686.0	20,916.8
Total	98,109.8	10,564.6	108,674.4

Years of school completed

a. How many cells does this contingency table have?
b. What is the size of the civilian labor force?
c. How many people in the civilian labor force have completed exactly 12 years of school?
d. How many in the civilian labor force are unemployed?
e. How many in the civilian labor force with 16 or more years of school are unemployed?

4.58 The following contingency table provides a cross classification of U.S. hospitals by type and number of beds. [SOURCE: American Hospital Association, *Hospital Statistics*.]

Number of beds

	6–24 B_1	25–74 B_2	75+ B_3	Total
General H_1	299	1894	3945	6138
Psychiatric H_2	17	121	378	516
Chronic H_3	0	7	40	47
Tuberculosis H_4	0	1	10	11
Other H_5	22	131	162	315
Total	338	2154	4535	7027

Type

a. How many hospitals have at least 75 beds?
b. How many psychiatric hospitals are there?
c. How many general hospitals have between 25 and 74 beds?

4.59 According to *Census of Agriculture*, published by the U.S. Bureau of the Census, a joint frequency distribution for the number of farms by acreage and tenure of operator, is as shown in the contingency table at the top of the first column on the next page. Frequencies are given in thousands.
a. Fill in the three empty cells.
b. How many cells does this contingency table have?
c. How many farms have under 50 acres?
d. How many farms are tenant operated?
e. How many farms are operated by part owners and have between 500 and 999 acres?
f. How many farms are not full-owner operated?
g. How many tenant-operated farms have at least 180 acres?

Tenure of operator

	Full owner T_1	Part owner T_2	Tenant T_3	Total
Under 50 A_1	532	74	84	690
50–179 A_2	563		94	814
180–499 A_3	262		87	596
500–999 A_4	57	128		215
1000+ A_5	36	107	18	161
Total	1450	713	313	2476

(left axis label: Acreage)

4.60 The contingency table shown below provides a joint frequency distribution for data on annual income level of families by type of family. Frequencies are in thousands of families. [SOURCE: U.S. Bureau of the Census, *Current Population Reports.*]

Type of family

	Married couple F_1	Husband only F_2	Wife only F_3	Total
Under $10,000 I_1	5,852	392	4,308	10,552
$10,000–$19,999 I_2	12,326		2,992	15,925
$20,000–$49,999 I_3	26,240			29,109
$50,000 and over I_4	5,213	107	111	5,431
Total	49,631	1,984	9,402	61,017

(left axis label: Income level)

a. Fill in the three empty cells.
b. How many cells does this contingency table have?

c. Find the number of families that make between $20,000 and $49,999.
d. How many families have only the husband present?
e. How many families have only the wife present and make under $10,000?
f. How many families make at least $20,000?
g. Determine the number of married couples that make less than $20,000.

4.61 Refer to Exercise 4.55.
a. For a randomly selected institution of higher education, describe each of the following events in words: T_2, R_3, and (T_1 & R_4).
b. Compute the probability of each event in part (a). Interpret your results in terms of percentages.
c. Construct a joint probability distribution similar to Table 4.6 on page 212.
d. Verify that the sum of each row and column of joint probabilities equals the marginal probability in that row or column. (*Note:* Rounding may cause slight deviations.)

4.62 Refer to Exercise 4.56. Suppose a vehicle (car or truck) in use is selected at random.
a. Describe each of the following events in words: A_3, V_1, and (A_3 & V_1).
b. Determine the probability of each event in part (a). Interpret your results in terms of percentages.
c. Construct a joint probability distribution similar to Table 4.6 on page 212.
d. Verify that the sum of each row and column of joint probabilities equals, up to rounding error, the corresponding marginal probability.

4.63 The contingency table in Exercise 4.57 cross classifies the civilian labor force according to employment status and educational level. Suppose a person in the civilian labor force is selected at random.
a. Describe each of the following events in words: E_1, S_5, and (E_1 & S_5).
b. Compute the probability of each event in part (a).
c. Compute $P(E_1 \text{ or } S_5)$
   i. using the contingency table and the f/N rule.
   ii. using the general addition rule and your results from part (b).
d. Construct a joint probability distribution.

4.64 Refer to Exercise 4.58. Suppose a U.S. hospital is selected at random.
a. Describe each of the following events in words: H_2, B_2, (H_2 & B_2), and (H_4 & B_1).
b. Compute the probability of each event in part (a).

c. Compute $P(H_2 \text{ or } B_2)$
   i. using the contingency table and the f/N rule.
   ii. using the general addition rule and your results from part (b).
d. Construct a joint probability distribution.

4.65 The contingency table in Exercise 4.59 cross classifies U.S. farms by acreage and tenure of operator. Suppose a U.S. farm is selected at random.

a. Use the letters in the margins of the contingency table to represent each of the following events.
   i. The farm obtained has between 180 and 499 acres, inclusive.
   ii. The farm obtained is part-owner operated.
   iii. The farm obtained is full-owner operated and has at least 1000 acres.
b. Compute the probability of each event in part (a).
c. Construct a *joint percentage distribution,* a table similar to a joint probability distribution except with percentages replacing probabilities.

4.66 Refer to Exercise 4.60. Suppose a family is selected at random.

a. Use the letters in the margins of the contingency table to represent each of the following events.
   i. The family obtained has only the wife present.
   ii. The family obtained makes at least $50,000.
   iii. The family obtained has only the husband present and makes between $10,000 and $19,999.
b. Determine the probability of each event in part (a).
c. Construct a *joint percentage distribution,* a table similar to a joint probability distribution except with percentages replacing probabilities.

4.67 Explain why the joint events in a contingency table are mutually exclusive.

4.68 This exercise supplies a proof of the fact that the sum of the joint probabilities in a row or column of a joint probability distribution equals the marginal probability in that row or column. Consider the following joint probability distribution.

	C_1	$\cdots$	C_n	$P(R_i)$
R_1	$P(R_1 \& C_1)$	$\cdots$	$P(R_1 \& C_n)$	$P(R_1)$
.	.	$\cdots$	.	.
.	.	$\cdots$	.	.
.	.	$\cdots$	.	.
R_m	$P(R_m \& C_1)$	$\cdots$	$P(R_m \& C_n)$	$P(R_m)$
$P(C_j)$	$P(C_1)$	$\cdots$	$P(C_n)$	1

a. Explain why we can write
$$R_1 = \big((R_1 \& C_1) \text{ or } \cdots \text{ or } (R_1 \& C_n)\big).$$

b. Why are the events $(R_1 \& C_1), \ldots, (R_1 \& C_n)$ mutually exclusive?

c. Why can we conclude from parts (a) and (b) that
$$P(R_1) = P(R_1 \& C_1) + \cdots + P(R_1 \& C_n)?$$

This equation shows that the first row of joint probabilities sums to the marginal probability at the end of that row. A similar argument applies to any other row or column.

4.5 CONDITIONAL PROBABILITY

In this section we will introduce the concept of conditional probability. The **conditional probability** of an event is the probability that the event occurs under the assumption that another event has occurred.

DEFINITION 4.7 **CONDITIONAL PROBABILITY**

Suppose A and B are events. Then the probability that event B occurs given that event A has occurred is called a ***conditional probability.*** It is denoted by the symbol $P(B \mid A)$, which is read "the probability of B given A."

Example 4.18   Illustrates Definition 4.7

When a balanced die is rolled once, six equally likely outcomes are possible, as displayed in Fig. 4.22.

FIGURE 4.22

Sample space for rolling a die once

Let

$$F = \text{event a 5 is rolled,}$$

$$O = \text{event the die comes up odd.}$$

Determine the following probabilities:

a. $P(F)$, the probability that a 5 is rolled.

b. $P(F \mid O)$, the conditional probability that a 5 is rolled given that the die comes up odd.

c. $P(O \mid (\text{not } F))$, the conditional probability that the die comes up odd given that a 5 is not rolled.

SOLUTION   a. To obtain $P(F)$, the probability that a 5 is rolled, we proceed as usual. From Fig. 4.22 we see that six outcomes are possible. Also, event F can occur in only one way: if the die comes up 5. Thus the probability that a 5 is rolled equals

$$P(F) = \frac{f}{N} = \frac{1}{6} = 0.167.$$

b. Given that the die comes up odd, that is, that event O has occurred, there are no longer six possible outcomes. There are only three, as shown in Fig. 4.23.

FIGURE 4.23

Event O

Therefore the conditional probability that a 5 is rolled given that the die comes up odd equals

$$P(F \mid O) = \frac{f}{N} = \frac{1}{3} = 0.333.$$

Comparing this last probability with the one that we obtained in part (a), we see that $P(F \mid O) \neq P(F)$; that is, the conditional probability that a 5 is rolled given that the die comes up odd is not the same as the (unconditional) probability that a 5 is rolled. In other words, knowing that the die comes up odd affects the probability that a 5 is rolled.

c. Given that a 5 is not rolled, that is, that event (not F) has occurred, the possible outcomes are the five shown in Fig. 4.24.

FIGURE 4.24
Event (not F)

Under these circumstances, event O (odd) can occur in two ways: if a 1 or a 3 is rolled. Thus the conditional probability that the die comes up odd given that a 5 is not rolled equals

$$P(O \mid (\text{not } F)) = \frac{f}{N} = \frac{2}{5} = 0.4.$$

■

Conditional probability plays a central role in the analysis of cross-classified data. In Section 4.4 we discussed contingency tables as a method for tabulating such data. Now we will learn how to obtain conditional probabilities for cross-classified data directly from a contingency table.

Example 4.19   Illustrates Definition 4.7

In Example 4.16 we presented a contingency table resulting from cross classifying the faculty at Arizona State University according to age and rank. We repeat that table here as Table 4.7.

TABLE 4.7

Contingency table for cross classification of ASU faculty by age and rank

Age	Rank				
	Full professor R_1	Associate professor R_2	Assistant professor R_3	Instructor R_4	Total
Under 30 A_1	2	3	57	6	68
30–39 A_2	52	170	163	17	402
40–49 A_3	156	125	61	6	348
50–59 A_4	145	68	36	4	253
60 & over A_5	75	15	3	0	93
Total	430	381	320	33	1164

Suppose a faculty member is selected at random.

a. Determine the (unconditional) probability that the faculty member selected is in his or her 50s.

b. Determine the (conditional) probability that the faculty member selected is in his or her 50s given that an assistant professor is selected.

c. Interpret the probabilities obtained in parts (a) and (b) in terms of percentages.

SOLUTION
a. Here we are to determine the probability that the faculty member selected is in his or her 50s (event A_4). From Table 4.7 we see that $N = 1164$, since the total number of faculty members is 1164. Also, $f = 253$, since 253 of the faculty members are in their 50s. Therefore

$$P(A_4) = \frac{f}{N} = \frac{253}{1164} = 0.217.$$

b. For this part we are to find the probability that the faculty member selected is in his or her 50s (event A_4) given that an assistant professor is selected (event R_3). In other words, we want to determine $P(A_4 \mid R_3)$. To obtain that probability, we restrict our attention to the assistant-professor column of Table 4.7. We have $N = 320$, since the total number of assistant professors is 320. Also, $f = 36$, since 36 of the assistant professors are in their 50s. Thus

$$P(A_4 \mid R_3) = \frac{f}{N} = \frac{36}{320} = 0.113.$$

c. $P(A_4) = 0.217$ indicates that 21.7% of the faculty are in their 50s; $P(A_4 \mid R_3) = 0.113$ indicates that 11.3% of the assistant professors are in their 50s. ∎

The Conditional-Probability Rule

In the preceding examples, we computed conditional probabilities *directly;* that is, we first obtained the new sample space determined by the *given event* (i.e., the event assumed to have occurred) and then, using the new sample space, we calculated probabilities in the usual manner. For instance, in Example 4.18(b) on page 216, we computed the conditional probability that a 5 is rolled given that the die comes up odd. To do that we first obtained the new sample space (in this case, 1, 3, 5) and then went on from there.

Sometimes, however, we cannot determine conditional probabilities directly but must instead compute them in terms of unconditional probabilities. To see how this can be done, we return to the situation of Example 4.19.

Example 4.20   Introduces the Conditional-Probability Rule

In Example 4.19(b) we determined the conditional probability that a randomly selected faculty member is in his or her 50s (event A_4) given that an assistant professor is selected (event R_3). To accomplish that we restricted our attention to the R_3 column of Table 4.7 on page 217 and obtained

$$P(A_4 \mid R_3) = \frac{36}{320} = 0.113.$$

This is a *direct* computation of the conditional probability $P(A_4 \mid R_3)$.

To determine $P(A_4 \mid R_3)$ using unconditional probabilities, we proceed as follows: First we note that the number 36 in the numerator of the above fraction is the number of assistant professors in their 50s, that is, the number of ways event $(R_3 \ \& \ A_4)$ can occur. Next we observe that the number 320 in the denominator of the above fraction is the total number of assistant professors, that is, the number of ways event R_3 can occur. Thus the numbers 36 and 320 are those used to compute the unconditional probabilities of events $(R_3 \ \& \ A_4)$ and R_3, respectively:

$$P(R_3 \ \& \ A_4) = \frac{36}{1164} = 0.031, \qquad P(R_3) = \frac{320}{1164} = 0.275.$$

So by arithmetic and the previous three probabilities, we see that

$$P(A_4 \mid R_3) = \frac{36}{320} = \frac{\frac{36}{1164}}{\frac{320}{1164}} = \frac{P(R_3 \ \& \ A_4)}{P(R_3)}.$$

Consequently, the conditional probability, $P(A_4 \mid R_3)$, can be obtained from the unconditional probabilities, $P(R_3 \ \& \ A_4)$ and $P(R_3)$, by using the formula

$$P(A_4 \mid R_3) = \frac{P(R_3 \ \& \ A_4)}{P(R_3)}.$$

That formula holds in general and is called the **conditional-probability rule.** ∎

FORMULA 4.4

THE CONDITIONAL-PROBABILITY RULE

If A and B are any two events, then

$$P(B \mid A) = \frac{P(A \ \& \ B)}{P(A)}.$$

In words, the conditional probability that event B occurs given that event A has occurred is equal to the joint probability of events A and B divided by the probability of event A.

Note: When applying the conditional-probability rule, remember that we divide by the probability of the *given event,* that is, the event on the right of the $\mid$.

For the faculty-member example, conditional probabilities can be obtained either directly or by employing the conditional-probability rule. However, as Examples 4.21 and 4.22 illustrate, the conditional-probability rule is sometimes the only way conditional probabilities can be determined.

Example 4.21 Illustrates Formula 4.4

The U.S. Bureau of the Census compiles data on the marital status of U.S. adults and publishes the results in *Current Population Reports.* Table 4.8 provides a joint probability distribution for the marital status of U.S. adults by sex.

TABLE 4.8
Joint probability distribution
of marital status vs. sex

Marital status

		Single M_1	Married M_2	Widowed M_3	Divorced M_4	$P(S_i)$
Sex	Male S_1	0.116	0.319	0.012	0.028	0.475
	Female S_2	0.093	0.325	0.066	0.041	0.525
	$P(M_j)$	0.209	0.644	0.078	0.069	1.000

Suppose a U.S. adult is selected at random.

a. Determine the probability that the adult selected is divorced, given that the adult selected is a male.

b. Determine the probability that the adult selected is a male, given that the adult selected is divorced.

SOLUTION Unlike our previous illustrations with contingency tables, we do not have the frequency data here, only the probability (relative-frequency) data. Because of that we cannot compute conditional probabilities directly; we must use the conditional-probability rule.

a. For this part we want to obtain the conditional probability that the adult selected is divorced, given that the adult selected is a male: $P(M_4 \mid S_1)$. Using the conditional-probability rule and Table 4.8, we get

$$P(M_4 \mid S_1) = \frac{P(S_1 \& M_4)}{P(S_1)} = \frac{0.028}{0.475} = 0.059.$$

In terms of percentages, this means that 5.9% of adult males are divorced.

b. Here we want to determine the conditional probability that the adult selected is a male, given that the adult selected is divorced: $P(S_1 \mid M_4)$. Referring to Table 4.8 and applying the conditional-probability rule, we see that

$$P(S_1 \mid M_4) = \frac{P(M_4 \& S_1)}{P(M_4)} = \frac{0.028}{0.069} = 0.406.$$

In other words, 40.6% of divorced adults are males. ∎

Example 4.22   Illustrates Formula 4.4

The National Center for Education Statistics publishes information about educational institutions in *Digest of Education Statistics*. According to that publication, 13.4% of all
students at all levels attend private institutions and 4.5% of all students attend private
colleges. What percentage of the students attending private schools are in college?

SOLUTION   To solve this problem, we first translate it into the language of probability. So suppose a
student is selected at random. Let

$$C = \text{event the student selected is in college,}$$

$$D = \text{event the student selected attends a private school.}$$

We need to determine $P(C \mid D)$.

From the percentage data supplied in the statement of the problem, we know that
$P(D) = 0.134$ and $P(D \& C) = 0.045$. Hence, by the conditional-probability rule,

$$P(C \mid D) = \frac{P(D \& C)}{P(D)} = \frac{0.045}{0.134} = 0.336.$$

Thus 33.6% of the students attending private schools are in college.   ■

EXERCISES 4.5

*For Exercises 4.69–4.74, compute conditional probabilities
directly; that is, do not use the conditional-probability rule.*

4.69   Suppose one card is selected at random from an
ordinary deck of 52 playing cards. Let

$$A = \text{event a face card is selected,}$$
$$B = \text{event a king is selected,}$$
$$C = \text{event a heart is selected.}$$

Determine the following probabilities and express your results
in words.

a. $P(B)$ b. $P(B \mid A)$ c. $P(B \mid C)$
d. $P\big(B \mid (\text{not } A)\big)$ e. $P(A)$ f. $P(A \mid B)$
g. $P(A \mid C)$ h. $P\big(A \mid (\text{not } B)\big)$

4.70   A balanced dime is tossed twice. The four possible equally likely outcomes are HH, HT, TH, TT. Let

$$A = \text{event the first toss is heads,}$$
$$B = \text{event the second toss is heads,}$$
$$C = \text{event at least one toss is heads.}$$

Determine the following probabilities and express your results
in words.

a. $P(B)$ b. $P(B \mid A)$ c. $P(B \mid C)$
d. $P(C)$ e. $P(C \mid A)$ f. $P\big(C \mid (\text{not } B)\big)$

4.71   The absentee records over the past year for the
employees of Cudahey Masonry are presented below in a
frequency distribution.

Days missed	0	1	2	3	4	5	6	7
No. of employees	4	2	14	10	16	18	10	6

a. Find the probability that a randomly selected employee
   missed exactly 3 days of work.
b. Find the conditional probability that a randomly selected
   employee missed exactly 3 days, given that the employee
   missed at least 1 day.
c. Find the conditional probability that a randomly selected
   employee missed at most 3 days, given that the employee
   missed at least 1 day.
d. Interpret the probabilities that you obtained in parts (a)–(c)
   in terms of percentages.

4.72   As reported by the U.S. Bureau of the Census in *Current Population Reports,* a frequency distribution for the population of the states in the United States and Washington, D.C., is as shown in the following table.

Population size (millions)	Frequency
Under 1	8
1–under 2	10
2–under 3	5
3–under 5	12
5–under 10	9
10 or over	7

Suppose a state (include Washington, D.C., as a state) is selected at random. Obtain the probability that the population of the state selected
a. is between 2 million and 3 million.
b. is between 2 million and 3 million, given that it is at least 1 million.
c. is less than 5 million, given that it is at least 1 million.
d. Interpret the probabilities that you obtained in parts (a)–(c) in terms of percentages.

4.73   The following contingency table cross classifies institutions of higher education in the United States by region and type. [SOURCE: U.S. National Center for Education Statistics, *Digest of Education Statistics.*]

Type

Region	Public T_1	Private T_2	Total
Northeast R_1	266	555	821
Midwest R_2	359	504	863
South R_3	533	502	1035
West R_4	313	242	555
Total	1471	1803	3274

Suppose an institution of higher education is selected at random. Determine the probability that the institution obtained
a. is in the Northeast.
b. is in the Northeast, given that it is a private school.

c. is a private school, given that it is in the Northeast.
d. Interpret the probabilities that you obtained in parts (a)–(c) in terms of percentages.

4.74   As reported by the Motor Vehicle Manufacturers Association of the United States in *Motor Vehicle Facts and Figures,* the number of cars and trucks in use by age is as follows. Frequencies are in millions.

Type

Age (yrs)	Car V_1	Truck V_2	Total
Under 3 A_1	21.5	3.2	24.7
3–5 A_2	29.9	2.5	32.4
6–8 A_3	22.2	2.0	24.2
9–11 A_4	17.9	1.8	19.7
12 & over A_5	15.4	3.6	19.0
Total	106.9	13.1	120.0

Suppose a vehicle is selected at random. What is the probability that the vehicle obtained
a. is under 3 years old?
b. is under 3 years old, given that it is a car?
c. is a car?
d. is a car, given that it is under 3 years old?
e. Interpret the probabilities that you obtained in parts (a)–(d) in terms of percentages.

4.75   According to *Census of Agriculture,* published by the U.S. Bureau of the Census, a contingency table for the number of farms, by acreage and tenure of operator, is as shown at the top of the first column on the following page. Frequencies are in thousands of farms. Suppose a U.S. farm is selected at random.
a. Find $P(T_3)$.
b. Find $P(T_3 \& A_3)$.
c. Find $P(A_3 \mid T_3)$ directly from the table.
d. Find $P(A_3 \mid T_3)$ using the conditional-probability rule and your results from parts (a) and (b).
e. State your results in parts (a)–(c) in words.

Tenure of operator

	Full owner T_1	Part owner T_2	Tenant T_3	Total
Under 50 A_1	532	74	84	690
50–179 A_2	563	157	94	814
180–499 A_3	262	247	87	596
500–999 A_4	57	128	30	215
1000+ A_5	36	107	18	161
Total	1450	713	313	2476

Acreage

·   **4.76**   The table below cross classifies U.S. hospitals according to type and number of beds. [SOURCE: American Hospital Association, *Hospital Statistics.*]

Number of beds

	6–24 B_1	25–74 B_2	75+ B_3	Total
General H_1	299	1894	3945	6138
Psychiatric H_2	17	121	378	516
Chronic H_3	0	7	40	47
Tuberculosis H_4	0	1	10	11
Other H_5	22	131	162	315
Total	338	2154	4535	7027

Type

Suppose a U.S. hospital is selected at random.
a. Find $P(H_1)$.
b. Find $P(H_1 \& B_3)$.
c. Compute $P(B_3 \mid H_1)$ directly from the table.
d. Compute $P(B_3 \mid H_1)$ using the conditional-probability rule and your results from parts (a) and (b).
e. State your results in parts (a)–(c) in words.

·   **4.77**   The table below provides a joint probability distribution for the members of the 102nd Congress by legislative group and political party. [SOURCE: U.S. Congress, Joint Committee on Printing, *Congressional Directory.*]

Group

	Rep C_1	Senator C_2	$P(P_i)$
Democratic P_1	0.500	0.105	0.605
Republican P_2	0.313	0.082	0.395
$P(C_j)$	0.813	0.187	1.000

Party

If a member of the 102nd Congress is selected at random, what is the probability that the member obtained
a. is a senator?
b. is a Republican senator?
c. is a Republican, given that he or she is a senator?
d. is a senator, given that he or she is a Republican?
e. Interpret each of your results in parts (a)–(d) in terms of percentages.

·   **4.78**   The National Center for Education Statistics publishes information on U.S. engineers and scientists in *Digest of Education Statistics*. The table below presents a joint probability distribution for engineers and scientists by highest degree obtained.

Type

	Engineer T_1	Scientist T_2	$P(D_i)$
Bachelors D_1	0.343	0.289	0.632
Masters D_2	0.098	0.146	0.244
Doctorate D_3	0.017	0.091	0.108
Other D_4	0.013	0.003	0.016
$P(T_j)$	0.471	0.529	1.000

Highest degree

A person is selected at random from among the engineers and scientists. Determine the probability that the person obtained
a. is an engineer.

b. has a doctorate.

c. is an engineer with a doctorate.

d. is an engineer, given the person has a doctorate.

e. has a doctorate, given the person is an engineer.

f. Interpret each of your results in parts (a)–(e) in terms of percentages.

·   **4.79**   According to the Census Bureau's *Current Population Reports,* 10.6% of U.S. families are African-American, and 3.5% are African-American and have incomes below the poverty level. What percentage of African-American families have incomes below the poverty level?

·   **4.80**   As reported by the Federal Bureau of Investigation in *Crime in the United States,* 4.9% of property crimes are committed in rural areas and 1.9% of property crimes are burglaries committed in rural areas. What percentage of property crimes committed in rural areas are burglaries?

· ·   **4.81**   Refer to Exercise 4.74.

a. Construct a joint probability distribution.

b. Determine the probability distribution of age for cars in use; that is, construct a table showing the conditional probabilities that a car in use is under 3 years old, 3–5 years old, 6–8 years old, and so on.

c. Determine the probability distribution of type for vehicles 3–5 years old.

d. The probability distributions in parts (b) and (c) are examples of **conditional-probability distributions.** Determine two other conditional-probability distributions for the type-versus-age data of motor vehicles in use.

· ·   **4.82**   Do you think the conditional probability of an event, *B,* given another event, *A,* must always be different than the unconditional probability of event *B*? That is, do you think it is always true that $P(B \mid A) \neq P(B)$? [*Hint:* Try to construct an example in which $P(B \mid A) = P(B)$.]

4.6   THE MULTIPLICATION RULE; INDEPENDENCE

The conditional-probability rule, Formula 4.4 on page 219, is used for computing conditional probabilities in terms of unconditional probabilities:

$$P(B \mid A) = \frac{P(A \ \& \ B)}{P(A)}.$$

Multiplying both sides of the equation by $P(A)$, we obtain a formula for computing joint probabilities in terms of marginal and conditional probabilities. That formula is called the **general multiplication rule** and is presented in Formula 4.5.

FORMULA 4.5   **THE GENERAL MULTIPLICATION RULE**

If *A* and *B* are any two events, then

$$P(A \ \& \ B) = P(A) \cdot P(B \mid A).$$

In words, the probability that both event *A* and event *B* occur equals the probability that event *A* occurs times the conditional probability that event *B* occurs given that event *A* has occurred.

The conditional-probability rule and the general multiplication rule are simply variations of each other. When the joint and marginal probabilities are known or easily determined directly, we can use the conditional-probability rule to obtain conditional probabilities. On the other hand, when the marginal and conditional probabilities are known or easily determined directly, we can use the general multiplication rule to obtain joint probabilities.

Example 4.23   Illustrates Formula 4.5

The U.S. Bureau of Labor Statistics collects data on U.S. workers and publishes the information in *Employment and Earnings.* According to that publication, 30.9% of all employed African-Americans are white-collar workers and 9.2% of all employed workers are African-Americans. Determine the probability that a randomly selected employed worker is an African-American white-collar worker.

SOLUTION   Let

$$A = \text{event the worker selected is African-American,}$$
$$W = \text{event the worker selected is a white-collar worker.}$$

Then the event that the worker selected is an African-American white-collar worker can be expressed as $(A \,\&\, W)$. Our problem is to determine $P(A \,\&\, W)$.

Since 9.2% of all employed workers are African-Americans, $P(A) = 0.092$; and since 30.9% of all employed African-Americans are white-collar workers, $P(W \mid A) = 0.309$. Applying the general multiplication rule, we get

$$P(A \,\&\, W) = P(A) \cdot P(W \mid A) = 0.092 \cdot 0.309 = 0.028.$$

Thus the probability is 0.028 that a randomly selected employed worker is an African-American white-collar worker. Expressed in percentages, 2.8% of all employed workers are African-American white-collar workers.   ∎

Another application of the general multiplication rule occurs when two or more members are selected from a population. Example 4.24 provides an illustration.

Example 4.24   Illustrates Formula 4.5

In Professor Weiss's introductory statistics class, the numbers of males and females are as shown in the frequency distribution in Table 4.9.

TABLE 4.9
Frequency distribution of males and females in Professor Weiss's introductory statistics class

Sex	Frequency
Male	17
Female	23
	40

Two students are selected at random from the class. The first student chosen is not returned to the class for possible reselection; that is, the sampling is without replacement. Find the probability that the first student selected is female and the second male.

SOLUTION   Let's use the following notation:

$$F1 = \text{event the first student selected is female,}$$
$$M2 = \text{event the second student selected is male.}$$

The problem is to determine the probability that the first student selected is female and the second student selected is male, $P(F1 \text{ \& } M2)$. By the general multiplication rule, Formula 4.5, we can write

$$P(F1 \text{ \& } M2) = P(F1) \cdot P(M2 \mid F1).$$

We now obtain the two probabilities on the right-hand side of the preceding equation. As we see from Table 4.9, 23 out of the 40 students are female. Therefore, by the f/N rule,

$$P(F1) = \frac{f}{N} = \frac{23}{40}.$$

Also, given that the first student selected is female, that is, that event $F1$ has occurred, there are 39 students remaining in the class, of which 17 are male and 22 are female. Thus the conditional probability that the second student selected is male given that the first student selected is female equals

$$P(M2 \mid F1) = \frac{f}{N} = \frac{17}{39}.$$

Applying the general multiplication rule, we can now conclude that

$$P(F1 \text{ \& } M2) = P(F1) \cdot P(M2 \mid F1) = \frac{23}{40} \cdot \frac{17}{39} = 0.251.$$

The probability is 0.251 that the first student selected is female and the second is male.

It is often helpful to draw a **tree diagram** when applying the general multiplication rule. A tree diagram for the present example is displayed in Fig. 4.25.

FIGURE 4.25
Tree diagram for
student-selection problem

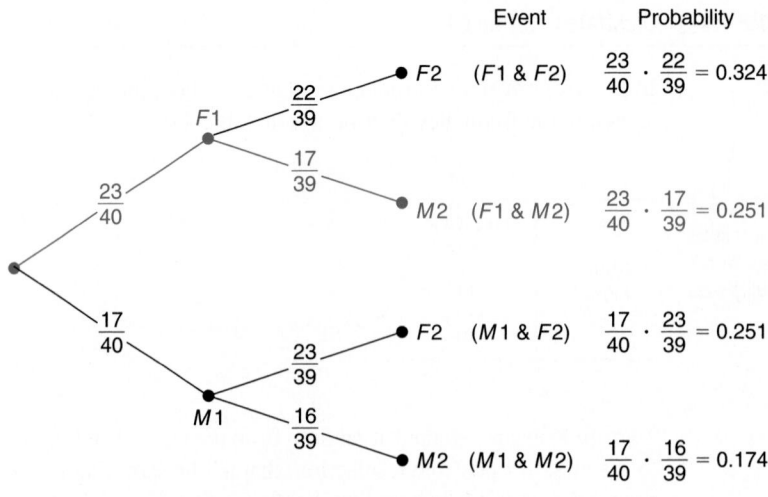

	Event	Probability
F2	(F1 & F2)	$\frac{23}{40} \cdot \frac{22}{39} = 0.324$
M2	(F1 & M2)	$\frac{23}{40} \cdot \frac{17}{39} = 0.251$
F2	(M1 & F2)	$\frac{17}{40} \cdot \frac{23}{39} = 0.251$
M2	(M1 & M2)	$\frac{17}{40} \cdot \frac{16}{39} = 0.174$

Each branch of the tree corresponds to one possibility for selecting two students at random from the class. For instance, the second branch of the tree, shown in color, corresponds to event ($F1$ & $M2$)—the event that the first student selected is female (event $F1$) and the second is male (event $M2$).

Starting from the left on that branch, the number $\frac{23}{40}$ is the probability that the first student selected is female, $P(F1)$, and the number $\frac{17}{39}$ is the conditional probability that the second student selected is male given that the first student selected is female, $P(M2 \mid F1)$. The product of those two probabilities is, by the general multiplication rule, the probability that the first student selected is female and the second is male, $P(F1$ & $M2)$. The second entry in the Probability column of Fig. 4.25 shows that this latter probability equals 0.251, as we discovered earlier in this example.   ∎

The general multiplication rule can be extended to more than two events. This will be explored in Exercise 4.104.

Independence

Two events are called *statistically independent* if the occurrence (or nonoccurrence) of one of the events does not affect the probability of the other event. A more formal definition is presented in Definition 4.8.

DEFINITION 4.8   **STATISTICAL INDEPENDENCE**

> Event B is said to be ***statistically independent*** of event A if the occurrence of event A does not affect the probability that event B occurs. In symbols,
>
> $$P(B \mid A) = P(B).$$
>
> This means that knowing whether event A has occurred provides no probabilistic information about the occurrence of event B.

For brevity we will usually omit the adjective *statistically* when discussing independence of events. Thus, for instance, we will use the term *independent* instead of *statistically independent*.

Example 4.25   Illustrates Definition 4.8

For the experiment of randomly selecting one card from a deck of 52 playing cards, let

$$A = \text{event a face card is selected,}$$
$$B = \text{event a king is selected,}$$
$$C = \text{event a heart is selected.}$$

a. Determine whether event B is independent of event A.

b. Determine whether event B is independent of event C.

SOLUTION   First we note that the unconditional probability that event B occurs equals

$$P(B) = \frac{f}{N} = \frac{4}{52} = \frac{1}{13} = 0.077.$$

a. To determine whether event B is independent of event A, we must compute $P(B \mid A)$ and compare it to $P(B)$. If those two probabilities are equal, then event B is independent of event A; otherwise, event B is not independent of event A. Now, given that event A has occurred, 12 outcomes are possible (four jacks, four queens, and four kings), and event B can occur in four ways out of those 12 possibilities. So

$$P(B \mid A) = \frac{f}{N} = \frac{4}{12} = 0.333.$$

This is not equal to $P(B)$. Thus the occurrence of event A affects the probability that event B occurs. In other words, event B is *not* independent of event A. This lack of independence stems from the fact that the percentage of kings among the face cards (33.3%) is not the same as the percentage of kings among all the cards (7.7%).

b. Here we need to compute $P(B \mid C)$ and compare it to $P(B)$. Given that event C has occurred, 13 outcomes are possible (the 13 hearts), and event B can occur in one way out of those 13 possibilities. Therefore

$$P(B \mid C) = \frac{f}{N} = \frac{1}{13} = 0.077.$$

This is equal to $P(B)$. Thus the occurrence of event C does not affect the probability that event B occurs. In other words, event B *is* independent of event C. This independence stems from the fact that the percentage of kings among the hearts is the same as the percentage of kings among all the cards; namely, 7.7%.   ■

Example 4.26   Illustrates Definition 4.8

The American Medical Association compiles information on U.S. physicians in *Physician Characteristics and Distribution in the U.S.* Table 4.10 provides a joint probability distribution for U.S. surgeons cross classified by specialty and base of practice.

Suppose a surgeon is selected at random. Is the event that the surgeon obtained is office-based independent of the event that the surgeon obtained is an orthopedist? That is, is event B_1 independent of event S_3?

SOLUTION   To solve this problem, we need to compare $P(B_1 \mid S_3)$ and $P(B_1)$. If those two probabilities are equal, then event B_1 is independent of event S_3; otherwise, event B_1 is not independent

TABLE 4.10

Surgeons by specialty
and base of practice

TABLE 4.10

Surgeons by specialty
and base of practice

		Base of practice			
		Office B_1	Hospital B_2	Other B_3	$P(S_i)$
Specialty	General surgery S_1	0.233	0.118	0.016	0.367
	Obstetrics/Gynecology S_2	0.233	0.065	0.011	0.309
	Orthopedics S_3	0.129	0.041	0.004	0.174
	Ophthalmology S_4	0.119	0.026	0.005	0.150
	$P(B_j)$	0.714	0.250	0.036	1.000

of event S_3. From Table 4.10 we find that $P(B_1) = 0.714$ and

$$P(B_1 \mid S_3) = \frac{P(S_3 \ \& \ B_1)}{P(S_3)} = \frac{0.129}{0.174} = 0.741.$$

Thus $P(B_1 \mid S_3) \neq P(B_1)$; so event B_1 is not independent of event S_3. That is, the event that the surgeon obtained is office-based is not independent of the event that the surgeon obtained is an orthopedist. This lack of independence results from the fact that the percentage of orthopedic surgeons who are office-based (74.1%) is not the same as the percentage of all surgeons who are office-based (71.4%). ∎

It can be shown that if event B is independent of event A, then it is also true that event A is independent of event B. (You are asked to verify this statement in Exercise 4.105.) So, in such cases, we often say that event A and event B are **independent,** or that A and B are **independent events.** If two events are not independent, then they are said to be **dependent events.** Hence in Example 4.26, S_3 and B_1 are dependent events.

The Special Multiplication Rule

Recall that the general multiplication rule, Formula 4.5 on page 224, states that for any two events, A and B,

$$P(A \ \& \ B) = P(A) \cdot P(B \mid A).$$

If A and B are independent events, then $P(B \mid A) = P(B)$. Thus for the special case of independent events, we can replace the term $P(B \mid A)$ in the general multiplication rule by the term $P(B)$. This yields the following rule.

| FORMULA 4.6 | **THE SPECIAL MULTIPLICATION RULE (FOR TWO INDEPENDENT EVENTS)** |

> If A and B are independent events, then
>
> $$P(A \ \& \ B) = P(A) \cdot P(B),$$
>
> and conversely, if $P(A \ \& \ B) = P(A) \cdot P(B)$, then A and B are independent events. In words, two events are independent if and only if their joint probability equals the product of their marginal probabilities.

In Examples 4.25 and 4.26, we used the definition of independence, Definition 4.8 on page 227, to decide whether two specified events are independent. If the two events are, say, A and B, this means determining whether $P(B \mid A) = P(B)$. Alternatively, we can decide whether event A and event B are independent by employing the special multiplication rule, that is, by determining whether $P(A \ \& \ B) = P(A) \cdot P(B)$.

The special multiplication rule is also used to compute joint probabilities when we know, or can reasonably assume, that two events are independent. Example 4.27 illustrates this point.

Example 4.27   Illustrates Formula 4.6

A roulette wheel contains 38 numbers, of which 18 are red, 18 are black, and 2 are green. When the roulette ball is spun, it is equally likely to land on any of the 38 numbers. In two plays at a roulette wheel, what is the probability that the ball will land on green the first time and on black the second time?

SOLUTION   First of all we note that it is reasonable to assume that outcomes on successive plays at the wheel are independent. (Why is this so?) Let

$$G1 = \text{event the ball lands on green the first time,}$$
$$B2 = \text{event the ball lands on black the second time.}$$

The problem is to determine $P(G1 \ \& \ B2)$. Because outcomes on successive plays at the wheel are independent, event $G1$ and event $B2$ are independent. So, by the special multiplication rule,

$$P(G1 \ \& \ B2) = P(G1) \cdot P(B2).$$

But $P(G1) = \frac{2}{38}$ and $P(B2) = \frac{18}{38}$. Consequently,

$$P(G1 \ \& \ B2) = P(G1) \cdot P(B2) = \frac{2}{38} \cdot \frac{18}{38} = 0.025.$$

In two plays at a roulette wheel, there is a 2.5% chance that the ball will land on green the first time and on black the second time.   ∎

The definition of independence for three or more events is more complicated than that for two events. Nevertheless, the special multiplication rule still holds. Specifically, we have the following formula.

FORMULA 4.7

THE SPECIAL MULTIPLICATION RULE

If events A, B, C, ... are independent, then

$$P(A \& B \& C \& \cdots) = P(A) \cdot P(B) \cdot P(C) \cdots .$$

Example 4.28 Illustrates Formula 4.7

In five plays at a roulette wheel, what is the probability that the ball will land on red all five times?

SOLUTION Let

$$R1 = \text{event the ball lands on red the first time,}$$
$$R2 = \text{event the ball lands on red the second time,}$$

and so forth. We need to compute $P(R1 \& R2 \& R3 \& R4 \& R5)$. Since outcomes on successive plays at the wheel are independent, the special multiplication rule, Formula 4.7, applies to give

$$P(R1 \& R2 \& R3 \& R4 \& R5) = P(R1) \cdot P(R2) \cdot P(R3) \cdot P(R4) \cdot P(R5)$$
$$= \frac{18}{38} \cdot \frac{18}{38} \cdot \frac{18}{38} \cdot \frac{18}{38} \cdot \frac{18}{38} = 0.024.$$

In five plays at a roulette wheel, there is a 2.4% chance that the ball will land on red all five times. ∎

Mutually Exclusive Versus Independent Events

It is important to realize that the terms *mutually exclusive* and *independent* refer to different concepts. Mutually exclusive events are those that cannot occur simultaneously. Independent events are those for which the occurrence of some does not affect the probabilities of occurrence of the others.

If two events are mutually exclusive, then the occurrence of one precludes the occurrence of the other; so the two events are certainly not independent. More generally, Exercise 4.109 shows that, except in trivial cases, it is not possible for two events to be both mutually exclusive and independent.

EXERCISES 4.6

4.83 According to the Census Bureau's *Current Population Reports,* 29.2% of all farm families make at least $25,000 per year, and 2.6% of all families are farm families. Use the general multiplication rule to find the probability that a randomly selected family is a farm family making at least $25,000 per year. Interpret your result in terms of percentages.

4.84 The National Center for Education Statistics states in *Digest of Education Statistics* that 43.9% of all public elementary schools have between 250 and 499 students. Moreover, 51.4% of all public schools are elementary schools. Use the general multiplication rule to determine the probability that a randomly selected public school is an elementary school with between 250 and 499 students. Interpret your result in terms of percentages.

4.85 Cards numbered 1, 2, 3, ..., 10 are placed in a box. The box is shaken and a blindfolded person selects two successive cards without replacement.
a. What is the probability that the first card selected is numbered 6?
b. Given that the first card is numbered 6, what is the probability that the second is numbered 9?
c. What is the probability of selecting first a 6 and then a 9?
d. What is the probability that both cards selected are numbered over 5?

4.86 A person has agreed to participate in an ESP experiment. He is asked to randomly pick two numbers between 1 and 6. The second number must be different from the first. Let

H = event the first number picked is a 3,
K = event the second number picked exceeds 4.

Determine
a. $P(H)$.   b. $P(K \mid H)$.   c. $P(H \& K)$.
Find the probability that both numbers picked are
d. less than 3.   e. greater than 3.

4.87 The table that follows provides a frequency distribution for the political-party affiliations of U.S. governors. [SOURCE: National Governors' Association, Washington, D.C., *Directory of Governors of the American States, Commonwealths & Territories.*]

Party	Frequency
Democratic	27
Republican	21
Independent	2

Suppose two governors are selected at random without replacement. Obtain the probability that
a. the first selected is a Republican and the second selected is a Democrat.
b. both governors selected are Republicans.
c. Draw a tree diagram for this problem similar to Fig. 4.25 on page 226.
d. What is the probability that the two governors selected are both Democrats, are both Republicans, or are both Independents?

4.88 A frequency distribution for the class of students in a midwestern high school is as follows.

Class	Frequency
Freshman	89
Sophomore	127
Junior	118
Senior	93

Suppose two students are randomly selected without replacement. Determine the probability that
a. the first student selected is a junior and the second selected is a senior.
b. both students selected are sophomores.
c. Draw a tree diagram for this problem similar to Fig. 4.25 on page 226.
d. What is the probability that one of the students selected is a freshman and the other student selected is a sophomore?

4.89 The U.S. National Center for Health Statistics compiles data on injuries and publishes the information in *Vital and Health Statistics.* A contingency table for injuries in the United States by circumstance and sex is as follows. Frequencies are in millions.

		Circumstance			
		Work C_1	Home C_2	Other C_3	Total
Sex	Male S_1	8.0	9.8	17.8	35.6
	Female S_2	1.3	11.6	12.9	25.8
	Total	9.3	21.4	30.7	61.4

a. Find $P(C_1)$.   b. Find $P(C_1 \mid S_2)$.
c. Are events C_1 and S_2 independent? Why?

d. Is the event that an injured person is male independent of the event that an injured person was hurt at home? Why?

4.90 A study conducted by the Census Bureau revealed the following data on the methods Americans use to get to work, by residence. The frequencies are in millions of workers.

Method		Residence: Urban R_1	Rural R_2	Total
	Automobile M_1	45.0	15.0	60.0
	Public trans. M_2	6.5	0.5	7.0
	Total	51.5	15.5	67.0

a. Find $P(M_1)$. b. Find $P(M_1 \mid R_2)$.
c. Are M_1 and R_2 independent events? Why?
d. Is the event that a worker resides in an urban area independent of the event that the worker uses an automobile to get to work? Justify your answer.

4.91 When a balanced dime is tossed three times, eight equally likely outcomes are possible:

HHH	HTH	THH	TTH
HHT	HTT	THT	TTT

Let

A = event the first toss is heads,
B = event the third toss is tails,
C = event the total number of heads is one.

a. Compute $P(A)$, $P(B)$, and $P(C)$.
b. Compute $P(B \mid A)$.
c. Are A and B independent events? Why?
d. Compute $P(C \mid A)$.
e. Are A and C independent events? Why?

4.92 When two balanced dice are rolled, 36 equally likely outcomes are possible, as seen in Fig. 4.2 on page 186. Let

A = event the blue die comes up even,
B = event the black die comes up odd,
C = event the sum of the dice is 10,
D = event the sum of the dice is even.

a. Compute $P(A)$, $P(B)$, $P(C)$, and $P(D)$.
b. Compute $P(B \mid A)$.
c. Are events A and B independent? Why?
d. Compute $P(C \mid A)$.
e. Are events A and C independent? Why?
f. Compute $P(D \mid A)$.
g. Are events A and D independent? Why?

4.93 The table that follows gives a joint probability distribution for the members of the 102nd Congress by political party and legislative group. [SOURCE: U.S. Congress, Joint Committee on Printing, *Congressional Directory.*]

Party		Group: Rep C_1	Senator C_2	$P(P_i)$
	Democratic P_1	0.500	0.105	0.605
	Republican P_2	0.313	0.082	0.395
	$P(C_j)$	0.813	0.187	1.000

a. Determine $P(P_1)$, $P(C_2)$, and $P(P_1 \& C_2)$.
b. Use the special multiplication rule to determine whether events P_1 and C_2 are independent.

4.94 The National Center for Education Statistics publishes information on U.S. engineers and scientists in *Digest of Education Statistics.* The table that follows provides a joint probability distribution for engineers and scientists by highest degree obtained.

Highest degree		Type: Engineer T_1	Scientist T_2	$P(D_i)$
	Bachelors D_1	0.343	0.289	0.632
	Masters D_2	0.098	0.146	0.244
	Doctorate D_3	0.017	0.091	0.108
	Other D_4	0.013	0.003	0.016
	$P(T_j)$	0.471	0.529	1.000

a. Determine $P(T_2)$, $P(D_3)$, and $P(T_2 \& D_3)$.
b. Are T_2 and D_3 independent events? Why?

4.95 Two cards are drawn at random from an ordinary deck of 52 cards. Determine the probability that both cards are aces if

a. the first card is replaced before the second card is drawn.
b. the first card is not replaced before the second card is drawn.

4.96 In *Yahtzee,* five balanced dice are rolled.

a. What is the probability of rolling all 2s? *(Hint:* Use the fact that the outcomes of different dice are independent, and apply the special multiplication rule.*)*
b. What is the probability that all the dice come up the same number? *(Hint:* Apply both the special addition rule and the special multiplication rule.*)*

4.97 Suppose E and F are independent events and $P(E) = \frac{1}{3}$ and $P(F) = \frac{1}{4}$. Find

a. $P(E \& F)$.   b. $P(E \text{ or } F)$.

4.98 A family has two portable computers. There is a 70% chance that a computer will run for over 60 operating hours without a change of batteries. New batteries are installed in each computer. Determine the probability that

a. both computers will run for over 60 hours without a change of batteries.
b. at least one of the two computers will run for over 60 hours without a change of batteries.

4.99 In a letter to the editor that appeared in the February 23, 1987, issue of *U.S. News and World Report,* a reader discussed the issue of space shuttle safety. Each "criticality 1" item must have 99.99% reliability, according to NASA standards. This means that the probability of failure for a "criticality 1" item is only 0.0001. Mission 25, the mission in which the Challenger exploded, had 748 "criticality 1" items. Determine the probability that

a. none of the "criticality 1" items would fail.
b. at least one "criticality 1" item would fail.
c. Interpret your result in part (b) in words.

4.100 A hardware manufacturer produces nuts and bolts. Each bolt produced is attached to a nut to make a single unit. It is known that 2% of the nuts produced and 3% of the bolts produced are defective in some way. A nut-bolt unit is considered defective if either the nut or the bolt has a defect. Determine the percentage of nondefective nut-bolt units.

4.101 As reported by the Chicago Title Insurance Company in *The Guarantor,* the probability is 0.768 that a home buyer will purchase a resale home. In the next four home purchases, find the probability that

a. the first three will be resales and the fourth a new home.

b. the first will be a resale, the second will be a new home, and the last two will be resales.
c. the first will be a resale, the next two will be new homes, and the last will be a resale.
d. exactly three of the four will be resales. [*Hint:* There are four distinct ways in which exactly three of the four home purchases are resales; two of the four ways were already considered in parts (a) and (b).]

4.102 The Federal Bureau of Investigation compiles information on violent crimes by type and publishes the results in *Crime in the United States.* Here is a probability distribution for the various types of violent crimes.

Violent crime	Probability
Murder	0.016
Forcible rape	0.064
Robbery	0.403
Aggravated assault	0.517

The table shows, for instance, that the probability is 0.016 that a violent crime will be a murder. Out of three violent crimes, find the probability that

a. the first two are robberies and the last a forcible rape.
b. the first is a murder, the second is a robbery, and the third is an aggravated assault.

4.103 The National Center for Health Statistics collects data on activity limitations. Results are published in *Vital and Health Statistics.* The data show that 13.6% of males and 14.4% of females have an activity limitation. Are sex and activity limitation statistically independent? Why?

4.104 For three events, say, A, B, and C, the general multiplication rule is

$$P(A \& B \& C) = P(A) \cdot P(B \mid A) \cdot P\left(C \mid (A \& B)\right).$$

a. Suppose three cards are randomly selected without replacement from an ordinary deck of 52. Find the probability that all three cards are hearts; that the first two cards are hearts and the third is a spade.
b. Provide a mathematical statement of the general multiplication rule for four events.

4.105 Prove that if $P(B \mid A) = P(B)$, then it is also true that $P(A \mid B) = P(A)$. This shows that if event B is independent of event A, then it is also true that event A is independent of event B.

· · · **4.106** Three events, say, A, B, and C, are said to be independent if

$$P(A \& B) = P(A) \cdot P(B),$$
$$P(A \& C) = P(A) \cdot P(C),$$
$$P(B \& C) = P(B) \cdot P(C),$$

and

$$P(A \& B \& C) = P(A) \cdot P(B) \cdot P(C).$$

What do you think is required for four events to be independent? Explain your definition in words.

· · · **4.107** Consider the experiment of rolling two balanced dice, one blue and one black. Let

A = event the blue die comes up even,
B = event the black die comes up even,
C = event the sum of the dice is even,
D = event the blue die comes up 1, 2, or 3,
E = event the blue die comes up 3, 4, or 5,
F = event the sum of the dice is 5.

Apply the definition of independence for three events, stated in Exercise 4.106, to solve each of the following problems.
a. Are A, B, and C independent events?
b. Show that $P(D \& E \& F) = P(D) \cdot P(E) \cdot P(F)$ but that D, E, and F are not independent events.

· · · **4.108** When a balanced coin is tossed four times, 16 equally likely outcomes are possible:

HHHH	THHH	THHT	THTT
HHHT	HHTT	THTH	TTHT
HHTH	HTHT	TTHH	TTTH
HTHH	HTTH	HTTT	TTTT

Let

A = event the first toss is heads,
B = event the second toss is tails,
C = event the last two tosses are heads.

Apply the definition of independence for three events, stated in Exercise 4.106, to show that A, B, and C are independent.

· · · **4.109** This exercise explores the relationship between the concepts of mutually exclusive events and independent events. Consider two events, A and B, neither of which is impossible; in other words, assume $P(A) > 0$ and $P(B) > 0$.
a. Show that if A and B are independent events, then they cannot be mutually exclusive. [*Hint:* Use the special multiplication rule to show that $P(A \& B) > 0$.]
b. Deduce from part (a) that if A and B are mutually exclusive events, then they cannot be independent.
c. Find an example of two events that are neither mutually exclusive nor independent.

4.7 BAYES'S RULE (OPTIONAL)

· · · · · ·

In this section we will discuss a rule of probability developed by Thomas Bayes, an eighteenth-century clergyman. This rule is aptly called Bayes's rule. One of the primary uses of Bayes's rule is to revise probabilities in accordance with newly acquired information. Such revised probabilities are actually conditional probabilities, and so in some sense we have already examined much of the material in this section. However, as we will see, Bayes's rule involves some new concepts and techniques.

The Rule of Total Probability

In preparation for Bayes's rule, we need to study another rule of probability called the rule of total probability. First we consider the concept of exhaustive events. Events A_1, A_2, ..., A_k are said to be **exhaustive** if at least one of them must occur.

For instance, the National Governors' Association classifies governors as Democrat, Republican, or Independent. Suppose a governor is selected at random; let E_1 denote the event that the governor selected is a Democrat, E_2 the event that the governor selected is a Republican, and E_3 the event that the governor selected is an Independent. Then events

E_1, E_2, and E_3 are exhaustive since at least one of them must occur when a governor is selected—the governor obtained must be a Democrat, Republican, or Independent.

The events E_1, E_2, and E_3 (Democrat, Republican, and Independent) are not only exhaustive, but are also mutually exclusive (why is this so?). In general, if events are both exhaustive and mutually exclusive, then *exactly one* of them must occur. This is true because at least one of the events must occur (since the events are exhaustive) and at most one of the events can occur (since the events are mutually exclusive).

An event and its complement are always mutually exclusive and exhaustive. Figure 4.26(a) portrays three events, A_1, A_2, and A_3, that are both mutually exclusive and exhaustive. In the figure the three events do not overlap, indicating that they are mutually exclusive; furthermore, they fill out the entire region enclosed by the heavy rectangle (i.e., the sample space), indicating that they are exhaustive.

FIGURE 4.26
(a) Three mutually exclusive
and exhaustive events
(b) An event B and three mutually
exclusive and exhaustive events

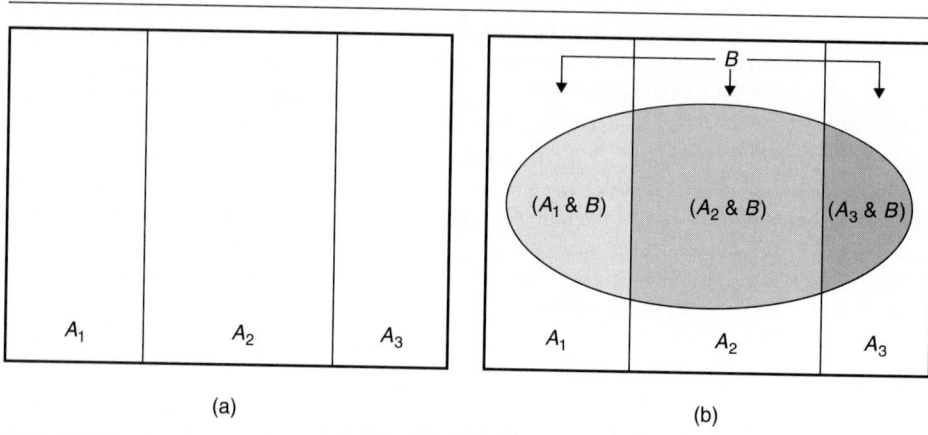

(a) (b)

Now consider, say, three mutually exclusive and exhaustive events, A_1, A_2, and A_3, and any event B, as portrayed in Fig. 4.26(b). As we see from Fig. 4.26(b), event B is comprised of the mutually exclusive events (A_1 & B), (A_2 & B), and (A_3 & B), shown in color. This reflects the fact that event B must occur in conjunction with exactly one of the events A_1, A_2, and A_3.

If we think of the colored regions in Fig. 4.26(b) as probabilities, then the total colored region is $P(B)$, and the three colored subregions are, from left to right, $P(A_1$ & $B)$, $P(A_2$ & $B)$, and $P(A_3$ & $B)$. Because events (A_1 & B), (A_2 & B), and (A_3 & B) are mutually exclusive, the total colored region equals the sum of the three colored subregions; in other words,

$$P(B) = P(A_1 \text{ \& } B) + P(A_2 \text{ \& } B) + P(A_3 \text{ \& } B).$$

Applying the general multiplication rule, Formula 4.5 on page 224, to each term on the right-hand side of this equation, we obtain

$$P(B) = P(A_1) \cdot P(B \mid A_1) + P(A_2) \cdot P(B \mid A_2) + P(A_3) \cdot P(B \mid A_3).$$

This formula holds in general and is called the **rule of total probability.**

FORMULA 4.8

THE RULE OF TOTAL PROBABILITY

Suppose events $A_1, A_2, \ldots, A_k$ are mutually exclusive and exhaustive; that is, exactly one of the events must occur. Then for any event B,

$$P(B) = \sum_{j=1}^{k} P(A_j) \cdot P(B \mid A_j).$$

Example 4.29   Illustrates Formula 4.8

The U.S. Bureau of the Census collects data on the resident population by age and region of residence. Results are published in *Current Population Reports.* In the first two columns of Table 4.11, we have provided a percentage distribution for region of residence; the third column displays the percentage of seniors (age 65 or over) in each region. The table shows, for instance, that 20.6% of U.S. residents live in the Northeast region and that 13.5% of residents living in the Northeast are seniors. Use Table 4.11 to determine the percentage of U.S. residents that are seniors; that is, find the probability that a randomly selected U.S. resident is a senior.

TABLE 4.11

Percentage distribution for region of residence, and percentage of seniors in each region

Region	Percentage of U.S. population	Percentage seniors
Northeast	20.6	13.5
Midwest	24.5	12.6
South	34.5	12.1
West	20.4	10.8
	100.0	

SOLUTION

To solve this problem, we first translate the information displayed in Table 4.11 into the language of probability. Suppose a U.S. resident is selected at random. Let

$$S = \text{event the resident selected is a senior,}$$

and

$$R_1 = \text{event the resident selected lives in the Northeast,}$$
$$R_2 = \text{event the resident selected lives in the Midwest,}$$
$$R_3 = \text{event the resident selected lives in the South,}$$
$$R_4 = \text{event the resident selected lives in the West.}$$

Then the percentages shown in the second and third columns of Table 4.11 translate into the probabilities displayed in Table 4.12.

TABLE 4.12

Probabilities derived
from Table 4.11

$P(R_1) = 0.206$	$P(S \mid R_1) = 0.135$
$P(R_2) = 0.245$	$P(S \mid R_2) = 0.126$
$P(R_3) = 0.345$	$P(S \mid R_3) = 0.121$
$P(R_4) = 0.204$	$P(S \mid R_4) = 0.108$

The problem is to determine the percentage of U.S. residents that are seniors, or, in terms of probability, $P(S)$. Because a U.S. resident must reside in exactly one of the four regions, events R_1, R_2, R_3, and R_4 are mutually exclusive and exhaustive. Therefore, by the rule of total probability applied to the event S, we have from Table 4.12 that

$$P(S) = \sum_{j=1}^{4} P(R_j) \cdot P(S \mid R_j)$$
$$= 0.206 \cdot 0.135 + 0.245 \cdot 0.126 + 0.345 \cdot 0.121 + 0.204 \cdot 0.108$$
$$= 0.122.$$

A tree diagram for this calculation is shown in Fig. 4.27 (in the figure, J represents the event that the resident selected is not a senior). We obtain $P(S)$ from the tree diagram by first multiplying the two probabilities on each branch of the tree that ends with S (the branches shown in color) and then summing all those products.

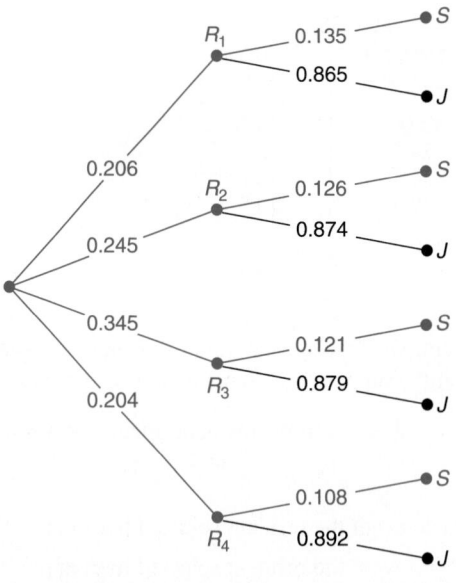

In any case, we see that the probability is 0.122 that a randomly selected U.S. resident is a senior. In other words, 12.2% of U.S. residents are seniors. ∎

Bayes's Rule

Using the rule of total probability, we can derive Bayes's rule. For simplicity let's consider three events, A_1, A_2, and A_3, that are mutually exclusive and exhaustive; and let B be any event. For Bayes's rule we assume the probabilities $P(A_1)$, $P(A_2)$, $P(A_3)$, $P(B \mid A_1)$, $P(B \mid A_2)$, and $P(B \mid A_3)$ are known. The problem is to use those six probabilities to determine the conditional probabilities $P(A_1 \mid B)$, $P(A_2 \mid B)$, and $P(A_3 \mid B)$.

We will show how to express $P(A_2 \mid B)$ in terms of the six known probabilities; $P(A_1 \mid B)$ and $P(A_3 \mid B)$ are handled similarly. First we apply the conditional probability rule, Formula 4.4 on page 219, to write

$$(1) \qquad P(A_2 \mid B) = \frac{P(B \ \& \ A_2)}{P(B)} = \frac{P(A_2 \ \& \ B)}{P(B)}.$$

Next we apply the general multiplication rule, Formula 4.5 on page 224, to the numerator of the fraction on the right, and the rule of total probability, Formula 4.8, to the denominator of the fraction on the right. This gives

$$P(A_2 \ \& \ B) = P(A_2) \cdot P(B \mid A_2)$$

and

$$P(B) = P(A_1) \cdot P(B \mid A_1) + P(A_2) \cdot P(B \mid A_2) + P(A_3) \cdot P(B \mid A_3).$$

Substituting these last two formulas into the fraction on the right of Equation (1), we obtain

$$P(A_2 \mid B) = \frac{P(A_2) \cdot P(B \mid A_2)}{P(A_1) \cdot P(B \mid A_1) + P(A_2) \cdot P(B \mid A_2) + P(A_3) \cdot P(B \mid A_3)}.$$

This formula holds in general and is called **Bayes's rule.**

FORMULA 4.9

BAYES'S RULE

Suppose events A_1, A_2, ..., A_k are mutually exclusive and exhaustive; that is, exactly one of the events must occur. Then for any event B,

$$P(A_i \mid B) = \frac{P(A_i) \cdot P(B \mid A_i)}{\sum_{j=1}^{k} P(A_j) \cdot P(B \mid A_j)},$$

where A_i can be any one of the events A_1, A_2, ..., A_k.

Example 4.30   Illustrates Formula 4.9

Recall that the first two columns of Table 4.11 on page 237 provide a percentage distribution for the region of residence of U.S. residents, and the third column displays the percentage of seniors in each region. Determine the percentage of U.S. seniors that live in the Northeast.

SOLUTION   Referring to the notation introduced at the beginning of the solution to Example 4.29, we see that in terms of probability the problem is to determine $P(R_1 \mid S)$, the probability that a U.S. resident lives in the Northeast given that the resident is a senior. To obtain that conditional probability, we apply Bayes's rule and Table 4.12 on page 238:

$$P(R_1 \mid S) = \frac{P(R_1) \cdot P(S \mid R_1)}{\sum_{j=1}^{4} P(R_j) \cdot P(S \mid R_j)}$$

$$= \frac{0.206 \cdot 0.135}{0.206 \cdot 0.135 + 0.245 \cdot 0.126 + 0.345 \cdot 0.121 + 0.204 \cdot 0.108}$$

$$= 0.227.$$

Thus 22.7% of U.S. seniors live in the Northeast.   ∎

Example 4.31   Illustrates Formula 4.9

According to the Arizona Chapter of the American Lung Association, 7.0% of the population has lung disease. Of those people having lung disease, 90.0% are smokers; and of those not having lung disease, 25.3% are smokers. Determine the probability that a randomly selected smoker has lung disease.

SOLUTION   Suppose a person is selected at random. Let

$$S = \text{event the person selected is a smoker,}$$

and

$$L_1 = \text{event the person selected has no lung disease,}$$
$$L_2 = \text{event the person selected has lung disease.}$$

Note that events L_1 and L_2 are complementary, which implies that they are mutually exclusive and exhaustive.

The data provided in the statement of the problem indicate that $P(L_2) = 0.070$, $P(S \mid L_2) = 0.900$, and $P(S \mid L_1) = 0.253$. Also, since $L_1 = (\text{not } L_2)$, we deduce that $P(L_1) = P(\text{not } L_2) = 1 - P(L_2) = 1 - 0.070 = 0.930$. We summarize this information in Table 4.13.

TABLE 4.13
Known probability information

$P(L_1) = 0.930$	$P(S \mid L_1) = 0.253$
$P(L_2) = 0.070$	$P(S \mid L_2) = 0.900$

The problem is to determine the probability that a randomly selected smoker has lung disease, $P(L_2 \mid S)$. Applying Bayes's rule to the probability data in Table 4.13, we obtain

$$P(L_2 \mid S) = \frac{P(L_2) \cdot P(S \mid L_2)}{P(L_1) \cdot P(S \mid L_1) + P(L_2) \cdot P(S \mid L_2)}$$

$$= \frac{0.070 \cdot 0.900}{0.930 \cdot 0.253 + 0.070 \cdot 0.900} = 0.211.$$

Thus the probability is 0.211 that a randomly selected smoker has lung disease; in other words, 21.1% of smokers have lung disease. ∎

We observe from Example 4.31 that the rate of lung disease among smokers (21.1%) is more than three times the rate among the general population (7.0%). Using arguments similar to those in Example 4.31, we can show that the probability is 0.010 that a randomly selected nonsmoker has lung disease; in other words, 1.0% of nonsmokers have lung disease. This implies that the rate of lung disease among smokers (21.1%) is more than 20 times that among nonsmokers (1.0%).[†]

We conclude this section by introducing some terminology that is often used in conjunction with Bayes's rule. To that end, let's return to Example 4.31. From the information provided, we know that the probability is 0.070 that a randomly selected person has lung disease: $P(L_2) = 0.070$. This probability does not take into consideration whether the person is a smoker. It is therefore called a **prior probability** since it represents the probability that the person selected has lung disease *before* knowing whether the person is a smoker.

Now suppose the person selected is found to be a smoker. On the basis of this additional information, we can revise the probability that the person has lung disease. This can be done by determining the conditional probability that the person selected has lung disease, given that the person selected is a smoker: $P(L_2 \mid S) = 0.211$ (from Example 4.31). This revised probability is called a **posterior probability** since it represents the probability that the person selected has lung disease *after* knowing that the person is a smoker.

EXERCISES 4.7

4.110 An article appearing in *The Arizona Republic* on February 14, 1993, reported on a study by researchers at Harvard University and the National Institute of Aging. The study compared the life spans of left- and right-handed people in the United States. According to the article, 9% of women and 13% of men are left-handed. Census Bureau data indicate that 51.2% of the people in the United States are women and 48.8% are men.

a. What percentage of people in the United States are left-handed?

b. What percentage of left-handed people in the United States are men?

4.111 In the first two columns of the following table, we have provided a percentage distribution for the religious affiliation of the voters in a hypothetical city. The data are based on the national distribution for religious preference found in *Emerging Trends*, published by the Princeton Religion Re-

search Center of Princeton, New Jersey. In the third column of the table, we show the percentage of Democrats in each religious group of voters. The table shows, for instance, that 28% of the voters in the city are Catholic and that 53% of the Catholic voters in the city are Democrats.

Religion	Percentage of voters	Percentage Democrats
Catholic	28	53
Jewish	2	61
Protestant	57	42
Other	4	58
None	9	67
	100	

a. What percentage of the voters are Democrats?

b. What percentage of the Democrats are Protestant?

[†] Since the study under consideration here is observational, we cannot conclude that smoking causes lung disease, but only that a strong positive association exists between smoking and lung disease.

4.112   Textbook editors must estimate the sales of new (first-edition) books. The records of one major publishing company indicate that 10% of all new books sell more than projected, 30% sell at (close to) projected, and 60% sell less than projected. Of those that sell more than projected, 70% are revised for a second edition, as are 50% of those that sell close to projected, and 20% of those that sell less than projected.

a. What percentage of books published by this publishing company go to a second edition?

b. What percentage of books published by this publishing company that go to a second edition sold less than projected in their first edition?

4.113   EDA Products, a manufacturer of customized metal fabrication items, produces forged tools for a major retailer. EDA currently uses three 50-ton forge presses to manufacture a particular model of slip joint pliers. Although each of the presses accounts for one-third of production, the presses produce defective units with varying percentages. In fact, recent quality-assurance tests indicate that Press 1 produces 1.1% defective units, Press 2 produces 0.8% defective units, and Press 3 produces 1.6% defective units.

a. What percentage of slip joint pliers produced by EDA are defective?

b. Of those slip joint pliers that are defective, what percentage are produced by Press 2?

4.114   The National Center for Health Statistics provides information on suicides by sex and method used. Data are published in *Vital Statistics of the United States*. In 1989 there were 30,232 suicides in the United States, of which 24,102 were males and 6,130 were females. The following table gives the relative-frequency distributions for method used by males and females who committed suicide in 1989.

Method used	Relative frequency for males	Relative frequency for females
Poisoning	0.133	0.364
Hanging/ strangulation	0.154	0.126
Firearms	0.651	0.408
Other	0.062	0.102

Suppose a 1989 suicide report is selected at random.

a. Determine the probability that a firearm was used for the suicide.

b. Find the prior probability that the person who committed suicide was a female.

c. Find the posterior probability that the person who committed suicide was a female, given that a firearm was used.

d. Interpret the probabilities that you obtained in parts (a)–(c) in terms of percentages.

4.115   A 1989 Gallup poll asked 1005 adults and 500 teenagers the question, "What is the nation's top problem?" The pie charts shown below were used to summarize the results of the survey.

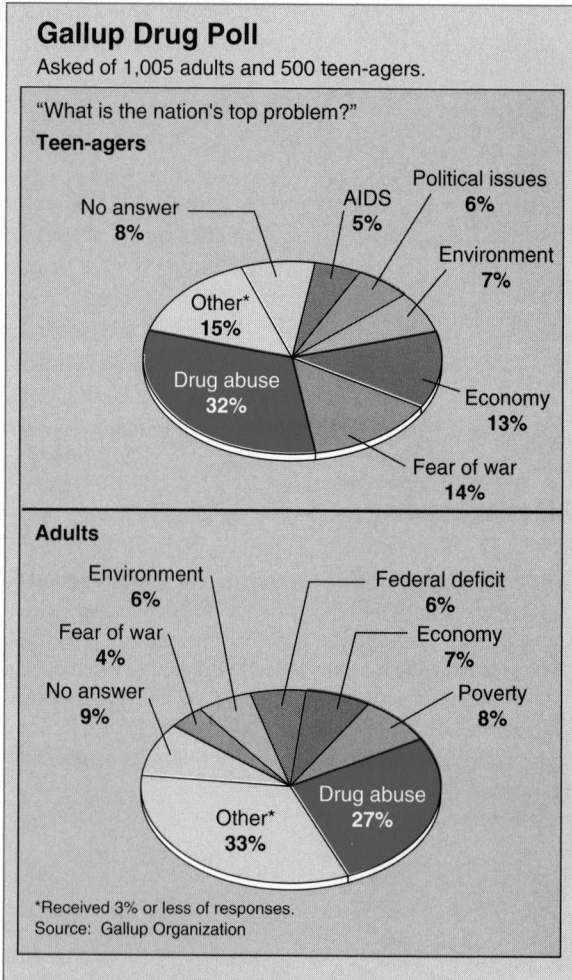

Gallup Drug Poll
Asked of 1,005 adults and 500 teen-agers.

"What is the nation's top problem?"
Teen-agers

No answer 8% — AIDS 5% — Political issues 6% — Environment 7% — Other* 15% — Drug abuse 32% — Economy 13% — Fear of war 14%

Adults

Environment 6% — Federal deficit 6% — Fear of war 4% — Economy 7% — No answer 9% — Poverty 8% — Other* 33% — Drug abuse 27%

*Received 3% or less of responses.
Source: Gallup Organization

Suppose a person who participated in the survey is selected at random.

a. Determine the probability that the person selected said that drug abuse is the nation's top problem.

b. Find the prior probability that the person selected is a teenager.

c. Find the posterior probability that the person selected is a teenager, given that the person selected said that drug abuse is the nation's top problem.

d. Interpret the probabilities that you obtained in parts (a)–(c) in terms of percentages.

4.116 At a grocery store, eggs come in cartons that hold a dozen eggs. Experience indicates that 78.5% of the cartons have no broken eggs, 19.2% have one broken egg, 2.2% have two broken eggs, and 0.1% have three broken eggs (the percentage of cartons with four or more broken eggs is negligible). One egg is selected at random from a carton and is found to be broken. What is the probability that this egg is the only broken one in the carton?

4.117 Medical tests are frequently used to decide whether a person has a particular disease. The **sensitivity** of a test is defined as the probability that a person having the disease will test positive; the **specificity** of a test is defined as the probability that a person not having the disease will test negative. A test for a certain disease has been used for many years. Experience with the test indicates that its sensitivity is 0.934 and its specificity is 0.968. Furthermore, it is known that roughly 1 in 500 people has the disease.

a. Interpret the sensitivity and specificity of this test in terms of percentages.

b. Determine the probability that a person testing positive actually has the disease.

c. Interpret your result from part (b) in terms of percentages.

4.118 Kress, Inc., markets two types of evaporative coolers, metal and fiberglass. The company has four sales districts. In the first two columns of the table shown at the top of the next column, we give the percentage distribution of total sales by district. The third column shows the percentage of sales in each district that are metal coolers. We see, for instance, that 45% of all sales occur in District I and that 35% of the coolers sold in District I are metal coolers.

a. What percentage of sales are metal coolers?

b. What percentage of metal-cooler sales occur in District II?

District	Percentage of total sales	Percentage metal
I	45	35
II	26	30
III	18	60
IV	11	22
	100	

4.119 *Bottom Line/Personal* newsletter interviewed Gerald Kushel, Ed.D., on the secrets of successful people. To study success, Kushel questioned 1200 people, among whom were lawyers, artists, teachers, and students. He found that 15% enjoy neither their jobs nor their personal lives, 80% enjoy their jobs but not their personal lives, and 4% enjoy both their jobs and their personal lives.

a. Determine the percentage of the people interviewed who enjoy their jobs.

b. What percentage of the people interviewed who enjoy their jobs also enjoy their personal lives?

4.120 Refer to Example 4.31 on page 240.

a. Determine the probability that a randomly selected non-smoker has lung disease.

b. Use part (a) and the result of Example 4.31 to compare the rates of lung disease for smokers and nonsmokers.

4.121 Suppose events $A_1, A_2, \ldots, A_k$ are mutually exclusive and exhaustive. Further suppose those events are equally likely to occur.

a. Show that for any event B,

$$P(B) = \frac{1}{k} \sum_{j=1}^{k} P(B \mid A_j).$$

Interpret this equation in words.

b. Show that for any event B,

$$P(A_i \mid B) = \frac{P(B \mid A_i)}{\sum_{j=1}^{k} P(B \mid A_j)},$$

for $i = 1, 2, \ldots, k$. Interpret this equation in words.

c. Use parts (a) and (b) to solve Exercise 4.113.

4.8 COUNTING RULES (OPTIONAL)

We often find it necessary to determine the number of ways something can happen—the number of possible outcomes for an experiment, the number of ways an event can occur, the number of ways a certain task can be performed, and so forth. Sometimes we can list the possibilities and then count them; but in most cases, the number of possibilities is so large that a direct listing is impractical.

Thus we need to develop techniques that do not rely on a direct listing for determining the number of ways something can happen. Such techniques are usually referred to as **counting rules.** In this section we will examine some widely used counting rules.

The Fundamental Counting Rule

One counting rule, called the **fundamental counting rule (FCR),** is basic to all the counting techniques we will discuss.[†] We introduce this rule in Example 4.32.

Example 4.32   Introduces the Fundamental Counting Rule

A builder of new homes in the metropolitan Phoenix area offers four models, called the Shalimar, Palacia, Valencia, and Monterey. Each model comes in a choice of three elevations, designated A, B, and C. How many choices are there for the selection of a home, including both model and elevation?

SOLUTION   We will first use a tree diagram (Fig. 4.28) to systematically obtain a direct listing of the possibilities. In the tree diagram, we have used S for Shalimar, P for Palacia, V for Valencia, and M for Monterey. Each branch of the tree corresponds to one possibility for model and elevation. For instance, the first branch of the tree, ending in SA, corresponds to the Shalimar model with the A elevation. The total number of possibilities can be obtained by counting the number of branches at the end of the tree. Thus we see that there are 12 choices for the selection of a home, including both model and elevation.

Although the tree-diagram approach for determining the number of possibilities is a direct listing, it provides us with a clue for obtaining the number of possibilities without resorting to a direct listing. Specifically, there are four possibilities for model, indicated by the four sub-branches emanating from the starting point of the tree; and to each possibility for model, there correspond three possibilities for elevation, indicated by the three sub-branches emanating from the end of each model sub-branch. Consequently, there are

$$\underbrace{3 + 3 + 3 + 3}_{4 \text{ times}} = 4 \cdot 3 = 12$$

possibilities altogether. So we see that the total number of possibilities can be obtained by *multiplying* the number of possibilities for the model by the number of possibilities for the elevation.

                                                                        ■

[†] The fundamental counting rule is also known as the **basic principle of counting** and the **multiplication rule.**

FIGURE 4.28

Tree diagram for model
and elevation possibilities

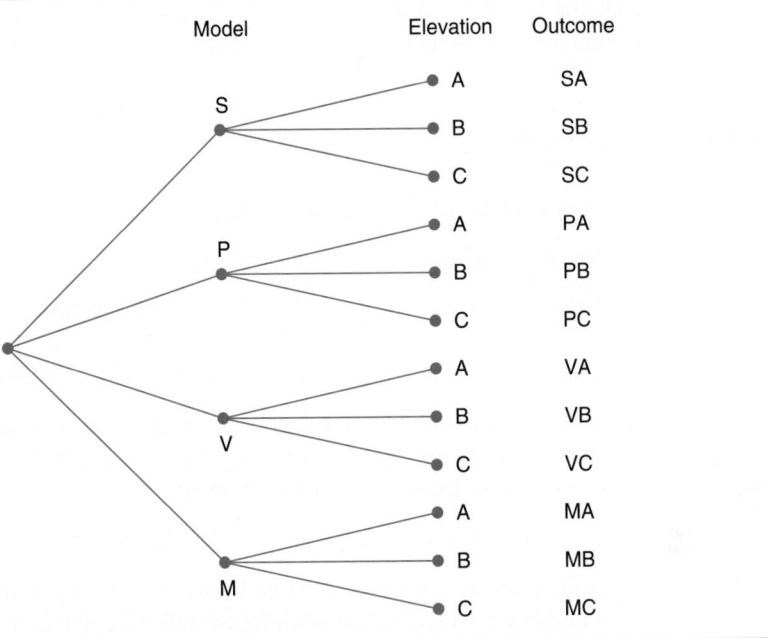

KEY FACT 4.3

THE FUNDAMENTAL COUNTING RULE (FCR)

Suppose two actions (choices, experiments) are to be performed in a definite order. Further suppose there are m_1 possibilities for the first action, and that corresponding to each of these possibilities there are m_2 possibilities for the second action. Then there are $m_1 \cdot m_2$ possibilities altogether for the two actions.

More generally, suppose r actions (choices, experiments) are to be performed in a definite order. Further suppose there are m_1 possibilities for the first action, and that corresponding to each of these possibilities there are m_2 possibilities for the second action, and that corresponding to each of these possibilities there are m_3 possibilities for the third action, and so on. Then there are $m_1 \cdot m_2 \cdots m_r$ possibilities altogether for the r actions.

In Example 4.32 there are two actions ($r = 2$), selecting a model and selecting an elevation. Because there are four possibilities for model, $m_1 = 4$; and because corresponding to each model there are three possibilities for elevation, $m_2 = 3$. Therefore by the FCR, the total number of possibilities, including both model and elevation, is

$$m_1 \cdot m_2 = 4 \cdot 3 = 12,$$

as we discovered in Example 4.32.

Because the number of possibilities in the model/elevation problem is quite small, it is not too difficult to determine the number by a direct listing, as we did in the tree diagram in Fig. 4.28. Nonetheless, it is still much easier to obtain the number of possibilities by applying the FCR. Moreover, in problems where the number of possibilities is large, a direct listing is not feasible and the FCR is the only practical way to proceed.

Example 4.33   Illustrates Key Fact 4.3

The license plates of Arizona consist of three letters followed by three digits.

a. How many different license plates are possible?

b. How many possibilities are there for license plates in which no letter or digit is repeated?

SOLUTION   For both parts (a) and (b), we will apply the FCR with six actions ($r = 6$).

a. There are 26 possibilities for the first letter, 26 for the second letter, and 26 for the third letter; and there are 10 possibilities for the first digit, 10 for the second digit, and 10 for the third digit. Hence by the FCR, there are

$$m_1 \cdot m_2 \cdot m_3 \cdot m_4 \cdot m_5 \cdot m_6 = 26 \cdot 26 \cdot 26 \cdot 10 \cdot 10 \cdot 10 = 17,576,000$$

possibilities altogether for different license plates. Obviously, it would not be practical to obtain the number of possibilities by a direct listing—the tree diagram would have 17,576,000 branches!

b. For this part there are again 26 possibilities for the first letter. But to each possibility for the first letter, there correspond 25 possibilities for the second letter because the second letter cannot be the same as the first. And to each possibility for the first two letters, there correspond 24 possibilities for the third letter because the third letter cannot be the same as either the first or the second. Similarly, there are 10 possibilities for the first digit, 9 for the second digit, and 8 for the third digit. So by the FCR, there are

$$m_1 \cdot m_2 \cdot m_3 \cdot m_4 \cdot m_5 \cdot m_6 = 26 \cdot 25 \cdot 24 \cdot 10 \cdot 9 \cdot 8 = 11,232,000$$

possibilities for license plates in which no letter or digit is repeated.   ∎

Factorials

Before we continue our presentation of counting rules, we need to discuss factorials. Factorials are used extensively in mathematics and its applications.

DEFINITION 4.9   **FACTORIALS**

> The product of the first k positive integers is called k *factorial* and is denoted by the symbol $k!$. Thus
>
> $$k! = k(k-1)\cdots 2 \cdot 1.$$
>
> We also define $0! = 1$.

Example 4.34   Illustrates Definition 4.9

Determine 3!, 4!, and 5!.

SOLUTION   Applying Definition 4.9, we obtain that $3! = 3 \cdot 2 \cdot 1 = 6$, $4! = 4 \cdot 3 \cdot 2 \cdot 1 = 24$, and $5! = 5 \cdot 4 \cdot 3 \cdot 2 \cdot 1 = 120$.   ■

Note, for instance, that $6! = 6 \cdot 5!$, $6! = 6 \cdot 5 \cdot 4!$, $6! = 6 \cdot 5 \cdot 4 \cdot 3!$, and so on. In general, if $j \leq k$, then $k! = k(k-1) \cdots (k-j+1)(k-j)!$.

Permutations

A **permutation** of r objects from a collection of m objects is any *ordered* arrangement of r of the m objects. The number of possible permutations of r objects that can be formed from a collection of m objects is denoted by $(m)_r$ or $_mP_r$.[†] Let's look at a simple example.

Example 4.35   Introduces Permutations

Consider the collection consisting of the five letters a, b, c, d, e.

a. List all possible permutations of three letters from this collection of five letters.

b. Use part (a) to determine the number of possible permutations of three letters that can be formed from the collection of five letters; that is, determine $(5)_3$.

c. Use the FCR to determine the number of possible permutations of three letters that can be formed from the collection of five letters; that is, determine $(5)_3$ using the FCR.

SOLUTION   a. For this part we need to list all ordered arrangements of three letters from the first five letters of the English alphabet. This is done in Table 4.14.

TABLE 4.14
Possible permutations of three letters from the collection of five letters

abc	abd	abe	acd	ace	ade	bcd	bce	bde	cde
acb	adb	aeb	adc	aec	aed	bdc	bec	bed	ced
bac	bad	bae	cad	cae	dae	cbd	cbe	dbe	dce
bca	bda	bea	cda	cea	dea	cdb	ceb	deb	dec
cab	dab	eab	dac	eac	ead	dbc	ebc	ebd	ecd
cba	dba	eba	dca	eca	eda	dcb	ecb	edb	edc

b. From Table 4.14 we see that there are 60 possible permutations of three letters from the collection of five letters; in other words, $(5)_3 = 60$.

[†] The notation $_mP_r$ is more common than $(m)_r$, but we prefer the latter since it is less cumbersome and has other pedagogical advantages.

c. Here we want to use the FCR to determine the number of possible permutations of three letters from the collection of five letters. There are five possibilities for the first letter, four possibilities for the second letter, and three possibilities for the third letter. Hence by the FCR, there are

$$m_1 \cdot m_2 \cdot m_3 = 5 \cdot 4 \cdot 3 = 60$$

possibilities altogether. So again we see that $(5)_3 = 60$. ∎

We can make two relevant observations from Example 4.35. First, it is generally tedious or impractical to list all possible permutations under consideration. Second, it is not necessary to list the possible permutations in order to determine how many there are—we can use the FCR to count them.

By studying part (c) of Example 4.35, we see that the FCR can be used to obtain a general formula for $(m)_r$, the number of possible permutations of r objects from a collection of m objects. There are m possibilities for the first object, $m - 1$ for the second object, $m - 2$ for the third object, and so on. Applying the FCR, we find that $(m)_r = m(m - 1) \cdots (m - r + 1)$. Multiplying the numerator and denominator of this last expression by $(m - r)!$, we get the equivalent expression $(m)_r = m!/(m - r)!$. We summarize this discussion in Formula 4.10.

FORMULA 4.10

PERMUTATIONS RULE

The number of possible permutations of r objects from a collection of m objects is given by the formula

$$(m)_r = \frac{m!}{(m - r)!}.$$

Example 4.36 Illustrates Formula 4.10

In an exacta wager at the race track, the bettor picks the two horses that he or she thinks will finish first and second, in a specified order. For a race with 12 entrants, determine the number of possible exacta wagers.

SOLUTION Selecting 2 horses from the 12 horses for an exacta wager is equivalent to specifying a permutation of 2 objects from a collection of 12 objects; the first object is the horse selected to come in first place and the second object is the horse selected to come in second place. Thus the number of possible exacta wagers is $(12)_2$—the number of possible permutations of 2 objects from a collection of 12 objects. Applying the permutations rule, Formula 4.10, with $m = 12$ and $r = 2$, we obtain

$$(12)_2 = \frac{12!}{(12 - 2)!} = \frac{12!}{10!} = \frac{12 \cdot 11 \cdot \cancel{10!}}{\cancel{10!}} = 12 \cdot 11 = 132.$$

In a 12-horse race, there are 132 possible exacta wagers. ∎

Example 4.37   Illustrates Formula 4.10

A student has 10 books to arrange on a shelf of a bookcase. In how many ways can the 10 books be arranged?

SOLUTION   Any particular arrangement of the 10 books on the shelf is a permutation of 10 objects from a collection of 10 objects. Thus for this problem we need to determine $(10)_{10}$, the number of possible permutations of 10 objects from a collection of 10 objects, more commonly expressed as the number of possible permutations of 10 objects among themselves. Applying the permutations rule, we get

$$(10)_{10} = \frac{10!}{(10-10)!} = \frac{10!}{0!} = \frac{10!}{1} = 10! = 3,628,800.$$

There are 3,628,800 ways to arrange the 10 books on the shelf. It doesn't seem possible that there could be this many ways, but there are!   ∎

Let's generalize Example 4.37 to find the number of possible permutations of m objects among themselves. Using the permutations rule, we conclude that

$$(m)_m = \frac{m!}{(m-m)!} = \frac{m!}{0!} = \frac{m!}{1} = m!.$$

Consequently, we have the following formula as a special case of the permutations rule.

FORMULA 4.11   **SPECIAL PERMUTATIONS RULE**

> The number of possible permutations of m objects among themselves is
>
> $$(m)_m = m!.$$

Combinations

A **combination** of r objects from a collection of m objects is any *unordered* arrangement of r of the m objects, in other words, any subset of r objects from the collection of m objects. Note that order matters in permutations, but not in combinations. The number of possible combinations of r objects that can be formed from a collection of m objects is denoted by $\binom{m}{r}$ or $_mC_r$. Let's return to the situation of Example 4.35.

Example 4.38   Introduces Combinations

Consider the collection consisting of the five letters a, b, c, d, e.

a. List all possible combinations of three letters from this collection of five letters.

b. Use part (a) to determine the number of possible combinations of three letters that can be formed from the collection of five letters; that is, determine $\binom{5}{3}$.

SOLUTION   a. For this part we need to list all unordered arrangements (subsets) of three letters from the first five letters in the English alphabet. This is done in Table 4.15.

TABLE 4.15
Combinations

| $\{a, b, c\}$ | $\{a, b, d\}$ | $\{a, b, e\}$ | $\{a, c, d\}$ | $\{a, c, e\}$ | $\{a, d, e\}$ | $\{b, c, d\}$ | $\{b, c, e\}$ | $\{b, d, e\}$ | $\{c, d, e\}$ |

b. From Table 4.15 we see that there are 10 possible combinations of three letters from the collection of five letters; in other words, $\binom{5}{3} = 10$.   ∎

In Example 4.38 we obtained the number of possible combinations by a direct listing. We can avoid resorting to a direct listing by deriving a formula for determining the number of possible combinations. To see how this is done, let's return once more to the English-letters example.

Look at the first combination in Table 4.15, $\{a, b, c\}$. By the special permutations rule, Formula 4.11, there are $3! = 6$ permutations of these three letters among themselves; namely, *abc, acb, bac, bca, cab,* and *cba.* These six permutations are the ones displayed in the first column of Table 4.14 on page 247. Similarly, there are $3! = 6$ permutations of the three letters in the second combination in Table 4.15, $\{a, b, d\}$. These six permutations are the ones displayed in the second column of Table 4.14. The same comments apply to the other eight combinations in Table 4.15.

Therefore we see that to each combination of three letters from the collection of five letters, there correspond $3!$ permutations of three letters from the collection of five letters. Moreover, any such permutation is accounted for in this way. Consequently, there must be $3!$ times as many permutations as combinations; or, equivalently, the number of possible combinations of three letters from the collection of five letters must equal the number of possible permutations of three letters from the collection of five letters divided by $3!$:

$$\binom{5}{3} = \frac{(5)_3}{3!} = \frac{5!/(5-3)!}{3!} = \frac{5!}{3!\,(5-3)!} = \frac{5 \cdot 4 \cdot \cancel{3!}}{\cancel{3!}\,2!} = \frac{5 \cdot 4}{2} = 10.$$

This is the number we obtained in Example 4.38 by a direct listing.

The same type of argument that we just gave holds in general. In other words, we have the following rule for determining the number of possible combinations.

FORMULA 4.12   **COMBINATIONS RULE**

> The number of possible combinations of r objects from a collection of m objects is given by the formula
>
> $$\binom{m}{r} = \frac{m!}{r!\,(m-r)!}.$$

Example 4.39 Illustrates Formula 4.12

In order to recruit new members, a compact-disc club advertises a special introductory offer: A new member agrees to buy 1 compact disc at regular club prices and receives free any 4 compact discs of his or her choice from a collection of 69 compact discs. How many possibilities does a new member have for the selection of the 4 free compact discs?

SOLUTION Any particular selection of 4 compact discs from 69 compact discs is a combination of 4 objects from a collection of 69 objects. Therefore, by the combinations rule, the number of possible selections equals

$$\binom{69}{4} = \frac{69!}{4!\,(69-4)!} = \frac{69!}{4!\,65!} = \frac{69 \cdot 68 \cdot 67 \cdot 66 \cdot \cancel{65!}}{4!\,\cancel{65!}} = 864,501.$$

There are 864,501 possibilities for the selection of 4 compact discs from the collection of 69 compact discs. ∎

Example 4.40 Illustrates Formula 4.12

An economics professor is teaching a junior-level course containing 42 students using a new teaching method. The professor wants to conduct in-depth interviews with the students to get feedback on the new teaching method, but does not want to interview all 42 of them. She decides to interview a sample of five students from the class. How many different samples are possible?

SOLUTION A sample of 5 students from the class of 42 students can be considered a combination of 5 objects from a collection of 42 objects. Consequently, by the combinations rule, the number of possible samples is

$$\binom{42}{5} = \frac{42!}{5!\,(42-5)!} = \frac{42!}{5!\,37!} = 850,668.$$

There are 850,668 different samples of 5 students that can be obtained from the population of 42 students in the class. ∎

Example 4.40 shows how to determine the number of possible samples of a specified size from a finite population. This is so important that we record it as the following formula.

FORMULA 4.13 **NUMBER OF POSSIBLE SAMPLES**

The number of possible samples of size n from a population of size N is

$$\binom{N}{n} = \frac{N!}{n!\,(N-n)!}.$$

Some Applications to Probability

Suppose an experiment has N equally likely possible outcomes. Then according to the f/N rule, the probability that a specified event occurs equals the number of ways, f, that the event can occur divided by the total number, N, of possible outcomes.

Although in the probability problems we have considered up to this point it has been easy to determine f and N, that is not always the case. We must often use counting rules to obtain the number of possible outcomes, N, and the number of ways that the specified event can occur, f. Examples 4.41 and 4.42 provide some typical ways in which counting rules are applied to solve probability problems.

Example 4.41 Applying Counting Rules to Probability

A drug known to have a 50% effectiveness rate in curing a certain disease is administered to seven people who have the disease. Determine the probability that exactly four of the seven people are cured.

SOLUTION Because the drug has a 50% effectiveness rate, the possible outcomes of the experiment (cure–noncure results) are equally likely. Thus we can apply the f/N rule to obtain the probability in question. First let's determine the number of possible outcomes, N. There are two possibilities for the first person (cured or not cured), two possibilities for the second person, and so on. Hence by the FCR, there are

$$\underbrace{2 \cdot 2 \cdots 2}_{7 \text{ times}} = 2^7 = 128$$

possibilities altogether. Thus $N = 128$.

Next we need to determine the number of ways, f, that the specified event can occur; that is, we must ascertain in how many outcomes exactly four of the seven people are cured. But this is just the number of ways that we can stipulate four people (the ones who will be cured) from seven people—the number of possible combinations of four objects from seven objects. Hence, by the combinations rule, Formula 4.12 on page 250, the number of ways that the specified event can occur is

$$\binom{7}{4} = \frac{7!}{4!\,(7-4)!} = \frac{7!}{4!\,3!} = 35.$$

Thus $f = 35$.

Consequently, by the f/N rule, the probability that exactly four of the seven people are cured is

$$\frac{f}{N} = \frac{35}{128} = 0.273.$$

∎

Example 4.42 Applying Counting Rules to Probability

The quality-assurance engineer of a television company inspects TVs in lots of 100. He selects 5 of the 100 TVs at random and inspects them thoroughly. Assuming that 6 of the 100 TVs in the current lot are actually defective, find the probability that exactly 2 of the 5 TVs selected by the engineer are defective.

SOLUTION Because the engineer makes his selection at random, each of the possible outcomes is equally likely. This means that we can apply the f/N rule to obtain the required probability.

First we determine the number of possible outcomes, N, for the experiment. This is the number of ways that 5 TVs can be selected from the 100 TVs—the number of possible combinations of 5 objects from a collection of 100 objects. Applying the combinations rule, we obtain

$$\binom{100}{5} = \frac{100!}{5!\,(100-5)!} = \frac{100!}{5!\,95!} = 75{,}287{,}520.$$

Thus $N = 75{,}287{,}520$.

Next we determine the number of ways, f, that the specified event can occur, that is, the number of outcomes in which exactly 2 of the 5 TVs selected are defective. To accomplish this, it is helpful to think of the 100 TVs as partitioned into two groups, namely, the defective TVs and the nondefective TVs, as shown in the top part of Fig. 4.29.

FIGURE 4.29
Calculating the number of outcomes in which exactly 2 of the 5 TVs selected are defective

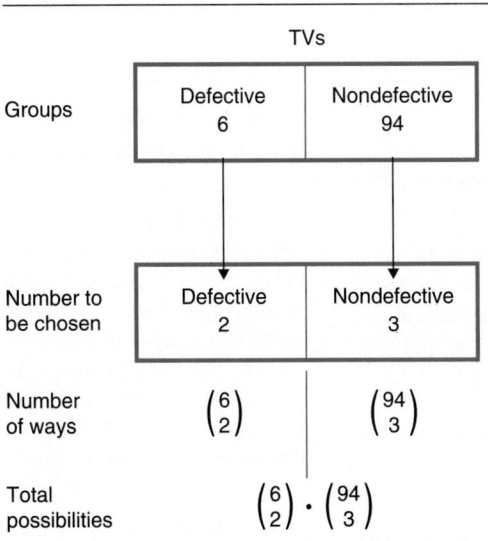

There are 6 TVs in the first group, of which 2 are to be selected. This can be done in

$$\binom{6}{2} = \frac{6!}{2!\,(6-2)!} = \frac{6!}{2!\,4!} = 15$$

ways. There are 94 TVs in the second group, of which 3 are to be selected. This can be done in

$$\binom{94}{3} = \frac{94!}{3!\,(94-3)!} = \frac{94!}{3!\,91!} = 134{,}044$$

ways. Consequently, by the FCR, there are a total of

$$\binom{6}{2} \cdot \binom{94}{3} = 15 \cdot 134{,}044 = 2{,}010{,}660$$

outcomes in which exactly 2 of the 5 TVs selected are defective. Thus $f = 2{,}010{,}660$. Figure 4.29 summarizes the calculations done in this paragraph.

Applying the f/N rule, we now conclude that the probability that exactly 2 of the 5 TVs selected are defective equals

$$\frac{f}{N} = \frac{2{,}010{,}660}{75{,}287{,}520} = 0.027.$$

In other words, there is a 2.7% chance that exactly 2 of the 5 TVs selected by the engineer will be defective.   ∎

EXERCISES 4.8

4.122 A zip code consists of five digits.
a. How many possible zip codes are there?
b. How many possible zip codes are there in which no digit appears more than once?

4.123 Telephone numbers in the United States consist of a three-digit area code followed by a seven-digit local number. Suppose neither the first digit of an area code nor the first digit of a local number can be a zero but that all other choices are acceptable.
a. How many different area codes are possible?
b. For a given area code, how many local telephone numbers are possible?
c. How many telephone numbers are possible?

4.124 Computerized testing systems are used extensively by professors. A physics professor needs to construct a five-question quiz, one question for each of five topics. The computerized testing system she uses provides 8 choices for the question on the first topic, 10 choices for the question on the second topic, 7 choices for the question on the third topic, 8 choices for the question on the fourth topic, and 6 choices

for the question on the fifth topic. How many possibilities are there for the five-question quiz?

4.125 *Scientific Computing & Automation* magazine offers free subscriptions to the scientific community. The magazine does ask, however, that a person answer six questions: primary title, type of facility, area of work, brand of computer used, type of operating system in use, and type of instruments in use. There are 6 choices given for the first question, 8 for the second, 5 for the third, 19 for the fourth, 16 for the fifth, and 14 for the sixth. How many possibilities are there for answering all six questions?

4.126 Determine the value of each of the following quantities.
a. $(7)_3$ b. $(5)_2$ c. $(8)_4$ d. $(6)_0$ e. $(9)_9$

4.127 Determine the value of each of the following quantities.
a. $(4)_3$ b. $(15)_4$ c. $(6)_2$ d. $(10)_0$ e. $(8)_8$

4.128 At a movie festival, a team of judges is to pick the first, second, and third place winners from the 18 films entered. How many possibilities are there?

4.129 Investment firms usually have a large selection of mutual funds from which an investor can choose. One such firm has 30 mutual funds. Suppose you plan to invest in four of these mutual funds, one during each quarter of next year. In how many different ways can you make these four investments?

4.130 The sales manager of a clothing company needs to assign seven salespeople to seven different territories. How many possibilities are there for the assignments?

4.131 An extrasensory-perception (ESP) experiment is conducted by a psychologist. For part of the experiment, the psychologist takes 10 cards, numbered 1–10, and shuffles them. Then she looks at the cards one at a time. While she looks at each card, the subject writes down the number he thinks is on the card.
a. How many possibilities are there for the order in which the subject writes down the numbers?
b. If the subject has no ESP and is just guessing each time, what is the probability that he writes down the numbers in the correct order (i.e., in the order that the cards are actually arranged)?

4.132 Determine the value of each of the following quantities.
a. $\binom{7}{3}$ b. $\binom{5}{2}$ c. $\binom{8}{4}$ d. $\binom{6}{0}$ e. $\binom{9}{9}$

4.133 Determine the value of each of the following quantities.
a. $\binom{4}{3}$ b. $\binom{15}{4}$ c. $\binom{6}{2}$ d. $\binom{10}{0}$ e. $\binom{8}{8}$

4.134 A poker hand consists of 5 cards dealt from an ordinary deck of 52 playing cards.
a. How many possible poker hands are there?
b. How many different hands are there consisting of three kings and two queens?
c. The hand in part (b) is an example of a full house: three cards of one denomination and two of another. How many different full houses are there?
d. Obtain the probability of being dealt a full house.

4.135 The U.S. Senate consists of 100 senators, two from each state. A committee consisting of five senators is to be formed.
a. How many different committees are possible?
b. How many are possible if no state may have more than one senator on the committee?

c. If the committee is selected at random from all 100 senators, what is the probability that no state will have both of its senators on the committee?

4.136 How many samples of size 5 are possible from a population of size 70?

4.137 How many samples of size 6 are possible from a population with 45 members?

4.138 Suppose you have a key ring with eight keys on it, one of which is your house key. Further suppose you get home after dark and can't see the keys on the key ring. You randomly try one key at a time, being careful not to mix the keys you have already tried with the ones you haven't. What is the probability that you get the right key
a. on the first try?
b. on the eighth try?
c. on or before the fifth try?

4.139 Refer to Example 4.42 on page 253. Determine the probability that
a. exactly one of the TVs selected is defective.
b. at most one of the TVs selected is defective.
c. at least one of the TVs selected is defective.

4.140 *The Birthday Problem.* A biology class has 38 students. Find the probability that at least two students in the class have the same birthday. For simplicity assume there are always 365 days in a year and that birth rates are constant throughout the year. *(Hint: First determine the probability that no two students have the same birthday and then apply the complementation rule.)*

4.141 The Arizona state lottery, *Lotto,* is played as follows: The player selects six numbers from the numbers 1–42 and buys a ticket for $1. There are six winning numbers, which are selected at random from the numbers 1–42. To win a prize, a *Lotto* ticket must contain three or more of the winning numbers. A ticket with exactly three winning numbers is paid $2. The prize for a ticket with exactly four, five, or six winning numbers depends on sales and on how many other tickets were sold that have exactly four, five, or six winning numbers, respectively. If you buy one *Lotto* ticket, determine the probability that
a. you win the jackpot; that is, your six numbers are the same as the six winning numbers.
b. your ticket contains exactly four winning numbers.
c. you don't win a prize.

4.142 A student takes a true-false test consisting of 15 questions. Assuming the student guesses at each question, find the probability that
a. the student gets at least one question correct.
b. the student gets a 60% or better on the exam.

4.143 According to the Center for Political Studies at the University of Michigan, Ann Arbor, roughly 50% of U.S. adults are Democrats. Suppose 10 U.S. adults are selected at random. Determine the approximate probability that
a. exactly five are Democrats.
b. at least eight are Democrats.

4.144 Refer to Exercise 4.140, but now assume the class consists of N students.

a. Determine the probability that at least two of the students have the same birthday.
b. This part presumes you have access to a computer or a programmable calculator. Use part (a) to construct a table giving the probability that at least two of the students in the class have the same birthday, for $N = 2, 3, \ldots, 70$.

4.145 Suppose a simple random sample of size n is to be taken without replacement from a population of size N.
a. Determine the probability that any particular sample of size n is the one selected.
b. Determine the probability that any specified member of the population is included in the sample.
c. Determine the probability that any k specified members of the population are included in the sample.

Chapter　Review

Key Terms

Formulas   Classical probability (f/N rule for equally likely outcomes), *184*

$$P(E) = \frac{f}{N}$$

(f = number of ways E can occur, N = total number of possible outcomes)

Special addition rule, *201*

$$P(A \text{ or } B \text{ or } C \text{ or } \cdots) = P(A) + P(B) + P(C) + \cdots$$

($A, B, C, \ldots$ mutually exclusive)

Complementation rule, *203*

$$P(E) = 1 - P(\text{not } E)$$

General addition rule, *204*

$$P(A \text{ or } B) = P(A) + P(B) - P(A \& B)$$

Conditional-probability rule, *219*

$$P(B \mid A) = \frac{P(A \& B)}{P(A)}$$

General multiplication rule, *224*

$$P(A \& B) = P(A) \cdot P(B \mid A)$$

Special multiplication rule, *231*

$$P(A \& B \& C \& \cdots) = P(A) \cdot P(B) \cdot P(C) \cdots$$

($A, B, C, \ldots$ independent)

Rule of total probability,* *237*

$$P(B) = \sum_{j=1}^{k} P(A_j) \cdot P(B \mid A_j)$$

($A_1, A_2, \ldots, A_k$ mutually exclusive and exhaustive)

Bayes's rule,* *239*

$$P(A_i \mid B) = \frac{P(A_i) \cdot P(B \mid A_i)}{\sum_{j=1}^{k} P(A_j) \cdot P(B \mid A_j)}$$

($A_1, A_2, \ldots, A_k$ mutually exclusive and exhaustive)

Factorial,* *246*

$$k! = k(k-1)\cdots 2 \cdot 1$$

(k a positive integer)

Permutations rule,* *248*

$$(m)_r = \frac{m!}{(m-r)!}$$

Special permutations rule,* *249*

$$(m)_m = m!$$

Combinations rule,* *250*

$$\binom{m}{r} = \frac{m!}{r!\,(m-r)!}$$

Number of possible samples,* *251*

$$\binom{N}{n} = \frac{N!}{n!\,(N-n)!}$$

(N = population size, n = sample size)

You Should Be Able To

1. use and understand the preceding formulas.
2. find and describe (not E), (A & B), and (A or B).
3. determine whether two or more events are mutually exclusive.
4. read and interpret contingency tables.
5. construct a joint probability distribution from a contingency table.
6. compute conditional probabilities both directly and by using the conditional-probability rule.
7. determine whether two events are independent.
8. determine whether two or more events are exhaustive.*

9. state and apply the rule of total probability.*
10. state and apply Bayes's rule.*
11. state and apply the fundamental counting rule (FCR).*
12. state and apply the permutations and combinations rules.*
13. apply counting rules to solve probability problems when appropriate.*

REVIEW TEST

1. The first two columns of Table 4.16 provide a frequency distribution for the adjusted gross incomes from 1989 federal individual income tax returns. Frequencies (number of returns) are in thousands. [SOURCE: U.S. Internal Revenue Service, *Statistics of Income, Individual Income Tax Returns.*]

TABLE 4.16 Adjusted gross incomes

Adjusted gross income	Number of returns	Event	Probability
Under $10,000	14,322	A	
$10,000–$19,999	21,702	B	
$20,000–$29,999	16,716	C	
$30,000–$39,999	12,047	D	
$40,000–$49,999	8,560	E	
$50,000–$99,999	12,960	F	
$100,000 & over	2,873	G	
	89,180		

Suppose a 1989 federal individual income tax return is selected at random.

a. Determine $P(A)$, the probability that the return selected shows an adjusted gross income under $10,000.
b. Find the probability that the return selected shows an adjusted gross income between $30,000 and $99,999.
c. Compute the probability of each of the seven events in the third column of Table 4.16 and record those probabilities in the fourth column.

2. Refer to Problem 1. Suppose a 1989 federal individual income tax return is selected at random. Let

H = event the return shows an adjusted gross income between $20,000 and $99,999,

I = event the return shows an adjusted gross income of at most $49,999,

J = event the return shows an adjusted gross income of at most $99,999,

K = event the return shows an adjusted gross income of at least $50,000.

Describe each of the following events in words and determine the number of outcomes (returns) that comprise each event.

a. (not J)   b. (H & I)   c. (H or K)
d. (H & K)

3. For the following groups of events from Problem 2, determine which are mutually exclusive.

a. H and I b. I and K c. H and (not J)
d. H, (not J), and K

4. Refer to Problems 1 and 2.
a. Use the second column of Table 4.16 and the f/N rule to compute the probability of each of the four events H, I, J, and K.
b. Express each of the events H, I, J, and K in terms of the mutually exclusive events in the third column of Table 4.16.
c. Compute the probability of each of the four events H, I, J, and K using your results from part (b), the special addition rule, and the fourth column of Table 4.16, which you completed in Problem 1(c).

5. Consider the events (not J), (H & I), (H or K), and (H & K) discussed in Problem 2.
a. Find the probability of each of those four events using the f/N rule. (Use the second column of Table 4.16.)
b. Compute $P(J)$ using the complementation rule and your result for $P(\text{not } J)$ from part (a).
c. In Problem 4(a) you found that $P(H) = 0.564$ and $P(K) = 0.178$; and in part (a) of this problem you found that $P(H \& K) = 0.145$. Using those probabilities and the general addition rule, find $P(H \text{ or } K)$. Compare your answer with the value for $P(H \text{ or } K)$ that you obtained in part (a).

6. The U.S. National Center for Education Statistics publishes information about school enrollment in *Digest of Education Statistics*. Table 4.17, shown at the top of the first column on the next page, provides a contingency table for enrollment in public and private schools by level. Frequencies are in thousands of students.

TABLE 4.17   Enrollment by level and type

Type

	Public T_1	Private T_2	Total
Elementary L_1	26,951	3,600	30,551
High school L_2	12,215	1,400	13,615
College L_3	9,612	2,562	12,174
Total	48,778	7,562	56,340

(Level)

a. How many cells are in this contingency table?
b. How many students are in high school?
c. How many students attend public schools?
d. How many students attend private colleges?

7. Refer to Problem 6. Suppose a student is selected at random.
   a. Describe each of the following events in words.
      i. L_3 ii. T_1 iii. (T_1 & L_3)
   b. Obtain the probability of each event in part (a) and interpret in terms of percentages.
   c. Construct a joint probability distribution corresponding to Table 4.17.
   d. Compute $P(T_1$ or $L_3)$ using
      i. Table 4.17 and the f/N rule.
      ii. the general addition rule and your results from part (b).

8. Refer to Problem 6. Suppose a student is selected at random.
   a. Find $P(L_3 \mid T_1)$ directly using Table 4.17 and the f/N rule. Interpret the probability you obtain in terms of percentages.
   b. Find $P(L_3 \mid T_1)$ using the conditional-probability rule and your results from Problem 7(b).

9. Refer to Problem 6. Suppose a student is selected at random.
   a. Using Table 4.17, find $P(T_2)$ and $P(T_2 \mid L_2)$.
   b. Are events L_2 and T_2 independent? Explain your answer in terms of percentages.
   c. Are events L_2 and T_2 mutually exclusive?
   d. Is the event that a student is in elementary school independent of the event that a student attends public school? Justify your answer.

10. During one year, the College of Public Programs at Arizona State University awarded the following number of masters degrees.

Type of degree	Frequency
Master of arts	3
Master of public administration	28
Master of science	19

Suppose two students who received such masters degrees are selected at random without replacement. Determine the probability that
   a. the first student selected received a master of arts and the second student selected a master of science.
   b. both students selected received a master of public administration.
   c. Construct a tree diagram for this problem similar to Fig. 4.25 on page 226.
   d. Obtain the probability that the two students selected received the same degree.

11. A manufacturer of electric water heaters knows that 25% of the water heaters made by his company last more than 10 years. Suppose four water heaters made by this company are selected at random. Determine the probability that
   a. all four last more than 10 years.
   b. the first three selected last more than 10 years and the fourth does not.
   c. exactly three of the four last more than 10 years.

12. Complete the following statements.
   a. A and B are mutually exclusive if . . .
   b. A and B are independent if . . .
   c. The general addition rule states . . .
   d. The special multiplication rule states . . .

13. Suppose A and B are events such that $P(A) = 0.4$, $P(B) = 0.5$, and $P(A$ & $B) = 0.2$.
   a. Are A and B mutually exclusive? Why or why not?
   b. Are A and B independent? Why or why not?

*14. A poll was taken in June 1989 by *The New York Times* to gauge the sentiment of Americans on the changes in the role of women over the last 20 years. One question asked was: "Many women have better jobs and more opportunities than they did 20 years ago. Do you think women have had to give up too much in the process, or not?" The relevant data are shown in the following chart.

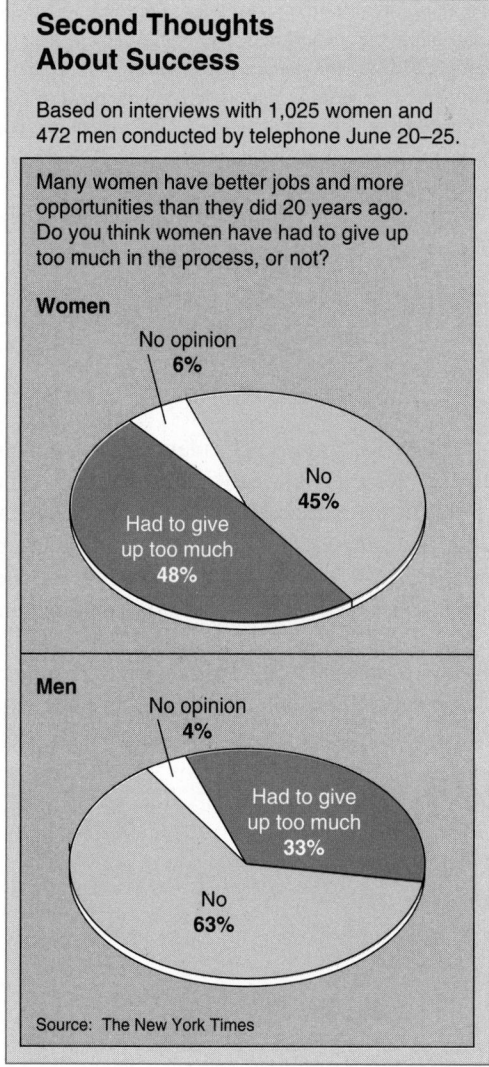

Second Thoughts About Success

Based on interviews with 1,025 women and 472 men conducted by telephone June 20–25.

Many women have better jobs and more opportunities than they did 20 years ago. Do you think women have had to give up too much in the process, or not?

Women

No opinion
6%

No
45%

Had to give
up too much
48%

Men

No opinion
4%

Had to give
up too much
33%

No
63%

Source: The New York Times

Suppose a person who participated in the survey is selected at random. Find the probability that

a. the person answered "no" to the question, given that the person selected is a woman.

b. the person answered "no" to the question.

c. the person is a woman.

d. the person is a woman, given that the person selected answered "no" to the question.

e. Interpret the probabilities obtained in parts (a)–(d) in terms of percentages.

f. Of the four probabilities in parts (a)–(d), which are prior and which are posterior?

*15. In Example 4.36 we considered exacta wagering in horse racing. Two similar wagers are the quinella and the trifecta. In a quinella wager, the bettor picks the two horses he or she believes will finish first and second, but not in a specified order. In a trifecta wager, the bettor picks the three horses he or she thinks will finish first, second, and third in a specified order. For a 12-horse race,

a. how many different quinella wagers are there?

b. how many different trifecta wagers are there?

c. Repeat parts (a) and (b) for an eight-horse race.

*16. A bridge hand consists of 13 cards dealt at random from an ordinary deck of 52 playing cards.

a. How many possible bridge hands are there?

b. Find the probability of being dealt a bridge hand that contains exactly two of the four aces.

c. Find the probability of being dealt an 8-4-1 distribution, that is, eight cards of one suit, four of another, and one of another.

d. Determine the probability of being dealt a 5-5-2-1 distribution.

e. Determine the probability of being dealt a hand void in a specified suit.

Using the Focus Database

Appendix B contains a printout of a database obtained by randomly selecting 500 Arizona State University sophomores. Seven variables are considered for each student: sex, high-school GPA, SAT math score, cumulative GPA, SAT verbal score, age, and total hours. Use Minitab or some other statistical software to solve the problems presented in parts (a)–(f) below.

a. Obtain a relative-frequency distribution for the sex data.

b. Using your answer from part (a), determine the probability that a randomly selected sophomore (from the sample of 500) is a female.

c. Consider the experiment of selecting a sophomore at random from the sample of 500 and observing the sex of the person obtained. Simulate that experiment 1000 times. *(Hint:* The simulation is equivalent to taking a random sample of size 1000 with replacement.*)*

d. Referring to the simulation performed in part (c), in approximately what percentage of the 1000 experiments would you expect a female to be selected? Compare that percentage to the actual percentage of the 1000 experiments in which a female was selected.

e. Obtain a contingency table that cross classifies the 500 sophomores by age and sex.

f. Obtain a joint probability distribution or joint percentage distribution of age and sex for the 500 sophomores. Interpret your results.

Case Study   *Lotto*—The Arizona State Lottery

The Arizona state lottery, called *Lotto,* is played as follows: The player selects six numbers from the numbers 1–42 and buys a ticket for $1. There are six winning numbers, which are selected at random from the numbers 1–42. To win a prize, a *Lotto* ticket must contain three or more of the winning numbers. A ticket with exactly three winning numbers is paid $2. The prize for a ticket with exactly four, five, or six winning numbers depends on sales and on how many other tickets were sold that have exactly four, five, or six winning numbers, respectively.

Table 4.18 displays the probabilities (to seven decimal places) for the number of winning numbers that a *Lotto* ticket contains.[†] The table shows, for instance, that the probability of winning the jackpot (i.e., having all six winning numbers) is approximately 0.0000002, or 1 in 5 million.

TABLE 4.18
Probabilities for *Lotto*

Number of winning numbers	Probability
0	0.3713060
1	0.4311941
2	0.1684352
3	0.0272219
4	0.0018014
5	0.0000412
6	0.0000002

Suppose we were to purchase one *Lotto* ticket per week. Let k be one of the numbers 0–6 and let p_k denote the probability that a ticket contains exactly k winning numbers. Then it can be shown that, on the average, it would take $1/p_k$ weeks until we obtained a ticket with exactly k winning numbers. For example, as we see from Table 4.18, $p_6 = 0.0000002$,

---

[†] These probabilities can be computed using the f/N rule or by employing a formula derived from the f/N rule. That formula is called the *hypergeometric probability formula.* The interested reader should refer to Exercise 5.68 after completing Section 5.4.

and so $1/p_6 = 5,000,000$. Consequently, if we purchased one *Lotto* ticket per week, then we should expect to wait approximately 5 million weeks, or roughly 96,154 years, before winning the jackpot.[†]

a. If you purchase one ticket, what is the probability that you win a prize?
b. If you purchase one ticket, what is the probability that you don't win a prize?
c. If you win a prize, what is the probability it is the $2 prize for exactly three winning numbers?
d. If you were to buy one ticket per week, approximately how long should you expect to wait before getting a ticket with exactly three winning numbers?
e. If you were to buy one ticket per week, approximately how long should you expect to wait before winning a prize?

[†] The probability 0.0000002 for winning the jackpot is approximate. Using a more exact value, we find that we should expect to wait 100,880.5 years before winning the jackpot.

James Bernoulli

James Bernoulli was born on December 27, 1654, in Basle, Switzerland. He was the first of the Bernoulli family of mathematicians; his younger brother John and various nephews and grandnephews were also renowned mathematicians. His father, Nicolaus Bernoulli (1623–1708), planned the ministry as James's career. James rebelled, however, finding mathematics much more interesting.

Although Bernoulli was schooled in theology, he studied mathematics on his own. He was especially fascinated with calculus. In a 1690 issue of the journal *Acta eruditorum,* Bernoulli used the word *integral* to describe the inverse of differential. The results of his studies of calculus and the catenary (the curve formed by a cord freely suspended between two fixed points) were soon applied to the building of suspension bridges.

Some of Bernoulli's most important work was published posthumously in *Ars Conjectandi* (The Art of Conjecturing) in 1713.

This book contains his theory of permutations and combinations, the Bernoulli numbers, and his writings on probability, which include the weak law of large numbers for Bernoulli trials. *Ars Conjectandi* has been regarded as the beginning of the theory of probability.

Both James and his younger brother John were highly accomplished mathematicians. But rather than collaborating in their work, they were most often competing. James would publish a question inviting solutions in a professional journal. John would reply in the same journal with a solution, only to find that an ensuing issue would contain another article by James, telling him that he was wrong. In their later years, they communicated only in this manner.

Bernoulli began lecturing in natural philosophy and mechanics at the University of Basle in 1682 and became a Professor of Mathematics there in 1687. He remained at the university until his death of a "slow fever" on August 10, 1705.

Chapter 5

Discrete Random Variables

· · · · ·

General Objective

In Chapter 4 we began our investigation of probability. We continue that investigation in this chapter by studying random variables. A **random variable** is a numerical quantity whose value depends on chance. For instance, suppose a person is to be selected at random. The number of CDs (compact discs) owned by the person chosen is an example of a random variable, as is the height of the person chosen. Both of these quantities are numerical and their values depend on chance, namely, on which person is selected.

There are two main types of random variables: discrete and continuous. A **discrete random variable** is a random variable whose possible values form a finite (or countably infinite) set of numbers, usually some collection of whole numbers. The number of CDs owned by a randomly selected person is an example of a discrete random variable. A **continuous random variable** is a random variable whose possible values form some interval of numbers. The height of a randomly selected person is an example of a continuous random variable.

As we will see, random variables play a key role in designing and implementing statistical-inference procedures. In this chapter we will study discrete random variables. Continuous random variables will be studied in several future chapters.

Chapter Outline

Case Study

Aces Wild on the Sixth at Oak Hill

A most amazing event occurred during the second round of the 1989 U.S. Open at Oak Hill in Pittsford, New York—four golfers made holes-in-one on the sixth hole. What are the chances of such a remarkable event occurring? As we will discover in the case study at the end of this chapter, the chances are extremely slight; in fact, they are less than the chances of winning the jackpot in the Arizona state lottery!

5.1   DISCRETE RANDOM VARIABLES; PROBABILITY DISTRIBUTIONS

In this section we will introduce discrete random variables and their probability distributions. We begin with the following illustration.

Example 5.1   Random Variables

Table 5.1 presents a grouped-data table for the number of cars owned by each of the families in a small city. The table shows, for instance, that 1796 of the 6487 families, or 27.7%, own three cars.

TABLE 5.1
Grouped-data table for
number of cars owned

Cars owned x	Frequency f	Relative frequency
0	27	0.004
1	1422	0.219
2	2865	0.442
3	1796	0.277
4	324	0.050
5	53	0.008
	6487	1.000

Since the "number of cars owned" varies from family to family, it is a *variable*.[†] Suppose now that a family is to be selected at random. Then the "number of cars owned" by the family chosen is called a *random variable* because its value depends on chance, namely, on which family is selected.   ■

DEFINITION 5.1   **RANDOM VARIABLE**

> A *random variable* is a numerical quantity whose value depends on chance.

The random variable in Example 5.1 is the "number of cars owned" by a randomly selected family. The possible values of that random variable form a finite set: the numbers 0, 1, 2, 3, 4, and 5 (see Table 5.1). Consequently, the random variable "number of cars owned" is called a discrete random variable.

DEFINITION 5.2   **DISCRETE RANDOM VARIABLE**

> A *discrete random variable* is a random variable whose possible values form a finite or countably infinite set of numbers.

[†] We discussed variables in detail in Section 2.1.

A discrete random variable usually involves a count of something, like the number of cars owned by a randomly selected family, the number of people waiting for a haircut in a barber shop, or the number of households in a sample that own a personal computer.

Random-Variable Notation

It is customary to denote random variables using letters such as x, y, and z. In doing so, we can develop useful notation for dealing with random variables.

Let's return to the situation of Example 5.1, where we considered the random variable "number of cars owned" by a randomly selected family. Suppose we let x denote that random variable. Then, for instance, we can represent the event that the family selected owns three cars by $\{x = 3\}$, read as "x equals three."

The probability that the family selected owns three cars can be expressed symbolically as $P(x = 3)$, read as "the probability that x equals three." When there is no question about which random variable is under consideration, we often write $P(3)$ instead of $P(x = 3)$.

We should clarify a point that is frequently confused. The notation $\{x = 3\}$ refers to the event that the family selected owns three cars, whereas the notation $P(x = 3)$ refers to the probability of that event.

Probability Distributions and Histograms

We next discuss probability distributions and probability histograms for discrete random variables. These are defined as follows.

DEFINITION 5.3 **PROBABILITY DISTRIBUTION AND PROBABILITY HISTOGRAM**

> ***Probability distribution:*** A listing of the possible values and corresponding probabilities of a discrete random variable; or a formula for the probabilities.
>
> ***Probability histogram:*** A graph of the probability distribution that displays the possible values of a discrete random variable on the horizontal axis and the probabilities of those values on the vertical axis. The probability of each value is represented by a vertical bar whose height is equal to the probability.

Example 5.2 Illustrates Definition 5.3

Refer to Example 5.1 and, as before, let x denote the number of cars owned by a randomly selected family.

a. Determine the probability distribution of the random variable x.

b. Construct a probability histogram for the random variable x.

SOLUTION a. We want to determine the probability of each of the possible values of the random variable x. To obtain, for instance, $P(x = 3)$, the probability that the family selected owns three cars, we apply the f/N rule. Referring to Table 5.1 on page 266, we see that

$$P(x = 3) = \frac{f}{N} = \frac{1796}{6487} = 0.277.$$

The other probabilities are found in the same way. Table 5.2 displays those probabilities and provides the probability distribution of the random variable x.

TABLE 5.2

Probability distribution of the random variable x, the number of cars owned by a randomly selected family

Cars owned x	Probability $P(x)$
0	0.004
1	0.219
2	0.442
3	0.277
4	0.050
5	0.008
	1.000

b. To construct a probability histogram for x, we plot its possible values on a horizontal axis and display the corresponding probabilities using vertical bars. Referring to Table 5.2, we get the probability histogram shown in Fig. 5.1.

FIGURE 5.1

Probability histogram for the random variable x, the number of cars owned by a randomly selected family

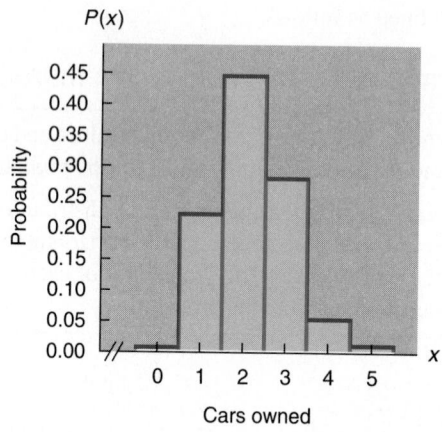

The probability histogram provides a quick and easy way to visualize how the probabilities of the random variable x are distributed. ∎

Notice that the probabilities in the second column of Table 5.2 sum to 1. This is always true for discrete random variables.

| KEY FACT 5.1 | **SUM OF THE PROBABILITIES OF A RANDOM VARIABLE** |

For any discrete random variable, x, the sum of the probabilities is equal to 1; in symbols, $\Sigma P(x) = 1$.

For an *integer-valued* random variable, the area of each histogram bar equals the corresponding probability. So for such a random variable, the total area of the histogram bars in its probability histogram equals 1.

Also observe that for the random variable x considered in Examples 5.1 and 5.2, its probabilities, given in the second column of Table 5.2, are identical to the relative frequencies shown in the third column of Table 5.1 on page 266. This is always the case when dealing with finite populations, as indicated in Key Fact 5.2.

| KEY FACT 5.2 | **PROBABILITIES FOR A RANDOM VARIABLE AS PERCENTAGES** |

Suppose a random variable x is defined to be the value of a randomly selected member from a finite population. Then probabilities for x are equal to relative frequencies (percentages). In other words, the probability distribution of the random variable x is identical to the relative-frequency distribution of the population values.

Examples 5.3 and 5.4 provide additional illustrations of random-variable notation and probability distributions.

Example 5.3   Random-Variable Notation and Probability Distributions

---

The U.S. National Center for Education Statistics compiles enrollment data on U.S. public schools and publishes the results in *Digest of Education Statistics*. Table 5.3 displays a frequency distribution for the enrollment by grade in public elementary schools ($0 =$ kindergarten, $1 =$ first grade, and so on). Frequencies are in thousands of students.

TABLE 5.3

Frequency distribution for enrollment by grade in U.S. public elementary schools

Grade y	Frequency f
0	2,700
1	2,951
2	2,786
3	2,812
4	2,926
5	3,131
6	3,180
7	3,184
8	3,062
	26,732

Suppose a student in elementary school is to be selected at random. Let y denote the grade of the student selected. Then y is a random variable whose possible values are $0, 1, 2, \ldots, 8$.

a. Use random-variable notation to represent the event that the student selected is in the fifth grade.

b. Determine $P(y = 5)$ and express the result in terms of percentages.

c. Find $P(7)$.

d. Determine the probability distribution of the random variable y.

SOLUTION   a. The event that the student selected is in the fifth grade can be represented as $\{y = 5\}$.

b. $P(y = 5)$ is the probability that the student selected is in the fifth grade. Using Table 5.3 and the f/N rule, we can obtain that probability:

$$P(y = 5) = \frac{f}{N} = \frac{3,131}{26,732} = 0.117.$$

In terms of percentages, this means that 11.7% of the students in elementary school are in the fifth grade.

c. $P(7)$ is the probability that the student selected is in the seventh grade. So again using Table 5.3 and the f/N rule, we get

$$P(7) = \frac{f}{N} = \frac{3,184}{26,732} = 0.119.$$

d. The probability distribution of y is obtained by computing $P(y)$ for $y = 0, 1, 2, \ldots, 8$. We have already done that for $y = 5$ and $y = 7$. The other probabilities are computed similarly and are displayed in Table 5.4.

TABLE 5.4

Probability distribution of the random variable y, the grade of a randomly selected elementary-school student

Grade y	Probability $P(y)$
0	0.101
1	0.110
2	0.104
3	0.105
4	0.109
5	0.117
6	0.119
7	0.119
8	0.115
	0.999 (rounding error)

Note: In Table 5.4 the sum of the probabilities is given as 0.999. As we know from Key Fact 5.1, the sum of the probabilities must be exactly 1. However, our computation is off a little since we rounded the probabilities for y to three decimal places.   ∎

Once we have the probability distribution of a discrete random variable, it is easy to determine any probability involving that random variable. The basic tool for accomplishing this is the special addition rule, Formula 4.1 on page 201.

Example 5.4   Random-Variable Notation and Probability Distributions

When a balanced dime is tossed three times, eight equally likely outcomes are possible, as shown in Table 5.5.

TABLE 5.5
Possible outcomes

HHH	HTH	THH	TTH
HHT	HTT	THT	TTT

Here, for instance, HHT means that the first two tosses are heads and the third is tails. Let x denote the total number of heads obtained in the three tosses. Then x is a random variable whose possible values are 0, 1, 2, and 3.

a. Use random-variable notation to represent the event that exactly two heads are tossed.

b. Determine $P(x = 2)$.

c. Find the probability distribution of the random variable x.

d. Use random-variable notation to represent the event that at most two heads are tossed.

e. Find $P(x \leq 2)$.

SOLUTION   a. The event that exactly two heads are tossed can be represented as $\{x = 2\}$.

b. $P(x = 2)$ is the probability that exactly two heads are tossed. We see from Table 5.5 that there are three ways to get a total of two heads and that there are eight possible outcomes altogether. Hence, by the f/N rule,

$$P(x = 2) = \frac{f}{N} = \frac{3}{8} = 0.375.$$

c. The remaining probabilities for x are computed as in part (b) and are shown in Table 5.6.

TABLE 5.6
Probability distribution of the random variable x, the number of heads obtained in three tosses of a balanced dime

No. of heads x	Probability $P(x)$
0	0.125
1	0.375
2	0.375
3	0.125
	1.000

d. The event that at most two heads are tossed can be represented as $\{x \leq 2\}$, read as "x is less than or equal to two."

e. $P(x \leq 2)$ is the probability that at most two heads are tossed. The event that at most two heads are tossed can be expressed as

$$\{x \leq 2\} = \big(\{x = 0\} \text{ or } \{x = 1\} \text{ or } \{x = 2\}\big).$$

Because the three events on the right are mutually exclusive (why is this so?), we can use the special addition rule and Table 5.6 to conclude that

$$P(x \leq 2) = P(x = 0) + P(x = 1) + P(x = 2)$$
$$= 0.125 + 0.375 + 0.375 = 0.875.$$

Thus the probability is 0.875 that at most two heads are tossed.   ∎

Interpretation of Probability Distributions

Recall that the relative-frequency interpretation of probability construes the probability of an event to be the relative frequency of its occurrence in a large number of (independent) repetitions of the experiment. Using that interpretation we can clarify the meaning of probability distributions. Example 5.5 shows how.

Example 5.5   Interpreting a Probability Distribution

Consider again the random variable x discussed in Example 5.4, the number of heads obtained in three tosses of a balanced dime. Suppose we repeat the experiment of observing the number of heads obtained in three tosses of a balanced dime a large number of times. Then the relative frequency of those times in which, say, no heads are obtained (i.e., $x = 0$) should be approximately equal to the probability of that event [i.e., $P(x = 0)$]. The same statement holds true for the other three possible values of the random variable x.

We used a computer to simulate 1000 observations of the number of heads obtained in three tosses of a balanced dime. Table 5.7 shows a frequency distribution and a relative-frequency distribution for the numbers of heads obtained in the 1000 observations. Thus, for instance, 136 of the 1000 observations resulted in no heads out of three tosses, giving a relative frequency of 0.136.

TABLE 5.7
Frequency and relative-frequency distributions for the numbers of heads obtained in three tosses of a balanced dime for 1000 observations

No. of heads x	Frequency f	Relative frequency
0	136	0.136
1	377	0.377
2	368	0.368
3	119	0.119
	1000	1.000

As expected, the relative frequencies in the third column of Table 5.7 are fairly close to the true probabilities in the second column of Table 5.6 on page 271. We can see this more easily if we compare the relative-frequency histogram for the simulation to the probability histogram of the random variable x. See Fig. 5.2.

FIGURE 5.2
(a) Relative-frequency histogram for the numbers of heads obtained in three tosses of a balanced dime for 1000 observations
(b) Probability histogram for the number of heads obtained in three tosses of a balanced dime

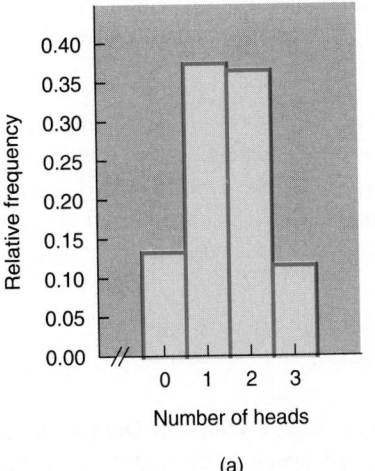
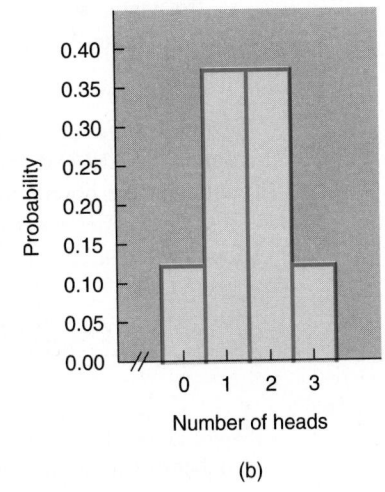

If we simulated, say, 10,000 observations instead of 1000, then the relative frequencies that would appear in the third column of Table 5.7 would most likely be very close to the true probabilities in the second column of Table 5.6.

In general, we can make the statement in Key Fact 5.3.

KEY FACT 5.3   **INTERPRETATION OF PROBABILITY DISTRIBUTIONS**

> The probability distribution of a random variable x can be given the following interpretation: In a large number of independent observations of x, a relative-frequency distribution of the observations will approximate the probability distribution of x, or equivalently, a relative-frequency histogram for the observations will approximate the probability histogram for x.

Using the Computer (Optional)

Minitab has a command called **Random** that will simulate observations of a specified random variable. In particular, by employing the **Discrete** subcommand of the Random command, we can simulate observations of any (finite-valued) discrete random variable. To illustrate, we return to the simulation discussed in Example 5.5.

Example 5.6   The Random Command and Discrete Subcommand

Table 5.7 on page 272 shows the results of simulating 1000 observations of the number of heads obtained in three tosses of a balanced dime. We performed the simulation by applying Minitab's Random command and Discrete subcommand. The details for doing this will now be explained.

To begin, we store the probability distribution shown in Table 5.6 on page 271 in columns named X and P(X). Then we proceed as follows.

Session commands: First we name an available column NUMHEADS. Next we type the command **RANDOM** followed by the number of observations required and the storage location for the observations; that is, we type RANDOM 1000 obs into 'NUMHEADS'; and press [Enter]. Then we use the **DISCRETE** subcommand to specify the probability distribution of the random variable to be simulated; that is, we type DISCRETE 'X' 'P(X)'. and press [Enter].

(or)

Menu commands: We choose **Calc ▸ Random Data ▸ Discrete...**, type 1000 for the number of rows to be generated, specify NUMHEADS in the **Store in column(s)** text box, specify X for **Values in**, specify 'P(X)' for **Probabilities in**, and then select **OK**.

As a consequence of the above commands, the numbers of heads obtained in three tosses of a balanced dime for 1000 observations are stored in NUMHEADS. To obtain the frequencies and relative frequencies for the numbers of heads, we apply Minitab's **Tally** command as follows.

Session commands: We type the command **TALLY** followed by the storage location of the data to be tallied; that is, we type TALLY 'NUMHEADS'; and press [Enter]. Then, in order to obtain the frequencies and relative frequencies, we use the subcommands **COUNTS** and **PERCENTS**. In other words, we type COUNTS;, press [Enter], type PERCENTS., and press [Enter] again.

(or)

Menu commands: We choose **Stat ▸ Tables ▸ Tally...**, specify NUMHEADS in the **Variables** text box, select **Counts** and **Percents** from the list of **Display** check boxes, and then select **OK**.

The computer output we actually got by performing the simulation and applying the Tally command is shown in Printout 5.1. Compare the table in Printout 5.1 to the frequency and relative-frequency distributions in Table 5.7.

PRINTOUT 5.1
Minitab output for the Random
command and Discrete
subcommand; and for the
Tally command and Counts
and Percents subcommands

```
MTB > RANDOM 1000 obs into 'NUMHEADS';
SUBC> DISCRETE 'X' 'P(X)'.
MTB > TALLY 'NUMHEADS';
SUBC> COUNTS;
SUBC> PERCENTS.


    NUMHEADS   COUNT  PERCENT
           0     136    13.60
           1     377    37.70
           2     368    36.80
           3     119    11.90
        N=  1000
```

If we performed the simulation again, it is virtually certain that we would obtain results different from (although similar to) those in Printout 5.1.   ■

EXERCISES 5.1

· **5.1** According to the Census Bureau publication *Current Population Reports,* a frequency distribution for the number of people per household in the United States is as follows. Frequencies are in millions of households.

No. of people	1	2	3	4	5	6	7†
Frequency	19.4	26.5	14.6	12.9	6.1	2.5	1.6

† Actually this is "7 or more," but we will consider it "7" for illustrative purposes.

Let x denote the number of people in a household selected at random.
a. What are the possible values of the random variable x?
b. Employ random-variable notation to represent the event that the household selected has exactly five people.
c. Determine $P(x = 5)$ and interpret your result in terms of percentages.
d. Find $P(3)$.
e. Obtain the probability distribution of x.
f. Construct a probability histogram for x.

· **5.2** The National Center for Education Statistics compiles enrollment data on U.S. public schools and reports the information in *Digest of Education Statistics.* The following table displays a frequency distribution for the enrollment

by grade in public secondary schools. Frequencies are in thousands of students.

Grade	9	10	11	12
Frequency	3290	3223	3041	2908

Suppose a student in public secondary school is to be selected at random. Let x denote the grade level of the student chosen.
a. What are the possible values of the random variable x?
b. Employ random-variable notation to represent the event that the student chosen is in the tenth grade.
c. Determine $P(x = 10)$ and interpret your result in terms of percentages.
d. Find $P(12)$.
e. Obtain the probability distribution of x.
f. Construct a probability histogram for x.

· **5.3** When two balanced dice are rolled, 36 equally likely outcomes are possible, as seen in Fig. 4.2 on page 186. Let y denote the sum of the dice.
a. What are the possible values of the random variable y?
b. Employ random-variable notation to represent the event that the sum of the dice is 7.
c. Find $P(y = 7)$.

d. Determine $P(11)$.

e. Find the probability distribution of y. Leave your probabilities in fraction form.

f. Construct a probability histogram for y.

• **5.4** When two balanced dice are rolled, 36 equally likely outcomes are possible, as seen in Fig. 4.2 on page 186. Let x denote the larger number showing on the two dice. For example, if the outcome is

then $x = 5$. If both dice come up the same number, then x equals that common value.

a. What are the possible values of the random variable x?

b. Employ random-variable notation to represent the event that the larger number is 4.

c. Find $P(x = 4)$.

d. Determine $P(2)$.

e. Find the probability distribution of x. Leave your probabilities in fraction form.

f. Construct a probability histogram for x.

• **5.5** Prescott National Bank has six tellers available to serve customers. The number of tellers busy with customers at, say, 1:00 P.M. varies from day to day, so it is a random variable, which we will call x. From past records the probability distribution of x is known to be as follows.

x	$P(x)$
0	0.03
1	0.05
2	0.08
3	0.15
4	0.21
5	0.26
6	0.22

The table indicates, for example, that the probability is 0.26 that exactly five of the tellers will be busy with customers at 1:00 P.M.; that is, about 26% of the time, exactly five tellers are busy with customers at 1:00 P.M. Use random-variable notation to represent each of the following events. At 1:00 P.M.,

a. exactly four tellers are busy.

b. at least two tellers are busy.

c. less than five tellers are busy.

d. at least two but less than five tellers are busy.

Use the special addition rule and the probability distribution to determine

e. $P(x = 4)$.   f. $P(x \geq 2)$.

g. $P(x < 5)$.   h. $P(2 \leq x < 5)$.

• **5.6** Based on past experience, a car salesperson knows that the number of cars she sells per week is a random variable, y, with a probability distribution as shown here.

y	$P(y)$
0	0.135
1	0.271
2	0.271
3	0.180
4	0.090
5	0.036
6	0.012
7	0.003
8	0.002

Suppose a week is selected at random. Use random-variable notation to represent each of the following events: The salesperson sells

a. exactly three cars.

b. at least three cars.

c. less than seven cars.

d. at least three but less than seven cars.

Use the special addition rule and the probability distribution to find

e. $P(y = 3)$.   f. $P(y \geq 3)$.

g. $P(y < 7)$.   h. $P(3 \leq y < 7)$.

• **5.7** Benny's Barber Shop in Cleveland has five chairs for waiting customers. Previous records indicate that the probability distribution for the number of customers waiting, y, is as follows.

y	0	1	2	3	4	5
$P(y)$	0.424	0.161	0.134	0.111	0.093	0.077

Represent each of the following events using random-variable notation.

a. Exactly two customers are waiting.

b. At most four customers are waiting.

c. More than one customer is waiting.

d. Between two and four customers, inclusive, are waiting.

Apply the special addition rule and the probability distribution to find

e. $P(y = 2)$.   f. $P(y \leq 4)$.   g. $P(y > 1)$.

h. $P(2 \leq y \leq 4)$.   i. $P(1 < y \leq 4)$.

5.8   A small company in Tulsa, Oklahoma, has a total of 10 employees; six are male and four are female. Three of the 10 employees are to be selected to represent the company at a small-business conference sponsored by the Small Business Administration. If the selection is done randomly, then the number of women chosen is a random variable, x, whose probability distribution is as follows.

x	0	1	2	3
$P(x)$	0.167	0.500	0.300	0.033

Use random-variable notation to express each of the following events. The number of women chosen is
a. exactly two.
b. at most two.
c. at least one.
d. either one or two.

Apply the special addition rule and the probability distribution to determine
e. $P(x = 2)$.   f. $P(x \le 2)$.
g. $P(x \ge 1)$.   h. $P(x = 1 \text{ or } 2)$.

5.9   Suppose that z is a random variable and that $P(z > 1.96) = 0.025$. Find $P(z \le 1.96)$. *(Hint:* Use the complementation rule, Formula 4.2 on page 203.)

5.10   (Computer exercise)   Refer to the probability distribution displayed in Table 5.6 on page 271.
a. Use Minitab or some other statistical software to repeat the simulation done in Example 5.6 on page 274.

b. Obtain a relative-frequency distribution for the numbers of heads in three tosses and compare it to the probability distribution in Table 5.6.
c. Obtain a relative-frequency histogram for the numbers of heads in three tosses and compare it to the probability histogram in Fig. 5.2(b) on page 273.
d. What do parts (b) and (c) illustrate?

5.11   (Computer exercise)   Refer to the probability distribution displayed in Table 5.2 on page 268.
a. Use Minitab or some other statistical software to simulate 2000 observations of the number of cars owned by a randomly selected family.
b. Obtain a relative-frequency distribution for the numbers of cars owned and compare it to the probability distribution in Table 5.2.
c. Obtain a relative-frequency histogram for the numbers of cars owned and compare it to the probability histogram in Fig. 5.1 on page 268.
d. What do parts (b) and (c) illustrate?

5.12   Suppose t and z are random variables.
a. If $P(t > 2.02) = 0.05$ and $P(t < -2.02) = 0.05$, obtain $P(-2.02 \le t \le 2.02)$.
b. If $P(-1.64 \le z \le 1.64) = 0.90$ and $P(z > 1.64) = P(z < -1.64)$, find $P(z > 1.64)$.

5.13   Let c and α be numbers, where $c > 0$ and $0 \le \alpha \le 1$. Also let x, y, and t denote random variables.
a. If $P(x > c) = \alpha$, determine $P(x \le c)$ in terms of α.
b. If $P(y > c) = \alpha/2$ and $P(y < -c) = P(y > c)$, find $P(-c \le y \le c)$ in terms of α.
c. If $P(-c \le t \le c) = 1 - \alpha$ and $P(t < -c) = P(t > c)$, find $P(t > c)$ in terms of α.

5.2   THE MEAN AND STANDARD DEVIATION OF A DISCRETE RANDOM VARIABLE

In this section we introduce the mean and standard deviation of a discrete random variable. As we will see, the mean and standard deviation of a discrete random variable are analogous to the mean and standard deviation of a population.

Mean of a Discrete Random Variable

We first introduce the mean of a discrete random variable. This is accomplished with the aid of Example 5.7.

Example 5.7   Introduces the Mean of a Discrete Random Variable

CJ^2 Business Services is a small company specializing in word processing. The company employs eight people. A grouped-data table for the weekly salaries of the eight employees is presented in Table 5.8.

TABLE 5.8
Grouped-data table for the weekly salaries of CJ^2 employees

Salary ($) x	Frequency f	Relative frequency
240	3	0.375
320	2	0.250
450	1	0.125
600	2	0.250
	8	1.000

a. Determine the population mean weekly salary of CJ^2 employees.

b. Suppose a CJ^2 employee is to be selected at random. Let x denote the weekly salary of the employee obtained. Express the formula for the mean weekly salary in terms of the probability distribution of the random variable x.

SOLUTION   a. The population mean weekly salary can be determined by employing Formula 3.4:

$$\mu = \frac{\Sigma xf}{N}.$$

This formula is used to compute the mean of a population when the data are in grouped form, as they are here. To apply the formula, we append an xf column to the frequency distribution displayed in the first two columns of Table 5.8, as shown in Table 5.9.

TABLE 5.9
Table for computing μ

x	f	xf
240	3	720
320	2	640
450	1	450
600	2	1200
	8	3010

Thus the mean weekly salary of the employees of CJ^2 is

$$\mu = \frac{\Sigma xf}{N} = \frac{3010}{8} = \$376.25.$$

b. For this part we are to express the formula for the mean weekly salary in terms of the probability distribution of the random variable x, where x is the weekly salary of a randomly selected CJ^2 employee. Because we are dealing here with a finite population, we know from Key Fact 5.2 on page 269 that probabilities for x are the same as relative frequencies. In other words, for each x value, the probability, $P(x)$, and the relative frequency, f/N, are equal.

Because probabilities and relative frequencies are the same here, we can express the formula for μ in terms of probabilities as follows:

$$\mu = \frac{\Sigma x f}{N} = \Sigma x \cdot \frac{f}{N} = \Sigma x P(x).$$

Since, expressed in this way, the formula for the mean, μ, involves the probability distribution of the random variable x, we make the following definition.   ■

DEFINITION 5.4   **MEAN OF A DISCRETE RANDOM VARIABLE**

> The **mean of a discrete random variable** x is defined by
>
> $$\mu_x = \Sigma x P(x).$$
>
> The terms **expected value** and **expectation** are commonly used in place of *mean*.

Note: μ has the subscript x to emphasize that it is the mean of the random variable x. If the random variable under consideration is called y, then we would write μ_y for its mean and express the formula in Definition 5.4 as $\mu_y = \Sigma y P(y)$.

As we learned in Chapter 3, the easiest way to compute descriptive measures is by using tables. The same is true for computing the mean of a discrete random variable, as Example 5.8 shows.

Example 5.8   Illustrates Definition 5.4

Suppose a CJ^2 employee is to be selected at random, and let x denote the weekly salary of the employee obtained. Find the mean, μ_x, of the random variable x.

SOLUTION   To compute μ_x, we must first determine the probability distribution of x. In this case, probabilities for x are equal to relative frequencies. So we can obtain the probability distribution of x simply by referring to the first and third columns of Table 5.8. The probability distribution of x is as shown in Table 5.10.

TABLE 5.10
Probability distribution of the random variable x, the weekly salary of a randomly selected CJ^2 employee

Salary x	Probability $P(x)$
240	0.375
320	0.250
450	0.125
600	0.250

From the printout we find that the probability of at most 2 arrivals between 6:00 P.M. and 7:00 P.M. is 0.0320.

c. Here we are to determine the probability of between 4 and 10 arrivals, inclusive. Minitab can be used in several ways to obtain that probability. One way is to instruct Minitab to print the individual probabilities for all the required x-values (in this case 4–10) for a Poisson distribution with parameter $\lambda = 6.9$. This is done as follows. First we store the integers between 4 and 10, inclusive, into a column named VALUES. Then we proceed in the following manner.

Session commands: We first type the command PDF followed by the storage location of the x-values whose probabilities are required; that is, we type PDF 'VALUES'; and press Enter. Then we type POISSON with lambda=6.9. and press Enter. Printout 5.7 displays these commands and the resulting output.

(or)

Menu commands: We choose **Calc ▸ Probability Distributions ▸ Poisson**. Next we select **Probability** to indicate that we want to obtain individual probabilities. To specify the appropriate Poisson distribution, we type 6.9 in the **Mean** text box. Next we select **Input column** and specify VALUES to tell Minitab that we want an individual probability for each of the numbers in that column. Finally, we select **OK**. The resulting output is portrayed in Printout 5.7.

PRINTOUT 5.7

Minitab output of individual
Poisson probabilities
for $x = 4$ to $x = 10$

```
MTB > PDF 'VALUES';
SUBC> POISSON with lambda=6.9.
       K          P( X = K)
      4.00           0.0952
      5.00           0.1314
      6.00           0.1511
      7.00           0.1489
      8.00           0.1284
      9.00           0.0985
     10.00           0.0679
```

The probability of between 4 and 10 arrivals is equal to the sum of the probabilities in the second column of Printout 5.7. Thus

$$P(4 \le x \le 10) = 0.0952 + 0.1314 + 0.1511 + 0.1489$$
$$+ 0.1284 + 0.0985 + 0.0679$$
$$= 0.8214.$$

Incidentally, more efficient ways exist of using Minitab to obtain this probability. See, for instance, Exercise 5.101. ∎

Now we can compute μ_x. We append an $xP(x)$ column to Table 5.10 and apply Definition 5.4. This is done in Table 5.11.

TABLE 5.11
Table for computing μ_x

x	$P(x)$	$xP(x)$
240	0.375	90.00
320	0.250	80.00
450	0.125	56.25
600	0.250	150.00
		376.25

Consequently, the mean of the random variable x is

$$\mu_x = \Sigma x P(x) = \$376.25.$$

 As expected, the mean, μ_x, of the random variable x equals the population mean weekly salary, μ, which we computed in Example 5.7(a) on page 278.   ∎

We have seen that if we let x denote the weekly salary of a randomly selected CJ[2] employee, then the mean of the random variable x is equal to the population mean weekly salary. More generally, we have Key Fact 5.4.

KEY FACT 5.4

MEAN OF A RANDOM VARIABLE AS A POPULATION MEAN

Suppose a random variable x is defined to be the value of a randomly selected member from a finite population. Then the mean of the random variable equals the population mean; that is, $\mu_x = \mu$.

The definition of the mean of a discrete random variable, presented in Definition 5.4, applies to any discrete random variable, not just to those associated with finite populations. Consider Example 5.9.

Example 5.9   Illustrates Definition 5.4

Prescott National Bank has six tellers available to serve customers. The number of tellers busy with customers at, say, 1:00 P.M. varies from day to day; it is a discrete random variable, which we will denote by x. On the basis of past records, it is known that the probability distribution of this random variable is as shown in the first two columns of Table 5.12. The table indicates, for instance, that the probability is 0.26 that exactly five tellers will be busy with customers at 1:00 P.M. Find the mean, μ_x, of the random variable x.

TABLE 5.12
Table for computing μ_x, the
mean number of tellers busy
with customers at 1:00 P.M.

No. busy x	Probability $P(x)$	$xP(x)$
0	0.03	0.00
1	0.05	0.05
2	0.08	0.16
3	0.15	0.45
4	0.21	0.84
5	0.26	1.30
6	0.22	1.32
		4.12

SOLUTION   Applying Definition 5.4 we obtain the mean of the random variable x by summing the $xP(x)$ values in the third column of Table 5.12. We find that $\mu_x = \Sigma x P(x) = 4.12$. The mean number of tellers busy with customers at 1:00 P.M. is 4.12.   ∎

Interpretation of the Mean of a Random Variable

As we know, the mean of a finite population is the arithmetic average of the population values. A similar interpretation holds for the mean of a random variable.

For instance, in Example 5.9 the random variable x denotes the number of tellers busy with customers at 1:00 P.M. The mean, μ_x, of that random variable is 4.12. Of course, we can never observe a day when 4.12 tellers are busy (with customers) at 1:00 P.M. The mean of 4.12 simply indicates that in a large number of observations, the average number of busy tellers at 1:00 P.M. will be about 4.12. This interpretation holds in all cases.

KEY FACT 5.5   **INTERPRETATION OF THE MEAN OF A RANDOM VARIABLE**

> The mean, μ_x, of a random variable x can be interpreted as follows: In a large number of independent observations of x, the average value of those observations will be approximately equal to μ_x. The larger the number of observations, the closer the average tends to be to μ_x.

We used a computer to simulate the number of busy tellers at 1:00 P.M., x, on 100 randomly selected days; that is, we obtained 100 independent observations of the random variable x. The data are displayed in Table 5.13.

TABLE 5.13
One hundred observations of the
random variable x, the number
of busy tellers at 1:00 P.M.

5	3	5	3	4	3	4	3	6	5	6	4	5	4	3	5	4	5	6	3
4	1	6	5	3	6	3	5	5	4	6	4	1	6	5	3	3	6	4	5
3	4	2	5	5	6	5	4	6	2	4	5	4	6	4	5	5	3	4	6
1	5	4	6	4	4	4	5	6	2	5	4	5	1	3	3	6	4	6	4
5	6	5	5	3	2	4	6	6	1	5	1	3	6	5	3	5	4	3	6

The average value of the 100 observations in Table 5.13 is 4.25. This value is quite close to the mean, $\mu_x = 4.12$, of the random variable x. If we made, say, 1000 observations instead of 100, then the average value of those 1000 observations would most likely be even closer to 4.12.

Figure 5.3(a) shows a plot of the average number of busy tellers versus the number of observations for the data in Table 5.13. The dashed line is at $\mu_x = 4.12$. In Fig. 5.3(b) is another such plot for a different simulation of the number of busy tellers at 1:00 P.M. on 100 randomly selected days. Both plots suggest that as the number of observations increases, the average number of busy tellers approaches the mean, $\mu_x = 4.12$, of the random variable x.

FIGURE 5.3

Graphs showing the average number of busy tellers versus the number of observations for two simulations of 100 observations each

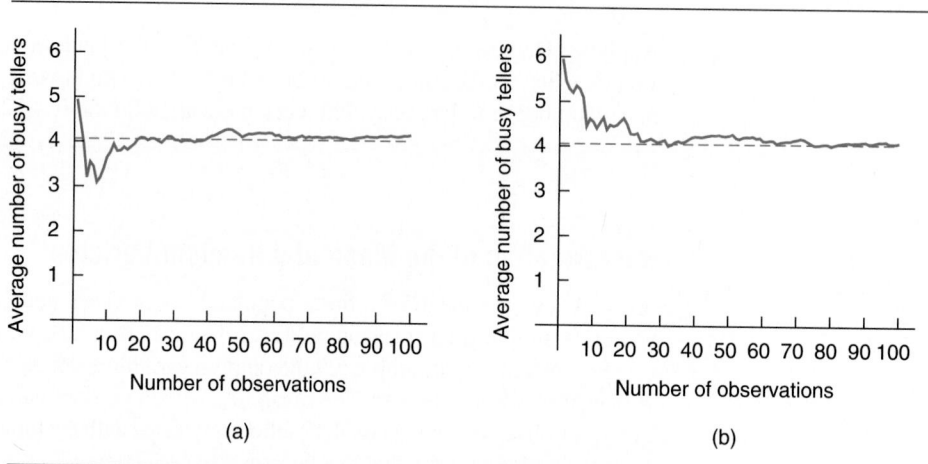

(a)   (b)

Standard Deviation of a Discrete Random Variable

We next examine the standard deviation of a discrete random variable. To begin, let's return to the salary data of Example 5.7.

Example 5.10   Introduces the Standard Deviation of a Discrete Random Variable

A grouped-data table for the weekly salaries of the eight employees of CJ^2 Business Services is presented in Table 5.8 on page 278.

a. Determine the population standard deviation of the weekly salaries.

b. Suppose a CJ^2 employee is to be selected at random. Let x denote the weekly salary of the employee obtained. Express the formula for the population standard deviation of the weekly salaries in terms of the probability distribution of the random variable x.

SOLUTION   a. The population standard deviation of the weekly salaries can be found by employing Formula 3.5:

$$\sigma = \sqrt{\frac{\Sigma(x - \mu)^2 f}{N}}.$$

This formula is used to compute the standard deviation of a population when the data are in grouped form, as they are here. To apply the formula, we first recall that the population mean weekly salary is $\mu = \$376.25$. Referring to the first two columns of Table 5.8, we construct Table 5.14.

TABLE 5.14
Table for computing σ

x	f	$x - \mu$	$(x - \mu)^2$	$(x - \mu)^2 f$
240	3	−136.25	18,564.0625	55,692.1875
320	2	−56.25	3,164.0625	6,328.1250
450	1	73.75	5,439.0625	5,439.0625
600	2	223.75	50,064.0625	100,128.1250
	8			167,587.5000

From the second and fifth columns of Table 5.14, we get

$$\sigma = \sqrt{\frac{\Sigma(x - \mu)^2 f}{N}} = \sqrt{\frac{167{,}587.5}{8}} = \$144.74.$$

Thus the population standard deviation of the weekly salaries is $\sigma = \$144.74$.

b. For this part we are to express the formula for the population standard deviation of the weekly salaries in terms of the probability distribution of the random variable x, where x is the weekly salary of a randomly selected CJ^2 employee. Because we are dealing here with a finite population, we know from Key Fact 5.2 on page 269 that probabilities for x are the same as relative frequencies. In other words, for each x value, the probability, $P(x)$, and the relative frequency, f/N, are equal.

Because probabilities and relative frequencies are the same here, we can express the formula for σ in terms of probabilities as follows:

$$\sigma = \sqrt{\frac{\Sigma(x - \mu)^2 f}{N}} = \sqrt{\Sigma(x - \mu)^2 \cdot \frac{f}{N}} = \sqrt{\Sigma(x - \mu)^2 P(x)}.$$

This last expression serves as the definition of the standard deviation of a discrete random variable. ∎

DEFINITION 5.5   **STANDARD DEVIATION OF A DISCRETE RANDOM VARIABLE**

The *standard deviation of a discrete random variable* x is defined by

$$\sigma_x = \sqrt{\Sigma(x - \mu_x)^2 P(x)}.$$

The variance of a population is σ^2, the square of the population standard deviation. Similarly, σ_x^2 is called the **variance of a discrete random variable** x.

Although we motivated Definition 5.5 using a discrete random variable associated with a finite population, the definition applies to any discrete random variable. In Example 5.11 we will illustrate the calculations involved in computing the standard deviation of a discrete random variable.

Example 5.11   Illustrates Definition 5.5

Suppose a CJ^2 employee is to be selected at random, and let x denote the weekly salary of the employee obtained. Determine the standard deviation, σ_x, of the random variable x.

SOLUTION   The probability distribution of the random variable x, displayed in Table 5.10, is repeated here in the first two columns of Table 5.15.

TABLE 5.15
Table for computing σ_x

Salary x	Probability $P(x)$	$x - \mu_x$	$(x - \mu_x)^2$	$(x - \mu_x)^2 P(x)$
240	0.375	−136.25	18,564.0625	6,961.5234375
320	0.250	−56.25	3,164.0625	791.0156250
450	0.125	73.75	5,439.0625	679.8828125
600	0.250	223.75	50,064.0625	12,516.0156250
				20,948.4375000

The mean of x was found in Example 5.8: $\mu_x = \$376.25$. To apply Definition 5.5, we need columns for $x - \mu_x$, $(x - \mu_x)^2$, and $(x - \mu_x)^2 P(x)$. These are given in the last three columns of Table 5.15.

The last column of Table 5.15 shows that $\Sigma(x - \mu_x)^2 P(x) = 20{,}948.4375$. Taking the square root, we get σ_x:

$$\sigma_x = \sqrt{\Sigma(x - \mu_x)^2 P(x)} = \sqrt{20{,}948.4375} = \$144.74.$$

Thus the standard deviation, σ_x, of the random variable x is $144.74. As expected, this equals the population standard deviation, σ, of the weekly salaries, which we computed in Example 5.10(a). ∎

We have seen that if we let x denote the weekly salary of a randomly selected CJ^2 employee, then the standard deviation of the random variable x is equal to the population standard deviation of the weekly salaries. More generally, we have Key Fact 5.6.

KEY FACT 5.6

STANDARD DEVIATION OF A RANDOM VARIABLE AS A POPULATION STANDARD DEVIATION

> Suppose a random variable x is defined to be the value of a randomly selected member from a finite population. Then the standard deviation of the random variable equals the population standard deviation; that is, $\sigma_x = \sigma$.

Shortcut Formula for σ_x

Example 5.11 makes it clear that the computations required to find σ_x are time-consuming and tedious, even in simple cases. To deal with that problem, we can use a computer or develop a shortcut formula. The latter approach is presented in Formula 5.1.

FORMULA 5.1

SHORTCUT FORMULA FOR σ_x

> The shortcut formula for computing the standard deviation of a discrete random variable x is
>
> $$\sigma_x = \sqrt{\Sigma x^2 P(x) - \mu_x^2}.$$

Example 5.12   Illustrates Formula 5.1

Let x be the weekly salary of a randomly selected CJ^2 employee. Use Formula 5.1 to determine the standard deviation, σ_x, of the random variable x.

SOLUTION   The probability distribution of x, first shown in Table 5.10, is repeated in the first two columns of Table 5.16. To apply the shortcut formula, we need columns for x^2 and $x^2 P(x)$, which are presented in the last two columns of Table 5.16.

TABLE 5.16
Table for computing σ_x
using the shortcut formula

x	$P(x)$	x^2	$x^2 P(x)$
240	0.375	57,600	21,600.0
320	0.250	102,400	25,600.0
450	0.125	202,500	25,312.5
600	0.250	360,000	90,000.0
			162,512.5

From the final column of Table 5.16, we see that $\Sigma x^2 P(x) = 162,512.5$. Recalling now that $\mu_x = \$376.25$, we apply Formula 5.1 to get

$$\sigma_x = \sqrt{\Sigma x^2 P(x) - \mu_x^2} = \sqrt{162,512.5 - (376.25)^2} = \sqrt{20,948.4375} = \$144.74.$$

This is the same result we obtained in Example 5.11. However, the computations done here using the shortcut formula are much easier.   ■

Interpretation of the Standard Deviation of a Random Variable

Recall that the standard deviation of a population is a measure of the dispersion of the population values; roughly speaking, it measures how far the population values are from the mean, on the average. A similar interpretation can be given to the standard deviation of a random variable.

KEY FACT 5.7   **INTERPRETATION OF THE STANDARD DEVIATION OF A RANDOM VARIABLE**

> The standard deviation, σ_x, of a random variable x measures the dispersion of the possible values of the random variable relative to the mean; the smaller the standard deviation, the more likely that an observed value of x will be close to the mean.

EXERCISES 5.2

·   **5.14** Suppose the random variables x and y represent the amount of return on two different investments. Further suppose the mean of x equals the mean of y, but the standard deviation of x is greater than the standard deviation of y.
a. On the average, is there a difference between the returns of the two investments? Explain your answer.
b. Which investment is more conservative? Why?

In Exercises 5.15–5.20, we have given the probability distributions of the random variables considered in Exercises 5.1–5.6. For each exercise,
a. *determine and interpret the mean of the random variable.*
b. *determine the standard deviation of the random variable using the defining formula, Definition 5.5 on page 284.*
c. *determine the standard deviation of the random variable using the shortcut formula, Formula 5.1 on page 285.*
d. *draw a probability histogram for the random variable; locate the mean; and show one, two, and three standard-deviation intervals.*

·   **5.15** The random variable x is the number of people in a randomly selected U.S. household. Its probability distribution follows.

x	1	2	3	4	5	6	7
$P(x)$	0.232	0.317	0.175	0.154	0.073	0.030	0.019

·   **5.16** The random variable x is the grade level of a secondary-school student selected at random. Its probability distribution is as follows.

x	9	10	11	12
$P(x)$	0.264	0.259	0.244	0.233

·   **5.17** The random variable y is the sum of the dice when two balanced dice are rolled. Its probability distribution is shown in the following table. Perform the required computations using fractions; that is, do not convert to decimals until the final answer is obtained.

y	2	3	4	5	6	7	8	9	10	11	12
$P(y)$	$\frac{1}{36}$	$\frac{1}{18}$	$\frac{1}{12}$	$\frac{1}{9}$	$\frac{5}{36}$	$\frac{1}{6}$	$\frac{5}{36}$	$\frac{1}{9}$	$\frac{1}{12}$	$\frac{1}{18}$	$\frac{1}{36}$

5.18   The random variable x is the larger number showing when two balanced dice are rolled. Its probability distribution follows.

x	1	2	3	4	5	6
$P(x)$	$\frac{1}{36}$	$\frac{1}{12}$	$\frac{5}{36}$	$\frac{7}{36}$	$\frac{1}{4}$	$\frac{11}{36}$

5.19   The random variable x is the number of tellers busy with customers at 1:00 P.M. at Prescott National Bank. Its probability distribution follows.

x	0	1	2	3	4	5	6
$P(x)$	0.03	0.05	0.08	0.15	0.21	0.26	0.22

5.20   The random variable y is the number of cars sold during a week by a car salesperson. Its probability distribution is as follows.

y	$P(y)$
0	0.135
1	0.271
2	0.271
3	0.180
4	0.090
5	0.036
6	0.012
7	0.003
8	0.002

5.21 (Computer exercise)   In Exercise 5.7 we gave the probability distribution for the number of customers waiting at Benny's Barber Shop in Cleveland. We repeat it below.

y	0	1	2	3	4	5
$P(y)$	0.424	0.161	0.134	0.111	0.093	0.077

a. Compute the mean number of customers waiting, μ_y.
b. In a large number of independent observations, roughly how many customers will be waiting, on the average?
c. Use Minitab or some other statistical software to simulate 100 observations of the number of customers waiting.

d. Obtain the mean of the observations in part (c), and compare the mean to μ_y.
e. What is part (d) illustrating?

5.22 (Computer exercise)   In Exercise 5.8 we displayed the probability distribution for the number of women chosen to represent a company at a small-business conference. We repeat it below.

x	0	1	2	3
$P(x)$	0.167	0.500	0.300	0.033

a. Compute μ_x.
b. On the average, how many women would you expect to be chosen?
c. Use Minitab or some other statistical software to simulate 75 observations of the number of women chosen.
d. Obtain the mean of the observations in part (c), and compare the mean to μ_x.
e. What is part (d) illustrating?

Expected value. As mentioned in Definition 5.4 on page 279, the mean of a random variable is also called its *expected value*. This terminology is especially useful in gambling and decision theory, as illustrated in Exercises 5.23 and 5.24.

5.23   A roulette wheel contains 38 numbers: 18 are red, 18 are black, and 2 are green. When the roulette ball is spun, it is equally likely to land on any of the 38 numbers. Suppose you bet $1 on red. Then if the ball lands on a red number, you win $1; otherwise you lose your $1. Let x be the amount you win on your $1 bet. Then x is a random variable. Its probability distribution is as follows.

x	1	-1
$P(x)$	0.474	0.526

a. Verify that the probability distribution shown in the table is correct.
b. Find the expected value of the random variable x.
c. On the average, how much will you lose per play? *(Hint: Refer to the interpretation of the mean of a random variable given in Key Fact 5.5 on page 281.)*
d. Approximately how much would you expect to lose if you bet $1 on red 100 times? 1000 times?
e. Do you think roulette is a profitable game to play?

5.24   An investor plans to invest $50,000 in one of four investments. The return on each investment depends on whether next year's economy is strong or weak. The following table summarizes the possible payoffs, in dollars, for the four investments.

Next year's economy

Investment		Strong	Weak
	Certificate of deposit	6,000	6,000
	Office complex	15,000	5,000
	Land speculation	33,000	−17,000
	Technical school	5,500	10,000

Let v, w, x, and y denote, respectively, the payoffs for the certificate of deposit, office complex, land speculation, and technical school. Then v, w, x, and y are random variables. Assume next year's economy has a 40% chance of being strong and a 60% chance of being weak.

a. Find the probability distribution of each of the four random variables, v, w, x, and y.

b. Determine the expected value of each of the four random variables.

c. Which investment has the best expected payoff? Which has the worst?

d. Which investment would you select? Explain your answer.

·   **5.25**   A factory manager collected data on the number of equipment breakdowns per day. From those data, she derived the probability distribution shown in the table below.

w	0	1	2
$P(w)$	0.80	0.15	0.05

a. Determine μ_w and σ_w.

b. How many breakdowns occur per day, on the average?

c. About how many breakdowns are expected per year, assuming 250 work days per year?

Properties of the mean and standard deviation of a random variable. Exercises 5.26 and 5.27 develop some important properties of the mean and standard deviation of a random variable. Two of these properties relate the mean and standard deviation of the sum of two random variables to the individual means and standard deviations, respectively; two others relate the mean and standard deviation of a constant times a random variable to the constant and the mean and standard deviation of the random variable, respectively. In developing

these properties, we will need the concept of **independent random variables.** Two discrete random variables, x and y, are said to be independent if

$$P\left(\{x = j\} \& \{y = k\}\right) = P(x = j)P(y = k)$$

for all values of j and k, that is, if the joint probability distribution of x and y equals the product of their marginal probability distributions. This is equivalent to requiring that events $\{x = j\}$ and $\{y = k\}$ are independent for all values of j and k. A similar definition holds for independence of more than two discrete random variables.

· ·   **5.26**   Refer to Exercise 5.25. Assume the number of breakdowns on different days are independent of one another. Let x and y denote the number of breakdowns on each of two consecutive days.

a. Complete the following joint probability distribution table.

		y			
		0	1	2	$P(x)$
x	0				
	1				
	2				
	$P(y)$				

Hint: To obtain the joint probability in the first row and third column, use the definition of independence for discrete random variables and the table in Exercise 5.25:

$$P\left(\{x = 0\} \& \{y = 2\}\right) = P(x = 0)P(y = 2)$$
$$= 0.80 \cdot 0.05 = 0.04.$$

b. Employ the joint probability distribution you obtained in part (a) to determine the probability distribution of the random variable $x + y$, the total number of breakdowns in two days; that is, complete the following table.

$x + y$	0 1 2 3 4
$P(x + y)$	

c. Use your answer from part (b) to determine μ_{x+y} and σ_{x+y}.

d. Use part (c) to verify that the following equations are true for this example:

$$\mu_{x+y} = \mu_x + \mu_y \quad \text{and} \quad \sigma_{x+y} = \sqrt{\sigma_x^2 + \sigma_y^2}.$$

(Note: The mean and standard deviation of x and y are the same as that of w in Exercise 5.25.)

e. The equations in part (d) hold in general: If x and y are random variables (discrete or continuous), then

$$\mu_{x+y} = \mu_x + \mu_y.$$

If, in addition, x and y are independent, then

$$\sigma_{x+y} = \sqrt{\sigma_x^2 + \sigma_y^2}.$$

Interpret these two equations in words.

· · **5.27** The factory manager in Exercise 5.25 estimates that each breakdown costs the company \$100 in repairs and loss of production. If w is the number of breakdowns in a day, then \$100w is the cost due to breakdowns for that day.

a. Referring to the probability distribution in Exercise 5.25, determine the probability distribution of the random variable 100w.
b. Determine the mean daily breakdown cost, μ_{100w}, using your answer from part (a).
c. What is the relationship between μ_{100w} and μ_w? *(Note: From Exercise 5.25, $\mu_w = 0.25$.)*
d. Find σ_{100w} using your answer from part (a).
e. What is the apparent relationship between σ_{100w} and σ_w? *(Note: From Exercise 5.25, $\sigma_w = 0.536$.)*
f. The results in parts (c) and (e) are true in general: If w is a random variable (discrete or continuous) and c is a constant, then

$$\mu_{cw} = c\mu_w \quad \text{and} \quad \sigma_{cw} = |c|\sigma_w.$$

Interpret these two equations in words.

5.3 BINOMIAL COEFFICIENTS; BERNOULLI TRIALS

· · · · · ·

Many problems in probability and statistics involve situations in which an experiment with two possible outcomes is repeated several times. Each repetition of the experiment is called a **trial.**

For example, consider testing the effectiveness of a drug. Several patients take the drug (the trials), and for each patient, the drug is either effective or not effective (the two possible outcomes). Or consider the weekly sales of a car salesperson. The salesperson has several customers during the week (the trials), and for each customer, the salesperson either makes a sale or does not make a sale (the two possible outcomes). Or consider taste tests for colas. A number of people taste two different colas (the trials), and for each person, the preference is either for the first cola or for the second cola (the two possible outcomes).

Analyzing repeated trials of an experiment having two possible outcomes requires knowledge of factorials, binomial coefficients, Bernoulli trials, and the binomial distribution. We will study the first three of these topics in this section and the fourth topic in Section 5.4.

Factorials

Factorials are defined in Definition 5.6.

DEFINITION 5.6 **FACTORIALS**

The product of the first k positive integers is called **k factorial** and is denoted by the symbol **$k!$**. Thus

$$k! = k(k-1)\cdots 2 \cdot 1.$$

We also define $0! = 1$.

Example 5.13   Illustrates Definition 5.6

Determine 3!, 4!, and 5!.

SOLUTION   Applying Definition 5.6 we find that $3! = 3 \cdot 2 \cdot 1 = 6$, $4! = 4 \cdot 3 \cdot 2 \cdot 1 = 24$, and $5! = 5 \cdot 4 \cdot 3 \cdot 2 \cdot 1 = 120$.   ∎

Note, for example, that $6! = 6 \cdot 5!$, $6! = 6 \cdot 5 \cdot 4!$, $6! = 6 \cdot 5 \cdot 4 \cdot 3!$, and so on. In general, if $j \leq k$, then $k! = k(k-1) \cdots (k-j+1)(k-j)!$.

Binomial Coefficients

You may have already encountered binomial coefficients in algebra when you studied the binomial expansion, the expansion of $(a+b)^n$. Here is the definition of binomial coefficients.[†]

DEFINITION 5.7   **BINOMIAL COEFFICIENTS**

If n is a positive integer and x is a nonnegative integer less than or equal to n, then the *binomial coefficient* $\binom{n}{x}$ is defined as

$$\binom{n}{x} = \frac{n!}{x!\,(n-x)!}.$$

Example 5.14   Illustrates Definition 5.7

Determine the value of each of the following binomial coefficients.

a. $\binom{6}{1}$ b. $\binom{5}{3}$ c. $\binom{7}{3}$ d. $\binom{4}{4}$

SOLUTION   We apply Definition 5.7.

a. $\binom{6}{1} = \frac{6!}{1!\,(6-1)!} = \frac{6!}{1!\,5!} = \frac{6 \cdot 5!}{1!\,5!} = \frac{6}{1} = 6$

b. $\binom{5}{3} = \frac{5!}{3!\,(5-3)!} = \frac{5!}{3!\,2!} = \frac{5 \cdot 4 \cdot 3!}{3!\,2!} = \frac{5 \cdot 4}{2} = 10$

[†] If you have read Section 4.8, you will recognize the binomial coefficient $\binom{n}{x}$ as the number of possible combinations of x objects from a collection of n objects.

c. $\dbinom{7}{3} = \dfrac{7!}{3!\,(7-3)!} = \dfrac{7!}{3!\,4!} = \dfrac{7 \cdot 6 \cdot 5 \cdot \cancel{4!}}{3!\,\cancel{4!}} = \dfrac{7 \cdot 6 \cdot 5}{6} = 35$

d. $\dbinom{4}{4} = \dfrac{4!}{4!\,(4-4)!} = \dfrac{4!}{4!\,0!} = \dfrac{\cancel{4!}}{\cancel{4!}\,0!} = \dfrac{1}{1} = 1$ ■

Bernoulli Trials; Assignment of Probabilities

We now discuss Bernoulli trials, so called in honor of the Swiss mathematician James Bernoulli (see the biography presented at the beginning of this chapter for more information on Bernoulli).

DEFINITION 5.8   **BERNOULLI TRIALS**

Repeated identical trials are called **Bernoulli trials** if the following three conditions are satisfied:

1. Each trial has two possible outcomes, denoted generically by *s,* for **success,** and *f,* for **failure.**
2. The trials are independent.
3. The probability of a success remains the same from trial to trial. We call that probability the **success probability** and denote it by the letter **p.**

Example 5.15   Illustrates Definition 5.8

A drug is known to be 80% effective in curing a certain disease. Suppose the drug is administered to four patients having the disease and that the cure-noncure results are recorded.

a. Formulate this process as a sequence of four Bernoulli trials.

b. Determine the possible outcomes of the four Bernoulli trials.

c. Determine the probability of each outcome in part (b).

SOLUTION   a. Each trial consists of administering the drug to one of the patients and has two possible outcomes: cure or noncure. The trials are independent (why is this so?). If we let a success, *s,* correspond to a cure, then the success probability is 0.8 (80%); that is, $p = 0.8$.

b. For this part we are to obtain the possible outcomes of the four Bernoulli trials, that is, the possible cure-noncure results for the four patients. The possible outcomes are shown in Table 5.17 (s = success = cure, f = failure = noncure).

ssss	*sfss*	*fsss*	*ffss*
sssf	*sfsf*	*fssf*	*ffsf*
ssfs	*sffs*	*fsfs*	*fffs*
ssff	*sfff*	*fsff*	*ffff*

For instance, *sssf* represents the outcome that the first three patients are cured and the fourth is not.

c. Here we are to determine the probability of each of the possible outcomes. As we see from Table 5.17, 16 outcomes are possible. However, because these 16 outcomes are not equally likely, we cannot use the f/N rule to determine their probabilities; instead we must proceed as follows. First of all, by part (a), the success probability equals 0.8:

$$P(s) = p = 0.8.$$

Therefore the failure probability is

$$P(f) = 1 - p = 1 - 0.8 = 0.2.$$

Now using the fact that the trials are independent, we can apply the special multiplication rule, Formula 4.7 on page 231, to obtain the probability of each of the 16 possible outcomes. For instance, the probability of the outcome *sssf* is

$$P(sssf) = P(s) \cdot P(s) \cdot P(s) \cdot P(f) = 0.8 \cdot 0.8 \cdot 0.8 \cdot 0.2 = 0.1024$$

and that for *fsfs* is

$$P(fsfs) = P(f) \cdot P(s) \cdot P(f) \cdot P(s) = 0.2 \cdot 0.8 \cdot 0.2 \cdot 0.8 = 0.0256.$$

Similar computations yield the probabilities of the other 14 possible outcomes. All 16 possible outcomes and their probabilities are shown in Table 5.18.

Outcome	Probability
ssss	$(0.8)(0.8)(0.8)(0.8) = 0.4096$
sssf	$(0.8)(0.8)(0.8)(0.2) = 0.1024$
ssfs	$(0.8)(0.8)(0.2)(0.8) = 0.1024$
ssff	$(0.8)(0.8)(0.2)(0.2) = 0.0256$
sfss	$(0.8)(0.2)(0.8)(0.8) = 0.1024$
sfsf	$(0.8)(0.2)(0.8)(0.2) = 0.0256$
sffs	$(0.8)(0.2)(0.2)(0.8) = 0.0256$
sfff	$(0.8)(0.2)(0.2)(0.2) = 0.0064$
fsss	$(0.2)(0.8)(0.8)(0.8) = 0.1024$
fssf	$(0.2)(0.8)(0.8)(0.2) = 0.0256$
fsfs	$(0.2)(0.8)(0.2)(0.8) = 0.0256$
fsff	$(0.2)(0.8)(0.2)(0.2) = 0.0064$
ffss	$(0.2)(0.2)(0.8)(0.8) = 0.0256$
ffsf	$(0.2)(0.2)(0.8)(0.2) = 0.0064$
fffs	$(0.2)(0.2)(0.2)(0.8) = 0.0064$
ffff	$(0.2)(0.2)(0.2)(0.2) = 0.0016$

Note that outcomes containing the same number of successes have the same probability. For instance, four outcomes contain exactly three successes: *sssf, ssfs, sfss,* and *fsss*. Each of those four outcomes has the same probability: 0.1024. This is because each probability is obtained by multiplying three success probabilities of 0.8 and one failure probability of 0.2.

A tree diagram is useful for organizing and summarizing the possible outcomes and their probabilities. Such a diagram is presented in Fig. 5.4. ■

FIGURE 5.4

Tree diagram corresponding to Table 5.18

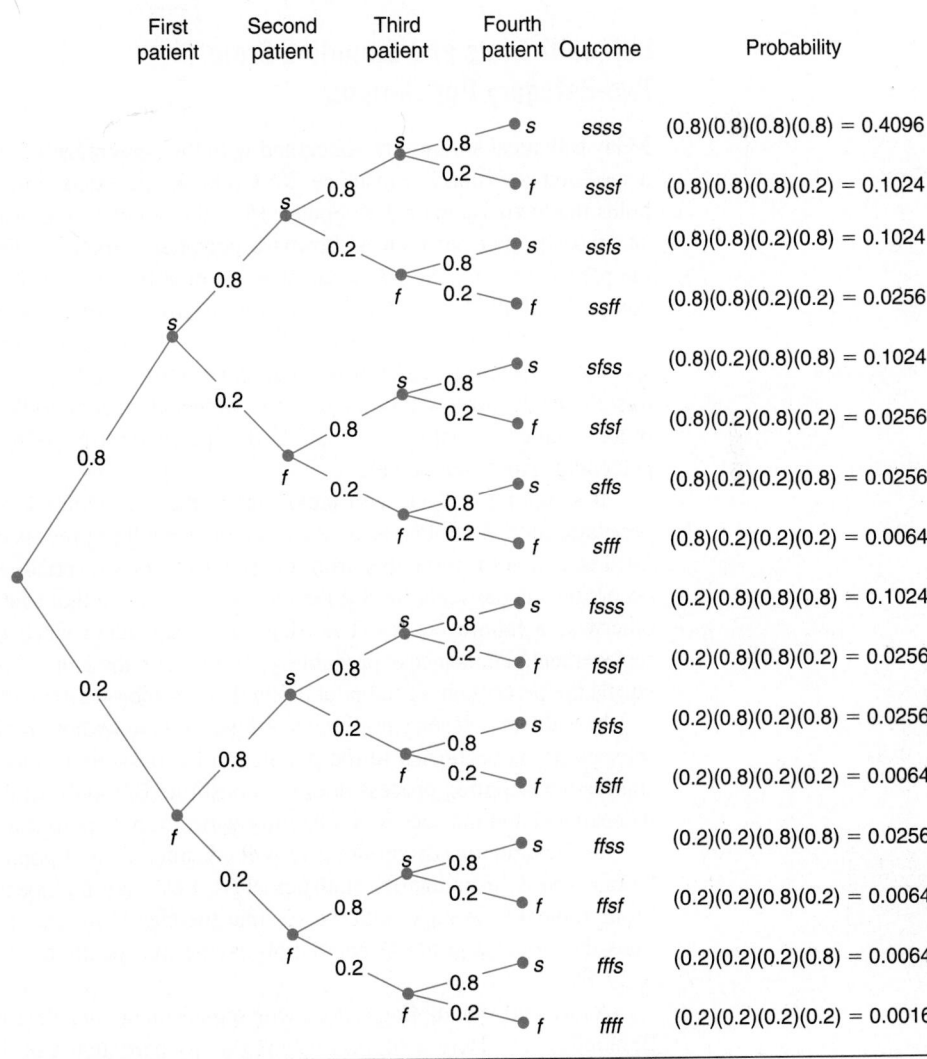

First patient	Second patient	Third patient	Fourth patient	Outcome	Probability
			s	ssss	(0.8)(0.8)(0.8)(0.8) = 0.4096
			f	sssf	(0.8)(0.8)(0.8)(0.2) = 0.1024
			s	ssfs	(0.8)(0.8)(0.2)(0.8) = 0.1024
			f	ssff	(0.8)(0.8)(0.2)(0.2) = 0.0256
			s	sfss	(0.8)(0.2)(0.8)(0.8) = 0.1024
			f	sfsf	(0.8)(0.2)(0.8)(0.2) = 0.0256
			s	sffs	(0.8)(0.2)(0.2)(0.8) = 0.0256
			f	sfff	(0.8)(0.2)(0.2)(0.2) = 0.0064
			s	fsss	(0.2)(0.8)(0.8)(0.8) = 0.1024
			f	fssf	(0.2)(0.8)(0.8)(0.2) = 0.0256
			s	fsfs	(0.2)(0.8)(0.2)(0.8) = 0.0256
			f	fsff	(0.2)(0.8)(0.2)(0.2) = 0.0064
			s	ffss	(0.2)(0.2)(0.8)(0.8) = 0.0256
			f	ffsf	(0.2)(0.2)(0.8)(0.2) = 0.0064
			s	fffs	(0.2)(0.2)(0.2)(0.8) = 0.0064
			f	ffff	(0.2)(0.2)(0.2)(0.2) = 0.0016

A success, *s,* in Bernoulli trials is often derived from a collection of outcomes. For example, as we mentioned earlier, a roulette wheel consists of 38 numbers, of which 18 are red, 18 are black, and 2 are green. When the roulette ball is spun, it is equally likely to land on any one of the 38 numbers. If we are interested in which number the ball lands on, then each play at the wheel has 38 possible outcomes (the 38 different numbers).

Suppose, however, that we are betting on red. Then we are interested only in whether the ball lands on a red number. From this point of view, each play at the wheel has only two possible outcomes—either the ball lands on a red number or it doesn't. Hence successive bets on red constitute a sequence of Bernoulli trials: Each trial consists of one play at the wheel and has two possible outcomes, red or not red; the trials are independent because successive plays at the wheel are independent; and if we let a success correspond to the ball landing on red, then the success probability is $\frac{18}{38}$ since 18 of the 38 numbers are red.

Bernoulli Trials and Sampling From Two-Category Populations

Many statistical studies are concerned with the percentage of a (finite) population that has a specified attribute. For instance, we might be interested in the percentage of U.S. households that own a personal computer. Here the population consists of all U.S. households, and the specified attribute is "owns a personal computer." Or we might want to know the percentage of U.S. businesses that are minority owned. In this case the population is composed of all U.S. businesses, and the specified attribute is "minority owned."

A population in which each member is classified as either having or not having a specified attribute is called a **two-category population.** Since most often the population under consideration will be quite large, we usually cannot determine the exact percentage of a two-category population that has the specified attribute; instead we must estimate the percentage from sample data.

If sampling is done with replacement, that is, members selected are returned to the population for possible reselection, then the sampling process constitutes Bernoulli trials: Each selection of a member from the population corresponds to a single trial. A success occurs on any particular trial if the member selected in that trial has the specified attribute; otherwise a failure occurs. The trials are independent since the sampling is done with replacement. The success probability, *p,* remains the same from trial to trial—it always equals the percentage of the population that has the specified attribute.

In reality, however, sampling is ordinarily done without replacement, that is, members selected are not returned to the population for possible reselection. Under these circumstances the sampling process does not constitute Bernoulli trials because the trials are not independent and the success probability varies from trial to trial.

To illustrate our discussion, we will consider a small population, the students in Professor Weiss's introductory statistics class. Let's say the specified attribute is "female." From Table 4.9 on page 225, we see that the class consists of 17 males and 23 females. Thus the percentage of the population having the specified attribute equals $\frac{23}{40} = 0.575$, or 57.5%.

As we said, if sampling is done with replacement, then the sampling process constitutes Bernoulli trials. Here a success occurs on any particular trial if the student selected is a female; the trials are independent since the sampling is done with replacement; and the success probability, *p,* always equals 0.575.

On the other hand, suppose the sampling is done without replacement. Then the sampling process does not constitute Bernoulli trials because the trials are not independent and the success probability varies from trial to trial. For instance, the probability of a success (female) on the first trial (selection) is $\frac{23}{40} = 0.575$. However, given that the first trial results in a success, the probability of a success on the second trial is $\frac{22}{39} = 0.564$.

So we see that sampling without replacement from a two-category population does not constitute Bernoulli trials. Nonetheless, if the size of the sample is small relative to the size of the population, then for all practical purposes the sampling process can be regarded as Bernoulli trials. The reason for this is as follows: If the sample size is small relative to the population size, then there is very little difference between sampling without replacement and sampling with replacement; and in the latter case, we do indeed have Bernoulli trials. We summarize the previous discussion in Key Fact 5.8.

KEY FACT 5.8 **SAMPLING FROM A TWO-CATEGORY POPULATION AS BERNOULLI TRIALS**

> Sampling without replacement from a two-category population can be regarded as Bernoulli trials provided the size of the sample is small relative to the size of the population. A rule of thumb is that Bernoulli trials may be assumed whenever the sample size does not exceed 5% of the population size.

EXERCISES 5.3

5.28 Find 1!, 2!, and 6!.

5.29 Compute 7!, 8!, and 9!.

5.30 Evaluate the following binomial coefficients.
a. $\binom{4}{1}$ b. $\binom{6}{2}$ c. $\binom{8}{3}$ d. $\binom{9}{6}$

5.31 Evaluate the following binomial coefficients.
a. $\binom{5}{2}$ b. $\binom{7}{4}$ c. $\binom{10}{3}$ d. $\binom{12}{5}$

5.32 Determine the value of each of the following binomial coefficients.
a. $\binom{3}{2}$ b. $\binom{6}{0}$ c. $\binom{6}{6}$ d. $\binom{7}{3}$

5.33 Determine the value of each of the following binomial coefficients.
a. $\binom{5}{3}$ b. $\binom{10}{0}$ c. $\binom{10}{10}$ d. $\binom{9}{5}$

5.34 In an ESP experiment, a person in one room randomly selects 1 of 10 cards numbered 1–10 and a person in another room tries to guess the number on the card chosen. This experiment is to be repeated three times, with each card chosen being replaced before the next selection. Assuming the

person guessing lacks ESP, the probability of a correct guess on any particular trial is $\frac{1}{10} = 0.1$. Suppose we consider a success, *s,* to be a correct guess. Then, for example, *ssf* represents a correct guess on the first two trials and an incorrect guess on the third.
a. Identify the success probability, *p.*
b. Construct a table similar to Table 5.18 on page 292 for the three guesses.
c. Draw a tree diagram for this problem similar to Fig. 5.4 on page 293.
d. List the outcomes containing exactly one success.
e. Find the probability of each outcome in part (d). Why are those probabilities all the same?

5.35 Based on data from the *Statistical Abstract of the United States,* the probability that a newborn baby will be a girl is about 0.487. Suppose we consider a success, *s,* in a given birth to be "a girl."
a. Identify the success probability, *p.*
b. Construct a table similar to Table 5.18 on page 292 for the next three births. Display the probabilities to three decimal places.

c. Draw a tree diagram for this problem similar to Fig. 5.4 on page 293.

d. List the outcomes in which exactly two of the three babies born are girls.

e. Find the probability of each outcome in part (d). Why are those probabilities all the same?

5.36 According to a survey by the Opinion Research Corporation, 60% of all clerical workers in the United States "like their jobs very much." Suppose four randomly selected clerical workers are to be asked whether they like their jobs very much. Let's consider an affirmative response to be a success, s.

a. What is the success probability, p?

b. Construct a table like Table 5.18 on page 292 for the possible responses of the four clerical workers. Display the probabilities to four decimal places.

c. Draw a tree diagram for this problem similar to Fig. 5.4 on page 293.

d. List the outcomes in which exactly two of the four clerical workers respond affirmatively.

e. Find the probability of each outcome in part (d). Why are those probabilities all the same?

5.37 The National Institute of Mental Health reports that 20% of adult Americans suffer from at least one psychiatric disorder. Suppose four randomly selected adult Americans are to be examined for psychiatric disorders.

a. If we let a success, s, correspond to an adult American having a psychiatric disorder, then what is the success probability, p? *(Note:* The use of the word *success* in Bernoulli trials need not conform to the ordinarily positive connotation of the word.)

b. Construct a table similar to Table 5.18 on page 292 for the four people examined. Display the probabilities to four decimal places.

c. Draw a tree diagram for this problem similar to Fig. 5.4 on page 293.

d. List the outcomes in which exactly three of the four people examined have a psychiatric disorder.

e. Find the probability of each outcome in part (d). Why are those probabilities all the same?

5.38 A phone-company study conducted in Phoenix revealed that the probability is 0.25 that a randomly selected phone call will last longer than the mean duration of 3.8 minutes. For any particular phone call, suppose we consider a success, s, to be that the call lasts at most 3.8 minutes.

a. What is the success probability, p?

b. Construct a table similar to Table 5.18 on page 292 for the possible success-failure results of three randomly selected calls. Display the probabilities to three decimal places.

c. List the outcomes in which exactly two of the three calls last at most 3.8 minutes.

d. Find the probability of each outcome in part (c). Why are those probabilities all the same?

5.39 As reported by the Chicago Title Insurance Company in *The Guarantor,* the probability is 0.768 that a home buyer will purchase a resale home. Let a success correspond to the purchase of a new (nonresale) home.

a. What is the success probability, p?

b. Construct a table similar to Table 5.18 on page 292 for the possible success-failure results of four randomly selected home purchases. Display the probabilities to three decimal places.

c. List the outcomes in which exactly two of the four purchases are new homes.

d. Find the probability of each outcome in part (c). Why are those probabilities all the same?

5.40 Show that the equality $k! = k(k-1)!$ holds for every positive integer k.

5.41 Establish the following equalities.

a. $\binom{n}{0} = 1$ 　　b. $\binom{n}{n} = 1$ 　　c. $\binom{n}{x} = \binom{n}{n-x}$

5.42 The binomial theorem states that

$$(a+b)^n = \binom{n}{0}a^n b^0 + \binom{n}{1}a^{n-1}b^1 + \cdots + \binom{n}{n}a^0 b^n.$$

a. Verify the binomial theorem for the case $n = 2$ by computing $(a+b)^2 = (a+b)(a+b)$ and showing that the result is identical to

$$\binom{2}{0}a^2 b^0 + \binom{2}{1}a^1 b^1 + \binom{2}{2}a^0 b^2.$$

b. Repeat part (a) for the case $n = 3$.

5.43 Refer to the discussion on Bernoulli trials and sampling from two-category populations given on pages 294 and 295. Consider the following frequency distribution for students in Professor Weiss's introductory statistics class.

Sex	Frequency
Male	17
Female	23

a. Suppose two students are to be selected at random. Find the probability that both students are male: if the selection is done with replacement (apply the special multiplication rule); if the selection is done without replacement (apply the general multiplication rule).

b. Compare the two answers obtained in part (a).

Suppose Professor Weiss's class had 10 times the students, but in the same proportions, that is, 170 males and 230 females.

c. Repeat parts (a) and (b) using this hypothetical distribution of students.

d. In which case is there less difference between sampling without and with replacement? Explain why this is so.

5.4 THE BINOMIAL DISTRIBUTION

Now that we have studied binomial coefficients and Bernoulli trials, we are ready to examine the binomial distribution. The **binomial distribution** is the probability distribution for the number of successes in a sequence of Bernoulli trials. We introduce this concept by returning to the drug experiment of Example 5.15.

Example 5.16   Introduces the Binomial Distribution

A drug is known to be 80% effective in curing a certain disease; that is, the probability is 0.80 that a person with the disease will be cured by the drug. If four patients with the disease are given the drug, find the probability that

a. exactly two are cured.   b. exactly three are cured.

SOLUTION   To begin, recall that we can regard the process of administering the drug to four patients as a sequence of four Bernoulli trials. If we let a success, s, correspond to a cure, and a failure, f, correspond to a noncure, then the success probability is $p = 0.8$ and the failure probability is $1 - p = 0.2$. The 16 possible cure-noncure outcomes and their probabilities are displayed in Table 5.18, which we repeat here as Table 5.19. Using this table, we can solve the problems posed in parts (a) and (b).

TABLE 5.19
Outcomes and probabilities
for drug experiment

Outcome	Probability
ssss	$(0.8)(0.8)(0.8)(0.8) = 0.4096$
sssf	$(0.8)(0.8)(0.8)(0.2) = 0.1024$
ssfs	$(0.8)(0.8)(0.2)(0.8) = 0.1024$
ssff	$(0.8)(0.8)(0.2)(0.2) = 0.0256$
sfss	$(0.8)(0.2)(0.8)(0.8) = 0.1024$
sfsf	$(0.8)(0.2)(0.8)(0.2) = 0.0256$
sffs	$(0.8)(0.2)(0.2)(0.8) = 0.0256$
sfff	$(0.8)(0.2)(0.2)(0.2) = 0.0064$
fsss	$(0.2)(0.8)(0.8)(0.8) = 0.1024$
fssf	$(0.2)(0.8)(0.8)(0.2) = 0.0256$
fsfs	$(0.2)(0.8)(0.2)(0.8) = 0.0256$
fsff	$(0.2)(0.8)(0.2)(0.2) = 0.0064$
ffss	$(0.2)(0.2)(0.8)(0.8) = 0.0256$
ffsf	$(0.2)(0.2)(0.8)(0.2) = 0.0064$
fffs	$(0.2)(0.2)(0.2)(0.8) = 0.0064$
ffff	$(0.2)(0.2)(0.2)(0.2) = 0.0016$

a. The problem here is to determine the probability that exactly two of the four patients are cured. As we see from Table 5.19, the event that exactly two of the four patients are cured consists of six outcomes: *ssff, sfsf, sffs, fssf, fsfs,* and *ffss.* Each of those six outcomes has the same probability, 0.0256, because each probability is obtained by multiplying two success probabilities of 0.8 and two failure probabilities of 0.2. Hence, by the special addition rule, we have

$$P(\text{Exactly two are cured})$$

$$= P(ssff) + P(sfsf) + P(sffs) + P(fssf) + P(fsfs) + P(ffss)$$

$$= 0.0256 + 0.0256 + 0.0256 + 0.0256 + 0.0256 + 0.0256$$

$$= 6 \cdot 0.0256 = 0.1536.$$

The probability is 0.1536 that exactly two of the four patients are cured.

b. For this part we need to determine the probability that exactly three of the four patients are cured. As we see from Table 5.19, the event that exactly three of the four patients are cured consists of four outcomes: *sssf, ssfs, sfss,* and *fsss.* Each of those four outcomes has the same probability, 0.1024, because each probability is obtained by multiplying three success probabilities of 0.8 and one failure probability of 0.2. So, by the special addition rule, we have

$$P(\text{Exactly three are cured})$$

$$= P(sssf) + P(ssfs) + P(sfss) + P(fsss)$$

$$= 0.1024 + 0.1024 + 0.1024 + 0.1024$$

$$= 4 \cdot 0.1024 = 0.4096.$$

The probability is 0.4096 that exactly three of the four patients are cured. ∎

Table 5.19 provides the 16 possible success-failure (cure-noncure) outcomes and their probabilities. However, in this and many other situations, it is not the success-failure outcomes but rather the *total number of successes* that are of interest. If we denote the total number of successes (cures) by x, then x is a random variable—its value depends on chance, namely, on how many patients are cured.

Now we can use random-variable notation to represent events and probabilities concisely. For instance, the event that exactly two of the four patients are cured can be written simply as $\{x = 2\}$. In part (a) of Example 5.16, we determined the probability of that event to be 0.1536. We can express this as $P(x = 2) = 0.1536$, or even more briefly as $P(2) = 0.1536$. This is recorded in the third row of Table 5.20.

In part (b) of Example 5.16, we found that the probability is 0.4096 that exactly three of the four patients are cured. Thus $P(3) = 0.4096$. This is recorded in the fourth row of Table 5.20. Using reasoning similar to that used in Example 5.16, we can determine the remaining probabilities for the random variable x, and hence obtain its probability distribution. Table 5.20 shows that probability distribution.

TABLE 5.20

Probability distribution for
the number of patients cured

Number cured x	Probability $P(x)$
0	0.0016
1	0.0256
2	0.1536
3	0.4096
4	0.4096

The Binomial Probability Formula

Table 5.20 displays the probability distribution of the random variable "number of patients cured" out of four by a drug. To obtain that probability distribution, we used a tabulation method (Table 5.19), which required a significant amount of work. But in most practical applications the work required would be considerably more and often prohibitive, because the number of trials is generally much larger than 4. For instance, if 20 patients instead of 4 were given the drug, there would be over 1 million possible outcomes. In that case the tabulation method would certainly not be feasible.

Fortunately, there is a relatively simple formula for obtaining binomial probabilities. The first step in developing that formula is the following key fact.

KEY FACT 5.9

**NUMBER OF OUTCOMES CONTAINING A
SPECIFIED NUMBER OF SUCCESSES**

> In n Bernoulli trials, the number of outcomes containing exactly x successes is equal to the binomial coefficient $\binom{n}{x}$; that is, the event of exactly x successes in n Bernoulli trials consists of $\binom{n}{x}$ outcomes.

We will not stop to prove Key Fact 5.9, but let's quickly check to see that it is consistent with the results obtained in Example 5.16. In part (a) of that example, we found from Table 5.19 that there are six outcomes in which exactly two of the four patients are cured: *ssff, sfsf, sffs, fssf, fsfs,* and *ffss.* In other words, the event of exactly two successes ($x = 2$) in four trials ($n = 4$) consists of six outcomes.

Using binomial coefficients we can determine that fact without resorting to a direct listing. Applying Key Fact 5.9, we have

$$\begin{bmatrix} \text{Number of outcomes} \\ \text{comprising the event} \\ \text{of exactly two cures} \end{bmatrix} = \binom{4}{2} = \frac{4!}{2!\,(4-2)!} = \frac{4!}{2!\,2!} = \frac{4 \cdot 3 \cdot 2!}{2!\,2!} = \frac{4 \cdot 3}{2} = 6.$$

The same approach holds for part (b) of Example 5.16:

$$\begin{bmatrix} \text{Number of outcomes} \\ \text{comprising the event} \\ \text{of exactly three cures} \end{bmatrix} = \binom{4}{3} = \frac{4!}{3!\,(4-3)!} = \frac{4!}{3!\,1!} = \frac{4 \cdot 3!}{3!\,1!} = \frac{4}{1} = 4.$$

We can now develop a probability formula for the number of successes in Bernoulli trials. We indicate briefly how that formula is derived by referring to Example 5.16. For instance, to determine the probability of exactly three cures, $P(x = 3)$, we reason as follows: Any particular outcome containing exactly three cures, for example, *ssfs,* has probability

$$\underset{\substack{\uparrow \\ \text{Probability} \\ \text{of a cure}}}{\overset{\overset{\text{Three cures}}{\downarrow}}{(0.8)^3}} \cdot \underset{\substack{\uparrow \\ \text{Probability} \\ \text{of a noncure}}}{\overset{\overset{\text{One noncure}}{\downarrow}}{(0.2)^1}} = 0.512 \cdot 0.2 = 0.1024,$$

obtained by multiplying three success probabilities of 0.8 and one failure probability of 0.2. Also, by Key Fact 5.9, the number of outcomes containing exactly three cures is equal to

$$\underset{\substack{\uparrow \\ \text{Number of cures}}}{\overset{\overset{\text{Number of trials}}{\downarrow}}{\binom{4}{3}}} = \frac{4!}{3!\,(4-3)!} = 4.$$

So, by the special addition rule, the probability of exactly three cures is

$$P(x = 3) = \binom{4}{3} \cdot (0.8)^3 (0.2)^1 = 4 \cdot 0.1024 = 0.4096.$$

Of course, this is the same result obtained in Example 5.16(b). However, this time we determined the probability quickly and easily—no tabulation and no listing were required. More important, the above reasoning applies to any sequence of Bernoulli trials and leads to the following formula.

FORMULA 5.2   **BINOMIAL PROBABILITY FORMULA**

> Suppose n Bernoulli trials are to be performed and the probability of success on any particular trial equals p. Let x denote the total number of successes in the n trials. Then the probability distribution of the random variable x is given by the formula
>
> $$P(x) = \binom{n}{x} p^x (1 - p)^{n-x}.$$
>
> The random variable x is called a ***binomial random variable*** and is said to have the ***binomial distribution*** with parameters n and p.

To determine a binomial probability formula in specific problems, it is useful to have a well-organized strategy, such as the one presented in Procedure 5.1.

PROCEDURE 5.1 **TO FIND A BINOMIAL PROBABILITY FORMULA**

ASSUMPTIONS

1. *n* identical trials are to be performed.
2. Two outcomes, success or failure, are possible for each trial.
3. The trials are independent.
4. The success probability, *p*, remains the same from trial to trial.

Step 1 Identify a success.

Step 2 Determine *p*, the success probability.

Step 3 Determine *n*, the number of trials.

Step 4 The binomial probability formula for the number of successes, *x*, is

$$P(x) = \binom{n}{x} p^x (1-p)^{n-x}.$$

Example 5.17 Illustrates Procedure 5.1

A new-car salesperson knows from past experience that she will make a sale to about 20% of her customers. Find the probability that in five (randomly selected) attempts, she makes a sale to

a. exactly three customers.

b. at most one customer.

c. at least one customer.

d. Determine the probability distribution of the number of sales in five attempts.

e. Construct a probability histogram for the number of sales in five attempts, and identify the skewness of the distribution.

SOLUTION Let *x* denote the number of sales in five attempts. To solve parts (a)–(d), we first apply Procedure 5.1.

Step 1 *Identify a success.*

In this problem a success is a sale to a customer.

Step 2 *Determine p, the success probability.*

This is the probability that the salesperson makes a sale to any particular customer, which is 20%. So $p = 0.2$.

Step 3 *Determine n, the number of trials.*

In this case the number of trials is the number of customers for which a sale is attempted, namely, five. Thus $n = 5$.

Step 4   *The binomial probability formula for the number of successes, x, is*

$$P(x) = \binom{n}{x} p^x (1-p)^{n-x}.$$

Since $n = 5$ and $p = 0.2$, the formula becomes

$$P(x) = \binom{5}{x} (0.2)^x (0.8)^{5-x}.$$

Now that we have applied Procedure 5.1, it is relatively easy to solve the problems posed in parts (a)–(d).

a. Here we want the probability of exactly three sales (successes). Applying the binomial probability formula with $x = 3$ yields

$$P(3) = \binom{5}{3} (0.2)^3 (0.8)^{5-3} = \frac{5!}{3!\,(5-3)!} (0.2)^3 (0.8)^2 = 0.051.$$

Consequently, the probability is 0.051 that the salesperson makes exactly three sales in five attempts.

b. The probability of at most one sale is

$$P(x \le 1) = P(0) + P(1) = \binom{5}{0} (0.2)^0 (0.8)^{5-0} + \binom{5}{1} (0.2)^1 (0.8)^{5-1}$$

$$= 0.328 + 0.410 = 0.738.$$

c. The probability of at least one sale is $P(x \ge 1)$. This can be obtained by first using the fact that $P(x \ge 1) = P(1) + P(2) + P(3) + P(4) + P(5)$ and then applying the binomial probability formula to calculate each of the five individual probabilities. However, it is easier to use the complementation rule:

$$P(x \ge 1) = 1 - P(x < 1) = 1 - P(x = 0)$$

$$= 1 - P(0) = 1 - \binom{5}{0} (0.2)^0 (0.8)^{5-0}$$

$$= 1 - 0.328 = 0.672.$$

The salesperson has a 67.2% chance of making at least one sale in five attempts.

d. Here we are to obtain the probability distribution of the number of sales, x, in five attempts. Thus we need to compute $P(x)$ for $x = 0, 1, 2, 3, 4,$ and 5 using the binomial probability formula. This has already been done for $x = 0, 1,$ and 3 in parts (a) and (b).

For $x = 5$ we have

$$P(5) = \binom{5}{5}(0.2)^5(0.8)^{5-5} = (0.2)^5 = 0.000,$$

to three decimal places [the exact value of $P(5)$ is 0.00032]. Similar computations yield $P(2) = 0.205$ and $P(4) = 0.006$. Therefore the probability distribution of the number of sales, x, is as shown in Table 5.21.

TABLE 5.21

Probability distribution of the random variable x, the number of sales in five attempts

Number of sales x	Probability $P(x)$
0	0.328
1	0.410
2	0.205
3	0.051
4	0.006
5	0.000

e. From part (d) we obtain the probability histogram for x, the number of sales in five attempts. This is shown in Fig. 5.5.

FIGURE 5.5

Probability histogram for the random variable x, the number of sales in five attempts

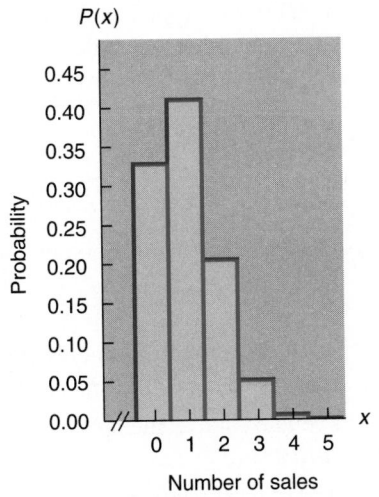

As we see from the probability histogram in Fig. 5.5, the probability distribution of the random variable x is right skewed.

Figure 5.5 shows that the distribution of the number of sales in five attempts is right skewed; this is because the success probability, $p = 0.2$, is less than 0.5. More generally, *a binomial distribution is right skewed if $p < 0.5$, is symmetric if $p = 0.5$, and is left skewed if $p > 0.5$.* Figure 5.6 illustrates these facts for binomial distributions with $n = 6$ in case $p = 0.25$, $p = 0.5$, and $p = 0.75$.

FIGURE 5.6
Probability histograms for
binomial distributions
with parameters $n = 6$
and (a) $p = 0.25$,
(b) $p = 0.5$, (c) $p = 0.75$

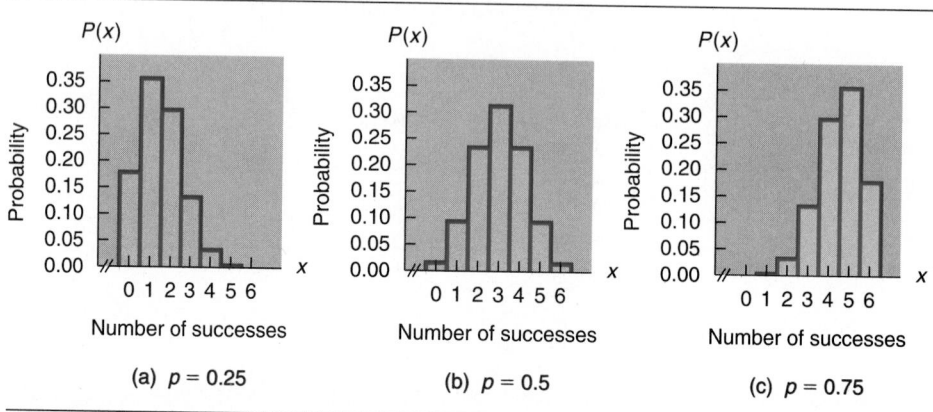

(a) $p = 0.25$ (b) $p = 0.5$ (c) $p = 0.75$

Binomial Probability Tables

Because of the importance of the binomial distribution, tables of binomial probabilities have been extensively compiled. Table I in Appendix A displays the number of trials, n, in the far left column; the number of successes, x, in the next column to the right; and the success probability, p, along the top. Example 5.18 illustrates how to use Table I.

Example 5.18   Using Binomial Probability Tables

Employ Table I in Appendix A to determine the probabilities computed in parts (a) and (b) of Example 5.17.

SOLUTION   The number of trials is $n = 5$ and the success probability is $p = 0.2$. The binomial distribution with those two parameters is displayed on the first page of Table I in Appendix A.

a. The probability of exactly three sales in five attempts is given as the fourth entry in the binomial distribution, which is 0.051.

b. To obtain the probability of at most one sale (0 or 1), we need to add the first two entries in the binomial distribution: $0.328 + 0.410 = 0.738$.

These are the same answers we obtained in parts (a) and (b) of Example 5.17, where we employed the binomial probability formula. However, by using Table I, we have reduced the computations required to almost nil.   ∎

As we have just observed, binomial probability tables eliminate most of the computations necessary in dealing with the binomial distribution. But such tables are of limited usefulness because they contain only a small number of different values of n and p. For example, Table I in Appendix A has only 11 different values of p and stops at $n = 20$.

Consequently, if we want to determine a binomial probability whose n or p parameter is not included in the table, then we must either use the binomial probability formula or statistical software. The latter method will be discussed at the end of this section.

The Binomial Distribution and Two-Category Populations

Recall that a two-category population is one in which each member is classified as either having or not having a specified attribute. Suppose a random sample of size n is to be taken from a (finite) two-category population in which the proportion of the population having the specified attribute equals p. Let x denote the number of members sampled that have the specified attribute.

If the sampling is with replacement, then the sampling process constitutes Bernoulli trials, and so the random variable x has the binomial distribution with parameters n and p. However, as we mentioned in Section 5.3, sampling is usually done without replacement. If the sampling is without replacement, then the sampling process does not constitute Bernoulli trials, and therefore x does not have a binomial distribution; rather, it has a **hypergeometric distribution.**

We will not present the hypergeometric probability formula here, because in practice a hypergeometric distribution can usually be approximated by a binomial distribution.[†] Specifically, Key Fact 5.8 on page 295 states that sampling without replacement from a two-category population can be regarded as Bernoulli trials provided the sample size does not exceed 5% of the population size. This implies that under those circumstances the probability distribution of the random variable x, which is a hypergeometric distribution, can be approximated by a binomial distribution. In summary, we have the following fact.

KEY FACT 5.10

BINOMIAL APPROXIMATION TO THE HYPERGEOMETRIC DISTRIBUTION

> Suppose a random sample of size n is to be taken without replacement from a two-category population in which the proportion of the population having the specified attribute equals p. Further suppose the sample size, n, does not exceed 5% of the population size, N. Let x denote the number of members sampled that have the specified attribute. Then the random variable x has approximately a binomial distribution with parameters n and p.

For example, according to the Census Bureau publication *Current Population Reports,* 74.7% of U.S. adults have completed high school. Suppose eight U.S. adults are to be

---

[†] See Exercise 5.68 for a discussion of the hypergeometric probability formula.

randomly selected without replacement, and let x denote the number of those sampled that have completed high school. Then, since the sample size does not exceed 5% of the population size (why is this so?), the random variable x has approximately a binomial distribution with parameters $n = 8$ and $p = 0.747$. Thus, probabilities for x are given approximately by the binomial probability formula, $P(x) = \binom{8}{x}(0.747)^x(0.253)^{8-x}$.

In practice, when sampling is done without replacement from a two-category population, the sample size does not exceed 5% of the population size. Consequently, a binomial distribution can generally be used to approximate the probability distribution of the number of members sampled that have the specified attribute.

Other Discrete Probability Distributions

The binomial distribution is the most important and most widely used discrete probability distribution. However, many other discrete probability distributions occur frequently in practice. We have already mentioned the hypergeometric distribution. Additional ones are the Poisson, discrete uniform, geometric, negative binomial, and multinomial distributions. We will examine the Poisson distribution in Section 5.6 and the geometric distribution in Exercise 5.69.

Using the Computer (Optional)

Minitab has a subcommand called **Binomial** that can be employed to determine binomial probabilities. We will illustrate the use of the Binomial subcommand by applying it in the car-salesperson example.

Example 5.19   The Binomial Subcommand

A new-car salesperson knows from past experience that she will make a sale to approximately 20% of her customers. Use Minitab to obtain the probability that in five attempts, she makes a sale to

a. exactly three customers.

b. at most one customer.

c. Determine the probability distribution of the number of sales in five attempts.

SOLUTION   In this problem a success is a sale to a customer. The probability of a success is the probability that the salesperson makes a sale to a customer, which is 20%; so $p = 0.2$. The number of trials is the number of attempts, namely, five; so $n = 5$. Consequently, the appropriate binomial distribution is the one with parameters $n = 5$ and $p = 0.2$.

a. To determine the probability of exactly three successes (sales), we use the PDF command and Binomial subcommand as follows.

Session commands: We type the command **PDF** followed by the number of successes for which the probability is required; that is, we type PDF 3; and press [Enter]. Then on the next line, we type the **BINOMIAL** subcommand followed by the parameters for the binomial distribution under consideration; that is, we type BINOMIAL with n=5 and p=0.2. and press [Enter]. These commands and the resulting output are shown in Printout 5.2.

(or)

Menu commands: We choose **Calc ▶ Probability Distributions ▶ Binomial**. Then we select **Probability** to indicate that we want to obtain an individual probability. To specify the appropriate binomial distribution, we first select **Number of trials** and type 5 and then select **Probability of success** and type 0.2. Next we select **Input constant** and type 3 to specify that we want the probability for the value 3, that is, $P(x = 3)$. Finally, we select **OK**. Printout 5.2 displays the output that results.

PRINTOUT 5.2
Minitab output for $P(x = 3)$

```
MTB > PDF 3;
SUBC> BINOMIAL with n=5 and p=0.2.
       K              P( X = K)
     3.00              0.0512
```

Printout 5.2 shows that the probability is 0.0512 that the salesperson makes exactly three sales in five attempts. Notice that Minitab uses K instead of *x,* and P(X = K) instead of $P(x)$.

b. For this part we want the probability of at most one sale, that is, the probability of one or fewer sales. Minitab can be used in several ways to find that probability. One way is to employ Minitab's capability for obtaining cumulative probabilities. A **cumulative probability** gives the probability that a random variable is less than or equal to a specified value. Here we want the cumulative probability for the value 1 (one success). We use the CDF command and Binomial subcommand as follows.

Session commands: We type the command **CDF** followed by the number of successes for which the cumulative probability is required; that is, we type CDF 1; and press [Enter]. Then on the next line, we type the **BINOMIAL** subcommand followed by the parameters for the binomial distribution under consideration; that is, we type BINOMIAL with n=5 and p=0.2. and press [Enter]. Printout 5.3 at the top of the next page displays these commands and the output that results.

(or)

Menu commands: We choose **Calc ▶ Probability Distributions ▶ Binomial**. Then we select **Cumulative probability** to indicate that we want to obtain a cumulative probability. To specify the appropriate binomial distribution, we first select **Number of trials** and type 5 and then select **Probability of success** and type 0.2. Next we select **Input constant** and type 1 to specify that we want the cumulative probability for the value 1, that is, $P(x \leq 1)$. Then we select **OK**. Printout 5.3 shows the resulting output.

PRINTOUT 5.3

Minitab output for $P(x \leq 1)$

```
MTB > CDF 1;
SUBC> BINOMIAL with n=5 and p=0.2.
     K  P( X LESS OR = K)
    1.00            0.7373
```

As we see from Printout 5.3, the probability is 0.7373 of at most one sale in five attempts. *Note:* If we round this probability to three decimal places, we get 0.737. This differs from the probability of 0.738 obtained in Examples 5.17(b) and 5.18(b). The discrepancy is due to rounding error.

c. Here we are to determine the probability distribution of the number of sales in five attempts. In other words, we want to obtain the binomial distribution with parameters $n = 5$ and $p = 0.2$. This can be accomplished in the following way. We begin by storing the possible values for the random variable x, namely, the integers 0–5, in a column named VALUES. Then we proceed as described below.

Session commands: We first type the command PDF followed by the storage location of the values for which we want probabilities; that is, we type PDF 'VALUES'; and press ⌨Enter. Then on the next line, we type the BINOMIAL subcommand followed by the parameters for the binomial distribution under consideration; that is, we type BINOMIAL with n=5 and p=0.2. and press ⌨Enter. Printout 5.4 shows these commands and the resulting output.

(or)

Menu commands: We choose **Calc ▶ Probability Distributions ▶ Binomial**. Then we select **Probability** to indicate that we want to obtain individual probabilities. To specify the appropriate binomial distribution, we first select **Number of trials** and type 5 and then select **Probability of success** and type 0.2. Next we select **Input column** and specify VALUES in order to tell Minitab that we want an individual probability for each of the numbers in that column. Finally, we select **OK**. The resulting output is portrayed in Printout 5.4.

PRINTOUT 5.4

Minitab output for the binomial distribution with parameters $n = 5$ and $p = 0.2$

```
MTB > PDF 'VALUES';
SUBC> BINOMIAL with n=5 and p=0.2.
       K          P( X = K)
      0.00          0.3277
      1.00          0.4096
      2.00          0.2048
      3.00          0.0512
      4.00          0.0064
      5.00          0.0003
```

Compare the table in Printout 5.4 to the binomial distribution displayed in Table 5.21 on page 303.

EXERCISES 5.4

5.44 This exercise uses the results obtained in Exercise 5.34 on page 295. If you have not done that exercise, you should do it before proceeding. In an ESP experiment, a person in one room randomly selects 1 of 10 cards numbered 1–10 and a person in another room tries to guess the number on the card chosen. This experiment is to be repeated three times, with each card chosen being replaced before the next selection. Assuming the person guessing lacks ESP, the probability of a correct guess on any particular trial is $\frac{1}{10} = 0.1$. Suppose we consider a success, s, to be a correct guess.
a. Use your results from Exercise 5.34 to determine the probability of exactly one correct guess in the three tries.
b. Obtain the probability in part (a) by applying the binomial probability formula. (Use Procedure 5.1 on page 301.)

5.45 This exercise uses the results obtained in Exercise 5.37 on page 296. If you have not done that exercise, you should do it before proceeding. According to the National Institute of Mental Health, 20% of adult Americans suffer from a psychiatric disorder. Suppose four randomly selected adult Americans are to be examined for psychiatric disorders.
a. Use your results from Exercise 5.37 to determine the probability that exactly three of the four people examined have a psychiatric disorder.
b. Obtain the probability in part (a) by applying the binomial probability formula. (Use Procedure 5.1 on page 301.)

Use Procedure 5.1 on page 301 to solve Exercises 5.46–5.49. Express each probability as a decimal rounded to three places.

5.46 According to the *Daily Racing Form,* the probability is approximately 0.67 that the favorite in a horse race will finish in the money (first, second, or third place). In the next five races, what is the probability that the favorite finishes in the money
a. exactly twice?
b. exactly four times?
c. at least four times?
d. between two and four times, inclusive?
e. Determine the probability distribution of the random variable x, the number of times the favorite finishes in the money in the next five races.
f. Identify the probability distribution of x as right skewed, symmetric, or left skewed without consulting its probability distribution or drawing its probability histogram.
g. Draw a probability histogram for x.

5.47 Based on data from the *Statistical Abstract of the United States,* the probability that a newborn baby will be a girl is roughly 0.487. In the next three births, what is the probability that
a. exactly one is a girl?
b. at most one is a girl?
c. at least one is a girl?
d. either one or two are girls?
e. Determine the probability distribution of the random variable x, the number of newborns in the next three births that are girls.
f. Identify the probability distribution of x as right skewed, symmetric, or left skewed without consulting its probability distribution or drawing its probability histogram.
g. Draw a probability histogram for x.

5.48 According to a survey by the Opinion Research Corporation, 60% of all clerical workers in the United States "like their jobs very much." Suppose four randomly selected clerical workers are to be asked whether they like their jobs very much. Find the probability that the number of affirmative responses is
a. exactly two.
b. at most two.
c. at least one.
d. between one and three, inclusive.
e. Obtain the probability distribution of the random variable x, the number of affirmative responses.
f. Identify the probability distribution of x as right skewed, symmetric, or left skewed without consulting its probability distribution or drawing its probability histogram.
g. Draw a probability histogram for x.

5.49 As reported by the A. C. Nielsen Company in *Nielsen Report on Television,* 85.5% of U.S. households have a color television set. If six households are randomly selected, what is the probability that the number having a color TV is
a. exactly four?
b. at least four?
c. at most five?
d. between two and five, inclusive?
e. Determine the probability distribution of the random variable x, the number of households out of six that have a color television set.
f. Identify the probability distribution of x as right skewed, symmetric, or left skewed without consulting its probability distribution or drawing its probability histogram.
g. Draw a probability histogram for x.

For Exercises 5.50–5.53, obtain the required probabilities using Procedure 5.1. If possible, use Table I in Appendix A to check your results.

5.50 According to *Current Population Reports,* a publication of the U.S. Bureau of the Census, 25% of U.S. children are not living with both parents. If 10 U.S. children are selected at random, determine the probability that the number not living with both parents is
a. exactly two.
b. at most two.
c. between three and six, inclusive.
d. either less than three or more than seven.

5.51 The U.S. Energy Information Administration collects data on appliances used by U.S. households and publishes the results in *Residential Energy Consumption Survey: Housing Characteristics.* According to that publication, 36.7% of U.S. households use an automatic dishwasher. If 12 U.S. households are randomly selected, determine the probability that the number using an automatic dishwasher is
a. exactly three.
b. at most three.
c. at least one.
d. between two and four, inclusive.
e. either less than three or more than 10.

5.52 The U.S. Bureau of the Census collects data on the educational attainment of Americans and states its findings in *Current Population Reports.* According to that publication, 74.7% of U.S. adults have completed high school. Suppose eight U.S. adults are selected at random. What is the probability that
a. exactly six have completed high school?
b. at least six have completed high school?
c. at least three have completed high school?
d. between four and seven, inclusive, have completed high school?

5.53 Surveys indicate that in 80% of Swedish couples, both partners work. If nine Swedish couples are selected at random, what is the probability that the number of working couples is
a. exactly seven?
b. at least seven?
c. at most nine?
d. between five and eight, inclusive?
e. either less than three or more than six?

In Exercises 5.54 and 5.55, use Table I in Appendix A to determine the required probabilities.

5.54 Studies show that 60% of U.S. families use physical aggression to resolve conflict. Suppose 20 families are selected at random. Find the probability that the number that use physical aggression to resolve conflict is
a. exactly 10.
b. between 10 and 15, inclusive.
c. over 75% of those surveyed.
d. less than 8.

5.55 According to the American Bankers Association, only 1 in 10 people are dissatisfied with their local bank. If 15 people are randomly selected, what is the probability that the number dissatisfied with their local bank is
a. exactly two?
b. at most two?
c. at least two?
d. between one and three, inclusive?

5.56 The case study at the end of Chapter 4 discusses the Arizona state lottery, *Lotto,* which is played as follows: The player selects six numbers from the numbers 1–42 and buys a ticket for $1. There are six winning numbers, which are selected at random from the numbers 1–42. To win a prize, a *Lotto* ticket must contain three or more of the winning numbers. Table 4.18 on page 262 displays the probabilities (to seven decimal places) for the number of winning numbers a *Lotto* ticket contains. Suppose you buy one *Lotto* ticket per week for a year. Determine the probability that you win a prize at least once in the 52 tries.

5.57 Following are two probability histograms of binomial distributions. For each one, specify whether the success probability, *p,* is less than, equal to, or greater than 0.5.

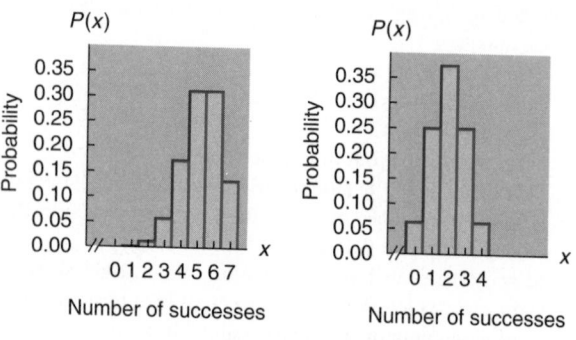

(a)

(b)

5.58 Following are two probability histograms of binomial distributions. For each one, specify whether the success probability, p, is less than, equal to, or greater than 0.5.

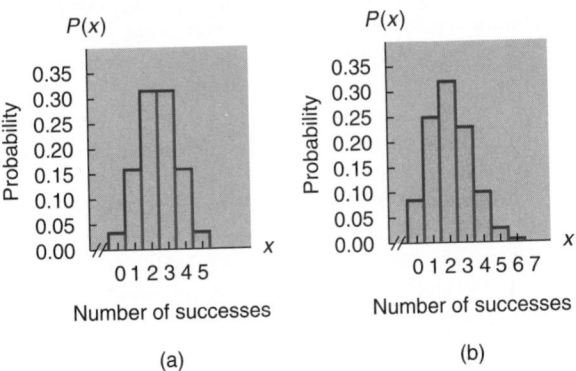

(a)   (b)

5.59 (Computer exercise) Use Minitab or some other statistical software to obtain the required probability or probabilities in parts (a)–(e) of Exercise 5.49.

5.60 (Computer exercise) Use Minitab or some other statistical software to obtain the required probability or probabilities in parts (a)–(e) of Exercise 5.48.

The printouts in Exercises 5.61–5.64, which have a slightly different appearance from those in Printouts 5.2–5.4, were obtained using Release 7 of Minitab's PC version.

5.61 (Computer exercise) Following is a Minitab printout obtained by applying the PDF command and Binomial subcommand.

```
BINOMIAL WITH N =    7  P = 0.340000
     K          P( X = K)
     0           0.0546
     1           0.1967
     2           0.3040
     3           0.2610
     4           0.1345
     5           0.0416
     6           0.0071
     7           0.0005
```

Use the printout to determine
a. the number of trials.
b. the success probability.
c. the probability of exactly three successes.
d. the probability of between two and five successes, inclusive.
e. the probability of at most four successes.
f. the probability of at least four successes.

5.62 (Computer exercise) The following printout was obtained from Minitab by employing the CDF command and Binomial subcommand.

```
BINOMIAL WITH N =    6  P = 0.590000
     K   P( X LESS OR = K)
     0           0.0048
     1           0.0458
     2           0.1933
     3           0.4764
     4           0.7819
     5           0.9578
     6           1.0000
```

Use the printout to find
a. the number of trials.
b. the success probability.
c. the probability of at most three successes.
d. the probability of at least three successes. *(Hint: Use the complementation rule and the printout.)*
e. the probability of exactly three successes. *(Hint: Use the special addition rule to express $P(x = 3)$ as the difference of two cumulative probabilities.)*

5.63 (Computer exercise) According to an article in *Reader's Digest*, 10% of people are left-handed. Suppose 10 people are to be selected at random. Let x denote the number of people out of the 10 who are left-handed. We used Minitab to obtain the probability distribution of the random variable x. The output is shown below. Notice that Minitab stops printing the probabilities after it encounters one that is zero to four decimal places.

```
BINOMIAL WITH N =   10  P = 0.100000
     K          P( X = K)
     0           0.3487
     1           0.3874
     2           0.1937
     3           0.0574
     4           0.0112
     5           0.0015
     6           0.0001
     7           0.0000
```

Employ the printout to obtain the probability that out of the 10 people chosen,
a. exactly one is left-handed.
b. at least one is left-handed.
c. at most one is left-handed.
d. more than one is left-handed.
e. between one and three, inclusive, are left-handed.

5.64 (Computer exercise) The U.S. National Center for Health Statistics reports that 27% of U.S. adults ages 20 to 74 have high cholesterol levels. Suppose eight U.S. adults are to be selected at random. Let x denote the number of those chosen who have high cholesterol levels. The following Minitab printout displays the probability distribution of the random variable x.

```
BINOMIAL WITH N =   8  P = 0.270000
    K         P( X = K)
    0          0.0806
    1          0.2386
    2          0.3089
    3          0.2285
    4          0.1056
    5          0.0313
    6          0.0058
    7          0.0006
    8          0.0000
```

From the printout, obtain the probability that out of the eight adults chosen, the number having high cholesterol levels is
a. exactly three.   b. at least three.
c. less than two.
d. between two and four, inclusive.
e. either less than two or more than five.

5.65 A sales representative for a tire manufacturer claims that the company's steel-belted radials last at least 35,000 miles. A tire dealer decides to check that claim by testing eight of the tires. If 75% or more of the eight tires he tests last at least 35,000 miles, he will purchase tires from the sales representative. If, in fact, 90% of the steel-belted radials produced by the manufacturer last at least 35,000 miles, what is the probability that the tire dealer will purchase tires from the sales representative?

5.66 From past experience the owner of a restaurant knows that, on average, 4% of the parties making reservations never show. How many reservations can the owner accept and still be at least 80% sure that all parties making a reservation will show?

5.67 Sickle cell anemia is an inherited blood disease that occurs primarily in blacks. In the United States, roughly 15 of every 10,000 black children have sickle cell anemia. The red blood cells of an affected person are abnormal; the result is severe chronic anemia (inability to carry the required amount of oxygen), which causes headaches, shortness of breath, jaundice, increased risk of pneumococcal pneumonia and gallstones, and other severe problems. Sickle cell anemia arises in children inheriting an abnormal type

of hemoglobin, called hemoglobin S, from both parents. If hemoglobin S is inherited from only one parent, then the person is said to have sickle cell trait and is generally free from symptoms. There is a 50% chance that a person having sickle cell trait will pass hemoglobin S to an offspring.
a. Obtain the probability that a child of two people having sickle cell trait will have sickle cell anemia.
b. If two people having sickle cell trait have five children, determine the probability that at least one of the children will have sickle cell anemia.
c. If two people having sickle cell trait have five children, obtain the probability distribution of the number of those children who will inherit sickle cell anemia.
d. Construct a probability histogram for the probability distribution in part (c).

5.68 In this exercise we will discuss the hypergeometric distribution in more detail. When sampling without replacement from a finite two-category population, the hypergeometric distribution is the exact probability distribution for the number of successes, that is, the number of members sampled that have the specified attribute. The hypergeometric probability formula is

$$P(x) = \frac{\binom{Np}{x}\binom{N(1-p)}{n-x}}{\binom{N}{n}}.$$

Here N is the population size, n is the sample size, and p is the proportion of successes in the population, that is, the proportion of the population having the specified attribute.

To illustrate, suppose a customer purchases 4 fuses from a shipment of 250, of which 94% are not defective. Let a success correspond to a fuse that is not defective.
a. Determine N, n, and p.
b. Use the hypergeometric probability formula to find the probability distribution of the number of nondefective fuses the customer gets.

Key Fact 5.10 states that a hypergeometric distribution can be approximated by a binomial distribution provided the sample size does not exceed 5% of the population size. In particular, we can use the binomial probability formula

$$P(x) = \binom{n}{x}p^x(1-p)^{n-x},$$

with $n = 4$ and $p = 0.94$, to approximate the probability distribution of the number of nondefective fuses that the customer gets.
c. Obtain the binomial distribution with parameters $n = 4$ and $p = 0.94$.

d. Compare the hypergeometric distribution from part (b) with the binomial distribution from part (c).

· ·   **5.69** Another important discrete probability distribution is called the geometric distribution. The **geometric distribution** is the probability distribution for the number of trials until the first success in Bernoulli trials. The geometric probability formula is

$$P(x) = p(1 - p)^{x-1},$$

where p denotes the success probability and x can be any positive integer. This formula gives the probability that the first success occurs on trial x.

To illustrate, let's again consider the Arizona state lottery, *Lotto*, as described in Exercise 5.56. Suppose you buy one *Lotto* ticket per week. Let x denote the number of weeks until you win a prize.
a. Find and interpret the probability formula for the random variable x. *(Note:* The appropriate success probability was obtained in the process of solving Exercise 5.56.)
b. Compute the probability that the number of weeks until you win a prize is exactly 3; at most 3; at least 3.

5.5 THE MEAN AND STANDARD DEVIATION OF A
· · · · · · BINOMIAL RANDOM VARIABLE

In Section 5.2 we learned how to obtain the mean and standard deviation of a discrete random variable. Definition 5.4 on page 279 provides the formula for computing the mean of a discrete random variable. Definition 5.5 on page 284 and Formula 5.1 on page 285 provide formulas for computing the standard deviation of a discrete random variable. Those formulas apply to any discrete random variable, and in particular to a binomial random variable. Consider Example 5.20.

Example 5.20   Calculating the Mean and Standard Deviation of a Binomial Random Variable

As reported by the U.S. National Center for Health Statistics in *Vital and Health Statistics*, approximately 60% of all eye operations are performed on females. Suppose three eye-operation patients are to be selected at random. Let x denote the number of patients out of the three chosen that are female. Then x is a binomial random variable with parameters $n = 3$ and $p = 0.6$.

a. Find the mean of the random variable x.

b. Find the standard deviation of the random variable x.

SOLUTION   To obtain the mean and standard deviation of the random variable x, we first need to determine its probability distribution. Since x has the binomial distribution with parameters $n = 3$ and $p = 0.6$, probabilities for x can be computed using the binomial probability formula,

$$P(x) = \binom{3}{x}(0.6)^x(0.4)^{3-x}.$$

Employing that formula we get the probability distribution of x, shown in the first two columns of Table 5.22.

TABLE 5.22
Table for computing μ_x and σ_x

x	$P(x)$	$xP(x)$	x^2	$x^2P(x)$
0	0.064	0.000	0	0.000
1	0.288	0.288	1	0.288
2	0.432	0.864	4	1.728
3	0.216	0.648	9	1.944
		1.800		3.960

a. The mean of the random variable x can now be obtained by applying Definition 5.4. We see from the third column of Table 5.22 that

$$\mu_x = \Sigma x P(x) = 1.8.$$

The mean of the random variable x is $\mu_x = 1.8$.

b. To determine the standard deviation of the random variable x, we will use the shortcut formula, Formula 5.1. From the fifth column of Table 5.22 and part (a), we conclude that

$$\sigma_x = \sqrt{\Sigma x^2 P(x) - \mu_x^2} = \sqrt{3.96 - (1.8)^2} = \sqrt{0.72} = 0.85.$$

The standard deviation of the random variable x is $\sigma_x = 0.85$.   ∎

In part (a) of Example 5.20 we found that if three eye-operation patients are selected at random and x denotes the number that are female, then the mean of the random variable x is $\mu_x = 1.8$. This indicates that, on the average, about 1.8 of every three eye-operation patients are female.

We could have guessed that result before doing any computations by using the following reasoning: Sixty percent of all eye-operation patients are female. Thus out of three eye-operation patients, we would expect roughly 60% to be female. Sixty percent of 3 is 1.8. Consequently, on the average, we would expect about 1.8 of every three eye-operation patients to be female.

This is the same result obtained for μ_x in part (a) of Example 5.20. However, here we did almost no calculations. We simply multiplied the number of trials, $n = 3$, by the success probability, $p = 0.6$.

It can be shown that this type of reasoning always works; that is, the mean of a binomial random variable is equal to the number of trials, n, times the success probability, p. Formula 5.3 states this formally.

FORMULA 5.3   **MEAN OF A BINOMIAL RANDOM VARIABLE**

Suppose x has the binomial distribution with parameters n and p. Then the mean of the random variable x can be obtained from the formula

$$\mu_x = np.$$

Formula 5.3 can be derived mathematically by substituting the binomial probability formula, $P(x) = \binom{n}{x} p^x (1-p)^{n-x}$, into the formula $\mu_x = \Sigma x P(x)$, and simplifying. The details are a little tricky, but the idea is not.

There is also a simple formula for obtaining the standard deviation of a binomial random variable. Although the formula is not intuitively clear like the formula for the mean, it is nonetheless derived in essentially the same way: Substitute the binomial probability formula into the formula $\sigma_x = \sqrt{\Sigma x^2 P(x) - \mu_x^2}$, and simplify. The result of doing that is presented in Formula 5.4.

FORMULA 5.4

STANDARD DEVIATION OF A BINOMIAL RANDOM VARIABLE

> Suppose x has the binomial distribution with parameters n and p. Then the standard deviation of the random variable x can be obtained from the formula
>
> $$\sigma_x = \sqrt{np(1-p)}.$$

We applied the general formulas $\mu_x = \Sigma x P(x)$ and $\sigma_x = \sqrt{\Sigma x^2 P(x) - \mu_x^2}$ in Example 5.20 to find the mean and standard deviation of a random variable x, the number of eye-operation patients out of three that are female. Since that random variable has a binomial distribution, we can instead use the special formulas $\mu_x = np$ and $\sigma_x = \sqrt{np(1-p)}$, applicable only to binomial random variables. This is done in Example 5.21.

Example 5.21   Illustrates Formulas 5.3 and 5.4

Refer to Example 5.20, and as before, let x denote the number of patients out of three randomly selected ones that are female.

a. Find the mean of the random variable x.

b. Find the standard deviation of the random variable x.

SOLUTION   First recall that the random variable x has the binomial distribution with parameters $n = 3$ and $p = 0.6$.

a. Applying Formula 5.3 we obtain the mean of x:

$$\mu_x = np = 3 \cdot 0.6 = 1.8.$$

b. Using Formula 5.4 we get the standard deviation of x:

$$\sigma_x = \sqrt{np(1-p)} = \sqrt{3 \cdot 0.6 \cdot 0.4} = \sqrt{0.72} = 0.85.$$

The values just obtained for μ_x and σ_x are the same as the ones that we found in Example 5.20. However, the calculations required here using Formulas 5.3 and 5.4 are much easier.   ■

As we have seen, the special formulas, Formulas 5.3 and 5.4, are real time-savers for computing the mean and standard deviation of a binomial random variable, even when the number of trials, n, is small. In most applications n is quite large, and in those cases the special formulas are indispensable.

Example 5.22   Illustrates Formulas 5.3 and 5.4

According to a spokesperson for Southwest Airlines, the no-show rate for reservations is about 16%; that is, the probability is 0.16 that a party making a reservation will not show up. Find the mean and standard deviation of the number of no-shows for a flight having 42 parties with reservations.

SOLUTION   Let x denote the number of no-shows out of the 42 reservations. We need to compute the mean and standard deviation of x. The random variable x has the binomial distribution with parameters $n = 42$ and $p = 0.16$. Thus by Formula 5.3,

$$\mu_x = np = 42 \cdot 0.16 = 6.72,$$

and by Formula 5.4,

$$\sigma_x = \sqrt{np(1 - p)} = \sqrt{42 \cdot 0.16 \cdot 0.84} = 2.38.$$

In particular, the airline can expect about 6.72 no-shows on the flight.   ■

EXERCISES 5.5

·   **5.70** In an ESP experiment, a person in one room randomly selects 1 of 10 cards numbered 1–10 and a person in another room tries to guess the number on the card chosen. This experiment is to be repeated three times, with each card chosen being replaced before the next selection. Assuming the person guessing lacks ESP, the probability of a correct guess on any particular trial is $\frac{1}{10} = 0.1$. Let x denote the number of correct guesses out of the three tries.
a. Find the mean and standard deviation of the random variable x by applying the general formulas, Definition 5.4 on page 279 and Formula 5.1 on page 285. (Use the method employed in Example 5.20.)
b. Find the mean and standard deviation of x using the special formulas, Formulas 5.3 and 5.4.
c. Compare the amount of work required in parts (a) and (b).

·   **5.71** According to the National Institute of Mental Health, 20% of all adult Americans suffer from a psychiatric disorder. Suppose four adult Americans are to be examined for psychiatric disorders, and let x denote the number having a psychiatric disorder.
a. Find the mean and standard deviation of the random variable x by applying the general formulas, Definition 5.4 on

page 279 and Formula 5.1 on page 285. (Use the method employed in Example 5.20.)
b. Find the mean and standard deviation of x using the special formulas, Formulas 5.3 and 5.4.
c. Compare the amount of work required in parts (a) and (b).

In each of Exercises 5.72–5.77, use the special formulas, Formulas 5.3 and 5.4, to obtain the mean and standard deviation of the binomial random variable in question. Interpret your answer for the mean.

·   **5.72** According to the *Daily Racing Form*, the probability is approximately 0.67 that the favorite in a horse race will finish in the money (first, second, or third place). Find the mean and standard deviation of the number of favorites finishing in the money in the next five races.

·   **5.73** As reported by the A. C. Nielsen Company in *Nielsen Report on Television*, 85.5% of U.S. households have a color television set. In a random sample of six households, determine the mean and standard deviation of the number having a color TV.

· **5.74** According to the Census Bureau's *Current Population Reports,* 25% of U.S. children are not living with both parents. Determine the mean and standard deviation of the number of children in a sample of 10 that are not living with both parents.

· **5.75** The U.S. Energy Information Administration collects data on appliances used in U.S. households and publishes its findings in *Residential Energy Consumption Survey: Housing Characteristics.* According to that publication, 36.7% of U.S. households use an automatic dishwasher. If 12 households are to be selected at random, find the mean and standard deviation of the number that use an automatic dishwasher.

· **5.76** An air-conditioner contractor knows from past experience that about 40% of the evaporative coolers he sells are fiberglass (as opposed to metal). Find the mean and standard deviation of the number of fiberglass units sold out of the next 20 evaporative-cooler sales.

· **5.77** A student takes a multiple-choice test having 50 questions. Each question has four possible answers. If the student guesses at each question, find the mean and standard deviation of the number of questions answered correctly.

· · **5.78** In Exercise 5.67 on page 312, we discussed sickle cell anemia. We found that the probability is 0.25 that a child of two people having sickle cell trait will have sickle cell anemia. If two people with sickle cell trait have five children, how many can they expect will have sickle cell anemia?

· · · **5.79** Suppose n Bernoulli trials are to be performed. Let x denote the number of successes. Also, let $\hat{p}$ denote the **proportion of successes;** that is, $\hat{p} = x/n$. Use Formulas 5.3 and 5.4 and the results of part (f) of Exercise 5.27 on page 289 to show that
a. $\mu_{\hat{p}} = p$.   b. $\sigma_{\hat{p}} = \sqrt{p(1-p)/n}$.
These formulas are required for making inferences about a population proportion.

5.6 THE POISSON DISTRIBUTION (OPTIONAL)

· · · · ·

Another important discrete probability distribution is the **Poisson distribution,** named after the French mathematician and physicist Simeon D. Poisson (1781–1840). The Poisson distribution is often used to model the frequency with which a specified event occurs during a particular period of time. For instance, we might employ the Poisson distribution when analyzing

- the number of patients arriving at an emergency room between 6:00 P.M. and 7:00 P.M.
- the number of telephone calls received per day at a switchboard.
- the number of alpha particles emitted per minute by a radioactive substance.

Additionally, the Poisson distribution might be used to describe the number of misprints in books, the number of defective teeth per person, or the number of bacterial colonies appearing on a petri dish smeared with a bacterial suspension.

The Poisson Probability Formula

A particular Poisson distribution is identified by one parameter, usually denoted by the Greek letter λ (lambda). As we will see, that parameter represents the mean of the distribution. Formula 5.5 provides the **Poisson probability formula,** the formula used to obtain probabilities for a random variable having a Poisson distribution.

FORMULA 5.5   **POISSON PROBABILITY FORMULA**

Probabilities for a random variable x having a Poisson distribution are given by the formula

$$P(x) = e^{-\lambda} \frac{\lambda^x}{x!}, \qquad x = 0, 1, 2, \ldots,$$

where λ is a positive real number and $e \approx 2.718$.[†] The random variable x is called a **Poisson random variable** and is said to have the **Poisson distribution** with parameter λ.

A Poisson random variable has a (countably) infinite number of possible values, namely, all nonnegative integers. Thus we cannot display all the probabilities for a Poisson random variable in a probability-distribution table.

Example 5.23   Illustrates Formula 5.5

Desert Samaritan Hospital, located in Mesa, Arizona, keeps records of emergency-room traffic. From those records we find that the number of patients arriving between 6:00 P.M. and 7:00 P.M. has a Poisson distribution with parameter $\lambda = 6.9$. Determine the probability that, on a given day, the number of patients arriving at the emergency room between 6:00 P.M. and 7:00 P.M. will be

a. exactly 4.

b. at most 2.

c. between 4 and 10, inclusive.

d. Obtain a table of probabilities for x, stopping when the probabilities become 0 to three decimal places.

e. Use part (d) to construct a (partial) probability histogram for the random variable x.

f. Identify the shape of the probability distribution of x.

SOLUTION   Let x denote the number of patients arriving between 6:00 P.M. and 7:00 P.M. Then the random variable x has a Poisson distribution with parameter $\lambda = 6.9$. Thus, by Formula 5.5, probabilities for x are given by the Poisson probability formula,

$$P(x) = e^{-6.9} \frac{(6.9)^x}{x!}.$$

Using this formula we can now solve the problems posed in parts (a)–(e).

a. Here we want the probability of exactly 4 arrivals. Applying the Poisson probability formula with $x = 4$ gives

$$P(4) = e^{-6.9} \frac{(6.9)^4}{4!} = e^{-6.9} \cdot \frac{2266.7121}{24} = 0.095.$$

[†] Most calculators have an e-key.

b. The probability of at most 2 arrivals is

$$P(x \leq 2) = P(0) + P(1) + P(2)$$

$$= e^{-6.9}\,\frac{(6.9)^0}{0!} + e^{-6.9}\,\frac{(6.9)^1}{1!} + e^{-6.9}\,\frac{(6.9)^2}{2!}$$

$$= e^{-6.9}\left(\frac{6.9^0}{0!} + \frac{6.9^1}{1!} + \frac{6.9^2}{2!}\right)$$

$$= e^{-6.9}\left(1 + 6.9 + 23.805\right) = e^{-6.9} \cdot 31.705 = 0.032.$$

c. The probability of between 4 and 10 arrivals, inclusive, is

$$P(4 \leq x \leq 10) = P(4) + P(5) + \cdots + P(10)$$

$$= e^{-6.9}\left(\frac{6.9^4}{4!} + \frac{6.9^5}{5!} + \cdots + \frac{6.9^{10}}{10!}\right) = 0.821.$$

d. Proceeding as in part (a), we obtain a partial probability distribution of the random variable x, as shown in Table 5.23.

TABLE 5.23

Partial probability distribution of the random variable x, the number of patients arriving at the emergency room between 6:00 P.M. and 7:00 P.M.

Number arriving x	Probability $P(x)$	Number arriving x	Probability $P(x)$
0	0.001	10	0.068
1	0.007	11	0.043
2	0.024	12	0.025
3	0.055	13	0.013
4	0.095	14	0.006
5	0.131	15	0.003
6	0.151	16	0.001
7	0.149	17	0.001
8	0.128	18	0.000
9	0.098		

e. Using Table 5.23 we obtain a partial probability histogram for the random variable x, as depicted in Fig. 5.7, shown at the top of the next page.

f. From Fig. 5.7 we see that the probability distribution of the random variable x is right skewed.   ■

In part (f) of Example 5.23 we found that the probability distribution of the number of patients arriving at the emergency room between 6:00 P.M. and 7:00 P.M. is right skewed. This is true for all Poisson distributions—the probability distribution of any Poisson random variable is right skewed.

FIGURE 5.7

Partial probability histogram
for the random variable x, the
number of patients arriving
at the emergency room
between 6:00 P.M. and 7:00 P.M.

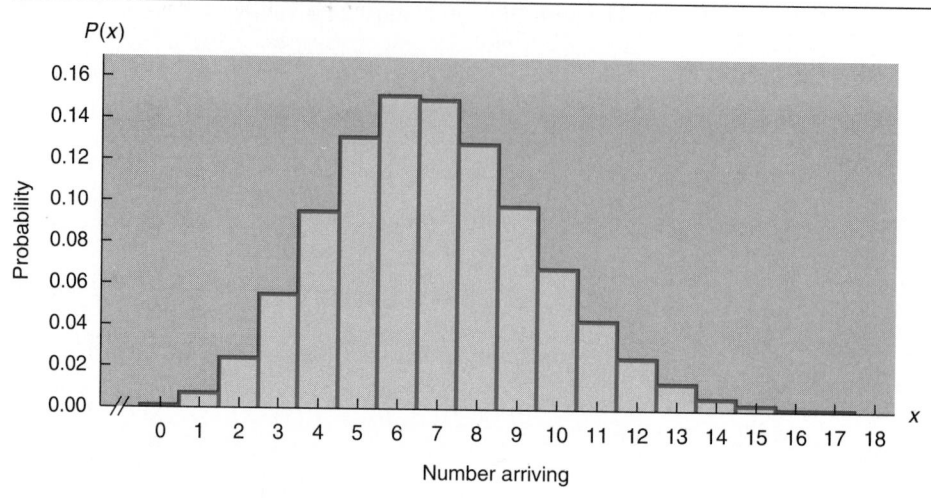

FIGURE 5.7

Partial probability histogram for the random variable x, the number of patients arriving at the emergency room between 6:00 P.M. and 7:00 P.M.

The Mean and Standard Deviation of a Poisson Random Variable

We can derive special formulas for the mean and standard deviation of a Poisson random variable. Those formulas are obtained by first substituting the Poisson probability formula, $P(x) = e^{-\lambda}\lambda^x/x!$, into the general formulas for the mean and standard deviation of a discrete random variable, $\mu_x = \Sigma x P(x)$ and $\sigma_x = \sqrt{\Sigma x^2 P(x) - \mu_x^2}$, and then simplifying. The results of doing that are presented in Formula 5.6.

FORMULA 5.6

MEAN AND STANDARD DEVIATION OF A POISSON RANDOM VARIABLE

> Suppose x has the Poisson distribution with parameter λ. Then the mean and standard deviation of the random variable x can be obtained from the formulas
>
> $$\mu_x = \lambda \qquad \text{and} \qquad \sigma_x = \sqrt{\lambda}.$$

Example 5.24   Illustrates Formula 5.6

Refer to Example 5.23 where x denotes the number of patients arriving at the emergency room of Desert Samaritan Hospital between 6:00 P.M. and 7:00 P.M.

a. Determine and interpret the mean of the random variable x.

b. Determine the standard deviation of x.

SOLUTION   From Example 5.23 we know that x has the Poisson distribution with parameter $\lambda = 6.9$. Thus we can apply Formula 5.6 to determine the mean and standard deviation of x.

a. The mean of the random variable x is $\mu_x = \lambda = 6.9$. In other words, on the average, 6.9 patients arrive at the emergency room between 6:00 P.M. and 7:00 P.M.

b. The standard deviation of x is $\sigma_x = \sqrt{\lambda} = \sqrt{6.9} = 2.6$.   ∎

Poisson Approximation to the Binomial Distribution

Recall that the binomial probability formula is

$$P(x) = \binom{n}{x} p^x (1 - p)^{n-x}.$$

This is the formula from which we obtain probabilities for the number of successes, x, in n Bernoulli trials with success probability p.

When n is large, the binomial probability formula can be awkward or impractical to use because of computational difficulties. Therefore, methods have been developed that permit us to approximate binomial probabilities using formulas that are easier to work with. One of those methods employs the Poisson probability formula and applies when n is large and p is small (as a rule of thumb, we will require that $n \geq 100$ and $np \leq 10$). In such cases we can use the Poisson distribution with parameter $\lambda = np$ to approximate the binomial distribution. Specifically, we have the following procedure.

PROCEDURE 5.2   **TO APPROXIMATE BINOMIAL PROBABILITIES USING A POISSON PROBABILITY FORMULA**

Step 1   Determine n, the number of trials, and p, the success probability.

Step 2   Check that $n \geq 100$ and $np \leq 10$. If they are not, the Poisson approximation should not be used.

Step 3   Use the Poisson probability formula

$$P(x) = e^{-np} \frac{(np)^x}{x!}$$

to approximate the required binomial probabilities.

Example 5.25   Illustrates Procedure 5.2

According to the *World Almanac*, the infant mortality rate in Sweden is 3.3 per 1000 live births. Use the Poisson approximation to determine the probability that out of 500 randomly selected live births there are

a. no infant deaths.

b. at most three infant deaths.

SOLUTION   We apply Procedure 5.2.

Step 1   *Determine n, the number of trials, and p, the success probability.*

We have $n = 500$ and $p = \frac{3.3}{1000} = 0.0033$.

Step 2   *Check that $n \geq 100$ and $np \leq 10$.*

From Step 1 we conclude that $n = 500$ and $np = 500 \cdot 0.0033 = 1.65$. Thus we see that $n \geq 100$ and $np \leq 10$.

Step 3   *Use the Poisson probability formula*

$$P(x) = e^{-np}\,\frac{(np)^x}{x!}$$

to approximate the required binomial probabilities.

As we noted in Step 2, $np = 1.65$. Thus in this case the appropriate Poisson probability formula is

$$P(x) = e^{-1.65}\,\frac{(1.65)^x}{x!}.$$

a. For this part we want the probability of no infant deaths in 500 live births. That probability is

$$P(0) = e^{-1.65}\,\frac{(1.65)^0}{0!} = 0.192,$$

approximately.

b. Here we want the probability of at most three infant deaths in 500 live births. That probability is

$$P(x \leq 3) = P(0) + P(1) + P(2) + P(3)$$

$$= e^{-1.65}\left(\frac{1.65^0}{0!} + \frac{1.65^1}{1!} + \frac{1.65^2}{2!} + \frac{1.65^3}{3!}\right) = 0.914,$$

approximately. ■

Referring to Example 5.25, we will now illustrate the accuracy of the Poisson approximation to the binomial distribution. We used a computer to obtain both the binomial distribution with parameters $n = 500$ and $p = 0.0033$ and the Poisson distribution with parameter $\lambda = np = 500 \cdot 0.0033 = 1.65$.

Table 5.24 shows both distributions and exhibits how well the Poisson approximates the binomial. The probabilities are displayed to four decimal places in order to present a clearer picture of the differences between the two distributions. Notice that we stopped listing the probabilities once they became 0 to four decimal places.

TABLE 5.24

Comparison of the binomial distribution with parameters $n = 500$ and $p = 0.0033$ to the Poisson distribution with parameter $\lambda = 1.65$

x	Binomial probability $P(x)$	Poisson approximation $P(x)$
0	0.1915	0.1920
1	0.3171	0.3169
2	0.2619	0.2614
3	0.1440	0.1438
4	0.0592	0.0593
5	0.0195	0.0196
6	0.0053	0.0054
7	0.0012	0.0013
8	0.0003	0.0003
9	0.0000	0.0000

For large n and small p, it is not always possible to use a computer instead of a Poisson approximation to determine a required binomial distribution—sometimes n is so large or p is so small that even a computer can't handle the computations to obtain a binomial distribution. Nonetheless, the Poisson approximation will still be easy to apply.

Using the Computer (Optional)

Minitab has a subcommand called **Poisson** that can be employed to determine Poisson probabilities. We will illustrate the use of the Poisson subcommand by applying it to solve the problems posed in parts (a)–(c) of Example 5.23.

Example 5.26   The Poisson Subcommand

Desert Samaritan Hospital, located in Mesa, Arizona, keeps records of emergency-room traffic. From those records we find that the number of patients arriving between 6:00 P.M. and 7:00 P.M. has a Poisson distribution with parameter $\lambda = 6.9$. Determine the probability that, on a given day, the number of patients arriving at the emergency room between 6:00 P.M. and 7:00 P.M. will be

a. exactly 4.

b. at most 2.

c. between 4 and 10, inclusive.

SOLUTION   a. To determine the probability of exactly 4 arrivals, we use the PDF command and Poisson subcommand as follows.

Session commands: We first type the command **PDF** followed by the number of arrivals for which the probability is required; that is, we type PDF 4; and press Enter. Then we type the **POISSON** subcommand followed by the parameter for the Poisson distribution under consideration; that is, we type POISSON with lambda=6.9. and press Enter. These commands and the resulting output are shown in Printout 5.5.

(or)

Menu commands: We choose **Calc ▸ Probability Distributions ▸ Poisson**. Then we select **Probability** to indicate that we want to obtain an individual probability. To specify the appropriate Poisson distribution, we type 6.9 in the **Mean** text box. Next we select **Input constant** and type 4 to specify that we want the probability for the value 4, that is, $P(x = 4)$. Finally, we select **OK**. Printout 5.5 displays the output that results.

PRINTOUT 5.5
Minitab output for $P(x = 4)$

```
MTB > PDF 4;
SUBC> POISSON with lambda=6.9.
       K          P( X = K)
      4.00          0.0952
```

Printout 5.5 shows that the probability is 0.0952 that exactly 4 patients will arrive at the emergency room between 6:00 P.M. and 7:00 P.M.

b. For this part we want the probability of at most 2 arrivals, that is, 2 or fewer. Since that probability is a cumulative probability (see page 307), the easiest way to obtain it using Minitab is to apply the CDF command and Poisson subcommand as follows.

Session commands: We first type the command **CDF** followed by the number of arrivals for which the cumulative probability is required; that is, we type CDF 2; and press Enter. Then we type the **POISSON** subcommand followed by the parameter for the Poisson distribution under consideration; that is, we type POISSON with lambda=6.9. and press Enter. Printout 5.6 displays these commands and the output that results.

(or)

Menu commands: We choose **Calc ▸ Probability Distributions ▸ Poisson**. Then we select **Cumulative probability** to indicate that we want to obtain a cumulative probability. To specify the appropriate Poisson distribution, we type 6.9 in the **Mean** text box. Next we select **Input constant** and type 2 to specify that we want the cumulative probability for the value 2, that is, $P(x \leq 2)$. Finally, we select **OK**. Printout 5.6 shows the resulting output.

PRINTOUT 5.6
Minitab output for $P(x \leq 2)$

```
MTB > CDF 2;
SUBC> POISSON with lambda=6.9.
       K  P( X LESS OR = K)
      2.00          0.0320
```

EXERCISES 5.6

5.80 Identify two uses of Poisson distributions.

5.81 Suppose x has a Poisson distribution with parameter $\lambda = 3$. Determine
a. $P(2)$. b. $P(x \leq 3)$.
c. $P(x > 0)$. *(Hint: Use the complementation rule.)*
d. the mean of the random variable x.
e. the standard deviation of the random variable x.

5.82 Suppose x has a Poisson distribution with parameter $\lambda = 4.7$. Find
a. $P(5)$. b. $P(x < 2)$.
c. $P(x \geq 3)$. *(Hint: Use the complementation rule.)*
d. the mean of the random variable x.
e. the standard deviation of the random variable x.

5.83 A 1910 article, "The Probability Variations in the Distribution of α Particles," appearing in *Philosophical Magazine,* describes the results of experiments with polonium. The experiments, conducted by Ernest Rutherford and Hans Geiger, indicate that the number of α (alpha) particles reaching a small screen during an 8-minute interval has a Poisson distribution with parameter $\lambda = 3.87$. Determine the probability that, during an 8-minute interval, the number of α particles reaching the screen will be
a. exactly four. b. at most one.
c. between two and five, inclusive.

5.84 A paper by L. F. Richardson, published in the *Journal of the Royal Statistical Society,* analyzed the distribution of wars in time. From the data we find that the number of wars that begin during a given calendar year has approximately a Poisson distribution with parameter $\lambda = 0.7$. If a calendar year is selected at random, find the probability that the number of wars that begin during that calendar year will be
a. zero. b. at most two.
c. between one and three, inclusive.

5.85 M. F. Driscoll and N. A. Weiss discussed the modeling and solution of motel-reservation-network problems in "An Application of Queuing Theory to Reservation Networks," which appeared in the journal *Management Science.* They defined a Type 1 call to be a call from a motel's computer terminal to the national reservation center. For a certain motel, the number of Type 1 calls per hour has a Poisson distribution with parameter $\lambda = 1.7$. Determine the probability that the number of Type 1 calls made from this motel during a period of 1 hour will be
a. exactly one. b. at most two.

c. at least two. *(Hint: Compute the probability of at most one Type 1 call and then apply the complementation rule.)*

5.86 The owner of a fast-food restaurant knows that, on the average, 2.4 cars (customers) use the drive-through window between 3:00 P.M. and 3:15 P.M. Assuming that the number of such cars has a Poisson distribution, find the probability that, between 3:00 P.M. and 3:15 P.M.,
a. exactly two cars will use the drive-through window.
b. at least three cars will use the drive-through window.

5.87 Refer to Exercise 5.83. Let x denote the number of α particles reaching the screen during an 8-minute interval.
a. Find and interpret the mean of the random variable x.
b. Determine the standard deviation of x.

5.88 Refer to Exercise 5.84. Let x denote the number of wars that begin during a randomly selected calendar year.
a. Find and interpret the mean of the random variable x.
b. Determine the standard deviation of x.

5.89 Refer to Exercise 5.85. Let x denote the number of Type 1 calls made by the motel during a 1-hour period.
a. Construct a table of probabilities for the random variable x. Compute the probabilities until they are 0 to three decimal places.
b. Draw a histogram of the probabilities in part (a).

5.90 Refer to Exercise 5.86. Let x denote the number of cars using the drive-through window between 3:00 P.M. and 3:15 P.M.
a. Construct a table of probabilities for the random variable x. Compute the probabilities until they are 0 to three decimal places.
b. Draw a histogram of the probabilities in part (a).

In each of Exercises 5.91–5.94, determine the required probabilities by using the Poisson approximation to the binomial distribution, Procedure 5.2 on page 321.

5.91 In a letter to the editor appearing in the February 23, 1987, issue of *U.S. News and World Report,* a reader discussed the issue of space-shuttle safety. Each "criticality 1" item must have a 99.99% reliability, according to NASA standards. This means that the probability of failure for a "criticality 1" item is only 0.0001. Mission 25, the mission in which the Challenger exploded on take-off, had 748 "criticality 1" items. Determine the probability that
a. none of the "criticality 1" items would fail.
b. at least one "criticality 1" item would fail.

5.92 An experienced and very accurate data-entry operator has a probability of 0.0002 of making an incorrect keystroke. Determine the probability that on a page containing 3680 characters the data-entry operator makes
a. no mistakes.   b. at most one mistake.
c. at least two mistakes.

5.93 The literacy rate of a population represents the percent of the population over 15 years old that can read and write. According to the *Reader's Digest Almanac and Yearbook,* the literacy rate in Russia is 99.8%. If 1000 residents of Russia are selected at random, find the probability that the number who are illiterate is
a. exactly two.   b. at most two.
c. at least two.

5.94 During the second round of the 1989 U.S. Open, a strange thing happened: Four golfers made holes-in-one on the sixth hole. According to the experts, the odds of a PGA golfer making a hole-in-one are 3708-1; that is, the probability is $\frac{1}{3709}$. There were 155 golfers participating in the second round. Determine the probability that at least 4 of the 155 golfers would get a hole-in-one on the sixth hole.

5.95 (Computer exercise) Use Minitab or some other statistical software to obtain the required probabilities in Exercise 5.83.

5.96 (Computer exercise) Use Minitab or some other statistical software to obtain the required probabilities in Exercise 5.84.

5.97 (Computer exercise) Consider the following Minitab printout showing a portion of a Poisson distribution.

```
POISSON WITH MEAN =    2.100
     K          P( X = K)
   0.00           0.1225
   1.00           0.2572
   2.00           0.2700
   3.00           0.1890
   4.00           0.0992
   5.00           0.0417
   6.00           0.0146
   7.00           0.0044
   8.00           0.0011
   9.00           0.0003
  10.00           0.0001
```

a. What is the parameter λ for this Poisson distribution?
b. Use the printout to obtain $P(x = 3)$.
c. Use the printout to obtain $P(2 \leq x \leq 5)$.

5.98 (Computer exercise) Consider the printout below showing part of a cumulative Poisson distribution.

```
POISSON WITH MEAN =    1.400
   K   P( X LESS OR = K)
  0.00           0.2466
  1.00           0.5918
  2.00           0.8335
  3.00           0.9463
  4.00           0.9857
  5.00           0.9968
  6.00           0.9994
  7.00           0.9999
  8.00           1.0000
```

a. Identify λ for this Poisson distribution.
b. Determine $P(x \leq 4)$, $P(x \geq 3)$, and $P(x = 2)$.

5.99 (Computer exercise) In Exercise 5.93 you used the Poisson approximation to the binomial distribution with parameters $n = 1000$ and $p = 0.002$. Use Minitab or some other statistical software to obtain both the binomial distribution and its approximating Poisson distribution. Construct a table similar to Table 5.24 on page 323 in order to compare the two distributions.

5.100 (Computer exercise) In Exercise 5.92 you employed the Poisson approximation to the binomial distribution with parameters $n = 3680$ and $p = 0.0002$. Use Minitab or some other statistical software to obtain both the binomial distribution and its approximating Poisson distribution. Construct a table similar to Table 5.24 on page 323 in order to compare the two distributions.

5.101 (Computer exercise) On page 325 we stated that there are more efficient ways to use Minitab to determine $P(4 \leq x \leq 10)$ than the method used in Example 5.26(c).
a. Explain how Minitab's CDF command and Poisson subcommand can be used to obtain $P(4 \leq x \leq 10)$ for a Poisson random variable with parameter $\lambda = 6.9$.
b. Use Minitab or some other statistical software to carry out part (a).

5.102 As we know, when a binomial distribution with parameters n and p is approximated by a Poisson distribution, the parameter used for the Poisson distribution is $\lambda = np$. Can you provide an intuitive explanation of why that is the appropriate parameter?

5.103 A Poisson distribution can also be used to approximate a binomial distribution when n is large and p is large (i.e., close to 1). Explain how this can be done and under what conditions the approximation should be used.

Chapter Review

Formulas

Mean of a discrete random variable x, *279*

$$\mu_x = \Sigma x P(x)$$

Standard deviation of a discrete random variable x, *284, 285*

$$\sigma_x = \sqrt{\Sigma(x - \mu_x)^2 P(x)} \qquad \text{or} \qquad \sigma_x = \sqrt{\Sigma x^2 P(x) - \mu_x^2}$$

Factorial, *289*

$$k! = k(k - 1) \cdots 2 \cdot 1$$

Binomial coefficient, *290*

$$\binom{n}{x} = \frac{n!}{x!\,(n - x)!}$$

Binomial probability formula, *300*

$$P(x) = \binom{n}{x} p^x (1 - p)^{n-x}$$

(n = number of trials, p = success probability)

Mean of a binomial random variable, *314*

$$\mu_x = np$$

(n = number of trials, p = success probability)

Standard deviation of a binomial random variable, *315*

$$\sigma_x = \sqrt{np(1 - p)}$$

(n = number of trials, p = success probability)

Poisson probability formula,* *318*

$$P(x) = e^{-\lambda}\, \frac{\lambda^x}{x!}$$

Mean of a Poisson random variable,* *320*

$$\mu_x = \lambda$$

Standard deviation of a Poisson random variable,* *320*

$$\sigma_x = \sqrt{\lambda}$$

You Should Be Able To

1. use and understand each of the preceding formulas.
2. determine the probability distribution of a discrete random variable.
3. describe events using random-variable notation, when appropriate.
4. interpret the probability distribution of a random variable in terms of long-run relative frequency.
5. find and interpret the mean and standard deviation of a discrete random variable.
6. define and apply the concept of Bernoulli trials.
7. assign probabilities to the outcomes in a sequence of Bernoulli trials.
8. apply Procedure 5.1 to obtain binomial probabilities.
9. use the binomial probability table, Table I in Appendix A.
10. compute the mean and standard deviation of a binomial random variable using the special formulas, Formulas 5.3 and 5.4.
11. obtain Poisson probabilities.*
12. compute the mean and standard deviation of a Poisson random variable.*
13. use the Poisson distribution to approximate binomial probabilities, when appropriate.*
14. use the Minitab commands covered in this chapter.*
15. interpret the output obtained from the application of the Minitab commands discussed in this chapter.*

REVIEW TEST

1. The table below provides a frequency distribution for the number of undergraduate students attending a large state university, by class level. For the class levels, we used the following coding: $1 =$ freshman, $2 =$ sophomore, $3 =$ junior, and $4 =$ senior.

Class level	1	2	3	4
No. of students	5,745	6,240	7,486	10,063

Suppose an undergraduate at this university is to be selected at random. Let x denote the class level of the student obtained.
a. What are the possible values of the random variable x?
b. Use random-variable notation to represent the event that the student selected is a junior (class-level 3).
c. Determine $P(x = 3)$ and interpret your result in terms of percentages.
d. Find $P(2)$.
e. Determine the probability distribution of the random variable x.
f. Construct a probability histogram for the random variable x.

*2. (Computer problem) Refer to the probability distribution obtained in Problem 1(e).
a. Use Minitab or some other statistical software to simulate 2500 observations of the class level of a randomly selected undergraduate at the university.
b. Construct a relative-frequency distribution for the 2500 class levels obtained in part (a) and compare it to the probability distribution obtained in Problem 1(e).
c. Construct a relative-frequency histogram for the 2500 class levels obtained in part (a) and compare it to the probability histogram obtained in Problem 1(f).
d. What do parts (b) and (c) illustrate?

3. An accounting office has six incoming telephone lines. The probability distribution of the number of busy lines, y, is as follows.

y	0	1	2	3	4	5	6
$P(y)$	0.052	0.154	0.232	0.240	0.174	0.105	0.043

Use random-variable notation to express each of the following events. The number of busy lines is
a. exactly four.   b. at least four.
c. between two and four, inclusive.
d. at least one.

Apply the special addition rule and the probability distribution to determine
e. $P(y = 4)$.   f. $P(y \geq 4)$.
g. $P(2 \leq y \leq 4)$.   h. $P(y \geq 1)$.

4. Refer to the probability distribution displayed in the table in Problem 3.
a. Find the mean of the random variable y.
b. On the average, how many lines are busy?
c. Compute the standard deviation of y using the defining formula.
d. Compute the standard deviation of y using the shortcut formula.
e. Construct a probability histogram for y, locate the mean, and show one, two, and three standard deviation intervals.

*5. (Computer problem) Refer to the probability distribution displayed in the table in Problem 3.
a. Use Minitab or some other statistical software to simulate 200 observations of the number of busy lines.
b. Obtain the mean of the observations in part (a), and compare the mean to μ_y, which was determined in Problem 4(a).
c. What is part (b) illustrating?

6. Fill in the blanks: A finite population has mean μ and standard deviation σ. Suppose one member of the population is to be selected at random. If x denotes the value of the member selected, then $\mu_x = \underline{\hspace{1cm}}$ and $\sigma_x = \underline{\hspace{1cm}}$.

7. Determine $0!$, $3!$, $4!$, and $7!$.

8. Determine the value of each of the following binomial coefficients.
a. $\binom{8}{3}$ b. $\binom{8}{5}$ c. $\binom{6}{6}$
d. $\binom{10}{2}$ e. $\binom{40}{4}$ f. $\binom{100}{0}$

9. The game of craps is played by rolling two balanced dice. A first roll of a sum of 7 or 11 wins; and a first roll of a sum of 2, 3, or 12 loses. To win with any other first sum, that sum must be repeated before a sum of 7 is thrown. It can be shown that the probability is 0.493 that a player wins a game of craps. Suppose we consider a win by a player to be a success, s.
a. What is the success probability, p?

b. Construct a table showing the possible win-lose results and their probabilities for three games of craps. Round each probability to three decimal places.

c. Draw a tree diagram.

d. List the outcomes in which the player wins exactly two out of three times.

e. Determine the probability of each of the outcomes in part (d). Explain why those probabilities are equal.

10. The nation of Surinam is located on the northern coast of South America. According to the *World Almanac,* 80% of the population is literate. Suppose four people from Surinam are to be selected at random. Use the binomial probability formula to find the probability that the number of literate people in the sample is

a. exactly three.

b. at most three.

c. at least three.

d. Determine the probability distribution of the random variable x, the number of literate Surinamese in a sample of four.

e. Without referring to the probability distribution obtained in part (d) or constructing a probability histogram, decide whether the probability distribution is right skewed, symmetric, or left skewed.

f. Draw a probability histogram for x.

11. Solve parts (a)–(d) of Problem 10 by employing Table I in Appendix A.

*12. (**Computer problem**) Use Minitab or some other statistical software to obtain the required probability or probabilities in parts (a)–(d) of Problem 10.

13. Following are two probability histograms of binomial distributions. For each one, specify whether the success probability, p, is less than, equal to, or greater than 0.5.

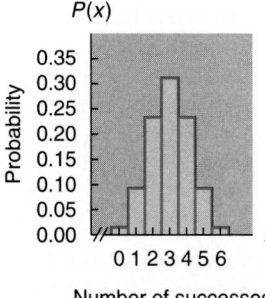

(a)   (b)

*14. (**Computer problem**) The following is a Minitab printout obtained by applying the PDF command and Binomial subcommand.

```
BINOMIAL WITH N =   5  P = 0.650000
    K          P( X = K)
    0           0.0053
    1           0.0488
    2           0.1811
    3           0.3364
    4           0.3124
    5           0.1160
```

Use the printout to determine

a. the number of trials.   b. the success probability.

c. the probability of exactly one success.

d. the probability of between one and three successes, inclusive.

e. the probability of at most one success.

f. the probability of at least one success.

*15. (**Computer problem**) The following printout was obtained by employing Minitab's CDF command and Binomial subcommand.

```
BINOMIAL WITH N =   8  P = 0.570000
   K  P( X LESS OR = K)
   0           0.0012
   1           0.0136
   2           0.0711
   3           0.2235
   4           0.4762
   5           0.7440
   6           0.9216
   7           0.9889
   8           1.0000
```

Use the printout to find

a. the number of trials.   b. the success probability.

c. the probability of at most four successes.

d. the probability of at least four successes. (*Hint:* Employ the printout and the complementation rule, Formula 4.2 on page 203.)

e. the probability of exactly four successes. (*Hint:* Use the special addition rule, Formula 4.1 on page 201, to express $P(x = 4)$ as the difference of two cumulative probabilities.)

*16. (**Computer problem**) In a 1987 statement, the Department of Agriculture reported that approximately 4 of every 10 chickens sold to consumers are contaminated by salmonella. Salmonella is a microorganism that can produce flu-like symptoms of fever, diarrhea, and vomiting within 12 to 36 hours after eating food contaminated by it. Suppose seven chickens are to be selected at random. Let x denote the number of those chosen that are contaminated by salmonella. The Minitab printout presented below provides the probability distribution of the random variable x.

```
BINOMIAL WITH N =   7  P = 0.400000
    K           P( X = K)
    0             0.0280
    1             0.1306
    2             0.2613
    3             0.2903
    4             0.1935
    5             0.0774
    6             0.0172
    7             0.0016
```

From the printout, determine the probability that of the seven chickens chosen, the number contaminated by salmonella is

a. exactly two.   b. at least two.

c. less than two.

d. between two and five, inclusive.

17. Refer to Problem 10. Determine the mean and standard deviation of the number of literate Surinamese in a sample of four.

18. Strictly speaking, why doesn't the sampling in Problem 10 constitute a sequence of four Bernoulli trials? Why is it permissible to use the binomial distribution?

*19. The number of telephone calls, x, received per hour by a small CPA firm has a Poisson distribution with parameter $\lambda = 5.2$. Find the probability that during a 1-hour period the firm receives

a. exactly two calls.

b. between four and six calls, inclusive.

c. at least one call.

d. Obtain a table of probabilities for x, stopping when the probabilities become 0 to three decimal places.

e. Use part (d) to construct a partial probability histogram for the random variable x.

f. Identify the shape of the probability distribution of x. Is this shape typical of Poisson distributions?

*20. Refer to Problem 19.

a. Find and interpret the mean of the random variable x.

b. Find the standard deviation of the random variable x.

*21. (Computer problem) Use Minitab or some other statistical software to obtain the probabilities required in parts (a)–(c) of Problem 19.

*22. (Computer problem) The following printout was obtained by applying Minitab's CDF command and Poisson subcommand.

```
POISSON WITH MEAN =   2.400
    K  P( X LESS OR = K)
    0           0.0907
    1           0.3084
    2           0.5697
    3           0.7787
    4           0.9041
    5           0.9643
    6           0.9884
    7           0.9967
    8           0.9991
    9           0.9998
   10           1.0000
```

Use the printout to

a. identify the parameter λ for the Poisson random variable, x, under consideration.

b. determine $P(x = 2)$.

c. determine $P(x \geq 2)$.

*23. The probability is approximately 0.00024 of being dealt four of a kind in a hand of five-card poker.

a. In 10,000 hands of five-card poker, roughly how many times would you expect to be dealt four of a kind (before the draw)?

b. Employ the Poisson approximation to the binomial distribution to determine the probability of being dealt four of a kind exactly twice in 10,000 hands of five-card poker.

c. Employ the Poisson approximation to the binomial distribution to determine the probability of being dealt four of a kind at least twice in 10,000 hands of five-card poker.

Using the Focus Database   Appendix B contains a printout of a database obtained by randomly selecting 500 Arizona State University sophomores. Seven variables are considered for each student: sex, high-school GPA, SAT math score, cumulative GPA, SAT verbal score, age, and total hours. Use Minitab or some other statistical software to solve the following problems.

a. Let x denote the age of a randomly selected sophomore from the sample of 500. Obtain the probability distribution of the random variable x.

b. Obtain a probability histogram or similar graphic for the random variable in part (a).

c. Determine the mean and standard deviation of the random variable x defined in part (a).

d. Consider the experiment of randomly selecting 10 sophomores with replacement from the sample of 500 and observing the number of those selected who are 21 years old. Simulate that experiment 2000 times.

e. Referring to the simulation from part (d), in approximately what percentage of the 2000 experiments would you expect exactly 3 of the 10 sophomores selected to be 21 years old? Compare that percentage to the actual percentage of the 2000 experiments in which exactly 3 of the 10 sophomores selected are 21 years old.

Case Study   **Aces Wild on the Sixth at Oak Hill**

On June 16, 1989, during the second round of the 1989 U.S. Open, four golfers—Doug Weaver, Mark Wiebe, Jerry Pate, and Nick Price—made holes-in-one on the sixth hole at Oak Hill in Pittsford, New York.

An article appeared the next day in the *Boston Globe* that discussed the remarkable event in detail. To quote the article, ". . . for perspective, consider this: This is the 89th U.S. Open, and through the thousands and thousands and thousands of rounds played in the previous 88, there had been only 17 holes-in-one. Yet on this dark Friday morning, there were four holes-in-one on the same hole in less than two hours. Four times into a cup $4\frac{1}{2}$ inches in diameter from 160 yards away."

The article also reported odds estimates obtained from several sources. These estimates varied considerably, from 1 in 10 million to 1 in 1,890,000,000,000,000 to 1 in 8.7 million to 1 in 332,000. We now ask you to compute the odds.

Here are the relevant data: According to the experts, the odds of a professional golfer making a hole-in-one are 3708-1; that is, the probability is $\frac{1}{3709}$. There were 155 golfers participating in the second round.

a. Determine the probability that at least 4 of the 155 golfers would get a hole-in-one on the sixth hole.

b. What assumptions did you make in solving part (a)? Do those assumptions seem reasonable?

c. Use Minitab or some other statistical software to solve part (a).

Carl Friedrich Gauss

Born on April 30, 1777, in Brunswick, Germany, the only son in a poor, semi-literate peasant family, Carl Friedrich Gauss taught himself to calculate before he could talk. At the age of 3, he pointed out an error in his father's calculations of wages. In addition to his arithmetic experimentation, he taught himself to read. At the age of 8, Gauss instantly solved the summing of all numbers from 1 to 100. His father was persuaded to allow him to stay in school and to study after school instead of working to help support the family.

Impressed by Gauss's brilliance, the Duke of Brunswick supported him monetarily from the ages of 14 to 30, permitting him to pursue his studies exclusively. Gauss conceived most of his mathematical discoveries by the time he was 17. He was granted his doctorate in absentia from the university at Helmstedt; his doctoral thesis developed the concept of complex numbers and proved the fundamental theorem of algebra, which had previously been only partially established. Shortly thereafter, Gauss published his theory of numbers, which has been said to be one of the most brilliant achievements in the history of mathematics.

Gauss made important discoveries in mathematics, physics, astronomy, and statistics. Two of his major contributions to statistics were the development of the least-squares method and fundamental work with the normal distribution, often called the *Gaussian distribution* in his honor.

In 1807 Gauss accepted the directorship of the observatory at the University of Göttingen, ending his dependence on the Duke of Brunswick. He remained there the rest of his life. In 1833 Gauss and a colleague, Wilhelm Weber, invented a working electric telegraph, 5 years before Samuel Morse. Gauss died in Göttingen in 1855.

Chapter 6

The Normal Distribution

General Objective

As we learned in Chapter 5, there are two main types of random variables: discrete and continuous. A **discrete random variable** is a random variable whose possible values form a finite (or countably infinite) set of numbers, usually some collection of whole numbers. Generally, a discrete random variable involves a count, like the number of radios owned by a randomly selected person. A **continuous random variable** is a random variable whose possible values form some interval of numbers. Typically, a continuous random variable involves a measurement, like the weight of a newborn baby.

The probability distribution of a continuous random variable is called a **continuous probability distribution.** In this chapter we will discuss the most important continuous probability distribution, the **normal distribution,** or so-called bell-shaped curve.

The normal distribution arises frequently, both in theory and in practice. It has been discovered, for instance, that many physical measurements have distributions that are bell-shaped. Thus it is often appropriate to use the normal distribution as the distribution of a population or random variable. The normal distribution is also applied frequently in inferential statistics. For example, under certain circumstances the normal distribution can be used to make inferences about the mean of a population.

Chapter Outline

6.1 The Standard Normal Curve

6.2 Normal Curves

6.3 Normally Distributed Populations

6.4 Normally Distributed Random Variables

6.5 Normal Probability Plots

6.6 The Normal Approximation to the Binomial Distribution

Case Study

Chest Sizes of Scottish Militiamen

An 1817 issue of the *Edinburgh Medical and Surgical Journal* included data that cross classified the members of 11 different local militia by chest circumference and height. Those data were used by the nineteenth-century astronomer, statistician, and sociologist Adolphe Quetelet in explaining his method for fitting a normal curve to data. (See the biography at the beginning of Chapter 2 for more information on Quetelet.) In the case study at the end of this chapter, we will examine the chest-circumference data in detail.

6.1   THE STANDARD NORMAL CURVE

In the world around us, we observe a variety of populations and random variables. Many are intrinsically different. But some—such as aptitude-test scores, heights of women, and wheat yield—share an important characteristic: the probabilities associated with them are equal, at least approximately, to areas under a **normal curve,** that is, a bell-shaped curve like the one shown in Fig. 6.1. In this section and the next, we will learn how to find areas under normal curves.

FIGURE 6.1
A normal curve

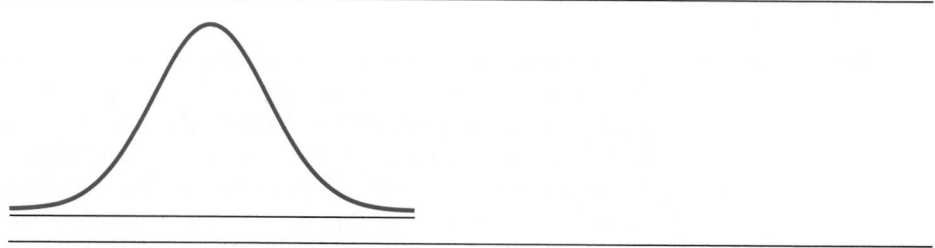

There are many, in fact infinitely many, normal curves. But, fortunately, there is a way to find areas under any normal curve once we know how to find areas under one particular normal curve—the **standard normal curve,** or **z-curve.** That curve is shown in Fig. 6.2.[†] Notice that the horizontal axis under the standard normal curve is labeled with the letter z.

Some of the more important properties of the standard normal curve are presented in Key Fact 6.1. Property 1 of Key Fact 6.1 is not unique to the standard normal curve. In fact, the total area under any curve representing a continuous probability distribution is equal to 1.

FIGURE 6.2
The standard normal curve

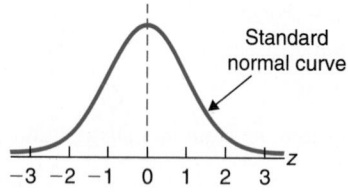

[†]   The equation of the standard normal curve is

$$y = \frac{1}{\sqrt{2\pi}} \, e^{-z^2/2},$$

where $e \approx 2.718$ and $\pi \approx 3.142$.

KEY FACT 6.1 | **BASIC PROPERTIES OF THE STANDARD NORMAL CURVE**

> *Property 1:* The total area under the standard normal curve is equal to 1.
>
> *Property 2:* The standard normal curve extends indefinitely in both directions, approaching, but never touching, the horizontal axis as it does so.
>
> *Property 3:* The standard normal curve is symmetric about 0; that is, the part of the curve to the left of the dashed line in Fig. 6.2 is the mirror image of the part of the curve to the right of it.
>
> *Property 4:* Most of the area under the standard normal curve lies between -3 and 3.

Using the Standard-Normal Table (Table II)

Because of the importance of areas under the standard normal curve, tables of those areas have been constructed. Such a table is Table II, which can be found inside the front cover of this book as well as in Appendix A.

A typical four-decimal-place number in the body of Table II gives the area under the standard normal curve that lies to the left of a specified value of z. The left page of Table II is for negative values of z, and the right page is for positive values of z.[†] Example 6.1 explains how to use Table II.

Example 6.1 | Finding the Area to the Left of a Specified Value

---

Determine the area under the standard normal curve that lies to the left of 1.23, as shown in Fig. 6.3(a).

FIGURE 6.3
Finding the area under
the standard normal
curve to the left of 1.23

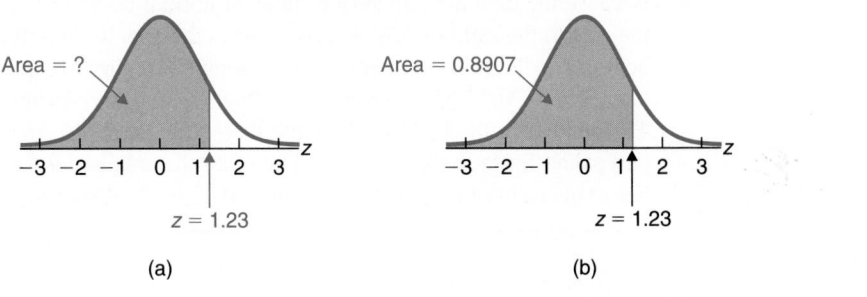

(a)   (b)

[†] Although z-values can be negative, areas must be positive. In fact, an area under the standard normal curve must always be between 0 and 1, inclusive.

SOLUTION   We use Table II, specifically the portion on the right page, since 1.23 is positive. First we go down the left-hand column, labeled *z,* to "1.2." Then we go across that row until we are under the "0.03" in the top row. The number in the body of the table there, 0.8907, is the area under the standard normal curve that lies to the left of 1.23. This area is shown in Fig. 6.3(b).   ∎

Finding the area under the standard normal curve that lies to the left of a specified value of *z* is one important use of Table II. Two other important uses of that table are finding the area to the right of a specified value of *z* and finding the area between two specified values of *z*. We illustrate these two uses in Examples 6.2 and 6.3, respectively.

Example 6.2   Finding the Area to the Right of a Specified Value

Determine the area under the standard normal curve that lies to the right of 0.76, as seen in Fig. 6.4(a).

FIGURE 6.4
Finding the area under
the standard normal
curve to the right of 0.76

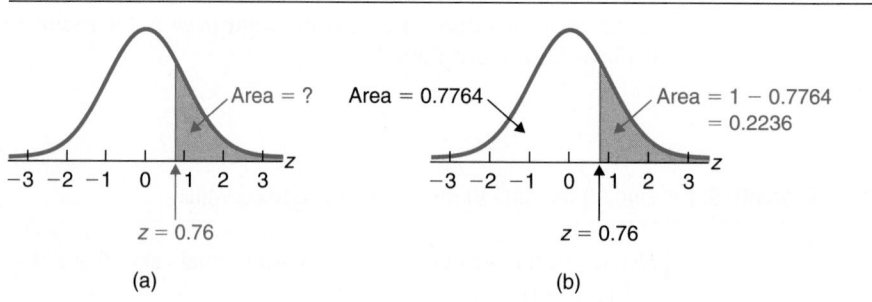

(a)   (b)

SOLUTION   Because the total area under the standard normal curve is 1 (Property 1 of Key Fact 6.1), the area to the right of 0.76 equals 1 minus the area to the left of 0.76. This latter area can be found in Table II, as explained in Example 6.1: First we go down the left-hand column, labeled *z,* to "0.7." Then we go across that row until we are under the "0.06" in the top row. The number in the body of the table there, 0.7764, is the area under the standard normal curve that lies to the left of 0.76. Therefore the area under the standard normal curve that lies to the right of 0.76 is $1 - 0.7764 = 0.2236$, as shown in Fig. 6.4(b).   ∎

Example 6.3   Finding the Area Between Two Specified Values

Determine the area under the standard normal curve that lies between −0.68 and 1.82, as seen in Fig. 6.5(a).

FIGURE 6.5
Finding the area under the
standard normal curve that
lies between −0.68 and 1.82

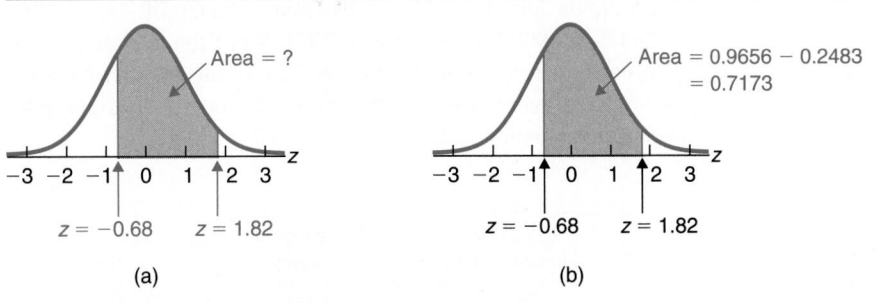

SOLUTION   The area under the standard normal curve that lies between −0.68 and 1.82 equals the area
to the left of 1.82 minus the area to the left of −0.68. Table II shows that the area to the left
of 1.82 is 0.9656 and that the area to the left of −0.68 is 0.2483 (this latter area is obtained
from the left page of Table II, since −0.68 is negative; notice that the second decimal places
displayed at the top of this half of Table II go from 0.09 to 0.00, not from 0.00 to 0.09).
Consequently, the area under the standard normal curve that lies between −0.68 and 1.82
is $0.9656 − 0.2483 = 0.7173$, as depicted in Fig. 6.5(b). ∎

The discussion in Examples 6.1–6.3 is summarized by the three graphs in Fig. 6.6.
Each graph shows how Table II, which gives areas to the left of a specified value of z,
can be used to obtain a required area. Obtaining the area to the left of a specified value
of z requires one table look-up, as seen in Fig. 6.6(a); obtaining the area to the right of
a specified value of z requires one table look-up and one subtraction (from 1), as seen in
Fig. 6.6(b); and obtaining the area between two specified values of z requires two table
look-ups and one subtraction, as seen in Fig. 6.6(c).

FIGURE 6.6
Using Table II to find the
area under the standard
normal curve that lies
(a) to the left of a specified value,
(b) to the right of a specified value,
(c) between two specified values

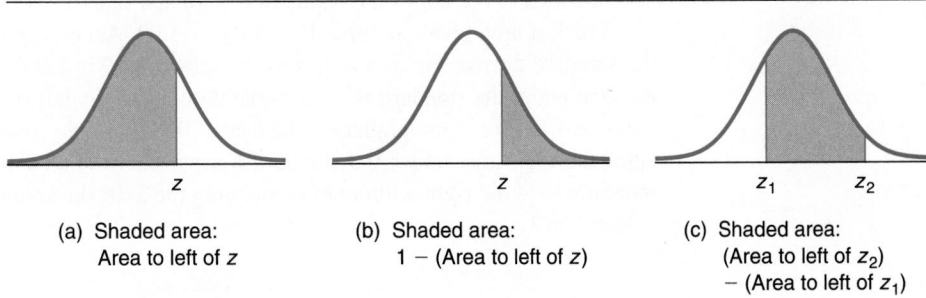

Some Important Areas

Property 4 of Key Fact 6.1 states that most of the area under the standard normal curve lies
between −3 and 3. Using Table II we can determine exactly how much area that is. The

area to the left of 3 is 0.9987; the area to the left of -3 is 0.0013. Thus the area between -3 and 3 is $0.9987 - 0.0013 = 0.9974$. Since the total area under the standard normal curve is equal to 1, we conclude that 99.74% of the area lies between -3 and 3.

Employing similar reasoning, we can obtain the useful information shown in Table 6.1. That information is depicted in Fig. 6.7.

TABLE 6.1
Some important areas under the standard normal curve

z	Area under curve between $-z$ and z	Percentage of total area
1	0.6826	68.26%
2	0.9544	95.44%
3	0.9974	99.74%

FIGURE 6.7
Percentage of area under the standard normal curve that lies between
(a) -1 and 1
(b) -2 and 2
(c) -3 and 3

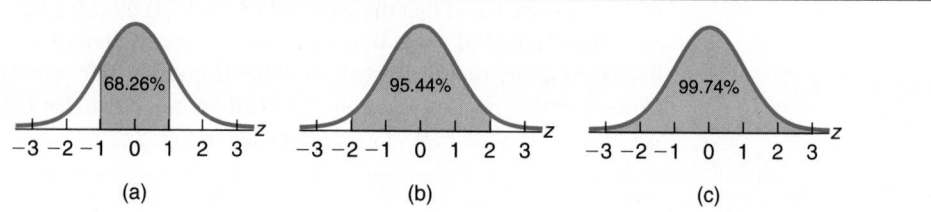

A Note Concerning Table II

The first area given in Table II is for $z = -3.90$. According to the table, the area under the standard normal curve that lies to the left of -3.90 is 0.0000. This does not mean that the area under the standard normal curve that lies to the left of -3.90 is exactly 0, but only that it is 0 to four decimal places (the area is 0.0000481 to seven decimal places). Indeed, since the standard normal curve extends indefinitely to the left without ever touching the axis, the area to the left of any value of z is greater than 0.

The last area given in Table II is for $z = 3.90$. According to the table, the area under the standard normal curve that lies to the left of 3.90 is 1.0000. This does not mean that the area under the standard normal curve that lies to the left of 3.90 is exactly 1, but only that it is 1 to four decimal places (the area is 0.9999519 to seven decimal places). Indeed, since the total area under the standard normal curve is exactly 1 and the curve extends indefinitely to the right without ever touching the axis, the area to the left of any value of z is less than 1.

Finding the z-Value for a Specified Area

Up to this point, we have used Table II to find areas under the standard normal curve to the left of a specified value of z, to the right of a specified value of z, and between two specified values of z. Now we will learn how to use Table II to find the z-value(s) corresponding to a specified area under the standard normal curve. Consider Example 6.4.

Example 6.4   Finding the *z*-Value Having a Specified Area to Its Left

Determine the *z*-value for which the area under the standard normal curve to the left of that value is 0.04, as seen in Fig. 6.8(a).

FIGURE 6.8

Finding the *z*-value
having area 0.04 to its left

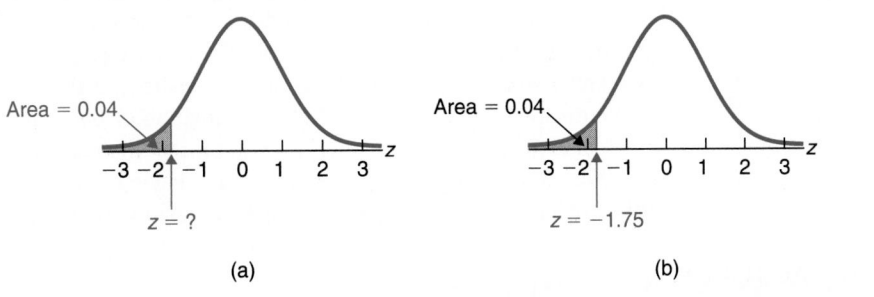

| | (a) | | (b) |

SOLUTION   We use Table II to obtain the *z*-value corresponding to the area 0.04. For ease of reference, we have reproduced a portion of Table II in Table 6.2.

TABLE 6.2

Areas under the
standard normal curve

	Second decimal place in *z*									
z	*0.09*	*0.08*	*0.07*	*0.06*	*0.05*	*0.04*	*0.03*	*0.02*	*0.01*	*0.00*
.	.	.	.	.	.	.	.	.	.	.
.	.	.	.	.	.	.	.	.	.	.
.	.	.	.	.	.	.	.	.	.	.
−1.9	0.0233	0.0239	0.0244	0.0250	0.0256	0.0262	0.0268	0.0274	0.0281	0.0287
−1.8	0.0294	0.0301	0.0307	0.0314	0.0322	0.0329	0.0336	0.0344	0.0351	0.0359
−1.7	0.0367	0.0375	0.0384	0.0392	0.0401	0.0409	0.0418	0.0427	0.0436	0.0446
−1.6	0.0455	0.0465	0.0475	0.0485	0.0495	0.0505	0.0516	0.0526	0.0537	0.0548
−1.5	0.0559	0.0571	0.0582	0.0594	0.0606	0.0618	0.0630	0.0643	0.0655	0.0668
.	.	.	.	.	.	.	.	.	.	.
.	.	.	.	.	.	.	.	.	.	.
.	.	.	.	.	.	.	.	.	.	.

We search the body of Table 6.2 (or Table II) for the area 0.04. Because we find no such area in the table, we use the area closest to 0.04, which is 0.0401. The *z*-value corresponding to that area is −1.75, as seen in Table 6.2. Figure 6.8(b) summarizes our results.   ∎

In Example 6.4, we were to determine the *z*-value having area 0.04 to its left. Because we were unable to find an area-entry of 0.04 in Table II, we selected the area closest to 0.04

and took the z-value corresponding to that area as an approximation of the required z-value. This illustrates the most typical case: when there is no area-entry in Table II exactly equal to the one desired, but there is one area-entry closest to the one desired. In this case, we take the z-value corresponding to the closest area-entry as an approximation of the required z-value.[†]

Two other cases are possible. One case is when there is an area-entry in Table II exactly equal to the one desired; nothing more need be said in this case. The other case is when there is no area-entry in Table II exactly equal to the one desired, but there are two area-entries equally closest to the one desired; in this case, we take the mean of the two corresponding z-values as an approximation of the required z-value. Both cases are illustrated in Example 6.5, which we will present momentarily.

We often need to determine the z-value having a specified area to its right. Because this problem occurs so frequently, a special notation is employed.

DEFINITION 6.1   **THE z_α NOTATION**

> The symbol z_α is used to denote the z-value having area α (alpha) to its right under the standard normal curve, as illustrated in Fig. 6.9. We read "z_α" as "z sub α" or more simply as "z α."

FIGURE 6.9
The z_α notation

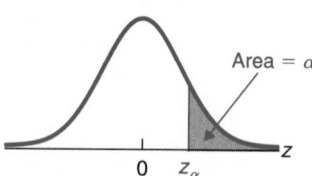

Example 6.5   Finding z_α

Use Table II to find

a. $z_{0.025}$.   b. $z_{0.05}$.

SOLUTION   a. $z_{0.025}$ is the z-value having area 0.025 to its right under the standard normal curve, as shown in Fig. 6.10(a). Because the area under the standard normal curve to the right of $z_{0.025}$ equals 0.025, the area to its left is $1 - 0.025 = 0.975$, as shown in Fig. 6.10(b). We search the body of Table II for the area 0.975 and find that such an area-entry does indeed exist. Its corresponding z-value, 1.96, is the required value of z; that is, $z_{0.025} = 1.96$, as seen in Fig. 6.10(b).

[†] There are more exact approaches, such as linear interpolation or using a computer. This latter approach will be discussed at the end of this section.

FIGURE 6.10

Finding $z_{0.025}$

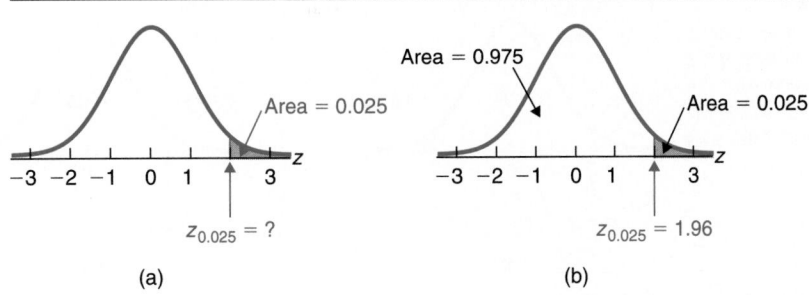

(a)   (b)

b. $z_{0.05}$ is the z-value having area 0.05 to its right under the standard normal curve, as shown in Fig. 6.11(a).

FIGURE 6.11

Finding $z_{0.05}$

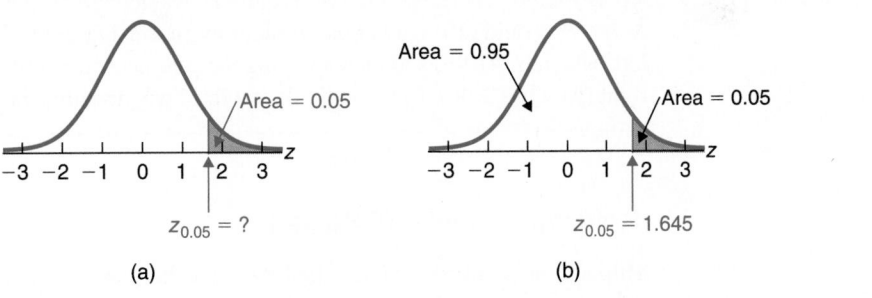

(a)   (b)

Because the area under the standard normal curve to the right of $z_{0.05}$ equals 0.05, the area to its left is $1 - 0.05 = 0.95$, as shown in Fig. 6.11(b). We search the body of Table II for the area 0.95, but find no such area. Instead we find that there are two areas closest to 0.95: 0.9495 and 0.9505. The z-values corresponding to those two areas are 1.64 and 1.65, respectively. Consequently, we take $z_{0.05}$ to be halfway between 1.64 and 1.65; that is, $z_{0.05} = 1.645$, approximately, as shown in Fig. 6.11(b). ∎

In Example 6.6 we show how to find the two z-values that divide the area under the standard normal curve into a specified middle area and two outside areas.

Example 6.6   Finding the z-Values for a Specified Area

Determine the two z-values that divide the area under the standard normal curve into a middle 0.95 area and two outside 0.025 areas, as depicted in Fig. 6.12(a).

FIGURE 6.12
Finding the two z-values
dividing the area under the
standard normal curve into
a middle 0.95 area and
two outside 0.025 areas

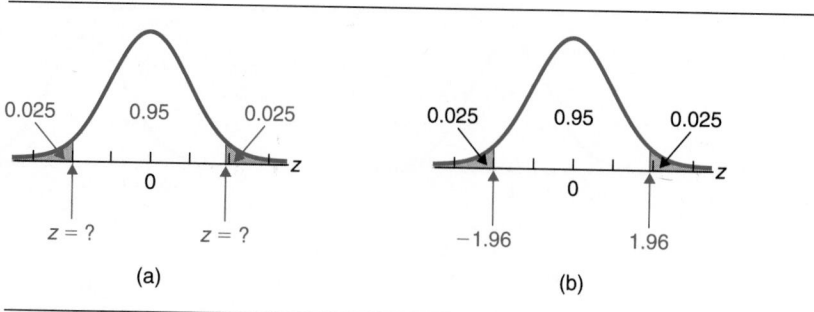

(a)   (b)

SOLUTION   This problem can be solved in several ways. Here is one way. As we see from Fig. 6.12(a), the area of the shaded region on the right-hand side is 0.025. This means that the z-value on the right is $z_{0.025}$. In Example 6.5(a) we found that $z_{0.025} = 1.96$. Thus the z-value on the right is 1.96. Since the standard normal curve is symmetric about 0, the z-value on the left is -1.96. Therefore the two required z-values are ± 1.96, as shown in Fig. 6.12(b). *Note:* We could also solve this problem by first using Table II to find the z-value on the left (which is -1.96) and then applying the symmetry property to obtain the z-value on the right (which is 1.96). Can you think of a third way to solve the problem?   ∎

Using the Computer (Optional)

Minitab has a subcommand called **Normal** that can be used in place of Table II. As a subcommand to **CDF** (cumulative distribution function), Normal yields the area under the standard normal curve to the left of a specified value of z; as a subcommand to **InvCDF** (inverse cumulative distribution function), Normal yields the z-value having a specified area to its left. Examples 6.7 and 6.8 illustrate these two uses of the Normal subcommand.

Example 6.7   The CDF Command and Normal Subcommand

Use Minitab to solve Example 6.1, that is, to determine the area under the standard normal curve that lies to the left of 1.23.

SOLUTION   We proceed as follows.

Session commands: We type the command **CDF** followed by the specified z-value; that is, we type CDF 1.23; and press Enter. Then we type the subcommand **NORMAL** followed by the numbers 0 and 1; that is, we type NORMAL 0 1. and press Enter.[†] Printout 6.1 shows these commands and the resulting output.

(or)

[†] The numbers 0 and 1 are the two parameters that distinguish the standard normal curve from other normal curves (Section 6.2 discusses parameters for normal curves). Actually, NORMAL 0 1 is the default for CDF.

Menu commands: We choose **Calc ▸ Probability Distributions ▸ Normal...**, select **Cumulative probability**, type 1.23 in the **Input constant** text box, type 0 in the **Mean** text box, type 1 in the **Standard deviation** text box, and then select **OK**.[†] The output that results is shown in Printout 6.1.

PRINTOUT 6.1

Minitab output for the CDF command and Normal subcommand

```
MTB > CDF 1.23;
SUBC> NORMAL 0 1.
    1.2300    0.8907
```

The first number in the last line of Printout 6.1 is the specified z-value, 1.23; the second number is the area under the standard normal curve to the left of the specified z-value. So we see that the area under the standard normal curve to the left of 1.23 is 0.8907. This is the area we obtained in Example 6.1 by employing Table II.   ∎

Minitab can also be used to find the area under the standard normal curve to the right of a specified value of z or between two specified values of z. For details on how to do this, refer to the *Minitab Supplement*.

Example 6.8   The InvCDF Command and Normal Subcommand

Use Minitab to solve Example 6.4, that is, to find the z-value having area 0.04 to its left.

SOLUTION   We proceed as follows.

Session commands: We type the command **INVCDF** followed by the specified area; that is, we type INVCDF 0.04; and press Enter. Then we type the subcommand NORMAL followed by the numbers 0 and 1; that is, we type NORMAL 0 1. and press Enter. Printout 6.2 shows these commands and the resulting output.

(or)

Menu commands: We choose **Calc ▸ Probability Distributions ▸ Normal...**, select **Inverse cumulative probability**, type 0.04 in the **Input constant** text box, type 0 in the **Mean** text box, type 1 in the **Standard deviation** text box, and then select **OK**. The output that results is shown in Printout 6.2.

PRINTOUT 6.2

Minitab output for the InvCDF command and Normal subcommand

```
MTB > INVCDF 0.04;
SUBC> NORMAL 0 1.
    0.0400    -1.7507
```

[†] See the footnote on the preceding page.

The first number in the last line of Printout 6.2 is the specified area, 0.04; the second number is the z-value having the specified area to its left. So we see that the z-value having area 0.04 to its left under the standard normal curve is -1.7507. This value differs slightly from and is more exact than the value of -1.75 that we obtained in Example 6.4 by employing Table II. ■

EXERCISES 6.1

6.1 Without consulting Table II, explain why the area under the standard normal curve that lies to the right of 0 is equal to 0.5.

6.2 According to Table II, the area under the standard normal curve that lies to the left of 1.96 is 0.975. Without further consulting Table II, determine the area under the standard normal curve that lies to the left of -1.96.

Use Table II to obtain the areas under the standard normal curve required in Exercises 6.3–6.10. Sketch a standard normal curve and shade the area of interest in each problem.

6.3 Determine the area under the standard normal curve that lies to the left of
a. 2.24. b. -1.56. c. 0. d. -4.

6.4 Determine the area under the standard normal curve that lies to the left of
a. -0.87. b. 3.56. c. 5.12.

6.5 Find the area under the standard normal curve that lies to the right of
a. -1.07. b. 0.6. c. 0. d. 4.2.

6.6 Find the area under the standard normal curve that lies to the right of
a. 2.02. b. -0.56. c. -4.

6.7 Determine the area under the standard normal curve that lies between
a. -2.18 and 1.44. b. -2 and -1.5.
c. 0.59 and 1.51. d. 1.1 and 4.2.

6.8 Determine the area under the standard normal curve that lies between
a. -0.88 and 2.24. b. -2.5 and -2.
c. 1.48 and 2.72. d. -5.1 and 1.

6.9 Find the area under the standard normal curve that lies
a. either to the left of -2.12 or to the right of 1.67.

b. either to the left of 0.63 or to the right of 1.54.

6.10 Find the area under the standard normal curve that lies
a. either to the left of -1 or to the right of 2.
b. either to the left of -2.51 or to the right of -1.

6.11 Use Table II to obtain the following shaded areas under the standard normal curve.
a.

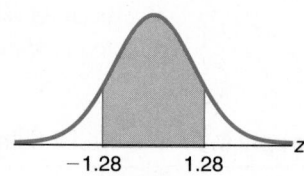

b.

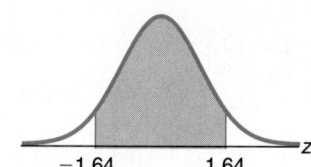

c.

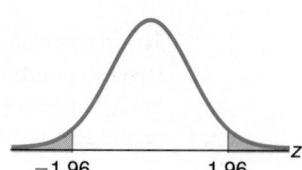

d.

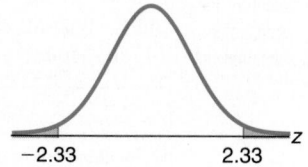

6.12   Use Table II to obtain the following shaded areas under the standard normal curve.

a.

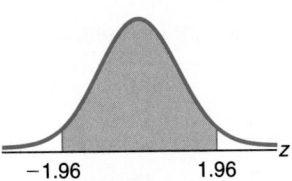

b.

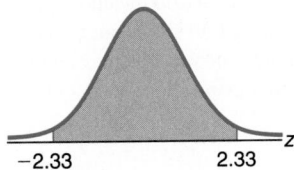

c.

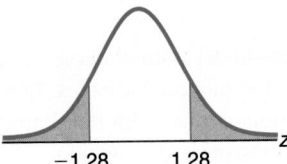

d.

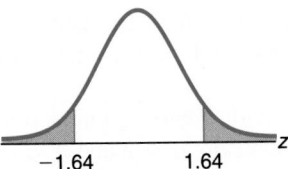

6.13   Verify that the information in Table 6.1 on page 340 is correct. Draw standard normal curves for the three entries and shade the appropriate areas.

6.14   Below is a standard normal curve. The total area under the curve is divided into eight regions.

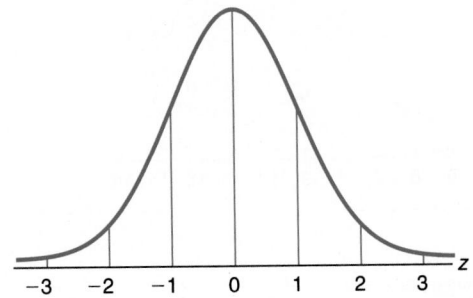

a. Determine the area of each region.
b. Complete the following table.

Region	Area	Percentage of total area
$-\infty$ to -3	0.0013	0.13
-3 to -2		
-2 to -1		
-1 to 0		
0 to 1	0.3413	34.13
1 to 2		
2 to 3		
3 to ∞		
	1.0000	100.00

In Exercises 6.15–6.26, employ Table II to obtain the required z-values. Illustrate your work using graphs.

6.15   Obtain the z-value for which the area under the standard normal curve to the left of that value is equal to 0.025.

6.16   Determine the z-value for which the area under the standard normal curve to the left of that value is 0.01.

6.17   Find the z-value having area 0.75 to its left under the standard normal curve.

6.18   Obtain the z-value having area 0.80 to its left under the standard normal curve.

6.19   Obtain the z-value having area 0.95 to its right; that is, find $z_{0.95}$.

6.20   Obtain the z-value having area 0.70 to its right; that is, find $z_{0.70}$.

6.21   Determine $z_{0.33}$; that is, find the z-value having area 0.33 to its right under the standard normal curve.

6.22   Determine $z_{0.015}$; that is, find the z-value having area 0.015 to its right under the standard normal curve.

6.23   Find the following z-values.
a. $z_{0.03}$ b. $z_{0.005}$

6.24   Obtain the following z-values.
a. $z_{0.20}$ b. $z_{0.06}$

6.25   Determine the two z-values that divide the area under the standard normal curve into a middle 0.90 area and two outside 0.05 areas.

·   **6.26**   Determine the two z-values that divide the area under the standard normal curve into a middle 0.99 area and two outside 0.005 areas.

·   **6.27 (Computer exercise)**   Use Minitab or some other statistical software to obtain the areas required in Exercise 6.3.

·   **6.28 (Computer exercise)**   Use Minitab or some other statistical software to obtain the areas required in Exercise 6.4.

·   **6.29 (Computer exercise)**   Use Minitab or some other statistical software to obtain the z-value required in Exercise 6.17.

·   **6.30 (Computer exercise)**   Use Minitab or some other statistical software to obtain the z-value required in Exercise 6.18.

·   **6.31**   Complete the following table.

$z_{0.10}$	$z_{0.05}$	$z_{0.025}$	$z_{0.01}$	$z_{0.005}$
1.28				

· ·   **6.32**   Let $0 < \alpha < 1$. Determine
a. the z-value having area α to its right in terms of z_α.
b. the z-value having area α to its left in terms of z_α.
c. the two z-values that divide the area under the curve into a middle $1 - \alpha$ area and two outside areas of $\alpha/2$.
d. Draw graphs to illustrate your results in parts (a)–(c).

6.2   NORMAL CURVES

Recall that it is important to be able to obtain areas under normal curves since those areas correspond to probabilities for many populations and random variables. In Section 6.1 we learned how to find areas under the standard normal curve. With that knowledge we can obtain areas under any normal curve, as we will discover in this section.

Each normal curve can be identified by two parameters, usually denoted by μ and σ. The parameter μ designates where the normal curve is centered, and the parameter σ indicates its spread or shape.[†] Two normal curves that have the same μ-parameter are centered at the same place; and two normal curves that have the same σ-parameter have the same shape.

Figure 6.13 shows three normal curves. The normal curve on the left has parameters $\mu = -2$ and $\sigma = 1$; the one in the center has parameters $\mu = 3$ and $\sigma = \frac{1}{2}$; and the one on the right has parameters $\mu = 8$ and $\sigma = 2$.

FIGURE 6.13
Three normal curves

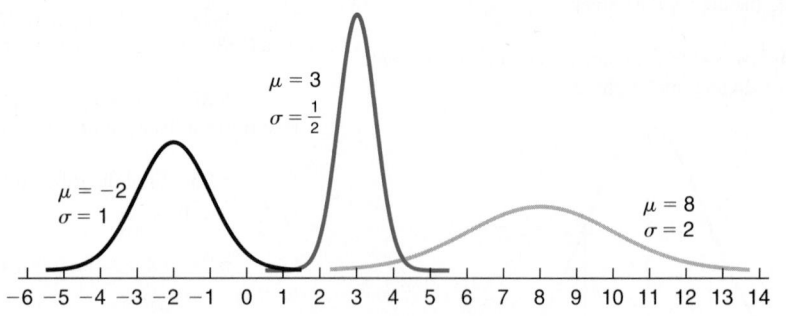

[†]   The equation of the normal curve with parameters μ and σ is $y = \frac{1}{\sqrt{2\pi}\sigma} e^{-(x-\mu)^2/2\sigma^2}$.

Notice that each of the normal curves in Fig. 6.13 is centered at its μ-parameter. Also notice that the larger the σ-parameter, the more the normal curve is spread out. Some of the most important properties of normal curves are presented in Key Fact 6.2.

KEY FACT 6.2

BASIC PROPERTIES OF NORMAL CURVES

Property 1: The total area under a normal curve is equal to 1.

Property 2: A normal curve extends indefinitely in both directions, approaching, but never touching, the horizontal axis as it does so.

Property 3: The normal curve with parameters μ and σ is symmetric about μ; that is, the part of the curve to the left of μ is the mirror image of the part of the curve to the right of μ.

Property 4: Most of the area under the normal curve with parameters μ and σ lies between $\mu - 3\sigma$ and $\mu + 3\sigma$.

Sketching Normal Curves

Although for this text it is not necessary to draw normal curves exactly, it is helpful to be able to sketch them. Using Key Fact 6.2, it is easy to quickly sketch any normal curve. Example 6.9 illustrates how this is done.

Example 6.9 Sketching Normal Curves

Sketch the normal curve with parameters

a. $\mu = 5$ and $\sigma = 2$.   b. $\mu = 0$ and $\sigma = 1$.

SOLUTION a. By Property 3 of Key Fact 6.2, the normal curve with parameters $\mu = 5$ and $\sigma = 2$ is symmetric about 5. Also, by Property 4, most of the area under that normal curve lies between

$$\mu - 3\sigma = 5 - 3 \cdot 2 = 5 - 6 = -1 \quad \text{and} \quad \mu + 3\sigma = 5 + 3 \cdot 2 = 5 + 6 = 11.$$

So we sketch this curve as shown in Fig. 6.14.

FIGURE 6.14
Sketch of normal curve with
parameters $\mu = 5$ and $\sigma = 2$

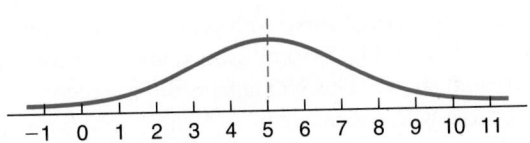

b. Next we sketch the normal curve with parameters $\mu = 0$ and $\sigma = 1$. By Property 3 of Key Fact 6.2, we know that this normal curve is symmetric about 0. From Property 4 we know that most of the area under this normal curve lies between

$$\mu - 3\sigma = 0 - 3 \cdot 1 = 0 - 3 = -3 \quad \text{and} \quad \mu + 3\sigma = 0 + 3 \cdot 1 = 0 + 3 = 3.$$

Thus we can sketch this curve as in Fig. 6.15.   ■

FIGURE 6.15

Sketch of normal curve with parameters $\mu = 0$ and $\sigma = 1$

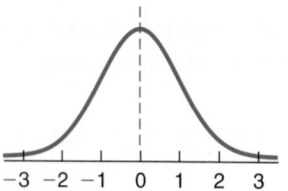

The curve in Fig. 6.15 should look familiar. It is the standard normal curve (see Fig. 6.2 on page 336).

KEY FACT 6.3

PARAMETERS FOR THE STANDARD NORMAL CURVE

The normal curve with parameters $\mu = 0$ and $\sigma = 1$ is the standard normal curve.

Finding Areas Under Normal Curves

We now proceed with the main business of this section—to learn how to find areas under normal curves. The idea is to relate any particular normal curve to the standard normal curve. Example 6.10 provides the details.

Example 6.10   Finding Areas Under a Normal Curve

Determine the area under the normal curve with parameters $\mu = 5$ and $\sigma = 2$ that lies

a. to the right of 7.   b. between 3 and 8.

SOLUTION   a. First we sketch the normal curve with parameters $\mu = 5$ and $\sigma = 2$ and shade the area to the right of 7, as in Fig. 6.16(a). Notice that we have labeled the horizontal axis in Fig. 6.16(a) with an "x." This will help us differentiate values on the horizontal axis of the normal curve with parameters $\mu = 5$ and $\sigma = 2$ from values on the horizontal axis of the standard normal curve.

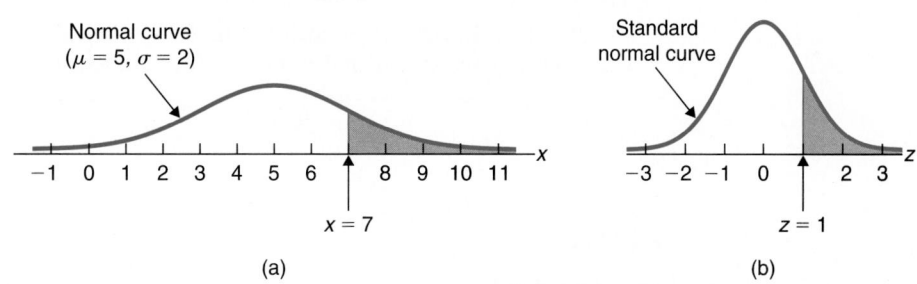

(a)   (b)

It can be shown mathematically (see Exercise 6.64) that the area under the normal curve with parameters $\mu = 5$ and $\sigma = 2$ that lies to the right of $x = 7$ is equal to the area under the standard normal curve that lies to the right of

$$z = \frac{x - \mu}{\sigma} = \frac{7 - 5}{2} = 1.$$

In other words, the shaded areas in Figs. 6.16(a) and 6.16(b) are equal.

By Table II the area shaded in Fig. 6.16(b) is $1 - 0.8413 = 0.1587$. Thus the area shaded in Fig. 6.16(a) is also 0.1587; that is, the area under the normal curve with parameters $\mu = 5$ and $\sigma = 2$ that lies to the right of 7 equals 0.1587.

b. For this part we need to determine the area under the normal curve with parameters $\mu = 5$ and $\sigma = 2$ that lies between 3 and 8, as shown in Fig. 6.17(a).

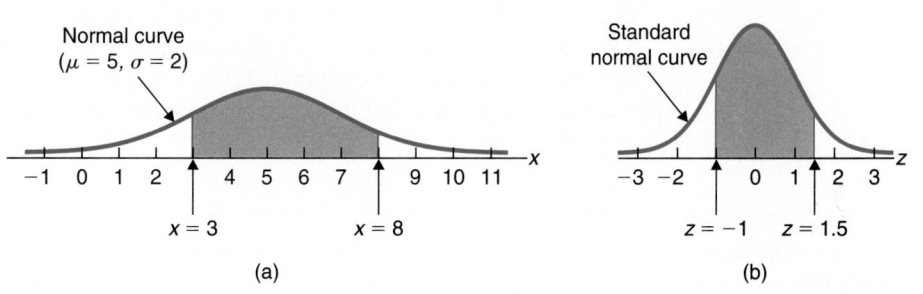

(a)   (b)

The area under the normal curve with parameters $\mu = 5$ and $\sigma = 2$ that lies between $x = 3$ and $x = 8$ is equal to the area under the standard normal curve that lies between

$$z = \frac{x - \mu}{\sigma} = \frac{3 - 5}{2} = -1$$

and

$$z = \frac{x - \mu}{\sigma} = \frac{8 - 5}{2} = 1.5.$$

This latter area is shown in Fig. 6.17(b).

By Table II, the shaded area in Fig. 6.17(b) equals $0.9332 - 0.1587 = 0.7745$. Consequently, the shaded area in Fig. 6.17(a) also equals 0.7745. ∎

z-Scores

As indicated in Example 6.10, determining areas under normal curves entails converting x-values to z-values by first subtracting μ and then dividing by σ:

$$z = \frac{x - \mu}{\sigma}.$$

That conversion process is often referred to as **standardizing.** It is the same process that we use to compute population z-scores (see Definition 3.15 on page 168). In applications, the parameters μ and σ will be the mean and standard deviation of a population or random variable. Hence we will refer to z-values obtained by standardizing as **z-scores.**

We now summarize the fundamental fact that permits us to determine areas under any normal curve by using the Table II values of areas under the standard normal curve.

KEY FACT 6.4

DETERMINING AREAS UNDER NORMAL CURVES

The area under the normal curve with parameters μ and σ that lies between a and b is equal to the area under the standard normal curve that lies between $(a - \mu)/\sigma$ and $(b - \mu)/\sigma$, as seen in Fig. 6.18. The area under the normal curve with parameters μ and σ that lies to the right (or left) of a particular x-value is found similarly, namely, by first converting to the z-score and then using Table II to find the area under the standard normal curve that lies to the right (or left) of the z-score.

FIGURE 6.18

Two-diagram approach for finding normal-curve areas

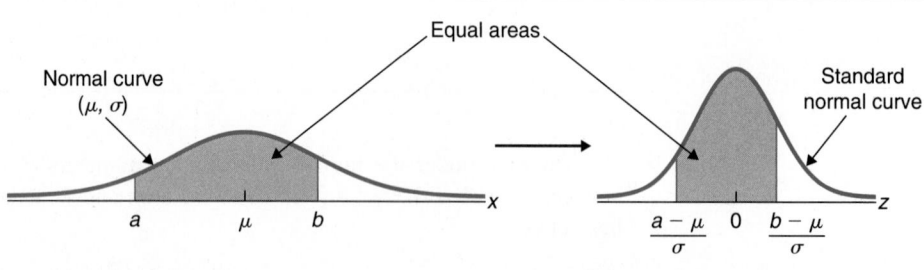

Note: The z-values in Table II are displayed to two decimal places. Therefore, when computing a z-score, always round the result to two decimal places.

A Streamlined Procedure for Finding Normal-Curve Areas

In Example 6.10 we drew two diagrams to solve each area problem—one for the normal curve with parameters $\mu = 5$ and $\sigma = 2$ and one for the standard normal curve. Generally, the two-diagram approach for finding normal-curve areas is as depicted in Fig. 6.18. Although that approach is helpful for explaining and understanding the necessary ideas, it is cumbersome to apply in practice.

Procedure 6.1 provides a one-diagram approach for finding normal-curve areas that is faster and simpler to use than the two-diagram approach. Figure 6.19 depicts the one-diagram approach.

PROCEDURE 6.1

TO DETERMINE AREAS UNDER THE NORMAL CURVE WITH PARAMETERS μ AND σ

Step 1 Sketch the normal curve with parameters μ and σ.

Step 2 Indicate on the graph the area to be determined.

Step 3 Compute the required z-scores and mark them on the graph beneath the x-values.

Step 4 Use Table II to obtain the required area.

FIGURE 6.19

One-diagram approach for finding normal-curve areas

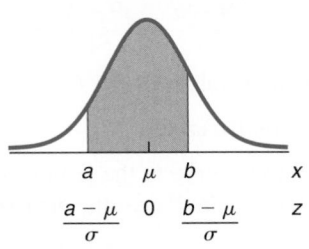

Example 6.11 Illustrates Procedure 6.1

Determine the area under the normal curve with parameters $\mu = 3$ and $\sigma = 4$ that lies between -6 and 9.

SOLUTION We apply Procedure 6.1.

Step 1 *Sketch the normal curve with parameters μ and σ.*

Here $\mu = 3$ and $\sigma = 4$. We have sketched that normal curve in Fig. 6.20. Notice that the tick marks are $\sigma = 4$ units apart.

FIGURE 6.20
One-diagram approach for finding
the area under the normal curve
with parameters $\mu = 3$ and
$\sigma = 4$ that lies between -6 and 9

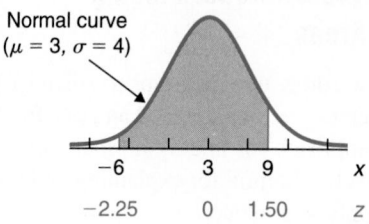

z-score computations: Area to the left of z:

$x = -6 \longrightarrow z = \dfrac{-6 - 3}{4} = -2.25$ 0.0122

$x = 9 \longrightarrow z = \dfrac{9 - 3}{4} = 1.50$ 0.9332

Shaded area = $0.9332 - 0.0122 = 0.9210$

Step 2   *Indicate on the graph the area to be determined.*

See the shaded area in Fig. 6.20.

Step 3   *Compute the required z-scores and mark them on the graph beneath the x-values.*

We need to obtain the z-scores for the x-values, -6 and 9:

$$x = -6 \quad \longrightarrow \quad z = \frac{-6 - \mu}{\sigma} = \frac{-6 - 3}{4} = -2.25,$$

$$x = 9 \quad \longrightarrow \quad z = \frac{9 - \mu}{\sigma} = \frac{9 - 3}{4} = 1.50.$$

These z-scores are marked beneath the x-values in Fig. 6.20.

Step 4   *Use Table II to obtain the required area.*

The area under the standard normal curve to the left of -2.25 is 0.0122 and that to the left of 1.50 is 0.9332. Hence the required area, the area shaded in Fig. 6.20, equals $0.9332 - 0.0122 = 0.9210$.

This four-step procedure can be accomplished quickly and easily by performing all the steps in a "picture," as in Fig. 6.20.   ■

Finding the *x*-Value for a Specified Area

The formula for converting x-values to z-scores is

$$z = \frac{x - \mu}{\sigma}.$$

As we said, this process is called standardizing. If we solve the preceding equation for x, we obtain a formula for converting z-scores to x-values:

$$x = \mu + z \cdot \sigma.$$

We call this latter process **destandardizing.**

  Procedure 6.1 provides a method for finding the area under a normal curve that lies between two specified values of x. Frequently, however, we need to perform the reverse procedure, that is, to find the x-values corresponding to a specified area under a normal curve. To accomplish this we employ Procedure 6.2.

PROCEDURE 6.2

TO FIND THE x-VALUE(S) CORRESPONDING TO A SPECIFIED AREA UNDER THE NORMAL CURVE WITH PARAMETERS μ AND σ

> **Step 1**   Sketch the normal curve with parameters μ and σ.
>
> **Step 2**   On the graph, indicate the specified area and the x-values to be found.
>
> **Step 3**   Determine the z-values corresponding to the specified area, as explained in Section 6.1.
>
> **Step 4**   Use the formula $x = \mu + z \cdot \sigma$ to destandardize the z-values found in Step 3 and hence obtain the required x-values.

Example 6.12   Illustrates Procedure 6.2

For the normal curve with parameters $\mu = 100$ and $\sigma = 16$, determine the x-value having area 0.04 to its left.

SOLUTION   We apply Procedure 6.2.

Step 1   *Sketch the normal curve with parameters μ and σ.*

  Here $\mu = 100$ and $\sigma = 16$. We have sketched that normal curve in Fig. 6.21(a).

FIGURE 6.21

Finding the x-value having area 0.04 to its left under the normal curve with parameters $\mu = 100$ and $\sigma = 16$

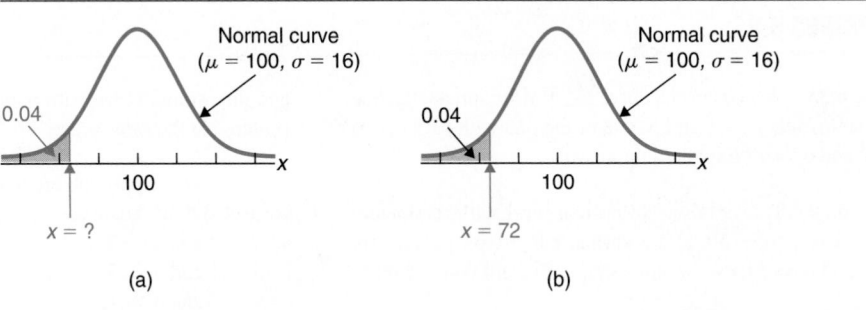

Step 2   *On the graph, indicate the specified area and the x-values to be found.*

  This is done in Fig. 6.21(a).

Step 3   *Determine the z-values corresponding to the specified area, as explained in Section 6.1.*

We need to determine the z-score having area 0.04 to its left, that is, the z-value having area 0.04 to its left under the standard normal curve. This was done earlier in Example 6.4 on page 341, where we found that $z = -1.75$.

Step 4   *Use the formula $x = \mu + z \cdot \sigma$ to destandardize the z-values found in Step 3 and hence obtain the required x-values.*

We want to obtain the x-value whose z-score is -1.75, the z-value found in Step 3. We have

$$z = -1.75 \quad \longrightarrow \quad x = \mu + z \cdot \sigma = 100 + (-1.75) \cdot 16 = 100 - 28 = 72.$$

Thus for the normal curve with parameters $\mu = 100$ and $\sigma = 16$, the x-value having area 0.04 to its left is 72, as seen in Fig. 6.21(b).   ∎

Using the Computer (Optional)

In Section 6.1 we learned how to apply Minitab's Normal subcommand to solve problems concerning the standard normal curve. As a subcommand to the CDF command, Normal yields the area under the standard normal curve to the left of a specified value of z; as a subcommand to the InvCDF command, Normal yields the z-value having a specified area to its left under the standard normal curve.

We can use those same two commands and the Normal subcommand to solve problems concerning any normal curve. The only difference is that in the process of applying the Normal subcommand, we type the two parameters of the normal curve under consideration instead of the parameters 0 and 1 (corresponding to the standard normal curve). See the *Minitab Supplement* for details.

EXERCISES 6.2

6.33   Which normal curve has a wider spread: the one with parameters $\mu = 1$ and $\sigma = 2$ or the one with parameters $\mu = 2$ and $\sigma = 1$? Explain your answer.

6.34   True or false: The normal curve with parameters $\mu = -4$ and $\sigma = 3$ and the normal curve with parameters $\mu = 6$ and $\sigma = 3$ have the same shape. Explain your answer.

6.35   True or false: The value of the parameter μ has no effect on the shape of a normal curve. Explain your answer.

6.36   Answer true or false and give a reason for your answer: The normal curve with parameters $\mu = -4$ and $\sigma = 3$

and the normal curve with parameters $\mu = 6$ and $\sigma = 3$ are centered at the same place.

6.37   Apply the procedure illustrated in Example 6.9 on page 349 to sketch the normal curve with parameters
a. $\mu = 3$ and $\sigma = 3$.
b. $\mu = 1$ and $\sigma = 3$.
c. $\mu = 3$ and $\sigma = 1$.

6.38   Sketch the normal curve with parameters
a. $\mu = -2$ and $\sigma = 2$.
b. $\mu = -2$ and $\sigma = \frac{1}{2}$.
c. $\mu = 0$ and $\sigma = 2$.

In Exercises 6.39–6.42, use the two-diagram procedure, portrayed in Fig. 6.18 on page 352, to find the designated areas.

· **6.39** Determine the area under the normal curve with parameters $\mu = 1$ and $\sigma = 2.5$ that lies
a. to the right of 0.   b. to the left of -1.5.
c. between -2 and 2.

· **6.40** Find the area under the normal curve with parameters $\mu = -1.5$ and $\sigma = 1$ that lies
a. between 0 and 1.4.   b. to the left of -1.5.
c. to the right of 1.

· **6.41** Find the area under the normal curve with parameters $\mu = 2$ and $\sigma = \frac{1}{2}$ that lies
a. to the left of 2.87.   b. between 1 and 1.5.
c. to the right of 2.75.

· **6.42** Determine the area under the normal curve with parameters $\mu = 3$ and $\sigma = 0.75$ that lies
a. to the left of 2.5.   b. between 2 and 4.
c. to the right of 1.5.

In Exercises 6.43–6.48, use Procedure 6.1 on page 353 to determine the required areas.

· **6.43** Find the area under the normal curve with parameters $\mu = 74$ and $\sigma = 2$ that lies
a. between 71 and 78.   b. to the right of 76.5.

· **6.44** Obtain the area under the normal curve with parameters $\mu = 7.3$ and $\sigma = 2$ that lies
a. between 4.3 and 11.8.   b. to the left of 9.94.

· **6.45** Determine the area under the normal curve with parameters $\mu = 335$ and $\sigma = 10$ that lies
a. to the left of 348.5.   b. between 340 and 350.

· **6.46** Find the area under the normal curve with parameters $\mu = 64.4$ and $\sigma = 2.4$ that lies
a. to the right of 70.   b. between 68 and 71.

· **6.47** Find the area under the normal curve with parameters $\mu = 40.9$ and $\sigma = 7.1$ that lies
a. to the left of 35.   b. between 25 and 30.

· **6.48** Determine the area under the normal curve with parameters $\mu = 15.6$ and $\sigma = 5.1$ that lies
a. to the left of 5.   b. between 3.42 and 10.

In Exercises 6.49–6.54, use Procedure 6.2 on page 355 to obtain the required x-values.

· **6.49** For the normal curve with parameters $\mu = 74$ and $\sigma = 2$, determine the

a. *x*-value having area 0.05 to its right.
b. *x*-value having area 0.10 to its left.
c. two *x*-values that divide the area under the curve into a middle 0.95 area and two outside 0.025 areas.

· **6.50** For the normal curve with parameters $\mu = 7.3$ and $\sigma = 2$, determine the
a. *x*-value having area 0.025 to its right.
b. *x*-value having area 0.01 to its left.
c. two *x*-values that divide the area under the curve into a middle 0.90 area and two outside 0.05 areas.

· **6.51** For the normal curve with parameters $\mu = 335$ and $\sigma = 10$, obtain the *x*-value having area 0.60
a. to its left.   b. to its right.

· **6.52** For the normal curve with parameters $\mu = 64.4$ and $\sigma = 2.4$, find the *x*-value having area 0.90
a. to its left.   b. to its right.

· **6.53** Find the three *x*-values that divide the area under the normal curve with parameters $\mu = 40.9$ and $\sigma = 7.1$ into four 0.25 areas.

· **6.54** Obtain the three *x*-values that divide the area under the normal curve with parameters $\mu = 15.6$ and $\sigma = 5.1$ into four 0.25 areas.

· · **6.55** **(Computer exercise)** Use Minitab or some other statistical software to obtain the areas required in Exercise 6.41.

· · **6.56** **(Computer exercise)** Use Minitab or some other statistical software to obtain the areas required in Exercise 6.42.

· · **6.57** **(Computer exercise)** Use Minitab or some other statistical software to obtain the *x*-values required in Exercise 6.49.

· · **6.58** **(Computer exercise)** Use Minitab or some other statistical software to obtain the *x*-values required in Exercise 6.50.

· · **6.59** Find the designated area under the normal curve with the specified parameters:
a. $\mu = 0$, $\sigma = 1$; area between -3 and 3.
b. $\mu = 4$, $\sigma = 2$; area between -2 and 10.
c. $\mu = \mu$, $\sigma = \sigma$; area between $\mu - 3\sigma$ and $\mu + 3\sigma$.

· · **6.60** Apply Procedure 6.1 on page 353 to solve the following problems.

a. Complete the following table for the normal curve with parameters $\mu = 4$ and $\sigma = 3$. Draw three graphs illustrating your results.

x-values	Corresponding z-scores	Area between x-values
1 & 7	−1 & 1	0.6826
−2 & 10	−2 & 2	
−5 & 13		

b. Complete the following table for the normal curve with parameters μ and σ. Draw three graphs illustrating your results.

x-values	Corresponding z-scores	Area between x-values
$\mu - \sigma$ & $\mu + \sigma$	−1 & 1	0.6826
$\mu - 2\sigma$ & $\mu + 2\sigma$		
$\mu - 3\sigma$ & $\mu + 3\sigma$		

· ·   **6.61**   Employ Procedure 6.1 on page 353 to solve the following problems.

a. Following is a sketch of the normal curve with parameters $\mu = 4$ and $\sigma = 3$. The area under the curve is divided into eight regions. We found the area of the region between 7 and 10 and recorded that area on the graph in color. Find and record the z-scores and areas for the remaining regions.

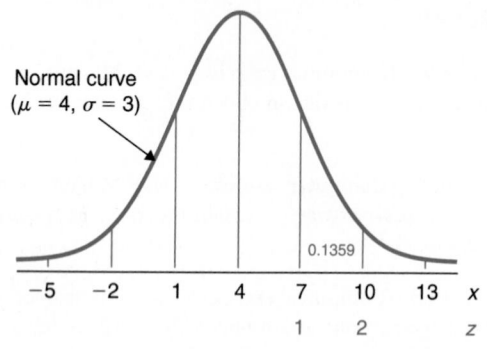

b. Repeat part (a) for the normal curve with parameters μ and σ.

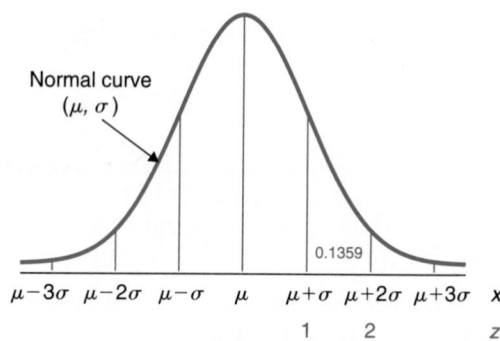

· ·   **6.62**   Let $0 < \alpha < 1$. For the normal curve with parameters μ and σ, determine

a. the x-value having area α to its right, in terms of z_α, μ, and σ.

b. the x-value having area α to its left, in terms of z_α, μ, and σ.

c. the two x-values that divide the area under the curve into a middle area of $1 - \alpha$ and two outside areas of $\alpha/2$.

· ·   **6.63**   For the normal curve with parameters μ and σ, obtain the three x-values that divide the area under the curve into four areas of 0.25.

· · ·   **6.64**   This exercise verifies Key Fact 6.4 (page 352). Knowledge of elementary calculus is required.

a. Referring to the footnote on page 348, show that the area under the normal curve with parameters μ and σ that lies between a and b equals

$$\int_a^b \frac{1}{\sqrt{2\pi}\sigma} e^{-(x-\mu)^2/2\sigma^2}\, dx.$$

b. Making the substitution $z = (x - \mu)/\sigma$, show that the integral in part (a) equals

$$\int_{(a-\mu)/\sigma}^{(b-\mu)/\sigma} \frac{1}{\sqrt{2\pi}} e^{-z^2/2}\, dz.$$

c. What area does the integral in part (b) equal? (*Hint:* See the footnote on page 336.)

6.3   NORMALLY DISTRIBUTED POPULATIONS

As we mentioned in Section 6.1, many populations have distributions that can be represented by normal curves. This means that a histogram for the population is bell-shaped and that percentages for the population are equal, at least approximately, to areas under a suitable normal curve. Example 6.13 should help make this idea clear.

Example 6.13 A Normally Distributed Population

A midwestern college has an enrollment of 3264 female students. Records show that the population mean height of these students is 64.4 inches and that the standard deviation is 2.4 inches: $\mu = 64.4$ inches and $\sigma = 2.4$ inches.

Frequency and relative-frequency distributions for this population of heights are presented in Table 6.3. The table shows, for instance, that 7.35% (0.0735) of the students are between 67 and 68 inches tall.

TABLE 6.3
Frequency and relative-frequency distributions for heights

Height (in.)	Frequency f	Relative frequency
56–under 57	3	0.0009
57–under 58	6	0.0018
58–under 59	26	0.0080
59–under 60	74	0.0227
60–under 61	147	0.0450
61–under 62	247	0.0757
62–under 63	382	0.1170
63–under 64	483	0.1480
64–under 65	559	0.1713
65–under 66	514	0.1575
66–under 67	359	0.1100
67–under 68	240	0.0735
68–under 69	122	0.0374
69–under 70	65	0.0199
70–under 71	24	0.0074
71–under 72	7	0.0021
72–under 73	5	0.0015
73–under 74	1	0.0003
	3264	1.0000

A relative-frequency histogram for the heights of the 3264 female students is included in Fig. 6.22, shown at the top of the following page. Notice that the histogram is bell-shaped. Because of this we can approximate percentages (relative frequencies, probabilities) for the population of heights by areas under a suitable normal curve. As you might expect, the appropriate normal curve is the one with parameters μ and σ, where μ is the mean of the population and σ is its standard deviation. Here $\mu = 64.4$ and $\sigma = 2.4$. We have superimposed the normal curve with parameters $\mu = 64.4$ and $\sigma = 2.4$ on the histogram in Fig. 6.22.

Now let's see how we can approximate percentages for this population of heights by areas under the normal curve with parameters $\mu = 64.4$ and $\sigma = 2.4$. To be specific, let's consider the percentage of female students who are between 67 and 68 inches tall. According to Table 6.3, the exact percentage is 7.35%, or 0.0735. Note that 0.0735 also equals the area of the cross-hatched bar in Fig. 6.22 because the bar has height 0.0735 and width 1.

FIGURE 6.22

Relative-frequency
histogram for heights with
superimposed normal curve

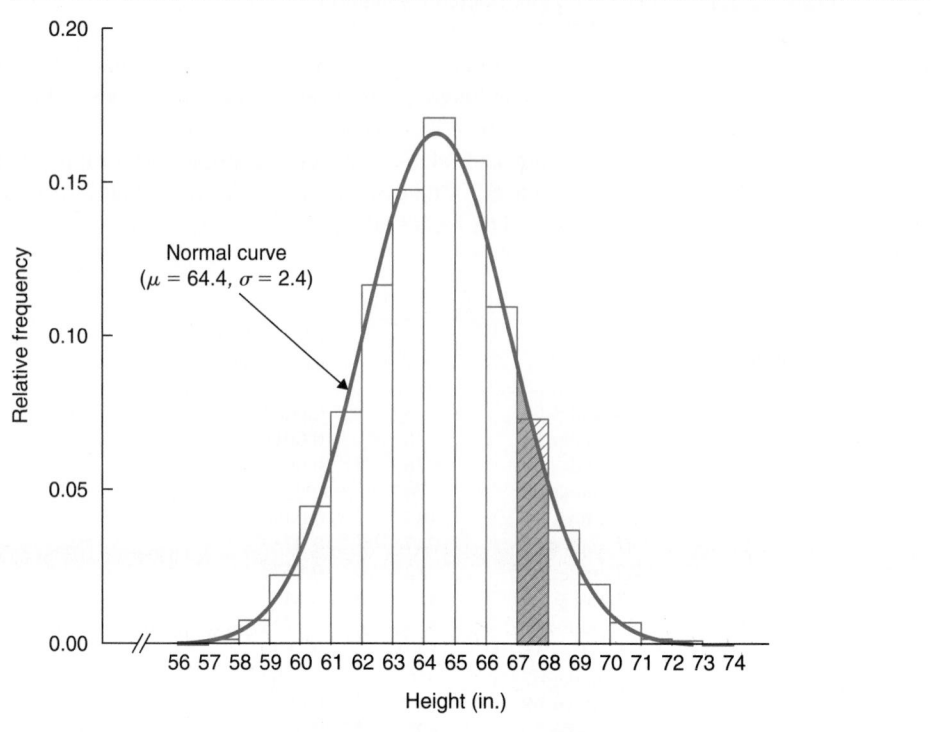

Look now at the area under the normal curve between 67 and 68, the area shaded in Fig. 6.22. Observe that this area is approximately equal to the area of the cross-hatched bar which, as we have noted, equals the percentage of students who are between 67 and 68 inches tall. Consequently, *we can approximate the percentage of students who are between 67 and 68 inches tall by the area under the normal curve between 67 and 68.*

Although the main point of this example has now been made, it is interesting to compare the numerical values under consideration here. We already know that the percentage of students who are between 67 and 68 inches tall is exactly 7.35%. To determine the area under the normal curve between 67 and 68, we apply Procedure 6.1 on page 353, as is done in Fig. 6.23.

The last line of Fig. 6.23 shows that the area under the normal curve between 67 and 68 is 0.0733, or 7.33%. Comparing this with the exact percentage of students who are between 67 and 68 inches tall, 7.35%, we see that the approximation by the area under the normal curve provides excellent results.   ∎

The important point of Example 6.13 is that for certain populations, percentages are approximately equal to areas under a suitable normal curve. Not all populations have that property, but if a population does, then we say it is a **normally distributed population.**

FIGURE 6.23

Determination of the area under the normal curve with parameters $\mu = 64.4$ and $\sigma = 2.4$ that lies between 67 and 68

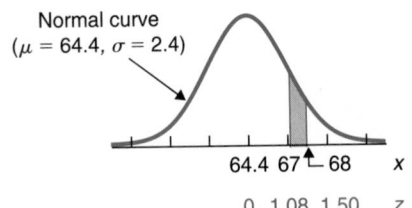

Normal curve
($\mu = 64.4$, $\sigma = 2.4$)

64.4 67 68 x

0 1.08 1.50 z

z-score computations:

Area to the left of z:

$x = 67 \longrightarrow z = \dfrac{67 - 64.4}{2.4} = 1.08$ 0.8599

$x = 68 \longrightarrow z = \dfrac{68 - 64.4}{2.4} = 1.50$ 0.9332

Shaded area $= 0.9332 - 0.8599 = 0.0733$

DEFINITION 6.2

NORMALLY DISTRIBUTED POPULATION

A population is said to be (approximately) **normally distributed** if percentages for the population are (approximately) equal to areas under a normal curve. If such a population has mean μ and standard deviation σ, then the appropriate normal curve is the one with parameters μ and σ.

Keep in mind what it means qualitatively for a population to be normally distributed, namely, that a histogram of the population is bell-shaped.

It is often known that a population is normally distributed because of past experience, previous statistical studies, or theoretical considerations. In such cases it is easy to obtain any required percentage. Example 6.14 illustrates this point.

Example 6.14 Finding Percentages for a Normally Distributed Population

In 1905 R. Pearl published the article "Biometrical Studies on Man. I. Variation and Correlation in Brain Weight" (*Biometrica*, Vol. 4, pp. 13–104). According to the study, brain weights of Swedish men are normally distributed with a mean of $\mu = 1.40$ kilograms (kg) and a standard deviation of $\sigma = 0.11$ kg. Obtain the percentage of Swedish men having brain weights between 1.50 kg and 1.70 kg.

SOLUTION Since the brain weights are normally distributed, percentages are equal to areas under a normal curve, namely, the normal curve whose parameters are the same as the mean and standard deviation of the population. Here $\mu = 1.40$ and $\sigma = 0.11$.

Thus to obtain the percentage of Swedish men having brain weights between 1.50 kg and 1.70 kg, we need to find the area under the normal curve with parameters $\mu = 1.40$ and $\sigma = 0.11$ that lies between 1.50 and 1.70. That area is computed in Fig. 6.24.

FIGURE 6.24

Determination of the area under
the normal curve with parameters
$\mu = 1.40$ and $\sigma = 0.11$ that
lies between 1.50 and 1.70

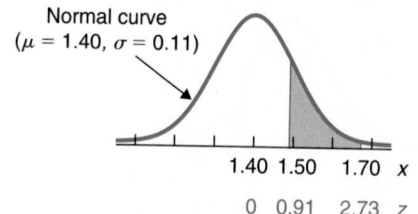

Normal curve
$(\mu = 1.40,\ \sigma = 0.11)$

| | 1.40 | 1.50 | 1.70 | x |
| | 0 | 0.91 | 2.73 | z |

z-score computations: Area to the left of z:

$x = 1.5 \longrightarrow z = \dfrac{1.50 - 1.40}{0.11} = 0.91$ 0.8186

$x = 1.7 \longrightarrow z = \dfrac{1.70 - 1.40}{0.11} = 2.73$ 0.9968

Shaded area $= 0.9968 - 0.8186 = 0.1782$

Consequently, we see that 17.82% (0.1782) of Swedish men have brain weights between 1.50 kg and 1.70 kg.   ∎

Population z-Scores

In Section 3.7 we defined the z-score of a population value. For ease of reference, we repeat
that definition here.

DEFINITION 6.3   **POPULATION z-SCORE**

> The **population z-score** for a value, x, is defined as the number of standard deviations
> that x is away from the mean. The population z-score is computed using the formula
>
> $$z = \frac{x - \mu}{\sigma},$$
>
> where μ and σ are, respectively, the mean and standard deviation of the population. A
> negative z-score indicates that the population value is smaller than the mean, whereas
> a positive z-score indicates that the population value is larger than the mean.

Thus, for example, a population value with a z-score of 2 is two standard deviations
above the mean, a population value with a z-score of -1.25 is 1.25 standard deviations
below the mean, and a population value with a z-score of 0 is equal to the mean.

Keep in mind, as we have already pointed out, that the process of computing a population z-score is identical to the process of standardizing. One implication of that fact is
the following: For a normally distributed population, Table II gives, for positive z, the percentage of population values that are less than z standard deviations above the mean; and
it gives, for negative z, the percentage of population values that are more than $-z$ standard
deviations below the mean.

For instance, Table II shows that for a normally distributed population, 98.3% of the population values are less than 2.12 standard deviations above the mean and that 5.82% of the population values are more than 1.57 $[-(-1.57)]$ standard deviations below the mean, as illustrated in Fig. 6.25.

FIGURE 6.25

Percentage of a normally distributed population that lies (a) less than 2.12 standard deviations above the mean, (b) more than 1.57 standard deviations below the mean

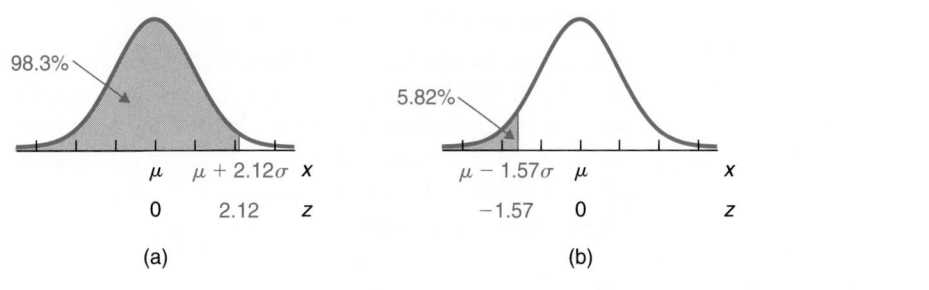

The Empirical Rule

In Chapter 3 we discussed Chebychev's rule. That rule provides estimates for the percentage of population values that lie within a specified number of standard deviations to either side of the mean. The power of Chebychev's rule is its generality—it holds for any population. But if we know something about the distribution of the population, then we can obtain better estimates than those provided by Chebychev's rule.

For a normally distributed population, we can use Table II to obtain the exact percentage of population values that lie within a specified number of standard deviations to either side of the mean. To illustrate, we again use the information from Example 6.14.

Example 6.15 Introduces the Empirical Rule

Refer to Example 6.14. Obtain the percentage of Swedish men having brain weights

a. within one standard deviation to either side of the mean.

b. within two standard deviations to either side of the mean.

c. within three standard deviations to either side of the mean.

SOLUTION Since the brain weights are normally distributed with mean $\mu = 1.40$ kg and standard deviation $\sigma = 0.11$ kg, percentages are equal to areas under the normal curve with parameters $\mu = 1.40$ and $\sigma = 0.11$. As we know, such areas are obtained by first converting to z-scores and then using the standard-normal table, Table II.

a. The z-score for a brain weight one standard deviation below the mean is -1 and that for a brain weight one standard deviation above the mean is 1. Thus the percentage of Swedish men having brain weights within one standard deviation to either side of the mean equals the area under the standard normal curve between -1 and 1, which is 0.6826, or 68.26%. (To obtain the area 0.6826, we can either use Table II in the usual way or simply refer to Table 6.1 on page 340.)

b. The z-score for a brain weight two standard deviations below the mean is -2 and that for a brain weight two standard deviations above the mean is 2. Therefore the percentage of Swedish men having brain weights within two standard deviations to either side of the mean equals the area under the standard normal curve between -2 and 2, which is 0.9544, or 95.44%.

c. The z-score for a brain weight three standard deviations below the mean is -3 and that for a brain weight three standard deviations above the mean is 3. Consequently, the percentage of Swedish men having brain weights within three standard deviations to either side of the mean equals the area under the standard normal curve between -3 and 3, which is 0.9974, or 99.74%.   ∎

The methods employed in Example 6.15 apply to any normally distributed population; we never used the information that the population consists of brain weights or that $\mu = 1.40$ kg and $\sigma = 0.11$ kg. All we used was the fact that the population is normally distributed. Thus we have the following rule.

KEY FACT 6.5

EMPIRICAL RULE FOR NORMALLY DISTRIBUTED POPULATIONS

For any normally distributed population:

Property 1: 68.26% of the population values lie within one standard deviation to either side of the mean, that is, between $\mu - \sigma$ and $\mu + \sigma$.

Property 2: 95.44% of the population values lie within two standard deviations to either side of the mean, that is, between $\mu - 2\sigma$ and $\mu + 2\sigma$.

Property 3: 99.74% of the population values lie within three standard deviations to either side of the mean, that is, between $\mu - 3\sigma$ and $\mu + 3\sigma$.

These properties are displayed graphically in Fig. 6.26.

FIGURE 6.26

Percentage of a normally distributed population that lies within (a) one, (b) two, and (c) three standard deviations to either side of the mean

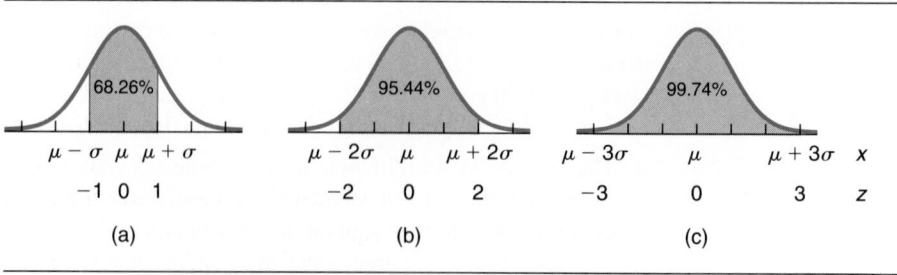

Example 6.16   Illustrates Key Fact 6.5

Intelligence quotients (IQs) measured on the Stanford Revision of the Binet-Simon Intelligence Scale are known to be approximately normally distributed with a mean of $\mu = 100$ and a standard deviation of $\sigma = 16$. Apply the empirical rule to make some observations about IQs.

SOLUTION   From Property 1 of the empirical rule, 68.26% of the population have IQs within one standard deviation to either side of the mean. One standard deviation below the mean is $\mu - \sigma = 100 - 16 = 84$; one standard deviation above the mean is $\mu + \sigma = 100 + 16 = 116$. So 68.26% of the population have IQs between 84 and 116, as seen in Fig. 6.27(a).

FIGURE 6.27
Graphical display of
empirical rule for IQs

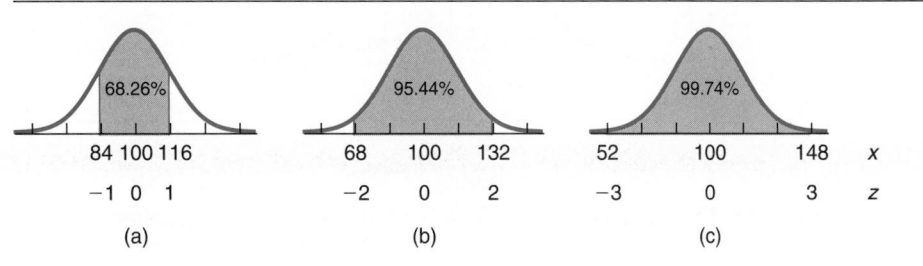

From Property 2 of the empirical rule, 95.44% of the population have IQs within two standard deviations to either side of the mean. Two standard deviations below the mean is $\mu - 2\sigma = 100 - 2 \cdot 16 = 100 - 32 = 68$; two standard deviations above the mean is $\mu + 2\sigma = 100 + 2 \cdot 16 = 100 + 32 = 132$. Therefore 95.44% of the population have IQs between 68 and 132, as seen in Fig. 6.27(b).

From Property 3 of the empirical rule, we find that 99.74% of the population have IQs between 52 $(= 100 - 3 \cdot 16)$ and 148 $(= 100 + 3 \cdot 16)$, as seen in Fig. 6.27(c). ■

The empirical rule allows us to obtain useful information about a normally distributed population quickly and easily, as illustrated in Example 6.16. We should point out, however, that similar facts are obtainable for any number of standard deviations. For instance, we can use Table II to conclude that for a normally distributed population, 86.64% of the population values lie within 1.5 standard deviations to either side of the mean.

Quartiles and Percentiles for Normally Distributed Populations

In Chapter 3 we discussed quartiles and percentiles. Recall that **quartiles** divide a population into quarters, or four equal parts. A population has three quartiles, denoted by Q_1, Q_2, and Q_3. The first quartile, Q_1, is the number that divides the bottom 25% of the population from the top 75%; the second quartile, Q_2, is the median, which, as we know, is the number that divides the bottom 50% of the population from the top 50%; and the third quartile, Q_3, is the number that divides the bottom 75% of the population from the top 25%.

For a normally distributed population, percentages are equal to areas under a normal curve. Thus the quartiles for a normally distributed population are the three x-values that divide the area under the normal curve into four 0.25 areas, as shown in Fig. 6.28.

FIGURE 6.28
Quartiles for a normally
distributed population

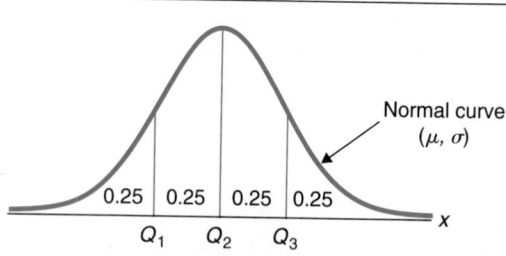

Example 6.17   Finding Quartiles for a Normally Distributed Population

Intelligence quotients are normally distributed with a mean of 100 and a standard deviation of 16. Determine the quartiles for IQs.

SOLUTION   Percentages for IQs are equal to areas under the normal curve with parameters $\mu = 100$ and $\sigma = 16$. Therefore to determine the quartiles, we need to find the three x-values that divide the area under that normal curve into four 0.25 areas, as shown in Fig. 6.29.

FIGURE 6.29
Finding the quartiles for IQs

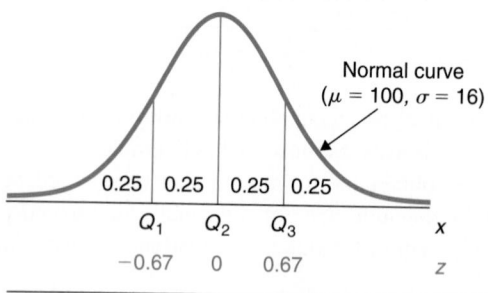

As we observe from Fig. 6.29, the first quartile, Q_1, is the x-value having area 0.25 to its left. So the z-score corresponding to Q_1 is the z-value having area 0.25 to its left under the standard normal curve. Using Table II we find that z-value to be -0.67, approximately. Similarly, we find that the z-scores corresponding to the second and third quartiles, Q_2 and Q_3, are 0 and 0.67, respectively. These three z-scores are displayed in Fig. 6.29.

To obtain the quartiles, we need only destandardize the three z-scores. Thus the first quartile is

$$Q_1 = \mu + z \cdot \sigma = 100 + (-0.67) \cdot 16 = 100 - 0.67 \cdot 16 = 100 - 10.72,$$

or $Q_1 = 89.28$. Similarly, we find that $Q_2 = 100$ and $Q_3 = 110.72$.   ■

Finally, let's discuss percentiles. As we learned in Chapter 3, the **percentiles** of a population divide it into hundredths, or 100 equal parts. A population has 99 percentiles, denoted by $P_1, P_2, \ldots, P_{99}$. The first percentile, P_1, is the number that divides the bottom 1% of the population from the top 99%; the second percentile, P_2, is the number that divides the bottom 2% of the population from the top 98%; and so forth.

For a normally distributed population, we can obtain percentiles by employing Table II, just as we did to find quartiles. Example 6.18 explains how to do this.

Example 6.18   Finding Percentiles for a Normally Distributed Population

Consider again the population of IQs, which is normally distributed with mean $\mu = 100$ and standard deviation $\sigma = 16$. Determine the 90th percentile, P_{90}.

SOLUTION   The 90th percentile, P_{90}, is the number that divides the bottom 90% of the population from the top 10%. Since percentages for IQs are equal to areas under the normal curve with parameters $\mu = 100$ and $\sigma = 16$, P_{90} is the x-value having area 0.90 to its left or, equivalently, area 0.10 to its right, under that normal curve, as seen in Fig. 6.30(a).

FIGURE 6.30

Finding the 90th percentile for IQs

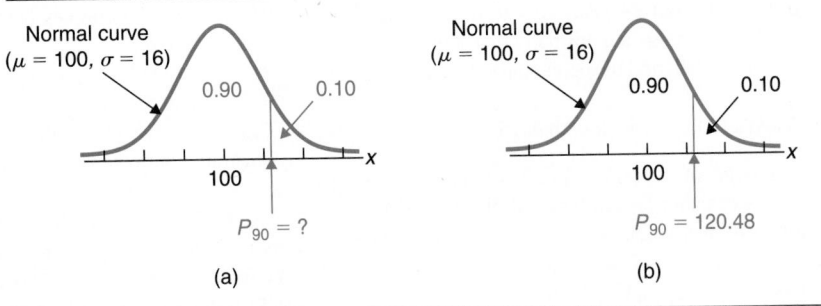

(a)   (b)

The z-score corresponding to P_{90} is therefore the z-value having area 0.90 to its left under the standard normal curve. From Table II we find that z-value to be 1.28, approximately. Now we destandardize the z-score to obtain the 90th percentile:

$$P_{90} = \mu + z \cdot \sigma = 100 + 1.28 \cdot 16 = 100 + 20.48 = 120.48.$$

Therefore 90% of the population have IQs below 120.48 and 10% have IQs above 120.48, as seen in Fig. 6.30(b).   ■

EXERCISES 6.3

A relative-frequency distribution for the heights of the 3264 female students attending a midwestern college is displayed in the first and third columns of Table 6.3 on page 359. That population of heights has a mean of 64.4 inches and a standard deviation of 2.4 inches. In each part of Exercises 6.65 and 6.66, first obtain the exact percentage from Table 6.3 and then find the corresponding area under the normal curve with parameters $\mu = 64.4$ and $\sigma = 2.4$. Compare your results.

6.65 The percentage of female students with heights
a. between 62 and 63 in.   b. between 65 and 70 in.

6.66 The percentage of female students with heights
a. between 71 and 72 in.   b. between 61 and 65 in.

6.67 The annual wages, excluding board, of U.S. farm laborers in 1926 are normally distributed with a mean of $\mu = \$586$ and a standard deviation of $\sigma = \$97$. In 1926 what percentage of U.S. farm laborers had an annual wage of
a. between $500 and $700?   b. at least $400?

6.68 The length of an adult yellow-bellied sapsucker is normally distributed with a mean of $\mu = 8.5$ in. and a standard deviation of $\sigma = 0.17$ in. Determine the percentage of yellow-bellied sapsuckers that are
a. between 8.3 and 9.0 in. long.   b. less than 8.8 in. long.

6.69 As reported by the U.S. National Center for Health Statistics in *Vital and Health Statistics,* males who are 6 ft tall and between 18 and 24 years of age have a mean weight of 175 lb. If the weights are normally distributed with a standard deviation of 14 lb, find the percentage of such males that weigh
a. between 190 and 210 lb.   b. less than 150 lb.

6.70 An issue of *Scientific American* reveals that the batting averages of major-league baseball players are approximately normally distributed with a mean of 0.270 and a standard deviation of 0.015. Determine the percentage of major-league baseball players having batting averages
a. between 0.225 and 0.250.   b. at least 0.300.

6.71 Refer to Exercise 6.67. How many standard deviations away from the mean is a 1926 U.S. farm laborer's annual wage of
a. $780?   b. $416.25?   c. $586?

6.72 Refer to Exercise 6.68. How many standard deviations away from the mean is the length of a yellow-bellied sapsucker
a. 8.7125 in. long?   b. 8.16 in. long?   c. 8.5 in. long?

6.73 Refer to Exercise 6.69, where we considered the weights of U.S. males who are 6 ft tall and between 18 and 24 years of age. Assuming the weights are normally distributed, determine the percentage of such males that have weights within
a. 1 standard deviation to either side of the mean.
b. 2 standard deviations to either side of the mean.
c. 3 standard deviations to either side of the mean.
d. 1.5 standard deviations to either side of the mean.

6.74 Refer to Exercise 6.70. Obtain the percentage of major-league baseball players who have batting averages within
a. 1 standard deviation to either side of the mean.
b. 2 standard deviations to either side of the mean.
c. 3 standard deviations to either side of the mean.
d. 2.5 standard deviations to either side of the mean.

6.75 The Department of Agriculture compiles information on food costs and publishes its findings in *Human Nutrition Information Service.* According to that document, the mean weekly food cost for a couple with two children 6–11 years old is $132. Presuming the costs are normally distributed with a standard deviation of $17.20, fill in the following blanks by applying the empirical rule.
a. 68.26% of such couples have weekly food costs between $____ and $____.
b. 95.44% of such couples have weekly food costs between $____ and $____.
c. 99.74% of such couples have weekly food costs between $____ and $____.
d. Draw graphs similar to the ones in Fig. 6.27 on page 365 to illustrate your results.

6.76 The A. C. Nielsen Company reports in *Nielsen Report on Television* that the mean weekly television viewing time for children age 2–5 years is 27.15 hours. Assuming the weekly television viewing times of these children are normally distributed with a standard deviation of 6.23 hours, fill in the following blanks by applying the empirical rule.
a. 68.26% of all such children watch between ____ and ____ hours of TV per week.
b. 95.44% of all such children watch between ____ and ____ hours of TV per week.
c. 99.74% of all such children watch between ____ and ____ hours of TV per week.
d. Draw graphs similar to the ones in Fig. 6.27 on page 365 to illustrate your results.

6.77 For a normally distributed population, fill in the following blanks.
a. ____% of the population values lie within 1.96 standard deviations to either side of the mean.
b. ____% of the population values lie within 1.64 standard deviations to either side of the mean.

6.78 For a normally distributed population, fill in the following blanks.
a. ____% of the population values lie within 1.28 standard deviations to either side of the mean.
b. ____% of the population values lie within 2.33 standard deviations to either side of the mean.

6.79 For a normally distributed population, fill in the following blanks.
a. 99% of the population values lie within _____ standard deviations to either side of the mean.
b. 80% of the population values lie within _____ standard deviations to either side of the mean.

6.80 For a normally distributed population, fill in the following blanks.
a. 95% of the population values lie within _____ standard deviations to either side of the mean.
b. 90% of the population values lie within _____ standard deviations to either side of the mean.

6.81 Refer to Exercise 6.67.
a. Determine the quartiles for the wages.
b. Obtain the 15th percentile.
c. Find the 98th percentile.

6.82 Refer to Exercise 6.68.
a. Determine the quartiles for the lengths.
b. Obtain the first decile (i.e., 10th percentile).
c. Find the 82nd percentile.

6.83 Refer to Exercise 6.75.
a. Determine the quartiles for the food costs.
b. Obtain the 3rd decile.
c. Find the 85th percentile.

6.84 Refer to Exercise 6.76.
a. Determine the quartiles for the viewing times.
b. Obtain the 45th percentile.
c. Find the 9th decile.

6.85 (Computer exercise) Use Minitab or some other statistical software to solve Exercise 6.81.

6.86 (Computer exercise) Use Minitab or some other statistical software to solve Exercise 6.82.

6.87 Verify Key Fact 6.5 on page 364, the empirical rule for normally distributed populations.

6.88 Derive general formulas for the quartiles of a normally distributed population; that is, express the quartiles, Q_1, Q_2, and Q_3, in terms of μ and σ. (*Hint:* See Fig. 6.28 on page 366.)

6.4 NORMALLY DISTRIBUTED RANDOM VARIABLES

In Section 6.3 we discussed normally distributed populations. As we know, a population is normally distributed if percentages for the population are equal to areas under a normal curve—the normal curve whose parameters are the same as the mean and standard deviation of the population.

Also of great importance are normally distributed random variables. The definition of a normally distributed random variable is as follows.

DEFINITION 6.4   **NORMALLY DISTRIBUTED RANDOM VARIABLE**

A random variable is said to be (approximately) **normally distributed** if probabilities for the random variable are (approximately) equal to areas under a normal curve. If such a random variable has mean μ_x and standard deviation σ_x, then the appropriate normal curve is the one with parameters μ_x and σ_x.

Normally distributed populations and normally distributed random variables are quite closely related. Indeed, if we let x denote the value of a randomly selected member from a normally distributed population, then x is a normally distributed random variable. However, as we will see later, normally distributed random variables arise in other contexts.

Example 6.19   Finding Probabilities for a Normally Distributed Random Variable

A local bottling plant fills bottles of soda for distribution in the surrounding area. Although the advertised content is 354 mL, the filling machine is actually set to a mean content of 356 mL. In fact, the amount of soda put into a bottle is normally distributed with a mean of 356 mL and a standard deviation of 1.63 mL. Find the probability that a randomly selected bottle of soda will contain less than the advertised content of 354 mL.

SOLUTION   Let x denote the content, in milliliters, of a randomly selected bottle of soda. Then x is a normally distributed random variable with mean $\mu_x = 356$ mL and standard deviation $\sigma_x = 1.63$ mL. Therefore probabilities for x are equal to areas under the normal curve with parameters $\mu = 356$ and $\sigma = 1.63$.

We want to determine the probability that a randomly selected bottle will contain less than 354 mL of soda, $P(x < 354)$. That probability is equal to the area under the normal curve with parameters $\mu = 356$ and $\sigma = 1.63$ that lies to the left of 354. This area is found in the usual way, as done in Fig. 6.31.

FIGURE 6.31

Determination of the area under the normal curve with parameters $\mu = 356$ and $\sigma = 1.63$ that lies to the left of 354

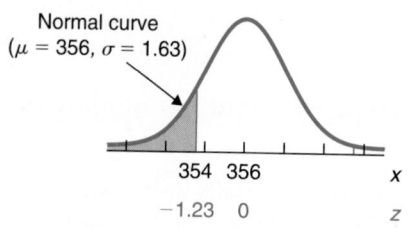

Normal curve
$(\mu = 356, \sigma = 1.63)$

354 356 x

−1.23   0 z

z-score computation:

Area to the left of z:

$$x = 354 \longrightarrow z = \frac{354 - 356}{1.63} = -1.23 \qquad 0.1093$$

Shaded area $= 0.1093$

Consequently, $P(x < 354) = 0.1093$. There is roughly an 11% chance that a randomly selected bottle will contain less than the advertised content of 354 mL of soda. In other words, approximately 11% of the bottles will contain less than the advertised content of 354 mL.

In Example 6.19 we found that approximately 11% of the bottles will contain less than the advertised content of 354 mL of soda if the filling machine is set to a mean content of 356 mL. What percentage of the bottles would contain less than the advertised content of 354 mL of soda if the filling machine were set to a mean content of 354 mL? Do you see why the filling machine is set to a mean higher than the advertised content?

Interpretation of Normally Distributed Random Variables

Key Fact 5.3 on page 273 provides an interpretation for the probability distribution of a random variable. Applied to a normally distributed random variable, it implies that in a large number of independent observations, a histogram of the observations will be roughly bell-shaped. To illustrate, let's return to the situation of Example 6.19.

Example 6.20   Interpreting a Normally Distributed Random Variable

Consider again the random variable x from Example 6.19, the content, in milliliters, of a randomly selected bottle of soda. We know that x is normally distributed with mean $\mu_x = 356$ mL and standard deviation $\sigma_x = 1.63$ mL. So if we observe the contents of a large number of bottles of soda, a histogram of those contents should be bell-shaped; more specifically, it should be shaped approximately like the normal curve with parameters $\mu = 356$ and $\sigma = 1.63$.

We used a computer to first simulate the contents of 1000 bottles of soda and then obtain a histogram for those contents. Printout 6.3 shows the histogram, which, as expected, is roughly bell-shaped. Compare the histogram in Printout 6.3 to the theoretical distribution in Fig. 6.31.   ■

PRINTOUT 6.3

Minitab histogram for the contents of 1000 bottles of soda

```
Histogram of CONTENTS    N = 1000
Each * represents 5 obs.

Midpoint    Count
     352       12   ***
     353       60   ************
     354      109   *********************
     355      201   *****************************************
     356      226   **********************************************
     357      208   ******************************************
     358      118   ************************
     359       47   **********
     360       18   ****
     361        1   *
```

The Empirical Rule for Normally Distributed Random Variables

The empirical rule for normally distributed populations, Key Fact 6.5 on page 364, provides useful information about percentages for normally distributed populations. Here are the corresponding probability statements for normally distributed random variables.

KEY FACT 6.6

EMPIRICAL RULE FOR NORMALLY DISTRIBUTED RANDOM VARIABLES

For any normally distributed random variable x:

Property 1: The probability is 0.6826 that an observed value of x will be within one standard deviation to either side of the mean:

$$P(\mu_x - \sigma_x < x < \mu_x + \sigma_x) = 0.6826.$$

Property 2: The probability is 0.9544 that an observed value of x will be within two standard deviations to either side of the mean:

$$P(\mu_x - 2\sigma_x < x < \mu_x + 2\sigma_x) = 0.9544.$$

Property 3: The probability is 0.9974 that an observed value of x will be within three standard deviations to either side of the mean:

$$P(\mu_x - 3\sigma_x < x < \mu_x + 3\sigma_x) = 0.9974.$$

These properties are displayed graphically in Fig. 6.32.

FIGURE 6.32

Probability that a normally distributed random variable will be within (a) one, (b) two, and (c) three standard deviations to either side of its mean

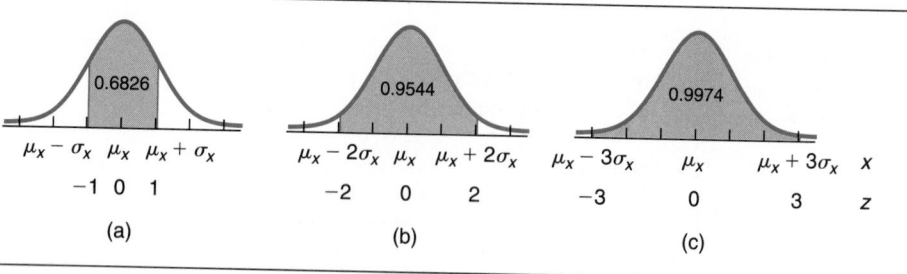

It is important to be able to interpret Properties 1–3 and similar statements in terms of percentages. For instance, Property 2 can be interpreted as follows: In a large number of independent observations, a normally distributed random variable will be within two standard deviations of its mean roughly 95.44% of the time.

Example 6.21   Illustrates Key Fact 6.6

In Example 6.19 we considered the contents of soda bottles filled in a bottling plant. The amount of soda put into a bottle is normally distributed with a mean of 356 mL and a standard deviation of 1.63 mL. If we let x denote the content of a randomly selected bottle, then x is a normally distributed random variable with mean $\mu_x = 356$ mL and standard deviation $\sigma_x = 1.63$ mL. Apply the empirical rule for normally distributed random variables to make some pertinent statements about the contents of the filled bottles.

SOLUTION   From Property 1 of the empirical rule, the probability is 0.6826 that an observed value of x will be within one standard deviation to either side of the mean. One standard deviation below the mean is $\mu_x - \sigma_x = 356 - 1.63 = 354.37$; one standard deviation above the mean is $\mu_x + \sigma_x = 356 + 1.63 = 357.63$. Thus the probability is 0.6826 that a randomly selected bottle will contain between 354.37 mL and 357.63 mL of soda. In other words, 68.26% of the bottles will contain between 354.37 mL and 357.63 mL of soda.

By Property 2 of the empirical rule, the probability is 0.9544 that an observed value of x will be within two standard deviations to either side of the mean. Two standard deviations below the mean is $\mu_x - 2\sigma_x = 356 - 2 \cdot 1.63 = 352.74$; two standard deviations above the mean is $\mu_x + 2\sigma_x = 356 + 2 \cdot 1.63 = 359.26$. So the probability is 0.9544 that a randomly selected bottle will contain between 352.74 mL and 359.26 mL of soda; that is, 95.44% of the bottles will contain between 352.74 mL and 359.26 mL of soda.

Applying Property 3 of the empirical rule, we conclude that the probability is 0.9974 that a randomly selected bottle will contain between 351.11 ($= 356 - 3 \cdot 1.63$) mL and 360.89 ($= 356 + 3 \cdot 1.63$) mL of soda. In other words, 99.74% of the bottles will contain between 351.11 mL and 360.89 mL of soda. It is very unlikely that a bottle will have a content outside that range.   ■

Statements similar to those in the empirical rule hold for any number of standard deviations. For example, using Table II we find that the probability is 0.3830 that an observed value of a normally distributed random variable will be within 0.5 standard deviation to either side of the mean. More generally, we have the following fact, which we will need in Chapter 8.

KEY FACT 6.7   **GENERAL EMPIRICAL RULE FOR NORMALLY DISTRIBUTED RANDOM VARIABLES**

> The probability is $1 - \alpha$ that an observed value of a normally distributed random variable, x, will be within $z_{\alpha/2}$ standard deviations to either side of the mean:
>
> $$P(\mu_x - z_{\alpha/2} \cdot \sigma_x < x < \mu_x + z_{\alpha/2} \cdot \sigma_x) = 1 - \alpha.$$
>
> Here, $0 < \alpha < 1$. This key fact is portrayed in Fig. 6.33.

FIGURE 6.33

General empirical rule

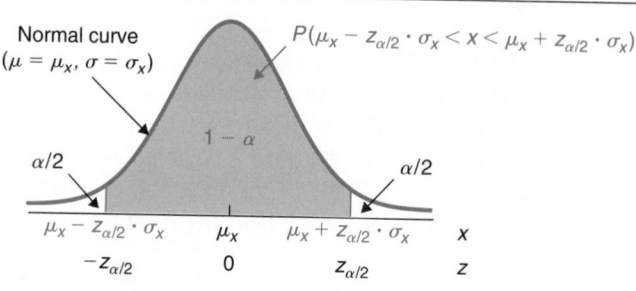

The general empirical rule, like the empirical rule, is simply a consequence of the fact that for a normally distributed random variable, x, probabilities are equal to areas under the normal curve with parameters μ_x and σ_x.

To illustrate the general empirical rule, suppose we let $\alpha = 0.10$. Then $1 - \alpha = 1 - 0.10 = 0.90$ and $z_{\alpha/2} = z_{0.10/2} = z_{0.05} = 1.645$ (from Table II). So, by the general empirical rule, the probability is 0.90 that an observed value of a normally distributed random variable will be within 1.645 standard deviations to either side of the mean. In symbols, if x is a normally distributed random variable, then

$$P(\mu_x - 1.645\sigma_x < x < \mu_x + 1.645\sigma_x) = 0.90.$$

For a concrete application of this equation, consider once again the contents of soda bottles filled in a bottling plant. Here x denotes the content of a randomly selected bottle of soda, and we know that x is normally distributed. Since $\mu_x = 356$ mL and $\sigma_x = 1.63$ mL, we have that $\mu_x - 1.645\sigma_x = 356 - 1.645 \cdot 1.63 = 353.32$ and $\mu_x + 1.645\sigma_x = 356 + 1.645 \cdot 1.63 = 358.68$. Hence the probability is 0.90 that a randomly selected bottle will contain between 353.32 mL and 358.68 mL of soda.

An Interpretation of the Standard Deviation of a Normally Distributed Random Variable

In Key Fact 5.7, we gave the following interpretation to the standard deviation, σ_x, of a random variable x: It measures the dispersion of the possible values of the random variable relative to the mean; the smaller the standard deviation, the more likely that an observed value of x will be close to the mean.

When the random variable is normally distributed, we can use the empirical rule to make this interpretation of σ_x more explicit. For example, by Property 2 of the empirical rule, there is a 95.44% chance that an observed value of a normally distributed random variable, x, will be within two standard deviations of the mean. If σ_x is small, then two standard deviations to either side of the mean will also be small, making it likely (a 95.44% chance) that an observed value of x will be close to the mean. If σ_x is large, then two standard deviations to either side of the mean will also be large, indicating that we cannot expect an observed value of x to be close to the mean, although it may be. A graphical summary of the discussion in this paragraph is presented in Fig. 6.34.

FIGURE 6.34

Illustration of Property 2
of the empirical rule for
(a) small σ_x and (b) large σ_x

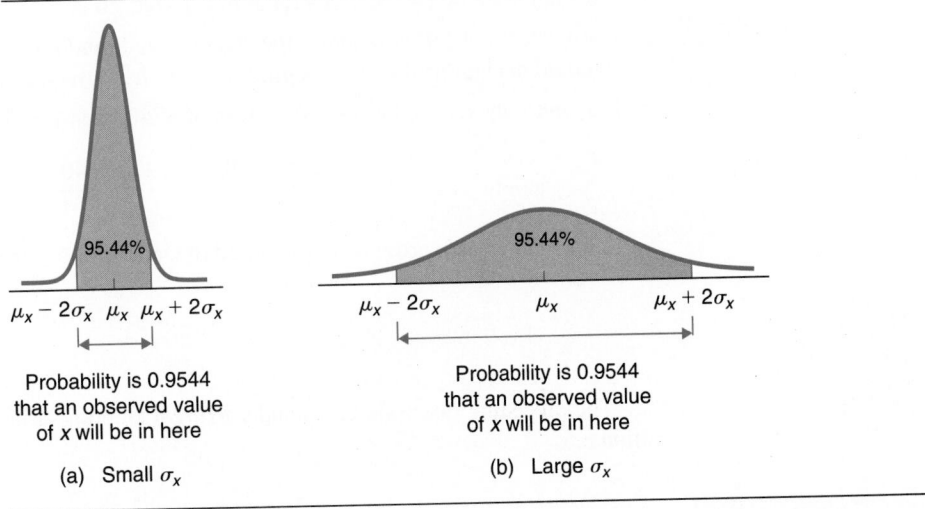

Probability is 0.9544
that an observed value
of x will be in here

(a) Small σ_x

Probability is 0.9544
that an observed value
of x will be in here

(b) Large σ_x

Standardizing a Random Variable

Recall that the process of standardizing involves converting x-values to z-values using the formula $z = (x - \mu)/\sigma$. Standardizing is employed both to determine z-scores and to obtain areas under normal curves.

It is also useful to consider the process of standardizing a random variable. A random variable, x, is standardized by first subtracting its mean and then dividing by its standard deviation. The resulting random variable is called the **standardized version** of x and is denoted by the letter z. We introduce this concept in Example 6.22.

Example 6.22 Standardizing a Random Variable

A one-person barber shop has a mean weekly gross income of \$640, with a standard deviation of \$50. Let x be the shop's gross income during a randomly selected week.

a. Determine the standardized version, z, of the random variable x.

b. If we select a week at random, then the random variable x tells us that week's gross income. What does the standardized random variable z tell us?

c. If a randomly selected week has a gross income of \$740, then how many standard deviations away from the mean is that week's gross income?

SOLUTION   We first note that x is a random variable with mean $\mu_x = 640$ and standard deviation $\sigma_x = 50$.

a. As we said, to obtain the standardized version, z, we first subtract μ_x from x and then divide by σ_x. Thus the standardized version of x is

$$z = \frac{x - \mu_x}{\sigma_x} = \frac{x - 640}{50}.$$

We emphasize that since x is a random variable, so is z.

b. If we select a week at random, the standardized random variable z tells us how many standard deviations that week's gross is away from the mean.

c. If a randomly selected week has a gross of $740, then $x = 740$, and consequently,

$$z = \frac{x - 640}{50} = \frac{740 - 640}{50} = 2.$$

Hence that week's gross is two standard deviations above the mean.   ∎

The preceding discussion on standardizing a random variable is summarized in Definition 6.5.

DEFINITION 6.5 **STANDARDIZED VERSION OF A RANDOM VARIABLE**

Let x be a random variable. Then we define the **standardized version** of x to be the random variable

$$z = \frac{x - \mu_x}{\sigma_x}.$$

The standardized random variable z gives the number of standard deviations that an observed value of x is away from the mean. It has mean 0 and standard deviation 1; that is, $\mu_z = 0$ and $\sigma_z = 1$.

Standardizing a Normally Distributed Random Variable

If a normally distributed random variable has mean 0 and standard deviation 1, then probabilities for that random variable are equal to areas under the normal curve with parameters $\mu = 0$ and $\sigma = 1$, the standard normal curve. Hence the following definition.

DEFINITION 6.6 **STANDARD NORMAL DISTRIBUTION**

A normally distributed random variable having mean 0 and standard deviation 1 is said to have the **standard normal distribution.**

We can form the standardized version of a random variable x regardless of whether x is normally distributed. If x is not normally distributed, then neither is z. But if x is normally distributed, then so is z; in fact, z then has the standard normal distribution.

| KEY FACT 6.8 | **STANDARDIZED NORMALLY DISTRIBUTED RANDOM VARIABLE** |

The standardized version of a normally distributed random variable x,

$$z = \frac{x - \mu_x}{\sigma_x},$$

has the standard normal distribution; that is, z is a normally distributed random variable with mean $\mu_z = 0$ and standard deviation $\sigma_z = 1$. Thus probabilities for the standardized random variable z are equal to areas under the standard normal curve.

Example 6.23   Illustrates Key Fact 6.8

Consider again the contents of soda bottles filled in a bottling plant. The amount of soda put into a bottle is normally distributed with a mean of 356 mL and a standard deviation of 1.63 mL. Let x denote the content of a randomly selected bottle of soda.

a. Obtain the standardized version, z, of the random variable x.

b. Determine the probability distribution of the standardized version, z, of the random variable x.

SOLUTION   a. The random variable x has mean $\mu_x = 356$ and standard deviation $\sigma_x = 1.63$. Thus the standardized version of x is the random variable

$$z = \frac{x - 356}{1.63}.$$

b. We know that x is normally distributed. Therefore, by Key Fact 6.8, the standardized random variable z, obtained in part (a), has the standard normal distribution.   ∎

Using the Computer (Optional)

Minitab's **Random** command, which we introduced in Section 5.1 for simulating discrete random variables, can also be used to simulate normally distributed random variables. To do so we use the **Normal** subcommand, as illustrated in Example 6.24.

Example 6.24   The Random Command and Normal Subcommand

Printout 6.3 on page 371 shows the result of simulating the contents of 1000 bottles of soda, where the amount of soda put into a bottle is normally distributed with a mean of 356 mL and a standard deviation of 1.63 mL. We performed the simulation by applying Minitab's Random command and Normal subcommand. The details for doing this will now be explained.

Session commands: First we name an available column CONTENTS. Next we type the command **RANDOM** followed by the number of observations required and the storage location for the observations; that is, we type RANDOM 1000 obs into 'CONTENTS'; and press Enter. Then we use the **NORMAL** subcommand to specify the probability distribution of the random variable to be simulated; that is, we type NORMAL with mu=356 and sigma=1.63. and press Enter.

(or)

Menu commands: We choose **Calc ▸ Random Data ▸ Normal...**, type 1000 for the number of rows to be generated, specify CONTENTS in the **Store in column(s)** text box, type 356 in the **Mean** text box, type 1.63 in the **Standard deviation** text box, and then select **OK**.

As a consequence of the above commands, the contents of 1000 bottles of soda are stored in CONTENTS. To obtain a histogram for the contents, we apply the Histogram command in the usual way, as explained in Section 2.3. The computer output we actually got by performing the simulation and applying the Histogram command is shown in Printout 6.3 on page 371. If we repeat the simulation, we will most likely obtain different results than those depicted in Printout 6.3; but the results should be similar.   ■

EXERCISES 6.4

6.89 As reported by the National Education Association in *Estimates of School Statistics,* the mean salary for secondary school teachers is $32,000. Assume the salaries are normally distributed with a standard deviation of $5,900. Let x be the salary, in thousands of dollars, of a randomly selected secondary school teacher. Find
a. $P(x < 28)$.   b. $P(35 \leq x \leq 44)$.
c. Interpret your results in parts (a) and (b).

6.90 According to *The World Almanac,* the mean travel time to work in New York State is 29 minutes. Let x be the time, in minutes, that it takes a randomly selected New Yorker to get to work on a randomly selected day. If the travel times are normally distributed with a standard deviation of 9.3 minutes, find
a. $P(x < 45)$.   b. $P(20 \leq x \leq 30)$.
c. Interpret your results in parts (a) and (b).

6.91 The lifetime of a certain brand of flashlight battery is normally distributed with a mean of 30 hours and a standard deviation of 5.6 hours. Let x be the lifetime of a randomly selected flashlight battery of this brand. Determine
a. $P(x > 20)$.   b. $P(15 \leq x \leq 45)$.
c. Interpret each of your results in parts (a) and (b) in terms of percentages.

6.92 A manufacturer of timepieces claims that the weekly error, in seconds, of the watches she makes has a normal distribution with a mean of 0 and a standard deviation of 1. Let x denote the amount of time, in seconds, that one of these watches is off at the end of a randomly selected week. Determine
a. $P(x < -1)$.   b. $P(x < -2$ or $x > 2)$.
c. Interpret your results in parts (a) and (b).

6.93 As reported by *Runner's World* magazine, the times of the finishers in the New York City 10-km run are normally distributed with a mean of 61 minutes and a standard deviation of 9 minutes. Let x be the time of a randomly selected finisher. Find
a. $P(x > 75)$.   b. $P(x < 50$ or $x > 70)$.
c. Interpret your results in parts (a) and (b).

6.94 The length of the western rattlesnake is normally distributed with a mean of 42 inches and a standard deviation of 2.04 inches. Let x be the length of one of these snakes selected at random. Determine
a. $P(x > 45)$.   b. $P(35 \leq x \leq 40)$.
c. Interpret each of your results in parts (a) and (b) in terms of percentages.

6.95 A company manufactures bolts approximately 10 mm in diameter to fit into a circular hole 10.4 mm in diameter. In reality, the diameters of the bolts produced are normally distributed with a mean of 10 mm and a standard deviation of 0.1 mm. Let x be the diameter of a randomly selected bolt. Apply the empirical rule for normally distributed random variables, Key Fact 6.6 on page 372, to make some pertinent statements about the diameters of the bolts produced. (See Example 6.21 on page 373 for a model.)

6.96 An electronics company administers a general aptitude test to all prospective employees. The test is designed to take roughly 1 hour to complete. Experience suggests that the completion times are normally distributed with a mean of 62.4 minutes and a standard deviation of 7.2 minutes. Let x denote the time it takes a randomly selected applicant to complete the test. Apply the empirical rule for normally distributed random variables, Key Fact 6.6 on page 372, to make some pertinent statements about the completion times for the aptitude test.

6.97 Refer to Exercise 6.91. Use the empirical rule to obtain some information about the lifetimes of the flashlight batteries.

6.98 Refer to Exercise 6.92. Use the empirical rule to obtain some information about the weekly error of one of the manufacturer's watches.

6.99 According to *Statistical Report,* published by the U.S. Bureau of Prisons, the mean time served by prisoners released from federal institutions for the first time is 16.3 months. Assume the standard deviation of the times served is 17.9 months. Let x be the time served by a randomly selected prisoner released for the first time from a federal institution.
a. Find the standardized version of the random variable x.
b. How many standard deviations from the mean is a prison time served of 20.3 months? 64.7 months? 4.2 months?
c. Is it necessary to assume that the times served are normally distributed in order to answer parts (a) and (b)?

6.100 The Health Insurance Association of America compiles data on room charges in nongovernment, short-term, general hospitals. Results are published in *Survey of Hospital Semi-Private Room Charges.* According to that publication, the average daily charge for a semi-private hospital room was $253 in 1988. Assume a standard deviation of $45. Let x be the daily semi-private room charge of a randomly selected hospital in 1988.
a. Find the standardized version of the random variable x.

b. How many standard deviations from the mean is a daily room charge of $348? $107? $192?
c. Must we assume that the daily semi-private room charges are normally distributed to answer parts (a) and (b)?

6.101 Refer to Exercise 6.93.
a. Find the standardized version, z, of the random variable x.
b. What does the random variable z represent?
c. Identify the probability distribution of z.

6.102 Refer to Exercise 6.94.
a. Find the standardized version, z, of the random variable x.
b. What does the random variable z represent?
c. Identify the probability distribution of z.

6.103 Let x be a normally distributed random variable. Employ the general empirical rule, Key Fact 6.7 on page 373, to fill in the following blanks.
a. The probability is 0.95 that an observed value of x will be within _____ standard deviations to either side of the mean. *(Hint:* Here $1 - \alpha = 0.95$. Solve for α.)
b. The probability is 0.99 that an observed value of x will be within _____ standard deviations to either side of the mean.
c. Express your results in parts (a) and (b) using formulas like those in the empirical rule, Key Fact 6.6 on page 372.

6.104 Let x be a normally distributed random variable. Employ the general empirical rule, Key Fact 6.7 on page 373, to fill in the following blanks.
a. The probability is 0.90 that an observed value of x will be within _____ standard deviations to either side of the mean.
b. The probability is 0.85 that an observed value of x will be within _____ standard deviations to either side of the mean.
c. Express your results in parts (a) and (b) using formulas like those in the empirical rule, Key Fact 6.6 on page 372.

6.105 (Computer exercise) Refer to the simulation of contents of soda bottles discussed in Example 6.20 on page 371.
a. Use Minitab or some other statistical software to repeat the simulation and to obtain a histogram of the results.
b. Compare the histogram you obtained in part (a) to the theoretical distribution in Fig. 6.31 on page 370.
c. What does part (b) illustrate?

6.106 (Computer exercise) Refer to Exercise 6.94, where x denotes the length of a randomly selected western rattlesnake.
a. Sketch the normal curve for the random variable x.
b. Use Minitab or some other statistical software to simulate 1500 independent observations of the random variable x and to obtain a histogram of the lengths obtained.

c. Compare the histogram you obtained in part (b) to the theoretical distribution in part (a).

d. What does part (c) illustrate?

· ·   **6.107**   This exercise concerns the filling of soda bottles discussed in Example 6.19 on page 370. Recall that although the advertised content is 354 mL, the amount of soda actually put into a bottle is normally distributed with a mean of 356 mL and a standard deviation of 1.63 mL. To what setting should the mean be changed in order to ensure that only 1% of the bottles will contain less than the advertised content of 354 mL?

· ·   **6.108**   Refer to Exercise 6.107. Assuming the mean of 356 mL remains unchanged, what standard deviation must the filling process have in order to ensure that only 1% of the bottles will contain less than the advertised content of 354 mL?

· ·   **6.109**   Verify the empirical rule for normally distributed random variables, Key Fact 6.6 on page 372.

· · ·   **6.110**   Provide a proof of the general empirical rule for normally distributed random variables, Key Fact 6.7 on page 373. (*Hint:* You will need to determine the z-scores for the x-values $\mu_x \pm z_{\alpha/2} \cdot \sigma_x$.)

6.5   NORMAL PROBABILITY PLOTS

We have now seen how to work with normally distributed populations and normally distributed random variables. For instance, we know how to determine percentages and obtain percentiles for a normally distributed population.

Another problem involves deciding whether a population or random variable is normally distributed, or at least approximately so. Such decisions often play a major role in subsequent analyses—from percentage or percentile calculations to statistical inferences.

As we learned in Key Fact 2.1 on page 88, the distribution of a random sample from a population will approximate the distribution of the population, with larger samples tending to provide better approximations. We can use that key fact to help decide whether a population is normally distributed—if a population is normally distributed, then a graph of a sample from the population should reflect that.

A histogram or stem-and-leaf diagram of a large sample from a normally distributed population should be roughly bell-shaped; for a very large sample, even moderate departures from a bell shape cast doubt on the normality of the population. On the other hand, for a small sample, say of size 15 or less, it is often difficult to ascertain a clear shape in a histogram or stem-and-leaf diagram, and in particular whether a histogram or stem-and-leaf diagram is bell-shaped. So in this latter case, a more sensitive graphical technique is required for assessing normality. Such a technique is provided by normal probability plots.

Essentially, a **normal probability plot** is a plot of the sample data versus the data we would expect to get by taking a sample of the same size from a standard normal distribution. These latter data are called **normal scores.** If the sample is from a normally distributed population, then the normal probability plot should be roughly linear (i.e., fall roughly in a straight line), and vice versa.

In employing the normal probability plot of a sample to assess the normality of a population, we must remember two things: that the decision of whether or not a normal probability plot is roughly linear is a subjective one, and that we are using a sample to make a judgment about a population. With these considerations in mind, we present the following guidelines for assessing normality.

KEY FACT 6.9   **GUIDELINES FOR ASSESSING NORMALITY USING NORMAL PROBABILITY PLOTS**

To assess the normality of a population, construct a normal probability plot for the sample data.

- If the plot is roughly linear, then accept as reasonable that the population is approximately normally distributed.

- If the plot shows systematic deviations from linearity (e.g., if it displays significant curvature), then conclude that the population is probably not approximately normally distributed.

These guidelines should be interpreted loosely for small samples, but can be interpreted rather strictly for large samples.

Key Fact 6.9 applies to random variables as well as to populations: To assess the normality of a random variable, construct a normal probability plot for the observations. If the plot is roughly linear, then accept as reasonable that the random variable is approximately normally distributed; if the plot shows systematic deviations from linearity, then conclude that the random variable is probably not approximately normally distributed.

In practice, normal probability plots are constructed by computer. However, to really understand normal probability plots, it is helpful to draw some by hand. To do so, we have provided in Table III of Appendix A the normal scores for sample sizes from 5 to 30. Example 6.25 explains how Table III can be used to obtain a normal probability plot.

Example 6.25   Constructing a Normal Probability Plot

The Internal Revenue Service publishes data on federal individual income tax returns in *Statistics of Income, Individual Income Tax Returns*. A sample of 12 returns revealed the adjusted gross incomes, in thousands of dollars, shown in Table 6.4.

TABLE 6.4
Adjusted gross incomes ($1000s)

9.7	93.1	33.0	21.2
81.4	51.1	43.5	10.6
12.8	7.8	18.1	12.7

a. Construct a normal probability plot for these data.

b. Assess the normality of adjusted gross incomes.

SOLUTION   a. To construct a normal probability plot, we first arrange the data in increasing order and obtain the normal scores from Table III. The ordered data are shown in the first column of Table 6.5. The normal scores are shown in the second column of Table 6.5; these were obtained from the $n = 12$ column of Table III.

TABLE 6.5

Ordered data and normal scores

Adjusted gross income	Normal score
7.8	−1.64
9.7	−1.11
10.6	−0.79
12.7	−0.53
12.8	−0.31
18.1	−0.10
21.2	0.10
33.0	0.31
43.5	0.53
51.1	0.79
81.4	1.11
93.1	1.64

Next we plot the points in Table 6.5, using the horizontal axis for the adjusted gross incomes and the vertical axis for the normal scores. For instance, the first point plotted has a horizontal coordinate of 7.8 and a vertical coordinate of −1.64. Figure 6.35 shows all 12 points from Table 6.5; this is the normal probability plot for the sample of adjusted gross incomes.

FIGURE 6.35

Normal probability plot for the sample of adjusted gross incomes

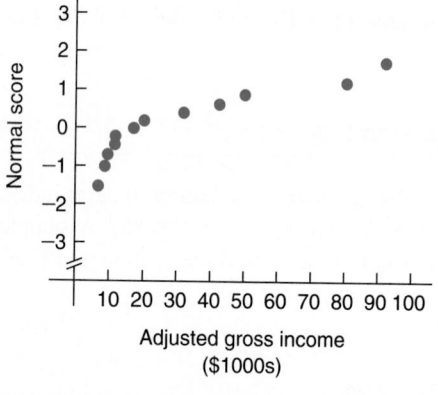

b. The normal probability plot in Fig. 6.35 displays significant curvature. Evidently, adjusted gross incomes are not approximately normally distributed. ■

It is helpful to know the shapes of normal probability plots for samples from commonly occurring distributions. Figure 6.36 displays normal (bell-shaped), uniform, right-skewed, and left-skewed distributions; it also shows smooth-curve representations of normal probability plots for samples from those distributions. But be aware that the shapes of these plots are based on ideal situations, namely, large samples from exact distributions. Use Fig. 6.36 only as a guide. Generally speaking, we must be careful about using a normal probability plot to infer the shape of a distribution, especially when the sample size is small.

FIGURE 6.36

Distribution shapes and normal probability plots for (a) normal, (b) uniform, (c) right-skewed, and (d) left-skewed distributions

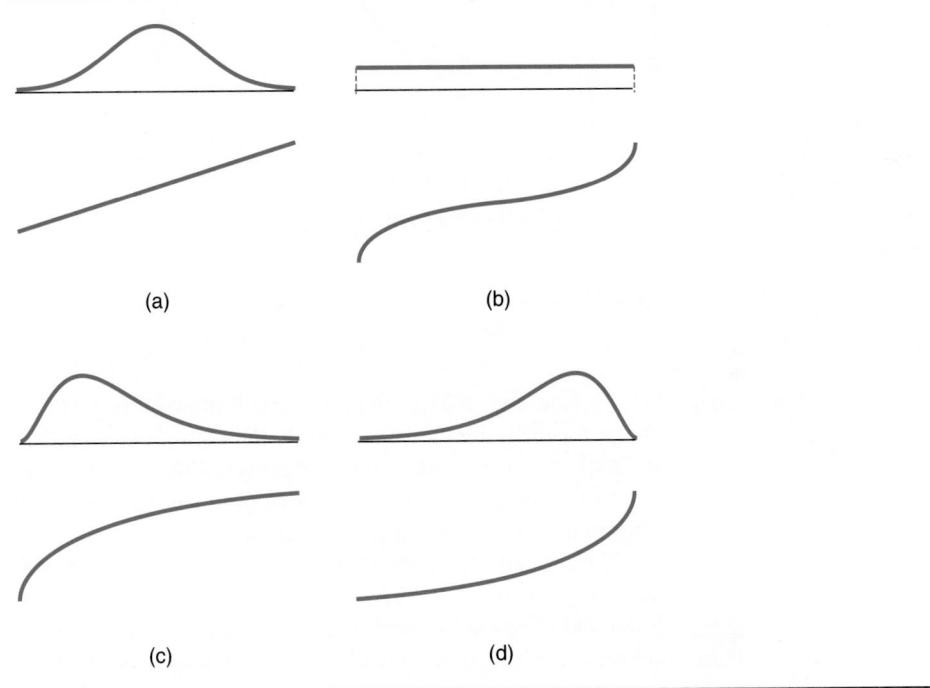

(a)   (b)

(c)   (d)

Detecting Outliers Using Normal Probability Plots

Recall that outliers are data values that fall well outside the overall pattern of a data set. We can use normal probability plots to detect outliers. Example 6.26 explains how.

Example 6.26   Using Normal Probability Plots to Detect Outliers

The U.S. Department of Agriculture publishes data on U.S. chicken consumption in *Food Consumption, Prices, and Expenditures*. Last year's chicken consumptions, in pounds, for 17 randomly selected people are displayed in Table 6.6. A normal probability plot for these data is presented in Fig. 6.37(a). Use the plot to discuss the distribution of chicken consumptions and to detect any outliers.

TABLE 6.6

Sample of last year's chicken consumptions (lbs)

47	39	62	49	50	70
59	53	55	0	65	63
53	51	50	72	45	

FIGURE 6.37

Normal probability plots
for chicken-consumptions:
(a) original data,
(b) data with outlier removed

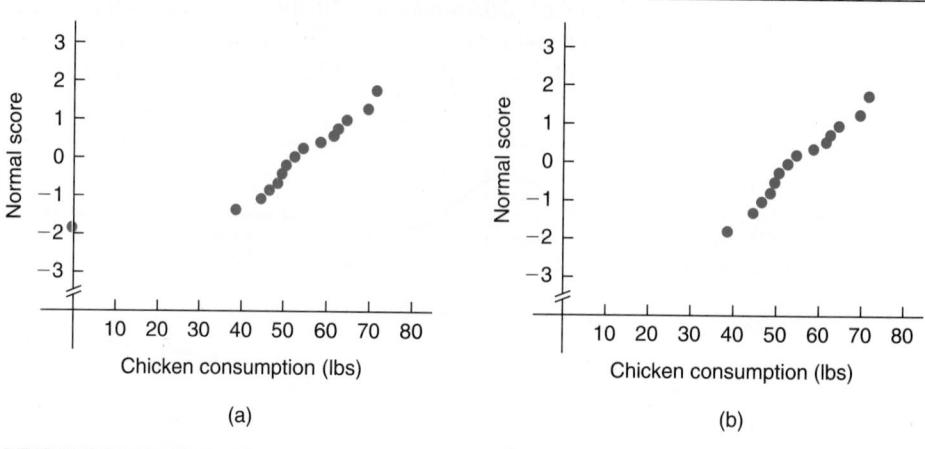

SOLUTION   We see from Fig. 6.37(a) that the normal probability plot falls roughly in a straight line except for the point corresponding to the consumption of 0 lb. That point on the normal probability plot falls well outside the overall pattern of the plot; hence 0 lb is an outlier. This outlier might be a recording error or due to a person in the sample who does not eat chicken for some reason (e.g., a vegetarian).

   If we remove the outlier 0 lb from the data set and draw a normal probability plot for the abridged data set, then, as we see from Fig. 6.37(b), this normal probability plot is quite linear and shows no outliers. It appears plausible that among people who eat chicken, the amounts they consume annually are approximately normally distributed.   ■

Note: There appears to be only 15 points (instead of the expected 17) in Fig. 6.37(a). This is because we used an averaging process to assign identical normal scores to the two 50s and identical normal scores to the two 53s. Thus there really are 17 points in the graph, but only 15 are distinguishable because there are two sets of two identical points. Minitab also uses the averaging process to deal with ties. Keep this in mind when you examine normal probability plots in figures and printouts. An alternative procedure for dealing with ties when obtaining normal scores is to treat equal data values as being slightly different; then no averaging process is required.

   In this section we have learned how a normal probability plot can be used as an aid for deciding whether a sample is from a population whose distribution is (approximately) normal. Although this visual assessment of normality is subjective, it will be sufficient for most of our analyses.[†]

[†] An objective method for assessing normality can be based on the correlation between the sample data and its normal scores. We will study that method in Section 14.6.

Using the Computer (Optional)

Minitab has a command called **NScores** that can be used to determine the normal scores for a data set. Following the application of the NScores command, we can employ Minitab's **Plot** command to obtain a normal probability plot. Example 6.27 shows how this is done.

Example 6.27   Using Minitab to Obtain a Normal Probability Plot

Use Minitab to obtain a normal probability plot for the adjusted gross incomes displayed in Table 6.4 on page 381.

SOLUTION   First we store the data from Table 6.4 in a column named AGI and name an available column NSCORES. Then we determine the normal scores as follows.

Session commands: We type the command **NSCORES** followed by the storage location of the data and the column in which the normal scores are to be stored; that is, we type `NSCORES 'AGI' put into 'NSCORES'` and press Enter.

(or)

Menu commands: We choose **Calc ▶ Functions and Statistics ▶ Functions...**, specify AGI in the **Input column** text box, specify NSCORES in the **Result in** text box, select **Normal scores**, and then select **OK**.

The normal scores for the adjusted gross incomes are now stored in NSCORES. Next we apply Minitab's Plot command to obtain the normal probability plot for the adjusted gross incomes, that is, a plot of the adjusted gross incomes versus the normal scores for that data.

Session commands: We type the command **PLOT** followed first by the storage location of the normal scores and then by the storage location of the sample data; that is, we type `PLOT 'NSCORES' versus 'AGI'` and press Enter.

(or)

Menu commands: We choose **Graph ▶ Scatter Plot...**, specify NSCORES in the **Vertical axis** text box, specify AGI in the **Horizontal axis** text box, select **Plotting symbol** and type * to indicate that an asterisk should be used as the symbol for plotting points,[†] select **High resolution** and disable it by pressing Spacebar,[‡] and then select **OK**.

The normal probability plot obtained by applying the above commands is shown in Printout 6.4. Compare that normal probability plot to the one we drew by hand in Fig. 6.35 on page 382. ∎

[†] We have chosen an asterisk as the plotting symbol only so the output will be consistent with the default plotting symbol used in session commands. As a consequence of this, your screen will display the subcommand `Symbol '*'` . between the second line and plot in Printout 6.4.

[‡] See the footnote on page 155.

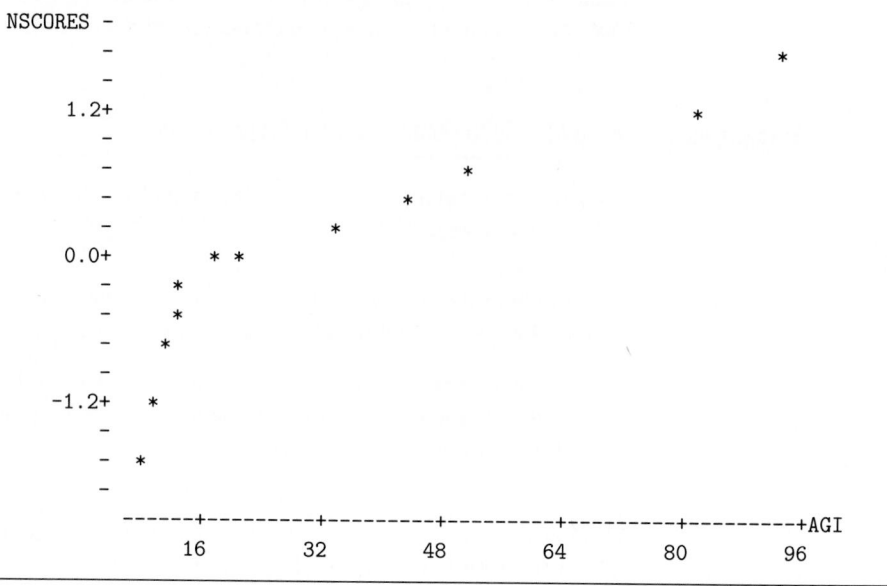

```
MTB > NSCORES 'AGI' put into 'NSCORES'
MTB > PLOT 'NSCORES' versus 'AGI'


NSCORES -
        -
        -
        -
   1.2+                                                               *
        -                                                    *
        -                                           *
        -                                      *
        -                                 *
        -                           *
   0.0+                   *   *
        -             *
        -             *
        -           *
        -
  -1.2+         *
        -
        -     *
        -
        ------+---------+---------+---------+---------+---------+---------+AGI
             16        32        48        64        80        96
```

EXERCISES 6.5

In each of Exercises 6.111–6.118,
a. use Table III in Appendix A to construct a normal probability plot for the data. For simplicity, treat equal data values as being slightly different when obtaining normal scores.
b. use part (a) to identify any outliers.
c. use part (a) to assess normality.

6.111   A sample of the final-exam scores in a large introductory statistics course is as follows.

88	67	64	76	86
85	82	39	75	34
90	63	89	90	84
81	96	100	70	96

6.112   As reported by the R. R. Bowker Company of New York in *Library Journal,* the mean annual subscription rate to law periodicals was $39.82 in 1987. A random sample of 12 of this year's law periodicals provided the following data on subscription rates.

$30	46	44	47
42	38	62	55
52	48	43	54

6.113   The U.S. Federal Highway Administration conducts studies on motor vehicle travel by type of vehicle. Results are published annually in *Highway Statistics*. A sample of 15 cars yields the following data on number of miles driven, in thousands, for last year.

10.2	10.3	8.9	12.7	8.3
9.2	13.7	7.7	3.3	10.6
11.8	6.6	8.6	5.7	12.0

6.114   The Bureau of Labor Statistics publishes information on average annual expenditures by consumers in

Consumer Expenditure Survey. In 1989 the mean amount spent by consumers on nonalcoholic beverages was $216. A random sample of 12 consumers yielded the following data, in dollars, on last year's expenditures on nonalcoholic beverages.

361	176	184	265
259	281	240	273
259	249	194	258

· **6.115** The U.S. Energy Information Administration reports figures on residential energy consumption and expenditures in *Residential Energy Consumption Survey: Consumption and Expenditures.* A sample of 18 households using electricity as their primary energy source yields the following data on one year's energy expenditures.

$1376	1452	1235	1480	1185	1327
1059	1400	1227	1102	1168	1070
949	1351	1259	1179	1393	1456

· **6.116** In January 1984 the U.S. Department of Agriculture reported in *Family Economic Review* that a typical U.S. family of four with an intermediate budget spent about $92 per week for food. A consumer researcher in Kansas suspected the median weekly cost was less in her state. She took a sample of 10 Kansas families of four, each with an intermediate budget, and obtained the weekly food costs shown in the following table.

$78	104	84	70	96
73	87	85	76	94

· **6.117** The U.S. National Center for Health Statistics compiles data on the length of stay by patients in short-term hospitals and publishes its findings in *Vital and Health Statistics.* A random sample of 21 patients yielded the following data on length of stay. The data are in days.

4	4	12	18	9	6	12
3	6	15	7	3	55	1
10	13	5	7	1	23	9

· **6.118** IQs measured on the Stanford Revision of the Binet-Simon Intelligence Scale have a mean of 100 points and a standard deviation of 16 points. Twenty-five randomly se-

lected people are given the IQ test; the results are shown in the following table.

91	96	106	116	97
102	96	124	115	121
95	111	105	101	86
88	129	112	82	98
104	118	127	66	102

In each of Exercises 6.119 and 6.120, we have drawn a normal probability plot for a large sample from a population. For each exercise, use Fig. 6.36 on page 383 to identify the approximate shape of the distribution of the population.

· **6.119**

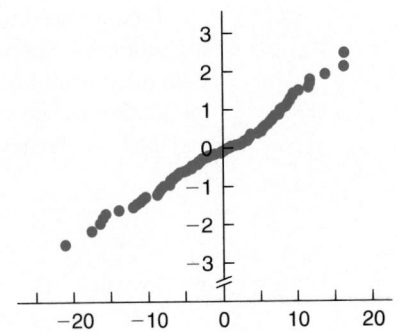

· **6.120**

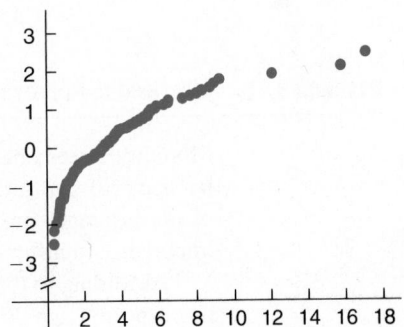

· **6.121 (Computer exercise)** Use Minitab or some other statistical software to obtain a normal probability plot for the exam scores in Exercise 6.111.

· **6.122 (Computer exercise)** Use Minitab or some other statistical software to obtain a normal probability plot for the subscription rates in Exercise 6.112.

·   **6.123 (Computer exercise)** Use Minitab or some other statistical software to obtain a normal probability plot for the data in Exercise 6.113 on number of miles driven.

·   **6.124 (Computer exercise)** Use Minitab or some other statistical software to obtain a normal probability plot for the expenditures in Exercise 6.114.

6.6   THE NORMAL APPROXIMATION TO THE BINOMIAL DISTRIBUTION

In this section we will discover that it is often possible to approximate binomial probabilities by areas under a normal curve. The development of the mathematical theory for doing this is credited to Abraham de Moivre (1667–1754) and Pierre-Simon Laplace (1749–1827). For more information on de Moivre and Laplace, see the biographies at the beginning of Chapters 11 and 7, respectively.

First we need to briefly review the binomial distribution, which we discussed in detail in Section 5.4. Suppose n identical independent success-failure experiments are performed, with the probability of success on any given trial being equal to p. Let x denote the total number of successes in the n trials. Then the probability distribution of the random variable x is given by the binomial probability formula,

$$P(x) = \binom{n}{x} p^x (1-p)^{n-x}.$$

We say that x has the binomial distribution with parameters n and p.

You might be wondering why we would use normal-curve areas to approximate binomial probabilities when we can obtain them exactly by employing the binomial probability formula. Example 6.28 explains why.

Example 6.28   The Need to Approximate Binomial Probabilities

Mortality tables enable actuaries to obtain the probability that a person at any particular age will live a specified number of years. This, in turn, permits the determination of life-insurance premiums, retirement pensions, annuity payments, and related items of importance to insurance companies and others.

According to the Department of Health and Human Services, there is an 80% chance that a person age 70 will be alive at age 75. Suppose 500 people age 70 are selected at random. Find the probability that

a. exactly 390 of them will be alive at age 75.

b. between 375 and 425 of them, inclusive, will be alive at age 75.

SOLUTION   Let x denote the number of people out of the 500 that will be alive at age 75. Then x has the binomial distribution with parameters $n = 500$ (the 500 people) and $p = 0.8$ (the probability a person age 70 will be alive at age 75). Thus probabilities for x can be determined exactly by using the binomial probability formula,

$$P(x) = \binom{500}{x}(0.8)^x(0.2)^{500-x}.$$

Let's apply that formula to the problems posed in parts (a) and (b).

a. Here we want the probability that exactly 390 of the 500 people will still be alive at age 75, $P(x = 390)$. The "answer" is

$$P(390) = \binom{500}{390}(0.8)^{390}(0.2)^{110}.$$

However, to actually obtain the numerical value of the expression on the right of the preceding equation is not easy, even with a calculator. In performing the computations, we must be careful to avoid such pitfalls as making roundoff errors and getting numbers so large or so small that they are outside the range of the calculator. Fortunately, as we will soon see, the computations can be sidestepped altogether by using normal-curve areas in a simple way.

b. For this part we need to determine the probability that between 375 and 425, inclusive, of the 500 people will be alive at age 75, $P(375 \le x \le 425)$. The "answer" is

$$P(375 \le x \le 425) = P(375) + P(376) + \cdots + P(425)$$

$$= \binom{500}{375}(0.8)^{375}(0.2)^{125} + \binom{500}{376}(0.8)^{376}(0.2)^{124}$$

$$+ \cdots + \binom{500}{425}(0.8)^{425}(0.2)^{75}.$$

We have the same computational difficulties as we did in part (a), except that here we must evaluate 51 complex expressions instead of 1. Thus, although in theory we can use the binomial probability formula to determine the answer, doing so in practice is another matter.

Surprising as it might seem, there is a way to use normal-curve areas to get the (approximate) answer—and it is easy! We will return to this problem momentarily and obtain the probabilities required in parts (a) and (b). ∎

Example 6.28 makes it clear why we often need to approximate binomial probabilities, even though the binomial probability formula is available for computing them exactly: *It is not practical to use the binomial probability formula when the number of trials, n, is very large.*

Under certain conditions on n and p, the distribution of a binomial random variable is (roughly) bell-shaped. In such cases we can approximate probabilities for the random variable by areas under a suitable normal curve. To show how this is done, we present Example 6.29. For this example it is actually easy to calculate binomial probabilities exactly using the binomial probability formula. However, for purposes of illustration, we will also show how normal-curve areas approximate the binomial probabilities.

Example 6.29   Approximating Binomial Probabilities by Areas Under a Normal Curve

---

A student is taking a true-false exam with 10 questions. Assuming the student guesses at all 10 questions, use the binomial probability formula to determine the exact probability that the student gets either 7 or 8 answers correct. Then approximate that probability by an area under a suitable normal curve.

SOLUTION   Let x be the number of correct guesses by the student. There are 10 questions, so $n = 10$. Since the student guesses at each question, the success probability, p, is 0.5. Hence x has the binomial distribution with parameters $n = 10$ and $p = 0.5$. In other words, probabilities for x are given by the binomial probability formula,

$$P(x) = \binom{10}{x}(0.5)^x(1 - 0.5)^{10-x}.$$

Applying that formula, we obtain the probability distribution of x, shown in Table 6.7.

TABLE 6.7

Probability distribution of the number of correct guesses by the student

Number correct x	Probability $P(x)$
0	0.0010
1	0.0098
2	0.0439
3	0.1172
4	0.2051
5	0.2461
6	0.2051
7	0.1172
8	0.0439
9	0.0098
10	0.0010

The problem is to determine the probability that the student gets either 7 or 8 questions correct, $P(x = 7 \text{ or } 8)$. From Table 6.7 we find that the exact probability is equal to

$$P(x = 7 \text{ or } 8) = P(7) + P(8) = 0.1172 + 0.0439 = 0.1611.$$

Let's now see how we can approximate the probability $P(x = 7 \text{ or } 8)$ by an area under a suitable normal curve. To begin, we refer to Table 6.7 in order to obtain the probability histogram for the random variable x, as seen in Fig. 6.38.

FIGURE 6.38

Probability histogram for x with superimposed normal curve

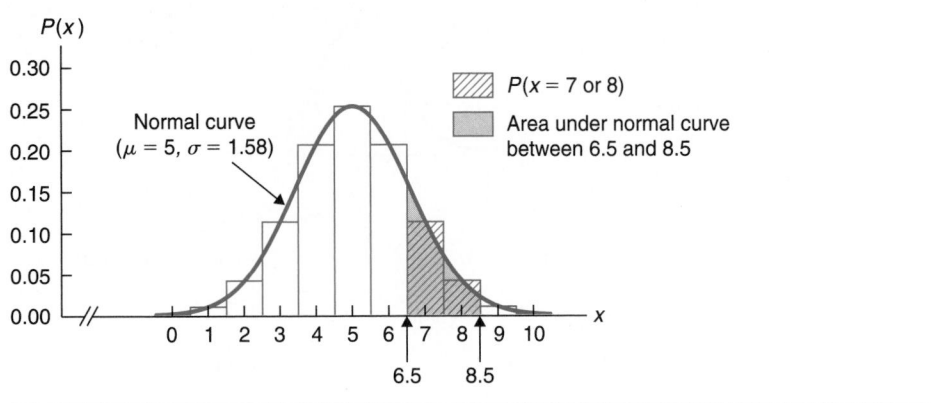

Because the probability histogram in Fig. 6.38 is bell-shaped, probabilities for x can be approximated by areas under a normal curve. As expected, the appropriate normal curve is the one whose parameters are the same as the mean and standard deviation of the random variable x. Since x has the binomial distribution with parameters $n = 10$ and $p = 0.5$, we can apply Formula 5.3 (page 314) and Formula 5.4 (page 315) to easily obtain the mean and standard deviation of x:

$$\mu_x = np = 10 \cdot 0.5 = 5$$

and

$$\sigma_x = \sqrt{np(1 - p)} = \sqrt{10 \cdot 0.5 \cdot (1 - 0.5)} = 1.58.$$

So the normal curve used here is the one with parameters $\mu = 5$ and $\sigma = 1.58$. That normal curve is superimposed on the probability histogram in Fig. 6.38.

The probability $P(x = 7 \text{ or } 8)$ is exactly equal to the combined area of the corresponding bars of the histogram, the cross-hatched area in Fig. 6.38. By examining the figure carefully, we observe that the cross-hatched area is approximately equal to the area under the normal curve between 6.5 and 8.5, the shaded area in Fig. 6.38. It should be clear from Fig. 6.38 why we consider the area under the normal curve between 6.5 and 8.5 instead of between 7 and 8. This is called the **correction for continuity** because it is a correction required as a result of approximating a discrete probability distribution by a continuous probability distribution.

Thus we see, at least qualitatively, that the probability $P(x = 7 \text{ or } 8)$ is approximately equal to the area under the normal curve with parameters $\mu = 5$ and $\sigma = 1.58$ that lies between 6.5 and 8.5. To compare these values quantitatively, we must obtain the normal-curve area. This is done in the usual way, as depicted in Fig. 6.39.

FIGURE 6.39

Determination of the area
under the normal curve with
parameters $\mu = 5$ and $\sigma = 1.58$
that lies between 6.5 and 8.5

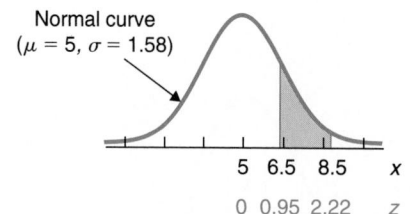

Normal curve
($\mu = 5$, $\sigma = 1.58$)

5   6.5   8.5 x

0   0.95   2.22 z

z-score computations: Area to the left of z:

$x = 6.5 \longrightarrow z = \dfrac{6.5 - 5}{1.58} = 0.95$ 0.8289

$x = 8.5 \longrightarrow z = \dfrac{8.5 - 5}{1.58} = 2.22$ 0.9868

Shaded area $= 0.9868 - 0.8289 = 0.1579$

The last line in Fig. 6.39 shows that the area under the normal curve between 6.5 and 8.5 is 0.1579. Comparing that area to the exact value of $P(x = 7$ or $8)$, which is 0.1611, we see that the normal-curve area provides an excellent approximation of the exact probability. ∎

As illustrated in Example 6.29, we can use normal-curve areas to approximate probabilities for binomial random variables that have bell-shaped distributions. Whether or not a particular binomial random variable has a bell-shaped distribution depends on its parameters, n and p. Figure 6.40 shows nine different binomial distributions.

As portrayed by Figs. 6.40(a) and 6.40(c), a binomial distribution with $p \neq 0.5$ is skewed. For small n, the skewness is enough to preclude using a normal approximation. However, as n increases, the skewness subsides and the binomial distribution becomes sufficiently bell-shaped to permit a normal approximation. On the other hand, as illustrated in Fig. 6.40(b), a binomial distribution with $p = 0.5$ is symmetric, regardless of the number of trials. Nonetheless, such a distribution will not be sufficiently bell-shaped to permit a normal approximation if n is too small.

The customary rule of thumb for using the normal approximation is that *both np and $n(1 - p)$ are at least 5*. This indicates, as suggested in Fig. 6.40, that the further the success probability is from 0.5 (in either direction), the larger the number of trials must be in order to employ the normal approximation.

A Procedure for Approximating Binomial Probabilities by Normal-Curve Areas

By examining carefully what we did in Example 6.29, we can write a general procedure for approximating binomial probabilities by areas under a normal curve. This is presented as Procedure 6.3.

FIGURE 6.40   Nine different binomial distributions

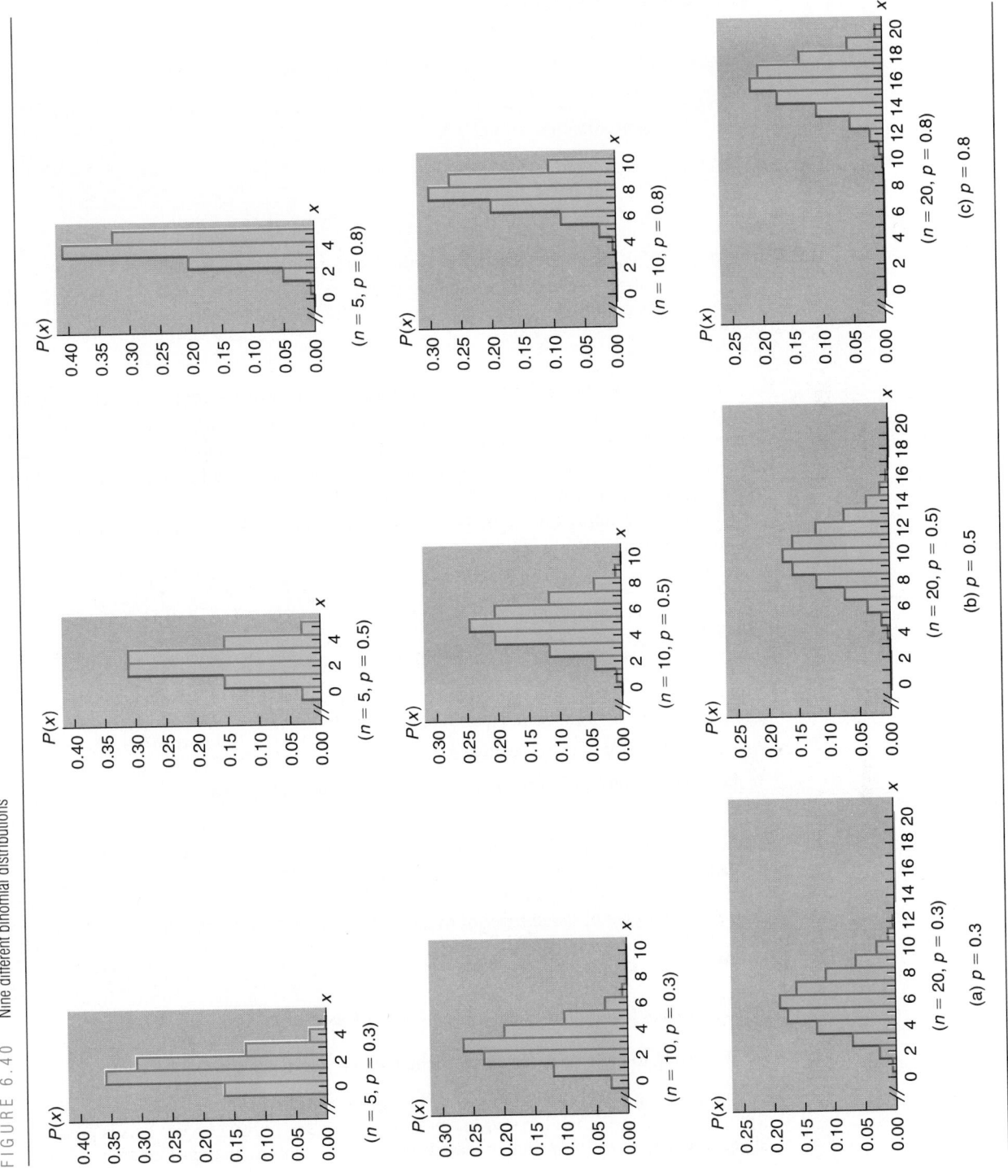

PROCEDURE 6.3	**TO APPROXIMATE BINOMIAL PROBABILITIES BY NORMAL-CURVE AREAS**

Step 1 Determine n, the number of trials, and p, the success probability.

Step 2 Check that both np and $n(1 - p)$ are at least 5. If they are not, the normal approximation should not be used.

Step 3 Find μ_x and σ_x using the formulas

$$\mu_x = np \quad \text{and} \quad \sigma_x = \sqrt{np(1 - p)}.$$

Step 4 Make the correction for continuity and find the required area under the normal curve with parameters μ_x and σ_x.

Step 4 of Procedure 6.3 requires us to make the correction for continuity. As illustrated in Example 6.29, this means the following: When using normal-curve areas to approximate the probability that an observed value of a binomial random variable will be between two integers, inclusive, subtract 0.5 from the smaller integer and add 0.5 to the larger integer before finding the area under the normal curve.

We will now apply Procedure 6.3 to solve the problems posed at the beginning of this section in Example 6.28.

Example 6.30   Illustrates Procedure 6.3

The probability is 0.80 that a person age 70 will be alive at age 75. Suppose 500 people age 70 are selected at random. Determine the probability that

a. exactly 390 of them will be alive at age 75.

b. between 375 and 425 of them, inclusive, will be alive at age 75.

SOLUTION   We will obtain the approximate values of the probabilities in parts (a) and (b) by applying Procedure 6.3.

Step 1   *Determine n, the number of trials, and p, the success probability.*

We have $n = 500$ and $p = 0.8$.

Step 2   *Check that both np and $n(1 - p)$ are at least 5.*

Referring to the values for n and p obtained in Step 1, we see that

$$np = 500 \cdot 0.8 = 400 \quad \text{and} \quad n(1 - p) = 500 \cdot 0.2 = 100.$$

Since both np and $n(1 - p)$ are at least 5, we can use the normal approximation.

Step 3   *Find μ_x and σ_x using the formulas*

$$\mu_x = np \quad \text{and} \quad \sigma_x = \sqrt{np(1 - p)}.$$

We have

$$\mu_x = np = 500 \cdot 0.8 = 400$$

and

$$\sigma_x = \sqrt{np(1 - p)} = \sqrt{500 \cdot 0.8 \cdot 0.2} = 8.94.$$

Step 4   *Make the correction for continuity and find the required area under the normal curve with parameters μ_x and σ_x.*

a. Here we want the probability that exactly 390 of the 500 people selected will be alive at age 75, that is, $P(x = 390)$. To make the correction for continuity, we subtract 0.5 from 390 and add 0.5 to 390. Thus we need to find the area under the normal curve with parameters $\mu = 400$ and $\sigma = 8.94$ that lies between 389.5 and 390.5. The required area is obtained in Fig. 6.41.

FIGURE 6.41

Determination of the area under the normal curve with parameters $\mu = 400$ and $\sigma = 8.94$ that lies between 389.5 and 390.5

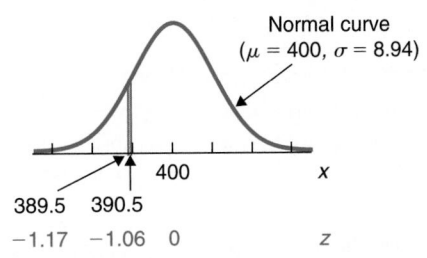

Normal curve
($\mu = 400$, $\sigma = 8.94$)

389.5   390.5   400   *x*

-1.17 -1.06 0   *z*

z-score computations: Area to the left of *z*:

$x = 389.5 \longrightarrow z = \dfrac{389.5 - 400}{8.94} = -1.17$ 0.1210

$x = 390.5 \longrightarrow z = \dfrac{390.5 - 400}{8.94} = -1.06$ 0.1446

Shaded area = 0.1446 − 0.1210 = 0.0236

So $P(x = 390) = 0.0236$, approximately. The probability is roughly 0.0236 that exactly 390 of the 500 people selected will be alive at age 75.

b. For this part we want the probability that between 375 and 425, inclusive, of the 500 people selected will be alive at age 75, $P(375 \le x \le 425)$. To make the correction for continuity, we subtract 0.5 from 375 and add 0.5 to 425. Hence we need to find the area under the normal curve with parameters $\mu = 400$ and $\sigma = 8.94$ that lies between 374.5 and 425.5. This area is determined in Fig. 6.42.

FIGURE 6.42

Determination of the area under
the normal curve with parameters
$\mu = 400$ and $\sigma = 8.94$ that
lies between 374.5 and 425.5

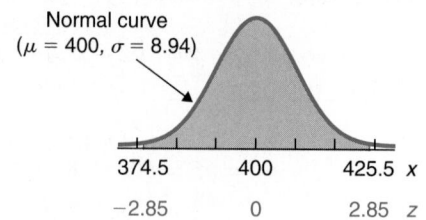

z-score computations:

$x = 374.5 \longrightarrow z = \dfrac{374.5 - 400}{8.94} = -2.85$ 　　Area to the left of z:　 0.0022

$x = 425.5 \longrightarrow z = \dfrac{425.5 - 400}{8.94} = 2.85$ 　　 0.9978

Shaded area = 0.9978 − 0.0022 = 0.9956

Consequently, $P(375 \leq x \leq 425) = 0.9956$, approximately. The probability that between 375 and 425 of the 500 people selected will be alive at age 75 is approximately equal to 0.9956. ■

EXERCISES 6.6

6.125 Why do we sometimes use normal-curve areas to approximate binomial probabilities even though we have a formula for computing them exactly?

6.126 The rule of thumb for using the normal approximation to the binomial is that both np and $n(1 - p)$ are at least 5. Why is this restriction necessary?

6.127 Refer to Example 6.29 on page 390.
a. Use Table 6.7 to find the exact probability that the student guesses correctly on
  i. 4 or 5 questions, $P(x = 4 \text{ or } 5)$.
  ii. between 3 and 7 questions, $P(3 \leq x \leq 7)$.
b. Apply Procedure 6.3 on page 394 to approximate the probabilities in part (a) by areas under a normal curve. Compare your answers.

6.128 Refer to Example 6.29 on page 390.
a. Use Table 6.7 to obtain the exact probability that the student guesses correctly on
  i. at most 5 questions, $P(0 \leq x \leq 5)$.
  ii. at least 6 questions, $P(6 \leq x \leq 10)$.

b. Apply Procedure 6.3 on page 394 to approximate the probabilities in part (a) by areas under a normal curve. Compare your answers.

6.129 If in Example 6.29 the true-false exam had 30 questions instead of 10, which normal curve would you use to approximate probabilities for the number of correct guesses?

6.130 If in Example 6.29 the true-false exam had 25 questions instead of 10, which normal curve would you use to approximate probabilities for the number of correct guesses?

In Exercises 6.131–6.138, apply Procedure 6.3 to approximate the required binomial probabilities.

6.131 According to the *Daily Racing Form,* the probability is 0.67 that the favorite in a horse race will finish in the money (first, second, or third place). Determine the probability that in the next 200 races the favorite will finish in the money
a. exactly 140 times.
b. between 120 and 130 times, inclusive.
c. at least 150 times.

6.132 As reported by a spokesperson for Southwest Airlines, the no-show rate for reservations is 16%. In other words, the probability is 0.16 that a party making a reservation will not show up. The next flight has 42 parties with reservations. What is the probability that
a. exactly 5 parties do not show up?
b. between 9 and 12, inclusive, do not show up?
c. at least 1 does not show up?
d. at most 2 do not show up?

6.133 The U.S. National Center for Health Statistics states in *Vital and Health Statistics* that 38.3% of all injuries in the United States occur at home. Out of 500 randomly selected injuries, what is the probability that the number occurring at home will be
a. exactly 200?
b. between 180 and 210, inclusive?
c. at most 225?

6.134 *The World Almanac* reports that the infant mortality rate in India is 139 per 1000 live births. Determine the probability that out of 1000 randomly selected live births in India, there are
a. exactly 139 infant deaths.
b. between 120 and 150 infant deaths, inclusive.
c. at most 130 infant deaths.

6.135 A mail-order firm receives orders on approximately 12% of its mailings. If 750 brochures are mailed, find the probability that the number resulting in orders will be
a. exactly 100.
b. either less than 80 or more than 105.
c. at least 14% of the number of brochures mailed.

6.136 Data obtained from the *Statistical Abstract of the United States* indicate that there is a 51.3% chance that a baby born in the United States will be male. Find the probability that out of the next 10,000 births, at least half will be male.

6.137 A roulette wheel consists of 38 numbers, of which 18 are red, 18 are black, and 2 are green. When the roulette ball is spun, it is equally likely to land on each of the 38 numbers. A gambler is playing roulette and bets $10 on red each time. If the ball lands on a red number, the gambler wins $10 from the house; otherwise, the gambler loses $10. What is the probability that the gambler will be ahead after
a. 100 bets?   b. 1000 bets?   c. 5000 bets?
(Hint: The gambler will be ahead after a series of bets if and only if she has won more than half of the bets.*)*

6.138 A brand of flashlight battery has normally distributed lifetimes with a mean of 30 hours and a standard deviation of 5 hours. A supermarket purchases 500 of these batteries from the manufacturer. What is the probability that at least 80% of them will last longer than 25 hours?

Chapter Review

Key Terms

CDF,* *344*
CDF,* *344*
continuous probability distribution, *335*
continuous random variable, *335*
correction for continuity, *391*
destandardizing, *355*
discrete random variable, *335*
empirical rule, *364, 372*
general empirical rule, *373*
InvCDF,* *344*
INVCDF,* *345*
Normal,* *344, 377*
NORMAL,* *344, 378*
normal curve, *336*
normal distribution, *335*

normal probability plot, *380*
normal scores, *380*
normally distributed population, *361*
normally distributed random
   variable, *369*
NScores,* *385*
NSCORES,* *385*
percentiles, *367*
Plot,* *385*
PLOT,* *385*
population z-score, *362*
quartiles, *365*
Random,* *377*
RANDOM,* *378*
standard normal curve, *336*

standard normal distribution, *376*
standardized normally distributed
   random variable, *377*
standardized version of a random
   variable, *376*

standardizing, *352*
z_α, *342*
z-curve, *336*
z-scores, *352*

Formulas z-score for an x-value, *352, 362*

$$z = \frac{x - \mu}{\sigma}$$

(μ and σ are the parameters for a normal curve or the mean and standard deviation of a population)

x-value for a z-score, *354*

$$x = \mu + z \cdot \sigma$$

(μ and σ are the parameters for a normal curve or the mean and standard deviation of a population)

Standardized version of a random variable x, *376*

$$z = \frac{x - \mu_x}{\sigma_x}$$

(μ_x = mean of x, σ_x = standard deviation of x)

You Should Be Able To

1. use and understand the preceding formulas.
2. find areas under the standard normal curve using Table II in Appendix A.
3. use Table II in Appendix A to find the z-value(s) corresponding to a specified area under the standard normal curve.
4. explain the meaning of the parameters μ and σ for a normal curve.
5. sketch a normal curve.
6. find areas under a normal curve.
7. find the x-value(s) corresponding to a specified area under a normal curve.
8. determine percentages for a normally distributed population.
9. obtain and interpret a population z-score.
10. state and apply the empirical rule for normally distributed populations.
11. find quartiles and percentiles for a normally distributed population.
12. determine probabilities for a normally distributed random variable.
13. state and apply the empirical rule for normally distributed random variables.
14. state and apply the general empirical rule for normally distributed random variables.
15. obtain and interpret the standardized version of a random variable.
16. explain how to assess the normality of a population or random variable using a normal probability plot.
17. construct a normal probability plot with the aid of Table III in Appendix A.
18. identify distribution shapes from normal probability plots.
19. detect outliers from normal probability plots.
20. approximate binomial probabilities by normal-curve areas, when appropriate.
21. use the Minitab commands covered in this chapter.*
22. interpret the output obtained from the application of the Minitab commands discussed in this chapter.*

REVIEW TEST

1. Identify two primary reasons for studying the normal distribution.

2. Determine and sketch the area under the standard normal curve that lies
   a. to the left of -3.02.
   b. to the right of 0.61.
   c. between 1.11 and 2.75.
   d. between -2.06 and 5.02.
   e. between -4.11 and -1.5.
   f. either to the left of 1 or to the right of 3.

*3. **(Computer problem)** Use Minitab or some other statistical software to find the areas required in Problem 2.

4. For the standard normal curve, find the z-value(s)
   a. having area 0.30 to its left.
   b. having area 0.10 to its right.
   c. $z_{0.025}$, $z_{0.05}$, $z_{0.01}$, and $z_{0.005}$.
   d. that divide the area under the curve into a middle 0.99 area and two outside 0.005 areas.

*5. **(Computer problem)** Use Minitab or some other statistical software to find the z-values required in Problem 4.

6. Sketch the normal curve with parameters
   a. $\mu = -1$ and $\sigma = 2$.   b. $\mu = 3$ and $\sigma = 2$.
   c. $\mu = -1$ and $\sigma = 0.5$.

7. Consider the normal curves with the following parameters: $\mu = 1.5$ and $\sigma = 3$; $\mu = 1.5$ and $\sigma = 6.2$; $\mu = -2.7$ and $\sigma = 3$; $\mu = 0$ and $\sigma = 1$.
   a. Which curve has the largest spread?
   b. Which curves are centered at the same place?
   c. Which curves have the same shape?
   d. Which curve is centered farthest to the left?
   e. Which curve is the standard normal curve?

8. Determine the area under the normal curve with parameters $\mu = -1$ and $\sigma = 2.5$ that lies
   a. between 2 and 6.   b. to the right of -5.6.
   c. to the right of 3.2.

*9. **(Computer problem)** Use Minitab or some other statistical software to find the areas required in Problem 8.

10. For the normal curve in Problem 8, obtain the x-value(s)
    a. having area 0.025 to its right.
    b. having area 0.05 to its left.
    c. having area 0.84 to its left.

d. that divide the area under the curve into a middle 0.90 area and two outside 0.05 areas.

*11. **(Computer problem)** Use Minitab or some other statistical software to find the x-values required in Problem 10.

12. Each year, thousands of college seniors take the Graduate Record Examination (GRE). The scores are transformed so they have a mean of $\mu = 500$ and a standard deviation of $\sigma = 100$. Furthermore, the scores are known to be approximately normally distributed. Determine the percentage of students that score
    a. between 350 and 625.   b. at least 375.
    c. above 750.

13. Refer to Problem 12.
    a. Obtain the quartiles for GRE scores. Interpret your results in words.
    b. Find the 99th percentile for GRE scores. Interpret your result in words.

14. Refer to Problem 12. How many standard deviations away from the mean is a score of
    a. 645?   b. 320?   c. 500?

15. Refer to Problem 12. Apply the empirical rule for normally distributed populations to fill in the blanks.
    a. Approximately 68.26% of the students who take the GRE score between _____ and _____ .
    b. Approximately 95.44% of the students who take the GRE score between _____ and _____ .
    c. Approximately 99.74% of the students who take the GRE score between _____ and _____ .

16. For a normally distributed population,
    a. _____% of the population values lie within 2.75 standard deviations to either side of the mean.
    b. 90% of the population values lie within _____ standard deviations to either side of the mean.

17. According to R. R. Paul, the mean gestation period of the Morgan mare is 339.6 days ("Foaling Date," *The Morgan Horse*, 33:40, 1973). The gestation periods are normally distributed with a standard deviation of 13.3 days. Let x denote the gestation period of a randomly selected Morgan mare.
    a. Determine the mean and standard deviation of the random variable x.
    b. Find $P(320 \le x \le 330)$.   c. Find $P(x > 370)$.
    d. Interpret your results in parts (b) and (c).

*18. **(Computer problem)** Refer to Problem 17, where x is the gestation period of a randomly selected Morgan mare.
   a. Sketch the normal curve for the random variable x.
   b. Use Minitab or some other statistical software to simulate 750 independent observations of the random variable x and to obtain a histogram of the results obtained.
   c. Compare the histogram you obtained in part (b) to the theoretical distribution in part (a).
   d. What does part (c) illustrate?

19. Refer to Problem 17. Apply the empirical rule for normally distributed random variables to fill in the blanks.
   a. The probability is 0.6826 that a Morgan mare will have a gestation period between _____ and _____ days.
   b. The probability is 0.9544 that a Morgan mare will have a gestation period between _____ and _____ days.
   c. The probability is 0.9974 that a Morgan mare will have a gestation period between _____ and _____ days.

20. Refer to Problem 17.
   a. Determine the standardized version, z, of the random variable x.
   b. What does the random variable z represent?
   c. How many standard deviations from the mean is a gestation period of 325 days? 368 days? 339.6 days?
   d. Identify the probability distribution of z.
   e. For the answer you gave in part (b), was it necessary to know that the gestation periods are normally distributed? What about for the answer you gave in part (d)?

21. Use the general empirical rule for normally distributed random variables to fill in the following blank: The probability is 0.98 that an observed value of a normally distributed random variable will be within _____ standard deviations to either side of the mean.

22. Each year, manufacturers perform mileage tests on new car models and submit the results to the Environmental Protection Agency (EPA). The EPA then tests the vehicles to determine whether the manufacturers are correct. In 1992 one company reported that a particular model equipped with a four-speed manual transmission averaged 29 mpg on the highway. Suppose the EPA tested 15 of the cars and obtained the following gas mileages.

27.3	31.2	29.4	31.6	28.6
30.9	29.7	28.5	27.8	27.3
25.9	28.8	28.9	27.8	27.6

   a. Use Table III to construct a normal probability plot for the data. For simplicity, treat equal data values as being slightly different when obtaining the normal scores.
   b. Use part (a) to identify any outliers.
   c. Use part (a) to assess normality.

*23. **(Computer problem)** Use Minitab or some other statistical software to obtain a normal probability plot for the gas mileages in Problem 22.

24. Acute rotavirus diarrhea is the leading cause of death among children under age 5, killing an estimated 4.5 million annually in developing countries. Scientists from Finland and Belgium have claimed that a new oral vaccine is 80% effective against rotavirus diarrhea. Assuming the claim is correct, find the probability that out of 1500 cases, the vaccine will be effective in
   a. exactly 1225 cases.   b. at least 1175 cases.
   c. between 1150 and 1250 cases, inclusive.

Using the Focus Database   Appendix B contains a printout of a database obtained by randomly selecting 500 Arizona State University (ASU) sophomores. Seven variables are considered for each student: sex, high-school GPA, SAT math score, cumulative GPA, SAT verbal score, age, and total hours. In this exercise we will investigate normality for the last six of those variables. Use Minitab or some other statistical software to solve the following problems.
   a. For each of the six data sets—high-school GPA, SAT math score, cumulative GPA, SAT verbal score, age, and total hours—in the Focus database, obtain a histogram and a dotplot.
   b. Based on part (a), which of the six variables for ASU sophomores—high-school GPA, SAT math score, cumulative GPA, SAT verbal score, age, and total hours—appear to be approximately normally distributed?
   c. For each of the six data sets—high-school GPA, SAT math score, cumulative GPA, SAT verbal score, age, and total hours—in the Focus database, obtain a normal probability plot.

d. Based on part (c), which of the six variables for ASU sophomores—high-school GPA, SAT math score, cumulative GPA, SAT verbal score, age, and total hours—appear to be approximately normally distributed?

Case Study

Chest Sizes of Scottish Militiamen

In 1817 the medical journal *Edinburgh Medical and Surgical Journal* contained data on chest circumference by height for 5732 Scottish militiamen.[†] The data were collected by an army contractor who was responsible for providing clothing for the militia. Table 6.8 displays a frequency distribution for the chest circumferences.

TABLE 6.8

Chest circumferences, to the nearest inch, of 5732 Scottish militiamen

Chest size	Frequency	Chest size	Frequency
33	3	41	935
34	19	42	646
35	81	43	313
36	189	44	168
37	409	45	50
38	753	46	18
39	1062	47	3
40	1082	48	1

In his book *Lettres à S.A.R. le Duc Régnant de Saxe-Cobourg et Gotha sur la théorie des probabilités appliquée aux sciences morales et politiques* (Brussels: Hayez, 1846), Adolphe Quetelet discussed a procedure for fitting a normal curve to the data on chest circumferences. His method was based on the binomial distribution. You will be asked to fit a normal curve to the data using the technique explained at the beginning of Section 6.3.

a. Construct a relative-frequency histogram for the chest-circumference data.
b. Find the population mean and population standard deviation of the data.
c. Identify the normal curve that should be used for the chest circumferences of the 5732 Scottish militiamen.
d. Use Table 6.8 to determine the percentage of militiamen in the survey having chest circumferences between 36 and 41 inches, inclusive. (*Note:* Since the circumferences were rounded to the nearest inch, you are actually determining the percentage of militiamen in the survey having chest circumferences between 35.5 and 41.5 inches.)
e. Use the normal curve you identified in part (c) to obtain an approximation to the percentage of militiamen in the survey having chest circumferences between 35.5 and 41.5 inches. Compare your answer to the exact percentage found in part (d).
f. Use Minitab or some other statistical software to solve parts (a), (b), (d), and (e).

[†] See "Statement of the Sizes of Men in Different Counties of Scotland, Taken from the Local Militia," 1817, *Edinburgh Medical and Surgical Journal, 13*, pp. 260–264.

Pierre-Simon Laplace

Pierre-Simon Laplace was born on March 23, 1749, at Beaumount-en-Auge, Normandy, France, the son of a peasant farmer. His early schooling was at the military academy at Beaumount, where he developed his mathematical abilities. At the age of 18, he went to Paris. Within two years he was recommended for a professorship at the École Militaire by the French mathematician and philosopher Jean d'Alembert. (It is said that Laplace examined and passed Napoleon Bonaparte there in 1785.) In 1773 Laplace was granted membership in the Academy of Sciences.

Laplace held various positions in public life: he was president of the Bureau des Longitudes, professor at the École Normale, Minister of the Interior under Napoleon for six weeks (at which time he was replaced by Napoleon's brother), and Chancellor of the Senate; he was also made a marquis.

His professional interests were also varied. He published several volumes on celestial mechanics (which the Scottish geologist and mathematician John Playfair said were "the highest point to which man has yet ascended in the scale of intellectual attainment"), a book entitled *Théorie analytique des probabilités* (Analytic Theory of Probability), and other works on physics and mathematics. Laplace's primary contribution to the field of probability and statistics was the remarkable and all-important central limit theorem, which appeared in an 1809 publication and was read to the Academy of Sciences on April 9, 1810.

Astronomy was Laplace's major work; approximately half of his publications were concerned with the solar system and its gravitational interactions. These interactions were so complex that even Sir Isaac Newton had concluded "divine intervention was periodically required to preserve the system in equilibrium." Laplace, however, proved that planets' average angular velocities are invariable and periodic, thus making the most important advance in physical astronomy since Newton.

When Laplace died in Paris on March 5, 1827, he was eulogized by the famous French mathematician and physicist Simeon Poisson as "the Newton of France."

Chapter 7

The Sampling Distribution of the Mean

. . . . .

General Objective

In the preceding chapters, we have studied sampling, descriptive statistics, probability, random variables, and the normal distribution. Now we will learn how those seemingly diverse topics can be integrated to lay the groundwork for inferential statistics.

We will first explain why probability distributions must be incorporated into the design of inferential studies. Next we will obtain formulas for the mean and standard deviation of all possible sample means for samples of a given size from a population. Following that we will investigate the sampling distribution of the mean. That concept sets the stage for two important statistical-inference procedures—using the mean, $\bar{x}$, of a sample from a population to estimate and draw conclusions about the mean, μ, of the entire population.

Chapter Outline

7.1   Sampling Error; the Need for Sampling Distributions

7.2   The Mean and Standard Deviation of $\bar{x}$

7.3   The Sampling Distribution of the Mean

Case Study

The Chesapeake and Ohio Freight Study

Can relatively small samples really provide results that are nearly as accurate as those obtained from a complete examination (census)? Although statisticians have shown mathematically that this is indeed the case, you might be more convinced by a study in which the results of a sample are compared with those of a census. The case study at the end of this chapter examines such a study.

7.1   SAMPLING ERROR; THE NEED FOR SAMPLING DISTRIBUTIONS

We have already discovered that using a sample to acquire information about a population is often preferable to conducting a census, in which data for the entire population are collected. Generally, sampling is less costly and can be done more quickly than a census; it is often the only practical way to gather information.

But now we need to deal with the following problem: Since a sample from a population provides data for only a portion of the entire population, it is unreasonable to expect the sample to yield perfectly accurate information about the population. Thus we should anticipate that a certain amount of error will result simply because we are sampling. This kind of error is called **sampling error.**

DEFINITION 7.1   **SAMPLING ERROR**

> *Sampling error* is the error resulting from using a sample, instead of a census, to estimate a population quantity.

Example 7.1   Sampling Error and the Need for Sampling Distributions

The Census Bureau publishes annual figures on the mean income of U.S. households in *Current Population Reports.* Actually, the Census Bureau reports the mean income of a sample of about 60,000 households out of a total of more than 94 million households. For instance, in 1990 the mean income of U.S. households was reported to be $37,403. This amount is really the sample mean income, $\bar{x}$, of the 60,000 households surveyed, not the true (population) mean income, μ, of all U.S. households.

We certainly cannot expect the mean income, $\bar{x}$, of the 60,000 households sampled by the Census Bureau to be exactly the same as the mean income, μ, of all U.S. households—some sampling error is to be anticipated. But how much sampling error should we expect; that is, how accurate are such estimates likely to be? Is it likely, for instance, that the sample mean household income reported by the Census Bureau will be within $1000 of the population mean household income?

 To answer such questions, we need to know the distribution of all possible sample means that could be obtained by sampling the incomes of 60,000 households. This type of distribution is called a **sampling distribution.**   ■

As we just mentioned in Example 7.1, to answer questions about the accuracy of estimating a population mean by a sample mean, we need to know the distribution of all possible sample means that could be obtained. That distribution is called the **sampling distribution of the mean.** In this chapter we will study the sampling distribution of the mean. In future chapters we will examine other sampling distributions, such as the sampling distribution of the standard deviation.

Introducing the sampling distribution of the mean with an example that is both real-istic and concrete is difficult because even for moderately large populations the number of possible samples is enormous, thus prohibiting an actual listing of the possibilities.[†] Consequently, we will use an unrealistically small population to introduce the sampling dis-tribution of the mean. Keep in mind, however, that populations are much larger in real-life applications.

Example 7.2   Introduces the Sampling Distribution of the Mean

Suppose the population of interest consists of the heights of the five starting players on a men's basketball team. The heights, in inches, are displayed in Table 7.1.

TABLE 7.1
Population of heights

76	78	79	81	86

a.   Determine the sampling distribution of the mean for random samples of two heights from the population of five heights.[‡]

b.   Make some observations about sampling error when the mean, $\bar{x}$, of a random sample of two heights is used to estimate the population mean, μ.

c.   Employ the sampling distribution of the mean obtained in part (a) to find the probability that the sampling error made in estimating the population mean, μ, by the mean, $\bar{x}$, of a random sample of two heights will be at most 1 inch; that is, for a random sample of size two, determine the probability that $\bar{x}$ will be within 1 inch of μ.

SOLUTION   For future reference we first compute the mean of the population of heights:

$$\mu = \frac{\Sigma x}{N} = \frac{76 + 78 + 79 + 81 + 86}{5} = 80.$$

a.   For this part we need to determine the distribution of all possible $\bar{x}$-values (sample means) for random samples of size two. The population under consideration here is so small that we can list the possibilities directly. There are 10 possible samples of size two. The first column of Table 7.2 displays the 10 possible samples, and the second column displays their means. As a visual aid, we have also drawn a dotplot of the $\bar{x}$-values in Fig. 7.1.

[†]   For example, the number of possible samples of size 50 from a population of size 10,000 is approximately 3×10^{135}, a 3 followed by 135 zeros.

[‡]   As we mentioned in Section 1.6, the statistical-inference techniques considered in this book are intended for use only with simple random sampling. Therefore, unless otherwise specified, when we say *sample* or *random sample*, we mean *simple random sample*. Furthermore, we assume sampling is without replacement unless explicitly stated to the contrary.

Sample	$\bar{x}$
76, 78	77.0
76, 79	77.5
76, 81	78.5
76, 86	81.0
78, 79	78.5
78, 81	79.5
78, 86	82.0
79, 81	80.0
79, 86	82.5
81, 86	83.5

FIGURE 7.1

Dotplot of possible $\bar{x}$-values for samples of size two ($n = 2$)

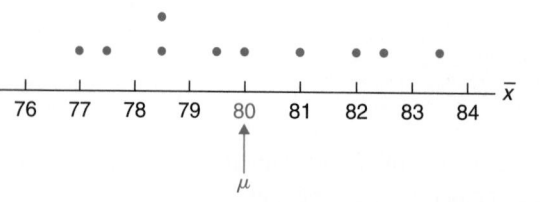

Note: From the second column of Table 7.2, we see that the value of $\bar{x}$ varies from sample to sample. For instance, the first sample has a mean of 77.0 and the second sample has a mean of 77.5. Thus when sampling from a population, the sample mean, $\bar{x}$, is a random variable because its value depends on chance, namely, on which sample is obtained.

Since we are dealing with random samples, each of the 10 possible samples shown in the first column of Table 7.2 has the same probability of being the one selected, a probability of $\frac{1}{10}$, or 0.1. Using that fact and the list of $\bar{x}$-values in the second column of Table 7.2, we obtain the probability distribution of the random variable $\bar{x}$ (the sampling distribution of the mean) for random samples of size two. This is shown in Table 7.3.

TABLE 7.3

Sampling distribution of the mean for random samples of size two

Sample mean $\bar{x}$	Probability $P(\bar{x})$	
77.0	0.1	
77.5	0.1	
78.5	0.2	*(Two of the 10 samples have $\bar{x} = 78.5$.)*
79.5	0.1	
80.0	0.1	
81.0	0.1	
82.0	0.1	
82.5	0.1	
83.5	0.1	

b. Using the results obtained in part (a), we can make some simple but significant observations about sampling error when the mean, $\bar{x}$, of a random sample of two heights is used to estimate the population mean, μ. For instance, it is unlikely that the mean of the sample selected will equal the population mean of 80. In fact, only 1 of the 10 samples has mean 80, the eighth sample in Table 7.2. Thus in this case the chances are only $\frac{1}{10}$ that $\bar{x}$ will equal μ; some sampling error is likely.

c. Here we want to obtain the probability that for a random sample of size two, the sampling error made in estimating μ by $\bar{x}$ will be at most 1 inch. Since $\mu = 80$ inches, we need to find $P(79 \le \bar{x} \le 81)$. Using the sampling distribution of the mean for samples of size two, Table 7.3, we see that

$$P(79 \le \bar{x} \le 81) = P(\bar{x} = 79.5, \ 80.0, \ \text{or} \ 81.0)$$
$$= P(79.5) + P(80.0) + P(81.0)$$
$$= 0.1 + 0.1 + 0.1 = 0.3.$$

If we take a random sample of two heights, there is a 30% chance that the mean of the sample selected will be within 1 inch of the population mean. ∎

Definition 7.2 summarizes the key points made so far in this section.

DEFINITION 7.2   **SAMPLING DISTRIBUTION OF THE MEAN**

When sampling from a population, the sample mean, $\bar{x}$, is a random variable because its value depends on chance, namely, on which sample is obtained. The probability distribution of the random variable $\bar{x}$ is called the *sampling distribution of the mean.*

We now use the height data from Example 7.2 to continue our discussion of the sampling distribution of the mean in Example 7.3. Up to this point, we have considered samples of size two ($n = 2$). If we consider samples of another size, say, of size four, then we obtain a new and different random variable $\bar{x}$.

Example 7.3   Further Illustrates the Sampling Distribution of the Mean

The heights of the five starting players on a men's basketball team are 76, 78, 79, 81, and 86 inches.

a. Determine the sampling distribution of the mean for random samples of four heights from the population of five heights.

b. Make some observations about sampling error when the mean, $\bar{x}$, of a random sample of four heights is used to estimate the population mean, μ.

c. Employ the sampling distribution of the mean obtained in part (a) to find the probability that the sampling error made in estimating the population mean, μ, by the mean, $\bar{x}$, of a random sample of four heights will be at most 1 inch; that is, for a random sample of size four, determine the probability that $\bar{x}$ will be within 1 inch of μ.

SOLUTION   a. There are five possible samples of size four. Those five samples and their means are listed in Table 7.4. A dotplot of the five possible $\bar{x}$-values is displayed in Fig. 7.2.

TABLE 7.4

Possible samples and their means for samples of four heights from the population of five heights

Sample	$\bar{x}$
76, 78, 79, 81	78.50
76, 78, 79, 86	79.75
76, 78, 81, 86	80.25
76, 79, 81, 86	80.50
78, 79, 81, 86	81.00

FIGURE 7.2

Dotplot of possible $\bar{x}$-values for samples of size four ($n = 4$)

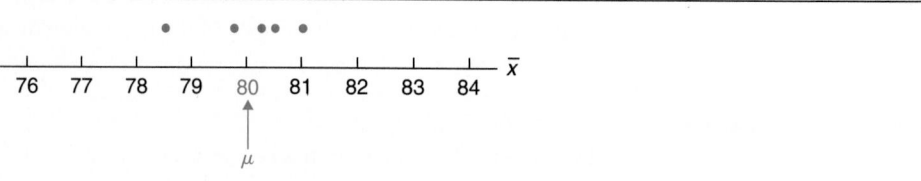

Since we are dealing with random samples, each of the five possible samples of size four has probability $\frac{1}{5} = 0.2$ of being the one selected. Applying that fact and Table 7.4, we can now tabulate the probability distribution of the random variable $\bar{x}$ (the sampling distribution of the mean) for random samples of size four, as shown in Table 7.5.

TABLE 7.5

Sampling distribution of the mean for random samples of size four

Sample mean $\bar{x}$	Probability $P(\bar{x})$
78.50	0.2
79.75	0.2
80.25	0.2
80.50	0.2
81.00	0.2

b. Using the results obtained in part (a), we observe that none of the samples of four heights has a mean equal to the population mean of 80. Thus when the mean, $\bar{x}$, of a random sample of four heights is used to estimate the population mean, μ, some sampling error is certain.

c. Here we want to obtain the probability that for a random sample of size four, the sampling error made in estimating μ by $\bar{x}$ will be at most 1 inch. Since $\mu = 80$ inches, we need to find $P(79 \leq \bar{x} \leq 81)$. Using the sampling distribution of the mean for samples

of size four, Table 7.5, we see that

$$P(79 \leq \bar{x} \leq 81) = P(\bar{x} = 79.75,\ 80.25,\ 80.50,\ \text{or}\ 81.00)$$
$$= P(79.75) + P(80.25) + P(80.50) + P(81.00)$$
$$= 0.2 + 0.2 + 0.2 + 0.2 = 0.8.$$

If we take a random sample of four heights, there is an 80% chance that the mean of the sample selected will be within 1 inch of the population mean. ■

We emphasize that to each sample size, there corresponds a different random variable $\bar{x}$ and hence a different sampling distribution of the mean.

Sample Size and Sampling Error

In Figs. 7.1 and 7.2, we drew dotplots of the possible $\bar{x}$-values for samples of sizes two and four, respectively. Those two dotplots, as well as one for samples of size three, are displayed together in Fig. 7.3.

FIGURE 7.3

Dotplots of possible $\bar{x}$-values for sample sizes two, three, and four

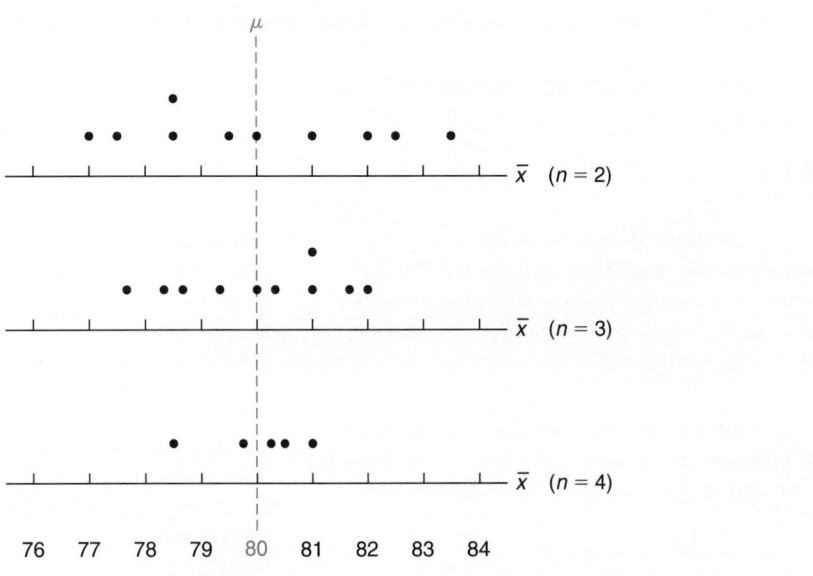

Figure 7.3 vividly depicts that the possible $\bar{x}$-values cluster closer around the population mean, μ, as the sample size increases. For example, for samples of size two, 3 out of 10, or 30%, of the possible $\bar{x}$-values lie within 1 inch of μ; for samples of size three, 5 out of 10, or 50%, of the possible $\bar{x}$-values lie within 1 inch of μ; and for samples of

size four, 4 out of 5, or 80%, of the possible $\bar{x}$-values lie within 1 inch of μ. More generally, we can make the following qualitative statement.

KEY FACT 7.1   **SAMPLE SIZE AND SAMPLING ERROR**

> The possible sample means cluster closer around the population mean as the sample size increases. Thus the larger the sample size, the smaller the sampling error tends to be in estimating a population mean, μ, by a sample mean, $\bar{x}$.

What We Do in Practice

We have used a population of five heights to illustrate and explain the importance of the sampling distribution of the mean. For that small population, it is easy to obtain the sampling distribution of the mean for any sample size by listing all of the possible samples.

However, as we have noted, the populations we deal with in practice are usually large. For such populations it is not feasible to obtain the sampling distribution of the mean by a direct listing. A more serious practical problem is that in reality we do not even know the population values—if we did, there would be no need to sample! Realistically then, we could not list the possible samples to determine the sampling distribution of the mean even if we were willing to expend the effort.

So, what can we do in the usual case of a large and unknown population? Fortunately, there exist mathematical relationships that allow us to determine, at least approximately, the sampling distribution of the mean for any specified sample size. We will discuss those relationships in Sections 7.2 and 7.3.

EXERCISES 7.1

Exercises 7.1–7.11 are intended solely to provide concrete illustrations of the sampling distribution of the mean. For that reason the populations considered are unrealistically small. (Note: Remember that sampling is assumed to be without replacement.)

7.1 As reported by the *World Almanac,* the annual salaries of the governors of the Great Lakes states are as follows. The salaries are in thousands of dollars, rounded to the nearest thousand.

76	58	64	82	60

a. Compute the population mean of these five salaries.
b. List the possible samples of size two and their means. Construct a table similar to Table 7.2 on page 406.
c. Draw a dotplot for the possible $\bar{x}$-values like the one in Fig. 7.1 on page 406.

d. Determine the probability distribution of the random variable $\bar{x}$ (the sampling distribution of the mean) for random samples of size two. Construct a table similar to Table 7.3 on page 406.
e. For a random sample of size two, find the probability that the sample mean, $\bar{x}$, will equal the population mean, μ; that is, determine $P(\bar{x} = \mu)$.
f. For a random sample of size two, obtain the probability that the mean of the sample selected will be within 4 (i.e., $4000) of the population mean. Interpret your result in terms of percentages.

7.2 Repeat parts (b)–(f) of Exercise 7.1 for samples of size three.

7.3 Repeat parts (b)–(f) of Exercise 7.1 for samples of size four.

7.4 This exercise requires that you have done Exercises 7.1–7.3.

a. Draw a graph similar to Fig. 7.3 on page 409 for sample sizes of two, three, and four from the population of five governors' salaries.
b. What does your graph in part (a) illustrate about the impact that increasing sample size has on sampling error?

· **7.5**   The lengths, in centimeters, of six bullfrogs in a small pond are as follows.

19	14	15	9	16	17

Consider these six lengths a population of interest.
a. Calculate the mean length, μ, of the six bullfrogs.
b. List the possible samples of size two and their means. (There are 15 possible samples of size two.)
c. Draw a dotplot for the possible $\bar{x}$-values.
d. Determine the sampling distribution of the mean for random samples of size two. Leave your probabilities in fraction form.
e. Obtain the probability that the sample mean, $\bar{x}$, of a random sample of size two will equal the population mean, μ.
f. For a random sample of size two, determine the probability that the mean of the sample selected will be within 1 cm of the population mean. Interpret your result in terms of percentages.

· **7.6**   Repeat parts (b)–(f) of Exercise 7.5 for samples of size three. (There are 20 possible samples.)

· **7.7**   Repeat parts (b)–(f) of Exercise 7.5 for samples of size four. (There are 15 possible samples.)

· **7.8**   Repeat parts (b)–(f) of Exercise 7.5 for samples of size five. (There are six possible samples.)

· **7.9**   Repeat parts (b)–(f) of Exercise 7.5 for samples of size six. What is the relationship between the only possible sample here and the population?

· **7.10**   Repeat parts (b)–(f) of Exercise 7.5 for samples of size one.

· **7.11**   What do the dotplots in parts (c) of Exercises 7.5–7.10 illustrate about the impact increasing sample size has on sampling error?

· · **7.12**   Suppose a sample is to be taken without replacement from a finite population of size N. If the sample size is the same as the population size, that is, $n = N$, then
a. how many possible samples are there?
b. what are the possible $\bar{x}$-values?
c. what is the relationship between the only possible sample and the population?

· · **7.13**   Suppose a random sample of size one is to be taken from a finite population of size N.
a. How many possible samples are there?
b. Identify the relation between the possible sample means and the population values.
c. What is the difference between taking a random sample of size one from a population and selecting a member at random from the population?

7.2   THE MEAN AND STANDARD DEVIATION OF $\bar{x}$

· · · · ·

In Section 7.1 we discussed the sampling distribution of the mean (the probability distribution of $\bar{x}$). That sampling distribution is used to make inferences about a population mean based on the mean of a sample from the population.

As we said earlier, it is generally not possible to know the sampling distribution of the mean exactly. Fortunately, however, we can often approximate that sampling distribution by a normal distribution; that is, under certain conditions the random variable $\bar{x}$ is approximately normally distributed.

Recall that a random variable is normally distributed if probabilities for the random variable are equal to areas under a normal curve. To obtain probabilities for such a random variable, we must know its mean and standard deviation because they are also the parameters for the appropriate normal curve.

Hence a first step in learning how to approximate the sampling distribution of the mean by a normal distribution is to study the mean and standard deviation of the random variable $\bar{x}$. That is what we will do in this section.

To begin, let's review the notation used for the mean and standard deviation of a random variable. The mean of a random variable is denoted by the Greek letter μ subscripted with the letter representing the random variable. So the mean of x is written as μ_x, the mean of y as μ_y, and so forth. In particular, then, the **mean of $\bar{x}$** is written as $\mu_{\bar{x}}$; similarly, the **standard deviation of $\bar{x}$** is written as $\sigma_{\bar{x}}$.

The Mean of $\bar{x}$

There is a simple relationship between the mean of the random variable $\bar{x}$ and the mean of the population being sampled. Namely, the mean of $\bar{x}$ is equal to the population mean: $\mu_{\bar{x}} = \mu$. In other words, for any particular sample size, the mean of all possible sample means equals the population mean. This equality holds regardless of the size of the sample. We illustrate the relationship $\mu_{\bar{x}} = \mu$ in Example 7.4.

Example 7.4　　The Mean of $\bar{x}$ Equals the Mean of the Population

---

The heights of the five starting players on a men's basketball team are 76, 78, 79, 81, and 86 inches. Let's regard these five heights as a population of interest.

a. Determine the population mean, μ.

b. Definition 5.4 provides the defining formula for the mean of a discrete random variable: $\mu_x = \Sigma x P(x)$. Use this formula to obtain the mean, $\mu_{\bar{x}}$, of the random variable $\bar{x}$ for samples of size two. Verify that the relation $\mu_{\bar{x}} = \mu$ holds.

c. Repeat part (b) for samples of size four.

SOLUTION　　a. To obtain the population mean, μ, we apply Definition 3.12 on page 161:

$$\mu = \frac{\Sigma x}{N} = \frac{76 + 78 + 79 + 81 + 86}{5} = 80.$$

Thus the mean height of the five players is 80 inches.

b. To obtain $\mu_{\bar{x}}$ we first need the probability distribution of $\bar{x}$ for samples of size two. That probability distribution was determined in Table 7.3 and is repeated here in the first two columns of Table 7.6.

Now we can compute $\mu_{\bar{x}}$ by proceeding in the usual way: We first append an $\bar{x}P(\bar{x})$ column to the probability distribution of $\bar{x}$ and then apply Definition 5.4, as is done in Table 7.6. We find that $\mu_{\bar{x}} = \Sigma \bar{x}P(\bar{x}) = 80$. By part (a), $\mu = 80$. So for samples of size two, we see that $\mu_{\bar{x}} = \mu$.

TABLE 7.6

Table for computing $\mu_{\bar{x}}$
for samples of size two

$\bar{x}$	$P(\bar{x})$	$\bar{x}P(\bar{x})$
77.0	0.1	7.70
77.5	0.1	7.75
78.5	0.2	15.70
79.5	0.1	7.95
80.0	0.1	8.00
81.0	0.1	8.10
82.0	0.1	8.20
82.5	0.1	8.25
83.5	0.1	8.35
		80.00

c. The probability distribution of $\bar{x}$ for samples of size four, determined in Table 7.5, is repeated here in the first two columns of Table 7.7. The third column of Table 7.7 displays the $\bar{x}P(\bar{x})$-values.

TABLE 7.7

Table for computing $\mu_{\bar{x}}$
for samples of size four

$\bar{x}$	$P(\bar{x})$	$\bar{x}P(\bar{x})$
78.50	0.2	15.70
79.75	0.2	15.95
80.25	0.2	16.05
80.50	0.2	16.10
81.00	0.2	16.20
		80.00

Applying Definition 5.4 we find from the third column of Table 7.7 that for samples of size four, $\mu_{\bar{x}} = 80$, which again is the same as μ.

Formula 7.1, which you are asked to verify in Exercise 7.32, summarizes the preceding discussion.

FORMULA 7.1   **MEAN OF THE RANDOM VARIABLE $\bar{x}$**

Suppose a random sample of size n is to be taken from a population with mean μ. Then

$$\mu_{\bar{x}} = \mu.$$

In other words, for each sample size, the mean of the random variable $\bar{x}$ is equal to the mean of the population.

The Standard Deviation of $\bar{x}$

Next we will investigate the standard deviation of the random variable $\bar{x}$. Our purpose is to discover any apparent relationship that $\sigma_{\bar{x}}$ has with the population standard deviation, σ. We once again use the population of heights.

Example 7.5 The Relation Between $\sigma_{\bar{x}}$ and σ

The heights of the five starting players on a men's basketball team are 76, 78, 79, 81, and 86 inches. We take those heights to be a population of interest.

a. Determine the population standard deviation, σ.

b. Formula 5.1 provides a shortcut formula for the standard deviation of a discrete random variable: $\sigma_x = \sqrt{\Sigma x^2 P(x) - \mu_x^2}$. Use this formula to obtain the standard deviation, $\sigma_{\bar{x}}$, of the random variable $\bar{x}$ for samples of size two. Indicate any apparent relationship between $\sigma_{\bar{x}}$ and σ.

c. Repeat part (b) for samples of sizes three and four.

d. Summarize and discuss the results obtained in parts (a)–(c).

SOLUTION

a. To obtain the population standard deviation, σ, we will apply the shortcut formula presented in Definition 3.13 on page 163. Recalling that $\mu = 80$, we have

$$\sigma = \sqrt{\frac{\Sigma x^2}{N} - \mu^2} = \sqrt{\frac{76^2 + 78^2 + 79^2 + 81^2 + 86^2}{5} - 80^2} = \sqrt{11.6} = 3.41.$$

Thus the standard deviation of the population of heights is 3.41 inches.

b. The probability distribution of $\bar{x}$ for samples of size two is repeated in the first two columns of Table 7.8. The third column displays the $\bar{x}^2 P(\bar{x})$-values. Those values are needed to compute $\sigma_{\bar{x}}$ using Formula 5.1.

TABLE 7.8
Table for computing $\sigma_{\bar{x}}$
for samples of size two

$\bar{x}$	$P(\bar{x})$	$\bar{x}^2 P(\bar{x})$
77.0	0.1	592.900
77.5	0.1	600.625
78.5	0.2	1232.450
79.5	0.1	632.025
80.0	0.1	640.000
81.0	0.1	656.100
82.0	0.1	672.400
82.5	0.1	680.625
83.5	0.1	697.225
		6404.350

Recalling that $\mu_{\bar{x}} = \mu = 80$, we find from the third column of Table 7.8 that

$$\sigma_{\bar{x}} = \sqrt{\Sigma \bar{x}^2 P(\bar{x}) - \mu_{\bar{x}}^2} = \sqrt{6404.350 - 80^2} = \sqrt{4.35} = 2.09.$$

Thus for samples of size two, the standard deviation of $\bar{x}$ is $\sigma_{\bar{x}} = 2.09$. Note that this is not the same as the population standard deviation, which is $\sigma = 3.41$. Also note that $\sigma_{\bar{x}}$ is smaller than σ.

c. Using the same procedure as in part (b), we can compute $\sigma_{\bar{x}}$ for samples of size three and size four. For samples of size three, $\sigma_{\bar{x}} = 1.39$; and for samples of size four, $\sigma_{\bar{x}} = 0.85$.

d. We summarize our results for $\sigma_{\bar{x}}$ in Table 7.9.

TABLE 7.9

The standard deviation of $\bar{x}$ for sample sizes two, three, and four

Sample size n	Standard deviation of $\bar{x}$ $\sigma_{\bar{x}}$
2	2.09
3	1.39
4	0.85

Table 7.9 suggests that the standard deviation of $\bar{x}$ gets smaller as the sample size gets larger. We could have predicted this phenomenon from the dotplots of the possible $\bar{x}$-values in Fig. 7.3 on page 409 and from the fact that the standard deviation of a random variable measures the spread of the possible values of the random variable. Figure 7.3 indicates graphically that the spread, and hence the standard deviation of the possible $\bar{x}$-values, decreases with increasing sample size. Table 7.9 indicates that same thing numerically. ∎

Example 7.5 provides evidence that the standard deviation of $\bar{x}$ gets smaller as the sample size increases. Our question now is whether there is a formula that relates the standard deviation of $\bar{x}$ to the sample size and standard deviation of the population. The answer is yes. In fact, two different formulas express the precise relationship.

When sampling is done without replacement from a finite population, as in Example 7.5, the appropriate formula is

$$\sigma_{\bar{x}} = \sqrt{\frac{N - n}{N - 1}} \cdot \frac{\sigma}{\sqrt{n}},$$

where, as usual, n denotes the sample size and N the population size.

When sampling is done with replacement from a finite population or when it is done from an infinite population, the appropriate formula is

$$\sigma_{\bar{x}} = \frac{\sigma}{\sqrt{n}}.$$

If the sample size is small relative to the size of the population, then there is very little difference in the values given by the two formulas (see Exercise 7.31). Generally, in inferential statistics the sample size *is* small relative to the size of the population. Therefore we will use the second formula exclusively in this book since it is simpler than the first.[†]

FORMULA 7.2   **STANDARD DEVIATION OF THE RANDOM VARIABLE $\bar{x}$**

> Suppose a random sample of size n is to be taken from a population with standard deviation σ. Then
>
> $$\sigma_{\bar{x}} \approx \frac{\sigma}{\sqrt{n}}.$$
>
> In other words, the standard deviation of the random variable $\bar{x}$ is equal, at least approximately, to the standard deviation of the population divided by the square root of the sample size.

Note: For simplicity, we will write $\sigma_{\bar{x}} = \sigma/\sqrt{n}$ instead of $\sigma_{\bar{x}} \approx \sigma/\sqrt{n}$, with the understanding that the equality may be only approximate.

Applying the Formulas

We have seen that there are simple formulas relating the mean and standard deviation of the random variable $\bar{x}$ to the mean and standard deviation of the population being sampled: $\mu_{\bar{x}} = \mu$ and $\sigma_{\bar{x}} = \sigma/\sqrt{n}$ (at least approximately). Those two formulas are applied in Example 7.6.

Example 7.6   Illustrates Formulas 7.1 and 7.2

As reported by the U.S. Bureau of the Census in *Current Housing Reports,* the mean livable square footage for single-family detached homes is 1742 square feet. The standard deviation is 568 square feet.

a. Suppose 25 single-family detached homes are to be selected at random. Let $\bar{x}$ denote the mean livable square footage of the homes obtained. Determine the mean, $\mu_{\bar{x}}$, and the standard deviation, $\sigma_{\bar{x}}$, of the random variable $\bar{x}$.

b. Repeat part (a) for a sample of size 500.

SOLUTION   We know that $\mu = 1742$ and $\sigma = 568$.

a. Applying Formula 7.1 (page 413), we get

$$\mu_{\bar{x}} = \mu = 1742.$$

Since $n = 25$, we conclude from Formula 7.2 that

$$\sigma_{\bar{x}} = \frac{\sigma}{\sqrt{n}} = \frac{568}{\sqrt{25}} = 113.6.$$

[†] Exercises 7.24–7.31 give you some practice with using and comparing both formulas.

In other words, for random samples of size 25, the mean of all possible sample means is 1742 square feet and the standard deviation is 113.6 square feet.

b. Proceeding as in part (a), we find that

$$\mu_{\bar{x}} = \mu = 1742$$

and since, here, $n = 500$,

$$\sigma_{\bar{x}} = \frac{\sigma}{\sqrt{n}} = \frac{568}{\sqrt{500}} = 25.4.$$

That is, for random samples of size 500, the mean of all possible sample means is 1742 square feet and the standard deviation is 25.4 square feet.   ■

Sample Size and Sampling Error (Revisited)

In Key Fact 7.1, we pointed out that the sampling error made in estimating a population mean, μ, by a sample mean, $\bar{x}$, tends to be smaller with increasing sample size. We can now make that statement more precise.

Recall from Key Fact 5.7 that the standard deviation of a random variable measures the dispersion of the possible values of the random variable relative to the mean. So the smaller the standard deviation, the more likely that an observed value of the random variable will be close to the mean. In particular, the smaller the value of $\sigma_{\bar{x}}$, the more likely that an observed value of $\bar{x}$ will be close to the mean, $\mu_{\bar{x}}$.

But as we now know, the mean of $\bar{x}$ is equal to the population mean: $\mu_{\bar{x}} = \mu$. Thus the smaller the value of $\sigma_{\bar{x}}$, the more likely that an observed value of $\bar{x}$ will be close to μ. In other words, the smaller the value of $\sigma_{\bar{x}}$, the greater the tendency for small sampling error.

KEY FACT 7.2   **THE STANDARD DEVIATION OF $\bar{x}$ AND SAMPLING ERROR**

> The standard deviation of $\bar{x}$ indicates the amount of sampling error to be expected when a population mean is estimated by the mean of a sample from the population; that is, the smaller the value of $\sigma_{\bar{x}}$, the smaller the sampling error tends to be. Hence $\sigma_{\bar{x}}$ is often referred to as the ***standard error of the mean.***[†]

We have seen that the standard deviation of $\bar{x}$ is equal, at least approximately, to the standard deviation of the population divided by the square root of the sample size: $\sigma_{\bar{x}} = \sigma/\sqrt{n}$. Since n appears in the denominator, the larger the sample size, the smaller the value of $\sigma_{\bar{x}}$. Combining this last fact with Key Fact 7.2, we can make the following fundamental statement.

[†] In general, the standard deviation of a statistic that is used to estimate a parameter is called the *standard error* of the statistic.

KEY FACT 7.3	SAMPLE SIZE AND SAMPLING ERROR

> When estimating a population mean by a sample mean, the larger the sample size, the greater the likelihood for small sampling error. In other words, larger sample sizes tend to produce more accurate estimates of the population mean.

The formula $\sigma_{\bar{x}} = \sigma/\sqrt{n}$ also implies that the standard deviation of $\bar{x}$ is directly proportional to the standard deviation of the population being sampled. Thus for any particular sample size, the smaller the standard deviation of the population, the smaller the sampling error tends to be.

EXERCISES 7.2

7.14 Why do we need to know the mean and standard deviation of the random variable $\bar{x}$ in order to use normal curve areas to approximate probabilities for $\bar{x}$?

7.15 As reported by the *World Almanac,* the annual salaries of the governors of the Great Lakes states are as follows. The salaries are in thousands of dollars, rounded to the nearest thousand.

76	58	64	82	60

a. Find the population mean, μ, of these five salaries.
b. Consider samples of size two without replacement. Find the mean, $\mu_{\bar{x}}$, of the random variable $\bar{x}$ by applying the formula $\mu_{\bar{x}} = \Sigma \bar{x} P(\bar{x})$. *(Note: The probability distribution of $\bar{x}$ for samples of size two was found in part (d) of Exercise 7.1.)*
c. Find $\mu_{\bar{x}}$ using only the result of part (a).

7.16 Repeat parts (b) and (c) of Exercise 7.15 for samples of size three. *(Note: The probability distribution of $\bar{x}$ for samples of size three was found in part (d) of Exercise 7.2.)*

7.17 Repeat parts (b) and (c) of Exercise 7.15 for samples of size four. *(Note: The probability distribution of $\bar{x}$ for samples of size four was found in part (d) of Exercise 7.3.)*

In Exercises 7.18–7.22, determine the mean and the standard deviation of $\bar{x}$ for each of the specified sample sizes by using the formulas $\mu_{\bar{x}} = \mu$ and $\sigma_{\bar{x}} = \sigma/\sqrt{n}$, Formulas 7.1 and 7.2.

7.18 According to the Census Bureau publication *Construction Reports,* the mean price of new mobile homes is $25,000. The standard deviation of the prices is $7200.
a. Suppose a random sample of 50 new mobile homes is to be

selected. Let $\bar{x}$ be the mean price of the homes obtained. Find $\mu_{\bar{x}}$ and $\sigma_{\bar{x}}$.
b. Repeat part (a) for a sample size of 100.

7.19 The U.S. National Center for Health Statistics compiles information on the length of stay by patients in short-stay hospitals. Data are published in *Vital and Health Statistics.* According to that publication, the average stay of female patients in short-stay hospitals is $\mu = 6.9$ days. The standard deviation is $\sigma = 4.3$ days.
a. Suppose 75 female patients are to be selected at random. Let $\bar{x}$ denote the mean hospital stay of the 75 patients obtained. Find $\mu_{\bar{x}}$ and $\sigma_{\bar{x}}$.
b. Repeat part (a) for a random sample of size 500.

7.20 The National Council of the Churches of Christ in the United States of America reports in *Yearbook of American and Canadian Churches* that the mean number of members per local church is 408. Assume a standard deviation of 116.7.
a. Suppose 100 local churches are to be selected at random. Let $\bar{x}$ denote the mean number of members for the 100 churches obtained. Determine the mean and standard deviation of $\bar{x}$.
b. Repeat part (a) with $n = 200$.

7.21 According to *Motor Vehicle Facts and Figures,* published by the Motor Vehicle Manufacturers Association of the United States, the mean age of cars in use is 7.4 years and the standard deviation is 2.6 years.
a. Let $\bar{x}$ denote the mean age of a random sample of 50 cars. Determine the mean and standard deviation of the random variable $\bar{x}$.
b. Repeat part (a) with $n = 200$.

7.22 The Bureau of Labor Statistics collects data on earnings of U.S. workers, by industry, and publishes the results in *Employment and Earnings*. That publication reveals that the average weekly earnings of workers in the trucking industry is $456. Suppose n such workers are to be selected at random. Let $\bar{x}$ denote the mean weekly earnings of the workers obtained. Assuming a population standard deviation of $63, find the mean and standard deviation of $\bar{x}$ if

a. $n = 100$.   b. $n = 200$.   c. $n = 400$.

7.23 Suppose you plan to use a sample mean, $\bar{x}$, to estimate a population mean, μ. To increase the likelihood of small sampling error, what must you do?

Finite population correction factor. For Exercises 7.24–7.31, recall the following fact from page 415: If sampling is done without replacement from a finite population of size N, then the standard deviation of $\bar{x}$ can be obtained from the formula

(1) $$\sigma_{\bar{x}} = \sqrt{\frac{N-n}{N-1}} \cdot \frac{\sigma}{\sqrt{n}},$$

where n is the sample size. The term $\sqrt{\frac{N-n}{N-1}}$ is referred to as the **finite population correction factor.** If the sample size is small relative to the population size, then we can ignore the finite population correction factor and use the simpler formula,

(2) $$\sigma_{\bar{x}} = \frac{\sigma}{\sqrt{n}}.$$

A rule of thumb is that the finite population correction factor can be ignored provided the sample size does not exceed 5% of the population size, that is, $n \leq 0.05N$.

7.24 Refer to Example 7.5 on page 414. We computed the standard deviation of the basketball players' heights to be $\sigma = 3.41$ inches. We also applied the formula $\sigma_{\bar{x}} = \sqrt{\sum \bar{x}^2 P(\bar{x}) - \mu_{\bar{x}}^2}$ to determine the standard deviation of $\bar{x}$ for samples of sizes two, three, and four. The results are summarized in Table 7.9 on page 415. Since the sampling is without replacement from a finite population, we can also use (1) to obtain $\sigma_{\bar{x}}$.

a. Use (1) to compute $\sigma_{\bar{x}}$ for samples of size
  i. two.   ii. three.   iii. four.
  Compare your results to those in Table 7.9.

b. Use the simpler formula, (2), to compute $\sigma_{\bar{x}}$ for samples of size
  i. two.   ii. three.   iii. four.
  Compare your results to the true values of $\sigma_{\bar{x}}$ displayed in Table 7.9. Why does (2) yield such poor results?

c. What percentage of the population size is a sample of size
  i. two?   ii. three?   iii. four?

7.25 Refer to Exercise 7.15.
a. Find the standard deviation, σ, of the governors' salaries.
b. Suppose we consider samples of size two without replacement. Find the standard deviation, $\sigma_{\bar{x}}$, of the random variable $\bar{x}$ by applying the formula $\sigma_{\bar{x}} = \sqrt{\sum \bar{x}^2 P(\bar{x}) - \mu_{\bar{x}}^2}$.
c. In part (b) which formula would be appropriate, (1) or (2)?
d. Use (1) to compute $\sigma_{\bar{x}}$ and compare your result with that from part (b).
e. Use (2) to compute $\sigma_{\bar{x}}$ and compare your result with that from part (b). Why does (2) yield such a poor approximation to the true value of $\sigma_{\bar{x}}$?

7.26 Refer to Exercise 7.16. Repeat parts (b)–(e) of Exercise 7.25 for samples of size three.

7.27 Refer to Exercise 7.17. Repeat parts (b)–(e) of Exercise 7.25 for samples of size four.

7.28 Repeat parts (b)–(e) of Exercise 7.25 for samples of size five.

7.29 Repeat parts (b)–(d) of Exercise 7.25 for samples of size one.

7.30 Suppose a random sample of size n is to be taken from a population of size N.
a. Which formula, (1) or (2), should be used to compute $\sigma_{\bar{x}}$ if sampling is done without replacement? with replacement?
b. Assume $n = 1$. Compute $\sigma_{\bar{x}}$ using both (1) and (2). Why do both formulas yield the same result? Explain in words why $\sigma_{\bar{x}} = \sigma$.
c. Assume $n = N$ and sampling is done without replacement. Compute $\sigma_{\bar{x}}$ using (1). Could you have guessed the answer without doing any computations? Why or why not?

7.31 Suppose a random sample of size n is to be taken without replacement from a population of size N.
a. Prove that if $n \leq 0.05N$, then

$$0.97 \leq \sqrt{\frac{N-n}{N-1}} \leq 1.$$

b. Use part (a) to explain why there is little difference in the values provided by (1) and (2) when the sample size does not exceed 5% of the population size.

7.32 Exercises 5.26(e) and 5.27(f) indicate that if x, y, and w are random variables and c is a constant, then $\mu_{x+y} = \mu_x + \mu_y$, $\mu_{cw} = c\mu_w$, and $\sigma_{cw} = |c|\sigma_w$; and if, in addition, x and y are independent, then $\sigma_{x+y} = \sqrt{\sigma_x^2 + \sigma_y^2}$. These four formulas can be used to derive some of the facts stated in this section. Suppose a random sample of size n is to be taken from a population with mean μ and standard deviation σ. Let x_1 denote the value of the first member selected, x_2 the value of the second member selected, and so on. Then each of the random variables, $x_1, x_2, \ldots, x_n$, has mean μ and standard deviation σ; that is, for $i = 1, 2, \ldots, n$, $\mu_{x_i} = \mu$ and $\sigma_{x_i} = \sigma$.

The sample mean is $\overline{x} = (x_1 + x_2 + \cdots + x_n)/n$. Employ the formulas at the beginning of this exercise to prove that
a. $\mu_{\overline{x}} = \mu$
and that if the sampling is done with replacement,
b. $\sigma_{\overline{x}} = \sigma/\sqrt{n}$.

7.3   THE SAMPLING DISTRIBUTION OF THE MEAN

Recall that the probability distribution of the random variable $\overline{x}$ is referred to as the sampling distribution of the mean. Also recall that we need to know that probability distribution in order to make inferences about the mean of a population based on the mean of a sample from the population.

In Section 7.2 we took the first step in describing the sampling distribution of the mean. There we discovered that the mean and standard deviation of the random variable $\overline{x}$ can be expressed in terms of the sample size and the mean and standard deviation of the population being sampled: $\mu_{\overline{x}} = \mu$ and $\sigma_{\overline{x}} = \sigma/\sqrt{n}$.

In this section we will take the second (and final) step in describing the sampling distribution of the mean. It is helpful to distinguish between the case in which the population being sampled is normally distributed and the case in which it may not be so. We first consider normally distributed populations.

Normally Distributed Populations

Suppose the population under consideration is normally distributed. Recall that, qualitatively, this means the histogram of the population is bell-shaped. More precisely, a population with mean μ and standard deviation σ is normally distributed if percentages for the population are equal to areas under the normal curve with parameters μ and σ.

Although it is by no means obvious, when sampling is done from a normally distributed population, the random variable $\overline{x}$ is also normally distributed. In other words, probabilities for $\overline{x}$ are equal to areas under the normal curve with parameters $\mu_{\overline{x}}$ and $\sigma_{\overline{x}}$. Combining this with the fact that $\mu_{\overline{x}} = \mu$ and $\sigma_{\overline{x}} = \sigma/\sqrt{n}$, we can now state the following fundamental result concerning the sampling distribution of the mean for normally distributed populations.

KEY FACT 7.4   **THE SAMPLING DISTRIBUTION OF THE MEAN
FOR NORMAL POPULATIONS**

> Suppose a random sample of size n is to be taken from a normally distributed population with mean μ and standard deviation σ. Then the random variable $\overline{x}$ is also normally distributed and has mean $\mu_{\overline{x}} = \mu$ and standard deviation $\sigma_{\overline{x}} = \sigma/\sqrt{n}$. In other words, if the population being sampled is normally distributed, then probabilities for $\overline{x}$ are equal to areas under the normal curve with parameters μ and $\sigma/\sqrt{n}$.

For convenience we will often use the phrase **normal population** as an abbreviation of "normally distributed population."

Example 7.7   Illustrates Key Fact 7.4

The U.S. Bureau of the Census compiles information on U.S. farm operators and publishes the data in *Census of Agriculture*. According to that document, the mean age of U.S. farm operators is 50 years and the standard deviation is 8 years. Assuming the ages are normally distributed, the normal curve for that population of ages is as displayed in Fig. 7.4(a). Find the sampling distribution of the mean for random samples of size

a. 3.   b. 10.

FIGURE 7.4

(a) Normal curve for ages of farm operators
(b) Sampling distribution of the mean for $n = 3$
(c) Sampling distribution of the mean for $n = 10$

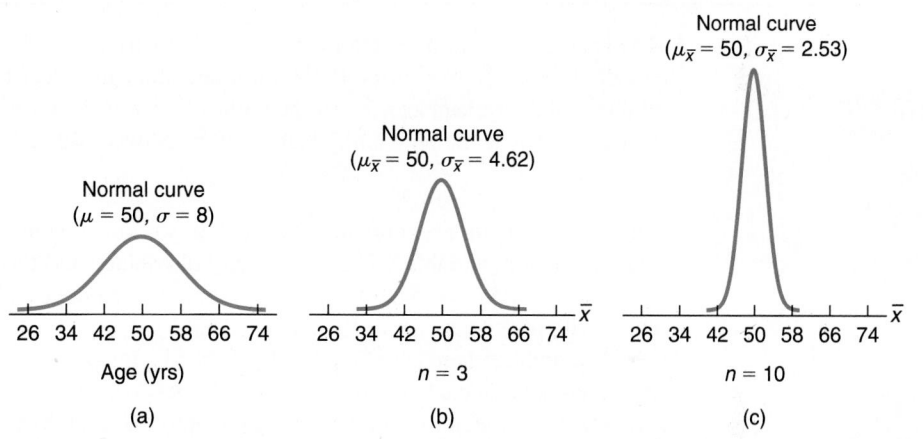

Normal curve
($\mu = 50, \sigma = 8$)

26 34 42 50 58 66 74
Age (yrs)
(a)

Normal curve
($\mu_{\bar{x}} = 50, \sigma_{\bar{x}} = 4.62$)

26 34 42 50 58 66 74
$n = 3$
(b)

Normal curve
($\mu_{\bar{x}} = 50, \sigma_{\bar{x}} = 2.53$)

26 34 42 50 58 66 74
$n = 10$
(c)

SOLUTION By assumption, the population of ages is normally distributed. Thus, by Key Fact 7.4, for any particular sample size n, the random variable $\bar{x}$ is also normally distributed and has mean $\mu_{\bar{x}} = \mu = 50$ and standard deviation $\sigma_{\bar{x}} = \sigma/\sqrt{n} = 8/\sqrt{n}$.

a. For samples of size 3, $\mu_{\bar{x}} = \mu = 50$ and $\sigma_{\bar{x}} = \sigma/\sqrt{n} = 8/\sqrt{3} = 4.62$. So for random samples of size 3, the sampling distribution of the mean is a normal distribution with $\mu_{\bar{x}} = 50$ and $\sigma_{\bar{x}} = 4.62$. The normal curve for $\bar{x}$ is shown in Fig. 7.4(b).

b. For samples of size 10, $\mu_{\bar{x}} = \mu = 50$ and $\sigma_{\bar{x}} = \sigma/\sqrt{n} = 8/\sqrt{10} = 2.53$. So for random samples of size 10, the sampling distribution of the mean is a normal distribution with $\mu_{\bar{x}} = 50$ and $\sigma_{\bar{x}} = 2.53$. The normal curve for $\bar{x}$ is shown in Fig. 7.4(c). ∎

The normal curves in Figs. 7.4(b) and 7.4(c) are drawn to scale so that we can compare them visually and observe two important things we already know. First, both curves are centered at the value of the population mean, $\mu = 50$, since $\mu_{\bar{x}} = \mu = 50$ regardless of the sample size, n. Second, the spread of the curve for $n = 10$ is less extensive than that of the curve for $n = 3$. This is because the standard deviation of the random variable $\bar{x}$ decreases with increasing sample size ($\sigma_{\bar{x}} = \sigma/\sqrt{n}$).

The two curves illustrate vividly that the sampling error made in estimating a population mean, μ, by a sample mean, $\bar{x}$, tends to be smaller for larger sample sizes; that is, the larger the sample size, the more likely that $\bar{x}$ will be close to μ.

We now illustrate how to determine probabilities for the random variable $\bar{x}$ when sampling from a normally distributed population. Since under those circumstances $\bar{x}$ is normally distributed, its probabilities are equal to areas under the normal curve with parameters $\mu_{\bar{x}} = \mu$ and $\sigma_{\bar{x}} = \sigma/\sqrt{n}$.

Example 7.8   Illustrates Key Fact 7.4

The ages of U.S. farm operators are normally distributed with a mean of 50 years and a standard deviation of 8 years. If 250 farm operators are selected at random, find the probability that their mean age, $\bar{x}$, will be within 1 year of the mean age of all farm operators; that is, determine the probability that $\bar{x}$ will be between 49 and 51 years.

SOLUTION   Since the population is normally distributed, we know from Key Fact 7.4 on page 420 that the random variable $\bar{x}$ is also normally distributed and that $\mu_{\bar{x}} = \mu = 50$ and $\sigma_{\bar{x}} = \sigma/\sqrt{n} = 8/\sqrt{250} = 0.51$. Therefore probabilities for $\bar{x}$ are equal to areas under the normal curve with parameters $\mu_{\bar{x}} = 50$ and $\sigma_{\bar{x}} = 0.51$.

The problem here is to find $P(49 \le \bar{x} \le 51)$, the probability that the mean age of the 250 randomly selected farm operators will be between 49 and 51 years. So we need to find the area under the normal curve with parameters 50 and 0.51 that lies between 49 and 51. That normal-curve area is found in the usual way, as depicted in Fig. 7.5.

FIGURE 7.5

Determination of the area under the normal curve with parameters $\mu_{\bar{x}} = 50$ and $\sigma_{\bar{x}} = 0.51$ that lies between 49 and 51

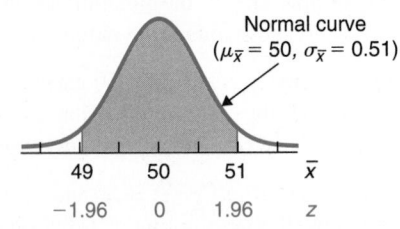

Normal curve
($\mu_{\bar{x}} = 50$, $\sigma_{\bar{x}} = 0.51$)

	49	50	51	$\bar{x}$
	−1.96	0	1.96	z

z-score computations: Area to the left of z:

$\bar{x} = 49 \longrightarrow z = \dfrac{49 - 50}{0.51} = -1.96$ 0.0250

$\bar{x} = 51 \longrightarrow z = \dfrac{51 - 50}{0.51} = 1.96$ 0.9750

Shaded area = $0.9750 - 0.0250 = 0.9500$

Therefore $P(49 \leq \bar{x} \leq 51) = 0.95$. We can interpret this result as follows: Suppose 250 farm operators are to be selected at random. Then there is a 95% chance that the mean age, $\bar{x}$, of the 250 farm operators obtained will be within 1 year of the mean age, $\mu = 50$, of all farm operators. We can already begin to see the power of sampling. ■

Qualitatively, Key Fact 7.4 implies that for any particular sample size, a histogram of all possible sample means will be bell-shaped when sampling from a normally distributed population. Alternatively, it implies that if we take a large number of random samples of a particular size from a normally distributed population, then a histogram of the sample means will be approximately bell-shaped.

To illustrate this last interpretation of Key Fact 7.4, we used Minitab. First we simulated the taking of 1000 random samples of size three from the normal population of ages considered in Examples 7.7 and 7.8. Next we determined the mean of each of the 1000 samples. And then we obtained a histogram of those 1000 sample means. Printout 7.1 displays the histogram.

PRINTOUT 7.1

Minitab histogram of the sample means for 1000 samples of size three from a normal population

```
Histogram of XBARS    N = 1000
Each * represents 5 obs.

Midpoint    Count
      36        2   *
      38        6   **
      40       12   ***
      42       46   *********
      44       75   ***************
      46      109   **********************
      48      134   **************************
      50      180   ************************************
      52      159   ********************************
      54      128   *************************
      56       83   *****************
      58       47   *********
      60       12   ***
      62        5   *
      64        2   *
```

As expected, the histogram of the 1000 sample means is roughly bell-shaped. Compare the histogram in Printout 7.1 to the theoretical sampling distribution shown in Fig. 7.4(b) on page 421.

General Populations; The Central Limit Theorem

According to Key Fact 7.4, if the population being sampled is normally distributed, then the random variable $\bar{x}$ is also normally distributed. What if the population being sampled is not normally distributed? Remarkably, the random variable $\bar{x}$ is still approximately normally distributed, provided only that the sample size is relatively large. This extraordinary fact is called the **central limit theorem.**

KEY FACT 7.5

THE CENTRAL LIMIT THEOREM

> For a relatively large sample size, the random variable $\bar{x}$ is approximately normally distributed, regardless of the population's distribution. The approximation becomes better and better with increasing sample size.

Generally speaking, the more nonnormal a population, the larger the sample size must be for a normal distribution to provide an adequate approximation to the probability distribution of $\bar{x}$. As a rule of thumb, we will consider sample sizes of 30 or more ($n \geq 30$) large enough. Thus we have the following statement regarding the sampling distribution of the mean.

KEY FACT 7.6

THE SAMPLING DISTRIBUTION OF THE MEAN FOR GENERAL POPULATIONS

> Suppose a random sample of size $n \geq 30$ is to be taken from a population with mean μ and standard deviation σ. Then regardless of the distribution of the population, the random variable $\bar{x}$ is approximately normally distributed and has mean $\mu_{\bar{x}} = \mu$ and standard deviation $\sigma_{\bar{x}} = \sigma/\sqrt{n}$. In other words, if the sample size, n, is 30 or more, then probabilities for $\bar{x}$ are approximately equal to areas under the normal curve with parameters μ and $\sigma/\sqrt{n}$.

Example 7.9   Illustrates Key Fact 7.6

According to the Census Bureau publication *Current Population Reports,* a frequency distribution for the number of people per household in the United States is as displayed in Table 7.10. Frequencies are in millions of households.

TABLE 7.10
Frequency distribution
for U.S. household size

No. of people	1	2	3	4	5	6	7
Frequency	19.4	26.5	14.6	12.9	6.1	2.5	1.6

Applying the formulas for the mean and standard deviation of grouped population data, Formulas 3.4 and 3.5 on page 166, to the data in Table 7.10, we find that the mean household size is 2.685 and the standard deviation is 1.47. Suppose 30 households are randomly selected. Determine the probability that the mean size of the households obtained will exceed the population mean size by more than 0.5.

SOLUTION   In Fig. 7.6 we have presented a relative-frequency histogram for the population of household sizes, obtained by consulting Table 7.10.

FIGURE 7.6

Relative-frequency histogram for household size

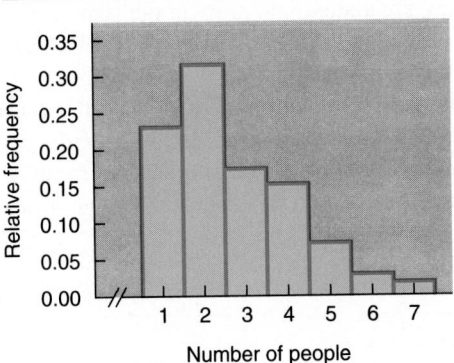

Number of people

Notice that the population is far from being normally distributed; it is clearly right skewed. Nonetheless, since the sample size is 30, it follows from Key Fact 7.6 that the random variable $\bar{x}$ is approximately normally distributed and has mean $\mu_{\bar{x}} = \mu = 2.685$ and standard deviation $\sigma_{\bar{x}} = \sigma/\sqrt{n} = 1.47/\sqrt{30} = 0.27$.

The problem is to determine the probability that $\bar{x}$ will exceed 2.685 by more than 0.5, $P(\bar{x} > 3.185)$, which we now know approximately equals the area under the normal curve with parameters 2.685 and 0.27 that lies to the right of 3.185. We compute that area in the usual way in Fig. 7.7.

FIGURE 7.7

Determination of the area under the normal curve with parameters $\mu_{\bar{x}} = 2.685$ and $\sigma_{\bar{x}} = 0.27$ that lies to the right of 3.185

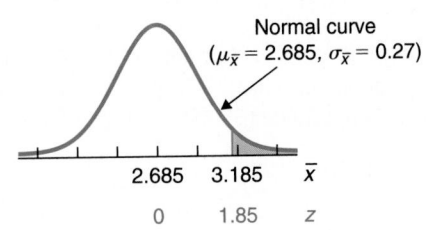

Normal curve
($\mu_{\bar{x}} = 2.685$, $\sigma_{\bar{x}} = 0.27$)

| | 2.685 | 3.185 | $\bar{x}$ |
| | 0 | 1.85 | z |

z-score computation: Area to the left of z:

$\bar{x} = 3.185 \longrightarrow z = \dfrac{3.185 - 2.685}{0.27} = 1.85$ 0.9678

Shaded area $= 1 - 0.9678 = 0.0322$

Consequently, we see that $P(\overline{x} > 3.185) = 0.0322$. There is only a 3.2% chance that the mean size of the 30 households selected will exceed the population mean size by more than 0.5.   ∎

Qualitatively, Key Fact 7.6 implies that for any population and any sample size of 30 or more, a histogram of all possible sample means will be roughly bell-shaped. Alternatively, it implies that for any population and any sample size of 30 or more, if we take a large number of random samples of that particular size, then a histogram of the sample means will be approximately bell-shaped.

To illustrate this last interpretation of Key Fact 7.6, we used Minitab. First we simulated the taking of 500 random samples of size 30 from the population of household sizes considered in Example 7.9. Next we determined the mean of each of the 500 samples. And then we obtained a histogram of those 500 sample means. Printout 7.2 displays the histogram.

PRINTOUT 7.2

Minitab histogram of the sample means for 500 samples of size 30 from a nonnormal population

```
Histogram of XBARS    N = 500
Each * represents 5 obs.

Midpoint    Count
     1.8        2   *
     2.0        0
     2.2       27   ******
     2.4       83   ****************
     2.6      145   *****************************
     2.8      139   ****************************
     3.0       64   *************
     3.2       33   *******
     3.4        4   *
     3.6        3   *
```

As expected, the histogram of the 500 sample means is roughly bell-shaped, even though the population itself is highly skewed. Compare the histogram in Printout 7.2 to the theoretical sampling distribution in Fig. 7.7 on page 425.

Some Visual Displays

In the preceding pages of this section, we have examined the sampling distribution of the mean. We discovered that if the population being sampled is normally distributed, then so is the random variable $\overline{x}$, regardless of the sample size. This fact is portrayed graphically in Fig. 7.8(a).

FIGURE 7.8
Population and sampling
distributions for (a) normal,
(b) reverse-J-shaped, and
(c) uniform populations

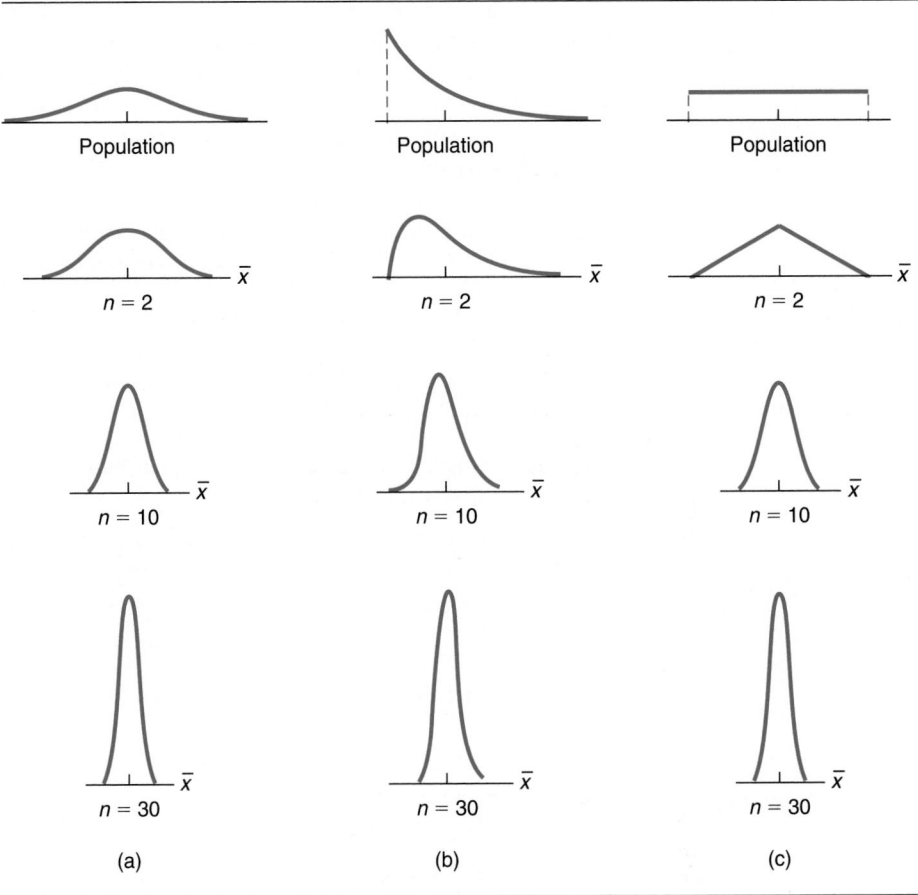

Additionally, the central limit theorem indicates that if the sample size is relatively large, then the random variable $\bar{x}$ is approximately normally distributed, regardless of the population's distribution. Figs. 7.8(b) and 7.8(c) provide a visual display of the central limit theorem for two nonnormal populations, one reverse J-shaped and the other uniform.

In both of these cases, we see the following: For samples of size 2, the random variable $\bar{x}$ is definitely not normally distributed; for samples of size 10, the random variable $\bar{x}$ is already somewhat normally distributed; and for samples of size 30, the random variable $\bar{x}$ is very close to being normally distributed.

Standardized Form

In Chapters 8 and 9 we will need to consider the standardized version of the random variable $\bar{x}$, that is,

$$z = \frac{\bar{x} - \mu_{\bar{x}}}{\sigma_{\bar{x}}} = \frac{\bar{x} - \mu}{\sigma/\sqrt{n}}.$$

We learned in Key Fact 6.8 that if we standardize a normally distributed random variable, then the resulting random variable has the standard normal distribution. Therefore we can summarize the results of Section 7.2 and this section as follows.

KEY FACT 7.7   **THE SAMPLING DISTRIBUTION OF THE MEAN (STANDARDIZED FORM)**

Suppose a random sample of size n is to be taken from a population with mean μ and standard deviation σ. Then the **standardized version of $\bar{x}$,**

$$z = \frac{\bar{x} - \mu}{\sigma/\sqrt{n}},$$

1. has exactly the standard normal distribution if the population itself is normally distributed, regardless of sample size;

2. has approximately the standard normal distribution if $n \geq 30$, regardless of how the population is distributed.

In other words, if either the population is normally distributed or the sample size is at least 30, then probabilities for the random variable

$$z = \frac{\bar{x} - \mu}{\sigma/\sqrt{n}}$$

are equal, at least approximately, to areas under the standard normal curve.

Example 7.10   Illustrates Key Fact 7.7

As reported by the Bureau of Labor Statistics in *Employment and Earnings,* the mean weekly earnings of workers in the trucking industry is $456. The standard deviation is $63. Suppose 81 workers in the trucking industry are to be selected at random. Let $\bar{x}$ denote the mean weekly earnings of the workers obtained.

a. Determine the standardized version of the random variable $\bar{x}$.

b. Identify the probability distribution of the standardized version of $\bar{x}$.

c. Find the probability that the standardized version of $\bar{x}$ will be between -2 and 2.

SOLUTION   a. Since $\mu = 456$, $\sigma = 63$, and $n = 81$, the standardized version of $\bar{x}$ is

$$z = \frac{\bar{x} - \mu_{\bar{x}}}{\sigma_{\bar{x}}} = \frac{\bar{x} - \mu}{\sigma/\sqrt{n}} = \frac{\bar{x} - 456}{63/\sqrt{81}} = \frac{\bar{x} - 456}{7}.$$

b. Because the sample size, $n = 81$, exceeds 30, the standardized version of $\bar{x}$,

$$z = \frac{\bar{x} - 456}{7},$$

has approximately the standard normal distribution. This is true regardless of how the population of weekly earnings is distributed, because $n \geq 30$. However, if that population *is* normally distributed, then $z = (\bar{x} - 456)/7$ has exactly, not just approximately, the standard normal distribution.

c. From part (b), the standardized version, z, of $\bar{x}$ has approximately the standard normal distribution. Therefore the probability that z will be between -2 and 2 is approximately equal to the area under the standard normal curve between -2 and 2. From Table II, that area is $0.9772 - 0.0228 = 0.9544$. Thus

$$P(-2 \le z \le 2) = P\left(-2 \le \frac{\bar{x} - 456}{7} \le 2\right) = 0.9544,$$

approximately. ∎

EXERCISES 7.3

7.33 A population has a mean of $\mu = 100$ and a standard deviation of $\sigma = 28$.
a. Determine the sampling distribution of the mean for random samples of size 49.
b. In answering part (a), what assumptions did you make about the distribution of the population?
c. Can you answer part (a) if the sample size is 16 instead of 49? Why or why not?

7.34 A population has a mean of $\mu = 35$ and a standard deviation of $\sigma = 42$.
a. If the population is normally distributed, determine the sampling distribution of the mean for random samples of size 9.
b. Can you answer part (a) if the distribution of the population is unknown? Explain your answer.
c. Can you answer part (a) if the distribution of the population is unknown but the sample size is 36 instead of 9? Why or why not?

7.35 Suppose a random sample of size n is to be taken from a normally distributed population with mean μ and standard deviation σ.
a. Identify the probability distribution of $\bar{x}$.
b. Does your answer to part (a) depend on how large the sample size is? Explain your answer.
c. What are the mean and the standard deviation of $\bar{x}$?

7.36 The length of the western rattlesnake is normally distributed with mean $\mu = 42$ inches and standard deviation $\sigma = 2.04$ inches.
a. Sketch the normal curve for this population.

b. Determine the sampling distribution of the mean for random samples of size four. Draw the normal curve for $\bar{x}$.
c. Repeat part (b) for samples of size eight.

7.37 As reported by the U.S. National Center for Health Statistics in *Vital and Health Statistics*, males who are 6 feet tall and between 18 and 24 years of age have a mean weight of 175 lb. Assume the weights are normally distributed and have a standard deviation of 14 lb.
a. Sketch the normal curve for this population.
b. Determine the sampling distribution of the mean for samples of size two and sketch the associated normal curve.
c. Repeat part (b) for samples of size nine.

7.38 Refer to Exercise 7.36. Suppose a random sample of 16 snakes is to be taken.
a. Find the probability that the mean length, $\bar{x}$, of the snakes obtained will be within 1 inch of the population mean of 42 inches, that is, between 41 and 43 inches.
b. Interpret your result in part (a) in terms of sampling error.
c. For samples of size 16, what percentage of the possible samples have means that lie within 1 inch of the population mean of 42 inches?
d. Repeat part (a) for a sample of size 50.

7.39 Refer to Exercise 7.37. Suppose nine 6-foot-tall males age 18–24 years are to be selected at random.
a. Determine the probability that the mean weight, $\bar{x}$, of the males obtained will be within 5 lb of the population mean weight of 175 lb, that is, between 170 and 180 lb.
b. Interpret your result in part (a) in terms of sampling error.
c. What percentage of all possible samples of size nine have means that lie between 170 and 180?
d. Repeat part (a) for a sample of size 30.

·   **7.40**   The *Arizona Regional Multiple Listing Service* publishes several books for the real-estate industry. One such book deals with the region termed the "Residential Southwest," composed of Mesa, Tempe, and Chandler, Arizona. This book indicates that the mean sale price of homes in that region is $77,500. Suppose you take a random sample of 250 homes that have recently sold.

a. Assuming a standard deviation of $57,000, find the probability that the mean sale price of the 250 homes you obtain will be within $5000 of the population mean sale price of $77,500.
b. Must you assume the sale prices are normally distributed to answer part (a)? Why or why not?
c. Sketch the normal curve for $\bar{x}$ for random samples of size 250.
d. Repeat part (a) for a sample size of 150.

·   **7.41**   According to *Current Population Reports,* published by the U.S. Bureau of the Census, the mean annual alimony income received by women is $4000. Assume a standard deviation of $7500. Suppose 100 women receiving alimony are selected at random.

a. Determine the probability that the mean annual alimony received by the 100 women will be within $500 of the population mean of $4000.
b. Must you assume the population of annual alimony payments is normally distributed to answer part (a)? Explain your answer.
c. Sketch the normal curve for $\bar{x}$ for random samples of size 100.
d. Repeat part (a) for a sample of size 1000.
e. For the alimony incomes considered here, why is it necessary to take such a large sample in order to be assured of relatively small sampling error?

·   **7.42**   An air-conditioning contractor is preparing to offer service contracts on the brand of compressor used in all of the units her company installs. Before she can work out the details, she must have an idea of how long those compressors last on the average. The contractor anticipated this need and has kept detailed records on the lifetimes of a random sample of 250 compressors. She plans to use the sample mean lifetime, $\bar{x}$, of those 250 compressors as her estimate for the population mean lifetime, μ, of all such compressors. If the lifetimes of this brand of compressor have a standard deviation of 40 months, what is the probability that the contractor's estimate will be within 5 months of the true mean?

·   **7.43**   An economist employed by the Department of Agriculture needs to estimate the mean weekly food cost for couples with two children 6–11 years old. The economist plans to randomly sample 500 such families and use the mean weekly food cost, $\bar{x}$, of those families as her estimate of the true mean weekly food cost, μ. If $\sigma = \$17.20$, determine the probability that the economist's estimate will be within $1 of the actual mean.

·   **7.44**   Refer to Fig. 7.8 on page 427.
a. Why are the four graphs in Fig. 7.8(a) all centered at the same place?
b. Why does the spread of the graphs diminish with increasing sample size? How does this fact affect the sampling error when estimating a population mean, μ, by a sample mean, $\bar{x}$?
c. Why are the graphs in Fig. 7.8(a) bell-shaped?
d. Why do the graphs in Figs. 7.8(b) and 7.8(c) become bell-shaped as the sample size increases?

·   **7.45**   IQs measured on the Stanford Revision of the Binet-Simon Intelligence Scale are known to be normally distributed with mean $\mu = 100$ and standard deviation $\sigma = 16$. Suppose nine people are to be selected at random, and let $\bar{x}$ denote the mean IQ of the people obtained.
a. Find the standardized version of the random variable $\bar{x}$.
b. Identify the probability distribution of the standardized version of $\bar{x}$.
c. Does your answer to part (b) depend on the fact that IQs are normally distributed? Why or why not?
d. Find the probability that the standardized version of $\bar{x}$ will be between -1.64 and 1.64.

·   **7.46**   The U.S. Energy Information Administration collects data on household vehicles and publishes the results in *Residential Transportation Energy Consumption Survey, Consumption Patterns of Household Vehicles.* This document states that the mean monthly fuel expenditure per household vehicle is $58.80. The standard deviation is $30.40. Suppose 50 household vehicles are to be selected at random, one for each of 50 randomly selected months. Let $\bar{x}$ denote the mean monthly fuel expenditure per vehicle selected.
a. Find the standardized version of the random variable $\bar{x}$.
b. Identify the probability distribution of the standardized version of $\bar{x}$.
c. Does your answer to part (b) depend on the fact that the sample size is at least 30? Why or why not?
d. Find the probability that the standardized version of $\bar{x}$ will be less than 1.96.

·   **7.47**   A brand of water-softener salt comes in packages marked "net weight 40 lb." The company claims that the bags contain an average of 40 lb of salt and that the standard deviation of the weights is 1.5 lb. Furthermore, it is known that the weights are normally distributed.

a. Obtain the probability that the weight of one randomly selected bag of water-softener salt will be 39 lb or less, if the company's claim is true.

b. Obtain the probability that the mean weight of 10 randomly selected bags of water-softener salt will be 39 lb or less, if the company's claim is true.

c. If you bought one bag of water-softener salt and it weighed 39 lb, would you consider this evidence that the company's claim is incorrect? Explain your answer.

d. If you bought 10 bags of water-softener salt and their mean weight was 39 lb, would you consider this evidence that the company's claim is incorrect? Explain your answer.

7.48 According to the Salt River Project, a supplier of electricity to the greater Phoenix area, the mean annual electric bill in 1984 was $852.31. Assume that for any particular year, annual electric bills are normally distributed and have a standard deviation of $204. At the end of 1986, an independent consumer agency wanted to determine whether the mean annual electric bill had increased over the 1984 figure of $852.31.

a. Determine the probability that the mean of a random sample of 25 annual electric bills for 1986 will be $875 or greater if the 1986 mean annual electric bill for all customers was still $852.31.

b. If the consumer agency takes a random sample of 25 annual electric bills for 1986 and finds that $\bar{x} = \$875$, does this provide evidence that the 1986 mean was greater than the 1984 mean of $852.31? [*Hint:* Refer to your answer from part (a).]

c. Repeat parts (a) and (b) for $n = 250$.

d. To answer part (c), is it necessary to assume that annual electric bills are normally distributed?

Note: Some statistical software will not have sufficient storage space to carry out Exercises 7.49 and 7.50.

7.49 (Computer exercise) Consider a normally distributed population with mean 100 and standard deviation 16.

a. Simulate the taking of 2000 random samples of size four from the population.

b. Find the sample mean of each of the 2000 samples.

c. Obtain the mean, the standard deviation, and a histogram of the 2000 sample means.

d. Theoretically, what are the mean, standard deviation, and distribution of all possible sample means for samples of size four from the population?

e. Compare your results from parts (c) and (d).

7.50 (Computer exercise) A population is said to be **exponentially distributed** if percentages for the population are equal to areas under the curve $y = e^{-x/\mu}/\mu$ for $x > 0$, where μ is the mean of the population. The standard deviation of such a population also equals μ. Consider now an exponentially distributed population with mean $\mu = 4$.

a. Sketch the exponential curve for this population. Note that the population is far from being normally distributed. What shape does it have?

b. Simulate the taking of 300 random samples of size four from the population.

c. Find the sample mean of each of the 300 samples.

d. Determine the mean and standard deviation of the 300 sample means.

e. Theoretically, what are the mean and the standard deviation of all possible sample means for samples of size four from the population? Compare these values to the ones you obtained in part (d).

f. Obtain a histogram of the 300 sample means. Is the histogram bell-shaped? Would you necessarily expect it to be?

g. Repeat parts (b)–(f) for a sample size of 40.

Chapter **Review**

Key Terms   central limit theorem, *424*
mean of $\bar{x}$ ($\mu_{\bar{x}}$), *412*
normal population, *421*
sampling distribution, *404*
sampling distribution of the mean, *407*

sampling error, *404*
standard deviation of $\bar{x}$ ($\sigma_{\bar{x}}$), *412*
standard error of the mean, *417*
standardized version of $\bar{x}$, *428*

Formulas   Mean of the random variable $\bar{x}$, *413*

$$\mu_{\bar{x}} = \mu$$

(μ = mean of the population)

Standard deviation of the random variable $\bar{x}$, *416*

$$\sigma_{\bar{x}} = \frac{\sigma}{\sqrt{n}}$$

(σ = standard deviation of the population, n = sample size)

Standardized version of the random variable $\bar{x}$, *428*

$$z = \frac{\bar{x} - \mu}{\sigma/\sqrt{n}}$$

You Should Be   1. use and understand the preceding formulas.
Able To   2. define sampling error and explain the need for sampling distributions.
3. find the mean and standard deviation of the random variable $\bar{x}$, given the mean and standard deviation of the population and the sample size.
4. state and apply the central limit theorem.
5. determine the sampling distribution of the mean when the population being sampled is normally distributed or when the sample size is at least 30.
6. find probabilities for $\bar{x}$ when sampling from a normally distributed population.
7. find probabilities for $\bar{x}$ when the sample size is at least 30.
8. obtain the standardized version of the random variable $\bar{x}$ and identify its probability distribution when the population being sampled is normally distributed or when the sample size is at least 30.

REVIEW TEST

1. In 1989 the U.S. Internal Revenue Service (IRS) sampled approximately 125,000 tax returns to obtain estimates of various parameters. Data were published in *Statistics of Income, Individual Income Tax Returns.* According to that document, the mean income tax per taxable return for the returns sampled was $\bar{x}$ = $4855.
   a. Explain the meaning of sampling error in this context.
   b. If the population mean income tax per taxable return in 1989 was μ = $4943, how much sampling error was made in estimating μ by $\bar{x}$?
   c. If the IRS had sampled 250,000 returns instead of 125,000, would the sampling error necessarily have been smaller? Explain your answer.
   d. In future surveys how can the IRS increase the likelihood of a small sampling error?

2. Consider these six salaries (in $1000s): 8, 12, 16, 20, 24, and 28.
   a. Calculate μ for this population of six salaries.

There are 15 possible samples of size four from the population of six salaries. They are listed in the first column of the following table.

Sample	$\bar{x}$
8, 12, 16, 20	14
8, 12, 16, 24	15
8, 12, 16, 28	16
8, 12, 20, 24	16
8, 12, 20, 28	17
8, 12, 24, 28	18
8, 16, 20, 24	
8, 16, 20, 28	
8, 16, 24, 28	
8, 20, 24, 28	
12, 16, 20, 24	
12, 16, 20, 28	
12, 16, 24, 28	
12, 20, 24, 28	
16, 20, 24, 28	

b. Complete the $\bar{x}$-column of the above table.
c. Complete the dotplot of the possible $\bar{x}$-values shown below; locate the value of μ on the graph.

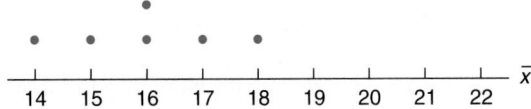

d. Determine the sampling distribution of the mean for samples of size four.
e. Obtain the probability that the mean of a random sample of four salaries will be within 1 (i.e., $1000) of the population mean.
f. Calculate the mean of all the $\bar{x}$-values.
g. Could you have obtained the answer to part (f) without performing the computations? Explain your answer.

3. As reported by the U.S. Department of Agriculture in *Crop Production*, the mean yield of cotton per acre for U.S. farms is 506 lb. The standard deviation is 237 lb.
   a. Suppose 25 one-acre plots of cotton are to be selected at random. Let $\bar{x}$ denote the mean yield of the 25 plots obtained. Find the mean and standard deviation of $\bar{x}$.
   b. Repeat part (a) for a sample size of 200.
   c. Suppose 500 one-acre plots of cotton are to be selected at random. Without doing any computations, answer the following question: Will the value of $\sigma_{\bar{x}}$ here be larger, smaller, or the same as the value of $\sigma_{\bar{x}}$ in part (b)? Why?

4. A population, which may or may not be normally distributed, has mean $\mu = 40$ and standard deviation $\sigma = 10$. Suppose 100 members of the population are to be selected at random. Decide whether each of the following statements is true or false or whether it is not possible to tell. Give a reason for each of your answers.
   a. There is roughly a 68.26% chance that the mean of the sample will be between 30 and 50.
   b. Approximately 68.26% of the population values lie between 30 and 50.
   c. There is roughly a 68.26% chance that the mean of the sample will be between 39 and 41.

5. Repeat Problem 4 under the assumption that the population is normally distributed.

6. The monthly rents for studio apartments in a large city are normally distributed with a mean of $385 and a standard deviation of $45.
   a. Sketch the normal curve for this population of monthly rents.

b. Find the sampling distribution of the mean for samples of size three. Draw a graph of the normal curve for $\bar{x}$.
   c. Repeat part (b) for samples of size nine.

7. Refer to Problem 6. Suppose three studio apartments are to be selected at random.
   a. Obtain the probability that the mean monthly rent of the three apartments chosen will be within $10 of the population mean of $385, that is, between $375 and $395.
   b. Interpret the probability you obtained in part (a) in terms of sampling error.
   c. What percentage of the possible samples of size three have means within $10 of the population mean?
   d. Repeat part (a) for samples of size 75.

8. The following figure shows the curve for a normally distributed population (in color). Superimposed are the curves for the sampling distribution of the mean for two different sample sizes.

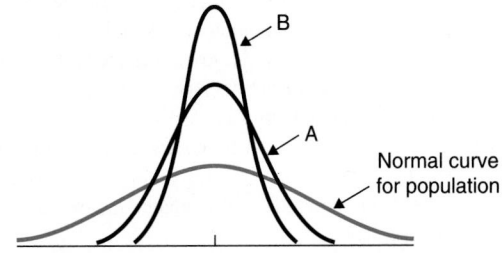

a. Explain why all three curves are centered at the same place.
   b. Which curve corresponds to the larger sample size? Explain your answer.
   c. Why is the spread of each curve different?
   d. Which of the two sampling-distribution curves corresponds to the sample size that will tend to produce less sampling error? Why?

9. The American Council of Life Insurance reports in *Life Insurance Fact Book* that the mean life insurance in force per covered family is $121,400. Assume a standard deviation of $50,900. Suppose 500 covered families are to be randomly selected.
   a. Determine the probability that the mean life insurance in force for the 500 families obtained will be within $2000 of the population mean of $121,400.
   b. Must you assume that the population of life-insurance amounts is normally distributed in order to answer part (a)? What if the sample size is 20 instead of 500?
   c. Repeat part (a) for a sample size of 5000.

10. The Social Security Administration compiles information on benefits to retired workers and publishes the data in *Social Security Bulletin.* According to that publication, the mean monthly benefit to retired workers is $603. Assume a standard deviation of $93. Suppose 125 retired workers are to be selected at random. Let $\bar{x}$ denote the mean monthly benefit received by the retired workers chosen.
    a. Determine the standardized version, z, of the random variable $\bar{x}$.
    b. Find the probability distribution of the standardized version of $\bar{x}$.
    c. Determine the probability that the standardized version of $\bar{x}$ will be either greater than 2.33 or less than -2.33.
    d. Could you answer part (b) if the sample size were 5 instead of 125? Why or why not?

11. A paint manufacturer in Pittsburgh claims his paint will last an average of 5 years. Assume paint life is normally distributed and has a standard deviation of 0.5 years.
    a. Suppose you paint one house with the paint and the paint lasts 4.5 years. Would you consider that evidence against the manufacturer's claim? *(Hint: Let x be the paint life for a randomly selected house painted with the paint. Compute $P(x \leq 4.5)$, assuming the manufacturer's claim is correct.)*
    b. Suppose you paint 10 houses with the paint and the paint lasts an average of 4.5 years for the 10 houses. Would you consider that evidence against the manufacturer's claim?
    c. Repeat part (b) if the paint lasts an average of 4.9 years for the 10 houses painted.

Note: Some statistical software will not have sufficient storage to carry out Problems 12 and 13.

*12. **(Computer problem)** Consider a normally distributed population with mean 70 and standard deviation 15.

    a. Simulate the taking of 500 random samples of size nine from the population.
    b. Determine the sample mean of each of the 500 samples obtained.
    c. Determine the mean, the standard deviation, and a histogram of the 500 sample means.
    d. Theoretically, what are the mean, standard deviation, and distribution of all possible sample means for samples of size nine from the population?
    e. Compare your results from parts (c) and (d).

*13. **(Computer problem)** A population is said to be **uniformly distributed** with parameters α and β if percentages for the population are equal to areas under the straight line $y = 1/(\beta - \alpha)$, for $\alpha < x < \beta$. The mean and standard deviation of such a population are $(\alpha + \beta)/2$ and $(\beta - \alpha)/\sqrt{12}$, respectively. Consider now a uniformly distributed population with parameters 0 and 1.
    a. Sketch the distribution of this population. Observe from your sketch that the population is far from being normally distributed.
    b. Simulate the taking of 400 random samples of size six from the population.
    c. Determine the sample mean of each of the 400 samples obtained.
    d. Determine the mean and standard deviation of the 400 sample means.
    e. Theoretically, what are the mean and the standard deviation of all possible sample means for samples of size six from the population? Compare these values to the ones you obtained in part (d).
    f. Obtain a histogram of the 400 sample means. Is the histogram bell-shaped? Would you necessarily expect it to be?
    g. Repeat parts (b)–(f) for a sample size of 35.

Using the Focus Database

Appendix B contains a printout of a database obtained by randomly selecting 500 Arizona State University sophomores. Seven variables are considered for each student: sex, high-school GPA, SAT math score, cumulative GPA, SAT verbal score, age, and total hours. Suppose that in addition to studying these seven variables, we wish to conduct extensive interviews on college life with 25 of the 500 sophomores. Use Minitab or some other statistical software to obtain a simple random sample of 25 of the 500 sophomores in the Focus database. *(Hint: Refer to the computer discussion in Example 1.10 on page 24.)*

Case Study

The Chesapeake and Ohio Freight Study

. . . .

When a freight shipment travels over several railroads, the freight charge is divided among them according to prearranged agreements. With each shipment of freight, a document called a *waybill* is issued that provides information on the goods, route, and total charges. From the waybill of any particular freight shipment, the amount due each railroad can be calculated.

For a large number of shipments, the computations required for allocating the shares properly among the railroads are time-consuming and costly. Consequently, if the division of total revenue among the railroads in question could be done accurately on the basis of a sample—as statisticians contend—then considerable savings could be realized in accounting and clerical costs.

To convince themselves of the validity of the sampling approach, officials of the Chesapeake and Ohio Railroad Company (C & O) undertook a study of freight shipments that had traveled over its Pere Marquette district and another railroad during a 6-month period. The total number of waybills for the 6-month period was known (22,984), as was the total freight revenue. The problem was to obtain an accurate estimate of the total revenue due C & O using as small a sample of waybills as possible.

Statistical theory was applied to determine the sample size required in order to obtain an estimate of the total revenue due C & O with a prescribed accuracy.[†] In all, 2072 of the 22,984 waybills, roughly 9%, were sampled. For each waybill in the sample, the necessary calculations were performed to find the amount of freight revenue for that shipment belonging to C & O. From those amounts the total revenue due C & O for all shipments was estimated to be $64,568.

Since all 22,984 waybills were available, a census could be taken to determine exactly the total revenue due C & O and thereby reveal the accuracy of the estimate obtained by sampling. The exact amount due C & O was found to be $64,651.

Consequently, we see that a sample of only 9% of the waybills yielded an estimate that was off by only $83, an error of roughly $\frac{1}{10}$ of 1%. Noting that, at the time, the cost of a complete examination was approximately $5000, whereas the cost of the sampling was only $1000, it is clear that sampling was preferable to a census—why spend $4000 to save an error of $83?

a. In the study described above, the $83 error was against C & O. Would it necessarily have to be that way?
b. In view of your answer to part (a), explain why the long-run average error would be even smaller than that found in the study.

[†] We will study sample-size determination in Section 8.3.

Inferential Statistics

*William
Gosset*

William Sealy Gosset was born in Canterbury, England, on June 13, 1876, the eldest son of Colonel Frederic Gosset and Agnes Sealy. He studied mathematics and chemistry at Winchester College and New College, Oxford, receiving a first-class degree in natural sciences in 1899.

After graduation, Gosset began work with Arthur Guinness and Sons, a brewery in Dublin, Ireland. Seeing the need for accurate statistical analyses of various brewing processes ranging from barley production to yeast fermentation, he pressed the firm to solicit mathematical advice. So in 1906 the brewery sent him to work under Karl Pearson at University College in London.

In the next few years, Gosset developed what has come to be called Student's t-distribution. This distribution has proved to be fundamental in statistical analyses involving normal distributions. In particular, Student's t-distribution is used in performing small-sample inferences for population means when the population being sampled is (approximately) normally distributed. Although the statistical theory for large samples had been completed in the early 1800s. no small-sample theory was available before Gosset.

Because Guinness's brewery prohibited its employees from publishing any of their research, Gosset published his contributions to statistical theory under the pseudonym "Student," thus the name "Student" in Student's t-distribution. Gosset remained with Guinness his entire life, moving to London in 1935 to take charge of a new brewery. But his tenure there was short-lived; he died in Beaconsfield, England, on October 16, 1937.

Chapter 8

Confidence Intervals for
One Population Mean

. . . . .

General Objective   We now begin our study of inferential statistics. In this chapter we will examine methods for estimating the mean of a population. As you might suspect, the statistic used to estimate a population mean, μ, is the sample mean, $\bar{x}$. Because of sampling error, we cannot expect $\bar{x}$ to be exactly equal to μ. Thus it is important to provide information about the accuracy of the estimate. This leads to the discussion of *confidence intervals,* the main topic of this chapter.

We will examine two procedures for obtaining confidence intervals for the mean of a population. The first procedure applies when the sample size is large; the second when the population being sampled is approximately normally distributed.

Chapter Outline

Case Study   *Zooplankton Nutrition in the Gulf of Mexico*

Zooplankton are tiny marine organisms that are the base of the food chain for many commercially fished animals. How do the nutrient-rich waters of the Mississippi River influence the nutritional quality of the zooplankton residing in the northern Gulf of Mexico? Two marine researchers set out to answer this question in the spring of 1993. We will examine their results in the case study at the end of this chapter.

8.1 ESTIMATING A POPULATION MEAN

A common problem in statistics is to obtain information about the mean, μ, of a population. For instance, we might be interested in

1. the mean tar content of a certain brand of cigarette,
2. the mean life of a newly developed steel-belted radial tire,
3. the mean gas mileage of a new-model car, or
4. the mean annual income of liberal-arts graduates.

If the population is small, we can ordinarily determine μ exactly by first taking a census and then computing μ from the population data. But, if the population is large, as is often the case in practice, then taking a census is generally impractical, extremely expensive, or impossible. Nonetheless we can usually obtain sufficiently accurate information about μ by taking a sample from the population. Let's look at an example.

Example 8.1 Using a Sample Mean to Estimate a Population Mean

The U.S. Bureau of the Census publishes annual price figures for new mobile homes in *Construction Reports*. The figures are obtained from sampling, not from a census. A random sample of 40 new mobile homes yields the prices, in thousands of dollars, shown in Table 8.1. Use the data to estimate the population mean price, μ, of all new mobile homes.

TABLE 8.1

Prices ($1000s) of 40 randomly selected new mobile homes

24.4	30.6	26.4	26.8	33.5	32.2	32.4	13.9
24.4	29.3	26.2	14.1	21.4	20.0	33.0	17.6
24.8	27.0	22.8	18.8	35.1	26.7	22.1	37.2
31.9	24.0	28.4	15.8	29.3	31.4	22.8	8.4
24.7	16.6	31.1	13.9	16.8	29.5	17.0	9.9

SOLUTION　As expected, we will estimate the mean price, μ, of all new mobile homes by the mean price, $\bar{x}$, of the 40 new mobile homes sampled. From Table 8.1 we find that

$$\bar{x} = \frac{\Sigma x}{n} = \frac{972.2}{40} = 24.31.$$

Therefore, based on the sample data, we estimate the mean price, μ, of all new mobile homes to be approximately $24.31 thousand, that is, $24,310. An estimate of this kind is called a **point estimate** for μ because it consists of a single number, or point. ∎

The term *point estimate* applies to the use of a statistic to estimate any parameter, not just a population mean. Thus we have Definition 8.1.

DEFINITION 8.1   **POINT ESTIMATE**

A *point estimate* for a parameter is the value of a statistic that is used to estimate the parameter.

As we learned in Chapter 7, it is unreasonable to expect that a sample mean, $\bar{x}$, will be exactly equal to the population mean, μ; some sampling error is to be anticipated. Therefore, in addition to reporting a point estimate for μ, we need to provide information that indicates the accuracy of the estimate. We can do this by giving a **confidence-interval estimate** for μ. With a confidence-interval estimate for μ, we use the mean of a sample to construct an interval of numbers and state how confident we are that μ lies in that interval. We will soon see how to obtain such confidence intervals.

The discussion in the preceding paragraph is summarized in Definition 8.2. Again, the terminology applies to any parameter, not only to a population mean.

DEFINITION 8.2   **CONFIDENCE-INTERVAL ESTIMATE; CONFIDENCE LEVEL**

A *confidence-interval estimate* for a parameter consists of an interval of numbers obtained from a point estimate of the parameter together with a percentage that specifies how confident we are that the parameter lies in the interval. The confidence percentage is called the *confidence level*.

We will now present a method for obtaining a confidence interval for a population mean when the sample size is large. Before proceeding, you should review the following two facts: (1) Key Fact 6.6 on page 372, which is the empirical rule for normally distributed random variables, and (2) Key Fact 7.6 on page 424, which gives the sampling distribution of the mean for general populations.

Example 8.2   Confidence Intervals

Suppose a random sample of size $n \geq 30$ is to be taken from a population with mean μ and standard deviation σ.

a. Determine the probability that the interval from

$$\bar{x} - 2 \cdot \frac{\sigma}{\sqrt{n}} \quad \text{to} \quad \bar{x} + 2 \cdot \frac{\sigma}{\sqrt{n}}$$

will contain the population mean, μ.

b. Interpret the result from part (a) in terms of percentages.

SOLUTION   a. First we note that since the sample size is at least 30, Key Fact 7.6 implies that the random variable $\bar{x}$ is approximately normally distributed with mean $\mu_{\bar{x}} = \mu$ and standard deviation $\sigma_{\bar{x}} = \sigma/\sqrt{n}$. Consequently, by Property 2 of Key Fact 6.6, the probability is approximately 0.9544 that an observed value of $\bar{x}$ will be within two standard deviations to either side of the mean:

$$P(\mu_{\bar{x}} - 2\sigma_{\bar{x}} < \bar{x} < \mu_{\bar{x}} + 2\sigma_{\bar{x}}) = 0.9544.$$

Since $\mu_{\bar{x}} = \mu$ and $\sigma_{\bar{x}} = \sigma/\sqrt{n}$, we can rewrite the preceding equation as

$$P\left(\mu - 2 \cdot \frac{\sigma}{\sqrt{n}} < \bar{x} < \mu + 2 \cdot \frac{\sigma}{\sqrt{n}}\right) = 0.9544.$$

In words, this equation states that the probability is 0.9544 that $\bar{x}$ will be within $2 \cdot \sigma/\sqrt{n}$ of μ. But saying that "$\bar{x}$ will be within $2 \cdot \sigma/\sqrt{n}$ of μ" is the same as saying that "μ will be within $2 \cdot \sigma/\sqrt{n}$ of $\bar{x}$." Therefore the preceding equation can be rewritten as

$$P\left(\bar{x} - 2 \cdot \frac{\sigma}{\sqrt{n}} < \mu < \bar{x} + 2 \cdot \frac{\sigma}{\sqrt{n}}\right) = 0.9544.$$

This last equation provides us with the answer to our question: The probability is 0.9544 that the interval from

$$\bar{x} - 2 \cdot \frac{\sigma}{\sqrt{n}} \quad \text{to} \quad \bar{x} + 2 \cdot \frac{\sigma}{\sqrt{n}}$$

will contain μ. It is important to realize that in this and similar statements, the random variable is $\bar{x}$, not μ. The population mean, μ, is a fixed number, although it may be unknown; the sample mean, $\bar{x}$, is a random variable—its value depends on chance, namely, on which sample is obtained.

b. Here we are to interpret the result obtained in part (a) in terms of percentages. One interpretation is that roughly 95.44% of all samples of size n have the property that the interval with endpoints $\bar{x} \pm 2\sigma/\sqrt{n}$ contains the population mean, μ. A second interpretation is that if we take a large number of random samples of size n, then approximately 95.44% of the samples obtained will have the property that the interval with endpoints $\bar{x} \pm 2\sigma/\sqrt{n}$ contains the population mean, μ. ∎

In Example 8.2, we could have just as well used Property 1 or Property 3 of Key Fact 6.6 instead of Property 2. By doing so we would have gotten a different confidence interval with a different confidence level. (See Exercises 8.11 and 8.12.)

Let's now apply the results of Example 8.2 to obtain a confidence interval for the mean price, μ, of all new mobile homes.

Example 8.3 Confidence Intervals

A random sample of 40 new mobile homes yields the price data shown in Table 8.1 on page 440. Obtain a 95.44% confidence interval for the mean price, μ, of all new mobile homes. Assume the population standard deviation of the prices is $7200.[†]

[†] We might know this from previous research or from a preliminary study of prices. The more usual case where σ is unknown will be considered in Section 8.2.

SOLUTION   Since the sample size, $n = 40$, is at least 30, we can apply the results of Example 8.2. From part (a) of that example, the probability is 0.9544 that the interval from

$$\bar{x} - 2 \cdot \frac{\sigma}{\sqrt{n}} \quad \text{to} \quad \bar{x} + 2 \cdot \frac{\sigma}{\sqrt{n}}$$

will contain μ. We know that $\sigma = 7.2$ (i.e., \$7.2 thousand). Also, from the data in Table 8.1, we find that $n = 40$ and $\bar{x} = 24.31$. Hence

$$\bar{x} - 2 \cdot \frac{\sigma}{\sqrt{n}} = 24.31 - 2 \cdot \frac{7.2}{\sqrt{40}} = 22.03$$

and

$$\bar{x} + 2 \cdot \frac{\sigma}{\sqrt{n}} = 24.31 + 2 \cdot \frac{7.2}{\sqrt{40}} = 26.59.$$

Thus our confidence interval for μ is from 22.03 to 26.59. Since in 95.44% of all possible samples of size 40, such intervals contain the population mean, μ, the confidence level is 95.44%. In other words, we can be 95.44% confident that the mean price, μ, of all new mobile homes is somewhere between \$22,030 and \$26,590. It is crucial to remember that this or any other 95.44% confidence interval may or may not contain the population mean, μ; but we can be 95.44% confident that it does. ∎

The confidence interval obtained depends on the value of $\bar{x}$, which in turn depends on the sample selected. For example, suppose the prices of the 40 new mobile homes sampled were as shown in Table 8.2 instead of as in Table 8.1.

TABLE 8.2

Prices (\$1000s) of another sample of 40 randomly selected new mobile homes

20.0	21.8	32.1	25.2	38.4	19.5	28.9	24.8
17.2	26.2	10.3	25.1	30.8	35.1	33.7	11.0
22.9	23.3	30.1	24.0	47.1	23.2	16.9	30.1
30.2	26.5	29.7	14.6	13.4	29.8	35.7	30.7
20.4	35.8	38.2	14.8	16.8	42.0	32.3	34.6

Then we would have $\bar{x} = 26.58$ so that

$$\bar{x} - 2 \cdot \frac{\sigma}{\sqrt{n}} = 26.58 - 2 \cdot \frac{7.2}{\sqrt{40}} = 24.30$$

and

$$\bar{x} + 2 \cdot \frac{\sigma}{\sqrt{n}} = 26.58 + 2 \cdot \frac{7.2}{\sqrt{40}} = 28.86.$$

Hence, in this case, the 95.44% confidence interval for μ would be the interval from 24.30 to 28.86. We could be 95.44% confident that the mean price, μ, of all new mobile homes is somewhere between \$24,300 and \$28,860.

Example 8.4 stresses the importance of interpreting a confidence interval correctly. It also illustrates something that we recently mentioned: the population mean, μ, may or may not lie in the confidence interval.

Example 8.4   Interpreting Confidence Intervals

Consider again the prices of new mobile homes. Suppose 40 new mobile homes are to be selected at random. Then, as we have seen, the probability is 0.9544 that the interval from

(1)
$$\bar{x} - 2 \cdot \frac{\sigma}{\sqrt{n}} \quad \text{to} \quad \bar{x} + 2 \cdot \frac{\sigma}{\sqrt{n}}$$

will contain the mean price, μ, of all new mobile homes. In other words, once the sample of 40 new mobile homes is obtained and their mean price, $\bar{x}$, is computed, the interval given by (1) will be a 95.44% confidence interval for μ.

To illustrate that the mean price, μ, of all new mobile homes may or may not lie in the 95.44% confidence interval obtained, we used a computer to simulate the sampling of 40 new mobile homes 20 times. For the simulation we assumed that $\mu = 25$ (i.e., $25 thousand). Of course in practice we don't know μ; we are assuming a value for μ to illustrate a point.

For each of the 20 samples of size 40, we did three things: computed the sample mean, $\bar{x}$; used (1) and the fact that $\sigma = 7.2$ to obtain the 95.44% confidence interval for μ; and noted whether the population mean, $\mu = 25$, actually lies in the confidence interval.

Figure 8.1 summarizes our results (values are in thousands of dollars). For each sample, we have drawn a graph on the right-hand side of Fig. 8.1. The dot represents the sample mean, $\bar{x}$, and the horizontal line represents the corresponding 95.44% confidence interval. As we see, the population mean, μ, lies in the confidence interval when and only when the horizontal line crosses the dashed line.

From Fig. 8.1 we observe that μ lies in the 95.44% confidence interval in 19 of the 20 samples, that is, in 95% of the samples. If instead of 20 samples, we simulated, say, 1000 samples, then we would most likely find that the percentage of those 1000 samples for which μ lies in the 95.44% confidence interval would be even closer to 95.44%. Consequently, we can be 95.44% confident that any computed 95.44% confidence interval will contain μ. ∎

Up to this point, we have only considered confidence intervals with a confidence level of 95.44%. To obtain such confidence intervals, we applied Key Fact 7.6 (the sampling distribution of the mean for general populations) and Property 2 of Key Fact 6.6 (the empirical rule for normally distributed random variables). If instead of using Key Fact 6.6, we use Key Fact 6.7 (the general empirical rule for normally distributed random variables), then we can obtain confidence intervals with any specified confidence level. We will do that in the next section.

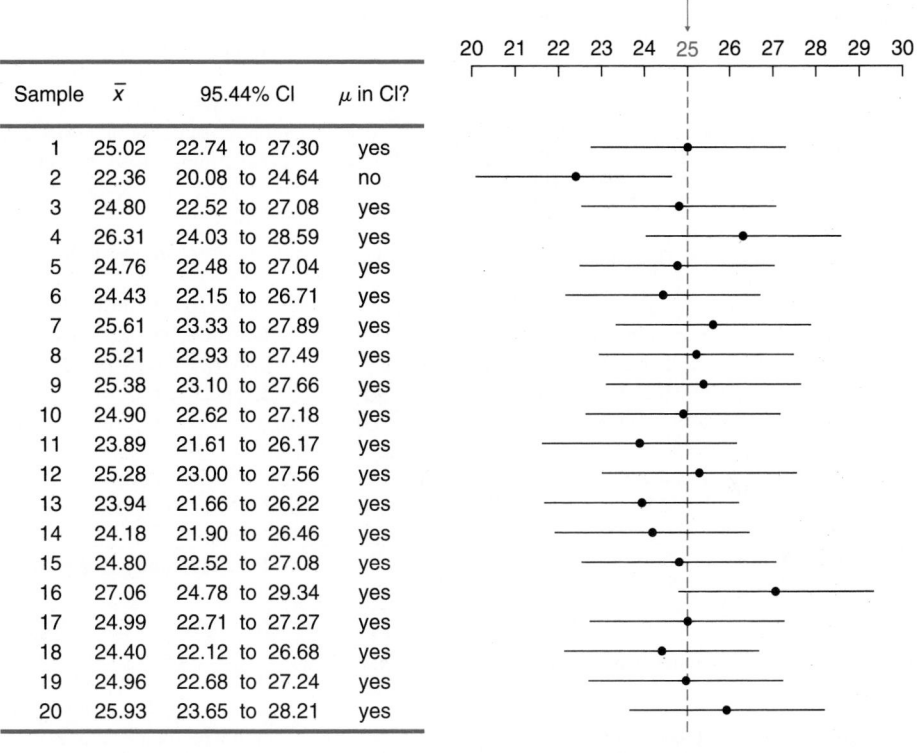

FIGURE 8.1

Confidence intervals for
20 random samples of size 40

Sample	x̄	95.44% CI	μ in CI?
1	25.02	22.74 to 27.30	yes
2	22.36	20.08 to 24.64	no
3	24.80	22.52 to 27.08	yes
4	26.31	24.03 to 28.59	yes
5	24.76	22.48 to 27.04	yes
6	24.43	22.15 to 26.71	yes
7	25.61	23.33 to 27.89	yes
8	25.21	22.93 to 27.49	yes
9	25.38	23.10 to 27.66	yes
10	24.90	22.62 to 27.18	yes
11	23.89	21.61 to 26.17	yes
12	25.28	23.00 to 27.56	yes
13	23.94	21.66 to 26.22	yes
14	24.18	21.90 to 26.46	yes
15	24.80	22.52 to 27.08	yes
16	27.06	24.78 to 29.34	yes
17	24.99	22.71 to 27.27	yes
18	24.40	22.12 to 26.68	yes
19	24.96	22.68 to 27.24	yes
20	25.93	23.65 to 28.21	yes

EXERCISES 8.1

8.1 The U.S. National Center for Health Statistics compiles natality (birth) statistics and publishes the results in *Vital Statistics of the United States*. A random sample of 35 babies yields the following birth weights, in pounds.

7.4	6.0	8.6	4.5	2.0	7.9	4.0
2.6	5.9	7.3	7.3	7.0	6.3	8.1
7.1	7.3	6.6	5.2	9.8	8.0	10.9
6.3	3.8	5.0	8.0	10.7	9.7	6.0
6.8	10.3	7.6	6.5	7.1	5.8	6.9

a. Use the data to obtain a point estimate for the population mean weight, μ, of all newborns. (*Note:* The sum of the data is 240.3 lb.)

b. Is it likely that your point estimate in part (a) is exactly equal to μ? Explain your answer.

8.2 An educational psychologist at a large university wants to estimate the mean IQ of the students in attendance. A random sample of 30 students gives the following data on IQs.

107	99	101	93	99	103
134	132	103	109	104	103
101	128	113	106	126	103
131	106	119	102	98	116
108	103	111	119	112	105

a. Use the data to obtain a point estimate for the mean IQ, μ, of all students attending the university. (*Note:* The sum of the data is 3294.)

b. Is it likely that your point estimate in part (a) is exactly equal to μ? Explain your answer.

8.3 The R. R. Bowker Company of New York collects data on books and periodicals. Sources for information are *Publishers Weekly, The Bowker Annual of Library and Book Trade Information,* and *Library Journal.* Forty randomly selected science books have the prices, to the nearest dollar, shown in the following table.

70	71	56	71	69	78	87	83
74	94	52	84	65	69	62	65
73	70	63	64	79	61	69	86
70	48	75	62	90	100	53	74
64	68	90	56	60	79	73	50

Obtain a point estimate for the mean price, μ, of all science books. *(Note: $\Sigma x = \$2827$.)*

8.4   *Trends in Television,* published by the Television Bureau of Advertising, contains information on the number of television sets owned by U.S. households. Fifty households are selected at random. The number of TV sets per household sampled is as follows.

1	1	1	2	6	3	3	4	2	4
3	2	1	5	2	1	3	6	2	2
3	1	1	4	3	2	2	2	2	3
0	3	1	2	1	2	3	1	1	3
3	2	1	2	1	1	3	1	5	1

Use the sample data to find a point estimate for the mean number of TV sets, μ, per U.S. household. *(Note: $\Sigma x = 114$.)*

8.5   Define the following terms.
a. point estimate   b. confidence-interval estimate
c. confidence level

8.6   Why is a confidence-interval estimate of a parameter more useful than a point estimate?

For Exercises 8.7–8.10, you may want to review Example 8.3, which begins on page 442.

8.7   Refer to Exercise 8.1. Assume the population standard deviation of weights for newborns is 1.9 lb.
a. Use the data displayed in Exercise 8.1 to determine a 95.44% confidence interval for the mean weight, μ, of all newborns.
b. Interpret your result in part (a).
c. Does the population mean birth weight, μ, lie in the confidence interval you obtained in part (a)? Explain your answer.

8.8   Refer to Exercise 8.2. Assume the standard deviation of IQs for all students attending the university is 12.
a. Use the data given in Exercise 8.2 to find a 95.44% confidence interval for the mean IQ, μ, of all students attending the university.

b. Interpret your result in part (a).
c. Does the mean IQ, μ, of all students attending the university lie in the confidence interval you obtained in part (a)? Explain your answer.

8.9   Refer to Exercise 8.3. Assume the standard deviation of prices for all science books is $15.
a. Determine a 95.44% confidence interval for the mean price of all science books based on the data displayed in Exercise 8.3.
b. Interpret your result in part (a).

8.10   Refer to Exercise 8.4. Assume $\sigma = 1.4$.
a. Obtain a 95.44% confidence interval for the mean number of TV sets per U.S. household using the data from Exercise 8.4.
b. Interpret your result in part (a).

8.11   Suppose a random sample of size $n \geq 30$ is to be taken from a population having mean μ and standard deviation σ.
a. Determine the probability that the interval from

$$\bar{x} - 3 \cdot \frac{\sigma}{\sqrt{n}} \quad \text{to} \quad \bar{x} + 3 \cdot \frac{\sigma}{\sqrt{n}}$$

will contain the population mean, μ. *(Hint: Refer to Example 8.2, which begins on page 441, and to the first paragraph following that example.)*
b. Interpret your result from part (a) in terms of percentages.
c. Apply your result from part (a) and the data in Exercise 8.1 to obtain a 99.74% confidence interval for the mean weight of all newborns. (Recall that $\sigma = 1.9$ lb.)

8.12   Suppose a random sample of size $n \geq 30$ is to be taken from a population having mean μ and standard deviation σ.
a. Determine the probability that the interval from

$$\bar{x} - 1 \cdot \frac{\sigma}{\sqrt{n}} \quad \text{to} \quad \bar{x} + 1 \cdot \frac{\sigma}{\sqrt{n}}$$

will contain the population mean, μ. *(Hint: Refer to Example 8.2, which starts on page 441, and to the first paragraph following that example.)*
b. Interpret your result from part (a) in terms of percentages.
c. Apply your result from part (a) and the data in Exercise 8.2 to obtain a 68.26% confidence interval for the mean IQ of all students attending the university. (Recall that $\sigma = 12$.)

8.13   *Employment and Earnings,* published by the U.S. Bureau of Labor Statistics, contains information on the ages of people 16 years old and over in the civilian labor force.

Fifty such people are randomly selected. Their ages are those displayed in the following table.

22	58	40	42	43	32	34	45	38	19
33	16	49	29	30	43	37	19	21	62
60	41	28	35	37	51	37	65	57	26
27	31	33	24	34	28	39	43	26	38
42	40	31	34	38	35	29	33	32	33

a. Use the data to obtain a point estimate for the mean age of all people in the civilian labor force. *(Note: $\Sigma x = 1819$.)*
b. Use the data to obtain a point estimate for the standard deviation, σ, of the ages of all people in the civilian labor force. *(Note: $\Sigma x^2 = 72,179$.)*
c. Obtain a 95.44% confidence interval for the mean age, μ, of all people in the civilian labor force. *(Hint: Use the point estimate obtained in part (b) in place of the unknown value of σ.)*

8.14 On page 442 we made the following statement: "... saying that '$\bar{x}$ will be within $2 \cdot \sigma/\sqrt{n}$ of μ' is the same as saying that 'μ will be within $2 \cdot \sigma/\sqrt{n}$ of $\bar{x}$.'" Mathematically this means that the two pairs of inequalities,

$$\mu - 2 \cdot \frac{\sigma}{\sqrt{n}} < \bar{x} < \mu + 2 \cdot \frac{\sigma}{\sqrt{n}}$$

and

$$\bar{x} - 2 \cdot \frac{\sigma}{\sqrt{n}} < \mu < \bar{x} + 2 \cdot \frac{\sigma}{\sqrt{n}},$$

are equivalent. Verify that by proving the following more general result: The inequalities

$$\mu - E < \bar{x} < \mu + E$$

and

$$\bar{x} - E < \mu < \bar{x} + E$$

are equivalent.

8.2 LARGE-SAMPLE CONFIDENCE INTERVALS FOR ONE POPULATION MEAN

In Section 8.1 we discovered how to find a large-sample confidence interval for a population mean, μ, with a confidence level of 95.44%. Now we will learn how to obtain a large-sample confidence interval for a population mean, μ, with any prescribed confidence level.

To begin, it will be helpful to introduce some general notation that is used with confidence intervals. Frequently, we want to write the confidence level in the form $1 - \alpha$, where α is a number between 0 and 1; that is, if the confidence level is expressed as a decimal, then α is the number that must be subtracted from 1 to get that confidence level. For example, if the confidence level is 0.9544 (i.e., 95.44%), then we have $0.9544 = 1 - 0.0456$ and, so, $\alpha = 0.0456$. Similarly, if the confidence level is 0.90, then $\alpha = 0.10$.

Next let's review the z_α-notation, which we introduced in Section 6.1. Recall that the symbol z_α is used to denote the z-value having area α to its right under the standard normal curve. So, for example, $z_{0.05}$ denotes the z-value having area 0.05 to its right under the standard normal curve, $z_{0.025}$ denotes the z-value having area 0.025 to its right under the standard normal curve, and $z_{\alpha/2}$ denotes the z-value having area $\alpha/2$ to its right under the standard normal curve.

Obtaining Large-Sample Confidence Intervals for a Population Mean

To obtain a large-sample confidence interval for a population mean with any specified confidence level, we proceed as in Example 8.2, with one exception. Instead of using Property 2 of Key Fact 6.6 (the empirical rule for normally distributed random variables), we use Key Fact 6.7 (the general empirical rule for normally distributed random variables).

Specifically, suppose a random sample of size $n \geq 30$ is to be taken from a population with mean μ and standard deviation σ. Then by Key Fact 7.6 on page 424, the random variable $\bar{x}$ is approximately normally distributed with mean $\mu_{\bar{x}} = \mu$ and standard deviation $\sigma_{\bar{x}} = \sigma/\sqrt{n}$. Consequently, by Key Fact 6.7 on page 373, the probability is approximately $1 - \alpha$ that an observed value of $\bar{x}$ will be within $z_{\alpha/2}$ standard deviations to either side of the mean:

$$P(\mu_{\bar{x}} - z_{\alpha/2} \cdot \sigma_{\bar{x}} < \bar{x} < \mu_{\bar{x}} + z_{\alpha/2} \cdot \sigma_{\bar{x}}) = 1 - \alpha.$$

Since $\mu_{\bar{x}} = \mu$ and $\sigma_{\bar{x}} = \sigma/\sqrt{n}$, we can express the above equation as

$$P\left(\mu - z_{\alpha/2} \cdot \frac{\sigma}{\sqrt{n}} < \bar{x} < \mu + z_{\alpha/2} \cdot \frac{\sigma}{\sqrt{n}}\right) = 1 - \alpha.$$

Using algebra, the preceding equation can be rewritten as

$$P\left(\bar{x} - z_{\alpha/2} \cdot \frac{\sigma}{\sqrt{n}} < \mu < \bar{x} + z_{\alpha/2} \cdot \frac{\sigma}{\sqrt{n}}\right) = 1 - \alpha.$$

This last equation shows that once the sample is taken, the interval from

$$\bar{x} - z_{\alpha/2} \cdot \frac{\sigma}{\sqrt{n}} \quad \text{to} \quad \bar{x} + z_{\alpha/2} \cdot \frac{\sigma}{\sqrt{n}}$$

will be a $(1 - \alpha)$-level confidence interval for μ. Thus we have the following procedure, which we sometimes refer to as the **one-sample z-interval procedure** or more briefly as the **z-interval procedure.**

PROCEDURE 8.1

THE ONE-SAMPLE z-INTERVAL PROCEDURE FOR A POPULATION MEAN, μ

ASSUMPTION

Large sample ($n \geq 30$)

Step 1   For a confidence level of $1 - \alpha$, use Table II to find $z_{\alpha/2}$.

Step 2   The confidence interval for μ is from

$$\bar{x} - z_{\alpha/2} \cdot \frac{\sigma}{\sqrt{n}} \quad \text{to} \quad \bar{x} + z_{\alpha/2} \cdot \frac{\sigma}{\sqrt{n}}$$

where $z_{\alpha/2}$ is found in Step 1, n is the sample size, and $\bar{x}$ is computed from the actual sample data obtained. If σ is unknown, use s in its place.

In Step 2 of Procedure 8.1 it is stated that if the population standard deviation, σ, is unknown, then the sample standard deviation, s, should be used in its place. This substitution is acceptable because for large samples the sample standard deviation is likely to be a good approximation of the population standard deviation.

Example 8.5   Illustrates Procedure 8.1

The U.S. Bureau of Labor Statistics collects information on the ages of people in the civilian labor force and publishes the results in *Employment and Earnings*. Fifty people in the civilian labor force are randomly selected; their ages are as displayed in Table 8.3. Find a 90% confidence interval for the mean age, μ, of all people in the civilian labor force.

TABLE 8.3
Ages of 50 randomly selected
people in the civilian labor force

22	58	40	42	43	32	34	45	38	19
33	16	49	29	30	43	37	19	21	62
60	41	28	35	37	51	37	65	57	26
27	31	33	24	34	28	39	43	26	38
42	40	31	34	38	35	29	33	32	33

SOLUTION   Since the sample size is 50, which exceeds 30, we can apply Procedure 8.1 to obtain the required confidence interval.

Step 1   *For a confidence level of* $1 - \alpha$, *use Table II to find* $z_{\alpha/2}$.

We want a 90% confidence interval, so the confidence level is $0.90 = 1 - 0.10$. This means that $\alpha = 0.10$. Consulting Table II we find that

$$z_{\alpha/2} = z_{0.10/2} = z_{0.05} = 1.645.$$

Step 2   *The confidence interval for* μ *is from*

$$\bar{x} - z_{\alpha/2} \cdot \frac{\sigma}{\sqrt{n}} \quad to \quad \bar{x} + z_{\alpha/2} \cdot \frac{\sigma}{\sqrt{n}}.$$

Since σ is unknown, we will use s in its place. We have $n = 50$ and, from Step 1, $z_{\alpha/2} = 1.645$. To compute $\bar{x}$ and s for the data in Table 8.3, we apply the usual formulas:

$$\bar{x} = \frac{\Sigma x}{n} = \frac{1819}{50} = 36.38$$

and

$$s = \sqrt{\frac{\Sigma x^2 - (\Sigma x)^2/n}{n - 1}} = \sqrt{\frac{72,179 - (1819)^2/50}{49}} = 11.07.$$

Consequently, a 90% confidence interval for μ is from

$$36.38 - 1.645 \cdot \frac{11.07}{\sqrt{50}} \quad to \quad 36.38 + 1.645 \cdot \frac{11.07}{\sqrt{50}}$$

 or 33.8 to 39.0. We can be 90% confident that the mean age, μ, of all people in the civilian labor force is somewhere between 33.8 years and 39.0 years.   ∎

Although strictly speaking the only condition required for using Procedure 8.1 is that the sample size be large, it is important to watch for outliers. Because the sample mean (and sample standard deviation) are not resistant to outliers, the presence of outliers can have an undue influence on a confidence interval for a population mean, especially when the sample size is only moderately large.

Therefore it is always a good idea to first look at the data using a histogram, stem-and-leaf diagram, or other graphical display. If outliers are detected, we must decide between retaining and removing them before obtaining a confidence interval.

Relation of the Confidence Level to the Length of the Confidence Interval

The *confidence level* of a confidence interval for a population mean, μ, signifies the confidence of the estimate; that is, it expresses the confidence we have that μ actually lies in the confidence interval. The *length* of the confidence interval indicates the precision of the estimate. Long confidence intervals indicate poor precision, whereas short confidence intervals indicate good precision.

How does the confidence level affect the length of the confidence interval? To answer this question, let's return to Example 8.5, where we found a 90% confidence interval for the mean age, μ, of all people in the civilian labor force. The confidence level there is 0.90 and the confidence interval we computed is from 33.8 to 39.0 years. If we change the confidence level from 0.90 to, say, 0.95, then $z_{\alpha/2}$ changes from $z_{0.10/2} = z_{0.05} = 1.645$ to $z_{0.05/2} = z_{0.025} = 1.96$. The resulting confidence interval, using the same sample data (Table 8.3), is then from

$$36.38 - 1.96 \cdot \frac{11.07}{\sqrt{50}} \quad \text{to} \quad 36.38 + 1.96 \cdot \frac{11.07}{\sqrt{50}}$$

or from 33.3 to 39.4 years. We picture both the 90% and 95% confidence intervals in Fig. 8.2.

FIGURE 8.2
90% and 95% confidence intervals for μ using the data in Table 8.3

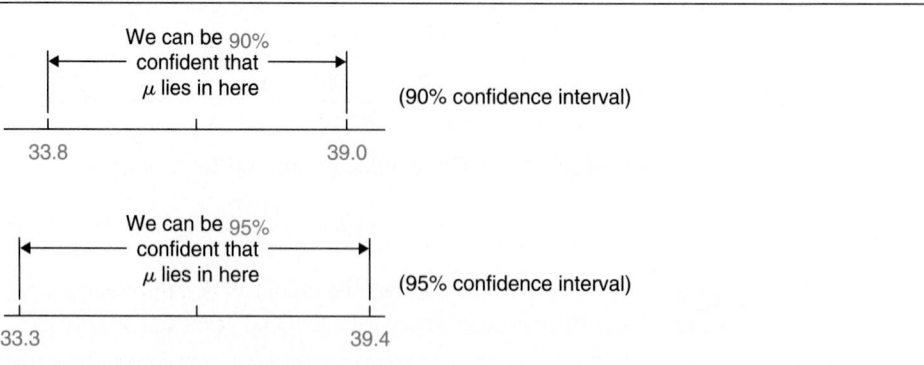

Thus, increasing the confidence level increases the length of the confidence interval. This makes sense: If we want to be more confident that μ lies in our confidence interval, then we must naturally have a more extensive interval. In summary we have the following important fact, whose verification is taken up in Exercise 8.34.

KEY FACT 8.1   **RELATION OF THE CONFIDENCE LEVEL TO THE LENGTH OF THE CONFIDENCE INTERVAL**

> For a fixed sample size, the greater the confidence level, the greater the length of the confidence interval.

Using the Computer (Optional)

Procedure 8.1 on page 448 provides a step-by-step method for obtaining a confidence interval for a population mean, μ, when the sample size is large ($n \geq 30$). Minitab has a command, called **ZInterval,** that will perform Procedure 8.1 for us.

In Example 8.5 we applied Procedure 8.1 to determine a confidence interval for the mean age of all people in the civilian labor force. Example 8.6 shows how Minitab's ZInterval command can be employed to obtain that confidence interval.

Example 8.6   The ZInterval Command

Table 8.3 on page 449 displays the ages of 50 randomly selected people in the civilian labor force. Use Minitab to determine a 90% confidence interval for the mean age, μ, of all people in the civilian labor force.

SOLUTION   First we store the age data in Table 8.3 in a column named AGES. To apply the ZInterval command, we must specify the population standard deviation, σ, which in this case is unknown. However, because the sample size is large ($n = 50$), we can use the sample standard deviation, s, in place of σ. We will employ Minitab to obtain s and store its value in K1.[†] This is accomplished in the following way.

Session commands: We type the command STDEV followed by the storage location of the sample data and the storage location for the sample standard deviation; that is, we type STDEV 'AGES' put into K1 and press Enter.

(or)

Menu commands: We choose **Calc ▸ Functions and Statistics ▸ Column Statistics...,** select **Standard deviation** from the **Statistic** list, specify AGES as the **Input variable,** select **Store result in** and type K1, and then select **OK.**

Now we are ready to employ the ZInterval command.

[†] In Minitab, storage locations for single numbers, called *constants,* are designated by a "K" followed by a number. So K1 denotes Constant 1, K2 denotes Constant 2, and so forth.

Session commands: We type the command **ZINTERVAL**, followed by the desired confidence level, the (estimated) value of σ, and the storage location of the sample data; that is, we type <u>ZINTERVAL 90% confidence, sigma=K1, data in 'AGES'</u> and press [Enter].

(or)

Menu commands: We choose **Stat ▶ Basic Statistics ▶ 1-Sample Z...**, specify AGES in the **Variables** text box, select **Confidence interval** and type <u>90</u> for **Level**, select **Sigma** and type <u>K1</u>, and then select **OK**.

The output resulting from applying either the above session commands or the above menu commands is shown in Printout 8.1.

PRINTOUT 8.1

Minitab output for the StDev and ZInterval commands

```
MTB > STDEV 'AGES' put into K1
   ST.DEV. =      11.069
MTB > ZINTERVAL 90% confidence, sigma=K1, data in 'AGES'

THE ASSUMED SIGMA =11.1

               N     MEAN    STDEV   SE MEAN    90.0 PERCENT C.I.
AGES          50    36.38    11.07    1.57   (   33.80,    38.96)
```

The fourth line of Printout 8.1 displays (to three significant digits) the value used for the population standard deviation, σ: THE ASSUMED SIGMA =11.1, which in this case is the sample standard deviation of the data in Table 8.3. Then we find the sample size, sample mean, sample standard deviation, and standard error of the mean. The final item is the confidence interval. Hence a 90% confidence interval for μ is from 33.80 to 38.96. We can be 90% confident that the mean age, μ, of all people in the civilian labor force is somewhere between 33.80 years and 38.96 years. ∎

EXERCISES 8.2

8.15 Find the confidence level and α for
a. a 90% confidence interval.
b. a 99% confidence interval.

8.16 Find the confidence level and α for
a. an 85% confidence interval.
b. a 95% confidence interval.

In Exercises 8.17–8.27, use Procedure 8.1 on page 448 to determine the required confidence intervals.

8.17 The Gallup Organization conducts annual national surveys on home gardening. Results are published by the National Association for Gardening in *National Gardening Survey*. A random sample is taken of 250 households with vegetable gardens. The average size of their gardens is 643 sq ft.
a. Determine a 90% confidence interval for the mean size, μ, of all household vegetable gardens in the United States. Assume $\sigma = 247$ sq ft.
b. Interpret your result in part (a).

8.18 A quality-control engineer in a bakery goods plant needs to estimate the mean weight, μ, of bags of potato chips that are packed by a machine. He knows from experience that $\sigma = 0.1$ oz for this machine. A random sample of 36 bags has a mean weight of 16.01 oz.

a. Find a 99% confidence interval for μ.

b. Interpret your result in part (a).

8.19   The Bureau of Labor Statistics collects data on employment and hourly earnings in the aircraft industry and publishes its findings in *Employment and Earnings*. Thirty people working in the aircraft industry are selected at random; their hourly earnings are as follows.

$15.20	5.94	12.28	14.99	5.99
16.42	15.71	4.01	8.90	9.25
13.77	18.22	9.18	11.60	13.20
14.05	15.17	13.57	12.52	14.49
9.43	8.85	18.82	5.83	14.82
17.02	11.84	12.16	7.73	9.30

a. Find a 95% confidence interval for the mean hourly earnings, μ, of all people employed in the aircraft industry. Assume the population standard deviation of the hourly earnings is $3.25. *(Note:* The sum of the data is $360.26.)*

b. Interpret your result in part (a).

8.20   A sociologist wants information on the number of children per farm family in her native state of Nebraska. Forty randomly selected farm families have the following number of children.

3	5	2	1	1	0	2	3
1	1	2	1	2	0	1	5
4	1	0	1	3	1	0	1
0	1	8	0	1	2	2	2
2	1	5	3	1	4	1	0

a. Find a 90% confidence interval for the mean number of children, μ, per farm family in Nebraska. Assume $\sigma = 1.95$. *(Note:* $\Sigma x = 74.)*

b. Interpret your result in part (a).

8.21   The U.S. National Center for Health Statistics estimates mean weights of Americans by age, height, and sex and publishes the results in *Vital and Health Statistics*. Forty U.S. women, 5 ft 4 in. tall and age 18–24, are randomly selected. Their weights, in pounds, are as follows.

140	136	147	138	143	122	115	125
136	152	130	134	150	153	148	132
116	159	128	136	134	126	120	146
131	167	145	132	138	137	115	145
154	139	139	147	123	154	127	116

a. Find a 90% confidence interval for the mean weight, μ, of all U.S. women 5 ft 4 in. tall and in the age group 18–24 years. *(Note:* $\bar{x} = 136.88$ and $s = 12.77$.)*

b. Interpret your result in part (a).

8.22   A research physician wants to estimate the average age of people with diabetes. She takes a sample of 35 diabetics and obtains the following ages.

48	41	57	83	41	55	59
61	38	48	79	75	77	7
54	23	47	56	79	68	61
64	45	53	82	68	38	70
10	60	83	76	21	65	47

a. Determine a 95% confidence interval for the mean age, μ, of people with diabetes. *(Note:* $\bar{x} = 55.40$ and $s = 19.88$.)*

b. Interpret your result in part (a).

8.23   The U.S. National Center for Education Statistics surveys colleges and universities to obtain data on the costs of attending an institution of higher education. Results are published in *Digest of Education Statistics*. A random sample is taken of 150 private 4-year colleges and universities. The mean tuition and fees for the schools selected is $10,424 and the standard deviation is $3241. Use these statistics to obtain a 90% confidence interval for the mean tuition and fees of all private 4-year colleges and universities.

8.24   A telephone company in the Southwest undertook a study on various aspects of phone usage. One item of interest was how long calls last. According to the media relations manager, the company randomly selected 15,000 local calls involving Phoenix residential customers. The mean duration of the calls sampled was 3.8 minutes and the standard deviation was 4.0 minutes. Use this information to obtain a 95% confidence interval for the mean duration of all phone calls made by Phoenix residential customers.

8.25   Refer to Exercise 8.21.

a. Find a 99% confidence interval for μ.

b. Why is the confidence interval you found in part (a) longer than the one in Exercise 8.21?

c. Draw a graph similar to Fig. 8.2 on page 450 that displays both confidence intervals.

8.26   Refer to Exercise 8.22.

a. Determine an 80% confidence interval for μ.

b. Why is the confidence interval you found in part (a) shorter than the one in Exercise 8.22?

c. Draw a graph similar to Fig. 8.2 on page 450 that displays both confidence intervals.

8.27 The U.S. Bureau of the Census compiles data on family size and presents its findings in *Current Population Reports*. Suppose 500 U.S. families are randomly selected in order to estimate the mean size, μ, of all U.S. families. Further suppose the results are as shown in the following frequency distribution.

Size	2	3	4	5	6	7	8	9
Frequency	198	118	101	59	12	3	8	1

a. Determine a 95% confidence interval for the mean size, μ, of all U.S. families.
b. Interpret your result in part (a).

In each of Exercises 8.28–8.31, use Minitab or some other statistical software to
a. *identify outliers, if any, for the specified data set.*
b. *construct the designated confidence interval, first removing data values you deem appropriate.*

8.28 (Computer exercise) The data set and confidence interval in part (a) of Exercise 8.20.

8.29 (Computer exercise) The data set and confidence interval in part (a) of Exercise 8.19.

8.30 (Computer exercise) The data set and confidence interval in part (a) of Exercise 8.22.

8.31 (Computer exercise) The data set and confidence interval in part (a) of Exercise 8.21.

8.32 (Computer exercise) The manufacturer of a new-model car, called the Orion, claims that a typical Orion gets 26 miles per gallon (mpg). An independent consumer group is skeptical of this claim and thinks the mean gas mileage of all Orions may be less than 26 mpg. The consumer group performs mileage tests on a random sample of Orions and obtains the following abridged Minitab output.

```
THE ASSUMED SIGMA =1.50

 N    MEAN   STDEV   SE MEAN    95.0 PERCENT C.I.
30   25.230  1.592    0.274    ( 24.692,  25.768)
```

Use the printout to determine
a. the (assumed) population standard deviation of the gas mileages of all Orions.
b. the standard deviation of the gas mileages of the Orions in the sample.

c. the number of Orions in the sample.
d. the mean gas mileage of the Orions in the sample.
e. a 95% confidence interval for the mean gas mileage of all Orions.
f. Does it appear that the consumer group's skepticism is justified? Explain your answer.

8.33 (Computer exercise) A cotton farmer is interested in a new brand of fertilizer. The farmer uses the new fertilizer on a random sample of 1-acre plots and records the yield, in pounds, for each plot. Below is an abridged Minitab printout obtained by applying the ZInterval command to the cotton-yield data.

```
THE ASSUMED SIGMA =20.8

 N    MEAN   STDEV   SE MEAN    90.0 PERCENT C.I.
80   625.16  21.47    2.33     ( 621.33,  628.99)
```

From the printout determine
a. the (assumed) population standard deviation of cotton yields for all 1-acre plots on the farm when the new fertilizer is used.
b. the standard deviation of cotton yields for the 1-acre plots in the sample.
c. the number of 1-acre plots in the sample.
d. the mean yield of cotton for the 1-acre plots in the sample.
e. a 90% confidence interval for the mean yield of cotton for all 1-acre plots on the farm when the new fertilizer is used.

8.34 In this exercise we will verify Key Fact 8.1, which states that for a fixed sample size, the greater the confidence level, the greater the length of the confidence interval. We will assume the population standard deviation, σ, is known, so that the form of the confidence interval is

$$\bar{x} - z_{\alpha/2} \cdot \frac{\sigma}{\sqrt{n}} \quad \text{to} \quad \bar{x} + z_{\alpha/2} \cdot \frac{\sigma}{\sqrt{n}}.$$

a. Show that the length of the confidence interval is

$$L = 2z_{\alpha/2} \cdot \frac{\sigma}{\sqrt{n}}.$$

b. Show that increasing the confidence level decreases the value of α.
c. Verify that decreasing the value of α increases the value of $z_{\alpha/2}$.
d. Deduce from parts (a)–(c) that for a fixed sample size, increasing the confidence level increases the length of the confidence interval.

8.3   SAMPLE-SIZE CONSIDERATIONS

In this section we will examine in detail how the sample size affects the precision of estimating a population mean by a sample mean. We have already seen that the larger the sample size, the greater the likelihood for small sampling error (Key Fact 7.3 on page 418). Now that we have studied confidence intervals, we can determine exactly how the sample size affects the accuracy of the estimate.

Example 8.7   Margin of Error

The U.S. Energy Information Administration surveys households to obtain data on monthly fuel expenditures for household vehicles. Results of those surveys are contained in *Residential Transportation Energy Consumption Survey, Consumption Patterns of Household Vehicles.* Thirty monthly fuel expenditures for household vehicles are selected at random; their mean is $58.56.

a. Find a 95% confidence interval for the mean monthly fuel expenditure, μ, for all household vehicles. Assume $\sigma = \$20.65$.

b. Discuss the precision with which $\bar{x}$ estimates μ.

SOLUTION   a. For a 95% confidence interval, $\alpha = 0.05$, and so $z_{\alpha/2} = z_{0.05/2} = z_{0.025} = 1.96$. Also we have $n = 30$, $\sigma = \$20.65$, and $\bar{x} = \$58.56$. Applying Procedure 8.1 on page 448, we find that a 95% confidence interval for μ is from

$$\bar{x} - z_{\alpha/2} \cdot \frac{\sigma}{\sqrt{n}} \quad \text{to} \quad \bar{x} + z_{\alpha/2} \cdot \frac{\sigma}{\sqrt{n}}$$

or

$$58.56 - 1.96 \cdot \frac{20.65}{\sqrt{30}} \quad \text{to} \quad 58.56 + 1.96 \cdot \frac{20.65}{\sqrt{30}}$$

or

$$58.56 - 7.39 \quad \text{to} \quad 58.56 + 7.39$$

or

$$51.17 \quad \text{to} \quad 65.95.$$

We can be 95% confident that the mean monthly fuel expenditure, μ, per household vehicle is somewhere between $51.17 and $65.95.

b. The confidence interval that we obtained in part (a) is relatively long and hence provides a rather wide range for the possible values of μ. In other words, the precision of the estimate is poor. To improve the precision, we need to decrease the length of the confidence interval.

As we learned in Section 8.2, one way to decrease the length of the confidence interval is to lower the confidence level from 95% to some lower level. But suppose we want to retain the same level of confidence and still narrow the confidence interval. How can we accomplish this? To answer that question, let's first look more closely at the confidence interval in part (a). That confidence interval is displayed graphically in Fig. 8.3.

FIGURE 8.3

95% confidence interval for the mean monthly fuel expenditure, μ, per household vehicle

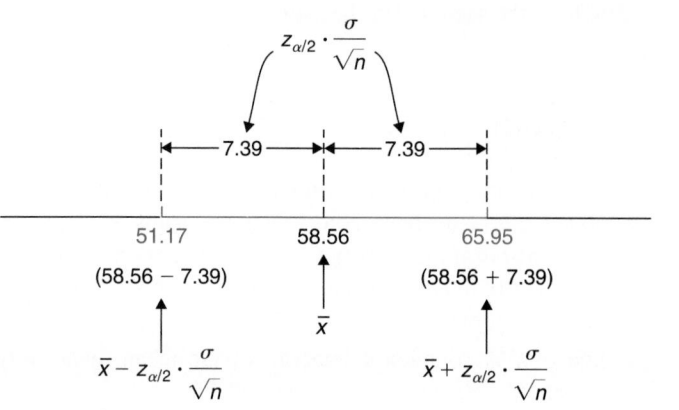

By studying Fig. 8.3 or referring to the computations done in part (a), we find that the length of the confidence interval is determined by the quantity,

$$E = z_{\alpha/2} \cdot \frac{\sigma}{\sqrt{n}},$$

which is half the length of the confidence interval. In this case, $E = 7.39$. That quantity is called the **margin of error,** also known as the **maximum error of the estimate.** We use this terminology because we can be 95% confident that μ lies in the confidence interval or, equivalently, that the margin of error in estimating μ by $\bar{x}$ is $7.39 (see Fig. 8.3). In newspapers and magazines, this is often expressed as "the poll has a margin of error of $7.39" or as "theoretically, in 95 out of 100 such polls the margin of error will be $7.39."

In any event we see that to narrow the confidence interval and thereby increase the precision of the estimate, we need only decrease the margin of error, E. Since the sample size, n, occurs in the denominator of the formula for E, we can decrease E by increasing the sample size. This makes sense, of course, because we expect to get more accurate information from larger samples. ■

In the preceding example, we introduced some concepts and terminology important in confidence-interval analysis. We summarize that discussion in Definition 8.3 and Key Fact 8.2.

DEFINITION 8.3

MARGIN OF ERROR FOR THE ESTIMATE OF μ

The **margin of error** for the estimate of μ is

$$E = z_{\alpha/2} \cdot \frac{\sigma}{\sqrt{n}}.$$

The margin of error is equal to half the length of the confidence interval, as illustrated in Fig. 8.4.

FIGURE 8.4
Margin of error, $E = z_{\alpha/2} \cdot \dfrac{\sigma}{\sqrt{n}}$

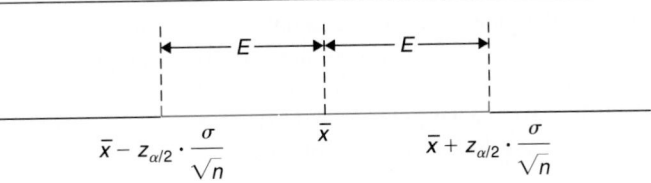

As we know, if the population standard deviation, σ, is unknown, then we use the sample standard deviation, s, in its place when computing a large-sample confidence interval for μ. This means that if σ is unknown, then the margin of error for the estimate of μ is $z_{\alpha/2} \cdot s/\sqrt{n}$ not $z_{\alpha/2} \cdot \sigma/\sqrt{n}$.

KEY FACT 8.2

MARGIN OF ERROR AND THE PRECISION OF THE ESTIMATE

The length of a confidence interval for a population mean, μ, and hence the precision with which $\bar{x}$ estimates μ, is determined by the margin of error, E. For a prescribed confidence level, we can increase the precision of the estimate by increasing the sample size, n.

Determining the Required Sample Size

The margin of error and confidence level of a confidence interval are often specified in advance. We must then determine the sample size required to meet the specifications. The formula for the required sample size can be obtained by solving for n in the formula for the margin of error, $E = z_{\alpha/2} \cdot \sigma/\sqrt{n}$. The result is Formula 8.1.

FORMULA 8.1

SAMPLE SIZE FOR ESTIMATING μ

The sample size required for a $(1 - \alpha)$-level confidence interval for μ with a specified margin of error, E, is given by the formula

$$n = \left(\frac{z_{\alpha/2} \cdot \sigma}{E} \right)^2,$$

rounded up to the nearest whole number.

Example 8.8   Illustrates Formula 8.1

Consider again the problem of estimating the mean monthly fuel expenditure, μ, per household vehicle.

a. Determine the sample size required to ensure that we can be 95% confident that the estimate, $\bar{x}$, is within $0.50 of μ. (Recall that $\sigma = \$20.65$.)

b. Find a 95% confidence interval for μ if a sample of the size determined in part (a) has a mean of $59.02.

SOLUTION   a. To determine the required sample size, we apply Formula 8.1. In doing so, we must identify σ, E, and $z_{\alpha/2}$. First note that, by assumption, $\sigma = \$20.65$, and that the margin of error, E, is specified at $0.50. Next note that since the confidence level is stipulated to be 0.95, we have $\alpha = 0.05$. Consulting Table II we find that $z_{\alpha/2} = z_{0.05/2} = z_{0.025} = 1.96$. Thus the required sample size is

$$ n = \left(\frac{z_{\alpha/2} \cdot \sigma}{E} \right)^2 = \left(\frac{1.96 \cdot 20.65}{0.5} \right)^2 = 6552.58. $$

Obviously, we cannot take a fractional sample size, so to be conservative we round up to 6553. Thus if 6553 monthly fuel expenditures are randomly selected, then we can be 95% confident that their mean, $\bar{x}$, is within $0.50 of the mean monthly fuel expenditure, μ, for all household vehicles. (By the way, the U.S. Energy Information Administration takes a sample of size 6841.)

b. For this part, we are to find a 95% confidence interval for μ if a sample of the size determined in part (a) has a mean of $59.02. Applying Procedure 8.1 with $\alpha = 0.05$, $\sigma = 20.65$, $\bar{x} = 59.02$, and $n = 6553$, we obtain the confidence interval:

$$ \bar{x} - z_{\alpha/2} \cdot \frac{\sigma}{\sqrt{n}} \quad \text{to} \quad \bar{x} + z_{\alpha/2} \cdot \frac{\sigma}{\sqrt{n}} $$

or

$$ 59.02 - 1.96 \cdot \frac{20.65}{\sqrt{6553}} \quad \text{to} \quad 59.02 + 1.96 \cdot \frac{20.65}{\sqrt{6553}} $$

or

$$ 59.02 - 0.50 \quad \text{to} \quad 59.02 + 0.50 $$

or

$$ 58.52 \quad \text{to} \quad 59.52. $$

We can be 95% confident that the mean monthly fuel expenditure, μ, per household vehicle is somewhere between $58.52 and $59.52.

The sample size, $n = 6553$, was determined in part (a) to guarantee a margin of error of $0.50 for a 95% confidence interval. Therefore, in view of Fig. 8.4 on page 457, the 95% confidence interval required here could be obtained simply by computing $\bar{x} \pm E = 59.02 \pm 0.50$. This gives the same confidence interval, 58.52 to 59.52, we found above, but with much less work. It is possible, however, that because the sample size, $n = 6553$, is a rounded value, the simpler method might have incorrectly yielded a slightly wider confidence interval. In practice, the simpler method is considered acceptable since, at worst, it provides a slightly conservative estimate.   ■

The formula for finding the required sample size, Formula 8.1, involves the population standard deviation, σ. If σ is unknown, which is usually the case, and we want to apply the formula, then we must first estimate σ by taking a preliminary sample of size 30 or more. The sample standard deviation, s, of the sample obtained provides an estimate of σ and can be used in place of σ in Formula 8.1.

EXERCISES 8.3

8.35   In Example 8.8 we considered the problem of estimating the mean monthly fuel expenditure, μ, per household vehicle. We mentioned that the U.S. Energy Information Administration takes a sample of size 6841. Determine their margin of error in estimating μ at the 95% level of confidence. (Recall that $\sigma = \$20.65$.)

8.36   In Exercise 8.20 of Section 8.2, you were asked to determine a 90% confidence interval, based on a sample of size 40, for the mean number of children, μ, per farm family in Nebraska. The 90% confidence interval for μ is from 1.3 to 2.4 children per farm family.
a. Determine the margin of error, E.
b. Explain the meaning of E in this context as far as the accuracy of the estimate is concerned.
c. Determine the sample size required to ensure that we can be 90% confident that our estimate, $\bar{x}$, is within 0.1 child of μ. (Recall that $\sigma = 1.95$.)
d. Find a 90% confidence interval for μ if a sample of the size determined in part (c) yields a mean of $\bar{x} = 1.9$ children.

8.37   In Exercise 8.19 of Section 8.2, you were asked to determine a 95% confidence interval, based on a sample of size 30, for the mean hourly earnings, μ, of people employed in the aircraft industry. The 95% confidence interval is from $10.85 to $13.17.
a. Determine the margin of error, E.
b. Explain the meaning of E in this context as far as the accuracy of the estimate is concerned.
c. Determine the sample size required to ensure that we can be 95% confident that our estimate, $\bar{x}$, is within $0.50 of μ. (Recall that $\sigma = \$3.25$.)

d. Find a 95% confidence interval for μ if a sample of the size determined in part (c) has a mean of $12.87.

8.38   In Exercise 8.22 of Section 8.2, you were asked to find a 95% confidence interval, based on a sample of size 35, for the mean age of people with diabetes. The required confidence interval is from 48.8 to 62.0 years.
a. Determine the margin of error.
b. Find the sample size required to have a margin of error of 0.5 year and a 95% confidence level. (*Note:* The sample standard deviation of the sample of 35 ages from Exercise 8.22 is 19.88 years.)
c. Why did you use the sample standard deviation, $s = 19.88$, in place of σ in your solution to part (b)?
d. Find a 95% confidence interval for the mean age, μ, of people with diabetes if a sample of the size determined in part (b) has a mean of 58.3 years and a standard deviation of 19.0 years.

8.39   In Exercise 8.21 of Section 8.2, you were given a sample of weights obtained from the random selection of 40 U.S. women, 5 ft 4 in. tall and age 18–24. Based on that data, a 90% confidence interval for the mean weight, μ, of all such women is from 133.6 to 140.2 lb.
a. Determine the margin of error.
b. Find the sample size required to have a margin of error of 2.0 lb and a 99% confidence level. (*Note:* $s = 12.77$ lb for the sample data in Exercise 8.21.)
c. Why did you use the sample standard deviation, $s = 12.77$, in place of σ in your solution to part (b)?
d. Obtain a 99% confidence interval for the mean weight, μ, of all U.S. women age 18–24 who are 5 ft 4 in. tall if a sample of the size determined in part (b) has a mean of 134.2 lb and a standard deviation of 13.0 lb.

8.40 Explain how to apply the formula

$$n = \left(\frac{z_{\alpha/2} \cdot \sigma}{E} \right)^2$$

if σ is unknown.

8.41 The U.S. Bureau of the Census estimates the mean value, μ, of the land and buildings per corporate farm. Those estimates are published in *Census of Agriculture.* Sup-

pose an estimate, $\bar{x}$, is obtained and the margin of error is $1000. Does this imply that the estimate is within $1000 of the true mean, μ? Explain your answer.

8.42 Use Formula 8.1 on page 457 to establish the following fact: For a fixed confidence level, it is necessary to (approximately) quadruple the sample size in order to halve the margin of error.

8.4 CONFIDENCE INTERVALS FOR ONE NORMAL POPULATION MEAN

In Section 8.2 we learned how to determine a confidence interval for a population mean, μ, when the sample size is large ($n \geq 30$). The procedure for obtaining a large-sample confidence interval for μ is based on Key Fact 7.6, the sampling distribution of the mean for general populations. That key fact states that for large samples, the random variable $\bar{x}$ is approximately normally distributed and has mean $\mu_{\bar{x}} = \mu$ and standard deviation $\sigma_{\bar{x}} = \sigma/\sqrt{n}$, or, equivalently, that the standardized random variable

$$(1) \qquad\qquad z = \frac{\bar{x} - \mu}{\sigma/\sqrt{n}}$$

has approximately the standard normal distribution when the sample size is large.

However, a large sample is often unavailable, extremely expensive, or undesirable. For example, tests of automobiles to analyze the impact of collisions frequently involve wrecking the cars. Clearly, it is more appropriate to employ a small sample in this and similar situations.

Several methods exist for obtaining a confidence interval for a population mean that do not require a large sample. One such method applies when the population being sampled is *normally distributed.* We will study that method in this section.

Student's *t*-Distribution

As we said, our next objective is to learn how to find a confidence interval for a population mean, μ, when the population being sampled is normally distributed. We will assume the population standard deviation, σ, is unknown, since that is usually the case in practice.[†]

To begin, we need to identify the probability distribution of the random variable obtained by replacing the unknown population standard deviation, σ, in Equation (1) by the sample standard deviation, s; that is, we need to identify the probability distribution of the random variable

$$(2) \qquad\qquad t = \frac{\bar{x} - \mu}{s/\sqrt{n}}$$

when the population being sampled is normally distributed.

[†] See page 474 for a discussion of confidence intervals for a normal population mean when σ is known.

The random variables z and t in Equations (1) and (2) have similar forms; the only computational difference is the use of σ in the formula for z and the use of s in the formula for t. However, their probability distributions are different. To get an idea of how they differ, we used Minitab to simulate each random variable. First we simulated the taking of 2000 random samples of size five from a normally distributed population with mean $\mu = 60$ and standard deviation $\sigma = 0.9$. (Any mean and standard deviation will do.) Next we obtained the sample mean and sample standard deviation of each of the 2000 samples. Then, for each of the 2000 samples, we determined the values of both the quantity z in Equation (1) and the quantity t in Equation (2). Finally, we obtained histograms for the 2000 values of z and the 2000 values of t. Printout 8.2, shown on the following page, displays both histograms.

From Key Fact 7.7 on page 428, we know that the random variable, z, in Equation (1) has the standard normal distribution. This is reflected in the first histogram in Printout 8.2. Comparing the two histograms in Printout 8.2, it appears that the distributions of the random variables z and t are different—the variation of the z-values is less than that of the t-values. In particular, then, we would conjecture that the random variable t does not have the standard normal distribution.

In 1908 William Gosset determined the probability distribution of the random variable t, now called **Student's t-distribution.** (See the biography on page 438 for more on Gosset and the story of the Student's t-distribution.) For brevity we will usually omit "Student's" and just use *t*-distribution.

As we know, probabilities for a random variable having a normal distribution are equal to areas under a normal curve. Probabilities for a random variable having a t-distribution are also equal to areas under a curve, aptly called a *t*-curve.

Actually, there are infinitely many t-curves; the one that is used depends on the sample size. If the sample size is n, then we identify the t-curve in question by saying that it is the t-curve with $n - 1$ **degrees of freedom.** The mathematical concepts involved in defining degrees of freedom are a bit complex. Therefore we define degrees of freedom simply as a number that identifies the appropriate t-curve or t-distribution. For convenience we usually write $df = n - 1$ to indicate $n - 1$ degrees of freedom.

Before further examining the t-distribution and t-curves, we summarize the preceding discussion in Key Fact 8.3.

KEY FACT 8.3 *t*-STATISTIC

Suppose a random sample of size n is to be taken from a normally distributed population with mean μ. Then the random variable

$$t = \frac{\bar{x} - \mu}{s/\sqrt{n}}$$

has the t-distribution with $n - 1$ degrees of freedom. Thus probabilities for that random variable are equal to areas under the t-curve with $df = n - 1$.

Although there is a different t-curve for each number of degrees of freedom, all t-curves are similar and resemble the standard normal curve. Figure 8.5 shows the standard normal curve and two t-curves.

PRINTOUT 8.2

Minitab histograms of z and t
for 2000 samples of size five
from a normal population

Histogram of z   N = 2000
Each * represents 10 obs.

```
Midpoint    Count
   -3.5       1   *
   -3.0       8   *
   -2.5      23   ***
   -2.0      52   ******
   -1.5     144   ***************
   -1.0     237   ************************
   -0.5     342   ***********************************
    0.0     399   ****************************************
    0.5     335   **********************************
    1.0     242   *************************
    1.5     130   *************
    2.0      60   ******
    2.5      24   ***
    3.0       3   *
```

Histogram of t   N = 2000
Each * represents 10 obs.

```
Midpoint    Count
   -7.0       1   *
   -6.5       0
   -6.0       0
   -5.5       1   *
   -5.0       2   *
   -4.5       6   *
   -4.0       7   *
   -3.5      11   **
   -3.0      13   **
   -2.5      46   *****
   -2.0      80   ********
   -1.5     130   *************
   -1.0     219   *********************
   -0.5     317   ********************************
    0.0     351   ************************************
    0.5     309   *******************************
    1.0     225   **********************
    1.5     138   **************
    2.0      68   *******
    2.5      30   ***
    3.0      21   ***
    3.5      11   **
    4.0       6   *
    4.5       3   *
    5.0       3   *
    5.5       1   *
    6.0       1   *
```

FIGURE 8.5

Standard normal curve
and two *t*-curves

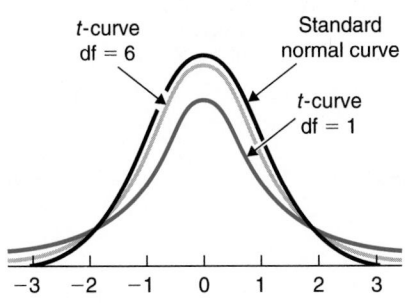

As illustrated by Fig. 8.5, *t*-curves have the following properties.

KEY FACT 8.4   **BASIC PROPERTIES OF *t*-CURVES**

> *Property 1:* The total area under a *t*-curve is equal to 1.
>
> *Property 2:* A *t*-curve extends indefinitely in both directions, approaching, but never touching, the horizontal axis as it does so.
>
> *Property 3:* A *t*-curve is symmetric about 0.
>
> *Property 4:* As the number of degrees of freedom gets larger, *t*-curves look increasingly like the standard normal curve.

Using the *t*-Table

In Section 6.1 we learned how to find areas under the standard normal curve using the standard-normal table, Table II. Now we will learn how to find areas under *t*-curves using the *t*-table, Table IV, which appears in Appendix A and on the page facing the inside back cover.

For our purposes, one of which is obtaining confidence intervals for a population mean, we do not need a complete *t*-table for each *t*-curve. Only certain areas will be important for us to know.

The two outside columns of the *t*-table, labeled df, display the number of degrees of freedom. As expected, the symbol t_α denotes the *t*-value having area α to its right under a *t*-curve. Thus the column headed $t_{0.10}$ contains *t*-values having area 0.10 to their right; the column headed $t_{0.05}$ contains *t*-values having area 0.05 to their right; and so on. We illustrate a use of Table IV in Example 8.9.

Example 8.9   Finding the *t*-Value Having a Specified Area to Its Right

For a *t*-curve with 13 degrees of freedom, determine $t_{0.05}$; that is, find the *t*-value having area 0.05 to its right, as shown in Fig. 8.6(a).

FIGURE 8.6

Finding the *t*-value having
area 0.05 to its right

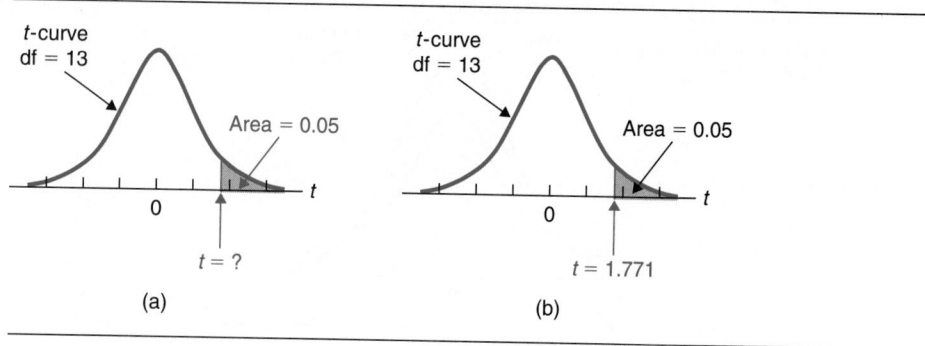

(a)   (b)

SOLUTION   To find the *t*-value in question, we use Table IV. For ease of reference, we have repeated a
portion of Table IV in Table 8.4.

TABLE 8.4

Values of t_α

df	$t_{0.10}$	$t_{0.05}$	$t_{0.025}$	$t_{0.01}$	$t_{0.005}$	df
.	.	.	.	.	.	.
.	.	.	.	.	.	.
.	.	.	.	.	.	.
12	1.356	1.782	2.179	2.681	3.055	12
13	1.350	1.771	2.160	2.650	3.012	13
14	1.345	1.761	2.145	2.624	2.977	14
15	1.341	1.753	2.131	2.602	2.947	15
.	.	.	.	.	.	.
.	.	.	.	.	.	.
.	.	.	.	.	.	.

Since the number of degrees of freedom is 13, we first go down the outside columns,
labeled df, to "13." Then we go across that row until we are under the column headed $t_{0.05}$.
The number in the body of the table there, 1.771, is the required *t*-value; that is, for a
t-curve with df $= 13$, the *t*-value having area 0.05 to its right is $t_{0.05} = 1.771$, as seen in
Fig. 8.6(b).   ∎

Confidence Intervals for a Normal Population Mean

Having discussed *t*-distributions and *t*-curves, we can now develop a procedure to obtain
a confidence interval for a population mean, μ, when the population being sampled is
normally distributed. The basis for the procedure is outlined below. For details and proofs,
see Exercise 8.68.

Suppose a random sample of size n is to be taken from a normally distributed population with mean μ. Then, by Key Fact 8.3 on page 461, the random variable

$$t = \frac{\bar{x} - \mu}{s/\sqrt{n}}$$

has the t-distribution with $n - 1$ degrees of freedom; that is, probabilities for that random variable are equal to areas under the t-curve with df $= n - 1$. Therefore

$$P\left(-t_{\alpha/2} < \frac{\bar{x} - \mu}{s/\sqrt{n}} < t_{\alpha/2}\right) = 1 - \alpha.$$

Using algebra we can rewrite the preceding equation as

$$P\left(\bar{x} - t_{\alpha/2} \cdot \frac{s}{\sqrt{n}} < \mu < \bar{x} + t_{\alpha/2} \cdot \frac{s}{\sqrt{n}}\right) = 1 - \alpha.$$

This last equation shows that once the sample is taken, the interval from

$$\bar{x} - t_{\alpha/2} \cdot \frac{s}{\sqrt{n}} \quad \text{to} \quad \bar{x} + t_{\alpha/2} \cdot \frac{s}{\sqrt{n}}$$

will be a $(1 - \alpha)$-level confidence interval for μ. Thus we have Procedure 8.2, which we refer to as the **one-sample t-interval procedure** or, more briefly, the **t-interval procedure.**

PROCEDURE 8.2

THE ONE-SAMPLE t-INTERVAL PROCEDURE FOR A POPULATION MEAN, μ

ASSUMPTION

Normal population

Step 1 For a confidence level of $1 - \alpha$, use Table IV to find $t_{\alpha/2}$ with df $= n - 1$, where n is the sample size.

Step 2 The confidence interval for μ is from

$$\bar{x} - t_{\alpha/2} \cdot \frac{s}{\sqrt{n}} \quad \text{to} \quad \bar{x} + t_{\alpha/2} \cdot \frac{s}{\sqrt{n}}$$

where $t_{\alpha/2}$ is found in Step 1 and $\bar{x}$ and s are computed from the actual sample data obtained.

The t-interval procedure applies for any sample size, large or small. The only condition for using it is that the population being sampled is normally distributed. Actually, the procedure works reasonably well even if the population being sampled is only approximately normally distributed. Procedures that are insensitive to departures from the assumptions on which they are based are called **robust.** Thus the t-interval procedure is robust to moderate deviations of the normality assumption.

Because of robustness, we need only be able to detect gross violations of the normality assumption. This we can usually do quite effectively with a normal probability plot: *Unless a normal probability plot for the sample reveals outliers or shows systematic and substantial deviations from linearity, we can apply the t-interval procedure to obtain a confidence interval for a population mean.*

KEY FACT 8.5

WHEN TO USE THE *t*-INTERVAL PROCEDURE (PROCEDURE 8.2)

- If a normal probability plot or some other graphical display indicates the presence of outliers or that the population is far from being normally distributed, Procedure 8.2 should not be used. Otherwise it is an appropriate method for obtaining a confidence interval for a population mean.

- If outliers are present in the sample data but their removal is justified and results in a data set that is approximately bell-shaped, then Procedure 8.2 can be used.

- When considering the application of Procedure 8.2 to a population that is only approximately normally distributed, keep in mind that the smaller the sample size, the smaller the deviation from normality should be. Thus for very small samples, the population should be close to normally distributed; for larger samples, the population can deviate more from being normally distributed.

Example 8.10   Illustrates Procedure 8.2

To estimate the mean gestation period of domestic dogs, 15 randomly selected dogs are observed during pregnancy. Their gestation periods, in days, are recorded in Table 8.5. Obtain a 95% confidence interval for the mean gestation period, μ, of the domestic dog.

TABLE 8.5
Gestation periods, in days, of 15 dogs

62.0	61.4	59.8	62.2	60.3
60.4	59.4	60.2	60.4	60.8
61.8	59.2	61.1	60.4	60.9

SOLUTION   First we construct a normal probability plot for the data in Table 8.5, as shown in Fig. 8.7.

FIGURE 8.7
Normal probability plot of the sample of gestation periods in Table 8.5

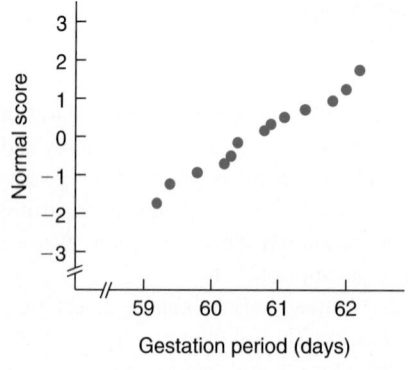

The normal probability plot in Fig. 8.7 shows no outliers and falls roughly in a straight line. Thus we can apply Procedure 8.2 to obtain the required confidence interval.

Step 1 *For a confidence level of $1-\alpha$, use Table IV to find $t_{\alpha/2}$ with df $= n-1$, where n is the sample size.*

The specified confidence level is 0.95, so $\alpha = 0.05$. Since $n = 15$, we have df $= n - 1 = 15 - 1 = 14$. Table IV shows that for df $= 14$,

$$t_{\alpha/2} = t_{0.05/2} = t_{0.025} = 2.145.$$

Step 2 *The confidence interval for μ is from*

$$\overline{x} - t_{\alpha/2} \cdot \frac{s}{\sqrt{n}} \quad to \quad \overline{x} + t_{\alpha/2} \cdot \frac{s}{\sqrt{n}}.$$

From Step 1, $t_{\alpha/2} = 2.145$. Applying the usual formulas for $\overline{x}$ and s to the data in Table 8.5, we find that $\overline{x} = 60.69$ and $s = 0.90$. Consequently, a 95% confidence interval for μ is from

$$60.69 - 2.145 \cdot \frac{0.90}{\sqrt{15}} \quad to \quad 60.69 + 2.145 \cdot \frac{0.90}{\sqrt{15}}$$

or 60.19 to 61.18. We can be 95% confident that the mean gestation period, μ, of the domestic dog is somewhere between 60.19 days and 61.18 days. ∎

Example 8.11 Illustrates Procedure 8.2

The U.S. Department of Agriculture publishes data on U.S. chicken consumption in *Food Consumption, Prices, and Expenditures.* Last year's chicken consumptions, in pounds, for 17 randomly selected people are displayed in Table 8.6. Use the data to obtain a 90% confidence interval for last year's mean chicken consumption, μ.

TABLE 8.6
Sample of last year's
chicken consumptions (lbs)

47	39	62	49	50	70
59	53	55	0	65	63
53	51	50	72	45	

SOLUTION A normal probability plot of the data in Table 8.6 is displayed in Fig. 8.8(a), shown on the following page. The plot reveals an outlier—the data value 0 lb. Thus it is inappropriate to apply Procedure 8.2 to the data in Table 8.6.

The outlier of 0 lb might be a recording error or be due to a person in the sample who does not eat chicken (e.g., a vegetarian). If we remove the outlier from the data set, the normal probability plot for the abridged data set shows no outliers and is quite linear, as we see from Fig. 8.8(b).

FIGURE 8.8

Normal probability plots
for chicken-consumptions:
(a) original data,
(b) data with outlier removed

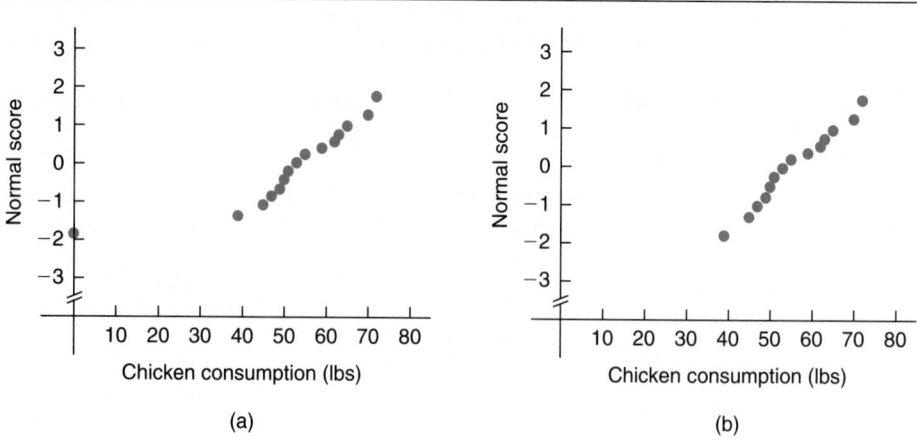

(a)

(b)

This means that if we are willing to take as our population of interest only people who eat chicken, then we can use Procedure 8.2 to obtain a confidence interval. Applying that procedure to the sample data with the outlier removed, we find that a 90% confidence interval is from 51.2 to 59.2. We can be 90% confident that last year's mean chicken consumption, among people who eat chicken, is somewhere between 51.2 lbs and 59.2 lbs. ∎

By taking our population of interest in Example 8.11 to consist only of people who eat chicken, we were justified in removing the outlier 0. Generally, an outlier should not be removed unless careful consideration indicates that it is appropriate to do so. Simply removing an outlier because it is an outlier is unacceptable statistical practice.

If in Example 8.11 we had been careless in our analysis by blindly finding a confidence interval without first looking at the data, then our result would have been invalid and misleading. This explains one reason for the following fundamental principle of data analysis, which applies to any inferential procedure.

KEY FACT 8.6   **A FUNDAMENTAL PRINCIPLE OF DATA ANALYSIS**

> Before performing a statistical-inference procedure, first look at the sample data. If any of the conditions required for using the procedure appear to be violated, do not apply the procedure. Instead use a different procedure that is appropriate.[†]

Although, as we have mentioned previously, graphical displays of small samples must be interpreted carefully, it is still far better to look at the data than to not. Even for very small samples, graphical displays can sometimes detect violations of assumptions required

---

[†] If you are unsure of one, consult a statistician.

for inferential procedures. Just remember to proceed cautiously when conducting graphical analyses of small samples, especially very small samples, say, of size 10 or less.

Which Procedure Should Be Used?

We have now learned two methods for obtaining a confidence interval for a population mean, μ. If the sample size is large, we can use the z-interval procedure (Procedure 8.1 on page 448), regardless of the distribution of the population being sampled. If the population being sampled is normally distributed, we can use the t-interval procedure (Procedure 8.2 on page 465), regardless of the sample size. Two questions may come to mind.

First, what if both conditions are satisfied? That is, suppose we want to obtain a large-sample confidence interval for the mean of a normally distributed population (with unknown σ). Then we can use the z-interval procedure, which is based on the fact that for large samples the random variable $(\bar{x} - \mu)/(\sigma/\sqrt{n})$ has *approximately* the standard normal distribution, regardless of the distribution of the population; or we can use the t-interval procedure, which is based on the fact that for a normal population the random variable $(\bar{x} - \mu)/(s/\sqrt{n})$ has *exactly* the t-distribution with df $= n - 1$, regardless of the size of the sample.

Strictly speaking, the t-interval procedure is the correct procedure to use for a normal population, because it is exact. Practically, however, both procedures yield essentially the same results if the sample size is large. This is because for large samples (and hence large df), there is little difference between the t-distribution and the standard normal distribution. Thus we stopped the t-table at df $= 29$ and then added a final row with df $= \infty$. This last row actually gives values of z_α.

In summary, to obtain a confidence interval for the mean of a normally distributed population, the correct procedure to use is the t-interval procedure. To apply that procedure when the sample size exceeds 30, determine $t_{\alpha/2}$ from the last row (the ∞ row) of the t-table, Table IV.

The second question that may come to mind is: What if neither the large-sample condition nor the normality condition are satisfied? That is, suppose we want to obtain a small-sample confidence interval for the mean of a population that is far from being normally distributed. Then neither the z-interval procedure nor the t-interval procedure is appropriate.

However, under certain conditions we can use a *nonparametric method*. Although most nonparametric methods have some conditions for their use, they do not require even approximate normality and can be applied regardless of sample size.

Example 8.12   Which Procedure Should Be Used?

The Internal Revenue Service publishes information on federal individual income tax returns in *Statistics of Income, Individual Income Tax Returns*. A sample of 12 returns from last year revealed the adjusted gross incomes in thousands of dollars shown in Table 8.7. Which procedure should be used to obtain a confidence interval for the mean adjusted gross income, μ, of all last year's individual income tax returns?

TABLE 8.7
Adjusted gross incomes ($1000)

9.7	93.1	33.0	21.2
81.4	51.1	43.5	10.6
12.8	7.8	18.1	12.7

SOLUTION   Since the sample size is small ($n = 12$), we cannot use the z-interval procedure, Procedure 8.1. Furthermore, a normal probability plot of the sample data, as shown in Fig. 6.35 on page 382, suggests that the population of adjusted gross incomes is far from being normally distributed; so the t-interval procedure, Procedure 8.2, also appears to be inappropriate. Consequently, neither Procedure 8.1 nor Procedure 8.2 should be used; instead some nonparametric confidence-interval procedure should be employed.   ■

Using the Computer (Optional)

We can use Minitab to obtain a confidence interval for the mean, μ, of a normally distributed population. The appropriate command is called **TInterval** (the "T" stands for t-distribution). In Example 8.10 we applied Procedure 8.2 to obtain a confidence interval for the mean gestation period of the domestic dog. Example 8.13 shows how to determine that confidence interval by employing Minitab's TInterval command.

Example 8.13   The TInterval Command

The gestation periods, in days, of 15 randomly selected dogs are displayed in Table 8.5 on page 466. Use Minitab to obtain a 95% confidence interval for the mean gestation period, μ, of the domestic dog.

SOLUTION   First we store the sample data from Table 8.5 in a column named GESTN. Next we use the method described on page 385 to have Minitab produce a normal probability plot of the data. This plot is displayed in Printout 8.3. *(Note:* The 3 in the printout indicates there are three points in that vicinity that are either the same or very close together.*)*

The normal probability plot in Printout 8.3 shows no outliers and is roughly linear. Hence we can apply TInterval to obtain the required confidence interval.

Session commands: We type the command **TINTERVAL**, followed by the confidence level that we require and the storage location of the sample data. In other words, we type TINTERVAL 95% confidence, data in 'GESTN' and press Enter. Printout 8.4 shows this command and the output that results.

(or)

Menu commands: We choose **Stat ▸ Basic Statistics ▸ 1-Sample t...**, specify GESTN in the **Variables** text box, select **Confidence interval** and type 95 for **Level**, and then select **OK**. The output obtained is displayed in Printout 8.4.

PRINTOUT 8.3

Minitab normal probability plot
for the gestation-period data

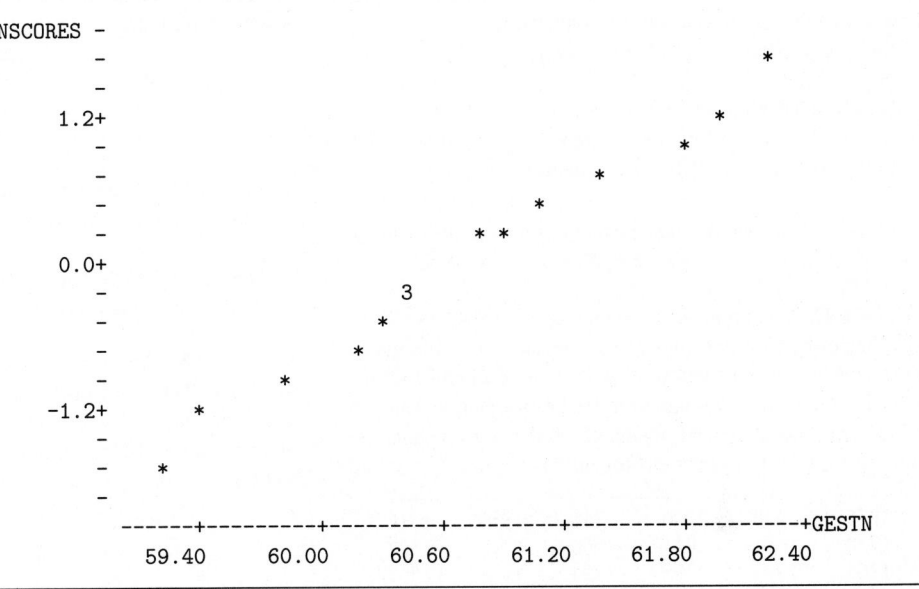

PRINTOUT 8.4

Minitab output for the
TInterval command

```
MTB > TINTERVAL 95% confidence, data in 'GESTN'

          N     MEAN    STDEV   SE MEAN    95.0 PERCENT C.I.
GESTN    15   60.687   0.898    0.232   ( 60.190,  61.184)
```

The output in Printout 8.4 provides the sample size, sample mean, sample standard deviation, estimated standard error of the mean ($s/\sqrt{n}$), and the confidence interval. This last item shows that a 95% confidence interval for μ is from 60.190 to 61.184. We can be 95% confident that the mean gestation period, μ, of the domestic dog is somewhere between 60.190 days and 61.184 days. ∎

EXERCISES 8.4

8.43 For a t-curve with df = 6, use Table IV to find the following t-values.

a. $t_{0.10}$ b. $t_{0.025}$ c. $t_{0.01}$

8.44 For a t-curve with df = 17, use Table IV to find the following t-values.

a. $t_{0.05}$ b. $t_{0.025}$ c. $t_{0.005}$

8.45 For a t-curve with df = 21, find the following t-values and illustrate your results graphically.

a. the t-value having area 0.10 to its right

b. $t_{0.01}$

c. the t-value having area 0.025 to its left (*Hint:* A t-curve is symmetric about 0.)

d. the two t-values that divide the area under the curve into a middle 0.90 area and two outside areas of 0.05

8.46   For a *t*-curve with df = 8, find the following *t*-values and illustrate your results graphically.

a. the *t*-value having area 0.05 to its right

b. $t_{0.10}$

c. the *t*-value having area 0.01 to its left

d. the two *t*-values that divide the area under the curve into a middle 0.95 area and two outside areas of 0.025

Normal probability plots indicate that it is reasonable to apply the t-interval procedure in each of Exercises 8.47–8.52.

8.47   According to the Salt River Project (SRP), a supplier of electricity to the greater Phoenix area, the mean annual electric bill in 1984 was $852.31. An economist wants to estimate the mean for last year. He takes a random sample of 18 SRP customers and obtains the following amounts, in dollars, for their last year's electric bills.

1875	1478	2206	1740	1830	1516
1738	1486	1941	1608	1794	1828
1264	1999	1798	1794	1598	1568

a. Determine a 95% confidence interval for last year's mean annual electric bill, μ, for all SRP customers. *(Note: The sample mean and sample standard deviation of the data are $\bar{x} = \$1725.61$ and $s = \$222.45$.)*

b. Does it appear that the mean annual electric bill has increased from the 1984 mean of $852.31? Explain your answer.

8.48   A city planner working on bikeways needs information about local bicycle commuters. She designs a questionnaire. One of the questions asks how many minutes it takes the rider to pedal from home to his or her destination. A sample of local bicycle commuters yields the following times.

22	19	24	31	29	29
21	15	27	23	37	31
30	26	16	26	12	
23	48	22	29	28	

a. Find a 90% confidence interval for the mean commuting time of all local bicycle commuters in the city. *(Note: The sample mean and sample standard deviation of the data are $\bar{x} = 25.82$ and $s = 7.71$.)*

b. Interpret your result in part (a).

8.49   As reported by the Department of Agriculture in *Crop Production*, the mean yield of oats for U.S. farms is 58.4 bushels per acre. A farmer wants to estimate his mean yield using a newly developed fertilizer. He uses the fertilizer on a random sample of 1-acre plots and obtains the following yields, in bushels.

67	65	55	57	58
61	61	61	64	62
62	60	62	60	67

a. Find a 99% confidence interval for the mean yield per acre, μ, that this farmer will get on his land with the new fertilizer. *(Note: $\bar{x} = 61.47, s = 3.38$.)*

b. Does it appear that the farmer can get a better mean yield than the national average by using the new fertilizer?

8.50   According to *Library Journal*, published by the R. R. Bowker Company of New York, the mean annual subscription rate for law periodicals was $29.66 in 1983. A random sample of 12 law periodicals yields the following annual subscription rates, to the nearest dollar, for this year.

30	46	44	47
42	38	62	55
52	48	43	54

a. Determine a 95% confidence interval for this year's mean annual subscription rate, μ, for all law periodicals. *(Note: $\bar{x} = 46.75, s = 8.44$.)*

b. Does your result from part (a) suggest an increase in the mean annual subscription rate over that in 1983?

8.51   *Physician's Handbook* provides statistics on heights and weights of children by age. The heights, in inches, of 20 randomly selected 6-year-old girls follow.

44	44	47	46	38
42	46	41	50	43
40	51	47	43	47
48	48	45	41	46

a. Obtain a 95% confidence interval for the mean height of all 6-year-old girls. *(Note: $\bar{x} = 44.85, s = 3.39$.)*

b. Interpret your result from part (a).

8.52   To estimate the mean length, μ, of western rattlesnakes, 10 such snakes are randomly selected. Their lengths, in inches, are as follows.

40.2	43.1	45.5	44.5	39.5
40.2	41.0	41.6	43.1	44.9

a. Find a 90% confidence interval for the mean length, μ, of all western rattlesnakes. *(Note: $\bar{x} = 42.36, s = 2.16$.)*

b. Interpret your result from part (a).

8.53 A random sample of size 60 is taken from a population. A normal probability plot displays extreme curvature but no outliers. Which procedure should be used to obtain a confidence interval for the mean of the population? Explain your answer.

8.54 A random sample of size 60 is taken from a population. A normal probability plot shows no outliers and is roughly linear. Which procedure should be used to obtain a confidence interval for the mean of the population? Explain your answer.

8.55 (Computer exercise) Refer to Exercise 8.47. Use Minitab or some other statistical software to
a. obtain a normal probability plot of the sample data.
b. determine the required confidence interval.
c. Justify the use of your procedure in part (b).

8.56 (Computer exercise) Refer to Exercise 8.48. Use Minitab or some other statistical software to
a. obtain a normal probability plot of the sample data.
b. determine the required confidence interval.
c. Justify the use of your procedure in part (b).

8.57 (Computer exercise) A manufacturer of tobacco products intends to begin marketing a new brand of cigarette. To do so, the manufacturer needs information about the mean tar content, μ. Obviously, he cannot test all such cigarettes, so he tests a sample of them. The company statistician knows that tar content is normally distributed, so she applies Minitab's TInterval command to the tar-content data. Following is the abridged Minitab output. Tar contents are in milligrams.

N	MEAN	STDEV	SE MEAN	95.0 PERCENT C.I.
25	10.985	0.604	0.121	(10.736, 11.235)

Use the printout to find
a. the mean tar content of the cigarettes tested.
b. the standard deviation of the tar contents.
c. the number of cigarettes tested.
d. a 95% confidence interval for the mean tar content, μ, of all cigarettes of the new brand.

8.58 (Computer exercise) For advertising purposes a tire company wants to estimate the mean life of a line of steel-belted radials. From past experience it is known that tire life is normally distributed. A random sample of steel-belted radials is tested, and the following abridged output is obtained from the application of Minitab's TInterval command. Data are in thousands of miles.

N	MEAN	STDEV	SE MEAN	90.0 PERCENT C.I.
16	39.469	1.849	0.462	(38.659, 40.280)

From the printout, determine
a. the mean life of the tires sampled.
b. the estimated standard error of the mean.
c. the number of tires in the sample.
d. a 90% confidence interval for the mean life, μ, of all tires in this line.

Note: Some statistical software will not have sufficient storage to carry out Exercises 8.59 and 8.60.

8.59 (Computer exercise) Consider a normally distributed population with mean 100 and standard deviation 16.
a. Simulate the taking of 1000 random samples of size 10 from the population.
b. Determine the sample mean and sample standard deviation of each of the 1000 samples.
c. For each of the 1000 samples, obtain the quantity

$$\frac{\bar{x} - 100}{s/\sqrt{10}}.$$

d. Obtain a histogram of the 1000 values found in part (c).
e. Theoretically, what is the distribution of the random variable in part (c)?
f. Compare your results from parts (d) and (e).

8.60 (Computer exercise) Consider a normally distributed population with mean 100 and standard deviation 16.
a. Simulate the taking of 1000 random samples of size 10 from the population.
b. Find the sample mean of each of the 1000 samples.
c. For each of the 1000 samples, obtain the quantity

$$\frac{\bar{x} - 100}{16/\sqrt{10}}.$$

d. Obtain a histogram of the 1000 values found in part (c).
e. Theoretically, what is the distribution of the random variable in part (c)?
f. Compare your results from parts (d) and (e).

8.61 Suppose a random sample of size n is to be taken from a normally distributed population with mean μ and standard deviation σ. Identify the probability distributions of the random variables in parts (a) and (b):

a. $\dfrac{\bar{x} - \mu}{\sigma/\sqrt{n}}$ b. $\dfrac{\bar{x} - \mu}{s/\sqrt{n}}$

c. How do we obtain probabilities for the random variable in part (a)?
d. How do we obtain probabilities for the random variable in part (b)?

· · **8.62** IQs measured on the Stanford Revision of the Binet-Simon Intelligence Scale are normally distributed with mean $\mu = 100$ and standard deviation $\sigma = 16$. Suppose 250 IQs are to be selected at random. Identify the probability distribution of each of the following random variables.

a. $\dfrac{\bar{x} - 100}{16/\sqrt{250}}$ b. $\dfrac{\bar{x} - 100}{s/\sqrt{250}}$

The random variable in part (b) has a t-distribution with df $= 249$; thus probabilities for that random variable are equal to areas under the t-curve with df $= 249$. But our t-table, Table IV, does not have df $= 249$. So, for instance,

c. how would you find the t-value having area 0.025 to its right for a t-curve with df $= 249$?

Confidence intervals for a normal population mean when σ is known. Suppose a random sample of size n is to be taken from a normally distributed population with mean μ and standard deviation σ. Then, by Key Fact 7.4 on page 420, the random variable $\bar{x}$ is normally distributed with mean $\mu_{\bar{x}} = \mu$ and standard deviation $\sigma_{\bar{x}} = \sigma/\sqrt{n}$. This is true regardless of the size of the sample. If we now use the same argument as we did on page 448 prior to Procedure 8.1, then we find that once the sample is taken, the interval from

(3) $\qquad \bar{x} - z_{\alpha/2} \cdot \dfrac{\sigma}{\sqrt{n}} \quad$ to $\quad \bar{x} + z_{\alpha/2} \cdot \dfrac{\sigma}{\sqrt{n}}$

will be a $(1 - \alpha)$-level confidence interval for μ. Use this fact to solve Exercises 8.63–8.65.

· · **8.63** A certain brand of water-softener salt comes in packages marked "net weight 40 lb." The weights are known to be normally distributed with a standard deviation of 1.5 lb. To ensure that the amount of salt being packaged does weigh roughly 40 lb, on the average, 15 bags are randomly selected and their contents carefully weighed. The results follow.

38.1	38.6	42.5	40.7	40.0
39.9	40.4	41.0	41.6	40.4
43.2	37.9	39.5	42.3	39.4

a. Use the data to find a 99% confidence interval for the mean net weight, μ, of all such packages of water-softener salt.
b. Does your result from part (a) conflict with the advertised weight of 40 lb? Explain your answer.

· · **8.64** An English professor wants to estimate the average study time per week for students in introductory English courses at her school. To accomplish that she randomly se-

lects 25 such students and records their weekly study times. The times, in hours, follow.

9	8	7	6	7
8	9	4	7	6
6	4	11	5	4
3	7	8	8	7
6	2	2	8	6

Assume weekly study times are normally distributed with a standard deviation of 2.0 hours.
a. Find a 95% confidence interval for the mean weekly study time, μ, of introductory English students at the school.
b. Interpret your result from part (a).

· · **8.65** Refer to Exercise 8.49.
a. Rework part (a) of Exercise 8.49 under the assumption that σ is known and equals 3.38 bushels.
b. Compare the confidence interval from part (a) of this exercise to the one found in Exercise 8.49.
c. What is the general effect on a confidence interval for a normal population mean if σ is known and hence Formula (3) instead of the t-interval procedure is used to obtain the confidence interval?

· · **8.66** Suppose a random sample of size $n \geq 30$ is to be taken from a normally distributed population in order to obtain a confidence interval for the mean, μ, of the population. Then we can use the z-interval procedure, which applies when the sample size is large, or the t-interval procedure, which applies when the population being sampled is normally distributed. Which is the better procedure to use? Why?

· · **8.67** Let $0 < \alpha < 1$. For a t-curve, determine
a. the t-value having area α to its right.
b. the t-value having area α to its left.
c. the two t-values that divide the area under the curve into a middle $1 - \alpha$ area and two outside areas of $\alpha/2$.
d. Draw graphs to illustrate your results in parts (a)–(c).

· · · **8.68** Suppose a random sample of size n is to be taken from a normally distributed population with mean μ.
a. Use Key Fact 8.3 on page 461 and part (c) of Exercise 8.67 to verify that

$$P\left(-t_{\alpha/2} < \frac{\bar{x} - \mu}{s/\sqrt{n}} < t_{\alpha/2}\right) = 1 - \alpha.$$

b. Use algebra to show that the result from part (a) can be expressed as

$$P\left(\bar{x} - t_{\alpha/2} \cdot \frac{s}{\sqrt{n}} < \mu < \bar{x} + t_{\alpha/2} \cdot \frac{s}{\sqrt{n}}\right) = 1 - \alpha.$$

c. Use the result from part (b) to explain why, once the sample is taken, the interval from

$$\bar{x} - t_{\alpha/2} \cdot \frac{s}{\sqrt{n}} \quad \text{to} \quad \bar{x} + t_{\alpha/2} \cdot \frac{s}{\sqrt{n}}$$

will be a $(1 - \alpha)$-level confidence interval for μ.

Chapter Review

Key Terms

Formulas

In the formulas below,

μ = population mean
σ = population standard deviation
$\bar{x}$ = sample mean

n = sample size
$\alpha = 1 -$ confidence level
s = sample standard deviation

z-interval for μ (large sample), *448*

$$\bar{x} - z_{\alpha/2} \cdot \frac{\sigma}{\sqrt{n}} \quad \text{to} \quad \bar{x} + z_{\alpha/2} \cdot \frac{\sigma}{\sqrt{n}}$$

(if σ is unknown, use s in its place)

Margin of error for the estimate of μ, *457*

$$E = z_{\alpha/2} \cdot \frac{\sigma}{\sqrt{n}}$$

(if σ is unknown, use s in its place)

Sample size for the estimate of μ, *457*

$$n = \left(\frac{z_{\alpha/2} \cdot \sigma}{E}\right)^2,$$

rounded up to the nearest whole number.

t-interval for μ (normal population), *465*

$$\bar{x} - t_{\alpha/2} \cdot \frac{s}{\sqrt{n}} \quad \text{to} \quad \bar{x} + t_{\alpha/2} \cdot \frac{s}{\sqrt{n}}$$

(df $= n - 1$)

You Should Be Able To

1. use and understand each of the preceding formulas.
2. use Table II to find $z_{\alpha/2}$ for any specified value of α.
3. find a large-sample confidence interval for a population mean, μ.
4. compute the margin of error for the estimate of μ.
5. understand the relationship between the sample size, standard deviation, confidence level, and margin of error for a confidence interval for μ.
6. determine the sample size required for a specified confidence level and margin of error for the estimate of μ.
7. use Table IV to find $t_{\alpha/2}$ for df $= n - 1$ and any specified value of α.
8. find a confidence interval for a population mean, μ, when the population being sampled is normally distributed.
9. decide which procedure to use when finding a confidence interval for a population mean, μ.
10. interpret a confidence interval for a population mean.
11. state and understand a fundamental principle of data analysis.
12. use the Minitab commands covered in this chapter.*
13. interpret the output obtained from the application of the Minitab commands discussed in this chapter.*

REVIEW TEST

1. Suppose a random sample of size $n \geq 30$ is to be taken from a population with mean μ and standard deviation σ.
   a. Identify the probability distribution of $\bar{x}$.
   b. Identify the probability distribution of

   $$\frac{\bar{x} - \mu}{\sigma/\sqrt{n}}.$$

   c. How do we find probabilities for the random variable in part (b)?

2. Use Table II to find the following z-values, and draw graphs to illustrate your work.
   a. $z_{0.063}$ b. $z_{0.40}$ c. $z_{0.007}$

3. Dr. Thomas Stanley of Georgia State University has surveyed millionaires since 1973. Among other things, Stanley obtains estimates for the mean age, μ, of all U.S. millionaires. Suppose 36 U.S. millionaires are randomly selected and their ages are as follows.

31	45	79	64	48	38	39	68	52
59	68	79	42	79	53	74	66	66
71	61	52	47	39	54	67	55	71
77	64	60	75	42	69	48	57	48

Determine a 95% confidence interval for the mean age, μ, of all U.S. millionaires. *(Note:* The sample mean and sample standard deviation of the data are 58.53 years and 13.36 years, respectively.*)*

4. From Problem 3 we know that "a 95% confidence interval for the mean age of all U.S. millionaires is from 54.2 years to 62.9 years." Decide which of the following provide a correct interpretation of the statement in quotes. Justify your answers.
   a. 95% of all U.S. millionaires are between the ages of 54.2 years and 62.9 years.
   b. There is a 95% chance that the mean age of all U.S. millionaires is between 54.2 years and 62.9 years.
   c. We can be 95% confident that the mean age of all U.S. millionaires is between 54.2 years and 62.9 years.
   d. The probability is 0.95 that the mean age of all U.S. millionaires is between 54.2 years and 62.9 years.

*5. **(Computer problem)** Use Minitab or some other statistical software to obtain the confidence interval required in Problem 3.

*6. **(Computer problem)** The following abridged Minitab output was obtained by applying the ZInterval command to the age data in Problem 3.

```
THE ASSUMED SIGMA =13.4

N    MEAN   STDEV   SE MEAN    95.0 PERCENT C.I.
36   58.53  13.36    2.23   (  54.16,  62.90)
```

From the printout, determine
   a. the (assumed) standard deviation, σ, of the ages of all U.S. millionaires.
   b. the sample standard deviation, s.

c. the number of millionaires sampled.

d. the mean age, $\bar{x}$, of the millionaires sampled.

e. a 95% confidence interval for the mean age, μ, of all U.S. millionaires.

7. A certain brand of hand-held computer runs on four 1.5-volt AAA batteries. The mean battery life of a sample of 50 such computers is found to be 60.1 hours.
   a. Assuming $\sigma = 4.3$ hours, obtain a 99% confidence interval for the mean battery life, μ, of this make of computer.
   b. Interpret your result from part (a).

8. Refer to Problem 7.
   a. Find the margin of error, E.
   b. Explain the meaning of E as far as the accuracy of the estimate is concerned.
   c. Determine the sample size required to have a margin of error of 0.5 hours and a 99% confidence level.
   d. Find a 99% confidence interval for μ if a sample of the size determined in part (c) yields a mean of 59.8 hours.

9. The figure below shows the standard normal curve and two t-curves. Which of the two t-curves has the larger degrees of freedom? Why?

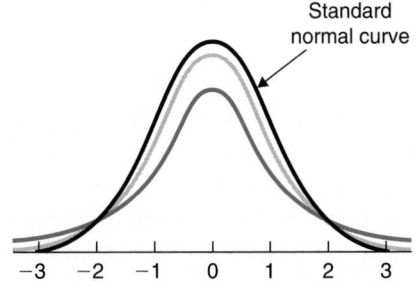

10. Suppose a random sample of size n is to be taken from a normally distributed population with mean μ and standard deviation σ.
    a. Identify the probability distribution of $\bar{x}$.
    b. Identify the probability distribution of

    $$\frac{\bar{x} - \mu}{s/\sqrt{n}}.$$

    c. How do we find probabilities for the random variable in part (b)?

11. For a t-curve with df $= 18$, obtain the following t-values and illustrate your results graphically.
    a. the t-value having area 0.025 to its right
    b. $t_{0.05}$

c. the t-value having area 0.10 to its left

d. the two t-values that divide the area under the curve into a middle 0.99 area and two outside 0.005 areas

12. The A. C. Nielsen Company publishes information on television viewing by Americans in *Nielsen Report on Television*. A random sample of 20 U.S. households yields the following daily viewing times in hours.

7.5	2.5	6.8	5.0	7.9
5.3	8.9	8.8	10.3	8.8
9.5	9.5	6.1	9.4	8.4
8.2	6.5	9.0	6.4	8.4

a. A normal probability plot of this sample data shows no outliers and is roughly linear. Given that fact, find a 95% confidence interval for the mean daily viewing time, μ, of all U.S. households. *(Note: $\bar{x} = 7.66$ and $s = 1.913$.)*

b. Interpret your result from part (a).

c. In 1988 the average U.S. household watched 7 hours and 6 minutes of TV per day. Does your result in part (a) provide evidence of an increase in average daily viewing time? Explain your answer.

*13. **(Computer problem)** Refer to Problem 12. Use Minitab or some other statistical software to
    a. obtain and interpret a normal probability plot of the sample data.
    b. find the required confidence interval.

*14. **(Computer problem)** The following abridged Minitab output was obtained by applying the TInterval command to the data in Problem 12.

```
N    MEAN   STDEV   SE MEAN    95.0 PERCENT C.I.
20   7.660  1.913   0.428   (   6.764,   8.556)
```

From the printout, determine
a. the mean daily viewing time of the households in the sample.
b. the standard deviation of the daily viewing times of the households in the sample.
c. the estimated standard error of the mean.
d. the number of households in the sample.
e. a 95% confidence interval for the mean daily viewing time, μ, of all U.S. households.

15. In each part of this problem, we have provided a scenario for a confidence interval. Decide whether the ap-

propriate method for obtaining the confidence interval is the z-interval procedure (Procedure 8.1), the t-interval procedure (Procedure 8.2), or neither.

a. A random sample of size 17 is taken from a population. A normal probability plot of the sample data is found to be very close to linear (straight line).

b. A random sample of size 50 is taken from a population. A normal probability plot of the sample data is found to be roughly linear.

c. A random sample of size 25 is taken from a population. A normal probability plot of the sample data shows three outliers but is otherwise roughly linear.

Furthermore, it is determined that the outliers are due to recording errors.

d. A random sample of size 30 is taken from a population. A normal probability plot of the sample data shows three outliers but is otherwise roughly linear. It is not clear whether it is reasonable to remove the outliers.

e. A random sample of size 128 is taken from a population. A normal probability plot of the sample data shows no outliers but significant curvature.

f. A random sample of size 13 is taken from a population. A normal probability plot of the sample data shows no outliers but significant curvature.

Using the Focus Database

The database printed in Appendix B contains data on 500 randomly selected Arizona State University sophomores. Among the variables considered are high-school GPA, SAT math score, cumulative GPA, SAT verbal score, age, and total hours. Use Minitab or some other statistical software to solve the following problems.

a. Obtain a 99% confidence interval for the mean high-school GPA of all Arizona State University sophomores.

b. Repeat part (a) for the other five variables listed above.

c. Justify the use of the confidence-interval procedures you employed in parts (a) and (b).

Case Study   **Zooplankton Nutrition in the Gulf of Mexico**

As the Mississippi River meets the Gulf of Mexico, the river's fresh water, with its high concentration of nutrient-laden sediment, mixes with gulf water. This results in a body of low salinity, nutrient-rich water (often referred to as the *plume*) that spreads westward along the Louisiana–Texas coast. The nutrients that are mixed into the gulf fuel the growth of microscopic marine algae, which in turn are eaten by tiny marine animals called zooplankton.

Dr. George McManus and Gregory Weiss, two marine scientists studying the dynamics of the lower food chain along the Gulf Coast, were interested in examining the effects of the nutrient-rich Mississippi River water on zooplankton nutrition. In the early spring of 1993, they collected zooplankton samples at three sites: near the mouth of the river, in the western part of the plume, and outside the plume (to the east).

The scientists used the lipid content of the zooplankton as an index of nutritional quality.[†] Specifically, they used the ratio of two kinds of lipid compounds, triacylglycerols (energy storage compounds) and sterols (cell-membrane components), to determine the nutritional state of each zooplankton. Table 8.8 displays the data collected at the three sites. The data are unitless, being the ratio of two weights.

[†] Lipids are biochemical compounds that, among other things, serve as energy storage and cell-membrane components.

	Mouth of river	Western plume	Outside plume
TABLE 8.8	1.89	1.79	1.36
Lipid contents (ratio of triacylglycerol to sterol) for zooplankton samples in three regions of the Gulf of Mexico	2.27	2.30	0.32
	1.14	2.52	0.34
	1.51	1.37	0.00
		0.93	0.11

a. Construct a normal probability plot for each of the three samples. *(Note:* The ordered normal scores for a sample of size four are -1.05, -0.30, 0.30, and 1.05.*)*

b. Identify outliers, if any, in each of the three samples.

c. For which samples might it be reasonable to use the t-interval procedure, Procedure 8.2 on page 465, to obtain a confidence interval for the mean lipid content in the region? Explain your answer.

d. For each sample for which it might be reasonable to use the t-interval procedure, determine a 95% confidence interval for the mean lipid content of the region. Interpret your results.

e. Use Minitab or some other statistical software to solve parts (a), (b), and (d).

*Jerzy
Neyman*

Jerzy Neyman was born on April 16, 1894, in Bendery, Russia. His father, Czeslaw, was a member of the Polish nobility, a lawyer, a judge, and an amateur archaeologist. Because the Russian authorities prohibited the family from living in Poland, Jerzy Neyman grew up in various cities in Russia, attending the university in Kharkov beginning in 1912. At Kharkov he was at first interested in physics, but because of his clumsiness in the laboratory, he decided to pursue mathematics.

After World War I, when Russia was at war with Poland over borders, Neyman was jailed as an enemy alien. In 1921, in a prisoner exchange, he went to Poland for the first time, and in 1924 he received his doctorate from the University of Warsaw. Between 1924 and 1934, Neyman worked with Karl Pearson (see biography, Chapter 12) and his son Egon Pearson, and held a position at the University of Kraków. In 1934 Neyman took a position in Karl Pearson's statistical laboratory at University College in London. He stayed in England, working with Egon Pearson, until 1938, at which time he accepted an offer to join the faculty at the University of California at Berkeley.

When the United States entered World War II, Neyman did war work, setting aside the development of a statistics program. After the war ended, Neyman organized a symposium to celebrate its end and "the return to theoretical research." That symposium, held in August 1945, and succeeding ones, held every 5 years until 1970, were instrumental in establishing Berkeley as a preeminent statistical center.

Neyman was a principal founder of the theory of modern statistics. His work on hypothesis testing, confidence intervals, and survey sampling transformed both the theory and the practice of statistics. His achievements were acknowledged by the receipt of many honors and awards, including election to the United States National Academy of Sciences, the Guy Medal in Gold of the Royal Statistical Society, and the United States National Medal of Science.

Neyman remained active until his death of heart failure on August 5, 1981, at the age of 87, in Oakland, California.

Chapter 9

Hypothesis Tests for One Population Mean

. . . . . .

General Objective

In Chapter 8 we examined methods for obtaining confidence intervals for one population mean. As we know, a confidence interval for a population mean, μ, is based on the statistic $\bar{x}$. Now we will learn how that statistic can be used to make decisions about hypothesized values of a population mean. For example, we might want to use the mean sentence, $\bar{x}$, of a sample of people imprisoned last year for drug offenses to decide whether last year's mean sentence, μ, for all such people exceeds the 1990 mean of 86.2 months. Statistical inferences of this kind are called *hypothesis tests.*

In this chapter we will study hypothesis tests for one population mean. Several different procedures will be considered: one applies when the sample size is large, another when the population being sampled is normally distributed, and still another when the population being sampled has a symmetric (but not necessarily normal) distribution. We will also examine the two different approaches to hypothesis testing, the classical approach and the *P*-value approach.

Chapter Outline

Case Study

Effects of Brewery Effluent on Soil

How does the application of brewery wastes affect soil composition and plant growth? Two researchers answered that question in an article appearing in the journal *Environmental Pollution.* We will examine that article in the case study at the end of this chapter.

9.1 THE NATURE OF HYPOTHESIS TESTING

We often use inferential statistics to make decisions or judgments about the value of a parameter, such as a population mean. For example, we might need to decide whether the mean weight, μ, of all bags of pretzels being packaged by a particular company differs from the advertised weight of 454 grams; or we might want to determine whether the mean age, μ, of all juveniles currently being held in public custody has decreased from the 1989 mean of 16 years.

One of the most commonly used methods for making such decisions or judgments is to perform a **hypothesis test.** A **hypothesis** is simply a statement that something is true. For instance, the statement "the mean weight of all bags of pretzels being packaged differs from the advertised weight of 454 grams" is a hypothesis.

Typically, a hypothesis test involves two hypotheses. One hypothesis is called the **null hypothesis,** the other the **alternative hypothesis** (or **research hypothesis**). These hypotheses are defined in Definition 9.1.

DEFINITION 9.1   **NULL AND ALTERNATIVE HYPOTHESES**

> *Null hypothesis:* A hypothesis to be tested. We use the symbol H_0 to stand for the null hypothesis.
>
> *Alternative hypothesis:* A hypothesis to be considered as an alternate to the null hypothesis. We use the symbol H_a to stand for the alternative hypothesis.

For example, in the pretzel-packaging illustration, the null hypothesis might be "the mean weight of all bags of pretzels being packaged equals the advertised weight of 454 grams," and the alternative hypothesis might be "the mean weight of all bags of pretzels being packaged differs from the advertised weight of 454 grams."

Originally, the word *null* in *null hypothesis* stood for "no difference" or "the difference is null." Over the years, however, *null hypothesis* has come to mean simply a hypothesis to be tested. The problem in a hypothesis test is to decide whether or not the null hypothesis should be rejected in favor of the alternative hypothesis.

Choosing the Hypotheses

The first step in setting up a hypothesis test is to decide what the null hypothesis and the alternative hypothesis should be. Below we offer some guidelines for choosing these two hypotheses. Although the guidelines refer specifically to hypothesis tests for one population mean, μ, they apply to any hypothesis test concerning one parameter.

Null hypothesis: In this book the null hypothesis for a hypothesis test concerning a population mean, μ, should always specify a single value for that parameter. This means that the null hypothesis should always be of the form $\mu = \mu_0$, where μ_0 is some number. In other words, there should be an equal sign (=) in the null hypothesis. We can therefore express the null hypothesis concisely as

$$H_0: \mu = \mu_0.$$

Alternative hypothesis: The choice of the alternative hypothesis depends on and should reflect the purpose of performing the hypothesis test. Three choices are possible for the alternative hypothesis.

1. If our primary concern is deciding whether a population mean, μ, is *different from* a specified value μ_0, then the alternative hypothesis should be $\mu \neq \mu_0$. In other words, there should be a not-equal sign ($\neq$) in the alternative hypothesis. We express such an alternative hypothesis as

$$H_a: \mu \neq \mu_0.$$

A hypothesis test whose alternative hypothesis is of this form is called a **two-tailed test.**

2. If our primary concern is deciding whether a population mean, μ, is *less than* a specified value μ_0, then the alternative hypothesis should be $\mu < \mu_0$. In other words, there should be a less-than sign ($<$) in the alternative hypothesis. We express such an alternative hypothesis as

$$H_a: \mu < \mu_0.$$

A hypothesis test whose alternative hypothesis is of this form is called a **left-tailed test.**

3. If our primary concern is deciding whether a population mean, μ, is *greater than* a specified value μ_0, then the alternative hypothesis should be $\mu > \mu_0$. In other words, there should be a greater-than sign ($>$) in the alternative hypothesis. We express such an alternative hypothesis as

$$H_a: \mu > \mu_0.$$

A hypothesis test whose alternative hypothesis is of this form is called a **right-tailed test.**

A hypothesis test is called a **one-tailed test** if it is either left-tailed or right-tailed, that is, if it is not two-tailed. In Section 9.2 we will explain the significance of the term *tailed*. But let's now consider Examples 9.1–9.3, which illustrate the preceding discussion.

Example 9.1   Choosing the Null and Alternative Hypotheses

A snack-food company produces a 454-gram bag of pretzels. Although the actual net weights deviate somewhat from 454 grams and vary from one bag to another, it is important to the company that the mean net weight of the bags be kept at 454 grams. Consequently, the quality-assurance department periodically performs a hypothesis test to decide whether or not the packaging machine is working properly, that is, to decide whether or not the mean net weight of all bags being packaged is 454 grams.

a. Determine the null hypothesis for the hypothesis test.

b. Determine the alternative hypothesis for the hypothesis test.

c. Classify the hypothesis test as two-tailed, left-tailed, or right-tailed.

SOLUTION   Let μ denote the (population) mean net weight of all bags being packaged.

a. As we said, the null hypothesis for a hypothesis test concerning a population mean, μ, should always specify a single value for that parameter. Thus the null hypothesis for this hypothesis test is that the packaging machine is working properly; that is, the mean net weight, μ, of all bags being packaged equals 454 grams. In symbols, H_0: $\mu = 454$ grams.

b. The alternative hypothesis for this hypothesis test is that the packaging machine is not working properly; that is, the mean net weight, μ, of all bags being packaged is *different from* 454 grams. In symbols, H_a: $\mu \neq 454$ grams.

c. This hypothesis test is two-tailed since a not-equal sign ($\neq$) appears in the alternative hypothesis.   ∎

Example 9.2   Choosing the Null and Alternative Hypotheses

The R. R. Bowker Company of New York collects information on the retail prices of books. Data are published in *Publishers Weekly*. In 1990 the mean retail price of all hardcover history books was $35.44. Suppose we want to perform a hypothesis test to decide whether this year's mean retail price of all hardcover history books has increased over the 1990 mean.

a. Determine the null hypothesis for the hypothesis test.

b. Determine the alternative hypothesis for the hypothesis test.

c. Classify the hypothesis test as two-tailed, left-tailed, or right-tailed.

SOLUTION   Let μ denote this year's mean retail price of all hardcover history books.

a. Again, the null hypothesis for a hypothesis test concerning a population mean, μ, should always specify a single value for that parameter. Thus the null hypothesis for this hypothesis test is that this year's mean retail price of all hardcover history books is the same as the 1990 mean of $35.44; that is, H_0: $\mu = \$35.44$.

b. We want to decide whether this year's mean retail price for hardcover history books has increased over the 1990 mean. So the alternative hypothesis for this hypothesis test is that this year's mean retail price of all hardcover history books is *greater than* $35.44; that is, H_a: $\mu > \$35.44$.

c. This hypothesis test is right-tailed since a greater-than sign ($>$) appears in the alternative hypothesis.   ∎

Example 9.3   Choosing the Null and Alternative Hypotheses

Calcium is the most abundant and one of the most important minerals in the body. It works with phosphorus to build and maintain bones and teeth. According to the Food and

Nutrition Board of the National Academy of Sciences, the recommended daily allowance (RDA) of calcium for adults is 800 milligrams (mg). A nutritionist thinks the average person with an income below the poverty level gets less than the RDA of 800 mg. She plans to perform a hypothesis test to determine whether her suspicion is correct.

a. Determine the null hypothesis for the hypothesis test.

b. Determine the alternative hypothesis for the hypothesis test.

c. Classify the hypothesis test as two-tailed, left-tailed, or right-tailed.

SOLUTION Let μ denote the mean calcium intake (per day) of all people with incomes below the poverty level.

a. The null hypothesis must specify a single value for the parameter μ. Hence the null hypothesis for this hypothesis test is that the mean calcium intake of all people with incomes below the poverty level is 800 mg per day; that is, H_0: $\mu = 800$ mg.

b. The nutritionist suspects that the mean calcium intake of all people with incomes below the poverty level is *less than* the RDA of 800 mg per day. Thus the alternative hypothesis for this hypothesis test is H_a: $\mu < 800$ mg.

c. This hypothesis test is left-tailed since a less-than sign ($<$) appears in the alternative hypothesis. ■

The Logic of Hypothesis Testing

We have now seen how to choose appropriate null and alternative hypotheses for a hypothesis test. The next question is this: How do we decide which of the two hypotheses is true; that is, how do we decide whether or not to reject the null hypothesis in favor of the alternative hypothesis?

Very roughly, the procedure for deciding is the following: Take a random sample from the population. If the sample data are consistent with the null hypothesis, then do not reject the null hypothesis; if the sample data are inconsistent with the null hypothesis, then reject the null hypothesis and conclude that the alternative hypothesis is true.

Of course, in practice we must have a precise criterion for deciding whether or not to reject the null hypothesis. Example 9.4 illustrates how such a criterion can be devised for a large-sample, two-tailed hypothesis test concerning a population mean, μ. The example also introduces the logic and some of the terminology of hypothesis testing. Several general procedures for performing hypothesis tests will be provided later in this chapter.

Example 9.4   Hypothesis Testing

Consider again the situation of Example 9.1. A company that produces snack foods uses a machine to package 454-gram bags of pretzels. To check whether the machine is working properly, the quality-assurance department takes a random sample of 50 bags of pretzels. The net weights, in grams, of the 50 bags of pretzels obtained are displayed in Table 9.1.

TABLE 9.1
Weights, in grams, of 50 randomly
selected bags of pretzels

464	450	450	456	452	433	446	446	450	447
442	438	452	447	460	450	453	456	446	433
448	450	439	452	459	454	456	454	452	449
463	449	447	466	446	447	450	449	457	464
468	447	433	464	469	457	454	451	453	443

Do the data provide sufficient evidence to conclude that the packaging machine is not working properly? Use the following steps to answer the question.

a. State the null and alternative hypotheses for the hypothesis test.

b. Discuss the logic behind carrying out the hypothesis test.

c. Obtain a precise criterion for deciding whether or not to reject the null hypothesis in favor of the alternative hypothesis.

d. Apply the criterion in part (c) to the sample data and state the conclusion.

SOLUTION   Let μ denote the mean net weight of all bags being packaged.

a. The null and alternative hypotheses for the hypothesis test were found in Example 9.1. They are

$$H_0: \mu = 454 \text{ grams (the machine is working properly)}$$
$$H_a: \mu \neq 454 \text{ grams (the machine is not working properly).}$$

b. Basically, the logic behind carrying out the hypothesis test is this: If the null hypothesis is true, that is, if $\mu = 454$ grams, then the mean weight, $\bar{x}$, of the sample of 50 bags of pretzels should be approximately equal to 454 grams. Of course, we cannot expect a sample mean to be exactly equal to a population mean—some sampling error is to be anticipated. However, if the sample mean weight, $\bar{x}$, differs too much from 454 grams, then we would be inclined to reject the null hypothesis and conclude that the alternative hypothesis is true.

   From Table 9.1 we can calculate the mean weight of the sample of 50 bags of pretzels:

$$\bar{x} = \frac{\Sigma x}{n} = \frac{22{,}561}{50} = 451.2 \text{ grams.}$$

   The question now is whether the difference of 2.8 grams between the sample mean of 451.2 grams and the hypothesized population mean of 454 grams can reasonably be attributed to sampling error or whether the difference is large enough to indicate that the population mean is not 454 grams. To answer this question, we need to know how likely it would be to get such a difference if in fact the null hypothesis, $\mu = 454$ grams, is true. That likelihood can be determined by using our knowledge of the sampling distribution of the mean, as explained in part (c).

c. For this part we are to obtain a precise criterion for deciding whether or not to reject the null hypothesis in favor of the alternative hypothesis. We will obtain that criterion by applying Key Fact 7.6 on page 424, the sampling distribution of the mean for general populations. Since the sample size here is large ($n = 50$), Key Fact 7.6 implies that the random variable $\bar{x}$ is approximately normally distributed and has mean $\mu_{\bar{x}} = \mu$ and standard deviation $\sigma_{\bar{x}} = \sigma/\sqrt{n}$.

Knowing that $\bar{x}$ is approximately normally distributed, we can, for instance, use Property 2 of Key Fact 6.6 on page 372, the empirical rule for normally distributed random variables, to deduce the following: The probability is (approximately) 0.9544 that $\bar{x}$ will be within two standard deviations of its mean, $\mu_{\bar{x}}$, which is the same as the population mean, μ. This probability is depicted in Fig. 9.1.

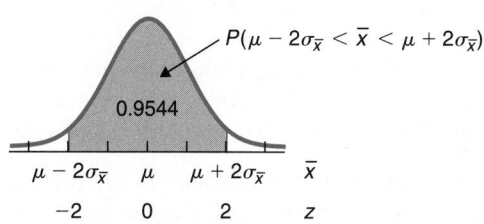

Thus the probability is only 0.0456 ($= 1 - 0.9544$) that $\bar{x}$ will be more than two standard deviations away from μ—it is quite unlikely that the sample mean will be more than two standard deviations away from the population mean. We can employ this fact as a basis for deciding whether or not to reject the null hypothesis, $\mu = 454$ grams.

Specifically, if the mean weight, $\bar{x}$, of the 50 bags of pretzels sampled is within two standard deviations of 454 grams, then we can reasonably attribute the difference between $\bar{x}$ and 454 grams to sampling error and thereby accept the null hypothesis. On the other hand, if $\bar{x}$ is more than two standard deviations away from 454 grams, then either (1) the null hypothesis, $\mu = 454$ grams, is true and an extremely unlikely event occurred, or (2) the null hypothesis is false and the alternative hypothesis, $\mu \neq 454$ grams, is true. Surely we would select (2) as the more reasonable conclusion.

In summary, then, we have obtained the following precise criterion for deciding whether or not to reject the null hypothesis:

> If the mean weight, $\bar{x}$, of the 50 bags of pretzels sampled is more than two standard deviations away from 454 grams, then reject the null hypothesis, $\mu = 454$ grams, and conclude that the alternative hypothesis, $\mu \neq 454$ grams, is true. Otherwise, do not reject the null hypothesis.

We can show this criterion graphically as in Fig. 9.2(a) at the top of the next page.

In order to see the implications of our decision criterion, if the null hypothesis, $\mu = 454$ grams, is in fact true, we superimpose on Fig. 9.2(a) the normal curve for $\bar{x}$ under that condition—the normal curve with parameters $\mu_{\bar{x}} = \mu = 454$ and $\sigma_{\bar{x}} = \sigma/\sqrt{n}$. This is done in Fig. 9.2(b).

FIGURE 9.2

(a) Criterion for deciding whether or not to reject the null hypothesis (b) Implications of the decision criterion if the null hypothesis is true

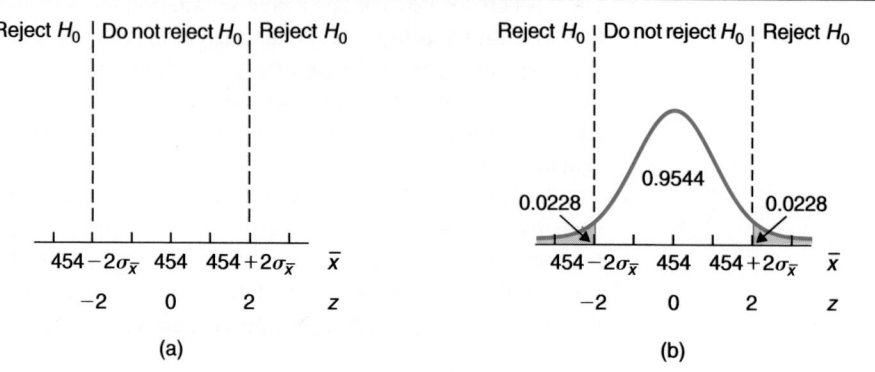

Figure 9.2(b) shows that, using our decision criterion, the probability is only 0.0456 ($= 1 - 0.9544 = 0.0228 + 0.0228$) of rejecting the null hypothesis if it is true. That probability is called the *significance level* of the hypothesis test.

d. Finally, we are to apply the criterion obtained in part (c) to the sample data and state the conclusion. In doing so, we must determine how many standard deviations the sample mean, $\bar{x}$, is away from 454 grams. This entails computing the z-score for $\bar{x}$, that is, finding the value of the standardized random variable,

$$z = \frac{\bar{x} - 454}{\sigma_{\bar{x}}} = \frac{\bar{x} - 454}{\sigma/\sqrt{n}}.$$

We will assume as known that $\sigma = 7.9$ grams. Now, $n = 50$, and from part (b), $\bar{x} = 451.2$ grams. Thus

$$z = \frac{\bar{x} - 454}{\sigma/\sqrt{n}} = \frac{451.2 - 454}{7.9/\sqrt{50}} = -2.51.$$

Consequently, the sample mean, $\bar{x}$, is 2.51 standard deviations below the null-hypothesis mean of 454 grams, as illustrated in Fig. 9.3.

FIGURE 9.3

Graph showing the number of standard deviations that $\bar{x}$ is away from the null hypothesis mean of 454 grams

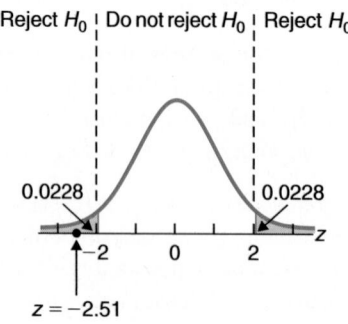

Since the mean weight, $\bar{x}$, of the 50 bags of pretzels sampled is more than two standard deviations away from 454 grams, we reject the null hypothesis, $\mu = 454$ grams, and conclude that the alternative hypothesis, $\mu \neq 454$ grams, is true. In other words, the data provide sufficient evidence to conclude that the packaging machine is not working properly. ∎

Example 9.4 contains all the elements of a hypothesis test. But don't worry too much about the details at this point. What you should understand now is how to choose the null and alternative hypotheses for a hypothesis test and the logic behind performing a hypothesis test.

EXERCISES 9.1

· **9.1** Explain the meaning of the term *hypothesis* as used in inferential statistics.

· **9.2** What role does the decision criterion play in a hypothesis test?

In each of Exercises 9.3–9.10, a hypothesis test will be proposed. For each hypothesis test,
a. *determine the null hypothesis.*
b. *determine the alternative hypothesis.*
c. *classify the hypothesis test as a two-tailed test, a left-tailed test, or a right-tailed test.*

· **9.3** According to the Census Bureau publication *Construction Reports,* the mean expenditure per residential property owner for maintenance and repairs in 1983 was $280. Suppose we want to perform a hypothesis test to decide whether last year's mean expenditure has increased over the 1983 mean of $280.

· **9.4** *The World Almanac, 1985,* reports that the mean travel time to work in 1980 for all South Dakota residents was 13 minutes. A transportation official wants to use a sample of this year's travel times for South Dakota residents to decide whether the mean travel time to work for all South Dakota residents has changed from the 1980 mean of 13 minutes.

· **9.5** The Food and Nutrition Board of the National Academy of Sciences states that the recommended daily allowance (RDA) of iron for adult females under the age of 51 is 18 mg. A hypothesis test is to be performed to decide whether adult females under the age of 51 are, on the average, getting less than the RDA of 18 mg of iron.

· **9.6** As reported by the U.S. Office of Juvenile Justice and Delinquency Prevention in *Children in Custody,* the mean age of all juveniles held in public custody in 1989 was 16.0 years. The ages of a random sample of juveniles currently being held in public custody are to be used to decide whether this year's mean age of all juveniles being held in public custody is less than the 1989 mean of 16.0 years.

· **9.7** A dog-food manufacturer sells "50-lb" bags of dog food. Seventy-five bags are randomly selected and carefully weighed. Using the weights obtained, a hypothesis test is to be performed in order to decide whether the mean weight of all bags of this dog food differs from the advertised weight of 50 lb.

· **9.8** The Health Insurance Association of America reports in *Survey of Hospital Semi-Private Room Charges* that the mean daily charge for a semi-private room in U.S. hospitals in 1988 was $253. In that same year, the semi-private room rates were obtained for a random sample of 30 Massachusetts hospitals. A hypothesis test was then performed to decide whether the mean semi-private room rate in Massachusetts hospitals exceeded the national mean.

· **9.9** The manufacturer of a new-model car, the Orion, claims that a typical car gets 26 miles per gallon (mpg). An independent consumer group is skeptical of this claim and thinks that the mean gas mileage of all Orions may very well be less than 26 mpg. To try to justify its contention, the consumer group plans to conduct mileage tests on 30 randomly selected Orions and use the results to perform a hypothesis test.

· **9.10**   A Louisiana cotton farmer has used a certain brand of fertilizer for the past 5 years. Based on experience, the farmer knows that the mean yield of cotton using this fertilizer is 623 lb/acre. Recently, a new fertilizer appeared on the market that will supposedly increase cotton yield. The farmer intends to try the new fertilizer on eighty 1-acre plots in order to decide for himself whether it increases the mean yield of cotton on his land.

For Exercises 9.11 and 9.12, use the same method as that in Example 9.4 on pages 485–489 to perform the required hypothesis tests.

· ·   **9.11**   The Radio Advertising Bureau of New York reports in *Radio Facts* that in 1990 the mean number of radios per U.S. household was 5.6. A random sample of 45 U.S. households taken this year yields the following data on number of radios owned.

4	10	4	7	4	4	5	10	6
8	6	9	7	5	4	5	6	9
7	5	3	4	9	5	4	4	7
8	4	9	8	5	9	1	3	2
8	6	4	4	4	10	7	9	3

Do the data provide sufficient evidence to conclude that this year's mean number of radios per U.S. household has changed from the 1990 mean of 5.6? Use the following steps to answer the question.
a. State the null and alternative hypotheses.
b. Discuss the logic behind carrying out the hypothesis test.
c. Obtain a precise criterion for deciding whether or not to reject the null hypothesis in favor of the alternative hypothesis.
d. Apply the criterion in part (c) to the sample data and state your conclusion. Assume the population standard deviation of this year's number of radios per U.S. household is 1.9.

· ·   **9.12**   The U.S. Energy Information Administration compiles data on energy consumption and publishes its findings in *Residential Energy Consumption Survey: Consumption and Expenditures*. In 1982 the mean energy consumed per U.S. household was 114 million BTU. For that same year, 50 randomly selected households in the South had the following energy consumptions, in millions of BTU.

130	55	45	64	155	66	60	80	102	62
58	101	75	111	151	139	81	55	66	90
97	77	51	67	125	50	136	55	83	91
54	86	100	78	93	113	111	104	96	113
96	87	129	109	69	94	99	97	83	97

Do the data provide sufficient evidence to conclude that in 1982 the mean energy consumed by southern households differed from that of all U.S. households? Use the following steps to answer the question.
a. State the null and alternative hypotheses.
b. Discuss the logic behind carrying out the hypothesis test.
c. Obtain a precise criterion for deciding whether or not to reject the null hypothesis in favor of the alternative hypothesis.
d. Apply the criterion in part (c) to the sample data and state your conclusion. Assume that in 1982 the standard deviation of energy consumptions of all southern households was 25 million BTU.

· ·   **9.13**   Refer to Example 9.4 on pages 485–489. Suppose in the second paragraph of the solution to part (c) we use Property 1 of Key Fact 6.6 (page 372) instead of Property 2.
a. Determine the resulting decision criterion and portray it graphically using a graph similar to the one in Fig. 9.2(a) on page 488.
b. Construct a graph similar to the one in Fig. 9.2(b) that shows the implications of the decision criterion in part (a) if in fact the null hypothesis is true.
c. Obtain the significance level of the hypothesis test.
d. Apply the criterion in part (a) to the sample data in Table 9.1 on page 486 and state your conclusion. *(Note: Recall that the sample mean of the data is 451.2 grams and that $\sigma = 7.9$ grams.)*

· ·   **9.14**   Refer to Example 9.4 on pages 485–489. Suppose in the second paragraph of the solution to part (c) we use Property 3 of Key Fact 6.6 (page 372) instead of Property 2.
a. Determine the resulting decision criterion and portray it graphically using a graph similar to the one in Fig. 9.2(a) on page 488.
b. Construct a graph similar to the one in Fig. 9.2(b) that shows the implications of the decision criterion in part (a) if in fact the null hypothesis is true.
c. Obtain the significance level of the hypothesis test.
d. Apply the criterion in part (a) to the sample data in Table 9.1 on page 486 and state your conclusion. *(Note: Recall that the sample mean of the data is 451.2 grams and that $\sigma = 7.9$ grams.)*

9.2   TERMS, ERRORS, AND HYPOTHESES

To fully understand the nature of hypothesis testing, we need to learn some additional terms and concepts. In this section we will define several more terms that are used in hypothesis testing, discuss the two types of errors that can be made in a hypothesis test, and interpret the possible conclusions for a hypothesis test.

Some Additional Terminology

To introduce some additional terminology used in hypothesis testing, we will refer to the pretzel-packaging hypothesis test of Example 9.4 on page 485. Recall that the null and alternative hypotheses for that hypothesis test are

$$H_0: \mu = 454 \text{ grams (the machine is working properly)}$$

$$H_a: \mu \neq 454 \text{ grams (the machine is not working properly)},$$

where μ is the mean net weight of all bags of pretzels being packaged.

As a basis for deciding whether to reject the null hypothesis, we employed, in part (d) of Example 9.4, the random variable

$$z = \frac{\bar{x} - 454}{\sigma/\sqrt{n}}.$$

That random variable is called the **test statistic** for the hypothesis test.

Figure 9.3 includes a graph portraying the criterion used to decide whether or not the null hypothesis should be rejected. For convenience we repeat that graph in Fig. 9.4.

FIGURE 9.4
Criterion used to decide whether
or not to reject the null hypothesis

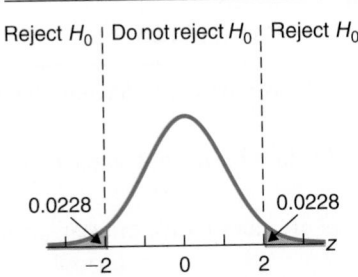

The set of values for the test statistic that leads us to reject the null hypothesis is called the **rejection region.** In this case the rejection region consists of all z-values that lie either to the left of -2 or to the right of 2, that part of the horizontal axis under the shaded areas in Fig. 9.4.

The set of values for the test statistic that leads us not to reject the null hypothesis is called the **nonrejection region,** or **acceptance region.** In this case the nonrejection region consists of all z-values that lie between -2 and 2, that part of the horizontal axis under the unshaded area in Fig. 9.4.

The values of the test statistic that separate the rejection and nonrejection regions are called the **critical values.** In this case the critical values are $z = \pm 2$, as we see from Fig. 9.4. We summarize the preceding discussion in Fig. 9.5.

FIGURE 9.5

Rejection region, nonrejection region, and critical values for pretzel-packaging illustration

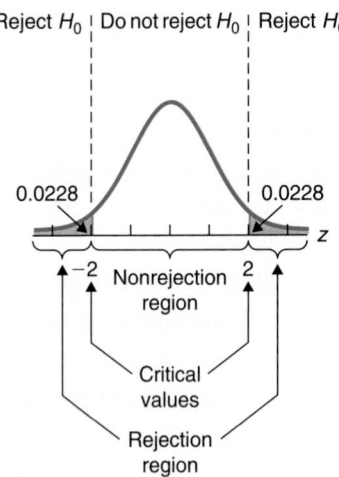

The terminology introduced so far in this section is presented formally in Definition 9.2. This terminology applies to any hypothesis test, not just to hypothesis tests for a population mean.

DEFINITION 9.2

TEST STATISTIC, REJECTION REGION,
NONREJECTION REGION, CRITICAL VALUES

> *Test statistic:* The statistic used as a basis for deciding whether the null hypothesis should be rejected.
>
> *Rejection region:* The set of values for the test statistic that leads to rejection of the null hypothesis.
>
> *Nonrejection region:* The set of values for the test statistic that leads to nonrejection of the null hypothesis.
>
> *Critical values:* The values of the test statistic that separate the rejection and nonrejection regions.

For a two-tailed test, as in the pretzel-packaging illustration, the null hypothesis will be rejected if the test statistic is either too small or too large. Consequently, the rejection region for such a test will consist of two parts, one on the left and one on the right, as illustrated in Fig. 9.5.

For a left-tailed test, as in Example 9.3 on page 484, the null hypothesis will be rejected only if the test statistic is too small. Thus the rejection region for such a test will consist of only one part, and that part will be on the left.

For a right-tailed test, as in Example 9.2 on page 484, the null hypothesis will be rejected only if the test statistic is too large. Hence the rejection region for such a test will consist of only one part, and that part will be on the right.

Table 9.2 summarizes the discussion in the preceding three paragraphs. A graphical summary is provided in Fig. 9.6. By examining Fig. 9.6, we can understand why the term *tailed* is used. Indeed, the rejection region is in both tails for a two-tailed test, in the left tail for a left-tailed test, and in the right tail for a right-tailed test.

TABLE 9.2

Rejection regions for two-tailed, left-tailed, and right-tailed tests

	Two-tailed test	Left-tailed test	Right-tailed test
Sign in H_a	$\neq$	$<$	$>$
Rejection region	Both sides	Left side	Right side

FIGURE 9.6

Graphical display of rejection regions for two-tailed, left-tailed, and right-tailed tests

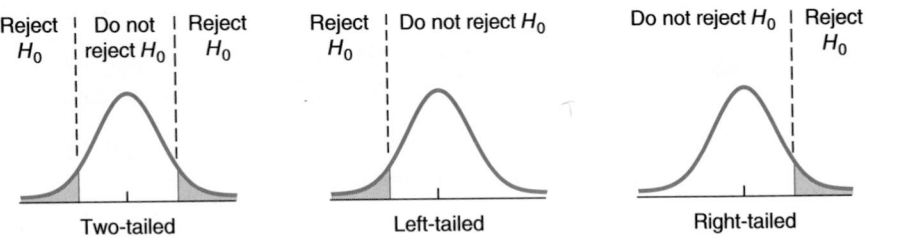

Type I and Type II Errors

Whenever statistical-inference methods are employed, it is always possible that the decision reached will be incorrect. This is because partial information, obtained from the sample, is used to draw conclusions about the entire population.

In hypothesis testing four outcomes are possible, two of which lead to incorrect decisions. The four possible outcomes are depicted in Table 9.3.

TABLE 9.3

The four possible outcomes for a hypothesis test

		H_0 is:	
		True	False
Decision:	Do not reject H_0	Correct decision	Type II error
	Reject H_0	Type I error	Correct decision

As we see from Table 9.3, an incorrect decision occurs if either a true null hypothesis is rejected or a false null hypothesis is not rejected. The first incorrect decision is called a **Type I error** and the second a **Type II error.**

| DEFINITION 9.3 | **TYPE I AND TYPE II ERRORS** |

> *Type I error:* Rejecting the null hypothesis when it is in fact true.
>
> *Type II error:* Not rejecting the null hypothesis when it is in fact false.

In a hypothesis test, it is presumed that the null and alternative hypotheses are exhaustive. In other words, if the null hypothesis is false, then the alternative hypothesis is true, and vice versa.

Example 9.5   Illustrates Definition 9.3

Consider once again the pretzel-packaging hypothesis test. The null and alternative hypotheses are

$$H_0: \mu = 454 \text{ grams (the machine is working properly)}$$
$$H_a: \mu \neq 454 \text{ grams (the machine is not working properly)},$$

where μ is the mean net weight of all bags of pretzels being packaged. Explain what each of the following would mean.

a. A Type I error

b. A Type II error

c. A correct decision

Recall from Example 9.4 that the results of sampling 50 bags of pretzels led to rejection of the null hypothesis, $\mu = 454$ grams, that is, to the conclusion that $\mu \neq 454$ grams. Classify that conclusion by error type or as a correct decision if in fact the mean net weight, μ, being packaged

d. is 454 grams.

e. is not 454 grams.

SOLUTION   a. A Type I error occurs when a true null hypothesis is rejected. In this case a Type I error would occur if in fact $\mu = 454$ grams but the results of the sampling lead to the conclusion that $\mu \neq 454$ grams.

b. A Type II error occurs when a false null hypothesis is not rejected. In this case a Type II error would occur if in fact $\mu \neq 454$ grams but the results of the sampling fail to lead to that conclusion.

c. A correct decision can occur in either of two ways. A correct decision occurs when a true null hypothesis is not rejected. In the present situation, this would happen if in fact $\mu = 454$ grams and the results of the sampling do not lead to the rejection of that fact. A correct decision also occurs when a false null hypothesis is rejected. In the present situation, this would happen if in fact $\mu \neq 454$ grams and the results of the sampling lead to that conclusion.

d. If in fact $\mu = 454$ grams, then the null hypothesis is true. So by rejecting the null hypothesis, $\mu = 454$ grams, a Type I error has been made—a true null hypothesis has been rejected.

e. If in fact $\mu \neq 454$ grams, then the null hypothesis is false. Consequently, by rejecting the null hypothesis, $\mu = 454$ grams, a correct decision has been made—a false null hypothesis has been rejected. ∎

Probabilities of Type I and Type II Errors

The probability of making a Type I error is the probability of rejecting a true null hypothesis. In other words, it is the probability that the test statistic will be in the rejection region if in fact the null hypothesis is true.

To illustrate, let's find the probability of making a Type I error in the pretzel-packaging hypothesis test of Example 9.4. Figure 9.2(b) on page 488 provides a graphical display of the criterion used to decide whether or not to reject the null hypothesis, $\mu = 454$ grams. It also shows the implications of that criterion when in fact the null hypothesis is true.

As we see from Fig. 9.2(b), the probability is $0.0456 (= 0.0228 + 0.0228)$ that the test statistic, z, will be in the rejection region if the null hypothesis is in fact true. Thus the probability of a Type I error for this hypothesis test is 0.0456. There is only a 4.56% chance of concluding that $\mu \neq 454$ grams when in fact $\mu = 454$ grams.

The probability of making a Type I error is called the **significance level** of the hypothesis test and is denoted by the Greek letter α (alpha). Hence the significance level of the pretzel-packaging hypothesis test is $\alpha = 0.0456$. We summarize the discussion of the probability of a Type I error in Definition 9.4.

DEFINITION 9.4   **SIGNIFICANCE LEVEL**

> The *significance level, α,* of a hypothesis test is defined as the probability of making a Type I error, that is, the probability of rejecting a true null hypothesis.

The probability of making a Type II error is the probability of not rejecting a false null hypothesis. In other words, it is the probability that the test statistic will be in the nonrejection region if in fact the null hypothesis is false. We use the Greek letter β (beta) to denote the probability of a Type II error. The probability, β, of a Type II error depends on the true value of μ. Exercise 9.34 deals briefly with the calculation of Type II error probabilities; a detailed discussion will be presented in Section 9.4.

Ideally, we would like both Type I and Type II errors to have small probabilities. Then the chances of making an incorrect decision would be small, regardless of whether the null hypothesis is true or the alternative hypothesis is true. As we will see, it is always possible to design a hypothesis test with any desired significance level (i.e., Type I error probability). Thus if it is important not to reject a true null hypothesis, then we should specify a small value for the significance level, α. However, in making our choice for α, we should keep the following fact in mind.

KEY FACT 9.1

RELATION BETWEEN TYPE I AND TYPE II ERROR PROBABILITIES

> For a fixed sample size, the smaller the Type I error probability, α, of rejecting a true null hypothesis, the larger the Type II error probability, β, of not rejecting a false null hypothesis; and vice versa.

Consequently, we must always assess the risks involved in committing both types of errors and use that assessment as a method for balancing the Type I and Type II error probabilities.

Possible Conclusions for a Hypothesis Test

As we know, the significance level, α, is the probability of making a Type I error, that is, of rejecting a true null hypothesis. Thus if the hypothesis test is performed at a small significance level (e.g., $\alpha = 0.05$), then it is unlikely that the null hypothesis will be rejected when it is in fact true. Since in this text we will generally specify a small significance level, we can make the following statement concerning a hypothesis test: If we do reject the null hypothesis in a hypothesis test, then we can be reasonably confident that the alternative hypothesis is true.

On the other hand, we will usually not know the probability, β, of making a Type II error, that is, of not rejecting a false null hypothesis. Therefore if we do not reject the null hypothesis in a hypothesis test, then we simply reserve judgment about which hypothesis is true. In other words, if we do not reject the null hypothesis, we conclude only that the data did not provide sufficient evidence to support the alternative hypothesis; we do not conclude that the data provided sufficient evidence to support the null hypothesis. Key Fact 9.2 summarizes this discussion.

KEY FACT 9.2

POSSIBLE CONCLUSIONS FOR A HYPOTHESIS TEST

> - If the null hypothesis is rejected, we conclude that the alternative hypothesis is probably true.
> - If the null hypothesis is not rejected, we conclude that the data do not provide sufficient evidence to support the alternative hypothesis.

When the null hypothesis is rejected in a hypothesis test performed at the significance level α, we frequently express that fact with the phrase "the test results are **statistically significant** at the α level." Similarly, when the null hypothesis is not rejected in a hypothesis test performed at the significance level α, we often express that fact with the phrase "the test results are not statistically significant at the α level."

EXERCISES 9.2

Exercises 9.15–9.20 contain graphs portraying the decision criterion for a hypothesis test for a population mean, μ. The null hypothesis for each test is H_0: $\mu = \mu_0$; the test statistic is

$$z = \frac{\overline{x} - \mu_0}{\sigma/\sqrt{n}}.$$

Also, the curve in each graph shows the implications of the decision criterion if in fact the null hypothesis is true. For each exercise, determine the

a. *rejection region.*   b. *nonrejection region.*
c. *critical value(s).*   d. *significance level.*
e. *Construct a graph similar to Fig. 9.5 on page 492 that depicts your results from parts (a)–(d).*
f. *Identify the hypothesis test as two-tailed, left-tailed, or right-tailed.*

9.15   A graphical display of the decision criterion is

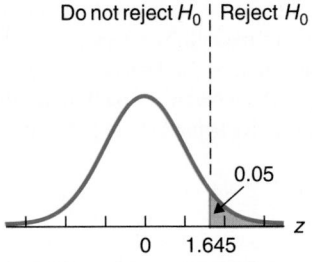

9.16   A graphical display of the decision criterion is

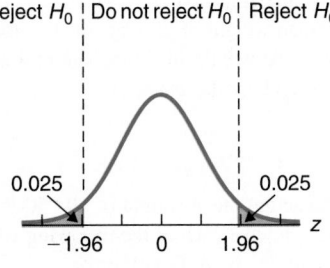

9.17   A graphical display of the decision criterion is

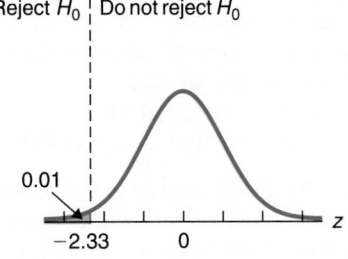

9.18   A graphical display of the decision criterion is

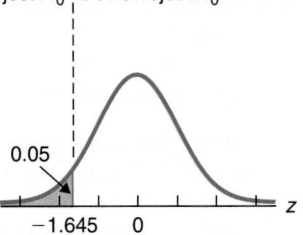

9.19   A graphical display of the decision criterion is

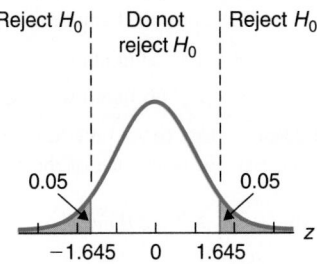

9.20   A graphical display of the decision criterion is

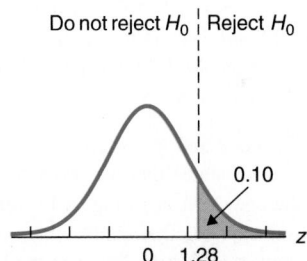

9.21   According to the Census Bureau publication *Construction Reports,* the mean expenditure per residential property owner for maintenance and repairs in 1983 was $280. Suppose we want to perform a hypothesis test to decide whether last year's mean amount spent for maintenance and repairs per residential property owner has increased over the 1983 mean. Then the null and alternative hypotheses are

$$H_0: \mu = \$280$$
$$H_a: \mu > \$280,$$

where μ is last year's mean amount spent for maintenance and repairs per residential property owner. Explain what each of the following would mean.
a. A Type I error   b. A Type II error

c. A correct decision

Now suppose the results of carrying out the hypothesis test lead to nonrejection of the null hypothesis. Classify that conclusion by error type or as a correct decision if in fact last year's mean amount spent for maintenance and repairs per residential property owner

d. is equal to the 1983 mean of $280.

e. is greater than the 1983 mean of $280.

9.22 *The World Almanac, 1985,* reports that the mean travel time to work in 1980 for all South Dakota residents was 13 minutes. A transportation official wants to use a sample of this year's travel times for South Dakota residents to decide whether the mean travel time to work for all South Dakota residents has changed from the 1980 mean. The null and alternative hypotheses for the hypothesis test are

$$H_0: \mu = 13 \text{ minutes}$$
$$H_a: \mu \neq 13 \text{ minutes},$$

where μ is this year's mean travel time to work for all South Dakota residents. Explain what each of the following would mean.

a. A Type I error   b. A Type II error

c. A correct decision

Now suppose the results of the sampling lead to nonrejection of the null hypothesis. Classify that conclusion by error type or as a correct decision if in fact this year's mean travel time to work, μ, for all South Dakota residents

d. has not changed from the 1980 mean of 13 minutes.

e. has changed from the 1980 mean of 13 minutes.

9.23 The Food and Nutrition Board of the National Academy of Sciences states that the RDA of iron for adult females under the age of 51 is 18 mg. A hypothesis test is to be performed to decide whether adult females under the age of 51 are, on the average, getting less than the RDA of 18 mg of iron. The null and alternative hypotheses for the hypothesis test are

$$H_0: \mu = 18 \text{ mg}$$
$$H_a: \mu < 18 \text{ mg},$$

where μ is the mean iron intake (per day) of all adult females under the age of 51. Explain what each of the following would mean.

a. A Type I error   b. A Type II error

c. A correct decision

Now suppose the results of carrying out the hypothesis test lead to rejection of the null hypothesis, $\mu = 18$ mg, that is, to the conclusion that $\mu < 18$ mg. Classify that conclusion by error type or as a correct decision if in fact the mean iron intake, μ, of all adult females under the age of 51

d. is not less than the RDA of 18 mg per day.

e. is less than the RDA of 18 mg per day.

9.24 As reported by the U.S. Office of Juvenile Justice and Delinquency Prevention in *Children in Custody,* the mean age of all juveniles held in public custody in 1989 was 16.0 years. The ages of a random sample of juveniles currently being held in public custody are to be used to decide whether this year's mean age of all juveniles being held in public custody is less than the 1989 mean. The null and alternative hypotheses for the hypothesis test are

$$H_0: \mu = 16.0 \text{ years}$$
$$H_a: \mu < 16.0 \text{ years},$$

where μ is this year's mean age of all juveniles being held in public custody. Explain what each of the following would mean.

a. A Type I error   b. A Type II error

c. A correct decision

Now suppose the results of carrying out the hypothesis test lead to rejection of the null hypothesis, $\mu = 16.0$ years, that is, to the conclusion $\mu < 16.0$ years. Classify that conclusion by error type or as a correct decision if in fact this year's mean age, μ, of all juveniles being held in public custody

d. is 16.0 years.

e. is less than 16.0 years.

9.25 A dog-food manufacturer sells "50-lb" bags of dog food. Seventy-five bags of this brand of dog food are randomly selected and carefully weighed. Using the weights obtained, a hypothesis test is to be performed in order to decide whether the mean weight of all bags of this dog food differs from the advertised weight of 50 lb. The null and alternative hypotheses for the hypothesis test are

$$H_0: \mu = 50 \text{ lb}$$
$$H_a: \mu \neq 50 \text{ lb},$$

where μ is the actual mean weight of all "50-lb" bags of this dog food. Explain what each of the following would mean.

a. A Type I error   b. A Type II error

c. A correct decision

Now suppose the weights of the 75 bags sampled lead to nonrejection of the null hypothesis. Classify that conclusion by error type or as a correct decision if in fact the mean weight, μ, of all "50-lb" bags of this dog food

d. equals the advertised weight of 50 lb.

e. does not equal the advertised weight of 50 lb.

9.26 The Health Insurance Association of America reports in *Survey of Hospital Semi-Private Room Charges* that the mean daily charge for a semi-private room in U.S. hospitals in 1988 was $253. That same year, the daily semi-private room charges were obtained for a sample of 30 Massachusetts hospitals. A hypothesis test was then performed to decide whether the mean daily semi-private room charge in Massachusetts hospitals exceeded the national mean. The null and alternative hypotheses for the hypothesis test performed are

$$H_0: \mu = \$253$$
$$H_a: \mu > \$253,$$

where μ is the 1988 mean daily semi-private room charge in Massachusetts hospitals. Explain what each of the following would mean.
a. A Type I error   b. A Type II error
c. A correct decision
Now suppose the results of the sampling led to rejection of the null hypothesis. Classify that conclusion by error type or as a correct decision if in fact the 1988 mean daily semi-private room charge in Massachusetts hospitals
d. did not exceed the national mean.
e. did exceed the national mean.

9.27 The manufacturer of a new-model car, called the Orion, claims that a typical car gets 26 mpg. An independent consumer group is skeptical of this claim and thinks that the mean gas mileage of all Orions may very well be less than 26 mpg. To try to justify its contention, the consumer group plans to conduct mileage tests on 30 randomly selected Orions and use the results to perform a hypothesis test. The null and alternative hypotheses are

$$H_0: \mu = 26 \text{ mpg}$$
$$H_a: \mu < 26 \text{ mpg},$$

where μ is the mean gas mileage of all Orions. Explain what each of the following would mean.
a. A Type I error   b. A Type II error
c. A correct decision
Now suppose the mileages of the 30 Orions tested lead to rejection of the null hypothesis. Classify that conclusion by error type or as a correct decision if in fact
d. the consumer group's conjecture is correct.
e. the manufacturer's claim is true.

9.28 A Louisiana cotton farmer has used a certain brand of fertilizer for the past 5 years. Based on experience, the farmer knows that the mean yield of cotton using this fertilizer is 623 lb/acre. Recently, a new brand of fertilizer appeared on the market that will supposedly increase cotton yield. The farmer intends to try the new fertilizer on eighty 1-acre plots in order to decide for himself whether it increases the mean yield

of cotton on his land. The null and alternative hypotheses are

$$H_0: \mu = 623 \text{ lb/acre}$$
$$H_a: \mu > 623 \text{ lb/acre},$$

where μ is the mean yield of cotton that the farmer will get on his land using the new fertilizer. Explain what each of the following would mean.
a. A Type I error   b. A Type II error
c. A correct decision
Now suppose the yields on the eighty 1-acre test plots lead to nonrejection of the null hypothesis. Classify that conclusion by error type or as a correct decision if in fact the new fertilizer
d. will increase the mean yield.
e. will not increase the mean yield.

9.29 Decide whether each of the following statements is true or false. Explain your answers.
a. If it is important not to reject a true null hypothesis, then the hypothesis test should be performed at a small significance level.
b. For a fixed sample size, decreasing the significance level of a hypothesis test results in an increase in the probability of making a Type II error.

9.30 Suppose we choose the significance level of a hypothesis test to be $\alpha = 0$.
a. What is the probability of a Type I error, that is, of rejecting a true null hypothesis?
b. What is the probability of a Type II error, that is, of not rejecting a false null hypothesis?

9.31 Identify an exercise in this section for which it is important to have
a. a small α probability.
b. a small β probability.
c. both α and β probabilities small.

9.32 Suppose you are performing a statistical test to decide whether a nuclear reactor should be approved for use. Further suppose that failing to reject the null hypothesis corresponds to approval. What property would you want the Type II error probability, β, to have?

9.33 In the U.S. court system, a defendant is assumed innocent until proven guilty. Suppose we regard a court trial as a hypothesis test with null and alternative hypotheses

$$H_0: \text{Defendant is innocent}$$
$$H_a: \text{Defendant is guilty.}$$

a. Explain the meaning of a Type I error.
b. Explain the meaning of a Type II error.

c. If you were the defendant, what kind of value would you want for α? Why?

d. If you were the prosecuting attorney, what kind of value would you want for β? Why?

e. What are the consequences to our court system if we make $\alpha = 0$? $\beta = 0$?

· ·   **9.34**   A brewer produces a 355 mL can of beer. To ensure that the machine is working properly, periodic checks are made. Specifically, the following hypothesis test is performed:

H_0: $\mu = 355$ mL (working properly)

H_a: $\mu \neq 355$ mL (not working properly),

where μ is the actual mean amount of beer being dispensed into the cans. For the test, 30 cans of beer are randomly selected and their contents carefully measured. Explain what each of the following would mean.

a. A Type I error

b. A Type II error

Suppose the hypothesis test is performed at the 5% significance level, and assume $\sigma = 5.9$ mL.

c. Find the critical values and draw a graph portraying the decision criterion.

d. Determine the probability, α, of a Type I error.

e. Determine the probability, β, of a Type II error if the true mean amount of beer being dispensed into the cans is 350 mL; 351 mL; 352 mL; 353 mL; 354 mL; 356 mL; 357 mL; 358 mL; 359 mL. *(Hint:* For each of the nine cases, you must compute the probability that the test statistic falls in the nonrejection region under the assumption that the true value of μ is as specified.)

f. Use your results from part (e) to draw a graph of the probability of a Type II error versus the true value of μ. What does the graph tell you?

9.3   LARGE-SAMPLE HYPOTHESIS TESTS FOR ONE POPULATION MEAN

· · · · · ·

In this section we will learn a procedure for performing a large-sample hypothesis test for a population mean, μ, at any prescribed significance level. The only preliminary topic we still need to discuss is how to obtain the critical value(s) for such a hypothesis test when the significance level, α, is specified in advance.

Obtaining the Critical Value(s) for a Specified Significance Level

Recall that the significance level, α, of a hypothesis test is the probability of making a Type I error, that is, the probability of rejecting a true null hypothesis. Equivalently, α is the probability that the test statistic will be in the rejection region if in fact the null hypothesis is true. Thus we have the following key fact, which holds for any hypothesis test.

KEY FACT 9.3   **OBTAINING THE CRITICAL VALUE(S) FOR A SPECIFIED SIGNIFICANCE LEVEL**

> Suppose a hypothesis test is to be performed at a specified significance level, α. Then the critical value(s) must be chosen so that if the null hypothesis is true, the probability is equal to α that the test statistic will be in the rejection region.

Let's apply Key Fact 9.3 to large-sample hypothesis tests for a population mean. As we have seen, the null hypothesis for a hypothesis test concerning one population mean, μ, is of the form H_0: $\mu = \mu_0$, where μ_0 is some number. If, as we are considering here, the sample size is large, then we use the random variable

(1) $$z = \frac{\bar{x} - \mu_0}{\sigma/\sqrt{n}}$$

as the test statistic for the hypothesis test. That test statistic tells us how many standard deviations the observed sample mean, $\bar{x}$, is away from μ_0.

From Key Fact 7.6 on page 424, the sampling distribution of the mean for general populations, we know that for large samples the random variable $\bar{x}$ is approximately normally distributed and has mean $\mu_{\bar{x}} = \mu$ and standard deviation $\sigma_{\bar{x}} = \sigma/\sqrt{n}$. Therefore if the null hypothesis, $\mu = \mu_0$, is true, then the test statistic in Equation (1) has approximately the standard normal distribution. In other words, if the null hypothesis is true, then probabilities for that test statistic are equal to areas under the standard normal curve.

Thus in view of Key Fact 9.3, we see that for a specified significance level, α, we need to choose the critical value(s) so that the area under the standard normal curve that lies above the rejection region is equal to α. Here is an example.

Example 9.6 Obtaining the Critical Values for a Specified Significance Level

Suppose we want to perform a large-sample hypothesis test for a population mean, μ, with null hypothesis H_0: $\mu = \mu_0$. If the hypothesis test is to be performed at the 5% significance level ($\alpha = 0.05$), find the critical value(s) for a

a. two-tailed test. b. left-tailed test. c. right-tailed test.

SOLUTION Since $\alpha = 0.05$, we need to choose the critical value(s) so that the area under the standard normal curve that lies above the rejection region is equal to 0.05.

a. For a two-tailed test, the rejection region is on both the left and right. So for a test with $\alpha = 0.05$, the critical values are the two z-values that divide the area under the standard normal curve into a middle 0.95 area and two outside areas of 0.025. In other words, the critical values are $\pm z_{0.025}$. Consulting Table II on the inside front cover or in Appendix A, we find that $\pm z_{0.025} = \pm 1.96$, as shown in Fig. 9.7(a).

FIGURE 9.7
Critical value(s) for a
hypothesis test at the
5% significance level if the test is
(a) two-tailed,
(b) left-tailed,
(c) right-tailed

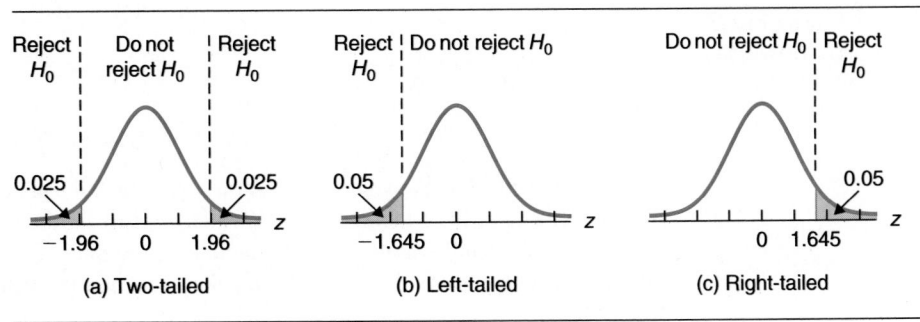

(a) Two-tailed   (b) Left-tailed   (c) Right-tailed

b. For a left-tailed test, the rejection region is on the left. So for a test with $\alpha = 0.05$, the critical value is the z-value having area 0.05 to its left under the standard normal curve. Since the standard normal curve is symmetric about 0, that z-value is the negative of the z-value having area 0.05 to its right. In other words, the critical value is $-z_{0.05}$. Consulting Table II, we find that $-z_{0.05} = -1.645$, as shown in Fig. 9.7(b).

c. For a right-tailed test, the rejection region is on the right. So for a test with $\alpha = 0.05$, the critical value is the z-value having area 0.05 to its right under the standard normal curve. In other words, the critical value is $z_{0.05}$. Consulting Table II, we find that $z_{0.05} = 1.645$, as shown in Fig. 9.7(c).   ■

By arguing as we did in Example 9.6, we can obtain the critical value(s) for any specified significance level, α. For a two-tailed test, the critical values are $\pm z_{\alpha/2}$; for a left-tailed test, the critical value is $-z_{\alpha}$; and for a right-tailed test, the critical value is z_{α}. These values are shown in Fig. 9.8.

FIGURE 9.8

Critical value(s) for a
hypothesis test at the
significance level α if the test is
(a) two-tailed,
(b) left-tailed,
(c) right-tailed

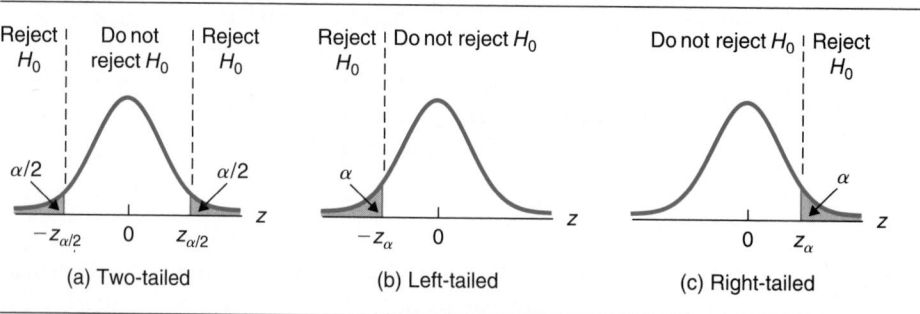

(a) Two-tailed   (b) Left-tailed   (c) Right-tailed

The most commonly used significance levels are 0.10, 0.05, and 0.01. If we consider both one-tailed and two-tailed tests, then these three significance levels give rise to five "tail areas." Using Table II we obtained the value of z_{α} corresponding to each of those five tail areas. Table 9.4 provides a summary.

TABLE 9.4

Some important values of z_{α}

$z_{0.10}$	$z_{0.05}$	$z_{0.025}$	$z_{0.01}$	$z_{0.005}$
1.28	1.645	1.96	2.33	2.575

Alternatively, the five values of z_{α} shown in Table 9.4 can be found in the last row (the row labeled ∞) of the t-table, Table IV, which appears on the page facing the inside back cover and in Appendix A. In Table IV, the five values of z_{α} are displayed to three decimal places. Can you explain the slight discrepancy between the values given for $z_{0.005}$ in the two tables?

Large-Sample Hypothesis Tests for a Population Mean

We now present Procedure 9.1, a simple method for performing a hypothesis test for a population mean when the sample size is large. The procedure is obtained by carefully studying what we have done in this and the previous two sections. We often refer to Procedure 9.1 as the **one-sample z-test** or more briefly as the **z-test.**

PROCEDURE 9.1

THE ONE-SAMPLE z-TEST FOR A POPULATION MEAN WITH NULL HYPOTHESIS $H_0: \mu = \mu_0$

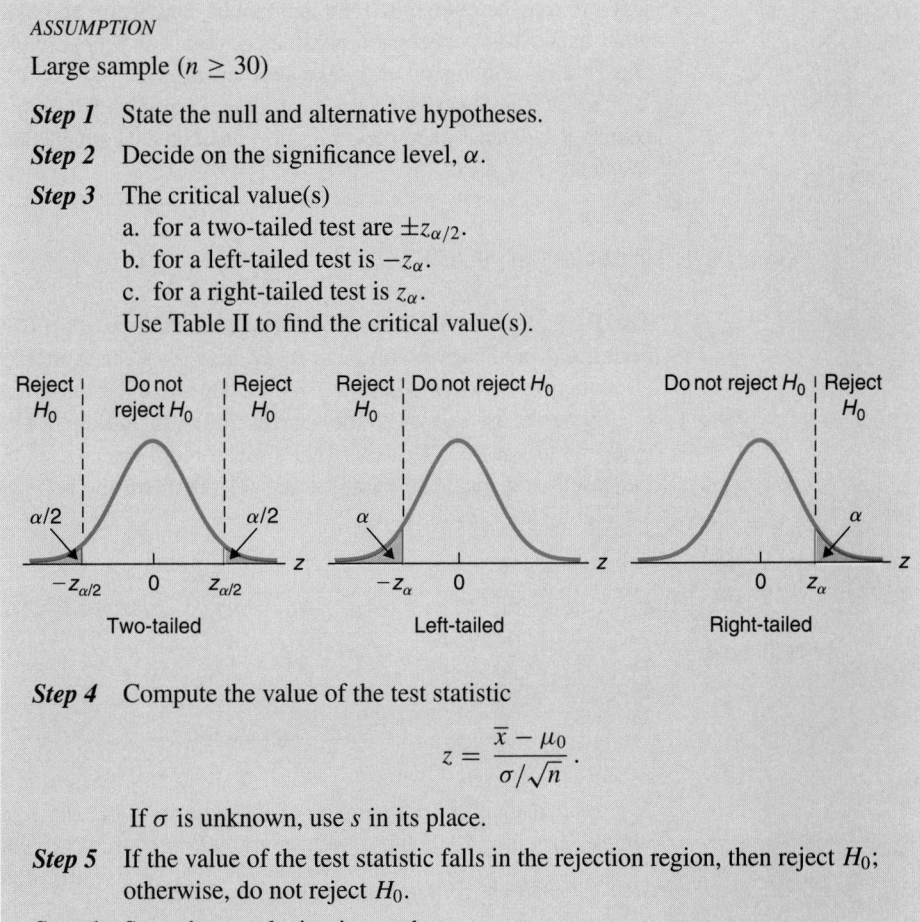

ASSUMPTION

Large sample ($n \geq 30$)

Step 1 State the null and alternative hypotheses.

Step 2 Decide on the significance level, α.

Step 3 The critical value(s)
  a. for a two-tailed test are $\pm z_{\alpha/2}$.
  b. for a left-tailed test is $-z_{\alpha}$.
  c. for a right-tailed test is z_{α}.
  Use Table II to find the critical value(s).

Reject H_0 | Do not reject H_0 | Reject H_0 Reject H_0 | Do not reject H_0 Do not reject H_0 | Reject H_0

$\alpha/2$ ... $\alpha/2$ α α

$-z_{\alpha/2}$ 0 $z_{\alpha/2}$ $-z_{\alpha}$ 0   0 z_{α}

Two-tailed   Left-tailed   Right-tailed

Step 4 Compute the value of the test statistic

$$z = \frac{\bar{x} - \mu_0}{\sigma/\sqrt{n}}.$$

  If σ is unknown, use s in its place.

Step 5 If the value of the test statistic falls in the rejection region, then reject H_0; otherwise, do not reject H_0.

Step 6 State the conclusion in words.

In Step 4 of Procedure 9.1 it is stated that if the population standard deviation, σ, is unknown, then the sample standard deviation, s, should be used in its place. This substitution is acceptable because for large samples the sample standard deviation is likely to be a good approximation of the population standard deviation.

Although strictly speaking the only condition required for using Procedure 9.1 is that the sample size be large, it is important to watch for outliers. Because the sample mean (and sample standard deviation) are not resistant to outliers, the presence of outliers can unduly influence a hypothesis test for a population mean, especially when the sample size is only moderately large.

Therefore, as we have mentioned several times before, it is always a good idea to first look at the data using a histogram, stem-and-leaf diagram, or other graphical display. If outliers are detected, we must try to decide whether to remove or retain them. In cases where it is not possible to decide, we can perform the hypothesis test twice—once with the outliers retained and once with them removed. If the conclusion remains the same either way, we may be content to take that as our conclusion and close the investigation. On the other hand, if the conclusion is affected, then it is probably wise either to make the more conservative conclusion or to take another sample.

We will now present three examples, Examples 9.7–9.9, to illustrate the use of Procedure 9.1. These examples will be considered again in Section 9.5 when we discuss *P*-values.

Example 9.7   Illustrates Procedure 9.1

The R. R. Bowker Company of New York collects information on the retail prices of books and publishes its findings in *Publishers Weekly*. In 1990 the mean retail price of all hardcover history books was $35.44. This year's retail prices for 40 randomly selected history books are shown, to the nearest dollar, in Table 9.5. Do the data provide sufficient evidence to conclude that this year's mean retail price of all hardcover history books has increased over the 1990 mean of $35.44? Perform the appropriate hypothesis test at the 1% significance level.

TABLE 9.5
This year's prices ($)
for 40 history books

42	44	40	33	57	39	37	46
39	40	55	34	27	31	40	38
46	32	35	38	43	39	33	48
40	32	42	30	48	43	52	34
25	35	39	43	29	41	33	28

SOLUTION   A graphical display (not shown) of the data in Table 9.5 reveals no outliers. Since the sample size, $n = 40$, is large, we can apply Procedure 9.1 to carry out the hypothesis test.

Step 1   *State the null and alternative hypotheses.*

Let μ denote this year's mean retail price of all hardcover history books. The null and alternative hypotheses were stated in Example 9.2. They are

$$H_0: \mu = \$35.44 \text{ (mean price has not increased)}$$
$$H_a: \mu > \$35.44 \text{ (mean price has increased)}.$$

Note that the hypothesis test is right-tailed since a greater-than sign ($>$) appears in the alternative hypothesis.

Step 2 *Decide on the significance level, α.*

We are to perform the test at the 1% significance level. Thus $\alpha = 0.01$.

Step 3 *The critical value for a right-tailed test is z_α.*

Since $\alpha = 0.01$, the critical value is $z_{0.01}$. From Table II (or Table 9.4 on page 502), we find that $z_{0.01} = 2.33$, as seen in Fig. 9.9.

FIGURE 9.9

Criterion for deciding whether or not to reject the null hypothesis

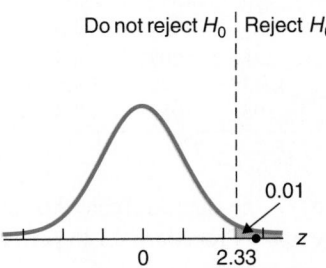

Step 4 *Compute the value of the test statistic*

$$z = \frac{\bar{x} - \mu_0}{\sigma/\sqrt{n}}.$$

Since σ is unknown, we will use s in its place. We have $\mu_0 = \$35.44$ and $n = 40$. Applying the usual formulas for computing $\bar{x}$ and s, we obtain for the data in Table 9.5 that $\bar{x} = \$38.75$ and $s = \$7.35$. Thus the value of the test statistic is

$$z = \frac{38.75 - 35.44}{7.35/\sqrt{40}} = 2.85.$$

This value of z is marked with a dot in Fig. 9.9.

Step 5 *If the value of the test statistic falls in the rejection region, reject H_0; otherwise, do not reject H_0.*

The value of the test statistic, found in Step 4, is $z = 2.85$. As we see from Fig. 9.9, this falls in the rejection region, and so we reject H_0.

Step 6 *State the conclusion in words.*

The test results are statistically significant at the 1% level; that is, at the 1% significance level, the data provide sufficient evidence to conclude that this year's mean retail price of all hardcover history books has increased over the 1990 mean of $35.44. ∎

Example 9.8   Illustrates Procedure 9.1

A nutritionist thinks the average person with an income below the poverty level gets less than the recommended daily allowance (RDA) of 800 mg of calcium. To test her conjecture, she obtains the daily intakes of calcium for a random sample of 35 people with incomes below the poverty level. Table 9.6 gives the sample data. At the 5% significance level, do the data provide sufficient evidence to conclude that the mean calcium intake of all people with incomes below the poverty level is less than the RDA of 800 mg?

TABLE 9.6

Daily calcium intakes, in milligrams, for a random sample of 35 people with incomes below the poverty level

879	1096	701	986	828	1077	703
555	422	997	473	702	508	530
513	720	944	673	574	707	864
1199	743	1325	655	1043	599	1008
705	180	287	542	893	1052	473

SOLUTION   A graphical display (not shown) of the data in Table 9.6 reveals no outliers. Since the sample size, $n = 35$, is large, we apply Procedure 9.1 to perform the hypothesis test.

Step 1   *State the null and alternative hypotheses.*

Let μ denote the mean calcium intake (per day) of all people with incomes below the poverty level. The null and alternative hypotheses were obtained in Example 9.3. They are

$$H_0: \mu = 800 \text{ mg (mean calcium intake is not less than the RDA)}$$
$$H_a: \mu < 800 \text{ mg (mean calcium intake is less than the RDA).}$$

Note that the hypothesis test is left-tailed since a less-than sign ($<$) appears in the alternative hypothesis.

Step 2   *Decide on the significance level, α.*

We are to perform the test at the 5% significance level. Thus $\alpha = 0.05$.

Step 3   *The critical value for a left-tailed test is $-z_\alpha$.*

Since $\alpha = 0.05$, the critical value is $-z_{0.05}$. From Table II (or Table 9.4 or Table IV), we find that $z_{0.05} = 1.645$. Hence the critical value is $-z_{0.05} = -1.645$, as seen in Fig. 9.10.

FIGURE 9.10

Criterion for deciding whether or not to reject the null hypothesis

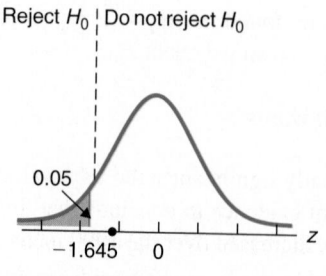

Reject H_0 $\mid$ Do not reject H_0

0.05

-1.645 0 z

Step 4   *Compute the value of the test statistic*

$$z = \frac{\bar{x} - \mu_0}{\sigma/\sqrt{n}}.$$

Since σ is unknown, we will use s in its place. We have $\mu_0 = 800$ mg and $n = 35$. From the data in Table 9.6, we find that $\bar{x} = 747.3$ mg and $s = 262.2$ mg. Thus the value of the test statistic is

$$z = \frac{747.3 - 800}{262.2/\sqrt{35}} = -1.19.$$

This value of z is marked with a dot in Fig. 9.10.

Step 5   *If the value of the test statistic falls in the rejection region, reject H_0; otherwise, do not reject H_0.*

The value of the test statistic, found in Step 4, is $z = -1.19$. As we see from Fig. 9.10, this does not fall in the rejection region, and so we do not reject H_0.

Step 6   *State the conclusion in words.*

The test results are not statistically significant at the 5% level; that is, at the 5% significance level, the sample of 35 calcium intakes does not provide sufficient evidence to conclude that the mean calcium intake, μ, of all people with incomes below the poverty level is less than the RDA of 800 mg.   ∎

Example 9.9   Illustrates Procedure 9.1

The general partner of a limited-partnership firm has told a potential investor that the mean monthly rent for three-bedroom apartments in the city is $587. To check this claim, the investor randomly selects 32 three-bedroom apartments in the city and determines their monthly rents. Table 9.7 displays the monthly rents, in dollars, for the 32 three-bedroom apartments obtained. Do the data suggest that the general partner's claim is incorrect? Perform the appropriate hypothesis test at the 0.05 level of significance.

TABLE 9.7
Monthly rents ($) for
32 three-bedroom apartments

289	560	726	643	586	657	565	676
656	577	663	729	745	597	669	626
450	669	603	545	661	610	604	598
507	675	609	503	589	521	595	472

SOLUTION   A frequency histogram for the data in Table 9.7 is displayed in Fig. 9.11. The histogram suggests that the first monthly rent, $289, is an outlier.

FIGURE 9.11
Frequency histogram for the
monthly rents in Table 9.7

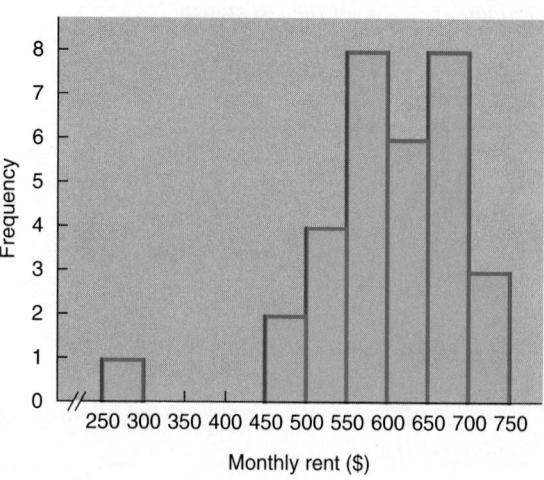

We will first apply Procedure 9.1 to the full data set in Table 9.7. Following that we will examine the effect on the test results when the outlier, $289, is removed.

Step 1   *State the null and alternative hypotheses.*

Let μ denote the mean monthly rent of all three-bedroom apartments in the city. Then the null and alternative hypotheses are

$$H_0: \mu = \$587 \text{ (general partner's claim is correct)}$$
$$H_a: \mu \neq \$587 \text{ (general partner's claim is not correct)}.$$

Note that the hypothesis test is two-tailed. (Why is this so?)

Step 2   *Decide on the significance level, α.*

We are to perform the hypothesis test at the 0.05 level of significance; so $\alpha = 0.05$.

Step 3   *The critical values for a two-tailed test are $\pm z_{\alpha/2}$.*

Since $\alpha = 0.05$, we obtain from Table II (or Table 9.4 or Table IV) that the critical values are $\pm z_{0.05/2} = \pm z_{0.025} = \pm 1.96$, as seen in Fig. 9.12.

FIGURE 9.12
Criterion for deciding whether or
not to reject the null hypothesis

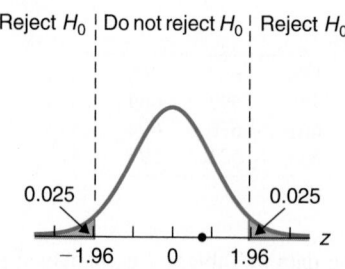

Step 4 *Compute the value of the test statistic*

$$z = \frac{\bar{x} - \mu_0}{\sigma/\sqrt{n}}.$$

Since σ is unknown, we will use s in its place. We have $\mu_0 = \$587$ and $n = 32$. From the data in Table 9.7, we find that $\bar{x} = \$599.22$ and $s = \$91.21$. Thus the value of the test statistic is

$$z = \frac{599.22 - 587}{91.21/\sqrt{32}} = 0.76.$$

This value of z is marked with a dot in Fig. 9.12.

Step 5 *If the value of the test statistic falls in the rejection region, reject H_0; otherwise, do not reject H_0.*

From Step 4 the value of the test statistic is $z = 0.76$. This does not fall in the rejection region, as we see from Fig. 9.12. Hence we do not reject H_0.

Step 6 *State the conclusion in words.*

The test results are not statistically significant at the 5% level; that is, at the 5% significance level, the data do not provide sufficient evidence to conclude that the mean monthly rent, μ, of all three-bedroom apartments in the city differs from the general partner's claim of \$587. This completes the hypothesis test.

Now recall that the first monthly rent, \$289, is an outlier. Although for this problem we don't actually know whether removing this outlier is justified (a commonly occurring situation), we can still remove it from the sample and assess the effect on the hypothesis test. Doing this, we find that the value of the test statistic for the abridged data is $z = 1.70$. This value still lies in the nonrejection region, although it is much closer to the rejection region than the value of test statistic for the unabridged data, $z = 0.76$. Hence, in this case, removing the outlier does not affect the conclusion of the hypothesis test. We can probably be content with accepting the general partner's claim. ■

The Classical Approach to Hypothesis Testing

Procedure 9.1 on page 503 provides a step-by-step method for performing a large-sample hypothesis test for a population mean. The steps used in that procedure are typical of those employed in the so-called **classical approach to hypothesis testing.**

We will usually provide procedures that are specific to a particular kind of inference (e.g., large-sample hypothesis test for a population mean). Nonetheless, it is a good idea to be aware of the elements common to all hypothesis-testing procedures that are based on the classical approach. Those elements are presented in Procedure 9.2.

PROCEDURE 9.2

**TO PERFORM A HYPOTHESIS TEST USING
THE CLASSICAL APPROACH**

Step 1   State the null and alternative hypotheses.

Step 2   Decide on the significance level, α.

Step 3   Determine the critical value(s).

Step 4   Compute the value of the test statistic.

Step 5   If the value of the test statistic falls in the rejection region, then reject H_0; otherwise, do not reject H_0.

Step 6   State the conclusion in words.

Statistical Significance Versus Practical Significance

Recall that the results of a hypothesis test are statistically significant if the null hypothesis is rejected at the chosen level of α. This means that the data provide sufficient evidence to conclude that the truth is different from that stated in the null hypothesis. It does not necessarily mean that the difference is important in any practical sense.

For instance, the manufacturer of a new-model car, the Orion, claims that a typical car gets 26 miles per gallon, that is, that the mean gas mileage of all Orions is $\mu = 26$ mpg. Suppose the mean gas mileage, $\bar{x}$, of a sample of 1000 Orions turns out to be 25.9 mpg, with a sample standard deviation of 1.5 mpg. The value of the test statistic for a test of

$$H_0: \mu = 26 \text{ mpg (mean gas mileage is 26 mpg)}$$

$$H_a: \mu < 26 \text{ mpg (mean gas mileage is less than 26 mpg)}$$

is $z = -2.11$. This is statistically significant at the 5% level (even at the 1.74% level). We can easily reject the null hypothesis, that is, the manufacturer's claim that the mean gas mileage of all Orions is 26 mpg.

Because the sample size, 1000, is so large, the sample mean, $\bar{x} = 25.9$ mpg, is probably nearly the same as the population mean. This indicates that the manufacturer's claim was rejected because μ is about 25.9 mpg instead of 26 mpg. From a practical point of view, the difference between 25.9 mpg and 26 mpg is not important. Therefore, in this case, the statistical significance is not practically significant. Remember: *Statistical significance does not necessarily imply practical significance!*

The Relation Between Hypothesis Tests and Confidence Intervals

Hypothesis tests and confidence intervals are closely related. Consider, for instance, a two-tailed hypothesis test for a population mean at the significance level α. It can be shown that the null hypothesis will be rejected if and only if the value μ_0 given for the mean in the null hypothesis lies outside the $(1 - \alpha)$-level confidence interval for μ. Exercises 9.50 and 9.51 discuss this relation between hypothesis tests and confidence intervals in greater detail.

EXERCISES 9.3

In each of Exercises 9.35–9.40, suppose a large-sample hypothesis test is to be performed for a population mean, μ, with null hypothesis H_0: $\mu = \mu_0$. Further suppose the test statistic used will be

$$z = \frac{\bar{x} - \mu_0}{\sigma/\sqrt{n}}.$$

For each exercise, obtain the required critical value(s) and draw a graph that illustrates your results.

9.35 A right-tailed test with $\alpha = 0.01$.

9.36 A left-tailed test with $\alpha = 0.10$.

9.37 A two-tailed test with $\alpha = 0.10$.

9.38 A right-tailed test with $\alpha = 0.05$.

9.39 A left-tailed test with $\alpha = 0.05$.

9.40 A two-tailed test with $\alpha = 0.01$.

In Exercises 9.41–9.48, apply Procedure 9.1 on page 503 to perform the required hypothesis tests. Comment on the practical significance of all hypothesis tests whose results are statistically significant.

9.41 According to the Census Bureau publication *Construction Reports*, the mean expenditure per residential property owner for maintenance and repairs in 1983 was $280. For last year, a random sample of 40 residential property owners revealed the following expenditures, in dollars.

158	110	185	136	167	84	437	420
230	57	942	287	259	123	712	170
531	347	505	148	111	442	145	38
49	188	107	134	223	36	360	253
321	821	89	202	759	151	1176	31

Do the data provide sufficient evidence to conclude that last year's mean amount spent for maintenance and repairs has increased over the 1983 mean of $280? Assume $\sigma = \$265$ and perform the hypothesis test at the 5% significance level. *(Note: The sum of the data is $11,644.)*

9.42 *The World Almanac, 1985*, reports that the mean travel time to work in 1980 for all South Dakota residents was 13 minutes. A transportation official obtained this year's travel times, in minutes, for a random sample of 35 South Dakota residents. Here are the data.

29	40	0	12	10	6	41
25	21	5	4	19	2	7
10	8	3	6	52	4	12
0	33	6	2	17	21	8
38	2	13	8	14	11	2

At the 5% significance level, do the data provide sufficient evidence to conclude that the mean travel time to work for all South Dakota residents has changed from the 1980 mean of 13 minutes? Assume $\sigma = 11.6$ minutes. *(Note: The sum of the data is 491 minutes.)*

9.43 The Food and Nutrition Board of the National Academy of Sciences states that the RDA of iron for adult females under the age of 51 is 18 mg. The following iron intakes, in milligrams, during a 24-hour period were obtained for 45 randomly selected adult females under the age of 51.

15.0	18.1	14.4	14.6	10.9	18.1	18.2	18.3	15.0
16.0	12.6	16.6	20.7	19.8	11.6	12.8	15.6	11.0
15.3	9.4	19.5	18.3	14.5	16.6	11.5	16.4	12.5
14.6	11.9	12.5	18.6	13.1	12.1	10.7	17.3	12.4
17.0	6.3	16.8	12.5	16.3	14.7	12.7	16.3	11.5

At the 1% significance level, do the data suggest that adult females under the age of 51 are, on the average, getting less than the RDA of 18 mg of iron? *(Note: $\bar{x} = 14.68$, $s = 3.08$.)*

9.44 As reported by the U.S. Office of Juvenile Justice and Delinquency Prevention in *Children in Custody*, the mean age of all juveniles held in public custody in 1989 was 16.0 years. The mean age of 250 randomly selected juveniles currently being held in public custody is 15.86 years, and the standard deviation of the ages is 1.01 years. Does it appear that the mean age, μ, of all juveniles being held in public custody this year is less than the 1989 mean of 16.0 years? Perform the appropriate hypothesis test using $\alpha = 0.10$.

9.45 A dog-food manufacturer sells "50-lb" bags of dog food. Suppose you randomly select 75 bags and find that $\bar{x} = 50.11$ lb and $s = 0.84$ lb.
a. Would you be inclined to believe that the actual mean weight, μ, of all "50-lb" bags of this dog food differs from the advertised weight of 50 lb? Perform your hypothesis test at the 5% significance level.
b. Repeat part (a) if the mean weight of the 75 bags is 50.21 lb instead of 50.11 lb.

9.46 The Health Insurance Association of America reports in *Survey of Hospital Semi-Private Room Charges* that the mean daily charge for a semi-private room in U.S. hospitals in 1988 was $253. In that same year, a random sample of 30 Massachusetts hospitals yielded a mean daily semi-private room charge of $260.68 with a standard deviation of $12.77. At the 5% significance level, do the data provide sufficient evidence to conclude that in 1988 the mean daily semi-private room charge in Massachusetts hospitals exceeded the national mean of $253?

9.47 The manufacturer of a new-model car, called the Orion, claims that a typical car gets 26 mpg. An independent consumer group is skeptical of this claim and thinks the mean gas mileage of all Orions may very well be less than 26 mpg. To try to justify its contention, the consumer group conducts mileage tests on 30 randomly selected Orions and obtains the following data.

25.3	25.1	29.6	24.6	26.0	26.0
26.3	23.6	26.0	25.4	26.1	23.8
25.1	24.1	25.8	26.4	23.4	24.8
22.6	26.6	25.1	26.6	28.0	23.3
23.8	25.4	26.2	25.1	25.3	21.5

At the 5% significance level, do the data support the consumer group's conjecture? *(Note: $\bar{x} = 25.23$, $s = 1.59$.)*

9.48 A Louisiana cotton farmer has used a certain brand of fertilizer for the past 5 years. Based on experience, the farmer knows that the mean yield of cotton using this fertilizer is 623 lb/acre. Recently, a new brand of fertilizer appeared on the market that will supposedly increase cotton yield. The farmer uses the new fertilizer on 80 of his 1-acre plots. Here are the resulting cotton yields, in pounds.

639	653	631	590	628	638	618	602
622	667	614	644	611	652	598	627
637	613	591	627	637	604	620	618
621	642	654	643	634	630	621	560
637	634	645	630	605	617	598	641
636	599	616	578	620	639	608	615
587	601	629	627	626	613	568	638
599	627	630	615	620	658	629	629
626	670	642	647	637	593	617	654
627	645	646	643	648	654	651	613

a. At the 10% significance level, do the data provide sufficient evidence to conclude that the new fertilizer increases the mean yield of cotton on the farmer's land? *(Note: $\bar{x} = 625.2$ lb/acre, $s = 21.5$ lb/acre.)*

b. If the new fertilizer costs more than the one the farmer presently uses, would you buy the new fertilizer if you were the farmer? Why or why not?

9.49 Explain why it is permissible to use a sample standard deviation, s, in place of an unknown population standard deviation, σ, when performing a large-sample hypothesis test for a population mean, μ.

9.50 In 1990, the average passenger vehicle was driven 10.3 thousand miles, as reported by the U.S. Federal Highway Administration in *Highway Statistics*. A random sample of 500 passenger vehicles had a mean of 10.1 thousand miles driven for last year and a standard deviation of 6.0 thousand miles. Let μ denote last year's mean distance driven for all passenger vehicles.

a. Perform the hypothesis test

$$H_0: \mu = 10.3 \text{ thousand miles}$$
$$H_a: \mu \neq 10.3 \text{ thousand miles}$$

at the 5% significance level.

b. Use Procedure 8.1 on page 448 to find a 95% confidence interval for μ.

c. Does the value of 10.3 thousand miles, hypothesized for the mean, μ, in the null hypothesis of part (a), lie within your confidence interval from part (b)?

d. Repeat parts (a)–(c) if the 500 passenger vehicles sampled were driven an average of 10.9 thousand miles last year.

e. Based on your observations in parts (a)–(d), complete the following statements concerning the relationship between a two-tailed hypothesis test,

$$H_0: \mu = \mu_0$$
$$H_a: \mu \neq \mu_0,$$

at the significance level α and a $(1 - \alpha)$-level confidence interval for μ:
   i. If μ_0 lies within the $(1 - \alpha)$-level confidence interval for μ, then the null hypothesis (*will, will not*) be rejected.
   ii. If μ_0 lies outside the $(1 - \alpha)$-level confidence interval for μ, then the null hypothesis (*will, will not*) be rejected.

9.51 The relationship between hypothesis tests and confidence intervals. In this exercise we will examine the relationship between a large-sample, two-tailed hypothesis test for a population mean and a large-sample confidence-interval estimate for a population mean.

a. Show that the inequalities

$$\overline{x} - z_{\alpha/2} \cdot \frac{\sigma}{\sqrt{n}} < \mu_0 < \overline{x} + z_{\alpha/2} \cdot \frac{\sigma}{\sqrt{n}}$$

are equivalent to

$$-z_{\alpha/2} < \frac{\overline{x} - \mu_0}{\sigma/\sqrt{n}} < z_{\alpha/2}.$$

b. Deduce the following fact from part (a): For a two-tailed

hypothesis test,

$$H_0: \mu = \mu_0$$
$$H_a: \mu \neq \mu_0,$$

at the significance level α, the null hypothesis will not be rejected if μ_0 lies within the $(1 - \alpha)$-level confidence interval for μ, and conversely, the null hypothesis will be rejected if μ_0 does not lie within the $(1-\alpha)$-level confidence interval for μ.

9.4 TYPE II ERROR PROBABILITIES; POWER (OPTIONAL)

As we learned in Section 9.2, hypothesis tests do not always yield correct conclusions; they have built-in margins of error. An important part of planning a study is to take an advance look at the types of errors that can be made and the effects those errors might have.

Recall that two types of errors are possible with hypothesis tests. One is a Type I error: rejecting a true null hypothesis. The other is a Type II error: not rejecting a false null hypothesis. Also recall that the probability of making a Type I error is called the significance level of the hypothesis test and is denoted by the Greek letter α; the probability of making a Type II error is denoted by the Greek letter β.

In this section we will learn how to compute Type II error probabilities. We will also investigate the concept of the power of a hypothesis test. Although we will limit the discussion to large-sample hypothesis tests for a population mean, μ, the ideas apply to any hypothesis test.

Computing Type II Error Probabilities

The probability, β, of a Type II error depends on the sample size, the significance level, and the true value of μ. Example 9.10 explains how to compute the probability of making a Type II error.

Example 9.10   Computing Type II Error Probabilities

The manufacturer of a new model car, the Orion, claims that a typical car gets 26 miles per gallon (mpg). An independent consumer group is skeptical of this claim and thinks the mean gas mileage of all Orions may very well be less than 26 mpg. The consumer group plans to perform the hypothesis test

$$H_0: \mu = 26 \text{ mpg (manufacturer's claim)}$$
$$H_a: \mu < 26 \text{ mpg (consumer group's conjecture)},$$

where μ is the mean gas mileage of all Orions.

Suppose the consumer group decides to use a significance level of 0.05 and a sample size of 30. Find the probability, β, of a Type II error if the true mean gas mileage, μ, is

a. 25.8 mpg. b. 25.0 mpg.

Assume the standard deviation of the gas mileages of all Orions is 1.4 mpg.[†]

SOLUTION The inference under consideration is a large-sample, left-tailed hypothesis test for a population mean at the 5% significance level. The test statistic is

$$(2) \qquad\qquad z = \frac{\bar{x} - \mu_0}{\sigma/\sqrt{n}},$$

and the critical value is $-z_\alpha = -z_{0.05} = -1.645$. Thus the decision criterion for the hypothesis test is this: If $z \leq -1.645$, reject H_0; otherwise, do not reject H_0.

It is somewhat simpler to compute Type II error probabilities if the decision criterion is expressed in terms of $\bar{x}$ instead of z. To do that, we first solve for $\bar{x}$ in Equation (2). The result is Equation (3):

$$(3) \qquad\qquad \bar{x} = \mu_0 + z \cdot \frac{\sigma}{\sqrt{n}}.$$

Next we use Equation (3) to determine the value of $\bar{x}$ corresponding to $z = -1.645$. Since $\mu_0 = 26$, $\sigma = 1.4$, and $n = 30$, we see that when $z = -1.645$,

$$\bar{x} = 26 - 1.645 \cdot \frac{1.4}{\sqrt{30}} = 25.6.$$

So the decision criterion can be expressed in terms of $\bar{x}$ as follows: If $\bar{x} \leq 25.6$ mpg, reject H_0; otherwise, do not reject H_0. This is portrayed graphically in Fig. 9.13.

FIGURE 9.13

Graphical display of decision criterion for the gas-mileage illustration ($\alpha = 0.05$, $n = 30$)

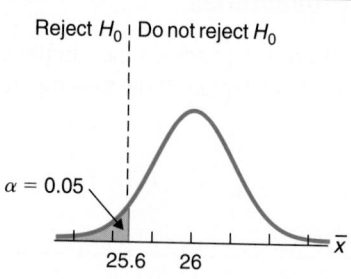

Reject H_0 ┊ Do not reject H_0

$\alpha = 0.05$

25.6 26 $\bar{x}$

a. For this part we want to determine the probability, β, of a Type II error if the true mean gas mileage of all Orions is 25.8 mpg; that is, we need to obtain the probability of not

[†] We are assuming σ is known for the sake of simplicity. If σ is unknown, which is usually the case, we must proceed somewhat differently.

rejecting the null hypothesis, $\mu = 26$ mpg, if in fact $\mu = 25.8$ mpg. According to the decision criterion, pictured in Fig. 9.13, the null hypothesis is not rejected if the sample mean gas mileage, $\bar{x}$, of the 30 Orions tested exceeds 25.6 mpg.

Consequently, we need to determine $P(\bar{x} > 25.6)$ given that the true mean gas mileage of all Orions is 25.8 mpg. Since the sample size is large ($n = 30$), the random variable $\bar{x}$ is approximately normally distributed and has a mean of $\mu_{\bar{x}} = \mu = 25.8$ and a standard deviation of $\sigma_{\bar{x}} = \sigma/\sqrt{n} = 1.4/\sqrt{30} = 0.26$. So $P(\bar{x} > 25.6)$ is equal to the area under the normal curve with parameters 25.8 and 0.26 that lies to the right of 25.6. That area is obtained in the usual manner, as shown in Fig. 9.14. *Note:* The curve we have drawn in Fig. 9.14 is not the curve based on the null hypothesis value of μ, which is 26 mpg, but rather the curve based on the true value of μ, which here is assumed to be 25.8 mpg.

FIGURE 9.14
Determination of the area under the normal curve with parameters $\mu_{\bar{x}} = 25.8$ and $\sigma_{\bar{x}} = 0.26$ that lies to the right of 25.6

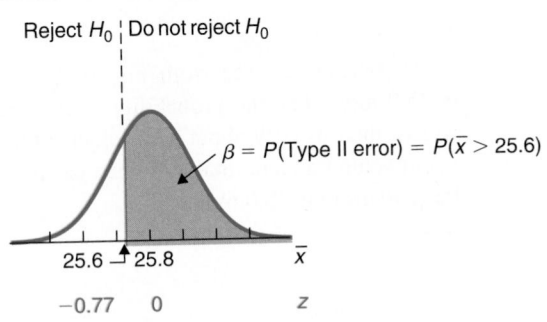

Therefore, as we see from Fig. 9.14, if the true mean gas mileage of all Orions is 25.8 mpg, then the probability of making a Type II error is $\beta = 0.7794$. In other words, there is roughly a 78% chance that the consumer group will fail to reject the manufacturer's claim that the mean gas mileage of all Orions is 26 mpg when, in fact, the true mean is 25.8 mpg. Although this is a rather high chance of error, we probably would not expect the hypothesis test to detect such a small difference in mean gas mileage (25.8 mpg as opposed to 26 mpg) with a sample size of only 30.

b. For this part we want to determine the probability, β, of a Type II error if the true mean gas mileage of all Orions is 25.0 mpg. As in part (a), this means we need to obtain $P(\bar{x} > 25.6)$, but this time assuming $\mu = 25.0$ mpg. Figure 9.15 shows the required computations.

FIGURE 9.15
Determination of the area under
the normal curve with parameters
$\mu_{\bar{x}} = 25.0$ and $\sigma_{\bar{x}} = 0.26$
that lies to the right of 25.6

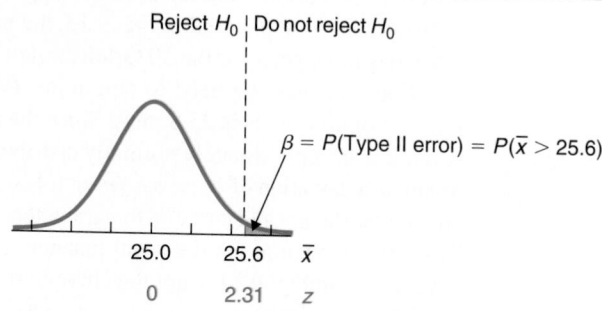

z-score computation:

Area to the left of z:

$$\bar{x} = 25.6 \longrightarrow z = \frac{25.6 - 25.0}{0.26} = 2.31 \qquad 0.9896$$

Shaded area $= 1 - 0.9896 = 0.0104$

Hence, as we see from Fig. 9.15, if the true mean gas mileage of all Orions is 25.0 mpg, then the probability of making a Type II error is $\beta = 0.0104$. In other words, there is only about a 1% chance that the consumer group will fail to reject the manufacturer's claim that the mean gas mileage of all Orions is 26 mpg when, in fact, the true mean is 25.0 mpg. ∎

By combining figures such as Figs. 9.14 and 9.15, we can better understand Type II error probabilities. In Fig. 9.16 we combined those two figures with two others. The Type II error probabilities for the two additional values of μ were obtained using the same techniques as those used in Example 9.10.

Figure 9.16 makes it clear that the farther the true value of μ is from the null hypothesis value of 26 mpg, the smaller the probability of making a Type II error. This is hardly surprising. We would expect it to be more likely for a false null hypothesis to be detected when the true value of μ is far from the null hypothesis value than when it is close to it.

Operating Characteristic Curves

Since in reality the true value of μ will be unknown, it is helpful to construct a table of Type II error probabilities for various values of μ. For the gas-mileage illustration, we have already obtained β when the true mean is 25.8 mpg, 25.6 mpg, 25.3 mpg, and 25.0 mpg (see Fig. 9.16). Similar calculations yield the β-values shown in Table 9.8.

Table 9.8 can be used as an aid for evaluating the overall effectiveness of the hypothesis test. We can also employ the table to obtain a visual display of that effectiveness. This is accomplished by plotting points of β versus μ and then connecting the points with a smooth curve, as shown in Fig. 9.17 at the top of page 518.

FIGURE 9.16

Type II error probabilities for
$\mu = 25.8, 25.6, 25.3$, and 25.0
$(\alpha = 0.05, n = 30)$

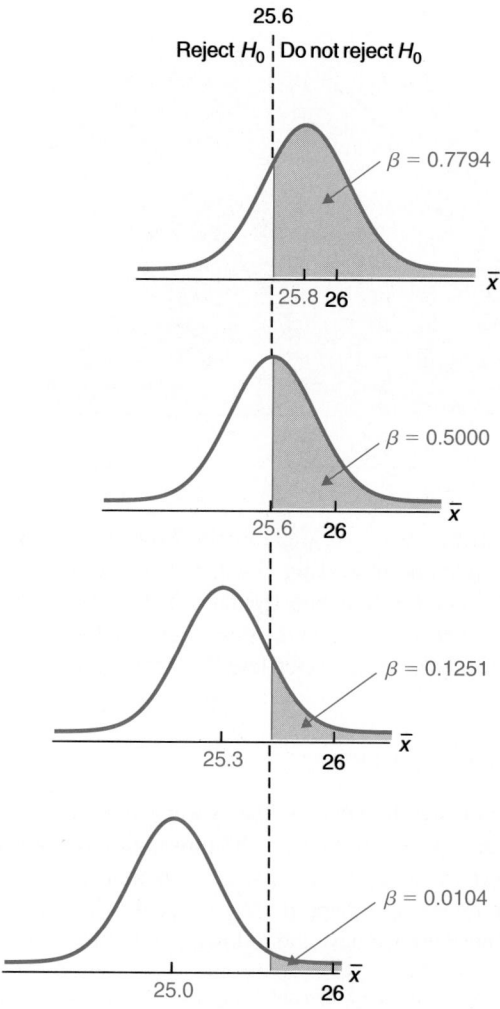

TABLE 9.8

Selected Type II error probabilities
for the gas-mileage illustration
$(\alpha = 0.05, n = 30)$

True mean μ	P(Type II error) β	True mean μ	P(Type II error) β
25.9	0.8749	25.3	0.1251
25.8	0.7794	25.2	0.0618
25.7	0.6480	25.1	0.0274
25.6	0.5000	25.0	0.0104
25.5	0.3520	24.9	0.0036
25.4	0.2206	24.8	0.0010

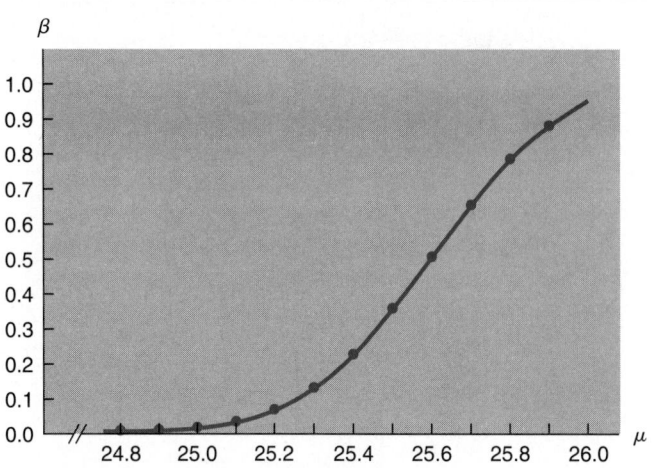

The curve in Fig. 9.17—a graph of the Type II error probability, β, versus the true value of μ—is called an **operating characteristic curve** or, more briefly, an **OC curve.** The closer an OC curve is to the horizontal axis, the better. (Why is this so?) Keep in mind that the OC curve depends on both the sample size and significance level. In other words, if either the sample size or significance level is changed, then the OC curve will also change.

Power and Power Curves

Closely related to the OC curve is the power curve. To discuss power curves, we first need to define the concept of power. The **power** of a hypothesis test is the probability of not making a Type II error. By the complementation rule, Formula 4.2 on page 203, the probability of not making a Type II error is equal to 1 minus the probability of making a Type II error. Therefore we have the following definition.

DEFINITION 9.5

POWER

The *power* of a hypothesis test is the probability of not making a Type II error, that is, the probability of rejecting a false null hypothesis. We have

$$\text{Power} = 1 - P(\text{Type II error}) = 1 - \beta.$$

The power of a hypothesis test is between 0 and 1 and measures the ability of the hypothesis test to detect a false null hypothesis. If the power is near 0, then the hypothesis test is not very good at detecting a false null hypothesis; if the power is near 1, then the hypothesis test is extremely good at detecting a false null hypothesis.

Once we know the Type II error probability, β, it is simple to obtain the power—just subtract β from 1. For the gas-mileage illustration, we used Table 9.8 to obtain the power of the hypothesis test for various values of μ. The results are presented in Table 9.9.

TABLE 9.9

Selected powers for the gas-mileage illustration ($\alpha = 0.05$, $n = 30$)

True mean μ	Power $1 - \beta$	True mean μ	Power $1 - \beta$
25.9	0.1251	25.3	0.8749
25.8	0.2206	25.2	0.9382
25.7	0.3520	25.1	0.9726
25.6	0.5000	25.0	0.9896
25.5	0.6480	24.9	0.9964
25.4	0.7794	24.8	0.9990

The **power curve** is a graph of the power, $1 - \beta$, versus the true value of μ. Using Table 9.9, we obtained the power curve for the gas-mileage illustration, as seen in Fig. 9.18.

FIGURE 9.18

Power curve for the gas-mileage illustration ($\alpha = 0.05$, $n = 30$)

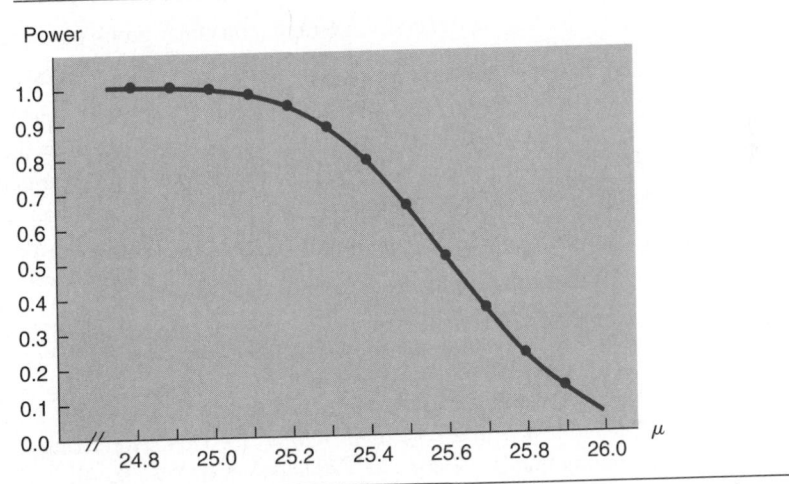

Usually there is no need to plot both an OC curve and a power curve since they provide essentially the same information. We have discussed both curves because they are both used extensively by researchers.

Sample-Size Considerations

Ideally, we would like both Type I and Type II errors to have small probabilities. For then the chances of making an incorrect decision would be small regardless of which hypothesis is actually true.

Because the probability of a Type I error is the significance level, α, we can control the size of the Type I error probability by specifying the appropriate significance level. For instance, if we want a small Type I error probability, then we can simply specify a small value for α. However, as we noted in Section 9.2, we must keep in mind that for a fixed sample size, the smaller the Type I error probability, α, of rejecting a true null hypothesis, the larger the Type II error probability, β, of not rejecting a false null hypothesis, and vice versa.

Nonetheless, there is a way to make both Type I and Type II error probabilities small: We can specify a small significance level, α, which makes the Type I error probability small, and we can use a large sample size, which makes the Type II error probabilities small for a fixed significance level. Example 9.11 provides an illustration.

Example 9.11   The Effect of Sample Size on Type II Error Probabilities

Consider again the gas-mileage illustration of Example 9.10. A consumer group wants to perform the hypothesis test

$$H_0\!: \mu = 26 \text{ mpg (manufacturer's claim)}$$
$$H_a\!: \mu < 26 \text{ mpg (consumer group's conjecture),}$$

where μ is the mean gas mileage of all Orions.

In Table 9.8 on page 517, we presented selected Type II error probabilities when $\alpha = 0.05$ and $n = 30$; and in Fig. 9.17 on page 518, we drew the corresponding OC curve. Now suppose the significance level is kept at 0.05 but the sample size is increased from 30 to 100.

a. Construct a table of Type II error probabilities similar to Table 9.8.

b. Use the table from part (a) to draw the OC curve.

c. Compare the Type II error probabilities for the sample sizes $n = 30$ and $n = 100$.

SOLUTION   As before, we first express the decision criterion in terms of $\bar{x}$. The critical value for a left-tailed test with $\alpha = 0.05$ is $-z_{0.05} = -1.645$. Referring to Equation (3) on page 514 and noting that $\mu_0 = 26$, $\sigma = 1.4$, and $n = 100$, we see that when $z = -1.645$,

$$\bar{x} = 26 - 1.645 \cdot \frac{1.4}{\sqrt{100}} = 25.8.$$

So the decision criterion can be expressed in terms of $\bar{x}$ as follows: If $\bar{x} \le 25.8$ mpg, reject H_0; otherwise, do not reject H_0. This is depicted in Fig. 9.19.

FIGURE 9.19

Graphical display of decision criterion for the gas-mileage illustration ($\alpha = 0.05$, $n = 100$)

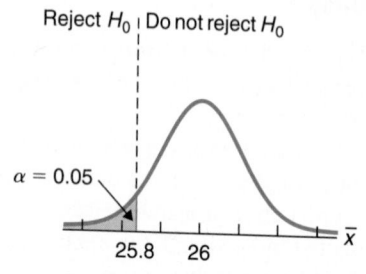

a. Now that the decision criterion has been expressed in terms of $\bar{x}$, Type II error probabilities can be obtained using the same techniques as in Example 9.10. We computed several Type II error probabilities and have displayed them in Table 9.10.

TABLE 9.10

Selected Type II error probabilities for the gas-mileage illustration ($\alpha = 0.05$, $n = 100$)

True mean μ	P(Type II error) β	True mean μ	P(Type II error) β
25.9	0.7611	25.3	0.0002
25.8	0.5000	25.2	0.0000†
25.7	0.2389	25.1	0.0000
25.6	0.0764	25.0	0.0000
25.5	0.0162	24.9	0.0000
25.4	0.0021	24.8	0.0000

† For $\mu \leq 25.2$, the β probabilities are zero to four decimal places.

b. Using Table 9.10, we can now draw the OC curve for the gas-mileage illustration when $n = 100$. This is shown in Fig. 9.20. For comparison purposes, we have also reproduced from Fig. 9.17 the OC curve for the sample size $n = 30$.

FIGURE 9.20

OC curves for the gas-mileage illustration when $n = 30$ and $n = 100$ ($\alpha = 0.05$)

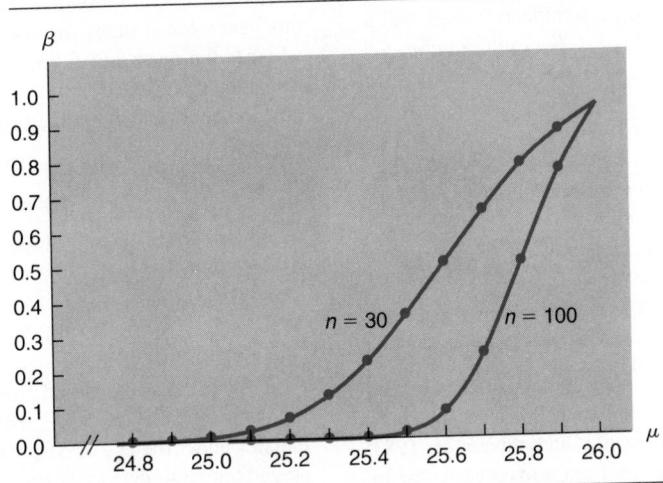

c. Comparing Tables 9.8 and 9.10, we see that each Type II error probability is smaller when $n = 100$ than when $n = 30$. Figure 9.20 displays that fact visually. ∎

In Example 9.11 we found that increasing the sample size, while keeping the significance level the same, reduced the Type II error probabilities. This is true in general, as indicated in Key Fact 9.4.

KEY FACT 9.4   **SAMPLE SIZE AND POWER**

> Increasing the sample size for a hypothesis test without changing the significance level decreases the Type II error probabilities. In other words, for a fixed significance level, increasing the sample size increases the power.

Key Fact 9.4 implies that by using a sufficiently large sample size, we can obtain a hypothesis test with as much power as we want. However, in practice we need to keep in mind that larger sample sizes tend to increase the cost of a study. Consequently, we must balance, among other things, the cost of a large sample against the cost of possible errors.

As we have discovered, power is useful for evaluating the overall effectiveness of a hypothesis-testing procedure. In addition, power can be used to compare different procedures. For example, a researcher might decide between two hypothesis-testing procedures on the basis of which test is more powerful for the situation under consideration.

EXERCISES 9.4

9.52 What does the power of a hypothesis test tell us?

In Exercises 9.53–9.58, we have given a hypothesis-testing situation and (i) a value for σ, (ii) a significance level, (iii) a sample size, and (iv) some values of μ. For each exercise,

a. *express the decision criterion for the hypothesis test in terms of $\bar{x}$.*

b. *determine the probability of a Type I error.*

c. *determine the probability of a Type II error for each of the given values of μ, and construct a table similar to Table 9.8 on page 517.*

d. *use the table obtained in part (c) to draw the OC curve.*

e. *use the table obtained in part (c) to construct a table of powers similar to Table 9.9 on page 519.*

f. *use the table obtained in part (e) to draw the power curve.*

9.53 According to the Census Bureau publication *Construction Reports,* the mean expenditure per residential property owner for maintenance and repairs in 1983 was \$280. Suppose we want to perform a hypothesis test to decide whether last year's mean amount spent for maintenance and repairs per residential property owner has increased over the 1983 mean of \$280. Then the null and alternative hypotheses are

$$H_0: \mu = \$280$$
$$H_a: \mu > \$280,$$

where μ is last year's mean amount spent for maintenance and repairs per residential property owner.

   i. $\sigma = 265$ ii. $\alpha = 0.05$ iii. $n = 40$
iv. $\mu = 305, 330, 355, 380, 405, 430, 455, 480$

9.54 *The World Almanac, 1985,* reports that the mean travel time to work in 1980 for all South Dakota residents was 13 minutes. A transportation official wants to use a sample of this year's travel times for South Dakota residents to decide whether the mean travel time to work for all South Dakota residents has changed from the 1980 mean of 13 minutes. The null and alternative hypotheses for the hypothesis test are

$$H_0: \mu = 13 \text{ minutes}$$
$$H_a: \mu \neq 13 \text{ minutes,}$$

where μ is this year's mean travel time to work for all South Dakota residents.

   i. $\sigma = 11.6$ ii. $\alpha = 0.05$ iii. $n = 35$
iv. $\mu = 3, 5, 7, 9, 11, 15, 17, 19, 21, 23$

9.55 The Food and Nutrition Board of the National Academy of Sciences states that the RDA of iron for adult females under the age of 51 is 18 mg. A hypothesis test is to be performed to decide whether adult females under the age of 51 are, on the average, getting less than the RDA of 18 mg of iron. The null and alternative hypotheses for the hypothesis test are

$$H_0: \mu = 18 \text{ mg}$$
$$H_a: \mu < 18 \text{ mg,}$$

where μ is the mean daily iron intake of all adult females under the age of 51.

   i. $\sigma = 3$ ii. $\alpha = 0.01$ iii. $n = 45$
iv. $\mu = 16.00, 16.25, 16.50, 16.75, 17.00, 17.25,$
          $17.50, 17.75$

9.56 As reported by the U.S. Office of Juvenile Justice and Delinquency Prevention in *Children in Custody,* the mean age of all juveniles held in public custody in 1989 was 16.0 years. The ages of a random sample of juveniles currently being held in public custody are to be used to decide whether this year's mean age of all juveniles being held in public custody is less than the 1989 mean. The null and alternative hypotheses for the hypothesis test are

$$H_0: \mu = 16.0 \text{ years}$$
$$H_a: \mu < 16.0 \text{ years},$$

where μ is this year's mean age of all juveniles being held in public custody.

   i. $\sigma = 1$ ii. $\alpha = 0.10$ iii. $n = 250$
iv. $\mu = 15.70, 15.75, 15.80, 15.85, 15.90, 15.95$

9.57 A dog-food manufacturer sells "50-lb" bags of dog food. Seventy-five bags of this brand of dog food are randomly selected and carefully weighed. Using the weights obtained, a hypothesis test is to be performed in order to decide whether the mean weight of all bags of this dog food differs from the advertised weight of 50 lb. The null and alternative hypotheses for the hypothesis test are

$$H_0: \mu = 50 \text{ lb}$$
$$H_a: \mu \neq 50 \text{ lb},$$

where μ is the actual mean weight of all "50-lb" bags of this dog food.

   i. $\sigma = 0.8$ ii. $\alpha = 0.05$ iii. $n = 75$
iv. $\mu = 49.5, 49.6, 49.7, 49.8, 49.9, 50.1, 50.2,$
      50.3, 50.4, 50.5

9.58 The Health Insurance Association of America reports in *Survey of Hospital Semi-Private Room Charges* that the mean daily charge for a semi-private room in U.S. hospitals in 1988 was $253. That same year, the daily semi-private room charges were obtained for a sample of 30 Massachusetts hos-

pitals. A hypothesis test was then performed to decide whether the mean daily semi-private room charge in Massachusetts hospitals exceeded the national mean. The null and alternative hypotheses for the hypothesis test performed are

$$H_0: \mu = \$253$$
$$H_a: \mu > \$253,$$

where μ is the 1988 mean daily semi-private room charge in Massachusetts hospitals.

   i. $\sigma = 13$ ii. $\alpha = 0.05$ iii. $n = 30$
iv. $\mu = 254, 255, 256, 257, 258, 259, 260, 261,$
      262, 263

9.59 Repeat parts (a)–(d) of Exercise 9.53 using a sample size of 80. Compare your OC curves for the two sample sizes and explain the principle being illustrated.

9.60 Repeat parts (a)–(d) of Exercise 9.54 using a sample size of 100. Compare your OC curves for the two sample sizes and explain the principle being illustrated.

9.61 Suppose you must choose between two procedures for performing a hypothesis test, say Procedure 1 and Procedure 2. Further suppose that for the same sample size and significance level, Procedure 1 has less power than Procedure 2. Which procedure would you choose? Why?

9.62 Consider a two-tailed hypothesis test for a population mean with null hypothesis $H_0: \mu = \mu_0$.
a. Draw the ideal OC curve.
b. Draw the ideal power curve.
c. Explain what your curves in parts (a) and (b) portray.

9.63 Consider a right-tailed hypothesis test for a population mean with null hypothesis $H_0: \mu = \mu_0$.
a. Draw the ideal OC curve.
b. Draw the ideal power curve.
c. Explain what your curves in parts (a) and (b) portray.

9.5 *P*-VALUES

In the classical approach to hypothesis testing (Procedure 9.2 on page 510), the significance level is specified in advance and then the conclusion is stated in terms of rejecting or not rejecting the null hypothesis. This approach has some disadvantages: it does not permit readers having access only to the conclusion of the hypothesis test to make their own evaluation (i.e., to select their own significance level); nor does it provide them with the information necessary to assess precisely the strength of the evidence against the null hypothesis.

To alleviate these problems, many researchers and most statistical software include the *P-value* of the hypothesis test in their reports. Roughly speaking, the **P-value** indicates the likelihood of observing the value obtained for the test statistic if the null hypothesis is true.

A large P-value indicates that it would not be unlikely to observe the value obtained for the test statistic if the null hypothesis is true; in other words, a large P-value does not provide evidence against the null hypothesis. On the other hand, a small P-value indicates that it would be unlikely to observe the value obtained for the test statistic if the null hypothesis is true; in other words, a small P-value provides evidence against the null hypothesis.

We now present the precise definition of the P-value of a hypothesis test. Following the definition we will consider several examples that illustrate P-values.

DEFINITION 9.6   **P-VALUE**

> The **P-value** of a hypothesis test is the probability of observing a value of the test statistic as inconsistent (or more) with the null hypothesis as the value of the test statistic actually observed. Therefore the smaller the P-value, the stronger the evidence against the null hypothesis. We often use the letter **P** to denote the P-value.

Note: The P-value is also frequently referred to as the **observed significance level** or the **probability value.**

If we are considering a large-sample hypothesis test for a population mean, μ, then the P-value is determined in the following way: For a two-tailed test, the P-value is the probability of observing a value of the test statistic, z, that is at least as large in magnitude as the value actually observed, as seen in Fig. 9.21(a). For a left-tailed test, the P-value is the probability of observing a value of the test statistic, z, that is as small as or smaller than the value actually observed, as seen in Fig. 9.21(b). For a right-tailed test, the P-value is the probability of observing a value of the test statistic, z, that is as large as or larger than the value actually observed, as seen in Fig. 9.21(c). All three probabilities are computed under the assumption that the null hypothesis is true, which implies that z has the standard normal distribution.

FIGURE 9.21

P-value for a large-sample hypothesis test for a population mean if the test is (a) two-tailed, (b) left-tailed, (c) right-tailed

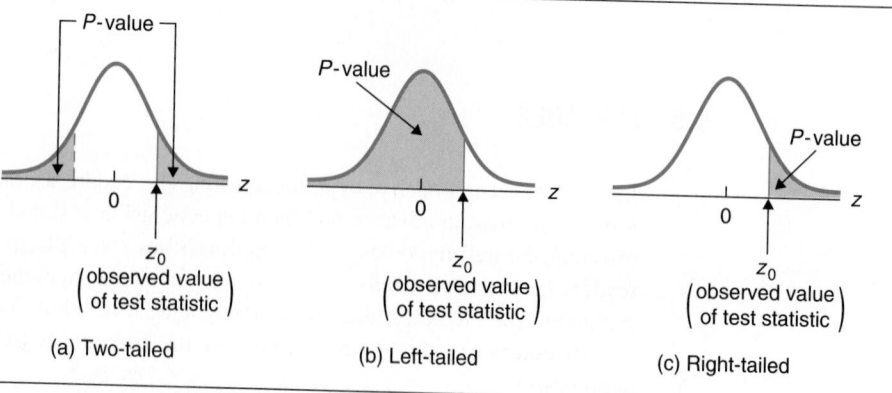

If we let z_0 denote the observed value of the test statistic, z, then we can obtain compact formulas for the *P*-value. Referring to Fig. 9.21, we see that the *P*-values for two-tailed, left-tailed, and right-tailed tests are given, respectively, by $P(|z| \geq |z_0|)$, $P(z \leq z_0)$, and $P(z \geq z_0)$. In terms of areas, this means that for a two-tailed test, the *P*-value is the area under the standard normal curve that lies either to the left of $-|z_0|$ or to the right of $|z_0|$, as in Fig. 9.21(a); for a left-tailed test, the *P*-value is the area that lies to the left of z_0, as in Fig. 9.21(b); and for a right-tailed test, the *P*-value is the area that lies to the right of z_0, as in Fig. 9.21(c).

P-values for other types of hypothesis tests are obtained similarly (see Exercise 9.89 for details). The easiest way to understand *P*-values is to look at several examples. In Examples 9.12 and 9.13, we will determine the *P*-values for two of the hypothesis tests we performed in Section 9.3.

Example 9.12   Illustrates Definition 9.6

Consider again the history-book hypothesis test of Example 9.7. The null and alternative hypotheses are

$$H_0: \mu = \$35.44 \text{ (mean price has not increased)}$$
$$H_a: \mu > \$35.44 \text{ (mean price has increased)},$$

where μ is this year's mean retail price of all hardcover history books. Note that the test is right-tailed since a greater-than sign ($>$) appears in the alternative hypothesis. Table 9.5 on page 504 displays this year's prices for 40 randomly selected history books. Using that data, we found the value of the test statistic to be 2.85. Determine and interpret the *P*-value of the hypothesis test.

SOLUTION   Since the test under consideration here is a large-sample, right-tailed hypothesis test for a population mean, the *P*-value is the probability of observing a value of z of 2.85 or greater, $P(z \geq 2.85)$, if the null hypothesis is true. That probability equals the area under the standard normal curve to the right of 2.85, the shaded area in Fig. 9.22. From Table II we find that area to be $1 - 0.9978 = 0.0022$.

FIGURE 9.22
P-value for the history-book hypothesis test

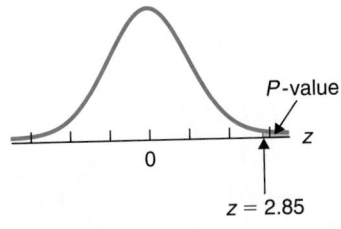

P-value

z

0

$z = 2.85$

Consequently, the *P*-value of this hypothesis test is 0.0022. This means there is only a 0.22% chance of observing a value of the test statistic as inconsistent (or more) with the null hypothesis as the value of the test statistic we actually observed; the data provide very strong evidence against the null hypothesis.

■

Example 9.13 Illustrates Definition 9.6

Consider again the monthly-rental hypothesis test of Example 9.9. The null and alternative hypotheses are

$$H_0: \mu = \$587 \text{ (general partner's claim is correct)}$$
$$H_a: \mu \neq \$587 \text{ (general partner's claim is not correct)},$$

where μ is the mean monthly rent of all three-bedroom apartments in the city. Note that the hypothesis test is two-tailed since a not-equal sign ($\neq$) appears in the alternative hypothesis. Table 9.7 on page 507 shows the monthly rents of a random sample of 32 three-bedroom apartments in the city. Recall that the first monthly rent, $289, is an outlier.

a. Obtain and interpret the P-value of the hypothesis test using the unabridged data (i.e., including the outlier).

b. Obtain and interpret the P-value of the hypothesis test using the abridged data (i.e., with the outlier removed).

c. Comment on the effect that removing the outlier has on the evidence against the null hypothesis.

SOLUTION a. For the unabridged data, the value of the test statistic was found in Example 9.9 to be 0.76. Therefore, since the test is a large-sample, two-tailed hypothesis test for a population mean, the P-value is the probability of observing a value of z of 0.76 or greater in magnitude if the null hypothesis is true. That probability, $P(|z| \geq 0.76)$, is depicted in Fig. 9.23(a) and equals 0.4472. There is roughly a 45% chance of observing a value of the test statistic as inconsistent (or more) with the null hypothesis as the value of the test statistic we actually observed; the unabridged data do not provide evidence against the null hypothesis.

FIGURE 9.23

P-value for the monthly-rents hypothesis test (a) including outlier, (b) with outlier removed

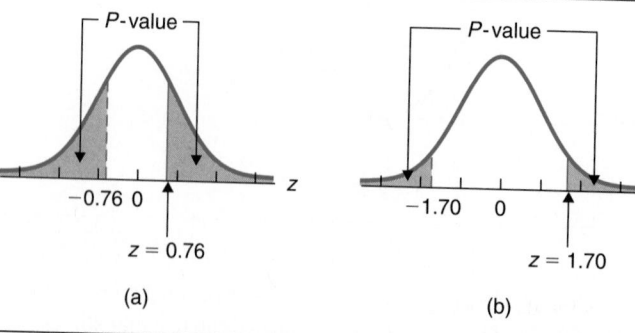

(a)

(b)

b. For the abridged data, the value of the test statistic was found in Example 9.9 to be 1.70. Thus in this case the *P*-value is $P(|z| \geq 1.70)$. That probability, depicted in Fig. 9.23(b), equals 0.0892. There is less than a 9% chance of observing a value of the test statistic as inconsistent (or more) with the null hypothesis as the value of the test statistic we actually observed; the abridged data provide moderate evidence against the null hypothesis.

c. From parts (a) and (b) we see that the strength of the evidence against the null hypothesis depends on whether the outlier is retained or removed. If the outlier is retained, there is virtually no evidence against the null hypothesis; if the outlier is removed, there is moderate evidence against the null hypothesis.   ∎

The *P*-Value Approach to Hypothesis Testing

The *P*-value can be interpreted as the *observed significance level* of a hypothesis test. To illustrate, suppose the value of the test statistic, *z*, for a large-sample, right-tailed hypothesis test for a population mean turns out to be 1.88. Then the *P*-value of the hypothesis test is 0.03 (actually 0.0301), as depicted by the shaded area in Fig. 9.24.

FIGURE 9.24

P-value as the observed significance level

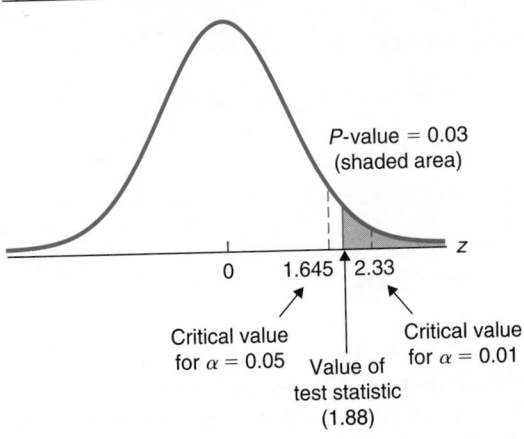

Then, as we see from Fig. 9.24, the null hypothesis would be rejected for a test at the 0.05 significance level but would not be rejected for a test at the 0.01 significance level. In fact, the figure makes it clear that the *P*-value is precisely the smallest significance level at which the null hypothesis would be rejected.

KEY FACT 9.5

P-VALUE AS THE OBSERVED SIGNIFICANCE LEVEL

The *P*-value of a hypothesis test is equal to the smallest significance level at which the null hypothesis can be rejected, that is, the smallest significance level for which the observed sample data results in rejecting H_0.

In view of Key Fact 9.5, we have the following criterion for deciding whether or not the null hypothesis should be rejected in favor of the alternative hypothesis.

KEY FACT 9.6

DECISION CRITERION FOR A HYPOTHESIS TEST USING THE *P*-VALUE

> If the *P*-value is less than or equal to the specified significance level, then reject the null hypothesis; otherwise, do not reject the null hypothesis.

Key Fact 9.6 provides us with a means to employ the *P*-value as a basis for performing a hypothesis test. A general method for doing that is presented in Procedure 9.3, which we refer to as the **P-value approach to hypothesis testing.**

PROCEDURE 9.3

TO PERFORM A HYPOTHESIS TEST USING THE *P*-VALUE APPROACH

Step 1   State the null and alternative hypotheses.

Step 2   Decide on the significance level, α.

Step 3   Compute the value of the test statistic.

Step 4   Determine the *P*-value.

Step 5   If $P \le \alpha$, then reject H_0; otherwise, do not reject H_0.

Step 6   State the conclusion in words.

In Example 9.8 on page 506, we performed a hypothesis test concerning the mean calcium intake, μ, of all people with incomes below the poverty level. To carry out that test, we used the classical approach to hypothesis testing. Now we will perform that same test using the *P*-value approach to hypothesis testing.

Example 9.14   Illustrates Procedure 9.3

A nutritionist thinks the average person with an income below the poverty level gets less than the recommended daily allowance (RDA) of 800 mg of calcium. To test her conjecture, she obtains the daily intakes of calcium for a random sample of 35 people with incomes below the poverty level. Table 9.11 displays the sample data. At the 5% significance level, do the data provide sufficient evidence to conclude that the mean calcium intake of all people with incomes below the poverty level is less than the RDA of 800 mg?

879	1096	701	986	828	1077	703
555	422	997	473	702	508	530
513	720	944	673	574	707	864
1199	743	1325	655	1043	599	1008
705	180	287	542	893	1052	473

SOLUTION   We will apply Procedure 9.3 to perform the hypothesis test.

Step 1   *State the null and alternative hypotheses.*

Let μ denote the mean calcium intake (per day) of all people with incomes below the poverty level. The null and alternative hypotheses are

$$H_0: \mu = 800 \text{ mg (mean calcium intake is not less than the RDA)}$$
$$H_a: \mu < 800 \text{ mg (mean calcium intake is less than the RDA).}$$

Note that the hypothesis test is left-tailed since a less-than sign ($<$) appears in the alternative hypothesis.

Step 2   *Decide on the significance level, α.*

We are to perform the test at the 5% significance level. Thus $\alpha = 0.05$.

Step 3   *Compute the value of the test statistic.*

Since the sample size, $n = 35$, is large, we use the test statistic

$$z = \frac{\bar{x} - \mu_0}{\sigma/\sqrt{n}}.$$

Because σ is unknown, we will use s in its place. We have $\mu_0 = 800$ mg and $n = 35$. From the data in Table 9.11, we find that $\bar{x} = 747.3$ mg and $s = 262.2$ mg. Thus the value of the test statistic is

$$z = \frac{747.3 - 800}{262.2/\sqrt{35}} = -1.19.$$

This value is shown in Fig. 9.25.

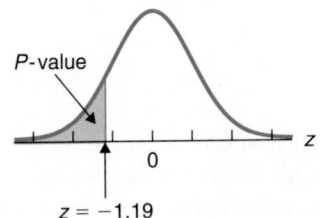

P-value

$z = -1.19$

Step 4   *Determine the P-value.*

Since the test is left-tailed, the P-value is the probability of observing a value of z of -1.19 or less, $P(z \leq -1.19)$, if the null hypothesis is true. That probability equals the shaded area in Fig. 9.25, which by Table II is 0.1170. Hence $P = 0.1170$.

Step 5   *If $P \leq \alpha$, then reject H_0; otherwise, do not reject H_0.*

From Step 4, $P = 0.1170$. Since this exceeds the specified significance level of $\alpha = 0.05$, we do not reject H_0.

Step 6   *State the conclusion in words.*

The test results are not statistically significant at the 5% level; that is, at the 5% significance level, the sample of 35 calcium intakes does not provide sufficient evidence to conclude that the mean calcium intake, μ, of all people with incomes below the poverty level is less than the RDA of 800 mg.   ■

Using the *P*-Value to Assess the Evidence Against H_0

One big advantage of the P-value is that it provides the actual significance of the hypothesis test—the smallest significance level at which the test results are statistically significant (i.e., at which the null hypothesis can be rejected). This allows us to assess significance at any level we desire. For instance, if the P-value of a hypothesis test is 0.03, then we know that the test results are statistically significant at any level larger than 0.03 (e.g., $\alpha = 0.05$) and are not statistically significant at any level smaller than 0.03 (e.g., $\alpha = 0.01$).

Knowing the actual significance of the hypothesis test also allows us to evaluate the strength of the evidence against the null hypothesis—the smaller the P-value, the stronger the evidence against the null hypothesis. Table 9.12 presents guidelines for interpreting the P-value of a hypothesis test. Keep in mind, though, that Table 9.12 is to be used only as a rule of thumb. There is no substitute for a careful analysis of the results of a hypothesis test within the context of the situation under consideration.

TABLE 9.12
Guidelines for using the
P-value to assess the evidence
against the null hypothesis

P-value	Evidence against H_0
$P > 0.10$	Weak or none
$0.05 < P \leq 0.10$	Moderate
$0.01 < P \leq 0.05$	Strong
$P \leq 0.01$	Very strong

Many researchers do not explicitly talk at all in terms of significance levels and critical values. Instead they simply obtain the P-value of the hypothesis test and use it to evaluate the strength of the evidence against the null hypothesis, as we did in Example 9.12 on page 525 and Example 9.13 on pages 526 and 527.

Using the Computer (Optional)

We can use Minitab to perform a large-sample hypothesis test for a population mean, μ. The appropriate command is called **ZTest.** To explain the details for using ZTest, we return once more to the hypothesis test concerning the mean calcium intake of all people with incomes below the poverty level.

Example 9.15   The ZTest Command

Use Minitab to perform the hypothesis test in Example 9.14 on page 528.

SOLUTION   Let μ denote the mean calcium intake (per day) of all people with incomes below the poverty level. The problem is to perform the hypothesis test,

$$H_0: \mu = 800 \text{ mg (mean calcium intake is not less than the RDA)}$$
$$H_a: \mu < 800 \text{ mg (mean calcium intake is less than the RDA),}$$

at the 5% significance level ($\alpha = 0.05$). Note that the hypothesis test is left-tailed since a less-than sign ($<$) appears in the alternative hypothesis.

First we store the sample data from Table 9.11 on page 529 in a column named CALCIUM. To apply the ZTest command, we must specify the population standard deviation, σ, which in this case is unknown. However, because the sample size is large ($n = 35$), we can use the sample standard deviation, s, in place of σ. We will employ Minitab to obtain s and store its value in K1. This is accomplished as follows.

Session commands: We type the command STDEV followed by the storage location of the sample data and the storage location for the sample standard deviation; that is, we type STDEV 'CALCIUM' put into K1 and press Enter.

(or)

Menu commands: We choose **Calc ▶ Functions and Statistics ▶ Column Statistics…**, select **Standard deviation** from the **Statistic** list, specify CALCIUM as the **Input variable**, select **Store result in** and type K1, and then select **OK**.

Now we are ready to employ the ZTest command.

Session commands: To specify whether a hypothesis test is two-tailed, left-tailed, or right-tailed, we use the **ALTERNATIVE** subcommand, setting it equal to 0, −1, and 1, respectively.[†] Table 9.13 lists the ALTERNATIVE subcommand settings. Those settings apply to any hypothesis test, not just to large-sample tests for means.

TABLE 9.13
Subcommands for specifying
an alternative hypothesis

Type of test	Subcommand
Two-tailed	ALTERNATIVE = 0
Left-tailed	ALTERNATIVE = −1
Right-tailed	ALTERNATIVE = 1

[†] Actually, a two-tailed test is the default, so we can omit the ALTERNATIVE subcommand for such a test.

To instruct Minitab to perform the required hypothesis test, we type the command **ZTEST** followed by the null hypothesis, the (estimated) value of σ, and the storage location of the sample data; that is, we type ZTEST of mu=800, sigma=K1, data in 'CALCIUM'; and press Enter. Additionally, because the test is left-tailed, we must type the subcommand ALTERNATIVE=-1., as indicated in Table 9.13.

(or)

Menu commands: We choose **Stat ▸ Basic Statistics ▸ 1-Sample Z...**, specify CALCIUM in the **Variables** text box, select **Test mean** and type 800, select **Alternative** and specify **less than**, select **Sigma** and type K1, and then select **OK**.

The results of applying either the above session commands or the above menu commands are shown in Printout 9.1.

PRINTOUT 9.1
Minitab output for the StDev and ZTest commands

```
MTB > STDEV 'CALCIUM' put into K1
   ST.DEV. =      262.23
MTB > ZTEST of mu=800, sigma=K1, data in 'CALCIUM';
SUBC> ALTERNATIVE=-1.

TEST OF MU = 800.000 VS MU L.T. 800.000
THE ASSUMED SIGMA = 262


               N     MEAN     STDEV   SE MEAN       Z    P VALUE
CALCIUM       35  747.314   262.227    44.325   -1.19      0.12
```

The output in Printout 9.1 displays a statement of the null and alternative hypotheses for the hypothesis test: TEST OF MU = 800.000 VS MU L.T. 800.000 (L.T. stands for "less than"). Next comes the value used for the population standard deviation, σ: THE ASSUMED SIGMA = 262. Then the output shows the sample size, sample mean, sample standard deviation, and standard error of the mean. The next-to-last entry, Z, gives the value of the test statistic, z. So $z = -1.19$.

The final entry shown in Printout 9.1 is P VALUE. This is really the only quantity we need in order to decide whether the null hypothesis should be rejected: If the P-value is less than or equal to the specified significance level, then we reject H_0; otherwise, we do not reject H_0.

From Printout 9.1 we see that the P-value for the hypothesis test equals 0.12. Since this exceeds the specified significance level of $\alpha = 0.05$, we do not reject H_0. The test results are not statistically significant at the 5% level; that is, at the 5% significance level, the data do not provide sufficient evidence to conclude that the mean calcium intake of all people with incomes below the poverty level is less than the RDA of 800 mg. ∎

EXERCISES 9.5

9.64 State two reasons why it is prudent to include the *P*-value when reporting the results of a hypothesis test.

In Exercises 9.65–9.70, we have given the value obtained for the test statistic

$$z = \frac{\bar{x} - \mu_0}{\sigma/\sqrt{n}}$$

in a large-sample hypothesis test concerning a population mean, μ. We have also specified whether the test is two-tailed, left-tailed, or right-tailed. Determine the P-value corresponding to each z-value.

9.65 Right-tailed test:
a. $z = 2.03$ b. $z = -0.31$

9.66 Left-tailed test:
a. $z = -1.84$ b. $z = 1.25$

9.67 Left-tailed test:
a. $z = -0.74$ b. $z = 1.16$

9.68 Two-tailed test:
a. $z = 3.08$ b. $z = -2.42$

9.69 Two-tailed test:
a. $z = -1.66$ b. $z = 0.52$

9.70 Right-tailed test:
a. $z = 1.24$ b. $z = -0.69$

In each of Exercises 9.41–9.48 of Section 9.3, you were asked to perform a hypothesis test for a population mean using Procedure 9.1, which is a classical approach to hypothesis testing. Now, in Exercises 9.71–9.78, you are asked to perform those same hypothesis tests using the P-value approach to hypothesis testing, Procedure 9.3 on page 528. In addition, use Table 9.12 on page 530 to assess the strength of the evidence against the null hypotheses.

9.71 According to the Census Bureau publication *Construction Reports,* the mean expenditure per residential property owner for maintenance and repairs in 1983 was $280. For last year, a random sample of 40 residential property owners revealed the following expenditures, in dollars.

158	110	185	136	167	84	437	420
230	57	942	287	259	123	712	170
531	347	505	148	111	442	145	38
49	188	107	134	223	36	360	253
321	821	89	202	759	151	1176	31

Do the data provide sufficient evidence to conclude that last year's mean amount spent for maintenance and repairs has increased over the 1983 mean of $280? Assume $\sigma = \$265$ and perform the hypothesis test at the 5% significance level. *(Note:* The sum of the data is $11,644.)

9.72 *The World Almanac, 1985,* reports that the mean travel time to work in 1980 for all South Dakota residents was 13 minutes. A transportation official obtained this year's travel times, in minutes, for a random sample of 35 South Dakota residents. Here are the data.

29	40	0	12	10	6	41
25	21	5	4	19	2	7
10	8	3	6	52	4	12
0	33	6	2	17	21	8
38	2	13	8	14	11	2

At the 5% significance level, do the data provide sufficient evidence to conclude that the mean travel time to work for all South Dakota residents has changed from the 1980 mean of 13 minutes? Assume $\sigma = 11.6$ minutes. *(Note:* The sum of the data is 491 minutes.)

9.73 The Food and Nutrition Board of the National Academy of Sciences states that the RDA of iron for adult females under the age of 51 is 18 mg. The following iron intakes, in milligrams, during a 24-hour period were obtained for 45 randomly selected adult females under the age of 51.

15.0	18.1	14.4	14.6	10.9	18.1	18.2	18.3	15.0
16.0	12.6	16.6	20.7	19.8	11.6	12.8	15.6	11.0
15.3	9.4	19.5	18.3	14.5	16.6	11.5	16.4	12.5
14.6	11.9	12.5	18.6	13.1	12.1	10.7	17.3	12.4
17.0	6.3	16.8	12.5	16.3	14.7	12.7	16.3	11.5

At the 1% significance level, do the data suggest that adult females under the age of 51 are, on the average, getting less than the RDA of 18 mg of iron? *(Note:* $\bar{x} = 14.68$, $s = 3.08$.)

9.74 As reported by the U.S. Office of Juvenile Justice and Delinquency Prevention in *Children in Custody,* the mean age of all juveniles held in public custody in 1989 was 16.0 years. The mean age of 250 randomly selected juveniles currently being held in public custody is 15.86 years, and the standard deviation of the ages is 1.01 years. Does it appear that the mean age, μ, of all juveniles being held in public custody this year is less than the 1989 mean of 16.0 years? Perform the appropriate hypothesis test using $\alpha = 0.10$.

9.75 A dog-food manufacturer sells "50-lb" bags of dog food. Suppose you randomly select 75 bags and find that $\bar{x} = 50.11$ lb and $s = 0.84$ lb.

a. Would you be inclined to believe that the actual mean weight, μ, of all "50-lb" bags of this dog food differs from the advertised weight of 50 lb? Perform your hypothesis test at the 5% significance level.

b. Repeat part (a) if the mean weight of the 75 bags is 50.21 lb instead of 50.11 lb.

9.76 The Health Insurance Association of America reports in *Survey of Hospital Semi-Private Room Charges* that the mean daily charge for a semi-private room in U.S. hospitals in 1988 was $253. In that same year, a random sample of 30 Massachusetts hospitals yielded a mean daily semi-private room charge of $260.68 with a standard deviation of $12.77. At the 5% significance level, do the data provide sufficient evidence to conclude that in 1988 the mean daily semi-private room charge in Massachusetts hospitals exceeded the national mean of $253?

9.77 The manufacturer of a new-model car, called the Orion, claims that a typical car gets 26 mpg. An independent consumer group is skeptical of this claim and thinks the mean gas mileage of all Orions may very well be less than 26 mpg. To try to justify its contention, the consumer group conducts mileage tests on 30 randomly selected Orions and obtains the following data.

25.3	25.1	29.6	24.6	26.0	26.0
26.3	23.6	26.0	25.4	26.1	23.8
25.1	24.1	25.8	26.4	23.4	24.8
22.6	26.6	25.1	26.6	28.0	23.3
23.8	25.4	26.2	25.1	25.3	21.5

At the 5% significance level, do the data support the consumer group's conjecture? *(Note: $\bar{x} = 25.23$, $s = 1.59$.)*

9.78 A Louisiana cotton farmer has used a certain brand of fertilizer for the past 5 years. Based on experience, the farmer knows that the mean yield of cotton using this fertilizer is 623 lb/acre. Recently, a new brand of fertilizer appeared on the market that will supposedly increase cotton yield. The farmer uses the new fertilizer on 80 of his 1-acre plots and obtains the cotton yields, in pounds, shown in the table at the top of the next column.

a. At the 10% significance level, do the data provide sufficient evidence to conclude that the new fertilizer increases the mean yield of cotton on the farmer's land? *(Note: $\bar{x} = 625.2$ lb/acre, $s = 21.5$ lb/acre.)*

639	653	631	590	628	638	618	602
622	667	614	644	611	652	598	627
637	613	591	627	637	604	620	618
621	642	654	643	634	630	621	560
637	634	645	630	605	617	598	641
636	599	616	578	620	639	608	615
587	601	629	627	626	613	568	638
599	627	630	615	620	658	629	629
626	670	642	647	637	593	617	654
627	645	646	643	648	654	651	613

b. If the new fertilizer costs more than the one the farmer presently uses, would you buy the new fertilizer if you were the farmer? Why or why not?

9.79 According to *Food Consumption, Prices, and Expenditures*, published by the U.S. Department of Agriculture, the mean consumption of beef per person in 1990 was 64 lb (boneless, trimmed weight). A random sample of 40 people taken this year yielded the following data, in pounds, on last year's beef consumptions.

77	65	57	54	68	79	56	0
50	49	51	56	56	78	63	72
0	62	74	61	61	60	56	37
76	77	67	67	62	89	56	75
69	73	75	62	8	71	20	47

a. Use the sample data to decide, at the 5% significance level, whether last year's mean beef consumption is less than the 1990 mean of 64 lb. *(Note: The mean and standard deviation of the sample data are 58.4 lb and 20.42 lb, respectively.)*

b. The above sample data contains four potential outliers: 0, 0, 8, and 20. Remove those four data values and repeat the hypothesis test in part (a). *(Note: The mean and standard deviation of the abridged sample data are 64.1 lb and 11.02 lb, respectively.)*

c. Compare your results in parts (a) and (b).

d. Assuming the four potential outliers are not recording errors, comment on the advisability of removing them from the sample data prior to performing the hypothesis test.

9.80 (Computer exercise) Refer to Exercise 9.79. Use Minitab or some other statistical software to

a. identify the possible and probable outliers.

b. perform the hypothesis test considered in part (a) of Exercise 9.79.

c. perform the hypothesis test considered in part (b) of Exercise 9.79.

d. Compare the results obtained in parts (b) and (c).

In each of Exercises 9.81–9.84, use Minitab or some other statistical software to

a. *identify possible and probable outliers, if any, for the specified data set.*

b. *perform the required hypothesis test using the unabridged sample data.*

c. *remove data values that you consider outliers, if any, and perform the required hypothesis test using the abridged sample data.*

d. *Comment on the effect that removing the outliers has on the hypothesis test.*

· **9.81 (Computer exercise)** The data set and hypothesis test in Exercise 9.71.

· **9.82 (Computer exercise)** The data set and hypothesis test in Exercise 9.72.

· **9.83 (Computer exercise)** The data set and hypothesis test in Exercise 9.73.

· **9.84 (Computer exercise)** The data set and hypothesis test in Exercise 9.78(a).

· **9.85 (Computer exercise)** The U.S. National Center for Health Statistics compiles information on the length of stay in hospitals by patients. Data are reported in *Vital and Health Statistics.* According to that publication, the mean hospital stay in 1989 was 6.5 days. A researcher thinks this year's mean will be less. She obtains the stays, in days, of a random sample of patients who were discharged from the hospital this year and applies Minitab's ZTest command to the data. Here is the abridged computer output.

```
TEST OF MU = 6.500 VS MU L.T.  6.500
THE ASSUMED SIGMA = 7.70
```

N	MEAN	STDEV	SE MEAN	Z	P VALUE
40	6.250	7.011	1.217	-0.21	0.42

Use the printout to determine the

a. null and alternative hypotheses for the researcher's hypothesis test.

b. (assumed) population standard deviation of this year's hospital stays.

c. standard deviation of the hospital stays in the sample obtained by the researcher.

d. number of patients sampled by the researcher.

e. mean hospital stay of the patients sampled.

f. value obtained for the test statistic, z.

g. *P*-value of the hypothesis test.

h. smallest significance level at which the null hypothesis can be rejected.

i. conclusion if the test is performed using $\alpha = 0.05$.

· **9.86 (Computer exercise)** A few years ago, the owner of a menswear store decided to advertise in local papers in an attempt to improve sales. His records showed that without advertising, average weekly sales had been $1700. Since he does not wish to continue spending money on advertising if sales have not increased, he decides to perform a hypothesis test. The owner determines the weekly sales for a random sample of weeks in which advertising was used and obtains the following abridged Minitab output.

```
TEST OF MU = 1700.000 VS MU G.T.  1700.000
THE ASSUMED SIGMA = 250
```

N	MEAN	STDEV	SE MEAN	Z	P VALUE
32	1801.469	283.586	44.194	2.30	0.011

Use the printout to determine the

a. null and alternative hypotheses for the test.

b. (assumed) population standard deviation of weekly sales with advertising.

c. standard deviation of the weekly sales in the sample obtained by the store owner.

d. number of weeks sampled by the store owner.

e. mean sales of the weeks sampled.

f. value obtained for the test statistic, z.

g. *P*-value of the hypothesis test.

h. conclusion if the test is performed with $\alpha = 0.05$.

· · **9.87** Suppose for a large-sample hypothesis test concerning a population mean, μ, the observed value of the test statistic, z, is z_0. If the test is right-tailed, then the *P*-value of the hypothesis test can be expressed as $P(z \geq z_0)$. Determine the corresponding expression for the *P*-value if the test is
a. left-tailed.   b. two-tailed.

· · · **9.88** The symbol $\Phi(z)$ is often used to denote the area under the standard normal curve that lies to the left of a specified value of z. Suppose for a large-sample hypothesis test concerning a population mean, μ, the observed value of the test statistic is z_0. Express the *P*-value of the hypothesis test in terms of Φ if the test is
a. left-tailed.   b. right-tailed.   c. two-tailed.

· · · **9.89 Obtaining the *P*-value.** This exercise discusses the general procedure for obtaining the *P*-value of a hypothesis test. Let x denote the test statistic for a hypothesis test and x_0 its observed value. Then the *P*-value of the hypothesis test equals

- $P(x \geq x_0)$ for a right-tailed test,
- $P(x \leq x_0)$ for a left-tailed test,
- $2 \cdot \min \{ P(x \leq x_0), P(x \geq x_0) \}$ for a two-tailed test,

where the probabilities are computed under the condition that the null hypothesis is true. Suppose we are considering a large-sample hypothesis test for a population mean using the test statistic z. Verify that the probability expressions listed at the left are equivalent to those obtained in Exercise 9.87.

9.6   HYPOTHESIS TESTS FOR ONE NORMAL POPULATION MEAN

In Section 9.3 we learned how to perform a hypothesis test for a population mean, μ, when the sample size is large. However, as we have mentioned, large samples are often unavailable, extremely expensive, or undesirable.

Several methods exist for performing a hypothesis test for a population mean that do not require a large sample. One such method applies when the population being sampled is *normally distributed*. That is the method we will study in this section. We will assume the population standard deviation, σ, is unknown, since that is usually the case in practice.[†]

To develop a hypothesis-testing procedure for a normal population mean, we begin by recalling Key Fact 8.3: Suppose a random sample of size n is to be taken from a normally distributed population with mean μ. Then the random variable

$$t = \frac{\bar{x} - \mu}{s/\sqrt{n}}$$

has the t-distribution with $n - 1$ degrees of freedom. In other words, probabilities for that random variable are equal to areas under the t-curve with df $= n - 1$.

Thus, when the population being sampled is normally distributed, we can perform a hypothesis test with null hypothesis H_0: $\mu = \mu_0$ by employing the random variable

$$t = \frac{\bar{x} - \mu_0}{s/\sqrt{n}}$$

as our test statistic and using the t-table, Table IV, to obtain the critical value(s). Specifically, we have the following procedure, which we often refer to as the **one-sample t-test** or more briefly as the **t-test.**

PROCEDURE 9.4   **THE ONE-SAMPLE t-TEST FOR A POPULATION MEAN WITH NULL HYPOTHESIS H_0: $\mu = \mu_0$**

ASSUMPTION

Normal population

Step 1   State the null and alternative hypotheses.

Step 2   Decide on the significance level, α.

[†] A discussion of hypothesis tests for a normal population mean when σ is known is presented in the exercises (see page 545).

Step 3 The critical value(s)

   a. for a two-tailed test are $\pm t_{\alpha/2}$,

   b. for a left-tailed test is $-t_\alpha$,

   c. for a right-tailed test is t_α,

   with df $= n - 1$. Use Table IV to find the critical value(s).

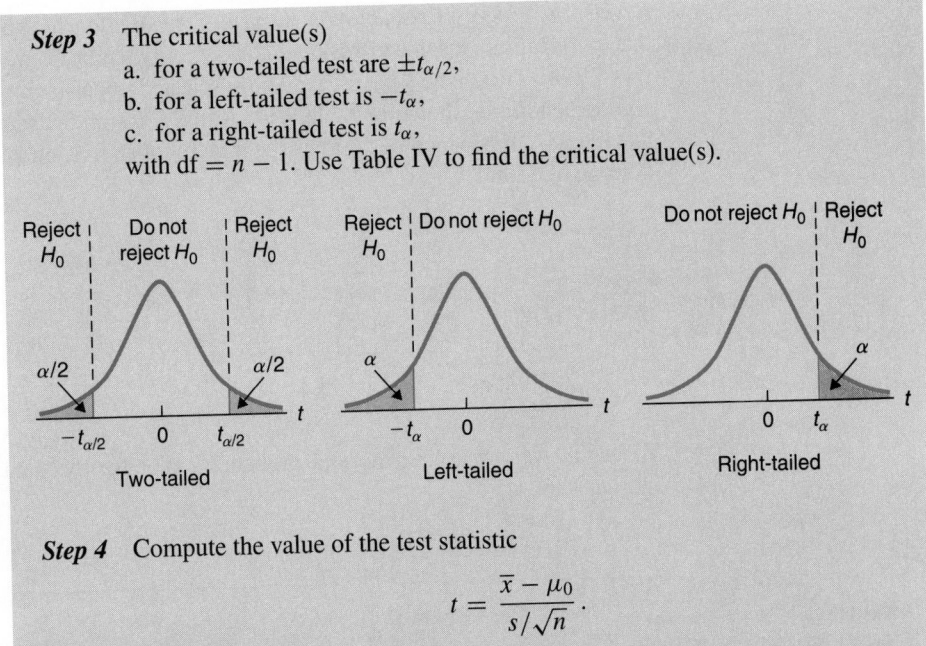

Step 4 Compute the value of the test statistic

$$t = \frac{\bar{x} - \mu_0}{s/\sqrt{n}}.$$

Step 5 If the value of the test statistic falls in the rejection region, then reject H_0; otherwise, do not reject H_0.

Step 6 State the conclusion in words.

As with the *t*-interval procedure (Procedure 8.2 on page 465), the *t*-test applies for any sample size, large or small. The only condition for using it is that the population being sampled is normally distributed; and in practice the *t*-test works reasonably well even if the population being sampled is only approximately normally distributed—it is robust to moderate deviations of the normality assumption.

Because of robustness, we need only be able to detect gross violations of the normality assumption. This we can usually do quite effectively with a normal probability plot: *Unless a normal probability plot for the sample reveals outliers or shows systematic and substantial deviations from linearity, we can apply the t-test to carry out a hypothesis test for a population mean.*

Additional guidelines for when to use the *t*-test can be found in Key Fact 8.5 on page 466. Although this key fact refers explicitly to the *t*-interval procedure (Procedure 8.2), it applies equally well to the *t*-test.

Example 9.16   Illustrates Procedure 9.4

The U.S. Energy Information Administration surveys households to obtain data on residential energy consumption and expenditures. Results of the surveys can be found in

Residential Energy Consumption Survey: Consumption and Expenditures. According to that publication, the mean residential energy expenditure of all U.S. families in 1987 was $1080. That same year, 15 randomly selected upper-income families reported the energy expenditures shown in Table 9.14. At the 5% significance level, do the data indicate that in 1987, upper-income families spent more for energy, on the average, than the national average of $1080?

TABLE 9.14

Energy expenditures ($) for 15 upper-income families

1211	1307	1184	1111	1747
1572	1478	865	1188	1326
1668	1250	1162	1308	1142

SOLUTION   To begin, we construct a normal probability plot for the data in Table 9.14, as shown in Fig. 9.26.

FIGURE 9.26

Normal probability plot for the sample of upper-income-family energy expenditures in Table 9.14

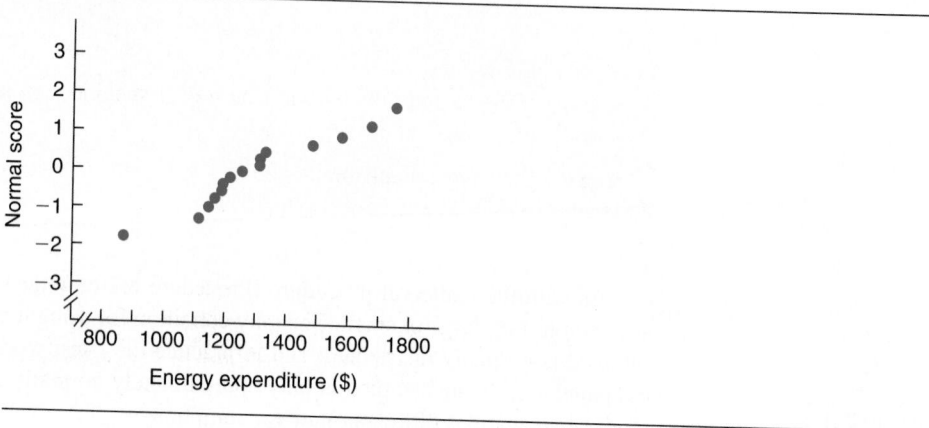

The normal probability plot in Fig. 9.26 reveals no outliers (although some might consider 865 a mild outlier). Whether the plot shows systematic deviations from linearity is a tough call; but the departure of the plot from linearity is probably not substantial enough to prevent the use of the *t*-test. So we'll go ahead and apply Procedure 9.4.

Step 1   *State the null and alternative hypotheses.*

Let μ denote the mean energy expenditure of all upper-income families in 1987. Then the null and alternative hypotheses are

H_0: $\mu = \$1080$ (mean was not greater than the national mean)

H_a: $\mu > \$1080$ (mean was greater than the national mean).

Note that the hypothesis test is right-tailed since a greater-than sign (>) appears in the alternative hypothesis.

Step 2   *Decide on the significance level, α.*

We are to perform the hypothesis test at the 5% significance level. Thus $\alpha = 0.05$.

Step 3   *The critical value for a right-tailed test is t_α, with $df = n - 1$.*

Here $n = 15$ and $\alpha = 0.05$. Table IV shows that for $df = 15 - 1 = 14$, $t_{0.05} = 1.761$, as seen in Fig. 9.27.

FIGURE 9.27

Criterion for deciding whether or not to reject the null hypothesis

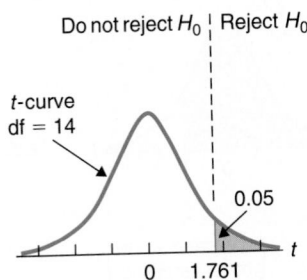

Step 4   *Compute the value of the test statistic*

$$t = \frac{\bar{x} - \mu_0}{s/\sqrt{n}}.$$

We have $\mu_0 = \$1080$ and $n = 15$. Furthermore, the mean and standard deviation of the sample data in Table 9.14 are \$1301.27 and \$231.00, respectively. Consequently, the value of the test statistic is

$$t = \frac{1301.27 - 1080}{231.00/\sqrt{15}} = 3.710.$$

Step 5   *If the value of the test statistic falls in the rejection region, reject H_0; otherwise, do not reject H_0.*

The value of the test statistic, found in Step 4, is $t = 3.710$. As we see from Fig. 9.27, this falls in the rejection region. Hence we reject H_0.

Step 6   *State the conclusion in words.*

The test results are statistically significant at the 5% level; that is, at the 5% significance level, the sample data provide sufficient evidence to conclude that in 1987, upper-income families spent more for energy, on the average, than the national average of \$1080.   ■

P-Values for a t-Test

We can also use the P-value approach to hypothesis testing to carry out a t-test. P-values for a t-test are obtained in a manner similar to that for a z-test.

If we let t_0 denote the observed value of the test statistic, t, then the P-values for two-tailed, left-tailed, and right-tailed tests are given, respectively, by $P(|t| \geq |t_0|)$, $P(t \leq t_0)$, and $P(t \geq t_0)$. In terms of areas, this means that for a two-tailed test, the P-value is the area under the t-curve that lies either to the left of $-|t_0|$ or to the right of $|t_0|$, as in Fig. 9.28(a); for a left-tailed test, the P-value is the area that lies to the left of t_0, as in Fig. 9.28(b); and for a right-tailed test, the P-value is the area that lies to the right of t_0, as in Fig. 9.28(c).

FIGURE 9.28
P-value for a t-test if the test is
(a) two-tailed,
(b) left-tailed,
(c) right-tailed

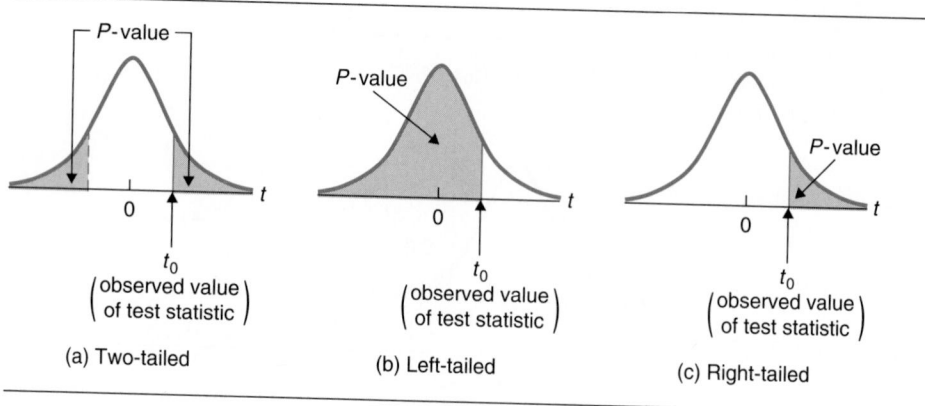

(a) Two-tailed   (b) Left-tailed   (c) Right-tailed

To obtain the exact P-value of a t-test, we need to use statistical software or a sophisticated calculator. However, we can use t-tables, such as Table IV, to estimate the P-value of a t-test; and an estimate of the P-value is often sufficient for deciding whether or not to reject the null hypothesis.

For instance, in the right-tailed t-test of Example 9.16, $\alpha = 0.05$, df $= 14$, and the value of the test statistic is $t = 3.710$. For df $= 14$, 3.710 is larger than any t-value in Table IV, where the largest value is $t_{0.005} = 2.977$ (which means that the area under the t-curve that lies to the right of 2.977 equals 0.005). This, in turn, implies that the area to the right of 3.710 is less than 0.005; in other words, for the t-test in Example 9.16, $P < 0.005$. Because the P-value is less than the designated significance level of 0.05, we can reject H_0.

Example 9.17 provides two more illustrations of how Table IV can be used to estimate the P-value of a t-test.

Example 9.17   Using Table IV to Estimate the P-Value of a t-Test

Use Table IV to estimate the P-value of each of the following t-tests.

a. Left-tailed test, $n = 12$, $t = -1.938$
b. Two-tailed test, $n = 25$, $t = -0.895$

SOLUTION   a. Because the test is left-tailed, the P-value is the area under the t-curve with df $= 12 - 1 = 11$ that lies to the left of -1.938, as seen in Fig. 9.29(a).

FIGURE 9.29

Estimating the P-value of a left-tailed t-test with sample size 12 and test statistic $t = -1.938$

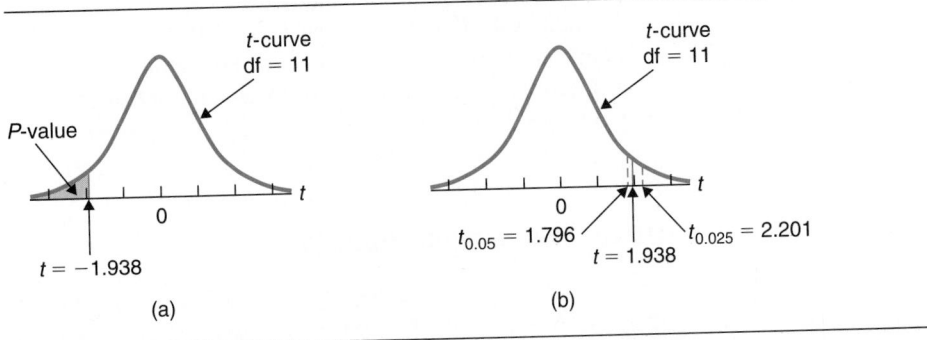

Since a t-curve is symmetric about 0, the area to the left of -1.938 equals the area to the right of 1.938; and we can use Table IV to estimate this latter area. Concentrating on the df $= 11$ row of Table IV, we search for the two t-values that straddle 1.938; they are $t_{0.05} = 1.796$ and $t_{0.025} = 2.201$. This implies that the area under the t-curve that lies to the right of 1.938 is somewhere between 0.025 and 0.05, as seen in Fig. 9.29(b).

Consequently, the area under the t-curve that lies to the left of -1.938 is also somewhere between 0.025 and 0.05; that is, $0.025 < P < 0.05$. So we can reject H_0 at any significance level of 0.05 or larger, and we cannot reject H_0 at any significance level of 0.025 or smaller. For significance levels between 0.025 and 0.05, Table IV is not sufficiently detailed for us to decide whether or not H_0 should be rejected.

b. Because the test is two-tailed, the P-value is the area under the t-curve with df $= 25 - 1 = 24$ that lies either to the left of -0.895 or to the right of 0.895, as seen in Fig. 9.30(a).

FIGURE 9.30

Estimating the P-value of a two-tailed t-test with sample size 25 and test statistic $t = -0.895$

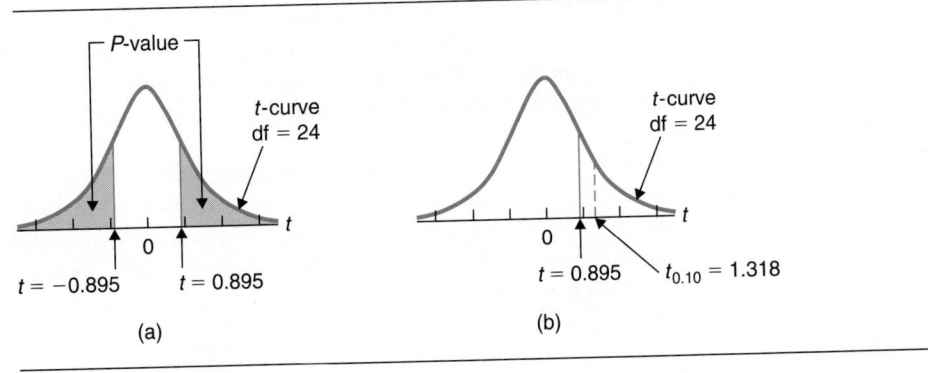

Since a t-curve is symmetric about 0, the area to the left of -0.895 and the area to the right of 0.895 are equal. Concentrating on the df $= 24$ row of Table IV, we find that 0.895 is smaller than any t-value in Table IV, the smallest being $t_{0.10} = 1.318$. This implies that the area under the t-curve that lies to the right of 0.895 is greater than 0.10, as seen in Fig. 9.30(b).

Consequently, the area under the t-curve that lies either to the left of -0.895 or to the right of 0.895 is greater than 0.20; that is, $P > 0.20$. So we cannot reject H_0 at any significance level of 0.20 or smaller. For significance levels larger than 0.20, Table IV is not sufficiently detailed for us to decide whether or not H_0 should be rejected.   ■

Using the Computer (Optional)

Procedure 9.4 provides a step-by-step method for performing a t-test for a population mean, μ. Minitab can also be used to carry out a t-test. Not surprisingly, the appropriate command is called **TTest**. Example 9.18 explains in detail how to apply TTest.

Example 9.18   The TTest Command

Use Minitab to perform the hypothesis test considered in Example 9.16.

SOLUTION   Let μ denote the mean energy expenditure of all upper-income families in 1987. The problem is to perform the hypothesis test

$$H_0: \mu = \$1080 \text{ (mean was not greater than the national mean)}$$
$$H_a: \mu > \$1080 \text{ (mean was greater than the national mean)}$$

at the 5% significance level ($\alpha = 0.05$). Note that the hypothesis test is right-tailed since a greater-than sign ($>$) appears in the alternative hypothesis.

First we store the sample data from Table 9.14 on page 538 in a column named ENERGY$. Next we use the method described on page 385 to have Minitab produce a normal probability plot of the data. This plot is displayed in Printout 9.2.

The plot contains no features that are sufficient to preclude the use of a t-test. So we'll apply TTest to carry out the hypothesis test.

Session commands: We type the command **TTEST** followed by the null hypothesis and the storage location of the sample data; in other words, we type TTEST of mu=1080, data in 'ENERGY$'; and press Enter. Because the test is right-tailed, we also type the subcommand ALTERNATIVE=1. and press Enter. These commands and the output obtained are displayed in Printout 9.3.

(or)

Menu commands: We choose **Stat ▸ Basic Statistics ▸ 1-Sample t...**, specify ENERGY$ in the **Variables** text box, select **Test mean** and type 1080, select **Alternative** and specify **greater than**, and then select **OK**. Printout 9.3 displays the output that results.

PRINTOUT 9.2
Minitab normal probability plot
for the energy-expenditure
data in Table 9.14

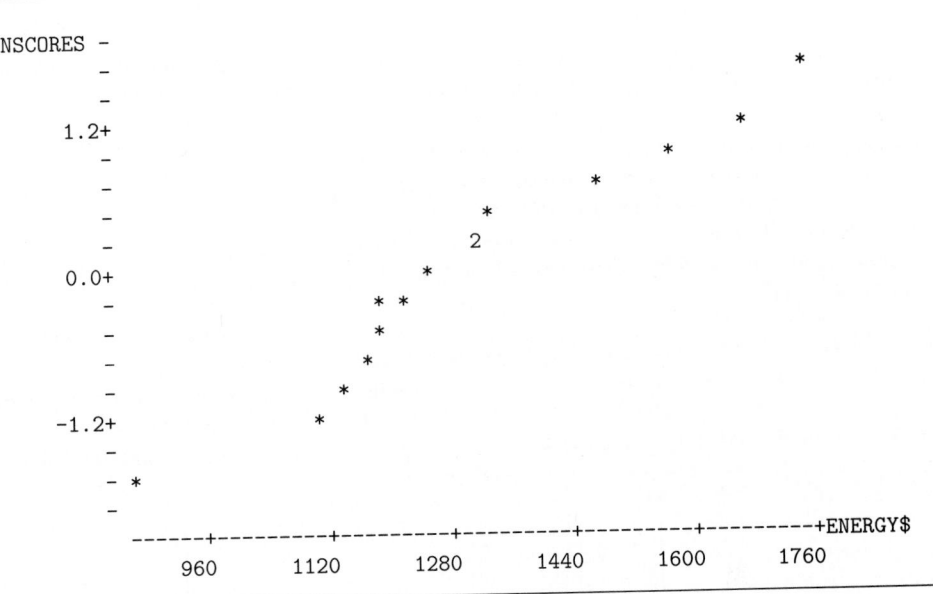

PRINTOUT 9.3
Minitab output for
the TTest command

```
MTB > TTEST of mu=1080, data in 'ENERGY$';
SUBC> ALTERNATIVE=1.

TEST OF MU = 1080.000 VS MU G.T. 1080.000

                N      MEAN     STDEV    SE MEAN      T    P VALUE
ENERGY$        15  1301.267   230.998    59.643    3.71    0.0012
```

On the third line of Printout 9.3, we find a statement of the null and alternative hypotheses: TEST OF MU = 1080.000 VS MU G.T. 1080.000 (G.T. stands for "greater than"). Then we find the sample size, sample mean, sample standard deviation, and estimated standard error of the mean. The next-to-last entry, T, shows the value of the test statistic, t. Hence we see that $t = 3.71$.

The final entry in Printout 9.3 displays the P-value of the hypothesis test, $P = 0.0012$. Since this is less than the specified significance level of $\alpha = 0.05$, we reject H_0. The test results are statistically significant at the 5% level; in other words, at the 5% significance level, the data provide sufficient evidence to conclude that in 1987, upper-income families spent more for energy, on the average, than the national average of $1080. ∎

EXERCISES 9.6

9.90 Is there any restriction on sample size for using the one-sample t-test, Procedure 9.4? Explain your answer.

Normal probability plots indicate that it is reasonable to use a t-test to carry out each of the hypothesis tests required in Exercises 9.91–9.96. Perform each t-test using either the classical approach or the P-value approach. Comment on the practical significance of all hypothesis tests whose results are statistically significant.

9.91 A paint manufacturer claims that the average drying time for its new latex paint is 2 hours. To test that claim, the drying times are obtained for 20 randomly selected cans of paint. Here are the drying times, in minutes.

123	109	115	121	130
127	106	120	116	136
131	128	139	110	133
122	133	119	135	109

Do the data provide sufficient evidence to conclude that the mean drying time is greater than the manufacturer's claim of 120 minutes? Use $\alpha = 0.05$. *(Note: The sample mean and sample standard deviation of the data are 123.1 minutes and 10.0 minutes, respectively.)*

9.92 The U.S. Energy Information Administration compiles data on household motor fuel expenditures and publishes the results in *Residential Transportation Energy Consumption Survey, Consumption Patterns of Household Vehicles*. According to that document, the mean annual motor fuel expenditure per U.S. household in 1988 was $998. That same year, 16 households within metropolitan areas had the following annual motor fuel expenditures, in dollars.

1071	1140	1724	1232
96	1040	1459	1218
848	1397	1144	585
241	1273	1391	319

At the 5% significance level, do the data provide sufficient evidence to conclude that the 1988 mean annual fuel expenditure for households within metropolitan areas differed from the national mean of $998? *(Note: The sample mean and sample standard deviation of the data are $1011.13 and $470.36.)*

9.93 A battery retailer has received a large shipment of automobile batteries from a supplier. The supplier claims that the batteries have a mean life of 36 months. Ten batteries, randomly sampled from the shipment, yielded the following lifetimes, in months.

27.6	28.7	34.7	29.0	22.9
29.6	29.4	30.2	36.5	34.7

Do the data indicate that the mean life of the supplier's batteries is less than the claimed 36 months? Perform the hypothesis test at the 1% significance level. *(Note: $\bar{x} = 30.33$, $s = 4.01$.)*

9.94 As reported by the College Entrance Examination Board in *National College-Bound Senior*, the mean verbal score on the Scholastic Aptitude Test in 1991 was 422 points out of a possible 800. A random sample of 25 verbal scores for last year yielded the following data.

338	488	344	370	307
483	352	377	492	550
373	295	426	554	488
412	477	438	471	414
486	281	428	508	607

At the 10% significance level, does it appear that last year's mean for verbal SAT scores is greater than the 1991 mean of 422 points? *(Note: $\bar{x} = 430.4$, $s = 85.5$.)*

9.95 According to *Food Cost Review*, published by the U.S. Department of Agriculture, the average retail price for oranges in 1983 was 38.5 cents per pound. Recently, 15 randomly selected markets reported the following prices for oranges, in cents per pound.

43.0	40.0	42.6	40.2	37.5
44.1	45.2	41.8	35.6	34.6
37.9	44.2	44.5	38.2	42.4

Can we conclude that the mean retail price for oranges now is different from the 1983 mean of 38.5 cents per pound? Use $\alpha = 0.05$. *(Note: $\bar{x} = 40.79$, $s = 3.38$.)*

9.96 Atlas Fishing Line produces a 10-lb test line. Twelve randomly selected spools are subjected to tensile-strength tests. The results follow.

9.8	10.2	9.8	9.4
9.7	9.7	10.1	10.1
9.8	9.6	9.1	9.7

Use the data to decide whether Atlas Fishing Line's 10-lb test line is not up to specifications. Perform the required hypothesis test at the 5% significance level. *(Note: $\bar{x} = 9.75$, $s = 0.31$.)*

9.97 (Computer exercise) Refer to Exercise 9.91. Use Minitab or some other statistical software to
a. obtain a normal probability plot of the data.
b. perform the required hypothesis test.
c. Justify the use of your procedure in part (b).

9.98 (Computer exercise) Refer to Exercise 9.92. Use Minitab or some other statistical software to
a. obtain a normal probability plot of the data.
b. perform the required hypothesis test.
c. Justify the use of your procedure in part (b).

9.99 (Computer exercise) An automobile manufacturer is experimenting with a new bumper that could reduce repair costs resulting from front-end collisions at low speeds. Experience with the presently used bumper indicates that at 10 mph the mean cost of repair resulting from a front-end collision is $550. The manufacturer equips a sample of cars with the new bumper and has these cars undergo front-end collisions at 10 mph. After the collisions the cars are repaired and the repair costs are recorded. The following Minitab printout was obtained by applying the TTest command to the repair-cost data.

```
TEST OF MU = 550.000 VS MU L.T. 550.000

 N    MEAN    STDEV   SE MEAN     T   P VALUE
15  496.667  68.624  17.719   -3.01   0.0047
```

Using the printout, determine the
a. null and alternative hypotheses for the manufacturer's hypothesis test.
b. sample standard deviation of the repair costs.
c. number of cars in the sample.
d. sample mean repair cost.
e. value obtained for the test statistic, t.
f. P-value of the hypothesis test.
g. smallest significance level at which the null hypothesis can be rejected.
h. conclusion if the hypothesis test is performed at the 5% significance level.

9.100 (Computer exercise) A company produces cans of stewed tomatoes with an advertised weight of 14 oz. Recently, the company hired a quality-control engineer. The engineer wants to check whether, on the average, the cans do contain 14 oz of stewed tomatoes. For an initial test, she takes a random sample of cans of stewed tomatoes, finds their net weights, and obtains the following Minitab printout by applying the TTest command.

```
TEST OF MU = 14.0000 VS MU N.E. 14.0000

 N    MEAN    STDEV   SE MEAN     T   P VALUE
20  13.9505  0.3242   0.0725  -0.68    0.50
```

From the printout, determine the
a. null and alternative hypotheses for the quality-control engineer's hypothesis test.
b. standard deviation of the net weights of the cans sampled by the quality-control engineer.
c. number of cans sampled.
d. mean net weight of the cans sampled.
e. value obtained for the test statistic, t.
f. P-value of the hypothesis test.
g. smallest significance level at which the null hypothesis can be rejected.
h. conclusion if the hypothesis test is performed at the 5% significance level.

9.101 A manufacturer of light bulbs produces a 60-watt bulb having a mean life of 1000 hours. The research and development (R&D) department has developed a new bulb that it claims will, on the average, outlast the present bulb. To try to justify its claim, R&D tests 10 new bulbs. The results show that the 10 bulbs tested have a mean life of 1050.2 hours and a standard deviation of 65.8 hours.
a. At the 1% significance level, do the data obtained by R&D support its claim? Assume bulb life is normally distributed.
b. Suppose in part (a) you had mistakenly concluded that the test statistic

$$\frac{\bar{x} - 1000}{s/\sqrt{10}}$$

has the standard normal distribution.
  i. What critical value would you have used?
  ii. What critical value did you actually use?
  iii. In general, does the mistaken use of a z critical value when a t critical value should have been used make it more or less likely that the null hypothesis will be rejected? Explain your answer.

Hypothesis tests for a normal population mean when σ is known. Suppose a random sample of size n is to be taken from a normally distributed population with mean μ and standard deviation σ. Then by Key Fact 7.4 on page 420 (the sampling distribution of the mean for normal populations), the random variable $\bar{x}$ is normally distributed with mean $\mu_{\bar{x}} = \mu$ and standard deviation $\sigma_{\bar{x}} = \sigma/\sqrt{n}$. Or, equivalently, the standardized random variable

$$z = \frac{\bar{x} - \mu}{\sigma/\sqrt{n}}$$

has the standard normal distribution. Consequently, if the population being sampled is normally distributed and σ is known, then we can perform a hypothesis test with null hypothesis H_0: $\mu = \mu_0$ by employing the random variable

(4) $$z = \frac{\bar{x} - \mu_0}{\sigma/\sqrt{n}}$$

as the test statistic and using the standard-normal table, Table II, to obtain the critical value(s). We will apply this fact in Exercises 9.102–9.106. Note that Exercise 9.102 is required for the other four exercises.

· ·   **9.102**   Suppose that we want to perform a hypothesis test for a normal population mean with null hypothesis H_0: $\mu = \mu_0$, and that the population standard deviation, σ, is known. Formulate a step-by-step procedure for the hypothesis test that uses the test statistic in Equation (4).

· ·   **9.103**   A consumer advocacy group suspects that Wheat Flakes cereal contains, on the average, less than the advertised weight of 15 oz per box. The following weights, in ounces, were obtained from a random sample of 40 boxes.

15.8	15.1	15.2	15.4	14.8	15.6	15.7	14.5
14.8	15.4	15.3	15.5	15.2	14.6	15.4	15.4
15.5	14.7	14.7	15.1	14.7	15.3	15.3	15.5
14.0	14.2	14.6	15.0	15.1	14.9	14.9	15.8
15.0	14.4	15.4	14.3	15.4	15.9	15.2	15.6

A normal probability plot for this sample data shows no outliers and is roughly linear and so, in particular, does not suggest that the population is far from being normally distributed. Assuming $\sigma = 0.5$ oz, use your procedure from Exercise 9.102 to decide whether the data provide sufficient evidence to conclude that the consumer group's conjecture is correct. Take $\alpha = 0.05$. *(Note: $\Sigma x = 604.2$ oz.)*

· ·   **9.104**   When performing a hypothesis test for the population mean, μ, of a normally distributed population, why can't we always use the random variable in Equation (4) as the test statistic?

· ·   **9.105**   Brown Swiss Dairy sells "half-gallon" cartons of milk. The contents of the cartons are known to be normally distributed with a standard deviation of 1.1 fluid oz. Suppose 15 randomly selected cartons have a mean content of 64.48 fluid oz.

a. Do the data provide sufficient evidence to infer that the cartons actually contain more milk, on the average, than 64 fluid oz? Perform the appropriate hypothesis test at the 0.05 level of significance using your procedure from Exercise 9.102.

b. Suppose in part (a) you had incorrectly concluded that the test statistic

$$\frac{\bar{x} - 64}{1.1/\sqrt{15}}$$

has the t-distribution with df $= 14$.
   i. What critical value would you have used?
   ii. What critical value did you actually use?
   iii. In general, does the mistaken use of a t critical value when a z critical value should have been used make it more or less likely that the null hypothesis will be rejected? Why?

· · ·   **9.106**   Refer to Exercise 9.105.

a. Suppose you mistakenly use the t-table (with df $= 14$) instead of the standard-normal table to obtain the critical value for the hypothesis test in part (a) of Exercise 9.105. What will be the significance level of the resulting test? Compare this with the desired significance level of 0.05.

b. More generally, suppose you are performing a hypothesis test at the significance level α for the mean of a normally distributed population. Further suppose the population standard deviation is known so that the appropriate hypothesis-testing procedure is the z-test of Exercise 9.102. If you mistakenly use the t-table instead of the standard-normal table to obtain the critical value(s), will the actual significance level of the resulting test be higher or lower than α? Justify your answer.

· · ·   **9.107**   Suppose you are performing a hypothesis test at the significance level α for the mean of a normally distributed population. Also suppose the population standard deviation is unknown so that the appropriate hypothesis-testing procedure is the t-test, Procedure 9.4 on page 536. If you mistakenly use the standard-normal table instead of the t-table to obtain the critical value(s), will the actual significance level of the resulting test be higher or lower than α? Why?

9.7 THE WILCOXON SIGNED-RANK TEST (OPTIONAL)

Up to this point, we have learned two methods for performing a hypothesis test for a population mean. One is the t-test, Procedure 9.4 on page 536; the other is the z-test, Procedure 9.1 on page 503. We can use the former procedure when the population being sampled is approximately normally distributed, the latter when the sample size is large.

In this section we will study a third method for performing a hypothesis test for a population mean—the **Wilcoxon signed-rank test.** This test, which is sometimes more appropriate than either the t-test or the z-test, is an example of a nonparametric method.

What Is a Nonparametric Method?

Recall that descriptive measures for a population, such as μ and σ, are called parameters. Technically, inferential methods concerned with parameters are called **parametric methods;** those that are not are called **nonparametric methods.** However, it has become common practice to refer to most methods that can be applied without assuming normality as nonparametric. Thus the term *nonparametric method* as used in contemporary statistics is somewhat of a misnomer.

Nonparametric methods have advantages beyond not requiring normality. They usually entail fewer and simpler computations than parametric methods and are resistant to outliers and other extreme values. On the other hand, parametric methods tend to give more accurate results (e.g., are more powerful) when the requirements for their use are met.

The Logic Behind the Wilcoxon Signed-Rank Test

The Wilcoxon signed-rank test assumes that the population under consideration has a symmetric distribution, but does not require that it be normal or have any other specific shape.[†] Thus, for example, the Wilcoxon signed-rank test applies to normal, triangular, uniform, and symmetric bimodal populations, but not to right-skewed or left-skewed populations.

Example 9.19 introduces and explains the reasoning behind the Wilcoxon signed-rank test. Following that example, we will present a step-by-step procedure for performing such a test.

Example 9.19   Introduces the Wilcoxon Signed-Rank Test

In January 1984 the Department of Agriculture estimated that a typical U.S. family of four with an intermediate budget would spend $92 per week for food. A consumer researcher in Kansas suspected that the mean weekly cost was less in her state. She took a random sample of 10 Kansas families of four with intermediate budgets and obtained the weekly food costs shown in Table 9.15. Do the data provide sufficient evidence to conclude that

---

[†] Recall that a distribution is symmetric if it can be divided into two pieces that are mirror images of each other.

in 1984 the mean weekly food cost for Kansas families of four with intermediate budgets was less than the national mean of $92?

TABLE 9.15
Sample of weekly food costs

$78	104	84	70	96
73	87	85	76	94

SOLUTION   Let μ denote the 1984 mean weekly food cost for all Kansas families of four with intermediate budgets. Then we want to perform the hypothesis test

$$H_0: \mu = \$92 \text{ (mean weekly food cost is not less than \$92)}$$

$$H_a: \mu < \$92 \text{ (mean weekly food cost is less than \$92).}$$

As we said, the Wilcoxon signed-rank test presumes that the population being sampled has a symmetric distribution. If the population of weekly food costs has a symmetric distribution, then a graphical display of the sample data should be roughly symmetric.

Figure 9.31 shows a stem-and-leaf diagram of the sample data in Table 9.15. The diagram is roughly symmetric and so does not reveal any obvious violations of the symmetry condition.[†] We will therefore apply the Wilcoxon signed-rank test to carry out the hypothesis test.

FIGURE 9.31
Stem-and-leaf diagram of sample data in Table 9.15

```
 7 | 3 0
 7 | 8 6
 8 | 4
 8 | 7 5
 9 | 4
 9 | 6
10 | 4
```

To begin, we rank the data in Table 9.15 according to distance and direction from the null hypothesis mean, $\mu_0 = \$92$. The steps for doing this are depicted in Table 9.16.

The absolute differences, $|D|$, displayed in the third column of Table 9.16 identify how far each data value is from $92; the ranks of those absolute differences, displayed in the fourth column, show which data values are closer to $92 and which are farther away; and the signed ranks, R, displayed in the last column, indicate additionally whether a data value is greater than $92 (+) or less than $92 (−). Figure 9.32 depicts this information for the second and third rows of Table 9.16.

[†] For ease in explaining the Wilcoxon signed-rank test, we have chosen an example in which the sample size is very small. This, however, makes it difficult to effectively check the symmetry condition. In general, we must proceed cautiously when dealing with very small samples.

TABLE 9.16

Steps for ranking the data
in Table 9.15 according to
distance and direction from
the null hypothesis mean

Cost ($) x	Difference $D = x - 92$	$\|D\|$	Rank of $\|D\|$	Signed rank R
78	−14	14	7	−7
73	−19	19	9	−9
104	12	12	6	6
87	−5	5	3	−3
84	−8	8	5	−5
85	−7	7	4	−4
70	−22	22	10	−10
76	−16	16	8	−8
96	4	4	2	2
94	2	2	1	1

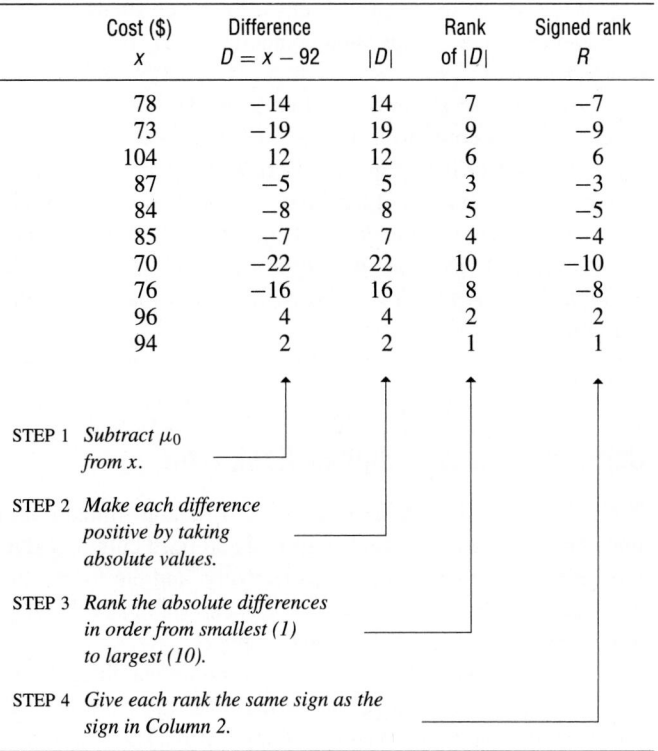

STEP 1   *Subtract μ_0*
from x.

STEP 2   *Make each difference*
positive by taking
absolute values.

STEP 3   *Rank the absolute differences*
in order from smallest (1)
to largest (10).

STEP 4   *Give each rank the same sign as the*
sign in Column 2.

FIGURE 9.32

Meaning of signed ranks
for two data values

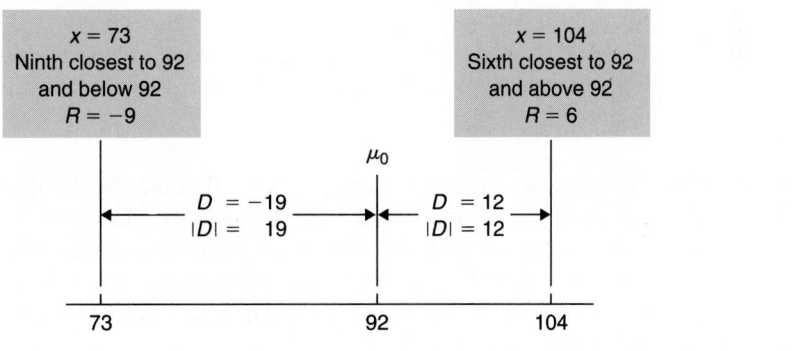

The reasoning behind the Wilcoxon signed-rank test is this: If the null hypothesis, $\mu = \$92$, is true, then because the distribution of weekly food costs is symmetric, we would expect the sum of the positive ranks and the sum of the negative ranks to be roughly the same in magnitude. Since the sample size here is 10, the sum of all the ranks must be $1 + 2 + \cdots + 10 = 55$; and half of 55 is 27.5. So if H_0 is true, we would expect the sum of the positive ranks (and the sum of the negative ranks) to be roughly 27.5. To put it another way, if the sum of the positive ranks is too much smaller than 27.5, then we would take this as evidence that H_0 is false and conclude that H_a is true, that is, that the mean weekly food cost is less than \$92.

From the last column of Table 9.16, we see that the sum of the positive ranks, denoted by W, is equal to $6 + 2 + 1 = 9$. This is quite a bit smaller than 27.5, the value we would expect if the mean is in fact \$92. Our question now is whether the difference between the observed and expected values of W can reasonably be attributed to sampling error or whether it indicates that the mean weekly food cost is actually less than \$92. To answer that question, we need a table of critical values for W, which we present as Table VI in Appendix A. We will discuss that table and then return to complete the hypothesis test.   ■

Using the Wilcoxon Signed-Rank Table

Table VI provides critical values for a Wilcoxon signed-rank test. The first column of the table gives the sample size; the second and third columns give the significance levels for a one-tailed and two-tailed test, respectively; and the fourth and fifth columns give the left-hand and right-hand critical values, respectively. Note that *a critical value from Table VI is to be included as part of the rejection region.*

The test statistic, W, is a discrete random variable. Because of this, it is generally not possible to perform a Wilcoxon signed-rank test at a significance level exactly equal to the one desired; usually we must be satisfied with a significance level close to the one desired.[†]

Example 9.20   Using the Wilcoxon Signed-Rank Table, Table VI

Refer to Example 9.19. Determine the critical value, rejection region, and nonrejection region for the hypothesis test if it is to be performed at a significance level as close to 0.05 as possible.

SOLUTION   The hypothesis test in Example 9.19 is left-tailed and the sample size is 10. To determine the critical value and the associated significance level, we first go down the sample-size column of Table VI to 10. Then we look for a significance level for a one-tailed test that is as close to 0.05 as possible, which in this case is $\alpha = 0.053$. Finally, we go across the row containing 0.053 to the W_ℓ column, where we find the corresponding left-hand critical value, 11.

Thus the critical value for the hypothesis test with $\alpha = 0.053$ is 11. The rejection region consists of all W-values less than or equal to 11, and the nonrejection region consists of all W-values greater than 11; in other words, we reject H_0 if $W \leq 11$ and do not reject H_0 if $W > 11$.   ■

[†] If we employ the P-value approach to hypothesis testing, then this problem does not arise. See the optional material on computer usage at the end of this section.

Performing the Wilcoxon Signed-Rank Test

We now present a step-by-step procedure for performing a Wilcoxon signed-rank test. For brevity, we will sometimes use the phrase *symmetric population* to indicate that a population has a symmetric distribution.

PROCEDURE 9.5

THE WILCOXON SIGNED-RANK TEST FOR A POPULATION MEAN WITH NULL HYPOTHESIS H_0: $\mu = \mu_0$

ASSUMPTION

Symmetric population

Step 1 State the null and alternative hypotheses.

Step 2 Decide on a significance level and use Table VI to find a significance level, α, as close as possible to the one desired.

Step 3 The critical value(s)
a. for a two-tailed test are W_ℓ and W_r.
b. for a left-tailed test is W_ℓ.
c. for a right-tailed test is W_r.
Use Table VI to find the critical value(s).

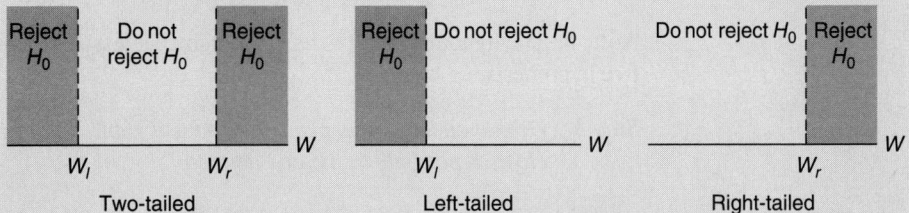

Step 4 Construct a work table of the following form:

Data value x	Difference $D = x - \mu_0$	$|D|$	Rank of $|D|$	Signed rank R
.	.	.	.	.
.	.	.	.	.
.	.	.	.	.

Step 5 Compute the value of the test statistic

$$W = \text{sum of the positive ranks.}$$

Step 6 If the value of the test statistic falls in the rejection region, then reject H_0; otherwise, do not reject H_0.

Step 7 State the conclusion in words.

Example 9.21   Illustrates Procedure 9.5

As an application of Procedure 9.5, let's complete the hypothesis test of Example 9.19. A random sample of 10 Kansas families of four with intermediate budgets yielded the data on 1984 weekly food costs shown in Table 9.17. Do the data provide sufficient evidence to conclude that in 1984, the mean weekly food cost for Kansas families of four with intermediate budgets was less than the national mean of $92? Apply the Wilcoxon signed-rank test with a significance level as close to 5% as possible.

TABLE 9.17
Sample of weekly food costs

$78	104	84	70	96
73	87	85	76	94

SOLUTION   We apply Procedure 9.5.

Step 1   *State the null and alternative hypotheses.*

Let μ denote the 1984 mean weekly food cost for all Kansas families of four with intermediate budgets. Then the null and alternative hypotheses are

$$H_0: \mu = \$92 \text{ (mean weekly food cost is not less than \$92)}$$
$$H_a: \mu < \$92 \text{ (mean weekly food cost is less than \$92)}.$$

Note that the hypothesis test is left-tailed since a less-than sign ($<$) appears in the alternative hypothesis.

Step 2   *Decide on a significance level and use Table VI to find a significance level, α, as close as possible to the one desired.*

The test is to be performed at a significance level as close to 0.05 as possible. Since $n = 10$ and the test is one-tailed, we see from Table VI that the closest possible significance level is $\alpha = 0.053$.

Step 3   *The critical value for a left-tailed test is W_ℓ.*

Referring again to Table VI with $n = 10$ and $\alpha = 0.053$, we find that the critical value is $W_\ell = 11$, as seen in Fig. 9.33.

FIGURE 9.33
Criterion for deciding whether or not to reject the null hypothesis

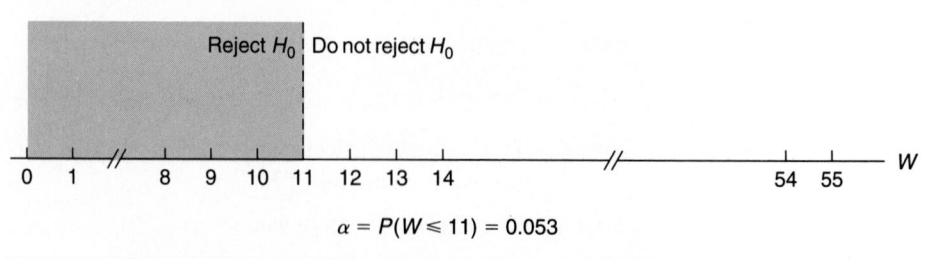

$$\alpha = P(W \leq 11) = 0.053$$

Step 4 *Construct a work table.*

We have already done this in Table 9.16 on page 549.

Step 5 *Compute the value of the test statistic*

$$W = \text{sum of the positive ranks.}$$

The last column of Table 9.16 shows that the sum of the positive ranks equals

$$W = 6 + 2 + 1 = 9.$$

Step 6 *If the value of the test statistic falls in the rejection region, reject H_0; otherwise, do not reject H_0.*

The value of the test statistic is $W = 9$, as found in Step 5. This falls in the rejection region, as we observe from Fig. 9.33. Thus we reject H_0.

Step 7 *State the conclusion in words.*

The test results are statistically significant at the 5.3% level; that is, at the 5.3% significance level, the data provide sufficient evidence to conclude that in 1984 the mean weekly food cost for Kansas families of four with intermediate budgets was less than the national mean of $92. ∎

As we mentioned earlier, one advantage of nonparametric methods is that they are resistant to outliers. We can illustrate that fact for the Wilcoxon signed-rank test by referring to Example 9.21. The stem-and-leaf diagram in Fig. 9.31 on page 548 shows that the sample data in Table 9.17 contain no outliers. The smallest data value, and also the farthest from the null hypothesis mean of 92, is 70. Replacing 70 by, say, 20, introduces an outlier but has absolutely no effect on the value of the test statistic and hence none on the hypothesis test itself. (Why is this so?)

The following points may be relevant when performing a Wilcoxon signed-rank test. Note them carefully.

- If a data value is equal to μ_0 (the value given for the mean in the null hypothesis), then that data value should be removed and the sample size reduced by 1.

- If two or more absolute differences are tied, each should be assigned the mean of the ranks they would have had if there were no ties. For example, if two absolute differences are tied for second place, each should be assigned rank $(2 + 3)/2 = 2.5$, and rank 4 should be assigned to the next largest absolute difference (which really is fourth!); if three absolute differences are tied for fifth place, each should be assigned rank $(5 + 6 + 7)/3 = 6$, and rank 8 should be assigned to the next largest absolute difference.

A Wilcoxon signed-rank test can be used to perform a hypothesis test for a population median, η, as well as for a population mean, μ. The reason is that the mean and median of a symmetric population are identical. To employ Procedure 9.5 to carry out a hypothesis test for a population median, simply replace μ by η and μ_0 by η_0. In some of the exercises at the end of this section, you will be asked to use the Wilcoxon signed-rank test to perform hypothesis tests for a population median.

Comparison of the Wilcoxon Signed-Rank Test and the *t*-Test

In Section 9.6 we learned how to perform a *t*-test (Procedure 9.4, page 536) for the mean of a normally distributed population. Since a normal population is necessarily symmetric, we can also use the Wilcoxon signed-rank test (Procedure 9.5, page 551) to perform such a hypothesis test.

So now the question is this: If we want to perform a hypothesis test for the mean of a population and we know the population is normally distributed, should we use the *t*-test or the Wilcoxon signed-rank test? As you might expect, we should use the *t*-test. For normal populations the *t*-test is more powerful than the Wilcoxon signed-rank test because it is designed expressly for normal populations; surprisingly, however, the *t*-test is not much more powerful than the Wilcoxon signed-rank test.

On the other hand, if the population being sampled has a symmetric distribution but is not normal, the Wilcoxon signed-rank test is usually more powerful than the *t*-test and is often considerably more powerful. In summary, we have the following key fact.

KEY FACT 9.7   **WILCOXON SIGNED-RANK TEST VERSUS THE *t*-TEST**

> Suppose a hypothesis test is to be performed for the mean, μ, of a population. When deciding between the *t*-test and the Wilcoxon signed-rank test, follow these guidelines.
>
> - If you are reasonably sure the population is normally distributed, use the *t*-test.
> - If you are not reasonably sure the population is normally distributed but are reasonably sure it is symmetric, use the Wilcoxon signed-rank test.

Using the Computer (Optional)

Procedure 9.5 on page 551 provides a step-by-step method for performing a Wilcoxon signed-rank test for a population mean, μ. Alternatively, we can use Minitab to carry out a Wilcoxon signed-rank test. The appropriate command is **WTest.**

As we said earlier, a Wilcoxon signed-rank test can be used to perform a hypothesis test for a population median, η, as well as for a population mean, μ. Minitab presents the output of the WTest command in terms of the median, but that output can be interpreted in terms of the mean: simply replace references to *median* by *mean*. With this in mind, we now consider Example 9.22, which shows how WTest is applied.

Example 9.22   The WTest Command

Use Minitab to perform the hypothesis test in Example 9.21.

SOLUTION   Let μ denote the 1984 mean weekly food cost for all Kansas families of four with intermediate budgets. The problem is to use the Wilcoxon signed-rank test to perform the hypothesis test

$$H_0: \mu = \$92 \text{ (mean weekly food cost is not less than \$92)}$$
$$H_a: \mu < \$92 \text{ (mean weekly food cost is less than \$92)}$$

at the 5% significance level ($\alpha = 0.05$). Note that the hypothesis test is left-tailed since a less-than sign ($<$) appears in the alternative hypothesis.

First we store the sample data from Table 9.17 on page 552 in a column named COSTS. Then we proceed as follows.

Session commands: We type **WTEST** followed by the null hypothesis and the storage location of the sample data; that is, we type <u>WTEST of mu=92, data in 'COSTS';</u> and press [Enter]. Since the test is left-tailed, we also type the subcommand <u>ALTERNATIVE=-1.</u> and press [Enter]. Printout 9.4 displays these commands and the output that results.

(or)

Menu commands: We choose **Stat ▸ Nonparametrics ▸ 1-Sample Wilcoxon...**, specify COSTS in the **Variables** text box, select **Test median** and type <u>92</u>, select **Alternative** and specify **less than**, and then select **OK**. The output obtained as a result of these commands is displayed in Printout 9.4.

PRINTOUT 9.4
Minitab output for
the WTest command

```
MTB > WTEST of mu=92, data in 'COSTS';
SUBC> ALTERNATIVE=-1.

TEST OF MEDIAN = 92.00 VERSUS MEDIAN L.T.  92.00

                N FOR   WILCOXON            ESTIMATED
          N     TEST   STATISTIC  P-VALUE    MEDIAN
COSTS    10      10         9.0    0.033      84.50
```

The third line of Printout 9.4 displays the null and alternative hypotheses for the hypothesis test: TEST OF MEDIAN = 92.00 VERSUS MEDIAN L.T. 92.00. (Remember, we can replace all references to *median* by *mean*.) Below that we find several entries.

The first entry, headed N, gives the sample size, which in this case is 10. The next entry, headed N FOR TEST, shows the number of data values in the sample that are not equal to the null hypothesis mean of 92. Recall that for a Wilcoxon signed-rank test, if a data value is equal to the null hypothesis mean, then that data value should be removed and the sample size reduced by 1. Since here the sample size is 10 and none of the data values in the sample equal 92, the sample size for the test (N FOR TEST) is also 10. The third entry, headed WILCOXON STATISTIC, provides the value of the test statistic, W; therefore we see that $W = 9$.

The final two entries in Printout 9.4, headed P-VALUE and ESTIMATED MEDIAN, give the P-value for the hypothesis test and a point estimate for the population median (or mean). Because the P-value of 0.033 is less than the designated significance level of $\alpha = 0.05$, we reject H_0. The test results are statistically significant at the 5% level; in other words, at the 5% significance level, the data provide sufficient evidence to conclude that in 1984, the mean weekly food cost for Kansas families of four with intermediate budgets was less than the national mean of $92. ∎

EXERCISES 9.7

9.108 What population assumption must be met in order to use the Wilcoxon signed-rank test?

9.109 We said on page 553 that if a data value is equal to μ_0 (the value given for the mean in the null hypothesis), then that data value should be removed and the sample size reduced by 1. Why do you think that is done?

9.110 Suppose you want to perform a hypothesis test for the mean, μ, of a population. For each of the following populations, decide whether you would use the t-test, the Wilcoxon signed-rank test, or neither.
a. Uniform population
b. Normal population
c. Right-skewed population

9.111 Suppose you want to perform a hypothesis test for the mean, μ, of a population. For each of the following populations, decide whether you would use the t-test, the Wilcoxon signed-rank test, or neither.
a. Triangular population
b. Normal population
c. Reverse-J-shaped population

9.112 The Wilcoxon signed-rank test can be used to perform a hypothesis test for a population median, η, as well as for a population mean, μ. Why is that so?

In each of Exercises 9.113–9.116, use the Wilcoxon signed-rank test, Procedure 9.5 on page 551, to perform the required hypothesis test.

9.113 In 1991 the median age of U.S. residents was 33.1 years, as reported by the Census Bureau in *Current Population Reports*. A random sample taken this year of 10 U.S. residents yielded the following ages, in years.

39	59	11	54	33
42	46	36	8	23

Do the data provide sufficient evidence to conclude that the median age of today's U.S. residents has increased over the 1991 median age of 33.1 years? Use a significance level as close to 0.05 as possible.

9.114 The Bureau of Labor Statistics publishes information on average annual expenditures by consumers in *Consumer Expenditure Survey*. In 1989 the mean amount spent by consumers on nonalcoholic beverages was $216. A random sample of 12 consumers yielded the following data, in dollars, on last year's expenditures on nonalcoholic beverages.

361	176	184	265
259	281	240	273
259	249	194	258

At a significance level as close to 0.05 as possible, do the data provide sufficient evidence to conclude that last year's mean amount spent by consumers on nonalcoholic beverages has increased over the 1989 mean of $216?

9.115 A chemist working for a pharmaceutical company has developed a new antacid tablet that she feels will relieve pain more quickly than the company's present tablet. Experience indicates that the present tablet requires an average of 12 minutes to take effect. The chemist records the following times, in minutes, for relief with the new tablet.

10.9	11.4	12.0	8.8	4.4
15.0	7.1	10.1	9.8	14.8
14.2	9.2	9.2	6.6	8.0

At a significance level as close to 5% as possible, does it appear that the new antacid tablet works faster?

9.116 The National Center for Health Statistics reports in *Vital Statistics of the United States* that the median birth weight of U.S. babies was 7.4 lb in 1990. A random sample of this year's births provided the following weights, in pounds.

8.6	7.4	5.3	13.8	7.8	5.7	9.2
8.8	8.2	9.2	5.6	6.0	11.6	7.2

Can we conclude that this year's median birth weight differs from that in 1990? Use a significance level as close to 5% as possible.

9.117 According to *Food Cost Review,* published by the U.S. Department of Agriculture, the mean retail price for oranges in 1983 was 38.5 cents per pound. Recently, a random sample of 15 markets reported the following prices for oranges, in cents per pound.

43.0	40.0	42.6	40.2	37.5
44.1	45.2	41.8	35.6	34.6
37.9	44.2	44.5	38.2	42.4

a. Can we conclude that the mean retail price for oranges now is different from the 1983 mean of 38.5 cents per pound? Perform a Wilcoxon signed-rank test at a significance level as close as possible to 0.05.
b. The hypothesis test considered in part (a) was done previously in Exercise 9.95 using a *t*-test. The assumption in that exercise is that retail prices for oranges are normally distributed (or at least approximately so). Assuming that, in fact, retail prices for oranges are normally distributed, why is it permissible to perform a Wilcoxon signed-rank test for the mean retail price of oranges?

9.118 Atlas Fishing Line produces a 10-lb test line. Twelve randomly selected spools are subjected to tensile-strength tests. The results follow.

9.8	10.2	9.8	9.4
9.7	9.7	10.1	10.1
9.8	9.6	9.1	9.7

a. Use the data to decide whether Atlas Fishing Line's 10-lb test line is not up to specifications. Perform a Wilcoxon signed-rank test for the mean tensile strength, μ, at a significance level as close to 0.05 as possible.
b. The hypothesis test of part (a) was done previously in Exercise 9.96 using a *t*-test. The assumption in that exercise is that tensile strengths are normally distributed (or at least approximately so). Assuming that, in fact, tensile strengths are normally distributed, why is it permissible to perform a Wilcoxon signed-rank test for the mean tensile strength?

9.119 A manufacturer of liquid soap produces a bottle with an advertised content of 310 mL. Sixteen bottles are randomly selected and found to have the following contents, in milliliters.

297	318	306	300
311	303	291	298
322	307	312	300
315	296	309	311

A normal probability plot of the data indicates that it is reasonable to assume the contents are normally distributed. Let μ denote the mean content of all bottles produced. To decide whether the mean content is less than advertised, perform the hypothesis test

$$H_0: \mu = 310 \text{ mL}$$
$$H_a: \mu < 310 \text{ mL}.$$

a. Use the *t*-test, Procedure 9.4 on page 536, with $\alpha = 0.05$.
b. Use the Wilcoxon signed-rank test with a significance level as close as possible to 0.05.
c. Assuming the mean content is in fact less than 310 mL, how do you explain the discrepancy between the two tests?

9.120 The U.S. Bureau of Justice Statistics reports in *Profile of Jail Inmates* that the median educational attainment of jail inmates was 10.2 years in 1978. Ten current inmates are randomly selected and found to have the following educational attainments, in years.

9	11	8	8	9
13	11	6	5	9

Assume the population of all such educational attainments has a symmetric, nonnormal distribution. At the 10.5% significance level, do the data provide sufficient evidence to conclude that this year's median educational attainment has changed from the 1978 median of 10.2 years?

a. Use the t-test, Procedure 9.4 on page 536. *(Note:* Employing statistical software, we determined that for df $= 9$, $t_{0.0525} = 1.802$.*)*
b. Use the Wilcoxon signed-rank test.
c. Presuming this year's median educational attainment has in fact changed from the 1978 median of 10.2 years, how do you explain the discrepancy between the two tests?

· **9.121 (Computer exercise)** Refer to Exercise 9.113. Use Minitab or some other statistical software to perform the required hypothesis test using the Wilcoxon signed-rank test.

· **9.122 (Computer exercise)** Refer to Exercise 9.114. Use Minitab or some other statistical software to perform the required hypothesis test using the Wilcoxon signed-rank test.

· **9.123 (Computer exercise)** The National Center for Health Statistics publishes data on the duration of marriages in *Vital Statistics of the United States.* In 1988 the median duration of a marriage was 7.1 years. A random sample is taken from last year's divorce certificates and the marriage durations are recorded. Then Minitab's WTest is applied. The resulting computer output is displayed in Printout 9.5. Using the computer output, determine
a. the null and alternative hypotheses.
b. the number of divorce certificates sampled.

c. the number of certificates in the sample for which the marriage lasted exactly 7.1 years.
d. a point estimate for last year's median marriage duration.
e. the P-value of the hypothesis test.
f. the smallest significance level at which the null hypothesis can be rejected.
g. the conclusion if the hypothesis test is performed at the 5% significance level.
h. Do you think it is appropriate to use the Wilcoxon signed-rank test to carry out this hypothesis test? Why?

· **9.124 (Computer exercise)** The Census Bureau estimates that the *U.S. Census Form* takes the average household 14 minutes to complete. A random sample of households is obtained and the time it takes each household to complete the form is recorded. Printout 9.6 shows the output resulting from applying Minitab's WTest to the completion-time data. Use the computer output to determine
a. the null and alternative hypotheses.
b. the number of households sampled.
c. the number of households in the sample that took exactly 14 minutes to complete the form.
d. the estimated population median completion time.
e. the P-value of the hypothesis test.
f. the smallest significance level at which the null hypothesis can be rejected.
g. the conclusion if the hypothesis test is performed at the 10% significance level.
h. Do you think it is appropriate to use the Wilcoxon signed-rank test to carry out this hypothesis test? Why?

PRINTOUT 9.5   Minitab output for Exercise 9.123

```
TEST OF MEDIAN = 7.100 VERSUS MEDIAN L.T. 7.100

              N FOR    WILCOXON              ESTIMATED
         N    TEST    STATISTIC   P-VALUE     MEDIAN
YEARS    50    50       696.0      0.716      7.531
```

PRINTOUT 9.6   Minitab output for Exercise 9.124

```
TEST OF MEDIAN = 14.00 VERSUS MEDIAN N.E. 14.00

                N FOR    WILCOXON              ESTIMATED
           N    TEST    STATISTIC   P-VALUE     MEDIAN
MINUTES    36    36       462.0      0.044      15.05
```

Large-sample Wilcoxon signed-rank test using a normal approximation. The table of critical values for the Wilcoxon signed-rank test, Table VI, stops at $n = 20$. For larger samples a normal approximation can be used. In fact, the normal approximation works well even for sample sizes as small as 10. Specifically, we have the following fact.

Suppose a random sample of size $n \geq 10$ is to be taken from a symmetric population. Then the random variable W is approximately normally distributed and has mean $\mu_W = n(n + 1)/4$ and standard deviation $\sigma_W = \sqrt{n(n + 1)(2n + 1)/24}$. In particular, the standardized random variable

$$(5) \qquad z = \frac{W - n(n + 1)/4}{\sqrt{n(n + 1)(2n + 1)/24}}$$

has approximately the standard normal distribution.

In Exercises 9.125–9.127, we will develop and apply a large-sample procedure for a Wilcoxon signed-rank test based on the previous fact.

· · **9.125** Formulate a large-sample hypothesis-testing procedure for a Wilcoxon signed-rank test that uses the test statistic in Equation (5).

· · **9.126** Refer to Exercise 9.116.
a. Use the procedure you formulated in Exercise 9.125 to perform the hypothesis test in Exercise 9.116 at the 4.8% significance level.
b. Compare your result in part (a) to the one you obtained in Exercise 9.116, where the normal approximation was not used.

· · **9.127** Refer to Exercise 9.115.
a. Use the procedure you formulated in Exercise 9.125 to perform the hypothesis test in Exercise 9.115 at the 5.2% significance level.

b. Compare your result in part (a) to the one you obtained in Exercise 9.115, where the normal approximation was not used.

· · · **9.128** In this exercise we will find the probability distribution of the random variable W when $n = 3$ and when $n = 4$. This will enable you to see how the critical values for the Wilcoxon signed-rank test are derived.
a. The rows of the following table give all possible signs for the signed ranks in a Wilcoxon signed-rank test with $n = 3$. For example, the first row covers the possibility that all three data values are greater than μ_0 and thus have positive sign ranks. There is an empty column for values of W. Fill it in. (*Hint:* The first entry is 6 and the last is 0.)

Rank			
1	2	3	W
+	+	+	
+	+	−	
+	−	+	
+	−	−	
−	+	+	
−	+	−	
−	−	+	
−	−	−	

b. If the null hypothesis, H_0: $\mu = \mu_0$, is true, what is the probability that a sample will match any particular row of the table? (*Hint:* The answer is the same for all rows.)
c. Use the answer from part (b) to find the probability distribution of the random variable W when $n = 3$.
d. Draw a probability histogram for the random variable W when $n = 3$.
e. Use your histogram from part (d) to obtain the critical value for a left-tailed Wilcoxon signed-rank test with a sample size of three and a significance level of 0.125.
f. Compare your critical value from part (e) with the critical value in Table VI.
g. Repeat parts (a)–(f) for a Wilcoxon signed-rank test with $n = 4$.

9.8 WHICH PROCEDURE SHOULD BE USED? (OPTIONAL)

In this chapter we have learned three procedures for performing a hypothesis test for one population mean, μ. We summarize those procedures in Table 9.18. Each row of the table gives the type of test, the condition required for using the test, the test statistic, and the procedure to use. In the table we have employed W-test as an abbreviation for Wilcoxon signed-rank test.

TABLE 9.18

Summary of hypothesis-testing procedures for one population mean, μ. The null hypothesis for all tests is H_0: $\mu = \mu_0$

Type	Assumption	Test statistic	Procedure to use
z-test	Large sample	$z = \dfrac{\bar{x} - \mu_0}{\sigma/\sqrt{n}}$	9.1 (page 503)
t-test	Normal population	$t = \dfrac{\bar{x} - \mu_0}{s/\sqrt{n}}$, df $= n - 1$	9.4 (page 536)
W-test	Symmetric population	$W = $ sum of positive ranks	9.5 (page 551)

In selecting the correct procedure, keep in mind that the best choice is the procedure expressly designed for the type of population under consideration, if such a procedure exists. For instance, if a large sample is taken from a normally distributed population with unknown σ, then all three procedures are applicable: The z-test applies since the sample size is large; the t-test applies since the population is normally distributed; and the W-test applies since the population has a symmetric distribution (because a normal population is symmetric). But the correct procedure is the t-test, because that test is designed specifically for a normal population.

The flowchart in Fig. 9.34 provides an organized strategy for choosing the correct hypothesis-testing procedure for a population mean. It has been constructed based on the criterion discussed in the preceding paragraph.

In practice we need to look at the sample data to ascertain distribution type before we can select the appropriate procedure. We recommend using a normal probability plot and either a stem-and-leaf diagram (for small or moderate-size samples) or a histogram (for moderate-size or large samples).

Example 9.23   Choosing the Correct Hypothesis-Testing Procedure

The *Consumer Expenditure Survey,* published by the U.S. Bureau of Labor Statistics, provides data on annual consumer expenditures for various items. In 1990 the mean expenditure for entertainment was $1422. A sample of 50 consumers yielded the data on last year's expenditures for entertainment shown in Table 9.19.

TABLE 9.19

Sample of last year's expenditures for entertainment

$1168	1945	1434	2457	2623	1628	1226	1883	1838	1286
1332	89	884	657	895	1524	1305	1661	1656	1453
1625	1971	3313	1397	1159	889	55	1549	1416	1832
1181	816	1654	1885	1419	1746	1450	217	1921	2028
1352	1282	1005	1088	1406	1501	1517	1397	1587	1704

Suppose we want to use the sample data in Table 9.19 to decide whether last year's mean expenditure for entertainment has changed from the 1990 mean. Then we want to

FIGURE 9.34

Flowchart for choosing the correct
hypothesis-testing procedure
for one population mean

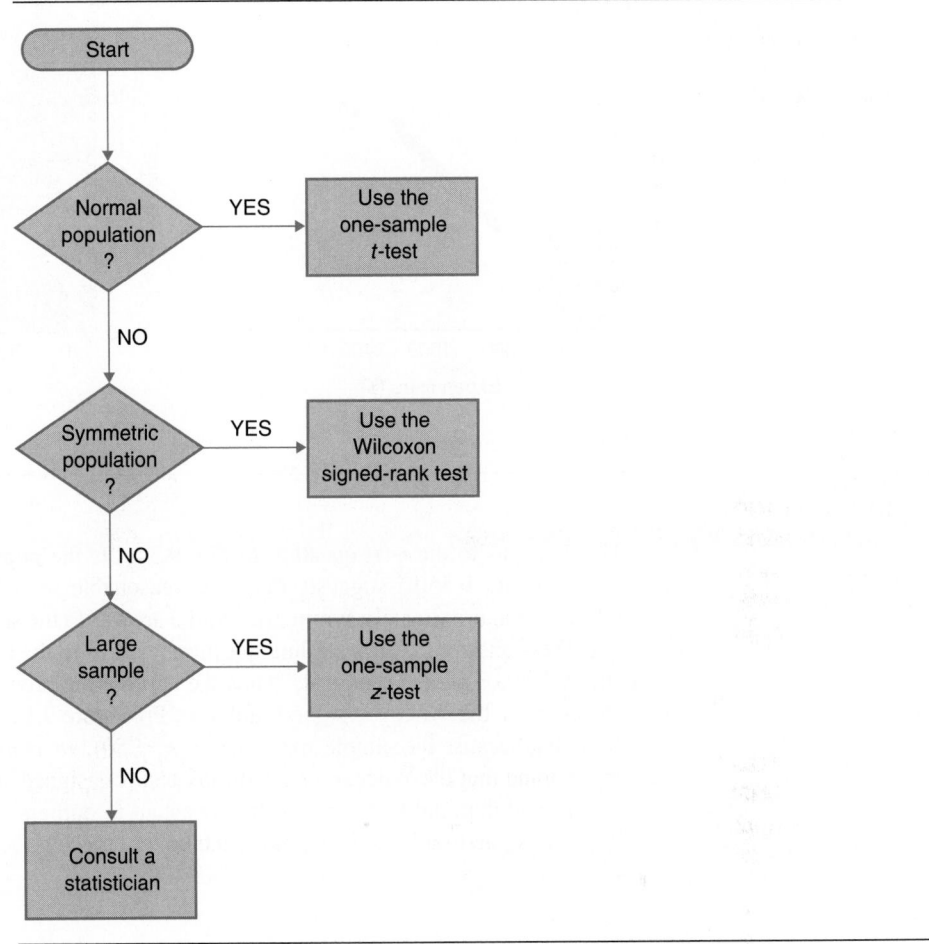

perform the hypothesis test

$$H_0: \mu = \$1422 \text{ (mean expenditure has not changed)}$$
$$H_a: \mu \neq \$1422 \text{ (mean expenditure has changed),}$$

where μ is last year's mean expenditure for entertainment. Which procedure should we use
to perform the hypothesis test?

SOLUTION   We begin by drawing a normal probability plot and histogram for the sample data in
Table 9.19, as seen in Figs. 9.35(a) and 9.35(b), respectively, at the top of the next page.

   Next we consult the flowchart in Fig. 9.34 and the graphs in Fig. 9.35 to decide which
procedure we should use. The first question we must answer in Fig. 9.34 is, "Is the population normal?" The normal probability plot in Fig. 9.35(a) shows substantial curvature
and/or outliers; so the answer to the first question is "No."

FIGURE 9.35

(a) Normal probability
plot and (b) histogram of
expenditure data in Table 9.19

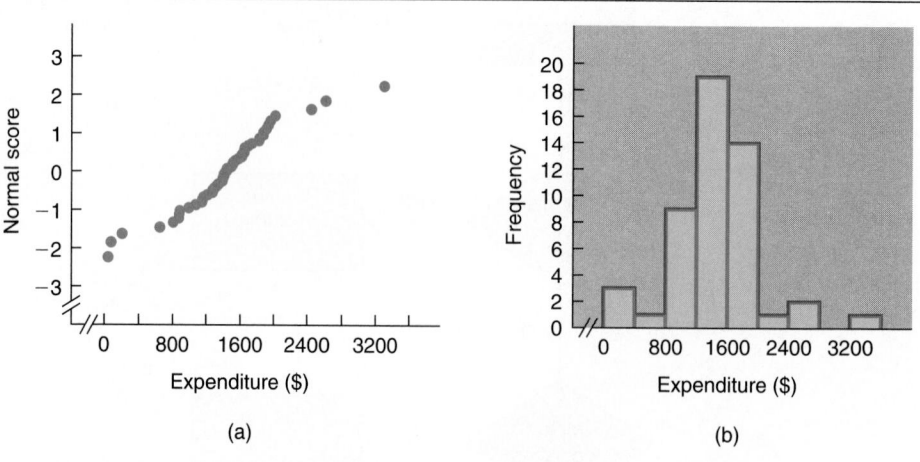

This leads us to the next question in Fig. 9.34: "Is the population symmetric?" The histogram in Fig. 9.35(b) suggests that it is reasonable to presume the population of expenditures is approximately symmetric; so the answer to the second question is "Yes."

The "Yes" answer to the preceding question leads us to the box in Fig. 9.34 that states "Use the Wilcoxon signed-rank test." Thus the appropriate procedure for carrying out the hypothesis test is the Wilcoxon signed-rank test, Procedure 9.5 on page 551.

Note that because the sample size is large ($n = 50$), we could also use the z-test. But keeping in mind that the Wilcoxon signed-rank test is designed specifically for symmetric populations and that, unlike the z-test, it is resistant to outliers and other extreme values, the Wilcoxon signed-rank test is definitely the better choice here. ■

EXERCISES 9.8

In each of Exercises 9.129–9.136, we have provided a normal probability plot and either a stem-and-leaf diagram or a frequency histogram for a set of sample data. The intent is to employ the sample data to perform a hypothesis test for the mean of the population from which the data were obtained. In each case consult the graphs provided and the flowchart in Fig. 9.34 to decide which procedure should be used.

9.129 The normal probability plot and stem-and-leaf diagram of the data are depicted in Fig. 9.36.

9.130 The normal probability plot and histogram of the data are depicted in Fig. 9.37.

9.131 The normal probability plot and histogram of the data are depicted in Fig. 9.38.

9.132 The normal probability plot and stem-and-leaf diagram of the data are depicted in Fig. 9.39 on page 564.

9.133 The normal probability plot and stem-and-leaf diagram of the data are depicted in Fig. 9.40 on page 564.

9.134 The normal probability plot and stem-and-leaf diagram of the data are depicted in Fig. 9.41 on page 564. *(Note:* The decimal parts of the data values were removed before the stem-and-leaf diagram was constructed.*)*

9.135 The normal probability plot and stem-and-leaf diagram of the data are depicted in Fig. 9.42 on page 565.

9.136 The normal probability plot and stem-and-leaf diagram of the data are depicted in Fig. 9.43 on page 565.

FIGURE 9.36   Normal probability plot and stem-and-leaf diagram for Exercise 9.129

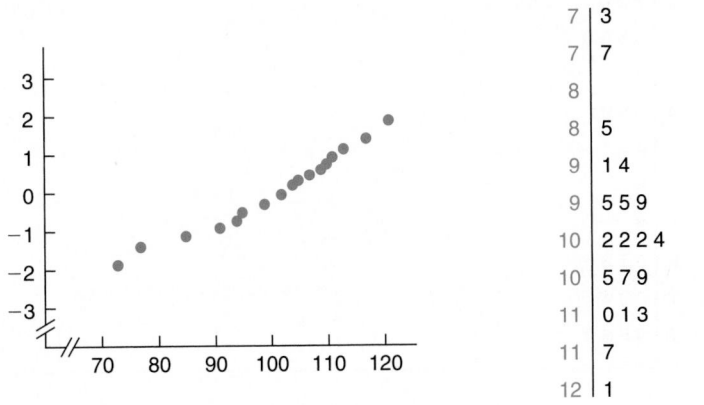

```
 7 | 3
 7 | 7
 8 |
 8 | 5
 9 | 1 4
 9 | 5 5 9
10 | 2 2 2 4
10 | 5 7 9
11 | 0 1 3
11 | 7
12 | 1
```

FIGURE 9.37   Normal probability plot and histogram for Exercise 9.130

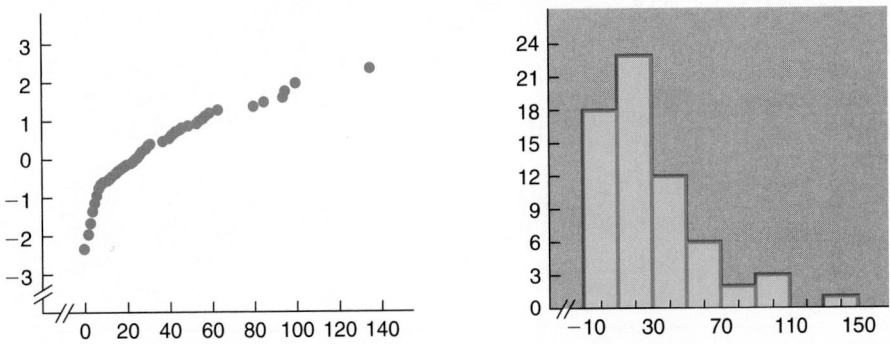

FIGURE 9.38   Normal probability plot and histogram for Exercise 9.131

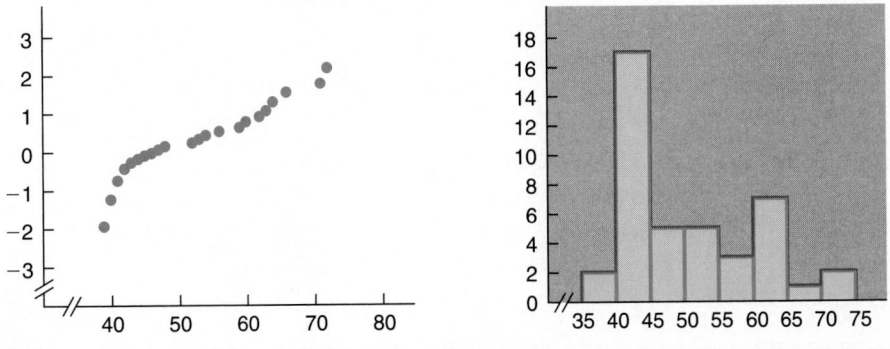

FIGURE 9.39   Normal probability plot and stem-and-leaf diagram for Exercise 9.132

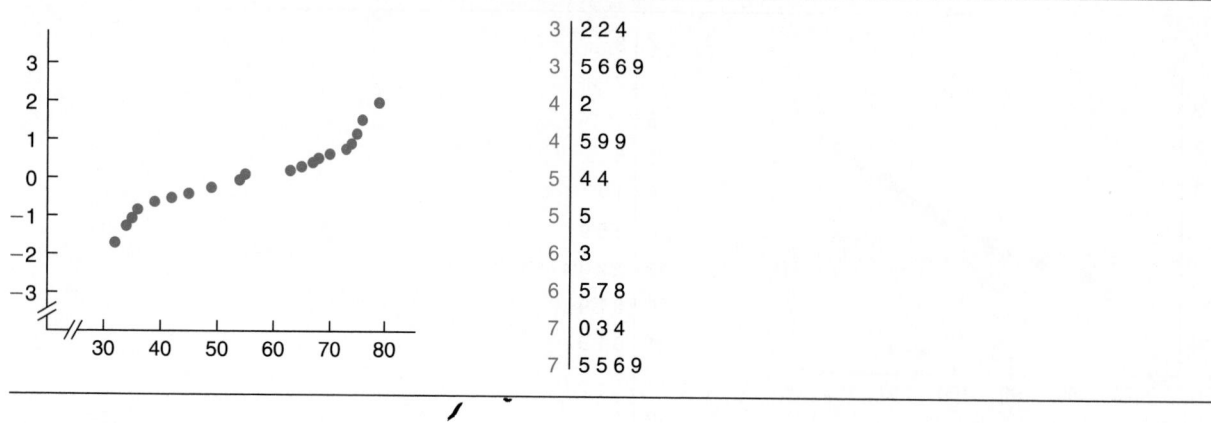

FIGURE 9.40   Normal probability plot and stem-and-leaf diagram for Exercise 9.133

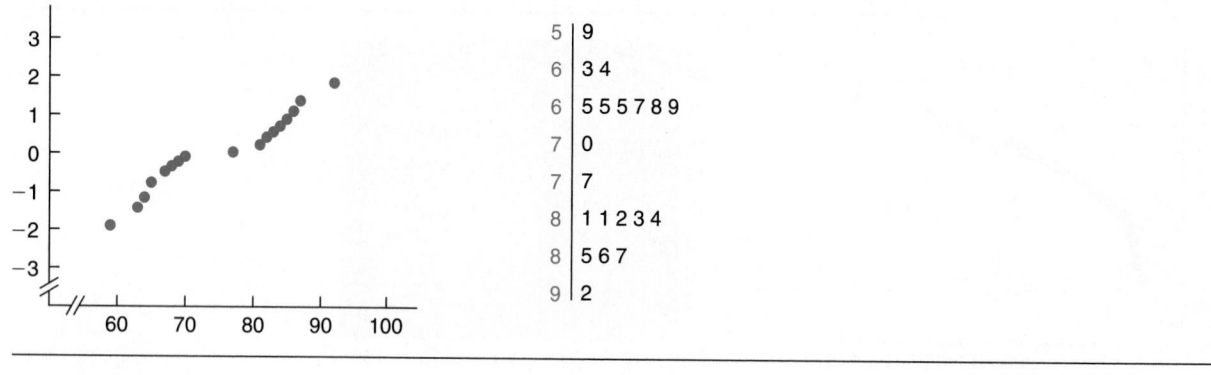

FIGURE 9.41   Normal probability plot and stem-and-leaf diagram for Exercise 9.134

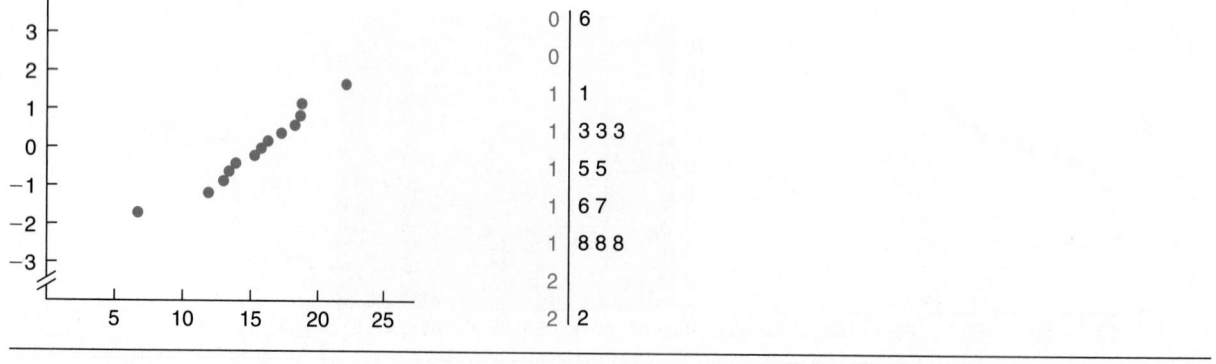

FIGURE 9.42   Normal probability plot and stem-and-leaf diagram for Exercise 9.135

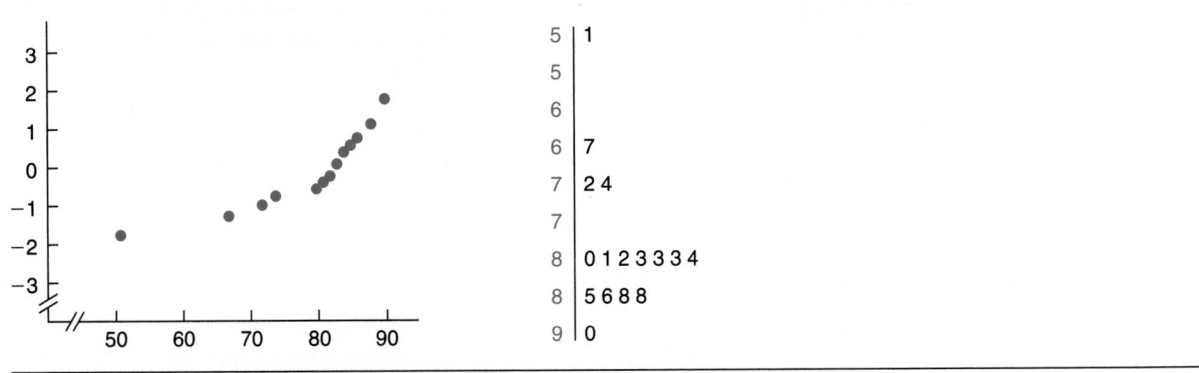

FIGURE 9.43   Normal probability plot and stem-and-leaf diagram for Exercise 9.136

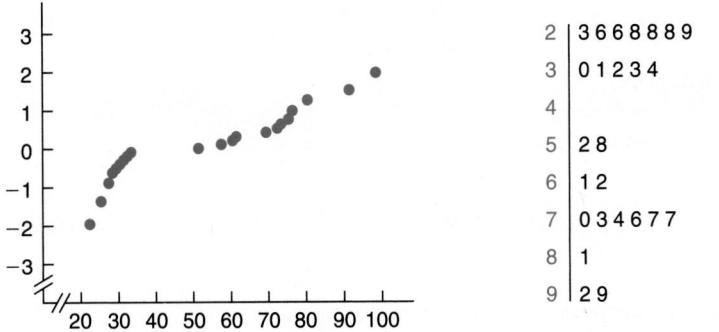

Chapter Review

Formulas

In the formulas below,

μ = population mean

σ = population standard deviation

$\overline{x}$ = sample mean

n = sample size

s = sample standard deviation

z-test statistic for H_0: $\mu = \mu_0$ (large sample), *503*

$$z = \frac{\overline{x} - \mu_0}{\sigma/\sqrt{n}}$$

(if σ is unknown, use s in its place)

t-test statistic for H_0: $\mu = \mu_0$ (normal population), *537*

$$t = \frac{\overline{x} - \mu_0}{s/\sqrt{n}}$$

with df = $n - 1$.

Wilcoxon signed-rank test statistic for H_0: $\mu = \mu_0$ (symmetric population),* *551*

$$W = \text{sum of the positive ranks}$$

You Should Be Able To

1. use and understand the preceding formulas.
2. define the terms associated with hypothesis testing.
3. choose the null and alternative hypotheses for a hypothesis test.
4. explain the logic behind hypothesis testing.
5. identify the test statistic, rejection region, nonrejection region, and critical value(s) for a hypothesis test.
6. define and apply the concepts of Type I and Type II errors.
7. state the possible conclusions for a hypothesis test.
8. obtain the critical value(s) for a specified significance level.
9. perform a large-sample hypothesis test for a population mean, μ.
10. state and apply the steps for performing a hypothesis test using the classical approach to hypothesis testing.
11. compute Type II error probabilities.*
12. draw an operating characteristic (OC) curve.*
13. calculate the power of a hypothesis test.*
14. draw a power curve.*
15. obtain the *P*-value of a hypothesis test.
16. state and apply the steps for performing a hypothesis test using the *P*-value approach to hypothesis testing.

17. perform a hypothesis test for a population mean, μ, when the population being sampled is normally distributed.
18. perform a hypothesis test for a population mean, μ, when the population being sampled has a symmetric distribution.*
19. decide which procedure should be used to perform a hypothesis test for a population mean.*
20. use the Minitab commands covered in this chapter.*
21. interpret the output obtained from the application of the Minitab commands discussed in this chapter.*

REVIEW TEST

1. The U.S. Department of Agriculture reports in *Food Consumption, Prices, and Expenditures* that the average American consumed 24.7 lb of cheese in 1990. There has been a steady increase in cheese consumption since 1960, when the average American ate only 8.3 lb of cheese annually. A researcher thinks the trend of increasing cheese consumption is still continuing. He wants to perform a hypothesis test to decide whether last year's mean cheese consumption is greater than the 1990 mean.
   a. Identify the null hypothesis for the researcher's hypothesis test.
   b. Identify the alternative hypothesis for the researcher's hypothesis test.
   c. Classify the hypothesis test as two-tailed, left-tailed, or right-tailed.

2. The following graph portrays the decision criterion for a hypothesis test concerning a population mean, μ. The null hypothesis for the hypothesis test is H_0: $\mu = \mu_0$, and the test statistic is

$$z = \frac{\bar{x} - \mu_0}{\sigma/\sqrt{n}}.$$

Also, the curve in the graph shows the implications of the decision criterion if in fact the null hypothesis is true.

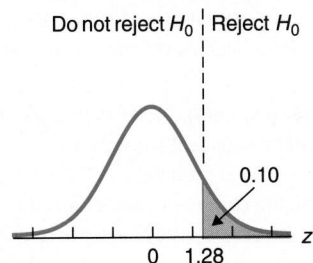

Do not reject H_0 | Reject H_0

0.10

0   1.28

Determine the
a. rejection region.
b. nonrejection region.
c. critical value(s).
d. significance level.

e. Construct a graph depicting the results you obtained in parts (a)–(d).
f. Classify the hypothesis test as two-tailed, left-tailed, or right-tailed.

3. The null and alternative hypotheses for the hypothesis test in Problem 1 are

H_0: $\mu = 24.7$ lb (mean has not increased)

H_a: $\mu > 24.7$ lb (mean has increased),

where μ is last year's mean cheese consumption for all Americans. Explain what each of the following would mean.
a. A Type I error   b. A Type II error
c. A correct decision.
Now suppose the results of carrying out the hypothesis test lead to rejection of the null hypothesis. Classify that decision by error type or as a correct decision if in fact last year's mean cheese consumption
d. has not increased over the 1990 mean of 24.7 lb.
e. has increased over the 1990 mean of 24.7 lb.

*4. Refer to Problem 1. Suppose the researcher decides to use a significance level of 0.10 and a sample size of 35. Assume as known that $\sigma = 6.9$ lb.
   a. Express the decision criterion for the hypothesis test in terms of $\bar{x}$.
   b. Determine the probability of a Type I error.
   c. Determine the probability of a Type II error for each of the following values of μ: 25.0, 25.5, 26.0, 26.5, 27.0, 27.5, 28.0, 28.5, 29.0, 29.5, 30.0. Construct a table similar to Table 9.8 on page 517.
   d. Use the table that you obtained in part (c) to draw the OC curve.
   e. Use the table that you obtained in part (c) to construct a table of powers similar to Table 9.9 on page 519.
   f. Use the table that you obtained in part (e) to draw the power curve.

Using a sample size of 100 instead of 35,
g. repeat part (a).   h. repeat part (b).
i. repeat part (c).   j. repeat part (d).
k. Compare your OC curves for the two sample sizes and explain the principle being illustrated.

5. Refer to Problem 1. Suppose the researcher randomly selects 35 Americans and obtains the following data, in pounds, on their cheese consumptions for last year.

37	20	24	29	33	27	31
25	24	23	27	13	19	38
17	33	27	35	23	36	15
30	19	27	24	35	24	17
28	20	18	22	27	28	28

a. At the 10% significance level, do the data provide sufficient evidence to conclude that last year's mean cheese consumption for all Americans has increased over the 1990 mean of 24.7 lb? Assume as known that $\sigma = 6.9$ lb. *(Note:* The sum of the data is 903 lb.*)*
b. Given the conclusion in part (a), if an error has been made, what type must it be? Explain your answer.

*6. **(Computer problem)** Use Minitab or some other statistical software to carry out the hypothesis test considered in part (a) of Problem 5.

*7. **(Computer problem)** The following abridged Minitab printout was obtained by applying the ZTest command to the data in Problem 5.

```
TEST OF MU = 24.700 VS MU G.T. 24.700
THE ASSUMED SIGMA = 6.90


N    MEAN   STDEV   SE MEAN    Z   P VALUE
35   25.800  6.480   1.166   0.94   0.17
```

Using the printout, determine
a. the null and alternative hypotheses.
b. the (assumed) population standard deviation of last year's cheese consumptions.
c. the sample standard deviation, s.
d. the number of people in the sample.
e. last year's mean cheese consumption, $\bar{x}$, for the people in the sample.
f. the value of the test statistic, z.
g. the P-value of the hypothesis test.
h. the smallest significance level at which the null hypothesis can be rejected.
i. the conclusion if the hypothesis test is performed at the 10% significance level.

8. According to *Crime in the United States,* a publication of the FBI, the mean value lost because of purse snatching in 1990 was $278. For this year, 40 randomly selected purse-snatching offenses have a mean value lost of $259 with a standard deviation of $84. Do the data provide sufficient evidence to conclude that the mean value lost because of purse snatching has decreased from the 1990 mean? Use $\alpha = 0.05$.

9. Refer to Problem 8.
a. Determine the P-value of the hypothesis test.
b. Perform the hypothesis test using the P-value approach to hypothesis testing.
c. Use Table 9.12 on page 530 to assess the strength of the evidence against the null hypothesis.

10. Each year, manufacturers perform mileage tests on new car models and submit the results to the Environmental Protection Agency (EPA). The EPA then tests the vehicles to determine whether the manufacturers are correct. In 1992 one company reported that a particular model equipped with a four-speed manual transmission averaged 29 mpg on the highway. Suppose the EPA tested 15 of the cars and obtained the following gas mileages.

27.3	31.2	29.4	31.6	28.6
30.9	29.7	28.5	27.8	27.3
25.9	28.8	28.9	27.8	27.6

A normal probability plot of the data shows no outliers and is roughly linear. What decision would you make regarding the company's report on the gas mileage of the car? Perform the required hypothesis test at the 5% significance level.
a. Use the classical approach to hypothesis testing. *(Note:* $\bar{x} = 28.75$, $s = 1.595$.*)*
b. Use the P-value approach to hypothesis testing.
c. Use Table 9.12 on page 530 to assess the strength of the evidence against the null hypothesis.

*11. **(Computer problem)** Refer to Problem 10. Use Minitab or some other statistical software to
a. obtain a normal probability plot of the data.
b. perform the required hypothesis test.
c. Justify the use of your procedure in part (b).

*12. **(Computer problem)** The following abridged Minitab printout was obtained by applying the TTest command to the data in Problem 10.

```
TEST OF MU = 29.000 VS MU N.E. 29.000

N    MEAN   STDEV  SE MEAN     T   P VALUE
15   28.753  1.595   0.412  -0.60    0.56
```

Employ the printout to determine

a. the null and alternative hypotheses.

b. the standard deviation of the mileages for the cars that were tested.

c. the number of cars tested.

d. the mean mileage of the cars tested.

e. the value of the test statistic, t.

f. the P-value of the hypothesis test.

g. the smallest significance level at which the null hypothesis can be rejected.

h. the conclusion if the hypothesis test is performed at the 5% significance level.

*13. Refer to Problem 10.

a. At a significance level as close to 5% as possible, what decision would you make regarding the company's report on the gas mileage of the car? Perform a Wilcoxon signed-rank test for the mean gas mileage, μ.

b. In performing the hypothesis test of part (a), what assumption are you making about the distribution of the gas mileages?

c. In Problem 10 we performed the hypothesis test in part (a) using the t-test. The assumption in that problem is that the gas mileages are normally distributed (or approximately so). Assuming that, in fact, the gas mileages are normally distributed, why is it permissible to perform a Wilcoxon signed-rank test for the mean gas mileage?

*14. (**Computer problem**) Use Minitab or some other statistical software to carry out the hypothesis test considered in part (a) of Problem 13.

*15. (**Computer problem**) The following abridged Minitab printout was obtained by applying the WTest command to the data in Problem 10.

```
TEST OF MEDIAN = 29.0 VS MEDIAN N.E. 29.0

      N FOR    WILCOXON            ESTIMATED
N     TEST    STATISTIC  P-VALUE     MEDIAN
15     15        48.5     0.532      28.65
```

Employ the output to determine

a. the null and alternative hypotheses.

b. the number of cars tested.

c. the number of cars tested whose gas mileage is exactly 29 mpg.

d. the value of the test statistic, W.

e. the P-value of the hypothesis test.

f. the smallest significance level at which the null hypothesis can be rejected.

g. the conclusion if the hypothesis test is performed at the 5% significance level.

h. a point estimate for the population median gas mileage.

*16. Refer to Problems 10 and 13. If in fact the gas mileages are normally distributed, which is the preferred procedure for performing the hypothesis test—the t-test or the Wilcoxon signed-rank test? Explain your answer.

In each of Problems 17 and 18, we have referenced a figure providing a normal probability plot and either a frequency histogram or a stem-and-leaf diagram for a set of sample data. The intent is to employ the sample data to perform a hypothesis test for the mean of the population from which the data were obtained. In each case, consult the graphs referenced and the flowchart in Fig. 9.34 on page 561 to decide which procedure should be used.

*17. Figure 9.44 on page 570.

*18. Figure 9.45 on page 570.

Using the Focus Database

The Focus database, which is printed in Appendix B, contains data on 500 randomly selected Arizona State University sophomores. Among the variables considered are SAT math score and SAT verbal score. Use Minitab or some other statistical software to solve the following problems.

a. In 1989 the mean SAT math score was 476 nationally. At the 5% significance level, do the Focus data on SAT math scores provide sufficient evidence to conclude that the mean SAT math score of Arizona State University sophomores exceeds the 1989 national mean?

b. In 1989 the mean SAT verbal score was 427 nationally. At the 5% significance level, do the Focus data on SAT verbal scores provide sufficient evidence to conclude that the mean SAT verbal score of Arizona State University sophomores exceeds the 1989 national mean?

c. Justify the use of the hypothesis-testing procedures you employed in parts (a) and (b).

FIGURE 9.44   Normal probability plot and histogram for Problem 17

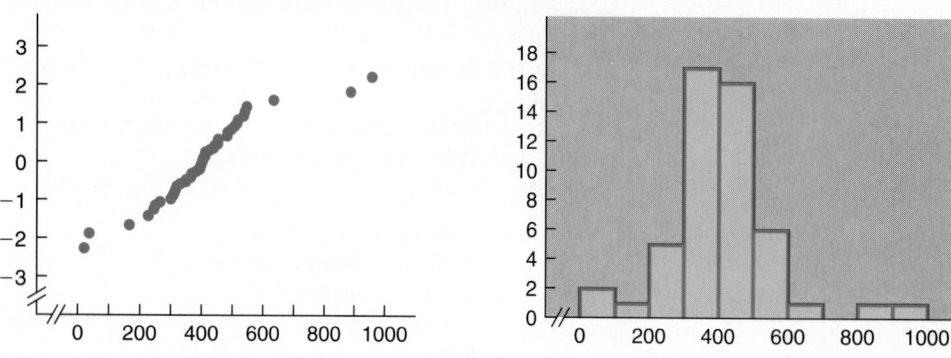

FIGURE 9.45   Normal probability plot and stem-and-leaf diagram for Problem 18

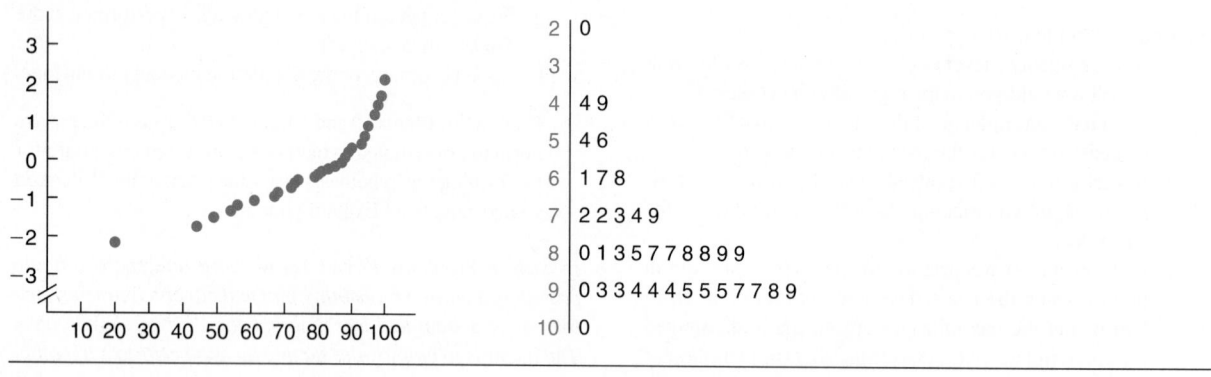

Case Study   **Effects of Brewery Effluent on Soil**

. . . . .

Because many industrial wastes contain nutrients that enhance crop growth, efforts are being made, for environmental purposes, to use such wastes on agricultural soils. In a 1984 issue of *Environmental Pollution (Series A)*, Mohammad Ajmal and Ahsan Ullah Khan reported their findings on experiments with brewery wastes used for agricultural purposes.[†]

Specifically, the researchers studied the physico-chemical properties of effluent from Mohan Meakin Breweries Ltd (MMBL), Ghazibad, UP, India, and "... its effects on the physico-chemical characteristics of agricultural soil, seed germination pattern, and the growth of two common crop plants." They assessed the impact using different concentrations of the effluent: 25%, 50%, 75%, and 100%.

[†] From "Effects of Brewery Effluent on Agricultural Soil and Crop Plants" by M. Ajmal and A. Khan, 1984, *Environmental Pollution (Series A), 33,* pp. 341–351.

Various chemical properties of the treated soil were measured, in particular, available nutrients. Table 9.20 shows the effects of the different dilutions of MMBL effluent on the available limestone, potassium, and phosphorus in the soil. A number not in parentheses is the mean of five observations; a number enclosed by parentheses is the sample standard deviation of the five observations. The data for limestone are in percentages and those for potassium and phosphorus are in milligrams per kilogram of soil.

TABLE 9.20

Effects of different dilutions of MMBL effluent on the available limestone, potassium, and phosphorus in the soil

		Treatment				
	Original	Control (water)	Effluent 100%	Effluent 75%	Effluent 50%	Effluent 25%
Limestone	2.30	2.80 (0.10)	2.50 (0.15)	2.60 (0.20)	3.00 (0.10)	2.95 (0.05)
Potassium	150	130 (1.5)	560 (2.0)	520 (1.0)	392 (1.5)	260 (3.0)
Phosphorus	530	560 (3.0)	810 (1.0)	690 (1.0)	630 (2.0)	590 (2.0)

a. Do the data provide sufficient evidence to conclude, at the 1% level of significance, that the mean available limestone in soil treated with 100% MMBL effluent exceeds that which is ordinarily found (original)?
b. Repeat part (a) for 50% MMBL effluent.
c. At the 1% significance level, do the data provide sufficient evidence to conclude that the mean available phosphorus in soil treated with 50% MMBL effluent exceeds that which is ordinarily found?
d. What assumptions are you making in solving parts (a)–(c)?
e. Use Minitab or some other statistical software to perform the hypothesis tests in parts (a)–(c). *(Hint: Since the raw data are unavailable, for each part you may first have to obtain a data set consisting of five numbers that has the same mean and standard deviation as the data set under consideration. That can be done by using the following fact: Let a and $b > 0$ be numbers and $x_1, x_2, \ldots, x_n$ be any data set having a nonzero standard deviation. Also let z_i be the sample z-score for x_i and $y_i = a + bz_i$. Then the data set $y_1, y_2, \ldots, y_n$ has mean a and sample standard deviation b.)*

Gertrude Cox

Gertrude Mary Cox was born on January 13, 1900, in Dayton, Iowa, the daughter of John and Emmaline Cox. She graduated from Perry High School, Perry, Iowa, in 1918. Between 1918 and 1925, she prepared to become a deaconess in the Methodist Episcopal Church.

In 1929 and 1931, Cox received a B.S. and an M.S., respectively, from Iowa State College in Ames. Her work there was directed by George W. Snedecor, and her degree was the first master's degree in statistics given by the department of mathematics at Iowa State.

From 1931 to 1933, Cox studied psychological statistics at the University of California at Berkeley. Snedecor meanwhile had established a new Statistical Laboratory at Iowa State, and in 1933 he asked her to be his assistant. This began her internationally influential career in statistics. Cox worked in the lab until becoming an Iowa State assistant professor in 1939.

In 1940 the committee in charge of filling a newly created position as head of the department of experimental statistics at North Carolina State College in Raleigh asked Snedecor for recommendations; he first named several male statisticians, then wrote, "... but if you would consider a woman for this position I would recommend Gertrude Cox of my staff." They did consider a woman and Cox accepted their offer.

In 1945 Cox organized and became director of the Institute of Statistics, which combined the teaching of statistics at the University of North Carolina and North Carolina State. Work conferences that Cox organized established the Institute as an international center for statistics. Cox also developed statistical programs throughout the South, referred to as "spreading the gospel according to St. Gertrude."

Cox's area of expertise was experimental design. She, with W. G. Cochran, wrote *Experimental Designs* (1950), recognized as the classic textbook on design and analysis of replicated experiments.

From 1960–1964, Cox was director of the Statistics Section of the Research Triangle Institute in Durham, North Carolina. She then retired, working only as a consultant. She died of leukemia on October 17, 1978, in Durham.

Inferences for Two Population Means

General Objective   In Chapters 8 and 9, we learned how to obtain confidence intervals and perform hypothesis tests for one population mean, μ. Frequently, however, inferential statistics is used to compare the means of two or more populations.

For instance, we might want to perform a hypothesis test to decide whether the mean age of buyers of new domestic cars is greater than the mean age of buyers of new imported cars; or we might want to find a confidence interval for the difference between the two mean ages. In this chapter we will study methods of making statistical inferences for two population means.

Chapter Outline   10.1   Large-Sample Inferences for Two Population Means Using Independent Samples

10.2   Inferences for the Means of Two Normal Populations Using Independent Samples (Standard Deviations Assumed Equal)

10.3   Inferences for the Means of Two Normal Populations Using Independent Samples (Standard Deviations Not Assumed Equal)

10.4   The Mann–Whitney Test (Optional)

10.5   Inferences for Two Population Means Using Paired Samples

10.6   Which Procedure Should Be Used? (Optional)

Case Study   *Breast Milk and IQ*
Does being fed breast milk as an infant affect a child's subsequent IQ? Several researchers attempted to answer that question for preterm (premature) babies in an article appearing in a 1992 issue of *The Lancet*. We will investigate a portion of that article in the case study at the end of this chapter.

10.1   LARGE-SAMPLE INFERENCES FOR TWO POPULATION MEANS USING INDEPENDENT SAMPLES

In this section we will examine methods of making statistical inferences for the means of two populations when dealing with large samples. The methods we will consider here require not only that the samples selected from the two populations be random but also that they be *independent*. Two samples are called **independent samples** if the sample selected from one of the populations has no effect or bearing on the sample selected from the other population.

We begin by discussing large-sample hypothesis tests for comparing the means of two populations using independent samples. Example 10.1 introduces the basic ideas involved in performing such tests.

Example 10.1   Introduces Hypothesis Tests for Comparing Two Population Means

Suppose we want to perform a hypothesis test to decide whether there is a difference between the mean salary of faculty teaching in public institutions and the mean salary of faculty teaching in private institutions. We can formulate the problem statistically in the following way: First, let the two populations in question be designated as Population 1 and Population 2:

> Population 1: All salaries of faculty teaching in public institutions.

> Population 2: All salaries of faculty teaching in private institutions.

Next, denote the mean salary of all faculty teaching in public institutions by μ_1 and the mean salary of all faculty teaching in private institutions by μ_2. Then the hypothesis test we want to perform can be stated as

> H_0: $\mu_1 = \mu_2$ (mean salaries are the same)
>
> H_a: $\mu_1 \neq \mu_2$ (mean salaries are different).

Roughly speaking, the hypothesis test can be carried out by proceeding in the following manner.

1. Take a random sample from each of the two populations.

2. Compute the mean, $\overline{x}_1$, of the sample from Population 1 and the mean, $\overline{x}_2$, of the sample from Population 2.

3. Reject the null hypothesis if $\overline{x}_1$ and $\overline{x}_2$ differ by too much; otherwise, do not reject the null hypothesis.

This process is pictured in Fig. 10.1.

FIGURE 10.1
Process for comparing
two population means

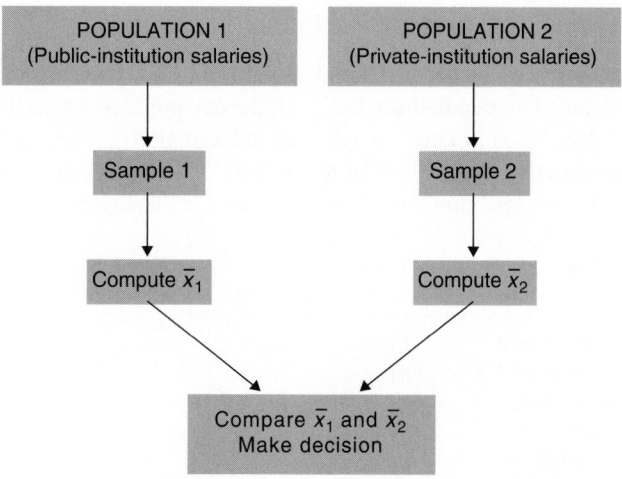

Suppose we take a random sample of 30 salaries from Population 1 and a random sample of 35 salaries from Population 2. Further suppose the salaries obtained are as depicted in Table 10.1, where they are given in thousands of dollars rounded to the nearest hundred.

TABLE 10.1
Annual salaries ($1000s)
for 30 faculty in public
institutions and 35 faculty
in private institutions

Sample 1 (public institutions)						Sample 2 (private institutions)						
34.6	37.5	54.3	52.4	40.0	54.3	35.7	57.1	38.3	86.2	48.5	46.4	35.0
32.8	76.4	41.2	31.2	24.2	52.8	59.0	46.7	58.9	59.3	37.8	44.3	76.7
30.7	38.1	50.8	33.0	51.7	70.9	31.7	38.3	43.7	35.6	43.8	34.5	26.5
42.7	62.7	31.4	38.3	63.5	23.9	50.2	50.5	34.6	42.9	24.2	22.8	28.8
37.8	41.8	37.6	56.3	47.4	39.8	42.7	75.8	47.6	47.6	50.7	55.9	71.2

The means of the two samples in Table 10.1 are

$$\bar{x}_1 = \frac{\Sigma x}{n_1} = \frac{1330.1}{30} = 44.34 \quad \text{and} \quad \bar{x}_2 = \frac{\Sigma x}{n_2} = \frac{1629.5}{35} = 46.56.$$

The question now is whether the difference of 2.22 ($2220) between these two sample means can reasonably be attributed to sampling error or whether the difference is large enough to indicate that the two populations have different means. To answer this question, we need to know how likely it would be to get such a difference in sample means if in fact the two populations have equal means. And that likelihood can be determined once we have obtained the probability distribution of the difference, $\bar{x}_1 - \bar{x}_2$, between two sample means—the **sampling distribution of the difference between two means.** Let's therefore examine that sampling distribution. After we have done that, we will complete the hypothesis test posed in this example. ■

The Sampling Distribution of the Difference Between Two Means (Large and Independent Samples)

First we need to become familiar with the notation for parameters and statistics when analyzing two populations. Let's call the two populations under consideration Population 1 and Population 2. Then, as indicated in Example 10.1, we use a subscript 1 when referring to parameters or statistics for Population 1 and a subscript 2 when referring to parameters or statistics for Population 2. This notation is displayed in Table 10.2.

TABLE 10.2
Notation for parameters
and statistics when
considering two populations

	Population 1	Population 2
Population mean	μ_1	μ_2
Population std. dev.	σ_1	σ_2
Sample mean	$\bar{x}_1$	$\bar{x}_2$
Sample std. dev.	s_1	s_2
Sample size	n_1	n_2

Next we present formulas that express the mean and standard deviation of the random variable $\bar{x}_1 - \bar{x}_2$ in terms of the population parameters and the sample sizes:

$$\mu_{\bar{x}_1 - \bar{x}_2} = \mu_1 - \mu_2 \quad \text{and} \quad \sigma_{\bar{x}_1 - \bar{x}_2} = \sqrt{(\sigma_1^2/n_1) + (\sigma_2^2/n_2)}.$$

The derivation of these two formulas is considered in Exercise 10.17.

Finally, by the central limit theorem (page 424), we know that the random variable $\bar{x}$ is approximately normally distributed for large samples. So if both n_1 and n_2 are large, the random variables $\bar{x}_1$ and $\bar{x}_2$ are both approximately normally distributed. From this we can prove mathematically that for independent samples, the random variable $\bar{x}_1 - \bar{x}_2$ is also approximately normally distributed.[†] Thus we have the following fact.

KEY FACT 10.1

THE SAMPLING DISTRIBUTION OF THE DIFFERENCE BETWEEN TWO MEANS (LARGE AND INDEPENDENT SAMPLES)

Suppose a random sample of size $n_1 \geq 30$ is to be taken from a population with mean μ_1 and standard deviation σ_1, and a random sample of size $n_2 \geq 30$ is to be taken from a population with mean μ_2 and standard deviation σ_2. Further suppose the two samples are to be selected independently. Then the random variable $\bar{x}_1 - \bar{x}_2$ is approximately normally distributed and has mean $\mu_{\bar{x}_1 - \bar{x}_2} = \mu_1 - \mu_2$ and standard deviation $\sigma_{\bar{x}_1 - \bar{x}_2} = \sqrt{(\sigma_1^2/n_1) + (\sigma_2^2/n_2)}$. Thus the standardized random variable,

$$z = \frac{(\bar{x}_1 - \bar{x}_2) - (\mu_1 - \mu_2)}{\sqrt{(\sigma_1^2/n_1) + (\sigma_2^2/n_2)}},$$

has approximately the standard normal distribution.

[†] Qualitatively, this means that for any two sample sizes n_1 and n_2, both large, a histogram of all possible differences of sample means, $\bar{x}_1 - \bar{x}_2$, will be roughly bell-shaped.

Large-Sample Hypothesis Tests for Two Population Means Using Independent Samples

Now that we have obtained the sampling distribution of the difference between two means when dealing with large and independent samples, we can develop a procedure for performing a hypothesis test. The null hypothesis will be

$$H_0: \mu_1 = \mu_2 \text{ (population means are equal).}$$

If the null hypothesis is true, then $\mu_1 - \mu_2 = 0$, and the standardized random variable in Key Fact 10.1 becomes

(1)
$$z = \frac{\bar{x}_1 - \bar{x}_2}{\sqrt{(\sigma_1^2/n_1) + (\sigma_2^2/n_2)}}.$$

This random variable can serve as the test statistic for a hypothesis test concerning two population means in the same way the random variable

$$z = \frac{\bar{x} - \mu_0}{\sigma/\sqrt{n}}$$

does for a hypothesis test concerning one population mean. The numerator of the test statistic for two means, given in Equation (1), measures the difference between the sample means. The denominator standardizes the test statistic so that the standard-normal table, Table II, can be used to obtain the critical value(s).

Let's now discuss the alternative hypothesis for a hypothesis test concerning the means of two populations. As we know, there are three possibilities for the alternative hypothesis:

1. *Two-tailed test.* If we want to decide whether the means of Population 1 and Population 2 are different, then the alternative hypothesis should be

$$H_a: \mu_1 \neq \mu_2.$$

If the null hypothesis, $H_0: \mu_1 = \mu_2$, is true, then we would expect the sample means $\bar{x}_1$ and $\bar{x}_2$ to be roughly equal (why?). Thus the null hypothesis is rejected in favor of the alternative hypothesis if $\bar{x}_1$ and $\bar{x}_2$ differ by too much, that is, if the test statistic in Equation (1) is either too small or too large. So the rejection region consists of two "tails" of the standard normal distribution, just as in two-tailed tests for one population mean.

2. *Left-tailed test.* If we want to decide whether the mean of Population 1 is smaller than the mean of Population 2, then the alternative hypothesis should be

$$H_a: \mu_1 < \mu_2.$$

In this case the null hypothesis, $H_0: \mu_1 = \mu_2$, is rejected in favor of the alternative hypothesis if $\bar{x}_1$ is too much smaller than $\bar{x}_2$, that is, if the test statistic in Equation (1) is too small. So the rejection region consists of a left "tail" of the standard normal distribution, just as in left-tailed tests for one population mean.

3. *Right-tailed test.* If we want to decide whether the mean of Population 1 is larger than the mean of Population 2, then the alternative hypothesis should be

$$H_a: \mu_1 > \mu_2.$$

In this case the null hypothesis, $H_0: \mu_1 = \mu_2$, is rejected in favor of the alternative hypothesis if $\bar{x}_1$ is too much larger than $\bar{x}_2$, that is, if the test statistic in Equation (1) is too large. So the rejection region consists of a right "tail" of the standard normal distribution, just as in right-tailed tests for one population mean.

With the above discussion in mind, we present Procedure 10.1.

PROCEDURE 10.1

THE TWO-SAMPLE z-TEST FOR TWO POPULATION MEANS WITH NULL HYPOTHESIS $H_0: \mu_1 = \mu_2$

ASSUMPTIONS

1. Independent samples
2. Large samples ($n_1 \geq 30$, $n_2 \geq 30$)

Step 1   State the null and alternative hypotheses.

Step 2   Decide on the significance level, α.

Step 3   The critical value(s)
        a. for a two-tailed test are $\pm z_{\alpha/2}$.
        b. for a left-tailed test is $-z_\alpha$.
        c. for a right-tailed test is z_α.
        Use Table II to find the critical value(s).

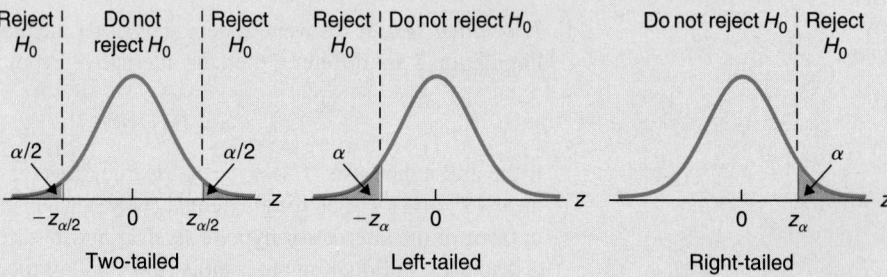

| Two-tailed | Left-tailed | Right-tailed |

Step 4   Compute the value of the test statistic

$$z = \frac{\bar{x}_1 - \bar{x}_2}{\sqrt{(\sigma_1^2/n_1) + (\sigma_2^2/n_2)}}.$$

If σ_1 is unknown, use s_1 in its place; if σ_2 is unknown, use s_2 in its place.

Step 5   If the value of the test statistic falls in the rejection region, then reject H_0; otherwise, do not reject H_0.

Step 6   State the conclusion in words.

Procedure 10.1 provides us with a method for performing a large-sample hypothesis test to compare the means of two populations using independent samples. For ease of reference, we will often refer to Procedure 10.1 as the **two-sample z-test.**

In Step 4 of Procedure 10.1 it is stated that if either population standard deviation is unknown, then the corresponding sample standard deviation should be used in its place. This substitution is acceptable because for large samples, the sample standard deviations are likely to be good approximations of the population standard deviations.

Although strictly speaking the only conditions required for using Procedure 10.1 are that the samples be independent and the sample sizes be large, it is important to watch for outliers. Because the sample mean (and sample standard deviation) are not resistant to outliers, the presence of outliers can unduly influence a hypothesis test for comparing two population means, especially when the sample sizes are only moderately large.

Therefore we should always first look at the sample data using histograms, stem-and-leaf diagrams, or other graphical displays. If outliers are detected, we must try to decide whether to remove or retain them.

Example 10.2   Illustrates Procedure 10.1

We now return to the salary problem posed in Example 10.1. Recall that we want to perform a hypothesis test to decide whether the mean salaries of faculty teaching in public and private institutions differ. Random samples of 30 faculty teaching in public institutions and 35 faculty teaching in private institutions yield the data in Table 10.1 on page 575. Do the data provide sufficient evidence to conclude that there is a difference in mean salaries for faculty teaching in public and private institutions? Perform the required hypothesis test at the 5% significance level.

SOLUTION   Preliminary data analyses of the two samples in Table 10.1 reveal no outliers. Since both sample sizes are large ($n_1 = 30$ and $n_2 = 35$) and the description of the problem implies that the samples are independent, we can employ Procedure 10.1 to conduct the required hypothesis test.

Step 1   *State the null and alternative hypotheses.*

The null and alternative hypotheses are

$$H_0: \mu_1 = \mu_2 \text{ (mean salaries are the same)}$$
$$H_a: \mu_1 \neq \mu_2 \text{ (mean salaries are different)},$$

where μ_1 and μ_2 are, respectively, the mean salaries of all faculty in public and private institutions. Note that the hypothesis test is two-tailed since a not-equal sign ($\neq$) appears in the alternative hypothesis.

Step 2   *Decide on the significance level, α.*

We are to perform the hypothesis test at the 5% significance level. So $\alpha = 0.05$.

Step 3   *The critical values for a two-tailed test are* $\pm z_{\alpha/2}$.

Since $\alpha = 0.05$, we find from Table II (or the last row of Table IV) that the critical values are $\pm z_{0.05/2} = \pm z_{0.025} = \pm 1.96$, as seen in Fig. 10.2.

FIGURE 10.2
Criterion for deciding whether or
not to reject the null hypothesis

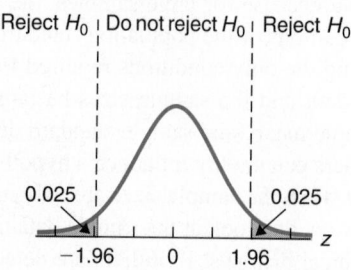

Reject H_0 ┊ Do not reject H_0 ┊ Reject H_0

0.025   0.025

−1.96   0   1.96 z

Step 4   *Compute the value of the test statistic*

$$z = \frac{\overline{x}_1 - \overline{x}_2}{\sqrt{(\sigma_1^2/n_1) + (\sigma_2^2/n_2)}}.$$

We have $n_1 = 30$ and $n_2 = 35$. The mean of each sample was obtained in Example 10.1: $\overline{x}_1 = 44.34$ and $\overline{x}_2 = 46.56$. Because the population standard deviations are unknown, we use the sample standard deviations in their place. We find in the usual way that $s_1 = 13.13$ and $s_2 = 14.94$. Thus the value of the test statistic is

$$z = \frac{44.34 - 46.56}{\sqrt{(13.13^2/30) + (14.94^2/35)}} = -0.64.$$

Step 5   *If the value of the test statistic falls in the rejection region, reject H_0; otherwise, do not reject H_0.*

From Step 4 we see that the value of the test statistic is $z = -0.64$, which does not fall in the rejection region. Hence we do not reject H_0.

Step 6   *State the conclusion in words.*

 The test results are not statistically significant at the 5% level; that is, at the 5% significance level, we have insufficient evidence to conclude that there is a difference in mean salaries for faculty in public and private institutions.   ∎

The *P*-value approach to hypothesis testing can also be used to carry out the hypothesis test in Example 10.2 and to assess more precisely the evidence against the null hypothesis. From Step 4 we see that the value of the test statistic is $z = -0.64$. Consulting Table II, we find that $P = 2 \cdot 0.2611 = 0.5222$. Because the *P*-value exceeds the significance level of 0.05, we cannot reject H_0. Furthermore, by referring to Table 9.12 on page 530, we see that the data provide no evidence against the null hypothesis of equal mean salaries.

Large-Sample Confidence Intervals for the Difference Between Two Population Means Using Independent Samples

We can also use Key Fact 10.1 on page 576 to derive Procedure 10.2, a confidence-interval procedure for the difference between two population means, when dealing with large and independent samples. (See Exercise 10.18 for details of the derivation.) For convenience we often refer to Procedure 10.2 as the **two-sample z-interval procedure.**

PROCEDURE 10.2

THE TWO-SAMPLE z-INTERVAL PROCEDURE FOR THE DIFFERENCE BETWEEN TWO POPULATION MEANS

ASSUMPTIONS

1. Independent samples
2. Large samples ($n_1 \geq 30$, $n_2 \geq 30$)

Step 1 For a confidence level of $1 - \alpha$, use Table II to find $z_{\alpha/2}$.

Step 2 The endpoints of the confidence interval for $\mu_1 - \mu_2$ are

$$(\bar{x}_1 - \bar{x}_2) \pm z_{\alpha/2} \cdot \sqrt{(\sigma_1^2/n_1) + (\sigma_2^2/n_2)}.$$

If σ_1 is unknown, use s_1 in its place; if σ_2 is unknown, use s_2 in its place.

Example 10.3 Illustrates Procedure 10.2

Consider once more the situation described in Example 10.1. Find a 95% confidence interval for the difference, $\mu_1 - \mu_2$, between the mean salaries of faculty teaching in public and private institutions.

SOLUTION We apply Procedure 10.2.

Step 1 *For a confidence level of $1 - \alpha$, use Table II to find $z_{\alpha/2}$.*

For a 95% confidence interval, the confidence level is $0.95 = 1 - 0.05$. So $\alpha = 0.05$. From Table II we find that $z_{\alpha/2} = z_{0.05/2} = z_{0.025} = 1.96$.

Step 2 *The endpoints of the confidence interval for $\mu_1 - \mu_2$ are*

$$(\bar{x}_1 - \bar{x}_2) \pm z_{\alpha/2} \cdot \sqrt{(\sigma_1^2/n_1) + (\sigma_2^2/n_2)}.$$

Because the population standard deviations are unknown, we will use the sample standard deviations in their place. From Step 1, $z_{\alpha/2} = 1.96$. Also, $n_1 = 30$, $n_2 = 35$, and from Example 10.2 we know that $\bar{x}_1 = 44.34$, $s_1 = 13.13$, $\bar{x}_2 = 46.56$, and $s_2 = 14.94$. Hence the endpoints of the confidence interval for $\mu_1 - \mu_2$ are

$$(44.34 - 46.56) \pm 1.96 \cdot \sqrt{(13.13^2/30) + (14.94^2/35)},$$

or -2.22 ± 6.82. Thus a 95% confidence interval for $\mu_1 - \mu_2$ is from -9.04 to 4.60. We can be 95% confident that the difference, $\mu_1 - \mu_2$, between the mean salaries of faculty teaching in public and private institutions is somewhere between $-\$9040$ and $\$4600$. ■

EXERCISES 10.1

10.1 Consider the quantities μ_1, σ_1, $\bar{x}_1$, s_1, μ_2, σ_2, $\bar{x}_2$, and s_2.
a. Which quantities represent parameters and which represent statistics?
b. Which quantities are fixed numbers and which are random variables?

In each of Exercises 10.2–10.7, perform the required hypothesis test using either the classical approach or the P-value approach.

10.2 Surveys are conducted by the Northwestern University Placement Center, Evanston, Illinois, on starting salaries for college graduates. Results of the surveys can be found in *The Northwestern Lindquist-Endicott Report.* In the following table, we have reproduced the starting annual salaries obtained from independent random samples of 32 accounting graduates and 35 liberal-arts graduates. The data are in thousands of dollars.

Accounting				Liberal arts				
28.9	24.4	27.3	24.9	26.0	22.8	27.3	26.6	27.0
27.4	29.3	30.8	26.6	24.7	29.8	24.3	23.7	24.0
28.0	26.0	25.7	27.0	25.1	27.1	25.6	25.0	24.7
30.2	29.9	28.8	29.3	26.8	26.4	23.1	25.5	
29.5	24.4	26.4	28.5	25.9	25.2	25.1	23.2	
28.9	29.4	29.8	28.2	24.3	28.7	26.0	25.2	
26.2	26.1	25.0	28.7	26.9	25.1	26.8	26.5	
25.5	26.5	29.4	30.6	27.4	22.3	22.3	26.5	

a. At the 5% significance level, can we conclude that accounting graduates have a higher mean starting salary than liberal-arts graduates? Assume the population standard deviations of starting salaries are 1.73 ($1730) for accounting graduates and 1.82 ($1820) for liberal-arts graduates. *(Note: The sum of the accounting data is 887.6, and the sum of the liberal-arts data is 892.9.)*
b. Identify the study as a designed experiment or an observational study. Explain your answer.
c. Interpret the results of the hypothesis test in light of your answer to part (b).

10.3 The U.S. National Center for Health Statistics compiles data on the length of stay by patients in short-term hospitals and publishes its findings in *Vital and Health Statistics.* Independent samples of 40 male patients and 35 female patients gave the following data on length of stay, in days.

Male					Female				
4	4	12	18	9	14	7	15	1	12
6	12	10	3	6	1	3	7	21	4
15	7	3	55	1	1	5	4	4	3
2	10	13	5	7	5	18	12	5	1
1	23	9	2	1	7	7	2	15	4
17	2	24	11	14	9	10	7	3	6
6	2	1	8	1	5	9	6	2	14
3	19	3	1	13					

At the 10% significance level, do the data provide sufficient evidence to conclude that, on the average, males stay in the hospital longer than females? Assume that $\sigma_1 = 7.5$ days and $\sigma_2 = 6.8$ days. *(Note: The sum of the male data is 363 days, and the sum of the female data is 249 days.)*

10.4 An agronomist wants to determine whether a larger corn crop can be obtained if sterilized males of an insect pest are introduced to control the pest population instead of using an insecticide. Eighty 1-acre plots are randomly divided into two groups of forty 1-acre plots. The insecticide is used on each 1-acre plot in the first group and the sterilized male insects on each 1-acre plot in the second group. The yields, in bushels, are as shown in the following table.

Insecticide					Sterilized males				
109	101	97	89	100	105	109	110	118	109
98	98	94	99	104	113	111	111	99	112
103	88	108	102	106	106	117	99	107	119
97	105	102	104	101	110	111	103	110	108
101	100	105	110	96	104	102	111	114	114
102	95	100	95	109	122	117	101	109	109
91	98	113	91	95	102	109	103	109	106
106	98	101	99	96	107	107	111	128	109

a. Do the data provide sufficient evidence to conclude that the use of sterilized male insects is more effective than the insecticide in controlling the insect pest? Use $\alpha = 0.01$. (*Note:* $\bar{x}_1 = 100.15$, $s_1 = 5.73$, $\bar{x}_2 = 109.53$, $s_2 = 6.06$.)

b. Is this a designed experiment or an observational study? Explain your answer.

c. Interpret the results of the hypothesis test in view of your answer to part (b).

10.5 The U.S. Energy Information Administration publishes data on residential energy consumption and expenditures in *Residential Energy Consumption Survey: Consumption and Expenditures.* Suppose you want to decide whether last year's mean annual fuel expenditure for households using natural gas is different from that for households using only electricity. At the 5% significance level, what conclusion would you draw given the data, in dollars, shown in the following table? (*Note:* $\bar{x}_1 = 1497.6$, $s_1 = 160.35$, $\bar{x}_2 = 1243.6$, $s_2 = 165.13$.)

Natural gas			Electricity			
2002	1456	1394	1376	1452	1235	1480
1541	1321	1338	1185	1327	1059	1400
1495	1526	1358	1227	1102	1168	1070
1801	1478	1376	1180	1221	1351	1014
1579	1375	1664	1461	1102	976	1394
1305	1458	1369	1379	987	1002	1532
1495	1507	1636	1450	1177	1150	
1698	1249	1377	1352	1266	1109	
1648	1557	1491	949	1351	1259	
1505	1355	1574	1179	1393	1456	

10.6 Researchers studying obesity wanted to compare the effectiveness of dieting with exercise to dieting without exercise. Seventy-three patients were randomly divided into two groups. Group 1, composed of 37 patients, was put on a program of dieting with exercise. Group 2, composed of 36 patients, dieted only. The results for weight loss, in pounds, after 2 months are summarized in the following table.

Diet-with-exercise group	Diet-only group
$\bar{x}_1 = 16.8$ lb	$\bar{x}_2 = 17.1$ lb
$s_1 = 3.5$ lb	$s_2 = 5.2$ lb

a. At the 0.05 significance level, determine whether there is a difference between the two treatments.

b. Identify the study as a designed experiment or an observational study. Explain your answer.

c. Interpret the results of the hypothesis test in light of your answer to part (b).

10.7 A regional sales manager chooses two similar offices in which to study the effectiveness of a new training program aimed at increasing sales. One office institutes the training program, and the other does not. The office that does not, Office 1, has 47 salespeople. For this office, the mean sales per person over the next month is $3197 with a standard deviation of $102. Office 2, the office that does institute the training program, has 51 salespeople. For this office, the mean sales per person over the next month is $3229 with a standard deviation of $107.

a. At the 5% significance level, does the training program appear to increase sales?

b. Repeat part (a) at the 10% significance level.

In Exercises 10.8–10.13, use Procedure 10.2 on page 581 to determine the required confidence interval.

10.8 Refer to Exercise 10.2.

a. Determine a 90% confidence interval for the difference, $\mu_1 - \mu_2$, between the mean starting salaries of accounting and liberal-arts graduates.

b. Interpret your result.

10.9 Refer to Exercise 10.3.

a. Determine an 80% confidence interval for the difference, $\mu_1 - \mu_2$, between the mean lengths of stay in short-term hospitals by males and females.

b. Interpret your result.

10.10 Refer to Exercise 10.4.

a. Determine a 98% confidence interval for the difference, $\mu_1 - \mu_2$, between the mean yields of corn when the insecticide is used to control the insect pest and when sterilized males are used.

b. Interpret your result.

10.11 Refer to Exercise 10.5.

a. Obtain a 95% confidence interval for the difference between last year's mean fuel expenditures for households using natural gas and those using only electricity.

b. Interpret your result.

10.12 Refer to Exercise 10.6.

a. Obtain a 95% confidence interval for the difference between the mean weight losses after 2 months using the diet-with-exercise method and the diet-only method.

b. Interpret your result.

10.13 Refer to Exercise 10.7.

a. Obtain a 90% confidence interval for the difference between the mean sales per person per month for those who do not participate in the training program and for those who do.

b. Repeat part (a) using an 80% confidence level.

10.14 (Computer exercise) To determine the sampling distribution of the difference between two means for large and independent samples, we need the fact that the difference between two independent normally distributed random variables is also normally distributed. In this exercise you are asked to perform a computer simulation to illustrate that fact.

a. Simulate the taking of 2000 observations from a normally distributed population having mean 100 and standard deviation 16.

b. Repeat part (a) for a normally distributed population having mean 120 and standard deviation 12.

c. Determine the difference between each pair of observations in parts (a) and (b); that is, if x_{1i} is the ith observation from part (a) and x_{2i} is the ith observation from part (b), compute $x_{1i} - x_{2i}$ for $i = 1, 2, \ldots, 2000$.

d. Obtain a histogram of the 2000 differences in part (c). Why is the histogram bell-shaped?

10.15 (Computer exercise) In this exercise you are to perform a computer simulation to illustrate the sampling distribution of the difference between two means for large and independent samples (Key Fact 10.1 on page 576).

a. In Exercise 7.50 on page 431, we defined an *exponentially distributed* population with mean μ and pointed out that the standard deviation of such a population also equals μ. Simulate the taking of 300 random samples of size 40 from an exponentially distributed population with $\mu = 4$. Then obtain the sample mean of each of the 300 samples.

b. In Problem 13 on page 434 of Chapter 7, we defined a *uniformly distributed* population with parameters α and β and pointed out that the mean and standard deviation of such a population are $(\alpha + \beta)/2$ and $(\beta - \alpha)/\sqrt{12}$, respectively. Simulate the taking of 300 random samples of size 35 from a uniformly distributed population with parameters $\alpha = 2$ and $\beta = 4$. Then obtain the sample mean of each of the 300 samples.

c. Find the difference, $\bar{x}_1 - \bar{x}_2$, for each of the 300 pairs of sample means obtained in parts (a) and (b).

d. Obtain the mean, the standard deviation, and a histogram of the 300 differences found in part (c).

e. Theoretically, what are the mean, standard deviation, and distribution of all possible differences, $\bar{x}_1 - \bar{x}_2$? (*Hint:* Refer to Key Fact 10.1 on page 576.)

f. Compare your results from parts (d) and (e).

10.16   The relationship between hypothesis tests and confidence intervals. Use the results that you obtained in Exercises 10.5, 10.6, 10.11, and 10.12 to complete the following statement: A hypothesis test of H_0: $\mu_1 = \mu_2$ versus H_a: $\mu_1 \neq \mu_2$ at the significance level α will lead to rejection of the null hypothesis if and only if the number _____ does not lie in the $(1 - \alpha)$-level confidence interval for $\mu_1 - \mu_2$.

10.17 In this exercise you will derive the formulas, given on page 576, for the mean and standard deviation of the random variable $\bar{x}_1 - \bar{x}_2$. Suppose independent random samples of sizes n_1 and n_2 are to be taken, respectively, from populations having means μ_1 and μ_2 and standard deviations σ_1 and σ_2.

a. Use the results of Exercise 5.26(e) and Exercise 5.27(f), both on page 289, to show that

$$\mu_{\bar{x}_1 - \bar{x}_2} = \mu_{\bar{x}_1} - \mu_{\bar{x}_2}$$

and

$$\sigma_{\bar{x}_1 - \bar{x}_2} = \sqrt{\sigma_{\bar{x}_1}^2 + \sigma_{\bar{x}_2}^2}.$$

b. Apply the formulas $\mu_{\bar{x}} = \mu$ and $\sigma_{\bar{x}} = \sigma/\sqrt{n}$ to the results in part (a) to derive the formulas

$$\mu_{\bar{x}_1 - \bar{x}_2} = \mu_1 - \mu_2$$

and

$$\sigma_{\bar{x}_1 - \bar{x}_2} = \sqrt{(\sigma_1^2/n_1) + (\sigma_2^2/n_2)}.$$

10.18 In this exercise you will verify Procedure 10.1 and Procedure 10.2. Suppose independent random samples of sizes $n_1 \geq 30$ and $n_2 \geq 30$ are to be taken, respectively, from populations with means μ_1 and μ_2 and standard deviations σ_1 and σ_2. Assume as known that the random variable $\bar{x}_1 - \bar{x}_2$ is approximately normally distributed.

a. Use the results of part (b) of Exercise 10.17 to show that the random variable

$$z = \frac{(\bar{x}_1 - \bar{x}_2) - (\mu_1 - \mu_2)}{\sqrt{(\sigma_1^2/n_1) + (\sigma_2^2/n_2)}}$$

has approximately the standard normal distribution. (This verifies Procedure 10.1.)

b. Use part (a) to show that

$$P\left(-z_{\alpha/2} < \frac{(\bar{x}_1 - \bar{x}_2) - (\mu_1 - \mu_2)}{\sqrt{(\sigma_1^2/n_1) + (\sigma_2^2/n_2)}} < z_{\alpha/2}\right) \approx 1 - \alpha.$$

c. Use part (b) to verify Procedure 10.2.

10.2   INFERENCES FOR THE MEANS OF TWO NORMAL POPULATIONS USING INDEPENDENT SAMPLES (STANDARD DEVIATIONS ASSUMED EQUAL)

In Section 10.1 we learned how to perform inferences (hypothesis tests and confidence intervals) to compare the means of two populations when the samples are chosen independently and the sample sizes are large. Several methods exist for performing such inferences that do not require large sample sizes.

Included among those methods are two that apply when the populations being sampled are normally distributed. One of these methods requires that the two populations have equal standard deviations; the other does not impose this restriction. We will study the first method in this section and the second in Section 10.3.

The Sampling Distribution of the Difference Between Two Means (Normal Populations and Independent Samples)

To begin, we need to determine the probability distribution of the random variable $\bar{x}_1 - \bar{x}_2$ when independent samples are taken from two normally distributed populations. From Key Fact 7.4 on page 420, we know that when sampling from a normally distributed population, the random variable $\bar{x}$ is normally distributed, regardless of the sample size, n.

Using that fact, it can be proved that when independent sampling is done from two normally distributed populations, the random variable $\bar{x}_1 - \bar{x}_2$ is also normally distributed, regardless of the sample sizes, n_1 and n_2.[†] Combining this last result with the formulas given on page 576 for the mean and standard deviation of $\bar{x}_1 - \bar{x}_2$, we obtain the following fact.

KEY FACT 10.2

THE SAMPLING DISTRIBUTION OF THE DIFFERENCE BETWEEN TWO MEANS (NORMAL POPULATIONS AND INDEPENDENT SAMPLES)

Suppose a random sample of size n_1 is to be taken from a normally distributed population with mean μ_1 and standard deviation σ_1, and a random sample of size n_2 is to be taken from a normally distributed population with mean μ_2 and standard deviation σ_2. Further suppose the two samples are to be selected independently. Then the random variable $\bar{x}_1 - \bar{x}_2$ is also normally distributed and has mean $\mu_{\bar{x}_1 - \bar{x}_2} = \mu_1 - \mu_2$ and standard deviation $\sigma_{\bar{x}_1 - \bar{x}_2} = \sqrt{(\sigma_1^2/n_1) + (\sigma_2^2/n_2)}$. Thus the standardized random variable,

$$z = \frac{(\bar{x}_1 - \bar{x}_2) - (\mu_1 - \mu_2)}{\sqrt{(\sigma_1^2/n_1) + (\sigma_2^2/n_2)}},$$

has the standard normal distribution.

[†]   Qualitatively, this means that for any two sample sizes n_1 and n_2, a histogram of all possible differences of sample means, $\bar{x}_1 - \bar{x}_2$, will be bell-shaped.

Hypothesis Tests for the Means of Two Normal Populations with Equal Standard Deviations Using Independent Samples

We will now develop a procedure for performing a hypothesis test to compare the means of two normally distributed populations with equal, but unknown, standard deviations using independent samples. Our immediate goal is to find a test statistic for such a test.

Let's use σ to denote the common standard deviation of the two populations. According to Key Fact 10.2, when independent samples are taken from two normally distributed populations, the random variable

$$z = \frac{(\bar{x}_1 - \bar{x}_2) - (\mu_1 - \mu_2)}{\sqrt{(\sigma_1^2/n_1) + (\sigma_2^2/n_2)}}$$

has the standard normal distribution. Replacing σ_1 and σ_2 in the above expression by their common value σ and using some algebra, we obtain the random variable

$$(2) \qquad z = \frac{(\bar{x}_1 - \bar{x}_2) - (\mu_1 - \mu_2)}{\sigma\sqrt{(1/n_1) + (1/n_2)}}.$$

Of course, this random variable cannot be used as a basis for obtaining the required test statistic since the common population standard deviation, σ, is unknown.

Consequently, we need to use sample information to estimate the unknown population standard deviation, σ, or equivalently, the population variance, σ^2. The best way to do this is to regard the sample variances, s_1^2 and s_2^2, as two estimates of σ^2 and then **pool** those estimates by weighting them according to sample size (actually by degrees of freedom). Thus our estimate for the common population variance, σ^2, is

$$s_p^2 = \frac{(n_1 - 1)s_1^2 + (n_2 - 1)s_2^2}{n_1 + n_2 - 2}$$

and hence for the common population standard deviation, σ, is

$$s_p = \sqrt{\frac{(n_1 - 1)s_1^2 + (n_2 - 1)s_2^2}{n_1 + n_2 - 2}}.$$

The subscript "p" stands for "pooled," and the quantity s_p is called the **pooled sample standard deviation.** In summary, we use the pooled sample standard deviation, s_p, as our estimate for the common, but unknown, standard deviation, σ, of the two populations.

Replacing the unknown population standard deviation, σ, in Equation (2) by its estimate, s_p, we get the random variable

$$\frac{(\bar{x}_1 - \bar{x}_2) - (\mu_1 - \mu_2)}{s_p\sqrt{(1/n_1) + (1/n_2)}},$$

which *can* be used as a basis for obtaining the required test statistic. However, unlike the random variable in Equation (2), this random variable does not have the standard normal distribution. But its distribution is one we are familiar with—a t-distribution.

KEY FACT 10.3

DISTRIBUTION OF POOLED-t STATISTIC

Suppose independent random samples of sizes n_1 and n_2 are to be taken from two normally distributed populations with means μ_1 and μ_2, respectively. Further suppose the standard deviations of the two populations are equal. Then the random variable

$$t = \frac{(\bar{x}_1 - \bar{x}_2) - (\mu_1 - \mu_2)}{s_p\sqrt{(1/n_1) + (1/n_2)}}$$

has the t-distribution with df $= n_1 + n_2 - 2$.

In view of Key Fact 10.3, we see that for a hypothesis test with null hypothesis H_0: $\mu_1 = \mu_2$, we can use the random variable

$$t = \frac{\bar{x}_1 - \bar{x}_2}{s_p\sqrt{(1/n_1) + (1/n_2)}}$$

as the test statistic and obtain the critical value(s) from the t-table, Table IV.

PROCEDURE 10.3

THE POOLED t-TEST FOR TWO POPULATION MEANS WITH NULL HYPOTHESIS H_0: $\mu_1 = \mu_2$

ASSUMPTIONS

1. Independent samples
2. Normal populations
3. Equal population standard deviations

Step 1   State the null and alternative hypotheses.

Step 2   Decide on the significance level, α.

Step 3   The critical value(s)
   a. for a two-tailed test are $\pm t_{\alpha/2}$,
   b. for a left-tailed test is $-t_\alpha$,
   c. for a right-tailed test is t_α,
   with df $= n_1 + n_2 - 2$. Use Table IV to find the critical value(s).

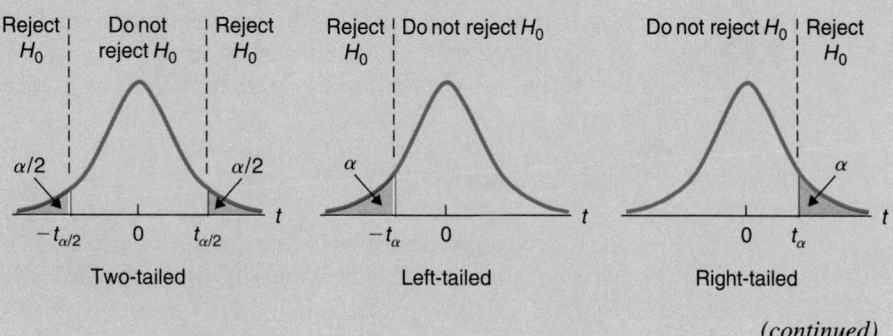

(continued)

Step 4   Compute the value of the test statistic

$$t = \frac{\overline{x}_1 - \overline{x}_2}{s_\mathrm{p}\sqrt{(1/n_1) + (1/n_2)}}$$

where

$$s_\mathrm{p} = \sqrt{\frac{(n_1 - 1)s_1^2 + (n_2 - 1)s_2^2}{n_1 + n_2 - 2}}.$$

Step 5   If the value of the test statistic falls in the rejection region, then reject H_0; otherwise, do not reject H_0.

Step 6   State the conclusion in words.

Before we apply Procedure 10.3, often called the **pooled t-test,** several comments are in order. In Step 4, the pooled sample standard deviation, s_p, is calculated. It always lies between the two sample standard deviations, s_1 and s_2. If you calculate s_p and it does not lie between s_1 and s_2, then you made an error.

Next let's discuss the three assumptions for the pooled t-test. The independent-samples assumption (Assumption 1) is essential; the samples must be independent or the procedure does not apply. As with our previous t-procedures, the pooled t-test is robust to moderate deviations of the normality assumption (Assumption 2). It is also reasonably robust to moderate deviations of the equal-standard-deviations assumption (Assumption 3), provided the sample sizes are roughly equal. We will have more to say about the robustness of the pooled t-test at the end of Section 10.3.

As before, we can check the normality assumption using normal probability plots. The equal-standard-deviations assumption is more difficult to check, especially when the sample sizes are small.[†] We recommend checking it by informally comparing the standard deviations of the two samples and by viewing together stem-and-leaf diagrams, histograms, or boxplots of the two samples (the same scales should be used for each pair of graphs).

The equal-standard-deviations assumption is sometimes checked by performing a formal hypothesis test, called an F-test for the equality of two standard deviations. We do not recommend this latter procedure because the F-test is extremely nonrobust to deviations from normality: unless the populations are very close to normally distributed, the F-test may yield unreliable results. As the noted statistician George E. P. Box remarked: "To make a preliminary test on variances [standard deviations] is rather like putting to sea in a rowing boat to find out whether conditions are sufficiently calm for an ocean liner to leave port!"

Example 10.4   Illustrates Procedure 10.3

The U.S. National Center for Health Statistics collects data on the daily intake of selected nutrients by race and income level. The results obtained are published in *Vital and Health*

[†] For this reason we generally suggest using the nonpooled t-test, which we will discuss in Section 10.3.

Statistics. Suppose we want to compare the mean protein intake of all people with incomes below the poverty level to that of all people with incomes above the poverty level.

The data in Table 10.3 display the protein intakes, in grams, over a 24-hour period for independent random samples of 15 people with incomes below the poverty level and 10 people with incomes above the poverty level. At the 5% significance level, do the data provide sufficient evidence to conclude that the mean protein intake of all people with incomes below the poverty level is less than the mean protein intake of all people with incomes above the poverty level?

TABLE 10.3

Samples of daily protein intakes (grams)

Below poverty level			Above poverty level	
51.4	49.7	72.0	86.0	69.0
76.7	65.8	55.0	59.7	80.2
73.7	62.1	79.7	68.6	78.1
66.2	75.8	65.4	98.6	69.8
65.5	62.0	73.3	87.7	77.2

SOLUTION First we present in Table 10.4 the required summary statistics for the two samples in Table 10.3. These statistics are obtained in the usual way.

TABLE 10.4

Summary statistics for the samples in Table 10.3

Below poverty level	Above poverty level
$\bar{x}_1 = 66.29$ g	$\bar{x}_2 = 77.49$ g
$s_1 = 9.17$ g	$s_2 = 11.34$ g
$n_1 = 15$	$n_2 = 10$

Next we check the three conditions required for using the pooled t-test. Since the samples are independent, Assumption 1 is satisfied. Normal probability plots (not shown) of the two samples in Table 10.3 reveal no outliers and are roughly linear; so we can consider Assumption 2 satisfied. From Table 10.4 we see that the sample standard deviations are 9.17 and 11.34; these are close enough for us to consider Assumption 3 satisfied.

The preceding paragraph suggests that the pooled t-test, Procedure 10.3, can be used to carry out the hypothesis test. We proceed as follows.

Step 1 *State the null and alternative hypotheses.*

Let μ_1 denote the mean protein intake (per day) of all people with incomes below the poverty level and μ_2 denote the mean protein intake (per day) of all people with incomes above the poverty level. Then the null and alternative hypotheses are

H_0: $\mu_1 = \mu_2$ (below-poverty mean is not less than the above-poverty mean)

H_a: $\mu_1 < \mu_2$ (below-poverty mean is less than the above-poverty mean).

Note that the hypothesis test is left-tailed since a less-than sign ($<$) appears in the alternative hypothesis.

Step 2   *Decide on the significance level, α.*

The hypothesis test is to be performed at the 5% significance level, so $\alpha = 0.05$.

Step 3   *The critical value for a left-tailed test is $-t_\alpha$ with $df = n_1 + n_2 - 2$.*

From Step 2, $\alpha = 0.05$. Also, we see from Table 10.4 that $n_1 = 15$ and $n_2 = 10$. Hence, df $= 15 + 10 - 2 = 23$. Consulting Table IV, we find that the critical value is $-t_\alpha = -t_{0.05} = -1.714$, as seen in Fig. 10.3.

FIGURE 10.3

Criterion for deciding whether or not to reject the null hypothesis

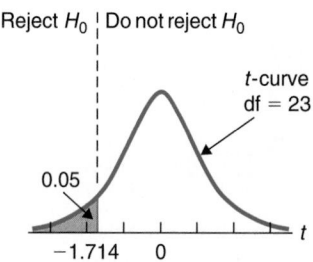

Reject H_0 | Do not reject H_0

t-curve
df = 23

0.05

−1.714   0

t

Step 4   *Compute the value of the test statistic*

$$t = \frac{\overline{x}_1 - \overline{x}_2}{s_p\sqrt{(1/n_1) + (1/n_2)}}$$

where

$$s_p = \sqrt{\frac{(n_1 - 1)s_1^2 + (n_2 - 1)s_2^2}{n_1 + n_2 - 2}}.$$

We first determine the pooled sample standard deviation, s_p. Referring to Table 10.4, we find that

$$s_p = \sqrt{\frac{(15 - 1) \cdot 9.17^2 + (10 - 1) \cdot 11.34^2}{15 + 10 - 2}} = 10.07 \text{ g}.$$

Referring again to Table 10.4, we obtain the value of the test statistic:

$$t = \frac{\overline{x}_1 - \overline{x}_2}{s_p\sqrt{(1/n_1) + (1/n_2)}} = \frac{66.29 - 77.49}{10.07\sqrt{(1/15) + (1/10)}} = -2.724.$$

Step 5   *If the value of the test statistic falls in the rejection region, reject H_0; otherwise, do not reject H_0.*

From Step 4 the value of the test statistic is $t = -2.724$, which falls in the rejection region (see Fig. 10.3). Thus we reject H_0.

Step 6   *State the conclusion in words.*

The test results are statistically significant at the 5% level. Evidently, the average person with an income below the poverty level consumes less protein than the average person with an income above the poverty level.   ■

The *P*-value approach to hypothesis testing can also be used to carry out the hypothesis test in Example 10.4 and to assess more precisely the evidence against the null hypothesis. From Step 4 we see that the value of the test statistic is $t = -2.724$. Recalling that df $= 23$, we find from Table IV that $0.005 < P < 0.01$. In particular, because the *P*-value is less than the significance level of 0.05, we can reject H_0. Furthermore, by referring to Table 9.12 on page 530, we see that the data provide very strong evidence against the null hypothesis of equal mean protein intakes and hence in support of the alternative hypothesis that $\mu_1 < \mu_2$.

Can we conclude from the results of the hypothesis test in Example 10.4 that, on the average, having an income below the poverty level *causes* a person to consume less protein? No! Because the study is observational, we can conclude only that an *association* exists between protein intake and poverty-level status.

Confidence Intervals for the Difference Between the Means of Two Normal Populations with Equal Standard Deviations Using Independent Samples

We can also use Key Fact 10.3 on page 587 to derive the following confidence-interval procedure for the difference between two means, which we often refer to as the **pooled *t*-interval procedure.**

PROCEDURE 10.4	**THE POOLED *t*-INTERVAL PROCEDURE FOR THE DIFFERENCE BETWEEN TWO POPULATION MEANS**

ASSUMPTIONS

1. Independent samples
2. Normal populations
3. Equal population standard deviations

Step 1   For a confidence level of $1 - \alpha$, use Table IV to find $t_{\alpha/2}$ with df $= n_1 + n_2 - 2$.

Step 2   The endpoints of the confidence interval for $\mu_1 - \mu_2$ are

$$(\bar{x}_1 - \bar{x}_2) \pm t_{\alpha/2} \cdot s_\text{p}\sqrt{(1/n_1) + (1/n_2)}.$$

Example 10.5   Illustrates Procedure 10.4

Refer to Example 10.4. Use the sample data displayed in Table 10.3 on page 589 to obtain a 95% confidence interval for the difference, $\mu_1 - \mu_2$, between the mean protein intake of all people with incomes below the poverty level and the mean protein intake of all people with incomes above the poverty level.

SOLUTION   We apply Procedure 10.4.

Step 1   *For a confidence level of $1 - \alpha$, use Table IV to find $t_{\alpha/2}$ with df $= n_1 + n_2 - 2$.*

For a 95% confidence interval, the confidence level is $0.95 = 1 - 0.05$; so $\alpha = 0.05$. From Table 10.4, $n_1 = 15$ and $n_2 = 10$; so df $= n_1 + n_2 - 2 = 15 + 10 - 2 = 23$. Consulting Table IV, we find that for df $= 23$, $t_{\alpha/2} = t_{0.05/2} = t_{0.025} = 2.069$.

Step 2   *The endpoints of the confidence interval for $\mu_1 - \mu_2$ are*

$$(\bar{x}_1 - \bar{x}_2) \pm t_{\alpha/2} \cdot s_p \sqrt{(1/n_1) + (1/n_2)}.$$

From Step 1, $t_{\alpha/2} = 2.069$. Also, $n_1 = 15$, $n_2 = 10$, and from Example 10.4 we know that $\bar{x}_1 = 66.29$ g, $\bar{x}_2 = 77.49$ g, and $s_p = 10.07$ g. Therefore the endpoints of the confidence interval for $\mu_1 - \mu_2$ are

$$(66.29 - 77.49) \pm 2.069 \cdot 10.07\sqrt{(1/15) + (1/10)},$$

or -11.20 ± 8.51. Consequently, a 95% confidence interval for $\mu_1 - \mu_2$ is from -19.71 to -2.69. We can be 95% confident that the difference, $\mu_1 - \mu_2$, between the mean protein intakes (per day) of all people with incomes below the poverty level and all people with incomes above the poverty level is somewhere between -19.71 g and -2.69 g. In particular, we can be 95% confident that the average person with an income above the poverty level gets at least 2.69 g more protein per day than the average person with an income below the poverty level.

■

Using the Computer (Optional)

Procedure 10.3 on page 587 gives a step-by-step method for performing a hypothesis test to compare the means of two normally distributed populations with equal standard deviations using independent samples. Procedure 10.4 on page 591 provides a step-by-step method for obtaining a confidence interval for the difference between the means of two normally distributed populations with equal standard deviations using independent samples.

Alternatively, we can employ Minitab to carry out both of those procedures simultaneously. To do so, we use Minitab's **TwoSample** command and **Pooled** subcommand. Example 10.6 explains how this is done.

Example 10.6   The TwoSample Command and Pooled Subcommand

Use Minitab to simultaneously perform the hypothesis test considered in Example 10.4 and obtain the confidence interval required in Example 10.5.

SOLUTION   Let μ_1 denote the mean protein intake (per day) of all people with incomes below the poverty level and μ_2 denote the mean protein intake (per day) of all people with incomes above the poverty level. The problem in Example 10.4 is to perform the hypothesis test

H_0: $\mu_1 = \mu_2$ (below-poverty mean is not less than the above-poverty mean)

H_a: $\mu_1 < \mu_2$ (below-poverty mean is less than the above-poverty mean)

at the 5% significance level; the problem in Example 10.5 is to obtain a 95% confidence interval for $\mu_1 - \mu_2$.

As we discovered in Example 10.4, it is reasonable to use pooled-t procedures here. Thus we can apply the TwoSample command and Pooled subcommand to simultaneously carry out the hypothesis test and obtain the required confidence interval. To begin, we enter the two sets of sample data from Table 10.3 on page 589 in columns named BELOWPOV and ABOVEPOV. Then we proceed in the following way.

Session commands: We type the command **TWOSAMPLE** followed by the desired confidence level for the confidence interval and the storage locations of the data; that is, we type TWOSAMPLE, 95% confidence, for 'BELOWPOV' vs 'ABOVEPOV'; and press Enter. Since the standard deviations of the two populations are equal, we want to use pooled-t procedures. To indicate that, we employ the **POOLED** subcommand; that is, we type POOLED; and press Enter. Finally, because the hypothesis test is left-tailed, we also type the subcommand ALTERNATIVE=-1. and press Enter. Printout 10.1 displays these commands and the resulting computer output.

(or)

Menu commands: We choose **Stat ▸ Basic Statistics ▸ 2-Sample t...**, select **Samples in different columns**, specify BELOWPOV for **First** and ABOVEPOV for **Second**, select **Alternative** and specify **less than**, select **Confidence level** and type 95, select **Assume equal variances** and press Spacebar, and then select **OK**. The output that results is shown in Printout 10.1.

PRINTOUT 10.1
Minitab output for the
TwoSample command
and Pooled subcommand

```
MTB > TWOSAMPLE, 95% confidence, for 'BELOWPOV' vs 'ABOVEPOV';
SUBC> POOLED;
SUBC> ALTERNATIVE=-1.

TWOSAMPLE T FOR BELOWPOV VS ABOVEPOV
              N      MEAN     STDEV    SE MEAN
BELOWPOV     15     66.29      9.17       2.4
ABOVEPOV     10     77.5      11.3        3.6

95 PCT CI FOR MU BELOWPOV - MU ABOVEPOV: (-19.7, -2.7)

TTEST MU BELOWPOV = MU ABOVEPOV (VS LT): T= -2.72  P=0.0060  DF=  23

POOLED STDEV =         10.1
```

The fourth line of Printout 10.1 describes the test being performed: TWOSAMPLE T FOR BELOWPOV VS ABOVEPOV. The next three lines display a table that gives the sample size, sample mean, sample standard deviation, and estimated standard error of the mean for each sample.

On the eighth line of Printout 10.1, we find the required 95% confidence interval for the difference between the two population means. Thus we can be 95% confident that the difference, $\mu_1 - \mu_2$, between the mean protein intake of all people with incomes below the poverty level and the mean protein intake of all people with incomes above the poverty level is somewhere between -19.7 g and -2.7 g per day.

The next-to-last line of Printout 10.1 provides a statement of the null and alternative hypotheses, followed by the value of the test statistic (T= -2.72), the P-value (P=0.0060), and the degrees of freedom (DF= 23). On the last line, we find the value of the pooled sample standard deviation, s_p.

Since the P-value of 0.006 is less than the specified significance level of $\alpha = 0.05$, we reject H_0. At the 5% significance level, the data provide sufficient evidence to conclude that the mean protein intake of all people with incomes below the poverty level is less than the mean protein intake of all people with incomes above the poverty level.   ■

EXERCISES 10.2

Preliminary data analyses indicate that it is reasonable to consider the assumptions for using pooled-t procedures satisfied in Exercises 10.19–10.24. For each exercise, perform the required hypothesis test using either the classical approach or the P-value approach.

10.19 A highway official wants to compare two brands of paint used for striping roads. Twenty locations are selected for paint stripes. The first brand is used on 10 locations randomly selected from the 20; the second brand is used on the remaining 10 locations. The number of months each stripe lasts is presented in the following table.

Brand A		Brand B	
35.6	36.1	37.2	36.4
37.0	35.8	39.7	37.5
34.9	34.9	37.2	40.5
36.0	37.8	38.8	38.2
36.6	36.5	37.7	36.6

a. Do the data provide sufficient evidence to conclude that there is a difference in mean lasting times for the two brands of paint? Perform the required hypothesis test at the 5% significance level. (*Note:* $\bar{x}_1 = 36.12$, $s_1 = 0.903$, $\bar{x}_2 = 37.98$, $s_2 = 1.331$.)

b. Is the study a designed experiment or an observational study? Explain your answer.

10.20 In a packing plant, a machine packs cartons with jars. A salesperson claims that the machine she is selling will pack faster. To test that claim, the times it takes each machine to pack 10 cartons are recorded. The results, in seconds, are shown in the following table.

New machine		Present machine	
42.0	41.0	42.7	43.6
41.3	41.8	43.8	43.3
42.4	42.8	42.5	43.5
43.2	42.3	43.1	41.7
41.8	42.7	44.0	44.1

Do the data provide sufficient evidence to conclude that, on the average, the new machine packs faster? Perform the required hypothesis test at the 5% level of significance. (*Note:* $\bar{x}_1 = 42.13$, $s_1 = 0.685$, $\bar{x}_2 = 43.23$, $s_2 = 0.750$.)

10.21 The U.S. Bureau of Labor Statistics conducts monthly surveys to estimate hourly earnings of nonsupervisory employees in various industry groups. Results of the surveys can be found in *Employment and Earnings*. Independent random samples of 14 mine workers and 17 construction workers yield the following statistics.

Mining	Construction
$\bar{x}_1 = \$13.93$	$\bar{x}_2 = \$14.42$
$s_1 = \$2.25$	$s_2 = \$2.36$

At the 5% significance level, do the data provide sufficient evidence to conclude that mine workers earn less on the average than construction workers?

10.22   Data on miles driven annually by U.S. households are compiled by the U.S. Energy Information Administration and published in *Residential Transportation Energy Consumption Survey, Consumption Patterns of Household Vehicles.* Independent samples of 15 midwestern households and 14 southern households provide the following statistics on the number of miles driven last year.

Midwest	South
$\bar{x}_1 = 16,229$ mi	$\bar{x}_2 = 17,689$ mi
$s_1 = 4,057$ mi	$s_2 = 4,420$ mi

At the 5% significance level, does there appear to be a difference in the average number of miles driven by midwestern and southern households?

10.23   *Vital and Health Statistics,* published by the National Center for Health Statistics, provides information on heights and weights of Americans, by age and sex. Independent samples of 10 males age 25–34 years and 15 males age 45–54 years yield the following heights, in inches.

25–34		45–54		
73.3	70.4	73.2	69.5	64.7
64.8	66.8	68.5	74.5	73.0
72.1	70.7	62.4	70.6	66.7
68.9	74.4	65.5	69.3	68.1
68.7	71.8	71.3	67.1	64.3

At the 5% significance level, do the data provide sufficient evidence to conclude that males in the age group 25–34 years are, on the average, taller than those who were in that age group 20 years ago? *(Note: $\bar{x}_1 = 70.19$, $s_1 = 2.951$, $\bar{x}_2 = 68.58$, $s_2 = 3.543$.)*

10.24   A supervisor records the time it takes each of two workers to perform an assembly-line task. Each worker

is observed on six randomly selected occasions. The times, to the nearest minute, are shown in the following table.

Frank			Marcia		
8	9	9	8	9	9
10	11	10	9	8	10

Do the data provide sufficient evidence to conclude, at the 10% significance level, that the mean times required to complete the assembly-line task differ for the two workers? *(Note: $\bar{x}_1 = 9.50$, $s_1 = 1.049$, $\bar{x}_2 = 8.83$, $s_2 = 0.753$.)*

In each of Exercises 10.25–10.30, apply Procedure 10.4 on page 591 to obtain the required confidence interval.

10.25   Refer to Exercise 10.19.
a. Determine a 95% confidence interval for the difference, $\mu_1 - \mu_2$, between the mean lasting times of Brand *A* and Brand *B* paints.
b. Interpret your result from part (a).

10.26   Refer to Exercise 10.20.
a. Determine a 90% confidence interval for the difference, $\mu_1 - \mu_2$, between the mean time it takes the new machine to pack 10 cartons and the mean time it takes the present machine to pack 10 cartons.
b. Interpret your result from part (a).

10.27   Refer to Exercise 10.21.
a. Determine a 90% confidence interval for the difference, $\mu_1 - \mu_2$, between the mean hourly earnings of nonsupervisory mine and construction workers.
b. Interpret your result from part (a).

10.28   Refer to Exercise 10.22.
a. Determine a 95% confidence interval for the difference, $\mu_1 - \mu_2$, between last year's mean numbers of miles driven by midwestern and southern households.
b. Interpret your result from part (a).

10.29   Refer to Exercise 10.23.
a. Obtain a 90% confidence interval for the difference between the mean height of males in the age group 25–34 years and the mean height of males in the age group 45–54 years.
b. Interpret your result from part (a).

10.30   Refer to Exercise 10.24.
a. Find a 90% confidence interval for the difference between the mean time it takes Frank to complete the assembly-line task and the mean time it takes Marcia to complete the assembly-line task.
b. Interpret your result from part (a).

10.31 (Computer exercise) Refer to Exercises 10.19 and 10.25. Use Minitab or some other statistical software to
a. obtain normal probability plots, stem-and-leaf diagrams, boxplots, and the standard deviations for the two samples.
b. perform the required hypothesis test and obtain the desired confidence interval.
c. Justify the use of your procedure in part (b).

10.32 (Computer exercise) Refer to Exercises 10.20 and 10.26. Use Minitab or some other statistical software to
a. obtain normal probability plots, stem-and-leaf diagrams, boxplots, and the standard deviations for the two samples.
b. perform the required hypothesis test and obtain the desired confidence interval.
c. Justify the use of your procedure in part (b).

10.33 (Computer exercise) The research and development (R&D) department of a light-bulb manufacturing company claims to have developed a new bulb that, on the average, will outlast the bulb currently produced. To try to justify the claim, R&D takes independent random samples of the current bulb and the new bulb. After the lifetimes (in hours) of the bulbs sampled are determined, Minitab's TwoSample command is applied to get the output shown in Printout 10.2. Use the output to obtain
a. the null and alternative hypotheses.
b. the sample standard deviations of the lifetimes for both the current bulb and the new bulb.
c. the number of current bulbs sampled and the number of new bulbs sampled.
d. the sample mean lifetimes for both the current bulb and the new bulb.
e. the value of the test statistic, t.
f. the P-value of the hypothesis test.
g. the smallest significance level at which the null hypothesis can be rejected.
h. the conclusion if the hypothesis test is performed at the 1% significance level.
i. a 99% confidence interval for the difference between the mean lifetimes of the current bulb and the new bulb.
j. the value of s_p.

10.34 (Computer exercise) A researcher in nutrition wants to compare the protein intakes of whites and blacks. He takes independent random samples of 10 whites and 12 blacks and determines their 24-hour protein intakes, in grams. Following that, he applies Minitab's TwoSample command to the resulting data and obtains the output shown in Printout 10.3. From the output, determine
a. the null and alternative hypotheses.
b. the standard deviations of the two samples.

c. the mean protein intakes (per day) of the whites sampled and the blacks sampled.
d. the value of the test statistic, t.
e. the P-value of the hypothesis test.
f. the smallest significance level at which the null hypothesis can be rejected.
g. the conclusion if the hypothesis test is performed at the 5% significance level.
h. a 95% confidence interval for the difference, $\mu_1 - \mu_2$, between the mean protein intakes of whites and blacks.
i. the value of s_p.

10.35 Suppose we want to perform a hypothesis test to compare the means of two normally distributed populations using independent samples. Further suppose the standard deviations of the two populations are known.
a. Which test statistic should be used to perform the hypothesis test? *(Hint: Refer to Key Fact 10.2 on page 585.)*
b. Which table should be consulted in order to obtain the critical value(s)?
c. Formulate a step-by-step procedure for the hypothesis test that uses the test statistic from part (a).

10.36 Let
$$z = \frac{(\bar{x}_1 - \bar{x}_2) - (\mu_1 - \mu_2)}{\sqrt{(\sigma_1^2/n_1) + (\sigma_2^2/n_2)}}.$$
Show that if $\sigma_1 = \sigma_2 = \sigma$, then we can rewrite the expression for z as
$$z = \frac{(\bar{x}_1 - \bar{x}_2) - (\mu_1 - \mu_2)}{\sigma\sqrt{(1/n_1) + (1/n_2)}}.$$
This verifies Equation (2) on page 586.

10.37 The formula given on page 586 for the pooled variance is
$$s_p^2 = \frac{(n_1 - 1)s_1^2 + (n_2 - 1)s_2^2}{n_1 + n_2 - 2}.$$
Show that if the sample sizes, n_1 and n_2, are equal, then s_p^2 is simply the mean of s_1^2 and s_2^2.

10.38 Suppose independent random samples of sizes n_1 and n_2 are to be taken from two normally distributed populations with means μ_1 and μ_2, respectively. Further suppose the standard deviations of the two populations are equal.
a. Use Key Fact 10.3 on page 587 to show that
$$P\left(-t_{\alpha/2} < \frac{(\bar{x}_1 - \bar{x}_2) - (\mu_1 - \mu_2)}{s_p\sqrt{(1/n_1) + (1/n_2)}} < t_{\alpha/2}\right) = 1 - \alpha.$$
b. Use part (a) to verify Procedure 10.4 on page 591.

PRINTOUT 10.2   Minitab output for Exercise 10.33

```
TWOSAMPLE T FOR CURRENT VS NEW
             N      MEAN     STDEV    SE MEAN
CURRENT     20    1023.2      55.3        12
NEW         10    1093.0      44.6        14


99 PCT CI FOR MU CURRENT - MU NEW: (-126, -14)


TTEST MU CURRENT = MU NEW (VS LT): T= -3.46  P=0.0009  DF=  28


POOLED STDEV =      52.1
```

PRINTOUT 10.3   Minitab output for Exercise 10.34

```
TWOSAMPLE T FOR WHITE VS BLACK
           N      MEAN     STDEV    SE MEAN
WHITE     10      77.4      16.4        5.2
BLACK     12      68.8      15.2        4.4


95 PCT CI FOR MU WHITE - MU BLACK: (-5.4, 22.8)


TTEST MU WHITE = MU BLACK (VS NE): T= 1.29  P=0.21  DF=  20


POOLED STDEV =      15.7
```

10.3 INFERENCES FOR THE MEANS OF TWO NORMAL POPULATIONS USING INDEPENDENT SAMPLES (STANDARD DEVIATIONS NOT ASSUMED EQUAL)

In Section 10.2 we examined methods of performing inferences to compare the means of two normally distributed populations using independent samples. The methods discussed there, called pooled t-procedures, require that the standard deviations of the two populations be equal.

In this section we will develop inferential procedures (hypothesis tests and confidence intervals) to compare the means of two normally distributed populations using independent samples that do not require the population standard deviations to be equal, even though they may be. As before, we will assume the population standard deviations are unknown, since that is usually the case in practice.

Hypothesis Tests for the Means of Two Normal Populations Using Independent Samples

Let's begin by finding a test statistic. From Key Fact 10.2 on page 585, we know that when independent samples are taken from two normally distributed populations, the random variable

$$z = \frac{(\bar{x}_1 - \bar{x}_2) - (\mu_1 - \mu_2)}{\sqrt{(\sigma_1^2/n_1) + (\sigma_2^2/n_2)}}$$

has the standard normal distribution. Since we are assuming the population standard deviations, σ_1 and σ_2, are unknown, we cannot use the preceding random variable as a basis for obtaining the required test statistic. We therefore replace σ_1 and σ_2 by their sample estimates, s_1 and s_2, and obtain the random variable

$$\frac{(\bar{x}_1 - \bar{x}_2) - (\mu_1 - \mu_2)}{\sqrt{(s_1^2/n_1) + (s_2^2/n_2)}},$$

which *can* be used as a basis for obtaining the required test statistic. This random variable does not have the standard normal distribution, but it does have approximately a t-distribution.

KEY FACT 10.4   **DISTRIBUTION OF NONPOOLED-t STATISTIC**

Suppose independent random samples of sizes n_1 and n_2 are to be taken from two normally distributed populations with means μ_1 and μ_2, respectively. Then the random variable

$$t = \frac{(\bar{x}_1 - \bar{x}_2) - (\mu_1 - \mu_2)}{\sqrt{(s_1^2/n_1) + (s_2^2/n_2)}}$$

has approximately the t-distribution with degrees of freedom Δ (delta), where

$$\Delta = \frac{\left[(s_1^2/n_1) + (s_2^2/n_2)\right]^2}{\dfrac{(s_1^2/n_1)^2}{n_1 - 1} + \dfrac{(s_2^2/n_2)^2}{n_2 - 1}},$$

rounded down to the nearest integer.

In view of Key Fact 10.4, we see that for a hypothesis test with null hypothesis $H_0\colon \mu_1 = \mu_2$, we can use the random variable

$$t = \frac{\bar{x}_1 - \bar{x}_2}{\sqrt{(s_1^2/n_1) + (s_2^2/n_2)}}$$

as the test statistic and obtain the critical value(s) from the t-table, Table IV. Specifically, we have the following procedure, which we often refer to as the **nonpooled t-test.**

PROCEDURE 10.5

THE NONPOOLED t-TEST FOR TWO POPULATION MEANS WITH NULL HYPOTHESIS H_0: $\mu_1 = \mu_2$

ASSUMPTIONS

1. Independent samples
2. Normal populations

Step 1 State the null and alternative hypotheses.

Step 2 Decide on the significance level, α.

Step 3 The critical value(s)
   a. for a two-tailed test are $\pm t_{\alpha/2}$,
   b. for a left-tailed test is $-t_\alpha$,
   c. for a right-tailed test is t_α,
   with df $= \Delta$, where

$$\Delta = \frac{\left[(s_1^2/n_1) + (s_2^2/n_2) \right]^2}{\dfrac{(s_1^2/n_1)^2}{n_1 - 1} + \dfrac{(s_2^2/n_2)^2}{n_2 - 1}},$$

rounded down to the nearest integer. Use Table IV to find the critical value(s).

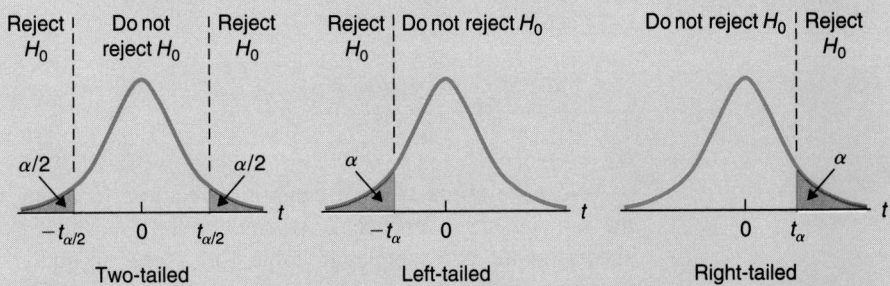

| Two-tailed | Left-tailed | Right-tailed |

Step 4 Compute the value of the test statistic

$$t = \frac{\bar{x}_1 - \bar{x}_2}{\sqrt{(s_1^2/n_1) + (s_2^2/n_2)}}.$$

Step 5 If the value of the test statistic falls in the rejection region, then reject H_0; otherwise, do not reject H_0.

Step 6 State the conclusion in words.

Regarding the assumptions for the nonpooled t-test: The independent-samples assumption (Assumption 1) is essential; the test is robust to moderate deviations of the normality assumption (Assumption 2).

Example 10.7   Illustrates Procedure 10.5

Costs to community hospitals per patient per day are reported by the American Hospital Association in *Hospital Statistics*. Independent random samples of 12 costs in Georgia and 15 costs in Illinois gave the data, in dollars, shown in Table 10.5. Do the data provide sufficient evidence to conclude that there is a difference between the mean costs to community hospitals per patient per day in Georgia and Illinois? Perform the appropriate hypothesis test at the 5% significance level.

TABLE 10.5
Samples of costs ($) to community hospitals per patient per day in Georgia and Illinois

Georgia				Illinois				
633	616	659	535	790	587	997	735	852
666	675	524	746	686	839	545	724	554
585	748	696	609	889	797	722	483	579

SOLUTION   First we present in Table 10.6 the required summary statistics for the two samples in Table 10.5. These statistics are obtained in the usual way.

TABLE 10.6
Summary statistics for the samples in Table 10.5

Georgia	Illinois
$\bar{x}_1 = 641.0$	$\bar{x}_2 = 718.6$
$s_1 = 72.2$	$s_2 = 146.6$
$n_1 = 12$	$n_2 = 15$

Next we check the two conditions required for using the nonpooled *t*-test. Since the samples are independent, Assumption 1 is satisfied. Normal probability plots (not shown) of the two samples in Table 10.5 reveal no outliers and are roughly linear; so we can consider Assumption 2 satisfied. We can therefore apply the nonpooled *t*-test, Procedure 10.5, to carry out the hypothesis test.

Step 1   State the null and alternative hypotheses.

Let μ_1 and μ_2 denote, respectively, the mean costs to community hospitals per patient per day in Georgia and Illinois. Then the null and alternative hypotheses are

$$H_0: \mu_1 = \mu_2 \text{ (mean costs are the same)}$$
$$H_a: \mu_1 \neq \mu_2 \text{ (mean costs are different)}.$$

Note that the hypothesis test is two-tailed since a not-equal sign ($\neq$) appears in the alternative hypothesis.

Step 2   Decide on the significance level, α.

The test is to be performed at the 5% significance level; thus $\alpha = 0.05$.

Step 3   *The critical values for a two-tailed test are $\pm t_{\alpha/2}$ with df $= \Delta$.*

From Step 2, $\alpha = 0.05$. Also, referring to Table 10.6 we find that

$$df = \frac{\left[(72.2^2/12) + (146.6^2/15)\right]^2}{\dfrac{\left(72.2^2/12\right)^2}{12 - 1} + \dfrac{\left(146.6^2/15\right)^2}{15 - 1}} = 21 \text{ (rounded down)}.$$

So the critical values are $\pm t_{\alpha/2} = \pm t_{0.05/2} = \pm t_{0.025} = \pm 2.080$, as seen in Fig. 10.4.

FIGURE 10.4

Criterion for deciding whether or not to reject the null hypothesis

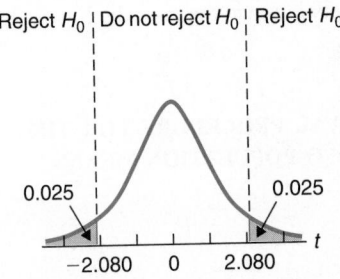

Reject H_0 | Do not reject H_0 | Reject H_0

0.025   0.025

-2.080 0   2.080 t

Step 4   *Compute the value of the test statistic*

$$t = \frac{\overline{x}_1 - \overline{x}_2}{\sqrt{(s_1^2/n_1) + (s_2^2/n_2)}}.$$

Referring again to Table 10.6, we obtain that

$$t = \frac{641.0 - 718.6}{\sqrt{(72.2^2/12) + (146.6^2/15)}} = -1.796.$$

Step 5   *If the value of the test statistic falls in the rejection region, reject H_0; otherwise, do not reject H_0.*

From Step 4 the value of the test statistic is $t = -1.796$, which, as we see from Fig. 10.4, does not fall in the rejection region. Thus we do not reject H_0.

Step 6   *State the conclusion in words.*

The test results are not statistically significant at the 5% level; that is, at the 5% significance level, the data do not provide sufficient evidence to conclude that there is a difference between the mean costs to community hospitals per patient per day in Georgia and Illinois. ∎

As usual we can also use the *P*-value approach to conduct the hypothesis test in Example 10.7. From Step 4 we know that the value of the test statistic is $t = -1.796$. Recalling that df $= 21$ and that the test is two-tailed, we find from Table IV that $0.05 < P < 0.10$. In particular, because the P-value exceeds the significance level of 0.05, we cannot reject H_0. But referring to Table 9.12 on page 530, we see that the data provide moderate evidence against the null hypothesis of equal mean costs; the evidence is just not strong enough to reject the null hypothesis at the 5% level.

Confidence Intervals for the Difference Between the Means of Two Normal Populations Using Independent Samples

We can also use Key Fact 10.4 on page 598 to derive the following confidence-interval procedure for the difference between two means, which we often refer to as the **nonpooled** **t-interval procedure.**

PROCEDURE 10.6   THE NONPOOLED t-INTERVAL PROCEDURE FOR THE DIFFERENCE BETWEEN TWO POPULATION MEANS

> *ASSUMPTIONS*
>
> 1. Independent samples
> 2. Normal populations
>
> **Step 1**   For a confidence level of $1 - \alpha$, use Table IV to find $t_{\alpha/2}$ with df $= \Delta$, where
>
> $$\Delta = \frac{\left[\left(s_1^2/n_1 \right) + \left(s_2^2/n_2 \right) \right]^2}{\dfrac{\left(s_1^2/n_1 \right)^2}{n_1 - 1} + \dfrac{\left(s_2^2/n_2 \right)^2}{n_2 - 1}},$$
>
> rounded down to the nearest integer.
>
> **Step 2**   The endpoints of the confidence interval for $\mu_1 - \mu_2$ are
>
> $$(\bar{x}_1 - \bar{x}_2) \pm t_{\alpha/2} \cdot \sqrt{\left(s_1^2/n_1 \right) + \left(s_2^2/n_2 \right)}.$$

Example 10.8   Illustrates Procedure 10.6

Refer to Example 10.7. Use the sample data in Table 10.5 on page 600 to obtain a 95% confidence interval for the difference, $\mu_1 - \mu_2$, between the mean costs to community hospitals per patient per day in Georgia and Illinois.

SOLUTION   We apply Procedure 10.6.

Step 1   *For a confidence level of $1 - \alpha$, use Table IV to find $t_{\alpha/2}$ with df $= \Delta$.*

For a 95% confidence interval, $\alpha = 0.05$. As we saw in Example 10.7, df $= 21$. Consulting Table IV, we find that for df $= 21$, $t_{\alpha/2} = t_{0.05/2} = t_{0.025} = 2.080$.

Step 2 *The endpoints of the confidence interval for $\mu_1 - \mu_2$ are*

$$(\bar{x}_1 - \bar{x}_2) \pm t_{\alpha/2} \cdot \sqrt{(s_1^2/n_1) + (s_2^2/n_2)}.$$

From Step 1, $t_{\alpha/2} = 2.080$. Referring to Table 10.6 on page 600, we conclude that the endpoints of the confidence interval for $\mu_1 - \mu_2$ are

$$(641.0 - 718.6) \pm 2.080 \cdot \sqrt{(72.2^2/12) + (146.6^2/15)}$$

or -167.5 to 12.3. We can be 95% confident that the difference, $\mu_1 - \mu_2$, between the mean costs to community hospitals per patient per day in Georgia and Illinois is somewhere between $-\$167.5$ and $\$12.3$. ∎

Pooled Versus Nonpooled

Suppose we want to perform a hypothesis test to compare the means, μ_1 and μ_2, of two normally distributed populations with unknown standard deviations. If we take independent samples, then two tests are candidates for the job: the pooled t-test (Procedure 10.3 of Section 10.2) or the nonpooled t-test (Procedure 10.5 of this section). As we have already mentioned, both tests are robust to moderate deviations from normality; they are very robust when the sample sizes are equal or nearly equal.

In theory, the pooled t-test requires that the population standard deviations, σ_1 and σ_2, be equal. What if the pooled t-test is employed when in fact the population standard deviations are not equal? The answer to this question depends on several factors. If the population standard deviations are unequal, but not too unequal, and the sample sizes, n_1 and n_2, are nearly the same, then using the pooled t-test will not cause serious difficulties. However, if the population standard deviations are quite different, then using the pooled t-test can result in a significantly larger Type I error probability than the one specified (i.e., α).

On the other hand, the nonpooled t-test does not require that the population standard deviations be equal; it applies whether or not they are equal. Then why use the pooled t-test at all? The reason is that if the population standard deviations are equal or nearly so, then, on the average, the pooled t-test is slightly more powerful; that is, there is a somewhat smaller probability of making a Type II error.

Thus we see that for a hypothesis test to compare the means of two normally distributed populations using independent samples, the pooled t-test should be used only when the two populations have nearly equal standard deviations; otherwise, the nonpooled t-test should be employed. Similar remarks apply to the pooled t-interval and nonpooled t-interval confidence-interval procedures.

KEY FACT 10.5   **CHOOSING BETWEEN A POOLED AND NONPOOLED PROCEDURE**

> Suppose you want to compare the means of two normally distributed populations using independent samples. If you are reasonably sure the populations have nearly equal standard deviations, use a pooled-t procedure; otherwise, use a nonpooled-t procedure.

Using the Computer (Optional)

Procedure 10.5 on page 599 gives a step-by-step method for performing a hypothesis test to compare the means of two normally distributed populations using independent samples. Procedure 10.6 on page 602 provides a step-by-step method for obtaining a confidence interval for the difference between the means of two normally distributed populations using independent samples.

Alternatively, we can employ Minitab to carry out both of those procedures simultaneously. The appropriate command is called **TwoSample** (without the POOLED subcommand for session commands and without the "Assume equal variances" specification for menu commands). Example 10.9 describes how to use the TwoSample command.

Example 10.9   The TwoSample Command

SOLUTION   Use Minitab to simultaneously perform the hypothesis test considered in Example 10.7 and obtain the confidence interval required in Example 10.8.

Let μ_1 and μ_2 denote, respectively, the mean costs to community hospitals per patient per day in Georgia and Illinois. The problem in Example 10.7 is to perform the hypothesis test

$$H_0: \mu_1 = \mu_2 \text{ (mean costs are the same)}$$
$$H_a: \mu_1 \neq \mu_2 \text{ (mean costs are different)}$$

at the 5% significance level; the problem in Example 10.8 is to obtain a 95% confidence interval for $\mu_1 - \mu_2$.

As we discovered in Example 10.7, it is reasonable to use nonpooled-t procedures here. Thus we can apply TwoSample to simultaneously carry out the hypothesis test and obtain the required confidence interval. To begin, we enter the two sets of sample data from Table 10.5 on page 600 in columns named GEORGIA and ILLINOIS. Then we proceed in the following way.

Session commands: We type the command **TWOSAMPLE** followed by the desired confidence level for the confidence interval and the storage locations of the data; that is, we type `TWOSAMPLE, 95% confidence, for 'GEORGIA' vs 'ILLINOIS'` and press $\boxed{\text{Enter}}$. Because the hypothesis test is two-tailed, we don't need to use the ALTERNATIVE subcommand. Printout 10.4 displays the TWOSAMPLE command and the resulting output.

(or)

Menu commands: We choose **Stat ▶ Basic Statistics ▶ 2-Sample t...**, select **Samples in different columns**, specify GEORGIA for **First** and ILLINOIS for **Second**, select **Alternative** and specify **not equal to**, select **Confidence level** and type 95, and then select **OK**. The output that results is shown in Printout 10.4.

For the hypothesis test, we refer to the last line of Printout 10.4. We see that the P-value is 0.087. Since this exceeds the designated significance level of $\alpha = 0.05$, we

PRINTOUT 10.4
Minitab output for the
TwoSample command

```
MTB > TWOSAMPLE, 95% confidence, for 'GEORGIA' vs 'ILLINOIS'

TWOSAMPLE T FOR GEORGIA VS ILLINOIS
                N      MEAN    STDEV   SE MEAN
GEORGIA    12     641.0     72.2       21
ILLINOIS   15      719       147        38


95 PCT CI FOR MU GEORGIA - MU ILLINOIS: ( -168,   12)

TTEST MU GEORGIA = MU ILLINOIS (VS NE): T= -1.80  P=0.087  DF=  21
```

do not reject H_0; at the 5% significance level, the data do not provide sufficient evidence to conclude that there is a difference between the mean costs to community hospitals per patient per day in Georgia and Illinois.

The required 95% confidence interval is found on the next-to-last line of Printout 10.4. We see that we can be 95% confident that the difference, $\mu_1 - \mu_2$, between the mean costs to community hospitals per patient per day in Georgia and Illinois is somewhere between −$168 and $12. ∎

EXERCISES 10.3

Preliminary data analyses indicate that it is reasonable to employ nonpooled-t procedures in Exercises 10.39–10.44. For each exercise, perform the required hypothesis test using either the classical approach or the P-value approach. (Do not assume the two populations under consideration have equal standard deviations, although they in fact may.)

10.39 Independent samples of 17 sophomores and 13 juniors attending a large state university gave the statistics below for cumulative grade point average (GPA).

Sophomores	Juniors
$\bar{x}_1 = 2.54$	$\bar{x}_2 = 2.68$
$s_1 = 0.52$	$s_2 = 0.31$

Can you conclude from these data that the mean GPAs of sophomores and juniors at the university differ? Use $\alpha = 0.05$.

10.40 According to *High School Profile Report,* in past years, college-bound males have outperformed college-bound females on the mathematics portion of tests given by

the American College Testing (ACT) Program. Independent samples of this year's scores yield the following statistics.

Males	Females
$\bar{x}_1 = 20.6$	$\bar{x}_2 = 19.4$
$s_1 = 6.5$	$s_2 = 4.1$
$n_1 = 15$	$n_2 = 15$

Does it appear that college-bound males are, on the average, still outperforming college-bound females on the mathematics portion of ACT tests? Use $\alpha = 0.05$.

10.41 The owner of a chain of car washes needs to decide between two brands of hot wax. One of the brands, Sureglow, costs less than the other brand, Mirror-sheen. So unless there is strong evidence that Mirror-sheen outlasts Sureglow, the owner will purchase Sureglow. With the cooperation of several local automobile dealers, 30 cars are selected to take part in a test. The 30 cars are randomly divided into two groups of 15 cars each. One group is waxed with Sureglow and the other with Mirror-sheen. The cars are then exposed to the same environmental conditions. In the following table, you will find the data obtained on effectiveness times, in days.

Sureglow			Mirror-sheen		
93	85	86	90	95	96
96	88	93	97	88	91
87	91	91	91	92	97
91	82	87	94	94	92
88	91	88	100	89	92

a. Do the data provide sufficient evidence to conclude that Mirror-sheen has a longer effectiveness time, on the average, than Sureglow? Perform the required hypothesis test at the 1% significance level. *(Note:* $\bar{x}_1 = 89.13$, $s_1 = 3.60$, $\bar{x}_2 = 93.20$, $s_2 = 3.34$.*)*
b. Do the data provide strong evidence that Mirror-sheen outlasts Sureglow? Explain your answer.
c. Identify the study as a designed experiment or an observational study. Explain your answer.

· **10.42**   The marketing manager of a firm that produces laundry products decides to test market a new laundry product in each of the firm's two sales regions. He wants to determine whether there will be a difference in mean sales per market per month between the two regions. Supermarkets from each region are independently and randomly selected to take part in the test—10 from Region 1 and 15 from Region 2. The following data give the number of cases sold in each store during the testing month.

Region 1		Region 2		
74	87	84	86	95
96	94	87	89	94
78	77	92	93	88
83	83	85	81	93
86	80	92	92	85

At the 10% significance level, does the test marketing reveal a difference in potential mean sales per market between the two regions? *(Note:* $\bar{x}_1 = 83.80$, $s_1 = 7.15$, $\bar{x}_2 = 89.07$, $s_2 = 4.27$.*)*

· **10.43**   The U.S. Department of Agriculture compiles information on acreage, production, and value of potatoes and publishes its findings in *Agricultural Statistics*. The yield of potatoes is measured in hundreds of pounds (cwt) per acre. A random sample of forty 1-acre plots of potatoes is taken from Idaho, and a random sample of thirty-two 1-acre plots of

potatoes is taken from Nevada. The yields of the plots sampled are displayed in the following table.

Idaho					Nevada			
229	267	326	309	231	283	254	328	292
283	344	310	258	316	315	336	378	314
241	281	218	284	311	312	328	272	307
254	217	267	299	266	348	233	354	400
264	264	290	312	298	341	313	309	308
305	244	303	299	285	340	300	316	268
308	260	204	291	242	259	276	271	362
329	315	246	322	293	340	339	300	333

At the 5% significance level, do the data provide sufficient evidence to conclude that Idaho has a smaller mean potato yield than Nevada? *(Note:* The sample mean and standard deviation of the Idaho data are 279.6 cwt and 34.6 cwt, respectively; the sample mean and standard deviation of the Nevada data are 313.4 cwt and 37.2 cwt, respectively.*)*

· **10.44**   A general contractor wants to compare the lifetimes of two major brands of electric water heaters, Eagle and National. Using independent samples, the contractor obtains the following data on lifetimes, in years.

Eagle				National				
6.9	7.3	7.8	7.4	7.6	7.6	9.4	9.0	6.4
7.2	6.6	6.2	8.2	8.8	6.2	8.2	4.7	9.2
7.6	5.7	5.5	6.9	6.3	7.8	6.9	10.4	9.1

At the 1% significance level, do the data provide sufficient evidence to conclude that, on the average, National water heaters outlast Eagle water heaters? *(Note:* $\bar{x}_1 = 6.94$, $s_1 = 0.82$, $\bar{x}_2 = 7.84$, $s_2 = 1.53$.*)*

In each of Exercises 10.45–10.50, apply Procedure 10.6 on page 602 to obtain the required confidence interval.

· **10.45**   Refer to Exercise 10.39.
a. Find a 95% confidence interval for the difference, $\mu_1 - \mu_2$, between the mean GPAs of sophomores and juniors at the university.
b. Interpret your result in words.

· **10.46**   Refer to Exercise 10.40.
a. Find a 90% confidence interval for the difference, $\mu_1 - \mu_2$, between this year's mean mathematics ACT scores for males and females.
b. Interpret your result in words.

· **10.47** Refer to Exercise 10.41.

a. Determine a 98% confidence interval for the difference, $\mu_1 - \mu_2$, between the mean effectiveness times of Sureglow and Mirror-sheen.

b. Interpret your result in words.

· **10.48** Refer to Exercise 10.42.

a. Determine a 90% confidence interval for the difference, $\mu_1 - \mu_2$, between potential mean monthly sales per market in Region 1 and Region 2.

b. Interpret your result in words.

· **10.49** Refer to Exercise 10.43.

a. Find a 90% confidence interval for the difference between the mean yields per acre of potatoes for Idaho and Nevada.

b. Interpret your result in words.

· **10.50** Refer to Exercise 10.44.

a. Find a 98% confidence interval for the difference between the mean lifetimes of Eagle and National water heaters.

b. Interpret your result in words.

· **10.51 (Computer exercise)** Refer to Exercises 10.43 and 10.49. Use Minitab or some other statistical software to

a. obtain normal probability plots for the two samples.

b. perform the required hypothesis test and obtain the desired confidence interval.

c. Justify the use of your procedure in part (b).

· **10.52 (Computer exercise)** Refer to Exercises 10.44 and 10.50. Use Minitab or some other statistical software to

a. obtain normal probability plots for the two samples.

b. perform the required hypothesis test and obtain the desired confidence interval.

c. Justify the use of your procedure in part (b).

· **10.53 (Computer exercise)** The National Center for Health Statistics compiles information on divorces in *Vital Statistics of the United States*. Independent random samples

are taken of divorced males and females to decide whether the mean age at the time of first divorce for males is greater than that for females. The output obtained by applying Minitab's TwoSample command to the data is displayed below in Printout 10.5. Use the printout to determine

a. the null and alternative hypotheses.

b. the standard deviation of the ages for each sample.

c. the number of males sampled and the number of females sampled.

d. the mean age at first divorce for both the males sampled and the females sampled.

e. the value of the test statistic, t.

f. the P-value of the hypothesis test.

g. the smallest significance level at which the null hypothesis can be rejected.

h. the conclusion if the hypothesis test is performed at the 5% significance level.

i. a 90% confidence interval for the difference between the mean ages at first divorce for males and females.

j. Do you think it's reasonable to apply TwoSample here? Explain your answer.

· **10.54 (Computer exercise)** A transportation official wants to compare the mean number of miles that cars were driven last year to the mean number of miles that trucks were driven last year. She takes independent random samples of cars and trucks and records the number of miles that each vehicle was driven last year. Then she applies Minitab's TwoSample command to the data and obtains the output displayed in Printout 10.6, shown at the top of the next page. Use the printout to determine

a. the null and alternative hypotheses.

b. the standard deviations of the number of miles driven last year for the cars sampled and for the trucks sampled.

c. the number of cars sampled and the number of trucks sampled.

d. the mean number of miles driven last year for both the cars sampled and the trucks sampled.

PRINTOUT 10.5   Minitab output for Exercise 10.53

```
TWOSAMPLE T FOR MALES VS FEMALES
              N      MEAN    STDEV   SE MEAN
MALES      11      35.69    4.62      1.4
FEMALES    12      32.81    7.25      2.1

90 PCT CI FOR MU MALES - MU FEMALES: (-1.5, 7.2)

TTEST MU MALES = MU FEMALES (VS GT): T= 1.14   P=0.13   DF= 18
```

PRINTOUT 10.6   Minitab output for Exercise 10.54

```
TWOSAMPLE T FOR CARS VS TRUCKS
              N      MEAN     STDEV    SE MEAN
CARS        15       9.31      2.78       0.72
TRUCKS      10      10.93      4.51        1.4


99 PCT CI FOR MU CARS - MU TRUCKS: (-6.44, 3.2)

TTEST MU CARS = MU TRUCKS (VS NE): T= -1.02  P=0.33  DF=  13
```

e. the value of the test statistic, t.
f. the P-value of the hypothesis test.
g. the smallest significance level at which the null hypothesis can be rejected.
h. the conclusion if the hypothesis test is performed at the 1% significance level.
i. a 99% confidence interval for the difference, $\mu_1 - \mu_2$, between the mean number of miles that cars were driven last year and the mean number of miles that trucks were driven last year.
j. Do you think it's reasonable to apply TwoSample here? Explain your answer.

· · · **10.55** We have presented two procedures for performing a hypothesis test to compare the means of two normally distributed populations using independent samples. One procedure is the pooled t-test, Procedure 10.3 on page 587, which applies when the standard deviations of the two popu-

lations are unknown but assumed equal. The test statistic used in that procedure is

$$t = \frac{\bar{x}_1 - \bar{x}_2}{s_p\sqrt{(1/n_1) + (1/n_2)}}.$$

The other procedure is the nonpooled t-test, Procedure 10.5 on page 599, which applies when the standard deviations of the two populations are unknown but not assumed equal. The test statistic used in that procedure is

$$t = \frac{\bar{x}_1 - \bar{x}_2}{\sqrt{(s_1^2/n_1) + (s_2^2/n_2)}}.$$

a. Show that if the sample sizes, n_1 and n_2, are equal, then the values of the two test statistics above will be identical. (*Hint:* Refer to Exercise 10.37 on page 596.)
b. Does part (a) imply that the two t-tests are equivalent when the sample sizes are equal? Explain your answer.

10.4 THE MANN–WHITNEY TEST (OPTIONAL)

We have now studied three procedures for performing a hypothesis test to compare the means of two populations using independent samples: the two-sample z-test, the pooled t-test, and the nonpooled t-test. The two-sample z-test applies when both sample sizes are large; the pooled t-test applies when the populations are normally distributed and have equal standard deviations; and the nonpooled t-test applies when the populations are normally distributed.

Recall that the standard deviation of a normal distribution determines its shape; so two normal populations having equal standard deviations have the same shape, and two normal populations having unequal standard deviations have different shapes. Consequently, the pooled t-test applies to two normal populations having the same shape, whereas the nonpooled t-test applies to any two normal populations, same shape or not.

Another procedure for performing a hypothesis test to compare the means of two populations using independent samples is the **Mann–Whitney test.** This nonparametric test, introduced by Wilcoxon and further developed by Mann and Whitney, is also commonly referred to as the **Wilcoxon rank-sum test** or the **Mann–Whitney–Wilcoxon test.** The Mann–Whitney test applies when the populations have the same shape, but does not require that they be normal or have any other specific shape. We can summarize the above discussion graphically as in Fig. 10.5.

FIGURE 10.5

Different population pairs and the appropriate procedures for comparing the means using independent samples

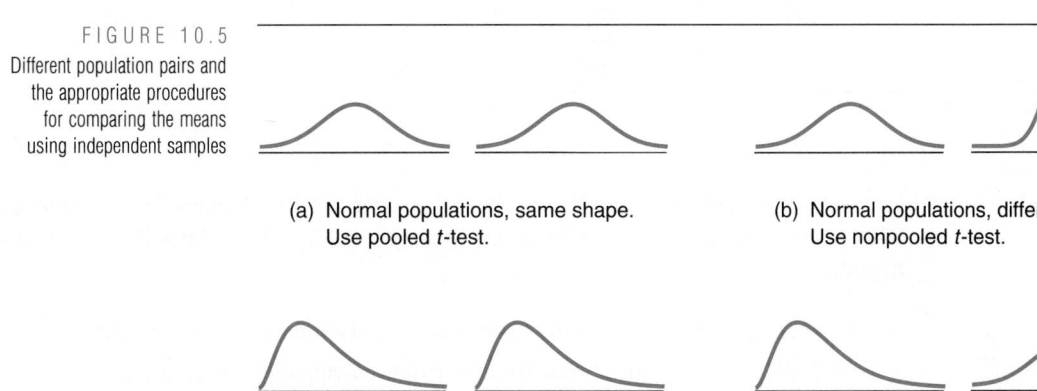

(a) Normal populations, same shape. Use pooled *t*-test.

(b) Normal populations, different shapes. Use nonpooled *t*-test.

(c) Nonnormal populations, same shape. Use Mann-Whitney test.

(d) Not both normal populations, different shapes. Use two-sample *z*-test for large samples; otherwise, consult a statistician.

Example 10.10 explains the reasoning behind the Mann–Whitney test.

Example 10.10   Introduces the Mann–Whitney Test

A nationwide shipping firm purchased a new computer system to keep track of the status of its current shipments, pickups, and deliveries. The system was linked to computer terminals in all regional offices, where office personnel could type in requests for information on the location of shipments and get answers immediately on display screens.

The company had to set up a training program to teach its staff how to use the computer terminals and decided to hire a technical writer to compose a short self-study manual for that purpose. The manual was designed so that a person could read it and be ready to use the computer terminal in 2 hours.

Some employees were able to apply the procedures outlined in the manual in very little time; other employees took considerably longer. Someone suggested that the reason for this difference in comprehension times might be that some employees had previous experience with computers whereas others did not. To test this suggestion, independent samples of employees with and without computer experience were randomly selected.

The times, in minutes, required for these employees to comprehend the manual are displayed in Table 10.7. At the 5% significance level, do the data provide sufficient evidence to conclude that the mean comprehension time for all employees without computer experience exceeds the mean comprehension time for all employees with computer experience?

TABLE 10.7
Times, in minutes, required to comprehend the self-study manual

Without experience	With experience
139	142
118	109
164	130
151	107
182	155
140	88
134	95
	104

SOLUTION Let μ_1 and μ_2 denote, respectively, the mean comprehension times for all employees without computer experience and with computer experience. Then the null and alternative hypotheses are

H_0: $\mu_1 = \mu_2$ (mean time for inexperienced employees is not greater)

H_a: $\mu_1 > \mu_2$ (mean time for inexperienced employees is greater).

As we said, the Mann–Whitney test presumes that the populations being sampled have the same shape. If the distributions of comprehension times for employees without and with computer experience have the same shape, then the distributions of the two samples in Table 10.7 should have roughly the same shape.

To check this, we constructed in Fig. 10.6 a **back-to-back stem-and-leaf diagram** of the two samples in Table 10.7. In a back-to-back stem-and-leaf diagram, the leaves for the first data set are on the left, the stems are in the middle, and the leaves for the second data set are on the right.

FIGURE 10.6
Back-to-back stem-and-leaf diagram of the two samples in Table 10.7

Without experience	stem	With experience
	8	8
	9	5
	10	9 7 4
8	11	
	12	
4 9	13	0
0	14	2
1	15	5
4	16	
	17	
2	18	

The stem-and-leaf diagrams in Fig. 10.6 have roughly the same shape and so do not reveal any obvious violations of the same-shape condition.[†] We will therefore apply the Mann–Whitney test to carry out the hypothesis test.

To apply the Mann–Whitney test, we first rank all the data from both samples combined. (It is very helpful to refer to Fig. 10.6 when ranking the data.) The results of ranking the data are depicted in Table 10.8, which shows, for instance, that the first employee without computer experience had the ninth shortest comprehension time among all 15 employees in the two samples combined.

TABLE 10.8

Results of ranking the combined data from Table 10.7

Without experience	Overall rank	With experience	Overall rank
139	9	142	11
118	6	109	5
164	14	130	7
151	12	107	4
182	15	155	13
140	10	88	1
134	8	95	2
		104	3

The idea behind the Mann–Whitney test is a simple one: If the sum of the ranks for the sample of employees without experience is too large, then we take this as evidence that the null hypothesis is false and conclude that the mean comprehension time for all employees without experience exceeds that for all employees with experience. From Table 10.8 we see that the sum of the ranks for the sample of employees without experience, denoted by M, is equal to $9 + 6 + 14 + 12 + 15 + 10 + 8 = 74$.

Of course, to decide whether this value of M is large enough to warrant rejecting the null hypothesis, we need a table of critical values for the random variable M, which can be found in Table VIII in Appendix A. We will discuss that table and then return to complete the hypothesis test.   ■

Using the Mann–Whitney Table

The test statistic, M, for a Mann–Whitney test is the sum of the ranks associated with the smaller sample size. For instance, in Example 10.10, the smaller sample size is the one for the employees without computer experience (see Table 10.7). Thus in that example the test statistic, M, is the sum of the ranks for the sample of employees without computer experience.

[†] For ease in explaining the Mann–Whitney test, we have chosen an example in which the sample sizes are very small. This, however, makes it difficult to effectively check the same-shape condition. In general, we must proceed cautiously when dealing with very small samples.

It is convenient to arrange things so that the sample size for Population 1 is less than or equal to that for Population 2. This can always be accomplished by interchanging the roles of the populations, if necessary. Therefore *we will assume that the populations have been designated so that the sample size for Population 1 is less than or equal to the sample size for Population 2; that is, $n_1 \leq n_2$.* Under that assumption the test statistic, M, for a Mann–Whitney test is the sum of the ranks for the sample from Population 1.

Tables VII and VIII in Appendix A provide critical values for a Mann–Whitney test. Table VII supplies critical values for a one-tailed test with $\alpha = 0.025$ or a two-tailed test with $\alpha = 0.05$; Table VIII supplies critical values for a one-tailed test with $\alpha = 0.05$ or a two-tailed test with $\alpha = 0.10$.[†]

The sample size for the sample from Population 1 is given along the top of each table and the sample size for the sample from Population 2 along the left side. The numbers in the columns headed M_ℓ and M_r are, respectively, left-hand and right-hand critical values. Note that *a critical value from Table VII or Table VIII is to be included as part of the rejection region.* Example 10.11 illustrates the use of the Mann–Whitney tables.

Example 10.11   Using the Mann–Whitney Tables

Determine the critical value, rejection region, and nonrejection region for the hypothesis test in Example 10.10.

SOLUTION   The hypothesis test in Example 10.10 is right-tailed (hence one-tailed) and is to be performed at the 5% significance level ($\alpha = 0.05$). Thus we use Table VIII. The sample sizes are $n_1 = 7$ and $n_2 = 8$. So to obtain the critical value for this hypothesis test, we do the following: First we locate the column labeled 7 ($n_1 = 7$) along the top of the table; then we go down that column until we are in the row labeled 8 ($n_2 = 8$) along the left side of the table. There we find two numbers, 41 and 71, which are the critical values for a left-tailed and right-tailed test, respectively.

Thus the critical value for this right-tailed hypothesis test is 71. The rejection region consists of all M-values greater than or equal to 71, and the nonrejection region consists of all M-values less than 71; in other words, we reject H_0 if $M \geq 71$ and do not reject H_0 if $M < 71$.   ∎

Performing the Mann–Whitney Test

We now present a step-by-step procedure for performing a Mann–Whitney test. The test can be used to compare two population medians as well as two population means. We will state the procedure in terms of population means. To employ the procedure for population medians, simply replace μ_1 by η_1 and μ_2 by η_2.

[†] Actually these are only approximate significance levels; but they are considered close enough in practice.

PROCEDURE 10.7

THE MANN–WHITNEY TEST FOR TWO POPULATION MEANS WITH NULL HYPOTHESIS H_0: $\mu_1 = \mu_2$

ASSUMPTIONS

1. Independent samples
2. Populations have same shape
3. $n_1 \le n_2$

Step 1 State the null and alternative hypotheses.

Step 2 Decide on the significance level, α.

Step 3 The critical value(s)
   a. for a two-tailed test are M_ℓ and M_r.
   b. for a left-tailed test is M_ℓ.
   c. for a right-tailed test is M_r.

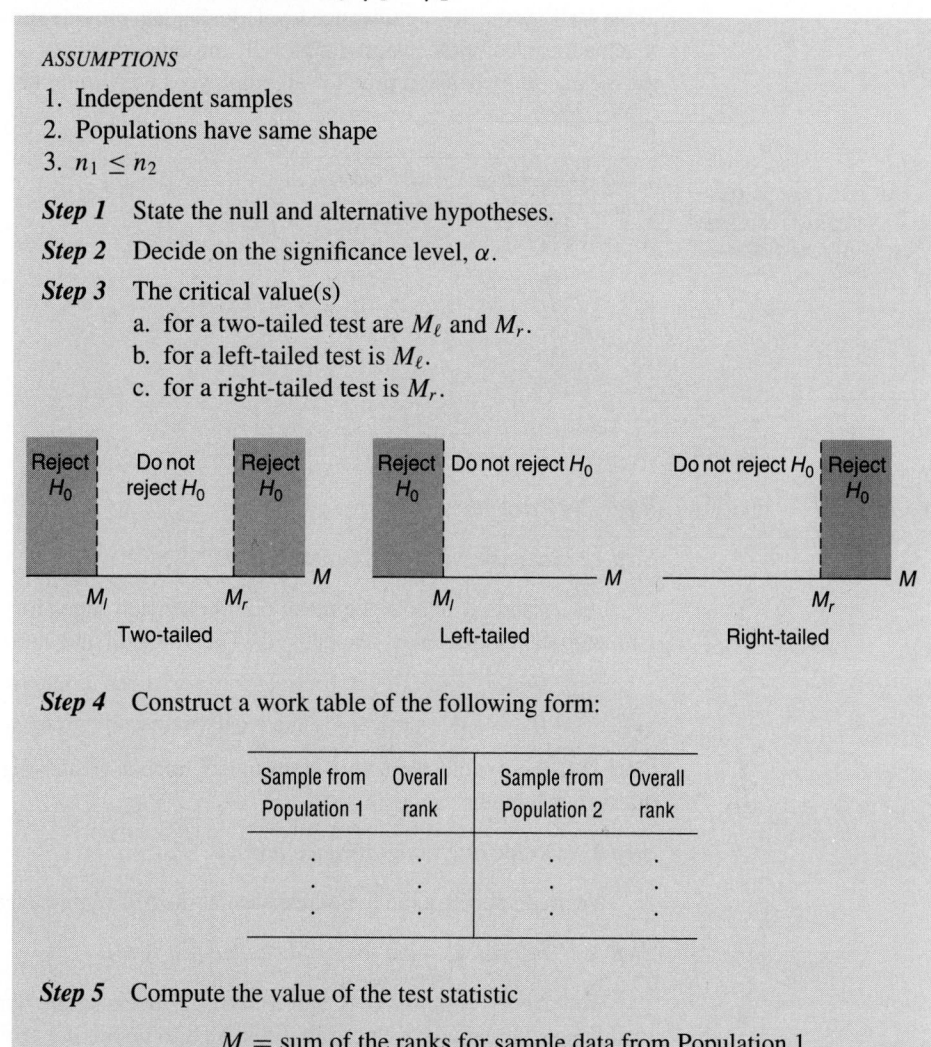

Step 4 Construct a work table of the following form:

Sample from Population 1	Overall rank	Sample from Population 2	Overall rank
.	.	.	.
.	.	.	.
.	.	.	.

Step 5 Compute the value of the test statistic

$$M = \text{sum of the ranks for sample data from Population 1.}$$

Step 6 If the value of the test statistic falls in the rejection region, then reject H_0; otherwise, do not reject H_0.

Step 7 State the conclusion in words.

Example 10.12   Illustrates Procedure 10.7

As an application of Procedure 10.7, let's complete the hypothesis test of Example 10.10. Independent samples of employees with and without computer experience were timed to see how long it would take them to comprehend a self-study manual that explained how to use a computer to track their company's products. The times, in minutes, are repeated in Table 10.9. At the 5% significance level, do the data provide sufficient evidence to conclude that the mean comprehension time for all employees without computer experience exceeds the mean comprehension time for all employees with computer experience?

TABLE 10.9

Times, in minutes, required to comprehend the self-study manual

Without experience	With experience
139	142
118	109
164	130
151	107
182	155
140	88
134	95
	104

SOLUTION

We apply Procedure 10.7.

Step 1   *State the null and alternative hypotheses.*

Let μ_1 and μ_2 denote the mean comprehension times for all employees without and with computer experience, respectively. Then the null and alternative hypotheses are

H_0: $\mu_1 = \mu_2$ (mean time for inexperienced employees is not greater)

H_a: $\mu_1 > \mu_2$ (mean time for inexperienced employees is greater).

Note that the hypothesis test is right-tailed since a greater-than sign ($>$) appears in the alternative hypothesis.

Step 2   *Decide on the significance level, α.*

We are to perform the hypothesis test at the 5% significance level; so $\alpha = 0.05$.

Step 3   *The critical value for a right-tailed test is M_r.*

We have $n_1 = 7$, $n_2 = 8$, and $\alpha = 0.05$. Because the hypothesis test is right-tailed (and hence one-tailed), we consult Table VIII to obtain the critical value. We find that the critical value is $M_r = 71$, as seen in Fig. 10.7.

FIGURE 10.7

Criterion for deciding whether or not to reject the null hypothesis

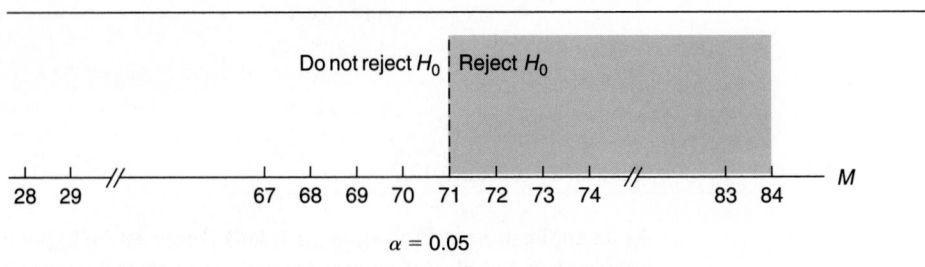

$\alpha = 0.05$

Step 4 *Construct a work table.*

We have already done this in Table 10.8 on page 611.

Step 5 *Compute the value of the test statistic*

$$M = \text{sum of the ranks for sample data from Population 1.}$$

Referring to the second column of Table 10.8, we find that

$$M = 9 + 6 + 14 + 12 + 15 + 10 + 8 = 74.$$

Step 6 *If the value of the test statistic falls in the rejection region, reject H_0; otherwise, do not reject H_0.*

From Step 5 the value of the test statistic is $M = 74$. Figure 10.7 shows that this falls in the rejection region. Thus we reject H_0.

Step 7 *State the conclusion in words.*

The test results are statistically significant at the 5% level; that is, at the 5% significance level, the data provide sufficient evidence to conclude that the mean comprehension time for all employees without computer experience exceeds the mean comprehension time for all employees with computer experience. Evidently, employees with computer experience can, on the average, comprehend the training manual more quickly than those without computer experience. ■

When there are ties in the sample data, ranks are assigned in the same way as in the Wilcoxon signed-rank test. Namely, *if two or more data values are tied, each is assigned the mean of the ranks they would have had if there were no ties.*

Comparison of the Mann–Whitney Test and the Pooled-*t* Test

In Section 10.2 we learned how to perform a pooled *t*-test (Procedure 10.3, page 587) to compare the means of two normally distributed populations having equal standard deviations using independent samples. Since two normal populations having equal standard deviations have the same shape, we can also use the Mann–Whitney test to perform such a hypothesis test.

So now the question is this: If we want to perform a hypothesis test to compare the means of two populations using independent samples and we know the populations are normally distributed and have equal standard deviations, should we use the pooled *t*-test or the Mann–Whitney test? As you might expect, we should use the pooled *t*-test; for normal populations the pooled *t*-test is more powerful than the Mann–Whitney test because it is designed expressly for normal populations. What might surprise you, however, is that for normal populations the pooled *t*-test is not much more powerful than the Mann–Whitney test.

On the other hand, if the populations being sampled have the same shape but are not normally distributed, then the Mann–Whitney test is usually more powerful than the pooled t-test, often considerably so. In summary, we have the following key fact.

KEY FACT 10.6

THE MANN–WHITNEY TEST VERSUS THE POOLED t-TEST

> Suppose a hypothesis test is to be performed to compare the means of two populations. When deciding between the pooled t-test and the Mann–Whitney test, follow these guidelines:
>
> - If you are reasonably sure the populations are normally distributed, use the pooled t-test.
>
> - If you are not reasonably sure the populations are normally distributed but are reasonably sure they have the same shape, use the Mann–Whitney test.

In this section we have discussed the Mann–Whitney test, which provides a nonparametric procedure for performing a hypothesis test to compare the means (or medians) of two populations having the same shape. There is a corresponding confidence-interval procedure, but we will not cover it here.

Using the Computer (Optional)

Procedure 10.7 on page 613 provides a step-by-step method for performing a Mann–Whitney test to compare the means of two populations having the same shape using independent samples. Alternatively, we can use Minitab to carry out a Mann–Whitney test. The appropriate command is **Mann-Whitney.**

As we said earlier, a Mann–Whitney test can be used to perform a hypothesis test to compare two population medians as well as two population means. Minitab presents the output of the Mann-Whitney command in terms of medians, but most of that output can be interpreted in terms of means. With this in mind, we now present Example 10.13, which shows how the Mann-Whitney command is applied.

Example 10.13   The Mann-Whitney Command

Use Minitab to perform the hypothesis test in Example 10.12.

SOLUTION   Let μ_1 and μ_2 denote the mean times for comprehension of the self-study manual for all employees without and with computer experience, respectively. The problem is to use the Mann–Whitney procedure to perform the hypothesis test

$$H_0: \mu_1 = \mu_2 \text{ (mean time for inexperienced employees is not greater)}$$

$$H_a: \mu_1 > \mu_2 \text{ (mean time for inexperienced employees is greater)}$$

at the 5% significance level. Note that the hypothesis test is right-tailed since a greater-than sign ($>$) appears in the alternative hypothesis.

First we store the two samples from Table 10.9 on page 614 in columns named WITHOUT and WITH. Then we proceed in the following manner.

Session commands: We type the command **MANNWHITNEY** followed by the storage locations of the data; that is, we type MANNWHITNEY for 'WITHOUT' vs 'WITH'; and press Enter. Because the test is right-tailed, we also type the subcommand ALTERNATIVE=1. and press Enter. Printout 10.7 displays these commands and the output that results.

(or)

Menu commands: We choose **Stat ▸ Nonparametrics ▸ Mann-Whitney...**, specify WITHOUT for **First Sample** and WITH for **Second Sample**, select **Alternative** and specify **greater than**, and then select **OK**. The output obtained is displayed in Printout 10.7.

PRINTOUT 10.7

Minitab output for the
Mann-Whitney command

```
MTB > MANNWHITNEY for 'WITHOUT' vs 'WITH';
SUBC> ALTERNATIVE=1.

Mann-Whitney Confidence Interval and Test

WITHOUT    N =   7    Median =      140.00
WITH       N =   8    Median =      108.00
Point estimate for ETA1-ETA2 is       31.50
95.7 pct c.i. for ETA1-ETA2 is (3.99,56.00)
W = 74.0
Test of ETA1 = ETA2  vs.  ETA1 g.t. ETA2 is significant at 0.0214
```

All references to the median in Printout 10.7 can be interpreted as references to the mean except for the ones in the fourth and fifth lines. The fourth and fifth lines give the sample sizes and sample medians for the two samples. On the next two lines we find, respectively, a point estimate and a confidence interval for the difference between the population medians, $\eta_1 - \eta_2$ (ETA1-ETA2). By default the confidence level is chosen to be as close to 95% as possible; but other specifications are allowed.

The next-to-last line of Printout 10.7 displays the value of the test statistic for a Mann–Whitney test (Minitab uses W instead of M); thus we see that the value of the test statistic is 74.0. Below the value of the test statistic is a statement of the null and alternative hypotheses (the ETAs can be replaced by MUs) and an approximate P-value; so $P = 0.0214$, approximately.[†] Since the (approximate) P-value is less than the designated significance level of $\alpha = 0.05$, we reject H_0. The test results are statistically significant at the 5% level; evidently, employees with computer experience can, on the average, comprehend the training manual more quickly than those without computer experience. ■

[†] Minitab employs a normal approximation with a continuity correction factor to obtain the approximate P-value. If there are ties in the data (there are none in this example), Minitab also prints an approximate P-value that adjusts for the ties. When there are ties, this latter approximation is usually closer to the actual P-value than the former approximation.

EXERCISES 10.4

10.56 What conditions are required for using the Mann–Whitney test?

In each of Exercises 10.57–10.60, use the Mann–Whitney test, Procedure 10.7 on page 613, to carry out the required hypothesis test.

10.57 A college chemistry instructor is concerned about the detrimental effects of poor mathematics background on his students. He randomly selects 15 students and divides them according to math background. Their semester averages turn out to be the following.

Fewer than two years of high-school algebra	Two or more years of high-school algebra
58 61	84 92 75
81 64	67 83 81
74 43	65 52 74

Do the data provide sufficient evidence to conclude that students with fewer than 2 years of high-school algebra have a lower mean semester average in this teacher's chemistry courses than do students with 2 or more years of high-school algebra? Perform the appropriate hypothesis test at the 5% significance level.

10.58 The lifetimes up to major breakdown for two brands of power lawn mowers are to be compared. Independent random samples yield the following lifetimes, in years.

Brand 1	2.3 3.7 5.9 6.8 3.5
Brand 2	1.9 3.8 6.4 5.6 4.9

Do the data suggest that the mean lifetimes of the two brands of power lawn mowers differ? Perform the required hypothesis test using a significance level of 0.05.

10.59 The National Center for Education Statistics surveys college and university libraries to obtain information on the number of volumes held. Independent random samples of public and private colleges and universities yield the following data on number of volumes held, in thousands.

Public	79	41	516	15	24	411	265
Private	139	603	113	27	67	500	

At the 5% significance level, can we conclude that the median number of volumes held by public colleges and universities is

less than that held by private colleges and universities? *(Note: The sample size for the public colleges and universities exceeds the sample size for the private colleges and universities.)*

10.60 The U.S. Bureau of Labor Statistics publishes data on weekly earnings of full-time wage and salary workers in *Employment and Earnings.* Independent random samples of male and female workers give the following data on weekly earnings, in dollars.

Male		Female	
726	2423	1894	2009
1690	188	410	191
377	217	326	174
207	618	190	997
		1261	228

At the 5% significance level, do the data provide sufficient evidence to conclude that the median weekly earnings of male full-time wage and salary workers exceeds the median weekly earnings of female full-time wage and salary workers?

10.61 A highway official wants to compare two brands of paint used for striping roads. Twenty locations are selected for paint stripes. The first brand is used on 10 locations randomly selected from the 20; the second brand is used on the remaining 10 locations. The number of months each stripe lasts is presented in the following table.

Brand A		Brand B	
35.6	36.1	37.2	36.4
37.0	35.8	39.7	37.5
34.9	34.9	37.2	40.5
36.0	37.8	38.8	38.2
36.6	36.5	37.7	36.6

a. Does there appear to be a difference in mean lasting times between the two paints? Perform a Mann–Whitney test using a significance level of $\alpha = 0.05$.
b. The hypothesis test in part (a) was done in Exercise 10.19 using the pooled t-test. The assumption in that exercise is that lasting times for both brands of paint are (approximately) normally distributed and have (approximately) equal standard deviations. Presuming that, in fact, lasting times for both brands of paint are normally distributed and have equal standard deviations, why is it permissible to perform a Mann–Whitney test to compare the means? Is it better in this case to use the pooled t-test or the Mann–Whitney test? Explain your answers.

10.62   In a packing plant, a machine packs cartons with jars. A salesperson claims that the machine she is selling will pack faster. To test that claim, the times it takes each machine to pack 10 cartons are recorded. The results, in seconds, are shown in the following table.

New machine		Present machine	
42.0	41.0	42.7	43.6
41.3	41.8	43.8	43.3
42.4	42.8	42.5	43.5
43.2	42.3	43.1	41.7
41.8	42.7	44.0	44.1

a. Do the data provide sufficient evidence to conclude that, on the average, the new machine packs faster? Perform a Mann–Whitney test at the 5% significance level.
b. The hypothesis test in part (a) was performed in Exercise 10.20 using the pooled t-test. The assumption in that exercise is that packing times for both machines are (approximately) normally distributed and have (approximately) equal standard deviations. Assuming that, in fact, packing times for both machines are normally distributed and have equal standard deviations, why is it permissible to perform a Mann–Whitney test to compare the means? Is it better in this case to use the pooled t-test or the Mann–Whitney test? Explain your answers.

10.63   Suppose you want to perform a hypothesis test to compare the means of two populations using independent samples. For each part below, decide whether you would use the two-sample z-test, the pooled t-test, the nonpooled t-test, the Mann–Whitney test, or none of these tests.
a. Preliminary data analyses of the samples suggest that both populations are normally distributed but do not have the same shape.
b. Preliminary data analyses of the samples suggest that the populations are not normally distributed but have the same shape.
c. Preliminary data analyses of the samples suggest that the populations are not normally distributed and do not have the same shape; both sample sizes are large.

10.64   Suppose you want to perform a hypothesis test to compare the means of two populations using independent samples. For each part below, decide whether you would use the two-sample z-test, the pooled t-test, the nonpooled t-test, the Mann–Whitney test, or none of these tests.

a. Preliminary data analyses of the samples suggest that both populations are normally distributed and have the same shape.
b. Preliminary data analyses of the samples suggest that the populations are not normally distributed and do not have the same shape; one of the sample sizes is large and the other is small.
c. Preliminary data analyses of the samples suggest that one of the populations is normally distributed and the other is not; both sample sizes are large.

10.65 (Computer exercise)   Refer to Exercise 10.59. Use Minitab or some other statistical software to perform the required hypothesis test using the Mann–Whitney procedure.

10.66 (Computer exercise)   Refer to Exercise 10.60. Use Minitab or some other statistical software to perform the required hypothesis test using the Mann–Whitney procedure.

10.67 (Computer exercise)   The research and development (R&D) department of a light-bulb manufacturing company claims to have developed a new bulb that, on the average, will outlast the bulb currently produced. To try to justify the claim, R&D takes independent random samples of the current bulb and the new bulb. After the lifetimes (in hours) of the bulbs sampled are determined, Minitab's Mann-Whitney command is applied to get the output shown in Printout 10.8 on the next page. Use the output to obtain
a. the null and alternative hypotheses.
b. the sample median lifetime for both the current bulb and the new bulb.
c. the number of current bulbs sampled and the number of new bulbs sampled.
d. the value of the Mann–Whitney test statistic.
e. the approximate P-value of the hypothesis test.
f. the smallest significance level at which the null hypothesis can be rejected.
g. the conclusion if the hypothesis test is performed at the 1% significance level.
h. a 95.5% confidence interval for the difference between the median (or mean) lifetimes of the current bulb and the new bulb.

10.68 (Computer exercise)   A researcher in nutrition wants to compare the protein intakes of whites and blacks. He takes independent random samples of 10 whites and 12 blacks and determines their 24-hour protein intakes, in grams. Following that, he applies Minitab's Mann-Whitney command to the resulting data and obtains the output shown in Printout 10.9 on the next page. From the output, determine

PRINTOUT 10.8   Minitab output for Exercise 10.67

```
Mann-Whitney Confidence Interval and Test

CURRENT    N =  20    Median =      1033.5
NEW        N =  10    Median =      1105.0
Point estimate for ETA1-ETA2 is      -72.0
95.5 pct c.i. for ETA1-ETA2 is (-109.0,-25.0)
W = 242.5
Test of ETA1 = ETA2  vs.  ETA1 l.t. ETA2 is significant at 0.0016
The test is significant at 0.0016 (adjusted for ties)
```

PRINTOUT 10.9   Minitab output for Exercise 10.68

```
Mann-Whitney Confidence Interval and Test

WHITE      N =  10    Median =        81.05
BLACK      N =  12    Median =        62.50
Point estimate for ETA1-ETA2 is        11.95
95.6 pct c.i. for ETA1-ETA2 is (-6.59,24.49)
W = 136.0
Test of ETA1 = ETA2  vs.  ETA1 n.e. ETA2 is significant at 0.1765

Cannot reject at alpha = 0.05
```

a. the null and alternative hypotheses.
b. the median protein intakes (per day) of the whites sampled and the blacks sampled.
c. the number of whites sampled and the number of blacks sampled.
d. the value of the Mann–Whitney test statistic.
e. the approximate P-value of the hypothesis test.
f. the smallest significance level at which the null hypothesis can be rejected.
g. the conclusion if the hypothesis test is performed at the 5% significance level.
h. a 95.6% confidence interval for the difference between the median (or mean) protein intakes per day of whites and blacks.

Large-sample Mann–Whitney test using the normal approximation. The tables of critical values for the Mann–Whitney test, Tables VII and VIII, stop at $n_1 = 10$ and

$n_2 = 10$. For larger samples, a normal approximation can be used. Specifically, we have the following result.

> Suppose independent random samples of sizes n_1 and n_2, both at least 10, are to be taken from two populations with means μ_1 and μ_2, respectively. Further suppose the distributions of the two populations have the same shape. Then, if $\mu_1 = \mu_2$, the random variable M is approximately normally distributed with mean $\mu_M = n_1(n_1 + n_2 + 1)/2$ and standard deviation $\sigma_M = \sqrt{n_1 n_2(n_1 + n_2 + 1)/12}$. In particular the standardized random variable
>
> $$(3) \qquad z = \frac{M - n_1(n_1 + n_2 + 1)/2}{\sqrt{n_1 n_2(n_1 + n_2 + 1)/12}}$$
>
> has approximately the standard normal distribution.

In Exercises 10.69–10.71 we will develop and apply a large-sample procedure for a Mann–Whitney test based on the preceding fact.

· · **10.69** Formulate a large-sample hypothesis-testing procedure for a Mann–Whitney test that uses the test statistic in Equation (3).

· · **10.70** Refer to Exercise 10.62.
a. Use your procedure from Exercise 10.69 to perform the hypothesis test.
b. Compare your result in part (a) to the one you obtained in Exercise 10.62(a), where the normal approximation was not used.

· · **10.71** Refer to Exercise 10.61.
a. Use your procedure from Exercise 10.69 to perform the hypothesis test.
b. Compare your result in part (a) to the one you obtained in Exercise 10.61(a), where the normal approximation was not used.

· · · **10.72** In this exercise you will obtain the probability distribution of the random variable M, when the sample sizes are $n_1 = 3$ and $n_2 = 3$. This will enable you to see how the critical values for the Mann–Whitney test are derived. We can display all possible ranks for the data by constructing the following table, in which the letter A stands for a member from Population 1 and the letter B stands for a member from Population 2.

Rank						
1	2	3	4	5	6	M
A	A	A	B	B	B	6
A	A	B	A	B	B	7
A	A	B	B	A	B	8
.	.	.	.	.	.	.
.	.	.	.	.	.	.
B	B	A	B	A	A	14
B	B	B	A	A	A	15

a. Complete the table. *(Hint: There are 20 rows.)*
b. Use your result from part (a) to find the probability distribution of the random variable M if $\mu_1 = \mu_2$. *(Hint: Each row of the table is equally likely.)*
c. If $\mu_1 = \mu_2$, draw a histogram for the probability distribution of M when $n_1 = 3$ and $n_2 = 3$.
d. Apply your result from part (b) to obtain the entries in Table VIII for M_ℓ and M_r when $n_1 = 3$ and $n_2 = 3$.

10.5 INFERENCES FOR TWO POPULATION MEANS USING PAIRED SAMPLES

· · · · · ·

Up to this point, the methods we have studied for comparing the means of two populations rely on independent samples. In this section we will examine methods for comparing the means of two populations using paired samples.

Suppose, for instance, we want to decide whether a newly developed gasoline additive increases gas mileage. Let μ_1 denote the mean gas mileage of all cars when the additive is used and μ_2 denote the mean gas mileage of all cars when the additive is not used. Then we want to perform the hypothesis test

H_0: $\mu_1 = \mu_2$ (mean gas mileage with additive is not greater)

H_a: $\mu_1 > \mu_2$ (mean gas mileage with additive is greater).

One method we can use to carry out the hypothesis test is this: Randomly and independently select two groups of, say, 10 cars each; have one group driven with the additive and the other group driven without the additive; and then apply a hypothesis-testing procedure to the two samples of mileages obtained. This method employs independent samples.

However, the following method for carrying out the hypothesis test may be more appropriate: Randomly select a single group of 10 cars; have each of the 10 cars driven both with and without the additive; and then apply a hypothesis-testing procedure (as will be described in this section) to the 10 *pairs* of mileages obtained. This method employs

paired samples. Each piece of sample data consists of a pair of numbers; in this case the gas mileage of a car both with and without the additive.

By pairing the samples, we can remove extraneous sources of variation, such as the variation due to cars and drivers. As a consequence, the sampling error made in estimating the difference between the population means will generally be smaller. This fact, in turn, makes it more likely that we will detect differences between the population means when such differences exist.

We will now use the gas-mileage illustration to explain the logic behind hypothesis tests that employ paired samples. In doing so we will also introduce some notation and terminology used in such hypothesis tests.

Example 10.14   Introduces Hypothesis Tests for Two Means Using Paired Samples

A major oil company has developed a new gasoline additive that is supposed to increase mileage. To test that hypothesis, 10 cars are randomly selected. Each car sampled is driven both with and without the additive. The resulting gas mileages, in miles per gallon (mpg), are displayed in the second and third columns of Table 10.10.

TABLE 10.10

Gas mileages, with and without additive, for 10 randomly selected cars

Car	With additive x_1	Without additive x_2	Paired difference $d = x_1 - x_2$
1	25.7	24.9	0.8
2	20.0	18.8	1.2
3	28.4	27.7	0.7
4	13.7	13.0	0.7
5	18.8	17.8	1.0
6	12.5	11.3	1.2
7	28.4	27.8	0.6
8	8.1	8.2	−0.1
9	23.1	23.1	0.0
10	10.4	9.9	0.5
			6.6

In the last column of Table 10.10, we have recorded the difference, d, between the gas mileages, with and without the additive, for each of the 10 cars sampled. Each difference is referred to as a **paired difference** since it is the difference of a pair of numbers. For instance, the first car got 25.7 mpg with the additive and 24.9 mpg without the additive, giving a paired difference of $d = 25.7 - 24.9 = 0.8$ mpg, an increase in gas mileage of 0.8 mpg with the additive.

We want to use the paired differences, d, in the last column of Table 10.10 to perform the hypotheses test

$$H_0: \mu_1 = \mu_2 \text{ (mean gas mileage with additive is not greater)}$$

$$H_a: \mu_1 > \mu_2 \text{ (mean gas mileage with additive is greater)},$$

where μ_1 denotes the mean gas mileage of all cars when the additive is used and μ_2 denotes the mean gas mileage of all cars when the additive is not used.

The logic behind performing the hypothesis test using the paired differences is as follows: If the null hypothesis is true, then the paired differences of the gas mileages for the cars sampled should average out to about zero. In other words, we would expect the sample mean, $\bar{d}$, of the paired differences in the final column of Table 10.10 to be roughly zero. To put it another way, if $\bar{d}$ is too much greater than zero, then we would take this as evidence that the null hypothesis is false and conclude that the mean gas mileage of all cars when the additive is used is greater than the mean gas mileage of all cars when the additive is not used (i.e., on the average, the additive improves gas mileage).

From the last column of Table 10.10, we find that the sample mean of the paired differences is

$$\bar{d} = \frac{\Sigma d}{n} = \frac{6.6}{10} = 0.66 \text{ mpg},$$

a mean increase in gas mileage of 0.66 mpg when the additive is used. The question now is whether this mean increase in gas mileage can reasonably be attributed to sampling error or whether it is large enough to indicate that, on the average, the additive improves gas mileage. To answer the question, we need to know the probability distribution of the random variable $\bar{d}$. We will discuss that probability distribution and then return to solve the gas-mileage problem.   ■

The Sampling Distribution of the Difference Between Two Means (Normal Differences and Paired Samples)

Consider two populations, Population 1 and Population 2, whose members can be paired. To each pair there corresponds a single number, obtained by subtracting the Population-2 value in the pair from the Population-1 value in the pair. Thus, from the two populations, we can form a single population consisting of the differences of all the pairs.[†] Denote the mean of that population of paired differences by μ_d. Then it can be shown that

(4) $$\mu_d = \mu_1 - \mu_2,$$

where μ_1 is the mean of Population 1 and μ_2 is the mean of Population 2. In other words, the mean of the population of paired differences is equal to the difference between the two population means. (See Exercise 10.98 for a detailed derivation.)

Now suppose a random sample of n pairs is to be taken from the two populations. Let $\bar{d}$ denote the sample mean paired difference of the pairs obtained, where each difference is computed by subtracting the second number in the pair from the first number in the pair ($d = x_1 - x_2$). Also let s_d denote the sample standard deviation of the paired differences.

[†] In Example 10.14, Population 1 and Population 2 consist, respectively, of the gas mileages of all cars with and without the additive. For any particular car, a pair is formed from the car's gas mileages with and without the additive. The difference of such a pair then represents the increase (or reduction) in gas mileage obtained for that car by using the additive.

We can think of the paired differences of the pairs sampled as a random sample from the population of all possible paired differences. If that population is normally distributed, then we can apply Key Fact 8.3 on page 461 and Equation (4) to obtain the following result.

KEY FACT 10.7   **DISTRIBUTION OF THE PAIRED-t STATISTIC**

Suppose a random sample of n pairs is to be taken from populations with means μ_1 and μ_2. Further suppose the population of all paired differences is normally distributed. Then the random variable

$$t = \frac{\overline{d} - (\mu_1 - \mu_2)}{s_d/\sqrt{n}}$$

has the t-distribution with df $= n - 1$.

Note: We will use the phrase **normal differences** as an abbreviation of "the population of paired differences is normally distributed."

Hypothesis Tests for Two Population Means Using Paired Samples

We can now present a hypothesis-testing procedure for comparing the means of two populations using paired samples when the population of all paired differences is normally distributed. In view of Key Fact 10.7, we see that for a hypothesis test with null hypothesis H_0: $\mu_1 = \mu_2$, we can use the random variable

$$t = \frac{\overline{d}}{s_d/\sqrt{n}}$$

as the test statistic and obtain the critical value(s) from the t-table, Table IV. Specifically, we have the following procedure, which we often refer to as the **paired t-test.**

PROCEDURE 10.8   **THE PAIRED t-TEST FOR TWO POPULATION MEANS WITH NULL HYPOTHESIS H_0: $\mu_1 = \mu_2$**

ASSUMPTIONS

1. Paired samples
2. Normal differences

Step 1   State the null and alternative hypotheses.

Step 2   Decide on the significance level, α.

Step 3   The critical value(s)
        a. for a two-tailed test are $\pm t_{\alpha/2}$,
        b. for a left-tailed test is $-t_\alpha$,
        c. for a right-tailed test is t_α,
        with df $= n - 1$. Use Table IV to find the critical value(s).

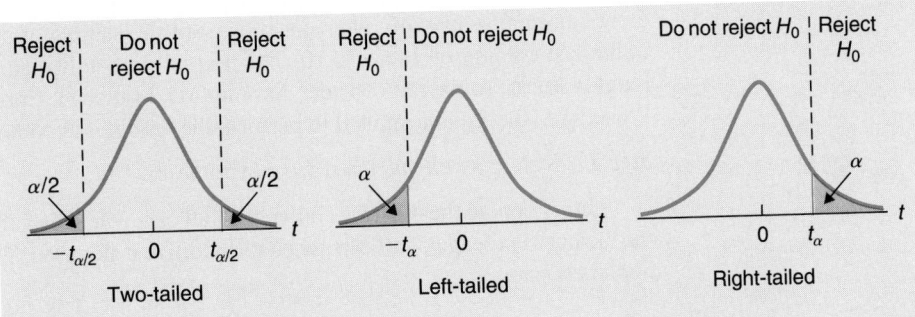

Step 4   Calculate the paired differences, $d = x_1 - x_2$, of the sample pairs.

Step 5   Compute the value of the test statistic

$$t = \frac{\bar{d}}{s_d / \sqrt{n}}.$$

Step 6   If the value of the test statistic falls in the rejection region, then reject H_0; otherwise, do not reject H_0.

Step 7   State the conclusion in words.

The paired t-test is simply a one-sample t-test, with null hypothesis H_0: $\mu = 0$, applied to a population of paired differences. So the paired t-test shares with the one-sample t-test the properties of robustness to moderate deviations of the normality assumption and sensitivity to outliers and other extreme values.

Regarding the two assumptions for the paired t-test: The paired-samples assumption (Assumption 1) is essential; the samples must be paired or the procedure does not apply. The normality assumption (Assumption 2) refers to the population of paired differences; the individual populations need not be normally distributed.

Example 10.15   Illustrates Procedure 10.8

We now return to the gas-mileage problem posed in Example 10.14. Recall that a major oil company has developed a new gasoline additive that is supposed to increase gas mileage. To test that hypothesis, the gas mileages are obtained, both with and without the additive, for each of 10 randomly selected cars. The results are displayed in the second and third columns of Table 10.10 on page 622.

Do the data provide sufficient evidence to conclude that, on the average, the gasoline additive improves gas mileage? Perform the appropriate hypothesis test at the 5% significance level.

SOLUTION   To begin, we check the two conditions required for using the paired t-test. We are dealing here with paired samples; each pair consists of the gas mileage of a car both with and without the additive. So Assumption 1 is satisfied. To check the normality condition, we drew a normal probability plot (not shown) of the sample of paired differences displayed in the last column of Table 10.10. The normal probability plot reveals no outliers and is roughly linear; so we can consider Assumption 2 satisfied. Consequently, the paired t-test, Procedure 10.8, can be applied to perform the required hypothesis test.

Step 1   *State the null and alternative hypotheses.*

Let μ_1 denote the mean gas mileage of all cars when the additive is used and μ_2 denote the mean gas mileage of all cars when the additive is not used. Then the null and alternative hypotheses are

$$H_0: \mu_1 = \mu_2 \text{ (mean gas mileage with additive is not greater)}$$
$$H_a: \mu_1 > \mu_2 \text{ (mean gas mileage with additive is greater)}.$$

Note that the hypothesis test is right-tailed since a greater-than sign ($>$) appears in the alternative hypothesis.

Step 2   *Decide on the significance level, α.*

The test is to be performed at the 5% significance level. Thus $\alpha = 0.05$.

Step 3   *The critical value for a right-tailed test is t_α with $df = n - 1$.*

From Step 2, $\alpha = 0.05$. Also, since there are 10 pairs in the sample, we have $df = n - 1 = 10 - 1 = 9$. So the critical value is $t_{0.05} = 1.833$, as seen in Fig. 10.8.

FIGURE 10.8
Criterion for deciding whether or not to reject the null hypothesis

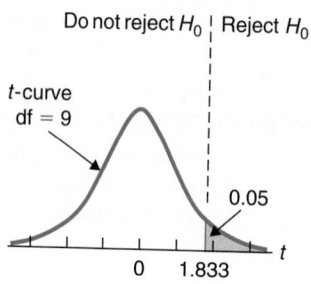

Do not reject H_0 | Reject H_0

t-curve
df = 9

0.05

0   1.833 t

Step 4   *Calculate the paired differences, $d = x_1 - x_2$, of the sample pairs.*

We have already done this in the last column of Table 10.10 on page 622.

Step 5   *Compute the value of the test statistic*

$$t = \frac{\overline{d}}{s_d/\sqrt{n}}.$$

We first need to determine the sample mean and sample standard deviation of the d-values in the last column of Table 10.10. This is accomplished in the usual manner. We find that $\overline{d} = 0.66$ and $s_d = 0.44$. Consequently, the value of the test statistic is

$$t = \frac{\overline{d}}{s_d/\sqrt{n}} = \frac{0.66}{0.44/\sqrt{10}} = 4.743.$$

Step 6   *If the value of the test statistic falls in the rejection region, reject H_0; otherwise, do not reject H_0.*

From Step 5, the value of the test statistic is $t = 4.743$, which falls in the rejection region. Hence we reject H_0.

Step 7   *State the conclusion in words.*

The test results are statistically significant at the 5% level; that is, at the 5% significance level, the data provide sufficient evidence to conclude that the mean gas mileage of all cars when the additive is used is greater than the mean gas mileage of all cars when the additive is not used. Evidently, the additive is effective in increasing gas mileage.   ■

The P-value approach to hypothesis testing can also be used to carry out the hypothesis test in Example 10.15 and to assess more precisely the evidence against the null hypothesis. From Step 5 we see that the value of the test statistic is $t = 4.743$. Recalling that df $= 9$, we find from Table IV that $P < 0.005$. In particular, because the P-value is less than the significance level of 0.05, we can reject H_0. Furthermore, by referring to Table 9.12 on page 530, we see that the data provide very strong evidence that, on the average, the additive increases gas mileage.

Confidence Intervals for the Difference Between Two Population Means Using Paired Samples

We can also use Key Fact 10.7 on page 624 to derive the following confidence-interval procedure for the difference between two population means, which we often refer to as the **paired t-interval procedure.**

PROCEDURE 10.9

THE PAIRED t-INTERVAL PROCEDURE FOR THE DIFFERENCE BETWEEN TWO POPULATION MEANS

ASSUMPTIONS

1. Paired samples
2. Normal differences

Step 1   For a confidence level of $1 - \alpha$, use Table IV to find $t_{\alpha/2}$ with df $= n - 1$.

Step 2   The endpoints of the confidence interval for $\mu_1 - \mu_2$ are

$$\bar{d} \pm t_{\alpha/2} \cdot \frac{s_d}{\sqrt{n}}.$$

Example 10.16   Illustrates Procedure 10.9

Consider again the gas-mileage illustration of Example 10.14. Use the sample data in Table 10.10 on page 622 to obtain a 90% confidence interval for the difference, $\mu_1 - \mu_2$, between the mean gas mileage of all cars when the additive is used and the mean gas mileage of all cars when the additive is not used.

SOLUTION   We apply Procedure 10.9.

Step 1   *For a confidence level of* $1 - \alpha$, *use Table IV to find* $t_{\alpha/2}$ *with* $df = n - 1$.

For a 90% confidence interval, we have $\alpha = 0.10$. From Table IV, we find that for $df = n - 1 = 10 - 1 = 9$, $t_{\alpha/2} = t_{0.10/2} = t_{0.05} = 1.833$.

Step 2   *The endpoints of the confidence interval for* $\mu_1 - \mu_2$ *are*

$$\bar{d} \pm t_{\alpha/2} \cdot \frac{s_d}{\sqrt{n}}.$$

From Step 1, $t_{\alpha/2} = 1.833$. Also, $n = 10$, and from Example 10.15 we know that $\bar{d} = 0.66$ and $s_d = 0.44$. So the endpoints of the confidence interval for $\mu_1 - \mu_2$ are

$$0.66 \pm 1.833 \cdot \frac{0.44}{\sqrt{10}}$$

or 0.40 to 0.92. We can be 90% confident that the difference, $\mu_1 - \mu_2$, between the mean gas mileage of all cars when the additive is used and the mean gas mileage of all cars when the additive is not used is somewhere between 0.40 mpg and 0.92 mpg. In particular we can be 90% confident that, on the average, the additive increases gas mileage by at least 0.40 mpg.

■

The paired-*t* procedures we have just discussed provide methods for comparing the means of two populations using paired samples. An assumption for using those procedures is that the population of all paired differences is normally distributed (normal differences). Methods for comparing the means of two populations using paired samples that do not require normal differences are also available. We now discuss two such methods.

Large-Sample Inferences for Two Population Means Using Paired Samples

If the samples are paired and the sample size is large, we can use the one-sample-*z* procedures to compare the means of the two populations under consideration, regardless of the distribution of the population of paired differences. For a hypothesis test, we apply the one-sample *z*-test (Procedure 9.1 on page 503) to the sample of paired differences, with null hypothesis H_0: $\mu = 0$; for a confidence interval, we apply the one-sample *z*-interval procedure (Procedure 8.1 on page 448) to the sample of paired differences.

In this context the one-sample z-test is called the **paired z-test,** and the one-sample z-interval procedure is called the **paired z-interval procedure.** We will examine the paired-z procedures in Exercises 10.94 and 10.95.

The Wilcoxon Paired-Sample Signed-Rank Test (Optional)

If the population of all paired differences is symmetric (but not necessarily normal), then we can use the Wilcoxon signed-rank test to perform a hypothesis test to compare the means of the two populations. Essentially we apply Procedure 9.5 on page 551, with null hypothesis H_0: $\mu = 0$, to the sample of paired differences.

In this context the Wilcoxon signed-rank test is called the **Wilcoxon paired-sample signed-rank test.** Procedure 10.10 provides the steps for performing a Wilcoxon paired-sample signed-rank test. We will use the phrase **symmetric differences** as an abbreviation of "the population of paired differences has a symmetric distribution."

PROCEDURE 10.10

THE WILCOXON PAIRED-SAMPLE SIGNED-RANK TEST FOR TWO POPULATION MEANS WITH NULL HYPOTHESIS H_0: $\mu_1 = \mu_2$

ASSUMPTIONS

1. Paired samples
2. Symmetric differences

Step 1 State the null and alternative hypotheses.

Step 2 Calculate the paired differences, $d = x_1 - x_2$, of the sample pairs.

Step 3 Discard all d-values equal to 0 and reduce the sample size accordingly.

Step 4 Decide on a significance level and use Table VI to find a significance level, α, as close as possible to the one desired.

Step 5 The critical value(s)
   a. for a two-tailed test are W_ℓ and W_r.
   b. for a left-tailed test is W_ℓ.
   c. for a right-tailed test is W_r.
   Use Table VI to find the critical value(s).

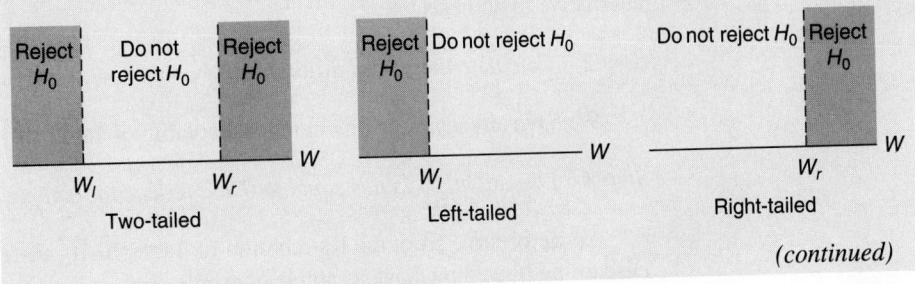

(continued)

Step 6   Construct a work table of the following form:

Paired difference d	$\lvert d \rvert$	Rank of $\lvert d \rvert$	Signed rank R
.	.	.	.
.	.	.	.
.	.	.	.

Step 7   Compute the value of the test statistic

$$W = \text{sum of the positive ranks.}$$

Step 8   If the value of the test statistic falls in the rejection region, then reject H_0; otherwise, do not reject H_0.

Step 9   State the conclusion in words.

Example 10.17   Illustrates Procedure 10.10

Use the Wilcoxon paired-sample signed-rank test to carry out the hypothesis test that we performed in Example 10.15 using the paired *t*-test.

SOLUTION   We apply Procedure 10.10.

Step 1   *State the null and alternative hypotheses.*

Let μ_1 denote the mean gas mileage of all cars when the additive is used and μ_2 denote the mean gas mileage of all cars when the additive is not used. Then the null and alternative hypotheses are

$$H_0\colon \mu_1 = \mu_2 \text{ (mean gas mileage with additive is not greater)}$$
$$H_a\colon \mu_1 > \mu_2 \text{ (mean gas mileage with additive is greater).}$$

Note that the hypothesis test is right-tailed since a greater-than sign ($>$) appears in the alternative hypothesis.

Step 2   *Calculate the paired differences, $d = x_1 - x_2$, of the sample pairs.*

We have already done this in the last column of Table 10.10 on page 622.

Step 3   *Discard all d-values equal to 0 and reduce the sample size accordingly.*

As we observe from the last column of Table 10.10, there is one *d*-value equal to 0. Discarding it, we now have a sample of size 9.

Step 4 *Decide on a significance level and use Table VI to find a significance level, α, as close as possible to the one desired.*

The test is to be performed at a significance level as close to 0.05 as possible. Since $n = 9$ and the test is one-tailed, we see from Table VI that the closest possible significance level is $\alpha = 0.049$.

Step 5 *The critical value for a right-tailed test is W_r.*

Referring again to Table VI with $n = 9$ and $\alpha = 0.049$ and recalling that the test is right-tailed, we find that the critical value is $W_r = 37$, as seen in Fig. 10.9.

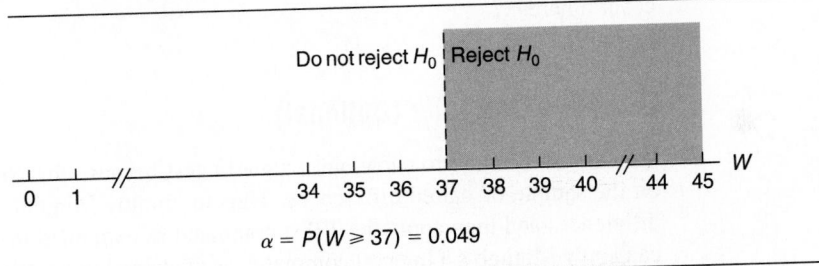

FIGURE 10.9

Criterion for deciding whether or not to reject the null hypothesis

$\alpha = P(W \geqslant 37) = 0.049$

Step 6 *Construct a work table.*

Paired difference d	$|d|$	Rank of $|d|$	Signed rank R
0.8	0.8	6	6
1.2	1.2	8.5	8.5
0.7	0.7	4.5	4.5
0.7	0.7	4.5	4.5
1.0	1.0	7	7
1.2	1.2	8.5	8.5
0.6	0.6	3	3
−0.1	0.1	1	−1
0.5	0.5	2	2

Step 7 *Compute the value of the test statistic*

$$W = \text{sum of the positive ranks.}$$

The last column of the work table shows that the sum of the positive ranks equals

$$W = 6 + 8.5 + 4.5 + 4.5 + 7 + 8.5 + 3 + 2 = 44.$$

Step 8 *If the value of the test statistic falls in the rejection region, reject H_0; otherwise, do not reject H_0.*

The value of the test statistic is $W = 44$, as found in Step 7. This falls in the rejection region, as we observe from Fig. 10.9. Thus we reject H_0.

Step 9   *State the conclusion in words.*

The test results are statistically significant at the 5% level; that is, at the 5% significance level, the data provide sufficient evidence to conclude that the mean gas mileage of all cars when the additive is used is greater than the mean gas mileage of all cars when the additive is not used. Evidently, the additive is effective in increasing gas mileage.   ∎

As we know, the Wilcoxon paired-sample signed-rank test provides a nonparametric procedure for performing a hypothesis test to compare the means of two populations using paired samples. There is a corresponding confidence-interval procedure, but we will not cover it here.

Using the Computer (Optional)

We can use Minitab to carry out a paired *t*-test because that procedure is simply a *t*-test on the sample of paired differences. Thus to employ Minitab, we first obtain the paired differences and then apply the TTest command as explained in Section 9.6. Similarly, we can apply Minitab's TInterval command, as explained in Section 8.4, to carry out a paired *t*-interval procedure. Exercises 10.86 and 10.87 give you practice with using Minitab to execute paired-*t* procedures.

Additionally, we can employ Minitab to carry out a Wilcoxon paired-sample signed-rank test because that procedure is simply a Wilcoxon signed-rank test on the sample of paired differences. Thus to employ Minitab, we first obtain the paired differences and then apply the WTest command as explained in Section 9.7. Exercises 10.90 and 10.91 give you practice with using Minitab to execute a Wilcoxon paired-sample signed-rank test.

EXERCISES 10.5

10.73 State the two conditions required for performing a paired-*t* procedure.

Preliminary data analyses indicate that it is reasonable to employ a paired t-test in Exercises 10.74–10.79. Perform each hypothesis test using either the classical approach or the P-value approach.

10.74 A pediatrician measured the blood cholesterol levels of her young patients. She was surprised to find that many of them had levels over 200 mg per 100 mL, indicating increased risk of artery disease. Ten such patients were randomly selected to take part in a nutritional program designed to lower blood cholesterol. Two months after the program started, the pediatrician measured the blood cholesterol levels of the 10 patients again. Here are the data.

Patient	Before program	After program
1	210	212
2	217	210
3	208	210
4	215	213
5	202	200
6	209	208
7	207	203
8	210	199
9	221	218
10	218	214

Do the data suggest that the nutritional program is, on the average, effective in reducing cholesterol levels? Perform the appropriate hypothesis test at the 1% level of significance.

10.75 An exercise physiologist wants to determine whether a certain type of running program will reduce heart rates. He measures the heart rates of 15 randomly selected people who are then placed on the running program. One year later the exercise physiologist again measures the heart rates of the 15 people. The heart rates, both before and after the running program, are displayed in the table below.

Person	Before program	After program
1	68	67
2	76	77
3	74	74
4	71	74
5	71	69
6	72	70
7	75	71
8	83	77
9	75	71
10	74	74
11	76	73
12	77	68
13	78	71
14	75	72
15	75	77

Do the data provide sufficient evidence to conclude that the running program will, on the average, reduce heart rates? Use $\alpha = 0.05$.

10.76 The A. C. Nielsen Company collects data on the TV viewing habits of Americans and publishes the information in *Nielsen Report on Television.* Twenty married couples are randomly selected. Their weekly viewing times, in hours, are as shown in the following table.

Husband	Wife	Husband	Wife	Husband	Wife
21	24	38	45	36	35
56	55	27	29	20	34
34	55	30	41	43	32
30	34	31	37	4	13
41	32	30	35	16	9
35	38	32	48	21	23
26	38	15	17		

At the 5% level of significance, does it appear that married men watch less TV, on the average, than married women? *(Note: $\bar{d} = -4.4$ and $s_d = 8.15$.)*

10.77 In Exercise 10.21 we performed a hypothesis test using independent samples to decide whether nonsupervisory mine workers earn a smaller average hourly wage than nonsupervisory construction workers. Now we will perform that same hypothesis test using paired samples. Nonsupervisory mine workers and nonsupervisory construction workers are paired by matching workers with similar experience and job classification. A random sample of 15 pairs yields the following paired differences for hourly wages (mining *minus* construction).

0.80	1.03	0.57	−0.63	1.20
−2.38	0.89	−2.16	−1.40	−0.67
−1.36	−0.05	−1.89	−1.23	0.47

Use these data to decide whether nonsupervisory mine workers earn a smaller average hourly wage than nonsupervisory construction workers. Perform the appropriate hypothesis test at the 5% significance level. *(Note: $\bar{d} = -0.454$, $s_d = 1.238$.)*

10.78 An algebra teacher wants to compare two methods of teaching college algebra. One is the lecture method and the other is the personalized system of instruction (PSI) method. Students are paired by matching those with similar mathematics background and performance. A random sample of 11 pairs is selected. From each pair, one student is randomly chosen to take the lecture course; the other student takes the PSI course. Both courses are taught by the algebra teacher. The final grades for the 11 pairs of students are as shown in the following table.

Lecture	66	93	36	84	60	66	80	73	74	83	52
PSI	67	93	35	85	64	57	79	70	67	79	50

Do the data provide sufficient evidence to conclude that there is a difference in mean student performance between the two instructional methods? Perform the required hypothesis test at the 5% significance level.

10.79 *Current Population Reports,* published by the U.S. Bureau of the Census, presents data on the ages of married people. Ten married couples are randomly selected and have the ages shown here.

Husband	54	21	32	78	70	33	68	32	54	52
Wife	53	22	33	74	64	35	67	28	41	44

Do the data suggest that the mean age of married men is greater than the mean age of married women? Perform the appropriate hypothesis test at the 5% significance level.

In each of Exercises 10.80–10.85, use Procedure 10.9 on page 627 to obtain the required confidence interval.

·   **10.80**   Refer to Exercise 10.74.
a. Determine a 98% confidence interval for the difference, $\mu_1 - \mu_2$, between the mean cholesterol levels of high-level patients before and after the nutritional program.
b. Interpret your result in part (a).

·   **10.81**   Refer to Exercise 10.75.
a. Determine a 90% confidence interval for the difference, $\mu_1 - \mu_2$, between the mean heart rates of all people before and after the running program.
b. Interpret your result in part (a).

·   **10.82**   Refer to Exercise 10.76.
a. Determine a 90% confidence interval for the difference, $\mu_1 - \mu_2$, between the mean weekly TV viewing times of married men and married women.
b. Interpret your result in part (a).

·   **10.83**   Refer to Exercise 10.77.
a. Determine a 90% confidence interval for the difference, $\mu_1 - \mu_2$, between the mean hourly earnings of nonsupervisory mine and construction workers.
b. Interpret your result in part (a).

·   **10.84**   Refer to Exercise 10.78.
a. Obtain a 95% confidence interval for the difference between mean student performance for the two instructional methods.
b. Interpret your result in part (a).

·   **10.85**   Refer to Exercise 10.79.
a. Find a 90% confidence interval for the difference between the mean ages of married men and married women.
b. Interpret your result in part (a).

·   **10.86 (Computer exercise)**   Refer to Exercises 10.78 and 10.84. Use Minitab or some other statistical software to
a. obtain a normal probability plot of the paired differences.
b. perform the required hypothesis test and obtain the desired confidence interval.
c. Justify the use of your procedure in part (b).

·   **10.87 (Computer exercise)**   Refer to Exercises 10.79 and 10.85. Use Minitab or some other statistical software to
a. obtain a normal probability plot of the paired differences.
b. perform the required hypothesis test and obtain the desired confidence interval.
c. Justify the use of your procedure in part (b).

Exercises 10.88–10.92 involve the optional material on the Wilcoxon paired-sample signed-rank test.

·   **10.88**   Refer to Exercise 10.74.
a. Use the Wilcoxon paired-sample signed-rank test to perform the hypothesis test.
b. Compare your results with those in Exercise 10.74.

·   **10.89**   Refer to Exercise 10.75.
a. Use the Wilcoxon paired-sample signed-rank test to perform the hypothesis test.
b. Compare your results with those in Exercise 10.75.

·   **10.90 (Computer exercise)**   Refer to Exercise 10.74. Use Minitab or some other statistical software to
a. obtain a normal probability plot, stem-and-leaf diagram, histogram, and boxplot of the paired differences.
b. perform the required hypothesis test using the paired t-test.
c. perform the required hypothesis test using the Wilcoxon paired-sample signed-rank test.
d. Compare your results from parts (b) and (c).
e. Based on your data analysis in part (a), which test do you think is more appropriate: the paired t-test or the Wilcoxon paired-sample signed-rank test? Explain your answer.

·   **10.91 (Computer exercise)**   Refer to Exercise 10.75. Use Minitab or some other statistical software to
a. obtain a normal probability plot, stem-and-leaf diagram, histogram, and boxplot of the paired differences.
b. perform the required hypothesis test using the paired t-test.
c. perform the required hypothesis test using the Wilcoxon paired-sample signed-rank test.
d. Compare your results from parts (b) and (c).
e. Based on your data analysis in part (a), which test do you think is more appropriate: the paired t-test or the Wilcoxon paired-sample signed-rank test? Explain your answer.

·   **10.92**   A hypothesis test is to be performed to compare the means of two populations using paired samples. The sample of 15 paired differences contains an outlier, but otherwise is roughly bell-shaped. Assuming it is not legitimate to remove the outlier, which is the better test to use—the paired t-test or the Wilcoxon paired-sample signed-rank test? Explain your answer.

· ·   **10.93**   This exercise shows what can happen when a hypothesis-testing procedure designed for use with independent samples is applied to perform a hypothesis test in which the samples are paired. In Example 10.15 on page 625, we applied the paired t-test, Procedure 10.8, to decide whether a gasoline additive is effective in increasing gas mileage. Specifically, if we let μ_1 and μ_2 denote, respectively, the mean gas mileage of all cars when the additive is and is not used, then the hypothesis test is

$$H_0: \mu_1 = \mu_2 \text{ (additive is not effective)}$$

$$H_a: \mu_1 > \mu_2 \text{ (additive is effective)}.$$

a. Apply the nonpooled t-test, Procedure 10.5 on page 599, to the sample data in the second and third columns of Table 10.10 on page 622 to perform the hypothesis test. Use $\alpha = 0.05$.
b. Why is it inappropriate to perform the hypothesis test the way you did in part (a)?
c. Compare your result in part (a) to the one obtained in Example 10.15.

Large-sample inferences for the means of two populations using paired samples. We will consider, in Exercises 10.94 and 10.95, large-sample inferences to compare the means of two populations using paired samples. The procedures used for such inferences are similar to those that apply when the population of paired differences is normally distributed. However, for large samples, (1) no assumptions need be made about the distribution of the population of paired differences, and (2) the standard-normal table, Table II, is used instead of the t-table, Table IV. The test statistic for a hypothesis test with null hypothesis $H_0: \mu_1 = \mu_2$ is

$$z = \frac{\overline{d}}{s_d / \sqrt{n}};$$

and the endpoints of a $(1 - \alpha)$-level confidence interval for $\mu_1 - \mu_2$ are

$$\overline{d} \pm z_{\alpha/2} \cdot \frac{s_d}{\sqrt{n}}.$$

These two inferential methods are often referred to as the **paired z-test** and **paired z-interval procedure,** respectively. Here σ_d denotes the standard deviation of the population of paired differences. If σ_d is unknown, which is generally the case, we use s_d in its place.

· · **10.94** A tire company has developed two new processes for making longer-wearing steel-belted radials. To compare the two processes, 50 tires made using Process A and 50 tires made using Process B are randomly selected. Each of the 50 Process-A tires is randomly assigned to either the front left or front right of one of 50 cars. If for a particular car, a Process-A tire gets assigned to be the left front tire, then a Process-B tire gets assigned to be the right front tire, and vice versa. The rear of each of the 50 cars is equipped with two tires currently manufactured by the tire company. The differences in tire life (Process-A tire life *minus* Process-B tire life) for the

50 pairs of tires are displayed below in thousands of miles, to the nearest hundred miles.

−0.8	−2.4	−0.2	1.6	−3.6
−0.6	−1.7	−1.4	−2.5	−1.8
−0.9	0.9	2.8	−2.7	−0.6
0.0	−2.0	1.8	−1.5	1.9
−4.2	−2.1	3.0	0.0	0.4
0.9	0.9	1.6	−2.7	0.9
1.0	−2.3	−0.4	0.5	2.4
−0.3	0.9	−0.6	0.8	3.6
−0.4	0.5	1.3	−4.0	2.4
−4.3	2.4	−0.5	1.5	3.1

a. At the 0.10 level of significance, does there appear to be a difference in the mean lifetimes of Process-A and Process-B tires? *(Note:* The sum of the data is −7.4, and the sum of the squares of the data is 197.90.*)*
b. Find a 90% confidence interval for the difference, $\mu_1 - \mu_2$, between the mean lifetime of Process-A tires and the mean lifetime of Process-B tires.

· · **10.95** In Example 10.2 (page 579), we performed a hypothesis test using independent samples to decide whether mean salaries differ for faculty teaching in public and private institutions. Now we will perform that same hypothesis test using paired samples. Pairs are formed by matching faculty in public and private institutions by rank and specialty. A random sample of 30 pairs yields the following annual salaries, in thousands of dollars.

Public	Private	Public	Private	Public	Private
42.1	39.8	70.4	69.3	25.1	26.4
36.1	46.4	39.5	38.7	27.2	27.9
33.9	37.5	24.4	29.8	26.8	28.6
30.6	34.2	45.6	47.2	25.7	29.0
57.3	68.8	25.7	37.3	32.9	31.9
39.6	38.9	33.9	36.6	24.9	25.3
43.2	46.4	36.7	32.2	28.9	36.2
31.7	32.0	37.3	37.7	36.5	38.2
29.6	29.8	20.4	20.7	52.7	51.0
48.0	54.9	55.6	56.4	61.5	60.7

a. Do the data provide sufficient evidence to conclude that mean salaries differ for faculty teaching in public and private institutions? Perform the required hypothesis test at the 5% significance level. *(Note:* $\overline{d} = -2.2$, $s_d = 3.968$.*)*
b. Compare your result in part (a) to the one obtained in Example 10.2.
c. Use the above data to find a 95% confidence interval for the difference between the mean salaries of faculty teaching in public and private institutions.

d. Compare your result in part (c) to the one obtained in Example 10.3 on page 582.

· ·   **10.96**   Suppose you want to perform a hypothesis test to compare the means of two populations using paired samples. For each part below, decide whether you would use the paired z-test, the paired t-test, the Wilcoxon paired-sample signed-rank test, or none of these tests.
a. Preliminary data analyses of the sample of paired differences suggest that the population of paired differences is approximately normally distributed.
b. Preliminary data analyses of the sample of paired differences suggest that the population of paired differences is highly skewed; the sample size is 20.
c. Preliminary data analyses of the sample of paired differences suggest that the population of paired differences has a symmetric bimodal distribution.

· ·   **10.97**   Suppose you want to perform a hypothesis test to compare the means of two populations using paired samples. For each part below, decide whether you would use the paired z-test, the paired t-test, the Wilcoxon paired-sample signed-rank test, or none of these tests.

a. Preliminary data analyses of the sample of paired differences suggest that the population of paired differences has a uniform distribution.
b. Preliminary data analyses of the sample of paired differences suggest that the population of paired differences is neither symmetric nor normally distributed; the sample size is 132.
c. Preliminary data analyses of the sample of paired differences suggest that the population of paired differences is very moderately skewed, but otherwise roughly normal.

· · ·   **10.98**   In Equation (4) on page 623, we presented a formula stating that the mean of the population of paired differences is equal to the difference between the two population means.
a. Suppose the two populations under consideration are finite, each of size N. Prove Equation (4).
b. In general, let (x_1, x_2) denote a randomly selected pair from the population of all pairs and let $d = x_1 - x_2$ denote the paired difference. Note that x_1, x_2, and d are random variables. Use the results of Exercises 5.26(e) and 5.27(f) on page 289 to prove Equation (4).

10.6   WHICH PROCEDURE SHOULD BE USED? (OPTIONAL)

· · · · ·

In this chapter we have learned several inferential procedures for comparing the means of two populations. Table 10.11 summarizes the hypothesis-testing procedures; a similar table can be constructed for confidence-interval procedures. Each row of Table 10.11 gives the type of test, the conditions required for using the test, the test statistic, and the procedure to use. In the table, we have employed "paired W-test" as an abbreviation for "Wilcoxon paired-sample signed-rank test."

In selecting the correct procedure, keep in mind that the best choice is the procedure expressly designed for the types of populations under consideration, if such a procedure exists. For instance, if large, independent samples are taken from two normally distributed populations with equal standard deviations, then the first four procedures are all applicable. (Why is this so?) But the correct procedure is the pooled t-test because that test is designed specifically for use with independent samples from two normally distributed populations having equal standard deviations. Similarly, if large, independent samples are taken from two nonnormal populations having the same shape, then both the two-sample z-test and the Mann–Whitney test are applicable. But the correct procedure is the Mann–Whitney test because it is designed specifically for use with independent samples from two populations having the same shape.

The flowchart in Fig. 10.10 on page 638 provides an organized strategy for choosing the correct hypothesis-testing procedure for comparing two population means. It has been constructed based on the criterion discussed in the preceding paragraph.

Type	Assumptions	Test statistic	Procedure to use
Two-sample z-test	1. Independent samples 2. Large samples	$z = \dfrac{\bar{x}_1 - \bar{x}_2}{\sqrt{(\sigma_1^2/n_1) + (\sigma_2^2/n_2)}}$	Procedure 10.1 (page 578)
Pooled t-test	1. Independent samples 2. Normal populations 3. Equal std. deviations	$t = \dfrac{\bar{x}_1 - \bar{x}_2}{s_p\sqrt{(1/n_1) + (1/n_2)}}$ † $(\text{df} = n_1 + n_2 - 2)$	Procedure 10.3 (page 587)
Nonpooled t-test	1. Independent samples 2. Normal populations	$t = \dfrac{\bar{x}_1 - \bar{x}_2}{\sqrt{(s_1^2/n_1) + (s_2^2/n_2)}}$ ‡	Procedure 10.5 (page 599)
Mann–Whitney test	1. Independent samples 2. Same shape populations 3. $n_1 \le n_2$	$M = $ sum of the ranks for sample data from Population 1	Procedure 10.7 (page 613)
Paired t-test	1. Paired samples 2. Normal differences	$t = \dfrac{\bar{d}}{s_d/\sqrt{n}}$ $(\text{df} = n - 1)$	Procedure 10.8 (page 624)
Paired W-test	1. Paired samples 2. Symmetric differences	$W = $ sum of positive ranks	Procedure 10.10 (page 629)
Paired z-test	1. Paired samples 2. Large sample	$z = \dfrac{\bar{d}}{\sigma_d/\sqrt{n}}$	Discussed in exercises (page 635)

† $s_p = \sqrt{\dfrac{(n_1 - 1)s_1^2 + (n_2 - 1)s_2^2}{n_1 + n_2 - 2}}$ ‡ $\text{df} = \dfrac{[(s_1^2/n_1) + (s_2^2/n_2)]^2}{\dfrac{(s_1^2/n_1)^2}{n_1 - 1} + \dfrac{(s_2^2/n_2)^2}{n_2 - 1}}$

In practice we need to look at the sample data before we can select the appropriate procedure. We recommend using normal probability plots and either stem-and-leaf diagrams
(for small or moderate-size samples) or histograms (for moderate-size or large samples);
boxplots can also be quite helpful, especially for moderate-size or large samples.

Example 10.18   Choosing the Correct Hypothesis-Testing Procedure

The *Consumer Expenditure Survey,* published by the U.S. Bureau of Labor Statistics,
provides data on annual consumer expenditures for various items, among which is the
amount spent on alcoholic beverages. Information is provided on amount spent by five
income classes: lowest 20%, second 20%, and so on. Independent random samples of

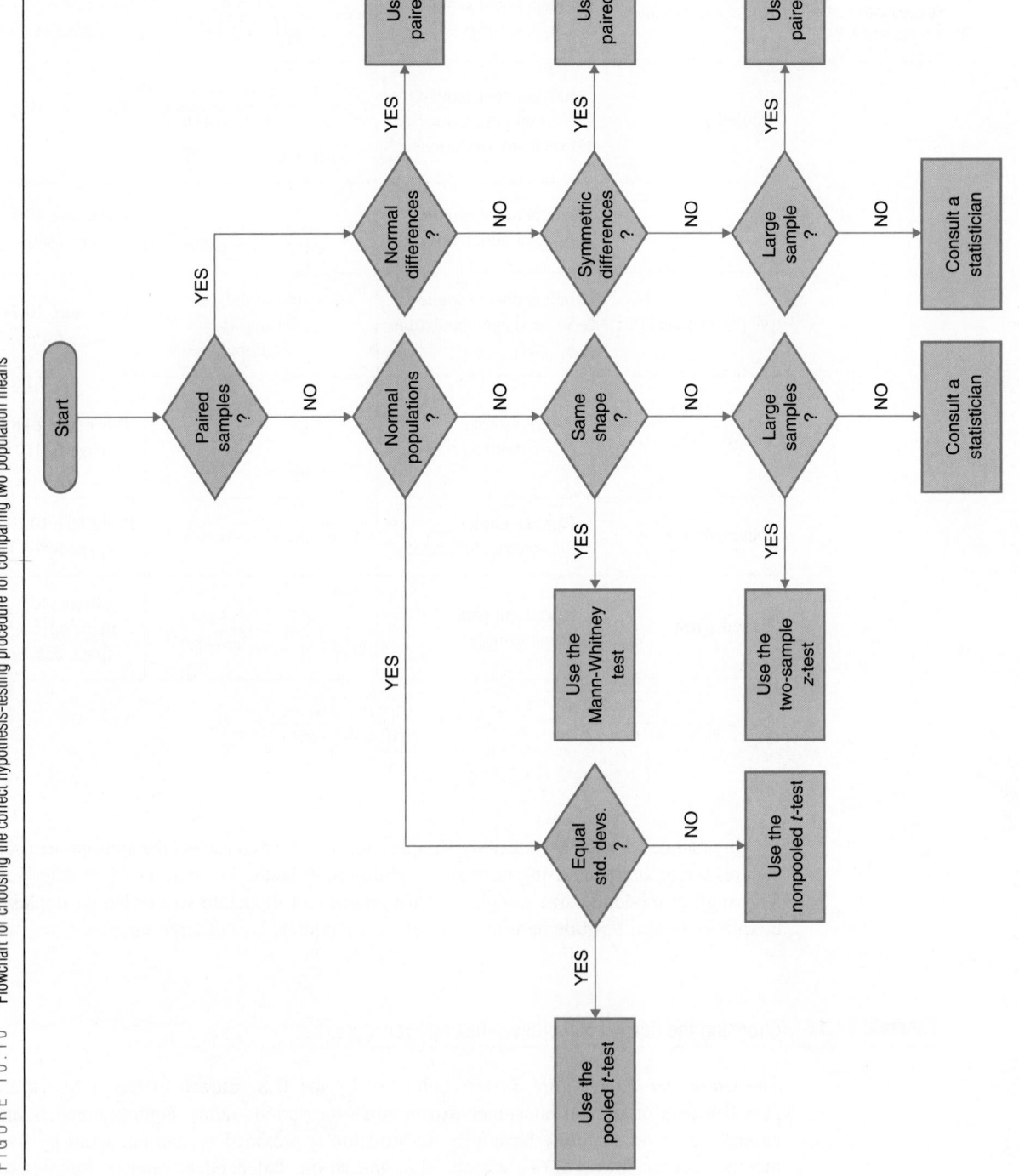

FIGURE 10.10   Flowchart for choosing the correct hypothesis-testing procedure for comparing two population means

15 consumers in the second-20% income class (Class 2) and 20 consumers in the third-20% income class (Class 3) yielded the data on last year's expenditures on alcoholic beverages shown in Table 10.12.

TABLE 10.12

Samples of last year's expenditures ($) on alcoholic beverages for Class 2 and Class 3 consumers

Class 2			Class 3			
314	123	178	399	502	351	637
53	263	228	353	248	34	303
136	193	254	250	178	20	98
207	191	302	390	409	160	267
185	77	218	585	524	184	231

Suppose we want to use the sample data in Table 10.12 to decide whether there is a difference between last year's mean expenditures on alcoholic beverages by Class 2 and Class 3 consumers. Let μ_1 and μ_2 denote, respectively, those mean expenditures. Then we want to perform the hypothesis test

$$H_0: \mu_1 = \mu_2 \text{ (mean expenditures are the same)}$$
$$H_a: \mu_1 \neq \mu_2 \text{ (mean expenditures are different)}.$$

Which procedure should we use to perform the hypothesis test?

SOLUTION   We consult the flowchart in Fig. 10.10. The first question we must answer is, "Are the samples paired?" From the information provided above, we know that the samples are independent, not paired. Thus the answer to the first question is "No."

This leads us to the question, "Are the populations normal?" To answer this question, we constructed the normal probability plots in Fig. 10.11. The plots are extremely straight and thereby indicate that it is certainly reasonable to assume that last year's expenditures on alcoholic beverages are approximately normally distributed for both Class 2 and Class 3 consumers. So the answer to the second question is "Yes."

FIGURE 10.11

Normal probability plots of the sample data for (a) Class 2 consumers (b) Class 3 consumers

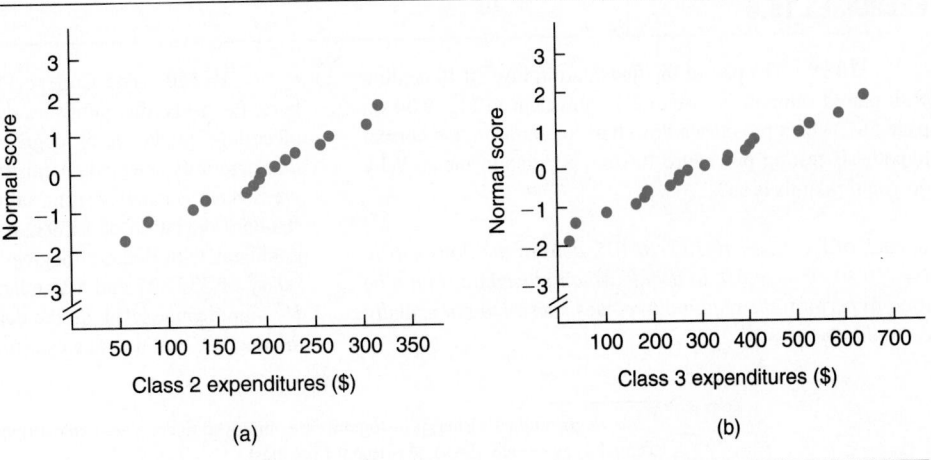

(a)   (b)

Next we must answer the question, "Are the population standard deviations equal?" The standard deviations of the two samples in Table 10.12 are $s_1 = 74.9$ and $s_2 = 173.1$, respectively. These statistics suggest that the population standard deviations are not equal. We can also see this by looking at a back-to-back stem-and-leaf diagram, histograms, or boxplots.[†] Figure 10.12 displays boxplots for the two samples in Table 10.12. The vast difference in the spreads of the two boxplots again suggest that last year's expenditures on alcoholic beverages for Class 2 and Class 3 consumers have different standard deviations. Thus the answer to the third question is "No."

FIGURE 10.12
Boxplots of the sample data for Class 2 and Class 3 consumers

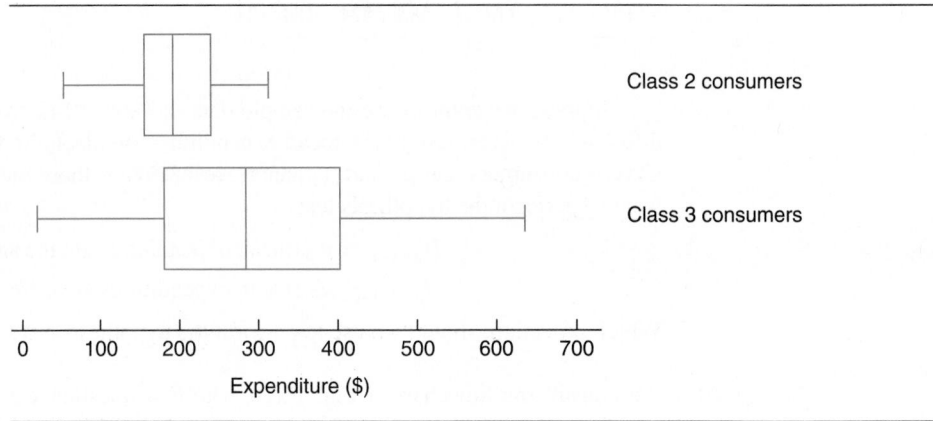

The "No" answer to the preceding question leads us to the box that states "Use the nonpooled t-test." In other words, we should use Procedure 10.5 on page 599 to carry out the hypothesis test.   ■

EXERCISES 10.6

10.99 The part of the flowchart in Fig. 10.10 dealing with paired samples is essentially equivalent to Fig. 9.34 on page 561, which provides a flowchart for choosing the correct hypothesis-testing procedure for one population mean. Why do you think this is so?

In each of Exercises 10.100–10.105, consult the flowchart in Fig. 10.10 on page 638 to decide which procedure should be used to perform the required hypothesis test. Do not actually carry out the test.

10.100 The College Placement Council of Bethlehem, Pennsylvania, publishes data on starting annual salaries of college graduates by degree and field of study. Seventy-five randomly selected chemistry graduates with doctoral degrees have a mean starting annual salary of $48,911 and a standard deviation of $3592; 100 randomly selected physics graduates with doctoral degrees have a mean starting annual salary of $39,825 and a standard deviation of $3241. At the 5% significance level, do the data provide sufficient evidence to conclude that the mean starting annual salary of chemistry

_____

[†] As we mentioned earlier, if histograms are employed to compare two distributions, they should be drawn using the same scale; the same is true for boxplots.

graduates with doctoral degrees exceeds that of physics graduates with doctoral degrees? *(Note:* Preliminary data analyses indicate that the two distributions of salaries do not have the same shape and that the starting annual salaries of physics graduates with doctoral degrees are not normally distributed.*)*

10.101 The research and development (R&D) department of a light-bulb company claims to have developed a new bulb that, on the average, will outlast the current bulb produced. In an initial attempt to justify its claim, R&D obtains the lifetimes of 20 currently produced bulbs and 10 newly developed bulbs. The lifetimes, in hours, follow.

Current bulb				New bulb	
1025	914	1015	1109	1099	1173
980	1057	1088	1015	1110	1174
1085	1065	1049	1043	1137	1100
943	981	967	1042	1213	1173
985	1051	1098	951	1077	1174

Do the data provide sufficient evidence to conclude that, on the average, the new bulb outlasts the current bulb? Use $\alpha = 0.01$. *(Note:* Preliminary data analyses indicate that it is reasonable to presume that the lifetimes of both bulbs are approximately normally distributed and have approximately equal standard deviations.*)*

10.102 Euromonitor Publications Limited, London, conducts surveys in various countries to obtain data on food consumption of major food commodities. Results of those surveys can be found in *European Marketing Data and Statistics.* Independent samples of 10 Germans and 15 Russians yielded the following fish consumptions, in kilograms, for last year.

Germans		Russians		
10	12	16	21	12
17	12	11	5	23
14	11	19	19	22
13	8	16	23	12
9	8	18	7	17

Do the data provide sufficient evidence to conclude that last year the average German consumed less fish than the average Russian? Use $\alpha = 0.05$. *(Note:* Preliminary data analyses indicate that it is reasonable to presume that last year's fish consumptions in both countries are approximately normally distributed but do not have equal standard deviations.*)*

10.103 The U.S. Federal Highway Administration compiles information on the number of miles driven by motor vehicles, by type of vehicle. Results are published annually in *Highway Statistics.* Independent random samples of 8 trucks and 10 cars yield the following data on number of miles, in thousands, driven last year.

Trucks		Cars	
8.0	11.0	14.7	2.8
34.9	17.2	13.9	1.9
14.5	36.6	3.5	3.7
11.4	13.3	16.6	12.3
		29.5	9.5

Can we conclude that for last year, the mean number of miles driven per truck differed from the mean number of miles driven per car? Use $\alpha = 0.05$. *(Note:* Preliminary data analyses suggest that it is reasonable to presume that the distributions of the number of miles driven for trucks and the number of miles driven for cars have roughly the same shape but that those distributions are right skewed.*)*

10.104 The U.S. Bureau of the Census conducts surveys to estimate average rents for housing units by region and publishes the statistics in *Current Housing Reports.* Renter-occupied housing units in the South and Midwest are paired according to square footage, quality of neighborhood, and so on. The monthly rents of 15 randomly selected pairs yield the following data, in dollars.

South	Midwest	South	Midwest	South	Midwest
648	564	199	119	324	232
155	205	237	285	150	74
258	329	419	422	519	545
987	933	78	111	354	262
505	495	127	57	750	794

At the 5% significance level, do the data provide sufficient evidence to conclude that the mean monthly rent for renter-occupied housing units in the South exceeds that for those in the Midwest? *(Note:* Preliminary data analyses suggest that it is reasonable to presume that the population of paired differences has a symmetric distribution.*)*

10.105 The effectiveness of two speed-reading programs is being compared. Ten pairs of people are randomly selected; each pair consists of people whose present reading speeds are essentially identical. From each pair, one person is randomly selected to take Program 1; the other person takes Program 2. After the speed-reading programs are completed, the reading speeds, in words per minute, for the 10 pairs of people are found to be the following.

Program 1	Program 2	Program 1	Program 2
1114	1032	996	1148
979	1074	1125	1076
910	959	1056	1094
1091	1091	1053	1096
996	1032	894	1012

Can we conclude that there is a difference in effectiveness of the two speed-reading programs? Use $\alpha = 0.10$. (*Note:* A normal probability plot of the paired differences of the samples suggests that it is reasonable to presume that the population of paired differences is approximately normally distributed.)

In each of Exercises 10.106–10.111, we have provided the type of sampling (independent or paired), sample size(s), and referenced a figure showing the results of preliminary data analyses on the samples. For independent samples, the graphs are for the two samples; for paired samples, the graphs are for the paired differences. The intent is to employ the sample data to perform a hypothesis test to compare the means of the two populations from which the data were obtained. In each case, use the information provided and the flowchart in Fig. 10.10 on page 638 to decide which procedure should be applied.

10.106   Paired; $n = 75$; Fig. 10.13.

10.107   Independent; $n_1 = 25$, $n_2 = 20$; Fig. 10.14.

10.108   Independent; $n_1 = 17$, $n_2 = 17$; Fig. 10.15.

10.109   Independent; $n_1 = 40$, $n_2 = 45$; Fig. 10.16.

10.110   Independent; $n_1 = 20$, $n_2 = 15$; Fig. 10.17.

10.111   Paired; $n = 18$; Fig. 10.18.

FIGURE 10.13   Results of preliminary data analyses of the data in Exercise 10.106

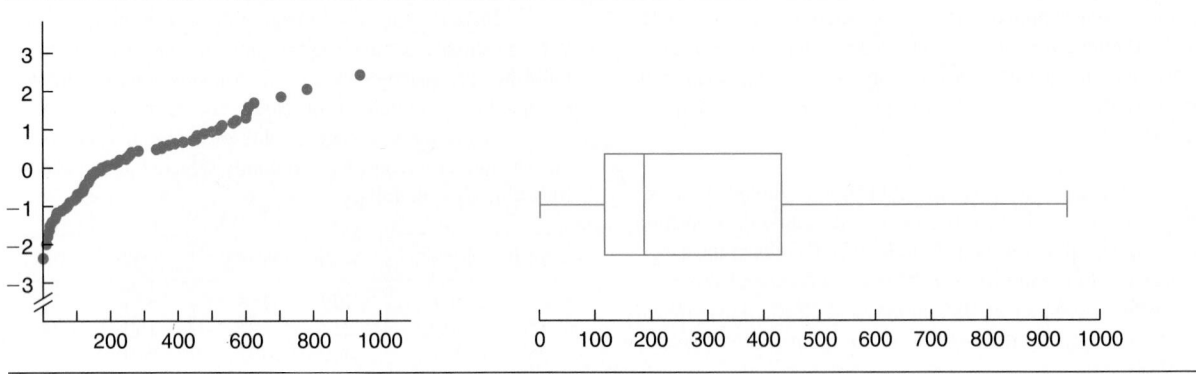

FIGURE 10.14   Results of preliminary data analyses of the data in Exercise 10.107

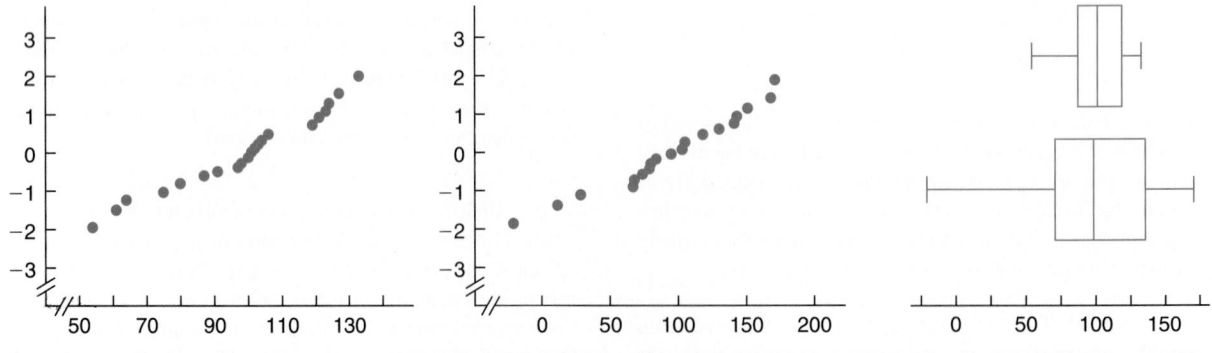

FIGURE 10.15   Results of preliminary data analyses of the data in Exercise 10.108

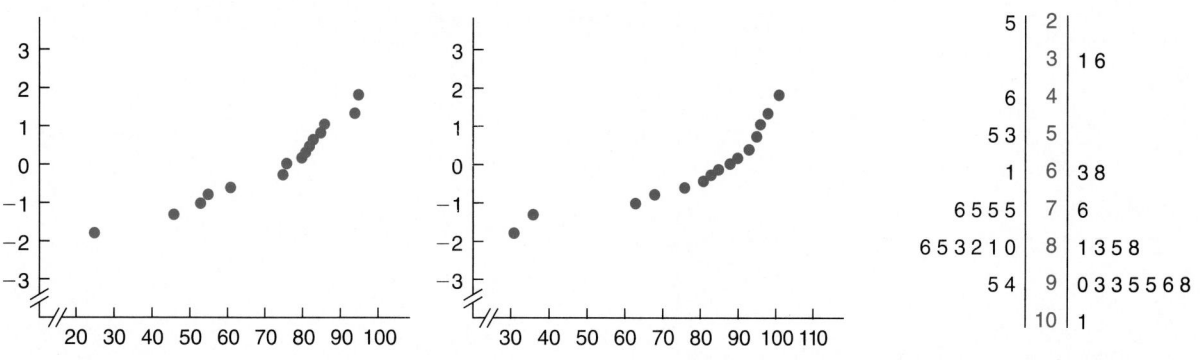

FIGURE 10.16   Results of preliminary data analyses of the data in Exercise 10.109

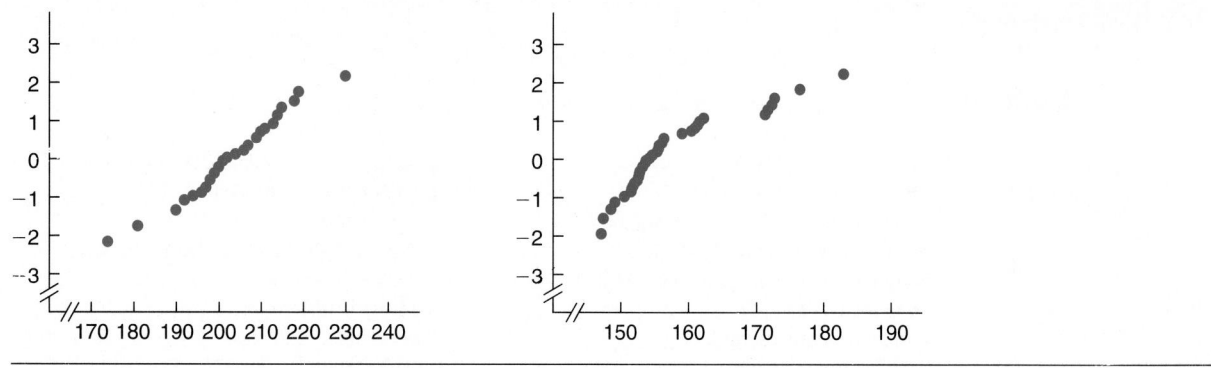

FIGURE 10.17   Results of preliminary data analyses of the data in Exercise 10.110

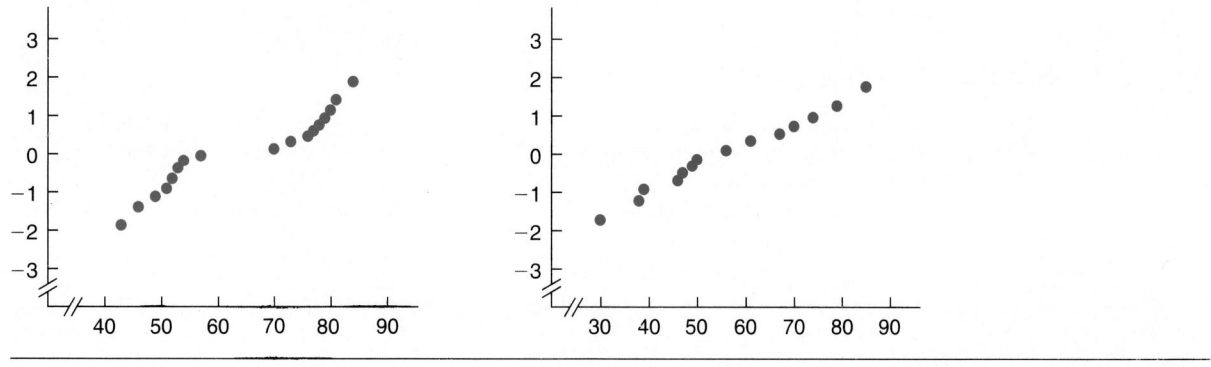

FIGURE 10.18   Results of preliminary data analyses of the data in Exercise 10.111

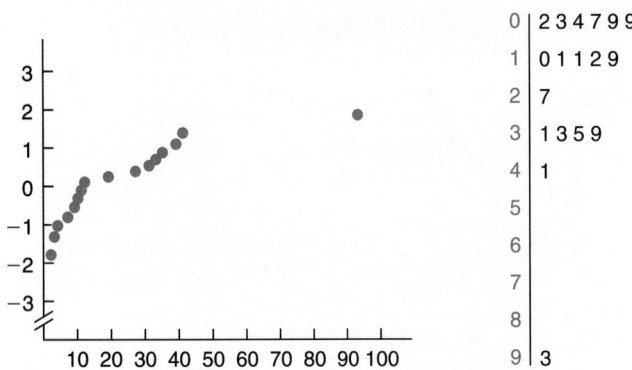

```
0 | 2 3 4 7 9 9
1 | 0 1 1 2 9
2 | 7
3 | 1 3 5 9
4 | 1
5 |
6 |
7 |
8 |
9 | 3
```

Chapter Review

Key Terms

back-to-back stem-and-leaf diagram,* *610*
independent samples, *574*
Mann–Whitney,* *616*
Mann–Whitney test,* *608*
Mann–Whitney–Wilcoxon test,* *609*
MANNWHITNEY,* *617*
nonpooled *t*-interval procedure, *602*
nonpooled *t*-test, *598*
normal differences, *624*
paired difference, *622*
paired samples, *622*
paired *t*-interval procedure, *627*
paired *t*-test, *624*
paired *z*-interval procedure, *629*
paired *z*-test, *629*
pool, *586*

Pooled,* *592*
POOLED,* *593*
pooled sample standard deviation (s_p), *586*
pooled *t*-interval procedure, *591*
pooled *t*-test, *588*
sampling distribution of the difference
    between two means, *575*
symmetric differences,* *629*
TwoSample,* *592, 604*
TWOSAMPLE,* *593, 604*
two-sample *z*-interval procedure, *581*
two-sample *z*-test, *579*
Wilcoxon paired-sample signed-rank
    test,* *629*
Wilcoxon rank-sum test,* *609*

Formulas

In the formulas below,

μ_1, μ_2 = population means
σ_1, σ_2 = population standard deviations
$\bar{x}_1, \bar{x}_2$ = sample means
s_1, s_2 = sample standard deviations
n_1, n_2 = sample sizes

$\bar{d}$ = sample mean of paired differences
s_d = sample standard deviation of
    paired differences
n = number of sample pairs

$$\Delta = \frac{\left[\left(s_1^2/n_1\right) + \left(s_2^2/n_2\right)\right]^2}{\dfrac{\left(s_1^2/n_1\right)^2}{n_1 - 1} + \dfrac{\left(s_2^2/n_2\right)^2}{n_2 - 1}} \text{, rounded down to the nearest integer}$$

Two-sample z-test statistic for H_0: $\mu_1 = \mu_2$ (independent and large samples), *578*

$$z = \frac{\overline{x}_1 - \overline{x}_2}{\sqrt{(\sigma_1^2/n_1) + (\sigma_2^2/n_2)}}$$

(if σ_1 is unknown, use s_1 in its place; if σ_2 is unknown, use s_2 in its place)

Two-sample z-interval for $\mu_1 - \mu_2$ (independent and large samples), *581*

$$(\overline{x}_1 - \overline{x}_2) \pm z_{\alpha/2} \cdot \sqrt{(\sigma_1^2/n_1) + (\sigma_2^2/n_2)}$$

(if σ_1 is unknown, use s_1 in its place; if σ_2 is unknown, use s_2 in its place)

Pooled sample standard deviation, *588*

$$s_\text{p} = \sqrt{\frac{(n_1 - 1)s_1^2 + (n_2 - 1)s_2^2}{n_1 + n_2 - 2}}$$

Pooled t-test statistic for H_0: $\mu_1 = \mu_2$ (independent samples, normal populations, and equal standard deviations), *588*

$$t = \frac{\overline{x}_1 - \overline{x}_2}{s_\text{p}\sqrt{(1/n_1) + (1/n_2)}}$$

with df $= n_1 + n_2 - 2$.

Pooled t-interval for $\mu_1 - \mu_2$ (independent samples, normal populations, and equal standard deviations), *591*

$$(\overline{x}_1 - \overline{x}_2) \pm t_{\alpha/2} \cdot s_\text{p}\sqrt{(1/n_1) + (1/n_2)}$$

with df $= n_1 + n_2 - 2$.

Nonpooled t-test statistic for H_0: $\mu_1 = \mu_2$ (independent samples and normal populations), *599*

$$t = \frac{\overline{x}_1 - \overline{x}_2}{\sqrt{(s_1^2/n_1) + (s_2^2/n_2)}}$$

with df $= \Delta$.

Nonpooled t-interval for $\mu_1 - \mu_2$ (independent samples and normal populations), *602*

$$(\overline{x}_1 - \overline{x}_2) \pm t_{\alpha/2} \cdot \sqrt{(s_1^2/n_1) + (s_2^2/n_2)}$$

with df $= \Delta$.

Mann–Whitney test statistic for H_0: $\mu_1 = \mu_2$ (independent samples, populations have same shape, and $n_1 \leq n_2$),* *613*

$$M = \text{sum of the ranks for sample data from Population 1}$$

Paired t-test statistic for H_0: $\mu_1 = \mu_2$ (paired samples and normal differences), *625*

$$t = \frac{\overline{d}}{s_d/\sqrt{n}}$$

with df $= n - 1$.

Paired t-interval for $\mu_1 - \mu_2$ (paired samples and normal differences), 627

$$\bar{d} \pm t_{\alpha/2} \cdot \frac{s_d}{\sqrt{n}}$$

with df $= n - 1$.

Wilcoxon paired-sample signed-rank test statistic for H_0: $\mu_1 = \mu_2$ (paired samples and symmetric differences),* 630

$$W = \text{sum of the positive ranks}$$

You Should Be Able To

1. use and understand the preceding formulas.
2. perform large-sample inferences to compare the means of two populations using independent samples.
3. perform inferences to compare the means of two normally distributed populations using independent samples when the population standard deviations are unknown but assumed equal.
4. perform inferences to compare the means of two normally distributed populations using independent samples when the population standard deviations are unknown but not assumed equal.
5. perform a hypothesis test to compare the means of two populations using independent samples when the populations have the same shape.*
6. perform inferences to compare the means of two populations using paired samples when the population of paired differences is normally distributed.
7. perform a hypothesis test to compare the means of two populations using paired samples when the population of paired differences has a symmetric distribution.*
8. use the Minitab commands covered in this chapter.*
9. interpret the output obtained from the application of the Minitab commands discussed in this chapter.*

REVIEW TEST

1. *Better Homes and Gardens* conducts semiannual surveys of housing in 100 cities across the United States. One item of considerable interest is the average price for resale homes. Fifty randomly selected resale home prices in Phoenix, Arizona, have a mean of $81,525 and a standard deviation of $29,670. Seventy-five randomly selected resale home prices in Flint, Michigan, have a mean of $47,603 and a standard deviation of $12,466. Do the data provide sufficient evidence to conclude that the mean resale price for homes in Phoenix, Arizona, exceeds the mean resale price for homes in Flint, Michigan? Perform a two-sample z-test at the 1% significance level.

2. Refer to Problem 1. Obtain a 98% confidence interval for the difference, $\mu_1 - \mu_2$, between the mean resale price of homes in Phoenix, Arizona, and the mean resale price of homes in Flint, Michigan. Use the two-sample z-interval procedure.

3. The effectiveness of two speed-reading programs is being compared. Ten pairs of people are randomly selected; each pair consists of people whose present reading speeds are essentially identical. From each pair, one person is randomly selected to take Program 1; the other person takes Program 2. After the speed-reading programs are completed, the reading speeds, in words per minute, for the 10 pairs of people are found to be the following.

Program 1	Program 2	Program 1	Program 2
1114	1032	996	1148
979	1074	1125	1076
910	959	1056	1094
1091	1091	1053	1096
996	1032	894	1012

At the 10% significance level, can we conclude that there is a difference in effectiveness of the two speed-reading programs? (*Note:* A normal probability plot of the paired differences of the sample data suggests that it is reasonable to presume that the population of paired differences is approximately normally distributed.)

4. Refer to Problem 3. Find a 90% confidence interval for the difference between the mean reading speeds of all people using Program 1 and all people using Program 2.

*5. **(Computer problem)** Refer to Problems 3 and 4. Use Minitab or some other statistical software to
   a. obtain a normal probability plot, stem-and-leaf diagram, and boxplot of the paired differences of the two samples.
   b. perform the required hypothesis test and obtain the desired confidence interval.
   c. Justify the use of your procedures in part (b).

6. A psychology professor teaching at a large university in the Northeast wants to know whether there is a difference between the mean IQs of male and female students in attendance. She randomly and independently selects 20 female students and 20 male students and has them take IQ tests. The resulting IQ data are as follows.

Female				Male			
116	126	121	127	112	117	115	114
128	120	112	124	104	132	120	129
131	107	105	114	110	108	126	114
114	114	121	109	113	110	103	116
120	130	117	113	121	112	107	125

Preliminary data analyses indicate that it is reasonable to presume that IQs of female and male students at the university are roughly normally distributed and have approximately equal standard deviations. Can we conclude that there is a difference between the mean IQs of male and female students at the university? Use $\alpha = 0.05$. (*Note:* $\bar{x}_1 = 118.45$, $s_1 = 7.61$, $\bar{x}_2 = 115.40$, $s_2 = 8.02$.)

7. Refer to Problem 6. Determine a 95% confidence interval for the difference between the mean IQs of female and male students at the university.

*8. **(Computer problem)** Refer to Problems 6 and 7. Use Minitab or some other statistical software to
   a. obtain normal probability plots, stem-and-leaf diagrams, boxplots, and the standard deviations of the two samples.
   b. perform the required hypothesis test and obtain the desired confidence interval.
   c. Justify the use of your procedures in part (b).

*9. **(Computer problem)** Printout 10.10 supplies the output obtained by applying the TwoSample command and Pooled subcommand to the data in Problem 6. From the output, determine
   a. the null and alternative hypotheses.
   b. the sample standard deviation of each of the two sets of IQ data.
   c. the mean IQ of the females sampled and the mean IQ of the males sampled.
   d. the value of the test statistic, t.
   e. the P-value of the hypothesis test.
   f. the smallest significance level at which the null hypothesis can be rejected.
   g. the conclusion if the hypothesis test is performed at the 5% significance level.
   h. a 95% confidence interval for the difference, $\mu_1 - \mu_2$, between the mean IQs of female and male students at the university.
   i. the value of s_p.

PRINTOUT 10.10   Minitab output for Problem 9

```
TWOSAMPLE T FOR FEMALE VS MALE
          N      MEAN     STDEV    SE MEAN
FEMALE   20     118.45     7.61      1.7
MALE     20     115.40     8.02      1.8

95 PCT CI FOR MU FEMALE - MU MALE: (-2.0, 8.1)

TTEST MU FEMALE = MU MALE (VS NE): T= 1.23  P=0.22  DF=  38

POOLED STDEV =      7.82
```

10. Euromonitor Publications Limited, London, conducts surveys in various countries to obtain data on food consumption of major food commodities. Results of those surveys can be found in *European Marketing Data and Statistics.* Independent samples of 10 Germans and 15 Russians yielded the following fish consumptions, in kilograms, for last year.

Germans		Russians		
10	12	16	21	12
17	12	11	5	23
14	11	19	19	22
13	8	16	23	12
9	8	18	7	17

Preliminary data analyses indicate that it is reasonable to presume that last year's fish consumptions in both countries are approximately normally distributed but do not have equal standard deviations. Do the data provide sufficient evidence to conclude, at the 5% significance level, that last year the average German consumed less fish than the average Russian? *(Note:* $\bar{x}_1 = 11.40$, $s_1 = 2.84$, $\bar{x}_2 = 16.07$, $s_2 = 5.61$.*)*

11. Refer to Problem 10. Find a 90% confidence interval for the difference, $\mu_1 - \mu_2$, between last year's mean fish consumptions by Germans and Russians.

*12. **(Computer problem)**   Refer to Problems 10 and 11. Use Minitab or some other statistical software to
a. obtain normal probability plots, stem-and-leaf diagrams, boxplots, and the standard deviations of the two samples.
b. perform the required hypothesis test and obtain the desired confidence interval.
c. Justify the use of your procedures in part (b).

*13. **(Computer problem)**   Printout 10.11 shows the Minitab output obtained by applying the TwoSample command to the data in Problem 10. Employ the output to determine
a. the null and alternative hypotheses.
b. the standard deviation of fish consumption for each of the two samples.
c. the number of Germans sampled and the number of Russians sampled.
d. the mean fish consumptions for the Germans and Russians sampled.
e. the value of the test statistic, t.
f. the P-value of the hypothesis test.
g. the smallest significance level at which the null hypothesis can be rejected.

h. the conclusion if the hypothesis test is performed at the 5% significance level.
i. a 90% confidence interval for the difference, $\mu_1 - \mu_2$, between last year's mean fish consumptions by Germans and Russians.

*14. The U.S. Federal Highway Administration compiles information on the number of miles driven by motor vehicles, by type of vehicle. Results are published annually in *Highway Statistics.* Independent random samples of 8 trucks and 10 cars yield the following data on number of miles, in thousands, driven last year.

Trucks		Cars	
8.0	11.0	14.7	2.8
34.9	17.2	13.9	1.9
14.5	36.6	3.5	3.7
11.4	13.3	16.6	12.3
		29.5	9.5

Can we conclude that for last year, the mean number of miles driven per truck differed from the mean number of miles driven per car? Use $\alpha = 0.05$. *(Note:* Preliminary data analyses suggest that it is reasonable to presume that the distributions of the number of miles driven for trucks and the number of miles driven for cars have roughly the same shape but that those distributions are right skewed.*)*

*15. **(Computer problem)**   Refer to Problem 14. Use Minitab or some other statistical software to
a. obtain normal probability plots, stem-and-leaf diagrams, boxplots, and the standard deviations of the two samples.
b. perform the required hypothesis test.
c. Justify the use of your procedure in part (b).

*16. **(Computer problem)**   Printout 10.12 shows the Minitab output obtained by applying the Mann-Whitney command to the data in Problem 14. Employ the output to determine
a. the null and alternative hypotheses.
b. the median number of miles driven by the trucks sampled and by the cars sampled.
c. the number of trucks sampled and the number of cars sampled.
d. the value of the Mann–Whitney test statistic.
e. the approximate P-value of the hypothesis test.
f. the smallest significance level at which the null hypothesis can be rejected.
g. the conclusion if the hypothesis test is performed at the 5% significance level.

h. a 95.4% confidence interval for the difference between last year's median (or mean) number of miles driven by trucks and cars.

*17. The U.S. Bureau of the Census conducts surveys to estimate average rents for housing units by region and publishes the statistics in *Current Housing Reports*. Renter-occupied housing units in the South and Midwest are paired according to square footage, quality of neighborhood, and so on. The monthly rents of 15 randomly selected pairs yield the following data, in dollars.

South	Midwest	South	Midwest	South	Midwest
648	564	199	119	324	232
155	205	237	285	150	74
258	329	419	422	519	545
987	933	78	111	354	262
505	495	127	57	750	794

Do the data provide sufficient evidence to conclude that the mean monthly rent for renter-occupied housing units in the South exceeds that for those in the Midwest? Perform the required hypothesis test at a significance level as close to 0.05 as possible. (*Note:* Preliminary data analyses suggest that it is reasonable to presume that the population of paired differences has approximately a symmetric distribution.)

*18. (**Computer problem**) Refer to Problem 17. Use Minitab or some other statistical software to
   a. obtain a normal probability plot, stem-and-leaf diagram, and boxplot of the paired differences of the two samples.
   b. perform the required hypothesis test.
   c. Justify the use of the procedure that you employed in solving part (b).

PRINTOUT 10.11   Minitab output for Problem 13

```
TWOSAMPLE T FOR GERMANS VS RUSSIANS
             N      MEAN    STDEV   SE MEAN
GERMANS     10     11.40    2.84     0.90
RUSSIANS    15     16.07    5.61     1.4

90 PCT CI FOR MU GERMANS - MU RUSSIANS: (-7.60, -1.7)

TTEST MU GERMANS = MU RUSSIANS (VS LT): T= -2.74  P=0.0062  DF=  21
```

PRINTOUT 10.12   Minitab output for Problem 16

```
Mann-Whitney Confidence Interval and Test

TRUCK      N =  8     Median =      13.90
CAR        N = 10     Median =      10.90
Point estimate for ETA1-ETA2 is       6.60
95.4 pct c.i. for ETA1-ETA2 is (-2.10,18.31)
W = 92.0
Test of ETA1 = ETA2  vs.  ETA1 n.e. ETA2 is significant at 0.1684

Cannot reject at alpha = 0.05
```

Using the Focus Database

The Focus database, which is printed in Appendix B, contains data on 500 randomly selected Arizona State University sophomores for seven different variables: sex, high-school GPA, SAT math score, cumulative GPA, SAT verbal score, age, and total hours. Use Minitab or some other statistical software to solve the following problems.

a. Obtain normal probability plots, boxplots, and the standard deviations of the SAT math scores of the male sophomores in the sample and the female sophomores in the sample.

b. At the 5% significance level, do the data provide sufficient evidence to conclude that male sophomores at Arizona State University have a higher mean SAT math score than female sophomores? Justify the use of the procedure you chose to carry out the hypothesis test.

c. Obtain a 90% confidence interval for the difference between the mean SAT math scores of male and female sophomores at Arizona State University.

d. Obtain normal probability plots, boxplots, and the standard deviations of the SAT verbal scores of the male sophomores in the sample and the female sophomores in the sample.

e. At the 5% significance level, do the data provide sufficient evidence to conclude that male and female sophomores at Arizona State University have different mean SAT verbal scores? Justify the use of the procedure you selected to perform the hypothesis test.

f. Obtain a 95% confidence interval for the difference between the mean SAT verbal scores of male and female sophomores at Arizona State University.

g. Obtain a normal probability plot, a histogram, and a boxplot for the paired differences of the SAT verbal and SAT math scores of the sophomores in the sample.

h. Do the data provide sufficient evidence to conclude that the mean SAT verbal score is less than the mean SAT math score for Arizona State University sophomores? Perform the required hypothesis test at the 0.01 significance level. Justify the use of the procedure you employed to conduct the hypothesis test.

i. Find a 98% confidence interval for the difference between the mean SAT math and verbal scores of Arizona State University sophomores.

j. Obtain normal probability plots, boxplots, and the standard deviations of the cumulative GPAs of the sophomores in the sample who are under 21 years of age and those who are 21 years of age or over.

k. Do the data provide sufficient evidence to conclude that for Arizona State University sophomores, there is a difference between the mean cumulative GPAs of those under 21 years of age and those 21 years of age or over? Perform the required hypothesis test at the 5% significance level. Justify the use of the procedure you chose to carry out the hypothesis test.

Case Study · · · · · ·

Breast Milk and IQ

Considerable controversy exists over whether neurodevelopment is affected long term by nutritional factors in early life. In a 1992 issue of *The Lancet,* five researchers summarized their findings on that question for preterm babies.[†] Their study was a continuation of work begun in January 1982.

Previously these researchers showed that a mother's decision to provide breast milk for preterm infants is associated with higher developmental scores for the children at age 18 months. Their 1992 article analyzed IQ data on the same children at age 7½–8 years.

[†] From "Breast Milk and Subsequent Intelligence Quotient in Children Born Preterm" by A. Lucas et al., 1992, *The Lancet, 339,* pp. 261–264.

IQ was measured for 300 children using an abbreviated form of the Weschler Intelligence Scale for Children (revised Anglicized version: WISC-R UK).

The mothers of the children in the study had chosen whether to provide their infants with breast milk within 72 hours of delivery; 90 did not and 210 did. Of those 210 who chose to provide their infants with breast milk, 193 succeeded and 17 did not. The children whose mothers declined to provide breast milk were designated by the researchers as Group I; those whose mothers had chosen but were unable to provide breast milk were designated as Group IIa; and those whose mothers had chosen and were able to provide breast milk were designated as Group IIb. Table 10.13 displays statistics for all three groups.

TABLE 10.13
IQ at 7½–8 years
in the three groups

Group	Sample size	Mean IQ	St. Dev.
I	90	92.8	15.2
IIa	17	94.8	19.0
IIb	193	103.7	15.3

Presuming that IQs are normally distributed in all three categories, solve each of the following problems.

a. Do the data provide sufficient evidence to conclude that for children age 7½–8 years who are born preterm, a difference exists in mean IQ between those whose mothers decline to provide breast milk and those whose mothers choose but are unable to provide breast milk? Perform the required hypothesis test at the 5% significance level.

b. Do the data provide sufficient evidence to conclude that for children age 7½–8 years who are born preterm, the mean IQ of those whose mothers decline to provide breast milk is less than that of those whose mothers choose and are able to provide breast milk? Perform the required hypothesis test at the 5% significance level.

c. Do the data provide sufficient evidence to conclude that for children age 7½–8 years who are born preterm, the mean IQ of those whose mothers choose but are unable to provide breast milk is less than that of those whose mothers choose and are able to provide breast milk? Perform the required hypothesis test at the 5% significance level.

d. Is this study observational, or is it a designed experiment? Explain your answer.

e. Based on your answers in parts (a)–(d), what conclusions would you draw?

f. Use Minitab or some other statistical software to perform the hypothesis tests in parts (a)–(c). *(Hint: You may want to refer to the hint on page 571.)*

The researchers also adjusted the data for such factors as social class, mother's education, and infant's sex, and still reached the same conclusions: "... preterm babies whose mothers provided breast milk had a substantial advantage in subsequent IQ at 7½–8 years over those who did not ...." However, the researchers emphasized that they could not exclude the possibility that their findings could be explained by differences in parental behavior or genetic potential between the groups.

*Abraham
de Moivre*

Abraham de Moivre was born in Vitry-le-Francois, France, on May 26, 1667, the son of a country surgeon. He was educated in the Catholic school in his village and at the Protestant Academy at Sedan. In 1684 he went to Paris to study under Jacques Ozanam.

In late 1685 de Moivre, a French Huguenot (Protestant), was imprisoned in Paris because of his religion. The duration of his incarceration is unclear, but de Moivre was probably jailed for roughly 1 to 3 years. In any case, upon his release he fled to London, where he began tutoring students in mathematics.

In London, de Moivre mastered Sir Isaac Newton's *Principia* and became a close friend of Newton and of Edmond Halley, an English astronomer (in whose honor, incidentally, Halley's Comet is named). In Newton's later years, he would refuse to take new students, saying, "Go to Mr. de Moivre; he knows these things better than I do."

De Moivre's contributions to mathematics range from the definition of statistical independence to analytical trigonometric formulas to his major discovery—the normal approximation to the binomial distribution, of monumental importance in its own right and precursor to the central limit theorem. The definition of statistical independence appeared in *The Doctrine of Chances,* published in 1718 and dedicated to Newton; the normal approximation to the binomial distribution was contained in a Latin pamphlet published in 1733.

De Moivre also did research on the analysis of mortality statistics and the theory of annuities. In 1725 the first edition of his *Annuities on Lives,* in which he derived annuity formulas and addressed other annuity problems, was published.

De Moivre was elected to the Royal Society in 1697, to the Berlin Academy of Sciences in 1735, and to the Paris Academy in 1754. But despite his obvious talents as a mathematician and his many champions, he was never able to obtain a position in any of England's universities. Instead he had to rely on his meager earnings as a tutor in mathematics and a consultant on gambling and insurance, supplemented by the sales of his books. De Moivre died in London on November 27, 1754.

Chapter 11

Inferences for Population Proportions

. . . . .

General Objective

In Chapters 8–10 we learned how to find confidence intervals and perform hypothesis tests for population means. Now we will discover how to conduct those inferences for population proportions. A *population proportion* is the proportion (percentage) of a population that has a specified attribute. For example, if the population under consideration consists of all Americans and the specified attribute is "retired," then the population proportion is the proportion of all Americans who are retired.

The first two sections of this chapter explain how to determine confidence intervals and perform hypothesis tests for one population proportion. The third section discusses how to perform hypothesis tests for comparing two population proportions and how to construct confidence intervals for the difference between two population proportions.

Chapter Outline

Case Study

Doctors Polled on AIDS Views

How do doctors feel about treating patients with AIDS? A study published in the *Journal of the American Medical Association* asked that and several other questions in order to identify barriers for physicians in treating HIV-infected patients. We will discuss that article in detail in the case study at the end of this chapter.

11.1   CONFIDENCE INTERVALS FOR ONE POPULATION PROPORTION

Many statistical studies are concerned with obtaining the proportion (percentage) of a population that has a specified attribute. For example, we might be interested in the percentage of U.S. adults who have health insurance or in the percentage of cars in the United States that are imports.

In most cases the population under consideration will be large, and hence it would be impractical (and probably impossible) to obtain the population proportion by taking a census; for instance, imagine trying to interview every U.S. adult in order to ascertain the proportion who are covered by health insurance. Thus we generally employ sampling and use the sample data to make inferences about the population proportion. Example 11.1 introduces proportion notation and terminology.

Example 11.1   Proportion Notation and Terminology

Many employers are concerned about employees who call in sick when in fact they are not ill. A 1991 survey commissioned by the Hilton Hotels Corporation investigated this issue. One question asked of the people taking part in the survey was whether they call in sick at least once a year when they simply need time to relax. For brevity we will use the phrase *play hooky* to refer to that practice. In the survey, 1010 randomly selected U.S. employees were polled. The proportion of the 1010 employees sampled who play hooky was used to estimate the proportion of all U.S. employees who play hooky.

We use the letter p to denote the proportion of all U.S. employees who play hooky; this is the **population proportion** and is the parameter whose value is to be estimated. The proportion of the 1010 U.S. employees sampled who play hooky is called a **sample proportion** and is designated by the symbol $\hat{p}$ (read "p hat"); this is the statistic that will be used to estimate the unknown population proportion, p. In summary, we use the following notation:

$$p = \text{population proportion}, \quad \hat{p} = \text{sample proportion}.$$

The population proportion, p, although unknown, is a fixed number. On the other hand, the sample proportion, $\hat{p}$, is a random variable; its value depends on chance, namely, on which employees are obtained as a result of the sampling. For instance, if 202 of the 1010 employees sampled play hooky, then

$$\hat{p} = \frac{202}{1010} = 0.2,$$

or 20.0%. But if 184 of the 1010 employees sampled play hooky, then

$$\hat{p} = \frac{184}{1010} = 0.182,$$

or 18.2%.

These last two calculations also indicate how a sample proportion is computed: Divide the number of employees sampled who play hooky, x, by the total number of employees sampled, n. In symbols, $\hat{p} = x/n$.   ∎

Example 11.1 introduced some notation and terminology we use when making inferences about a population proportion. In general, we have the following definitions.

DEFINITION 11.1

POPULATION PROPORTION AND SAMPLE PROPORTION

Consider a population in which each member is classified as either having or not having a specified attribute, a so-called *two-category population.* Then we use the following notation and terminology.

Population proportion, p: The proportion (percentage) of the entire population that has the specified attribute.

Sample proportion, $\hat{p}$: The proportion (percentage) of a sample from the population that has the specified attribute.

In Example 11.1 the specified attribute is "plays hooky"; the population proportion, p, is the proportion of all U.S. employees who play hooky; and a sample proportion, $\hat{p}$, is the proportion of employees sampled who play hooky. As we saw in Example 11.1, a sample proportion is computed using the following formula.

FORMULA 11.1

SAMPLE PROPORTION

A sample proportion, $\hat{p}$, is computed using the formula

$$\hat{p} = \frac{x}{n},$$

where x denotes the number of members sampled that have the specified attribute and n denotes the sample size.

Note: For convenience we will sometimes refer to x—the number of members sampled that have the specified attribute—as the **number of successes.**

Before proceeding, let's draw some parallels between proportions and means. Table 11.1 shows the correspondence between the notation for means and the notation for proportions.

TABLE 11.1

Correspondence between notations for means and proportions

	Parameter	Statistic
Means	μ	$\bar{x}$
Proportions	p	$\hat{p}$

As we know, a sample mean, $\bar{x}$, can be used to make inferences about a population mean, μ. Similarly, a sample proportion, $\hat{p}$, can be used to make inferences about a population proportion, p.

The Sampling Distribution of the Proportion

We have seen that to make inferences about a population mean, μ, we must know the sampling distribution of the mean, that is, the probability distribution of the random variable $\bar{x}$. The same is true for proportions: To make inferences about a population proportion, p, we need to know the **sampling distribution of the proportion,** that is, the probability distribution of the random variable $\hat{p}$.

The sampling distribution of the proportion can be derived from our knowledge of the sampling distribution of the mean, since a proportion can always be regarded as a mean. (See Exercise 11.23 for details.)

KEY FACT 11.1   **THE SAMPLING DISTRIBUTION OF THE PROPORTION**

> Suppose a large random sample of size n is to be taken from a two-category population with population proportion p. Then the random variable $\hat{p}$ is approximately normally distributed and has mean $\mu_{\hat{p}} = p$ and standard deviation $\sigma_{\hat{p}} = \sqrt{p(1-p)/n}$. In other words, probabilities for $\hat{p}$ are approximately equal to areas under the normal curve with parameters p and $\sqrt{p(1-p)/n}$.

The accuracy of the normal approximation depends on n and p. If p is close to 0.5, then the approximation is quite accurate even for moderate n. The farther p is from 0.5, the larger n needs to be for the approximation to be accurate. As a rule of thumb, we use the normal approximation when *both np and $n(1-p)$ are at least 5.*[†] This is what we will mean in this chapter when we say that n is large.

Qualitatively, Key Fact 11.1 means that for any two-category population and any particular sample size, a histogram of all possible sample proportions will be roughly bell-shaped, if the sample size is sufficiently large. Alternatively, it means that if we take a large number of random samples of a particular size, then a histogram of the sample proportions will be approximately bell-shaped, provided the sample size is large enough.

To illustrate the second interpretation, let's return to the situation in Example 11.1. Suppose, in reality, 19.1% of all U.S. employees play hooky, that is, the population proportion is $p = 0.191$. Using Minitab we simulated taking 2000 random samples of size 1010 from the population of all U.S. employees, determined the sample proportion for each of the 2000 samples, and obtained a histogram of those 2000 sample proportions. Printout 11.1 displays the histogram.

As expected, the histogram of the 2000 sample proportions is roughly bell-shaped. Compare the histogram in Printout 11.1 to the theoretical sampling distribution, which is approximately a normal distribution with parameters 0.191 and 0.012.

[†] Another commonly-used rule of thumb is that both np and $n(1-p)$ are at least 10; still another is that $np(1-p)$ is at least 25. Our rule of thumb, which is less conservative than either of these two, has been chosen so as to be consistent with the conditions required for performing a chi-square goodness-of-fit test (to be discussed in Section 12.2).

PRINTOUT 11.1

Minitab histogram of
sample proportions for
2000 samples of size 1010

```
Histogram of PHAT    N = 2000
Each * represents 10 obs.

Midpoint    Count
   0.155        4  *
   0.160       20  **
   0.165       33  ****
   0.170       75  ********
   0.175      134  **************
   0.180      218  **********************
   0.185      255  **************************
   0.190      314  ********************************
   0.195      327  *********************************
   0.200      259  **************************
   0.205      173  *****************
   0.210       90  *********
   0.215       60  ******
   0.220       22  ***
   0.225       12  **
   0.230        4  *
```

Large-Sample Confidence Intervals
for a Population Proportion

We now present Procedure 11.1, a step-by-step method for obtaining a confidence interval for a population proportion. Basically, the procedure is a special case of Procedure 8.1 on page 448, which provides a method for determining a large-sample confidence interval for a population mean. We often refer to Procedure 11.1 as the **one-sample z-interval procedure** for a population proportion or, more simply, as the **z-interval procedure** for a population proportion.

PROCEDURE 11.1

**THE ONE-SAMPLE z-INTERVAL PROCEDURE
FOR A POPULATION PROPORTION, p**

ASSUMPTION

The number of successes, x, and the number of failures, $n - x$, are both at least 5.

Step 1 For a confidence level of $1 - \alpha$, use Table II to find $z_{\alpha/2}$.

Step 2 The confidence interval for p is from

$$\hat{p} - z_{\alpha/2} \cdot \sqrt{\hat{p}(1 - \hat{p})/n} \quad \text{to} \quad \hat{p} + z_{\alpha/2} \cdot \sqrt{\hat{p}(1 - \hat{p})/n},$$

where $z_{\alpha/2}$ is found in Step 1, n is the sample size, and $\hat{p} = x/n$ is the sample proportion.

The assumption for using Procedure 11.1 is that "the number of successes, x, and the number of failures, $n-x$, are both at least 5." This can be restated as "both $n\hat{p}$ and $n(1-\hat{p})$ are at least 5," which, for unknown p, corresponds to the rule of thumb given on page 656 for using the normal approximation.

Example 11.2   Illustrates Procedure 11.1

Consider again the situation in Example 11.1. A poll was taken of 1010 U.S. employees. The employees sampled were asked whether they "play hooky," that is, call in sick at least once a year when they simply need time to relax; 202 responded "yes." Use these data to obtain a 95% confidence interval for the proportion, p, of all U.S. employees who play hooky.

SOLUTION   We will apply Procedure 11.1, but first we need to check that the condition for its use is satisfied. The attribute in question is "plays hooky," the sample size is 1010, and the number of employees sampled who play hooky is 202. So $x = 202$ and $n-x = 1010-202 = 808$, both of which are at least 5. Consequently, the condition for using Procedure 11.1 is met.

Step 1   *For a confidence level of $1-\alpha$, use Table II to find $z_{\alpha/2}$.*

We want a 95% confidence interval. This means $\alpha = 0.05$ and hence $z_{\alpha/2} = z_{0.05/2} = z_{0.025} = 1.96$.

Step 2   *The confidence interval for p is from*

$$\hat{p} - z_{\alpha/2} \cdot \sqrt{\hat{p}(1-\hat{p})/n} \quad \text{to} \quad \hat{p} + z_{\alpha/2} \cdot \sqrt{\hat{p}(1-\hat{p})/n}.$$

We have $n = 1010$ and, from Step 1, $z_{\alpha/2} = 1.96$. Also, because 202 of the 1010 employees sampled play hooky, $\hat{p} = x/n = 202/1010 = 0.2$. Thus a 95% confidence interval for p is from

$$0.2 - 1.96 \cdot \sqrt{(0.2)(1-0.2)/1010} \quad \text{to} \quad 0.2 + 1.96 \cdot \sqrt{(0.2)(1-0.2)/1010}$$

or

$$0.2 - 0.025 \quad \text{to} \quad 0.2 + 0.025$$

 or 0.175 to 0.225. We can be 95% confident that the percentage of all U.S. employees who play hooky is somewhere between 17.5% and 22.5%.   ∎

Margin of Error

In Section 8.3 we discussed the margin of error in estimating a population mean by a sample mean. In general the margin of error of an estimator represents the precision with which it estimates the parameter in question. Referring to the confidence-interval formula in Step 2 of Procedure 11.1, we see that the margin of error in estimating a population proportion by a sample proportion is $z_{\alpha/2} \cdot \sqrt{\hat{p}(1-\hat{p})/n}$.

DEFINITION 11.2

MARGIN OF ERROR FOR THE ESTIMATE OF p

> The *margin of error* for the estimate of p is
> $$E = z_{\alpha/2} \cdot \sqrt{\hat{p}(1 - \hat{p})/n}.$$
> The margin of error is equal to half the length of the confidence interval. It represents the precision with which the sample proportion, $\hat{p}$, estimates the population proportion, p, at the specified confidence level.

In Example 11.2 the margin of error is
$$E = z_{\alpha/2} \cdot \sqrt{\hat{p}(1 - \hat{p})/n} = 1.96 \cdot \sqrt{(0.2)(1 - 0.2)/1010} = 0.025,$$

which can also be obtained by taking one-half the length of the confidence interval: $(0.225 - 0.175)/2 = 0.025$. Consequently, we can be 95% confident that the error in estimating the proportion, p, of all U.S. employees who play hooky by the proportion, 0.2, of those in the sample who play hooky is at most 0.025, that is, plus or minus 2.5 percentage points.

As we have seen, given a confidence interval, we can find the margin of error by taking half the length of the confidence interval. On the other hand, given the sample proportion, $\hat{p}$, and the margin of error, E, we can determine the confidence interval—its endpoints are $\hat{p} \pm E$.

Most newspaper and magazine polls provide the sample proportion and the margin of error associated with a 95% confidence interval. For example, a 1993 survey of U.S. women conducted by Gallup for the CNBC cable network stated, "...36% of those polled believe their gender will hurt them; the margin of error for the poll is plus or minus 4 percentage points." Translated into our terminology, $\hat{p} = 0.36$ and $E = 0.04$. Thus the confidence interval has endpoints $\hat{p} \pm E = 0.36 \pm 0.04$; we can be 95% confident that the percentage of all U.S. women who believe their gender will hurt them is somewhere between 32% and 40%.

Determining the Required Sample Size

The margin of error and confidence level of a confidence interval are often specified in advance. We must then determine the sample size required to meet the specifications. If we solve for n in the formula for the margin of error, we obtain

(1) $$n = \hat{p}(1 - \hat{p}) \left(\frac{z_{\alpha/2}}{E} \right)^2.$$

This formula cannot be used to obtain the required sample size because the sample proportion, $\hat{p}$, is not known prior to sampling.

There are two ways around this problem. To begin, let's examine a graph of $\hat{p}(1 - \hat{p})$ versus $\hat{p}$, as shown in Fig. 11.1. As we see from the graph, the largest $\hat{p}(1 - \hat{p})$ can be is 0.25, which happens when $\hat{p} = 0.5$. The farther $\hat{p}$ is from 0.5, the smaller the value of $\hat{p}(1 - \hat{p})$.

FIGURE 11.1

Graph of $\hat{p}(1 - \hat{p})$ versus $\hat{p}$

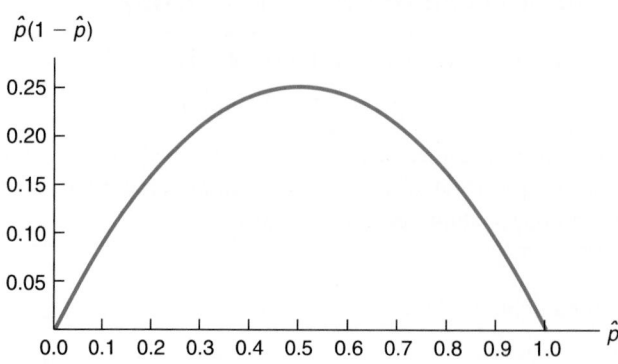

Since the largest possible value of $\hat{p}(1-\hat{p})$ equals 0.25, the most conservative approach for determining sample size is to use that value in Equation (1). The sample size thus obtained will generally be larger than necessary and the margin of error less than required. Nonetheless, this approach guarantees that the specifications will be met or bettered.

On the other hand, because sampling tends to be expensive, it is usually best not to take a larger sample than necessary. If we can make an educated guess for the observed value of $\hat{p}$, say from a previous study or theoretical considerations, then we can use that guess to obtain a more realistic sample size. In this same vein, if we have in mind a likely range for the observed value of $\hat{p}$, then in view of Fig. 11.1 we should take as our educated guess for $\hat{p}$ the value in the range closest to 0.5. But in either case we should be aware that if the observed value of $\hat{p}$ is closer to 0.5 than is our educated guess, then the margin of error will be larger than desired. (Why is this so?)

FORMULA 11.2   **SAMPLE SIZE FOR ESTIMATING p**

A $(1 - \alpha)$-level confidence interval for a population proportion having a margin of error of at most E can be obtained by choosing

$$n = 0.25 \left(\frac{z_{\alpha/2}}{E} \right)^2,$$

rounded up to the nearest whole number. If we can make an educated guess, $\hat{p}_g$ (g for guess), for the observed value of $\hat{p}$, then we should instead choose

$$n = \hat{p}_g(1 - \hat{p}_g) \left(\frac{z_{\alpha/2}}{E} \right)^2,$$

rounded up to the nearest whole number.

Example 11.3   Illustrates Formula 11.2

Consider once more the problem of estimating the proportion, p, of all U.S. employees who play hooky.

a. Obtain a sample size that will ensure a margin of error of at most 0.01 for a 95% confidence interval.

b. Find a 95% confidence interval for p if for a sample of the size determined in part (a), the proportion of those who play hooky is 0.194.

c. Determine the margin of error for the estimate in part (b) and compare it to the margin of error specified in part (a).

d. Repeat parts (a)–(c) if it is deemed reasonable to presume that the proportion of those sampled who play hooky will be somewhere between 0.1 and 0.3.

e. Compare the results obtained in parts (a)–(c) with those obtained in part (d).

SOLUTION   a. Here we apply the first displayed equation in Formula 11.2. In doing so, we must identify $z_{\alpha/2}$ and the margin of error, E. The confidence level is stipulated to be 0.95, so $z_{\alpha/2} = z_{0.05/2} = z_{0.025} = 1.96$; the margin of error is specified at 0.01. Thus a sample size that will ensure a margin of error of at most 0.01 for a 95% confidence interval is

$$n = 0.25 \left(\frac{z_{\alpha/2}}{E} \right)^2 = 0.25 \left(\frac{1.96}{0.01} \right)^2 = 9604.$$

If we take a sample of 9604 U.S. employees, then the margin of error for our estimate of the proportion of all U.S. employees who play hooky will be at most 0.01, that is, at most plus or minus 1 percentage point.

b. Applying Procedure 11.1 on page 657 with $\alpha = 0.05$, $n = 9604$, and $\hat{p} = 0.194$, we find that a 95% confidence interval for p has endpoints

$$0.194 \pm 1.96 \cdot \sqrt{(0.194)(1 - 0.194)/9604},$$

or 0.194 ± 0.008, or 0.186 to 0.202. We can be 95% confident that the percentage of all U.S. employees who play hooky is somewhere between 18.6% and 20.2%.

c. The margin of error for the estimate in part (b) is 0.008. Not surprisingly, this is less than the margin of error of 0.01 specified in part (a).

d. If it is reasonable to presume that the proportion of those sampled who play hooky will be somewhere between 0.1 and 0.3, then we should obtain the sample size using the second displayed equation in Formula 11.2 with $\hat{p}_g = 0.3$:

$$n = \hat{p}_g(1 - \hat{p}_g) \left(\frac{z_{\alpha/2}}{E} \right)^2 = (0.3)(1 - 0.3) \left(\frac{1.96}{0.01} \right)^2 = 8068 \text{ (rounded up)}.$$

Applying Procedure 11.1 with $\alpha = 0.05$, $n = 8068$, and $\hat{p} = 0.194$, we find that a 95% confidence interval for p has endpoints

$$0.194 \pm 1.96 \cdot \sqrt{(0.194)(1 - 0.194)/8068},$$

or 0.194 ± 0.009, or 0.185 to 0.203. We can be 95% confident that the percentage of all U.S. employees who play hooky is somewhere between 18.5% and 20.3%. The margin of error for the estimate is 0.009.

e. By employing the guess for $\hat{p}$ in part (d), we reduced the required sample size by more than 1500 (from 9604 to 8068). Moreover, only $\frac{1}{10}$ of 1% of precision was lost—the margin of error rose from 0.008 to 0.009. The risk of using the guess 0.3 for $\hat{p}$ is that if the observed value of $\hat{p}$ had turned out to be larger than 0.3 (but smaller than 0.7), then the achieved margin of error would have exceeded the specified 0.01. ∎

Using the Computer (Optional)

On page 656 we used Minitab to illustrate the sampling distribution of the proportion. Specifically, assuming 19.1% of all U.S. employees play hooky, we simulated taking 2000 random samples of size 1010 from the population of all U.S. employees; determined, for each of the 2000 samples, the proportion of employees sampled who play hooky; and obtained a histogram of those 2000 sample proportions.

We will now explain how this was done. To begin, we apply the **Random** command and the **Binomial** subcommand as follows.[†]

Session commands: We first name an available column HOOKY. Next we type the command **RANDOM** followed by the number of samples required and the storage location for the numbers of employees sampled who play hooky; that is, we type RANDOM 2000 obs into 'HOOKY'; and press Enter. Then we use the **BINOMIAL** subcommand to specify the probability distribution of the random variable to be simulated; that is, we type BINOMIAL with n=1010 and p=0.191. and press Enter.

(or)

Menu commands: We choose **Calc ▸ Random Data ▸ Binomial...**, type 2000 for the number of rows to be generated, specify HOOKY in the **Store in column(s)** text box, type 1010 in the **Number of trials** text box, type 0.191 in the **Probability of success** text box, and then select **OK**.

As a consequence of the above commands, 2000 independent observations of the number of employees who play hooky out of 1010 randomly selected employees are stored in HOOKY. To obtain the sample proportions, $\hat{p}$, for the 2000 samples of size 1010, we must divide each of the 2000 numbers in HOOKY by 1010. This can be done using Minitab's **Let** command as follows.

Session commands: First we name an available column PHAT. Then we type the command **LET** followed by the column in which the sample proportions are to be stored, an equal sign, and the arithmetic operation we want to perform; that is, we type LET 'PHAT'= 'HOOKY'/1010 and press Enter.

(or)

Menu commands: We choose **Calc ▸ Functions and Statistics ▸ General Expressions...**, specify PHAT in the **New/modified variable** text box, type 'HOOKY'/1010 in the **Expression** text box, and then select **OK**.

[†] Strictly speaking we should be simulating a hypergeometric distribution, not a binomial distribution. However, as we know, the binomial distribution adequately approximates the hypergeometric distribution when the sample size is small relative to the population size, which is certainly true here.

As a consequence of the above commands, the sample proportions for 2000 samples of size 1010 are stored in PHAT. To obtain a histogram of the sample proportions, we apply the Histogram command in the usual way, as explained in Section 2.3.

The computer output we actually got by performing the simulation, obtaining the sample proportions, and applying the Histogram command is shown in Printout 11.1 on page 657. Of course, every time we repeat the simulation, we will almost certainly get somewhat different results than those depicted in Printout 11.1. But in every case the histogram will be approximately bell-shaped. ∎

EXERCISES 11.1

· **11.1** Is a population proportion, p, a parameter or a statistic? What about a sample proportion, $\hat{p}$? Why?

In each of Exercises 11.2–11.7, apply Procedure 11.1 on page 657 to obtain the required confidence interval. Be sure to check the condition for using that procedure.

· **11.2** Studies are performed to estimate the percentage of the nation's 10 million asthmatics who are allergic to sulfites. In a recent survey, 38 of 500 randomly selected U.S. asthmatics were found to be allergic to sulfites.

a. Find a 95% confidence interval for the proportion of all U.S. asthmatics who are allergic to sulfites.

b. Interpret your result from part (a).

· **11.3** A *Reader's Digest/Gallup Survey* on the drinking habits of Americans estimated the percentage of adults across the country who drink beer, wine, or hard liquor, at least occasionally. Of the 1516 adults interviewed, 985 said they drank.

a. Determine a 95% confidence interval for the proportion, p, of all Americans who drink beer, wine, or hard liquor, at least occasionally.

b. Interpret your result from part (a).

· **11.4** A Gallup Poll asked public-school teachers nationwide to grade their fellow educators' performance. The poll found that 634 of the 813 teachers surveyed gave their fellow educators an A or B grade.

a. Find a 90% confidence interval for the percentage of all public-school teachers who would give their fellow educators an A or B grade for performance.

b. Interpret your result from part (a).

· **11.5** A Gallup Poll estimated the support among Americans for "right to die" laws. For the survey, 1528 adults were asked whether they favor voluntary withholding of life-support systems from the terminally ill. The results: 1238 said yes.

a. Find a 99% confidence interval for the percentage of all adult Americans who are in favor of "right to die" laws.

b. Interpret your result from part (a).

· **11.6** Dr. Charles Kuntzleman directed a study to estimate the percentage of schoolchildren who have at least one of the significant factors contributing to heart disease. As reported by *The Arizona Republic*, the 3-year study involved nearly 24,000 schoolchildren and concluded that 98% of the schoolchildren examined had at least one of the significant factors contributing to heart disease.

a. Use the data to obtain a 99% confidence interval for the proportion, p, of all schoolchildren who have at least one of the significant factors contributing to heart disease.

b. Interpret your result in terms of percentages.

· **11.7** Real Estate Research Corporation, a Los Angeles based research group, publishes an annual report entitled *Emerging Trends in Real Estate*.

a. Two-hundred randomly selected real-estate experts were asked whether they believe next year will be a good time to buy real estate; and 78% replied in the affirmative. Determine a 90% confidence interval for the proportion of all real-estate experts who believe next year will be a good time to buy real estate.

b. Interpret your result from part (a).

· **11.8** Suppose you have been hired to estimate the percentage of adults in your state who are literate. You take a random sample of 100 adults and find that 96 of them are literate. You then obtain a 95% confidence interval as follows:

$$0.96 \pm 1.96 \cdot \sqrt{(0.96)(0.04)/100},$$

or 0.922 to 0.998. From this you conclude that we can be 95% confident that the percentage of all adults in your state who are literate is somewhere between 92.2% and 99.8%. Is there anything wrong with your reasoning?

11.9   Suppose I have been commissioned to estimate the infant mortality rate in Norway. From a random sample of 500 live births, I find that 0.8% of them resulted in infant deaths. I then construct a 90% confidence interval for the infant mortality rate in Norway:

$$0.008 \pm 1.645 \cdot \sqrt{(0.008)(0.992)/500},$$

or 0.001 to 0.015. Then I conclude, "We can be 90% confident that the proportion of infant deaths in Norway is somewhere between 0.001 and 0.015." How did I do?

11.10   In April 1993, *Parade Magazine* conducted a nationwide survey of U.S. adults. The magazine reported that 82% of those polled said they have a "positive attitude" toward the police; the margin of error was plus or minus 1.5 percentage points (for a 0.95 confidence level). Use this information to obtain a 95% confidence interval for the percentage of all U.S. adults who would say they have a "positive attitude" toward the police.

11.11   A study of alternative-medicine use by Americans, directed by Dr. David M. Eisenberg of Boston's Beth Israel Hospital, appeared in a 1993 issue of the *New England Journal of Medicine.* The study revealed that 34% of the Americans surveyed used at least one unconventional therapy in 1990. The margin of error was plus or minus 2.4 percentage points (for a 0.95 confidence level). Use this information to obtain a 95% confidence interval for the percentage of all Americans who used at least one unconventional therapy in 1990.

11.12   Refer to Exercise 11.2, where you considered the problem of estimating the proportion, p, of all U.S. asthmatics who are allergic to sulfites.
a. Determine the margin of error for the estimate of p.
b. Obtain a sample size that will ensure a margin of error of at most 0.01 for a 95% confidence interval without making a guess for the observed value of $\hat{p}$.
c. Find a 95% confidence interval for p if for a sample of the size determined in part (b), the proportion of asthmatics sampled who are allergic to sulfites is 0.071.
d. Determine the margin of error for the estimate in part (c) and compare it to the margin of error specified in part (b).
e. Repeat parts (b)–(d) if you deem it reasonable to presume that the proportion of asthmatics sampled who are allergic to sulfites will be at most 0.10.
f. Compare the results you obtained in parts (b)–(d) with those obtained in part (e).

11.13   Refer to Exercise 11.3, where you found a 95% confidence interval for the proportion, p, of U.S. adults who drink alcoholic beverages.
a. Determine the margin of error for the estimate of p.
b. Obtain a sample size that will ensure a margin of error of at most 0.02 for a 95% confidence interval without making a guess for the observed value of $\hat{p}$.
c. Find a 95% confidence interval for p if for a sample of the size determined in part (b), 63% of those sampled drink alcoholic beverages.
d. Determine the margin of error for the estimate in part (c) and compare it to the margin of error specified in part (b).
e. Repeat parts (b)–(d) if you deem it reasonable to presume that the percentage of adults sampled who drink alcoholic beverages will be at least 60%.
f. Compare the results you obtained in parts (b)–(d) with those obtained in part (e).

11.14   Refer to Exercise 11.4, where you obtained a 90% confidence interval for the percentage of all public-school teachers who would give their fellow educators an A or B grade for performance.
a. Find the margin of error for the estimate of the percentage.
b. Obtain a sample size that will ensure a margin of error of at most 1.5 percentage points for a 90% confidence interval without making a guess for the observed value of $\hat{p}$.
c. Find a 90% confidence interval for p if for a sample of the size determined in part (b), 79.2% of the public-school teachers sampled would give their fellow educators an A or B grade for performance.
d. Determine the margin of error for the estimate in part (c) and compare it to the margin of error specified in part (b).
e. Repeat parts (b)–(d) if you deem it reasonable to presume that the percentage of public-school teachers sampled who would give their fellow educators an A or B grade for performance will be between 70% and 85%.
f. Compare the results you obtained in parts (b)–(d) with those obtained in part (e).

11.15   Refer to Exercise 11.5, where you determined a 99% confidence interval for the percentage of adult Americans who are in favor of "right to die" laws.
a. Find the margin of error for the estimate of the percentage.
b. Obtain a sample size that will ensure a margin of error of at most 1 percentage point for a 99% confidence interval without making a guess for the observed value of $\hat{p}$.
c. Find a 99% confidence interval for p if for a sample of the size determined in part (b), 82.5% of the adult Americans sampled are in favor of "right to die" laws.
d. Determine the margin of error for the estimate in part (c) and compare it to the margin of error specified in part (b).

e. Repeat parts (b)–(d) if you deem it reasonable to presume that the percentage of adult Americans sampled who are in favor of "right to die" laws will be between 75% and 90%.

f. Compare the results you obtained in parts (b)–(d) with those obtained in part (e).

11.16 A company manufactures goods that are sold exclusively by mail order. The director of market research needs to test market a new product. She plans to send out brochures to a random sample of households and use the proportion of orders obtained as an estimate of the true proportion, known as the *product response rate*. The results of the market research will be employed as a primary source for advance production planning. Consequently, the director wants the figures she presents to be as accurate as possible. Specifically, she wants to be 95% confident that the estimate of the product response rate will be accurate to within 1%.

a. Without making any assumptions, determine the sample size required.

b. Historically, product response rates for products sold by this company have ranged from 0.5% to 4.9%. If the director is willing to assume that the sample product response rate for this product will also fall in that range, determine the required sample size.

c. Compare the results from parts (a) and (b).

d. Discuss the possible consequences if the assumption made in part (b) turns out to be incorrect.

11.17 On February 19, 1993, the results of two polls on President Clinton's budget plan appeared in *The Arizona Republic*. A CNN/*USA Today* poll stated that 79% of those who saw Clinton's speech supported his plan; the margin of error was plus or minus 5 percentage points. An ABC/*Washington Post* poll stated that 74% of those who saw Clinton's speech supported his plan; the margin of error was plus or minus 5 percentage points. Is it possible that both of these polls were correct in their conclusions?

11.18 (Computer exercise) According to the *National Health Interview Survey,* published by the National Center for Health Statistics, 57.8% of last year's dental implants were done on people age 18–44 years. Consider the random selection of 300 people who had dental implants last year, where each person chosen is classified as either being in the 18–44 year age group (the specified attribute) or not. Use Minitab or some other statistical software to

a. simulate the taking of 2000 such samples.

b. obtain the sample proportion of each of the 2000 samples.

c. obtain the mean, the standard deviation, and a histogram of the 2000 sample proportions.

d. Theoretically, what are the mean, standard deviation, and distribution of all possible sample proportions for samples of size 300 from the population?

e. Compare your results from parts (c) and (d).

11.19 (Computer exercise) The U.S. Bureau of the Census reports that 14% of U.S. residents speak a language other than English at home. Consider the random selection of 500 U.S. residents, where each person obtained is classified as either speaking a language other than English at home (the specified attribute) or speaking only English at home. Use Minitab or some other statistical software to

a. simulate the taking of 1000 such samples.

b. obtain the sample proportion of each of the 1000 samples.

c. obtain the mean, the standard deviation, and a histogram of the 1000 sample proportions.

d. Theoretically, what are the mean, standard deviation, and distribution of all possible sample proportions for samples of size 500 from the population?

e. Compare your results from parts (c) and (d).

Exact confidence intervals for p. Procedure 11.1 provides a method for obtaining confidence intervals for a population proportion, p; it applies only when the number of successes and the number of failures are both at least 5, and even then the resulting confidence interval is only an approximation to the true confidence interval. Using more advanced methods, we can derive a procedure for determining exact confidence intervals for p. That procedure applies in all cases and can be implemented easily using any statistical software that provides a program for inverse cumulative probabilities of the beta distribution. We will show how to implement the procedure using the standard version of Minitab.

Suppose a random sample of size n is to be taken from a two-category population with population proportion p. Let x denote the number of members sampled having the specified attribute. We will assume $1 \leq x \leq n - 1$ since that is almost always the case. Then the left and right endpoints of a $(1 - \alpha)$-level confidence interval for p can be obtained, respectively, using the following Minitab commands:

```
MTB > INVCDF K1;
SUBC> BETA K2 K3.
```
and
```
MTB > INVCDF K4;
SUBC> BETA K5 K6.
```

In the commands, the constants are defined as follows: K1 = $\alpha/2$, K2 = x, K3 = $n - x + 1$, K4 = $1 - \alpha/2$, K5 = $x + 1$, and K6 = $n - x$. Exercises 11.20 and 11.21 ask you to apply this confidence-interval procedure.

· ·   **11.20 (Computer exercise)** Refer to Exercise 11.2.

a. Use Minitab or some other statistical software to apply the above procedure to obtain the confidence interval required in part (a) of Exercise 11.2.

b. Compare your answer to the one you obtained in Exercise 11.2(a).

· ·   **11.21 (Computer exercise)** Refer to Exercise 11.3.

a. Use Minitab or some other statistical software to apply the above procedure to obtain the confidence interval required in part (a) of Exercise 11.3.

b. Compare your answer to the one you obtained in Exercise 11.3(a).

Exercises 11.22 and 11.23 provide two different ways of verifying Key Fact 11.1 on page 656.

· · ·   **11.22** Formula 11.1 states that a sample proportion is computed using the formula $\hat{p} = x/n$, where x denotes the number of successes and n the sample size. The random variable x has, at least approximately, the binomial distribution with parameters n and p, where p is the population proportion. In Section 6.6 we learned that for large n, a binomial random variable is approximately normally distributed. From this it follows that the random variable $\hat{p}$ is also approximately normally distributed. So to complete the verification of Key Fact 11.1, we need only show that the mean and standard deviation of the random variable $\hat{p}$ are given by

$$\mu_{\hat{p}} = p \quad \text{and} \quad \sigma_{\hat{p}} = \sqrt{p(1-p)/n}.$$

Establish these two formulas. *(Hint: First apply part (f) of Exercise 5.27 on page 289 to Formula 11.1. Then use Formulas 5.3 and 5.4 on pages 314 and 315, respectively.)*

· · ·   **11.23** Consider a (finite) two-category population in which the proportion of members having the specified attribute is equal to p. We can think of such a population as consisting of 1s and 0s. A member of the population is a "1" if it has the specified attribute and is a "0" otherwise.

a. If the size of the population is N, how many 1s are in the population?

b. Use part (a) and Definition 3.12 on page 161 to show that the mean of this population of 1s and 0s is p, that is, $\mu = p$.

c. Use part (b) and the shortcut formula given in Definition 3.13 on page 163 to show that the standard deviation of this population of 1s and 0s is $\sqrt{p(1-p)}$; that is, $\sigma = \sqrt{p(1-p)}$.

d. Suppose a random sample of size n is to be taken from this population of 1s and 0s. Show that $\bar{x} = \hat{p}$.

e. Use parts (b)–(d) and the sampling distribution of the mean for general populations, Key Fact 7.6 on page 424, to verify Key Fact 11.1.

· · ·   **11.24** In this exercise we will derive Procedure 11.1 on page 657, which provides a method for obtaining a confidence interval for a population proportion. Suppose a large random sample of size n is to be taken from a two-category population with population proportion p.

a. Apply Key Fact 11.1 on page 656 (the sampling distribution of the proportion) and Key Fact 6.7 on page 373 (the general empirical rule for normally distributed random variables) to deduce that the probability is approximately $1 - \alpha$ that the sample proportion, $\hat{p}$, will be in the interval with endpoints

$$p \pm z_{\alpha/2} \cdot \sqrt{p(1-p)/n}.$$

b. Use part (a) to show that the probability is approximately $1 - \alpha$ that the interval with endpoints

$$\hat{p} \pm z_{\alpha/2} \cdot \sqrt{p(1-p)/n}$$

will contain p.

c. Deduce from part (b) that the probability is approximately $1 - \alpha$ that the interval with endpoints

$$\hat{p} \pm z_{\alpha/2} \cdot \sqrt{\hat{p}(1-\hat{p})/n}$$

will contain p.

d. Use part (c) to explain why once the sample is taken, the interval with endpoints

$$\hat{p} \pm z_{\alpha/2} \cdot \sqrt{\hat{p}(1-\hat{p})/n}$$

will be a $(1-\alpha)$-level confidence interval for p.

11.2   HYPOTHESIS TESTS FOR ONE POPULATION PROPORTION

· · · · ·

In Section 11.1 we discovered how to obtain confidence intervals for a population proportion. Now we will learn how to perform hypothesis tests for a population proportion. (Exercise 11.35 shows that the procedure we will develop is actually a special case of the one-sample z-test for a population mean, Procedure 9.1 on page 503.)

From Key Fact 11.1 on page 656 (the sampling distribution of the proportion) we can deduce that for large n, the standardized random variable

$$z = \frac{\hat{p} - p}{\sqrt{p(1-p)/n}}$$

has approximately the standard normal distribution. Consequently, to perform a large-sample hypothesis test with null hypothesis H_0: $p = p_0$, we can use the random variable

$$z = \frac{\hat{p} - p_0}{\sqrt{p_0(1-p_0)/n}}$$

as the test statistic and obtain the critical value(s) from the standard-normal table, Table II. Procedure 11.2 supplies the details. We often refer to that procedure as the **one-sample z-test** for a population proportion or, more briefly, as the **z-test** for a population proportion.

PROCEDURE 11.2

THE ONE-SAMPLE z-TEST FOR A POPULATION PROPORTION WITH NULL HYPOTHESIS H_0: $p = p_0$

ASSUMPTION

Both np_0 and $n(1 - p_0)$ are at least 5.

Step 1   State the null and alternative hypotheses.

Step 2   Decide on the significance level, α.

Step 3   The critical value(s)
   a. for a two-tailed test are $\pm z_{\alpha/2}$.
   b. for a left-tailed test is $-z_\alpha$.
   c. for a right-tailed test is z_α.
   Use Table II to find the critical value(s).

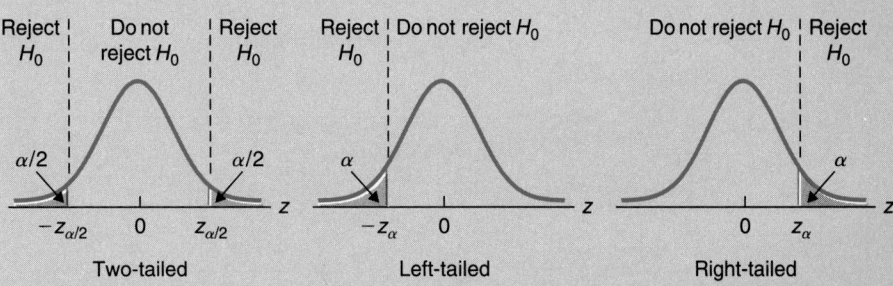

Step 4   Compute the value of the test statistic

$$z = \frac{\hat{p} - p_0}{\sqrt{p_0(1-p_0)/n}}.$$

Step 5   If the value of the test statistic falls in the rejection region, then reject H_0; otherwise, do not reject H_0.

Step 6   State the conclusion in words.

Example 11.4   Illustrates Procedure 11.2

One of the more controversial issues in the United States is gun control; there are many avid proponents and opponents of banning handgun sales. In an April 1993 survey, Louis Harris of LH Research polled 1250 U.S. adults regarding their view on banning handgun sales. The results were that 650 of those sampled favored a ban. Do the data provide sufficient evidence to conclude that a majority of U.S. adults (i.e., more than 50%) favor banning handgun sales? Perform the required hypothesis test at the 5% significance level.

SOLUTION   We will apply Procedure 11.2 to perform the required hypothesis test. But first we must check that the condition for its use is met. We have $n = 1250$ and $p_0 = 0.50$ (50%). Therefore

$$np_0 = 1250 \cdot 0.50 = 625 \quad \text{and} \quad n(1 - p_0) = 1250 \cdot (1 - 0.50) = 625.$$

Since both np_0 and $n(1 - p_0)$ are at least 5, we can employ Procedure 11.2.

Step 1   *State the null and alternative hypotheses.*

$$H_0\text{: } p = 0.50 \text{ (it is not true that a majority favor a ban)}$$
$$H_a\text{: } p > 0.50 \text{ (a majority favor a ban).}$$

Note that the hypothesis test is right-tailed since a greater-than sign ($>$) appears in the alternative hypothesis.

Step 2   *Decide on the significance level, α.*

We are to perform the test at the 5% significance level. So $\alpha = 0.05$.

Step 3   *The critical value for a right-tailed test is z_α.*

Since $\alpha = 0.05$, the critical value is $z_{0.05} = 1.645$, as seen in Fig. 11.2.

FIGURE 11.2
Criterion for deciding whether or not to reject the null hypothesis

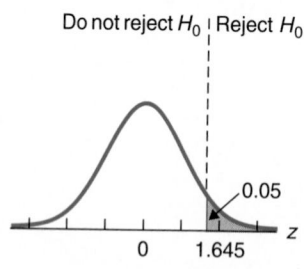

Do not reject H_0 | Reject H_0

0.05

0   1.645   z

Step 4   *Compute the value of the test statistic*

$$z = \frac{\hat{p} - p_0}{\sqrt{p_0(1 - p_0)/n}}.$$

We have $n = 1250$ and $p_0 = 0.50$. The number of U.S. adults surveyed who favor banning handgun sales is 650. Therefore the proportion of those surveyed who favor a ban is $\hat{p} = x/n = 650/1250 = 0.520$ (52.0%). Consequently, the value of the test statistic is

$$z = \frac{0.520 - 0.50}{\sqrt{(0.50)(1 - 0.50)/1250}} = 1.41.$$

Step 5 *If the value of the test statistic falls in the rejection region, reject H_0; otherwise, do not reject H_0.*

From Step 4, the value of the test statistic is $z = 1.41$, which as we see from Fig. 11.2 does not fall in the rejection region. Thus we do not reject H_0.

Step 6 *State the conclusion in words.*

The test results are not statistically significant at the 5% level; that is, at the 5% significance level, the data do not provide sufficient evidence to conclude that a majority of U.S. adults favor banning handgun sales. ■

Example 11.4 also provides a good illustration of how statistical results are sometimes misstated. The newspaper article we read that featured the survey had the headline "Fear prompts 52% in U.S. to back pistol-sale ban, poll says." In fact, the poll says no such thing. It says only that 52% of those *sampled* back a pistol-sale ban; and as we have seen, at the 5% significance level, those data do not provide sufficient evidence to conclude that even a majority of U.S. adults back a pistol-sale ban.

By using the P-value approach to hypothesis testing, we can assess more precisely the evidence against the null hypothesis. From Step 4 of Example 11.4, we know that the value of the test statistic is $z = 1.41$. Consulting Table II, we find that $P = 0.0793$. So we cannot reject H_0 at the 5% significance level, although we can reject it at the 8% significance level or for that matter at any level greater than or equal to 7.93%. Moreover, according to Table 9.12 on page 530, the data do provide moderate (but not strong) evidence against the null hypothesis and thereby for the alternative hypothesis that a majority of U.S. adults favor banning handgun sales.

EXERCISES 11.2

In Exercises 11.25–11.34, perform each hypothesis test using either the classical approach or the P-value approach. Comment on the practical significance of all tests whose results are statistically significant.

· **11.25** In December 1991, *The Arizona Republic* conducted a telephone poll of 758 Arizona adults who celebrate Christmas. The question asked was, "In your family, do you open presents on Christmas Eve or Christmas Day?" Of those surveyed, 394 said they wait until Christmas Day. At the 5% significance level, do the data provide sufficient evidence

to conclude that a majority of Arizona families who celebrate Christmas wait until Christmas Day to open their presents?

· **11.26** The Motor Vehicle Manufacturers Association of the United States reports in *Motor Vehicle Facts and Figures* that in 1962, foreign cars made up 4.8% of all U.S. car sales; in 1983 the percentage was 27.8%; and in 1990 the percentage was 25.8%. From a sample of 500 of this year's car sales, it is found that 128 are imports. Do the data suggest, at the 5% significance level, that the percentage of foreign-car sales for this year will differ from the 1990 figure of 25.8%?

11.27 In a 1993 study conducted by the American Management Association, 630 randomly selected major U.S. firms were polled on their drug-testing policies. According to the report, "...85% of the firms surveyed now test employees, applicants, or both." At the 5% significance level, do the data provide sufficient evidence to conclude that the percentage of major U.S. firms that drug-tested in 1993 exceeds the 1992 figure of 74%?

11.28 According to the Arizona Real Estate Commission, in 1982 only 10% of people holding real-estate licenses were active in the industry. An independent agency has been asked by the commission to determine whether this year's percentage is higher. A random sample of 150 licensed people reveals that 24 are currently active. Perform the appropriate hypothesis test at the 1% significance level.

11.29 The American Medical Association reports in *Physician Characteristics and Distribution in the U.S.* that in 1975, 9.0% of all physicians were women; in 1980, 11.6% were women; in 1981, 12.2% were women; in 1990, 16.9% were women. For this year, 25 of 125 randomly selected physicians are women. Do the data indicate, at the 10% significance level, that the percentage of female physicians is higher now than in 1990?

11.30 A 1985 study, conducted by the National Center for Health Statistics and published in *Vital and Health Statistics,* revealed that 31.8% of young adults (20–24 years old) smoked cigarettes. In 1988, 27.6% of 1283 randomly selected young adults smoked cigarettes. Do the data provide sufficient evidence to conclude that the percentage of young adult smokers declined between 1985 and 1988? Use a significance level of 0.05.

11.31 Of the 38 numbers on a roulette wheel, 18 are red, 18 are black, and 2 are green. If the wheel is balanced, the probability of the ball landing on red is $\frac{18}{38} = 0.474$. A gambler has been studying a roulette wheel. If the wheel is out of balance, then he can improve his odds of winning. The gambler observes 200 spins of the wheel and finds that the ball lands on red 93 times. At the 10% significance level, do the data provide sufficient evidence to conclude that the ball is not landing on red the correct percentage of the time for a balanced wheel?

11.32 A direct-mail firm wants to conduct a test-market offering for one of its new products. A random sample of 800 households is chosen to receive advertising material describing the new product. It is decided that additional advertising and promotion will occur only if the sample results

provide strong evidence that the actual (population) response rate, p, will exceed 6.5%. What decision will be made if 70 of the 800 households make a purchase? Use $\alpha = 0.01$.

11.33 In 1990, 10.7% of all U.S. families earned incomes below the poverty level, as reported by the Census Bureau in *Current Population Reports.* During that same year, 23 of 250 randomly selected northeastern families earned incomes below the poverty level. Does the study provide sufficient evidence to conclude that in 1990, the proportion of northeastern families earning incomes below the poverty level was less than the national proportion? Use $\alpha = 0.05$.

11.34 Polls are taken regularly to determine the public's view on how the president is handling his job. A CNN/*USA Today* poll, conducted by Gallup in July 1993, stated that 48% of 1002 randomly selected U.S. adults disapproved of Clinton's handling of the job. At the 5% significance level, do these data provide sufficient evidence to conclude that less than half of all U.S. adults disapproved of Clinton's handling of his job as president?

· · · 11.35 The test statistic used in Procedure 11.2 is

$$(2) \qquad z = \frac{\hat{p} - p_0}{\sqrt{p_0(1 - p_0)/n}} .$$

In this exercise we will show that this test statistic is the familiar

$$(3) \qquad z = \frac{\bar{x} - \mu_0}{\sigma/\sqrt{n}} ,$$

when the latter is specialized to the situation of proportions. Thus we will show that Procedure 11.2 is a special case of Procedure 9.1 on page 503, the z-test for a population mean.

Consider a (finite) two-category population in which the proportion of members having the specified attribute is equal to p. We can think of such a population as consisting of 1s and 0s. A member of the population is a "1" if it has the specified attribute and is a "0" otherwise.

a. If the size of the population is N, how many 1s are in the population?

b. Use part (a) and Definition 3.12 on page 161 to show that the mean of this population of 1s and 0s is p.

c. Use part (b) and the shortcut formula in Definition 3.13 on page 163 to show that the standard deviation of this population of 1s and 0s is $\sqrt{p(1-p)}$.

d. Suppose a random sample of size n is to be taken from this population of 1s and 0s. Show that $\bar{x} = \hat{p}$.

e. Deduce from parts (b)–(d) that the test statistic in Equation (3) becomes the one in Equation (2) when specialized to the situation of proportions.

11.3 INFERENCES FOR TWO POPULATION PROPORTIONS USING INDEPENDENT SAMPLES

· · · · · ·

In Sections 11.1 and 11.2 we studied inferences for one population proportion, p. Now we will examine methods for performing inferences to compare the proportions, p_1 and p_2, of two populations. We begin by discussing hypothesis testing.

Example 11.5 Introduces Hypothesis Tests for Two Population Proportions

A 1993 Associated Press article reported on a national smoking study conducted by the Centers for Disease Control and Prevention. About 43,000 U.S. adults were surveyed individually, in person. The following data are based on the results obtained in that survey.

Independent random samples of 20,650 U.S. males and 22,352 U.S. females were obtained in order to compare the percentage of males who smoke cigarettes to the percentage of females who smoke cigarettes. Of the males sampled, 5803 were cigarette smokers; of the females sampled, 5253 were cigarette smokers. Do the data provide sufficient evidence to conclude that the percentage of all U.S. males who smoke cigarettes exceeds the percentage of all U.S. females who smoke cigarettes?

SOLUTION For this problem the specified attribute is "smokes cigarettes." Let p_1 denote the proportion of all U.S. males who smoke cigarettes, and let p_2 denote the proportion of all U.S. females who smoke cigarettes. Then we want to perform the hypothesis test

$$H_0: p_1 = p_2 \text{ (percentage of male smokers is not higher)}$$
$$H_a: p_1 > p_2 \text{ (percentage of male smokers is higher)}.$$

Roughly speaking, the hypothesis test can be carried out as follows:

1. Compute the proportion of the males sampled who smoke cigarettes, $\hat{p}_1$, and the proportion of the females sampled who smoke cigarettes, $\hat{p}_2$.

2. If $\hat{p}_1$ is too much larger than $\hat{p}_2$, then reject H_0; otherwise, do not reject H_0.

The first step is easy. Since 5803 of the 20,650 males sampled and 5253 of the 22,352 females sampled were found to be smokers,

$$\hat{p}_1 = \frac{x_1}{n_1} = \frac{5803}{20,650} = 0.281 \ (28.1\%)$$

and

$$\hat{p}_2 = \frac{x_2}{n_2} = \frac{5253}{22,352} = 0.235 \ (23.5\%).$$

For the second step, we need to decide whether the sample proportion $\hat{p}_1 = 0.281$ exceeds the sample proportion $\hat{p}_2 = 0.235$ by a sufficient amount to warrant rejecting the null hypothesis in favor of the alternative hypothesis. In other words, we need to decide whether the difference between the two sample proportions can reasonably be attributed to sampling error or whether it indicates that the percentage of U.S. males who smoke cigarettes exceeds the percentage of U.S. females who smoke cigarettes.

To make that decision, we need to know the probability distribution of the difference, $\hat{p}_1 - \hat{p}_2$, between two sample proportions—the **sampling distribution of the difference between two proportions.** We will discuss that sampling distribution and then complete the hypothesis test. ∎

The Sampling Distribution of the Difference Between Two Proportions (Large and Independent Samples)

To begin our discussion of the sampling distribution of the difference between two proportions, we summarize the required notation in Table 11.2.

TABLE 11.2
Notation for parameters and statistics when considering two two-category populations

	Population 1	Population 2
Population proportion	p_1	p_2
Sample size	n_1	n_2
Number of successes	x_1	x_2
Sample proportion	$\hat{p}_1$	$\hat{p}_2$

Recall that the *number of successes* refers to the number of members sampled that have the specified attribute. Consequently, the sample proportions are computed from the formulas

$$\hat{p}_1 = \frac{x_1}{n_1} \quad \text{and} \quad \hat{p}_2 = \frac{x_2}{n_2}.$$

Next we present formulas that relate the mean and standard deviation of the random variable $\hat{p}_1 - \hat{p}_2$ to the population proportions, p_1 and p_2, and the sample sizes, n_1 and n_2. We have

$$\mu_{\hat{p}_1 - \hat{p}_2} = p_1 - p_2 \quad \text{and} \quad \sigma_{\hat{p}_1 - \hat{p}_2} = \sqrt{p_1(1 - p_1)/n_1 + p_2(1 - p_2)/n_2}.$$

These two formulas are derived in almost the same way as the formulas for the mean and standard deviation of the random variable $\bar{x}_1 - \bar{x}_2$. See Exercise 11.52 for details.

Finally, as we know from Key Fact 11.1 on page 656 (the sampling distribution of the proportion), the random variables $\hat{p}_1$ and $\hat{p}_2$ are both approximately normally distributed for large sample sizes. From this it can be shown that for independent samples, the random variable $\hat{p}_1 - \hat{p}_2$ is also approximately normally distributed. Hence we can state the following fact.

THE SAMPLING DISTRIBUTION OF THE DIFFERENCE BETWEEN TWO PROPORTIONS (LARGE AND INDEPENDENT SAMPLES)

Suppose a random sample of size n_1 is to be taken from a two-category population with population proportion p_1, and a random sample of size n_2 is to be taken from a two-category population with population proportion p_2. Further suppose the two samples are to be selected independently. Then for large samples, the random variable $\hat{p}_1 - \hat{p}_2$ is approximately normally distributed and has mean $\mu_{\hat{p}_1 - \hat{p}_2} = p_1 - p_2$ and standard deviation $\sigma_{\hat{p}_1 - \hat{p}_2} = \sqrt{p_1(1 - p_1)/n_1 + p_2(1 - p_2)/n_2}$. Thus the standardized random variable,

$$z = \frac{(\hat{p}_1 - \hat{p}_2) - (p_1 - p_2)}{\sqrt{p_1(1 - p_1)/n_1 + p_2(1 - p_2)/n_2}},$$

has approximately the standard normal distribution.

Key Fact 11.2 provides the necessary theory for deriving inferential procedures to compare the proportions of two two-category populations.

Large-Sample Hypothesis Tests for Two Population Proportions Using Independent Samples

We will now develop a hypothesis-testing procedure for comparing the proportions of two two-category populations. Our immediate goal is to employ Key Fact 11.2 to identify a random variable that can be used as the test statistic.

The null hypothesis for a hypothesis test to compare the proportions of two two-category populations will be

$$H_0: p_1 = p_2 \text{ (population proportions are equal)}.$$

If the null hypothesis is true, then $p_1 - p_2 = 0$, and so the standardized random variable in Key Fact 11.2 becomes

$$z = \frac{\hat{p}_1 - \hat{p}_2}{\sqrt{p(1 - p)/n_1 + p(1 - p)/n_2}},$$

where p denotes the common value of p_1 and p_2. Factoring $p(1-p)$ out of the denominator of the preceding expression yields the random variable

(4)
$$z = \frac{\hat{p}_1 - \hat{p}_2}{\sqrt{p(1 - p)}\sqrt{(1/n_1) + (1/n_2)}}.$$

We cannot use this random variable as the test statistic since p is unknown. Consequently, we must estimate p using sample information. The best estimate of p is obtained by pooling the data to get the proportion of successes in both samples combined; that is, we estimate p by

$$\hat{p}_p = \frac{x_1 + x_2}{n_1 + n_2}.$$

We call $\hat{p}_p$ the **pooled sample proportion.**

Replacing p in Equation (4) by its estimate, $\hat{p}_p$, yields the random variable

$$\frac{\hat{p}_1 - \hat{p}_2}{\sqrt{\hat{p}_p(1 - \hat{p}_p)}\sqrt{(1/n_1) + (1/n_2)}}.$$

This random variable *can* be used as the test statistic and, like the random variable in Equation (4), has approximately the standard normal distribution for large samples if the null hypothesis is true. Therefore we have the following procedure, which for ease of reference we sometimes refer to as the **two-sample z-test** for two population proportions.

PROCEDURE 11.3   **THE TWO-SAMPLE z-TEST FOR TWO POPULATION PROPORTIONS WITH NULL HYPOTHESIS $H_0\colon p_1 = p_2$**

ASSUMPTIONS

1. Independent samples
2. $x_1, n_1 - x_1, x_2$, and $n_2 - x_2$ are all at least 5

Step 1   State the null and alternative hypotheses.

Step 2   Decide on the significance level, α.

Step 3   The critical value(s)
  a. for a two-tailed test are $\pm z_{\alpha/2}$.
  b. for a left-tailed test is $-z_\alpha$.
  c. for a right-tailed test is z_α.
  Use Table II to find the critical value(s).

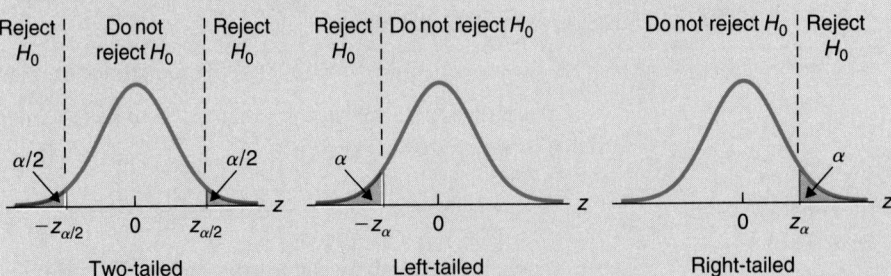

Step 4   Compute the value of the test statistic

$$z = \frac{\hat{p}_1 - \hat{p}_2}{\sqrt{\hat{p}_p(1 - \hat{p}_p)}\sqrt{(1/n_1) + (1/n_2)}},$$

where $\hat{p}_p = (x_1 + x_2)/(n_1 + n_2)$.

Step 5   If the value of the test statistic falls in the rejection region, then reject H_0; otherwise, do not reject H_0.

Step 6   State the conclusion in words.

Example 11.6 Illustrates Procedure 11.3

We now return to the problem posed in Example 11.5. Independent random samples of 20,650 U.S. males and 22,352 U.S. females were obtained in order to compare the percentage of males who smoke cigarettes to the percentage of females who smoke cigarettes. Of the males sampled, 5803 were cigarette smokers; of the females sampled, 5253 were cigarette smokers. At the 5% significance level, do the data provide sufficient evidence to conclude that the percentage of all U.S. males who smoke cigarettes exceeds the percentage of all U.S. females who smoke cigarettes?

SOLUTION   We apply Procedure 11.3, noting first that both assumptions for its use are satisfied.

Step 1   *State the null and alternative hypotheses.*

Let p_1 and p_2 denote, respectively, the proportions of all U.S. males and all U.S. females who smoke cigarettes. Then the null and alternative hypotheses are

$$H_0: p_1 = p_2 \text{ (percentage of male smokers is not higher)}$$
$$H_a: p_1 > p_2 \text{ (percentage of male smokers is higher).}$$

Note that the hypothesis test is right-tailed since a greater-than sign ($>$) appears in the alternative hypothesis.

Step 2   *Decide on the significance level, α.*

The test is to be performed at the 5% significance level; so $\alpha = 0.05$.

Step 3   *The critical value for a right-tailed test is z_α.*

Since $\alpha = 0.05$, the critical value is $z_{0.05} = 1.645$, as seen in Fig. 11.3.

FIGURE 11.3
Criterion for deciding whether or not to reject the null hypothesis

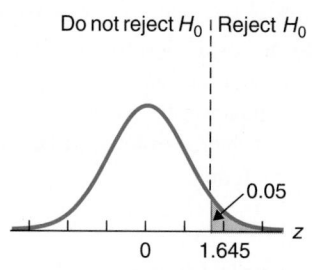

Do not reject H_0 ⋮ Reject H_0

0.05

0   1.645   z

Step 4   *Compute the value of the test statistic*

$$z = \frac{\hat{p}_1 - \hat{p}_2}{\sqrt{\hat{p}_p(1 - \hat{p}_p)}\sqrt{(1/n_1) + (1/n_2)}},$$

where $\hat{p}_p = (x_1 + x_2)/(n_1 + n_2)$.

We first obtain $\hat{p}_1$, $\hat{p}_2$, and $\hat{p}_{\mathrm{p}}$. Since 5803 of the 20,650 males sampled and 5253 of the 22,352 females sampled were found to be cigarette smokers, we have $x_1 = 5803$, $n_1 = 20,650$, $x_2 = 5253$, and $n_2 = 22,352$. Therefore

$$\hat{p}_1 = \frac{x_1}{n_1} = \frac{5803}{20,650} = 0.281, \qquad \hat{p}_2 = \frac{x_2}{n_2} = \frac{5253}{22,352} = 0.235,$$

and

$$\hat{p}_{\mathrm{p}} = \frac{x_1 + x_2}{n_1 + n_2} = \frac{5803 + 5253}{20,650 + 22,352} = \frac{11,056}{43,002} = 0.257.$$

Consequently, the value of the test statistic is

$$z = \frac{\hat{p}_1 - \hat{p}_2}{\sqrt{\hat{p}_{\mathrm{p}}(1 - \hat{p}_{\mathrm{p}})}\sqrt{(1/n_1) + (1/n_2)}}$$

$$= \frac{0.281 - 0.235}{\sqrt{(0.257)(1 - 0.257)}\sqrt{(1/20,650) + (1/22,352)}} = 10.91.$$

Step 5   *If the value of the test statistic falls in the rejection region, reject H_0; otherwise, do not reject H_0.*

From Step 4 the value of the test statistic is $z = 10.91$, which, as we see from Fig. 11.3, falls in the rejection region. Thus we reject H_0.

Step 6   *State the conclusion in words.*

The test results are statistically significant at the 5% level; that is, at the 5% significance level, the data provide sufficient evidence to conclude that the percentage of U.S. males who smoke cigarettes is greater than the percentage of U.S. females who smoke cigarettes.   ∎

We can also carry out the hypothesis test in Example 11.6 using the P-value approach. From Step 4 we see that the value of the test statistic is $z = 10.91$. Consulting Table II, we find that the P-value for the hypothesis test is 0 to four decimal places. Since the P-value is less than 0.05, we can reject the null hypothesis at the designated 5% significance level. Moreover, by referring to Table 9.12 on page 530, we see that the data provide very strong evidence against the null hypothesis and hence in favor of the alternative hypothesis that a higher percentage of males smoke than females.

Large-Sample Confidence Intervals for the Difference Between Two Population Proportions Using Independent Samples

Key Fact 11.2 on page 673 can also be used to derive the following confidence-interval procedure for the difference between two population proportions. We sometimes refer to this procedure as the **two-sample z-interval procedure** for the difference between two population proportions. See Exercise 11.53 for details of the derivation.

| PROCEDURE 11.4 | **THE TWO-SAMPLE z-INTERVAL PROCEDURE FOR THE DIFFERENCE BETWEEN TWO POPULATION PROPORTIONS** |

ASSUMPTIONS

1. Independent samples
2. $x_1, n_1 - x_1, x_2,$ and $n_2 - x_2$ are all at least 5

Step 1 For a confidence level of $1 - \alpha$, use Table II to find $z_{\alpha/2}$.

Step 2 The endpoints of the confidence interval for $p_1 - p_2$ are

$$(\hat{p}_1 - \hat{p}_2) \pm z_{\alpha/2} \cdot \sqrt{\hat{p}_1(1 - \hat{p}_1)/n_1 + \hat{p}_2(1 - \hat{p}_2)/n_2}.$$

Example 11.7 Illustrates Procedure 11.4

Refer to the study on smoking considered in Examples 11.5 and 11.6. Use the sample data to obtain a 90% confidence interval for the difference, $p_1 - p_2$, between the proportions of U.S. males and U.S. females who smoke cigarettes.

SOLUTION We apply Procedure 11.4, noting first that both conditions for its use are met.

Step 1 *For a confidence level of $1 - \alpha$, use Table II to find $z_{\alpha/2}$.*

For a 90% confidence interval, $\alpha = 0.10$. Consulting Table II, we find that $z_{\alpha/2} = z_{0.10/2} = z_{0.05} = 1.645$.

Step 2 *The endpoints of the confidence interval for $p_1 - p_2$ are*

$$(\hat{p}_1 - \hat{p}_2) \pm z_{\alpha/2} \cdot \sqrt{\hat{p}_1(1 - \hat{p}_1)/n_1 + \hat{p}_2(1 - \hat{p}_2)/n_2}.$$

From Step 1, $z_{\alpha/2} = 1.645$. Referring to Example 11.6, we find that $\hat{p}_1 = 0.281$, $n_1 = 20{,}650$, $\hat{p}_2 = 0.235$, and $n_2 = 22{,}352$. Therefore the endpoints of the confidence interval for $p_1 - p_2$ are

$$(0.281 - 0.235) \pm 1.645 \cdot \sqrt{(0.281)(1 - 0.281)/20{,}650 + (0.235)(1 - 0.235)/22{,}352}$$

or 0.046 ± 0.007. Thus a 90% confidence interval for $p_1 - p_2$ is from 0.039 to 0.053. We can be 90% confident that the difference between the proportions of U.S. males and U.S. females who smoke cigarettes is somewhere between 0.039 and 0.053. In other words, we can be 90% confident that the percentage of U.S. males who smoke cigarettes exceeds the percentage of U.S. females who smoke cigarettes by at least 3.9 percentage points but by no more than 5.3 percentage points. ∎

Margin of Error and Sample Size

The **margin of error** in estimating the difference between two population proportions can be obtained by referring to Step 2 of Procedure 11.4. And from the formula for the margin of error, we can determine the sample sizes required to obtain a confidence interval with a specified confidence level and margin of error. Formula 11.3 supplies the formulas.

FORMULA 11.3   **MARGIN OF ERROR AND SAMPLE SIZE FOR ESTIMATING $p_1 - p_2$**

> The margin of error for the estimate of $p_1 - p_2$ is
>
> $$E = z_{\alpha/2} \cdot \sqrt{\hat{p}_1(1 - \hat{p}_1)/n_1 + \hat{p}_2(1 - \hat{p}_2)/n_2}.$$
>
> The margin of error is equal to half the length of the confidence interval and represents the precision with which the difference between the sample proportions, $\hat{p}_1 - \hat{p}_2$, estimates the difference between the population proportions, $p_1 - p_2$, at the specified confidence level.
>
> A $(1 - \alpha)$-level confidence interval for the difference between two population proportions having a margin of error of at most E can be obtained by choosing
>
> $$n_1 = n_2 = 0.5 \left(\frac{z_{\alpha/2}}{E} \right)^2,$$
>
> rounded up to the nearest whole number. If we can make educated guesses, $\hat{p}_{1g}$ and $\hat{p}_{2g}$, for the observed values of $\hat{p}_1$ and $\hat{p}_2$, then we should instead choose
>
> $$n_1 = n_2 = \left(\hat{p}_{1g}(1 - \hat{p}_{1g}) + \hat{p}_{2g}(1 - \hat{p}_{2g}) \right) \left(\frac{z_{\alpha/2}}{E} \right)^2,$$
>
> rounded up to the nearest whole number.

The second displayed formula in Formula 11.3 provides sample sizes that ensure we will obtain a $(1 - \alpha)$-level confidence interval with a margin of error of at most E, but it may yield sample sizes that are unnecessarily large. The third displayed formula in Formula 11.3 yields smaller sample sizes, but should not be used unless the guesses for the sample proportions are considered reasonably accurate. If likely ranges for the observed values of the two sample proportions are known, then the values in the ranges closest to 0.5 should be taken as the educated guesses. For further discussion of these ideas and for applications of Formula 11.3, see Exercise 11.50.

EXERCISES 11.3

11.36   Consider a hypothesis test for two population proportions with null hypothesis H_0: $p_1 = p_2$. What parameter is being estimated by the pooled sample proportion, $\hat{p}_p$?

11.37   Of the quantities $p_1, p_2, x_1, x_2, \hat{p}_1, \hat{p}_2$, and $\hat{p}_p$,
a. which represent parameters and which represent statistics?
b. which are fixed numbers and which are random variables?

For each of Exercises 11.38–11.43, perform the required hypothesis test using either the classical approach or the P-value approach.

11.38   Roughly 450,000 vasectomies are performed each year in the United States. In this surgical procedure for contraception, the tube carrying sperm from the testicles is cut.

Several studies have been conducted to analyze the relationship between vasectomies and prostate cancer. One such study appeared in a February 1993 issue of the *Journal of the American Medical Association*. The following problem is based on data presented in that journal article. Of 21,300 men who have not had a vasectomy, 69 were found to have prostate cancer; and of 22,000 men who have had a vasectomy, 113 were found to have prostate cancer.

a. At the 1% significance level, do the data provide sufficient evidence to conclude that men who have had a vasectomy are at greater risk of having prostate cancer?

b. Is this a designed experiment or an observational study? Explain your answer.

c. In view of your answers to parts (a) and (b), would you say that it is reasonable to conclude that having a vasectomy causes an increased risk of prostate cancer? Why?

11.39 For several years, evidence has been mounting that folic acid reduces major birth defects. In December 1992, *The Arizona Republic* reported on a Hungarian study that provides the strongest evidence yet. The results of the study, directed by Dr. Andrew E. Czeizel and Dr. Istvan Dudas of the National Institute of Hygiene in Budapest, were published in the *New England Journal of Medicine*. For the study, the doctors enrolled 4753 women prior to conception. The women were divided randomly into two groups. One group, consisting of 2701 women, took daily multivitamins containing 0.8 mg of folic acid; the other group, consisting of 2052 women, received only trace elements. Major birth defects occurred in 35 cases where the women took folic acid and in 47 cases where the women did not.

a. At the 1% significance level, do the data provide sufficient evidence to conclude that women who take folic acid are at lesser risk of having children with major birth defects?

b. Is this a designed experiment or an observational study? Explain your answer.

c. In view of your answers to parts (a) and (b), would you say that it is reasonable to conclude that taking folic acid causes a reduction in major birth defects? Explain your answer.

11.40 The Organization for Economic Cooperation and Development, Paris, France, summarizes data on labor-force participation rates in *Labour Force Statistics*. From independent samples of 300 U.S. women and 250 Canadian women, it is found that 184 of the U.S. women and 148 of the Canadian women are in their respective labor forces. At the 5% significance level, do the data suggest that there is a difference between the labor-force participation rates of U.S. and Canadian women?

11.41 Annual surveys are performed by the U.S. Bureau of the Census to obtain estimates of the percentage of the voting-age population that has registered to vote. The information from those surveys is published in *Current Population Reports*. Independent random samples were taken of 400 employed people and 450 unemployed people. It was found that 262 of the employed people and 224 of the unemployed people had registered to vote. Can we conclude, at the 5% significance level, that the percentage of employed workers who have registered to vote exceeds the percentage of unemployed workers who have registered to vote?

11.42 Suppose we tell you that the percentage of adult U.S. males who are married exceeds the percentage of adult U.S. females who are married. Further suppose that to check this claim, you randomly select 550 adult U.S. males and 575 adult U.S. females. You find that 367 of the males and 353 of the females are married. Do your data provide sufficient evidence, at the 5% significance level, to support our claim? Explain your answer.

11.43 Information on U.S. physicians and dentists are compiled, respectively, by the American Medical Association and the American Dental Association. A random sample of 300 U.S. dentists contains 77 who practice in the Northeast; a random sample of 300 U.S. physicians contains 82 who practice in the Northeast. Do these data suggest that the percentage of U.S. dentists who practice in the Northeast is smaller than the percentage of U.S. physicians who practice in the Northeast? Use $\alpha = 0.10$.

11.44 Refer to Exercise 11.38.

a. Determine a 98% confidence interval for the difference between the prostate cancer rates of males who have not had a vasectomy and males who have.

b. Interpret your result from part (a).

11.45 Refer to Exercise 11.39.

a. Determine a 98% confidence interval for the difference between the rates of major birth defects for babies born to women who have taken folic acid and those born to women who have not taken folic acid.

b. Interpret your result from part (a).

11.46 Refer to Exercise 11.40.

a. Find a 95% confidence interval for the difference, $p_1 - p_2$, between the labor-force participation rates of U.S. and Canadian women.

b. Interpret your result from part (a).

11.47   Refer to Exercise 11.41.

a. Find a 90% confidence interval for the difference, $p_1 - p_2$, between the proportions of employed and unemployed workers who have registered to vote.

b. Interpret your result from part (a).

11.48   Refer to Exercise 11.42.

a. Determine a 90% confidence interval for the difference between the proportions of adult U.S. males and adult U.S. females who are married.

b. Interpret your result from part (a).

11.49   Refer to Exercise 11.43.

a. Determine an 80% confidence interval for the difference between the proportion of U.S. dentists who practice in the Northeast and the proportion of U.S. physicians who practice in the Northeast.

b. Interpret your result from part (a).

11.50   In this exercise we will apply Formula 11.3 on page 678 to the study on smoking considered in Examples 11.5–11.7.

a. Obtain the margin of error for the estimate of the difference between the proportions of male and female smokers by taking half the length of the confidence interval found in Example 11.7 on page 677. Interpret your result in words.

b. Obtain the margin of error for the estimate of the difference between the proportions of male and female smokers by applying the first displayed formula in Formula 11.3.

c. Obtain the common sample size that will ensure a margin of error of at most 0.01 for a 90% confidence interval without making a guess for the observed values of the sample proportions.

d. Find a 90% confidence interval for $p_1 - p_2$ if for samples of the size determined in part (c), 27.9% of the males and 24.0% of the females smoke cigarettes.

e. Determine the margin of error for the estimate in part (d) and compare it to the required margin of error specified in part (c).

f. Repeat parts (c)–(e) if it is deemed reasonable to presume that at most 30% of the males sampled and at most 25% of the females sampled will be smokers.

g. Compare the results obtained in parts (c)–(e) with those obtained in part (f).

11.51 **(Computer exercise)**   In this exercise you will perform a computer simulation to illustrate the sampling distribution of the difference between two proportions for large and independent samples, Key Fact 11.2 on page 673. Euromonitor Publications Limited, London, compiles data from various countries on the percentage of households that own selected appliances. Results are published in *European Marketing Data and Statistics*. According to that document, 25% of Italian households and 41% of Swiss households own VCRs. Consider the independent random selection of 300 Italian households and 200 Swiss households, where each household chosen is classified as either owning a VCR (the specified attribute) or not. Use Minitab or some other statistical software to

a. simulate taking 1000 such samples from each country. *(Hint:* Simulate appropriate binomial random variables.*)*

b. determine the sample proportion for each of the 1000 samples from each country.

c. obtain the difference between each of the 1000 pairs of sample proportions found in part (b).

d. Obtain the mean, the standard deviation, and a histogram of the 1000 differences found in part (c).

e. Theoretically, what are the mean, standard deviation, and distribution of all possible differences, $\hat{p}_1 - \hat{p}_2$?

f. Compare your results from parts (d) and (e).

11.52   In this exercise we will establish the formulas presented on page 672 for the mean and standard deviation of the random variable $\hat{p}_1 - \hat{p}_2$.

a. Employ the results of Exercises 5.26(e) and 5.27(f), both on page 289, to show that

$$\mu_{\hat{p}_1 - \hat{p}_2} = \mu_{\hat{p}_1} - \mu_{\hat{p}_2}$$

and

$$\sigma_{\hat{p}_1 - \hat{p}_2} = \sqrt{\sigma_{\hat{p}_1}^2 + \sigma_{\hat{p}_2}^2}.$$

b. Use the formulas $\mu_{\hat{p}} = p$ and $\sigma_{\hat{p}} = \sqrt{p(1-p)/n}$ from Section 11.1 and the results from part (a) to derive the formulas

$$\mu_{\hat{p}_1 - \hat{p}_2} = p_1 - p_2$$

and

$$\sigma_{\hat{p}_1 - \hat{p}_2} = \sqrt{p_1(1 - p_1)/n_1 + p_2(1 - p_2)/n_2}.$$

11.53   This exercise justifies Procedure 11.4.

a. Use Key Fact 11.2 on page 673 to explain why the random variable

$$\frac{(\hat{p}_1 - \hat{p}_2) - (p_1 - p_2)}{\sqrt{\hat{p}_1(1 - \hat{p}_1)/n_1 + \hat{p}_2(1 - \hat{p}_2)/n_2}}$$

has approximately the standard normal distribution for large samples.

b. Use part (a) to derive the confidence-interval formula in Step 2 of Procedure 11.4 on page 677.

Chapter Review

Key Terms

Formulas

In the formulas below,

p = population proportion
$\hat{p}$ = sample proportion
x = number of successes
n = sample size

p_1, p_2 = population proportions
$\hat{p}_1, \hat{p}_2$ = sample proportions
x_1, x_2 = numbers of successes
n_1, n_2 = sample sizes

Sample proportion, 655

$$\hat{p} = \frac{x}{n}$$

One-sample z-interval for p, 657

$$\hat{p} \pm z_{\alpha/2} \cdot \sqrt{\hat{p}(1 - \hat{p})/n}$$

(*Assumption:* both x and $n - x$ are at least 5)

Margin of error for the estimate of p, 659

$$E = z_{\alpha/2} \cdot \sqrt{\hat{p}(1 - \hat{p})/n}$$

Sample size for estimating p, 660

$$n = 0.25 \left(\frac{z_{\alpha/2}}{E} \right)^2 \quad \text{or} \quad n = \hat{p}_g(1 - \hat{p}_g) \left(\frac{z_{\alpha/2}}{E} \right)^2$$

rounded up to the nearest whole number (g = "educated guess")

One-sample z-test statistic for H_0: $p = p_0$, 667

$$z = \frac{\hat{p} - p_0}{\sqrt{p_0(1 - p_0)/n}}$$

(*Assumption:* both np_0 and $n(1 - p_0)$ are at least 5)

Pooled sample proportion, *674*

$$\hat{p}_p = \frac{x_1 + x_2}{n_1 + n_2}$$

Two-sample z-test statistic for H_0: $p_1 = p_2$, *674*

$$z = \frac{\hat{p}_1 - \hat{p}_2}{\sqrt{\hat{p}_p(1 - \hat{p}_p)}\sqrt{(1/n_1) + (1/n_2)}}$$

(*Assumptions:* independent samples; x_1, $n_1 - x_1$, x_2, $n_2 - x_2$ are all at least 5)

Two-sample z-interval for $p_1 - p_2$, *677*

$$(\hat{p}_1 - \hat{p}_2) \pm z_{\alpha/2} \cdot \sqrt{\hat{p}_1(1 - \hat{p}_1)/n_1 + \hat{p}_2(1 - \hat{p}_2)/n_2}$$

(*Assumptions:* independent samples; x_1, $n_1 - x_1$, x_2, $n_2 - x_2$ are all at least 5)

Margin of error for the estimate of $p_1 - p_2$, *678*

$$E = z_{\alpha/2} \cdot \sqrt{\hat{p}_1(1 - \hat{p}_1)/n_1 + \hat{p}_2(1 - \hat{p}_2)/n_2}$$

Sample size for estimating $p_1 - p_2$, *678*

$$n_1 = n_2 = 0.5 \left(\frac{z_{\alpha/2}}{E} \right)^2 \quad \text{or} \quad n_1 = n_2 = \left(\hat{p}_{1g}(1 - \hat{p}_{1g}) + \hat{p}_{2g}(1 - \hat{p}_{2g}) \right) \left(\frac{z_{\alpha/2}}{E} \right)^2$$

rounded up to the nearest whole number (g = "educated guess")

You Should Be Able To

1. use and understand each of the preceding formulas.
2. find a large-sample confidence interval for a population proportion, p.
3. compute the margin of error for the estimate of p.
4. understand the relationship between the sample size, confidence level, and margin of error for a confidence interval for p.
5. determine the sample size required for a specified confidence level and margin of error for the estimate of p.
6. perform a large-sample hypothesis test for a population proportion, p.
7. perform large-sample inferences (hypothesis tests and confidence intervals) to compare the proportions, p_1 and p_2, of two two-category populations.
8. understand the relationship between the sample sizes, confidence level, and margin of error for a confidence interval for $p_1 - p_2$.
9. determine the sample sizes required for a specified confidence level and margin of error for the estimate of $p_1 - p_2$.
10. use the Minitab commands covered in this chapter.*

REVIEW TEST

1. Suppose a large random sample of size n is to be taken from a two-category population having population proportion p.
   a. Identify the approximate probability distribution of the random variable $\hat{p}$.
   b. How do we determine probabilities for the random variable in part (a)?

*2. **(Computer problem)** According to a study by the Alan Guttmacher Institute, reported on in the *New York Times,* approximately 20% of all Americans are infected with a viral sexually transmitted disease (STD), such as herpes or hepatitis B. Consider the random selection of 350 Americans, where each person chosen is classified as either being infected with a viral STD (the specified attribute) or not. Use Minitab or some other statistical software to
   a. simulate the taking of 1500 such samples.
   b. find the sample proportion of each of the 1500 samples.
   c. determine the mean, the standard deviation, and a histogram of the 1500 sample proportions.
   d. Theoretically, what are the mean, standard deviation, and distribution of all possible sample proportions for samples of size 350 from the population?
   e. Compare your results from parts (c) and (d).

3. Between June 11 and 15, 1993, the Times Mirror Center for People and the Press interviewed 1006 adults concerning their views on media treatment of then recently inaugurated President Clinton. According to an article in the *Los Angeles Times,* "...a notable 43 percent of Americans [surveyed (433 of those sampled)] said that they thought news organizations were criticizing the Clinton administration unfairly."
   a. Find a 95% confidence interval for the proportion, p, of all Americans who thought news organizations were criticizing the Clinton administration unfairly.
   b. Interpret your result from part (a) in words.

4. Refer to Problem 3.
   a. Find the margin of error for the estimate of p.
   b. Obtain a sample size that will ensure a margin of error of at most 0.02 for a 95% confidence interval without making a guess for the observed value of $\hat{p}$.
   c. Find a 95% confidence interval for p if for a sample of the size determined in part (b), 41.6% of those surveyed thought news organizations were criticizing the Clinton administration unfairly.

   d. Determine the margin of error for the estimate in part (c) and compare it to the required margin of error specified in part (b).
   e. Repeat parts (b)–(d) if it is deemed reasonable to presume that the percentage of those surveyed who thought news organizations were criticizing the Clinton administration unfairly will be at most 45%.
   f. Compare the results obtained in parts (b)–(d) with those obtained in part (e).

5. A poll of Americans conducted by Gallup for the CNBC cable network found that only 17% of those surveyed felt they had achieved the American dream. The poll was taken by phone May 21–23, 1993, and had a margin of error of plus or minus 4 percentage points (for a 0.95 confidence level). Use this information to obtain a 95% confidence interval for the percentage of all Americans who feel they have achieved the American dream.

6. In the April, 18, 1993, issue of *Parade Magazine,* the editors reported on a national survey on law and order. One question asked of the 2512 U.S. adults taking part was whether they believed that juries "almost always" convict the guilty and free the innocent. Only 578 said they did. Do the data provide sufficient evidence to conclude that less than one in four Americans believe that juries "almost always" convict the guilty and free the innocent. Perform the appropriate hypothesis test at the 5% significance level using the
   a. classical approach to hypothesis testing.
   b. *P*-value approach to hypothesis testing.
   c. Assess the strength of the evidence against the null hypothesis. (Refer to Table 9.12 on page 530.)

7. An article published in the September 1992 issue of the *Annals of Epidemiology* discussed the relationship between height and breast cancer. The study by the National Cancer Institute, which took 5 years and involved more than 1500 women having breast cancer and 2000 women not having breast cancer, revealed that there is a trend between height and breast cancer: "...taller women have a 50 to 80 percent greater risk of getting breast cancer than women who are closer to 5 feet tall." But Christine Swanson, a nutritionist who was involved with the study, added that "...height may be associated with the culprit, ... but no one really knows" the exact relationship between height and breast-cancer risk.
   a. Classify this study as either an observational study or a designed experiment. Explain your answer.

b. Interpret the statement made by Christine Swanson in view of your answer to part (a).

8. State and local governments often poll their constituents about their views on the economy. In February 1992 and again in January 1993, O'Neil Associates asked 600 Maricopa County (Arizona) residents whether they thought the state's economy would improve over the next 2 years. In the 1992 poll, 48% said yes; in the 1993 poll, 60% said yes. At the 1% significance level, do the data provide sufficient evidence to conclude that the percentage of Maricopa County residents who thought the state's economy would improve over the next 2 years was less during the time of the 1992 poll than during the time of the 1993 poll? Perform the required hypothesis test using the
   a. classical approach to hypothesis testing.
   b. *P*-value approach to hypothesis testing.
   c. Assess the strength of the evidence against the null hypothesis. (Refer to Table 9.12 on page 530.)

9. Refer to Problem 8.
   a. Determine a 98% confidence interval for the difference, $p_1 - p_2$, between the proportions of Maricopa County residents who thought that the state's economy would improve over the next 2 years during the time of the 1992 poll and during the time of the 1993 poll.
   b. Interpret your result from part (a).

10. Refer to Problems 8 and 9.
   a. Take half the length of the confidence interval found in Problem 9(a) to obtain the margin of error for the estimate of the difference, $p_1 - p_2$, between the two proportions. Interpret your result in words.
   b. Solve part (a) by applying the first displayed formula in Formula 11.3 on page 678.
   c. Obtain the common sample size that will ensure a margin of error of at most 0.03 for a 98% confidence interval without making a guess for the observed values of the sample proportions.
   d. Find a 98% confidence interval for $p_1 - p_2$ if for samples of the size determined in part (c), the sample proportions are 0.475 and 0.603, respectively.
   e. Determine the margin of error for the estimate in part (d) and compare it to the required margin of error specified in part (c).

Using the Focus Database

The database printed in Appendix B contains data on 500 randomly selected Arizona State University sophomores. Among the variables considered are high-school GPA, SAT math score, cumulative GPA, SAT verbal score, age, and total hours. Use Minitab or some other statistical software to (help) solve the following problems.
   a. Obtain a 95% confidence interval for the percentage of all Arizona State University sophomores whose cumulative GPAs are at least 3.00. Interpret your result.
   b. Referring to part (a), determine the margin of error for the estimate. Interpret your answer.
   c. Do the data provide sufficient evidence to conclude that more than 20% of Arizona State University sophomores score 600 or over on the SAT math? Perform the required hypothesis test at the 5% level of significance.
   d. Can we conclude that a difference exists between the proportions of male and female sophomores at Arizona State University who score 500 or over on the SAT verbal? Perform the required hypothesis test at the 5% significance level.
   e. Determine a 95% confidence interval for the difference between the proportions of male and female sophomores at Arizona State University who score 500 or over on the SAT verbal.

Case Study Doctors Polled on AIDS Views

In the November 27, 1991, *Journal of the American Medical Association,* a study was published whose aim was to identify barriers for physicians in treating patients infected with HIV (human immunodeficiency virus), the virus responsible for AIDS.

According to Barbara Gerbert, lead author and head of behavioral sciences in the School of Dentistry at the University of California at San Francisco, those barriers include fear of homosexuals (who constitute two-thirds of all AIDS cases) and aversion to treating intravenous-drug users. Indeed, 35% of the doctors surveyed said they "would feel nervous among a group of homosexuals" and 55% indicated they would be uncomfortable treating IV-drug users.

The survey of 1121 randomly selected general-care physicians, taken in 1990, further revealed that 75% had treated at least one patient with HIV virus and that 68% believed they had a responsibility to treat people with HIV infection (although half said they would not, given the choice). An editorial accompanying the study pointed out that if 68% feel responsible to treat HIV-infected patients, then almost one-third "perceive no ethical difficulty with denying medical care" to them.

Practical barriers for physicians in treating HIV-infected patients were also identified: such treatment is "very time-consuming ... and over 80% of physicians feel they don't have enough knowledge to do it." As a matter of fact, 84% of those surveyed said that treating the disease is burdensome on a doctor's time and 83% revealed that they need to know more about AIDS.

a. Obtain a 95% confidence interval for the percentage of all general-care physicians who do not feel a responsibility to treat people with HIV infection.
b. Interpret your result from part (a).
c. Find the margin of error for the estimate in part (a). Interpret your answer.

**Karl
Pearson**

Karl Pearson was born on March 27, 1857, in London, the second son of William Pearson, a prominent lawyer, and his wife, Fanny Smith. Karl Pearson's early education took place at home. At the age of 9, he was sent to University College School in London, where he remained for the next 7 years. Because of ill health, Pearson was then privately tutored for a year. He received a scholarship at King's College, Cambridge, in 1875. There he earned a B.A. (with honors) in mathematics in 1879 and an M.A. in law in 1882. He then studied physics and metaphysics in Heidelberg, Germany.

In addition to his expertise in mathematics, law, physics, and metaphysics, Pearson was competent in literature and knowledgeable about German history, folklore, and philosophy. He was also considered somewhat of a political radical because of his interest in the ideas of Karl Marx and the rights of women.

In 1884 Pearson was appointed Goldsmid professor of applied mathematics and mechanics at University College; from 1891–1894 he was also a lecturer in geometry at Gresham College, London.

In 1911 he gave up the Goldsmid chair to become the first Galton professor of Eugenics at University College. Pearson was elected to the Royal Society, a prestigious association of scientists, in 1896 and awarded the society's Darwin Medal in 1898.

Pearson really began his pioneering work in statistics in 1893, mainly through an association with Walter Weldon (a zoology professor at University College), Francis Edgeworth (a professor of logic at University College), and Sir Francis Galton (see the Chapter 14 biography). An analysis of published data on roulette wheels at Monte Carlo led to Pearson's discovery of the chi-square goodness-of-fit test. He also coined the term "standard deviations," introduced his amazingly diverse skew curves, and developed the most widely used measure of correlation, the correlation coefficient.

Pearson, Weldon, and Galton co-founded the statistical journal *Biometrika,* of which Pearson was editor (1901–1936) and a major contributor. Pearson retired from University College in 1933. He died in London on April 27, 1936.

Chapter 12

Chi-Square Procedures

. . . . .

General Objective

The statistical-inference techniques presented so far have dealt exclusively with hypothesis tests and confidence intervals for means and proportions. In this chapter we will consider three other widely used inferential procedures. These three procedures are often referred to as **chi-square procedures** because they rely on a continuous probability distribution called the *chi-square distribution.*

First we will study the chi-square goodness-of-fit test. This test is used to make inferences about the percentage distribution of a population or the probability distribution of a random variable. For instance, we could apply the chi-square goodness-of-fit test to a sample of university students to decide whether the political-preference distribution of all university students differs from that of the population as a whole.

Next we will examine the chi-square independence test. This test is employed to decide whether an association exists between two characteristics of a population. For example, we could apply the chi-square independence test to a sample of adults to decide whether an association exists between annual income and educational level.

Finally, we will discuss chi-square inferences for a population standard deviation. We could use these inferential methods, for instance, to perform a hypothesis test or find a confidence interval for the standard deviation of the amounts of coffee being dispensed by a coffee machine.

Chapter Outline

12.1 The Chi-Square Distribution

12.2 Chi-Square Goodness-of-Fit Test

12.3 Chi-Square Independence Test

12.4 Inferences for a Population Standard Deviation

Case Study

Time On or Off the Job: Which Do Americans Enjoy More?

Do Americans enjoy their hours on the job more than their time spent off the job? Do preferences depend on sex? age? amount of education? A May 1993 CNN/*USA Today* poll conducted by The Gallup Organization attempted to answer these and other questions. We will investigate the results of that poll in the case study at the end of this chapter.

12.1   THE CHI-SQUARE DISTRIBUTION

The statistical-inference procedures discussed in this chapter all rely on a class of continuous probability distributions called **chi-square distributions.** Probabilities for a random variable having a chi-square distribution are equal to areas under a curve, aptly called a **chi-square (χ^2) curve.**

Actually there are infinitely many χ^2-curves, and we identify the χ^2-curve in question by stating its number of degrees of freedom, just as we did for t-curves. Figure 12.1 shows three different χ^2-curves. This figure illustrates some basic properties of χ^2-curves, which are presented in Key Fact 12.1.

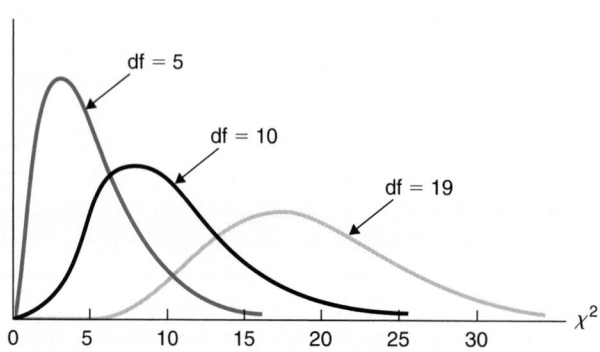

FIGURE 12.1
χ^2-curves for df = 5, 10, and 19

KEY FACT 12.1   **BASIC PROPERTIES OF χ^2-CURVES**

> *Property 1:* The total area under a χ^2-curve is equal to 1.
>
> *Property 2:* A χ^2-curve starts at 0 on the horizontal axis and extends indefinitely to the right, approaching, but never touching, the horizontal axis as it does so.
>
> *Property 3:* A χ^2-curve is not symmetric but is skewed to the right.
>
> *Property 4:* As the number of degrees of freedom becomes larger, χ^2-curves look increasingly like normal curves.

Using the χ^2-Table

To perform a hypothesis test or to obtain a confidence interval that is based on a chi-square distribution, we must be able to determine the χ^2-value that corresponds to a specified area under a χ^2-curve. Table V in Appendix A provides χ^2-values corresponding to several areas for various degrees of freedom.

The χ^2-table (Table V) is similar to the t-table (Table IV). The two outside columns of Table V, labeled df, display the number of degrees of freedom. As expected, the symbol χ_α^2 denotes the χ^2-value having area α to its right under a χ^2-curve. Thus the column headed $\chi^2_{0.995}$ contains χ^2-values having area 0.995 to their right; the column headed $\chi^2_{0.99}$ contains χ^2-values having area 0.99 to their right; and so on. Examples 12.1–12.3 explain how to use Table V.

Example 12.1 Finding the χ^2-Value Having a Specified Area to Its Right

For a χ^2-curve with 12 degrees of freedom, find $\chi^2_{0.025}$; that is, find the χ^2-value having area 0.025 to its right, as shown in Fig. 12.2(a).

FIGURE 12.2
Finding the χ^2-value having area 0.025 to its right

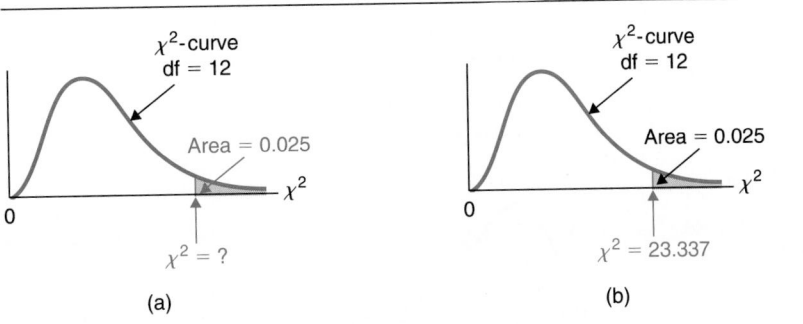

(a)   (b)

SOLUTION   To find this χ^2-value, we use Table V. Since the number of degrees of freedom is 12, we first go down the outside columns, labeled df, to "12." Then we go across that row until we are under the column headed $\chi^2_{0.025}$. The number in the body of the table there, 23.337, is the required χ^2-value; that is, for a χ^2-curve with df = 12, the χ^2-value having area 0.025 to its right is $\chi^2_{0.025} = 23.337$. Figure 12.2(b) shows this value.   ∎

Example 12.2 Finding the χ^2-Value Having a Specified Area to Its Left

Determine the χ^2-value having area 0.05 to its left for a χ^2-curve with df = 7, as seen in Fig. 12.3(a).

FIGURE 12.3
Finding the χ^2-value having area 0.05 to its left

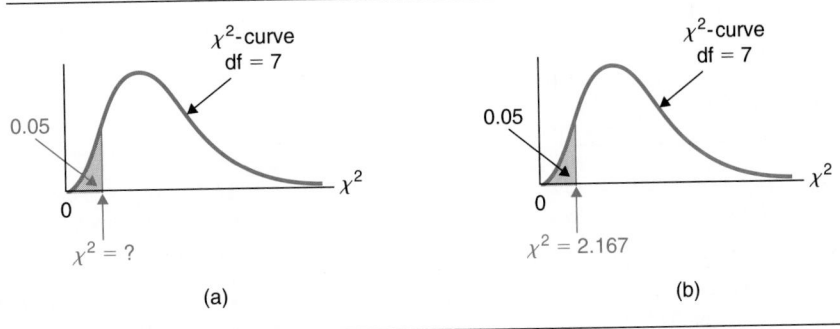

(a)   (b)

SOLUTION   Since the total area under a χ^2-curve is equal to 1 (Property 1 of Key Fact 12.1), the unshaded area in Fig. 12.3(a) must equal $1 - 0.05 = 0.95$. Thus the area to the right of the required χ^2-value is 0.95. This means that the required χ^2-value is $\chi^2_{0.95}$. From Table V we find that for df $= 7$, $\chi^2_{0.95} = 2.167$. Consequently, for a χ^2-curve with df $= 7$, the χ^2-value having area 0.05 to its left is 2.167, as shown in Fig. 12.3(b). ∎

Example 12.3   Finding the χ^2-Values for a Specified Area

For a χ^2-curve with df $= 20$, determine the two χ^2-values that divide the area under the curve into a middle 0.95 area and two outside 0.025 areas, as shown in Fig. 12.4(a).

FIGURE 12.4
Finding the two χ^2-values
dividing the area under the
curve into a middle 0.95 area
and two outside 0.025 areas

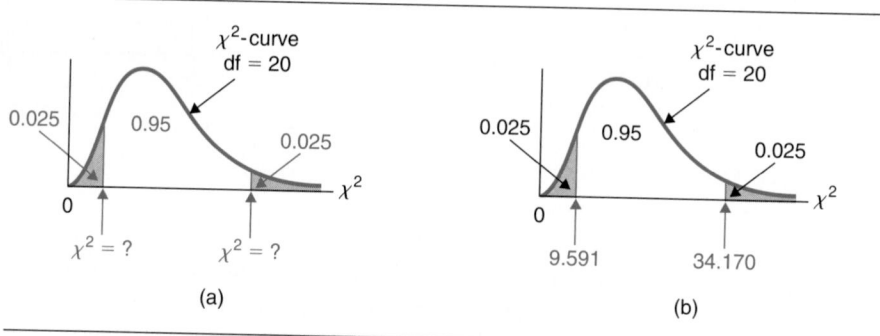

(a)   (b)

SOLUTION   First we obtain the χ^2-value on the right in Fig. 12.4(a). From the figure, we see that the shaded area on the right is 0.025. This means that the χ^2-value on the right is $\chi^2_{0.025}$. Consulting Table V, we find that for df $= 20$, $\chi^2_{0.025} = 34.170$.

Next we obtain the χ^2-value on the left in Fig. 12.4(a). Because the area to the left of that χ^2-value is 0.025, the area to its right is $1 - 0.025 = 0.975$. Hence the χ^2-value on the left is $\chi^2_{0.975}$, which, by Table V, equals 9.591 for df $= 20$.

Consequently, for a χ^2-curve with df $= 20$, the two χ^2-values that divide the area under the curve into a middle 0.95 area and two outside 0.025 areas are 9.591 and 34.170, as shown in Fig. 12.4(b). ∎

EXERCISES 12.1

In each of Exercises 12.1–12.8, use Table V to find the required χ^2-values. Illustrate your work graphically.

12.1   For a χ^2-curve with 19 degrees of freedom, find the χ^2-value having area
a. 0.025 to its right.   b. 0.95 to its right.

12.2   For a χ^2-curve with 22 degrees of freedom, find the χ^2-value having area
a. 0.01 to its right.   b. 0.995 to its right.

12.3   For a χ^2-curve with df $= 10$, determine
a. $\chi^2_{0.05}$.   b. $\chi^2_{0.975}$.

12.4 For a χ^2-curve with df $= 4$, determine
a. $\chi^2_{0.005}$.   b. $\chi^2_{0.99}$.

12.5 Consider a χ^2-curve with df $= 8$. Obtain the χ^2-value having area
a. 0.01 to its left.   b. 0.95 to its left.

12.6 Consider a χ^2-curve with df $= 16$. Obtain the χ^2-value having area
a. 0.025 to its left.   b. 0.975 to its left.

12.7 Determine the two χ^2-values that divide the area under the curve into a middle 0.95 area and two outside 0.025 areas for a χ^2-curve with
a. df $= 5$.   b. df $= 26$.

12.8 Determine the two χ^2-values that divide the area under the curve into a middle 0.90 area and two outside 0.05 areas for a χ^2-curve with
a. df $= 11$.   b. df $= 28$.

12.2 CHI-SQUARE GOODNESS-OF-FIT TEST

The first chi-square procedure we will discuss is the **chi-square goodness-of-fit test.** This procedure can be used to perform hypothesis tests about the percentage distribution of a population or the probability distribution of a random variable. Example 12.4 explains the reasoning behind the chi-square goodness-of-fit test.

Example 12.4   Introduces the Chi-Square Goodness-of-Fit Test

The American Medical Association compiles information on physicians and publishes its findings in *Physician Characteristics and Distribution in the U.S.* Physicians can be broadly classified into four categories: general practice, medical, surgical, and other.[†] Table 12.1 provides percentage and probability distributions for U.S. physicians in 1986 based on those four categories. The table shows, for instance, that in 1986, 18.0% of all U.S. physicians were in general practice, or equivalently, that the probability is 0.180 that a randomly selected 1986 U.S. physician was in general practice.

TABLE 12.1
Specialty distribution of
U.S. physicians, 1986

Specialty	Percentage	Probability, p
General practice	18.0	0.180
Medical	33.9	0.339
Surgical	27.0	0.270
Other	21.1	0.211
	100.0	1.000

A researcher wants to know whether this year's specialty distribution of U.S. physicians has changed from the 1986 distribution. To begin the study, he randomly selects

[†] The "medical" category includes internal medicine and pediatrics; the "surgical" category includes general surgery, obstetrics and gynecology, orthopedic surgery, and ophthalmology; the "other" category includes psychiatry, anesthesiology, pathology, and radiology.

500 U.S. physicians who are currently practicing medicine. Table 12.2 provides frequency and percentage distributions for the specialties of the 500 physicians.

Specialty	Frequency	Percentage
General practice	80	16.0
Medical	162	32.4
Surgical	156	31.2
Other	102	20.4
	500	100.0

The researcher plans to use the data in Table 12.2 to perform the hypothesis test

H_0: The current specialty distribution of U.S. physicians
is the same as the 1986 distribution.

H_a: The current specialty distribution of U.S. physicians
is different from the 1986 distribution.

The basic idea behind the hypothesis test is this: Compare the observed frequencies in the second column of Table 12.2 to the frequencies that would be expected if the current specialty distribution is the same as the 1986 specialty distribution. If the observed and expected frequencies match up fairly well, then do not reject the null hypothesis; otherwise, reject the null hypothesis.

To transform this idea into a precise procedure for performing the hypothesis test, we need to answer two questions: (1) What frequencies should we expect from a random sample of 500 U.S. physicians who are currently practicing medicine if the current specialty distribution is the same as the 1986 specialty distribution? (2) How do we decide whether the observed frequencies match up reasonably well with those that we would expect?

The first question is easy to answer. If the current specialty distribution is the same as the 1986 specialty distribution, then, for instance, 33.9% of all current U.S. physicians are in the medical category (see Table 12.1). Therefore in a sample of 500 U.S. physicians who are currently practicing medicine, we would expect about 33.9% of the 500, or 169.5, to be in the medical category.

In general, each expected frequency, denoted by the letter E, is computed using the formula

$$E = np,$$

where n is the sample size (in this case, 500) and p is the appropriate probability from the third column of Table 12.1. For instance, the expected frequency for the medical category is

$$E = np = 500 \cdot 0.339 = 169.5,$$

as we have already seen. The expected frequencies for all four specialties are calculated in Table 12.3.

Specialty	Probability p	Expected frequency $np = E$
General practice	0.180	$500 \cdot 0.180 = \;\;90.0$
Medical	0.339	$500 \cdot 0.339 = 169.5$
Surgical	0.270	$500 \cdot 0.270 = 135.0$
Other	0.211	$500 \cdot 0.211 = 105.5$

The third column of Table 12.3 provides the answer to our first question. It gives the frequencies we would expect if the current specialty distribution of U.S. physicians is the same as the 1986 specialty distribution.

The second question, whether the observed frequencies match up reasonably well with the expected frequencies, is harder to answer. We need to calculate a number that measures how good the fit is.

The second column of Table 12.4 repeats the **observed frequencies** from the second column of Table 12.2, that is, the frequencies actually obtained from a random sample of 500 U.S. physicians who are currently practicing medicine. The third column of Table 12.4 lists the **expected frequencies** from the third column of Table 12.3, that is, the frequencies we would expect from a random sample of 500 U.S. physicians who are currently practicing medicine if the current specialty distribution is the same as the 1986 specialty distribution.

Specialty	Observed frequency O	Expected frequency E	Difference $O - E$	Square of difference $(O - E)^2$	$(O - E)^2/E$
General practice	80	90.0	-10.0	100.00	1.111
Medical	162	169.5	-7.5	56.25	0.332
Surgical	156	135.0	21.0	441.00	3.267
Other	102	105.5	-3.5	12.25	0.116
	500	500.0	0		4.826

To measure how well the observed and expected frequencies match up, it is logical to look at the differences $O - E$, shown in the fourth column of Table 12.4. As we see, summing these differences to obtain a "total difference" is not very useful since the sum is 0. Instead, each difference is squared (fifth column of Table 12.4) and then divided by the corresponding expected frequency. This gives the values $(O - E)^2/E$ shown in the sixth column of Table 12.4. The sum of those values,

$$\Sigma(O - E)^2/E = 4.826,$$

is the statistic used to measure how well (or poorly) the observed and expected frequencies match up.

If the null hypothesis is true, then the observed and expected frequencies should be roughly equal, resulting in a small value of the test statistic, $\Sigma(O - E)^2/E$. In other words, large values of $\Sigma(O - E)^2/E$ provide evidence against the null hypothesis.

As we saw in Table 12.4, $\Sigma(O - E)^2/E = 4.826$. Can this value be reasonably attributed to sampling error, or is it large enough to indicate that the null hypothesis is false? To answer this question, we need to know the probability distribution of the test statistic, $\Sigma(O - E)^2/E$, if the null hypothesis is true. That probability distribution is identified in Key Fact 12.2.   ∎

KEY FACT 12.2   **TEST STATISTIC FOR A GOODNESS-OF-FIT TEST**

Consider a chi-square goodness-of-fit test in which the null hypothesis specifies the percentage distribution of a population or the probability distribution of a random variable. Suppose the sample size is large. Then, if the null hypothesis is true, the random variable,

$$\chi^2 = \Sigma(O - E)^2/E,$$

has approximately a chi-square distribution. The number of degrees of freedom is one less than the number of categories in the distribution.

Because Example 12.4 specified four categories (the four specialties), the number of degrees of freedom in that example is $4 - 1 = 3$.

Procedure for the Chi-Square Goodness-of-Fit Test

We can now state a procedure for performing a chi-square goodness-of-fit test. The null hypothesis is that a population (or random variable) has a specified distribution, and the alternative hypothesis is that the population (or random variable) has a distribution different from the one specified in the null hypothesis. Since the null hypothesis will be rejected only when the test statistic is too large, the rejection region is always on the right; that is, the hypothesis test is always right-tailed.

PROCEDURE 12.1   **THE CHI-SQUARE GOODNESS-OF-FIT TEST**

ASSUMPTIONS
1. All expected frequencies are at least 1.
2. At most 20% of the expected frequencies are less than 5.

Step 1   State the null and alternative hypotheses.

Step 2   Calculate the expected frequencies using the formula $E = np$, where n denotes the sample size and p denotes the probability specified for the category in the null hypothesis.

Step 3   Check whether the expected frequencies satisfy Assumptions 1 and 2. If they do not, this procedure should not be used.

Step 4   Decide on the significance level, α.

Step 5   The critical value is χ^2_α with df $= k - 1$, where k is the number of categories in the distribution. Use Table V to find the critical value.

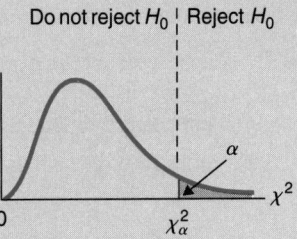

Do not reject H_0 | Reject H_0

Step 6   Compute the value of the test statistic

$$\chi^2 = \Sigma(O - E)^2/E,$$

where O and E represent observed and expected frequencies, respectively.

Step 7   If the value of the test statistic falls in the rejection region, then reject H_0; otherwise, do not reject H_0.

Step 8   State the conclusion in words.

Regarding Assumptions 1 and 2, many texts give the rule that all expected frequencies should be at least 5. Research by the noted statistician W. G. Cochran shows that the "rule of 5" is too restrictive.

Example 12.5   Illustrates Procedure 12.1

Let's return to the hypothesis test of Example 12.4. A researcher wants to know whether the specialty distribution of U.S. physicians currently in practice is different from the 1986 specialty distribution. The 1986 specialty distribution is displayed in Table 12.1 on page 691.

A random sample of 500 U.S. physicians currently in practice yields the frequency distribution shown in the first two columns of Table 12.2 on page 692. At the 5% significance level, do the data provide sufficient evidence to conclude that the current specialty distribution of U.S. physicians is different from the 1986 specialty distribution?

SOLUTION   We apply Procedure 12.1.

Step 1   *State the null and alternative hypotheses.*

The null and alternative hypotheses are

H_0: The current specialty distribution of U.S. physicians is the same as the 1986 distribution.

H_a: The current specialty distribution of U.S. physicians is different from the 1986 distribution.

Step 2 *Calculate the expected frequencies using the formula $E = np$, where n denotes the sample size and p denotes the probability specified for the category in the null hypothesis.*

The calculations are summarized in Table 12.3 on page 693.

Step 3 *Check whether the expected frequencies satisfy Assumptions 1 and 2.*

1. All expected frequencies are at least 1? Yes (see Table 12.3).
2. At most 20% of the expected frequencies are less than 5? Yes, in fact, none of the expected frequencies are less than 5 (see Table 12.3).

Step 4 *Decide on the significance level, α.*

We are to perform the test at the 5% significance level. Thus $\alpha = 0.05$.

Step 5 *The critical value is χ^2_α with $df = k - 1$, where k is the number of categories in the distribution.*

From Step 4, $\alpha = 0.05$. Also, there are four categories (specialties), so $k = 4$. Referring to Table V, we find that for df $= k - 1 = 4 - 1 = 3$, $\chi^2_{0.05} = 7.815$, as shown in Fig. 12.5.

FIGURE 12.5
Criterion for deciding whether or not to reject the null hypothesis

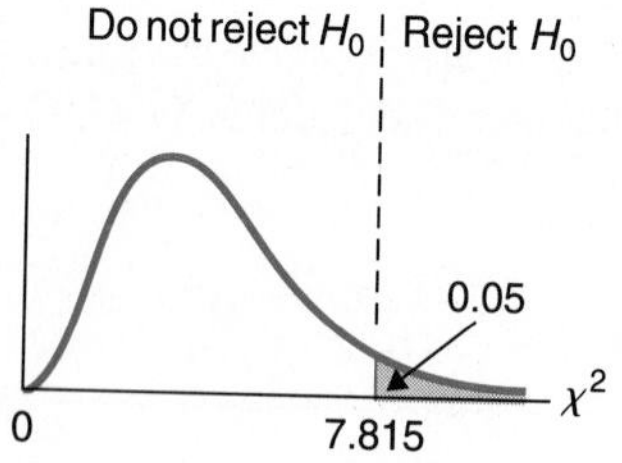

Step 6 *Compute the value of the test statistic*

$$\chi^2 = \Sigma(O - E)^2/E,$$

where O and E represent observed and expected frequencies, respectively.

Using the observed frequencies in the second column of Table 12.2 and the expected frequencies in the third column of Table 12.3, we can compute the value of the test statistic. This is done in Table 12.4 on page 693. From the last column of that table, we see that the value of the test statistic is

$$\chi^2 = \Sigma(O - E)^2/E = 4.826.$$

Step 7 *If the value of the test statistic falls in the rejection region, reject H_0; otherwise, do not reject H_0.*

From Step 6, the value of the test statistic is $\chi^2 = 4.826$. Since this does not fall in the rejection region, shown in Fig. 12.5, we do not reject H_0.

Step 8 *State the conclusion in words.*

The test results are not statistically significant at the 5% level; that is, at the 5% significance level, the data do not provide sufficient evidence to conclude that the current specialty distribution of U.S. physicians differs from the 1986 specialty distribution. ∎

The P-value approach to hypothesis testing can also be used to carry out the hypothesis test in Example 12.5. From Step 6 we see that the value of the test statistic is $\chi^2 = 4.826$. Recalling that df $= 3$, we find from Table V that $P > 0.05$. In particular, because the P-value exceeds the significance level of 0.05, we cannot reject H_0.

Before leaving this section, we need to make three additional points. In Table 12.4 on page 693, we calculated the sum of the observed frequencies, the sum of the expected frequencies, and the sum of their differences. Strictly speaking, those sums are not needed. However, they serve as a check for computational errors. The sums of the observed and expected frequencies, ΣO and ΣE, should both equal the sample size, n, which in this case is 500. The sum of the differences, $\Sigma (O - E)$, should equal 0. Exercise 12.20 asks you to verify these facts.

The chi-square goodness-of-fit test provides a method for performing a hypothesis test about the percentage distribution of a population in which each member is classified into one of k categories. If the number of categories is two, that is, $k = 2$, then the population is a two-category population. In that case the chi-square goodness-of-fit test is equivalent to the one-sample z-test for a population proportion (Procedure 11.2 on page 667). Exercise 12.19 compares the two tests.

Finally, the chi-square goodness-of-fit test can be used to decide whether a population is normally distributed. We do not recommend this procedure because a test based on normal scores, which we will present in Section 14.6, is far more powerful.

EXERCISES 12.2

12.9 Why do you think the term *goodness of fit* is used to describe the type of hypothesis test considered in this section?

12.10 Are the observed frequencies random variables? What about the expected frequencies? Explain your answers.

In Exercises 12.11–12.18, perform the required hypothesis test using either the classical approach or the P-value approach.

12.11 As reported by the Census Bureau in *U.S. Census of Population,* the 1983 distribution of the U.S. resident population, by region, is as follows.

Region	Northeast	Midwest	South	West
Percentage	21.2	25.2	34.0	19.6

A random sample of 500 current U.S. residents yields the following data.

Region	Northeast	Midwest	South	West
Frequency	97	121	176	106

At the 5% significance level, do the data suggest that the current distribution of the U.S. resident population by region is different from the 1983 distribution?

12.12   According to the Census Bureau publication *Current Population Reports,* the marital-status distribution of the U.S. adult population is as shown in the following table.

Marital status	Single	Married	Widowed	Divorced
Percentage	21.5	63.9	7.7	6.9

A random sample of 750 U.S. males, 25–29 years old, yielded the following frequency distribution.

Marital status	Single	Married	Widowed	Divorced
Frequency	289	408	0	53

At the 1% significance level, does it appear that the marital-status distribution of all 25–29-year-old U.S. males is different from that of the U.S. adult population as a whole?

12.13   The table that follows provides a relative-frequency distribution for the number of years of school completed by U.S. residents, 25 years old and over. [SOURCE: U.S. Bureau of the Census, *U.S. Census of Population.*]

Years	8 or less	9–11	12	13–15	16 or more
Rel. freq.	0.183	0.153	0.346	0.157	0.161

A random sample of 300 Tennessee residents, 25 years old and over, gave the following statistics.

Years	8 or less	9–11	12	13–15	16 or more
Frequency	83	48	95	36	38

Do the data provide sufficient evidence to conclude that the distribution of the number of years of school completed by Tennessee residents is different from the national distribution? Use $\alpha = 0.01$.

12.14   According to *Current Housing Reports,* published by the Census Bureau, the primary-heating-fuel distribution of all occupied housing units is as follows.

Primary heating fuel	Percentage
Natural gas	56.7
Fuel oil, kerosene	14.3
Electricity	16.0
Liquid propane gas	4.5
Wood	6.7
Other	1.8

A random sample of 250 occupied housing units built after 1974 yields the following frequency distribution.

Primary heating fuel	Frequency
Natural gas	91
Fuel oil, kerosene	16
Electricity	110
Liquid propane gas	14
Wood	17
Other	2

Do the data provide sufficient evidence to conclude that the primary-heating-fuel distribution of occupied housing units built after 1974 differs from that of all occupied housing units? Use $\alpha = 0.05$.

12.15   A gambler thinks a die may be loaded; that is, he thinks the six numbers may not be equally likely. To test his suspicion, he rolls the die 150 times and obtains the results shown in the following table.

Number	1	2	3	4	5	6
Frequency	23	26	23	21	31	26

Do the data provide sufficient evidence to conclude that the die is loaded? Perform the hypothesis test at the 0.05 level of significance.

12.16   A roulette wheel contains 18 red numbers, 18 black numbers, and 2 green numbers. The table below shows the frequency with which the ball landed on each color in 200 trials.

Color	Red	Black	Green
Frequency	88	102	10

At the 5% significance level, do the data suggest that the wheel is out of balance?

12.17   The FBI compiles data on violent crime and reports the information in *Crime in the United States.* According to that publication, a percentage distribution for the types of violent crime for the entire United States is as shown in the following table.

Violent crime	Percentage
Murder	1.6
Forcible rape	6.4
Robbery	40.3
Aggravated assault	51.7

A random sample of 600 violent-crime reports from the state of New Jersey yields the following statistics.

Violent crime	Frequency
Murder	6
Forcible rape	33
Robbery	292
Aggravated assault	269

At the 1% significance level, do the data provide sufficient evidence to conclude that the distribution of types of violent crime in New Jersey is different from the nation as a whole?

· **12.18** *Current Population Reports,* released by the Census Bureau, contains the following information on the 1983 money income of U.S. households.

Income level	Percentage
Under $5,000	9.2
$5,000–$9,999	13.7
$10,000–$14,999	13.0
$15,000–$19,999	12.0
$20,000–$24,999	10.8
$25,000–$34,999	17.0
$35,000–$49,999	14.0
$50,000 and over	10.3

A random sample of 825 U.S. household incomes for last year gave the following frequency distribution. The income levels are in 1983 constant dollars; that is, they are adjusted for inflation to 1983 levels.

Income level	Frequency
Under $5,000	70
$5,000–$9,999	117
$10,000–$14,999	92
$15,000–$19,999	111
$20,000–$24,999	101
$25,000–$34,999	140
$35,000–$49,999	111
$50,000 and over	83

At the 5% significance level, do the data provide sufficient evidence to conclude that last year's income-level distribution for U.S. households has changed from the 1983 distribution?

· · **12.19** Consider a two-category population having population proportion p. Suppose we want to perform a hypothesis test to decide whether p is different from the value p_0. Discuss the method for performing such a test using

a. the one-sample z-test for a population proportion (Procedure 11.2 on page 667).

b. the chi-square goodness-of-fit test (Procedure 12.1).

On page 697 we stated that the hypothesis tests in parts (a) and (b) are *equivalent.* What do you think that means?

· · · **12.20** On page 697 we mentioned that the sums of the observed and expected frequencies, ΣO and ΣE, always equal the sample size, n, and that the sum of the differences, $\Sigma(O - E)$, always equals 0.

a. Explain, in words, why $\Sigma O = n$.

b. Show mathematically that $\Sigma E = n$.

c. Prove that $\Sigma(O - E) = 0$.

12.3 CHI-SQUARE INDEPENDENCE TEST

· · · · ·

The next chi-square procedure we will study is the **chi-square independence test.** To begin, we need to define what it means for two characteristics of a population to be *statistically independent.* Roughly, it means that knowledge of the category to which a member of the population belongs for one of the characteristics imparts no information about the other characteristic. A more precise definition is given in Definition 12.1.

DEFINITION 12.1 | **STATISTICAL INDEPENDENCE FOR TWO CHARACTERISTICS OF A POPULATION**

> Two characteristics of a population are called *statistically independent* (or *nonassociated*) if within the categories of one of the characteristics, the distributions of the other characteristic are identical. If two characteristics of a population are not statistically independent, then we say that they are *statistically dependent* (or *associated*).

For instance, consider the two characteristics, marital status and sex. If marital-status distributions (i.e., percentages of single, married, widowed, divorced) are identical within the sex categories (i.e., are identical for males and females), then marital status and sex are statistically independent; otherwise, marital status and sex are statistically dependent.

Suppose, for the sake of argument, that marital-status distributions are identical within the sex categories, making marital status and sex statistically independent. Then those identical marital-status distributions for males and females would also be the same as the marital-status distribution of the entire population (i.e., of males and females combined). Furthermore, the reverse would also hold: sex distributions (i.e., percentages of males and females) within the four marital-status categories would be identical to each other and to the sex distribution of the entire population.

To generalize, consider two characteristics of a population, X and Y, that are statistically independent; say, that within the categories of X, the distributions of Y are identical. Then those identical distributions of Y are the same as the distribution of Y for the entire population. Furthermore, the reverse also holds: within the categories of Y, the distributions of X are identical to each other and to the distribution of X for the entire population.

Example 12.6   Illustrates Definition 12.1

The American Medical Association compiles information on U.S. physicians in *Physician Characteristics and Distribution in the U.S.* Table 12.5 provides a contingency table for U.S. surgeons cross classified by specialty and base of practice. Use the contingency table to determine whether the characteristics specialty and base of practice are statistically independent for U.S. surgeons.

TABLE 12.5
Joint frequency distribution of U.S. surgeons by specialty and base of practice

Specialty	Base of practice			
	Office B_1	Hospital B_2	Other B_3	Total
General surgery S_1	24,128	12,225	1,658	38,011
Obstetrics/Gynecology S_2	24,150	6,734	1,140	32,024
Orthopedics S_3	13,364	4,248	414	18,026
Ophthalmology S_4	12,328	2,694	518	15,540
Total	73,970	25,901	3,730	103,601

SOLUTION   To determine whether specialty and base of practice are statistically independent, we will obtain the specialty distribution within each base-of-practice category. If those specialty distributions are identical, then the characteristics specialty and base of practice are statistically independent; otherwise, they are statistically dependent.

Dividing each column in Table 12.5 by its column total, we get the relative-frequency distributions of specialty within the base-of-practice categories. These are shown in the columns of Table 12.6.

TABLE 12.6

Specialty distributions within the base-of-practice categories

Base of practice

Specialty		Office B_1	Hospital B_2	Other B_3	Total
General surgery	S_1	0.326	0.472	0.445	0.367
Obstetrics/Gynecology	S_2	0.326	0.260	0.306	0.309
Orthopedics	S_3	0.181	0.164	0.111	0.174
Ophthalmology	S_4	0.167	0.104	0.139	0.150
Total		1.000	1.000	1.000	1.000

Because the columns in Table 12.6 are not the same, the specialty distributions within the base-of-practice categories are not identical. Hence, for U.S. surgeons, specialty and base of practice are not statistically independent; in other words, they are statistically dependent. ■

Thus, knowing a surgeon's base of practice provides information about his or her specialty, and vice versa. For instance, as we see from Table 12.6, with no knowledge of a surgeon's base of practice, there is a 15.0% chance that the surgeon is an ophthalmologist. However, if we know the surgeon's base of practice is an office, then there is a 16.7% chance that the surgeon is an ophthalmologist.

A **segmented bar graph** is helpful for understanding the concept of statistical independence. Figure 12.6, at the top of the next page, depicts such a graph for the specialty distributions within the base-of-practice categories and the overall specialty distribution for U.S. surgeons.

If specialty and base of practice were statistically independent, then the three bars showing the specialty distributions within the bases of practice, as well as the overall specialty distribution (fourth bar), would be identical. The fact that specialty and base of practice are not statistically independent is reflected in the segmented bar graph by nonidentical bars.

FIGURE 12.6

Segmented bar graph for the
specialty distributions within the
base-of-practice categories and
the overall specialty distribution

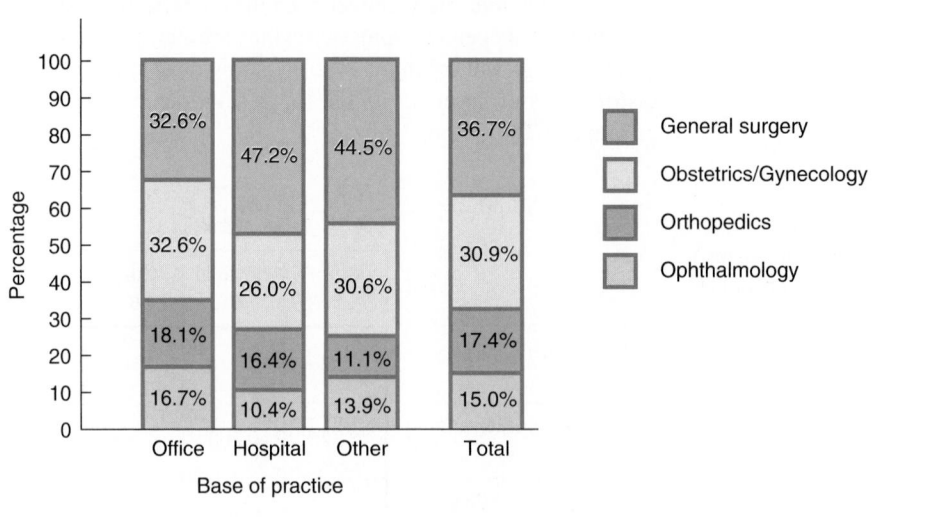

The concept of statistical independence for two characteristics of a population is closely related to the concept of statistical independence of events, which we studied in Section 4.6. In fact, it can be shown that two characteristics of a population are statistically independent if and only if the events corresponding to the categories of the two characteristics are statistically independent.

For instance, consider again the characteristics specialty and base of practice discussed in Example 12.6. Suppose a U.S. surgeon is selected at random. Let

$$S_1 = \text{event the surgeon obtained is a general surgeon,}$$

and so on (see the letters labeling the rows and columns of Table 12.5 on page 700). Then specialty and base of practice are statistically independent if and only if the events S_1, S_2, S_3, S_4 and B_1, B_2, B_3 are statistically independent, that is, event S_i and event B_j are statistically independent for all i and j. Since, as we know from Example 12.6, specialty and base of practice are statistically dependent, it follows that for some i and j, event S_i and event B_j are statistically dependent.

The Logic Behind the Chi-Square Independence Test

If we have data for an entire population, then we can always determine for sure whether two characteristics of the population are statistically independent by proceeding as in Example 12.6. However, because in most cases, data for an entire population are not available, we must usually employ inferential methods to decide whether two characteristics are statistically independent.

One of the most commonly employed procedures for making such decisions is the chi-square independence test. Example 12.7 introduces and explains the reasoning behind that hypothesis-testing procedure.

Example 12.7   Introduces the Chi-Square Independence Test

A national survey was conducted to obtain information on the alcohol consumption patterns of U.S. adults by marital status. A random sample of 1772 residents, 18 years old and over, yielded the data displayed in Table 12.7.[†] The table shows, for instance, that of the 1772 adults sampled, 1173 are married, 590 abstain, and 411 are married and abstain.

TABLE 12.7

Contingency table of marital status versus alcohol consumption for 1772 randomly selected U.S. adults

Drinks per month

Marital status	Abstain	1–60	Over 60	Total
Single	67	213	74	354
Married	411	633	129	1173
Widowed	85	51	7	143
Divorced	27	60	15	102
Total	590	957	225	1772

We want to use the sample data to decide whether there is an association between marital status and alcohol consumption; that is, we want to perform the hypothesis test

H_0: Marital status and alcohol consumption are statistically independent.

H_a: Marital status and alcohol consumption are statistically dependent.

The idea behind the chi-square independence test is to compare the observed frequencies in Table 12.7 with the frequencies that would be expected if the null hypothesis of statistical independence is true. The test statistic employed to make the comparison is the same one used for the goodness-of-fit test:

$$\chi^2 = \Sigma(O - E)^2/E,$$

where O represents observed frequency and E represents expected frequency.

We will now develop a formula for computing the expected frequencies. Consider, for instance, the cell of Table 12.7 corresponding to "Married *and* Abstain," the cell in the

---

[†] Adapted from "Alcohol Use and Alcohol Problems among U.S. Adults: Results of the 1979 National Survey" by W. B. Clark and L. Midanik. In National Institute on Alcohol Abuse and Alcoholism, *Alcohol and Health Monograph No. 1, Alcohol Consumption and Related Problems.* DHHS Pub. No. (ADM) 82–1190, 1982.

second row and first column of the table. To begin, note that the population proportion of all adults who abstain can be estimated by the sample proportion of the 1772 adults sampled who abstain, that is, by

Number sampled who abstain ↘

$$\frac{590}{1772} = 0.333 \ (33.3\%).$$

Total number sampled ↗

If marital status and alcohol consumption are statistically independent, that is, if H_0 is true, then the proportion of married adults who abstain is the same as the proportion of all adults who abstain. Thus, if H_0 is true, the sample proportion, $\frac{590}{1772}$, or 33.3%, is also an estimate of the population proportion of married adults who abstain.

A total of 1173 of the adults sampled are married, and as we have just seen, if H_0 is true, then approximately $\frac{590}{1772}$, or 33.3%, of married adults abstain. Therefore, if H_0 is true, we would expect roughly

$$\frac{590}{1772} \cdot 1173 = 390.6$$

of the adults in the survey to be married adults who abstain.

It is useful to rewrite the left-hand side of the above expected-frequency computation in a slightly different way. By using algebra and referring to Table 12.7, we obtain the following equalities:

$$\text{Expected frequency} = \frac{590}{1772} \cdot 1173 = \frac{1173 \cdot 590}{1772} = \frac{(\text{Row total}) \cdot (\text{Column total})}{\text{Sample size}}$$

If we let R denote "Row total" and C denote "Column total," then we can express this equation compactly as

$$E = \frac{R \cdot C}{n},$$

where, as usual, E denotes expected frequency and n denotes sample size.

Using this simple formula, we can obtain the expected frequencies for all 12 cells in Table 12.7. We have already done that for the cell in the second row and first column. For the cell in the upper right-hand corner of the table, we get

$$E = \frac{R \cdot C}{n} = \frac{354 \cdot 225}{1772} = 44.9.$$

Similar computations give the expected frequencies for the remaining cells.

In Table 12.8 we have modified Table 12.7 by placing the expected frequency for each cell beneath the corresponding observed frequency. Thus, for instance, Table 12.8 shows that the observed frequency of the adults sampled who are single and have over 60 drinks per month is 74, whereas if marital status and alcohol consumption are statistically independent, then the expected frequency is 44.9.

TABLE 12.8

Observed and expected frequencies for marital status versus alcohol consumption; expected frequencies are printed below the observed frequencies

Drinks per month

Marital status		Abstain	1–60	Over 60	Total
	Single	67 117.9	213 191.2	74 44.9	354
	Married	411 390.6	633 633.5	129 148.9	1173
	Widowed	85 47.6	51 77.2	7 18.2	143
	Divorced	27 34.0	60 55.1	15 13.0	102
	Total	590	957	225	1772

If the null hypothesis of statistical independence is true, then the observed and expected frequencies should be roughly equal, resulting in a relatively small value of the test statistic, $\chi^2 = \Sigma(O - E)^2/E$. Consequently, if χ^2 is too large, we will reject the null hypothesis of statistical independence in favor of the alternative hypothesis of statistical dependence. From Table 12.8 we find that

$$
\begin{aligned}
\chi^2 &= \Sigma(O - E)^2/E \\
&= (67 - 117.9)^2/117.9 + (213 - 191.2)^2/191.2 + (74 - 44.9)^2/44.9 \\
&\quad + (411 - 390.6)^2/390.6 + (633 - 633.5)^2/633.5 + (129 - 148.9)^2/148.9 \\
&\quad + (85 - 47.6)^2/47.6 + (51 - 77.2)^2/77.2 + (7 - 18.2)^2/18.2 \\
&\quad + (27 - 34.0)^2/34.0 + (60 - 55.1)^2/55.1 + (15 - 13.0)^2/13.0 \\
&= 21.952 + 2.489 + 18.776 + 1.070 + 0.000 + 2.670 \\
&\quad + 29.358 + 8.908 + 6.856 + 1.427 + 0.438 + 0.324 \\
&= 94.269.^{\dagger}
\end{aligned}
$$

Can this value be reasonably attributed to sampling error, or is it large enough to indicate that marital status and alcohol consumption are statistically dependent? Before we can answer that question, we must know the probability distribution of the test statistic $\chi^2 = \Sigma(O - E)^2/E$. ∎

[†] Although we have displayed the expected frequencies to one decimal place and the chi-square subtotals to three decimal places, the calculations were done using full calculator accuracy.

KEY FACT 12.3   **TEST STATISTIC FOR AN INDEPENDENCE TEST**

Consider a chi-square independence test in which the null hypothesis is that two characteristics of a population are statistically independent. Suppose the sample size is large. Then, if the null hypothesis of statistical independence is true, the random variable

$$\chi^2 = \Sigma (O - E)^2 / E$$

has approximately a chi-square distribution with df $= (r - 1)(c - 1)$. Here r and c are the number of rows and columns in the contingency table; that is, r is the number of categories for the characteristic displayed along the side of the contingency table, and c is the number of categories for the characteristic displayed along the top of the contingency table.

Procedure for the Chi-Square Independence Test

Procedure 12.2 provides a step-by-step method for performing a chi-square independence test. The null hypothesis is that the two characteristics under consideration are statistically independent, and the alternative hypothesis is that they are statistically dependent. Note that the hypothesis test is always right-tailed. (Why is this so?)

PROCEDURE 12.2   **THE CHI-SQUARE INDEPENDENCE TEST**

ASSUMPTIONS

1. All expected frequencies are at least 1.
2. At most 20% of the expected frequencies are less than 5.

Step 1   State the null and alternative hypotheses.

Step 2   Calculate the expected frequencies using the formula

$$E = \frac{R \cdot C}{n},$$

where $R =$ row total, $C =$ column total, and $n =$ sample size. Place each expected frequency below its corresponding observed frequency in the contingency table.

Step 3   Check whether the expected frequencies satisfy Assumptions 1 and 2. If they do not, this procedure should not be used.

Step 4   Decide on the significance level, α.

Step 5   The critical value is χ^2_α with df $= (r - 1)(c - 1)$, where r and c are, respectively, the number of rows and columns in the contingency table. Use Table V to find the critical value.

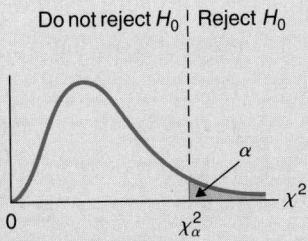

Step 6 Compute the value of the test statistic

$$\chi^2 = \Sigma(O - E)^2/E,$$

where O and E represent observed and expected frequencies, respectively.

Step 7 If the value of the test statistic falls in the rejection region, then reject H_0; otherwise, do not reject H_0.

Step 8 State the conclusion in words.

Example 12.8 Illustrates Procedure 12.2

Recall that a random sample of 1772 U.S. adults yielded the data on marital status versus alcohol consumption displayed in Table 12.7 on page 703. At the 5% significance level, do the data provide sufficient evidence to conclude that marital status and alcohol consumption are statistically dependent?

SOLUTION   We employ Procedure 12.2.

Step 1 *State the null and alternative hypotheses.*

The null and alternative hypotheses are

H_0: Marital status and alcohol consumption are statistically independent.

H_a: Marital status and alcohol consumption are statistically dependent.

Step 2 *Calculate the expected frequencies using the formula*

$$E = \frac{R \cdot C}{n},$$

where $R = $ row total, $C = $ column total, and $n = $ sample size. Place each expected frequency below its corresponding observed frequency in the contingency table.

We did this earlier in Table 12.8 on page 705.

Step 3 *Check whether the expected frequencies satisfy Assumptions 1 and 2.*

1. All expected frequencies are at least 1? Yes (see Table 12.8).
2. At most 20% of the expected frequencies are less than 5? Yes, in fact, none of the expected frequencies are less than 5 (see Table 12.8).

Step 4   *Decide on the significance level, α.*

The test is to be performed at the 5% significance level. Hence $\alpha = 0.05$.

Step 5   *The critical value is χ_α^2 with df $= (r - 1)(c - 1)$, where r and c are, respectively, the number of rows and columns in the contingency table.*

The number of rows, r, in the contingency table is the number of marital-status categories, which is four; so $r = 4$. The number of columns, c, in the contingency table is the number of drinks-per-month categories, which is three; so $c = 3$. Thus df $= (r-1)(c-1) = (4-1)(3-1) = 3 \cdot 2 = 6$. Since $\alpha = 0.05$, we find from Table V that the critical value is $\chi_{0.05}^2 = 12.592$, as seen in Fig. 12.7.

FIGURE 12.7

Criterion for deciding whether or not to reject the null hypothesis

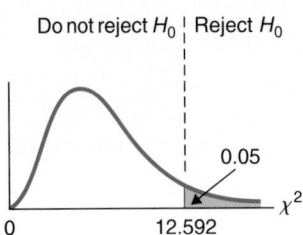

Step 6   *Compute the value of the test statistic*

$$\chi^2 = \Sigma(O - E)^2/E,$$

where O and E represent observed and expected frequencies, respectively.

The observed and expected frequencies are displayed in Table 12.8. Using these we compute the value of the test statistic:

$$\chi^2 = (67 - 117.9)^2/117.9 + (213 - 191.2)^2/191.2 + \cdots + (15 - 13.0)^2/13.0$$
$$= 21.952 + 2.489 + \cdots + 0.324 = 94.269.$$

Step 7   *If the value of the test statistic falls in the rejection region, reject H_0; otherwise, do not reject H_0.*

From Step 6, the value of the test statistic is $\chi^2 = 94.269$, which falls in the rejection region (see Fig. 12.7). Thus we reject H_0.

Step 8   *State the conclusion in words.*

The test results are statistically significant at the 5% level; that is, at the 5% significance level, the data provide sufficient evidence to conclude that marital status and alcohol consumption are statistically dependent.   ∎

The P-value approach to hypothesis testing can also be used to carry out the hypothesis test in Example 12.8 and to assess more precisely the evidence against the null hypothesis.

From Step 6 we see that the value of the test statistic is $\chi^2 = 94.269$. Recalling that df $= 6$, we find from Table V that $P < 0.005$. In particular, because the P-value is less than the significance level of 0.05, we can reject H_0. Furthermore, by referring to Table 9.12 on page 530, we see that the data provide very strong evidence against the null hypothesis of statistical independence and hence in favor of the alternative hypothesis that marital status and alcohol consumption are statistically dependent.

Concerning the Assumptions

In Procedure 12.2 we made two assumptions about expected frequencies:

1. All expected frequencies are at least 1.
2. At most 20% of the expected frequencies are less than 5.

What can we do if one or both of these assumptions are violated? Three approaches are possible. We can combine rows or columns to increase the expected frequencies in those cells in which they are too small; we can eliminate certain rows or columns in which the small expected frequencies occur; or we can increase the sample size. Exercises 12.35 and 12.36 give you some practice in making such modifications.

Statistical Dependence and Causation

The chi-square independence test is used to decide whether an association exists between two characteristics of a population; specifically, the null hypothesis is that the two characteristics are statistically independent, and the alternative hypothesis is that they are statistically dependent. If the null hypothesis is rejected, we can conclude that the two characteristics are associated, but not that they are causally related.

For instance, in Example 12.8 we rejected the null hypothesis of statistical independence for marital status and alcohol consumption. This means that knowing the marital status of a person provides statistical information about the alcohol consumption of that person, and vice versa. But it does not necessarily mean, for example, that being single causes a person to drink more.

Deciding on Statistical Dependence with Population Data

The chi-square independence test is an inferential procedure and as such should be applied only to sample data—not to data for an entire population. How then should we proceed if we have data for an entire population and want to decide whether two characteristics of the population are statistically dependent?

Two approaches are possible. One approach is to proceed as described in Example 12.6 on page 700. The other approach is to compute the expected frequencies as in the chi-square independence test, that is, to use the formula $E = R \cdot C/N$, where N is the size of the population. If the expected and observed frequencies are identical for each cell, then the two characteristics are statistically independent; otherwise, they are statistically dependent.

Using the Computer (Optional)

Procedure 12.2 gives a step-by-step method for performing a chi-square independence test. Alternatively, we can use Minitab's **ChiSquare** command to carry out such a test. Example 12.9 shows how ChiSquare is applied.

Example 12.9   The ChiSquare Command

Use Minitab to carry out the hypothesis test considered in Example 12.8.

SOLUTION   We want to perform the hypothesis test

H_0: Marital status and alcohol consumption are statistically independent.

H_a: Marital status and alcohol consumption are statistically dependent.

at the 5% significance level.

To employ Minitab, we first store the cell data appearing in the first three columns of Table 12.7 on page 703 in columns named ABSTAIN, 1-60, and OVER 60. Then we proceed in the following manner.

Session commands: We type **CHISQUARE** followed by the storage locations of the sample data; that is, we type CHISQUARE on 'ABSTAIN' '1-60' 'OVER 60' and press Enter. Printout 12.1 displays this command and the output that results.

(or)

Menu commands: We choose **Stat ▸ Tables ▸ Chisquare Test...**, specify ABSTAIN, 1-60, and OVER 60 for **Columns containing the table**, and then select **OK**. The resulting output is shown in Printout 12.1.

The first part of the output provides a table of the observed and expected frequencies. This is Minitab's version of Table 12.8 on page 705. After the table we find the value of the test statistic, $\chi^2 = \Sigma(O - E)^2/E$, including the cell-by-cell subtotals. Thus the value of the test statistic is $\chi^2 = 94.269$. The final item given in the output is the number of degrees of freedom, df = 6.

To complete the hypothesis test, we need only obtain the critical value and compare it to the value of the test statistic. The critical value for a chi-square independence test is χ^2_α with df $= (r - 1)(c - 1)$. Because the hypothesis test is to be performed at the 5% significance level, we have $\alpha = 0.05$. Also, the last line of Printout 12.1 shows that df = 6. Consulting Table V, we find that for df = 6, $\chi^2_{0.05} = 12.592$.

Since the value of the test statistic is $\chi^2 = 94.269$, which exceeds the critical value of 12.592, we reject H_0. In other words, at the 5% significance level, the data provide sufficient evidence to conclude that marital status and alcohol consumption are statistically dependent.   ■

PRINTOUT 12.1
Minitab output for the ChiSquare command

```
MTB > CHISQUARE on 'ABSTAIN' '1-60' 'OVER 60'

Expected counts are printed below observed counts

          ABSTAIN      1-60  OVER 60     Total
     1         67       213       74       354
           117.87    191.18    44.95

     2        411       633      129      1173
           390.56    633.50   148.94

     3         85        51        7       143
            47.61     77.23    18.16

     4         27        60       15       102
            33.96     55.09    12.95

Total         590       957      225      1772

ChiSq =  21.952 +  2.489 + 18.776 +
          1.070 +  0.000 +  2.670 +
         29.358 +  8.908 +  6.856 +
          1.427 +  0.438 +  0.324 = 94.269
df = 6
```

EXERCISES 12.3

• **12.21** Solve Example 12.6 on page 700 by obtaining the base-of-practice distribution within each specialty category. Draw a segmented bar graph illustrating your work.

• **12.22** Refer to the contingency table given in Exercise 4.74 on page 222.

a. Obtain the age distribution within each type category.
b. Are age and type statistically independent?
c. Construct a segmented bar graph to illustrate your results in parts (a) and (b).
d. Without doing any further calculations, answer true or false to the following statement and give a reason for your answer: The type distributions within the five age categories are not identical.

In Exercises 12.23–12.30, perform a chi-square independence test using either the classical approach or the P-value approach, provided the conditions for using this test are met.

• **12.23** The U.S. Bureau of the Census compiles data on money earnings of people by educational attainment, sex, and age and publishes the information in *Current Population Reports.* A study to decide whether an association exists between annual income and educational level yielded the sample data shown in the following contingency table.

Annual income	Years of schooling: 0–8	9–12	Over 12	Total
Under \$10,000	19	35	10	64
\$10,000–\$24,999	20	67	33	120
\$25,000–\$49,999	10	75	67	152
\$50,000 and over	3	35	78	116
Total	52	212	188	452

At the 1% significance level, do the data provide sufficient evidence to conclude that annual income and educational level are statistically dependent?

12.24 The Gallup Organization conducts periodic surveys to gauge the support by U.S. adults for regional primary elections. The question asked is, "It has been proposed that four individual primaries be held in different weeks of June during presidential election years. Does this sound like a good idea or a poor idea?" Here is a contingency table for responses by political affiliation, adapted from the results of a Gallup Poll appearing in *The Arizona Republic*.

Response

Political affiliation	Good idea	Poor idea	No opinion	Total
Republican	266	266	186	
Democrat	308	250	176	
Independent	28	27	21	
Total				1528

a. Fill in the row and column totals.
b. At the 5% level of significance, do the data suggest that the feelings of adults on the issue of regional primaries are dependent on political affiliation?

12.25 The U.S. Bureau of Labor Statistics gathers data on occupations of employed workers by sex and publishes its findings in *Employment and Earnings*. A random sample of 83 employed workers yields the following data.

Sex

Occupation type	Male	Female	Total
Managerial/ Professional	14	10	24
Technical sales Administrative	9	16	25
Service	4	6	10
Other	20	4	24
Total	47	36	83

Do the data provide sufficient evidence to conclude that there is an association between the characteristics sex and occupation type for employed workers? Use $\alpha = 0.01$.

12.26 In 1989 roughly 58 million Americans suffered injuries, as reported by the National Center for Health Statistics in *Vital and Health Statistics*. More males (31.7 million) were injured than females (26.3 million). Those statistics do not tell us whether males and females tend to be injured in similar circumstances. One set of categories commonly used for accident circumstance is "while at work," "home," "motor vehicle," and "other." In order to decide whether there is an association between accident circumstance and sex, a safety official in a large city took a random sample of accident reports. He obtained the following data.

Sex

Circumstance	Male	Female	Total
While at work	18	4	
Home	26	28	
Motor vehicle	4	6	
Other	36	24	
Total			

a. Fill in the row and column totals.
b. Determine the sample size.
c. Do the data provide sufficient evidence to conclude that in this city, accident circumstance and sex are statistically dependent? Perform the required hypothesis test at the 5% significance level.

12.27 The FBI compiles information on arrests for violent crimes by the type of crime committed and the age of the person arrested. Results are published in *Crime in the United States*. To decide whether there is an association between the type of violent crime committed and the age of the person arrested, 750 arrest records are randomly selected. The data obtained are displayed in the following table.

Age

Type of violent crime	18–24	25–44	45+	Total
Murder	11	16	4	
Forcible rape	21	26	4	
Robbery	128	92	6	
Aggravated assault	162	234	46	
Total				750

a. Fill in the row and column totals.
b. Is there evidence that an association exists between the type of violent crime committed and the age of the person arrested? Perform the required hypothesis test at the 1% significance level.

12.28 The National Center for Health Statistics reports information on people with acute medical conditions in *Vital and Health Statistics.* Acute conditions are counted only if they are medically attended or caused at least 1 day of restricted activity. A random sample of 376 acute conditions yielded the data in the contingency table below. We have used the following coding for the columns of the contingency table: *A* stands for infective and parasitic diseases, *B* stands for respiratory diseases, *C* stands for digestive-system diseases, and *D* stands for injuries.

Type of condition

Family income	A	B	C	D	Total
Under $15,000	4	24	2	8	
$15,000–$19,999	7	36	4	11	
$20,000–$29,999	8	35	4	9	
$30,000–$39,999	12	59	5	17	
$40,000 and over	19	81	6	25	
Total					376

a. Fill in the row and column totals.
b. Do the data provide sufficient evidence to conclude that type of acute condition and family income are associated? Perform the appropriate hypothesis test using a significance level of 0.05.

12.29 *Statistics of Income Bulletin,* an IRS publication, contains data on top wealthholders by marital status. A random sample of 487 top wealthholders yielded the contingency table shown at the top of the next column. At the 5% significance level, do the data provide sufficient evidence to conclude that for top wealthholders, net worth and marital status are statistically dependent?

Marital status

Net worth	Married	Single/ Divorced	Widowed	Total
$100,000–$249,999	227	54	63	344
$250,000–$499,999	60	15	22	97
$500,000–$999,999	20	4	7	31
$1,000,000 or more	10	2	3	15
Total	317	75	95	487

12.30 The American Bar Foundation publishes information on lawyer characteristics in *The Lawyer Statistical Report.* The following contingency table cross classifies 307 randomly selected lawyers by status in practice and size of city practicing in.

Size of city

Status in practice	Less than 250,000	250,000–499,999	500,000 or more	Total
Government	12	4	14	30
Judicial	8	1	2	11
Private practice	122	31	69	222
Salaried	19	7	18	44
Total	161	43	103	307

Do the data provide sufficient evidence to conclude that the characteristics size of city and status in practice are statistically dependent? Use $\alpha = 0.05$.

12.31 (Computer exercise) Use Minitab or some other statistical software to perform the hypothesis test in Exercise 12.23.

12.32 (Computer exercise) Use Minitab or some other statistical software to perform the hypothesis test in Exercise 12.24(b).

·   **12.33 (Computer exercise)** A study was conducted at Arizona State University to decide whether an association exists between grade and study time for intermediate algebra students who complete the course with a grade of C or better. Data were collected for a random sample of 422 students and compiled in a contingency table with categories for the number of hours studied per week across the top and categories for the grade earned along the side. For the number of hours studied, there were four categories: 0–3, 4–6, 7–9, and 10 or more. For the grade earned, there were three categories: A, B, and C. Printout 12.2 was obtained by applying Minitab's ChiSquare command to the data. Determine

a. the number of students in the sample who studied at least 10 hours per week.
b. the number of students in the sample who received a grade of B in the class.
c. the observed number of students in the sample who studied at least 10 hours per week and received a grade of B in the class.
d. the expected number of students in the sample who studied at least 10 hours per week and received a grade of B in the class if grade and study time are statistically independent.
e. the chi-square subtotal, $(O - E)^2/E$, for the "at least 10 hours per week *and* grade of B" cell.
f. the number of degrees of freedom.
g. the null and alternative hypotheses for a hypothesis test to decide whether there is an association between grade and study time for intermediate algebra students at Arizona State University.
h. the value of the test statistic, χ^2.
i. the conclusion if the hypothesis test is performed at the 5% significance level. *(Note:* Consult Table V.*)*

·   **12.34 (Computer exercise)** Surveys are performed by the Book Industry Study Group to obtain information on characteristics of book readers. A *book reader* is defined as a person who read one or more books in the 6 months prior to the survey; a *non–book reader* is defined as a person who read newspapers or magazines but no books in the 6 months prior to the survey; and a *nonreader* is defined as a person who did not read a book, newspaper, or magazine in the 6 months prior to the survey. Printout 12.3 was obtained by applying Minitab's ChiSquare command to data from a random sample of people 16 years old and over. Across the top of the contingency table are the reader-classification categories: book reader, non–book reader, and nonreader (labeled C1, C2, and C3). Along the side of the contingency table are household-income categories: less than $15,000, $15,000–$24,999, $25,000–$39,999, and $40,000 or over (labeled 1–4). Determine

a. the total number of people in the sample.
b. the number of people in the sample with a household income between $25,000 and $39,999.
c. the number of people in the sample who are book readers.
d. the observed number of people in the sample who are book readers with a household income between $25,000 and $39,999.
e. the expected number of people in the sample who are book readers with a household income between $25,000 and $39,999 if the characteristics household income and reader classification are statistically independent.
f. the number of degrees of freedom.
g. the null and alternative hypotheses for a hypothesis test to decide whether the characteristics household income and reader classification are statistically dependent.
h. the value of the test statistic, χ^2.
i. the conclusion if the hypothesis test is performed at the 1% significance level. *(Note:* Consult Table V.*)*

· ·   **12.35** In Exercise 12.29 it was not possible to perform the chi-square independence test because the assumptions regarding expected frequencies were not met. As mentioned on page 709, we can try three approaches to remedy the situation: (1) combine rows or columns; (2) eliminate rows or columns; or (3) increase the sample size.

a. Combine the last two rows of the contingency table in Exercise 12.29 to form a new contingency table.
b. Use the table obtained in part (a) to perform the hypothesis test required in Exercise 12.29, if possible.
c. Eliminate the last row of the contingency table in Exercise 12.29 to form a new contingency table.
d. Use the table obtained in part (c) to perform the hypothesis test required in Exercise 12.29, if possible.

· ·   **12.36** In Exercise 12.30 it was not possible to perform the chi-square independence test because the assumptions regarding expected frequencies were not met. As mentioned on page 709, we can try three approaches to remedy the situation: (1) combine rows or columns; (2) eliminate rows or columns; or (3) increase the sample size.

a. Combine the first two rows of the contingency table in Exercise 12.30 to form a new contingency table.
b. Use the table obtained in part (a) to perform the hypothesis test required in Exercise 12.30, if possible.
c. Eliminate the second row of the contingency table in Exercise 12.30 to form a new contingency table.
d. Use the table obtained in part (c) to perform the hypothesis test required in Exercise 12.30, if possible.

PRINTOUT 12.2   Minitab output for Exercise 12.33

Expected counts are printed below observed counts

	C1	C2	C3	C4	Total
1	48	51	34	13	146
	37.02	59.51	37.71	11.76	
2	37	75	43	9	164
	41.58	66.84	42.36	13.21	
3	22	46	32	12	112
	28.40	45.65	28.93	9.02	
Total	107	172	109	34	422

ChiSq = 3.257 + 1.216 + 0.365 + 0.130 +
         0.505 + 0.995 + 0.010 + 1.343 +
         1.441 + 0.003 + 0.326 + 0.982 = 10.574

df = 6

PRINTOUT 12.3   Minitab output for Exercise 12.34

Expected counts are printed below observed counts

	C1	C2	C3	Total
1	173	267	55	495
	247.33	217.88	29.79	
2	168	130	19	317
	158.39	139.53	19.08	
3	160	144	9	313
	156.39	137.77	18.84	
4	213	88	3	304
	151.89	133.81	18.30	
Total	714	629	86	1429

ChiSq = 22.337 + 11.072 + 21.334 +
         0.583 + 0.651 + 0.000 +
         0.083 + 0.281 + 5.137 +
        24.583 + 15.684 + 12.787 = 114.534

df = 6

· ·   **12.37** The following contingency table provides a joint frequency distribution for the population of all U.S. males on active military duty by classification and race. Frequencies are in thousands. [SOURCE: U.S. Department of Defense, unpublished data.]

Race

	White	Black	Other	Total
Officer	2,401	145	176	2,722
Enlisted	11,085	3,392	1,944	16,421
Total	13,486	3,537	2,120	19,143

Class

Suppose a U.S. male on active duty is selected at random. Let W denote the event that a white man is selected and A the event that an officer is selected.

a. Obtain $P(W)$, $P(A)$, and $P(W \& A)$.

b. Determine whether the events "white" and "officer" are statistically independent by employing the special multiplication rule, Formula 4.6 on page 230.

c. Determine whether the observed and expected frequencies for the cell in the upper left-hand corner of the contingency table are equal.

d. Prove that the checks you performed in parts (b) and (c) are equivalent.

e. Are classification and race statistically independent for U.S. males on active military duty? Explain your answer.

12.4 INFERENCES FOR A POPULATION STANDARD DEVIATION

Recall that the standard deviation, σ, of a population is a measure of dispersion (spread, variation) of the population values. A population with a great deal of variation will have a large standard deviation, whereas one with little variation will have a small standard deviation, as illustrated in Fig. 12.8.

FIGURE 12.8
Populations having, respectively, large and small standard deviations

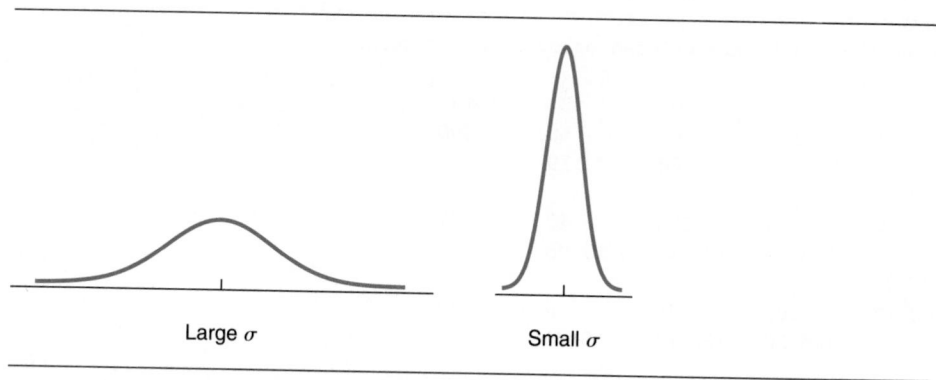

Large σ Small σ

Suppose we want to obtain information about the standard deviation of a population. If the population is small, then we can ordinarily determine σ exactly by first taking a census and then computing σ from the population data. However, if the population is large, which is usually the case, then a census is generally not feasible, and we must therefore use inferential methods to obtain the required information about σ.

In this section we will learn how to perform hypothesis tests and determine confidence intervals for the standard deviation, σ, of a normally distributed population. Consider Example 12.10.

Example 12.10 Introduces Hypothesis Tests for a Population Standard Deviation

A hardware manufacturer produces 10-mm bolts. The manufacturer knows that the diameters of the bolts produced vary somewhat from 10 mm and also from each other. But even if he is willing to accept some variation in bolt diameters, he cannot tolerate too much variation. For if the variation is too large, then too many of the bolts produced will be unusable.

Thus the manufacturer must make sure that the standard deviation, σ, of the bolt diameters is not unduly large. Since in this case it is not possible to obtain the value of σ exactly (why not?), inferential methods must be employed.

Let's suppose it has been determined that an acceptable standard deviation for the bolt diameters is one that is less than 0.09 mm.[†] Knowing this, the manufacturer can decide whether or not there is too much variation in the diameters of the bolts being produced by performing the hypothesis test

$$H_0: \sigma = 0.09 \text{ mm (too much variation)}$$

$$H_a: \sigma < 0.09 \text{ mm (not too much variation)}.$$

If the null hypothesis can be rejected, then the manufacturer can be confident that the variation in bolt diameters is acceptable.[‡]

The basic idea for carrying out the hypothesis test is as follows:

1. Take a random sample of bolts.

2. Compute the standard deviation, s, of the diameters of the bolts sampled.

3. If s is too much smaller than 0.09 mm, reject the null hypothesis in favor of the alternative hypothesis; otherwise, do not reject the null hypothesis.

The manufacturer takes a random sample of 12 bolts and carefully measures their diameters. Table 12.9 displays the diameters, in millimeters.

TABLE 12.9
Diameters (in millimeters) of
12 randomly selected bolts

10.05	10.00	10.02	9.97
10.07	10.03	9.98	10.10
9.95	9.99	10.00	10.08

The sample standard deviation of the bolt diameters in Table 12.9 is

$$s = \sqrt{\frac{\Sigma x^2 - (\Sigma x)^2/n}{n-1}} = \sqrt{\frac{1204.8290 - (120.24)^2/12}{11}} = 0.047 \text{ mm}.$$

Is this value of s too much smaller than 0.09 mm, so that the null hypothesis should be rejected, or can the difference between $s = 0.047$ mm and the null hypothesis value of

---

[†] See Exercise 12.55 for an explanation of how that information could be obtained.

[‡] We could also choose the alternative hypothesis to be $H_a: \sigma > 0.09$ mm. Then rejection of the null hypothesis would indicate that there is definitely too much variation in bolt diameters.

$\sigma = 0.09$ mm be attributed to sampling error? To answer that question, we need to know the probability distribution of s—the **sampling distribution of the standard deviation.** We will therefore discuss that sampling distribution and then return to complete the hypothesis test posed in this example.   ∎

The Sampling Distribution of the Standard Deviation

Recall that to perform a hypothesis test for the mean, μ, of a normally distributed population, we use the random variable

$$t = \frac{\bar{x} - \mu_0}{s/\sqrt{n}}$$

as the test statistic, not simply the random variable $\bar{x}$. Similarly, when performing a hypothesis test for the standard deviation, σ, of a normally distributed population, we do not employ the random variable s as the test statistic. Rather, we use the random variable

$$\chi^2 = \frac{n-1}{\sigma_0^2} s^2.$$

That random variable has a familiar probability distribution.

KEY FACT 12.4

THE SAMPLING DISTRIBUTION OF THE STANDARD DEVIATION[†]

Suppose a random sample of size n is to be taken from a normally distributed population with standard deviation σ. Then the random variable

$$\chi^2 = \frac{n-1}{\sigma^2} s^2$$

has the chi-square distribution with $n-1$ degrees of freedom. Thus probabilities for that random variable are equal to areas under the χ^2-curve with df $= n-1$.

To illustrate Key Fact 12.4, we used Minitab. First we simulated the taking of 1000 random samples of size 12 from a normally distributed population having mean $\mu = 10$ and standard deviation $\sigma = 0.05$ (any sample size, mean, and standard deviation will do). Next we obtained the sample standard deviation, s, of each of the 1000 samples. Then for each of the 1000 samples, we determined the value of

$$\chi^2 = \frac{n-1}{\sigma^2} s^2 = \frac{12-1}{0.05^2} s^2.$$

Finally, we obtained a histogram for the 1000 values of χ^2. Printout 12.4 shows the histogram, which, as expected, is shaped like a χ^2-curve (with df $= 11$).

[†] Strictly speaking, the sampling distribution presented here is not the sampling distribution of the standard deviation but is the sampling distribution of a function of the standard deviation.

PRINTOUT 12.4

Minitab histogram of χ^2 for
1000 samples of size 12
from a normal population

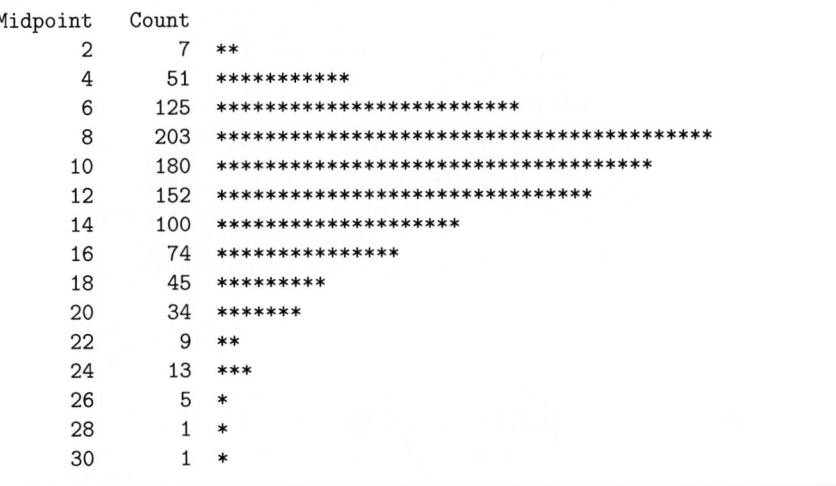

```
Histogram of chisq    N = 1000
Each * represents 5 obs.

Midpoint   Count
     2        7   **
     4       51   **********
     6      125   *************************
     8      203   *****************************************
    10      180   ************************************
    12      152   ******************************
    14      100   ********************
    16       74   ***************
    18       45   *********
    20       34   *******
    22        9   **
    24       13   ***
    26        5   *
    28        1   *
    30        1   *
```

Hypothesis Tests for a Population Standard Deviation

Now that we know the sampling distribution of the standard deviation, we can state a step-by-step method for performing a hypothesis test for a population standard deviation. The method is presented in Procedure 12.3 and is sometimes referred to as the χ^2-**test** for a population standard deviation.

PROCEDURE 12.3

**THE χ^2-TEST FOR A POPULATION STANDARD DEVIATION
WITH NULL HYPOTHESIS H_0: $\sigma = \sigma_0$**

ASSUMPTION

Normal population

Step 1 State the null and alternative hypotheses.

Step 2 Decide on the significance level, α.

Step 3 The critical value(s)

    a. for a two-tailed test are $\chi^2_{1-\alpha/2}$ and $\chi^2_{\alpha/2}$,

    b. for a left-tailed test is $\chi^2_{1-\alpha}$,

    c. for a right-tailed test is χ^2_{α},

    with df $= n - 1$. Use Table V to find the critical value(s).

(continued)

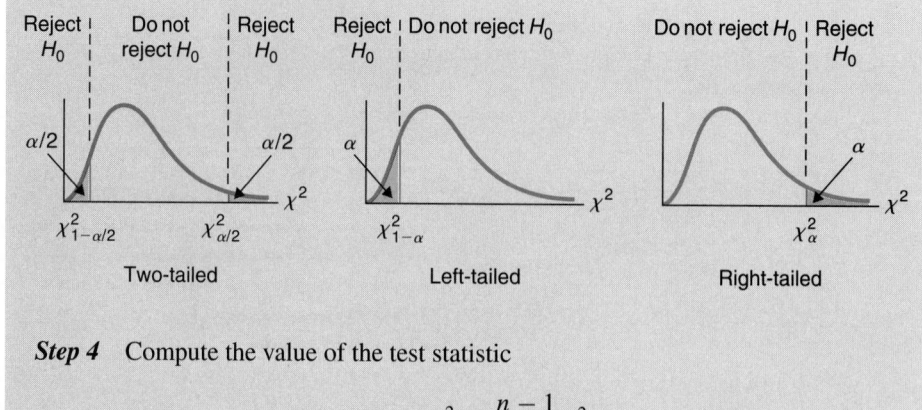

Two-tailed   Left-tailed   Right-tailed

Step 4   Compute the value of the test statistic

$$\chi^2 = \frac{n-1}{\sigma_0^2}\, s^2.$$

Step 5   If the value of the test statistic falls in the rejection region, then reject H_0; otherwise, do not reject H_0.

Step 6   State the conclusion in words.

Unlike the t-tests for one and two population means, the χ^2-test for a population standard deviation is not robust to departures from the normality assumption. In fact, it is so nonrobust that many statisticians advise against using it unless there is considerable evidence that the population is normally distributed or very nearly so. Thus, before applying Procedure 12.3, we should construct a normal probability plot; if the plot creates any doubt about the normality of the population, then we should not use the procedure.

Example 12.11   Illustrates Procedure 12.3

We can now complete the hypothesis test proposed in Example 12.10. Recall that a hardware manufacturer needs to decide whether the standard deviation of bolt diameters is less than 0.09 mm. He randomly samples 12 bolts and measures their diameters. The results are shown in Table 12.9 on page 717. At the 5% significance level, do the data provide sufficient evidence to conclude that the standard deviation, σ, of the diameters of all 10-mm bolts produced by the manufacturer is less than 0.09 mm?

SOLUTION   To begin, we construct a normal probability plot for the data in Table 12.9, as shown in Fig. 12.9. As we see from Fig. 12.9, the normal probability plot is quite linear. This and other data on bolt diameters collected previously by the manufacturer make it reasonable to presume that the diameters of 10-mm bolts produced by this manufacturer are normally distributed. Thus we will apply Procedure 12.3 to perform the required hypothesis test.

FIGURE 12.9

Normal probability plot
for the sample of bolt
diameters in Table 12.9

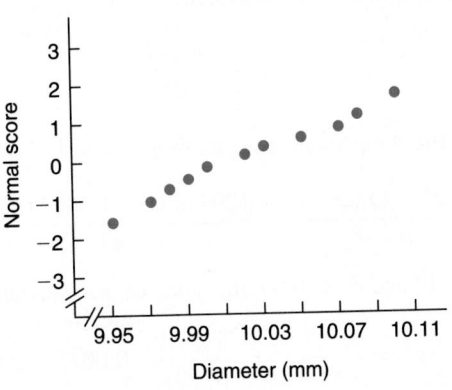

Step 1 *State the null and alternative hypotheses.*

The null and alternative hypotheses are

$$H_0: \sigma = 0.09 \text{ mm (too much variation)}$$
$$H_a: \sigma < 0.09 \text{ mm (not too much variation).}$$

Note that the hypothesis test is left-tailed since a less-than sign ($<$) appears in the alternative hypothesis.

Step 2 *Decide on the significance level, α.*

The test is to be performed at the 5% level of significance. Thus $\alpha = 0.05$.

Step 3 *The critical value for a left-tailed test is $\chi^2_{1-\alpha}$, with df $= n - 1$.*

We have $\alpha = 0.05$. Also, $n = 12$, and so df $= 12 - 1 = 11$. Consulting Table V, we find that the critical value is $\chi^2_{1-\alpha} = \chi^2_{1-0.05} = \chi^2_{0.95} = 4.575$, as seen in Fig. 12.10.

FIGURE 12.10

Criterion for deciding whether or
not to reject the null hypothesis

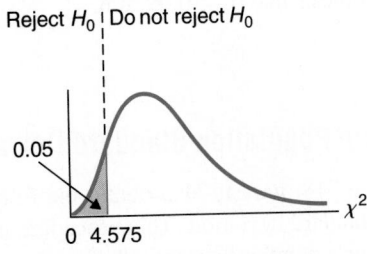

Step 4   *Compute the value of the test statistic*

$$\chi^2 = \frac{n-1}{\sigma_0^2} s^2.$$

First we obtain the sample variance, s^2. From Table 12.9 we find that

$$s^2 = \frac{\Sigma x^2 - (\Sigma x)^2/n}{n-1} = \frac{1204.8290 - (120.24)^2/12}{11} = 0.0022.$$

Therefore since $n = 12$ and $\sigma_0 = 0.09$, the value of the test statistic is

$$\chi^2 = \frac{n-1}{\sigma_0^2} s^2 = \frac{12-1}{(0.09)^2} \cdot 0.0022 = 2.988.$$

Step 5   *If the value of the test statistic falls in the rejection region, reject H_0; otherwise, do not reject H_0.*

From Step 4 the value of the test statistic is $\chi^2 = 2.988$. This falls in the rejection region, as we see by referring to Fig. 12.10. Thus we reject H_0.

Step 6   *State the conclusion in words.*

The test results are statistically significant at the 5% level; that is, at the 5% significance level, the data provide sufficient evidence to conclude that the standard deviation, σ, of the diameters of all 10-mm bolts produced by the manufacturer is less than 0.09 mm. Evidently, the variation in bolt diameters is not too large.   ■

The P-value approach to hypothesis testing can also be used to carry out the hypothesis test in Example 12.11 and to assess more precisely the evidence against the null hypothesis. From Step 4 we see that the value of the test statistic is $\chi^2 = 2.988$. Recalling that df = 11, we find from Table V that $0.005 < P < 0.01$. In particular, because the P-value is less than the significance level of 0.05, we can reject H_0. Furthermore, by referring to Table 9.12 on page 530, we see that the data provide very strong evidence against the null hypothesis, that is, in support of the hypothesis that $\sigma < 0.09$ mm.

Confidence Intervals for a Population Standard Deviation

Using Key Fact 12.4 on page 718, we can also obtain the following confidence-interval procedure for a population standard deviation. This procedure is sometimes referred to as the χ^2-**interval procedure** for a population standard deviation.

PROCEDURE 12.4

THE χ^2-INTERVAL PROCEDURE FOR A POPULATION STANDARD DEVIATION, σ

ASSUMPTION

Normal population

Step 1 For a confidence level of $1 - \alpha$, use Table V to find $\chi^2_{1-\alpha/2}$ and $\chi^2_{\alpha/2}$ with df $= n - 1$.

Step 2 The confidence interval for σ is from

$$\sqrt{\frac{n-1}{\chi^2_{\alpha/2}}} \cdot s \quad \text{to} \quad \sqrt{\frac{n-1}{\chi^2_{1-\alpha/2}}} \cdot s$$

where $\chi^2_{1-\alpha/2}$ and $\chi^2_{\alpha/2}$ are found in Step 1, n is the sample size, and s is computed from the actual sample data obtained.

Like the χ^2-test for one population standard deviation, the χ^2-interval procedure is not at all robust to deviations from the normality assumption; using it on data from a nonnormal population can result in misleading information. The χ^2-interval procedure should not be used unless there is considerable evidence that the population is normally distributed or very nearly so.

Example 12.12 Illustrates Procedure 12.4

Use the sample data in Table 12.9 on page 717 to determine a 95% confidence interval for the standard deviation, σ, of the diameters of all 10-mm bolts produced by the manufacturer.

SOLUTION As we discovered in Example 12.11, it is reasonable to presume that the diameters of 10-mm bolts produced by the manufacturer are normally distributed. Thus we can apply Procedure 12.4 to obtain the required confidence interval.

Step 1 For a confidence level of $1 - \alpha$, use Table V to find $\chi^2_{1-\alpha/2}$ and $\chi^2_{\alpha/2}$ with df $= n - 1$.

We want a 95% confidence interval; so the confidence level is $0.95 = 1 - 0.05$. This means that $\alpha = 0.05$. Also, since $n = 12$, df $= 12 - 1 = 11$. Referring to Table V, we find that

$$\chi^2_{1-\alpha/2} = \chi^2_{1-0.05/2} = \chi^2_{0.975} = 3.816$$

and

$$\chi^2_{\alpha/2} = \chi^2_{0.05/2} = \chi^2_{0.025} = 21.920.$$

Step 2   *The confidence interval for σ is from*

$$\sqrt{\frac{n-1}{\chi^2_{\alpha/2}}} \cdot s \quad to \quad \sqrt{\frac{n-1}{\chi^2_{1-\alpha/2}}} \cdot s.$$

We have $n = 12$, and from Step 1, $\chi^2_{1-\alpha/2} = 3.816$ and $\chi^2_{\alpha/2} = 21.920$. Also, we found in Example 12.10 that $s = 0.047$ mm. Thus a 95% confidence interval for σ is from

$$\sqrt{\frac{12-1}{21.920}} \cdot 0.047 \quad to \quad \sqrt{\frac{12-1}{3.816}} \cdot 0.047,$$

or 0.033 to 0.080. We can be 95% confident that the standard deviation, σ, of the diameters of all 10-mm bolts produced by the manufacturer is somewhere between 0.033 mm and 0.080 mm.   ∎

EXERCISES 12.4

12.38   Give two situations in which it would be important to make an inference about a population standard deviation.

12.39   In using chi-square procedures to make inferences about a population standard deviation, why is it important that the population being sampled is normally distributed or nearly so?

12.40   What does σ measure?

Preliminary data analyses and other information indicate that it is reasonable to assume the data sets in Exercises 12.41–12.46 are samples from normally distributed populations. For each exercise, perform the required hypothesis test using either the classical approach or the P-value approach.

12.41   Each year, thousands of high-school students bound for college take the Scholastic Aptitude Test (SAT). The test, provided by the Educational Testing Service of Princeton, New Jersey, measures the verbal and mathematical abilities of prospective college students. Scores are reported on a scale that ranges from a low of 200 to a high of 800; this scale was introduced in 1941. At that time the standard deviation of scores was 100 points. A random sample of 25 verbal scores for last year yields the following data.

346	496	352	378	315
491	360	385	500	558
381	303	434	562	496
420	485	446	479	422
494	289	436	516	615

Do the data provide sufficient evidence to conclude that the standard deviation, σ, of last year's verbal scores is different from the 1941 standard deviation of 100? Perform the required hypothesis test at the 5% significance level. *(Note:* The sample standard deviation of the data is 85.5.*)*

12.42   A company produces cans of stewed tomatoes with an advertised weight of 14 oz. Recently, the company hired a quality-control engineer, who has performed some statistical analyses and is satisfied that, on the average, the cans do contain 14 oz of stewed tomatoes. However, she must also be concerned about the variability of the weights. Specifically, in order to ensure that the vast majority of the cans have weights within 10% of the advertised weight, the standard deviation, σ, of the weights must be less than 0.47 oz. The weights, in ounces, of 10 randomly selected cans are as shown in the following table.

13.85	13.95	13.90	13.49	14.17
14.33	14.03	13.48	14.27	14.19

Do the data provide sufficient evidence to conclude that the standard deviation of the weights is less than 0.47 oz? Perform the hypothesis test using $\alpha = 0.05$. *(Note:* The sample standard deviation of the data is 0.298 oz.*)*

12.43   A manufacturer of watches claims that the weekly error made by the watches she produces has a standard deviation of about 1 second. To test that claim, a random sample of 20 watches are set to the correct time. After 1 week, the error made by each watch is recorded. The results, in seconds, are as follows.

0.6	2.3	2.0	−2.1	−1.4
−0.5	1.5	−0.3	0.4	0.6
0.4	−2.2	0.7	0.5	−1.3
−2.0	2.6	−0.8	1.0	−0.6

Can we conclude that the standard deviation, σ, of the weekly errors exceeds the 1-second claim made by the manufacturer? Use $\alpha = 0.01$. *(Note: $\Sigma x = 1.4$ and $\Sigma x^2 = 39.32$.)*

12.44 The designer of a 100-point aptitude test has attempted to construct the test so that the standard deviation of scores is 10 points. For a preliminary documentation, 30 randomly selected people are given the test. Their scores are as follows.

83	43	29	37	64	69
79	67	85	61	52	55
70	65	43	64	60	35
76	46	35	43	51	52
61	50	41	87	62	59

Can we conclude that the standard deviation of all scores on the aptitude test is not 10 points? Use $\alpha = 0.10$. *(Note: $\Sigma x = 1724$ and $\Sigma x^2 = 106,172$.)*

12.45 A coffee machine is supposed to dispense 6 fluid oz of coffee into a paper cup. In reality, the amounts dispensed vary from cup to cup. However, if the machine is working properly, then most of the cups will contain within 10% of the advertised 6 fluid oz. This means that the standard deviation of the amounts dispensed should be less than 0.2 fluid oz. A random sample of 15 cups provided the following data, in fluid ounces.

5.90	5.82	6.20	6.09	5.93
6.18	5.99	5.79	6.28	6.16
6.00	5.85	6.13	6.09	6.18

At the 5% significance level, do the data provide sufficient evidence to conclude that the standard deviation of the amounts being dispensed is less than 0.2 fluid oz? *(Note: $s = 0.154$.)*

12.46 Gas-mileage estimates for cars and light-duty trucks are determined and published by the Environmental Protection Agency (EPA). According to the EPA, "... the mileages obtained by most drivers will be within plus or minus 15 percent of the [EPA] estimates. ..." The mileage estimate given for one 1993 model is 23 mpg on the highway. If the EPA claim is true, then the standard deviation of mileages

should be about $0.15 \cdot 23/3 = 1.15$ mpg. A random sample of 12 cars of this model yields the following highway mileages.

24.1	23.3	22.5	23.2
22.3	21.1	21.4	23.4
23.5	22.8	24.5	24.3

At the 5% significance level, do the data suggest that the standard deviation of highway mileages for all 1993 cars of this model is different from 1.15 mpg? *(Note: $s = 1.071$.)*

In each of Exercises 12.47–12.52, use Procedure 12.4 on page 723 to obtain the required confidence interval.

12.47 Refer to Exercise 12.41. Obtain a 95% confidence interval for the standard deviation, σ, of last year's verbal SAT scores.

12.48 Refer to Exercise 12.42. Find a 90% confidence interval for the standard deviation, σ, of the weights of all cans of stewed tomatoes produced by the company.

12.49 Refer to Exercise 12.43. Obtain a 98% confidence interval for the standard deviation, σ, of the weekly errors of the watches manufactured.

12.50 Refer to Exercise 12.44. Obtain a 90% confidence interval for the standard deviation, σ, of all scores on the aptitude test.

12.51 Refer to Exercise 12.45. Find a 90% confidence interval for the standard deviation, σ, of the amounts of coffee being dispensed.

12.52 Refer to Exercise 12.46. Find a 95% confidence interval for the standard deviation of highway gas mileages for all 1993 cars of the model in question.

12.53 Refer to Exercise 12.45. Why is it important that the standard deviation, σ, of the amounts of coffee being dispensed not be too large?

12.54 Refer to Exercise 12.46. Why is it useful to know the standard deviation of the gas mileages as well as the mean of the gas mileages?

12.55 In the bolt-manufacturing problem of Example 12.10, we assumed it had been determined that an acceptable standard deviation, σ, for the bolt diameters is one that is less than 0.09 mm. We will now see how such information might be obtained. Let's suppose the manufacturer has set the tolerance specifications for the 10-mm bolts at ±0.3 mm; that is, a bolt's diameter is considered satisfactory if it is between 9.7 mm and 10.3 mm. Further suppose the manufacturer has decided that less than 0.1% (1 out of 1000) of the bolts produced should be defective.

a. Let x denote the diameter of a randomly selected bolt. Show that the manufacturer's production criteria can be expressed mathematically as

$$P(9.7 \leq x \leq 10.3) > 0.999.$$

b. Draw a normal-curve picture that illustrates the equation $P(9.7 \leq x \leq 10.3) = 0.999$. Include an x-axis and a z-axis. Assume $\mu = 10$ mm.

c. Deduce from your picture in part (b) that the manufacturer's production criteria are equivalent to the condition that $0.3/\sigma > z_{0.0005}$.

d. Use part (c) to conclude that the manufacturer's production criteria are equivalent to requiring that the standard deviation of bolt diameters be less than 0.09 mm, that is, $\sigma < 0.09$ mm.

· ·   **12.56 (Computer exercise)** Consider a normally distributed population having mean 100 and standard deviation 16.

a. Simulate the taking of 2000 random samples of size four from the population.

b. Determine the sample standard deviation of each of the 2000 samples.

c. For each of the 2000 samples, obtain the quantity

$$\frac{n-1}{\sigma^2} s^2 = \frac{4-1}{16^2} s^2.$$

d. Obtain a histogram of the 2000 values found in part (c).

e. Theoretically, what is the distribution of the random variable in part (c)?

f. Compare your results from parts (d) and (e).

· · ·   **12.57**   This exercise justifies Procedure 12.4.

a. Use Key Fact 12.4 on page 718 to show that

$$P\left(\chi^2_{1-\alpha/2} < \frac{n-1}{\sigma^2} s^2 < \chi^2_{\alpha/2}\right) = 1 - \alpha.$$

b. Deduce from part (a) that

$$P\left(\frac{n-1}{\chi^2_{\alpha/2}} \cdot s^2 < \sigma^2 < \frac{n-1}{\chi^2_{1-\alpha/2}} \cdot s^2\right) = 1 - \alpha.$$

c. Prove that the result in part (b) implies that once the sample is taken, the interval from

$$\sqrt{\frac{n-1}{\chi^2_{\alpha/2}}} \cdot s \quad \text{to} \quad \sqrt{\frac{n-1}{\chi^2_{1-\alpha/2}}} \cdot s$$

will be a $(1-\alpha)$-level confidence interval for the population standard deviation, σ.

Chapter   Review

Key Terms

associated, *699*
χ^2-interval procedure, *722*
χ^2-test, *719*
χ^2_α, *688*
chi-square (χ^2) curve, *688*
chi-square distributions, *688*
chi-square goodness-of-fit test, *691*
chi-square independence test, *699*
chi-square procedures, *687*
ChiSquare,* *710*

CHISQUARE,* *710*
expected frequencies, *693*
nonassociated, *699*
observed frequencies, *693*
sampling distribution of the standard
    deviation, *718*
segmented bar graph, *701*
statistically dependent, *699*
statistically independent, *699*

Formulas

In the formulas below,

O = observed frequency
E = expected frequency
k = number of categories in a probability
    or percentage distribution
p = probability or proportion
n = sample size

R = row total in a contingency table
C = column total in a contingency table
r = number of rows in a contingency table
c = number of columns in a contingency table
σ = population standard deviation
s = sample standard deviation

Expected frequencies for a chi-square goodness-of-fit test, *694*

$$E = np$$

Test statistic for a chi-square goodness-of-fit test, *695*

$$\chi^2 = \Sigma(O - E)^2/E$$

with df $= k - 1$.

Expected frequencies for a chi-square independence test, *706*

$$E = \frac{R \cdot C}{n}$$

Test statistic for a chi-square independence test, *707*

$$\chi^2 = \Sigma(O - E)^2/E$$

with df $= (r - 1)(c - 1)$.

χ^2-test statistic for H_0: $\sigma = \sigma_0$ (normal population), *720*

$$\chi^2 = \frac{n - 1}{\sigma_0^2} s^2$$

with df $= n - 1$.

χ^2-interval for σ (normal population), *723*

$$\sqrt{\frac{n - 1}{\chi_{\alpha/2}^2}} \cdot s \quad \text{to} \quad \sqrt{\frac{n - 1}{\chi_{1-\alpha/2}^2}} \cdot s$$

with df $= n - 1$.

You Should Be Able To

1. use and understand the preceding formulas.
2. use the chi-square table, Table V.
3. explain the reasoning behind the chi-square goodness-of-fit test.
4. perform a chi-square goodness-of-fit test for the percentage distribution of a population or the probability distribution of a random variable.
5. decide whether two characteristics of a population are statistically dependent, given the data for the entire population.
6. explain the reasoning behind the chi-square independence test.
7. perform a chi-square independence test to decide whether two characteristics of a population are statistically dependent.
8. perform a hypothesis test for the standard deviation, σ, of a normally distributed population.
9. obtain a confidence interval for the standard deviation, σ, of a normally distributed population.
10. use the Minitab commands covered in this chapter.*
11. interpret the output obtained from the application of the Minitab commands discussed in this chapter.*

REVIEW TEST

1. Consider a χ^2-curve with 17 degrees of freedom. Use Table V to determine
   a. $\chi^2_{0.99}$.   b. $\chi^2_{0.01}$.
   c. the χ^2-value having area 0.05 to its right.
   d. the χ^2-value having area 0.05 to its left.
   e. the two χ^2-values that divide the area under the curve into a middle 0.95 area and two outside 0.025 areas.

2. The Census Bureau and Department of Housing and Urban Development publish information on characteristics of new, privately owned, one-family homes in *Characteristics of New Housing.* A percentage distribution for the number of bedrooms in homes completed in 1987 follows.

Number of bedrooms	Percentage
2 or less	19
3	58
4 or more	23

A project called *Homestyle 1988,* directed by the Impulse Research Corporation of Santa Monica, California, was designed to provide builders with a profile of the next generation of home buyers in Arizona, California, and Nevada. Researchers conducted telephone interviews with 150 randomly selected potential home buyers between the ages of 24 and 35. They obtained the data on preferences for the number of bedrooms, as shown in the following frequency distribution.

Number of bedrooms	Frequency
2 or less	47
3	93
4 or more	10

Do the data provide sufficient evidence to conclude that the actual preference distribution for the number of bedrooms is different from the distribution for the number of bedrooms in homes completed in 1987? Perform the required hypothesis test at the 5% significance level.

3. According to *Current Population Reports,* published by the U.S. Bureau of the Census, in 1991 the distribution of the U.S. resident population by age and region of residence is as shown in the following contingency table. The frequencies are in millions. Also, NE and MW stand for Northeast and Midwest, respectively.

Region of residence

Age	NE	MW	South	West	Total
Under 5	3.7	4.5	6.5	4.5	19.2
5–17	8.5	11.3	16.0	10.1	45.9
18–44	21.9	25.3	37.3	24.1	108.6
45–64	9.9	11.2	16.2	9.4	46.7
65 & over	7.0	7.9	10.9	5.9	31.7
Total	51.0	60.2	86.9	54.0	252.1

a. Determine the age distribution within each region of residence.
b. Are age and region of residence associated?
c. Construct a segmented bar graph to illustrate your results in parts (a) and (b).
d. Answer true or false without doing any further calculations: The region-of-residence distributions within the five age groups are identical.

4. The Gallup Organization asked 1528 adults the following question: "The New Jersey Supreme Court recently ruled that all life-sustaining medical treatment may be withheld or withdrawn from terminally ill patients, provided that is what the patients want or would want if they were able to express their wishes. Would you like to see such a ruling in the state in which you live, or not?" The following contingency table presents a modified version of the results, which cross classify response by educational level.

Response

Educational level	Favor	Oppose	No opinion	Total
College grad	264	17	6	287
Some college	205	26	7	238
High-school grad	461	81	34	576
Non–high-school grad	290	81	56	427
Total	1220	205	103	1528

Can we conclude from the data that response and educational level are statistically dependent? Perform the required hypothesis test at the 1% significance level.

*5. **(Computer problem)** Use Minitab or some other statistical software to perform the hypothesis test in Problem 4.

*6. **(Computer problem)** Printout 12.5 shows the output obtained by applying Minitab's ChiSquare command to the cell data in the columns of the contingency table in Problem 4. Use the printout to determine

   a. the number of people sampled who would be opposed to the New Jersey Supreme Court ruling in their state.

   b. the number of people sampled who had some college. *(Note:* The row labeled "2" in Printout 12.5 corresponds to the "Some college" category, as we can see from the contingency table in Problem 4.*)*

   c. the observed number of people sampled who had some college and would be opposed to such a ruling in their state.

   d. the expected number of people sampled who had some college and would be opposed to such a ruling in their state if response and educational level are statistically independent.

   e. the chi-square subtotal, $(O - E)^2/E$, for the "Some college *and* Oppose" cell.

   f. the number of degrees of freedom.

   g. the value of the test statistic, χ^2.

   h. the conclusion if the hypothesis test concerning statistical independence of response and educational level is performed at the 1% significance level. *(Note:* Consult Table V.*)*

7. IQs measured on the Stanford Revision of the Binet-Simon Intelligence Scale are supposed to have a standard deviation of 16 points. Twenty-five randomly selected people were given the IQ test; here are the data that were obtained.

91	96	106	116	97
102	96	124	115	121
95	111	105	101	86
88	129	112	82	98
104	118	127	66	102

PRINTOUT 12.5   Minitab output for Problem 6

```
Expected counts are printed below observed counts

         FAVOR    OPPOSE   NOOPIN    Total
   1       264       17        6      287
        229.15    38.50    19.35

   2       205       26        7      238
        190.03    31.93    16.04

   3       461       81       34      576
        459.90    77.28    38.83

   4       290       81       56      427
        340.93    57.29    28.78

Total     1220      205      103     1528

ChiSq =   5.300 + 12.010 +  9.207 +
          1.180 +  1.102 +  5.097 +
          0.003 +  0.179 +  0.600 +
          7.608 +  9.815 + 25.735 = 77.837

df = 6
```

Preliminary data analyses and other information indicate that it is reasonable to presume that IQs measured on the Stanford Revision of the Binet-Simon Intelligence Scale are normally distributed.

a. Do the data provide sufficient evidence to conclude that IQs measured on this scale have a standard deviation different from 16 points? Perform the required hypothesis test at the 10% significance level. *(Note: $s = 15.006$.)*

b. How crucial is the normality assumption for the hypothesis test you performed in part (a)?

8. Refer to Problem 7. Determine a 90% confidence interval for the standard deviation of IQs.

Using the Focus Database

The Focus database, which is printed in Appendix B, contains data on 500 randomly selected Arizona State University sophomores for seven different variables, among which are sex, high-school GPA, and cumulative GPA. Use Minitab or some other statistical software to perform the chi-square independence tests below. Employ the coding scheme 0, 1, 2, 3, and 4, respectively, for grade point averages less than 1, at least 1 but less than 2, at least 2 but less than 3, at least 3 but less than 4, and (exactly) 4.

a. At the 5% significance level, do the data provide sufficient evidence to conclude that sex and high-school GPA are statistically dependent?

b. Repeat part (a) for sex and cumulative GPA.

Case Study

Time On or Off the Job: Which Do Americans Enjoy More?

In May 1993 a CNN/*USA Today* poll conducted by The Gallup Organization asked a sample of employed Americans the following question: "Which do you enjoy more, the hours when you are on your job, or the hours when you are not on your job?" The responses to this question were cross tabulated against several characteristics, among which were sex, age, type of community, amount of education, income, and type of employer.

Table 12.10 summarizes the poll's results. Each number represents the frequency of people in the category specified.

a. At the 5% significance level, do the data provide sufficient evidence to conclude that sex and response to the question ("Which do you enjoy more, the hours when you are on your job, or the hours when you are not on your job?") are statistically dependent?

b. At the 5% significance level, do the data provide sufficient evidence to conclude that age and response to the question are statistically dependent?

c. At the 5% significance level, do the data provide sufficient evidence to conclude that type of community and response to the question are statistically dependent?

d. At the 5% significance level, do the data provide sufficient evidence to conclude that amount of education and response to the question are statistically dependent?

e. At the 5% significance level, do the data provide sufficient evidence to conclude that income and response to the question are statistically dependent?

f. At the 5% significance level, do the data provide sufficient evidence to conclude that type of employer and response to the question are statistically dependent?

g. Use Minitab or some other statistical software to perform the hypothesis tests in parts (a)–(f).

TABLE 12.10

Joint frequency distributions obtained from a sample of employed Americans responding to the question about which they enjoy more, time on or off the job

	On the job	Off the job	Don't know
Male	77	263	28
Female	77	215	27
18–29 years	33	136	5
30–49 years	77	274	25
50–64 years	35	56	19
65 and older	9	11	5
Urban	62	197	14
Suburban	47	171	25
Rural	42	109	15
Postgraduate	23	51	10
College graduate	41	126	18
Some college	20	108	5
No college	68	192	21
Under $20,000	40	80	11
$20,000–$29,999	32	116	7
$30,000–$49,999	41	131	8
$50,000 and over	34	138	22
Private	67	326	27
Government	23	82	11
Self	61	69	14

Regression, Correlation, and ANOVA

Adrien Legendre

Adrien-Marie Legendre was born in Paris, France, on September 18, 1752, the son of a moderately wealthy family. He studied at the Collège Mazarin and received degrees in mathematics and physics in 1770 at the age of 18.

Although Legendre's financial assets were sufficient to allow him to devote himself to research, he took a position teaching mathematics at the École Militaire in Paris from 1775 to 1780. In March 1783 he was elected to the Academie des Sciences in Paris, and in 1787 he was assigned to a project undertaken jointly by the observatories at Paris and at Greenwich, England. At this time he became a fellow of the Royal Society.

As a result of the French Revolution, which began in 1789, Legendre lost his "small fortune" and was forced to find work. He held various positions during the early 1790s, among which were commissioner of astronomical operations for the Academie des Sciences, professor of pure mathematics at the Institut de Marat, and head of the National Executive Commission of Public Instruction.

During this same period, Legendre wrote a geometry book that became the major text used in elementary geometry courses for nearly a century.

Legendre's major contribution to statistics was the publication, in 1805, of the first statement and the first application of the most widely used, nontrivial technique of statistics—the method of least squares. Stigler writes in *The History of Statistics*, "[Legendre's] presentation . . . must be counted as one of the clearest and most elegant introductions of a new statistical method in the history of statistics." Because Gauss also claimed the method of least squares, there was strife between the two men. Although evidence shows that Gauss was not successful in any communication of the method prior to 1805, his development of the method was crucial to its usefulness.

In 1813 Legendre was appointed Chief of the Bureau des Longitudes, where he remained until his death, following a long illness, in Paris on January 10, 1833.

Chapter 13

Descriptive Methods in Regression and Correlation

. . . . .

General Objective We frequently want to know whether two or more variables are related, and if they are, how they are related. For instance, is there a relationship between SAT scores and college GPA? If these variables are related, how are they related? Or assume the president of a large corporation knows sales tend to increase as advertising expenditures increase but needs to know how strong that tendency is and how she can predict the approximate sales that will result from various advertising expenditures.

Some commonly used methods for examining the relationship between two or more variables and for making predictions are *linear regression* and *correlation*. Descriptive methods in linear regression and correlation will be discussed in this chapter. Inferential methods in linear regression and correlation will be discussed in Chapter 14.

Chapter Outline

Case Study *Fat Consumption and Prostate Cancer*

Is there a relationship between nutrition and cancer? Many studies indicate that the answer is yes. In the case study at the end of this chapter, we will investigate how fat consumption is related to prostate cancer among nations in the world.

13.1   LINEAR EQUATIONS WITH ONE INDEPENDENT VARIABLE

To understand linear regression, we first need to review linear equations with one independent variable. The general form of a **linear equation** with one independent variable can be written as

$$y = b_0 + b_1 x,$$

where b_0 and b_1 are constants (fixed numbers), x is the independent variable, and y is the dependent variable.[†]

The graph of a linear equation with one independent variable is a **straight line;** furthermore, any nonvertical straight line can be represented by such an equation. Three examples of linear equations with one independent variable are $y = 4 + 0.2x$, $y = -1.5 - 2x$, and $y = -3.4 + 1.8x$. In Fig. 13.1, we have drawn the straight-line graphs of these three linear equations.

FIGURE 13.1
Straight-line graphs of
three linear equations

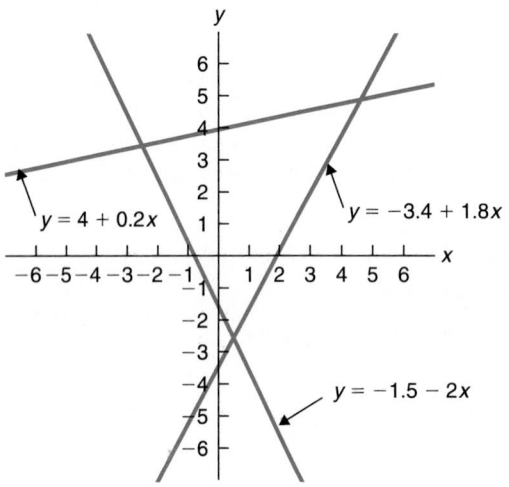

Linear equations with one independent variable occur frequently in applications of mathematics to many different fields, including the management, life, and social sciences, as well as the physical and mathematical sciences. In Examples 13.1 and 13.2, we illustrate the use of linear equations in a simple business application.

[†] You may be familiar with the form $y = mx + b$ instead of the form $y = b_0 + b_1 x$. In statistics the latter form is preferred because it allows a smoother transition to multiple regression, in which there is more than one independent variable. We will discuss multiple regression in Section 13.5.

Example 13.1   **Linear Equations**

CJ^2 Business Services does word processing as one of its basic functions. Its rate is \$20/hr plus a \$25 disk charge. The total cost to a customer depends, of course, on the number of hours it takes to complete the job. Find the equation that expresses the total cost in terms of the number of hours required to complete the job.

SOLUTION   Let x denote the number of hours required to complete the job and y denote the total cost to the customer. Since the rate for word processing is \$20/hr, a job that takes x hours will cost \$20x plus the \$25 disk charge. Therefore the total cost, y, of a job that takes x hours is $y = 25 + 20x$.   ■

The equation, $y = 25 + 20x$, for the total cost of a word-processing job is a linear equation; here $b_0 = 25$ and $b_1 = 20$. Using the equation, we can find the exact cost for a job once we know the number of hours required. For instance, a job that takes 5 hours will cost $y = 25 + 20 \cdot 5 = \$125$; a job that takes 7.5 hours will cost $y = 25 + 20 \cdot 7.5 = \$175$. Table 13.1 displays these and a few other cost illustrations.

TABLE 13.1
Times and costs for five
word-processing jobs

Time (hrs) x	5.0	7.5	15.0	20.0	22.5
Cost (\$) y	125	175	325	425	475

As we have already mentioned, a linear equation, such as $y = 25 + 20x$, has a straight-line graph. We can obtain the graph of $y = 25 + 20x$ by plotting the points in Table 13.1 and connecting them with a straight line, which is done in Fig. 13.2.

FIGURE 13.2
Graph of $y = 25 + 20x$, obtained
from the points in Table 13.1

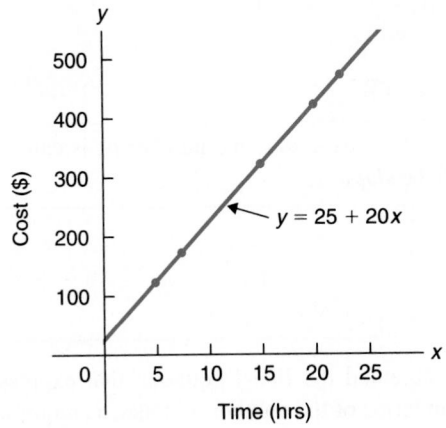

The graph in Fig. 13.2 is useful for quickly estimating cost. For example, a glance at the graph shows that a 10-hour job will cost somewhere between $200 and $300. The exact cost is $y = 25 + 20 \cdot 10 = \$225$.

Intercept and Slope

For a linear equation $y = b_0 + b_1x$, the numbers b_0 and b_1 have an important geometric interpretation. The number b_0 is the y-value at which the straight-line graph of the linear equation intersects the y-axis. The number b_1 measures the steepness of the straight line; more precisely, b_1 indicates how much the y-value on the straight line increases (or decreases) when the x-value increases by 1 unit. Figure 13.3 illustrates these relationships.

FIGURE 13.3
Graph of $y = b_0 + b_1x$

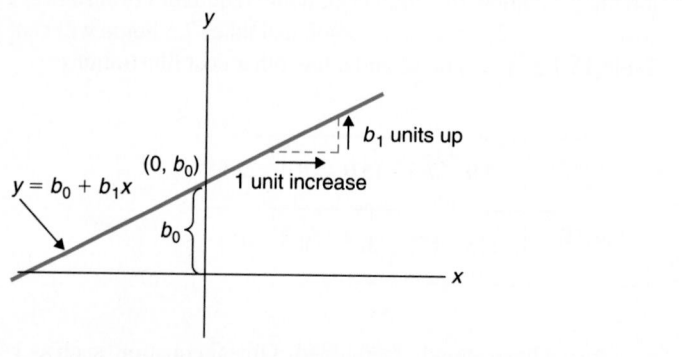

Because of the geometric interpretation, described above, of the numbers b_0 and b_1, they are given special names that reflect that interpretation. These special names are identified in Definition 13.1.

DEFINITION 13.1 y-INTERCEPT AND SLOPE

For a linear equation $y = b_0 + b_1x$, the number b_0 is called the **y-intercept** and the number b_1 is called the **slope.**

Example 13.2 y-Intercept and Slope

In Example 13.1 we obtained the linear equation that expresses the total cost, y, of a word-processing job in terms of the number of hours, x, required to complete the job. The equation is $y = 25 + 20x$.

a. Find the y-intercept and slope of that linear equation.

b. Interpret the y-intercept and slope in terms of the graph of the equation.

c. Interpret the y-intercept and slope in terms of word-processing costs.

SOLUTION   a. The y-intercept for the equation is $b_0 = 25$ and the slope is $b_1 = 20$.

b. The y-intercept $b_0 = 25$ is the y-value at which the straight line $y = 25 + 20x$ intersects the y-axis. The slope $b_1 = 20$ indicates that the y-value increases by 20 units for every increase in x of 1 unit. See Fig. 13.4.

FIGURE 13.4
Graph of
$y = 25 + 20x$

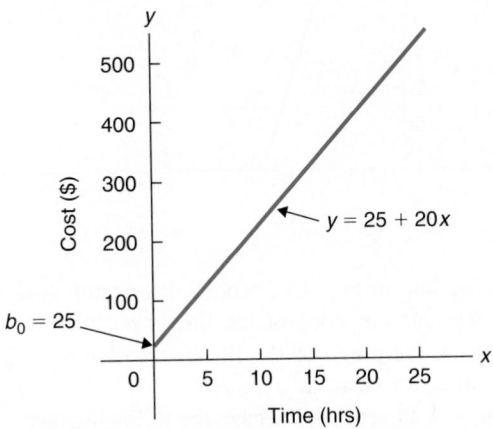

c. In terms of word-processing costs, the y-intercept $b_0 = 25$ represents the total cost of a job that takes 0 hours. In other words, the y-intercept of $25 is a fixed cost that is always charged no matter how long the job takes. The slope $b_1 = 20$ represents the fact that the cost per hour is $20; it is the amount that the total cost, y, goes up for every increase of 1 hour in the time, x, required to complete the job. ∎

A straight line is determined by any two distinct points that lie on the line. This means that the straight-line graph of a linear equation, $y = b_0 + b_1 x$, can be obtained by first substituting two different x-values into the equation to get two distinct points and then connecting those two points with a straight line.

For example, to graph the linear equation $y = 5 - 3x$, we can use the x-values 1 and 3 (or any other two x-values). The y-values corresponding to those two x-values are $y = 5 - 3 \cdot 1 = 2$ and $y = 5 - 3 \cdot 3 = -4$, respectively. Consequently, the graph of the linear equation $y = 5 - 3x$ is the straight line that passes through the two points $(1, 2)$ and $(3, -4)$, as depicted in Fig. 13.5.

FIGURE 13.5
Graph of
$y = 5 - 3x$

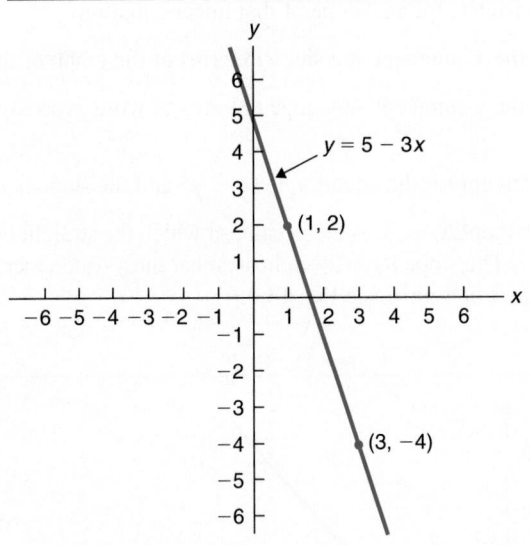

Notice that the line in Fig. 13.5 slopes downward—the y-values decrease as x increases. This is because the slope of the line is negative: $b_1 = -3 < 0$. Now look at the line in Fig. 13.4, the graph of the linear equation $y = 25 + 20x$. That line slopes upward—the y-values increase as x increases. This is because the slope of the line is positive: $b_1 = 20 > 0$. In general, we have the following fact.

KEY FACT 13.1   **GRAPHICAL INTERPRETATION OF SLOPE**

The straight-line graph of the linear equation $y = b_0 + b_1 x$ slopes upward if $b_1 > 0$, slopes downward if $b_1 < 0$, and is horizontal if $b_1 = 0$, as shown in Fig. 13.6.

FIGURE 13.6
Graphical interpretation of slope

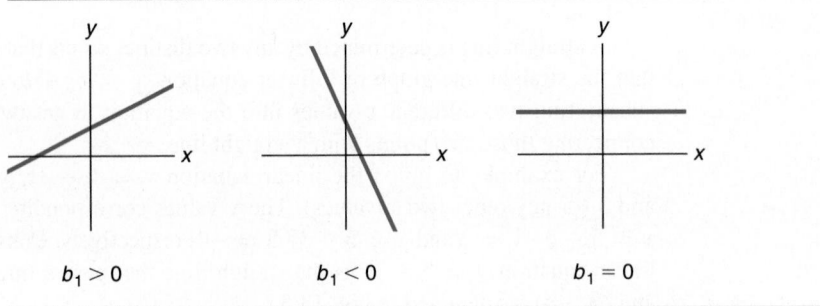

EXERCISES 13.1

13.1 On January 8, 1990, the Avis Rent-A-Car rate for renting a midsize car was $41.88 per day plus $0.33 per mile. For a 1-day rental, let x denote the number of miles driven and y denote the total cost.
a. Obtain the equation that expresses y in terms of x.
b. Find b_0 and b_1.
c. Construct a table similar to Table 13.1 on page 737 for the x-values 50, 100, and 250 miles.
d. Draw the graph of the equation that you obtained in part (a) by plotting the points from part (c) and connecting them with a straight line.
e. Apply the graph from part (d) to visually estimate the cost of driving the car 150 miles. Then calculate that cost exactly using the equation from part (a).

13.2 Encore Air Conditioning charges $36 per hour plus a $30 service charge. Let x denote the number of hours it takes for a job and y denote the total cost to the customer.
a. Obtain the equation that expresses y in terms of x.
b. Find b_0 and b_1.
c. Construct a table similar to Table 13.1 on page 737 for the x-values 0.5, 1, and 2.25 hours.
d. Draw the graph of the equation that you obtained in part (a) by plotting the points from part (c) and connecting them with a straight line.
e. Apply the graph from part (d) to visually estimate the cost of a job that takes 1.75 hours. Then calculate that cost exactly using the equation from part (a).

13.3 The most commonly used scales for measuring temperature are the Fahrenheit and Celsius scales. If we let y denote Fahrenheit temperature and x denote Celsius temperature, then we can express the relationship between those two scales with the linear equation $y = 32 + 1.8x$.
a. Determine b_0 and b_1.
b. Find the Fahrenheit temperatures corresponding to the following Celsius temperatures: $-40°$, $0°$, $20°$, and $100°$.
c. Graph the linear equation $y = 32 + 1.8x$ using the four points found in part (b).
d. Apply the graph obtained in part (c) to visually estimate the Fahrenheit temperature corresponding to a Celsius temperature of $28°$. Then calculate that temperature exactly by employing the linear equation $y = 32 + 1.8x$.

13.4 A ball is thrown straight up in the air with an initial velocity of 64 ft per second. According to the laws of physics, if we let y denote the velocity of the ball after x seconds, then $y = 64 - 32x$.
a. Determine b_0 and b_1 for this linear equation.

b. Find the velocity of the ball after 1, 2, 3, and 4 seconds.
c. Graph the linear equation $y = 64 - 32x$ using the four points obtained in part (b).
d. Use the graph from part (c) to visually estimate the velocity of the ball after 1.5 seconds. Then calculate that velocity exactly by employing the linear equation $y = 64 - 32x$.

In each of Exercises 13.5–13.8,
a. *determine the y-intercept and slope of the specified linear equation.*
b. *explain what the y-intercept and slope represent in terms of the graph of the equation.*
c. *explain what the y-intercept and slope represent in terms relating to the application.*

13.5 $y = 41.88 + 0.33x$ (from Exercise 13.1)

13.6 $y = 30 + 36x$ (from Exercise 13.2)

13.7 $y = 32 + 1.8x$ (from Exercise 13.3)

13.8 $y = 64 - 32x$ (from Exercise 13.4)

In each of Exercises 13.9–13.18, you are given a linear equation. For each exercise,
a. *find the y-intercept and slope.*
b. *determine whether the line slopes upward, slopes downward, or is horizontal, without graphing the equation.*
c. *graph the equation using two points.*

13.9 $y = 3 + 4x$

13.10 $y = -1 + 2x$

13.11 $y = 6 - 7x$

13.12 $y = -8 - 4x$

13.13 $y = 0.5x - 2$

13.14 $y = -0.75x - 5$

13.15 $y = 2$

13.16 $y = -3x$

13.17 $y = 1.5x$

13.18 $y = -3$

In each of Exercises 13.19–13.26, we have identified the y-intercept and slope of a straight line. For each exercise,

a. determine whether the line slopes upward, slopes downward, or is horizontal, without graphing the equation.
b. find the equation of the line.
c. graph the equation using two points.

·   **13.19** $b_0 = 5, b_1 = 2$

·   **13.20** $b_0 = -3, b_1 = 4$

·   **13.21** $b_0 = -2, b_1 = -3$

·   **13.22** $b_0 = 0.4, b_1 = 1$

·   **13.23** $b_0 = 0, b_1 = -0.5$

·   **13.24** $b_0 = -1.5, b_1 = 0$

·   **13.25** $b_0 = 3, b_1 = 0$

·   **13.26** $b_0 = 0, b_1 = 3$

· ·   **13.27**   On page 736 we stated that any nonvertical straight line can be described by an equation of the form $y = b_0 + b_1x$.

a. Why can't a vertical straight line be expressed in this form?
b. What is the form of the equation of a vertical straight line?
c. Does a vertical straight line have a slope? Explain your answer.

13.2   THE REGRESSION EQUATION

· · · · ·

In Examples 13.1 and 13.2, we discussed the linear equation $y = 25 + 20x$, which expresses the total cost, y, of a word-processing job in terms of the time in hours, x, required to complete the job. Given the amount of time required, x, we can use the equation to determine the *exact* cost of the job, y.

Real-life applications are not usually as simple as the word-processing example, in which one variable (cost) can be predicted exactly in terms of another variable (time required). More often than not, we must be content with rough predictions. For instance, we cannot predict the exact price, y, of a Nissan Z just by knowing its age, x. Indeed, even for a fixed age, say, 3 years old, the price of a Nissan Z varies from car to car. We must be satisfied with making a rough prediction for the price of a 3-year-old Nissan Z or with an estimate of the mean price of all 3-year-old Nissan Zs.

Table 13.2 displays data on age and price for a sample of 11 Nissan Zs. The data were obtained from the *Asian Import* edition of the *Auto Trader* magazine. Ages are in years; prices are in hundreds of dollars, rounded to the nearest hundred dollars.

TABLE 13.2

Age and price data for a sample of 11 Nissan Zs

Car	Age (yrs) x	Price ($100s) y
1	5	85
2	4	103
3	6	70
4	5	82
5	5	89
6	5	98
7	6	66
8	6	95
9	2	169
10	7	70
11	7	48

It is useful to plot the data so we can visualize any apparent relationships between age and price. Such a plot is called a **scatter diagram.** The scatter diagram for the data in Table 13.2 is depicted in Fig. 13.7.

FIGURE 13.7

Scatter diagram for the age and price data of Nissan Zs from Table 13.2

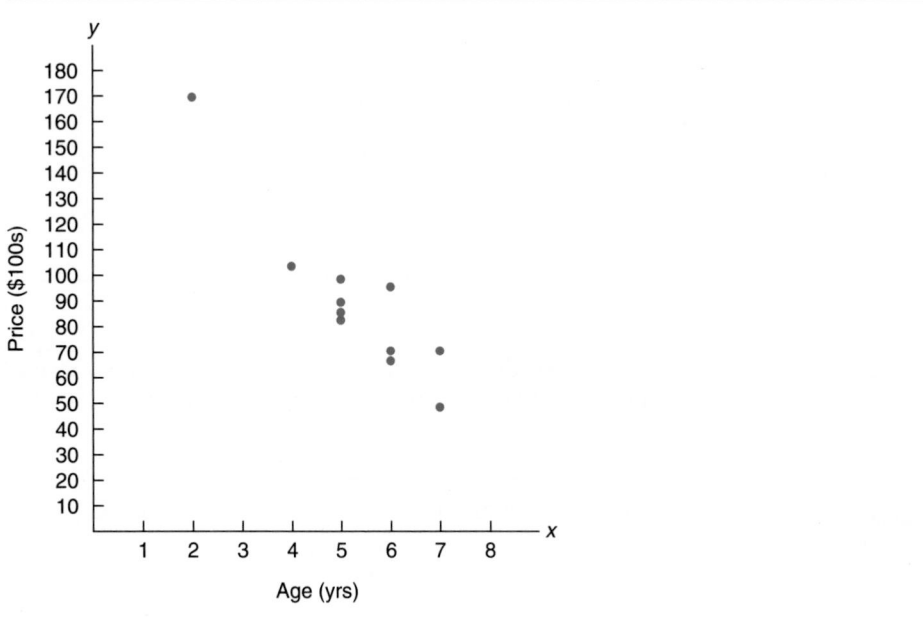

Although it is clear from the scatter diagram that the data points do not lie on a straight line, it appears that they are clustered about a straight line. We would like to fit a straight line to the data points; then we could use that line to predict the price of a Nissan Z given its age.

Since we could draw many different straight lines through the cluster of data points, we need a method to choose the "best" line. The method employed is called the **least-squares criterion.** It is based on an analysis of the errors made in using a straight line to fit the data points. To introduce the least-squares criterion, we will use a very simple data set. We will return to the Nissan Z data shortly.

Example 13.3   Introduces the Least-Squares Criterion

Let's consider the problem of fitting a straight line to the four data points displayed in Table 13.3. A scatter diagram for those data is pictured in Fig. 13.8.

TABLE 13.3
Four data points

x	1	1	2	4
y	1	2	2	6

FIGURE 13.8

Scatter diagram for the
data points in Table 13.3

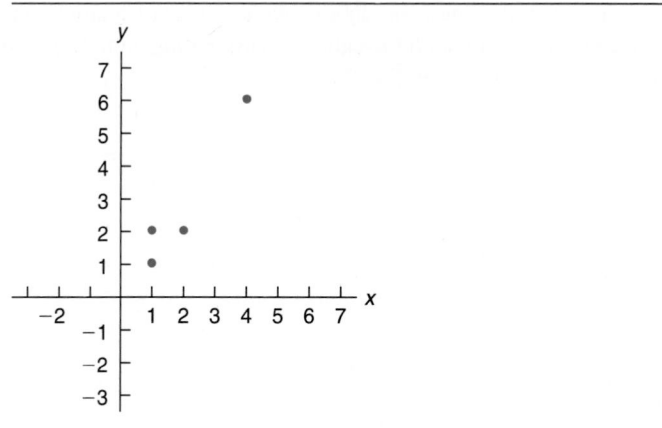

It is possible to fit (infinitely) many straight lines to the data points in Table 13.3.
Figures 13.9(a) and 13.9(b) show two possibilities.

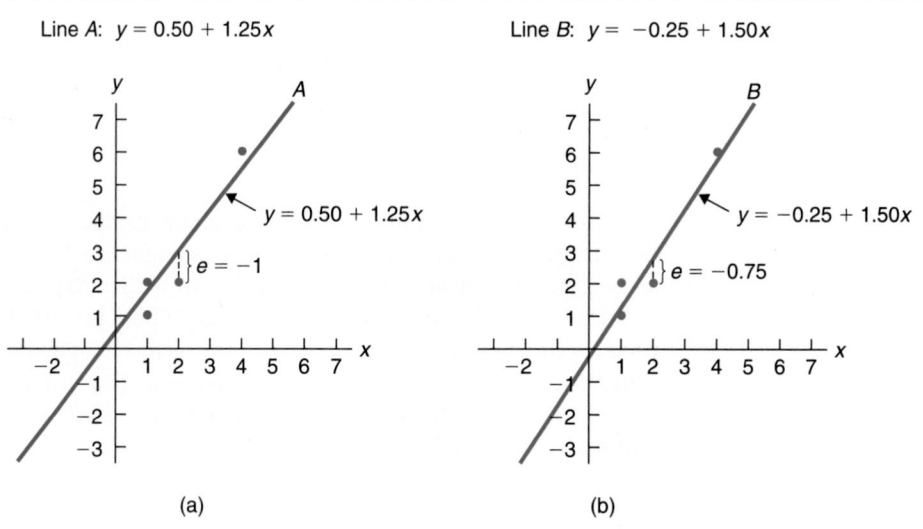

To avoid confusion, we use $\hat{y}$ to denote the y-value predicted by a straight line for a
value of x. For instance, the y-value predicted by Line A for $x = 2$ is

$$\hat{y} = 0.50 + 1.25 \cdot 2 = 3,$$

and the y-value predicted by Line B for $x = 2$ is

$$\hat{y} = -0.25 + 1.50 \cdot 2 = 2.75.$$

To measure quantitatively how well a line fits the data, we first look at the errors, e, made in using the line to predict the y-values of the data points. For instance, as we have just seen, Line A predicts a y-value of $\hat{y} = 3$ when $x = 2$. The actual y-value for $x = 2$ is $y = 2$ (see Table 13.3). Thus the error made in using Line A to predict the y-value of the data point $(2, 2)$ is

$$e = y - \hat{y} = 2 - 3 = -1,$$

as seen in Fig. 13.9(a).

The fourth column of Table 13.4(a) shows the errors made by Line A for all four data points; the fourth column of Table 13.4(b) shows that for Line B.

TABLE 13.4

Determining how well the data points in Table 13.3 are fit by (a) Line A, (b) Line B

Line A: $y = 0.50 + 1.25x$

x	y	$\hat{y}$	e	e^2
1	1	1.75	−0.75	0.5625
1	2	1.75	0.25	0.0625
2	2	3.00	−1.00	1.0000
4	6	5.50	0.50	0.2500
				1.8750

(a)

Line B: $y = -0.25 + 1.50x$

x	y	$\hat{y}$	e	e^2
1	1	1.25	−0.25	0.0625
1	2	1.25	0.75	0.5625
2	2	2.75	−0.75	0.5625
4	6	5.75	0.25	0.0625
				1.2500

(b)

To decide which line, Line A or Line B, fits the data better, we first compute the sum of the squared errors, Σe^2. This is done in the final columns of Tables 13.4(a) and 13.4(b). The line having the smaller sum of squared errors, in this case Line B, is the one that fits the data better. And among all straight lines, the least-squares criterion is that the line having the smallest sum of squared errors is the one that fits the data best. ∎

With the preceding example in mind, we can now state the least-squares criterion for the straight line that best fits a set of data points. Following that, we present the terminology used for the best-fitting line.

KEY FACT 13.2

LEAST-SQUARES CRITERION

The straight line that best fits a set of data points is the one for which the sum of squared errors is smallest.

DEFINITION 13.2

REGRESSION LINE AND REGRESSION EQUATION

Regression line: The straight line that best fits a set of data points according to the least-squares criterion.
Regression equation: The equation of the regression line.

Although the least-squares criterion tells us what property the regression line for a set of data points must have, it does not tell us how to find that line. We will soon provide formulas for obtaining the regression equation (equation of the regression line). However, before we do, we need to introduce some notation that will be used throughout our study of regression and correlation.

DEFINITION 13.3

NOTATION USED IN REGRESSION AND CORRELATION

We define S_{xx}, S_{xy}, and S_{yy} by $S_{xx} = \Sigma(x - \bar{x})^2$, $S_{xy} = \Sigma(x - \bar{x})(y - \bar{y})$, and $S_{yy} = \Sigma(y - \bar{y})^2$. These three quantities are most easily computed by using the following shortcut formulas:

$$S_{xx} = \Sigma x^2 - (\Sigma x)^2/n,$$

$$S_{xy} = \Sigma xy - (\Sigma x)(\Sigma y)/n,$$

$$S_{yy} = \Sigma y^2 - (\Sigma y)^2/n.$$

Note: It can be shown mathematically that the shortcut formulas for S_{xx}, S_{xy}, and S_{yy} are equivalent to the defining formulas.

We can now present the formulas that permit us to determine the regression line for a set of data points. These formulas can be derived using elementary calculus (see Exercise 13.52).

FORMULA 13.1

REGRESSION EQUATION

The regression equation (equation of the regression line) for a set of n data points is $\hat{y} = b_0 + b_1 x$, where

$$b_1 = \frac{S_{xy}}{S_{xx}} \quad \text{and} \quad b_0 = \frac{1}{n}(\Sigma y - b_1 \Sigma x) = \bar{y} - b_1\bar{x}.$$

Example 13.4   Illustrates Formula 13.1

Table 13.2 displays data on age and price for a sample of 11 Nissan Zs. We repeat that data in the first two columns of Table 13.5.

a. Determine the regression equation for the data.

b. Graph the regression equation and the data points.

c. Describe the apparent relationship between age and price of Nissan Zs.

d. What does the slope of the regression line represent in terms of prices for Nissan Zs?

e. Use the regression equation to predict the price of a 3-year-old Nissan Z and a 4-year-old Nissan Z.

SOLUTION a. To determine the regression equation, we need to compute b_1 and b_0 using Formula 13.1. To that end it is convenient to construct a table of values for x (age), y (price), xy, x^2, and their sums. This is done in Table 13.5.

TABLE 13.5
Table for computing the regression equation for the Nissan Z data

Age (yrs) x	Price ($100s) y	xy	x^2
5	85	425	25
4	103	412	16
6	70	420	36
5	82	410	25
5	89	445	25
5	98	490	25
6	66	396	36
6	95	570	36
2	169	338	4
7	70	490	49
7	48	336	49
58	975	4732	326

The slope of the regression line is, therefore,

$$b_1 = \frac{S_{xy}}{S_{xx}} = \frac{\Sigma xy - (\Sigma x)(\Sigma y)/n}{\Sigma x^2 - (\Sigma x)^2/n} = \frac{4732 - (58)(975)/11}{326 - (58)^2/11} = -20.26.$$

The y-intercept is

$$b_0 = \frac{1}{n}(\Sigma y - b_1 \Sigma x) = \frac{1}{11}\left[975 - (-20.26) \cdot 58\right] = 195.47.$$

Thus the regression equation is $\hat{y} = 195.47 - 20.26x$. *Note:* The usual warnings about rounding apply. When computing the slope, b_1, of the regression line, do not round until the computation is finished. Moreover, when computing the y-intercept, b_0, do not use the rounded value of b_1; instead, keep full calculator accuracy.

b. To graph the regression equation, we need to substitute two different x-values into the regression equation to obtain two distinct points. Let's use the x-values 2 and 8. The corresponding y-values are

$$\hat{y} = 195.47 - 20.26 \cdot 2 = 154.95 \quad \text{and} \quad \hat{y} = 195.47 - 20.26 \cdot 8 = 33.39.$$

Thus the regression line passes through the two points $(2, 154.95)$ and $(8, 33.39)$. In Fig. 13.10 we have plotted these two points using hollow dots. Drawing a straight line through the two hollow dots yields the regression line, the graph of the regression equation.

FIGURE 13.10

Regression line and data
points for Nissan Z data

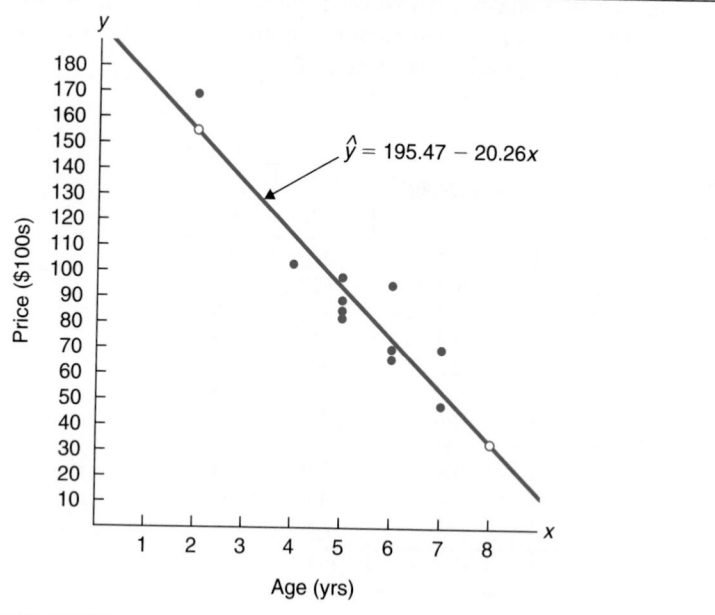

Also included in Fig. 13.10 are the data points from Table 13.2 (page 742). As we know, the regression line in Fig. 13.10 is the straight line that best fits the data points according to the least-squares criterion; that is, it is the straight line for which the sum of squared errors is smallest.

c. Here we are to describe the apparent relationship between age and price of Nissan Zs. Since the slope of the regression line is negative, we see that price tends to decrease as age increases—no particular surprise.

d. For this part, we are to interpret the slope of the regression line in terms of the prices for Nissan Zs. To begin, recall that x represents age, in years, and y represents price, in hundreds of dollars. The slope of -20.26, or $-\$2026$, indicates that Nissan Zs depreciate an estimated $2026 per year, at least in the 2- to 7-year-old range.

e. Finally, we are to use the regression equation, $\hat{y} = 195.47 - 20.26x$, to predict the price of a 3-year-old Nissan Z and a 4-year-old Nissan Z. For a 3-year-old Nissan Z, we have $x = 3$, and so the predicted price is

$$\hat{y} = 195.47 - 20.26 \cdot 3 = 134.69,$$

or $13,469. Similarly, the price the regression equation predicts for a 4-year-old Nissan Z is

$$\hat{y} = 195.47 - 20.26 \cdot 4 = 114.43,$$

or $11,443. Questions concerning the accuracy and reliability of such predictions will be discussed later in this chapter and also in Chapter 14.

Extrapolation

If a scatter diagram indicates a linear relationship between two variables, then it is reasonable to use the regression equation to make predictions for x-values within the range of the x-values in the sample data; but not necessarily for x-values outside that range, because the linear relationship between the variables may not hold there. Using the regression equation to make predictions for x-values outside the range of the x-values in the sample data is called **extrapolation.** Grossly incorrect predictions can result from extrapolation.

The Nissan Z example provides an excellent illustration of where extrapolation can lead to grossly incorrect predictions. The regression equation is $\hat{y} = 195.47 - 20.26x$, and the x-values of the sample points used in computing that regression equation range from $x = 2$ to $x = 7$, that is, from 2 to 7 years old.

Suppose that we extrapolate by using the regression equation to predict the price of an 11-year-old Nissan Z. The predicted price is $\hat{y} = 195.47 - 20.26 \cdot 11 = -27.39$, or $-\$2739$. Clearly, this is ridiculous—no one is going to pay us \$2739 to take away their 11-year-old Nissan Z. In fact, by consulting the classified ads, we find that a more reasonable prediction for the price of an 11-year-old Nissan Z is roughly \$3000.

Therefore, although the relationship between age and price of Nissan Zs appears to be linear in the range from $x = 2$ to $x = 7$, it is definitely not so in the range from $x = 2$ to $x = 11$. Figure 13.11 summarizes the discussion on extrapolation as it applies to age and price of Nissan Zs.

FIGURE 13.11

Extrapolation in the Nissan-Z example

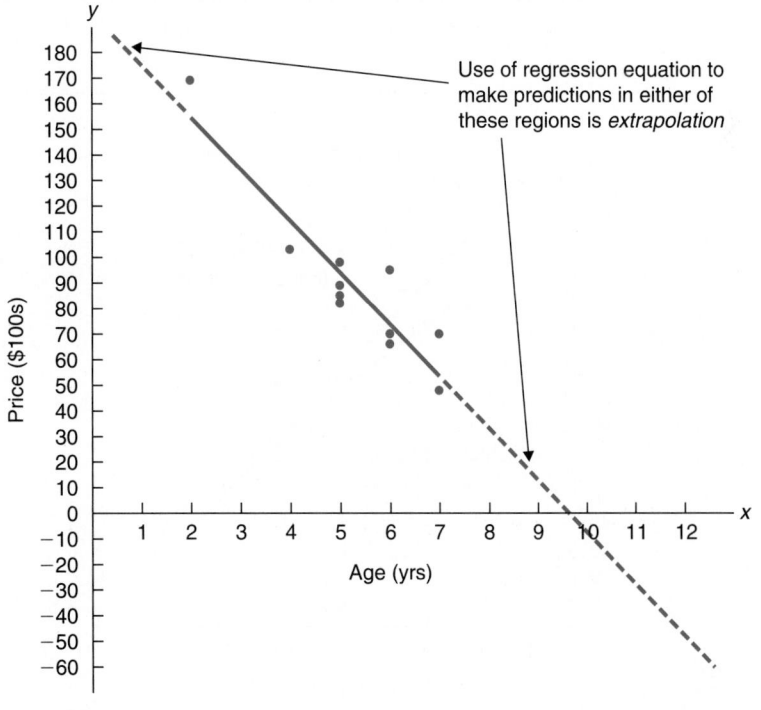

Use of regression equation to make predictions in either of these regions is *extrapolation*

Outliers and Influential Observations

Recall that an outlier is an observation that lies outside the overall pattern of the data. In the context of regression, an **outlier** is a data point that lies far from the regression line, relative to the other data points. Figure 13.10 shows that the Nissan Z data has no outliers.

An outlier can sometimes have a significant effect on a regression analysis. Thus, as usual, it is important to identify outliers and to remove them from the analysis if appropriate (e.g., if the outlier is found to be a measurement or recording error).

We must also watch for influential observations. In regression analysis, an **influential observation** is a data point whose removal causes the regression equation (and line) to change considerably. A data point that is separated in the x-direction from the other data points is often an influential observation because it "pulls" the regression line toward itself and is not counteracted by any other data points.

As with an outlier, you should try to determine the reason for an influential observation. If it is discovered that an influential observation is due to a measurement or recording error or that for some other reason it clearly does not belong in the data set, then it can be removed without further ado. However, if no explanation for the influential observation is apparent, then the decision whether or not to retain it in the data set can often be difficult and calls for a judgment by the researcher.

For the Nissan Z data, Fig. 13.10 (or Table 13.5) shows that the data point $(2, 169)$ is potentially an influential observation since its x-value, $x = 2$, is separated from the x-values of the other data points. We removed that data point and recalculated the regression equation; the result is $\hat{y} = 160.33 - 14.24x$. As we see from Fig. 13.12, this equation differs markedly from the regression equation, $\hat{y} = 195.47 - 20.26x$, which we obtained using the full data set. So the data point $(2, 169)$ is indeed an influential observation.

FIGURE 13.12

Regression lines with and without the influential observation removed

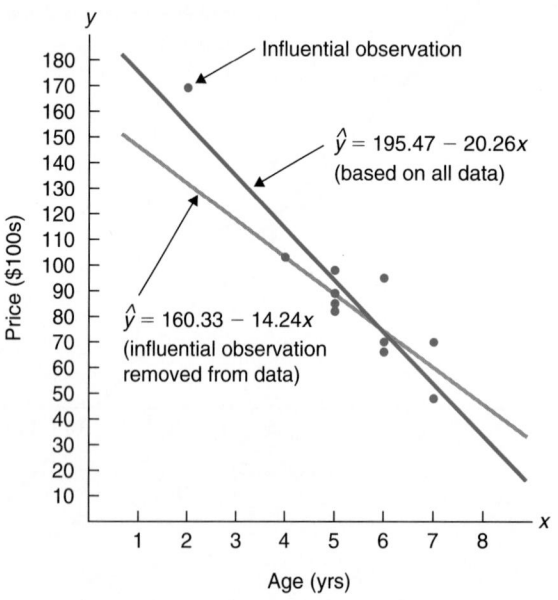

The influential observation (2, 169) is not a recording error, but a legitimate data point. Nonetheless it may be advisable either to remove it, thus limiting the analysis to Nissan Zs between 4 and 7 years old, or to obtain additional data on 2-year-old (and 3-year-old) Nissan Zs so that the regression analysis is not so dependent on one data point.

We added data for one 2-year-old and three 3-year-old Nissan Zs and obtained the regression equation $\hat{y} = 193.63 - 19.93x$. This regression equation differs very little from our original one, $\hat{y} = 195.47 - 20.26x$. So we could justify using the original regression equation to analyze the relationship between age and price of Nissan Zs between 2 and 7 years of age, even though the corresponding data set contains an influential observation.

An outlier may or may not be an influential observation; and an influential observation may or may not be an outlier. Many statistical software packages identify potential outliers and influential observations.

Predictor and Response Variables

For a linear equation $y = b_0 + b_1x$, x is the independent variable and y is the dependent variable. In the context of regression analysis, we call y the **response variable** and x the **predictor variable** or **explanatory variable** (because it is used to predict or explain the values of the response variable). For the Nissan Z example, "age" is the predictor variable and "price" is the response variable.

A Warning on the Use of Linear Regression

The idea behind finding a regression line is based on the assumption that the data points are scattered about a straight line.[†] But in some cases, data points may be scattered about a curve instead of a straight line, as in Fig. 13.13(a). Unfortunately, the formulas for b_0 and b_1 will still work for this data set and fit an inappropriate straight line. Indeed, the data, which really follow a curve, would be fitted by the straight line shown in Fig. 13.13(b).

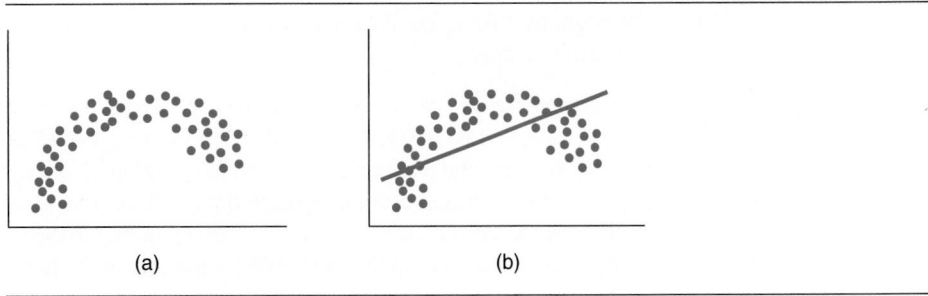

FIGURE 13.13

(a) Data points scattered about a curve (b) Inappropriate straight line fit to the data points

(a)   (b)

[†] We will discuss this assumption in detail and make it more precise in Section 14.1.

This procedure is misleading. For instance, it would lead us to predict that y-values in Fig. 13.13(a) will keep increasing when they have actually begun to decrease. Key Fact 13.3 summarizes the criterion for finding a regression line.

KEY FACT 13.3   **CRITERION FOR FINDING A REGRESSION LINE**

> Before finding a regression line for a set of data points, draw a scatter diagram. If the data points do not appear to be scattered about a straight line, do not determine a regression line.

Techniques are available for fitting curves to data points showing a curved pattern, like the data points in Fig. 13.13(a). Those techniques, referred to as *curvilinear regression,* will be discussed briefly in Section 13.5.

Using the Computer (Optional)

As we just mentioned, the idea behind finding a regression line is based on the assumption that the data points are scattered about a straight line. Consequently, before determining a regression line, we need to look at a scatter diagram of the data. We should then proceed with a regression analysis only when the data points appear to be scattered about a straight line.

In Section 6.5 we introduced Minitab's **Plot** command in connection with obtaining normal probability plots. Now we will explain how to use that command to obtain scatter diagrams. Consider Example 13.5.

Example 13.5   The Plot Command

Use Minitab to obtain a scatter diagram for the age and price data of Nissan Zs displayed in Table 13.2 on page 742.

SOLUTION   We begin by storing the data in columns named AGE and PRICE. Then we proceed in the following manner.

Session commands: We type the command **PLOT** followed by the storage locations of the sample data; that is, we type PLOT 'PRICE' versus 'AGE' and press Enter. Printout 13.1 shows this command and the resulting scatter diagram. Notice that the variable corresponding to the first storage location typed in the PLOT command, in this case PRICE, is plotted on the vertical axis and that the variable corresponding to the second storage location typed in the PLOT command, in this case AGE, is plotted on the horizontal axis.

(or)

Menu commands: First we choose **Graph ▸ Scatter Plot...** and specify PRICE for **Vertical axis** and AGE for **Horizontal axis**. Next we select **Plotting symbol** and type * to indicate that an asterisk should be used as the symbol for plotting points.[†] Then if high resolution is enabled, we select **High resolution** and press [Spacebar] to disable it.[‡] Finally, we select **OK**. The resulting scatter diagram is displayed in Printout 13.1.

PRINTOUT 13.1

Minitab output for the Plot command

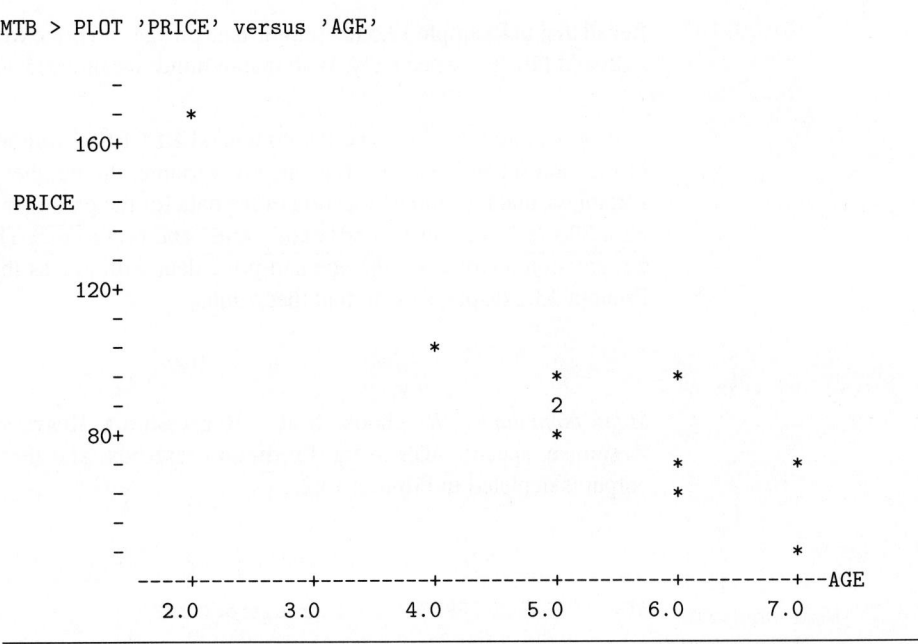

The plot in Printout 13.1 is Minitab's version of the scatter diagram we drew by hand in Fig. 13.7 on page 743. In general, the data points are plotted using asterisks; however, if two or more data points are the same or very close together, then Minitab prints the number of those points instead of asterisks. For example, the 2 plotted above the 5.0 indicates there are two data points in that vicinity that are either the same or very close together. In this case they are the points (5, 85) and (5, 89).

From the scatter diagram in Printout 13.1, it appears that the data points are scattered about a straight line. So it is reasonable to find a regression line for these data. ■

Formula 13.1 on page 746 provides the formulas to determine the regression equation for a set of data points. Alternatively, we can use Minitab's **Regress** command.

[†] We have chosen an asterisk as the plotting symbol only so the output will be consistent with the default plotting symbol used in session commands.

[‡] See the footnote on page 155.

Example 13.6   The Regress Command

Use Minitab to obtain the regression equation for the age and price data of Nissan Zs displayed in Table 13.2 on page 742.

SOLUTION   Recall that in Example 13.5 the sample data on age and price were stored in columns named AGE and PRICE, respectively. With that in mind, we proceed as follows.

Session commands: We type the command **REGRESS** followed by the storage location of the data for the response (dependent) variable, the number of predictors (independent variables), and the storage location of the data for the predictor variable(s); that is, we type REGRESS 'PRICE' on 1 predictor 'AGE' and press Enter. This tells Minitab to perform a regression analysis on the age and price data with age as the single predictor variable. Printout 13.2 displays the output that results.

(or)

Menu commands: We choose **Stat ▸ Regression ▸ Regression...**, specify PRICE for **Response**, specify AGE in the **Predictors** text box, and then select **OK**. The resulting output is depicted in Printout 13.2.

PRINTOUT 13.2
Minitab output for the
Regress command

```
MTB > REGRESS 'PRICE' on 1 predictor 'AGE'

The regression equation is
PRICE = 195 - 20.3 AGE

Predictor       Coef       Stdev     t-ratio       p
Constant      195.47       15.24       12.83   0.000
AGE           -20.261       2.800       -7.24   0.000

s = 12.58       R-sq = 85.3%     R-sq(adj) = 83.7%

Analysis of Variance

SOURCE       DF          SS          MS        F       p
Regression    1      8285.0      8285.0    52.38   0.000
Error         9      1423.5       158.2
Total        10      9708.5

Unusual Observations
Obs.    AGE     PRICE       Fit Stdev.Fit  Residual   St.Resid
  9    2.00    169.00    154.95      9.92     14.05       1.82 X

X denotes an obs. whose X value gives it large influence.
```

From the third line of Printout 13.2, we find that the required regression equation is PRICE = 195 - 20.3 AGE; in other words, the regression equation is $\hat{y} = 195 - 20.3x$. Next we find a table that provides information about the y-intercept, b_0, and slope, b_1, of the regression line. In particular, the entries 195.47 and -20.261 under the column headed Coef are the values of b_0 and b_1, respectively.

Near the bottom of Printout 13.2 we find information on Unusual Observations, which may be either potential outliers or influential observations. Minitab labels potential outliers with an R and potential influential observations with an X. Here we see that there is one potential influential observation, namely, the data point (2, 169). As we discovered on page 750, this data point is in fact an influential observation since its removal causes a drastic change in the regression equation.

Potential outliers detected by Minitab are identified as data points having large standardized residuals, namely, standardized residuals with magnitude greater than 2. (The *residual* of a data point is the difference between the observed and predicted y-values, $y - \hat{y}$; the *standardized residual* is obtained by dividing the residual by the standard deviation of the residual.) A standardized residual with magnitude greater than 3 or relatively many standardized residuals with magnitudes greater than 2 is cause for concern because least-squares regression is not resistant to outliers. Evidently, there are no potential outliers in the Nissan Z data.

We have discussed only a small portion of the information provided by the Regress output in Printout 13.2. Later sections will address other aspects. ∎

There are actually three forms for the output of the Regress command. The session command **BRIEF**, followed by an integer between 1 and 3, controls the amount of output for all subsequent Regress commands; the larger the integer, the more extensive the output. The default is 2. In future illustrations of the Regress command, we will sometimes use settings for BRIEF other than the default.

EXERCISES 13.2

In each of Exercises 13.28 and 13.29,
a. *graph each linear equation and the data points.*
b. *construct tables for x, y, $\hat{y}$, e, and e^2 similar to Table 13.4 on page 745.*
c. *determine which line fits the set of data points better according to the least-squares criterion.*

13.28   Line A: $y = 1.5 + 0.5x$
Line B: $y = 1.125 + 0.375x$

Data points

x	1	1	5	5
y	1	3	2	4

13.29   Line A: $y = 3 - 0.6x$
Line B: $y = 4 - x$

Data points

x	0	2	2	5	6
y	4	2	0	-2	1

For Exercises 13.30–13.37, be sure to save your worksheets. You will need them in later sections.

13.30   Refer to Exercise 13.28.
a. Find the regression equation for the data points.
b. Graph the regression equation and the data points.

13.31   Refer to Exercise 13.29.
a. Find the regression equation for the data points.
b. Graph the regression equation and the data points.

13.32   Ten Corvettes between 1 and 6 years old were randomly selected from the classified ads of *The Arizona Republic*. The following data were obtained, where x denotes age, in years, and y denotes price, in hundreds of dollars.

x	6	6	6	2	2	5	4	5	1	4
y	125	115	130	260	219	150	190	163	260	160

a. Determine the regression equation for the data.
b. Graph the regression equation and the data points.
c. Describe the apparent relationship between age and price for Corvettes.
d. What does the slope of the regression line represent in terms of Corvette prices?
e. Use the regression equation obtained in part (a) to predict the price of a 2-year-old Corvette; a 3-year-old Corvette.
f. Identify the predictor and response variables.
g. Identify outliers and potential influential observations.

13.33   The National Center for Health Statistics publishes data on heights and weights in *Vital and Health Statistics*. A random sample of 11 males age 18–24 years gave the following data, where x denotes height, in inches, and y denotes weight, in pounds.

x	65	67	71	71	66	75	67	70	71	69	69
y	175	133	185	163	126	198	153	163	159	151	155

a. Determine the regression equation for the data.
b. Graph the regression equation and the data points.
c. Describe the apparent relationship between height and weight for 18–24-year-old males.
d. What does the slope of the regression line represent in terms of weights of 18–24-year-old males?
e. Use the regression equation determined in part (a) to predict the weight of an 18–24-year-old male who is 67 inches tall; 73 inches tall.
f. Identify the predictor and response variables.
g. Identify outliers and potential influential observations.

13.34   Hanna Properties specializes in custom-home resales in the Equestrian Estates, an exclusive subdivision in Phoenix, Arizona. A random sample of nine custom homes currently listed for sale provided the following information on size and price. Here x denotes size, in hundreds of square feet, rounded to the nearest hundred, and y denotes price, in thousands of dollars, rounded to the nearest thousand.

x	26	27	33	29	29	34	30	40	22
y	235	249	267	269	295	345	415	475	195

a. Determine the regression equation for the data.
b. Graph the regression equation and the data points.
c. Describe the apparent relationship between square footage and price for custom homes in the Equestrian Estates.
d. What does the slope of the regression line represent in terms of sizes and prices of custom homes in the Equestrian Estates?
e. Use the regression equation determined in part (a) to predict the price of a custom home in the Equestrian Estates that has 2600 sq ft.
f. Identify the predictor and response variables.
g. Identify outliers and potential influential observations.

13.35   An article read by a physician indicated that the maximum heart rate an individual can reach during intensive exercise decreases with age. The physician decided to do his own study. Ten randomly selected people performed exercise tests and recorded their peak heart rates. The results are shown in the following table, where x denotes age, in years, and y denotes peak heart rate.

x	30	38	41	38	29	39	46	41	42	24
y	186	183	171	177	191	177	175	176	171	196

a. Determine the regression equation for the data.
b. Graph the regression equation and the data points.
c. Describe the apparent relationship between age and peak heart rate.
d. What does the slope of the regression line represent in terms of age and peak heart rate?
e. Use the regression equation to predict the peak heart rate of a 28-year-old person.
f. Identify the predictor and response variables.
g. Identify outliers and potential influential observations.

13.36   An instructor asked a random sample of eight students to record their study times in a beginning calculus course. She then made a table for total hours studied, x, over 2 weeks, and test score, y, at the end of the 2 weeks. Here are the results.

x	10	15	12	20	8	16	14	22
y	92	81	84	74	85	80	84	80

a. Determine the regression equation for the data.
b. Graph the regression equation and the data points.
c. Describe the apparent relationship between study time and test score. Does it surprise you?
d. What does the slope of the regression line represent in terms of study time and test score?
e. Use the regression equation to predict the test score of a student who studies for 15 hours.
f. Identify the predictor and response variables.
g. Identify outliers and potential influential observations.

•   **13.37** An economist is interested in the relationship between the disposable income of a family and the amount of money spent annually on food. For a preliminary study, the economist takes a random sample of eight middle-income families of the same size (father, mother, two children). The results are as follows, where x denotes disposable income, in thousands of dollars, and y denotes food expenditure, in hundreds of dollars.

x	30	36	27	20	16	24	19	25
y	55	60	42	40	37	26	39	43

a. Determine the regression equation for the data.
b. Graph the regression equation and the data points.
c. Describe the apparent relationship between disposable income and annual food expenditure.
d. What does the slope of the regression line represent in terms of disposable income and annual food expenditure?
e. Use the regression equation to predict the annual food expenditure of a family with a disposable income of $25,000.
f. Identify the predictor and response variables.
g. Identify outliers and potential influential observations.

•   **13.38** For which of the following sets of data points is it reasonable to determine a regression line?

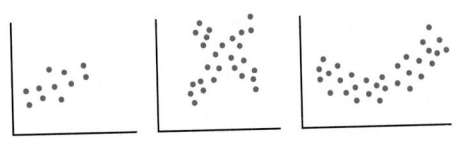

•   **13.39** For which of the following sets of data points is it reasonable to determine a regression line?

•   **13.40** In Exercise 13.32 you determined a regression equation that can be used to predict the price of a Corvette given its age.
a. Should that regression equation be used to predict the price of a 4-year-old Corvette? a 10-year-old Corvette? Explain your answers.
b. For which ages is it reasonable to use the regression equation to predict price?

•   **13.41** In Exercise 13.33 you determined a regression equation that relates the variables height and weight for 18–24-year-old males.
a. Should that regression equation be used to predict the weight of an 18–24-year-old male who is 68 inches tall? 60 inches tall? Explain your answers.
b. For which heights is it reasonable to use the regression equation to predict weight?

•   **13.42 (Computer exercise)** Refer to the data in Exercise 13.36. Use Minitab or some other statistical software to
a. obtain a scatter diagram for the data.
b. determine the regression equation for the data.
c. identify potential outliers and influential observations.
d. Justify the use of your procedure in part (b).

•   **13.43 (Computer exercise)** Refer to the data in Exercise 13.37. Use Minitab or some other statistical software to
a. obtain a scatter diagram for the data.
b. determine the regression equation for the data.
c. identify potential outliers and influential observations.
d. Justify the use of your procedure in part (b).

•   **13.44 (Computer exercise)** The Energy Information Administration publishes data on energy consumption by family income in *Residential Energy Consumption Survey: Consumption and Expenditures*. We applied Minitab's Regress command to data on family income and last year's energy consumption from a random sample of 25 families. Printout 13.3 at the top of the next page shows the output. The income data are in thousands of dollars, and the energy-consumption data are in millions of BTU.
a. Consult the printout to obtain the regression equation for the data.
b. Predict last year's energy consumption for a family with an income of $42,000.
c. Identify potential outliers and influential observations.

•   **13.45 (Computer exercise)** Greene and Touchstone conducted a study on the relationship between the estriol levels of pregnant women and the birth weights of their children. Their findings, "Urinary Tract Estriol: An Index of Placental

PRINTOUT 13.3   Minitab output for Exercise 13.44 (and Exercises 13.64, 14.18, and 14.34)

```
The regression equation is
CONSUMPT = 82.0 + 0.931 INCOME


Predictor      Coef      Stdev     t-ratio      p
Constant     82.036      2.054      39.94     0.000
INCOME       0.93051    0.05727     16.25     0.000


s = 5.375      R-sq = 92.0%     R-sq(adj) = 91.6%

Analysis of Variance

SOURCE         DF         SS         MS        F         p
Regression      1       7626.6     7626.6    264.02    0.000
Error          23        664.4      28.9
Total          24       8291.0


Unusual Observations
Obs.   INCOME  CONSUMPT      Fit Stdev.Fit  Residual   St.Resid
  3     15.0     81.69     95.99    1.40     -14.31      -2.76R
 19      5.0     97.07     86.69    1.82      10.38       2.05R


R denotes an obs. with a large st. resid.
```

PRINTOUT 13.4   Minitab output for Exercise 13.45 (and Exercises 13.65, 14.19, and 14.35)

```
The regression equation is
WEIGHT = 2152 + 60.8 ESTRIOL


Predictor      Coef      Stdev     t-ratio      p
Constant     2152.3      262.0       8.21     0.000
ESTRIOL       60.82      14.68       4.14     0.000


s = 382.1      R-sq = 37.2%     R-sq(adj) = 35.0%

Analysis of Variance

SOURCE         DF         SS          MS        F         p
Regression      1      2505745     2505745    17.16    0.000
Error          29      4234255      146009
Total          30      6740000


Unusual Observations
Obs. ESTRIOL    WEIGHT      Fit Stdev.Fit  Residual   St.Resid
 14     24.0    2800.0    3612.0   120.8     -812.0      -2.24R


R denotes an obs. with a large st. resid.
```

Function," were published in the *American Journal of Obstetrics and Gynecology*. Printout 13.4 shows the output that results by applying Minitab's Regress command to the data obtained by Greene and Touchstone. The estriol levels are in milligrams per 24 hours, and the birth weights are in grams.

a. Consult the printout to determine the regression equation for the data.
b. Use the regression equation to predict the birth weight of the child of a pregnant woman with an estriol level of 17 mg/24 hr.
c. Identify potential outliers and influential observations.

· · **13.46** The negative relation between study time and grade found in Exercise 13.36 has been discovered by many investigators, and it has puzzled them. Can you think of a possible explanation for it?

Sample covariance: The *sample covariance*, s_{xy}, of a sample of n data points is defined by

$$(1) \qquad s_{xy} = \frac{\Sigma(x - \bar{x})(y - \bar{y})}{n - 1}.$$

· · **13.47** Determine the sample covariance of the data points in Exercise 13.29.

· · **13.48** Determine the sample covariance of the data points in Exercise 13.28.

Determining the regression equation using the sample covariance. The sample covariance, defined in Equation (1), can be used as an alternate method for obtaining the slope and y-intercept of the regression line for a set of data points. The formulas are

$$(2) \qquad b_1 = s_{xy}/s_x^2 \quad \text{and} \quad b_0 = \bar{y} - b_1\bar{x},$$

where s_x denotes the standard deviation of the x-values.

· · **13.49** Use the equations in (2) to find the regression equation for the data points in Exercise 13.29. Compare your answer to the one you obtained in part (a) of Exercise 13.31.

· · **13.50** Use the equations in (2) to find the regression equation for the data points in Exercise 13.28. Compare your answer to the one you obtained in part (a) of Exercise 13.30.

· · · **13.51** Prove that the equations in (2) are equivalent to the ones given in Formula 13.1 on page 746.

· · · **13.52** In this exercise we will derive the equations in Formula 13.1 on page 746 for the slope and y-intercept of the regression line. The derivation requires elementary calculus. To begin, recall that according to the least-squares criterion, the regression line is the straight line for which the sum of squared errors is smallest.

a. Show that the regression line, $\hat{y} = b_0 + b_1 x$, is the straight line for which b_0 and b_1 minimize the function $f(b_0, b_1) = \Sigma[y - (b_0 + b_1 x)]^2$.
b. Use elementary calculus to find the values of b_0 and b_1 that minimize $f(b_0, b_1)$ and show that these values give the equations in Formula 13.1. [*Hint:* Compute the partial derivatives of $f(b_0, b_1)$, set them equal to zero, and solve for b_0 and b_1.]

13.3 THE COEFFICIENT OF DETERMINATION

In Example 13.4 we determined the regression equation, $\hat{y} = 195.47 - 20.26x$, for data on age and price of a sample of 11 Nissan Zs. Here x represents age, in years, and $\hat{y}$ predicted price, in hundreds of dollars. We can apply the regression equation to predict the price of a Nissan Z of a particular age, x. For instance, we predict that a 4-year-old Nissan Z will cost roughly $\hat{y} = 195.47 - 20.26 \cdot 4 = 114.43$, or \$11,443. But how valuable are such predictions? Is the regression equation useful for predicting price, or could we do just as well by ignoring age?

We can evaluate the utility of a regression equation for making predictions in several ways. One method is to measure the reduction in the errors made in prediction by using the regression equation instead of simply predicting the mean of the observed y-values, $\bar{y}$. To illustrate, we return to the Nissan Z data.

Example 13.7   Introduces the Coefficient of Determination

The age and price data for a sample of 11 Nissan Zs are repeated in the first two columns of Table 13.6 (next page). Ages are in years and prices are in hundreds of dollars.

One way we can employ this information to predict the price of a Nissan Z is to ignore age and simply use the mean price, $\bar{y}$, of the 11 Nissan Zs sampled. In other words, just use

$$\bar{y} = \frac{\Sigma y}{n} = \frac{975}{11} = 88.64 \ (\$8864)$$

as the predicted price for a Nissan Z, regardless of its age. Figure 13.14 graphically portrays the errors made when the mean price, $\bar{y} = 88.64$, is used as the predicted price for each of the 11 Nissan Zs sampled.

FIGURE 13.14

Errors made when the mean price is used for prediction

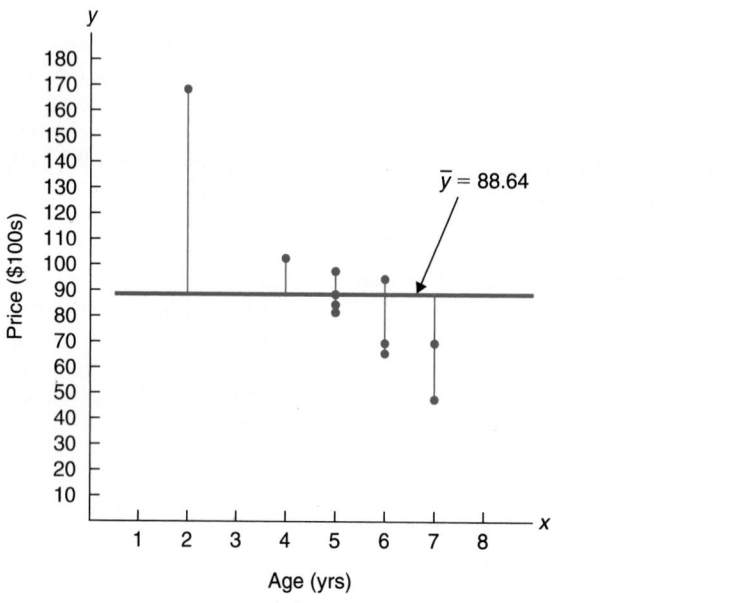

To obtain a quantitative measure of the total error made, we compute the sum of the squared errors. The required computations are shown in Table 13.6.[†]

From the final column of Table 13.6, we see that the total squared error is 9708.5 when the mean price, $\bar{y} = 88.64$, is used as the predicted price of each of the 11 Nissan Zs sampled. That total squared error is called the **total sum of squares,** *SST.* Thus for the Nissan Z data, the total sum of squares is

$$SST = \Sigma(y - \bar{y})^2 = 9708.5.$$

[†] Values in Table 13.6 and all other tables in this section are displayed to various numbers of decimal places, but computations are done using full calculator accuracy.

Age (yrs) x	Price ($100s) y	$y - \bar{y}$	$(y - \bar{y})^2$
5	85	−3.64	13.2
4	103	14.36	206.3
6	70	−18.64	347.3
5	82	−6.64	44.0
5	89	0.36	0.1
5	98	9.36	87.7
6	66	−22.64	512.4
6	95	6.36	40.5
2	169	80.36	6458.3
7	70	−18.64	347.3
7	48	−40.64	1651.3
	975		9708.5

If age is useful for predicting price, then we should obtain a reduction in the total squared error by using the regression equation, $\hat{y} = 195.47 - 20.26x$, instead of the mean price, $\bar{y}$, to make the price predictions. Figure 13.15 graphically portrays the errors made when the regression equation is used to predict the price of each of the 11 Nissan Zs sampled.

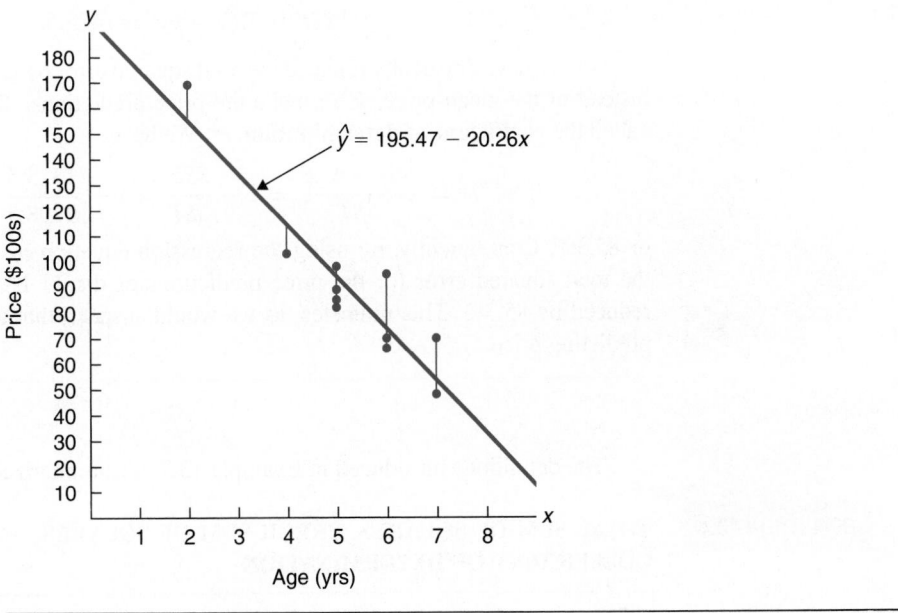

We will employ Table 13.7 to compute the total squared error made when the regression equation is used to predict the price of each of the 11 Nissan Zs sampled. Each predicted price, $\hat{y}$, is obtained by substituting the age of the Nissan Z in question into the regression equation $\hat{y} = 195.47 - 20.26x$.

TABLE 13.7
Table for computing *SSE*
for the Nissan Z data

Age (yrs) x	Price ($100s) y	$\hat{y}$	$y - \hat{y}$	$(y - \hat{y})^2$
5	85	94.16	−9.16	83.9
4	103	114.42	−11.42	130.5
6	70	73.90	−3.90	15.2
5	82	94.16	−12.16	147.9
5	89	94.16	−5.16	26.6
5	98	94.16	3.84	14.7
6	66	73.90	−7.90	62.4
6	95	73.90	21.10	445.2
2	169	154.95	14.05	197.5
7	70	53.64	16.36	267.7
7	48	53.64	−5.64	31.8
				1423.5

From the final column of Table 13.7, we see that the total squared error is 1423.5 when the regression equation is used to predict the price of each of the 11 Nissan Zs sampled. That total squared error is called the **error sum of squares, *SSE*.** Thus for the Nissan Z data, the error sum of squares is

$$SSE = \Sigma(y - \hat{y})^2 = 1423.5.$$

So, we have drastically reduced the total squared error by using the regression equation instead of the mean price, $\overline{y}$, to make the price predictions. The percentage reduction is called the **coefficient of determination, r^2.** We have

$$r^2 = \frac{SST - SSE}{SST} = 1 - \frac{SSE}{SST} = 1 - \frac{1423.5}{9708.5} = 0.853,$$

or 85.3%. Consequently, by using the regression equation instead of the mean price, $\overline{y}$, the total squared error for the price predictions of the 11 Nissan Zs sampled has been reduced by 85.3%. This indicates, as we would suspect, that age is extremely useful for predicting price.

 ■

The definitions introduced in Example 13.7 are summarized in Definition 13.4.

DEFINITION 13.4

TOTAL SUM OF SQUARES, ERROR SUM OF SQUARES, COEFFICIENT OF DETERMINATION

> *Total sum of squares:* $SST = \Sigma(y - \overline{y})^2$
> *Error sum of squares:* $SSE = \Sigma(y - \hat{y})^2$
> *Coefficient of determination:* $r^2 = 1 - SSE/SST$

As we discovered in Example 13.7, the coefficient of determination, r^2, is a descriptive measure of the usefulness of the regression equation for making predictions. Specifically, r^2 gives the percentage reduction obtained in the total squared error by using the regression equation instead of the sample mean, $\bar{y}$, to predict the observed y-values.

Explained Variation

Another way to interpret the coefficient of determination is as the percentage of the total variation in the observed y-values that is explained by the regression line, so-called **explained variation.** To see why this interpretation also applies, we return to the Nissan Z data.

Example 13.8 Introduces Explained Variation

The scatter diagram for the age and price data of 11 Nissan Zs, first seen in Fig. 13.7, is included in Fig. 13.16. Also shown in Fig. 13.16 is the regression line for the data, $\hat{y} = 195.47 - 20.26x$.

FIGURE 13.16

Scatter diagram and regression line for Nissan Z data

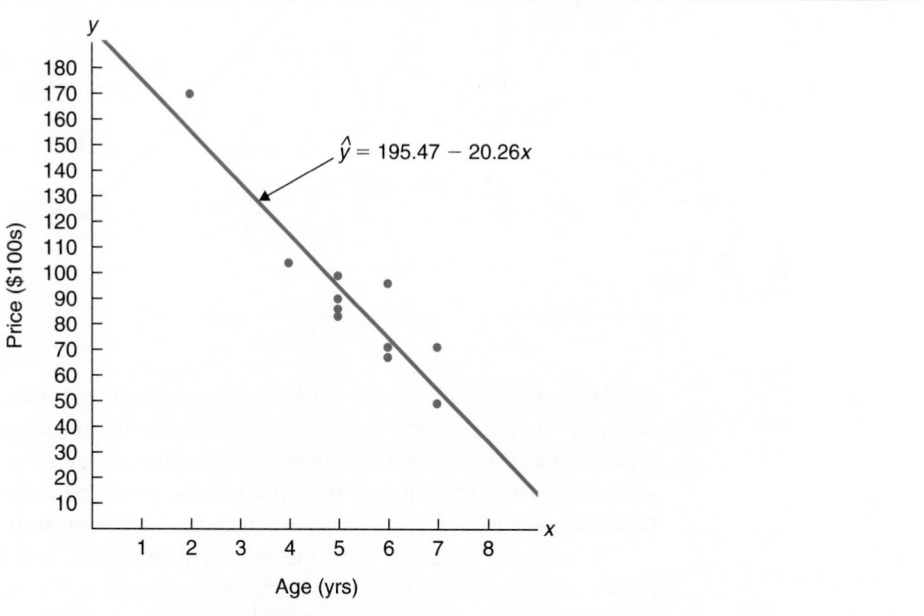

$\hat{y} = 195.47 - 20.26x$

As we can see from the scatter diagram in Fig. 13.16, the prices of the 11 Nissan Zs sampled vary widely, ranging from a low of 48 ($4800) to a high of 169 ($16,900). But as we can also see from the regression line in Fig. 13.16, much of the price variation is "explained" by age; that is, the regression line predicts a good portion of the type of variation found in the prices.

To describe quantitatively how much of the total variation in the observed (sampled) prices is explained by age, we proceed in the following manner. First, we use the total sum of squares, *SST*, as the measure of the total variation in the observed prices. From page 760 we have

$$SST = \Sigma(y - \bar{y})^2 = 9708.5.$$

Now let's look at a particular observed price, say, $y = 98$, corresponding to the data point $(5, 98)$. In Fig. 13.17 we have blown up a portion of Fig. 13.16 showing only the data point $(5, 98)$.

FIGURE 13.17

Blow-up of a portion of
Fig. 13.16 showing only
the data point (5, 98)

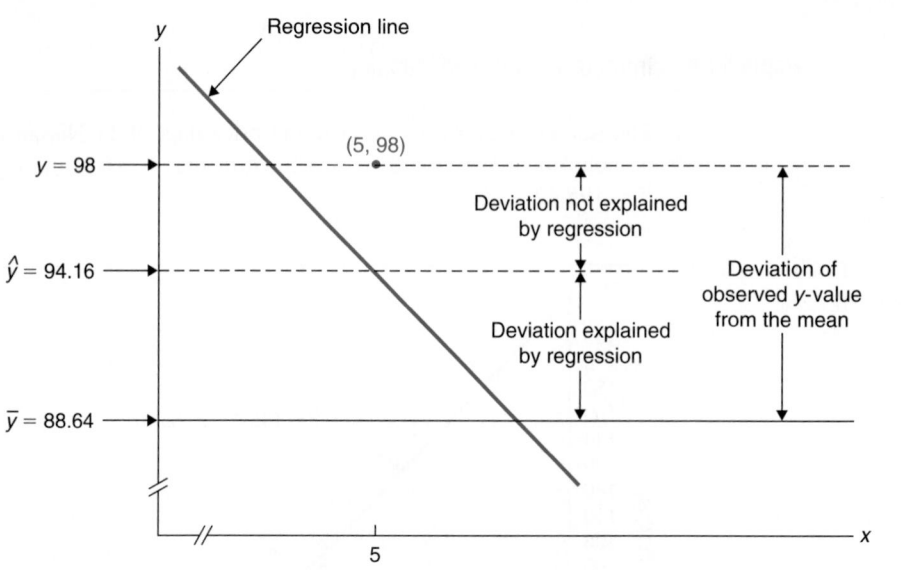

From Fig. 13.17 we see that the deviation of an observed price from the mean price, $y - \bar{y}$, can be decomposed into two parts: the deviation that is explained by the regression line, $\hat{y} - \bar{y}$, and the remaining unexplained deviation, $y - \hat{y}$. Therefore the total amount of variation (squared deviation) in the observed prices that is explained by the regression line is $\Sigma(\hat{y} - \bar{y})^2$. This is called the **regression sum of squares, *SSR*.**

To compute *SSR*, we need the predicted prices, $\hat{y}$, and the mean of the observed prices, $\bar{y}$. The predicted prices, which are displayed in the third column of Table 13.7, are repeated in the third column of Table 13.8.

Recalling that $\bar{y} = 88.64$, we construct the fourth column of Table 13.8 and then obtain the regression sum of squares, *SSR*, from the fifth column:

$$SSR = \Sigma(\hat{y} - \bar{y})^2 = 8285.0.$$

This is the amount of variation in the observed prices that is explained by the regression line. Therefore the percentage of the total variation in the observed prices that is explained

TABLE 13.8
Table for computing *SSR*
for the Nissan Z data

Age (yrs) x	Price ($100s) y	$\hat{y}$	$\hat{y} - \overline{y}$	$(\hat{y} - \overline{y})^2$
5	85	94.16	5.53	30.5
4	103	114.42	25.79	665.0
6	70	73.90	−14.74	217.1
5	82	94.16	5.53	30.5
5	89	94.16	5.53	30.5
5	98	94.16	5.53	30.5
6	66	73.90	−14.74	217.1
6	95	73.90	−14.74	217.1
2	169	154.95	66.31	4397.0
7	70	53.64	−35.00	1224.8
7	48	53.64	−35.00	1224.8
				8285.0

by the regression line is

$$\frac{SSR}{SST} = \frac{8285.0}{9708.5} = 0.853,$$

or 85.3%. This large percentage implies that age is quite useful for predicting price. ∎

In Example 13.8 we used data on age and price of Nissan Zs to introduce the regression sum of squares. As we discovered, that quantity is defined and interpreted as follows.

DEFINITION 13.5

REGRESSION SUM OF SQUARES

> The *regression sum of squares, SSR,* is defined by $SSR = \Sigma(\hat{y} - \overline{y})^2$ and represents the amount of variation in the observed *y*-values that is explained by the regression.

The Regression Identity

For the Nissan Z data, we have determined that $SST = 9708.5$, $SSR = 8285.0$, and $SSE = 1423.5$. Since $9708.5 = 8285.0 + 1423.5$, we see that $SST = SSR + SSE$. This equation is always true and is called the **regression identity.** We state that identity formally in the following key fact.

KEY FACT 13.4

REGRESSION IDENTITY

> The total sum of squares equals the regression sum of squares plus the error sum of squares; that is, $SST = SSR + SSE$.

You may have noted that for the Nissan Z data, the percentage of the total variation in the observed prices that is explained by the regression, SSR/SST, is equal to the coefficient of determination, r^2, which we computed on page 762. This is a consequence of the regression identity and hence is true not only for the Nissan Z data but for any set of data points (see Exercise 13.70). In summary, we have the following key fact.

KEY FACT 13.5   **INTERPRETATION OF THE COEFFICIENT OF DETERMINATION**

> The coefficient of determination, r^2, is defined by
>
> $$r^2 = 1 - \frac{SSE}{SST}$$
>
> and equals the percentage reduction obtained in the total squared error by using the regression equation instead of the sample mean, $\overline{y}$, to predict the observed y-values.
>   The coefficient of determination can also be computed as
>
> $$r^2 = \frac{SSR}{SST}.$$
>
> Thus it also equals the percentage of the total variation in the observed y-values that is explained by the regression.
>   In any case, r^2 always lies between 0 and 1 and is a descriptive measure of the utility of the regression equation for making predictions. Values of r^2 near 0 indicate that the regression equation is not very useful for making predictions, whereas values of r^2 near 1 indicate that the regression equation is extremely useful for making predictions.

Shortcut Formulas for the Sums of Squares

Computing the three sums of squares, *SST, SSR,* and *SSE,* using the defining formulas is time-consuming and can lead to significant roundoff error unless full accuracy is retained. For those reasons we usually employ shortcut formulas or a computer to compute the sums of squares. The shortcut formulas are presented in Formula 13.2. Exercise 13.71 considers the derivation of these formulas.

FORMULA 13.2   **SHORTCUT FORMULAS FOR THE SUMS OF SQUARES**

> The three sums of squares, *SST, SSR,* and *SSE,* can be computed using the following shortcut formulas:
>
> *Total sum of squares:* $SST = S_{yy}$
>
> *Regression sum of squares:* $SSR = S_{xy}^2/S_{xx}$
>
> *Error sum of squares:* $SSE = S_{yy} - S_{xy}^2/S_{xx}$
>
> The formulas for S_{yy}, S_{xy}, and S_{xx} are given in Definition 13.3 on page 746.

Example 13.9   Illustrates Formula 13.2

The age and price data for a sample of 11 Nissan Zs are repeated in the first two columns of Table 13.9. Use the shortcut formulas in Formula 13.2 to determine the three sums of squares, *SST, SSR,* and *SSE.*

SOLUTION   To apply the shortcut formulas, we will need a table of values for x (age), y (price), xy, x^2, y^2, and their sums. This is presented in Table 13.9.

TABLE 13.9
Table for computing the three sums of squares for the Nissan Z data using the shortcut formulas

Age (yrs) x	Price ($100s) y	xy	x^2	y^2
5	85	425	25	7,225
4	103	412	16	10,609
6	70	420	36	4,900
5	82	410	25	6,724
5	89	445	25	7,921
5	98	490	25	9,604
6	66	396	36	4,356
6	95	570	36	9,025
2	169	338	4	28,561
7	70	490	49	4,900
7	48	336	49	2,304
58	975	4732	326	96,129

Using the last row of Table 13.9 and Formula 13.2, we can now obtain the three sums of squares for the Nissan Z data. The total sum of squares equals

$$SST = S_{yy} = \Sigma y^2 - (\Sigma y)^2/n = 96,129 - (975)^2/11 = 9708.5;$$

the regression sum of squares equals

$$SSR = \frac{S_{xy}^2}{S_{xx}} = \frac{\left[\Sigma xy - (\Sigma x)(\Sigma y)/n\right]^2}{\Sigma x^2 - (\Sigma x)^2/n} = \frac{\left[4732 - (58)(975)/11\right]^2}{326 - (58)^2/11} = 8285.0;$$

and, using the two preceding results, we see that the error sum of squares equals

$$SSE = S_{yy} - \frac{S_{xy}^2}{S_{xx}} = 9708.5 - 8285.0 = 1423.5.$$

The values obtained here for the three sums of squares by using the shortcut formulas are, of course, the same as the values we found earlier by using the defining formulas. However, when the shortcut formulas are employed, the computations are much simpler and less subject to roundoff error.   ∎

Using the Computer (Optional)

In Section 13.2 we learned that Minitab's Regress command can be used to obtain the regression equation for a set of data points. The output that results from applying that command contains more than the regression equation. It also provides, among other statistics, the coefficient of determination, r^2, and the three sums of squares, SST, SSR, and SSE. Let's return once again to the Nissan Z data.

Example 13.10   The Regress Command

Printout 13.2 on page 754 shows the output obtained by applying the Regress command to the data on age and price for a sample of 11 Nissan Zs. Use the printout to find

a. the coefficient of determination, r^2.

b. the three sums of squares, SST, SSR, and SSE.

SOLUTION   a. The coefficient of determination, r^2, is displayed as the second entry in the seventh line of Printout 13.2: R-sq = 85.3%. In other words, $r^2 = 0.853$.

b. To obtain the three sums of squares, we use the table in Printout 13.2 entitled Analysis of Variance. Specifically, the values of the three sums of squares can be found in the column headed SS. The entries in the column headed SOURCE identify the sums of squares. Hence the first entry in the SS column is the regression sum of squares, SSR, the second is the error sum of squares, SSE, and the third is the total sum of squares, SST. So we see that $SSR = 8285.0$, $SSE = 1423.5$, and $SST = 9708.5$.   ∎

EXERCISES 13.3

13.53   In this section we introduced a descriptive measure of the utility of the regression equation for making predictions.
a. Identify the term and symbol for that descriptive measure.
b. Provide two different interpretations of that descriptive measure.

In Exercises 13.54 and 13.55, we have repeated the data sets from Exercises 13.28 and 13.29, respectively. We have also provided the regression equations for those data sets, which were found in Exercises 13.30 and 13.31, respectively. For each exercise,
a. compute SST, SSR, and SSE using the defining formulas.
b. verify the regression identity, $SST = SSR + SSE$.
c. compute the coefficient of determination by using both the definition $r^2 = 1 - SSE/SST$ and the formula $r^2 = SSR/SST$.

d. determine the percentage reduction obtained in the total squared error by using the regression equation instead of the sample mean, $\bar{y}$, to predict the observed y-values.
e. determine the percentage of the total variation in the observed y-values that is explained by the regression.
f. state how useful the regression equation appears to be for making predictions. (The answer for this part is subjective.)

13.54   The data from Exercise 13.28:

Data points

x	1	1	5	5
y	1	3	2	4

Regression equation is $\hat{y} = 1.75 + 0.25x$.

13.55 The data from Exercise 13.29:

Data points

x	0	2	2	5	6
y	4	2	0	-2	1

Regression equation is $\hat{y} = 2.875 - 0.625x$.

For each of Exercises 13.56–13.61,
a. *compute SST, SSR, and SSE using the shortcut formulas in Formula 13.2 on page 766.*
b. *compute the coefficient of determination, r^2.*
c. *determine the percentage of the total variation in the observed y-values that is explained by the regression, and interpret your result.*
d. *state how useful the regression equation appears to be for making predictions.*

13.56 The age and price data for Corvettes from Exercise 13.32:

x	6	6	6	2	2	5	4	5	1	4
y	125	115	130	260	219	150	190	163	260	160

13.57 The height and weight data for males age 18–24 years from Exercise 13.33:

x	65	67	71	71	66	75	67	70	71	69	69
y	175	133	185	163	126	198	153	163	159	151	155

13.58 The size and price data for custom homes from Exercise 13.34:

x	26	27	33	29	29	34	30	40	22
y	235	249	267	269	295	345	415	475	195

13.59 The data on age and peak heart rate from Exercise 13.35:

x	30	38	41	38	29	39	46	41	42	24
y	186	183	171	177	191	177	175	176	171	196

13.60 The data on study time and test score from Exercise 13.36:

x	10	15	12	20	8	16	14	22
y	92	81	84	74	85	80	84	80

13.61 The data on disposable income and annual food expenditure from Exercise 13.37:

x	30	36	27	20	16	24	19	25
y	55	60	42	40	37	26	39	43

13.62 (Computer exercise) Use Minitab or some other statistical software to obtain the coefficient of determination, r^2, and the three sums of squares, SST, SSR, and SSE, for the data in Exercise 13.60.

13.63 (Computer exercise) Use Minitab or some other statistical software to obtain the coefficient of determination, r^2, and the three sums of squares, SST, SSR, and SSE, for the data in Exercise 13.61.

13.64 (Computer exercise) The Energy Information Administration publishes data on energy consumption by family income in *Residential Energy Consumption Survey: Consumption and Expenditures*. We applied Minitab's Regress command to data on family income and last year's energy consumption from a random sample of 25 families. The income data are in thousands of dollars, and the energy-consumption data are in millions of BTU. Printout 13.3 on page 758 shows the computer output. Use the printout to find
a. the coefficient of determination for the data.
b. the regression sum of squares, the error sum of squares, and the total sum of squares.
c. the percentage of the total variation in the observed energy consumptions that is explained by family income.

13.65 (Computer exercise) Greene and Touchstone conducted a study on the relationship between the estriol levels of pregnant women and the birth weights of their children. Their findings, "Urinary Tract Estriol: An Index of Placental Function," were published in the *American Journal of Obstetrics and Gynecology*. Printout 13.4 on page 758 shows the computer output that results by applying Minitab's Regress command to the data obtained by Greene and Touchstone. The estriol levels are in milligrams per 24 hours, and the birth weights are in grams. Using the printout, determine
a. the coefficient of determination for the data.
b. the regression sum of squares, the error sum of squares, and the total sum of squares.
c. the percentage of the total variation in the observed birth weights that is explained by estriol level.

· ·   **13.66**   Suppose $r^2 = 1$ for a data set. What can you say about

a. *SSE?*   b. *SSR?*

c. the utility of the regression equation for making predictions?

· ·   **13.67**   Suppose $r^2 = 0$ for a data set. What can you say about

a. *SSE?*   b. *SSR?*

c. the utility of the regression equation for making predictions?

· ·   **13.68**   This exercise shows that the coefficient of determination always lies between 0 and 1.

a. Explain why the quantities *SST, SSR,* and *SSE* are nonnegative numbers.

b. Verify that $SSE \leq SST$. *(Hint: Use part (a) and the regression identity.)*

c. Prove that $0 \leq r^2 \leq 1$.

· · ·   **13.69**   This exercise provides a proof of the regression identity, $SST = SSR + SSE$. It uses some of the results obtained in the course of deriving the formulas for b_0 and b_1 in Exercise 13.52 of Section 13.2.

a. Show that

$$\Sigma(y - \overline{y})^2 = \Sigma(y - \hat{y})^2 + \Sigma(\hat{y} - \overline{y})^2$$
$$+ 2\Sigma(y - \hat{y})(\hat{y} - \overline{y}).$$

(Hint: Write $(y - \overline{y})^2$ as $[(y - \hat{y}) + (\hat{y} - \overline{y})]^2$ and apply the binomial formula.)

b. Verify that

$$\Sigma(y - \hat{y})(\hat{y} - \overline{y}) = \Sigma\hat{y}(y - \hat{y}) - \overline{y}\Sigma(y - \hat{y}).$$

c. In Exercise 13.52 (page 759), we found that $\Sigma(y - \hat{y}) = 0$ and $\Sigma x(y - \hat{y}) = 0$. Use those results and part (b) to show that $\Sigma(y - \hat{y})(\hat{y} - \overline{y}) = 0$.

d. Deduce the regression identity from the results obtained in parts (a) and (c).

· · ·   **13.70**   Prove that the percentage of the total variation in the observed *y*-values that is explained by the regression equals the coefficient of determination; that is, prove that $SSR/SST = r^2$. *(Hint: First apply the regression identity and then refer to the definition of r^2.)*

· · ·   **13.71**   In this exercise we will derive the shortcut formulas, given in Formula 13.2 on page 766, for the three sums of squares.

a. Obtain the shortcut formula for the total sum of squares, $SST = S_{yy}$, by showing that

$$\Sigma(y - \overline{y})^2 = \Sigma y^2 - (\Sigma y)^2/n.$$

(Hint: Expand the square on the left of the above equation and then apply properties of summation.)

b. Show that

$$\Sigma(\hat{y} - \overline{y})^2 = b_1^2 \Sigma(x - \overline{x})^2.$$

(Hint: Substitute $b_0 + b_1 x$ for $\hat{y}$ and then use the formula for b_0 given in Formula 13.1 on page 746.)

c. Apply the results of part (b) and the formula for b_1 in Formula 13.1 to obtain the shortcut formula for the regression sum of squares, $SSR = S_{xy}^2/S_{xx}$.

d. Use the regression identity and the shortcut formulas for *SST* and *SSR* to obtain the shortcut formula for the error sum of squares, $SSE = S_{yy} - S_{xy}^2/S_{xx}$.

13.4   LINEAR CORRELATION

We often hear statements pertaining to the correlation or lack of correlation between two variables: "There is a positive correlation between advertising expenditures and sales" or "IQ and alcohol consumption are uncorrelated." In this section we will explain the meaning of such statements.

Several statistics can be employed to measure the correlation between two variables. The one most commonly used is the **linear correlation coefficient, r,** also called the **Pearson product moment correlation coefficient.** The linear correlation coefficient is a descriptive measure of the strength of the linear (straight-line) relationship between two variables.

Formula 13.3 provides a shortcut formula for the linear correlation coefficient. (The defining formula and its equivalence to Formula 13.3 are discussed in the exercises.) Figure 13.18 depicts various degrees of linear correlation.

FORMULA 13.3

LINEAR CORRELATION COEFFICIENT

The *linear correlation coefficient, r,* of n data points can be computed using the formula

$$r = \frac{S_{xy}}{\sqrt{S_{xx}S_{yy}}}.$$

The formulas for S_{xx}, S_{xy}, and S_{yy} are given in Definition 13.3 on page 746.

FIGURE 13.18
Various degrees of linear correlation

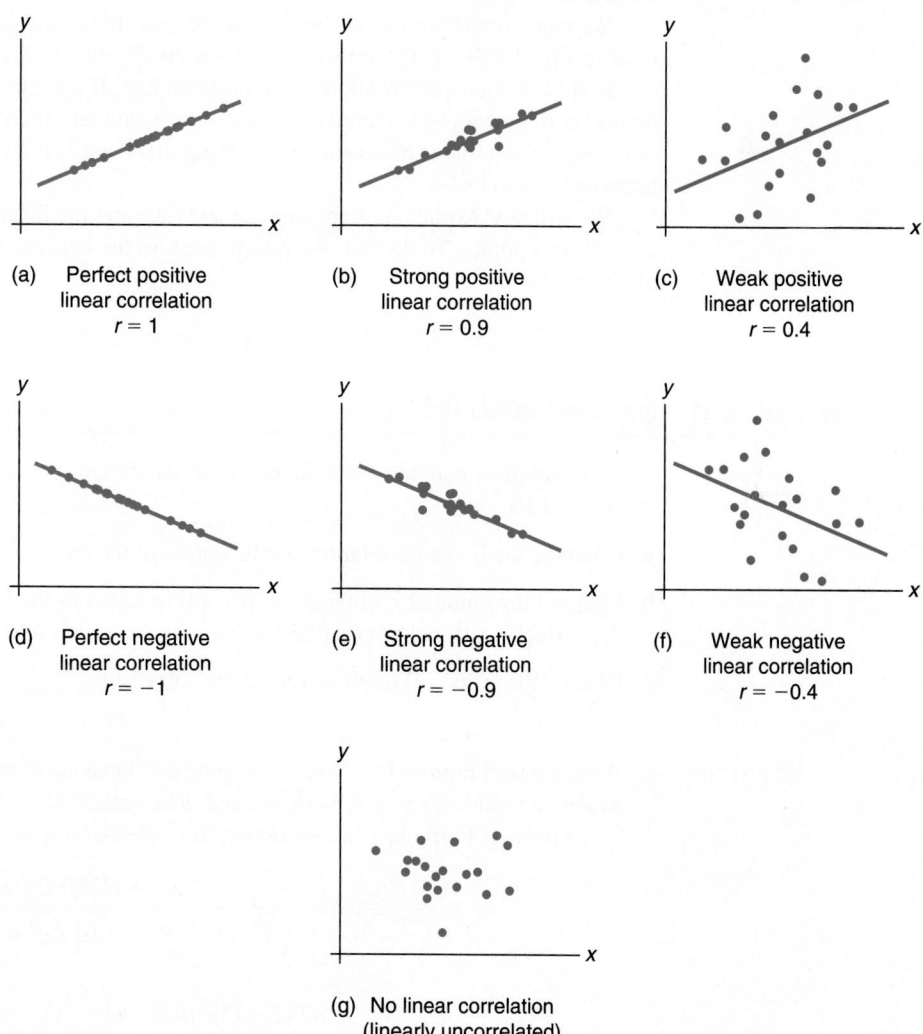

(a) Perfect positive
linear correlation
$r = 1$

(b) Strong positive
linear correlation
$r = 0.9$

(c) Weak positive
linear correlation
$r = 0.4$

(d) Perfect negative
linear correlation
$r = -1$

(e) Strong negative
linear correlation
$r = -0.9$

(f) Weak negative
linear correlation
$r = -0.4$

(g) No linear correlation
(linearly uncorrelated)
$r = 0$

The linear correlation coefficient, r, always lies between -1 and 1. Values of r close to -1 or 1 indicate a strong linear relationship between the variables and that the variable x is a good linear predictor of the variable y (i.e., the regression equation is extremely useful for making predictions). On the other hand, values of r near 0 indicate a weak linear relationship between the variables and that the variable x is a poor linear predictor of the variable y (i.e., the regression equation is not very useful for making predictions).

Positive values of r suggest that the variables are **positively linearly correlated,** meaning that y tends to increase linearly as x increases, with the tendency being greater the closer that r is to 1. Negative values of r suggest that the variables are **negatively linearly correlated,** meaning that y tends to decrease linearly as x increases, with the tendency being greater the closer that r is to -1. The sign of r is the same as the sign of the slope of the regression line.

We can summarize the discussion in the preceding two paragraphs as follows (also refer to Fig. 13.18): If the linear correlation coefficient, r, is close to ± 1, then the data points are clustered closely about the regression line. If r is farther from ± 1, then the data points are more widely scattered about the regression line. And if r is near 0, then the slope of the regression line is also near 0, indicating that there is probably no linear relationship between the variables.

We will now explain how to compute and interpret the linear correlation coefficient of a set of data points. To do that, we return again to the data on age and price for a sample of Nissan Zs.

Example 13.11   Illustrates Formula 13.3

The age and price data for a sample of 11 Nissan Zs are repeated in the first two columns of Table 13.10.

a. Compute the linear correlation coefficient, r, of the data.

b. Interpret the value of r obtained in part (a) in terms of the linear relationship between the variables age and price of Nissan Zs.

c. Discuss the graphical implications of the value of r.

SOLUTION   a. We see from Formula 13.3 that to compute the linear correlation coefficient, r, we need a table of values for x, y, xy, x^2, y^2, and their sums. This is presented in Table 13.10.

Applying Formula 13.3, we obtain from the last row of Table 13.10 that

$$r = \frac{S_{xy}}{\sqrt{S_{xx}S_{yy}}} = \frac{\Sigma xy - (\Sigma x)(\Sigma y)/n}{\sqrt{\left[\Sigma x^2 - (\Sigma x)^2/n\right]\left[\Sigma y^2 - (\Sigma y)^2/n\right]}}$$

$$= \frac{4732 - (58)(975)/11}{\sqrt{\left[326 - (58)^2/11\right]\left[96,129 - (975)^2/11\right]}} = -0.924.$$

TABLE 13.10

Table for computing the linear
correlation coefficient of
the data on age and price for
a sample of 11 Nissan Zs

Age (yrs) x	Price ($100s) y	xy	x^2	y^2
5	85	425	25	7,225
4	103	412	16	10,609
6	70	420	36	4,900
5	82	410	25	6,724
5	89	445	25	7,921
5	98	490	25	9,604
6	66	396	36	4,356
6	95	570	36	9,025
2	169	338	4	28,561
7	70	490	49	4,900
7	48	336	49	2,304
58	975	4732	326	96,129

b. The linear correlation coefficient, $r = -0.924$, suggests that there is a strong negative linear correlation between age and price of Nissan Zs. In particular, it indicates that as age increases there is a strong tendency for price to decrease, which is not surprising. It also implies that the regression equation, $\hat{y} = 195.47 - 20.26x$, is extremely useful for making predictions.

c. Since the correlation coefficient, $r = -0.924$, is quite close to -1, the data points should be clustered rather closely about the regression line. Figure 13.16 on page 763 shows that to be the case. ∎

Relationship Between the Correlation Coefficient and the Coefficient of Determination

In Section 13.3 we discussed the coefficient of determination, r^2, a descriptive measure of the utility of the regression equation for making predictions. Now we have introduced the linear correlation coefficient, r, as a descriptive measure of the strength of the linear relationship between two variables.

We expect the strength of the linear relationship to also indicate the usefulness of the regression equation for making predictions. In other words, there should be a relationship between the linear correlation coefficient and the coefficient of determination, and there is. The relationship is precisely the one suggested by the notation.

KEY FACT 13.6

RELATIONSHIP BETWEEN THE CORRELATION COEFFICIENT AND THE COEFFICIENT OF DETERMINATION

The coefficient of determination is the square of the linear correlation coefficient.

In Example 13.11 we found that the linear correlation coefficient for the data on age and price of a sample of 11 Nissan Zs is $r = -0.924$. From this and Key Fact 13.6, we can easily obtain the coefficient of determination: $r^2 = (-0.924)^2 = 0.854$. As predicted, this is the same value (except for roundoff error) as the one we found for r^2 in Example 13.7 on page 762 by using the defining formula, $r^2 = 1 - SSE/SST$.

In general, we can compute the coefficient of determination for a set of data points either by using the defining formula, $r^2 = 1 - SSE/SST$, or by first obtaining the linear correlation coefficient and then squaring the result.

A Warning on the Use of the Linear Correlation Coefficient

As we mentioned in Section 13.2, an assumption for finding the regression line for a set of data points is that the data points are scattered about a straight line. That same assumption applies to the linear correlation coefficient: The linear correlation coefficient, r, is used to describe the strength of the *linear* relationship between two variables. It should be employed as a descriptive measure only when a scatter diagram indicates that the data points are scattered about a straight line.

Correlation Is Not Causation

Two variables may have a high correlation without being causally related. For example, Table 13.11 displays data on total pari-mutuel turnover (money wagered) at U.S. racetracks and college enrollment for five randomly selected years.

TABLE 13.11

Pari-mutuel turnover and college enrollment for five randomly selected years

Pari-mutuel turnover ($millions) x	College enrollment (thousands) y
5,977	8,581
7,862	11,185
10,029	11,260
11,677	12,372
11,888	12,426

SOURCES: National Association of State Racing Commissioners and U.S. National Center for Education Statistics.

The linear correlation coefficient of the data points in Table 13.11 is $r = 0.931$, suggesting a strong positive linear correlation between pari-mutuel wagering and college enrollment. But this does not mean that a causal relationship exists between the two variables, such as that when people go to racetracks they are somehow inspired to go to college. On the contrary, we can only infer that the two variables have a strong tendency to increase (or decrease) simultaneously and that total pari-mutuel turnover is a good predictor of college enrollment.

Two variables may be strongly correlated because they are both associated with other variables, called **lurking variables,** that cause changes in the two variables under consideration. For example, a study showed that teachers' salaries and the dollar amount of liquor sales are positively linearly correlated. A possible explanation for this curious fact might be that both of the variables, teachers' salaries and liquor sales, are tied to other variables, such as the rate of inflation, that pull them along together.

Using the Computer (Optional)

Minitab has a command called **Correlation** that can be used to determine the linear correlation coefficient, r, of a set of data points. Example 13.12 shows how to apply the Correlation command.

Example 13.12   The Correlation Command

The data on age and price for a sample of 11 Nissan Zs are displayed in the first two columns of Table 13.10 on page 773. Use Minitab to determine the linear correlation coefficient of the data.

SOLUTION   Recall that we have previously stored the age and price data in columns named AGE and PRICE. With that in mind, we proceed as follows.

Session commands: We type the command **CORRELATION** followed by the storage locations of the sample data; that is, we type CORRELATION of 'AGE' and 'PRICE' and press Enter. This command and the resulting output are shown in Printout 13.5.

(or)

Menu commands: We choose **Stat ▸ Basic Statistics ▸ Correlation...**, specify AGE and PRICE in the **Variables** text box, and then select **OK**. Printout 13.5 shows the output that results.

PRINTOUT 13.5

Minitab output for the
Correlation command

```
MTB > CORRELATION of 'AGE' and 'PRICE'

Correlation of AGE and PRICE = -0.924
```

From Printout 13.5 we see that the linear correlation coefficient for the age and price data is -0.924; that is, $r = -0.924$. Evidently, there is a strong negative linear correlation between age and price of Nissan Zs. ∎

EXERCISES 13.4

In Exercises 13.72–13.79, we have repeated data from exercises in Section 13.2. For each exercise,

a. *compute the linear correlation coefficient, r.*
b. *interpret the value of r in terms of the linear relationship between the two variables in question.*
c. *discuss the graphical interpretation of the value of r and check that it is consistent with the graph you obtained in the corresponding exercise in Section 13.2.*
d. *square r and compare the result with the value of the coefficient of determination you obtained in the corresponding exercise in Section 13.3.*

13.72   The data for Exercise 13.30:

Data points

x	1	1	5	5
y	1	3	2	4

13.73   The data for Exercise 13.31:

Data points

x	0	2	2	5	6
y	4	2	0	−2	1

13.74   The age and price data for a random sample of 10 Corvettes (from Exercise 13.32):

x	6	6	6	2	2	5	4	5	1	4
y	125	115	130	260	219	150	190	163	260	160

13.75   The height and weight data for 11 randomly selected males age 18–24 years (from Exercise 13.33):

x	65	67	71	71	66	75	67	70	71	69	69
y	175	133	185	163	126	198	153	163	159	151	155

13.76   The size and price data for a random sample of nine custom homes in the Equestrian Estates (from Exercise 13.34):

x	26	27	33	29	29	34	30	40	22
y	235	249	267	269	295	345	415	475	195

13.77   The data on age and peak heart rate for a random sample of 10 people (from Exercise 13.35):

x	30	38	41	38	29	39	46	41	42	24
y	186	183	171	177	191	177	175	176	171	196

13.78   The study-time and test-score data for a random sample of eight students in a beginning calculus course (from Exercise 13.36):

x	10	15	12	20	8	16	14	22
y	92	81	84	74	85	80	84	80

13.79   The data on disposable income and annual food expenditure for eight randomly selected middle-income families (from Exercise 13.37):

x	30	36	27	20	16	24	19	25
y	55	60	42	40	37	26	39	43

13.80   We took a sample of 10 students from an introductory statistics class and obtained the following data, where *x* denotes height, in inches, and *y* denotes final-exam score.

x	71	68	71	65	66	68	68	64	62	65
y	87	96	66	71	71	55	83	67	86	60

a. What sort of value of *r* would you expect to find for these data? Explain your answer.
b. Compute *r*.

13.81   Consider the following set of data points.

x	−3	−2	−1	0	1	2	3
y	9	4	1	0	1	4	9

a. Compute the linear correlation coefficient, *r*.
b. Can you conclude from your result in part (a) that the variables *x* and *y* are unrelated? Explain your answer.
c. Draw a scatter diagram for the data.
d. Is it appropriate to use the linear correlation coefficient as a descriptive measure for the data? Why or why not?
e. Show that the data are related by the equation $y = x^2$, and graph that equation along with the data points.

13.82 Determine whether r is positive, negative, or zero for each of the following data sets.

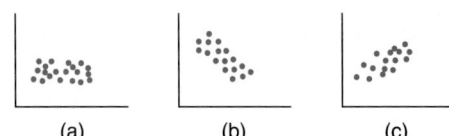

(a)   (b)   (c)

13.83 (Computer exercise) Use Minitab or some other statistical software to obtain the linear correlation coefficient of the data in Exercise 13.79.

13.84 (Computer exercise) Use Minitab or some other statistical software to obtain the linear correlation coefficient of the data in Exercise 13.78.

13.85 A Knight-Ridder News Service article appearing in the *Wichita Eagle* on November 27, 1992, discussed a study on the relationship between country music and suicide. The results of the study, coauthored by sociologist John Gundlach, appeared in the September 1992 issue of *Social Forces.* According to the article, ". . . analysis of 49 metropolitan areas shows that the greater the airtime devoted to country music, the greater the white suicide rate." (Suicide rates in the black population were found to be uncorrelated with the amount of country-music airtime.)

a. Use the terminology introduced in this section to describe the statement quoted above.

b. One of the conclusions stated in the journal article was that country music "nurtures a suicidal mood" by dwelling on marital status and alienation from work. Do you think this conclusion is warranted solely on the basis of the positive correlation found between airtime devoted to country music and white suicide rate? Explain your answer.

Defining formula for the linear correlation coefficient. In Exercises 13.47–13.51, we examined the concept of the sample covariance. Recall that the *sample covariance, s_{xy},* of a sample of n data points is

(3) $$s_{xy} = \frac{\Sigma(x - \bar{x})(y - \bar{y})}{n - 1}.$$

The defining formula for the *linear correlation coefficient* is

(4) $$r = \frac{s_{xy}}{s_x s_y},$$

where s_x and s_y are the sample standard deviations of the x-values and y-values, respectively.

13.86 Use Equation (4) to compute r for the set of data points in Exercise 13.72, and compare your answer with the value obtained for r in that exercise.

13.87 Use Equation (4) to compute r for the set of data points in Exercise 13.73, and compare your answer with the value obtained for r in that exercise.

13.88 In this exercise we will discuss the interpretation of the sample covariance.

a. Consider the following data set.

x	1	2	3	4	5	6
y	1	3	4	6	6	7

For these data, $\bar{x} = 3.5$ and $\bar{y} = 4.5$. Below we have drawn a coordinate system with a second set of axes passing through the point (3.5, 4.5). Construct a scatter diagram for the data points on the coordinate system provided.

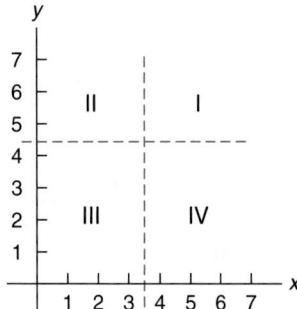

b. The dashed lines divide the above coordinate system into four regions, which we have labeled I, II, III, and IV. For a point (x, y) in Region I, $x - \bar{x}$ and $y - \bar{y}$ are both positive, so $(x - \bar{x})(y - \bar{y})$ is also positive. Fill in the remainder of the following table by using similar reasoning.

Region	Sign of $(x - \bar{x})(y - \bar{y})$
I	+
II	
III	
IV	

c. Without performing any calculations, determine whether the sample covariance of the data points in part (a) is positive or negative. [*Hint:* Use the graph from part (a) and the table from part (b).]

d. Use a similar graphing procedure to decide whether the sample covariance is positive or negative for the following data points.

x	1	2	3	4	5	6
y	7	6	6	4	3	1

e. The following graphs show two data sets, one scattered about a straight line with positive slope and the other scattered about a straight line with negative slope.

Complete the following statements.

i. For data scattered about a straight line with positive slope, the covariance is _____ .

ii. For data scattered about a straight line with negative slope, the covariance is _____ .

· · **13.89** Exercise 13.68 on page 770 shows that the coefficient of determination always lies between 0 and 1. Use that result and Key Fact 13.6 on page 773 to deduce that the linear correlation coefficient always lies between -1 and 1; that is, $-1 \le r \le 1$.

· · · **13.90** Show that for the linear correlation coefficient, the defining formula, Equation (4) on page 777, and the short-cut formula, Formula 13.3 on page 771, are equivalent. *(Hint: Use the defining formulas for S_{xx}, S_{xy}, and S_{yy} given in Definition 13.3 on page 746.)*

· · · **13.91** This exercise verifies Key Fact 13.6; namely, that the square of the linear correlation coefficient is equal to the coefficient of determination. *(Hint: Rewrite the defining formula for the coefficient of determination in terms of the shortcut formulas for SSE and SST.)*

13.5 MULTIPLE REGRESSION (OPTIONAL)

· · · · ·

So far we have considered linear regression analysis with one predictor variable. That type of linear regression analysis is called **simple linear regression.** Now we will study **multiple regression,** which applies when there is more than one predictor variable.

In simple linear regression, where there is only one predictor variable, x, the regression equation is of the form

$$\hat{y} = b_0 + b_1 x.$$

For multiple regression with two predictor variables, x_1 and x_2, the regression equation is of the form

$$\hat{y} = b_0 + b_1 x_1 + b_2 x_2.$$

And, in general, for multiple regression with k predictor variables, $x_1, x_2, \ldots, x_k$, the regression equation is of the form

$$\hat{y} = b_0 + b_1 x_1 + \cdots + b_k x_k.$$

Basically, multiple regression uses the same principles and techniques as simple linear regression: The regression equation is obtained by employing the least-squares criterion; the three sums of squares, *SST, SSR,* and *SSE,* and the coefficient of determination are defined in the same way; and so forth.

In multiple regression the **coefficient of determination** is denoted by R^2 instead of r^2, but its interpretation is the same. Thus the coefficient of determination, R^2, equals the percentage of the total variation in the observed y-values that is explained by the regression. It is a descriptive measure of the utility of the regression equation for making predictions: Values of R^2 near 0 indicate that the regression equation is not very useful for making predictions, whereas values of R^2 near 1 indicate that the regression equation is extremely useful for making predictions.

In Formula 13.1 on page 746, we presented formulas that can be used to obtain the regression equation in simple linear regression. Similar formulas exist for multiple regression but they are much more complicated, especially when there are many predictor variables. Because of this and other computational difficulties, multiple regression is almost always done by computer. Consequently, in this section we will use computer output to examine multiple regression.

Example 13.13   Multiple Regression

The age and price data for a sample of 11 Nissan Zs are repeated in the second and fourth columns of Table 13.12. Those data have been used to illustrate simple linear regression throughout this chapter.

In Example 13.7 on page 762, we found that the coefficient of determination for the age and price data is $r^2 = 0.853$. This means that 85.3% of the total variation in the price data is explained by age. Perhaps by using some additional predictor variables we could explain more of the variation in the price data and thereby obtain a regression equation that is a better predictor of price.

For instance, in addition to "age," we might include "number of miles driven" as a predictor variable. The third column of Table 13.12 displays the number of miles, in thousands, that each of the 11 Nissan Zs has been driven.

TABLE 13.12
Data on age, miles driven, and price for 11 Nissan Zs

Car	Age (yrs) x_1	Miles (thous) x_2	Price ($100s) y
1	5	57	85
2	4	40	103
3	6	77	70
4	5	60	82
5	5	49	89
6	5	47	98
7	6	58	66
8	6	39	95
9	2	8	169
10	7	69	70
11	7	89	48

We employed Minitab to perform a multiple regression analysis on the data in Table 13.12, with price as the response variable and age and miles as the predictor variables. The results are displayed in Printout 13.6.

PRINTOUT 13.6
Minitab output for multiple
regression of Nissan Z
data in Table 13.12

```
The regression equation is
PRICE = 183 - 9.50 AGE - 0.821 MILES

Predictor        Coef       Stdev     t-ratio        p
Constant       183.04       11.35       16.13    0.000
AGE            -9.504        3.874      -2.45     0.040
MILES          -0.8215       0.2552     -3.22     0.012


s = 8.805        R-sq = 93.6%      R-sq(adj) = 92.0%

Analysis of Variance

SOURCE        DF          SS          MS         F         p
Regression     2        9088.3      4544.2     58.61     0.000
Error          8         620.2        77.5
Total         10        9708.5
```

a. Use the output in Printout 13.6 to obtain the regression equation for price in terms of age and miles.

b. Apply the regression equation to predict the price of a Nissan Z that is 5 years old and has been driven 52,000 miles.

c. Use the output in Printout 13.6 to find the coefficient of determination, R^2.

d. Which regression equation better explains the variation in the price data: the multiple regression equation, using both age and miles as predictor variables; or the simple linear regression equation, using only age as a predictor variable?

SOLUTION
a. The regression equation is displayed in the second line of the computer output in Printout 13.6: PRICE = 183 - 9.50 AGE - 0.821 MILES. In other words, the regression equation is $\hat{y} = 183 - 9.50x_1 - 0.821x_2$, where x_1 denotes age, in years, x_2 denotes miles driven, in thousands, and $\hat{y}$ denotes predicted price, in hundreds of dollars.

b. For a 5-year-old Nissan Z with 52,000 miles, we have $x_1 = 5$ and $x_2 = 52$. The predicted price for such a car is

$$\hat{y} = 183 - 9.50 \cdot 5 - 0.821 \cdot 52 = 92.81,$$

or $9281.

c. The coefficient of determination is displayed as the second entry in the seventh line of Printout 13.6: R-sq = 93.6%; that is, $R^2 = 0.936$.

d. From part (c) we can conclude that 93.6% of the total variation in the observed prices is explained by age and miles driven. On the other hand, we have seen that only 85.3% of the total variation in the observed prices is explained by age alone. Thus the multiple regression equation provides a much better explanation of the variation in the price data than does the simple linear regression equation. As a consequence, we would expect to be able to make better price predictions using both age and miles driven than just age alone.   ∎

Curvilinear Regression

As we know, determining a regression line for a set of data points is based on the assumption that the data points are scattered about a straight line. If the data points are not scattered about a straight line but follow a curve, then a regression line should not be determined. Instead, a curve should be fit to the data points using a multiple regression technique called **curvilinear regression.** We will consider an application of curvilinear regression in Exercise 13.99.

Using the Computer (Optional)

In Section 13.2 we learned how Minitab's **Regress** command can be used to carry out a simple linear regression. That command works for multiple regression as well. Example 13.14 provides the details.

Example 13.14	The Regress Command

Use Minitab to obtain the output shown in Printout 13.6.

SOLUTION   For our purposes here, we need only the briefest computer output; therefore we begin by typing BRIEF 1 at the Minitab prompt MTB >. Next we store the age, miles, and price data from Table 13.12 on page 779 in columns named AGE, MILES, and PRICE, respectively. Then we do the following.

Session commands: We type the command **REGRESS** followed by the storage location of the data for the response variable, the number of predictors, and the storage locations of the data for the predictor variables; that is, we type REGRESS 'PRICE' on 2 predictors 'AGE' and 'MILES' and press [Enter]. Printout 13.6 displays the output that results.

(or)

Menu commands: We choose **Stat ▸ Regression ▸ Regression...**, specify PRICE for **Response**, specify AGE and MILES in the **Predictors** text box, and then select **OK**. The resulting output is depicted in Printout 13.6.   ∎

EXERCISES 13.5

13.92 Why is it often preferable to use more than one predictor variable in a regression analysis?

13.93 A household-appliance manufacturer wants to analyze the relationship between total sales and the company's three primary means of advertising. The first three columns of the following table provide the expenditures on advertising, by type, for each of 10 randomly selected sales periods. The fourth column contains the total sales. All data are in millions of dollars.

Television x_1	Magazines x_2	Radio x_3	Sales y
8.3	4.4	6.1	361.1
6.3	4.2	4.9	344.0
9.9	5.9	6.3	377.9
9.4	3.3	6.1	371.5
10.4	2.7	5.2	365.4
9.0	3.5	5.1	364.5
9.2	4.1	6.0	372.9
10.6	4.8	6.4	379.4
9.3	4.2	5.5	362.6
10.5	6.0	5.9	387.5

We used Minitab to perform a multiple regression analysis on the data using the variables television, magazine, and radio advertising expenditures as predictor variables for sales. The computer output is shown in Printout 13.7.

a. Use the computer output to obtain the regression equation for sales in terms of television, magazine, and radio advertising expenditures.

b. Apply the regression equation to predict total sales if the amounts spent on television, magazine, and radio advertising are $9.5 million, $4.3 million, and $5.2 million, respectively.

c. Use the computer output to obtain the coefficient of determination, R^2. Interpret your result.

13.94 The data on age and price for 10 Corvettes from Exercise 13.32 are repeated in the first and third columns of the table shown at the top of the next column. The second column of the table displays the number of miles, in thousands, that each of the 10 Corvettes has been driven. In Exercise 13.56 we found that the coefficient of determination for the age and price data is $r^2 = 0.937$. This means that 93.7% of the total variation in the price data is explained by age.

Age (yrs) x_1	Miles (thous) x_2	Price ($100s) y
6	36	125
6	36	115
6	36	130
2	22	260
2	5	219
5	31	150
4	22	190
5	39	163
1	9	260
4	27	160

We used Minitab to perform a multiple regression analysis on the data, with age and miles as predictor variables for price. The computer output is shown in Printout 13.8.

a. Use the output to obtain the regression equation for price in terms of age and miles.

b. Apply the regression equation to predict the price of a 4-year-old Corvette that has been driven 28,000 miles.

c. Use the computer output to find the coefficient of determination, R^2.

d. Which regression equation better explains the variation in the price data: the multiple regression equation, using both age and miles as predictor variables; or the simple linear regression equation, using only age as a predictor variable?

13.95 Graduation rates and what influences them have become a concern in U.S. colleges and universities. *U.S. News and World Report*'s "1992 College Guide" provides data on graduation rates for colleges and universities as a function of the percentage of freshmen in the top 10% of their high-school class, total spending per student, and student-to-faculty ratio. (Here *graduation rate* refers to the percentage of entering freshmen, attending full time, that graduate within 5 years.) Printout 13.9 at the top of page 785 shows the results of a multiple regression analysis for a sample of 44 schools.

a. Determine the regression equation.

b. Apply the regression equation to predict the graduation rate at a school where 70% of the freshmen were in the top 10% of their high-school class, the total spending per student is $15,500, and the student-to-faculty ratio is 18 to 1. (The values of the predictor variables are 70, 15,500, and 18.)

c. Find the coefficient of determination.

d. How useful do the variables percentage of freshmen in the top 10% of their high-school class, total spending per student, and student-to-faculty ratio appear to be for predicting graduation rates at colleges and universities?

The regression equation is
SALES = 266 + 6.73 TV + 3.26 MAG + 4.51 RADIO

Predictor	Coef	Stdev	t-ratio	p
Constant	266.23	16.34	16.29	0.000
TV	6.727	1.344	5.01	0.002
MAG	3.257	1.642	1.98	0.095
RADIO	4.507	3.703	1.22	0.269

s = 4.418   R-sq = 91.1%   R-sq(adj) = 86.6%

Analysis of Variance

SOURCE	DF	SS	MS	F	p
Regression	3	1194.53	398.18	20.40	0.002
Error	6	117.11	19.52		
Total	9	1311.64			

The regression equation is
PRICE = 287 - 37.4 AGE + 1.64 MILES

Predictor	Coef	Stdev	t-ratio	p
Constant	287.362	9.943	28.90	0.000
AGE	-37.375	5.174	-7.22	0.000
MILES	1.6378	0.8116	2.02	0.083

s = 12.11   R-sq = 96.0%   R-sq(adj) = 94.9%

Analysis of Variance

SOURCE	DF	SS	MS	F	p
Regression	2	24655	12328	84.07	0.000
Error	7	1026	147		
Total	9	25682			

13.96 Hanna Properties specializes in custom-home resales in the Equestrian Estates, an exclusive subdivision in Phoenix, Arizona. Thirty-three properties were randomly selected, and data on the following variables were obtained: square footage, number of bedrooms, number of bathrooms, number of days on the market, and selling price (in thousands of dollars). Then Minitab was used to perform a regression analysis for selling price in terms of the other four variables. The resulting computer output is displayed in Printout 13.10.

a. Use the computer output to obtain the regression equation.
b. Apply the regression equation to find the predicted selling price for a home in the Equestrian Estates that has 3200 sq ft, 4 bedrooms, and 3 bathrooms, and has been on the market for 60 days.
c. Use the computer output to obtain the coefficient of determination, R^2. Interpret your result.

13.97 (Computer exercise) Refer to Exercise 13.93. Use Minitab or some other statistical software to obtain computer output similar to that in Printout 13.7 on page 783.

13.98 (Computer exercise) Refer to Exercise 13.94. Use Minitab or some other statistical software to obtain computer output similar to that in Printout 13.8 on page 783.

13.99 Curvilinear regression. This exercise illustrates how to fit a curve to a set of data points using multiple regression. We stated earlier that although the relationship between age and price of Nissan Zs appears to be linear in the age range from 2 to 7 years, it is definitely not so in the age range from 2 to 11 years. From *Auto Trader* we obtained the data on age and price for a sample of 31 Nissan Zs shown in the table at the top of the next column. (Ages are in years, prices are in hundreds of dollars.) Below the table is a scatter diagram of the data.

As you can see from the scatter diagram, the data points are not clustered about a straight line but instead follow a curve. This means we should not determine a regression line but instead should try to fit a curve to the data. From the curvature of the scatter diagram, it appears that a parabola might be an appropriate curve to fit to the data. To fit a parabola to the data, we need a regression equation of the form

$$\hat{y} = b_0 + b_1 x + b_2 x^2.$$

If we let $x_1 = x$ and $x_2 = x^2$, then the above equation becomes

$$\hat{y} = b_0 + b_1 x_1 + b_2 x_2,$$

which is a multiple regression equation with two predictor variables, age and age^2 (square of the age).

Age x	Price y	Age x	Price y	Age x	Price y	Age x	Price y
5	85	4	103	10	25	3	135
6	70	4	100	5	82	9	44
4	90	6	75	10	35	9	36
2	150	3	140	5	89	11	33
5	98	6	66	6	95	1	180
6	95	2	169	4	65	5	80
3	129	6	60	6	82	5	105
4	115	8	50	9	42		

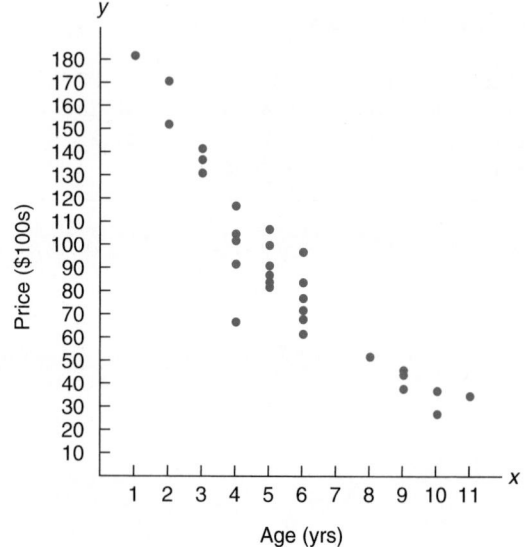

We used Minitab to perform a multiple regression analysis with price as the response variable and age and age^2 as the predictor variables. Printout 13.11, shown at the top of page 786, displays the computer output.

a. Use the computer output to obtain the regression equation, that is, the equation of the parabola that best fits the data points in the scatter diagram.
b. Graph the parabola you obtained in part (a) on the scatter diagram.
c. Apply the regression equation to predict the price of a 9-year-old Nissan Z.
d. What percentage of the total variation in the observed prices is explained by the variables age and age^2, that is, by the parabolic regression?

13.100 (Computer exercise) Refer to Exercise 13.99. Use Minitab or some other statistical software to obtain computer output similar to that in Printout 13.11 on page 786.

PRINTOUT 13.9   Minitab output for Exercise 13.95

The regression equation is
GRAD = 52.4 + 0.210 TOP10%HS +0.000220 EXPENDIT - 0.740 STU:FAC

Predictor	Coef	Stdev	t-ratio	p
Constant	52.36	15.98	3.28	0.002
TOP10%HS	0.2101	0.1134	1.85	0.071
EXPENDIT	0.0002197	0.0004401	0.50	0.620
STU:FAC	-0.7404	0.7480	-0.99	0.328

s = 13.34   R-sq = 16.6%   R-sq(adj) = 10.4%

Analysis of Variance

SOURCE	DF	SS	MS	F	p
Regression	3	1421.6	473.9	2.66	0.061
Error	40	7121.2	178.0		
Total	43	8542.8			

PRINTOUT 13.10   Minitab output for Exercise 13.96

The regression equation is
SELL$ = - 212 + 0.0754 SQFT + 20.9 BEDROOMS + 62.1 BATHS - 0.0899 DAYS

Predictor	Coef	Stdev	t-ratio	p
Constant	-211.95	73.05	-2.90	0.007
SQFT	0.07542	0.02345	3.22	0.003
BEDROOMS	20.94	19.99	1.05	0.304
BATHS	62.11	21.62	2.87	0.008
DAYS	-0.08986	0.07624	-1.18	0.248

s = 39.48   R-sq = 84.0%   R-sq(adj) = 81.7%

Analysis of Variance

SOURCE	DF	SS	MS	F	p
Regression	4	229575	57394	36.83	0.000
Error	28	43633	1558		
Total	32	273209			

PRINTOUT 13.11   Minitab output for Exercise 13.99

```
The regression equation is
PRICE = 209 - 30.8 AGE + 1.33 AGESQ

Predictor        Coef       Stdev     t-ratio        p
Constant       209.44       11.48       18.24    0.000
AGE           -30.776        4.056      -7.59    0.000
AGESQ           1.3297       0.3219      4.13    0.000

s = 12.81        R-sq = 90.3%     R-sq(adj) = 89.6%

Analysis of Variance

SOURCE         DF         SS          MS        F         p
Regression      2       42895       21448    130.70    0.000
Error          28        4595         164
Total          30       47490
```

Chapter Review

Key Terms

Formulas

In the formulas below,

SSE = error sum of squares
SST = total sum of squares
SSR = regression sum of squares

n = sample size (number of data points)
r^2 = coefficient of determination
r = linear correlation coefficient

S_{xx}, S_{xy}, and S_{yy}, 746

$$S_{xx} = \Sigma(x - \bar{x})^2 = \Sigma x^2 - (\Sigma x)^2/n$$
$$S_{xy} = \Sigma(x - \bar{x})(y - \bar{y}) = \Sigma xy - (\Sigma x)(\Sigma y)/n$$
$$S_{yy} = \Sigma(y - \bar{y})^2 = \Sigma y^2 - (\Sigma y)^2/n$$

Regression equation, 746

$$\hat{y} = b_0 + b_1 x,$$

where

$$b_1 = \frac{S_{xy}}{S_{xx}} \quad \text{and} \quad b_0 = \frac{1}{n}(\Sigma y - b_1 \Sigma x) = \bar{y} - b_1 \bar{x}$$

Total sum of squares, 762, 766

$$SST = \Sigma(y - \bar{y})^2 = S_{yy}$$

Regression sum of squares, 765, 766

$$SSR = \Sigma(\hat{y} - \bar{y})^2 = S_{xy}^2/S_{xx}$$

Error sum of squares, 762, 766

$$SSE = \Sigma(y - \hat{y})^2 = S_{yy} - S_{xy}^2/S_{xx}$$

Regression identity, 765

$$SST = SSR + SSE$$

Coefficient of determination, 762, 766

$$r^2 = 1 - \frac{SSE}{SST} = \frac{SSR}{SST}$$

Linear correlation coefficient, 771

$$r = \frac{S_{xy}}{\sqrt{S_{xx}S_{yy}}}$$

Multiple regression equation,* 778

$$\hat{y} = b_0 + b_1 x_1 + \cdots + b_k x_k$$

(k = number of predictor variables)

You Should Be Able To

1. use and understand the preceding formulas.
2. apply the concepts related to linear equations with one independent variable.
3. explain the least-squares criterion.
4. obtain and graph the regression equation for a set of data points, interpret the slope of the regression line, and use the regression equation to make predictions.
5. identify outliers and influential observations.
6. calculate and interpret the three sums of squares, SST, SSE, and SSR, and the coefficient of determination, r^2.
7. determine and interpret the linear correlation coefficient, r.

8. obtain the regression equation and coefficient of determination for a multiple regression from computer output.*
9. apply a multiple regression equation to make predictions.*
10. use the Minitab commands covered in this chapter.*
11. interpret the output obtained from the application of the Minitab commands discussed in this chapter.*

REVIEW TEST

1. A small company has purchased a microcomputer system for $7200 and plans to depreciate the value of the equipment by $1200 per year for 6 years. Let x denote the age of the equipment, in years, and y denote the value of the equipment, in hundreds of dollars.
   a. Determine the equation that expresses y in terms of x.
   b. Find the y-intercept, b_0, and slope, b_1, of the linear equation in part (a).
   c. Without graphing the equation in part (a), decide whether the line slopes upward, slopes downward, or is horizontal.
   d. Find the value of the computer equipment after 2 years; after 5 years.
   e. Obtain the graph of the equation in part (a) by plotting the points from part (d) and connecting them with a straight line.
   f. Use the graph from part (e) to visually estimate the value of the equipment after 4 years. Then calculate that value exactly using the equation from part (a).

2. The director of a large mathematics course hires upper-division science students to grade papers. On each grading day, he records the number of papers graded and the total amount of money paid to the graders. The following table gives data for 12 randomly selected days from last semester.

Papers (100s) x	Cost ($) y
16	234
16	220
18	258
22	298
19	273
16	227
18	246
15	210
19	265
17	250
15	223
18	251

a. Draw a scatter diagram of the data.
b. Is it reasonable to find a regression line for the data? Explain your answer.
c. Determine the regression equation for the data and draw its graph on the scatter diagram you drew in part (a).
d. Describe the apparent relationship between the two variables, number of papers graded and cost of grading.
e. What does the slope of the regression line represent in terms of paper-grading costs?
f. Use the regression equation to predict the cost of grading 1600 papers.
g. Identify outliers and potential influential observations.

3. Refer to Problem 2.
   a. Find SST, SSR, and SSE using the shortcut formulas.
   b. Calculate the coefficient of determination using the defining formula.
   c. Find the percentage reduction obtained in the total squared error by using the regression equation instead of the sample mean, $\bar{y}$, to predict the observed costs.
   d. Obtain the percentage of the total variation in the observed costs that is explained by the number of papers graded (i.e., by the regression line).
   e. State how useful the regression equation appears to be for making predictions.

*4. **(Computer problem)** Refer to Problem 2. Use Minitab or some other statistical software to
   a. obtain a scatter diagram for the data.
   b. determine the regression equation for the data.
   c. find the coefficient of determination, r^2, and the three sums of squares, SST, SSR, and SSE.
   d. identify potential outliers and influential observations.
   e. obtain the regression equation with the potential influential observation removed.
   f. Is the potential influential observation actually an influential observation; that is, does its removal markedly change the regression equation? *(Hint: Plot both regression lines on the same graph.)*

*5. (**Computer problem**) Printout 13.12 shows the Minitab output obtained by applying the Regress command to the data in Problem 2. Employ the output to determine the
   a. regression equation.
   b. coefficient of determination.
   c. regression sum of squares, error sum of squares, and total sum of squares.
   d. potential outliers and influential observations, if any.

6. Refer to Problem 2.
   a. Compute the linear correlation coefficient, r.
   b. Interpret the value of r in terms of the linear relationship between the number of papers graded and the cost of grading.
   c. Discuss the graphical implications of the value of the linear correlation coefficient, r.
   d. Use your answer from part (a) to obtain the coefficient of determination.

*7. (**Computer problem**) Use Minitab or some other statistical software to obtain the linear correlation coefficient of the data in Problem 2.

*8. The Census Bureau collects data on income by educational attainment, sex, and age. Results are published in *Current Population Reports*. From a random sample of 75 males between the ages of 25 and 50, all of whom have at least a ninth-grade education, data were collected on age, number of years of school completed, and annual income. Then Minitab was used to perform a multiple regression analysis for annual income (in thousands of dollars) with the variables age and number of years of school completed as predictor variables. The resulting output is shown in Printout 13.13 at the top of the next page.
   a. Use the computer output to obtain the regression equation for annual income in terms of age and number of years of school completed.
   b. Apply the regression equation to predict the annual income of a male who is 32 years old and has completed exactly 4 years of college (i.e., 16 years of school).
   c. Find and interpret the coefficient of determination, R^2.

*9. (**Computer problem**) Refer to Problem 8. Using your statistical software, explain how you would obtain computer output similar to that shown in Printout 13.13.

PRINTOUT 13.12   Minitab output for Problem 5

```
The regression equation is
COST = 35.8 + 12.1 PAPERS


Predictor       Coef       Stdev     t-ratio       p
Constant       35.80       17.06        2.10     0.062
PAPERS       12.0835      0.9738       12.41     0.000


s = 6.526      R-sq = 93.9%      R-sq(adj) = 93.3%


Analysis of Variance

SOURCE        DF          SS          MS         F         p
Regression     1       6558.3      6558.3    153.97     0.000
Error         10        425.9        42.6
Total         11       6984.3


Unusual Observations
Obs.  PAPERS      COST      Fit Stdev.Fit  Residual   St.Resid
  4     22.0    298.00    301.63     4.84     -3.63      -0.83 X

X denotes an obs. whose X value gives it large influence.
```

PRINTOUT 13.13   Minitab output for Problem 8

The regression equation is
INCOME = - 40.9 + 0.772 AGE + 3.11 EDUC

Predictor	Coef	Stdev	t-ratio	p
Constant	-40.855	5.511	-7.41	0.000
AGE	0.7719	0.1165	6.62	0.000
EDUC	3.1052	0.2250	13.80	0.000

s = 7.886   R-sq = 76.6%   R-sq(adj) = 75.9%

Analysis of Variance

SOURCE	DF	SS	MS	F	p
Regression	2	14657.2	7328.6	117.84	0.000
Error	72	4477.6	62.2		
Total	74	19134.7			

Using the Focus Database

Appendix B contains a printout of a database obtained by randomly selecting 500 Arizona State University sophomores. Seven variables are considered for each student: sex, high-school GPA, SAT math score, cumulative GPA, SAT verbal score, age, and total hours. For these database exercises, you should eliminate all cases (students) in which one or more of the four variables, cumulative GPA, high-school GPA, SAT math score, and SAT verbal score, equal 0.

First we will perform a correlation analysis to choose the best predictor of cumulative GPA from among the variables high-school GPA, SAT math score, and SAT verbal score.

a. Determine the correlation coefficient between the cumulative GPA data and each of the three data sets, high-school GPA, SAT math score, and SAT verbal score.

b. Among the variables high-school GPA, SAT math score, and SAT verbal score, identify the one that appears to be the best predictor of cumulative GPA for Arizona State University sophomores. Explain your answer.

Next we will perform a regression analysis on cumulative GPA using the predictor variable identified in part (b).

c. Obtain the regression equation for cumulative GPA using the predictor variable identified in part (b).

d. Find the coefficient of determination and interpret your answer.

e. Determine and interpret the three sums of squares, *SSR, SSE,* and *SST.*

Finally, for those who have studied the section on multiple regression (Section 13.5), we will perform a multiple regression analysis.

f. Obtain the regression equation for cumulative GPA using the three predictor variables, high-school GPA, SAT math score, and SAT verbal score.

g. Find the coefficient of determination and interpret your answer.

h. Which regression equation better explains the variation in cumulative GPA: the multiple regression equation obtained in part (f) or the simple linear regression equation obtained in part (c)? Explain your answer.

Case Study **Fat Consumption and Prostate Cancer**

Researchers have asked if there is a relationship between nutrition and cancer—and many studies have shown that there is. In fact, one of the conclusions of a 1980 study appearing in the journal *Advances in Cancer Research* was that "... none of the risk factors for cancer is probably more significant than diet and nutrition."[†]

Prostate cancer is one of the most virulent forms of cancer. Generally, it has spread before being detected and is usually fatal. One dietary factor that has been studied for its relationship with prostate cancer is fat consumption. The data in Table 13.13 were obtained from a graph in John Robbins's book *Diet for a New America* (Walpole, NH: Stillpoint Publishing, 1987).[‡]

TABLE 13.13
Fat consumption and prostate cancer death rates

Country	Dietary fat (grams/day)	Death rate (per 100,000)	Country	Dietary fat (grams/day)	Death rate (per 100,000)
El Salvador	38	0.9	Spain	97	10.1
Philippines	29	1.3	Portugal	73	11.4
Japan	42	1.6	Finland	112	11.1
Mexico	57	4.5	Hungary	100	13.1
Greece	96	4.8	United Kingdom	143	12.4
Colombia	47	5.4	Germany	134	12.9
Bulgaria	67	5.5	Canada	142	13.4
Yugoslavia	72	5.6	Austria	119	13.9
Poland	93	6.4	France	137	14.4
Panama	58	7.8	Netherlands	152	14.4
Israel	95	8.4	Australia	129	15.1
Romania	67	8.8	Denmark	156	15.9
Venezuela	62	9.0	United States	147	16.3
Czechoslovakia	96	9.1	Norway	133	16.8
Italy	86	9.4	Sweden	132	18.4

a. Draw a scatter diagram for the data on dietary fat and prostate cancer death rate. What does the scatter diagram tell you?

b. Does it appear reasonable to obtain a regression equation for the data? Explain your answer.

c. Find the regression equation for the data using dietary fat as the predictor variable.

d. Interpret the slope of the regression line.

e. Compute the correlation coefficient of the data and interpret your result.

f. Identify outliers and potential influential observations, if any.

g. Use Minitab or some other statistical software to solve parts (a), (c), (e) and (f).

[†] See "Nutrition and Its Relationship to Cancer" by B. Reddy et al., 1980, *Advances in Cancer Research, 32,* pp. 237–345.

[‡] The graph was adapted from data in the article cited in the preceding footnote.

Sir Francis Galton

Francis Galton was born on February 16, 1822, into a wealthy Quaker family of bankers and gunsmiths on his father's side and as a cousin of Charles Darwin on his mother's side. Although his IQ was estimated to be roughly 200, his formal education was unfinished.

He began training in medicine in Birmingham and London, but quit when, in his words, "A passion for travel seized me as if I had been a migratory bird." After a tour through Germany and southeastern Europe, he went to Trinity College in Cambridge to study mathematics. He left Cambridge in his third year, broken down from overwork. Recovering quickly, he resumed his medical studies in London; however, before he was finished, his father died, leaving him, at 22 years old, "a sufficient fortune to make me independent of the medical profession."

Galton held no professional or academic positions; nearly all of his experiments were conducted at his home or performed by friends. He was curious about almost everything and carried out research in fields including meteorology, biology, psychology, statistics, and genetics.

The origination of the concepts of regression and correlation, developed by Galton as tools for measuring the influence of heredity, are summed up in his work *Natural Inheritance.* He discovered regression during experiments with sweet-pea seeds to determine the law of inheritance of size. His other great discovery, correlation, was made while applying his techniques to the problem of measuring the degree of association between the sizes of two different body organs of an individual.

In his later years, Galton was associated with Karl Pearson, who became his champion and an extender of his ideas. Pearson was the first holder of the chair of eugenics at University College in London, which Galton had endowed in his will. Galton was knighted in 1909. He died in Haslemere, Surrey, England, in 1911.

Chapter 14

Inferential Methods in Regression and Correlation

. . . . .

General Objective In Chapter 13 we examined descriptive methods in regression and correlation. We discovered how to determine the regression equation for a set of data points and how to use that equation to make predictions. We also learned how to compute and interpret the coefficient of determination and the linear correlation coefficient for a set of data points.

In this chapter we will consider inferential methods in regression and correlation. For example, we will see how the regression equation can be used to obtain a confidence interval for the mean price of all Nissan Zs of any particular age; and how the linear correlation coefficient, r, can be used to decide whether there is a negative correlation between age and price of Nissan Zs.

Chapter Outline

Case Study *Fat Consumption and Prostate Cancer*

In the case study at the end of Chapter 13, we examined data on fat consumption and prostate cancer death rate for various nations of the world. The regression and correlation analyses done there were descriptive. At the end of this chapter, we will return to those data to make regression and correlation inferences.

14.1   THE REGRESSION MODEL; ANALYSIS OF RESIDUALS

To perform statistical inferences in regression and correlation, the variables under consideration must satisfy certain conditions. In this section we will discuss those conditions and examine methods for checking whether they hold.

The Regression Model

Let's return to the Nissan Z illustration used throughout Chapter 13. In Table 14.1 we have reproduced the data on age and price for a sample of 11 Nissan Zs. Ages are in years, and prices are in hundreds of dollars, rounded to the nearest hundred dollars.

TABLE 14.1
Age and price data for Nissan Zs

Age, x	5	4	6	5	5	5	6	6	2	7	7
Price, y	85	103	70	82	89	98	66	95	169	70	48

On page 747 we found that the regression equation for these data is $\hat{y} = 195.47 - 20.26x$. As we know, the regression equation can be used to predict the price of a Nissan Z given its age. However, we cannot expect such predictions to be completely accurate since prices vary even for Nissan Zs of the same age. For example, the sample data of Table 14.1 include four 5-year-old Nissan Zs. Their prices are $8500, $8200, $8900, and $9800. (The predicted price for a 5-year-old Nissan Z is $\hat{y} = 195.47 - 20.26 \cdot 5 = 94.17$, or $9417.)

This variation in price for Nissan Zs of the same age should be expected because these cars would have different mileages, interior conditions, paint quality, and so on. Thus we see that each age (x-value) has an entire population of corresponding prices (y-values), namely, the prices of all Nissan Zs of that age. There is a population of prices for 2-year-old Nissan Zs, another population of prices for 3-year-old Nissan Zs, and so on.

With the preceding discussion in mind, we now state the conditions required for using inferential methods in regression analysis.

KEY FACT 14.1

ASSUMPTIONS FOR REGRESSION INFERENCES

1. ***Population regression line:*** There is a straight line, $y = \beta_0 + \beta_1 x$, such that for each x-value, the mean of the corresponding population of y-values lies on that straight line. We refer to the straight line as the ***population regression line*** and to its equation as the ***population regression equation.***

2. ***Equal standard deviations:*** The standard deviation, σ, of the population of y-values corresponding to a particular x-value is the same, regardless of the x-value.

3. ***Normality:*** For each x-value, the corresponding population of y-values is normally distributed.

Assumptions 1, 2, and 3 require that there exist constants, β_0, β_1, and σ, such that for each x-value, the corresponding population of y-values is normally distributed with mean $\beta_0 + \beta_1 x$ and standard deviation σ. These assumptions are often referred to as the **regression model.** They can be expressed symbolically as

$$y = \beta_0 + \beta_1 x + \epsilon,$$

where ϵ (epsilon) represents a normally distributed random variable having mean 0 and standard deviation σ.

β_0, β_1, and σ are generally unknown and hence must be estimated from sample data. Point estimates for the y-intercept, β_0, and slope, β_1, of the population regression line are provided, respectively, by the y-intercept, b_0, and slope, b_1, of a sample regression line. A point estimate for σ will be given in the next subsection.

The inferential procedures in regression and correlation are robust to moderate deviations of the assumptions for regression inferences. In other words, the inferential procedures will work reasonably well provided the variables under consideration don't violate any of the assumptions too badly.

Example 14.1   Assumptions for Regression Inferences

Consider again the variables age and price of Nissan Zs.

a. Discuss what it would mean for the assumptions for regression inferences to be satisfied with age as the predictor variable for price.

b. Display the assumptions graphically.

SOLUTION

a. For the assumptions for regression inferences to be satisfied, it would mean there are constants, β_0, β_1, and σ, such that for each age, x, the prices of all Nissan Zs of that age are normally distributed with mean $\beta_0 + \beta_1 x$ and standard deviation σ. Thus it would mean that the prices of all 2-year-old Nissan Zs ($x = 2$) are normally distributed with mean $\beta_0 + \beta_1 \cdot 2$ and standard deviation σ; the prices of all 3-year-old Nissan Zs ($x = 3$) are normally distributed with mean $\beta_0 + \beta_1 \cdot 3$ and standard deviation σ; and so on.

b. To display the assumptions for regression inferences graphically, let's first consider Assumption 1. This assumption requires that for each age, the mean price of all Nissan Zs of that age lies on the straight line $y = \beta_0 + \beta_1 x$, as shown in Fig. 14.1 at the top of the next page.

Because the population regression line is usually not known, one of the main reasons for obtaining a sample regression line is to estimate the population regression line. Of course, a sample regression line, such as $\hat{y} = 195.47 - 20.26x$, ordinarily will not be the same as the population regression line, just as a sample mean, $\bar{x}$, generally will not equal the population mean, μ. We picture this situation in Fig. 14.2 (next page).

The solid line in Fig. 14.2 is the population regression line. The dashed line is a sample regression line, which is the best approximation that can be made to the population regression line by using the sample data in Table 14.1. A different sample of Nissan Zs would almost certainly yield a different sample regression line.

FIGURE 14.1

Population regression line

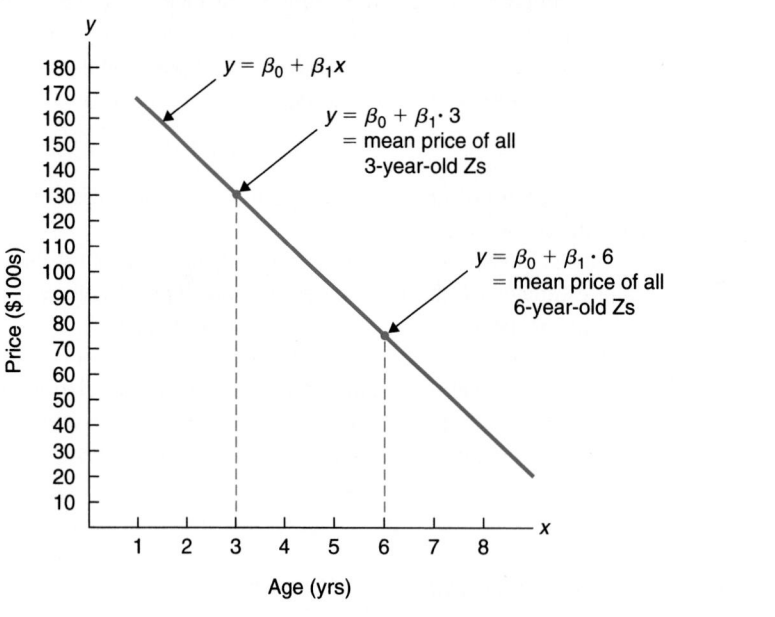

FIGURE 14.2

Population regression line
and sample regression line

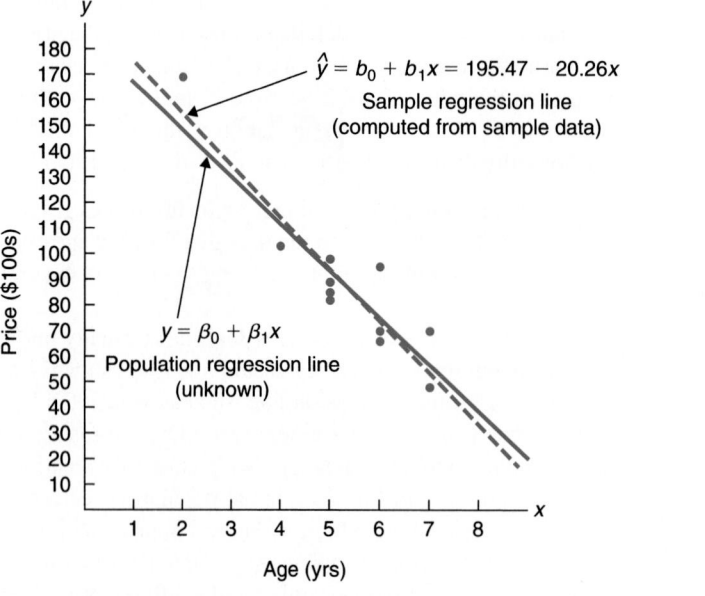

Next we will present graphs that depict Assumptions 2 and 3. Those assumptions require that the price distributions for the various ages of Nissan Zs are all normally distributed with the same standard deviation, σ. Figure 14.3 shows this for the price distributions of 2-year-old, 5-year-old, and 7-year-old Nissan Zs.

FIGURE 14.3

Price distributions for 2-year-old, 5-year-old, and 7-year-old Nissan Zs under Assumptions 2 and 3. (The means shown for the three normal distributions reflect Assumption 1.)

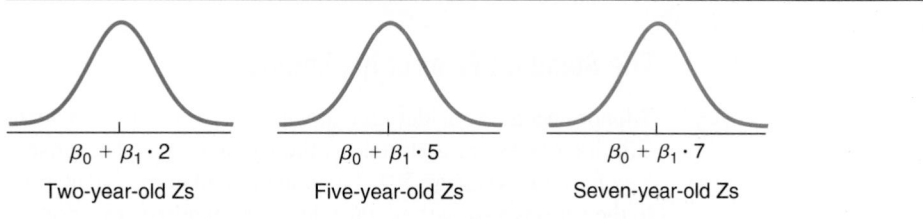

| $\beta_0 + \beta_1 \cdot 2$ | $\beta_0 + \beta_1 \cdot 5$ | $\beta_0 + \beta_1 \cdot 7$ |
| Two-year-old Zs | Five-year-old Zs | Seven-year-old Zs |

Notice that the shapes of the three normal curves in Fig. 14.3 are identical. This is because normal distributions having the same standard deviation have the same shape, and, under Assumptions 2 and 3, the price distributions for the various ages are all normally distributed and have equal standard deviations.

All three assumptions for regression inferences as they pertain to the variables age and price of Nissan Zs can be portrayed graphically by combining Figs. 14.1 and 14.3 into a three-dimensional graph. This is done in Fig. 14.4.

FIGURE 14.4 Graphical portrayal of the assumptions for regression inferences for age and price of Nissan Zs

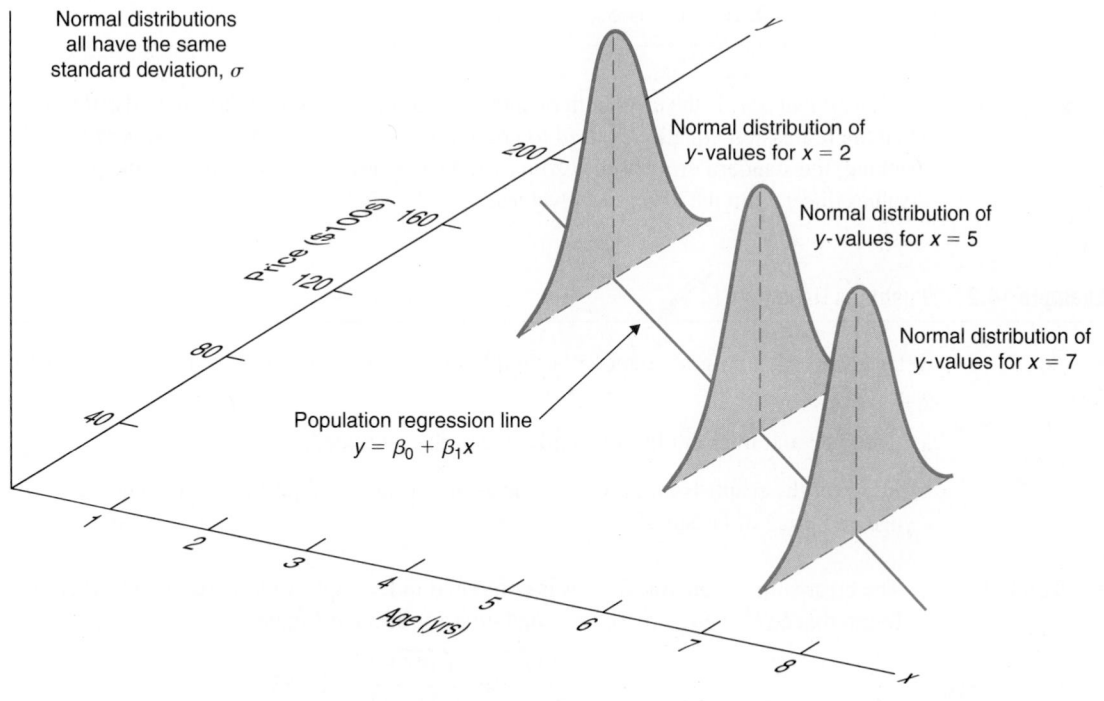

Figure 14.4 depicts the various price distributions for Nissan Zs under the assumption that the conditions for regression inferences hold, that is, that Assumptions 1, 2, and 3 are true for the variables age (x) and price (y) of Nissan Zs. Whether this is actually the case remains to be seen.   ∎

The Standard Error of the Estimate

Suppose we are considering two variables, x and y, for which the assumptions for regression inferences are met; that is, the variables x and y satisfy Assumptions 1, 2, and 3 in Key Fact 14.1 on page 794. Then, in particular, the populations of y-values corresponding to the various x-values all have the same standard deviation, σ.

As we mentioned earlier, the common population standard deviation, σ, is usually unknown and must be estimated from sample data. The statistic used to obtain a point estimate for σ is called the **standard error of the estimate** or the **residual standard deviation** and is defined as follows.

DEFINITION 14.1   **STANDARD ERROR OF THE ESTIMATE**

The *standard error of the estimate, s_e,* is defined by

$$s_e = \sqrt{\frac{SSE}{n-2}},$$

where $SSE = \Sigma(y - \hat{y})^2 = S_{yy} - S_{xy}^2/S_{xx}$.

Recall that SSE is the error sum of squares and represents the total squared error made when the regression equation is used to predict the observed y-values. Thus, very roughly speaking, the standard error of the estimate indicates how much, on average, the predicted y-values differ from the observed y-values.

Example 14.2   Illustrates Definition 14.1

Refer to the age and price data for a sample of 11 Nissan Zs displayed in Table 14.1 on page 794.

a. Compute and interpret the standard error of the estimate, s_e.

b. Interpret the result from part (a) if the assumptions for regression inferences hold for age and price of Nissan Zs.

SOLUTION   a. The error sum of squares, SSE, was computed in Example 13.9 on page 767, where we found that $SSE = 1423.5$. So the standard error of the estimate is

$$s_e = \sqrt{\frac{SSE}{n-2}} = \sqrt{\frac{1423.5}{11-2}} = 12.58.$$

As a rough estimate, we can say that, on the average, the predicted price of a Nissan Z in the sample differs from the observed price by $1258.

b. Presuming that the variables age (x) and price (y) for Nissan Zs satisfy the assumptions for regression inferences, the standard error of the estimate, $s_e = 12.58$, or $1258, provides an estimate for the (common) population standard deviation, σ, of prices for all Nissan Zs of any particular age. ∎

Analysis of Residuals

Now that we have examined the assumptions for regression inferences, we need to discuss how the sample data can be used to decide whether it is reasonable to presume that those assumptions are met. The method for deciding relies on an analysis of the errors made in using the regression equation to predict the observed y-values, that is, on the differences between the observed and predicted y-values, $y - \hat{y}$. Each such difference is called a **residual,** denoted generically by the letter **e.** Thus,

$$\text{Residual} = e = y - \hat{y}.$$

Figure 14.5 provides a graphical representation for the residual of a single data point.

FIGURE 14.5
Residual, e, of a data point

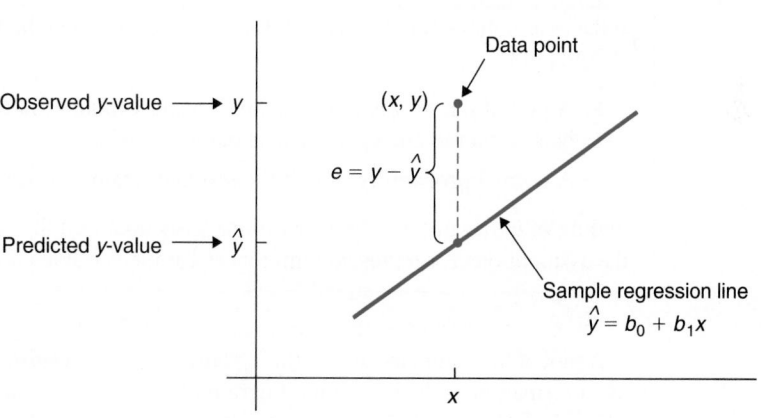

The residuals of all 11 data points for the Nissan Z data are given in the fourth column of Table 13.7 on page 762 and are pictured by the vertical lines in Fig. 13.15 on page 761.

We can express the standard error of the estimate, s_e, in terms of the residuals. Referring to Definition 14.1, we find that the standard error of the estimate can be written as

$$s_e = \sqrt{\frac{SSE}{n-2}} = \sqrt{\frac{\Sigma(y - \hat{y})^2}{n-2}} = \sqrt{\frac{\Sigma e^2}{n-2}}.$$

It can be shown that the sum of the residuals is always 0 and hence that $\bar{e} = 0$. Consequently, we see that the standard error of the estimate is essentially equal to the standard deviation of the residuals.[†] This explains why the standard error of the estimate is sometimes called the residual standard deviation.

The reason we can analyze the residuals to check whether the assumptions for regression inferences are met is that those assumptions can be translated into conditions on the residuals. To see how this is done, consider a sample of data points obtained from two variables that satisfy the assumptions for regression inferences.

In view of Assumption 1, the data points should be scattered about the (sample) regression line, which means that the residuals should be scattered about the x-axis; in view of Assumption 2, the variation of the observed y-values should remain approximately constant from one x-value to the next, which means the residuals should fall roughly in a horizontal band; and in view of Assumption 3, for each x-value the distribution of the corresponding observed y-values should be approximately bell-shaped, which implies that the horizontal band should be centered and symmetric about the x-axis.

Furthermore, considering Assumptions 1–3 simultaneously, we see that the residuals can be regarded as a random sample from a normal distribution having mean 0 and standard deviation σ. Thus a normal probability plot of the residuals should be roughly linear.

In summary, we have the following criteria for deciding whether the assumptions for regression inferences are met by the two variables under consideration.

KEY FACT 14.2   **RESIDUAL ANALYSIS FOR THE REGRESSION MODEL**

> If the assumptions for regression inferences are met, then the following two conditions should hold.
>
> 1. A plot of the residuals against the x-values should fall roughly in a horizontal band centered and symmetric about the x-axis.
>
> 2. A normal probability plot of the residuals should be roughly linear.
>
> Failure of either of these two conditions casts doubt on the validity of one or more of the assumptions for regression inferences for the variables under consideration.

A plot of the residuals against the x-values, called a **residual plot,** provides roughly the same information as does a scatter diagram of the y-values against the x-values. However, a residual plot makes it easier to spot patterns such as curvature and nonconstant standard deviation.

Figure 14.6(a) shows a residual plot in which the linearity and constant-standard-deviation assumptions appear to be met; Fig. 14.6(b) shows a residual plot in which the relation between the variables appears to be curved instead of linear; and Fig. 14.6(c) shows a residual plot in which the standard deviation of the population of y-values appears to increase as x increases instead of remaining constant.

[†] The exact standard deviation of the residuals is obtained by dividing by $n - 1$ instead of $n - 2$.

FIGURE 14.6

Residual plots indicating
(a) no violation of linearity or
constant standard deviation,
(b) violation of linearity, and
(c) violation of constant
standard deviation

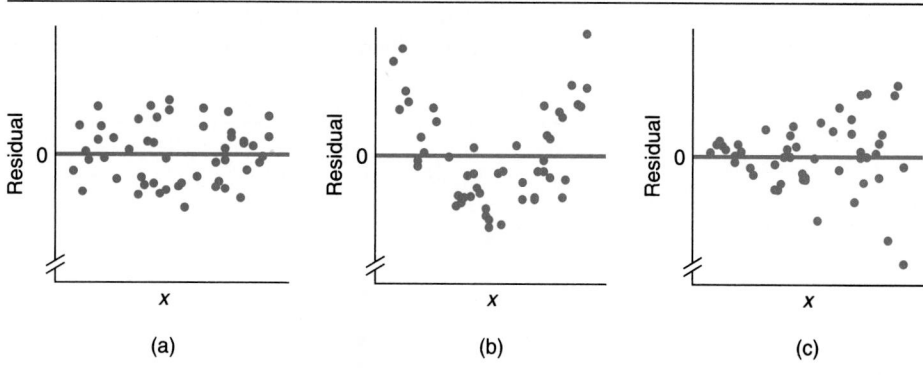

In our previous data analyses, we have seen that it is often difficult to decide on the appropriateness of a model when dealing with small samples, say, of size 20 or less. The same holds true in regression: For small samples we must be more liberal in allowing moderate departures from the idealized patterns when analyzing residual plots and normal probability plots to ascertain whether the assumptions for regression inferences are met. There are no definite rules—just judgment based on experience.

Example 14.3   Analysis of Residuals

Perform a residual analysis to decide whether it is reasonable to consider the assumptions for regression inferences met by the variables age (x) and price (y) of Nissan Zs.

SOLUTION   We apply the criteria in Key Fact 14.2. The ages and residuals for the Nissan Z data are displayed, respectively, in the first and fourth columns of Table 13.7 on page 762. We repeat that information here in Table 14.2.

TABLE 14.2

Age and residual
data for Nissan Zs

Age x	Residual e
5	−9.16
4	−11.42
6	−3.90
5	−12.16
5	−5.16
5	3.84
6	−7.90
6	21.10
2	14.05
7	16.36
7	−5.64

Figure 14.7(a) shows a plot of the residuals, *e,* against the ages, *x,* and Fig. 14.7(b) shows a normal probability plot for the residuals.

FIGURE 14.7
(a) Residual plot
(b) Normal probability
plot for residuals

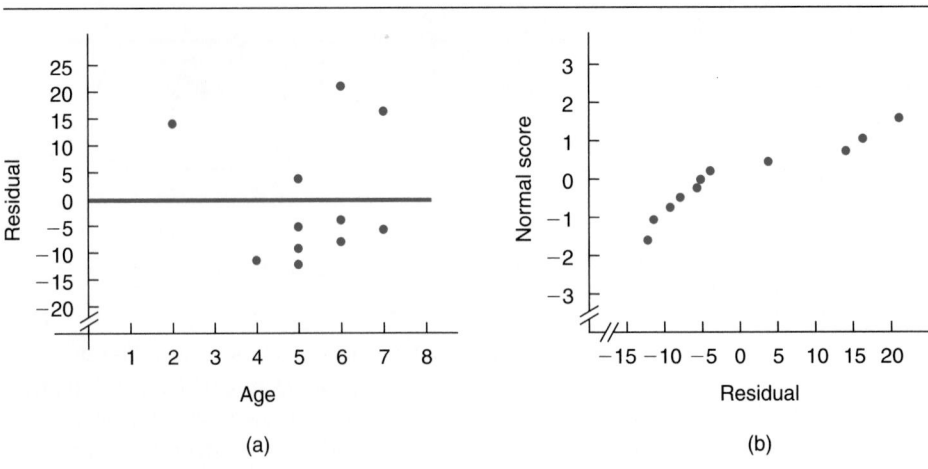

(a)

(b)

Taking into account the small sample size, we can say that the residuals fall roughly in a horizontal band centered and symmetric about the *x*-axis. We can also say that the normal probability plot for the residuals is (very) roughly linear, although the departure from linearity is sufficient for some concern.[†] Therefore, based on the sample data, there are no obvious violations of the assumptions for regression inferences for the variables age and price of Nissan Zs (in the age range from 2 to 7 years). ∎

Using the Computer (Optional)

In Section 13.2 we learned that Minitab's **Regress** command can be used to obtain the (sample) regression equation for a set of data points. The output resulting from the application of that command displays the standard error of the estimate, s_e, as well as the regression equation and several other statistics.

Example 14.4   Obtaining the Standard Error of the Estimate Using the Regress Command

Printout 13.2 on page 754 shows the output obtained by applying the Regress command to the age and price data for a sample of 11 Nissan Zs. Use the output to determine the standard error of the estimate.

[†] Recall, though, that the inferential procedures in regression and correlation are robust to moderate deviations of the assumptions for regression inferences.

SOLUTION   The standard error of the estimate, s_e, is the first entry in the seventh line of Printout 13.2: s = 12.58 (Minitab uses s instead of s_e to denote the standard error of the estimate). Thus for the Nissan Z data, $s_e = 12.58$.   ∎

We can also use the Regress command to obtain and store residuals. This is accomplished by employing the **Residuals** subcommand. Once we have the residuals, we can apply other Minitab commands to obtain a residual plot and a normal probability plot of the residuals. Example 14.5 provides the details.

Example 14.5   Obtaining Residuals and Associated Plots

Use Minitab to obtain the residuals, a residual plot, and a normal probability plot of the residuals for the age and price data of Nissan Zs.

SOLUTION   To obtain the residuals, we begin by storing the age and price data from Table 14.1 on page 794 in columns named AGE and PRICE, respectively. Then we do the following.

Session commands: First we name an available column RESI1.[†] Next we type the command **REGRESS** followed by the storage location of the data for the response variable, the number of predictors, and the storage location of the data for the predictor variable(s); that is, we type REGRESS 'PRICE' on 1 predictor 'AGE'; and press Enter. Then we type the subcommand **RESIDUALS** followed by the storage location for the residuals; that is, we type RESIDUALS into 'RESI1'. and press Enter.

(or)

Menu commands: We choose **Stat ▸ Regression ▸ Regression...**, specify PRICE for **Response**, specify AGE in the **Predictors** text box, select **Residuals** from the **Storage** check-box list, and then select **OK**.

The output obtained by employing either the above session commands or the above menu commands is the same as that obtained when we use only the Regress command. But now the residuals are stored in RESI1 and are available for analysis.

In particular, we can obtain a residual plot of the residuals in RESI1 against the ages in AGE by using the Plot command. The resulting output is shown in Printout 14.1 at the top of the next page. Compare Minitab's residual plot to the one we obtained by hand in Fig. 14.7(a).

We can also obtain a normal probability plot for the residuals stored in RESI1 by first applying the NScores command and then the Plot command, as explained in Section 6.5 on page 385. The resulting output is shown in Printout 14.2. Compare Minitab's normal probability plot to the one we obtained by hand in Fig. 14.7(b).

[†] We are choosing the name RESI1 for the storage location of the residuals because that is the name Minitab will select if menu commands are used and no residuals have been stored previously.

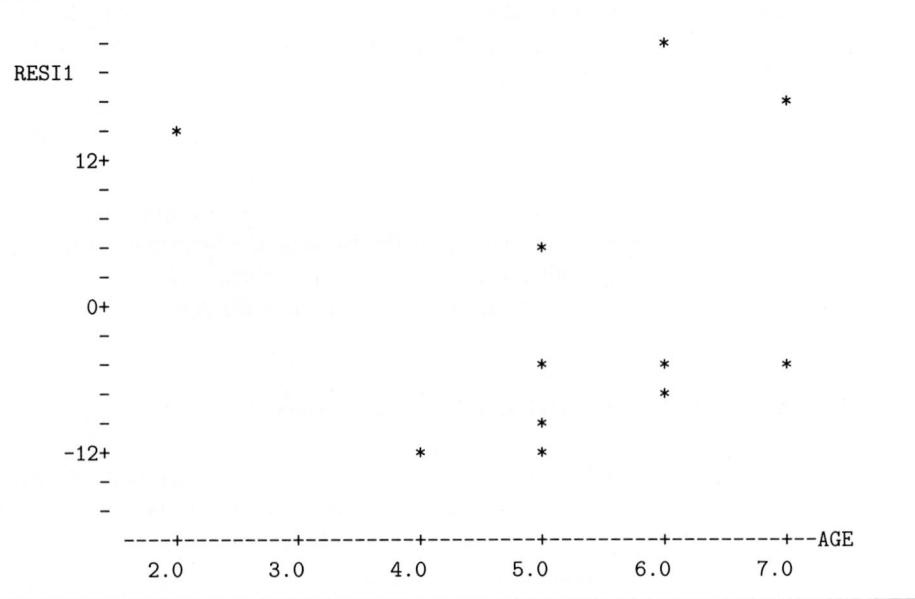

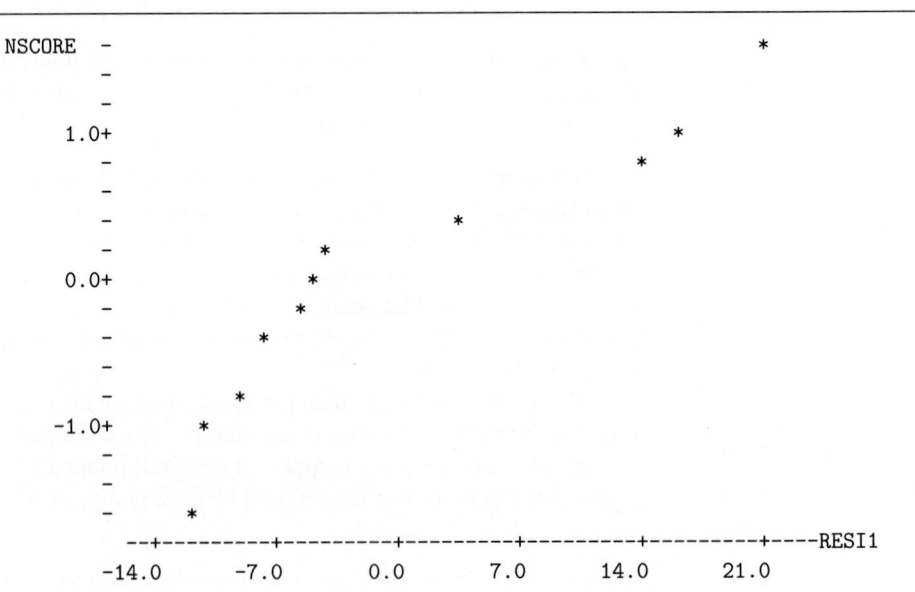

EXERCISES 14.1

14.1   State the three conditions required for making regression inferences.

In Exercises 14.2–14.7, we have repeated the information from Exercises 13.32–13.37. For each exercise, discuss what it would mean for the assumptions for regression inferences to be satisfied by the variables under consideration.

14.2   Ten Corvettes between 1 and 6 years old were randomly selected from the classified ads of *The Arizona Republic*. The following data were obtained, where x denotes age, in years, and y denotes price, in hundreds of dollars.

x	6	6	6	2	2	5	4	5	1	4
y	125	115	130	260	219	150	190	163	260	160

14.3   The National Center for Health Statistics publishes data on heights and weights in *Vital and Health Statistics*. A random sample of 11 males age 18–24 years gave the following data, where x denotes height, in inches, and y denotes weight, in pounds.

x	65	67	71	71	66	75	67	70	71	69	69
y	175	133	185	163	126	198	153	163	159	151	155

14.4   Hanna Properties specializes in custom-home resales in the Equestrian Estates, an exclusive subdivision in Phoenix, Arizona. A random sample of nine custom homes currently listed for sale provided the following information on size and price. Here x denotes size, in hundreds of square feet, rounded to the nearest hundred, and y denotes price, in thousands of dollars, rounded to the nearest thousand.

x	26	27	33	29	29	34	30	40	22
y	235	249	267	269	295	345	415	475	195

14.5   An article read by a physician indicated that the maximum heart rate an individual can reach during intensive exercise decreases with age. The physician decided to do his own study. Ten randomly selected people performed exercise tests and recorded their peak heart rates. The results are shown in the following table, where x denotes age, in years, and y denotes peak heart rate.

x	30	38	41	38	29	39	46	41	42	24
y	186	183	171	177	191	177	175	176	171	196

14.6   An instructor asked a random sample of eight students to record their study times in a beginning calculus course. She then made a table for total hours studied, x, over 2 weeks, and test score, y, at the end of the 2 weeks. Here are the results.

x	10	15	12	20	8	16	14	22
y	92	81	84	74	85	80	84	80

14.7   An economist is interested in the relationship between the disposable income of a family and the amount of money spent annually on food. For a preliminary study, the economist takes a random sample of eight middle-income families of the same size (father, mother, two children). The results are as follows, where x denotes disposable income, in thousands of dollars, and y denotes food expenditure, in hundreds of dollars.

x	30	36	27	20	16	24	19	25
y	55	60	42	40	37	26	39	43

For each of Exercises 14.8–14.13,
a. compute and interpret the standard error of the estimate, s_e.
b. interpret the result from part (a) if the assumptions for regression inferences hold.
c. obtain a residual plot and a normal probability plot of the residuals.
d. decide whether it is reasonable to consider the assumptions for regression inferences met by the variables in question. (The answer here is subjective, especially in view of the extremely small sample sizes.)

14.8   The age and price data for 10 Corvettes from Exercise 14.2.

14.9   The height and weight data for males age 18–24 years from Exercise 14.3.

14.10   The size and price data for custom homes from Exercise 14.4.

14.11   The data on age and peak heart rate from Exercise 14.5.

· **14.12** The data on study time and test score from Exercise 14.6.

· **14.13** The data on disposable income and annual food expenditure from Exercise 14.7.

· **14.14** Figure 14.8 shows three residual plots and a normal probability plot of residuals. For each part, decide whether the graph suggests violation of one or more of the assumptions for regression inferences. Provide a detailed explanation for your answers.

· **14.15** Figure 14.9 shows three residual plots and a normal probability plot of residuals. For each part, decide whether the graph suggests violation of one or more of the

assumptions for regression inferences. Provide a detailed explanation for your answers.

· **14.16 (Computer exercise)** Refer to Exercise 14.6. Use Minitab or some other statistical software to
a. determine the standard error of the estimate, s_e.
b. obtain a residual plot and a normal probability plot of the residuals.

· **14.17 (Computer exercise)** Refer to Exercise 14.7. Use Minitab or some other statistical software to
a. determine the standard error of the estimate, s_e.
b. obtain a residual plot and a normal probability plot of the residuals.

FIGURE 14.8   Plots for Exercise 14.14

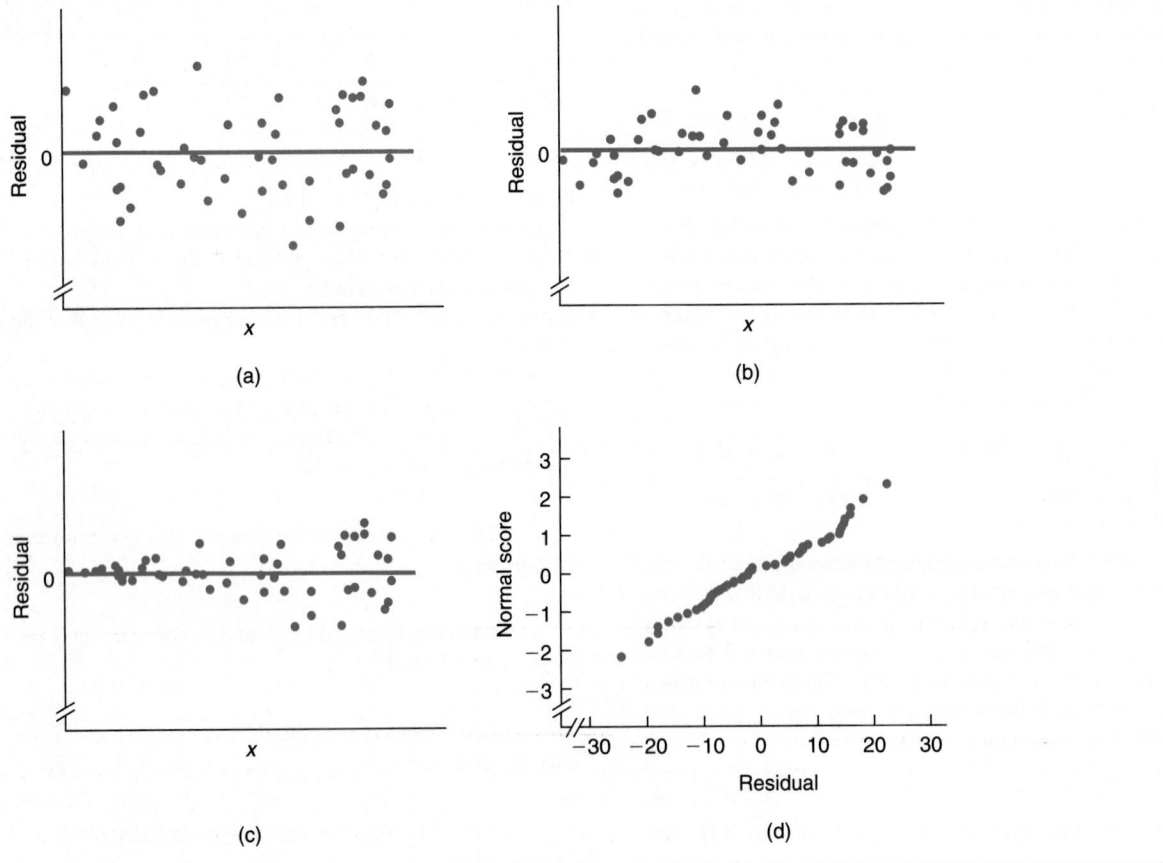

(a)

(b)

(c)

(d)

FIGURE 14.9   Plots for Exercise 14.15

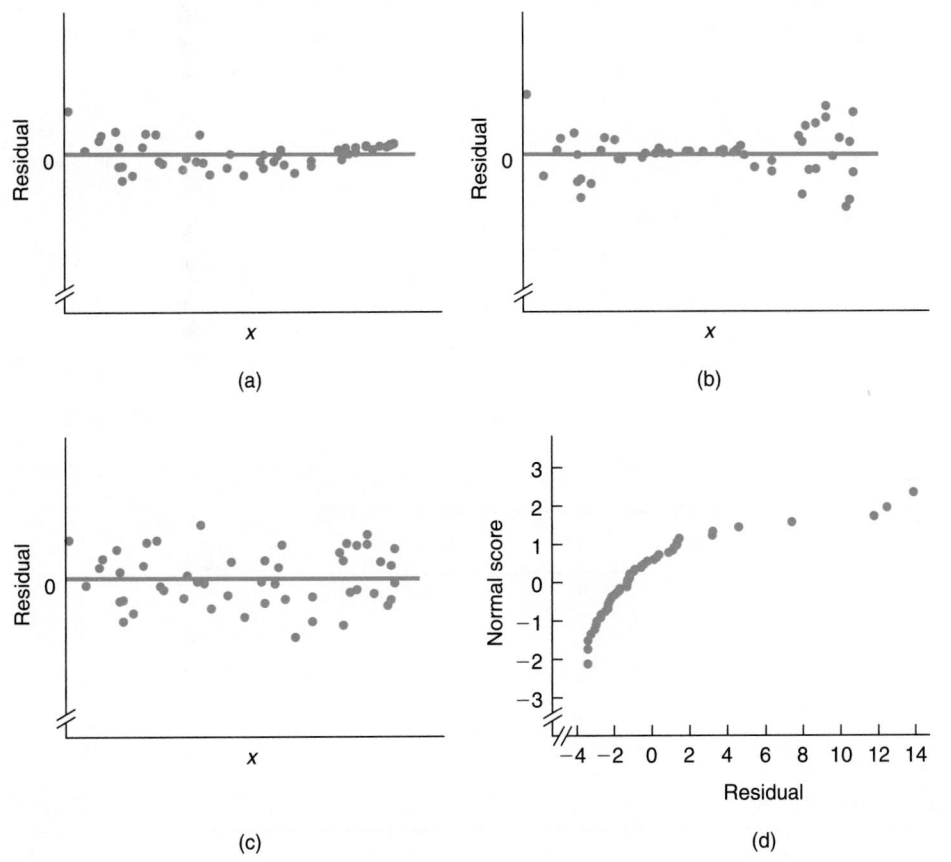

(a)

(b)

(c)

(d)

14.18 (Computer exercise) The U.S. Energy Information Administration compiles and publishes data on energy consumption by family income in *Residential Energy Consumption Survey: Consumption and Expenditures*. We applied Minitab's Regress command to data on family income and last year's energy consumption from a random sample of 25 families.

a. Printout 13.3 on page 758 displays the computer output. The income data are in thousands of dollars and the energy-consumption data are in millions of BTU. Employ the output to determine the standard error of the estimate.

b. We also used Minitab to obtain a residual plot (Printout 14.3 at the top of the next page) and a normal probability plot of the residuals (Printout 14.4 on the next page). Do these plots suggest violations of any of the assumptions for regression inferences? Explain your answer.

14.19 (Computer exercise) Greene and Touchstone conducted a study on the relationship between the estriol levels of pregnant women and the birth weights of their children. Their findings, "Urinary Tract Estriol: An Index of Placental Function," were published in the *American Journal of Obstetrics and Gynecology*.

a. Printout 13.4 on page 758 shows the output that results by applying Minitab's Regress command to the data obtained by Greene and Touchstone. The estriol levels are in milligrams per 24 hours and the birth weights are in grams. Use the output to find the standard error of the estimate.

b. We also used Minitab to obtain a residual plot (Printout 14.5 at the top of page 809) and a normal probability plot of the residuals (Printout 14.6 on page 809). Do these plots suggest violations of any of the assumptions for regression inferences? Explain your answer.

PRINTOUT 14.3   Minitab output for Exercise 14.18(b)

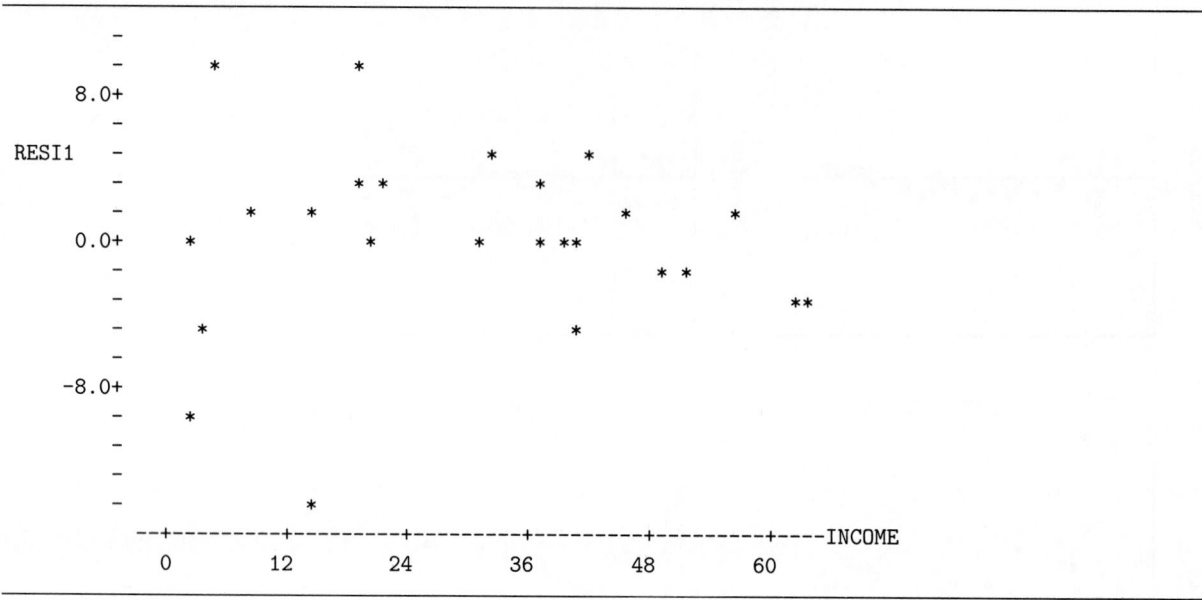

PRINTOUT 14.4   Minitab output for Exercise 14.18(b)

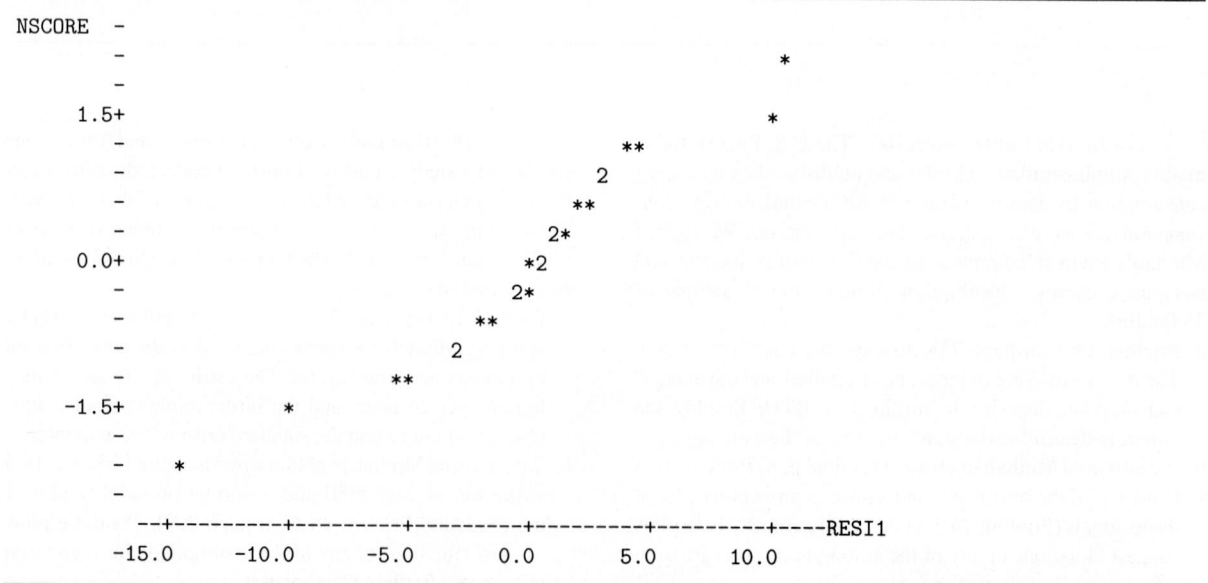

PRINTOUT 14.5   Minitab output for Exercise 14.19(b)

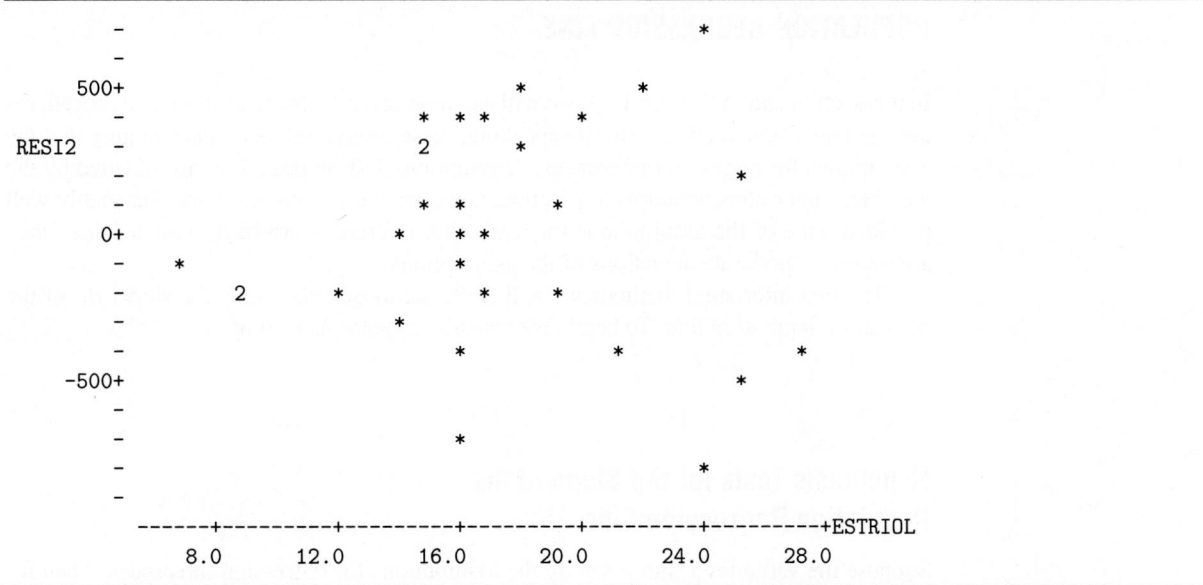

PRINTOUT 14.6   Minitab output for Exercise 14.19(b)

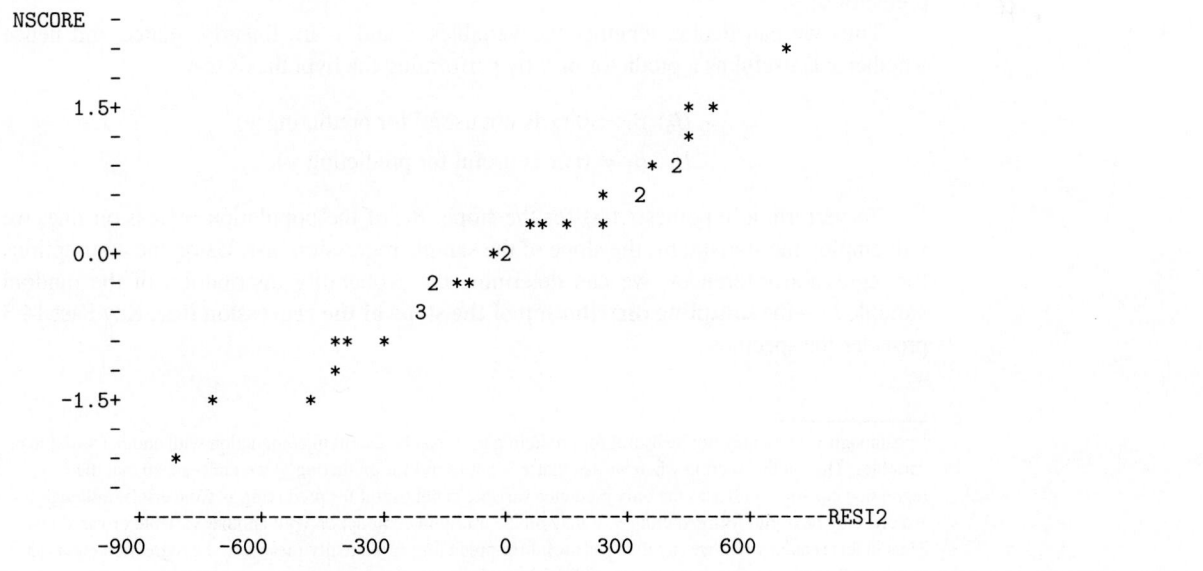

14.2   INFERENCES FOR THE SLOPE OF THE POPULATION REGRESSION LINE

In this section and in Section 14.3, we will examine several statistical-inference procedures used in regression analysis. Strictly speaking, these inferential techniques require that the assumptions for regression inferences, Assumptions 1–3 on page 794, are satisfied by the variables under consideration. In practice, however, the techniques work reasonably well provided none of the assumptions for regression inferences are badly violated (i.e., they are robust to moderate deviations of the assumptions).

The first inferential methods we will study are those concerning the slope, β_1, of the population regression line. To begin, we consider hypothesis testing.

Hypothesis Tests for the Slope of the Population Regression Line

Suppose the variables x and y satisfy the assumptions for regression inferences. Then for each x-value, the corresponding population of y-values is normally distributed with mean $\beta_0 + \beta_1 x$ and standard deviation σ.

Of particular interest is whether the slope, β_1, of the population regression line equals 0. If $\beta_1 = 0$, then for each x-value, the corresponding population of y-values is normally distributed with mean β_0 ($= \beta_0 + 0 \cdot x$) and standard deviation σ; and neither of those two parameters involves x. Consequently, we see that if $\beta_1 = 0$, then x provides no information about the distribution of the population of y-values. This implies that there is no linear relationship between the variables x and y, and therefore that x is useless as a predictor of y.[†]

Thus we can decide whether the variables x and y are linearly related and hence whether x is useful as a predictor of y by performing the hypothesis test

$$H_0: \beta_1 = 0 \ (x \text{ is not useful for predicting } y)$$

$$H_a: \beta_1 \neq 0 \ (x \text{ is useful for predicting } y).$$

To perform a hypothesis test for the slope, β_1, of the population regression line, we will employ the statistic b_1, the slope of the sample regression line. Using the assumptions for regression inferences, we can determine the probability distribution of the random variable b_1—the **sampling distribution of the slope of the regression line.** Key Fact 14.3 provides the specifics.

[†]   Although x alone may not be useful for predicting y, it may be useful in conjunction with another variable or variables. Thus in this section when we say that x is not useful for predicting y, we really mean that the regression equation with x as the only predictor variable is not useful for predicting y. Conversely, although x alone may be useful for predicting y, it may not be useful in conjunction with another variable or variables. Thus in this section when we say that x is useful for predicting y, we really mean that the regression equation with x as the only predictor variable is useful for predicting y.

KEY FACT 14.3

THE SAMPLING DISTRIBUTION OF THE SLOPE OF THE REGRESSION LINE

Suppose the variables x and y satisfy the assumptions for regression inferences. Then the random variable b_1 is normally distributed and has mean $\mu_{b_1} = \beta_1$ and standard deviation $\sigma_{b_1} = \sigma/\sqrt{S_{xx}}$. Thus the standardized random variable

$$z = \frac{b_1 - \beta_1}{\sigma/\sqrt{S_{xx}}}$$

has the standard normal distribution.

The standardized random variable in Key Fact 14.3 cannot be used as a basis for obtaining the required test statistic since the common population standard deviation, σ, is generally unknown. We therefore replace σ by its sample estimate s_e, the standard error of the estimate. As we might suspect, the resulting random variable has a t-distribution.

KEY FACT 14.4

t-DISTRIBUTION FOR INFERENCES FOR β_1

Suppose the variables x and y satisfy the assumptions for regression inferences. Then the random variable

$$t = \frac{b_1 - \beta_1}{s_e/\sqrt{S_{xx}}}$$

has the t-distribution with df $= n - 2$.

In view of Key Fact 14.4, we see that for a hypothesis test with null hypothesis H_0: $\beta_1 = 0$, we can use the random variable

$$t = \frac{b_1}{s_e/\sqrt{S_{xx}}}$$

as the test statistic and obtain the critical values from the t-table, Table IV. Specifically, we have the following procedure.

PROCEDURE 14.1

TO PERFORM A HYPOTHESIS TEST TO DECIDE WHETHER THE SLOPE OF A POPULATION REGRESSION LINE IS NOT 0 AND HENCE WHETHER x IS USEFUL AS A PREDICTOR OF y

ASSUMPTIONS

Assumptions 1–3 for regression inferences

Step 1 State the null and alternative hypotheses.

Step 2 Decide on the significance level, α.

(continued)

Step 3   The critical values are $\pm t_{\alpha/2}$, with df $= n - 2$. Use Table IV to find the critical values.

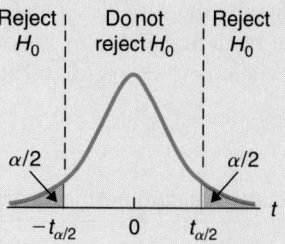

Reject H_0 | Do not reject H_0 | Reject H_0

$\alpha/2$ $\alpha/2$

$-t_{\alpha/2}$ 0 $t_{\alpha/2}$ t

Step 4   Compute the value of the test statistic

$$t = \frac{b_1}{s_e/\sqrt{S_{xx}}}.$$

Step 5   If the value of the test statistic falls in the rejection region, then reject H_0; otherwise, do not reject H_0.

Step 6   State the conclusion in words.

Example 14.6   Illustrates Procedure 14.1

The data on age and price for a sample of 11 Nissan Zs are displayed in Table 14.1 on page 794. At the 5% significance level, do the data provide sufficient evidence to conclude that age is useful as a predictor of price for Nissan Zs?

SOLUTION   As we discovered in Example 14.3, it is reasonable to consider the assumptions for regression inferences satisfied by the variables age and price for Nissan Zs, at least for Nissan Zs between 2 and 7 years old. Therefore we will apply Procedure 14.1 to carry out the required hypothesis test.

Step 1   *State the null and alternative hypotheses.*

Let β_1 denote the slope of the population regression line that relates price to age for Nissan Zs. Then the null and alternative hypotheses are

H_0: $\beta_1 = 0$ (age is not useful for predicting price)

H_a: $\beta_1 \neq 0$ (age is useful for predicting price).

Step 2   *Decide on the significance level, α.*

We are to perform the hypothesis test at the 5% significance level; so $\alpha = 0.05$.

Step 3   *The critical values are $\pm t_{\alpha/2}$, with df $= n - 2$.*

From Step 2, $\alpha = 0.05$. Also, $n = 11$; so df $= n - 2 = 11 - 2 = 9$. Using Table IV, we find that the critical values are $\pm t_{\alpha/2} = \pm t_{0.05/2} = \pm t_{0.025} = \pm 2.262$, as seen in Fig. 14.10.

FIGURE 14.10
Criterion for deciding whether or not to reject the null hypothesis

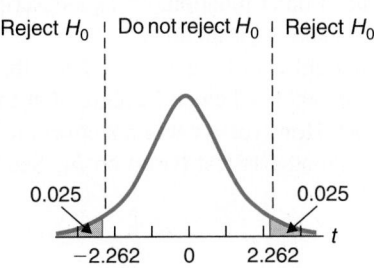

Step 4   *Compute the value of the test statistic*

$$t = \frac{b_1}{s_e/\sqrt{S_{xx}}}.$$

In Example 13.4 on page 747, we found that $b_1 = -20.26$, $\Sigma x^2 = 326$, and $\Sigma x = 58$. Also, in Example 14.2 on page 798, we determined that $s_e = 12.58$. Therefore since $n = 11$, the value of the test statistic is

$$t = \frac{b_1}{s_e/\sqrt{S_{xx}}} = \frac{b_1}{s_e/\sqrt{\Sigma x^2 - (\Sigma x)^2/n}} = \frac{-20.26}{12.58/\sqrt{326 - (58)^2/11}} = -7.235.$$

Step 5   *If the value of the test statistic falls in the rejection region, reject H_0; otherwise, do not reject H_0.*

The value of the test statistic, found in Step 4, is $t = -7.235$. Since this falls in the rejection region, we reject H_0.

Step 6   *State the conclusion in words.*

The test results are statistically significant at the 5% level; that is, at the 5% significance level, the data provide sufficient evidence to conclude that the slope of the population regression line is not 0 and hence that age is useful as a predictor of price for Nissan Zs.   ■

The P-value approach to hypothesis testing can also be used to carry out the hypothesis test in Example 14.6 and to assess more precisely the evidence against the null hypothesis. From Step 4 we see that the value of the test statistic is $t = -7.235$. Recalling that df $= 9$, we find from Table IV that $P < 0.005$. In particular, because the P-value is less than the designated significance level of 0.05, we can reject H_0. Furthermore, by referring to Table 9.12 on page 530, we see that the data provide very strong evidence against the null hypothesis and hence in favor of the alternative hypothesis that age is useful as a predictor of price for Nissan Zs.

Procedure 14.1, which is based on the statistic b_1, is used for performing a hypothesis test to decide whether the slope of the population regression line is not 0 or, equivalently, to decide whether the regression equation is useful for making predictions. In Section 13.3

we introduced the coefficient of determination, r^2, as a descriptive measure of the utility of the regression equation for making predictions.

This suggests that we should also be able to employ the statistic r^2 as a basis for performing a hypothesis test to decide whether the regression equation is useful for making predictions, and indeed we can. However, we will not cover the hypothesis test based on r^2 since it is equivalent to the hypothesis test based on b_1. See Exercise 14.37 for further discussion of this matter.

Confidence Intervals for the Slope of the Population Regression Line

Recall that the slope of a straight line represents the change in y resulting from an increase in x by 1 unit. Also recall that the population regression line, whose slope is β_1, gives the means of the populations of y-values corresponding to the various x-values. Therefore β_1 represents the change in the mean of the population of y-values for every increase in the value of x by 1 unit.

For example, consider the variables age (x) and price (y) of Nissan Zs. In this case, β_1 is the amount that the mean price decreases for every increase in age by 1 year. In other words, β_1 is the mean yearly depreciation of Nissan Zs.

Consequently, we see that it is worthwhile to obtain an estimate for the slope, β_1, of the population regression line. As we know, a point estimate for β_1 is provided by b_1. To determine a confidence-interval estimate for β_1, we apply Key Fact 14.4 on page 811 to get the following procedure.

PROCEDURE 14.2 **TO FIND A CONFIDENCE INTERVAL FOR THE SLOPE OF A POPULATION REGRESSION LINE**

ASSUMPTIONS

Assumptions 1–3 for regression inferences

Step 1 For a confidence level of $1 - \alpha$, use Table IV to find $t_{\alpha/2}$ with df $= n - 2$.

Step 2 The endpoints of the confidence interval for β_1 are

$$b_1 \pm t_{\alpha/2} \cdot \frac{s_e}{\sqrt{S_{xx}}}.$$

Example 14.7 Illustrates Procedure 14.2

Use the data in Table 14.1 on page 794 to obtain a 95% confidence interval for the slope, β_1, of the population regression line that relates price to age for Nissan Zs.

SOLUTION We apply Procedure 14.2.

Step 1   *For a confidence level of* $1 - \alpha$, *use Table IV to find* $t_{\alpha/2}$ *with* $df = n - 2$.

For a 95% confidence interval, $\alpha = 0.05$. Since $n = 11$, df $= 11 - 2 = 9$. Using Table IV, we find that $t_{\alpha/2} = t_{0.05/2} = t_{0.025} = 2.262$.

Step 2   *The endpoints of the confidence interval for* β_1 *are*

$$b_1 \pm t_{\alpha/2} \cdot \frac{s_e}{\sqrt{S_{xx}}}.$$

From Example 13.4, $b_1 = -20.26$, $\Sigma x^2 = 326$, and $\Sigma x = 58$. Also, from Example 14.2, $s_e = 12.58$. Hence the endpoints of the confidence interval for β_1 are

$$-20.26 \pm 2.262 \cdot \frac{12.58}{\sqrt{326 - (58)^2/11}}$$

or -20.26 ± 6.33 or -26.59 to -13.93. We can be 95% confident that the slope, β_1, of the population regression line is somewhere between -26.59 and -13.93. In other words, we can be 95% confident that the yearly decrease in mean price for Nissan Zs is somewhere between $1393 and $2659.   ∎

Using the Computer (Optional)

Procedure 14.1 on page 811 provides a step-by-step method for performing a hypothesis test to decide whether the slope, β_1, of a population regression line is not 0. Alternatively, we can use Minitab to carry out such a hypothesis test. In fact, the output obtained by applying Minitab's Regress command contains all the information we need. To illustrate, we return again to the Nissan Z example.

Example 14.8   The Regress Command

In Example 13.6 we applied the Regress command to the data on age and price for a sample of 11 Nissan Zs. The resulting computer output is shown in Printout 13.2 on page 754. Use that output to perform the hypothesis test considered in Example 14.6 on page 812.

SOLUTION   Let β_1 denote the slope of the population regression line that relates price to age for Nissan Zs. The problem is to perform the hypothesis test

$$H_0: \beta_1 = 0 \text{ (age is not useful for predicting price)}$$
$$H_a: \beta_1 \neq 0 \text{ (age is useful for predicting price)}$$

at the 5% significance level.

Refer to the sixth line of Printout 13.2, the line labeled AGE. The second entry in that line, which is under the column headed Coef, displays the slope, b_1, of the sample regression line; hence $b_1 = -20.261$. The third entry in that line, which is under the column headed Stdev, shows the estimated standard deviation of b_1, $s_e/\sqrt{S_{xx}}$; so

$s_e/\sqrt{S_{xx}} = 2.800$. The fourth entry in that line, which is under the column headed t-ratio, provides the value of the test statistic,

$$t = \frac{b_1}{s_e/\sqrt{S_{xx}}}.$$

Thus we see that $t = -7.24$.

The final entry in the line labeled AGE, under the column headed p, gives the P-value for the hypothesis test, which is 0.000 (to three decimal places). Since this is less than the specified significance level of $\alpha = 0.05$, we reject H_0. In other words, the data provide sufficient evidence to conclude that the slope of the population regression line is not 0 and hence that age is useful as a predictor of price for Nissan Zs.   ∎

EXERCISES 14.2

In Exercises 14.20–14.25, we have repeated the information from Exercises 13.32–13.37. Presuming that the assumptions for regression inferences are met, perform the required hypothesis tests using either the classical approach or the P-value approach. (Note: You previously obtained the sample regression equations in Exercises 13.32–13.37 and the standard errors of the estimate in Exercises 14.8–14.13.)

•   **14.20**   Ten Corvettes between 1 and 6 years old were randomly selected from the classified ads of *The Arizona Republic*. The following data were obtained, where x denotes age, in years, and y denotes price, in hundreds of dollars.

x	6	6	6	2	2	5	4	5	1	4
y	125	115	130	260	219	150	190	163	260	160

At the 10% significance level, do the data provide sufficient evidence to conclude that the slope of the population regression line is not 0 and hence that age is useful as a predictor of price for Corvettes?

•   **14.21**   The National Center for Health Statistics publishes data on heights and weights in *Vital and Health Statistics*. A random sample of 11 males age 18–24 years gave the following data, where x denotes height, in inches, and y denotes weight, in pounds.

x	65	67	71	71	66	75	67	70	71	69	69
y	175	133	185	163	126	198	153	163	159	151	155

Do the data provide sufficient evidence to conclude that the slope of the population regression line is not 0 and hence that

height is useful as a predictor of weight for 18–24-year-old males? Use $\alpha = 0.10$.

•   **14.22**   Hanna Properties specializes in custom-home resales in the Equestrian Estates, an exclusive subdivision in Phoenix, Arizona. A random sample of nine custom homes currently listed for sale provided the following information on size and price. Here x denotes size, in hundreds of square feet, rounded to the nearest hundred, and y denotes price, in thousands of dollars, rounded to the nearest thousand.

x	26	27	33	29	29	34	30	40	22
y	235	249	267	269	295	345	415	475	195

Do the data suggest that size is useful as a predictor of price for custom homes in the Equestrian Estates? Perform the required hypothesis test at the 0.01 level of significance.

•   **14.23**   An article read by a physician indicated that the maximum heart rate an individual can reach during intensive exercise decreases with age. The physician decided to do his own study. Ten randomly selected people performed exercise tests and recorded their peak heart rates. The results are shown in the following table, where x denotes age, in years, and y denotes peak heart rate.

x	30	38	41	38	29	39	46	41	42	24
y	186	183	171	177	191	177	175	176	171	196

At the 5% significance level, do the data provide sufficient evidence to conclude that age is useful as a predictor of peak heart rate?

14.24 An instructor asked a random sample of eight students to record their study times in a beginning calculus course. She then made a table for total hours studied, x, over 2 weeks, and test score, y, at the end of the 2 weeks. Here are the results.

x	10	15	12	20	8	16	14	22
y	92	81	84	74	85	80	84	80

Do the data provide sufficient evidence to conclude that study time is useful as a predictor of test score in beginning calculus courses? Use $\alpha = 0.05$.

14.25 An economist is interested in the relationship between the disposable income of a family and the amount of money spent annually on food. For a preliminary study, the economist takes a random sample of eight middle-income families of the same size (father, mother, two children). The results are as follows, where x denotes disposable income, in thousands of dollars, and y denotes food expenditure, in hundreds of dollars.

x	30	36	27	20	16	24	19	25
y	55	60	42	40	37	26	39	43

Do the data provide sufficient evidence to conclude that disposable income is useful as a predictor of annual food expenditure for middle-income families with a father, mother, and two children? Use $\alpha = 0.01$.

In each of Exercises 14.26–14.31, apply Procedure 14.2 on page 814 to obtain the required confidence interval.

14.26 Refer to Exercise 14.20.
a. Find a 90% confidence interval for the slope, β_1, of the population regression line that relates price to age for Corvettes.
b. Interpret your result from part (a).

14.27 Refer to Exercise 14.21.
a. Obtain a 90% confidence interval for the slope, β_1, of the population regression line that relates weight to height for males age 18–24.
b. Interpret your result from part (a).

14.28 Refer to Exercise 14.22.
a. Find a 99% confidence interval for the slope of the population regression line that relates price to size for custom homes in the Equestrian Estates.
b. Interpret your result from part (a).

14.29 Refer to Exercise 14.23.
a. Find a 95% confidence interval for the slope of the population regression line that relates peak heart rate to age.
b. Interpret your result from part (a).

14.30 Refer to Exercise 14.24.
a. Obtain a 95% confidence interval for the slope, β_1, of the population regression line that relates test score to study time in beginning calculus courses.
b. Interpret your result from part (a).

14.31 Refer to Exercise 14.25.
a. Obtain a 99% confidence interval for the slope, β_1, of the population regression line that relates annual food expenditure to disposable income for middle-income families with a father, mother, and two children.
b. Interpret your result from part (a).

14.32 (Computer exercise) Use Minitab or some other statistical software to perform the hypothesis test in Exercise 14.24.

14.33 (Computer exercise) Use Minitab or some other statistical software to perform the hypothesis test in Exercise 14.25.

14.34 (Computer exercise) The U.S. Energy Information Administration compiles and publishes data on energy consumption by family income in *Residential Energy Consumption Survey: Consumption and Expenditures.* We applied Minitab's Regress command to data on family income and last year's energy consumption from a random sample of 25 families. The income data are in thousands of dollars and the energy-consumption data are in millions of BTU. Printout 13.3 on page 758 shows the computer output. Using the printout,
a. find the slope of the sample regression line.
b. determine the estimated standard deviation of the random variable b_1.
c. obtain the value of the test statistic, t, for a hypothesis test to decide whether the slope of the population regression line is not 0.
d. determine the P-value for the hypothesis test.
e. decide whether family income is useful for predicting energy consumption. Use $\alpha = 0.01$.
f. obtain a 99% confidence interval for the slope of the population regression line and interpret your result. *(Note: You will need to use Table IV to determine $t_{\alpha/2}$, but everything else that is required to obtain the confidence interval can be found in the printout.)*

g. In performing the inferences that you did in this exercise, what assumptions are you making? How would you check those assumptions?

14.35 (Computer exercise) Greene and Touchstone conducted a study on the relationship between the estriol levels of pregnant women and the birth weights of their children. Their findings, "Urinary Tract Estriol: An Index of Placental Function," were published in the *American Journal of Obstetrics and Gynecology*. Printout 13.4 on page 758 shows the computer output that results by applying Minitab's Regress command to the 31 pairs of data obtained by Greene and Touchstone. The estriol levels are in milligrams per 24 hours and the birth weights are in grams. Using the printout,

a. find the slope of the sample regression line.
b. determine the estimated standard deviation of the random variable b_1.
c. obtain the value of the test statistic, t, for a hypothesis test to decide whether the slope of the population regression line is not 0.
d. determine the P-value for the hypothesis test.
e. decide, at the 5% significance level, whether estriol level is useful for predicting birth weight.
f. obtain a 95% confidence interval for the slope of the population regression line and interpret your result. *(Note:* You will need to use Table IV to determine $t_{\alpha/2}$, but everything else that is required to obtain the confidence interval can be found in the printout.)

g. In performing the inferences that you did in this exercise, what assumptions are you making? How would you check those assumptions?

14.36 This exercise justifies Procedure 14.2. Suppose the variables x and y satisfy the assumptions for regression inferences.

a. Use Key Fact 14.4 on page 811 to show that

$$P\left(-t_{\alpha/2} < \frac{b_1 - \beta_1}{s_e/\sqrt{S_{xx}}} < t_{\alpha/2}\right) = 1 - \alpha.$$

b. Use part (a) to show that the probability is $1 - \alpha$ that the interval with endpoints $b_1 \pm t_{\alpha/2} \cdot s_e/\sqrt{S_{xx}}$ will contain β_1.
c. Deduce from part (b) that once the sample is taken, the interval with endpoints $b_1 \pm t_{\alpha/2} \cdot s_e/\sqrt{S_{xx}}$ will be a $(1 - \alpha)$-level confidence interval for β_1.

14.37 On page 814 we mentioned that the coefficient of determination, r^2, can be employed as a basis for performing a hypothesis test to decide whether the regression equation is useful for making predictions. The test statistic, based on r^2, for performing such a hypothesis test is

$$F = \frac{(n-2)r^2}{1 - r^2}.$$

Show that the test statistic F is the square of the test statistic t used in Procedure 14.1 on page 812. *(Hint:* Express both statistics in terms of the quantities S_{xx}, S_{xy}, and S_{yy}.)

14.3 ESTIMATION AND PREDICTION

In this section we will learn how a (sample) regression equation can be used to make two important inferences. One inference estimates the mean of the population of y-values corresponding to a particular x-value, and the other inference predicts an individual y-value corresponding to a particular x-value.

We will use the Nissan Z example to illustrate the pertinent ideas. In doing so, we will consider the assumptions for regression inferences, Assumptions 1–3 on page 794, satisfied by the variables age and price for Nissan Zs. Example 14.3 on page 801 shows it is not unreasonable to do that.

Example 14.9 Estimating Means in Regression

The data on age and price for a sample of 11 Nissan Zs are displayed in Table 14.1 on page 794. Use that data to obtain an estimate for the mean price of all 3-year-old Nissan Zs.

SOLUTION By Assumption 1 of the assumptions for regression inferences, the population regression equation gives the mean prices for the various ages of Nissan Zs. Thus the mean price of all 3-year-old Nissan Zs is exactly equal to $\beta_0 + \beta_1 \cdot 3$. Since β_0 and β_1 are unknown, we will estimate the mean price of all 3-year-old Nissan Zs, $\beta_0 + \beta_1 \cdot 3$, by the corresponding value, $b_0 + b_1 \cdot 3$, on the sample regression line.

Recall that the sample regression equation for the data in Table 14.1 is $\hat{y} = 195.47 - 20.26x$. Thus our estimate for the mean price of all 3-year-old Nissan Zs is

$$\hat{y} = 195.47 - 20.26 \cdot 3 = 134.69,$$

or $13,469. ■

Note: The estimate for the mean price of all 3-year-old Nissan Zs is the same as the predicted price of a 3-year-old Nissan Z. Both are obtained by substituting $x = 3$ into the sample regression equation.

The estimate of $13,469 for the mean price of all 3-year-old Nissan Zs is a point estimate. As we know, it would be more informative if we had some idea of how accurate that point estimate is; in other words, it would be better to provide a confidence-interval estimate for the mean price of all 3-year-old Nissan Zs. We will now see how to obtain such confidence-interval estimates.

Confidence Intervals for Means in Regression

To develop a confidence-interval procedure for means in regression, we must first identify the probability distribution of the random variable $\hat{y}$. This is presented in Key Fact 14.5.

KEY FACT 14.5   **PROBABILITY DISTRIBUTION OF $\hat{y}_p$**

Suppose the variables x and y satisfy the assumptions for regression inferences. Let x_p denote a particular value of the predictor variable, x, and let $\hat{y}_p = b_0 + b_1 x_p$. Then the random variable $\hat{y}_p$ is normally distributed and has mean $\mu_{\hat{y}_p} = \beta_0 + \beta_1 x_p$ and standard deviation

$$\sigma_{\hat{y}_p} = \sigma \sqrt{\frac{1}{n} + \frac{(x_p - \Sigma x/n)^2}{S_{xx}}}.$$

If we standardize the random variable $\hat{y}_p$, then the resulting random variable has the standard normal distribution. However, because the standardized random variable contains the unknown parameter σ, we cannot employ that random variable to find a confidence-interval formula. So we replace σ by its estimate s_e, the standard error of the estimate. The resulting random variable has a t-distribution.

KEY FACT 14.6   **t-DISTRIBUTION FOR CONFIDENCE INTERVALS IN REGRESSION**

> Suppose the variables x and y satisfy the assumptions for regression inferences. Let x_p denote a particular value of the predictor variable, x, and let $\hat{y}_p = b_0 + b_1 x_p$. Then the random variable
>
> $$t = \frac{\hat{y}_p - (\beta_0 + \beta_1 x_p)}{s_e \sqrt{\dfrac{1}{n} + \dfrac{(x_p - \Sigma x/n)^2}{S_{xx}}}}$$
>
> has the t-distribution with df $= n - 2$.

Recalling that $\beta_0 + \beta_1 x_p$ is the mean of the population of y-values corresponding to x_p, we can use Key Fact 14.6 to derive the following confidence-interval procedure for means in regression.

PROCEDURE 14.3   **TO FIND A CONFIDENCE INTERVAL FOR THE MEAN OF THE POPULATION OF y-VALUES CORRESPONDING TO A PARTICULAR x-VALUE, x_p**

ASSUMPTIONS

Assumptions 1–3 for regression inferences

Step 1 For a confidence level of $1 - \alpha$, use Table IV to find $t_{\alpha/2}$ with df $= n - 2$.

Step 2 Compute the point estimate, $\hat{y}_p = b_0 + b_1 x_p$, for the mean of the population of y-values corresponding to x_p.

Step 3 The endpoints of the confidence interval for the mean are

$$\hat{y}_p \pm t_{\alpha/2} \cdot s_e \sqrt{\frac{1}{n} + \frac{(x_p - \Sigma x/n)^2}{S_{xx}}}.$$

Example 14.10   Illustrates Procedure 14.3

Use the sample data in Table 14.1 on page 794 to obtain a 95% confidence interval for the mean price of all 3-year-old Nissan Zs.

SOLUTION   We apply Procedure 14.3.

Step 1 *For a confidence level of $1 - \alpha$, use Table IV to find $t_{\alpha/2}$ with df $= n - 2$.*

We want a 95% confidence interval, which means $\alpha = 0.05$. Since $n = 11$, df $= 11 - 2 = 9$. Consulting Table IV, we find that $t_{\alpha/2} = t_{0.05/2} = t_{0.025} = 2.262$.

Step 2 *Compute the point estimate, $\hat{y}_p = b_0 + b_1 x_p$, for the mean of the population of y-values corresponding to x_p.*

From Example 13.4, the sample regression equation for the data in Table 14.1 is $\hat{y} = 195.47 - 20.26x$. Here we want $x = x_p = 3$ (3-year-old Nissan Zs). So

$$\hat{y}_p = 195.47 - 20.26 \cdot 3 = 134.69.$$

Step 3 *The endpoints of the confidence interval for the mean are*

$$\hat{y}_p \pm t_{\alpha/2} \cdot s_e \sqrt{\frac{1}{n} + \frac{(x_p - \Sigma x/n)^2}{S_{xx}}}.$$

In Example 13.4 we found that $\Sigma x = 58$ and $\Sigma x^2 = 326$; and in Example 14.2 we determined that $s_e = 12.58$. Also, from Step 1, $t_{\alpha/2} = 2.262$, and from Step 2, $\hat{y}_p = 134.69$. Hence the endpoints of the confidence interval for the mean are

$$134.69 \pm 2.262 \cdot 12.58 \sqrt{\frac{1}{11} + \frac{(3 - 58/11)^2}{326 - (58)^2/11}}$$

 or 134.69 ± 16.76 or 117.93 to 151.45. We can be 95% confident that the mean price of all 3-year-old Nissan Zs is somewhere between \$11,793 and \$15,145. ∎

Prediction Intervals

A primary use of a sample regression equation is for making predictions. The regression equation for the Nissan Z data in Table 14.1 is $\hat{y} = 195.47 - 20.26x$. Thus, for example, the predicted price for a 3-year-old Nissan Z is

$$\hat{y} = 195.47 - 20.26 \cdot 3 = 134.69,$$

or \$13,469. However, since the prices of such cars vary, it makes more sense to find a **prediction interval** for the price of a 3-year-old Nissan Z than to give a single predicted value.

Prediction intervals are similar to confidence intervals. The term *confidence* is usually reserved for interval estimates of parameters, such as the mean price of all 3-year-old Nissan Zs. The term *prediction* is used for interval estimates of random variables, such as the price of a randomly selected 3-year-old Nissan Z.

The procedure for obtaining prediction intervals is similar to the one for obtaining confidence intervals. The prediction-interval procedure is based on the following fact.

KEY FACT 14.7

PROBABILITY DISTRIBUTION OF $y_p - \hat{y}_p$

Suppose the variables x and y satisfy the assumptions for regression inferences. Let x_p denote a particular value of the predictor variable, x, and y_p denote the value of a randomly selected member from the population of y-values corresponding to x_p. Then the random variable $y_p - \hat{y}_p$ is normally distributed and has mean $\mu_{y_p - \hat{y}_p} = 0$ and standard deviation

$$\sigma_{y_p - \hat{y}_p} = \sigma \sqrt{1 + \frac{1}{n} + \frac{(x_p - \Sigma x/n)^2}{S_{xx}}}.$$

If we standardize the random variable $y_p - \hat{y}_p$, then the resulting random variable has the standard normal distribution. However, since the standardized random variable contains the unknown parameter σ, we cannot use it to develop a prediction-interval formula. So we replace σ by its estimate s_e, the standard error of the estimate. The resulting random variable has a t-distribution.

KEY FACT 14.8   **t-DISTRIBUTION FOR PREDICTION INTERVALS IN REGRESSION**

Suppose the variables x and y satisfy the assumptions for regression inferences. Let x_p denote a particular value of the predictor variable, x, and y_p denote the value of a randomly selected member from the population of y-values corresponding to x_p. Then the random variable

$$t = \frac{y_p - \hat{y}_p}{s_e \sqrt{1 + \dfrac{1}{n} + \dfrac{(x_p - \Sigma x/n)^2}{S_{xx}}}}$$

has the t-distribution with df $= n - 2$.

Using Key Fact 14.8, we can derive the following prediction-interval procedure for a population y-value corresponding to a particular x-value.

PROCEDURE 14.4   **TO FIND A PREDICTION INTERVAL FOR A POPULATION y-VALUE CORRESPONDING TO A PARTICULAR x-VALUE, x_p**

ASSUMPTIONS

Assumptions 1–3 for regression inferences

Step 1   For a prediction level of $1 - \alpha$, use Table IV to find $t_{\alpha/2}$ with df $= n - 2$.

Step 2   Compute the predicted y-value, $\hat{y}_p = b_0 + b_1 x_p$.

Step 3   The endpoints of the prediction interval for the y-value are

$$\hat{y}_p \pm t_{\alpha/2} \cdot s_e \sqrt{1 + \frac{1}{n} + \frac{(x_p - \Sigma x/n)^2}{S_{xx}}}.$$

Example 14.11   Illustrates Procedure 14.4

Using the sample data in Table 14.1 on page 794, obtain a 95% prediction interval for the price of a randomly selected 3-year-old Nissan Z.

SOLUTION   We apply Procedure 14.4.

Step 1 *For a prediction level of $1 - \alpha$, use Table IV to find $t_{\alpha/2}$ with $df = n - 2$.*

We want a 95% prediction interval; so $\alpha = 0.05$. Also, since $n = 11$, $df = 11 - 2 = 9$. Consulting Table IV, we find that $t_{\alpha/2} = t_{0.05/2} = t_{0.025} = 2.262$.

Step 2 *Compute the predicted y-value, $\hat{y}_p = b_0 + b_1 x_p$.*

The sample regression equation for the data in Table 14.1 is $\hat{y} = 195.47 - 20.26x$. Thus the predicted y-value for a 3-year-old Nissan Z is

$$\hat{y}_p = 195.47 - 20.26 \cdot 3 = 134.69.$$

Step 3 *The endpoints of the prediction interval for the y-value are*

$$\hat{y}_p \pm t_{\alpha/2} \cdot s_e \sqrt{1 + \frac{1}{n} + \frac{(x_p - \Sigma x/n)^2}{S_{xx}}}.$$

From Example 13.4, $\Sigma x = 58$ and $\Sigma x^2 = 326$; and from Example 14.2, $s_e = 12.58$. Also, $n = 11$, $t_{\alpha/2} = 2.262$, $x_p = 3$, and $\hat{y}_p = 134.69$. Consequently, the endpoints of the prediction interval are

$$134.69 \pm 2.262 \cdot 12.58 \sqrt{1 + \frac{1}{11} + \frac{(3 - 58/11)^2}{326 - (58)^2/11}}$$

or 134.69 ± 33.02 or 101.67 to 167.71. We can be 95% certain that the price of a randomly selected 3-year-old Nissan Z will be somewhere between \$10,167 and \$16,771. ∎

We have just seen that a 95% prediction interval for the price of a randomly selected 3-year-old Nissan Z is from \$10,167 to \$16,771; and in Example 14.10 we found that a 95% confidence interval for the mean price of all 3-year-old Nissan Zs is from \$11,793 to \$15,145. We picture both intervals in Fig. 14.11.

FIGURE 14.11
Prediction and confidence intervals for 3-year-old Nissan Zs

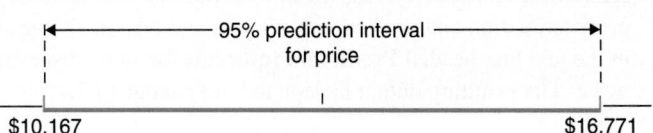

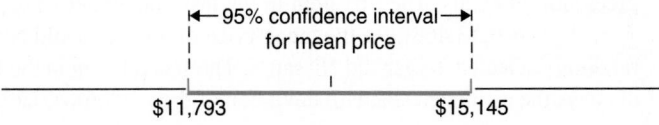

Notice that the prediction interval is wider than the confidence interval. This is to be expected for the following reason: The error in the estimate of the mean price of all

3-year-old Nissan Zs is due only to the fact that the population regression line is being estimated by a sample regression line. On the other hand, the error in the prediction of the price of a randomly selected 3-year-old Nissan Z is due to the previously mentioned error in estimating the mean price plus the variation in prices of 3-year-old Nissan Zs.

Using the Computer (Optional)

We can apply Minitab to obtain confidence intervals and prediction intervals in regression. This is accomplished by employing the **Regress** command and **Predict** subcommand, as described in Example 14.12.

Example 14.12   The Regress Command and Predict Subcommand

Use Minitab to simultaneously obtain a 95% confidence interval for the mean price of all 3-year-old Nissan Zs and a 95% prediction interval for the price of a randomly selected 3-year-old Nissan Z.

SOLUTION   We begin by storing the age and price data from Table 14.1 on page 794 in columns named AGE and PRICE, respectively. Then we do the following.

Session commands: We type **REGRESS** followed by the storage location of the data for the response variable, the number of predictors, and the storage location of the data for the predictor variable(s); that is, we type REGRESS 'PRICE' on 1 predictor 'AGE'; and press Enter. Then we type the subcommand **PREDICT** followed by the age for which we want price estimation and prediction; that is, we type PREDICT 3. and press Enter. Printout 14.7 shows these commands and the resulting output.

(or)

Menu commands: We choose **Stat ▸ Regression ▸ Regression...**, specify PRICE for **Response**, and specify AGE in the **Predictors** text box. Next, to indicate that we want price estimation and prediction for 3-year-old Nissan Zs, we select **Options...** and type 3 in the text box headed **Prediction intervals for new observations**. Finally, we select **OK** twice. The resulting output is depicted in Printout 14.7.

The output in Printout 14.7 is the same as that in Printout 13.2 on page 754 except for the last two lines. Those two lines contain the information about the confidence and prediction intervals. The first item in the last line, under Fit, provides the point estimate, $\hat{y}_p = 134.68$ ($13,468), for the mean price of all 3-year-old Nissan Zs or for the price of a randomly selected 3-year-old Nissan Z. The second item in the last line, under Stdev.Fit, displays the estimated standard deviation of the random variable, $\hat{y}_p = b_0 + b_1 x_p$, which is

$$s_e \sqrt{\frac{1}{n} + \frac{(x_p - \Sigma x/n)^2}{S_{xx}}}.$$

Here, of course, $x_p = 3$.

PRINTOUT 14.7

Minitab output for the
Regress command and
Predict subcommand

```
MTB > REGRESS 'PRICE' on 1 predictor 'AGE';
SUBC> PREDICT 3.

The regression equation is
PRICE = 195 - 20.3 AGE

Predictor        Coef      Stdev    t-ratio        p
Constant        195.47     15.24      12.83    0.000
AGE             -20.261     2.800     -7.24    0.000

s = 12.58      R-sq = 85.3%     R-sq(adj) = 83.7%

Analysis of Variance

SOURCE         DF        SS         MS        F        p
Regression      1      8285.0     8285.0    52.38    0.000
Error           9      1423.5      158.2
Total          10      9708.5

Unusual Observations
Obs.     AGE      PRICE       Fit Stdev.Fit  Residual   St.Resid
  9      2.00    169.00    154.95     9.92      14.05       1.82 X

X denotes an obs. whose X value gives it large influence.

    Fit   Stdev.Fit       95% C.I.          95% P.I.
 134.68        7.41   ( 117.92, 151.44)  ( 101.66, 167.71)
```

The third item in the last line of Printout 14.7, under 95% C.I., shows the required confidence interval. Hence a 95% confidence interval for the mean price of all 3-year-old Nissan Zs is from 117.92 to 151.44. We can be 95% confident that the mean price of all 3-year-old Nissan Zs is somewhere between $11,792 and $15,144.

The final item in the last line, under 95% P.I., gives the required prediction interval. So a 95% prediction interval for the price of a randomly selected 3-year-old Nissan Z is from 101.66 to 167.71. We can be 95% certain that the price of a randomly selected 3-year-old Nissan Z will be somewhere between $10,166 and $16,771. ∎

In solving Example 14.12, we did not specify either the confidence level or the prediction level. The reason is that Minitab automatically uses levels of 95%, and those were the levels we wanted. If levels other than 95% are required, then we must proceed somewhat differently. Refer to the *Minitab Supplement* for details.

EXERCISES 14.3

In Exercises 14.38–14.43, we have repeated the information from Exercises 13.32–13.37. Presuming that the assumptions for regression inferences are met, determine the required confidence and prediction intervals. (Note: You previously obtained the sample regression equations in Exercises 13.32–13.37 and the standard errors of the estimate in Exercises 14.8–14.13.)

14.38 Ten Corvettes between 1 and 6 years old were randomly selected from the classified ads of *The Arizona Republic*. The following data were obtained, where x denotes age, in years, and y denotes price, in hundreds of dollars.

x	6	6	6	2	2	5	4	5	1	4
y	125	115	130	260	219	150	190	163	260	160

a. Obtain a point estimate for the mean price of all 4-year-old Corvettes.
b. Determine a 90% confidence interval for the mean price of all 4-year-old Corvettes.
c. Find the predicted price of a randomly selected 4-year-old Corvette.
d. Determine a 90% prediction interval for the price of a randomly selected 4-year-old Corvette.
e. Draw graphs similar to the ones in Fig. 14.11 on page 823 showing both the 90% confidence interval from part (b) and the 90% prediction interval from part (d).
f. Why is the prediction interval wider than the confidence interval?

14.39 The National Center for Health Statistics publishes data on heights and weights in *Vital and Health Statistics*. A random sample of 11 males age 18–24 years gave the following data, where x denotes height, in inches, and y denotes weight, in pounds.

x	65	67	71	71	66	75	67	70	71	69	69
y	175	133	185	163	126	198	153	163	159	151	155

a. Obtain a point estimate for the mean weight of all 18–24-year-old males who are 70 inches tall.
b. Find a 90% confidence interval for the mean weight of all 18–24-year-old males who are 70 inches tall.
c. Find the predicted weight of a randomly selected 18–24-year-old male who is 70 inches tall.
d. Determine a 90% prediction interval for the weight of a randomly selected 18–24-year-old male who is 70 inches tall.

e. Draw graphs similar to the ones in Fig. 14.11 on page 823 showing both the 90% confidence interval from part (b) and the 90% prediction interval from part (d).
f. Why is the prediction interval wider than the confidence interval?

14.40 Hanna Properties specializes in custom-home resales in the Equestrian Estates, an exclusive subdivision in Phoenix, Arizona. A random sample of nine custom homes currently listed for sale provided the following information on size and price. Here x denotes size, in hundreds of square feet, rounded to the nearest hundred, and y denotes price, in thousands of dollars, rounded to the nearest thousand.

x	26	27	33	29	29	34	30	40	22
y	235	249	267	269	295	345	415	475	195

a. Obtain a point estimate for the mean price of all 2800-sq-ft Equestrian Estate homes.
b. Find a 99% confidence interval for the mean price of all 2800-sq-ft Equestrian Estate homes.
c. Find the predicted price of a randomly selected 2800-sq-ft Equestrian Estate home.
d. Determine a 99% prediction interval for the price of a randomly selected 2800-sq-ft Equestrian Estate home.

14.41 An article read by a physician indicated that the maximum heart rate an individual can reach during intensive exercise decreases with age. The physician decided to do his own study. Ten randomly selected people performed exercise tests and recorded their peak heart rates. The results are shown in the following table, where x denotes age, in years, and y denotes peak heart rate.

x	30	38	41	38	29	39	46	41	42	24
y	186	183	171	177	191	177	175	176	171	196

a. Obtain a point estimate for the mean peak heart rate of all 40-year-olds.
b. Find a 95% confidence interval for the mean peak heart rate of all 40-year-olds.
c. Determine the predicted peak heart rate of a randomly selected 40-year-old.
d. Find a 95% prediction interval for the peak heart rate of a randomly selected 40-year-old.

14.42 An instructor asked a random sample of eight students to record their study times in a beginning calculus course. She then made a table for total hours studied, x, over 2 weeks, and test score, y, at the end of the 2 weeks. Here are the results.

x	10	15	12	20	8	16	14	22
y	92	81	84	74	85	80	84	80

a. Obtain a 95% confidence interval for the mean test score of all beginning calculus students who study for 15 hours.
b. Obtain a 95% prediction interval for the test score of a randomly selected beginning calculus student who studies for 15 hours.

14.43 An economist is interested in the relationship between the disposable income of a family and the amount of money spent annually on food. For a preliminary study, the economist takes a random sample of eight middle-income families of the same size (father, mother, two children). The results are as follows, where x denotes disposable income, in thousands of dollars, and y denotes food expenditure, in hundreds of dollars.

x	30	36	27	20	16	24	19	25
y	55	60	42	40	37	26	39	43

a. Determine a 99% confidence interval for the mean annual food expenditure of all middle-income families consisting of a father, mother, and two children that have a disposable income of $25,000.
b. Find a 99% prediction interval for the annual food expenditure of a randomly selected middle-income family consisting of a father, mother, and two children that has a disposable income of $25,000.

14.44 (Computer exercise) Use Minitab or some other statistical software to obtain the confidence and prediction intervals required in Exercise 14.42.

14.45 (Computer exercise) Refer to the data in Exercise 14.43. Use Minitab or some other statistical software to
a. determine a 95% confidence interval for the mean annual food expenditure of all middle-income families consisting of a father, mother, and two children that have a disposable income of $25,000.

b. determine a 95% prediction interval for the annual food expenditure of a randomly selected middle-income family consisting of a father, mother, and two children that has a disposable income of $25,000.
c. In Exercise 14.43 you were asked to obtain 99% confidence and prediction intervals, whereas in this exercise you are asked to obtain 95% confidence and prediction intervals. Why didn't we ask you to obtain 99% confidence and prediction intervals in this exercise also?

14.46 (Computer exercise) The U.S. Energy Information Administration compiles and publishes data on energy consumption by family income in *Residential Energy Consumption Survey: Consumption and Expenditures.* We applied Minitab's Regress command to data on family income and last year's energy consumption from a random sample of 25 families. The income data are in thousands of dollars and the energy-consumption data are in millions of BTU. Printout 14.8 on the following page shows (abridged) computer output. The confidence and prediction intervals are for an income level of $40,000.
a. Determine a point estimate for last year's mean energy consumption of all families with an annual income of $40,000.
b. Find a 95% confidence interval for last year's mean energy consumption of all families with an annual income of $40,000.
c. Obtain a 95% prediction interval for last year's energy consumption by a randomly selected family with an annual income of $40,000.

14.47 (Computer exercise) Greene and Touchstone conducted a study on the relationship between the estriol levels of pregnant women and the birth weights of their children. Their findings, "Urinary Tract Estriol: An Index of Placental Function," were published in the *American Journal of Obstetrics and Gynecology.* Printout 14.9 on the following page shows (abridged) output that results by applying Minitab's Regress command to the 31 pairs of data obtained by Greene and Touchstone. Estriol levels are in milligrams per 24 hours and birth weights are in grams. The confidence and prediction intervals are for an estriol level of 15 mg/24 hours.
a. Obtain a point estimate for the mean birth weight of all babies whose mothers have an estriol level of 15 mg/24 hours.
b. Determine a 95% confidence interval for the mean birth weight of all babies whose mothers have an estriol level of 15 mg/24 hours.
c. Find a 95% prediction interval for the birth weight of a randomly selected baby whose mother has an estriol level of 15 mg/24 hours.

The regression equation is
CONSUMPT = 82.0 + 0.931 INCOME

Predictor	Coef	Stdev	t-ratio	p
Constant	82.036	2.054	39.94	0.000
INCOME	0.93051	0.05727	16.25	0.000

s = 5.375   R-sq = 92.0%   R-sq(adj) = 91.6%

Analysis of Variance

SOURCE	DF	SS	MS	F	p
Regression	1	7626.6	7626.6	264.02	0.000
Error	23	664.4	28.9		
Total	24	8291.0			

Fit	Stdev.Fit	95% C.I.	95% P.I.
119.26	1.20	(116.77, 121.75)	(107.86, 130.65)

The regression equation is
WEIGHT = 2152 + 60.8 ESTRIOL

Predictor	Coef	Stdev	t-ratio	p
Constant	2152.3	262.0	8.21	0.000
ESTRIOL	60.82	14.68	4.14	0.000

s = 382.1   R-sq = 37.2%   R-sq(adj) = 35.0%

Analysis of Variance

SOURCE	DF	SS	MS	F	p
Regression	1	2505745	2505745	17.16	0.000
Error	29	4234255	146009		
Total	30	6740000			

Fit	Stdev.Fit	95% C.I.	95% P.I.
3064.6	76.0	(2909.1, 3220.1)	(2267.6, 3861.6)

14.4 INFERENCES IN CORRELATION

Frequently, we want to decide whether two variables are linearly correlated, that is, whether there is a linear relationship between the two variables. As we learned in Section 14.2, we can make that decision by performing a hypothesis test for the slope, β_1, of the population regression line.

Alternatively, we can perform a hypothesis test for the **population linear correlation coefficient,** ρ (rho). The population linear correlation coefficient, ρ, measures the linear correlation between the population of all data points in the same way that the sample linear correlation coefficient, r, measures the linear correlation between a sample of data points. Thus it is ρ that actually describes the strength of the linear relationship between two variables; r is only an estimate of ρ.

The population linear correlation coefficient, ρ, lies between -1 and 1. Values of ρ near -1 or 1 indicate a strong linear relationship between the variables, whereas values of ρ near 0 indicate a weak linear relationship between the variables.

If $\rho > 0$, the variables are **positively linearly correlated,** meaning that y tends to increase linearly as x increases, with the tendency being greater the closer ρ is to 1. If $\rho < 0$, the variables are **negatively linearly correlated,** meaning that y tends to decrease linearly as x increases, with the tendency being greater the closer ρ is to -1. If $\rho = 0$, the variables are **linearly uncorrelated,** meaning that there is no linear relationship between the variables.

Because the sample linear correlation coefficient, r, is an estimate of the population linear correlation coefficient, ρ, we can use r as a basis for performing a hypothesis test for ρ. For a test with null hypothesis H_0: $\rho = 0$ (i.e., the variables are linearly uncorrelated), we will employ the following fact.

KEY FACT 14.9   **t-DISTRIBUTION FOR A CORRELATION TEST**

> Suppose the variables x and y satisfy the assumptions for regression inferences. Then if $\rho = 0$, the random variable
>
> $$t = \frac{r}{\sqrt{\dfrac{1 - r^2}{n - 2}}}$$
>
> has the t-distribution with df $= n - 2$.

In view of Key Fact 14.9, we see that for a hypothesis test with null hypothesis H_0: $\rho = 0$, we can use the random variable

$$t = \frac{r}{\sqrt{\dfrac{1 - r^2}{n - 2}}}$$

as the test statistic and obtain the critical values from the t-table, Table IV. Specifically, we have the following procedure.

PROCEDURE 14.5

TO PERFORM A HYPOTHESIS TEST FOR A POPULATION LINEAR CORRELATION COEFFICIENT WITH NULL HYPOTHESIS $H_0: \rho = 0$

ASSUMPTIONS

Assumptions 1–3 for regression inferences

Step 1 State the null and alternative hypotheses.

Step 2 Decide on the significance level, α.

Step 3 The critical value(s)
  a. for a two-tailed test are $\pm t_{\alpha/2}$,
  b. for a left-tailed test is $-t_\alpha$,
  c. for a right-tailed test is t_α,
  with df $= n - 2$. Use Table IV to find the critical value(s).

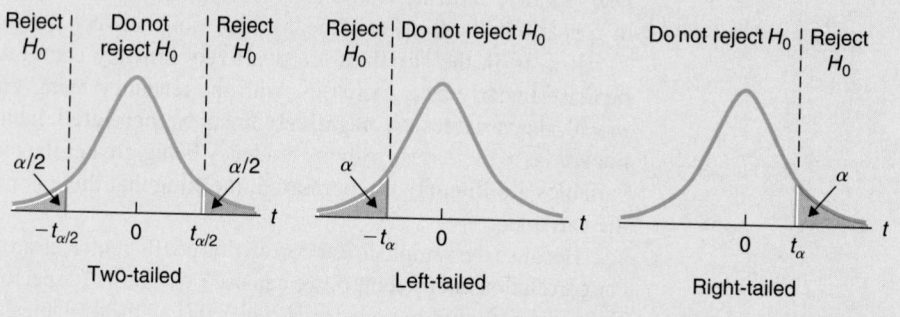

Step 4 Compute the value of the test statistic

$$t = \frac{r}{\sqrt{\dfrac{1 - r^2}{n - 2}}}.$$

Step 5 If the value of the test statistic falls in the rejection region, then reject H_0; otherwise, do not reject H_0.

Step 6 State the conclusion in words.

Example 14.13   Illustrates Procedure 14.5

Consider once more the age and price data for a sample of 11 Nissan Zs (Table 14.1 on page 794). At the 5% significance level, do the data provide sufficient evidence to conclude that age and price of Nissan Zs are negatively linearly correlated?

SOLUTION   As we discovered in Example 14.3 on page 801, it is not unreasonable to consider the assumptions for regression inferences satisfied by the variables age and price for Nissan Zs, at least for Nissan Zs between 2 and 7 years old. Therefore we will apply Procedure 14.5 to carry out the required hypothesis test.

Step 1 *State the null and alternative hypotheses.*

Let ρ denote the population linear correlation coefficient for the variables age and price of Nissan Zs. Then the null and alternative hypotheses are

$$H_0: \rho = 0 \text{ (age and price are linearly uncorrelated)}$$

$$H_a: \rho < 0 \text{ (age and price are negatively linearly correlated)}.$$

Note that the hypothesis test is left-tailed since a less-than sign ($<$) appears in the alternative hypothesis.

Step 2 *Decide on the significance level, α.*

We are to use $\alpha = 0.05$.

Step 3 *The critical value for a left-tailed test is $-t_\alpha$, with $df = n - 2$.*

We have $n = 11$, so df $= 9$. Also, $\alpha = 0.05$. Consulting Table IV, we find that for df $= 9$, $t_{0.05} = 1.833$. Thus the critical value is $-t_{0.05} = -1.833$, as seen in Fig. 14.12.

FIGURE 14.12

Criterion for deciding whether or not to reject the null hypothesis

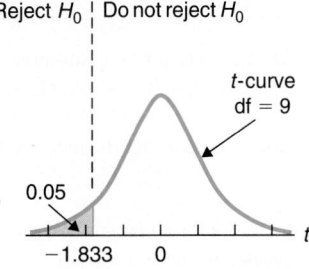

Step 4 *Compute the value of the test statistic*

$$t = \frac{r}{\sqrt{\dfrac{1 - r^2}{n - 2}}}.$$

We have already computed the sample linear correlation coefficient, r, for the age and price data in Table 14.1. This was done in Example 13.11 on page 772, where we found that $r = -0.924$. Therefore the value of the test statistic is

$$t = \frac{-0.924}{\sqrt{\dfrac{1 - (-0.924)^2}{11 - 2}}} = -7.249.$$

Step 5 *If the value of the test statistic falls in the rejection region, reject H_0; otherwise, do not reject H_0.*

The value of the test statistic, found in Step 4, is $t = -7.249$. A glance at Fig. 14.12 shows that this falls in the rejection region. Hence we reject H_0.

Step 6   *State the conclusion in words.*

The test results are statistically significant at the 5% level; that is, at the 5% significance level, the data provide sufficient evidence to conclude that age and price of Nissan Zs are negatively linearly correlated. Prices for Nissan Zs tend to decrease linearly with increasing age, at least for Nissan Zs between 2 and 7 years old.   ∎

As usual, we can also carry out the hypothesis test using the *P*-value approach to hypothesis testing. From Step 4 we see that the value of the test statistic is $t = -7.249$. Recalling that df $= 9$, we find from Table IV that $P < 0.005$. In particular, because the *P*-value is less than the specified significance level of 0.05, we can reject H_0. Furthermore, by referring to Table 9.12 on page 530, we see that the data provide very strong evidence against the null hypothesis and hence in favor of the alternative hypothesis that age and price of Nissan Zs are negatively linearly correlated.

EXERCISES 14.4

In Exercises 14.48–14.53, we have repeated the information from Exercises 13.32–13.37. Presuming that the assumptions for regression inferences are met, perform the required hypothesis test using either the classical approach or the P-value approach. (Note: You previously obtained the sample linear correlation coefficients in Exercises 13.74–13.79.)

14.48   Ten Corvettes between 1 and 6 years old were randomly selected from the classified ads of *The Arizona Republic*. The following data were obtained, where *x* denotes age, in years, and *y* denotes price, in hundreds of dollars.

x	6	6	6	2	2	5	4	5	1	4
y	125	115	130	260	219	150	190	163	260	160

At the 5% significance level, do the data provide sufficient evidence to conclude that age and price of Corvettes are negatively linearly correlated?

14.49   The National Center for Health Statistics publishes data on heights and weights in *Vital and Health Statistics*. A random sample of 11 males age 18–24 years gave the following data, where *x* denotes height, in inches, and *y* denotes weight, in pounds.

x	65	67	71	71	66	75	67	70	71	69	69
y	175	133	185	163	126	198	153	163	159	151	155

Do the data provide sufficient evidence to conclude that the variables height and weight are positively linearly correlated for 18–24-year-old males? Perform the required hypothesis test at the 5% significance level.

14.50   Hanna Properties specializes in custom-home resales in the Equestrian Estates, an exclusive subdivision in Phoenix, Arizona. A random sample of nine custom homes currently listed for sale provided the following information on size and price. Here *x* denotes size, in hundreds of square feet, rounded to the nearest hundred, and *y* denotes price, in thousands of dollars, rounded to the nearest thousand.

x	26	27	33	29	29	34	30	40	22
y	235	249	267	269	295	345	415	475	195

Do the data provide sufficient evidence to conclude that for custom homes in the Equestrian Estates, size and price are positively linearly correlated? Perform the required hypothesis test at the 0.5% significance level.

14.51   An article read by a physician indicated that the maximum heart rate an individual can reach during intensive exercise decreases with age. The physician decided to do his own study. Ten randomly selected people performed exercise tests and recorded their peak heart rates. The results are shown in the following table, where x denotes age, in years, and y denotes peak heart rate.

x	30	38	41	38	29	39	46	41	42	24
y	186	183	171	177	191	177	175	176	171	196

At the 2.5% significance level, do the data provide sufficient evidence to conclude that age and peak heart rate are negatively linearly correlated?

14.52 An instructor asked a random sample of eight students to record their study times in a beginning calculus course. She then made a table for total hours studied, x, over 2 weeks, and test score, y, at the end of the 2 weeks. Here are the results.

x	10	15	12	20	8	16	14	22
y	92	81	84	74	85	80	84	80

Do the data provide sufficient evidence to conclude that in beginning calculus courses, study time and test score are linearly correlated? Perform the hypothesis test using a significance level of 0.05.

14.53 An economist is interested in the relationship between the disposable income of a family and the amount of money spent annually on food. For a preliminary study, the economist takes a random sample of eight middle-income families of the same size (father, mother, two children). The results are as follows, where x denotes disposable income, in thousands of dollars, and y denotes food expenditure, in hundreds of dollars.

x	30	36	27	20	16	24	19	25
y	55	60	42	40	37	26	39	43

Do the data provide sufficient evidence to conclude that family disposable income and annual food expenditure are linearly correlated for middle-income families with a father, mother,

and two children? Perform the required hypothesis test at the 1% significance level.

14.54 We took a sample of 10 students from an introductory statistics class and obtained the following data, where x denotes height, in inches, and y denotes final-exam score.

x	71	68	71	65	66	68	68	64	62	65
y	87	96	66	71	71	55	83	67	86	60

Do the data provide sufficient evidence to conclude that for students in introductory statistics courses, height and final-exam score are linearly correlated? Perform the required hypothesis test at the 5% significance level.

14.55 Is the population linear correlation coefficient, ρ, a random variable? What about the sample linear correlation coefficient, r? Explain your answers.

14.56 Procedure 14.1 on page 811 employs the random variable

$$t = \frac{b_1}{s_e / \sqrt{S_{xx}}}$$

as the test statistic for a hypothesis test with null hypothesis $H_0: \beta_1 = 0$. Show that this test statistic is equal to the one used in Procedure 14.5 on page 830, that is, to

$$t = \frac{r}{\sqrt{\dfrac{1 - r^2}{n - 2}}}.$$

(*Hint:* Express both statistics in terms of the quantities S_{xx}, S_{xy}, and S_{yy}.) This result proves that the hypothesis tests

$$H_0: \beta_1 = 0$$
$$H_a: \beta_1 \neq 0$$

and

$$H_0: \rho = 0$$
$$H_a: \rho \neq 0$$

are equivalent.

14.5 MULTIPLE REGRESSION (OPTIONAL)

· · · · ·

Up to this point, we have considered inferential methods in regression analysis only for simple linear regression, that is, regression analysis with one predictor variable. Now we will study the corresponding methods in multiple regression, where more than one predictor variable is present.

The Multiple Regression Model

As in simple linear regression, to perform statistical inferences in multiple regression, the variables under consideration must satisfy certain conditions. For multiple regression with k predictor variables, those conditions are as follows.

KEY FACT 14.10

ASSUMPTIONS FOR MULTIPLE REGRESSION INFERENCES

1. **Population regression equation:** For each set of values, $x_1, x_2, \ldots, x_k$, of the predictor variables, the mean of the corresponding population of y-values is $\beta_0 + \beta_1 x_1 + \cdots + \beta_k x_k$. The equation

$$y = \beta_0 + \beta_1 x_1 + \cdots + \beta_k x_k$$

is called the *population regression equation.*

2. **Equal standard deviations:** The standard deviation, σ, of the population of y-values corresponding to a particular set of values, $x_1, x_2, \ldots, x_k$, of the predictor variables is the same, regardless of $x_1, x_2, \ldots, x_k$.

3. **Normality:** For each set of values, $x_1, x_2, \ldots, x_k$, of the predictor variables, the corresponding population of y-values is normally distributed.

Assumptions 1, 2, and 3 require that there exist constants, $\beta_0, \beta_1, \ldots, \beta_k$, and σ, such that for each set of values, $x_1, x_2, \ldots, x_k$, of the predictor variables, the corresponding population of y-values is normally distributed with mean $\beta_0 + \beta_1 x_1 + \cdots + \beta_k x_k$ and standard deviation σ. These assumptions are often referred to as the **multiple regression model.** They can be expressed symbolically as

$$y = \beta_0 + \beta_1 x_1 + \cdots + \beta_k x_k + \epsilon,$$

where ϵ (epsilon) represents a normally distributed random variable having mean 0 and standard deviation σ.

Example 14.14   Assumptions for Multiple Regression Inferences

Consider the variables age, miles driven, and price for Nissan Zs, where age is in years, miles driven is in thousands, and price is in hundreds of dollars. Discuss what it would mean for the assumptions for multiple regression inferences to be satisfied with age and miles driven as predictor variables for price.

SOLUTION   Here we have $k = 2$ since there are two predictor variables, age and miles driven. For the assumptions for multiple regression inferences to be satisfied, it would mean there are constants, $\beta_0, \beta_1, \beta_2$, and σ, such that for each age, x_1, and number of miles driven, x_2, the prices of all Nissan Zs of that age that have been driven that number of miles are normally distributed with mean $\beta_0 + \beta_1 x_1 + \beta_2 x_2$ and standard deviation σ.

Consequently, it would mean that the prices of all Nissan Zs that are 2 years old ($x_1 = 2$) and have been driven 15,000 miles ($x_2 = 15$) are normally distributed with mean

$\beta_0 + \beta_1 \cdot 2 + \beta_2 \cdot 15$ and standard deviation σ; the prices of all Nissan Zs that are 3 years old ($x_1 = 3$) and have been driven 32,000 miles ($x_2 = 32$) are normally distributed with mean $\beta_0 + \beta_1 \cdot 3 + \beta_2 \cdot 32$ and standard deviation σ; and so on. ∎

The interpretation of the coefficients $\beta_1, \beta_2, \ldots, \beta_k$ is similar to that in simple linear regression. Specifically, for each j, the coefficient β_j of the predictor variable x_j represents the change in the mean of the population of y-values for every increase in x_j by 1 unit, with all other predictor variables held fixed.

For instance, consider the situation in Example 14.14. The population regression equation is $y = \beta_0 + \beta_1 x_1 + \beta_2 x_2$, where x_1 denotes age in years, x_2 denotes miles driven in thousands, and y denotes price in hundreds of dollars. So, for any fixed number of miles driven, β_1 is the amount in hundreds of dollars that the mean price changes for every increase in age by 1 year. Similarly, for any fixed age, β_2 is the amount in hundreds of dollars that the mean price changes for every increase in the number of miles driven by 1 thousand.

When we determine a sample regression equation, $\hat{y} = b_0 + b_1 x_1 + \cdots + b_k x_k$, we obtain the best estimate, based on the sample data, of the unknown population regression equation, $y = \beta_0 + \beta_1 x_1 + \cdots + \beta_k x_k$. In fact, for each j, b_j is the best estimate of β_j.

As we mentioned in Section 13.5, because of computational complexity, multiple regression is now almost always done by computer. Consequently, in this section we will present computer output as a means for examining multiple regression.

Example 14.15   Obtaining a Sample Regression Equation

In Table 13.12 we presented data on age, miles driven, and price for a sample of 11 Nissan Zs. Those data are repeated here in Table 14.3.

TABLE 14.3
Data on age, miles driven, and price for 11 Nissan Zs

Car	Age (yrs) x_1	Miles (thous) x_2	Price ($100s) y
1	5	57	85
2	4	40	103
3	6	77	70
4	5	60	82
5	5	49	89
6	5	47	98
7	6	58	66
8	6	39	95
9	2	8	169
10	7	69	70
11	7	89	48

We employed Minitab to perform a multiple regression analysis on the data in Table 14.3, with age and miles driven as predictor variables for price. In doing so, we instructed Minitab to provide the most detailed output available, as seen in Printout 14.10. (The last two lines of Printout 14.10 do not result from instructing Minitab to provide the most detailed output available but rather from a special command, discussed on page 844.)

a. Use Printout 14.10 to obtain the sample regression equation.

b. Presuming that the assumptions for multiple regression inferences are met with age and miles driven as predictor variables for price, interpret the result obtained in part (a).

The regression equation is
PRICE = 183 - 9.50 AGE - 0.821 MILES

Predictor	Coef	Stdev	t-ratio	p
Constant	183.04	11.35	16.13	0.000
AGE	-9.504	3.874	-2.45	0.040
MILES	-0.8215	0.2552	-3.22	0.012

s = 8.805   R-sq = 93.6%   R-sq(adj) = 92.0%

Analysis of Variance

SOURCE	DF	SS	MS	F	p
Regression	2	9088.3	4544.2	58.61	0.000
Error	8	620.2	77.5		
Total	10	9708.5			

SOURCE	DF	SEQ SS
AGE	1	8285.0
MILES	1	803.3

Obs.	AGE	PRICE	Fit	Stdev.Fit	Residual	St.Resid
1	5.00	85.00	88.69	3.20	-3.69	-0.45
2	4.00	103.00	112.16	3.71	-9.16	-1.15
3	6.00	70.00	62.76	4.59	7.24	0.96
4	5.00	82.00	86.22	3.66	-4.22	-0.53
5	5.00	89.00	95.26	2.73	-6.26	-0.75
6	5.00	98.00	96.90	2.84	1.10	0.13
7	6.00	66.00	78.36	3.32	-12.36	-1.52
8	6.00	95.00	93.97	6.93	1.03	0.19
9	2.00	169.00	157.45	6.99	11.55	2.15R
10	7.00	70.00	59.82	4.71	10.18	1.37
11	7.00	48.00	43.39	5.35	4.61	0.66

R denotes an obs. with a large st. resid.

Fit	Stdev.Fit	95% C.I.	95% P.I.
92.80	2.74	(86.47, 99.12)	(71.53, 114.07)

SOLUTION   a. The sample regression equation is displayed in the second line of the computer output in Printout 14.10: PRICE = 183 - 9.50 AGE - 0.821 MILES. That is, the sample regression equation is $\hat{y} = 183 - 9.50x_1 - 0.821x_2$, where x_1 denotes age in years, x_2 denotes miles driven in thousands, and $\hat{y}$ denotes predicted price in hundreds of dollars.

b. Based on the sample data in Table 14.3, the equation obtained in part (a) is the best estimate of the unknown population regression equation, $y = \beta_0 + \beta_1 x_1 + \beta_2 x_2$. ■

The Standard Error of the Estimate

Assumption 2 for inferences in multiple regression requires that the standard deviations of the various populations of y-values be the same, regardless of the values of the predictor variables. As in simple linear regression, the statistic used to estimate the common population standard deviation, σ, is called the **standard error of the estimate** and is defined as follows.

DEFINITION 14.2   **STANDARD ERROR OF THE ESTIMATE**

> For a multiple regression analysis with k predictor variables, the **standard error of the estimate, s_e,** is defined by
>
> $$s_e = \sqrt{\frac{SSE}{n - (k + 1)}},$$
>
> where $SSE = \Sigma(y - \hat{y})^2$.

Using $k = 1$ (one predictor variable) in Definition 14.2, we get $s_e = \sqrt{SSE/(n - 2)}$, which is the definition of the standard error of the estimate in simple linear regression. Thus Definition 14.2 generalizes the standard error of the estimate to cover both simple and multiple regression.

Example 14.16   Illustrates Definition 14.2

Consider again the Nissan Z data in Table 14.3 on page 835.

a. Use Printout 14.10 to obtain the standard error of the estimate.

b. Presuming that the assumptions for multiple regression inferences are met with age and miles driven as predictor variables for price, interpret the result in part (a).

SOLUTION   a. The standard error of the estimate is the first entry in the seventh line of Printout 14.10: s = 8.805 (Minitab uses s instead of s_e to denote the standard error of the estimate). Thus for the Nissan Z data in Table 14.3, $s_e = 8.805$.

b. Based on the sample data in Table 14.3, our best estimate for the (common) population standard deviation, σ, of prices for all Nissan Zs of any particular age and number of miles driven is $880.5. ∎

Analysis of Residuals

Now that we have examined the assumptions for multiple regression inferences, we need to discuss how the sample data can be used to decide whether it is reasonable to presume that those assumptions are met. As in simple linear regression, the method for deciding relies on an analysis of the residuals, $e = y - \hat{y}$, the differences between the observed and predicted y-values.

KEY FACT 14.11 **RESIDUAL ANALYSIS FOR THE MULTIPLE REGRESSION MODEL**

> If the assumptions for multiple regression inferences are met, then the following three conditions should hold.
>
> 1. A plot of the residuals against each predictor variable should fall roughly in a horizontal band centered and symmetric about the horizontal axis.
>
> 2. A plot of the residuals against the predicted y-values should fall roughly in a horizontal band centered and symmetric about the horizontal axis.
>
> 3. A normal probability plot of the residuals should be roughly linear.
>
> Failure of any of these three conditions casts doubt on the validity of one or more of the assumptions for multiple regression inferences.

Example 14.17   Analysis of Residuals

Perform a residual analysis on the data in Table 14.3 on page 835, with age and miles driven as predictor variables for price.

SOLUTION   We apply Key Fact 14.11. The predicted y-values and residuals for the Nissan Z data can be found, respectively, in the fourth and sixth columns of the table near the bottom of Printout 14.10 on page 836. (Minitab uses the term Fit instead of *predicted y-value*.)

From the residual and age data, we obtain the residual plot in Fig. 14.13(a); from the residual and miles data, we obtain the residual plot in Fig. 14.13(b); and from the residual and predicted-price data, we obtain the residual plot in Fig. 14.13(c). Keeping in mind the small sample size, we can say that all three plots fall roughly in a horizontal band centered and symmetric about the horizontal axis. Finally, Fig. 14.13(d) shows a normal probability plot of the residuals, which is quite linear.

Thus our residual analysis uncovers no obvious violations of the assumptions for regression inferences for the variables under consideration. ∎

FIGURE 14.13

Residual plots and normal
probability plot of residuals for
the Nissan Z data in Table 14.3

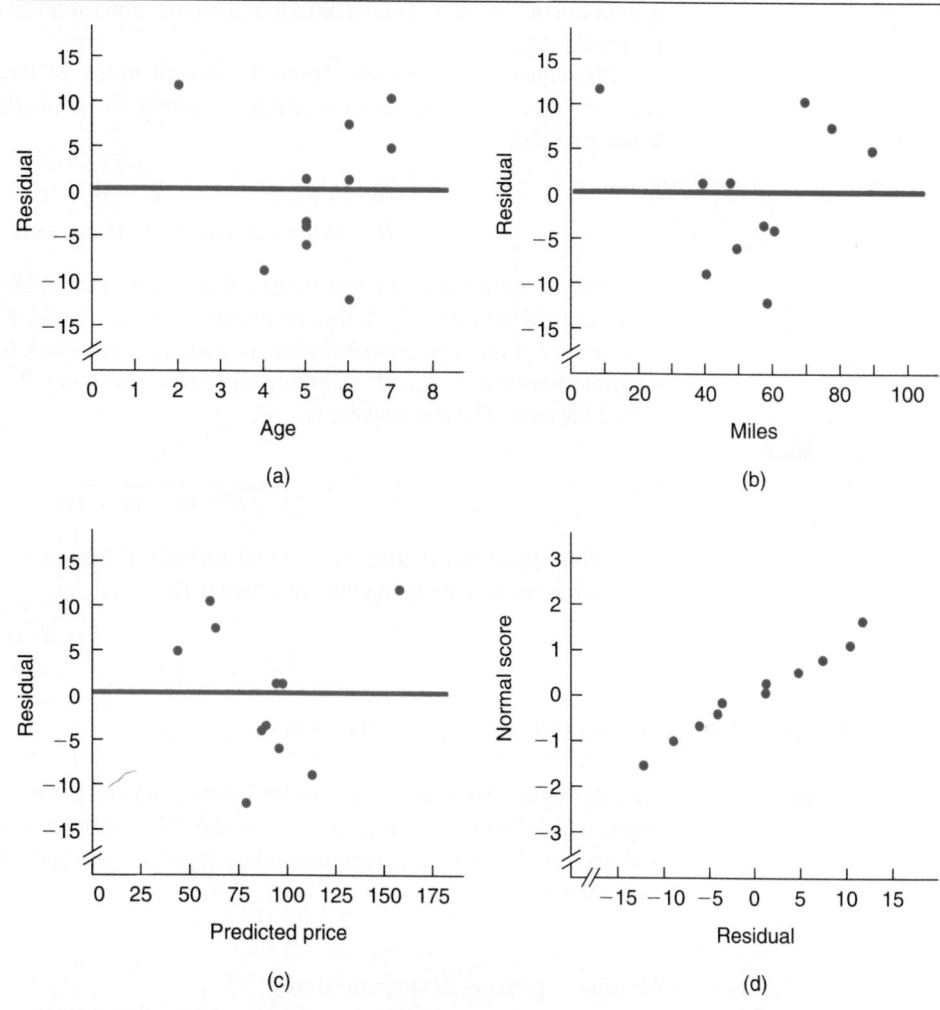

Inferences Concerning the Utility of the Regression

Suppose the variables $x_1, x_2, \ldots, x_k$ and y satisfy the assumptions for multiple regression inferences. Our first question is whether the regression as a whole is useful for making predictions, that is, whether the variables $x_1, x_2, \ldots, x_k$ taken together are useful for predicting y. The answer to that question depends on the parameters $\beta_1, \beta_2, \ldots, \beta_k$, the coefficients of $x_1, x_2, \ldots, x_k$ in the population regression equation.

If $\beta_1, \beta_2, \ldots, \beta_k$ are all 0, then for each set of values, $x_1, x_2, \ldots, x_k$, of the predictor variables, the corresponding population of y-values is normally distributed with mean β_0 $(= \beta_0 + 0 \cdot x_1 + 0 \cdot x_2 + \cdots + 0 \cdot x_k)$ and standard deviation σ; and neither of those two parameters involves the predictor variables. So we see that if $\beta_1 = \beta_2 = \cdots = \beta_k = 0$, then the predictor variables provide no information about the distribution of the population of y-values. This implies that the predictor variables taken together are useless for predicting y.

Consequently, we can decide on the overall utility of the regression—whether the variables $x_1, x_2, \ldots, x_k$ taken together are useful for predicting y—by performing the hypothesis test

$$H_0: \beta_1 = \beta_2 = \cdots = \beta_k = 0$$

$$H_a: \text{At least one of the } \beta \text{s is not zero.}$$

Now we must identify a test statistic that can be used to perform this hypothesis test. As we learned in Section 13.5, the coefficient of determination, R^2, is a descriptive measure of the utility of the regression equation for making predictions. So it seems reasonable that we should be able to use R^2 or some expression involving R^2 as a test statistic. This is indeed the case. The test statistic is

$$(1) \qquad\qquad F = \frac{R^2/k}{(1 - R^2)/[n - (k + 1)]}.$$

If the null hypothesis is true, the random variable F has an F-*distribution,* a probability distribution we will study in detail in Chapter 15.

Example 14.18   Inferences Concerning the Utility of the Regression

Consider again the data on age, miles driven, and price for a sample of 11 Nissan Zs, displayed in Table 14.3 on page 835. At the 5% significance level, do the data provide sufficient evidence to conclude that taken together, age and miles driven are useful for predicting price?

SOLUTION   We want to perform the hypothesis test

$$H_0: \beta_1 = \beta_2 = 0$$

$$H_a: \text{At least one of } \beta_1 \text{ and } \beta_2 \text{ is not zero,}$$

with $\alpha = 0.05$. As we just learned, the test statistic for this hypothesis test is given by Equation (1). Here $k = 2$ and $n = 11$. Look at the tenth line of the computer output in Printout 14.10 on page 836, the line whose first entry is `Regression`. The fifth entry in that line displays the value of the test statistic, F; thus $F = 58.61$.

The final entry in that line gives the P-value for the hypothesis test; so $P = 0.000$ (to three decimal places). In particular, because the P-value is less than the specified significance level of 0.05, we can reject H_0. Furthermore, by referring to Table 9.12 on page 530, we see that the data provide very strong evidence against the null hypothesis and hence in favor of the alternative hypothesis that taken together, age and miles driven are useful for predicting price.   ■

Inferences Concerning the Utility of a Particular Predictor Variable

To decide whether a particular predictor variable, say x_j, is useful for predicting y, we proceed as we did in simple linear regression. Namely, we perform the hypothesis test

$$H_0: \beta_j = 0 \ (x_j \text{ is not useful for predicting } y)$$

$$H_a: \beta_j \neq 0 \ (x_j \text{ is useful for predicting } y).$$

Rejection of the null hypothesis indicates that x_j is useful as a predictor of y. Nonrejection of the null hypothesis suggests that x_j may not be useful as a predictor of y and that it may be worthwhile to do a regression analysis with the variable x_j omitted.

The test statistic for this hypothesis test is essentially the same as the one used in simple linear regression:

$$t = \frac{b_j}{s_{b_j}},$$

where s_{b_j} is the estimated standard deviation of the random variable b_j. However, in multiple regression, df $= n - (k + 1)$ and the formula for s_{b_j} is much more complicated than in simple linear regression.

You must be careful when interpreting the conclusion of a hypothesis test for a β-parameter: Although x_j may not be useful for predicting y in conjunction with the other predictor variables under consideration, it may be useful when employed as the only predictor variable or with some collection of predictor variables. Thus in this section when we say that x_j is not useful for predicting y, we really mean that in the regression with $x_1, x_2, \ldots, x_k$ as the predictor variables, x_j is not useful for predicting y.

Likewise, although x_j may be useful for predicting y in conjunction with the other predictor variables under consideration, it may not be useful when employed as the only predictor variable or with some collection of predictor variables. Thus in this section when we say that x_j is useful for predicting y, we really mean that in the regression with $x_1, x_2, \ldots, x_k$ as the predictor variables, x_j is useful for predicting y.

Example 14.19 Inferences Concerning the Utility of a Particular Predictor Variable

---

Consider again the data on age, miles driven, and price for a sample of 11 Nissan Zs, presented in Table 14.3. At the 5% significance level, do the data provide sufficient evidence to conclude that in conjunction with age, number of miles driven is useful for predicting price?

SOLUTION We want to perform the hypothesis test

$$H_0: \beta_2 = 0 \text{ (miles driven is not useful for predicting price)}$$

$$H_a: \beta_2 \neq 0 \text{ (miles driven is useful for predicting price)},$$

with $\alpha = 0.05$.

Look at the sixth line of the computer output in Printout 14.10 on page 836, the line whose first entry is MILES. The second entry in that line gives the coefficient b_2 of the sample regression line; so $b_2 = -0.8215$. The third entry in that line shows the estimated standard deviation of b_2, that is, s_{b_2}; so $s_{b_2} = 0.2552$. The fourth entry in that line displays the value of the test statistic $t = b_2/s_{b_2}$; so $t = -3.22$.

The final entry in that line gives the P-value for the hypothesis test; thus $P = 0.012$. In particular, because the P-value is less than the specified significance level of 0.05, we can reject H_0. Furthermore, by referring to Table 9.12 on page 530, we see that the data provide strong evidence against the null hypothesis. Evidently, number of miles driven is useful for predicting price, or more precisely, in conjunction with age, number of miles driven is useful for predicting price.   ■

Confidence Intervals for Means

To determine a point estimate or confidence-interval estimate for the mean of the population of y-values corresponding to particular values, $x_{1p}, x_{2p}, \ldots, x_{kp}$, of the predictor variables, we proceed as in simple linear regression. A point estimate for the mean is obtained by substituting the predictor-variable values into the sample regression equation:

$$\hat{y}_p = b_0 + b_1 x_{1p} + b_2 x_{2p} + \cdots + b_k x_{kp}.$$

And the endpoints of a $(1-\alpha)$-level confidence interval for the mean are found by applying the formula

$$\hat{y}_p \pm t_{\alpha/2} \cdot s_{\hat{y}_p},$$

where $t_{\alpha/2}$ is computed for the t-curve with df $= n - (k + 1)$ and $s_{\hat{y}_p}$ is the estimated standard deviation of the random variable $\hat{y}_p$. In practice, a computer is almost always used to implement these formulas and obtain the required point estimate or confidence interval.

Example 14.20   Point Estimates and Confidence Intervals for Means

Refer to the data shown in Table 14.3 on age, miles driven, and price for a sample of 11 Nissan Zs.

a. Determine a point estimate for the mean price of all Nissan Zs that are 5 years old and have been driven 52,000 miles.

b. Obtain a 95% confidence interval for the mean price of all Nissan Zs that are 5 years old and have been driven 52,000 miles.

SOLUTION   Both of these problems can be solved by referring to the last two lines of Printout 14.10 on page 836. Those two lines, which provide information on price for Nissan Zs that are 5 years old and have been driven 52,000 miles, were obtained using a special computer command. (If that age and that number of miles driven are replaced in the command by different values, the last two lines of the computer output will change accordingly.)

a. The required point estimate for the mean is the first item in the last line of Printout 14.10, the item headed Fit. Based on the sample data, our point estimate for the mean price of all Nissan Zs that are 5 years old and have been driven 52,000 miles is 92.80, or $9280.

b. The required confidence interval is the third item in the last line of Printout 14.10, the item headed 95% C.I. We can be 95% confident that the mean price of all Nissan Zs that are 5 years old and have been driven 52,000 miles is somewhere between $8647 and $9912. ∎

Prediction Intervals

To determine the predicted value or a prediction interval for a randomly selected member of the population of y-values corresponding to particular values, $x_{1p}, x_{2p}, \ldots, x_{kp}$, of the predictor variables, we again proceed as in simple linear regression. Like a point estimate for the mean, the predicted y-value is obtained by substituting the predictor-variable values into the sample regression equation:

$$\hat{y}_p = b_0 + b_1 x_{1p} + b_2 x_{2p} + \cdots + b_k x_{kp}.$$

And the endpoints of a $(1 - \alpha)$-level prediction interval are found by applying the formula

$$\hat{y}_p \pm t_{\alpha/2} \cdot s_{y_p - \hat{y}_p},$$

where $t_{\alpha/2}$ is computed for the t-curve with df $= n - (k + 1)$ and $s_{y_p - \hat{y}_p}$ is the estimated standard deviation of the random variable $y_p - \hat{y}_p$.

Example 14.21 Predicted Values and Prediction Intervals

Refer to the data shown in Table 14.3 on age, miles driven, and price for a sample of 11 Nissan Zs.

a. Find the predicted price of a randomly selected Nissan Z that is 5 years old and has been driven 52,000 miles.

b. Obtain a 95% prediction interval for the price of a randomly selected Nissan Z that is 5 years old and has been driven 52,000 miles.

SOLUTION The necessary information can be found in the last two lines of Printout 14.10.

a. As we know, the predicted price is the same as the point estimate for the mean price, which we found in part (a) of Example 14.20. Thus, based on the sample data, the predicted price of a randomly selected Nissan Z that is 5 years old and has been driven 52,000 miles is 92.80, or $9280.

b. The required prediction interval is the final item in the last line of Printout 14.10, the item headed 95% P.I. We can be 95% certain that the price of a randomly selected Nissan Z that is 5 years old and has been driven 52,000 miles will be somewhere between $7153 and $11,407. ∎

Using the Computer (Optional)

Using Minitab's **Regress** command and **Predict** subcommand, we can obtain residuals, predicted values, confidence intervals, and prediction intervals. The method for doing this is explained in Example 14.22.

Example 14.22   The Regress Command and Predict Subcommand

Table 14.3 on page 835 gives data on age, miles driven, and price for a sample of 11 Nissan Zs. Explain how to obtain the Minitab output in Printout 14.10 on page 836 that in addition to the usual regression output includes the table giving fits, residuals, and so on, and provides the information in the last two lines about Nissan Zs that are 5 years old and have been driven 52,000 miles.

SOLUTION   We mentioned earlier that there are three forms for the output of the Regress command. The session command BRIEF followed by an integer between 1 and 3 controls the amount of output for all subsequent Regress commands; the larger the integer, the more extensive the output. To obtain a table giving fits, residuals, and so forth, we type BRIEF 3 at the MTB > prompt before issuing the Regress command. BRIEF 3 remains in effect until we issue another BRIEF command or exit or restart Minitab.

The method for obtaining the last two lines of Printout 14.10 has nothing to do with the setting for BRIEF; rather it relies on the way the Regress command is implemented, specifically, on the use of the Predict subcommand.

To begin, we store the age, miles, and price data from Table 14.3 in columns named AGE, MILES, and PRICE, respectively. Then we do the following.

Session commands: We type the command **REGRESS** followed by the storage location of the data for the response variable, the number of predictors, and the storage locations of the data for the predictor variables; that is, we type REGRESS 'PRICE' on 2 predictors 'AGE' and 'MILES'; and press ⌈Enter⌉. Then we type the subcommand **PREDICT** followed by the values of the predictor variables for which we want price estimation and prediction; that is, we type PREDICT 5 52 and press ⌈Enter⌉.

(or)

Menu commands: We choose **Stat ▸ Regression ▸ Regression...**, specify PRICE for **Response**, and specify AGE and MILES in the **Predictors** text box. Next, to indicate that we want price estimation and prediction for 5-year-old Nissan Zs that have been driven 52,000 miles, we select **Options...** and type 5 52 in the text box headed **Prediction intervals for new observations**. Finally, we select **OK** twice.   ■

By the way, the plots necessary for performing a residual analysis can be easily obtained from Minitab. First we use the Regress command to store the residuals and predicted *y*-values (fits) in columns, and then we apply the Plot and NScores commands as required.

EXERCISES 14.5

14.57 A household-appliance manufacturer wants to analyze the relationship between total sales and the company's three primary means of advertising. The first three columns of the following table provide the expenditures on advertising, by type, for each of 10 randomly selected sales periods. The fourth column contains total sales. All data are in millions of dollars.

Television x_1	Magazines x_2	Radio x_3	Sales y
8.3	4.4	6.1	361.1
6.3	4.2	4.9	344.0
9.9	5.9	6.3	377.9
9.4	3.3	6.1	371.5
10.4	2.7	5.2	365.4
9.0	3.5	5.1	364.5
9.2	4.1	6.0	372.9
10.6	4.8	6.4	379.4
9.3	4.2	5.5	362.6
10.5	6.0	5.9	387.5

Explain what it would mean for the assumptions for multiple regression inferences to be satisfied with television, magazine, and radio advertising expenditures as predictor variables for sales.

14.58 Ten Corvettes between 1 and 6 years old were randomly selected from the classified ads of *The Arizona Republic*. The following data were obtained on age, miles driven, and price.

Age (yrs) x_1	Miles (thous) x_2	Price ($100s) y
6	36	125
6	36	115
6	36	130
2	22	260
2	5	219
5	31	150
4	22	190
5	39	163
1	9	260
4	27	160

Explain what it would mean for the assumptions for multiple regression inferences to be satisfied with age and miles driven as predictor variables for price.

14.59 Refer to Exercise 14.57. We used Minitab to perform a regression analysis on the data with television, magazine, and radio advertising expenditures as predictor variables for sales. Printout 14.11 at the top of the next page displays the resulting computer output. The last two lines of the output give information about sales when the amounts spent on television, magazine, and radio advertising are $9.5 million, $4.3 million, and $5.2 million, respectively. Presuming the variables under consideration satisfy the assumptions for multiple regression inferences, use Printout 14.11 to solve the following problems.
a. Find and interpret the sample regression equation.
b. Obtain and interpret the standard error of the estimate, s_e.
c. At the 5% significance level, do the data provide sufficient evidence to conclude that taken together, television, magazine, and radio advertising expenditures are useful for predicting sales?
d. At the 5% significance level, do the data provide sufficient evidence to conclude that television advertising expenditure is useful for predicting sales? Be precise in your conclusion.
e. Repeat part (d) for radio advertising expenditure.
f. Obtain a point estimate for mean sales when the amounts spent on television, magazine, and radio advertising are $9.5 million, $4.3 million, and $5.2 million.
g. Find a 95% confidence interval for mean sales when the amounts spent on television, magazine, and radio advertising are $9.5 million, $4.3 million, and $5.2 million.
h. Determine the predicted sales if the amounts spent on television, magazine, and radio advertising are $9.5 million, $4.3 million, and $5.2 million.
i. Determine a 95% prediction interval for sales if the amounts spent on television, magazine, and radio advertising are $9.5 million, $4.3 million, and $5.2 million.

14.60 Refer to Exercise 14.58. We used Minitab to perform a multiple regression analysis on the data, with age and miles driven as predictor variables for price. Printout 14.12 on the next page shows the computer output obtained. The last two lines of the output provide information about price for Corvettes that are 4 years old and have been driven 28,000 miles. Presuming that the variables under consideration satisfy the assumptions for multiple regression inferences, employ Printout 14.12 to solve the following problems.
a. Find and interpret the sample regression equation.
b. Obtain and interpret the standard error of the estimate, s_e.
c. At the 5% significance level, do the data provide sufficient evidence to conclude that taken together, age and miles driven are useful for predicting price?

PRINTOUT 14.11 Minitab output for Exercise 14.59

```
The regression equation is
SALES = 266 + 6.73 TV + 3.26 MAG + 4.51 RADIO

Predictor      Coef      Stdev    t-ratio        p
Constant     266.23      16.34      16.29    0.000
TV            6.727      1.344       5.01    0.002
MAG           3.257      1.642       1.98    0.095
RADIO         4.507      3.703       1.22    0.269

s = 4.418      R-sq = 91.1%     R-sq(adj) = 86.6%

Analysis of Variance

SOURCE       DF         SS         MS        F        p
Regression    3     1194.53     398.18    20.40    0.002
Error         6      117.11      19.52
Total         9     1311.64


    Fit  Stdev.Fit       95% C.I.          95% P.I.
 367.58       2.59   ( 361.24, 373.92)  ( 355.05, 380.12)
```

PRINTOUT 14.12 Minitab output for Exercise 14.60

```
The regression equation is
PRICE = 287 - 37.4 AGE + 1.64 MILES

Predictor      Coef      Stdev    t-ratio        p
Constant     287.362     9.943      28.90    0.000
AGE          -37.375     5.174      -7.22    0.000
MILES         1.6378     0.8116      2.02    0.083

s = 12.11      R-sq = 96.0%     R-sq(adj) = 94.9%

Analysis of Variance

SOURCE       DF         SS         MS        F        p
Regression    2       24655      12328    84.07    0.000
Error         7        1026        147
Total         9       25682


    Fit  Stdev.Fit       95% C.I.          95% P.I.
 183.72       4.26   ( 173.65, 193.79)  ( 153.36, 214.08)
```

d. At the 5% significance level, do the data provide sufficient evidence to conclude that age is useful for predicting price? Be precise in your conclusion.

e. Repeat part (d) for miles driven. What if the hypothesis test is performed at the 10% significance level?

f. Obtain a point estimate for the mean price of all Corvettes that are 4 years old and have been driven 28,000 miles.

g. Obtain a 95% confidence interval for the mean price of all Corvettes that are 4 years old and have been driven 28,000 miles.

h. Find the predicted price of a randomly selected Corvette that is 4 years old and has been driven 28,000 miles.

i. Find a 95% prediction interval for the price of a randomly selected Corvette that is 4 years old and has been driven 28,000 miles.

14.61 Graduation rates and what influences them have become a concern in U.S. colleges and universities. *U.S. News and World Report*'s "1992 College Guide" provides data on graduation rates for colleges and universities as a function of the percentage of freshmen in the top 10% of their high-school class, total spending per student, and student-to-

faculty ratio. (Here *graduation rate* refers to the percentage of entering freshmen, attending full-time, that graduate within 5 years.) Printout 14.13 shows the results of a multiple regression analysis for a sample of 44 schools. In the last two lines of the printout, you will find information about graduation rate at schools where 70% of the freshmen were in the top 10% of their high-school class, the total spending per student is $15,500, and the student-to-faculty ratio is 18 to 1. Solve the following problems.

a. At the 5% significance level, do the data provide sufficient evidence to conclude that taken together, percentage of freshmen in the top 10% of their high-school class, total spending per student, and student-to-faculty ratio are useful for predicting graduation rate?

b. At the 5% significance level, do the data provide sufficient evidence to conclude that percentage of freshmen in the top 10% of their high-school class is useful for predicting graduation rate? Be precise in your conclusion.

c. Repeat part (b) for each of the variables total spending per student and student-to-faculty ratio.

d. Do you think another regression analysis is called for? If so, which predictor variable(s) might you include? Explain your answers.

PRINTOUT 14.13   Minitab output for Exercise 14.61

```
The regression equation is
GRAD = 52.4 + 0.210 TOP10%HS +0.000220 EXPENDIT - 0.740 STU:FAC

Predictor        Coef        Stdev     t-ratio        p
Constant        52.36        15.98        3.28      0.002
TOP10%HS        0.2101       0.1134       1.85      0.071
EXPENDIT      0.0002197    0.0004401      0.50      0.620
STU:FAC        -0.7404       0.7480      -0.99      0.328

s = 13.34      R-sq = 16.6%     R-sq(adj) = 10.4%

Analysis of Variance

SOURCE        DF        SS          MS         F        p
Regression     3      1421.6       473.9      2.66     0.061
Error         40      7121.2       178.0
Total         43      8542.8


    Fit   Stdev.Fit        95% C.I.          95% P.I.
   57.14       4.16    ( 48.73,  65.55)   ( 28.89,  85.40)
```

e. Determine a 95% confidence interval for the mean graduation rate of all schools where 70% of the freshmen were in the top 10% of their high-school class, the total spending per student is $15,500, and the student-to-faculty ratio is 18 to 1.

f. Determine a 95% prediction interval for the graduation rate at a randomly selected school where 70% of the freshmen were in the top 10% of their high-school class, the total spending per student is $15,500, and the student-to-faculty ratio is 18 to 1.

g. In solving parts (a)–(f), what assumptions are you making?

14.62   Hanna Properties specializes in custom-home resales in the Equestrian Estates. Thirty-three properties were randomly selected and data were obtained on square footage, number of bedrooms, number of bathrooms, number of days on the market, and selling price (in thousands of dollars). Then a regression analysis was performed for selling price in terms of the other four variables. Printout 14.14 shows the resulting output. The last two lines of the output contain information on selling price for homes in the Equestrian Estates that have 3200 sq ft, 4 bedrooms, and 3 bathrooms, and that remain on the market for 60 days. Solve the following problems.

a. Do the data provide sufficient evidence to conclude that taken together, square footage, number of bedrooms, number of bathrooms, and number of days on the market are useful for predicting selling price? Perform the required hypothesis test at the 1% significance level.

b. At the 1% significance level, do the data provide sufficient evidence to conclude that square footage is useful for predicting selling price? Be precise in your conclusion.

c. Repeat part (b) for the variable, number of bedrooms. Explain why number of bedrooms might not be useful for predicting selling price of homes in the Equestrian Estates.

d. Suppose you decide to run another regression analysis on the data, this time with fewer predictor variables. Which predictor variables would you include? Why?

e. Obtain a 95% confidence interval for the mean selling price of all homes in the Equestrian Estates that have 3200 sq ft, 4 bedrooms, and 3 bathrooms, and that remain on the market for 60 days.

f. Determine a 95% prediction interval for the selling price of a randomly selected home in the Equestrian Estates that has 3200 sq ft, 4 bedrooms, and 3 bathrooms, and that remains on the market for 60 days.

g. In solving parts (a)–(f), what assumptions are you making?

PRINTOUT 14.14   Minitab output for Exercise 14.62

The regression equation is
SELL$ = - 212 + 0.0754 SQFT + 20.9 BEDROOMS + 62.1 BATHS - 0.0899 DAYS

Predictor	Coef	Stdev	t-ratio	p
Constant	-211.95	73.05	-2.90	0.007
SQFT	0.07542	0.02345	3.22	0.003
BEDROOMS	20.94	19.99	1.05	0.304
BATHS	62.11	21.62	2.87	0.008
DAYS	-0.08986	0.07624	-1.18	0.248

s = 39.48   R-sq = 84.0%   R-sq(adj) = 81.7%

Analysis of Variance

SOURCE	DF	SS	MS	F	p
Regression	4	229575	57394	36.83	0.000
Error	28	43633	1558		
Total	32	273209			

Fit	Stdev.Fit	95% C.I.	95% P.I.
294.12	8.28	(277.15, 311.08)	(211.48, 376.76)

14.63 (Computer exercise) Refer to Exercise 14.57. Use Minitab or some other statistical software to
a. obtain computer output similar to that in Printout 14.11.
b. perform a residual analysis.
c. Does your analysis in part (b) reveal any violations of the assumptions for multiple regression inferences? Explain your answer.

14.64 (Computer exercise) Refer to Exercise 14.58. Use Minitab or some other statistical software to
a. obtain computer output similar to that in Printout 14.12.
b. perform a residual analysis.
c. Does your analysis in part (b) reveal any violations of the assumptions for multiple regression inferences? Explain your answer.

Individual confidence intervals for the β parameters. A $(1 - \alpha)$-level confidence interval for β_j has endpoints

$$(2) \qquad\qquad b_j \pm t_{\alpha/2} \cdot s_{b_j},$$

where $t_{\alpha/2}$ is computed for the t-curve with df $= n - (k + 1)$ and s_{b_j} is the estimated standard deviation of the random variable b_j. Here n denotes the sample size and k the number of predictor variables. We will apply this confidence-interval formula in Exercises 14.65 and 14.66.

14.65 Refer to Exercise 14.57 and Printout 14.11.
a. Use (2) to find a 95% confidence interval for the coefficient β_1 of the predictor variable x_1 (television advertising expenditure).
b. Interpret your result from part (a) in words.
c. Repeat part (a) for the coefficient β_3 of the predictor variable x_3 (radio advertising expenditure).

14.66 Refer to Exercise 14.58 and Printout 14.12.
a. Use (2) to determine a 95% confidence interval for the coefficient β_1 of the predictor variable x_1 (age).
b. Interpret your result from part (a) in words.
c. Repeat part (a) for the coefficient β_2 of the predictor variable x_2 (miles driven).

14.67 Curvilinear regression. For this exercise, refer to Exercise 13.99 on page 784. There we used Minitab to perform a curvilinear regression analysis for the price of Nissan Zs with age and age^2 as predictor variables. Printout 14.15 shows the resulting output, with additional information about price of 9-year-old Nissan Zs displayed in the last two lines.
a. Explain what it would mean for the assumptions for multiple regression inferences to be met.
b. Obtain the standard error of the estimate, s_e.

PRINTOUT 14.15   Minitab output for Exercise 14.67

```
The regression equation is
PRICE = 209 - 30.8 AGE + 1.33 AGESQ

Predictor       Coef        Stdev     t-ratio        p
Constant      209.44        11.48       18.24    0.000
AGE          -30.776         4.056      -7.59    0.000
AGESQ          1.3297        0.3219      4.13    0.000

s = 12.81      R-sq = 90.3%      R-sq(adj) = 89.6%

Analysis of Variance

SOURCE        DF         SS          MS        F         p
Regression     2      42895       21448    130.70    0.000
Error         28       4595         164
Total         30      47490


    Fit   Stdev.Fit       95% C.I.         95% P.I.
  40.16        3.97   ( 32.02,  48.30)  ( 12.68,  67.64)
```

c. At the 1% significance level, do the data provide sufficient evidence to conclude that taken together, age and age^2 are useful for predicting price?

d. At the 1% significance level, do the data provide sufficient evidence to conclude that age is useful for predicting price?

e. Repeat part (d) for the predictor variable age^2.

f. Using age and age^2 as the predictor variables, determine a point estimate and a 95% confidence interval for the mean price of all 9-year-old Nissan Zs.

g. Using age and age^2 as the predictor variables, determine the predicted price and a 95% prediction interval for a randomly selected 9-year-old Nissan Z.

· ·   **14.68 (Computer exercise)**   Refer to Exercise 14.67. Use Minitab or some other statistical software to

a. obtain computer output similar to that in Printout 14.15 shown on the previous page.

b. perform a residual analysis.

c. Does your analysis in part (b) reveal any violations of the assumptions for multiple regression inferences? Explain your answer.

d. Remove the outlier and repeat parts (a)–(c).

e. Repeat parts (c)–(g) of Exercise 14.67 using the computer output that you obtained in part (d) of this exercise. Compare your results.

14.6   TESTING FOR NORMALITY (OPTIONAL)

·   ·   ·   ·   ·

As we know, several descriptive methods are available for assessing normality of a population from sample data. One of the most commonly used methods is the normal probability plot, a plot of the normal scores against the sample data.

If the sample is from a normally distributed population, then the normal probability plot should be roughly linear. So we can assess normality as follows: If the normal probability plot is roughly linear, then we accept as reasonable that the population is normally distributed; if the normal probability plot is not roughly linear (e.g., if it shows curvature), then we conclude that the population is probably not normally distributed.

This visual assessment of normality is subjective because what constitutes "roughly linear" is a matter of opinion. To overcome this difficulty, we can perform a hypothesis test for normality based on the linear correlation coefficient: If the population under consideration is normally distributed, then the correlation between the sample data and its normal scores should be near 1 (because the normal probability plot should be roughly linear).[†]

Thus to perform a hypothesis test for normality, we compute the linear correlation coefficient between the sample data and its normal scores. If the correlation is too much smaller than 1, then we reject the null hypothesis that the population is normally distributed in favor of the alternative hypothesis that the population is not normally distributed. Of course, we need a table of critical values to decide what is "too much smaller than 1." This is provided by Table XIV in Appendix A.

We will use the letter w to denote a generic normal score and, as usual, the letter x to denote a generic data value. For this special context, we will use R_p instead of r to denote the linear correlation coefficient. Thus, in view of Formula 13.3 on page 771, the correlation between the sample data and its normal scores can be written symbolically as

$$R_p = \frac{S_{xw}}{\sqrt{S_{xx}S_{ww}}},$$

where $S_{xw} = \Sigma xw - (\Sigma x)(\Sigma w)/n$, $S_{xx} = \Sigma x^2 - (\Sigma x)^2/n$, and $S_{ww} = \Sigma w^2 - (\Sigma w)^2/n$.

[†] Since large normal scores are associated with large data values and vice versa, the correlation between the sample data and its normal scores cannot be negative.

However, because the sum of the normal scores for a data set always equals 0, we can simplify the preceding displayed equation to

$$R_p = \frac{\Sigma xw}{\sqrt{S_{xx}\,\Sigma w^2}}$$

and use this as our test statistic for a correlation test for normality. In a correlation test for normality, the null hypothesis is that the population is normally distributed, and the alternative hypothesis is that the population is not normally distributed.

PROCEDURE 14.6

THE CORRELATION TEST FOR NORMALITY WITH NULL HYPOTHESIS
H_0: **THE POPULATION IS NORMALLY DISTRIBUTED**

Step 1 State the null and alternative hypotheses.

Step 2 Decide on the significance level, α.

Step 3 The critical value is R_p^*. Use Table XIV to find the critical value.

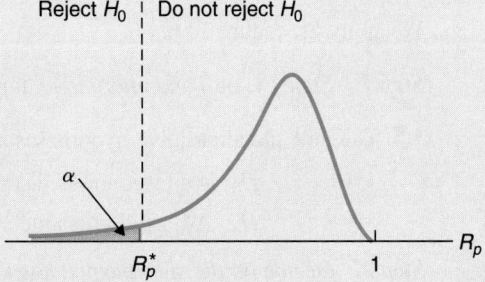

Step 4 Compute the value of the test statistic

$$R_p = \frac{\Sigma xw}{\sqrt{S_{xx}\,\Sigma w^2}},$$

where x and w denote, respectively, data values and normal scores.

Step 5 If the value of the test statistic falls in the rejection region, then reject H_0; otherwise, do not reject H_0.

Step 6 State the conclusion in words.

In addition to the correlation test for normality (Procedure 14.6), several other tests for normality exist. However, the correlation test is one of the most powerful.

In Example 6.25 on page 381, we considered data on adjusted gross incomes for a sample of 12 federal individual income tax returns. We obtained the normal scores for the data (Table 6.5) and drew a normal probability plot (Fig. 6.35). Because the normal probability plot shows significant curvature, we concluded that adjusted gross incomes are probably not normally distributed. This is a subjective conclusion based on a graph. Now we will employ Procedure 14.6 so that we can make an objective conclusion.

Example 14.23   Illustrates Procedure 14.6

The Internal Revenue Service publishes data on federal individual income tax returns in *Statistics of Income, Individual Income Tax Returns.* A random sample of 12 returns yielded the adjusted gross incomes, in thousands of dollars, shown in Table 14.4. At the 5% significance level, do the data provide sufficient evidence to conclude that adjusted gross incomes are not normally distributed?

TABLE 14.4
Adjusted gross incomes ($1000s)

9.7	93.1	33.0	21.2
81.4	51.1	43.5	10.6
12.8	7.8	18.1	12.7

SOLUTION   We apply Procedure 14.6.

Step 1   *State the null and alternative hypotheses.*

The null and alternative hypotheses are

H_0: Adjusted gross incomes are normally distributed.

H_a: Adjusted gross incomes are not normally distributed.

Step 2   *Decide on the significance level, α.*

We are to perform the hypothesis test at the 5% significance level; so $\alpha = 0.05$.

Step 3   *The critical value is R_p^*.*

We have $\alpha = 0.05$ and $n = 12$. Consulting Table XIV, we find that the critical value is $R_p^* = 0.927$, as seen in Fig. 14.14.

FIGURE 14.14

Criterion for deciding whether or not to reject the null hypothesis

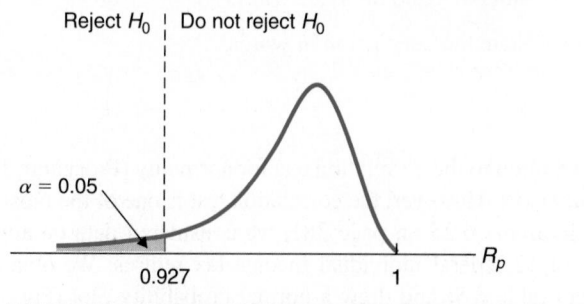

Step 4 *Compute the value of the test statistic*

$$R_p = \frac{\Sigma x w}{\sqrt{S_{xx} \Sigma w^2}}.$$

To compute the value of the test statistic, we need a table for the data (x-values), the normal scores (w-values), xw, x^2, and w^2. The normal scores for the adjusted gross incomes in Table 14.4 are obtained by consulting Table III in Appendix A. From the data and normal scores, we obtain Table 14.5.

TABLE 14.5
Table for computing R_p

Adjusted gross income x	Normal score w	xw	x^2	w^2
7.8	−1.64	−12.792	60.84	2.6896
9.7	−1.11	−10.767	94.09	1.2321
10.6	−0.79	−8.374	112.36	0.6241
12.7	−0.53	−6.731	161.29	0.2809
12.8	−0.31	−3.968	163.84	0.0961
18.1	−0.10	−1.810	327.61	0.0100
21.2	0.10	2.120	449.44	0.0100
33.0	0.31	10.230	1,089.00	0.0961
43.5	0.53	23.055	1,892.25	0.2809
51.1	0.79	40.369	2,611.21	0.6241
81.4	1.11	90.354	6,625.96	1.2321
93.1	1.64	152.684	8,667.61	2.6896
395.0	0.00	274.370	22,255.50	9.8656

Referring to Table 14.5, we find that

$$R_p = \frac{\Sigma x w}{\sqrt{S_{xx} \Sigma w^2}} = \frac{\Sigma x w}{\sqrt{\left[\Sigma x^2 - (\Sigma x)^2/n\right]\left[\Sigma w^2\right]}}$$

$$= \frac{274.370}{\sqrt{\left[22{,}255.50 - (395.0)^2/12\right] \cdot 9.8656}} = 0.908.$$

Step 5 *If the value of the test statistic falls in the rejection region, reject H_0; otherwise, do not reject H_0.*

From Step 4 the value of the test statistic is $R_p = 0.908$, which, as we see from Fig. 14.14, falls in the rejection region. Hence we reject H_0.

Step 6 *State the conclusion in words.*

The test results are statistically significant at the 5% level; that is, at the 5% significance level, the data provide sufficient evidence to conclude that adjusted gross incomes are not normally distributed.

The P-value approach to hypothesis testing can also be used to carry out the hypothesis test in Example 14.23 and to assess more precisely the evidence against the null hypothesis. From Step 4 we see that the value of the test statistic is $R_p = 0.908$. Recalling that $n = 12$, we find from Table XIV that $0.01 < P < 0.05$. In particular, because the P-value is less than the specified significance level of 0.05, we can reject H_0. Furthermore, by referring to Table 9.12 on page 530, we see that the data provide strong evidence against the null hypothesis of normality.

Using the Computer (Optional)

In Example 14.23 we went through the details of applying Procedure 14.6 so you could see exactly how a correlation test for normality works. But generally, correlation tests for normality are carried out by computer.

To employ Minitab to perform a correlation test for normality, we use two commands discussed earlier: NScores and Correlation. First we apply the NScores command to obtain the normal scores for the data; next we apply the Correlation command to obtain the correlation between the data and the normal scores; and then we compare that correlation to the appropriate critical value obtained from Table XIV. Exercises 14.77 and 14.78 ask you to carry out a correlation test for normality by computer.

EXERCISES 14.6

14.69 If you examine Procedure 14.6, you will notice that a correlation test for normality is always left-tailed. Explain in words why this is so.

14.70 Suppose you perform a correlation test for normality at the 1% significance level. Further suppose you reject the null hypothesis that the population is normally distributed. Can you be confident in stating that the population from which the sample was drawn is not normally distributed? Explain your answer.

In Exercises 14.71–14.76, perform a correlation test for normality using either the classical approach or the P-value approach. (Note: You obtained the normal scores for these data sets in Exercises 6.111–6.116.)

14.71 A sample of final-exam scores in a large introductory statistics course is as follows.

88	67	64	76	86
85	82	39	75	34
90	63	89	90	84
81	96	100	70	96

At the 5% significance level, do the data provide sufficient evidence to conclude that final-exam scores in this introductory statistics class are not normally distributed?

14.72 As reported by the R. R. Bowker Company of New York in *Library Journal,* the mean annual subscription rate to law periodicals was $39.82 in 1987. A random sample of 12 of this year's law periodicals provided the following data on subscription rates.

$30	46	44	47
42	38	62	55
52	48	43	54

Do the data provide sufficient evidence to conclude that this year's subscription rates to law periodicals are not normally distributed? Perform the required hypothesis test at the 5% significance level.

14.73 The U.S. Federal Highway Administration conducts studies on motor vehicle travel by type of vehicle. Results are published annually in *Highway Statistics.* A sample of 15 cars yields the following data on number of miles driven, in thousands, for last year.

10.2	10.3	8.9	12.7	8.3
9.2	13.7	7.7	3.3	10.6
11.8	6.6	8.6	5.7	12.0

Can we conclude from the data that last year's distribution for the number of miles cars were driven is not normal? Use $\alpha = 0.10$.

> **14.74** The Bureau of Labor Statistics publishes information on average annual expenditures by consumers in *Consumer Expenditure Survey*. In 1989 the mean amount spent by consumers on nonalcoholic beverages was $216. A random sample of 12 consumers yielded the following data, in dollars, on last year's expenditures on nonalcoholic beverages.

361	176	184	265
259	281	240	273
259	249	194	258

Can we conclude from the data that last year's distribution for amounts spent by consumers on nonalcoholic beverages is not normal? Perform the required hypothesis test at the 10% level of significance.

> **14.75** The U.S. Energy Information Administration reports figures on residential energy consumption and expenditures in *Residential Energy Consumption Survey: Consumption and Expenditures*. A sample of 18 households using electricity as their primary energy source yields the following data on one year's energy expenditures.

$1376	1452	1235	1480	1185	1327
1059	1400	1227	1102	1168	1070
949	1351	1259	1179	1393	1456

At the 1% significance level, do the data provide sufficient evidence to conclude that for the year in question, energy expenditures for households using electricity as their primary energy source are not normally distributed?

> **14.76** In January 1984 the U.S. Department of Agriculture reported in *Family Economic Review* that a typical U.S. family of four with an intermediate budget would spend about $92 per week for food. A consumer researcher in Kansas suspected the median weekly cost was less in her state. She took a sample of 10 Kansas families of four, each with an intermediate budget, and obtained the following weekly food costs.

$78	104	84	70	96
73	87	85	76	94

Based on these data, can we conclude that in 1984, weekly food costs for Kansas families of four with an intermediate budget were not normally distributed? Use $\alpha = 0.01$.

> **14.77 (Computer exercise)** Use Minitab or some other statistical software to perform the hypothesis test required in Exercise 14.71.

> **14.78 (Computer exercise)** Use Minitab or some other statistical software to perform the hypothesis test required in Exercise 14.72.

Chapter Review

Key Terms

Formulas   In the following formulas,

β_0 = y-intercept of population regression line b_0 = y-intercept of sample regression line
β_1 = slope of population regression line b_1 = slope of sample regression line
s_e = standard error of the estimate $\hat{y}_p = b_0 + b_1 x_p$
SSE = error sum of squares ρ = population linear correlation coefficient
n = sample size r = sample linear correlation coefficient

Note: You may also wish to refer to the formulas given in the Chapter Review in Chapter 13 on pages 786–787.

Population regression equation, *794*

$$y = \beta_0 + \beta_1 x$$

Standard error of the estimate, *798*

$$s_e = \sqrt{\frac{SSE}{n-2}}$$

Test statistic for H_0: $\beta_1 = 0$, *812*

$$t = \frac{b_1}{s_e/\sqrt{S_{xx}}}$$

with df = $n - 2$.

Confidence interval for β_1, *814*

$$b_1 \pm t_{\alpha/2} \cdot \frac{s_e}{\sqrt{S_{xx}}}$$

(df = $n - 2$)

Confidence interval for the mean of the population of y-values corresponding to x_p, *820*

$$\hat{y}_p \pm t_{\alpha/2} \cdot s_e \sqrt{\frac{1}{n} + \frac{(x_p - \Sigma x/n)^2}{S_{xx}}}$$

(df = $n - 2$)

Prediction interval for a population y-value corresponding to x_p, *822*

$$\hat{y}_p \pm t_{\alpha/2} \cdot s_e \sqrt{1 + \frac{1}{n} + \frac{(x_p - \Sigma x/n)^2}{S_{xx}}}$$

(df = $n - 2$)

Test statistic for H_0: $\rho = 0$, *830*

$$t = \frac{r}{\sqrt{\dfrac{1 - r^2}{n - 2}}}$$

with df = $n - 2$.

Population regression equation in multiple regression,* *834*

$$y = \beta_0 + \beta_1 x_1 + \cdots + \beta_k x_k$$

(k = number of predictor variables)

Standard error of the estimate in multiple regression,* *837*

$$s_e = \sqrt{\frac{SSE}{n - (k + 1)}}$$

(k = number of predictor variables)

Test statistic for a correlation test for normality,* *851*

$$R_p = \frac{\Sigma x w}{\sqrt{S_{xx} \Sigma w^2}}$$

(x = data value, w = normal score)

You Should Be Able To

1. use and understand the preceding formulas.
2. state the assumptions for regression inferences.
3. determine the standard error of the estimate.
4. perform a residual analysis to check the assumptions for regression inferences.
5. perform a hypothesis test to decide whether the slope, β_1, of the population regression line is not 0 and hence whether x is useful for predicting y.
6. obtain a confidence interval for β_1.
7. determine a point estimate and a confidence interval for the mean of the population of y-values corresponding to a particular x-value.
8. determine a predicted value and a prediction interval for a population y-value corresponding to a particular x-value.
9. perform a hypothesis test for a population linear correlation coefficient.
10. state the assumptions for multiple regression inferences.*
11. determine the standard error of the estimate from computer output.*
12. use computer output to perform a residual analysis to check the assumptions for multiple regression inferences.*
13. use computer output to decide on the overall utility of a multiple regression.*
14. use computer output to decide on the utility of a particular predictor variable in a multiple regression.*
15. use computer output to obtain confidence and prediction intervals corresponding to particular values of the predictor variables.*
16. perform a correlation test for normality.*
17. use the Minitab commands covered in this chapter.*
18. interpret the output obtained from the application of the Minitab commands discussed in this chapter.*

REVIEW TEST

1. The director of a large mathematics course hires upper-division science students to grade papers. On each grading day, he records the number of papers graded and the total amount of money paid to the graders. The following table provides data for 12 randomly selected days from last semester.

Papers (100s)	Cost ($)
x	y
16	234
16	220
18	258
22	298
19	273
16	227
18	246
15	210
19	265
17	250
15	223
18	251

Discuss what it would mean for the assumptions for regression inferences to be satisfied by the variables number of papers graded and cost.

2. Refer to Problem 1.
   a. Determine the regression equation for the data.
   b. Find and interpret the standard error of the estimate.
   c. Presuming that the assumptions for regression inferences are met, interpret your answer from part (b).

3. Refer to Problems 1 and 2. Perform a residual analysis to decide whether it is reasonable to consider the assumptions for regression inferences met by the variables number of papers graded and cost.

In Problems 4–9, presume that the variables number of papers graded (x) and cost (y) satisfy the assumptions for regression inferences.

4. Refer to Problems 1 and 2.
   a. At the 5% significance level, do the data provide sufficient evidence to conclude that number of papers graded is useful as a predictor of cost?
   b. Determine a 95% confidence interval for the slope, β_1, of the population regression line that relates cost to number of papers graded. Interpret your result.

5. Refer to Problems 1 and 2.
   a. Find a point estimate for the mean cost of grading 1600 papers.
   b. Determine a 95% confidence interval for the mean cost of grading 1600 papers.
   c. Find the predicted cost of grading 1600 papers.
   d. Obtain a 95% prediction interval for the cost of grading 1600 papers.
   e. Explain why the prediction interval in part (d) is wider than the confidence interval in part (b).

*6. **(Computer problem)** Refer to Problem 1. Use Minitab or some other statistical software to
   a. obtain the sample regression equation.
   b. determine the standard error of the estimate.

*7. **(Computer problem)** Use Minitab or some other statistical software to carry out the residual analysis required in Problem 3.

*8. **(Computer problem)** Use Minitab or some other statistical software to
   a. carry out the hypothesis test in Problem 4(a).
   b. obtain the confidence interval in Problem 5(b) and the prediction interval in Problem 5(d).

*9. **(Computer problem)** Printout 14.16 displays the computer output that results from applying Minitab's Regress command to the data displayed in Problem 1. The confidence and prediction intervals are for 1600 papers. Use the output to
   a. determine the sample regression equation.
   b. obtain the standard error of the estimate, s_e.
   c. find the slope of the sample regression line.
   d. determine the estimated standard deviation of the slope of the sample regression line.
   e. obtain the value of the test statistic, t, for a hypothesis test to decide whether the slope of the population regression line is not 0.
   f. determine the P-value for the hypothesis test referred to in part (e).
   g. decide whether number of papers graded is useful for predicting cost. Use $\alpha = 0.05$.
   h. determine a 95% confidence interval for the slope of the population regression line and interpret your result. (*Note:* You will need to use Table IV to find $t_{\alpha/2}$, but everything else that is required to obtain the confidence interval can be found in Printout 14.16.)

PRINTOUT 14.16   Minitab output for Problem 9

```
The regression equation is
COST = 35.8 + 12.1 PAPERS


Predictor        Coef        Stdev     t-ratio        p
Constant        35.80        17.06        2.10    0.062
PAPERS        12.0835       0.9738       12.41    0.000


s = 6.526        R-sq = 93.9%     R-sq(adj) = 93.3%

Analysis of Variance

SOURCE          DF          SS          MS        F        p
Regression       1       6558.3      6558.3    153.97    0.000
Error           10        425.9        42.6
Total           11       6984.3


Unusual Observations
Obs.  PAPERS      COST       Fit Stdev.Fit  Residual   St.Resid
  4     22.0    298.00    301.63      4.84     -3.63      -0.83 X

X denotes an obs. whose X value gives it large influence.


    Fit  Stdev.Fit        95% C.I.          95% P.I.
 229.13       2.34   ( 223.93, 234.34)  ( 213.68, 244.58)
```

i. obtain a point estimate for the mean cost of grading 1600 papers.

j. determine a 95% confidence interval for the mean cost of grading 1600 papers.

k. find the predicted cost of grading 1600 papers.

l. obtain a 95% prediction interval for the cost of grading 1600 papers.

10. Refer to Problem 1. At the 2.5% significance level, do the data provide sufficient evidence to conclude that the variables number of papers graded and cost are positively linearly correlated?

*11. The U.S. Bureau of the Census collects data on income by educational attainment, sex, and age and publishes the results in *Current Population Reports*. From a random sample of 75 males between the ages of 25 and 50, all of whom have at least a ninth-grade education, data were obtained on age, number of years of school completed, and annual income. Minitab was then used to perform a multiple regression analysis for annual income (in thousands of dollars), with age and number of years of school

completed as predictor variables. The resulting computer output is displayed in Printout 14.17 at the top of the next page. In the last two lines of the output you will find information on annual income of males who are 32 years old and have completed exactly 4 years of college (i.e., 16 years of school). Presuming that the assumptions for regression inferences are met, use Printout 14.17 to solve the following problems.

a. Obtain and interpret the sample regression equation.

b. Find and interpret the standard error of the estimate.

c. At the 5% significance level, do the data provide sufficient evidence to conclude that taken together, age and number of years of school completed are useful for predicting annual income for males (the type of males under consideration)?

d. At the 5% significance level, do the data provide sufficient evidence to conclude that age is useful as a predictor of annual income for males? Be precise in your conclusion.

e. Repeat part (d) for the predictor variable, number of years of school completed.

PRINTOUT 14.17   Minitab output for Problems 11 and 12

```
The regression equation is
INCOME = - 40.9 + 0.772 AGE + 3.11 EDUC

Predictor      Coef      Stdev     t-ratio       p
Constant    -40.855      5.511      -7.41     0.000
AGE           0.7719     0.1165      6.62     0.000
EDUC          3.1052     0.2250     13.80     0.000

s = 7.886        R-sq = 76.6%     R-sq(adj) = 75.9%

Analysis of Variance

SOURCE       DF         SS         MS         F         p
Regression    2     14657.2     7328.6    117.84     0.000
Error        72      4477.6       62.2
Total        74     19134.7


    Fit   Stdev.Fit       95% C.I.         95% P.I.
 33.528       1.096   ( 31.342, 35.714)  ( 17.653, 49.403)
```

*12. Refer to Problem 11 and to the computer output in Printout 14.17.
  a. Find a point estimate for the mean annual income of all males who are 32 years old and have completed exactly 4 years of college (i.e., 16 years of school).
  b. Obtain a 95% confidence interval for the mean annual income of all males who are 32 years old and have completed exactly 4 years of college.
  c. Determine the predicted annual income of a randomly selected male who is 32 years old and has completed exactly 4 years of college.
  d. Find a 95% prediction interval for the annual income of a randomly selected male who is 32 years old and has completed exactly 4 years of college.

*13. Refer to Problem 11. Figures 14.15(a), (b), and (c) display, respectively, plots of residuals against age, residuals against education, and residuals against predicted income; Fig. 14.15(d) shows a normal probability plot of the residuals. Do these graphs suggest any violations of the assumptions for multiple regression inferences for the variables under consideration? Explain your answer.

*14. (Computer problem)   Refer to Problem 11. Using your statistical software, explain how you would obtain computer output similar to that shown in Printout 14.17.

*15. Each year manufacturers perform mileage tests on new car models and submit the results to the Environmental Protection Agency (EPA). The EPA then tests the vehicles to determine whether the manufacturers' tests are accurate. In 1992 one company reported that a particular model equipped with a four-speed manual transmission averaged 29 mpg on the highway. Suppose the EPA tested 15 of the cars and obtained the following gas mileages.

27.3	31.2	29.4	31.6	28.6
30.9	29.7	28.5	27.8	27.3
25.9	28.8	28.9	27.8	27.6

At the 5% significance level, do the data provide sufficient evidence to conclude that gas mileages for this model are not normally distributed?

*16. (Computer problem)   Use Minitab or some other statistical software to carry out the correlation test for normality in Problem 15.

FIGURE 14.15   Residual plots and normal probability plot for Problem 13

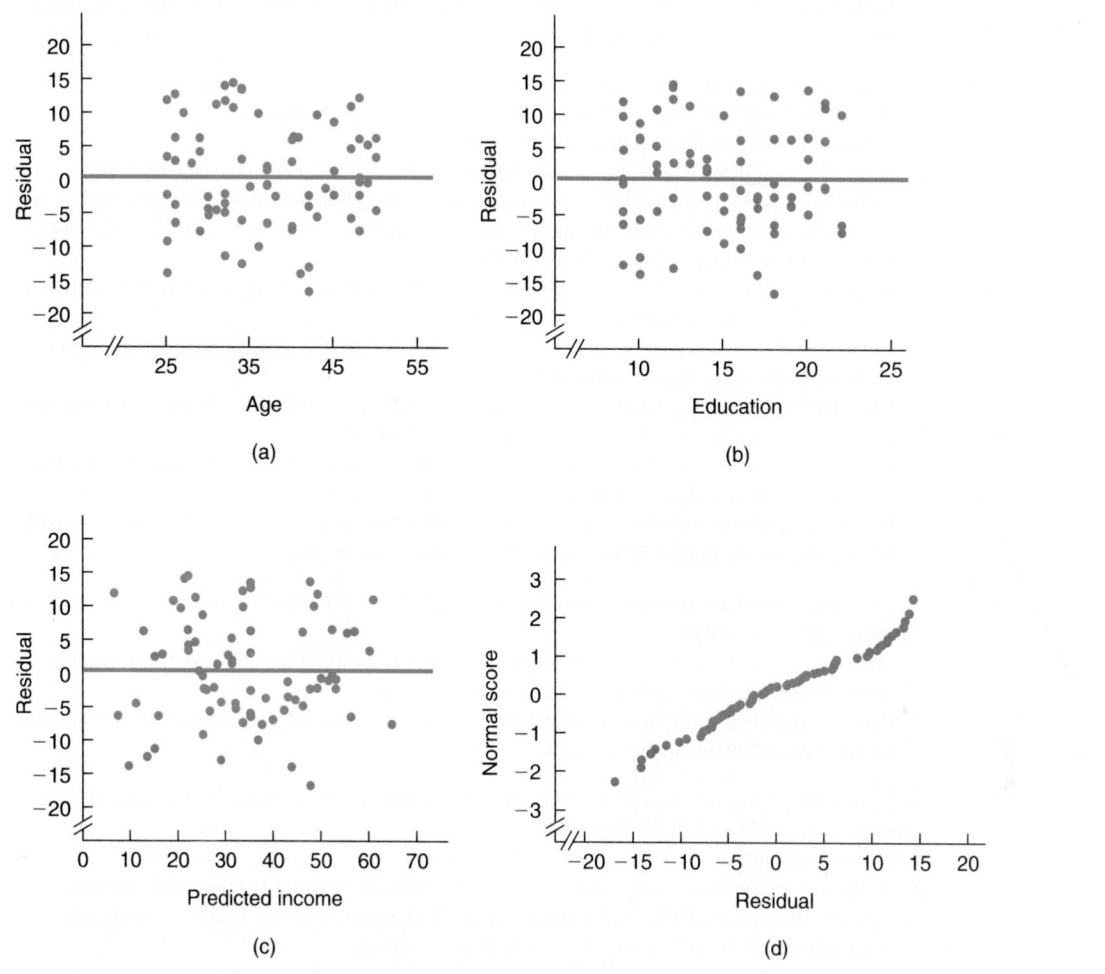

(a)

(b)

(c)

(d)

Using the Focus Database   Consider once again the database, printed in Appendix B, obtained by randomly selecting 500 Arizona State University sophomores. Seven variables are considered for each student: sex, high-school GPA, SAT math score, cumulative GPA, SAT verbal score, age, and total hours. For these database exercises, you should eliminate all cases (students) in which one or more of the four variables, cumulative GPA, high-school GPA, SAT math score, and SAT verbal score, equal 0.

First we will consider a regression analysis on cumulative GPA using high-school GPA as the predictor variable.

a. Determine the sample regression equation.
b. Perform a residual analysis to decide whether it appears reasonable to consider the assumptions for regression inferences satisfied.

Presuming now that the assumptions for regression inferences hold for the variables high-school GPA (x) and cumulative GPA (y), solve the following problems.

c. Obtain and interpret the standard error of the estimate.
d. At the 5% significance level, do the data provide sufficient evidence to conclude that high-school GPA is useful for predicting cumulative GPA of sophomores at Arizona State University?
e. Determine a point estimate for the mean cumulative GPA of all sophomores at Arizona State University who had a high-school GPA of 3.0.
f. Find a 95% confidence interval for the mean cumulative GPA of all sophomores at Arizona State University who had a high-school GPA of 3.0.
g. Determine the predicted cumulative GPA of a randomly selected sophomore at Arizona State University who had a high-school GPA of 3.0.
h. Obtain a 95% prediction interval for the cumulative GPA of a randomly selected sophomore at Arizona State University who had a high-school GPA of 3.0.
i. Repeat parts (a)–(h) with SAT math score instead of high-school GPA as the predictor variable. For the estimation and prediction, use an SAT math score of 500.
j. Repeat parts (a)–(h) with SAT verbal score instead of high-school GPA as the predictor variable. For the estimation and prediction, use an SAT verbal score of 450.

For those who have studied the section on multiple regression (Section 14.5), we will perform a multiple regression analysis.

k. Obtain the sample regression equation for cumulative GPA using the three predictor variables high-school GPA, SAT math score, and SAT verbal score.
l. Perform a residual analysis to decide whether it appears reasonable to consider the assumptions for multiple regression inferences met.

Presuming now that the assumptions for regression inferences hold for the variables under consideration, solve the following problems.

m. Obtain and interpret the standard error of the estimate.
n. At the 5% significance level, do the data provide sufficient evidence to conclude that taken together, high-school GPA, SAT math score, and SAT verbal score are useful for predicting cumulative GPA of sophomores at Arizona State University?
o. At the 5% significance level, do the data provide sufficient evidence to conclude that high-school GPA is useful for predicting cumulative GPA of sophomores at Arizona State University? Compare your answer with the one obtained in part (d). Must the two answers agree? Why or why not?
p. Determine a point estimate for the mean cumulative GPA of all sophomores at Arizona State University who had a high-school GPA of 3.0, an SAT math score of 500, and an SAT verbal score of 450.
q. Find a 95% confidence interval for the mean cumulative GPA of all sophomores at Arizona State University who had a high-school GPA of 3.0, an SAT math score of 500, and an SAT verbal score of 450.
r. Determine the predicted cumulative GPA of a randomly selected sophomore at Arizona State University who had a high-school GPA of 3.0, an SAT math score of 500, and an SAT verbal score of 450.
s. Obtain a 95% prediction interval for the cumulative GPA of a randomly selected sophomore at Arizona State University who had a high-school GPA of 3.0, an SAT math score of 500, and an SAT verbal score of 450.

Case Study   **Fat Consumption and Prostate Cancer**

·   ·   ·   ·

The case study in Chapter 13 presented data on fat consumption and prostate cancer death rate for various nations of the world (see Table 13.13 on page 791). There we used those data to perform some descriptive regression and correlation analyses. Now we will employ those same data to carry out several inferential procedures in regression and correlation.

a. Obtain the sample regression equation using fat consumption as the predictor variable for prostate cancer death rate.
b. Perform a residual analysis to decide whether it appears reasonable to consider the assumptions for regression inferences satisfied by the variables fat consumption (x) and prostate cancer death rate (y).
c. Obtain and interpret the standard error of the estimate.
d. At the 5% significance level, do the data provide sufficient evidence to conclude that fat consumption is useful for predicting prostate cancer death rate for nations of the world?
e. Find a point estimate for the mean prostate cancer death rate for nations with a fat consumption of 140 g per day.
f. Obtain a 95% confidence interval for the mean prostate cancer death rate for nations with a fat consumption of 140 g per day. Interpret your answer.
g. Determine the predicted prostate cancer death rate of a randomly selected nation with a fat consumption of 140 g per day.
h. Find a 95% prediction interval for the prostate cancer death rate of a randomly selected nation with a fat consumption of 140 g per day. Interpret your answer.
i. At the 5% significance level, do the data provide sufficient evidence to conclude that fat consumption and prostate cancer death rate are positively correlated?
j. Use Minitab or some other statistical software to solve parts (a)–(i).

Sir Ronald Fisher

Ronald Fisher was born on February 17, 1890, in London, England, a surviving twin in a family of eight children; his father was a prominent auctioneer. Fisher graduated from Cambridge in 1912, having studied mathematics and physics.

From 1912 to 1919, Fisher worked at an investment house, did farm chores in Canada, and taught high school. In 1919 he took a position as a statistician at Rothamsted Experimental Station in Harpenden, West Hertford, England. His charge was to sort and reassess a 66-year accumulation of data on manurial field trials and weather records.

Fisher's work at Rothamsted during the next 15 years earned him the reputation as the leading statistician of his day and as a top-ranking geneticist. It was there, in 1925, that he published *Statistics for Research Workers,* a book that remained in print for 50 years. Fisher made important contributions to analysis of variance, or ANOVA, exact tests of significance for small samples, and maximum-likelihood solutions. He developed experimental designs to address issues in biological research, such as small samples, variable materials, and fluctuating environments.

Fisher has been described as "slight, bearded, eloquent, reactionary, and quirkish; genial to his disciples and hostile to his dissenters." He was also a prolific writer— over a span of 50 years, he wrote an average of one paper every two months!

In 1933 Fisher became Galton professor of Eugenics at University College in London; in 1943, Balfour professor of genetics at Cambridge. In 1952 he was knighted. Fisher "retired" in 1959, moved to Australia, and spent the last three years of his life working at the Division of Mathematical Statistics of the Commonwealth Scientific and Industrial Research Organization. He died in 1962 in Adelaide, Australia.

Analysis of Variance (ANOVA)

· · · · ·

In Chapter 10 we studied inferential methods for comparing the means of two populations. Now we will study **analysis of variance,** or **ANOVA,** which provides methods for comparing the means of more than two populations. Just as there are several different procedures for comparing the means of two populations, there are several different ANOVA procedures.

First we will examine one-way analysis of variance, the simplest type of ANOVA. One-way analysis of variance is the generalization of the pooled-t procedure (Section 10.2) to more than two populations. We use one-way ANOVA to compare the means of populations that are classified in one way, that is, according to one factor.

Next we will study the Kruskal–Wallis test. This hypothesis-testing procedure is a generalization of the Mann–Whitney test to more than two populations and provides a nonparametric alternative to one-way ANOVA. Then we will consider two-way analysis of variance, which furnishes a method for comparing the means of populations that are classified in two ways, that is, according to two factors.

Chapter Outline

Case Study *Heavy Drinking Among College Students*

Alcohol abuse, a major national problem, is also a concern on college campuses. A recent study revealed that 81.5% of college students drink alcohol; 19% are heavy drinkers. What situations are associated with excessive drinking by college students? Are there situational contexts that differentiate heavy drinkers from light and moderate drinkers? A researcher examined these and other questions in a 1993 issue of the *Journal of Counseling Psychology.* We will discuss her findings in the case study at the end of this chapter.

15.1　THE *F*-DISTRIBUTION

Analysis-of-variance procedures utilize a class of continuous probability distributions called **F-distributions,** named in honor of Sir Ronald Fisher. (See the biography at the beginning of this chapter for more on Fisher.)

Probabilities for a random variable having an *F*-distribution are equal to areas under a curve called an **F-curve.** Recall that a *t*-distribution depends on the number of degrees of freedom, df. An *F*-distribution also depends on the number of degrees of freedom but has two numbers of degrees of freedom instead of one. Figure 15.1 depicts two different *F*-curves; one has df $= (10, 2)$, and the other has df $= (9, 50)$.

FIGURE 15.1

Two different *F*-curves

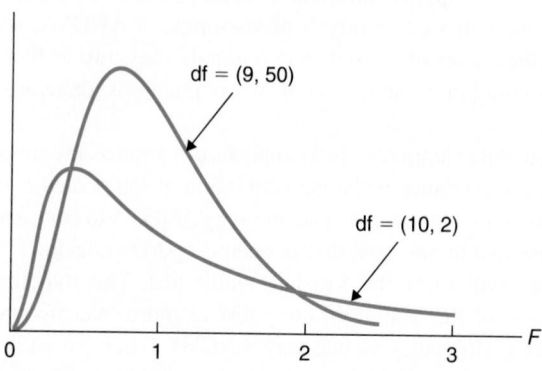

The first number of degrees of freedom for an *F*-curve is called the **degrees of freedom for the numerator** and the second the **degrees of freedom for the denominator.** (The reason for this terminology will become clear in Section 15.2.) Thus for the *F*-curve in Fig. 15.1 with df $= (10, 2)$, we have

$$df = (10, 2)$$

Degrees of freedom ↗　↖ Degrees of freedom
for the numerator　　for the denominator

Some basic properties of *F*-curves are presented in Key Fact 15.1.

KEY FACT 15.1

BASIC PROPERTIES OF *F*-CURVES

Property 1: The total area under an *F*-curve is equal to 1.

Property 2: An *F*-curve starts at 0 on the horizontal axis and extends indefinitely to the right, approaching but never touching the horizontal axis as it does so.

Property 3: An *F*-curve is not symmetric but is skewed to the right.

Using the *F*-Tables

ANOVA tests are right-tailed. Therefore we must be able to determine *F*-values having specified areas to their right. Because *F*-curves have two numbers of degrees of freedom, we need an entire table for each area.[†]

The symbol F_α is used to denote the *F*-value having area α to its right. Tables for $F_{0.01}$ and $F_{0.05}$ are presented, respectively, in Tables X and XI in Appendix A. Let's consider Table XI. The values of $F_{0.05}$ are displayed inside the table and the degrees of freedom for the numerator and denominator on the top and sides, respectively. Example 15.1 shows how to use Table XI.

Example 15.1   Finding the *F*-Value Having a Specified Area to Its Right

For an *F*-curve with df $= (4, 12)$, find $F_{0.05}$; that is, find the *F*-value having area 0.05 to its right, as shown in Fig. 15.2(a).

FIGURE 15.2

Finding the *F*-value having area 0.05 to its right

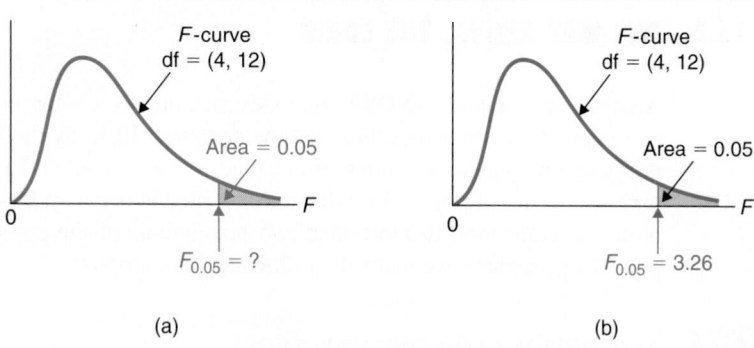

(a)   (b)

SOLUTION   To obtain the *F*-value in question, we use Table XI. In this case the degrees of freedom for the numerator is 4 and the degrees of freedom for the denominator is 12. Thus we first go down the outside columns to the row labeled "12." Then we go across that row until we are under the column headed "4." The number in the body of the table there, 3.26, is the required *F*-value; that is, for an *F*-curve with df $= (4, 12)$, the *F*-value having area 0.05 to its right is 3.26: $F_{0.05} = 3.26$, as seen in Fig. 15.2(b).

 ∎

EXERCISES 15.1

·   **15.1** An *F*-curve has df $= (12, 7)$. What is the number of degrees of freedom for the
a. numerator?   b. denominator?

·   **15.2** An *F*-curve has df $= (8, 19)$. What is the number of degrees of freedom for the
a. denominator?   b. numerator?

[†] Statistical software packages are a preferred alternative for obtaining required *F*-values.

In Exercises 15.3–15.8, use Tables X and XI in Appendix A to find the required F-values. Illustrate your work with graphs similar to Fig. 15.2 on page 867.

•   **15.3**   For an F-curve with df $= (24, 40)$, find the F-value having
a. area 0.05 to its right.   b. area 0.01 to its right.

•   **15.4**   For an F-curve with df $= (12, 5)$, find the F-value having
a. area 0.01 to its right.   b. area 0.05 to its right.

•   **15.5**   For an F-curve with df $= (20, 21)$, find
a. $F_{0.01}$.   b. $F_{0.05}$.

•   **15.6**   For an F-curve with df $= (6, 10)$, find
a. $F_{0.05}$.   b. $F_{0.01}$.

•   **15.7**   An F-curve has df $= (30, 15)$. Determine the F-value having
a. area 0.95 to its left.   b. area 0.99 to its left.

•   **15.8**   An F-curve has df $= (9, 8)$. Determine the F-value having
a. area 0.99 to its left.   b. area 0.95 to its left.

• •   **15.9**   Refer to Table XI in Appendix A. Because of space restrictions, the numbers of degrees of freedom are not consecutive. For instance, the degrees of freedom for the numerator skips from 24 to 30. If you had only Table XI to work with and you needed to find $F_{0.05}$ for df $= (25, 20)$, how would you do it?

15.2   ONE-WAY ANOVA: THE LOGIC

• • • • •

Analysis of variance (ANOVA) provides methods for comparing the means of more than two populations. In this section and the next we will study the simplest kind of ANOVA, **one-way analysis of variance.** It is called *one-way* analysis of variance because it compares the means of populations that are classified in one way. One-way analysis of variance is the generalization to more than two populations of the pooled-*t* procedure. As in the pooled-*t* procedure, we make the following assumptions.

KEY FACT 15.2

ASSUMPTIONS FOR ONE-WAY ANOVA

> 1. *Independent samples:*   The samples taken from the populations under consideration are independent of one another.
> 2. *Normal populations:*   The populations under consideration are normally distributed.
> 3. *Equal standard deviations:*   The standard deviations of the populations under consideration are equal.

One-way ANOVA has the same robustness properties as the pooled-*t* procedure. The independent-samples assumption (Assumption 1) is essential; the samples must be independent or the procedure does not apply. One-way ANOVA is robust to moderate deviations of the normality assumption (Assumption 2). It is also reasonably robust to moderate deviations of the equal-standard-deviations assumption (Assumption 3), provided the sample sizes are roughly equal.

Generally, normal probability plots are effective in detecting gross violations of the normality assumption. The equal-standard-deviations assumption is usually more difficult to check. As a rule of thumb, we consider that assumption satisfied if *the ratio of the largest to the smallest sample standard deviation is less than 2.* For convenience, we'll call this rule of thumb the **rule of 2.**

Additionally, the normality and equal-standard-deviations assumptions can be assessed by performing a residual analysis, similar to what we did in regression. In ANOVA the **residual** of a data value is the difference between the data value and the mean of the sample containing it. If the normality and equal-standard-deviations assumptions are met, then a normal probability plot of (all) the residuals should be roughly linear and a plot of the residuals against the sample means should fall roughly in a horizontal band centered and symmetric about the horizontal axis.

The Logic Behind One-Way ANOVA

The reason for the word *variance* in *analysis of variance* is that the procedure for comparing the means involves analyzing the variation in the sample data. To see how this works, suppose independent random samples are taken from two populations, Populations 1 and 2, having means μ_1 and μ_2. Further suppose the means of the two samples are $\bar{x}_1 = 20$ and $\bar{x}_2 = 25$. Can we reasonably conclude from these statistics that $\mu_1 \neq \mu_2$, that is, that the population means are different? To answer this question, we must consider the variation within the samples.

Suppose, for instance, the samples are as depicted in Table 15.1 and Fig. 15.3.

TABLE 15.1

Samples from Populations 1 and 2

Sample from Population 1	21	37	11	20	8	23
Sample from Population 2	24	31	29	40	9	17

FIGURE 15.3

Dotplots for samples in Table 15.1

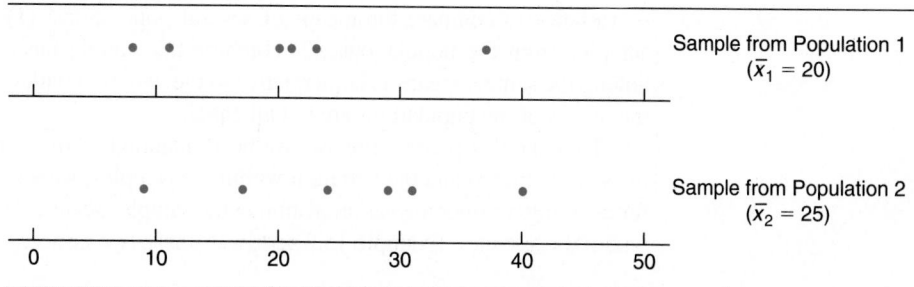

For these two samples, $\bar{x}_1 = 20$ and $\bar{x}_2 = 25$. But we cannot infer that $\mu_1 \neq \mu_2$ because it is not clear whether the difference between the sample means is due to a difference between the population means or due to the variation within the populations. In other words, because the variation between the sample means is not large relative to the variation within the samples, we cannot conclude that $\mu_1 \neq \mu_2$.

On the other hand, suppose the samples are as depicted in Table 15.2 and Fig. 15.4.

TABLE 15.2

Samples from Populations 1 and 2

Sample from Population 1	21	21	20	18	20	20
Sample from Population 2	25	28	25	24	24	24

FIGURE 15.4

Dotplots for samples in Table 15.2

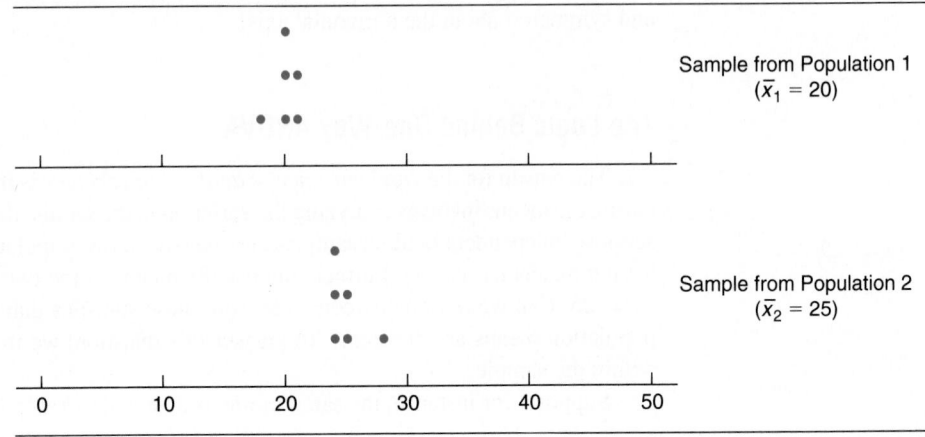

Again, for these two samples, $\bar{x}_1 = 20$ and $\bar{x}_2 = 25$. But this time we *can* infer that $\mu_1 \neq \mu_2$ because it seems clear that the difference between the sample means is due to a difference between the population means and not due to the variation within the populations. In other words, because the variation between the sample means is large relative to the variation within the samples, we can conclude that $\mu_1 \neq \mu_2$.

The preceding two illustrations reveal the basic idea for performing a one-way analysis of variance to compare the means of several populations: (1) take independent random samples from the populations; (2) compute the sample means; and (3) if the variation among the sample means is large relative to the variation within the samples, conclude that the means of the populations are not all equal.

To make this process precise, we need quantitative measures of the variation among the sample means and the variation within the samples; we also need an objective method for deciding whether the variation among the sample means is large relative to the variation within the samples. Example 15.2 addresses these two issues.

Example 15.2   Introduces One-Way ANOVA

The U.S. Energy Information Administration gathers data on residential energy consumption and expenditures and publishes its findings in *Residential Energy Consumption Survey: Consumption and Expenditures*. A researcher wants to know whether there is a difference in mean annual energy consumptions among households in the four regions of the United

States. Let μ_1, μ_2, μ_3, and μ_4 denote last year's mean energy consumptions for households in the Northeast, Midwest, South, and West, respectively. Then the hypotheses to be tested are

H_0: $\mu_1 = \mu_2 = \mu_3 = \mu_4$ (mean energy consumptions are all equal)

H_a: Not all the means are equal.

The basic strategy for carrying out this hypothesis test follows the three steps just mentioned:

1. Take independent random samples of last year's energy consumptions for households in the four regions.

2. Compute the means, $\bar{x}_1$, $\bar{x}_2$, $\bar{x}_3$, and $\bar{x}_4$, of the four samples.

3. Reject the null hypothesis if the variation among the sample means is large relative to the variation within the samples; otherwise, do not reject the null hypothesis.

This process is depicted in Fig. 15.5.

FIGURE 15.5

Process for comparing four population means

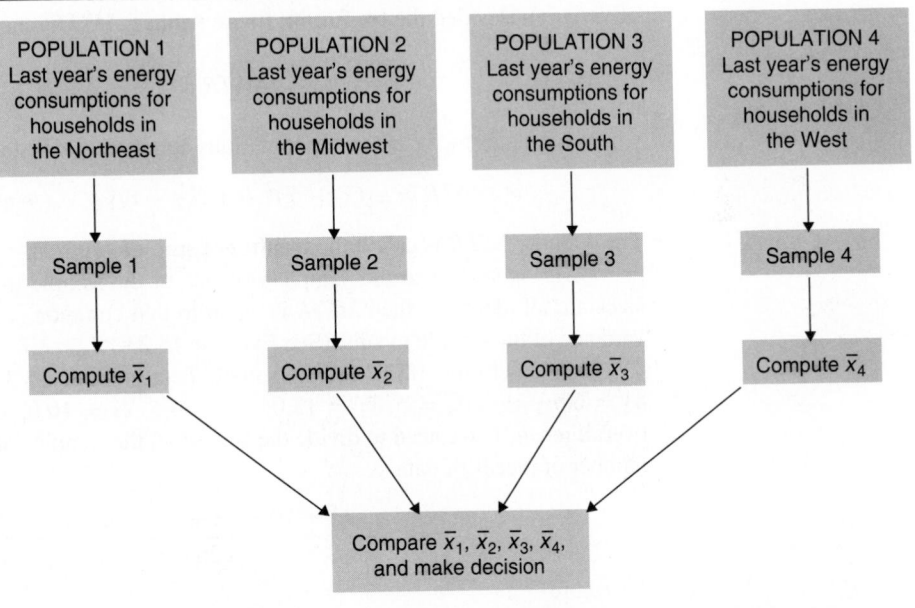

Steps 1 and 2 entail obtaining the sample data and computing the sample means. Suppose the results of those steps are as shown in Table 15.3 at the top of the next page, where the data are displayed to the nearest 10 million BTU.

Step 3 involves comparing the variation among the four sample means, shown at the bottom of Table 15.3, to the variation within the samples. Let's first consider the variation among the sample means.

TABLE 15.3

Samples and their means
of last year's energy
consumptions for households
in the four U.S. regions

Northeast	Midwest	South	West
15	17	11	10
10	12	7	12
13	18	9	8
14	13	13	7
13	15		9
	12		
13.0	14.5	10.0	9.2

← *Means*

In hypothesis tests for two population means, we measure the variation between the two sample means by calculating their difference, $\bar{x}_1 - \bar{x}_2$. When more than two populations are involved, as in this problem, we cannot measure the variation among the sample means simply by taking a difference. However, we can measure that variation by computing the standard deviation or variance of the sample means or, for that matter, by computing any descriptive statistic that measures the variation among the sample means.

In one-way ANOVA, we measure the variation among the sample means by a weighted average of their squared deviations about the mean, $\bar{x}$, of all the sample data. That measure of variation is called the **treatment mean square, MSTR,** and is defined by

$$MSTR = \frac{SSTR}{k-1},$$

where k denotes the number of populations being sampled and

$$SSTR = n_1(\bar{x}_1 - \bar{x})^2 + n_2(\bar{x}_2 - \bar{x})^2 + \cdots + n_k(\bar{x}_k - \bar{x})^2.$$

The quantity **SSTR** is called the **treatment sum of squares.**

MSTR is similar to the sample variance of the sample means. In fact, if the sample sizes are all identical, then MSTR is equal to that common sample size times the sample variance of the sample means. (See Exercise 15.23.)

Let's determine MSTR for the sample data in Table 15.3. We have $k = 4$, $n_1 = 5$, $n_2 = 6$, $n_3 = 4$, $n_4 = 5$, $\bar{x}_1 = 13.0$, $\bar{x}_2 = 14.5$, $\bar{x}_3 = 10.0$, and $\bar{x}_4 = 9.2$. To obtain the overall mean, $\bar{x}$, we need to divide the sum of all the sample data in Table 15.3 by the total number of pieces of data:

$$\bar{x} = \frac{\Sigma x}{n} = \frac{15 + 10 + 13 + \cdots + 7 + 9}{20} = \frac{238}{20} = 11.9.$$

Therefore

$$SSTR = n_1(\bar{x}_1 - \bar{x})^2 + n_2(\bar{x}_2 - \bar{x})^2 + n_3(\bar{x}_3 - \bar{x})^2 + n_4(\bar{x}_4 - \bar{x})^2$$
$$= 5(13.0 - 11.9)^2 + 6(14.5 - 11.9)^2 + 4(10.0 - 11.9)^2 + 5(9.2 - 11.9)^2$$
$$= 97.5,$$

and so

$$MSTR = \frac{SSTR}{k-1} = \frac{97.5}{4-1} = 32.5.$$

This is our measure of variation among the four sample means shown at the bottom of Table 15.3.

Next we must obtain a measure of variation within the samples. This latter measure is the pooled estimate of the common population variance, σ^2. It is called the **error mean square, *MSE*,** and is defined by

$$MSE = \frac{SSE}{n-k},$$

where n denotes the total number of pieces of sample data and

$$SSE = (n_1 - 1)s_1^2 + (n_2 - 1)s_2^2 + \cdots + (n_k - 1)s_k^2.$$

The quantity ***SSE*** is called the **error sum of squares.**[†‡]

For the sample data in Table 15.3, we have $k = 4$, $n_1 = 5$, $n_2 = 6$, $n_3 = 4$, $n_4 = 5$, and $n = 20$. Computing the sample variance for each of the four data sets in Table 15.3, we find that $s_1^2 = 3.5$, $s_2^2 = 6.7$, $s_3^2 = 6.\overline{6}$, and $s_4^2 = 3.7$. Consequently,

$$SSE = (n_1 - 1)s_1^2 + (n_2 - 1)s_2^2 + (n_3 - 1)s_3^2 + (n_4 - 1)s_4^2$$
$$= (5 - 1) \cdot 3.5 + (6 - 1) \cdot 6.7 + (4 - 1) \cdot 6.\overline{6} + (5 - 1) \cdot 3.7 = 82.3,$$

and so

$$MSE = \frac{SSE}{n-k} = \frac{82.3}{20 - 4} = 5.144.$$

This is our measure of variation within the samples.

Finally, we must compare the variation among the sample means, *MSTR,* to the variation within the samples, *MSE*. To accomplish that, we use the statistic $F = MSTR/MSE$, which we refer to as the ***F*-statistic.** Large values of F indicate that the variation among the sample means is large relative to the variation within the samples and hence that the null hypothesis of equal population means should be rejected.

For the energy-consumption data, we have seen that $MSTR = 32.5$ and $MSE = 5.144$. Thus the value of the F-statistic is

$$F = \frac{MSTR}{MSE} = \frac{32.5}{5.144} = 6.32.$$

Is this value of F large enough to conclude that the null hypothesis of equal population means is false? To answer that question, we need to know the probability distribution of the random variable F. We will discuss that in the next section and then return to complete the hypothesis test considered in this example. ∎

[†] The terms **treatment** and **error** arose from the fact that many ANOVA techniques were first developed to analyze agricultural experiments. In any case the treatments refer to the different populations and the errors pertain to the variation within the populations.

[‡] For two populations (i.e., $k = 2$), *MSE* is the pooled variance, defined in Section 10.2. See Exercise 15.20.

EXERCISES 15.2

15.10 State the three assumptions required for one-way ANOVA. How crucial are these assumptions?

15.11 One-way ANOVA is a procedure for comparing the means of several populations. It is the generalization of what procedure for comparing the means of two populations?

15.12 If we define $s = \sqrt{MSE}$, then of which parameter is s an estimate?

15.13 Explain the reason for the word *variance* in the phrase *analysis of variance*.

15.14 The null and alternative hypotheses for a one-way ANOVA test are

$$H_0: \mu_1 = \mu_2 = \cdots = \mu_k$$
$$H_a: \text{ Not all means are equal.}$$

Suppose in reality that the null hypothesis is false. Does this mean that no two of the populations have the same mean? If not, what does it mean?

15.15 In one-way ANOVA, identify the statistic used
a. as a measure of variation among the sample means.
b. as a measure of variation within the samples.
c. to compare the variation among the sample means to the variation within the samples.

15.16 Explain in your own words the logic behind one-way ANOVA.

15.17 What is the significance of the term *one-way* in *one-way ANOVA*?

15.18 Figure 15.6 shows side-by-side boxplots of independent samples from three normally distributed populations having equal standard deviations. Based on these boxplots, would you be inclined to reject the null hypothesis of equal population means? Explain your answer.

15.19 Figure 15.7 shows side-by-side boxplots of independent samples from three normally distributed populations having equal standard deviations. Based on these boxplots, would you be inclined to reject the null hypothesis of equal population means? Explain your answer.

FIGURE 15.6   Side-by-side boxplots for Exercise 15.18

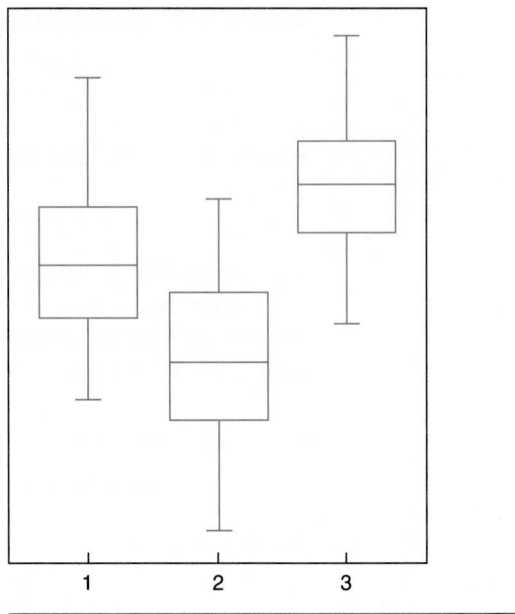

FIGURE 15.7   Side-by-side boxplots for Exercise 15.19

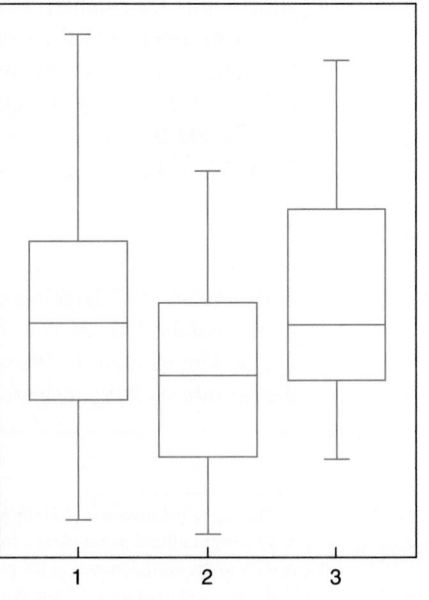

· ·   **15.20**   Show that for two populations (i.e., $k = 2$), $MSE = s_p^2$, where s_p^2 is the pooled variance defined in Section 10.2 on page 586. Conclude that $\sqrt{MSE}$ is the pooled sample standard deviation, s_p.

· ·   **15.21**   Consider two normally distributed populations having means μ_1 and μ_2 and equal standard deviations. Suppose we want to perform a hypothesis test to decide whether the populations have different means, that is, whether $\mu_1 \neq \mu_2$. If independent samples are used, identify two hypothesis-testing procedures that can be employed to carry out the test.

· · ·   **15.22**   Recall that $\bar{x}$ is the mean of all n pieces of sample data.
a. Show that $\bar{x}$ is a weighted average of the k sample means, weighted according to sample size; that is,

$$\bar{x} = \frac{n_1\bar{x}_1 + n_2\bar{x}_2 + \cdots + n_k\bar{x}_k}{n_1 + n_2 + \cdots + n_k}.$$

b. Prove that if the sample sizes are all equal, then $\bar{x}$ is simply the mean of the sample means.

· · ·   **15.23**   Show that if the sample sizes are all equal, say to m, then $MSTR$ equals m times the sample variance of the k sample means. *(Hint:* Use part (b) of Exercise 15.22.)

15.3   ONE-WAY ANOVA: THE PROCEDURE

In this section we will present a step-by-step procedure for performing a one-way ANOVA to compare the means of several populations. To begin, we need to identify the probability distribution of the random variable $F = MSTR/MSE$, introduced at the end of the preceding section.

KEY FACT 15.3

TEST STATISTIC FOR ONE-WAY ANOVA

Suppose independent random samples of sizes $n_1, n_2, \ldots, n_k$ are to be taken from k normally distributed populations with means $\mu_1, \mu_2, \ldots, \mu_k$, respectively. Further suppose the standard deviations of the k populations are equal. Then if $\mu_1 = \mu_2 = \cdots = \mu_k$, the random variable

$$F = \frac{MSTR}{MSE}$$

has the F-distribution with df $= (k - 1, n - k)$, where n denotes the total number of pieces of data.

We have now covered all the elements required to formulate a procedure for performing a one-way analysis of variance. However, it will be helpful to consider two additional concepts before presenting that procedure.

One-Way ANOVA Identity

First we define another sum of squares. That sum of squares provides a measure of total variation among all the sample data. It is called the **total sum of squares, SST,** and is defined by

$$SST = \Sigma(x - \bar{x})^2,$$

where the sum extends over all n pieces of sample data. If we divide SST by $n - 1$, then we get the sample variance of all the data. So *SST* really is a measure of total variation. For the energy-consumption data in Table 15.3 on page 872, $\bar{x} = 11.9$, and therefore

$$SST = \Sigma(x - \bar{x})^2 = (15 - 11.9)^2 + (10 - 11.9)^2 + \cdots + (9 - 11.9)^2$$

$$= 9.61 + 3.61 + \cdots + 8.41 = 179.8.$$

In Section 15.2 we found that for the energy-consumption data, $SSTR = 97.5$ and $SSE = 82.3$. Since $179.8 = 97.5 + 82.3$, we see that $SST = SSTR + SSE$. This equation is always true and is called the **one-way ANOVA identity.**

KEY FACT 15.4   **ONE-WAY ANOVA IDENTITY**

> The total sum of squares equals the treatment sum of squares plus the error sum of squares; that is, $SST = SSTR + SSE$.

The one-way ANOVA identity shows that we can partition the total variation in the data into a component representing variation among the sample means and a component representing variation within the samples. We can picture this partitioning as in Fig. 15.8.

FIGURE 15.8
Partitioning of the total sum of squares into the treatment sum of squares and the error sum of squares

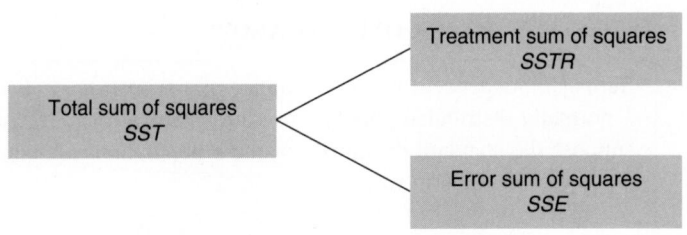

One-Way ANOVA Tables

Next we discuss **one-way ANOVA tables.** These tables are useful for organizing and summarizing the quantities required for performing a one-way analysis of variance. The general format of a one-way ANOVA table is shown in Table 15.4.

TABLE 15.4
ANOVA table format for a one-way analysis of variance

Source	df	SS	MS = SS/df	F-statistic
Treatment	$k - 1$	$SSTR$	$MSTR = \dfrac{SSTR}{k - 1}$	$F = \dfrac{MSTR}{MSE}$
Error	$n - k$	SSE	$MSE = \dfrac{SSE}{n - k}$	
Total	$n - 1$	SST		

For the energy-consumption data in Table 15.3, we have already computed all quantities appearing in the one-way ANOVA table. Table 15.5 displays the one-way ANOVA table for that data.

Source	df	SS	MS = SS/df	F-statistic
Treatment	3	97.5	32.500	6.32
Error	16	82.3	5.144	
Total	19	179.8		

The One-Way ANOVA Procedure

In performing a one-way ANOVA, we need to obtain the three sums of squares, *SST*, *SSTR*, and *SSE*. This can be accomplished using the defining formulas introduced earlier. Generally, however, when calculating by hand from the raw data, shortcut formulas are more accurate and easier to use. Both the defining formulas and their shortcut equivalents are presented in Formula 15.1.

FORMULA 15.1

SUMS OF SQUARES IN ONE-WAY ANOVA

For a one-way ANOVA of k population means, we have the following defining and shortcut formulas for the three sums of squares.

Sum of square	Defining formula	Shortcut formula
Total, *SST*	$\Sigma(x - \bar{x})^2$	$\Sigma x^2 - (\Sigma x)^2/n$
Treatment, *SSTR*	$\Sigma n_j(\bar{x}_j - \bar{x})^2$	$\Sigma(T_j^2/n_j) - (\Sigma x)^2/n$
Error, *SSE*	$\Sigma(n_j - 1)s_j^2$	$SST - SSTR$

Here

n = total number of pieces of data

$\bar{x}$ = mean of all n pieces of data

and, for $j = 1, 2, \ldots, k$,

n_j = size of sample from Population j

$\bar{x}_j$ = mean of sample from Population j

s_j^2 = variance of sample from Population j

T_j = sum of sample data from Population j

Keep the following facts in mind when using Formula 15.1:

- Only two of the three sums of squares need ever be calculated; the remaining one can always be determined from the other two by employing the one-way ANOVA identity, Key Fact 15.4 on page 876.

- Summations involving no subscripted variables are over all n pieces of sample data; those involving subscripts are over the k populations.

- When using the shortcut formulas, it is most efficient to compute the sum of all n pieces of sample data by employing the formula $\Sigma x = T_1 + T_2 + \cdots + T_k$.

We now present a step-by-step method that can be used to perform a one-way analysis of variance. Note that the hypothesis test is always right-tailed since the null hypothesis is rejected only when the test statistic, F, is too large.

PROCEDURE 15.1   **THE ONE-WAY ANOVA TEST FOR k POPULATION MEANS WITH NULL HYPOTHESIS $H_0: \mu_1 = \mu_2 = \cdots = \mu_k$**

ASSUMPTIONS

1. Independent samples
2. Normal populations
3. Equal population standard deviations

Step 1   State the null and alternative hypotheses.

Step 2   Decide on the significance level, α.

Step 3   The critical value is F_α, with df $= (k - 1, n - k)$, where n is the total number of pieces of data.

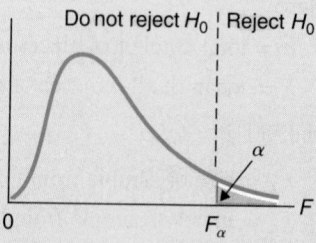

Step 4   Obtain the three sums of squares, *SST*, *SSTR*, and *SSE*.

Step 5 Construct the one-way ANOVA table:

Source	df	SS	MS = SS/df	F-statistic
Treatment	$k-1$	SSTR	$MSTR = \dfrac{SSTR}{k-1}$	$F = \dfrac{MSTR}{MSE}$
Error	$n-k$	SSE	$MSE = \dfrac{SSE}{n-k}$	
Total	$n-1$	SST		

Step 6 If the value of the F-statistic falls in the rejection region, then reject H_0; otherwise, do not reject H_0.

Step 7 State the conclusion in words.

Example 15.3 Illustrates Procedure 15.1

Recall that independent random samples of households in the four U.S. regions yielded the data on last year's energy consumptions shown in Table 15.6. At the 5% significance level, do the data provide sufficient evidence to conclude that there is a difference in last year's mean energy consumptions among households in the four U.S. regions?

TABLE 15.6
Samples of last year's energy consumptions for households in the four U.S. regions

Northeast	Midwest	South	West
15	17	11	10
10	12	7	12
13	18	9	8
14	13	13	7
13	15		9
	12		

SOLUTION First we check the three conditions required for performing a one-way ANOVA. Since the samples are independent, Assumption 1 is satisfied. Normal probability plots (not shown) of the four samples in Table 15.6 reveal no outliers and are roughly linear, thus indicating no gross violations of the normality assumption; so we can consider Assumption 2 satisfied. The sample standard deviations of the four samples in Table 15.6 are, respectively, 1.87, 2.59, 2.58, and 1.92. We see that the ratio of the largest to the smallest standard deviation is $\frac{2.59}{1.87} = 1.39 < 2$. Therefore, by the rule of 2, we can consider Assumption 3 satisfied. A residual analysis further attests to it being reasonable to consider Assumptions 2 and 3 satisfied.

The preceding paragraph suggests that the one-way ANOVA procedure can be used to carry out the hypothesis test. We proceed as follows.

Step 1 *State the null and alternative hypotheses.*

Let μ_1, μ_2, μ_3, and μ_4 denote last year's mean energy consumptions for households in the Northeast, Midwest, South, and West, respectively. Then the null and alternative hypotheses are

$$H_0: \mu_1 = \mu_2 = \mu_3 = \mu_4 \text{ (mean energy consumptions are equal)}$$
$$H_a: \text{Not all the means are equal.}$$

Step 2 *Decide on the significance level, α.*

We are to perform the test at the 5% significance level; thus $\alpha = 0.05$.

Step 3 *The critical value is F_α, with df $= (k - 1, n - k)$, where n is the total number of pieces of data.*

From Step 2, $\alpha = 0.05$. Also, as we see from Table 15.6, the number of populations under consideration is four ($k = 4$) and the total number of pieces of data is 20 ($n = 20$). Hence df $= (k - 1, n - k) = (4 - 1, 20 - 4) = (3, 16)$. Consulting Table XI, we find that the critical value is $F_\alpha = F_{0.05} = 3.24$, as seen in Fig. 15.9.

FIGURE 15.9
Criterion for deciding whether or not to reject the null hypothesis

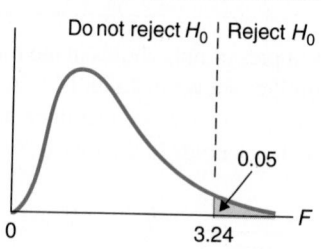

Do not reject H_0 | Reject H_0

0.05

0   3.24 F

Step 4 *Obtain the three sums of squares, SST, SSTR, and SSE.*

Although we obtained these earlier (pages 876, 872, and 873) using the defining formulas, we will compute them again to illustrate the shortcut formulas. Referring to Formula 15.1 on page 877 and Table 15.6 on page 879, we find that

$$k = 4$$

$n_1 = 5$	$n_2 = 6$	$n_3 = 4$	$n_4 = 5$
$T_1 = 65$	$T_2 = 87$	$T_3 = 40$	$T_4 = 46$

and

$$n = 5 + 6 + 4 + 5 = 20$$
$$\Sigma x = 65 + 87 + 40 + 46 = 238.$$

Summing the squares of all the data in Table 15.6 yields

$$\Sigma x^2 = 15^2 + 10^2 + 13^2 + \cdots + 7^2 + 9^2 = 3012.$$

Consequently,

$$SST = \Sigma x^2 - (\Sigma x)^2/n = 3012 - (238)^2/20 = 3012 - 2832.2 = 179.8,$$

$$
\begin{aligned}
SSTR &= \Sigma(T_j^2/n_j) - (\Sigma x)^2/n \\
&= 65^2/5 + 87^2/6 + 40^2/4 + 46^2/5 - (238)^2/20 \\
&= 2929.7 - 2832.2 = 97.5,
\end{aligned}
$$

and

$$SSE = SST - SSTR = 179.8 - 97.5 = 82.3.$$

Step 5   *Construct the one-way ANOVA table.*

The one-way ANOVA table for the energy-consumption data was constructed earlier in Table 15.5 on page 877.

Step 6   *If the value of the F-statistic falls in the rejection region, then reject H_0; otherwise, do not reject H_0.*

As we see from Table 15.5, $F = 6.32$. A glance at Fig. 15.9 shows that this value falls in the rejection region. Thus we reject H_0.

Step 7   *State the conclusion in words.*

The test results are statistically significant at the 5% level; that is, at the 5% significance level, the data provide sufficient evidence to conclude that last year's mean energy consumptions for households in the four U.S. regions are not all equal—at least two of the regions have different mean energy consumptions.   ■

The P-value approach to hypothesis testing can also be used to carry out the hypothesis test in Example 15.3 and to assess more precisely the evidence against the null hypothesis. From Step 6 we know that the value of the test statistic is $F = 6.32$. Consulting Tables X and XI and recalling that df $= (3, 16)$, we find that $P < 0.01$. In particular, because the P-value is less than the significance level of 0.05, we can reject H_0. Furthermore, by referring to Table 9.12 on page 530, we see that the data provide very strong evidence against the null hypothesis of equal mean energy consumptions.

The results of the one-way ANOVA in Example 15.3 show that (at the 5% significance level) last year's mean energy consumptions for households in the four U.S. regions are not all the same. But the analysis does not tell us which means are different, which mean is largest, or, more generally, the relationship among the four means. Such questions are answered using techniques called *multiple comparisons*. Section 15.4 discusses one widely-used method for performing multiple comparisons.

Using the Computer (Optional)

Procedure 15.1 provides a step-by-step method for performing a one-way ANOVA. Alternatively, we can apply Minitab to carry out such a hypothesis test by employing the **AOVOneway** command. Example 15.4 shows how this is done.

Example 15.4   The AOVOneway Command

Use Minitab to perform the hypothesis test in Example 15.3.

SOLUTION   Let μ_1, μ_2, μ_3, and μ_4 denote last year's mean energy consumptions for households in the Northeast, Midwest, South, and West, respectively. We want to perform the hypothesis test

H_0: $\mu_1 = \mu_2 = \mu_3 = \mu_4$ (mean energy consumptions are equal)

H_a: Not all the means are equal.

at the 5% significance level.

To employ Minitab, we first store the four samples in Table 15.6 on page 879 in columns named NRTHEAST, MIDWEST, SOUTH, and WEST. Then we proceed in the following manner.

Session commands: We type **AOVONEWAY** followed by the storage locations of the data; that is, we type AOVONEWAY for 'NRTHEAST' 'MIDWEST' 'SOUTH' 'WEST' and press [Enter]. Printout 15.1 displays this command and the resulting output.

(or)

Menu commands: We choose **Stat ▸ ANOVA ▸ Oneway (Unstacked)...**, specify NRTHEAST, MIDWEST, SOUTH, and WEST for **Responses (in separate columns)**, and then select **OK**. The output obtained is shown in Printout 15.1.

PRINTOUT 15.1
Minitab output for the
AOVOneway command

```
MTB > AOVONEWAY for 'NRTHEAST' 'MIDWEST' 'SOUTH' 'WEST'

ANALYSIS OF VARIANCE
SOURCE      DF        SS        MS        F        p
FACTOR       3      97.50     32.50     6.32    0.005
ERROR       16      82.30      5.14
TOTAL       19     179.80

                                INDIVIDUAL 95 PCT CI'S FOR MEAN
                                BASED ON POOLED STDEV
LEVEL       N       MEAN      STDEV   -------+---------+---------+---------
NRTHEAST    5     13.000      1.871                 (------*-------)
MIDWEST     6     14.500      2.588                      (-----*------)
SOUTH       4     10.000      2.582     (-------*-------)
WEST        5      9.200      1.924   (-------*------)
                                        -------+---------+---------+---------
POOLED STDEV =     2.268                 9.0      12.0      15.0
```

The first part of the computer output in Printout 15.1 displays a one-way ANOVA table. This is Minitab's version of the one-way ANOVA table in Table 15.5 on page 877. Note that Minitab uses the term "Factor" instead of "Treatment."

To the right of the one-way ANOVA table in Printout 15.1, under the column headed p, is the P-value; so $P = 0.005$. Because this is less than the specified significance level of 0.05, we reject H_0. The data provide sufficient evidence to conclude that last year's mean energy consumptions for households in the four U.S. regions are not all the same. ∎

Let's examine some of the other items shown in Printout 15.1. Below the ANOVA table is another table that provides the sample sizes, sample means, and sample standard deviations of the four samples. Beneath that table we find the item POOLED STDEV = 2.268, the pooled estimate of the common standard deviation, σ, of the four populations.

Finally, the lower right side of the computer output depicts individual 95% confidence intervals for the means of the four populations under consideration. The formula used to obtain those confidence intervals is presented in the exercises. (See page 886.)

EXERCISES 15.3

15.24 In each of parts (a)–(c), we have given the notation for one of the three sums of squares. For each sum of squares, state its name and the source of variation it represents.
a. *SSE* b. *SSTR* c. *SST*

15.25 State the one-way ANOVA identity and interpret its meaning with regard to partitioning the total variation in the data.

15.26 True or false: If you know any two of the three sums of squares, *SST, SSTR,* and *SSE,* you can determine the remaining one. Explain your answer.

In each of Exercises 15.27–15.29, construct the one-way ANOVA table for the data. Compute SSTR and SSE using the defining formulas given in Formula 15.1 on page 877.

15.27 The times required by three workers to perform an assembly-line task were recorded on five randomly selected occasions. Here are the times, to the nearest minute.

Hank	Joseph	Susan
8	8	10
10	9	9
9	9	10
11	8	11
10	10	9

(Note: $\bar{x}_1 = 9.6$, $\bar{x}_2 = 8.8$, $\bar{x}_3 = 9.8$, $s_1^2 = 1.3$, $s_2^2 = 0.7$, $s_3^2 = 0.7$, and $\bar{x} = 9.4$.)

15.28 The U.S. Bureau of the Census collects data on monthly rents of newly completed apartments and publishes the results in *Current Housing Reports.* Independent random samples of monthly rents for newly completed apartments in the four U.S. regions yield the following data, in dollars.

Northeast	Midwest	South	West
470	508	428	379
363	386	167	366
413	337	398	444
646	452	573	280
709	359		623
	244		

(Note: $\bar{x}_1 = 520.2$, $\bar{x}_2 = 381.0$, $\bar{x}_3 = 391.5$, $\bar{x}_4 = 418.4$, $s_1^2 = 22{,}548.7$, $s_2^2 = 8{,}476.8$, $s_3^2 = 28{,}239.0$, $s_4^2 = 16{,}492.3$, and $\bar{x} = 427.25$.)

15.29 Data on Scholastic Aptitude Test (SAT) scores are published by the College Entrance Examination Board in *National College-Bound Senior.* SAT scores for randomly selected students from each of four high-school rank categories are displayed in the following table.

Top tenth	Second tenth	Second fifth	Third fifth
528	514	649	372
586	457	506	440
680	521	556	495
718	370	413	321
	532	470	424
			330

(Note: $\bar{x}_1 = 628.0, \bar{x}_2 = 478.8, \bar{x}_3 = 518.8, \bar{x}_4 = 397.0, s_1^2 = 7522.667, s_2^2 = 4540.700, s_3^2 = 8018.700, s_4^2 = 4614.400$, and $\bar{x} = 494.1$.)

In Exercises 15.30 and 15.31, we have presented two partially completed one-way ANOVA tables. Fill in the missing entries.

15.30

Source	df	SS	MS = SS/df	F-statistic
Treatment		2.124	0.708	0.75
Error	20			
Total				

15.31

Source	df	SS	MS = SS/df	F-statistic
Treatment	2		21.652	
Error		84.400		
Total	14			

Preliminary data analyses indicate that it is reasonable to consider the assumptions for one-way ANOVA satisfied in Exercises 15.32–15.38. For each exercise, perform the required hypothesis test using either the classical approach or the P-value approach.

15.32 In Section 15.2 we considered two hypothetical examples in order to explain the logic behind one-way ANOVA. We will examine those examples further in this exercise.

a. Refer to Table 15.1 on page 869. Perform a one-way ANOVA on the data and compare your conclusion to the informal one made in the text. Use $\alpha = 0.05$.

b. Repeat part (a) for the data displayed in Table 15.2 on page 870.

15.33 Four brands of flashlight batteries are to be compared by testing each brand in five flashlights. Twenty flashlights are randomly selected and divided randomly into four groups of five flashlights each. Then each group of flashlights uses a different brand of battery. The lifetimes of the batteries to the nearest hour are as follows.

Brand A	Brand B	Brand C	Brand D
42	28	24	20
30	36	36	32
39	31	28	38
28	32	28	28
29	27	33	25

At the 5% significance level, does there appear to be a difference in mean lifetime among the four brands of batteries?

15.34 A chain of convenience stores wanted to test three different advertising policies:

Policy 1: No advertising.
Policy 2: Advertise in neighborhoods with circulars.
Policy 3: Use circulars and advertise in newspapers.

Eighteen stores were randomly selected and divided randomly into three groups of six stores. Each group used one of the three policies. Following the implementation of the policies, sales figures were obtained for each of the stores during a 1-month period. The figures are displayed, in thousands of dollars, in the following table.

Policy 1	Policy 2	Policy 3
22	21	29
20	25	24
26	25	31
21	20	32
24	22	26
22	26	27

Do the data provide evidence of a difference in mean monthly sales among the three policies? Perform the required hypothesis test at the 1% significance level.

15.35 The Bureau of Labor Statistics publishes data on weekly earnings of nonsupervisory workers in *Employment and Earnings*. The following data, in dollars, were obtained from random samples of (full and part-time) workers in five service-producing industries.

Transp. and Pub. util.	Wholesale trade	Retail trade	Finance, Insurance, Real estate	Services
467	402	208	424	364
507	347	136	378	376
468	327	118	460	383
512	396	246	346	299
559	380	133		336
490		227		273

Do the data provide sufficient evidence to conclude that a difference exists in mean weekly earnings among nonsupervisory workers in the five industries? Perform the required hypothesis test using $\alpha = 0.05$. (*Note:* $T_1 = 3003$, $T_2 = 1852$, $T_3 = 1068$, $T_4 = 1608$, $T_5 = 2031$, and $\Sigma x^2 = 3,755,826$.)

15.36 Manufacturers of golf balls always seem to be claiming that their ball goes the farthest. A writer for a sports magazine decided to conduct an impartial test. She randomly selected 20 golf professionals and then randomly assigned four golfers to each of five brands. Each golfer drove the assigned brand of ball. The driving distances, in yards, are displayed in the following table.

Brand 1	Brand 2	Brand 3	Brand 4	Brand 5
286	279	270	284	281
276	277	262	271	293
281	284	277	269	276
274	288	280	275	292

Do the data provide sufficient evidence to conclude that a difference exists in mean driving distances among the five brands of golf ball? Perform the required hypothesis test at the 5% significance level. (*Note:* $T_1 = 1117$, $T_2 = 1128$, $T_3 = 1089$, $T_4 = 1099$, $T_5 = 1142$, and $\Sigma x^2 = 1,555,185$.)

Journal articles and other sources frequently provide only summary statistics (means, standard deviations, and sample sizes) when publishing ANOVA results. Exercises 15.37 and 15.38 give you practice in working with such data. (Note: To obtain the mean of all the sample data, use the formula provided in Exercise 15.22 on page 875.)

15.37 The U.S. Bureau of Prisons publishes data in *Statistical Report* on the times served by prisoners released from federal institutions for the first time. Independent random samples of released prisoners for five different offense categories yielded the following information on time served, in months.

Offense	n_j	$\bar{x}_j$	s_j
Counterfeiting	15	14.5	4.5
Drug laws	17	18.4	3.8
Firearms	12	18.2	4.5
Forgery	10	15.6	3.6
Fraud	11	11.5	4.7

At the 1% significance level, do the data provide sufficient evidence to conclude that a difference exists in mean time served by prisoners among the five offense groups?

15.38 Data are collected by the Northwestern University Placement Center on starting salaries of college graduates, by major. Findings are reported in *The Northwestern Lindquist-Endicott Report*. Independent samples of college graduates in marketing, statistics, economics, and computer science provided the information on annual starting salaries, in thousands of dollars, shown in the following table.

Major	n_j	$\bar{x}_j$	s_j
Marketing	35	26.6	2.9
Statistics	25	29.2	2.7
Economics	30	26.4	3.0
Computer science	34	30.9	3.5

Do the data imply that a difference exists in mean annual starting salaries among the four majors? Use $\alpha = 0.05$.

15.39 (Computer exercise) Refer to Exercise 15.35. Use Minitab or some other statistical software to
a. obtain normal probability plots and the standard deviations of the samples.
b. perform a residual analysis.
c. perform the required hypothesis test.
d. Justify the use of your procedure in part (c).

15.40 (Computer exercise) Refer to Exercise 15.36. Use Minitab or some other statistical software to
a. obtain normal probability plots and the standard deviations of the samples.
b. perform a residual analysis.
c. perform the required hypothesis test.
d. Justify the use of your procedure in part (c).

15.41 (Computer exercise) The U.S. Bureau of the Census collects data on income by educational attainment, sex, and age. Results are published in *Current Population Reports*. Independent random samples were taken of women from three categories of educational attainment: elementary school, secondary school, and college (4-year degree). Then Minitab was applied to perform a one-way ANOVA on the annual incomes of the women sampled. Printout 15.2 displays the resulting computer output, where the data are in thousands of dollars. Determine

a. the three sums of squares, *SSTR, SSE,* and *SST.*
b. the treatment mean square, *MSTR,* and the error mean square, *MSE.*
c. the value of the test statistic, *F.*
d. the null and alternative hypotheses.
e. the *P*-value for the hypothesis test.
f. the conclusion if the hypothesis test is performed at the 1% significance level.
g. the sample size, sample mean, and sample standard deviation for each of the three samples.
h. a 95% confidence interval for the mean annual income of all women whose educational attainment is at the secondary-school level.

15.42 (Computer exercise) The Motor Vehicle Manufacturers Association of the United States conducts surveys on the costs of owning and operating a motor vehicle. Data are published in *Motor Vehicle Facts and Figures* and include costs for gas and oil, tires, maintenance, insurance, license and registration, and depreciation. Independent random samples of owners of large, intermediate, and compact cars were taken to obtain information on annual insurance premiums. Then Minitab's AOVOneway command was applied to the resulting data. Printout 15.3 displays the output generated by Minitab. Determine

a. the three sums of squares, *SSTR, SSE,* and *SST.*
b. the treatment mean square, *MSTR,* and the error mean square, *MSE.*
c. the value of the test statistic, *F.*
d. the null and alternative hypotheses.
e. the *P*-value for the hypothesis test.
f. the conclusion if the hypothesis test is performed at the 5% significance level.
g. the sample size, sample mean, and sample standard deviation for each of the three samples.
h. a 95% confidence interval for the mean annual insurance premium of all compact-car owners.

Confidence intervals for means and differences between means in one-way ANOVA. Suppose independent random samples of sizes $n_1, n_2, \ldots, n_k$ are to be taken from k normally distributed populations with means $\mu_1, \mu_2, \ldots, \mu_k$, respectively. Further suppose the standard deviations of the k populations are equal. Let $s = \sqrt{MSE}$. Then

- A $(1 - \alpha)$-level confidence interval for any particular population mean, say μ_i, has endpoints
$$\bar{x}_i \pm t_{\alpha/2} \cdot \frac{s}{\sqrt{n_i}} \, .$$
- A $(1 - \alpha)$-level confidence interval for the difference between any two particular population means, say μ_i and μ_j, has endpoints
$$(\bar{x}_i - \bar{x}_j) \pm t_{\alpha/2} \cdot s\sqrt{(1/n_i) + (1/n_j)}.$$

In both formulas, df $= n - k$. We will apply these confidence-interval formulas in Exercises 15.43 and 15.44.

15.43 Refer to Exercise 15.29.
a. Find a 90% confidence interval for the mean SAT score of all students ranked in the second fifth of their high-school class.
b. Find a 90% confidence interval for the difference between the mean SAT scores of students ranked in the top tenth and third fifth of their high-school class.
c. What assumptions are made in solving parts (a) and (b)?

15.44 Refer to Exercise 15.28.
a. Find a 99% confidence interval for the mean monthly rent of newly completed apartments in the Midwest.
b. Find a 99% confidence interval for the difference between the mean monthly rents of newly completed apartments in the Northeast and South.
c. What assumptions are made in solving parts (a) and (b)?

15.45 Refer to Exercise 15.43. Suppose you have obtained a 90% confidence interval for each of the two differences, $\mu_1 - \mu_2$ and $\mu_1 - \mu_3$. Can you be 90% confident of both results simultaneously, that is, that both differences are contained in their corresponding confidence intervals?

15.46 In this exercise we will derive the shortcut formulas, given in Formula 15.1 on page 877, for the three sums of squares, *SST, SSTR,* and *SSE.*
a. Show that $\Sigma(x - \bar{x})^2 = \Sigma x^2 - (\Sigma x)^2/n$, thus establishing the shortcut formula for *SST.*
b. Show that for each j,
$$n_j \left(T_j/n_j - \Sigma x/n\right)^2 = T_j^2/n_j - 2T_j \Sigma x/n + n_j(\Sigma x)^2/n^2.$$
c. Use part (b) to show that
$$\Sigma n_j \left(T_j/n_j - \Sigma x/n\right)^2 = \Sigma(T_j^2/n_j) - (\Sigma x)^2/n$$
(Hint: Recall that $\Sigma T_j = \Sigma x$ and $\Sigma n_j = n$.)
d. Conclude from part (c) and the defining formula for *SSTR* that the shortcut formula $SSTR = \Sigma(T_j^2/n_j) - (\Sigma x)^2/n$ is valid.
e. Why does $SSE = SST - SSTR$?

PRINTOUT 15.2   Minitab output for Exercise 15.41

```
ANALYSIS OF VARIANCE
SOURCE     DF      SS        MS        F         P
FACTOR      2    6545.7    3272.9    69.65    0.000
ERROR     154    7236.0      47.0
TOTAL     156   13781.7
                                INDIVIDUAL 95 PCT CI'S FOR MEAN
                                BASED ON POOLED STDEV
LEVEL       N     MEAN     STDEV    ------+---------+---------+---------+
ELEMENTA   40   11.099     6.720    (--*---)
HIGHSCHO   62   16.587     7.324               (--*--)
COLLEGE    55   27.190     6.388                           (--*--)
                                   ------+---------+---------+---------+
POOLED STDEV =   6.855             12.0      18.0      24.0      30.0
```

PRINTOUT 15.3   Minitab output for Exercise 15.42

```
ANALYSIS OF VARIANCE
SOURCE     DF      SS        MS        F         P
FACTOR      2    36763     18382     0.55     0.581
ERROR      74   2483155    33556
TOTAL      76   2519918
                                INDIVIDUAL 95 PCT CI'S FOR MEAN
                                BASED ON POOLED STDEV
LEVEL       N     MEAN     STDEV    ----------+---------+---------+------
LARGE      20    902.4    152.5          (-------------*------------)
INTERMED   30    851.7    169.2     (-----------*----------)
COMPACT    27    890.5    215.8       (----------*-----------)
                                   ----------+---------+---------+------
POOLED STDEV =   183.2              840       900       960
```

15.4 MULTIPLE COMPARISONS (OPTIONAL)

Suppose we perform a one-way ANOVA and reject the null hypothesis. Then we can conclude that the means of the populations under consideration are not all the same. Once we make that conclusion, we may also want to know which means are different, which mean is largest, or, more generally, the relation among all the means. Methods for dealing with these problems are called **multiple comparisons.**

Several multiple-comparison methods are available. In this book, we will discuss the **Tukey multiple-comparison method.** Other commonly used multiple-comparison methods are the Bonferroni method, the Fisher method, and the Scheffé method.

One approach for implementing multiple comparisons is to obtain confidence intervals for the differences between all possible pairs of population means. Two means are declared different if the confidence interval for their difference does not contain 0. (If a confidence interval for the difference between two population means does not contain 0, then we can reject the null hypothesis that the two means are equal in favor of the alternative hypothesis that the two means are different; and vice versa. See Exercise 10.16 on page 584.)

In multiple comparisons it is important to distinguish between the individual confidence level and the family confidence level. The **individual confidence level** is the confidence we have that *any particular* confidence interval contains the difference between the corresponding population means; the **family confidence level** is the confidence we have that *all* the confidence intervals contain the differences between the corresponding population means. It is at the family confidence level that we can be confident in the truth of our conclusions when comparing all the population means simultaneously.

The Studentized-Range Distribution

The probability distribution upon which the Tukey multiple-comparison method is based is the **studentized range distribution,** which for brevity we refer to as the **q-distribution.** There are infinitely many q-distributions; a particular one is identified by two parameters, which we denote by κ (kappa) and ν (nu).

Probabilities for a random variable having a q-distribution are equal to areas under a curve, aptly called a **q-curve.** The symbol **q_α** is used to denote the q-value having area α to its right. Values of $q_{0.01}$ and $q_{0.05}$ are presented in Tables XII and XIII in Appendix A, respectively. Example 15.5 explains how to use Table XIII.

Example 15.5 Finding the q-Value Having a Specified Area to Its Right

For the q-curve with parameters $\kappa = 4$ and $\nu = 16$, find $q_{0.05}$; that is, find the q-value having area 0.05 to its right, as shown in Fig. 15.10(a).

FIGURE 15.10
Finding the q-value having area 0.05 to its right

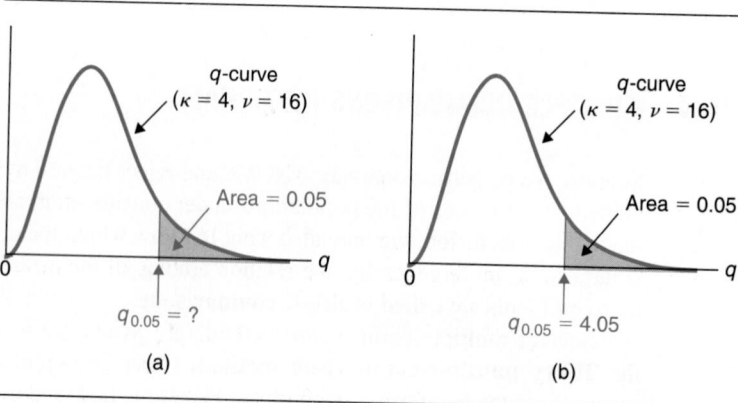

(a) (b)

SOLUTION   To obtain the q-value in question, we use Table XIII. In this case $\kappa = 4$ and $\nu = 16$. Thus we first go down the outside columns to the row labeled "16." Then we go across that row until we are under the column headed "4." The number in the body of the table there, 4.05, is the required q-value; that is, for the q-curve with parameters $\kappa = 4$ and $\nu = 16$, the q-value having area 0.05 to its right is 4.05, as seen in Fig. 15.10(b).   ∎

The Tukey Multiple-Comparison Method

The formulas used in the Tukey multiple-comparison method for obtaining confidence intervals for the differences between means are similar to the pooled-t confidence-interval formula (Procedure 10.4 on page 591). The essential difference is that in the Tukey multiple-comparison method we consult a q-table instead of a t-table.

PROCEDURE 15.2

THE TUKEY MULTIPLE-COMPARISON METHOD FOR COMPARING k POPULATION MEANS

ASSUMPTIONS

1. Independent samples
2. Normal populations
3. Equal population standard deviations

Step 1   Decide on the family confidence level, $1 - \alpha$.

Step 2   Find q_α for the q-curve with parameters $\kappa = k$ and $\nu = n - k$, where n is the total number of pieces of data.

Step 3   Obtain the endpoints of the confidence interval for $\mu_i - \mu_j$:

$$(\bar{x}_i - \bar{x}_j) \pm \frac{q_\alpha}{\sqrt{2}} \cdot s\sqrt{(1/n_i) + (1/n_j)},$$

where $s = \sqrt{MSE}$. Do this for all possible pairs of means with $i < j$.

Step 4   Declare two population means different if the confidence interval for their difference does not contain 0; otherwise, do not declare the two population means different.

Step 5   Summarize the results in Step 4 by ranking the sample means from smallest to largest and by connecting with lines those whose population means were not declared different.

In Example 15.3 on page 879, we conducted a one-way ANOVA to decide whether a difference exists in last year's mean energy consumptions among the four U.S. regions. Specifically, we performed the hypothesis test

H_0: $\mu_1 = \mu_2 = \mu_3 = \mu_4$ (mean energy consumptions are equal)

H_a: Not all the means are equal.

at the 5% significance level, where μ_1, μ_2, μ_3, and μ_4 denote last year's mean energy consumptions for households in the Northeast, Midwest, South, and West, respectively.

The test results were statistically significant; that is, we rejected H_0. Thus we can conclude at the 5% significance level that for last year, at least two of the regions have different mean energy consumptions. The Tukey multiple-comparison method allows us to elaborate on this conclusion.

Example 15.6   Illustrates Procedure 15.2

Apply the Tukey multiple-comparison method to the energy-consumption data, repeated here in Table 15.7.

TABLE 15.7
Samples of last year's energy consumptions for households in the four U.S. regions

Northeast	Midwest	South	West
15	17	11	10
10	12	7	12
13	18	9	8
14	13	13	7
13	15		9
	12		

SOLUTION   We apply Procedure 15.2.

Step 1   *Decide on the family confidence level, $1 - \alpha$.*

As we have done previously in this illustration, we will use $\alpha = 0.05$; so the family confidence level is 0.95 (95%).

Step 2   *Find q_α for the q-curve with parameters $\kappa = k$ and $\nu = n - k$, where n is the total number of pieces of data.*

As we see from Table 15.7, $\kappa = k = 4$ and $\nu = n - k = 20 - 4 = 16$. Consulting Table XIII, we find that $q_\alpha = q_{0.05} = 4.05$.

Step 3   *Obtain the endpoints of the confidence interval for $\mu_i - \mu_j$:*

$$(\bar{x}_i - \bar{x}_j) \pm \frac{q_\alpha}{\sqrt{2}} \cdot s\sqrt{(1/n_i) + (1/n_j)},$$

where $s = \sqrt{MSE}$. Do this for all possible pairs of means with $i < j$.

To begin, it is helpful to construct a table giving the sample means and sample sizes. Referring to Table 15.7, we obtain Table 15.8.

TABLE 15.8
Sample means and sample sizes for the energy-consumption data

j	1	2	3	4
$\bar{x}_j$	13.0	14.5	10.0	9.2
n_j	5	6	4	5

From Step 2, $q_\alpha = 4.05$. Also, on page 873 we found that $MSE = 5.144$ for the energy-consumption data. Now we are ready to obtain the required confidence intervals. The endpoints of the confidence interval for $\mu_1 - \mu_2$ are

$$(13.0 - 14.5) \pm \frac{4.05}{\sqrt{2}} \cdot \sqrt{5.144}\sqrt{(1/5) + (1/6)},$$

or -5.43 to 2.43. Likewise, the endpoints of the confidence interval for $\mu_1 - \mu_3$ are

$$(13.0 - 10.0) \pm \frac{4.05}{\sqrt{2}} \cdot \sqrt{5.144}\sqrt{(1/5) + (1/4)},$$

or -1.36 to 7.36. Proceeding in the same way, we obtain the remaining confidence intervals. All six confidence intervals are displayed in Table 15.9.

TABLE 15.9
Simultaneous 95% confidence intervals for the differences between the energy-consumption means

	1	2	3
2	$(-5.43, 2.43)$		
3	$(-1.36, 7.36)$	$(0.31, 8.69)$	
4	$(-0.31, 7.91)$	$(1.37, 9.23)$	$(-3.56, 5.16)$

Each entry in Table 15.9 is the confidence interval for the difference between the mean labeled by the column and the mean labeled by the row. For instance, the entry in the column labeled 2 and the row labeled 4 is $(1.37, 9.23)$; so the confidence interval for $\mu_2 - \mu_4$ is from 1.37 to 9.23.

Step 4 *Declare two population means different if the confidence interval for their difference does not contain 0; otherwise, do not declare the two population means different.*

Referring to Table 15.9, we declare means μ_2 and μ_3 different and means μ_2 and μ_4 different; all other pairs of means are not declared different.

Step 5 *Summarize the results in Step 4 by ranking the sample means from smallest to largest and by connecting with lines those whose population means were not declared different.*

In view of Table 15.8, Step 4, and the numbering used to represent the U.S. regions (shown parenthetically), we obtain the following diagram:

West (4)	South (3)	Northeast (1)	Midwest (2)
9.2	10.0	13.0	14.5

Interpreting this diagram, we conclude that last year's mean energy consumption in the Midwest exceeds that in the West and South, and that no other means can be declared different. *All of this* can be said with 95% confidence, the family confidence level. ■

As we see from Example 15.6, multiple comparisons require extensive computations, even when the sample sizes and the number of samples are quite small. This explains why multiple comparisons are almost always done by computer.

Using the Computer (Optional)

Minitab has a subcommand called **Tukey** that can be used to perform the Tukey multiple-comparison procedure. Although this subcommand is not available with the AOVOneway command (discussed in Section 15.3), it is available with the **Oneway** command, an alternative to the AOVOneway command for carrying out a one-way ANOVA.

An essential difference between the AOVOneway and Oneway commands is how the samples are stored. For AOVOneway, the samples are stored in different columns (*unstacked data*); for Oneway, the samples are stored in a single column (*stacked data*), with another column holding numbers that identify which data values belong to which samples.

Example 15.7   The Oneway Command and Tukey Subcommand

Use Minitab to perform the multiple comparison required in Example 15.6.

SOLUTION   First we store all 20 energy consumptions from Table 15.7 on page 890 in a column named ENERGY. Then we store numbers identifying which data values belong to which samples in a column named REGION. For instance, suppose we store the sample data for the Northeast in the first five rows of ENERGY, that for the Midwest in the next six rows of ENERGY, and so forth. Further suppose we use the numbers 1, 2, 3, and 4 to represent Northeast, Midwest, South, and West, respectively. Then we would store 1s in the first five rows of REGION, 2s in the next six rows of REGION, and so on.

Now we are ready to instruct Minitab to perform the multiple comparison. We proceed in the following manner.

Session commands: We type the command **ONEWAY** followed by the storage location of the sample data and the storage location of the identifying numbers; that is, we type ONEWAY for 'ENERGY', region in 'REGION'; and press Enter. Then, on the next line, we type the subcommand **TUKEY** followed by the required family significance level (i.e., 1 − family confidence level); that is, we type TUKEY 0.05. and press Enter. Printout 15.4 shows these commands and the resulting output.

(or)

Menu commands: We choose **Stat ▸ ANOVA ▸ Oneway...**, specify ENERGY in the **Response** text box, specify REGION in the **Factor** text box, select **Comparisons...**, select the **Tukey's, family error rate** check box and type 0.05,[†] and then select **OK** twice. The resulting output is displayed in Printout 15.4.

[†] This number is the desired family significance level (= 1 − family confidence level). Some versions of Minitab will not accept the significance level as a decimal, but require it in percentage form. In those versions, type 5 instead of 0.05.

PRINTOUT 15.4
Minitab output for the
Oneway command and
Tukey subcommand

```
MTB > ONEWAY for 'ENERGY', region in 'REGION';
SUBC> TUKEY 0.05.

ANALYSIS OF VARIANCE ON ENERGY
SOURCE     DF        SS        MS        F        P
REGION      3     97.50     32.50     6.32    0.005
ERROR      16     82.30      5.14
TOTAL      19    179.80
                                      INDIVIDUAL 95% CI'S FOR MEAN
                                      BASED ON POOLED STDEV

LEVEL       N      MEAN     STDEV   -------+---------+---------+---------
   1        5    13.000     1.871               (------*-------)
   2        6    14.500     2.588                 (-----*------)
   3        4    10.000     2.582     (-------*-------)
   4        5     9.200     1.924   (-------*------)
                                    -------+---------+---------+---------
POOLED STDEV =     2.268            9.0      12.0      15.0

Tukey's pairwise comparisons

    Family error rate = 0.0500
Individual error rate = 0.0113

Critical value = 4.05

Intervals for (column level mean) - (row level mean)

              1         2         3

    2      -5.433
            2.433

    3      -1.357     0.307
            7.357     8.693

    4      -0.308     1.367    -3.557
            7.908     9.233     5.157
```

The portion of the output in Printout 15.4 from the one-way ANOVA table up to but not including Tukey's pairwise comparisons is essentially identical to that obtained using the AOVOneway command (Printout 15.1 on page 882). Then comes the multiple comparison. The first and second items give the family error rate and individual error rate, which in our terminology are 1 minus the family confidence level and 1 minus the individual confidence level, respectively. Therefore we see that the family confidence level is 0.95 (as specified) and the individual confidence level is 0.9887. The next item in the printout, Critical value = 4.05, is q_α; so $q_\alpha = q_{0.05} = 4.05$.

The final item in Printout 15.4 is a table providing the endpoints of the confidence intervals for the differences between the means. Compare this table with the one we obtained by hand in Table 15.9 on page 891. ■

EXERCISES 15.4

15.47 Explain why the family confidence level is smaller than the individual confidence level for multiple comparisons involving three or more means.

15.48 In a Tukey multiple comparison, the parameter ν for the q-curve equals one of the degrees of freedom in a one-way ANOVA. Which one?

In Exercises 15.49–15.54, we have repeated the information from Exercises 15.33–15.38. For each exercise, use Procedure 15.2 on page 889 to perform a multiple comparison.

15.49 Four brands of flashlight batteries are to be compared by testing each brand in five flashlights. Twenty flashlights are randomly selected and divided randomly into four groups of five flashlights each. Then each group of flashlights uses a different brand of battery. The lifetimes of the batteries to the nearest hour are as follows.

Brand A	Brand B	Brand C	Brand D
42	28	24	20
30	36	36	32
39	31	28	38
28	32	28	28
29	27	33	25

Perform a Tukey multiple comparison using a family confidence level of 0.95.

15.50 A chain of convenience stores wanted to test three different advertising policies:

Policy 1: No advertising.
Policy 2: Advertise in neighborhoods with circulars.
Policy 3: Use circulars and advertise in newspapers.

Eighteen stores were randomly selected and divided randomly into three groups of six stores. Each group used one of the three policies. Following the implementation of the policies, sales figures were obtained for each of the stores during a

1-month period. The figures are displayed, in thousands of dollars, in the following table.

Policy 1	Policy 2	Policy 3
22	21	29
20	25	24
26	25	31
21	20	32
24	22	26
22	26	27

Perform a Tukey multiple comparison using a family confidence level of 0.99.

15.51 The Bureau of Labor Statistics publishes data on weekly earnings of nonsupervisory workers in *Employment and Earnings*. The following data, in dollars, were obtained from random samples of (full and part-time) workers in five service-producing industries.

Transp. and Pub. util.	Wholesale trade	Retail trade	Finance, Insurance, Real estate	Services
467	402	208	424	364
507	347	136	378	376
468	327	118	460	383
512	396	246	346	299
559	380	133		336
490		227		273

Conduct a Tukey multiple comparison at the 95% family confidence level. (*Hint:* For the q-value, use the mean of the two closest q-values in Table XIII.)

15.52 Manufacturers of golf balls always seem to be claiming that their ball goes the farthest. A writer for a sports magazine decided to conduct an impartial test. She randomly selected 20 golf professionals and then randomly assigned four golfers to each of five brands. Each golfer drove the assigned brand of ball. The driving distances, in yards, are displayed in the following table.

Brand 1	Brand 2	Brand 3	Brand 4	Brand 5
286	279	270	284	281
276	277	262	271	293
281	284	277	269	276
274	288	280	275	292

Conduct a Tukey multiple comparison at the 95% family confidence level.

· **15.53** The U.S. Bureau of Prisons publishes data in *Statistical Report* on the times served by prisoners released from federal institutions for the first time. Independent random samples of released prisoners for five different offense categories yielded the following information on time served, in months.

Offense	n_j	$\overline{x}_j$	s_j
Counterfeiting	15	14.5	4.5
Drug laws	17	18.4	3.8
Firearms	12	18.2	4.5
Forgery	10	15.6	3.6
Fraud	11	11.5	4.7

Using a family confidence level of 99%, perform a Tukey multiple comparison.

· **15.54** Data are collected by the Northwestern University Placement Center on starting salaries of college graduates, by major. Findings are reported in *The Northwestern Lindquist-Endicott Report*. Independent samples of college graduates in marketing, statistics, economics, and computer science provided the information on annual starting salaries, in thousands of dollars, shown in the following table.

Major	n_j	$\overline{x}_j$	s_j
Marketing	35	26.6	2.9
Statistics	25	29.2	2.7
Economics	30	26.4	3.0
Computer science	34	30.9	3.5

Using a family confidence level of 95%, perform a Tukey multiple comparison.

· **15.55 (Computer exercise)** Use Minitab or some other statistical software to perform the multiple comparison required in Exercise 15.51.

· **15.56 (Computer exercise)** Use Minitab or some other statistical software to perform the multiple comparison required in Exercise 15.52.

· **15.57 (Computer exercise)** The U.S. Bureau of the Census collects data on income by educational attainment, sex, and age. Results are published in *Current Population Reports*. Independent random samples were taken of women from three categories of educational attainment: elementary school, secondary school, and college (4-year degree). Then Minitab's Oneway command and Tukey subcommand were applied to the annual incomes of the women sampled. Printout 15.5 on the next page displays the portion of the computer output needed for a multiple comparison. The data are in thousands of dollars. Determine

a. the family confidence level and interpret your answer.
b. the individual confidence level and interpret your answer.
c. $q_{0.01}$.
d. the (simultaneous) confidence interval for the difference, $\mu_2 - \mu_3$, between the mean annual incomes of women whose educational attainments are at the secondary and college levels.
e. the (simultaneous) confidence interval for the difference, $\mu_3 - \mu_1$, between the mean annual incomes of women whose educational attainments are at the college and elementary levels.
f. which pairs of means should be declared different.
g. Summarize the results of the Tukey multiple comparison in words.

· **15.58 (Computer exercise)** Surveys are conducted by the Motor Vehicle Manufacturers Association of the United States on the costs of owning and operating a motor vehicle. Data are published in *Motor Vehicle Facts and Figures* and include costs for gas and oil, tires, maintenance, insurance, license and registration, and depreciation. Independent random samples of owners of large, intermediate, and compact cars were taken to obtain information on annual insurance premiums. Then Minitab's Oneway command and Tukey subcommand were applied to the resulting data. Printout 15.6 on the next page displays the portion of the output required for a multiple comparison. Determine

a. the family confidence level and interpret your answer.
b. the individual confidence level and interpret your answer.
c. $q_{0.05}$.
d. the (simultaneous) confidence interval for the difference, $\mu_1 - \mu_3$, between the mean annual insurance premiums of large and compact cars.
e. the (simultaneous) confidence interval for the difference, $\mu_3 - \mu_2$, between the mean annual insurance premiums of compact and intermediate cars.

f. which pairs of means should be declared different.

g. Summarize the results of the Tukey multiple comparison in words.

· ·   **15.59**   Explain precisely why the family confidence level is the appropriate level for comparing all population means simultaneously.

· ·   **15.60**   In Step 3 of Procedure 15.2, we obtain confidence intervals only when $i < j$.

a. Explain how to determine the remaining confidence intervals from those.

b. Apply your result in part (a) and Table 15.9 on page 891 to determine the remaining six confidence intervals for the differences between the energy-consumption means.

PRINTOUT 15.5   Minitab output for Exercise 15.57

```
Tukey's pairwise comparisons

    Family error rate = 0.0100
Individual error rate = 0.00359

Critical value = 4.18

Intervals for (column level mean) - (row level mean)

              1          2

    2      -9.596
           -1.378

    3     -20.301    -14.356
          -11.881     -6.851
```

PRINTOUT 15.6   Minitab output for Exercise 15.58

```
Tukey's pairwise comparisons

    Family error rate = 0.0500
Individual error rate = 0.0193

Critical value = 3.38

Intervals for (column level mean) - (row level mean)

              1          2

    2        -76
             177

    3       -117       -155
             141         77
```

15.5   THE KRUSKAL–WALLIS TEST (OPTIONAL)

In this section we will examine the **Kruskal–Wallis test,** a nonparametric alternative to the one-way ANOVA procedure discussed in Section 15.3. The Kruskal–Wallis test applies when the populations have the same shape, but it does not require that they be normal or have any other specific shape.

Like the Mann–Whitney test, the Kruskal–Wallis test is based on ranks. When ties occur, ranks are assigned in the same way as in the Mann–Whitney test: *If two or more data values are tied, each is assigned the mean of the ranks they would have had if there were no ties.* The Kruskal–Wallis test is introduced in Example 15.8.

Example 15.8   Introduces the Kruskal–Wallis Test

The U.S. Federal Highway Administration conducts annual surveys on motor vehicle travel by type of vehicle and publishes its findings in *Highway Statistics Summary.* Independent random samples of cars, buses, and trucks provided the data on number of miles driven last year, in thousands, shown in Table 15.10. At the 5% significance level, do the data provide sufficient evidence to conclude that a difference exists in last year's mean number of miles driven among cars, buses, and trucks?

TABLE 15.10

Number of miles driven (1000s) last year for independent samples of cars, buses, and trucks

Cars	Buses	Trucks
19.2	1.3	11.6
12.5	7.3	24.0
1.5	7.3	8.2
6.1	7.0	10.6
33.5	12.8	10.0
7.6	18.9	2.3
11.3		44.1
6.3		1.5
8.8		13.0
0.4		

SOLUTION   Let μ_1, μ_2, and μ_3 denote last year's mean number of miles driven for cars, buses, and trucks, respectively. Then the null and alternative hypotheses are

H_0: $\mu_1 = \mu_2 = \mu_3$ (mean numbers of miles driven are equal)

H_a: Not all the means are equal.

Preliminary data analyses (not shown) suggest extreme departures from normality for the populations being sampled but that it is not unreasonable to presume that the populations have roughly the same shape. Thus although the one-way ANOVA test of Section 15.3 is probably inappropriate, the Kruskal–Wallis procedure appears suitable.

To apply the Kruskal–Wallis test, we first rank the data from all three samples combined. The results of this ranking appear in Table 15.11.

TABLE 15.11
Results of ranking the combined
data from Table 15.10

Cars	Rank	Buses	Rank	Trucks	Rank
19.2	22	1.3	2	11.6	17
12.5	18	7.3	9.5	24.0	23
1.5	3.5	7.3	9.5	8.2	12
6.1	6	7.0	8	10.6	15
33.5	24	12.8	19	10.0	14
7.6	11	18.9	21	2.3	5
11.3	16			44.1	25
6.3	7			1.5	3.5
8.8	13			13.0	20
0.4	1				
	12.15		11.50		14.94	← *Mean ranks*

The idea behind the Kruskal–Wallis test is simple: If the null hypothesis of equal population means is true, then the means of the ranks for the three samples should be roughly equal. Put another way, if the variation among the mean ranks for the three samples is too large, then we have evidence against the null hypothesis.

To measure the variation among the mean ranks, we use the treatment sum of squares, *SSTR,* computed for the ranks. To decide whether that quantity is too large, we compare it to the variance of all the ranks, which can be expressed as $SST/(n-1)$, where SST is the total sum of squares for the ranks and n is the total number of pieces of data.[†] More precisely, the test statistic for a Kruskal–Wallis test, denoted by H, is

$$H = \frac{SSTR}{SST/(n-1)}.$$

Large values of H indicate that the variation among the mean ranks is large (relative to the variance of all the ranks) and hence that the null hypothesis of equal population means should be rejected.

For the ranks in Table 15.11, we find that $SSTR = 54.753$, $SST = 1299$, and $n = 25$. Thus the value of the test statistic is

$$H = \frac{SSTR}{SST/(n-1)} = \frac{54.753}{1299/24} = 1.012.$$

Is this value of H large enough to conclude that the null hypothesis of equal population means is false? To answer this question, we need to know the probability distribution of the random variable H. We will first discuss that and then finish the hypothesis test considered in this example.   ■

[†] Recall from Sections 15.2 and 15.3 that the treatment sum of squares, *SSTR,* is a measure of variation among means and that the total sum of squares, *SST*, is a measure of variation among all the data. The defining and shortcut formulas for *SSTR* and *SST* are given in Formula 15.1 on page 877. For the Kruskal–Wallis test, we apply those formulas to the ranks of the sample data, not to the sample data themselves.

KEY FACT 15.5

TEST STATISTIC FOR A KRUSKAL–WALLIS TEST

Suppose independent random samples of sizes $n_1, n_2, \ldots, n_k$ are to be taken from k populations having means $\mu_1, \mu_2, \ldots, \mu_k$, respectively. Further suppose the populations have the same shape. Then if $\mu_1 = \mu_2 = \cdots = \mu_k$, the random variable

$$H = \frac{SSTR}{SST/(n-1)}$$

has approximately a chi-square distribution with df $= k - 1$.

The usual rule of thumb for using the chi-square distribution as an approximation of the true distribution of H is that all sample sizes should be 5 or greater. We will adopt this rule of thumb even though some statisticians consider it too restrictive and consider the chi-square approximation adequate unless $k = 3$ and none of the sample sizes exceeds 5.

When computing the test statistic H by hand from the raw data, it is generally easier to use the shortcut formula

$$H = \frac{12}{n(n+1)} \sum \frac{R_j^2}{n_j} - 3(n+1),$$

where R_1 denotes the sum of the ranks for the sample data from Population 1, R_2 denotes the sum of the ranks for the sample data from Population 2, and so on. Strictly speaking, the shortcut formula for H is equivalent to the defining formula for H if no ties occur. In practice, however, the shortcut formula provides a sufficiently accurate approximation unless the number of ties is relatively large.

We now present a step-by-step procedure for performing a Kruskal–Wallis test. The test can be used to compare several population medians as well as several population means. We state the procedure in terms of population means. To employ the procedure for population medians, simply replace μ_1 by η_1, μ_2 by η_2, and so on.

PROCEDURE 15.3

THE KRUSKAL–WALLIS TEST FOR k POPULATION MEANS WITH NULL HYPOTHESIS $H_0: \mu_1 = \mu_2 = \cdots = \mu_k$

ASSUMPTIONS

1. Independent samples
2. Populations have same shape
3. All sample sizes are 5 or greater

Step 1   State the null and alternative hypotheses.

Step 2   Decide on the significance level, α.

(continued)

Step 3   The critical value is χ_α^2 with df $= k - 1$. Use Table V to determine the critical value.

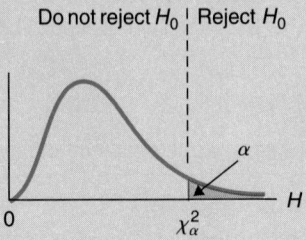

Do not reject H_0 | Reject H_0

α

0 χ_α^2 H

Step 4   Construct a work table of the following form:

Sample from Population 1	Overall rank	Sample from Population 2	Overall rank	...	Sample from Population k	Overall rank
.	.	.	.	.	.	.
.	.	.	.	.	.	.
.	.	.	.	.	.	.

Step 5   Compute the value of the test statistic

$$H = \frac{12}{n(n+1)} \sum \frac{R_j^2}{n_j} - 3(n+1),$$

where n denotes the total number of pieces of data and $R_1, R_2, \ldots, R_k$ denote, respectively, the sums of the ranks for the sample data from Populations $1, 2, \ldots, k$.

Step 6   If the value of the test statistic falls in the rejection region, then reject H_0; otherwise, do not reject H_0.

Step 7   State the conclusion in words.

Example 15.9   Illustrates Procedure 15.3

We now complete the hypothesis test introduced in Example 15.8. Independent random samples of cars, buses, and trucks provided the data on number of miles driven last year, in thousands, shown in Table 15.10 on page 897. At the 5% significance level, do the data provide sufficient evidence to conclude that a difference exists in last year's mean number of miles driven among cars, buses, and trucks?

SOLUTION   We apply Procedure 15.3.

Step 1 *State the null and alternative hypotheses.*

Let μ_1, μ_2, and μ_3 denote last year's mean number of miles driven for cars, buses, and trucks, respectively. Then the null and alternative hypotheses are

H_0: $\mu_1 = \mu_2 = \mu_3$ (mean numbers of miles driven are equal)

H_a: Not all the means are equal.

Step 2 *Decide on the significance level,* α.

We are to perform the hypothesis test at the 5% significance level; so $\alpha = 0.05$.

Step 3 *The critical value is* χ_α^2 *with df* $= k - 1$.

We have $k = 3$, the three types of vehicles; so df $= 3 - 1 = 2$. Consulting Table V, we find that the critical value is $\chi_{0.05}^2 = 5.991$, as seen in Fig. 15.11.

FIGURE 15.11

Criterion for deciding whether or not to reject the null hypothesis

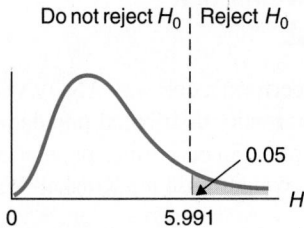

Do not reject H_0 | Reject H_0

0.05

0   5.991

H

Step 4 *Construct a work table.*

We have already done this in Table 15.11 on page 898. (Ignore the bottom row of that table now.)

Step 5 *Compute the value of the test statistic*

$$ H = \frac{12}{n(n+1)} \sum \frac{R_j^2}{n_j} - 3(n+1). $$

We have $n = 10 + 6 + 9 = 25$. Summing the second, fourth, and sixth columns of Table 15.11, we find that $R_1 = 121.5$, $R_2 = 69.0$, and $R_3 = 134.5$. Thus the value of the test statistic is

$$ H = \frac{12}{25(25+1)} \left(\frac{121.5^2}{10} + \frac{69.0^2}{6} + \frac{134.5^2}{9} \right) - 3(25+1) = 1.011. $$

Step 6 *If the value of the test statistic falls in the rejection region, reject* H_0*; otherwise, do not reject* H_0.

From Step 5 the value of the test statistic is $H = 1.011$. Figure 15.11 shows that this does not fall in the rejection region. Thus we do not reject H_0.

Step 7   *State the conclusion in words.*

The test results are not statistically significant at the 5% level; that is, at the 5% significance level, the data do not provide sufficient evidence to conclude that a difference exists in last year's mean number of miles driven among cars, buses, and trucks.   ■

The P-value approach to hypothesis testing can also be used to carry out the hypothesis test in Example 15.9. From Step 5 we see that the value of the test statistic is $H = 1.011$. Recalling that df $= 2$ and that H has approximately a chi-square distribution, we find from Table V that $P > 0.05$. In particular, because the P-value is greater than the specified significance level of 0.05, we cannot reject H_0.

Comparison of the Kruskal–Wallis Test and the One-Way ANOVA Test

In Section 15.3 we learned how to perform a one-way ANOVA (Procedure 15.1, page 878) to compare the means of several normally distributed populations having equal standard deviations using independent samples. Since normal populations having equal standard deviations have the same shape, we can also use the Kruskal–Wallis test to perform such a hypothesis test.

So now the question is this: If we want to perform a hypothesis test to compare the means of several populations using independent samples and we know the populations are normally distributed and have equal standard deviations, should we use the one-way ANOVA test or the Kruskal–Wallis test? As you might expect, we should use the one-way ANOVA test; for normal populations the one-way ANOVA test is more powerful than the Kruskal–Wallis test because it is designed expressly for normal populations. What might surprise you, however, is that for normal populations the one-way ANOVA test is not much more powerful than the Kruskal–Wallis test.

On the other hand, if the populations being sampled have the same shape but are not normally distributed, then the Kruskal–Wallis test is usually more powerful than the one-way ANOVA test, often considerably so. In summary, we have the following key fact.

KEY FACT 15.6

THE KRUSKAL–WALLIS TEST VERSUS THE ONE-WAY ANOVA TEST

Suppose a hypothesis test is to be performed to compare the means of several populations. When deciding between the one-way ANOVA test and the Kruskal–Wallis test, follow these guidelines:

- If you are reasonably sure the populations are normally distributed, use the one-way ANOVA test.

- If you are not reasonably sure the populations are normally distributed but are reasonably sure they have the same shape, use the Kruskal–Wallis test.

As we have seen, the Kruskal–Wallis test is a nonparametric procedure for performing a hypothesis test to compare the means (or medians) of several populations having the same shape. A corresponding Kruskal–Wallis multiple-comparison procedure is available, but we will not cover it here.

Using the Computer (Optional)

Procedure 15.3 provides a step-by-step method for performing a Kruskal–Wallis test. Alternatively, we can use Minitab to carry out a Kruskal–Wallis test. The appropriate command is **Kruskal-Wallis.**

Example 15.10   The Kruskal-Wallis Command

Use Minitab to perform the hypothesis test in Example 15.9.

SOLUTION   Let μ_1, μ_2, and μ_3 denote last year's mean number of miles driven for cars, buses, and trucks, respectively. The problem is to use the Kruskal–Wallis procedure to perform the hypothesis test

H_0: $\mu_1 = \mu_2 = \mu_3$ (mean numbers of miles driven are equal)

H_a: Not all the means are equal.

at the 5% significance level.

First we store all 25 mileages from Table 15.10 on page 897 in a column named MILES. Then we store numbers identifying which mileages belong to which samples in a column named VEHICLE. For instance, suppose we store the mileages for the cars in the first ten rows of MILES, for the buses in the next six rows of MILES, and for the trucks in the next nine rows of MILES. Further suppose we use the numbers 1, 2, and 3 to represent cars, buses, and trucks, respectively. Then we would store 1s in the first ten rows of VEHICLE, 2s in the next six rows of VEHICLE, and 3s in the next nine rows of VEHICLE.

Now we are ready to instruct Minitab to perform the Kruskal–Wallis test. We proceed in the following manner.

Session commands: We type the command **KRUSKAL-WALLIS** followed by the storage location of the sample data and the storage location of the identifying numbers; that is, we type KRUSKAL-WALLIS for 'MILES', type in 'VEHICLE' and press Enter. Printout 15.7 on the next page shows this command and the resulting output.

(or)

Menu commands: We choose **Stat ▶ Nonparametrics ▶ Kruskal-Wallis...**, specify MILES in the **Response** text box, specify VEHICLE in the **Factor** text box, and then select **OK**. The resulting output is displayed in Printout 15.7.

```
MTB > KRUSKAL-WALLIS for 'MILES', type in 'VEHICLE'

LEVEL    NOBS    MEDIAN   AVE. RANK    Z VALUE
   1      10      8.200     12.1       -0.47
   2       6      7.300     11.5       -0.57
   3       9     10.600     14.9        0.99
OVERALL   25                13.0

H = 1.01   d.f. = 2   p = 0.603
H = 1.01   d.f. = 2   p = 0.603  (adjusted for ties)
```

The first column of the output in Printout 15.7, headed LEVEL, gives the numbers used to represent the three types of vehicles. The next three columns give the sizes, medians, and mean ranks for the three samples.

The fifth column, headed Z VALUE, displays population z-scores for the sample mean ranks. This provides a standardized measure of how much each sample mean rank differs from its expected mean rank if the null hypothesis of equal population means is true; each expected mean rank equals the mean rank of all the data values, which in this case is 13. Thus if the null hypothesis is true, each sample mean rank should not differ too much from 13, or equivalently, each z-value should not be too far from 0.

In the second to last line of Printout 15.7, we find the value of the Kruskal–Wallis test statistic, the degrees of freedom, and the P-value for the hypothesis test. The last line provides the same information, but with an adjustment for ties; in this case, the unadjusted and adjusted statistics are the same. Since the P-value of 0.603 exceeds the specified significance level of 0.05, we do not reject H_0. The test results are not statistically significant at the 5% level. ∎

EXERCISES 15.5

15.61 What conditions are required for using the Kruskal–Wallis test?

In Exercises 15.62–15.65, perform a Kruskal–Wallis test using either the classical approach or the P-value approach.

15.62 The U.S. Bureau of Labor Statistics conducts surveys on consumer expenditures for various types of entertainment and publishes its findings in *Consumer Expenditure Survey*. Independent samples yielded the following data, in dollars, on last year's expenditures for three categories.

Fees and admissions	TV, radio, and sound equipment	Other equipment and services
173	100	0
112	1748	251
22	396	1293
495	0	31
111	470	75
1203	0	1024
609	562	1629
300		102
		1238

At the 5% significance level, do the data provide sufficient evidence to conclude that a difference exists among last year's mean expenditures in the three categories?

15.63 Indications are that Americans have become more aware of the dangers of excessive fat intake in their diets. The U.S. Department of Agriculture publishes data on annual consumption of selected beverages in *Food Consumption, Prices, and Expenditures.* Independent random samples of low-fat-milk consumptions for 1970, 1980, and 1990 revealed the following data, in gallons.

1970	1980	1990
5.1	9.4	11.9
4.7	9.9	15.6
2.6	10.0	11.2
3.4	8.6	13.7
3.2	5.4	15.9
9.1	11.0	12.4
5.6	10.1	13.1
2.3		9.7
		9.9

Do the data provide sufficient evidence to conclude that there is a difference in mean (per capita) consumption of low-fat milk for the years 1970, 1980, and 1990? Use $\alpha = 0.01$.

15.64 Information on characteristics of new-car buyers appeared in *1990 Buyers of New Cars,* a publication of Newsweek, Inc. Independent random samples of new-car buyers yielded the following data on age of purchaser (in years) by type of vehicle purchased.

Domestic	Asian	European
41	78	72
42	42	42
51	51	58
47	45	39
33	21	67
83	24	39
35	21	45
69	39	27
50	45	33
60	30	55

Do the data provide sufficient evidence to conclude that a difference exists in the median ages of new-car buyers among the three types of vehicles? Use $\alpha = 0.05$.

15.65 The U.S. Bureau of the Census publishes information on newly completed apartments in *Current Housing Reports.* Independent random samples of newly completed apartments in the four U.S. regions revealed the following data on asking rents, in dollars.

Northeast	Midwest	South	West
723	584	521	1247
569	435	339	518
392	1012	645	882
389	449	696	1062
287	740	894	526
509	1082	350	569
908	578	418	705
507	435	882	631

At the 5% significance level, do the data provide sufficient evidence to conclude that a difference exists among the median asking rents in the four U.S. regions?

15.66 A chain of convenience stores wanted to test three different advertising policies:

Policy 1: No advertising.
Policy 2: Advertise in neighborhoods with circulars.
Policy 3: Use circulars and advertise in newspapers.

Eighteen stores were randomly selected and randomly divided into three groups of six stores. Each group used one of the three policies. Following the implementation of the policies, sales figures were obtained for each of the stores during a 1-month period. The figures are displayed, in thousands of dollars, in the following table.

Policy 1	Policy 2	Policy 3
22	21	29
20	25	24
26	25	31
21	20	32
24	22	26
22	26	27

a. Do the data provide sufficient evidence to conclude that there is a difference in mean monthly sales among the three policies? Perform a Kruskal–Wallis test at the 1% significance level.
b. The hypothesis test in part (a) was done in Exercise 15.34 using the one-way ANOVA test. The assumption in that exercise is that for the three policies, monthly sales are (approximately) normally distributed and have (approximately) equal standard deviations. Presuming that, in fact, monthly sales for all three policies are normally distributed and have equal standard deviations, why is it permissible to perform a Kruskal–Wallis test to compare the means? Is it better in this case to use the one-way ANOVA test or the Kruskal–Wallis test? Explain your answers.

15.67 The Bureau of Labor Statistics publishes data on weekly earnings of nonsupervisory workers in *Employment and Earnings*. The following data, in dollars, were obtained from random samples of (full and part-time) workers in five service-producing industries.

Transp. and Pub. util.	Wholesale trade	Retail trade	Finance, Insurance, Real estate	Services
467	402	208	424	364
507	347	136	378	376
468	327	118	460	383
512	396	246	346	299
559	380	133		336
490		227		273

a. Do the data provide sufficient evidence to conclude that a difference exists in mean weekly earnings among nonsupervisory workers in the five industries? Perform a Kruskal–Wallis test at the 5% significance level. (*Note:* Although $n_4 < 5$, most statisticians would consider it reasonable to conduct a Kruskal–Wallis test on these data.)

b. The hypothesis test in part (a) was done in Exercise 15.35 using the one-way ANOVA test. The assumption in that exercise is that weekly earnings in the five industries are (approximately) normally distributed and have (approximately) equal standard deviations. Presuming that, in fact, weekly earnings in the five industries are normally distributed and have equal standard deviations, why is it permissible to perform a Kruskal–Wallis test to compare the means? Is it better in this case to use the one-way ANOVA test or the Kruskal–Wallis test? Explain your answers.

15.68 Suppose you want to perform a hypothesis test to compare the means of four populations using independent samples of size 20 each. For each part below, decide whether you would use the one-way ANOVA test, the Kruskal–Wallis test, or neither of these tests.

a. Preliminary data analyses of the samples suggest that the populations are not normally distributed but have the same shape.

b. Preliminary data analyses of the samples suggest that all four populations are normally distributed and have the same shape.

c. Preliminary data analyses of the samples suggest that the populations are far from being normally distributed and have quite different shapes.

15.69 (Computer exercise) Refer to Exercise 15.65. Use Minitab or some other statistical software to perform the required hypothesis test using the Kruskal–Wallis procedure.

15.70 (Computer exercise) Refer to Exercise 15.64. Use Minitab or some other statistical software to perform the required hypothesis test using the Kruskal–Wallis procedure.

15.71 (Computer exercise) The U.S. Bureau of the Census collects data on income by educational attainment, sex, and age. Results are published in *Current Population Reports*. Independent random samples were taken of women from three categories of educational attainment: elementary school, secondary school, and college (4-year degree). Then Minitab was applied to perform a Kruskal–Wallis test on the annual incomes of the women sampled. Printout 15.8 displays the resulting computer output, where the data are in thousands of dollars. Determine

a. the value of the test statistic, H.
b. the null and alternative hypotheses.
c. the P-value for the hypothesis test.
d. the conclusion if the hypothesis test is performed at the 1% significance level.
e. the sample size, sample median, and mean rank for each of the three samples.
f. Compare the result of the Kruskal–Wallis test to that of the one-way ANOVA test done on the same data in Exercise 15.41.

15.72 (Computer exercise) The Motor Vehicle Manufacturers Association of the United States conducts surveys on the costs of owning and operating a motor vehicle. Data are published in *Motor Vehicle Facts and Figures* and include costs for gas and oil, tires, maintenance, insurance, license and registration, and depreciation. Independent random samples of owners of large, intermediate, and compact cars were taken to obtain information on annual insurance premiums. Then Minitab's Kruskal-Wallis command was applied to the resulting data. Printout 15.9 displays the output generated by Minitab. Determine

a. the value of the test statistic, H.
b. the null and alternative hypotheses.
c. the P-value for the hypothesis test.
d. the conclusion if the hypothesis test is performed at the 5% significance level.
e. the sample size, sample median, and mean rank for each of the three samples.
f. Compare the result of the Kruskal–Wallis test to that of the one-way ANOVA test done on the same data in Exercise 15.42.

PRINTOUT 15.8   Minitab output for Exercise 15.71

LEVEL	NOBS	MEDIAN	AVE. RANK	Z VALUE
1	40	11.56	41.3	-6.08
2	62	17.60	67.5	-2.57
3	55	27.65	119.5	8.19
OVERALL	157		79.0	

$H = 75.08$ d.f. $= 2$ p $= 0.000$

PRINTOUT 15.9   Minitab output for Exercise 15.72

LEVEL	NOBS	MEDIAN	AVE. RANK	Z VALUE
1	20	895.2	42.0	0.71
2	30	821.8	35.3	-1.17
3	27	902.2	40.9	0.54
OVERALL	77		39.0	

$H = 1.40$ d.f. $= 2$ p $= 0.497$

· · · **15.73** This exercise asks you to derive the shortcut formula for the Kruskal–Wallis test statistic, H. In doing so, formulas for the sum and the sum of squares of the first n integers will be needed; they are

(1) $1 + 2 + \cdots + n = n(n+1)/2,$

(2) $1^2 + 2^2 + \cdots + n^2 = n(n+1)(2n+1)/6.$

a. Show that the treatment sum of squares for the ranks can be expressed as

$$SSTR = \Sigma(R_j^2/n_j) - n(n+1)^2/4.$$

[*Hint:* Use the shortcut formula for *SSTR* given in Formula 15.1 on page 877 and Equation (1).]

b. Assuming no ties, show that the total sum of squares for the ranks can be expressed as

$$SST = n(n+1)(n-1)/12.$$

[*Hint:* Use the shortcut formula for *SST* given in Formula 15.1 and Equations (1) and (2).]

c. Use parts (a) and (b) to obtain the shortcut formula for the Kruskal–Wallis test statistic.

15.6 TWO-WAY ANOVA: THE LOGIC

· · · · · ·

So far in this chapter, we have been studying one-way analysis of variance. We can consider one-way ANOVA a method for comparing the means of populations classified according to one factor. For example, in Section 15.3 we compared last year's mean energy consumptions of households in the four U.S. regions (Northeast, Midwest, South, and West). Here the factor is "region." One-way ANOVA permits us to analyze the effect that factor has on mean energy consumption.

Now we will study **two-way analysis of variance,** which provides methods for comparing the means of populations classified according to two factors, or more to the point, methods for simultaneously analyzing the effects of two factors on the mean of some variable, called the **response variable.** For example, suppose we want to consider the effects of "region" and "home-type" (the two factors) on energy consumption (the response variable). Two-way ANOVA permits us to determine simultaneously whether region affects mean energy consumption, whether home-type affects mean energy consumption, and whether region and home-type interact in their effect on mean energy consumption (e.g., whether the effect of home-type on mean energy consumption depends on region).

In ANOVA, the various categories of each factor are called the **levels** of the factor. For the energy-consumption example, the factor "region" has four levels: Northeast, Midwest, South, and West; and the factor "home-type" has five levels: single-family detached, single-family attached, two- to four-unit building, five or more unit building, and mobile home.[†] From the four regions and five home types, we get $4 \cdot 5 = 20$ populations: Northeast single-family detached, Northeast single-family attached, ..., West mobile home. Table 15.12 provides a schematic of the 20 populations, numbered 1 to 20.

TABLE 15.12
Schematic of the 20 populations for a two-way ANOVA with factors region and home-type

Home-type

		Single-family detached	Single-family attached	Two- to four-unit building	Five or more unit building	Mobile home
Region	Northeast	Population 1	Population 2	Population 3	Population 4	Population 5
	Midwest	Population 6	Population 7	Population 8	Population 9	Population 10
	South	Population 11	Population 12	Population 13	Population 14	Population 15
	West	Population 16	Population 17	Population 18	Population 19	Population 20

Generically, we refer to one factor as Factor A and the other as Factor B; the number of levels of Factor A is denoted by a and the number of levels of Factor B by b. For the energy-consumption example, if we let "region" be Factor A and "home-type" be Factor B, then $a = 4$ and $b = 5$.

Each combination of one level of one factor and one level of the other factor determines a population.[‡] We can picture the general layout for a two-way ANOVA as in Table 15.13. Such an ANOVA is called an $a \times b$ **ANOVA** since it involves two factors, one with a levels and the other with b levels.

[†] These home-type categories are the ones used by the U.S. Energy Information Administration.

[‡] In designed experiments the term *population* is usually replaced by *treatment.*

		Factor B			
		Level 1	Level 2	$\cdots$	Level b
Factor A	Level 1	Population 1	Population 2	$\cdots$	Population b
	Level 2	Population $b+1$	Population $b+2$	$\cdots$	Population $2b$
	$\vdots$	$\vdots$	$\vdots$	$\vdots$	$\vdots$
	Level a	Population $(a-1)b+1$	Population $(a-1)b+2$	$\cdots$	Population ab

The effects of each factor considered separately are called **main effects.** If the means for the levels of Factor A are identical (i.e., if Factor A has, on the average, no effect on the response variable), then we say that there is *no main effect due to Factor A;* otherwise, we say that there is *a main effect due to Factor A.* If the means for the levels of Factor B are identical (i.e., if Factor B has, on the average, no effect on the response variable), then we say that there is *no main effect due to Factor B;* otherwise, we say that there is *a main effect due to Factor B.*

Roughly speaking, the two factors are said to **interact** if the effect of one factor on the mean of the response variable depends on the level of the other factor. More precisely, if for any two levels of one of the factors, the difference between the means of the response variable is the same for all levels of the other factor, then we say that the two factors *do not interact;* otherwise, we say that the two factors *interact.* If the two factors interact, then there is said to be **interaction** between the factors. As we will see, when interaction is present, the main effects must be interpreted carefully.

The Logic Behind Two-Way ANOVA

In two-way ANOVA, independent random samples are taken from the ab populations shown in Table 15.13. We can use the resulting sample data to perform three hypothesis tests: one for Factor A main effects, one for Factor B main effects, and one for interaction. All three tests are based on the same logic used in one-way ANOVA: Compare the variation for the effect in question to the variation within samples.

As in one-way ANOVA, the total variation among all the sample data (total sum of squares, *SST*) is partitioned into a component representing variation among all ab sample means (treatment sum of squares, *SSTR*) and a component representing variation within the samples (error sum of squares, *SSE*). This is illustrated in Fig. 15.8 on page 876 and in the first part of Fig. 15.12.

FIGURE 15.12   Partitioning of the total sum of squares in two-way ANOVA

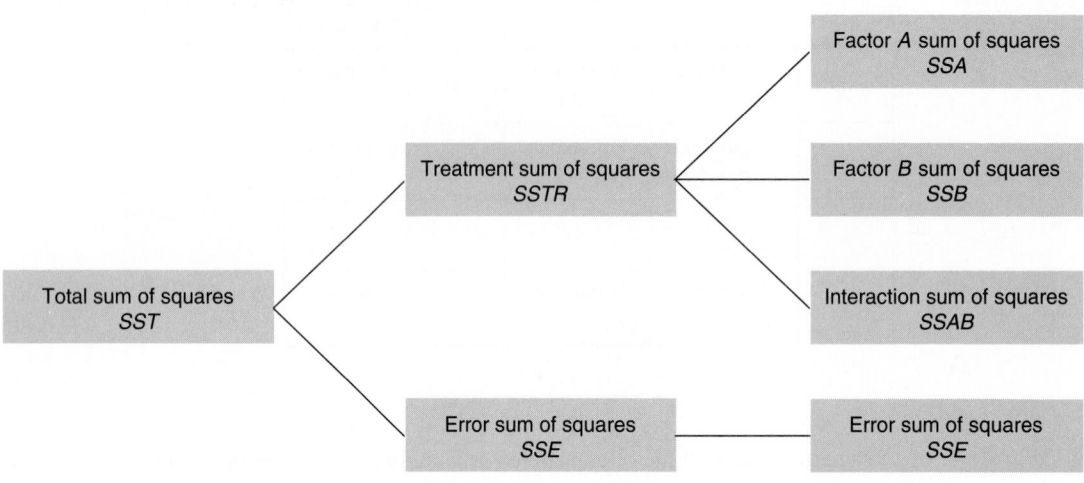

But in two-way ANOVA, the variation among all ab sample means, *SSTR*, is further partitioned into a component representing variation among the sample means for the different levels of Factor A (**Factor A sum of squares, SSA**), a component representing variation among the sample means for the different levels of Factor B (**Factor B sum of squares, SSB**), and a component representing variation due to interaction (**interaction sum of squares, $SSAB$**). This further partitioning is shown in the second part of Fig. 15.12 and is stated formally in Key Fact 15.7.

KEY FACT 15.7   **TWO-WAY ANOVA IDENTITY**

> The total sum of squares equals the Factor A sum of squares plus the Factor B sum of squares plus the interaction sum of squares plus the error sum of squares; that is,
>
> $$SST = SSA + SSB + SSAB + SSE.$$

As in one-way ANOVA, we measure variation using mean squares. That is, we measure the variation among the sample means for the different levels of Factor A using the **Factor A mean square, MSA,** the variation among the sample means for the different levels of Factor B using the **Factor B mean square, MSB,** the variation due to interaction using the **interaction mean square, $MSAB$,** and the variation within samples using the **error mean square, MSE.** Each mean square equals the ratio of the corresponding sum of squares to its degrees of freedom:

$$MSA = \frac{SSA}{a-1}, \quad MSB = \frac{SSB}{b-1}, \quad MSAB = \frac{SSAB}{(a-1)(b-1)}, \quad MSE = \frac{SSE}{n-ab},$$

where n denotes the total number of pieces of data.

Each test statistic in a two-way ANOVA compares the mean square for the effect in question to the error mean square in the same way as in one-way ANOVA. Thus the test statistics for Factor A main effects, Factor B main effects, and interaction are, respectively,

$$F_A = \frac{MSA}{MSE}, \qquad F_B = \frac{MSB}{MSE}, \qquad \text{and} \qquad F_{AB} = \frac{MSAB}{MSE}.$$

A **two-way ANOVA table** summarizes the quantities required for performing a two-way ANOVA. The general format of a two-way ANOVA table is shown in Table 15.14.

TABLE 15.14
ANOVA table format for a two-way analysis of variance

Source	df	SS	MS = SS/df	F-statistic
Factor A	$a - 1$	SSA	$MSA = \dfrac{SSA}{a-1}$	$F_A = \dfrac{MSA}{MSE}$
Factor B	$b - 1$	SSB	$MSB = \dfrac{SSB}{b-1}$	$F_B = \dfrac{MSB}{MSE}$
Interaction	$(a-1)(b-1)$	SSAB	$MSAB = \dfrac{SSAB}{(a-1)(b-1)}$	$F_{AB} = \dfrac{MSAB}{MSE}$
Error	$n - ab$	SSE	$MSE = \dfrac{SSE}{n-ab}$	
Total	$n - 1$	SST		

Although the formulas for the sums of squares in two-way ANOVA are similar to those in one-way ANOVA, we will not present them here because two-way ANOVA is almost always done by computer. Instead we will concentrate on interpreting computer printouts obtained for two-way ANOVAs.

The Elements of Two-Way ANOVA

In this book we will consider only **balanced** two-way ANOVA, that is, where all sample sizes are equal. We use the letter m to denote the common sample size. Because ab populations are being sampled, the total number of pieces of data will be mab, that is, $n = mab$. Key Fact 15.8 on the following page presents the elements of a balanced two-way ANOVA.

Note that the assumptions for a two-way ANOVA are identical to those for a one-way ANOVA. Moreover, two-way ANOVA has the same robustness properties as one-way ANOVA: The independent-samples assumption (Assumption 1) is essential; the samples must be independent or the procedure does not apply. Two-way ANOVA is robust to moderate deviations of the normality assumption (Assumption 2). It is also reasonably robust to moderate deviations of the equal-standard-deviations assumption (Assumption 3). Generally speaking, we can use the same diagnostics for checking the assumptions for a two-way ANOVA as we do for a one-way ANOVA.

KEY FACT 15.8   **ELEMENTS OF A BALANCED TWO-WAY ANOVA**

ASSUMPTIONS

1. Independent samples
2. Normal populations
3. Equal population standard deviations

Notation:

a = number of levels of Factor A
b = number of levels of Factor B
m = common sample size
n = total number of pieces of data = mab

Test for interaction. The null and alternative hypotheses are

H_0: The two factors do not interact.

H_a: The two factors interact.

The test statistic is

$$F_{AB} = \frac{MSAB}{MSE},$$

which has an F-distribution with df = $\big((a-1)(b-1), n-ab\big)$ if the null hypothesis is true.

Test for Factor A main effects. The null and alternative hypotheses are

H_0: There is no main effect due to Factor A.

H_a: There is a main effect due to Factor A.

The test statistic is

$$F_A = \frac{MSA}{MSE},$$

which has an F-distribution with df = $(a-1, n-ab)$ if the null hypothesis is true.

Test for Factor B main effects. The null and alternative hypotheses are

H_0: There is no main effect due to Factor B.

H_a: There is a main effect due to Factor B.

The test statistic is

$$F_B = \frac{MSB}{MSE},$$

which has an F-distribution with df = $(b-1, n-ab)$ if the null hypothesis is true.

We always test for interaction first. The two tests for main effects must be interpreted much more carefully when interaction is present (i.e., when the null hypothesis of no interaction is rejected) than when it is not.

In Examples 15.11 and 15.12, we will consider 2×2 ANOVAs because they are the simplest two-way ANOVAs to perform and interpret. Two-way ANOVAs in which one or both factors have more than two levels will be examined in Section 15.7.

Example 15.11   Two-Way ANOVA

The College Placement Council conducts surveys on salary offers to candidates for degrees and publishes its findings in *A Study of Beginning Offers.* Suppose we want to study the effects of major and degree on mean salary offer to civil- and mechanical-engineering majors receiving either a bachelor's or a master's degree. Then there are two factors: major and degree. Each factor has two levels: the two levels for major are civil engineering and mechanical engineering; the two for degree are bachelor's and master's. Thus we need to perform a 2×2 ANOVA.

Independent random samples of five candidates each were obtained from the four populations under consideration. The salary offers, in thousands of dollars, are displayed in Table 15.15. Perform a two-way ANOVA on these data. Conduct each hypothesis test at the 5% significance level.[†]

TABLE 15.15

Samples of salary offers ($1000s)

| | | Degree | |
		Bachelor's	Master's
Major	Civil	28.6	36.1
		29.5	31.4
		29.2	37.9
		26.2	39.9
		33.5	32.8
	Mech.	37.4	45.1
		29.8	41.6
		33.0	42.1
		37.8	36.5
		36.0	36.1

SOLUTION   We used Minitab to perform a two-way ANOVA on the data in Table 15.15. The results are displayed in Printout 15.10.

[†] If we conduct any three hypothesis tests, each at the significance level α, then the overall (family) significance level for all three tests is at most 3α. For the three tests in a two-way ANOVA, we can state something slightly stronger. If we conduct each of the three tests at the significance level α, then the overall significance level for all three tests is at most $1 - (1 - \alpha)^3$. This result, a special case of the *Kimball inequality,* utilizes the fact that for the three F-statistics in a two-way ANOVA, the numerators are independent random variables and the denominators are identical. Thus if we conduct each of the three tests in a two-way ANOVA at the 5% significance level, then the overall significance level for all three tests is at most $1 - (1 - 0.05)^3 = 0.143$, or 14.3%.

PRINTOUT 15.10

Minitab output for a two-way
ANOVA on the data in Table 15.15

```
Factor      Type Levels Values
MAJOR       fixed    2     1    2
DEGREE      fixed    2     1    2


Analysis of Variance for SALARY

Source         DF        SS       MS      F      P
MAJOR           1     126.50   126.50  11.08  0.004
DEGREE          1     171.11   171.11  14.98  0.001
MAJOR*DEGREE    1       0.68     0.68   0.06  0.810
Error          16     182.74    11.42
Total          19     481.04
```

The first portion of the computer output in Printout 15.10 provides information on the factors and the levels. Then comes the two-way ANOVA table. First we perform the test for interaction of the two factors, major and degree. The appropriate information is provided in the row labeled MAJOR*DEGREE. Under the column labeled F in that row, we find the value of the test statistic F_{AB} (0.06), obtained by dividing $MSAB$ (0.68) by MSE (11.42). The P-value for the hypothesis test is shown in the column labeled P; so $P = 0.810$. Since this exceeds the specified significance level of 0.05, we do not reject the null hypothesis of no interaction.

Next we perform the test for Factor A main effects to decide whether major affects mean salary offer. The appropriate information is in the row labeled MAJOR. Under the column labeled F in that row, we find the value of the test statistic F_A (11.08), obtained by dividing MSA (126.50) by MSE (11.42). The P-value for the hypothesis test is shown in the column labeled P; so $P = 0.004$. Since this is less than the specified significance level of 0.05, we reject the null hypothesis of no main effect due to major. Evidently, major affects mean salary offer.

Finally, we perform the test for Factor B main effects to decide whether degree affects mean salary offer. The appropriate information is provided in the row labeled DEGREE. Under the column labeled F in that row, we find the value of the test statistic F_B (14.98), obtained by dividing MSB (171.11) by MSE (11.42). The P-value for the hypothesis test is shown in the column labeled P; so $P = 0.001$. Since this is less than the specified significance level of 0.05, we reject the null hypothesis of no main effect due to degree. Evidently, degree affects mean salary offer.

We can better interpret the results of the two-way ANOVA by looking at the sample means. The cells of Table 15.16 display the sample means of the four samples in Table 15.15, which are called the **cell means;** the margins of Table 15.16 give the overall sample means for the four levels, which are appropriately called **marginal means.** Thus, for instance, we see that the mean salary offer to the 5 mechanical-engineering majors with a bachelor's degree is 34.800 ($34,800), and the mean salary offer to the 10 master's degree candidates is 37.950 ($37,950).

TABLE 15.16

Sample means of
salary offers ($1000s)

	Degree		
	Bachelor's	Master's	All
Civil	29.400	35.620	32.510
Mech.	34.800	40.280	37.540
All	32.100	37.950	35.025

Major

It is also helpful to graph the cell means. This is done in Fig. 15.13, where we first plotted the cell means against degree and then connected the means for each major with lines.

FIGURE 15.13

Plot of the cell
means in Table 15.16

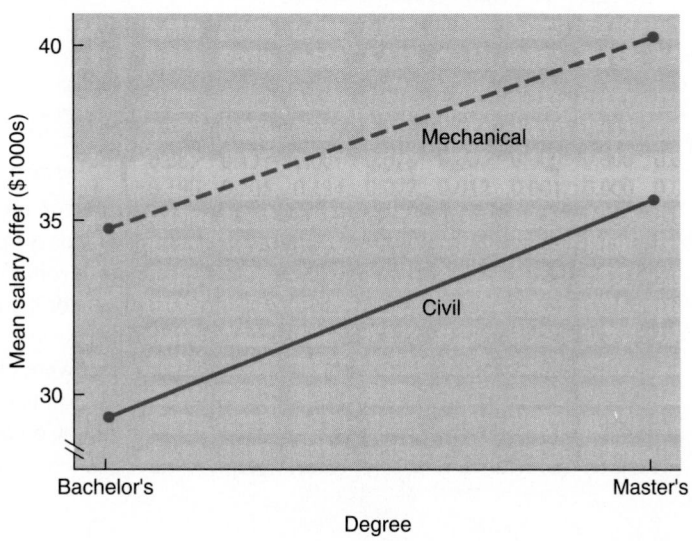

If two factors do not interact, then the lines in their cell-mean plot should be roughly parallel. We cannot expect the lines to be exactly parallel because of sample variability, but too much deviation from parallel indicates that the two factors interact. In our salary-offer example, we can see from Fig. 15.13 that the lines are nearly parallel, suggesting that major and degree probably do not interact. Our failure to reject the null hypothesis of no interaction demonstrates this more formally.

Table 15.16 shows that for the sample data, mechanical-engineering majors receive, on the average, higher salary offers than civil-engineering majors: $37,540 versus $32,510, respectively. As we have seen from our hypothesis test for Factor A main effects, this difference in sample means is sufficiently large to warrant rejecting the null hypothesis of no main effect due to major; that is, we have statistical evidence of a difference between the (population) mean salary offers to civil-engineering and mechanical-engineering majors. And because major has only two levels here, we can conclude more specifically that mechanical-engineering majors are offered more on the average than civil-engineering majors.

The relevance of noninteraction is that the interpretation of this statistical significance is straightforward. Not only are mechanical-engineering majors offered more on the average than civil-engineering majors, but this is also true within each of the two levels for degree: candidates for a bachelor's degree in mechanical engineering are offered more on the average than candidates for a bachelor's degree in civil engineering; candidates for a master's degree in mechanical engineering are offered more on the average than candidates for a master's degree in civil engineering; and in both cases the difference is approximately the same. (These facts are reflected by the sample information, as seen in Table 15.16 and Fig. 15.13.)

Remarks similar to those in the preceding two paragraphs apply to Factor B, degree. We leave the details to the reader. ∎

Example 15.12   Two-Way ANOVA

Suppose we want to study the effects of major and degree on mean salary offer to accounting and marketing majors receiving either a bachelor's or a master's degree. Then there are two factors: major and degree. Each factor has two levels: the two levels for major are accounting and marketing; the two for degree are bachelor's and master's. Thus we need to perform a 2×2 ANOVA.

Independent random samples of 10 candidates each were obtained from the four populations under consideration. The salary offers, in thousands of dollars, are displayed in Table 15.17. Perform a two-way ANOVA on these data. Conduct each hypothesis test at the 5% significance level.

TABLE 15.17
Samples of salary offers ($1000s)

		Degree			
		Bachelor's		Master's	
Major	Acct'g	25.0	25.9	27.4	37.3
		23.8	29.9	29.9	33.1
		28.7	22.3	33.3	31.4
		32.2	28.1	29.7	28.8
		30.8	25.4	31.5	30.6
	Mrkt'g	24.5	18.9	41.2	34.9
		23.5	22.0	43.5	36.8
		23.2	31.0	44.0	40.8
		23.6	22.8	37.8	41.7
		20.8	28.7	37.4	36.9

SOLUTION We used Minitab to perform a two-way ANOVA on the data in Table 15.17. Printout 15.11 shows the results.

PRINTOUT 15.11
Minitab output for a two-way
ANOVA on the data in Table 15.17

```
Factor      Type Levels Values
MAJOR       fixed     2   1    2
DEGREE      fixed     2   1    2


Analysis of Variance for SALARY

Source          DF       SS       MS      F      P
MAJOR            1     59.78    59.78   5.89  0.020
DEGREE           1    969.24   969.24  95.45  0.000
MAJOR*DEGREE     1    331.20   331.20  32.62  0.000
Error           36    365.57    10.15
Total           39   1725.79
```

As before, we first perform the test for interaction of the two factors, major and degree. From the last entry in the row labeled MAJOR*DEGREE, we find that the P-value for the hypothesis test is 0.000. Since this is less than the specified significance level of 0.05, we reject the null hypothesis of no interaction. Evidently, major and degree interact in their effect on mean salary offer for these populations.

The two tests for main effects are also statistically significant at the 5% significance level ($P = 0.020$ for major and $P = 0.000$ for degree). But because of interaction, we must interpret these results carefully.

Table 15.18 displays a table of cell and marginal means; Fig. 15.14, shown on the following page, provides a plot of the cell means.

TABLE 15.18
Sample means of
salary offers ($1000s)

	Degree		
Major	Bachelor's	Master's	All
Acct'g	27.210	31.300	29.255
Mrkt'g	23.900	39.500	31.700
All	25.555	35.400	30.478

As we see from the cell-mean plot in Fig. 15.14, the two lines are far from parallel, suggesting that the two factors interact. Our rejection of the null hypothesis of no interaction in favor of the alternative hypothesis of interaction demonstrates this more formally.

FIGURE 15.14

Plot of the cell
means in Table 15.18

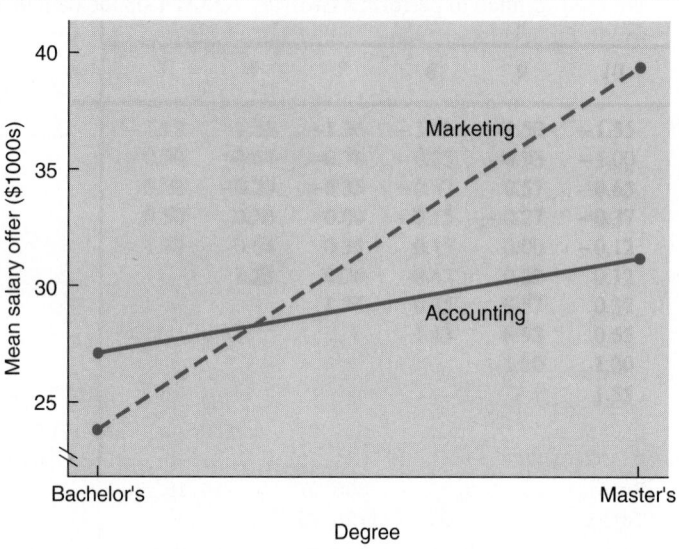

To interpret the test result of a main effect due to major, we consult Table 15.18. We observe that for the sample data, accounting majors receive lower salary offers on the average than marketing majors ($29,255 vs. $31,700). The test for Factor *A* main effects indicates that the difference between these two sample means is sufficiently large to warrant rejecting the null hypothesis of no main effect due to major. In other words, the data provide sufficient evidence to conclude that a difference exists between the (population) mean salary offers to accounting and marketing majors; and since major has only two levels here, we can conclude more specifically that accounting majors receive *lower* salary offers on the average than marketing majors.

However, because of the interaction between major and degree, this conclusion is misleading without further elaboration. If we look at the cell means in Table 15.18 or consult Fig. 15.14, we find that for the sample data, with a bachelor's degree, accounting majors receive *higher* salary offers on the average than marketing majors ($27,210 vs. $23,900); but that with a master's degree, accounting majors receive *lower* salary offers on the average than marketing majors ($31,300 vs. $39,500). Formal analyses using multiple comparisons indicate that this relationship holds for the population means as well.

To interpret the test result of a main effect due to degree, we again consult Table 15.18. We observe that for the sample data, candidates for a bachelor's degree receive lower salary offers on the average than candidates for a master's degree ($25,555 vs. $35,400). The test for Factor *B* main effects indicates that the difference between these two sample means is sufficiently large to warrant rejecting the null hypothesis of no main effect due to degree. In other words, the data provide sufficient evidence to conclude that a difference exists between the (population) mean salary offers to candidates for bachelor's and master's degrees; and since degree has only two levels here, we can conclude more specifically that candidates for a bachelor's degree receive lower salary offers on the average than candidates for a master's degree.

Furthermore, if we look at the cell means in Table 15.18 or consult Fig. 15.14, we find that for the sample data, this same relationship holds true within each major. Formal analyses using multiple comparisons indicate that this relationship holds for the population means as well. Consequently, although major and degree interact in their effect on mean salary offer, it is not necessary in this case to qualify the test result of a main effect due to degree. ∎

Example 15.12 illustrates why we must carefully interpret the test results for main effects when the test for interaction is statistically significant. In such cases we must look at the individual cell means to more accurately interpret the test results for the main effects. Formal analyses should be done using multiple comparisons.

EXERCISES 15.6

15.74 Consider a 3 × 4 ANOVA.
a. Identify the number of levels for Factor A.
b. Identify the number of levels for Factor B.
c. Altogether, how many populations are being sampled?
d. Construct a table similar to Table 15.13 on page 909 for this two-way ANOVA.

15.75 Consider a 4 × 2 ANOVA.
a. Identify the number of levels for Factor A.
b. Identify the number of levels for Factor B.
c. Altogether, how many populations are being sampled?
d. Construct a table similar to Table 15.13 on page 909 for this two-way ANOVA.

15.76 State the two-way ANOVA identity and interpret its meaning with regard to partitioning the total variation in the data.

15.77 In two-way ANOVA, identify the statistic that is used as a
a. measure of variation among the sample means for the different levels of Factor A.
b. measure of variation among the sample means for the different levels of Factor B.
c. measure of variation due to interaction.
d. measure of variation within the samples.

In Exercises 15.78 and 15.79, we have presented partially completed two-way ANOVA tables for a balanced ANOVA. For each exercise,

a. *fill in the missing entries.*
b. *determine the number of levels for each factor.*
c. *determine the common sample size.*

15.78

Source	df	SS	MS = SS/df	F-statistic
Factor A	3	605.272		
Factor B			1145.679	
Interaction	6			7.90
Error	24		10.943	
Total				

15.79

Source	df	SS	MS = SS/df	F-statistic
Factor A		7.174	3.587	
Factor B			1510.517	154.02
Interaction		50.070		
Error	27			
Total		3343.066		

15.80 State the three assumptions required for a two-way ANOVA. How crucial are these assumptions?

15.81 State the null and alternative hypotheses for a hypothesis test for
a. interaction.
b. Factor A main effects.
c. Factor B main effects.

15.82 Identify, give the degrees of freedom for, and interpret the test statistic used in a hypothesis test for
a. interaction.
b. Factor A main effects.
c. Factor B main effects.

15.83 In a two-way ANOVA, why is the test for interaction done first?

15.84 Each graph in Fig. 15.15 shows a plot of the cell means for a 2 × 2 ANOVA. For each part, using only the graphs, conjecture on the results of the three hypothesis tests. (*Note:* For convenience we have denoted the two levels of Factor A as A_1 and A_2.)

15.85 Each graph in Fig. 15.16 shows a plot of the cell means for a 2 × 2 ANOVA. For each part, using only the graphs, conjecture on the results of the three hypothesis tests. (*Note:* For convenience we have denoted the two levels of Factor A as A_1 and A_2.)

15.86 Refer to Exercise 15.84. For each part, interpret the results of the three tests.

15.87 Refer to Exercise 15.85. For each part, interpret the results of the three tests.

15.88 Referring to Exercise 15.84, for which part(s) is it reasonable to say that
a. Factor A does not affect the mean of the response variable.
b. Factor B does not affect the mean of the response variable.
c. neither factor affects the mean of the response variable.

15.89 Referring to Exercise 15.85, for which part(s) is it reasonable to say that
a. Factor A does not affect the mean of the response variable.
b. Factor B does not affect the mean of the response variable.
c. neither factor affects the mean of the response variable.

FIGURE 15.15   Sample-mean plots for Exercise 15.84

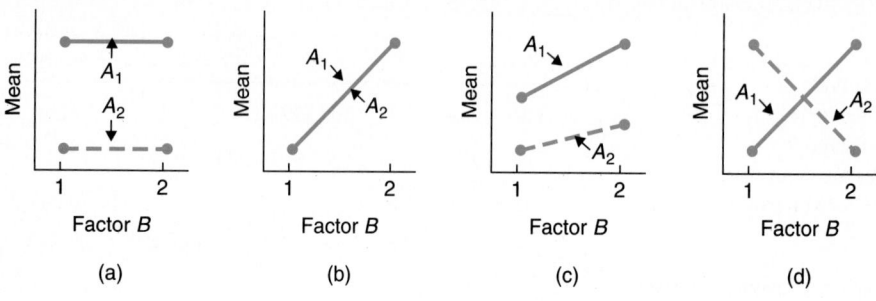

FIGURE 15.16   Sample-mean plots for Exercise 15.85

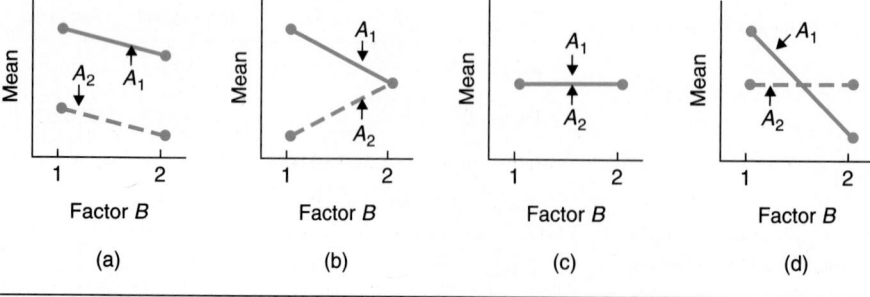

15.7 TWO-WAY ANOVA: THE PROCEDURE

As we have seen, when the test for interaction is statistically significant (i.e., the null hypothesis of no interaction is rejected), the interpretation of the tests for main effects is generally complex and requires using multiple comparisons. Consequently, in a first course in statistics, we usually conduct the tests for main effects only when the test for interaction is not statistically significant. With this in mind, we now present the following procedure for performing a two-way ANOVA.

PROCEDURE 15.4

THE TWO-WAY ANOVA PROCEDURE

ASSUMPTIONS

1. Independent samples
2. Normal populations
3. Equal population standard deviations

Step 1 Decide on the (individual) significance level.

Step 2 Obtain the computer output for the two-way ANOVA.

Step 3 Construct a table of sample means and a plot of the cell means.

Step 4 State the null and alternative hypotheses for the test for interaction, and use the computer output from Step 2 to carry out the test. Interpret the result using the table and plot obtained in Step 3.

Step 5 If the test for interaction is statistically significant, terminate the analysis; otherwise, proceed to Step 6.

Step 6 State the null and alternative hypotheses for the test for Factor *A* main effects, and use the computer output from Step 2 to carry out the test. Interpret the result using the table and plot obtained in Step 3.

Step 7 State the null and alternative hypotheses for the test for Factor *B* main effects, and use the computer output from Step 2 to carry out the test. Interpret the result using the table and plot obtained in Step 3.

Example 15.13 Illustrates Procedure 15.4

Golden Torch Cactus, a columnar cactus native to Argentina, is regarded as having excellent landscape potential. Feldman and Crosswhite, two researchers at the Boyce Thompson Southwestern Arboretum, conducted a thorough investigation of the optimal method for producing these cacti. They investigated the effects of two factors—hydrophilic polymer (with or without) and irrigation regime—on several variables: number of cuttings, length of cuttings, weight of cuttings, overall weight increase, and production per gram of original.

Hydrophilic polymers are used as soil additives to keep moisture in the root zone. For this study the researchers chose Broadleaf P-4 polyacrylamide, P4 for short. Five irrigation regimes were employed: none, light, medium, heavy, and very heavy. During the hottest 6 months, these were, respectively, no irrigation, once monthly, once biweekly, once weekly, and twice weekly; during the cooler 6 months, these were, respectively, no irrigation, once bimonthly, once monthly, once biweekly, and once weekly.

For the experiment, each treatment [choice of polymer (with or without) and irrigation regime] was applied to four cacti. In this example we will examine the results for weight gain. The weight gains, in grams, are displayed in Table 15.19. Perform a two-way ANOVA on these data. Conduct each hypothesis test at the 5% significance level.

TABLE 15.19
Weight gains (grams)

		Irrigation regime				
		None	Light	Medium	Heavy	Very heavy
Polymer	Without P4	683	867	1499	1445	2733
		344	1265	1672	2343	2079
		791	1387	1546	1938	1981
		756	1084	1457	1668	1896
	With P4	503	638	971	1814	2196
		315	1212	1361	2085	1794
		528	1748	1048	1497	1398
		455	884	1900	2188	2125

SOLUTION   We apply Procedure 15.4.

Step 1   *Decide on the (individual) significance level.*

The tests are to be performed at the 5% significance level; so $\alpha = 0.05$.

Step 2   *Obtain the computer output for the two-way ANOVA.*

We used Minitab to perform a two-way ANOVA on the data in Table 15.19. Printout 15.12 displays the results.

PRINTOUT 15.12
Minitab output for a two-way
ANOVA on the data in Table 15.19

```
Factor      Type Levels Values
POLYMER     fixed    2    1    2
WATER       fixed    5    1    2    3    4    5


Analysis of Variance for WTGAIN

Source         DF        SS        MS      F      P
POLYMER         1    192377    192377   1.85  0.184
WATER           4  11301236   2825309  27.10  0.000
POLYMER*WATER   4    161433     40358   0.39  0.816
Error          30   3127825    104261
Total          39  14782871
```

Step 3 *Construct a table of sample means and a plot of the cell means.*

Table 15.20 displays a table of sample means. Figure 15.17 provides a plot of the cell means.

TABLE 15.20

Sample means of weight gains (grams)

		None	Light	Medium	Heavy	Very heavy	All
Polymer	Without P4	643.5	1150.7	1543.5	1848.5	2172.2	1471.7
	With P4	450.3	1120.5	1320.0	1896.0	1878.2	1333.0
	All	546.9	1135.6	1431.7	1872.3	2025.2	1402.3

Irrigation regime

FIGURE 15.17

Plot of the cell means in Table 15.20

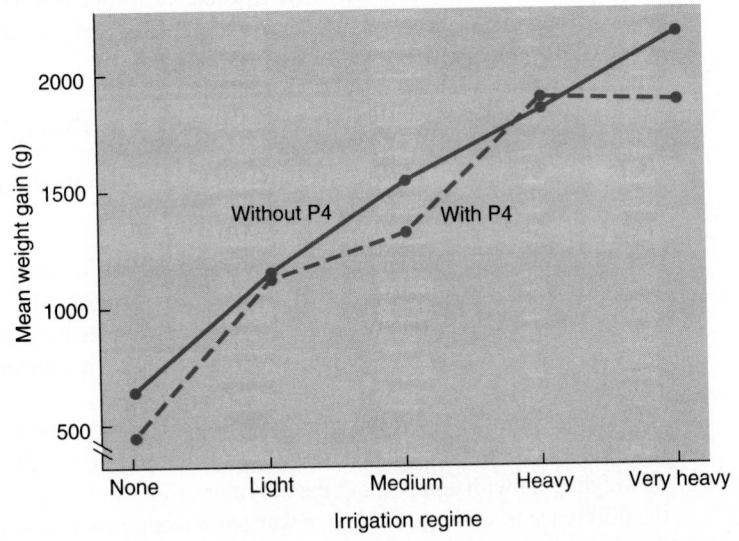

Step 4 *State the null and alternative hypotheses for the test for interaction, and use the computer output from Step 2 to carry out the test. Interpret the result using the table and plot obtained in Step 3.*

The null and alternative hypotheses for the test for interaction are

H_0: Polymer and irrigation regime do not interact in their effect on mean weight gain.

H_a: Polymer and irrigation regime interact in their effect on mean weight gain.

Referring to Printout 15.12, we see from the row labeled POLYMER*WATER that the P-value for the hypothesis test is 0.816. Since this exceeds the specified significance level of 0.05, we do not reject the null hypothesis. The data do not provide sufficient evidence to conclude that polymer and irrigation regime interact in their effect on mean weight gain of Golden Torch cacti.

Although the two polygonal lines in Fig. 15.17 are not exactly parallel (and in fact cross), the formal test for interaction that we just did shows that for the data in Table 15.19, the polygonal lines are not nearly far enough from parallel to warrant rejecting the null hypothesis of no interaction. In other words, we can reasonably attribute the deviation from parallel to sampling error.

Step 5   *If the test for interaction is statistically significant, terminate the analysis; otherwise, proceed to Step 6.*

As we discovered in Step 4, the test for interaction is not statistically significant. Consequently, we proceed to Step 6.

Step 6   *State the null and alternative hypotheses for the test for Factor A main effects, and use the computer output from Step 2 to carry out the test. Interpret the result using the table and plot obtained in Step 3.*

The null and alternative hypotheses for the test for Factor A main effects are

H_0: Polymer does not affect mean weight gain.

H_a: Polymer affects mean weight gain.

Referring again to Printout 15.12 on page 922, we find from the row labeled POLYMER that the test for a main effect due to polymer is not statistically significant at the 5% level ($P = 0.184$). The data do not provide sufficient evidence to conclude that there is a difference in mean weight gains for Golden Torch cacti grown with and without the hydrophilic polymer.

To interpret the test result of no main effect due to polymer, we consult Table 15.20 or Fig. 15.17, both on page 923. We observe that for the sample data, the marginal means of weight gain with and without the polymer differ (1333.0 g vs. 1471.7 g). But evidently the difference is not large enough to warrant rejecting the null hypothesis of no main effect due to polymer.

Step 7   *State the null and alternative hypotheses for the test for Factor B main effects, and use the computer output from Step 2 to carry out the test. Interpret the result using the table and plot obtained in Step 3.*

The null and alternative hypotheses for the test for Factor B main effects are

H_0: Irrigation regime does not affect mean weight gain.

H_a: Irrigation regime affects mean weight gain.

Referring once more to Printout 15.12, we observe from the row labeled WATER that the test for a main effect due to irrigation regime is statistically significant at the 5% level ($P = 0.000$). Evidently, irrigation regime affects mean weight gain of Golden Torch cacti.

To interpret the test result of a main effect due to irrigation regime, we again consult Table 15.20 or Fig. 15.17. We observe that for the sample data, considerable variation exists among the marginal means of weight gain for the five irrigation regimes. The test we just did shows that this variation in marginal means is sufficient to warrant rejecting the null hypothesis of no main effect due to irrigation regime. Descriptively, the greater the irrigation frequency, the greater the mean weight gain; but to be precise, a multiple comparison should be performed to determine which means can be declared different.[†]   ∎

We have barely scratched the surface of ANOVA; entire books and courses are devoted to its study. But partitioning the total sum of squares into several components representing different sources of variation, as illustrated by the one-way ANOVA identity and the two-way ANOVA identity, is the essential idea behind all ANOVA procedures.

Using the Computer (Optional)

Minitab has several commands that can be used to perform a balanced two-way ANOVA. Here we will discuss the **ANOVA** command.[‡] To use the ANOVA command to perform a two-way ANOVA, all the sample data are stored in a single column; then two other columns hold numbers that identify the appropriate levels of the two factors for each data value.

Example 15.14   The ANOVA Command

Use Minitab to obtain the output shown in Printout 15.12 on page 922.

SOLUTION   First we store all 40 weight gains from Table 15.19 in a column named WTGAIN. Then we store numbers in columns named POLYMER and WATER that identify the levels of the two factors for each data value. For instance, consider the third observation for "with P4" and "heavy" irrigation, which is 1497 g. In the same row that contains this observation in WTGAIN, we store the number 2 (the "with P4" level for polymer) in POLYMER and the number 4 (the "heavy" level for irrigation regime) in WATER.

Now we are ready to instruct Minitab to perform the two-way ANOVA. We proceed in the following manner.

Session commands: We type the command **ANOVA** followed by the storage location of the sample data, an equal sign, and the storage locations of the levels for the two factors

---

[†] The multiple comparison Feldman and Crosswhite performed indicates that the mean weight gain for no irrigation is smallest, the mean for light irrigation is less than that for heavy and very heavy irrigation, the mean for medium irrigation is less than that for very heavy irrigation, and no other means can be declared different.

[‡] The ANOVA command can be used to perform a balanced ANOVA with any number of factors.

separated by a vertical bar; that is, we type `ANOVA 'WTGAIN' = 'POLYMER' | 'WATER'` and press `Enter`. Printout 15.12 on page 922 shows the resulting output.

(or)

Menu commands: We choose **Stat ▶ ANOVA ▶ Balanced ANOVA...**, specify WTGAIN for **Responses**, specify POLYMER | WATER in the **Model** text box, and then select **OK**. The resulting output is depicted in Printout 15.12 on page 922.   ∎

If we use the **Means** subcommand to the ANOVA command, the output will also display the cell and marginal means. For the cactus example, session-command users would type the subcommand `MEANS 'POLYMER' | 'WATER'.`; menu-command users would select **Options...**, specify POLYMER | WATER in the **Display means for (list of terms)** text box, and then select **OK** twice.

EXERCISES 15.7

In Exercises 15.90–15.95, we have presented studies conducted using two-way ANOVA and have displayed corresponding (abridged) computer output that contains a two-way ANOVA table. For each exercise, perform a two-way ANOVA at the 5% individual significance level. (Note: Preliminary data analyses indicate that it is reasonable to consider the assumptions for performing a two-way ANOVA met in each case.)

15.90 The U.S. Bureau of the Census publishes data on money income of households by race and region in *Current Population Reports*. Independent samples of households yielded the following data on annual household income, in thousands of dollars. Printout 15.13 shows computer output obtained for a two-way ANOVA on these data.

Region

Race		Northeast	Midwest	South	West
White		58.9	47.1	36.2	34.4
		46.2	6.2	49.2	36.2
		48.1	52.9	22.3	56.3
		24.9	45.8	30.8	32.7
		38.1	36.5	45.3	45.8
Black		34.3	18.5	32.2	29.3
		15.4	15.4	34.0	26.4
		50.7	27.5	24.4	49.4
		29.9	24.6	12.6	20.7
		9.1	40.0	11.8	25.2

15.91 Another variable Feldman and Crosswhite investigated in their study of Golden Torch cacti was total length of cuttings at the end of 16 months. Here are the data, in millimeters. Printout 15.14 shows computer output obtained for a two-way ANOVA on these data.

Irrigation regime

Polymer		None	Light	Medium	Heavy	Very heavy
W/o P4		415	572	705	821	968
		360	722	839	942	862
		508	805	751	732	748
		487	638	733	759	809
With P4		410	518	488	874	831
		347	684	748	923	539
		373	676	541	659	492
		440	647	859	926	696

15.92 The R. R. Bowker Company collects data on retail prices of books and periodicals. Information about book prices are published in *Publishers Weekly* and in *Bowker Annual: Library and Book Trade Almanac*. Independent random samples were taken to compare retail book prices for five subjects (art, law, medicine, science, and technology) and three years (1985, 1990, and 1991). The table in the first column at the bottom of the following page displays the data, in dollars.

PRINTOUT 15.13   Minitab output for Exercise 15.90

Analysis of Variance for INCOME

Source	DF	SS	MS	F	P
RACE	1	1722.7	1722.7	10.47	0.003
REGION	3	255.9	85.3	0.52	0.673
RACE*REGION	3	27.1	9.0	0.05	0.983
Error	32	5264.7	164.5		
Total	39	7270.4			

PRINTOUT 15.14   Minitab output for Exercise 15.91

Analysis of Variance for LENGTH

Source	DF	SS	MS	F	P
POLYMER	1	56626	56626	4.99	0.033
WATER	4	769047	192262	16.93	0.000
POLYMER*WATER	4	61153	15288	1.35	0.276
Error	30	340669	11356		
Total	39	1227496			

Printout 15.15 on the next page shows computer output obtained for a two-way ANOVA on these data and also supplies a table of cell and marginal means.

Year

Subject	1985		1990		1991	
Art	32	28	48	36	49	36
	26	39	47	39	42	44
	39	47	38	44	61	37
Law	35	31	57	56	58	67
	52	48	68	60	49	56
	39	46	64	61	52	77
Medicine	58	37	68	75	70	72
	45	44	67	75	61	77
	33	47	71	77	69	71
Science	49	55	75	77	82	75
	49	48	81	68	86	75
	61	44	66	76	88	73
Technology	63	40	70	90	84	71
	60	58	85	71	73	78
	36	43	76	70	71	73

15.93 The U.S. Bureau of the Census conducts surveys to obtain information on state and local government employment. Results of the surveys are published in *Public Employment*. Independent random samples yielded the following data on earnings, in dollars, during October 1991. Printout 15.16 on the next page shows computer output obtained for a two-way ANOVA on these data and also supplies a table of cell and marginal means.

Region

Government		Northeast	Midwest	South	West
	State	3140	2951	1978	2982
		2196	2774	2501	2596
		2716	2317	1698	2907
		3538	2032	2646	2345
		2849	2870	2297	3480
	Local	2208	2892	1651	2534
		3237	2707	1317	3040
		3403	2761	2514	2524
		2561	1704	2504	3461
		2521	1911	2273	2726

PRINTOUT 15.15   Minitab output for Exercise 15.92

Analysis of Variance for PRICE

Source	DF	SS	MS	F	P
SUBJECT	4	9387.3	2346.8	42.29	0.000
YEAR	2	8953.8	4476.9	80.68	0.000
SUBJECT*YEAR	8	1169.5	146.2	2.63	0.013
Error	75	4161.8	55.5		
Total	89	23672.5			

ROWS: SUBJECT   COLUMNS: YEAR

	1	2	3	ALL
1	35.167	42.000	44.833	40.667
2	41.833	61.000	59.833	54.222
3	44.000	72.167	70.000	62.056
4	51.000	73.833	79.833	68.222
5	50.000	77.000	75.000	67.333
ALL	44.400	65.200	65.900	58.500

CELL CONTENTS --
       PRICE:MEAN

PRINTOUT 15.16   Minitab output for Exercise 15.93

Analysis of Variance for EARNING

Source	DF	SS	MS	F	P
GOVNMNT	1	139712	139712	0.64	0.430
REGION	3	3473223	1157741	5.30	0.004
GOVNMNT*REGION	3	54286	18095	0.08	0.969
Error	32	6985533	218298		
Total	39	10652754			

ROWS: GOVNMNT   COLUMNS: REGION

	1	2	3	4	ALL
1	2887.8	2588.8	2224.0	2862.0	2640.6
2	2786.0	2395.0	2051.8	2857.0	2522.4
ALL	2836.9	2491.9	2137.9	2859.5	2581.6

CELL CONTENTS --
       EARNING:MEAN

15.94 The U.S. National Center for Health Statistics collects data on length of stay in noninstitutional, short-stay hospitals by sex and age. Results are published in *Vital and Health Statistics*. Independent random samples of Americans were taken to compare the lengths of stay for males and females in four age groups (15–24 years, 25–34 years, 35–44 years, and 45–64 years). Then Minitab was applied to perform a two-way ANOVA on the data. Printout 15.17 on the next page shows the resulting computer output and also supplies a table of cell and marginal means. The length-of-stay data are in days.

15.95 The College Placement Council conducts surveys on salary offers to candidates for degrees and publishes its findings in *A Study of Beginning Offers*. To study the effects of major and degree for physics and computer-science (CS) majors receiving either a bachelor's or a master's degree, independent random samples of seven candidates each were obtained from the four populations under consideration. Then Minitab was applied to perform a two-way ANOVA on the data. Printout 15.18 on the next page shows the resulting computer output and also supplies a table of cell and marginal means. The salary data are in thousands of dollars.

15.96 (Computer exercise) Refer to Exercise 15.90. Use Minitab or some other statistical software to obtain computer output similar to that in Printout 15.13.

15.97 (Computer exercise) Refer to Exercise 15.91. Use Minitab or some other statistical software to obtain computer output similar to that in Printout 15.14.

15.98 (Computer exercise) Refer to Exercise 15.92. Use Minitab or some other statistical software to perform a residual analysis on the data.

15.99 (Computer exercise) Refer to Exercise 15.93. Use Minitab or some other statistical software to perform a residual analysis on the data.

Multiple comparisons in two-way ANOVA. This material presumes that you have studied the optional section on multiple comparisons, Section 15.4. There are two types of multiple comparisons in two-way ANOVA. One type compares the means of all ab populations (treatments); the other compares the level means of either factor. Here we will discuss the

Tukey multiple-comparison method. The procedure is basically identical to that given in Procedure 15.2 on page 889; the essential difference is that for comparisons of the level means of either factor, $v = n - ab$, just as it is for comparisons of the treatment means. Because we are considering only balanced two-way ANOVA, a simplification occurs in the expression for the margin of error. Here now are the formulas for Tukey multiple comparisons in a balanced two-way ANOVA. In the formulas, $s = \sqrt{MSE}$, m denotes the common sample size, $\bar{x}_i$ the ith treatment sample mean (cell mean), $\overline{A}_i$ the ith Factor A level sample mean (row mean), and $\overline{B}_i$ the ith Factor B level sample mean (column mean).

- Simultaneous $(1 - \alpha)$-level confidence intervals for differences between treatment means are given by the formula

$$(\bar{x}_i - \bar{x}_j) \pm q_\alpha \cdot s/\sqrt{m},$$

where q_α is obtained for a q-curve with parameters $\kappa = ab$ and $v = n - ab$.

- Simultaneous $(1 - \alpha)$-level confidence intervals for differences between Factor A level means are given by the formula

$$(\overline{A}_i - \overline{A}_j) \pm q_\alpha \cdot s/\sqrt{bm},$$

where q_α is obtained for a q-curve with parameters $\kappa = a$ and $v = n - ab$.

- Simultaneous $(1 - \alpha)$-level confidence intervals for differences between Factor B level means are given by the formula

$$(\overline{B}_i - \overline{B}_j) \pm q_\alpha \cdot s/\sqrt{am},$$

where q_α is obtained for a q-curve with parameters $\kappa = b$ and $v = n - ab$.

15.100 Refer to Exercise 15.91. Using a family significance level of 0.05, perform a Tukey multiple comparison for the
a. level means of irrigation regime.
b. level means of polymer.

15.101 Refer to Exercise 15.95. Perform a Tukey multiple comparison for the four treatment means. Use a family significance level of 0.05.

PRINTOUT 15.17　Minitab output for Exercise 15.94

Analysis of Variance for STAY

Source	DF	SS	MS	F	P
SEX	1	34.225	34.225	15.92	0.000
AGE	3	16.475	5.492	2.55	0.073
SEX*AGE	3	10.475	3.492	1.62	0.203
Error	32	68.800	2.150		
Total	39	129.975			

ROWS: SEX　　COLUMNS: AGE

	1	2	3	4	ALL
1	6.0000	6.6000	6.4000	6.6000	6.4000
2	3.4000	3.6000	5.0000	6.2000	4.5500
ALL	4.7000	5.1000	5.7000	6.4000	5.4750

CELL CONTENTS --
　　　STAY:MEAN

PRINTOUT 15.18　Minitab output for Exercise 15.95

Analysis of Variance for SALARY

Source	DF	SS	MS	F	P
MAJOR	1	137.73	137.73	12.12	0.002
DEGREE	1	223.46	223.46	19.66	0.000
MAJOR*DEGREE	1	62.10	62.10	5.46	0.028
Error	24	272.83	11.37		
Total	27	696.12			

ROWS: MAJOR　　COLUMNS: DEGREE

	1	2	ALL
1	29.029	31.700	30.364
2	30.486	39.114	34.800
ALL	29.757	35.407	32.582

CELL CONTENTS --
　　　SALARY:MEAN

Chapter Review

Key Terms

$a \times b$ ANOVA, *908*
analysis of variance (ANOVA), *865*
ANOVA,* *925*
ANOVA,* *925*
AOVOneway,* *882*
AOVONEWAY,* *882*
balanced, *911*
cell means, *914*
degrees of freedom for the denominator, *866*
degrees of freedom for the numerator, *866*
error, *873*
error mean square (*MSE*), *873, 910*
error sum of squares (*SSE*), *873*
F_α, *867*
F-curve, *866*
F-distribution, *866*
F-statistic, *873*
Factor *A* mean square (*MSA*), *910*
Factor *A* sum of squares (*SSA*), *910*
Factor *B* mean square (*MSB*), *910*
Factor *B* sum of squares (*SSB*), *910*
family confidence level,* *888*
individual confidence level,* *888*
interact, *909*
interaction, *909*
interaction mean square (*MSAB*), *910*
interaction sum of squares (*SSAB*), *910*
Kruskal-Wallis,* *903*
KRUSKAL-WALLIS,* *903*
Kruskal–Wallis test,* *897*

levels, *908*
main effects, *909*
marginal means, *914*
Means,* *926*
multiple comparisons, *887*
one-way analysis of variance, *868*
one-way ANOVA identity, *876*
one-way ANOVA tables, *876*
Oneway,* *892*
ONEWAY,* *892*
q_α,* *888*
q-curve,* *888*
q-distribution,* *888*
residual, *869*
response variable, *908*
rule of 2, *869*
studentized range distribution,* *888*
test for Factor *A* main effects, *912*
test for Factor *B* main effects, *912*
test for interaction, *912*
total sum of squares (*SST*), *875*
treatment, *873*
treatment mean square (*MSTR*), *872*
treatment sum of squares (*SSTR*), *872*
Tukey,* *892*
TUKEY,* *892*
Tukey multiple-comparison method,* *887*
two-way analysis of variance, *908*
two-way ANOVA identity, *910*
two-way ANOVA tables, *911*

Formulas

In the formulas below,

k = number of populations
n = total number of pieces of data
$\overline{x}$ = mean of all n pieces of data
n_j = size of sample from Population j
$\overline{x}_j$ = mean of sample from Population j
s_j^2 = variance of sample from Population j

T_j = sum of sample data from Population j
SST = total sum of squares
$SSTR$ = treatment sum of squares
SSE = error sum of squares
$MSTR$ = treatment mean square
MSE = error mean square

Defining formulas for sums of squares in one-way ANOVA, *877*

$$SST = \Sigma(x - \overline{x})^2$$

$$SSTR = \Sigma n_j(\overline{x}_j - \overline{x})^2$$

$$SSE = \Sigma(n_j - 1)s_j^2$$

One-way ANOVA identity, *876*

$$SST = SSTR + SSE$$

Shortcut formulas for sums of squares in one-way ANOVA, *877*

$$SST = \Sigma x^2 - (\Sigma x)^2/n$$
$$SSTR = \Sigma(T_j^2/n_j) - (\Sigma x)^2/n$$
$$SSE = SST - SSTR$$

Mean squares in one-way ANOVA, *872, 873*

$$MSTR = \frac{SSTR}{k-1}, \qquad MSE = \frac{SSE}{n-k}$$

Test statistic for one-way ANOVA, *875*

$$F = \frac{MSTR}{MSE}$$

with df $= (k-1, n-k)$.

Confidence interval for $\mu_i - \mu_j$ in the Tukey multiple-comparison method,* *889*

$$(\bar{x}_i - \bar{x}_j) \pm \frac{q_\alpha}{\sqrt{2}} \cdot s\sqrt{(1/n_i) + (1/n_j)},$$

where $s = \sqrt{MSE}$ and q_α is obtained for a q-curve with parameters k and $n-k$.

Test statistic for a Kruskal–Wallis test,* *899, 900*

$$H = \frac{SSTR}{SST/(n-1)}$$

or

$$H = \frac{12}{n(n+1)} \sum \frac{R_j^2}{n_j} - 3(n+1),$$

where *SSTR* and *SST* are computed for the ranks of the data, and R_j denotes the sum of the ranks for the sample data from Population j. The test statistic H has approximately a chi-square distribution with df $= k - 1$.

In the formulas below,

SST = total sum of squares
SSA = Factor A sum of squares
SSB = Factor B sum of squares
$SSAB$ = interaction sum of squares
SSE = error sum of squares
MSA = Factor A mean square
MSB = Factor B mean square

$MSAB$ = interaction mean square
MSE = error mean square
a = number of levels of Factor A
b = number of levels of Factor B
m = common sample size
n = total number of pieces of data = mab

Two-way ANOVA identity, *910*

$$SST = SSA + SSB + SSAB + SSE$$

Mean squares in two-way ANOVA, *910*

$$MSA = \frac{SSA}{a-1}, \quad MSB = \frac{SSB}{b-1}, \quad MSAB = \frac{SSAB}{(a-1)(b-1)}, \quad MSE = \frac{SSE}{n-ab}$$

Test statistics for Factor A main effects, Factor B main effects, and interaction, *912*

$$F_A = \frac{MSA}{MSE}, \qquad F_B = \frac{MSB}{MSE}, \qquad \text{and} \qquad F_{AB} = \frac{MSAB}{MSE}$$

with df = $(a-1, n-ab)$, df = $(b-1, n-ab)$, and df = $\big((a-1)(b-1), n-ab\big)$, respectively.

You Should Be Able To

1. use and understand the preceding formulas.
2. use the F-tables, Tables X and XI in Appendix A.
3. explain the essential ideas behind a one-way analysis of variance.
4. state and check the assumptions required for a one-way ANOVA.
5. compute the sums of squares for a one-way ANOVA using the defining formulas.
6. compute the sums of squares for a one-way ANOVA using the shortcut formulas.
7. compute the mean squares and the F-statistic for a one-way ANOVA.
8. construct a one-way ANOVA table.
9. perform a one-way ANOVA test.
10. use the q-tables, Tables XII and XIII in Appendix A.*
11. perform a multiple comparison using the Tukey method.*
12. perform a Kruskal–Wallis test.*
13. explain the essential ideas behind a two-way analysis of variance.
14. state and check the assumptions required for a two-way ANOVA.
15. construct a two-way ANOVA table, given the sums of squares.
16. perform a two-way ANOVA.
17. use the Minitab commands covered in this chapter.*
18. interpret the output obtained from the application of the Minitab commands discussed in this chapter.*

REVIEW TEST

1. Consider an F-curve with df $= (24, 5)$.
   a. Identify the number of degrees of freedom for the numerator.
   b. Identify the number of degrees of freedom for the denominator.
   c. Determine $F_{0.05}$.
   d. Find the F-value having area 0.01 to its right.
   e. Find the F-value having area 0.05 to its right.

2. In one-way ANOVA, identify a statistic that measures
   a. the variation among the sample means.
   b. the variation within the samples.

3. In one-way ANOVA,
   a. list and interpret the three sums of squares.
   b. state the one-way ANOVA identity and interpret its meaning with regard to partitioning the total variation in the data.

4. The Federal Bureau of Investigation conducts surveys to obtain information on the value of losses due to various types of robberies. Results of the surveys are published in *Population-at-Risk Rates and Crime Indicators*. Independent random samples of reports for three types of robberies—highway, gas station, and convenience store—gave the following data, in dollars, on value of losses.

Highway	Gas station	Convenience store
608	636	476
652	533	553
495	512	512
744	451	338
680	291	234
	618	246

   a. Obtain the sample mean and sample variance of each of the three data sets.
   b. Determine *MSTR* and *MSE*. Use the defining formulas to compute *SSTR* and *SSE*.
   c. What is *MSTR* measuring?
   d. What is *MSE* measuring?

5. Refer to Problem 4.
   a. Suppose we want to perform a one-way ANOVA to compare the mean losses for the three types of robberies. What conditions are necessary? How crucial are those conditions?
   b. Obtain normal probability plots for each of the three samples.

   c. Apply the rule of 2.
   d. Perform a residual analysis.
   e. Does it seem reasonable to consider the assumptions for one-way ANOVA met? Provide a detailed explanation for your answers.

6. Refer to Problem 4. At the 5% significance level, do the data provide sufficient evidence to conclude that a difference in mean losses exists among the three types of robberies? Use one-way ANOVA to perform the required hypothesis test.

*7. (Computer problem) Use Minitab or some other statistical software to carry out the one-way ANOVA test in Problem 6.

*8. (Computer problem) In Problem 6 you performed a one-way ANOVA on the data in Problem 4 to decide whether a difference exists in mean losses among three types of robberies. Printout 15.19 shows the output obtained by applying Minitab's AOVOneway command to the data in Problem 4. Use the output to determine
   a. the three sums of squares, *SSTR, SSE,* and *SST*.
   b. the treatment mean square and the error mean square.
   c. the value of the test statistic, F.
   d. the P-value for the hypothesis test.
   e. the conclusion, if the hypothesis test is performed at the 5% significance level.
   f. the size, mean, and standard deviation for each of the three samples.
   g. an individual 95% confidence interval for the mean loss due to convenience-store robberies.

*9. This problem concerns multiple comparisons.
   a. Give reasons for performing a multiple comparison.
   b. Explain the difference between the family confidence level and the individual confidence level.
   c. Which confidence level is appropriate for multiple comparisons, family or individual? Why?
   d. Identify the probability distribution used with the Tukey multiple-comparison method.

*10. Apply the Tukey multiple-comparison method to the data in Problem 4. Use a family confidence level of 0.95.

*11. (Computer problem) Use Minitab or some other statistical software to carry out the multiple comparison required in Problem 10.

*12. **(Computer problem)** We applied Minitab's Oneway command and Tukey subcommand to the data in Problem 4. Printout 15.20 displays the portion of the output required for a multiple comparison. (The remaining portion of the output is essentially the same as that shown in Printout 15.19.) Use the output to determine

a. the family confidence level.

b. the individual confidence level.

c. $q_{0.05}$.

d. the (simultaneous) confidence interval for the difference, $\mu_2 - \mu_3$, between the mean losses for gas-station and convenience-store robberies.

e. the (simultaneous) confidence interval for the difference, $\mu_2 - \mu_1$, between the mean losses for gas-station and highway robberies.

f. which means should be declared different.

g. Summarize the results of the Tukey multiple comparison in words.

PRINTOUT 15.19   Minitab output for Problem 8

```
ANALYSIS OF VARIANCE
SOURCE      DF       SS       MS        F        p
FACTOR       2    160601    80301     5.34    0.019
ERROR       14    210540    15039
TOTAL       16    371142
                                  INDIVIDUAL 95% CI'S FOR MEAN
                                  BASED ON POOLED STDEV
  LEVEL      N      MEAN    STDEV   -+---------+---------+---------+-----
  HIGHWAY    5     635.8     92.9                      (------*-------)
  GASSTATI   6     506.8    126.1             (------*------)
  CONVENIE   6     393.2    139.0   (------*------)
                                   -+---------+---------+---------+-----
  POOLED STDEV =   122.6           300       450       600       750
```

PRINTOUT 15.20   Minitab output for Problem 12

```
Tukey's pairwise comparisons

    Family error rate = 0.0500
Individual error rate = 0.0203

Critical value = 3.70

Intervals for (column level mean) - (row level mean)

               1         2

  2          -65
            323

  3           48       -72
            437       299
```

*13. Refer to the data in Problem 4.

a. At the 5% significance level, do the data provide sufficient evidence to conclude that a difference in mean losses exists among the three types of robberies? Use the Kruskal–Wallis procedure to perform the required hypothesis test.

b. The hypothesis test in part (a) was done in Problem 6 using the one-way ANOVA procedure. The assumption in that exercise is that for the three types of robberies, losses are (approximately) normally distributed and have (approximately) equal standard deviations. Presuming that, in fact, losses for all three types of robberies are normally distributed and have equal standard deviations, why is it permissible to perform a Kruskal–Wallis test to compare the means? Is it better in this case to use the one-way ANOVA test or the Kruskal–Wallis test? Explain your answers.

*14. **(Computer problem)** Use Minitab or some other statistical software to carry out the Kruskal–Wallis test required in Problem 13(a).

*15. **(Computer problem)** In Problem 13(a) you performed a Kruskal–Wallis test on the data in Problem 4 to decide whether a difference exists in mean losses among three types of robberies. Printout 15.21 shows the output obtained by applying Minitab's Kruskal-Wallis command to the data in Problem 4. Use the output to determine

a. the value of the test statistic, H.

b. the null and alternative hypotheses.

c. the P-value for the hypothesis test.

d. the conclusion if the hypothesis test is performed at the 5% significance level.

e. the sample size, sample median, and mean rank for each of the three samples.

f. Compare the result of the Kruskal–Wallis test to that of the one-way ANOVA test done on the same data in Problem 8.

16. For a two-way ANOVA, define the following concepts:
a. levels of Factor B
b. no interaction
c. a main effect due to Factor A

17. The U.S. Federal Highway Administration publishes data on domestic motor fuel consumption by type of vehicle in *Highway Statistics*. Independent samples gave the following data on fuel consumption, in hundreds of gallons.

		Year			
		1988	1989	1990	1991
Vehicle	Cars	3.51	4.05	4.95	4.50
		6.46	4.32	5.31	8.44
		5.80	4.57	6.34	5.07
		5.31	6.03	2.08	3.44
		4.38	6.46	6.46	3.29
	Buses	13.89	13.97	10.61	10.56
		14.02	13.06	14.36	15.29
		16.04	16.68	15.43	14.95
		15.65	15.08	15.62	13.88
		15.20	17.09	15.40	13.80
	Trucks	13.99	11.73	14.47	13.37
		16.41	12.07	12.81	9.35
		11.90	12.47	12.01	13.46
		13.61	15.25	15.19	13.01
		11.33	14.91	10.04	14.25

For a two-way ANOVA, let Factor A be "vehicle" and Factor B be "year." Given that $SSA = 1045.698$, $SSB = 5.757$, $SSAB = 2.914$, and $SSE = 146.023$, construct the two-way ANOVA table for the data.

PRINTOUT 15.21 Minitab output for Problem 15

```
LEVEL    NOBS    MEDIAN    AVE. RANK    Z VALUE
  1       5       652.0       13.4        2.32
  2       6       522.5        8.9       -0.05
  3       6       407.0        5.4       -2.16
OVERALL  17                    9.0

H = 6.82   d.f. = 2   p = 0.033
H = 6.83   d.f. = 2   p = 0.033 (adjusted for ties)
```

18. Refer to Problem 17. Printout 15.22 shows computer output obtained for a two-way ANOVA on these data and also provides a table of cell and marginal means. Perform a two-way ANOVA at the 5% individual significance level by employing Procedure 15.4 on page 921. *(Note: Preliminary data analyses indicate that it is reasonable*

to consider the assumptions for performing a two-way ANOVA satisfied.)*

*19. **(Computer problem)** Refer to the data in Problem 17. Use Minitab or some other statistical software to obtain computer output similar to that in Printout 15.22.

PRINTOUT 15.22   Minitab output for Problem 18

Analysis of Variance for CONSUMPT

Source	DF	SS	MS	F	P
VEHICLE	2	1045.698	522.849	171.87	0.000
YEAR	3	5.757	1.919	0.63	0.599
VEHICLE*YEAR	6	2.914	0.486	0.16	0.986
Error	48	146.023	3.042		
Total	59	1200.392			

ROWS: VEHICLE   COLUMNS: YEAR

	1	2	3	4	ALL
1	5.092	5.086	5.028	4.948	5.038
2	14.960	15.176	14.284	13.696	14.529
3	13.448	13.286	12.904	12.688	13.082
ALL	11.167	11.183	10.739	10.444	10.883

CELL CONTENTS --
    CONSUMPT:MEAN

Using the Focus Database   Consider again the database, printed in Appendix B, obtained by randomly selecting 500 Arizona State University sophomores. Seven variables are considered for each student: sex, high-school GPA, SAT math score, cumulative GPA, SAT verbal score, age, and total hours. For these database exercises, you should eliminate all cases (students) in which one or more of the four variables, cumulative GPA, high-school GPA, SAT math score, and SAT verbal score, equal 0.

a. Conduct a one-way ANOVA at the 5% significance level to test for differences among mean cumulative GPAs of Arizona State University sophomores in the three high-school GPA categories: under 2, 2–2.99, and 3 or over. Also perform a Tukey multiple comparison using a family confidence level of 95%.

b. Conduct a one-way ANOVA at the 5% significance level to test for differences among mean cumulative GPAs of Arizona State University sophomores in the five SAT math score categories: under 400, 400–499, 500–599, 600–699, and 700 or over. Also perform a Tukey multiple comparison using a family confidence level of 95%.

c. Conduct a one-way ANOVA at the 5% significance level to test for differences among mean cumulative GPAs of Arizona State University sophomores in the four SAT verbal score categories: under 400, 400–499, 500–599, and 600–699. (None of the students under consideration have an SAT verbal score of 700 or above.) Also perform a Tukey multiple comparison using a family confidence level of 95%.

d. Conduct a one-way ANOVA at the 5% significance level to test for differences among mean cumulative GPAs of Arizona State University sophomores in the three age categories: under 20, 20, and over 20. Also perform a Tukey multiple comparison using a family confidence level of 95%.

Case Study   Heavy Drinking Among College Students

Excessive drinking causes a multitude of problems for society in general and for college students in particular. According to a study by T. M. O'Hare, published in a 1990 issue of *Journal of Studies on Alcohol,* 81.5% of college students drink alcohol. Moreover, 37% of college students are moderate drinkers and 19% are heavy drinkers.

Recently, Professor Kate Carey of Syracuse University surveyed 78 college students from an introductory psychology class, all of whom were regular drinkers of alcohol. Her purpose was to identify interpersonal and intrapersonal situations associated with excessive drinking among college students and to detect those situations that differentiate heavy drinkers from light and moderate ones. She published her findings in a 1993 issue of the *Journal of Counseling Psychology.*[†]

To quantify drinking patterns, Carey used the time-line follow-back procedure (TLFB). The TLFB is a structured interview that is known to provide reliable estimates of daily drinking. In this case each student filled in each day of a blank calendar covering the previous month with the number of standard drink equivalents (SDEs) consumed on that day. One SDE is defined to be 1 fluid ounce of hard liquor, 12 fluid ounces of beer, or 4 fluid ounces of wine. Based on the results of the TLFB, the students were divided into three categories according to average quantity of alcohol consumed per drinking day: light drinkers ($\leq$ 3 SDEs), moderate drinkers (4–6 SDEs), and heavy drinkers (> 6 SDEs).

To assess the frequency of excessive drinking in intrapersonal and interpersonal situations, Carey employed the short form of the Inventory of Drinking Situations (IDS). This form consists of 42 items, each of which is rated on a 4-point scale ranging from (1) never drink heavily in that type of situation to (4) almost always drink heavily in that type of situation. The 42 items are divided into eight subscales, three interpersonal and five intrapersonal, as displayed in the first column of Table 15.21. A subscale score represents the average rating for the items constituting the subscale.

Table 15.21 provides the sample size, sample mean, and sample standard deviation for each drinking category and each situational context (subscale). The table shows, for example, that the mean score of all light drinkers in the sample on the "conflict with others" subscale is 1.23; and the standard deviation of the scores is 0.27.

[†] From "Situational Determinants of Heavy Drinking Among College Students" by Kate B. Carey, 1993, *Journal of Counseling Psychology, 40,* pp. 217–220.

TABLE 15.21

Means and standard deviations
for IDS scores, by situational
context and drinker classification

IDS subscale	Light drinkers ($n_1 = 16$)		Moderate drinkers ($n_2 = 47$)		Heavy drinkers ($n_3 = 15$)	
	$\bar{x}_1$	s_1	$\bar{x}_2$	s_2	$\bar{x}_3$	s_3
Interpersonal situations						
Conflict with others	1.23	0.27	1.53	0.49	1.79	0.49
Social pressure to drink	2.64	0.80	2.91	0.55	3.51	0.51
Pleasant times with others	2.21	0.67	2.53	0.51	3.03	0.38
Intrapersonal situations						
Unpleasant emotions	1.22	0.35	1.61	0.69	1.68	0.46
Physical discomfort	1.03	0.08	1.19	0.29	1.40	0.32
Pleasant emotions	2.09	0.73	2.61	0.58	3.03	0.30
Testing personal control	1.52	0.74	1.56	0.56	1.53	0.48
Urges and temptations	1.80	0.56	1.96	0.51	2.33	0.58

a. For each of the eight IDS subscales, perform a (separate) one-way ANOVA to decide whether a difference exists in mean IDS scores among the three drinker categories. Use $\alpha = 0.05$. (*Note:* Use the formula given in Exercise 15.22(a) on page 875 to obtain $\bar{x}$ for each ANOVA.)

b. For those one-way ANOVAs in part (a) that are statistically significant, carry out a Tukey multiple comparison using a family confidence level of 0.95.

c. Based on the data in Table 15.21, are there any of the eight ANOVAs that perhaps should not have been carried out? Explain your answer. (*Hint:* Rule of 2.)

d. Use Minitab or some other statistical software to perform the analyses required in parts (a) and (b). (*Hint:* Refer to the hint given in part (e) on page 571.)

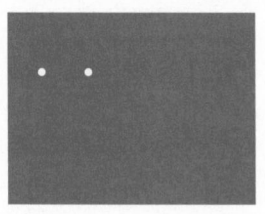

Appendixes

Statistical Tables

. . . . .

TABLE I

Binomial probabilities:

$$\binom{n}{x} p^x (1-p)^{n-x}$$

| | | | | | | | p | | | | | |
n	x	0.1	0.2	0.25	0.3	0.4	0.5	0.6	0.7	0.75	0.8	0.9
1	0	0.900	0.800	0.750	0.700	0.600	0.500	0.400	0.300	0.250	0.200	0.100
	1	0.100	0.200	0.250	0.300	0.400	0.500	0.600	0.700	0.750	0.800	0.900
2	0	0.810	0.640	0.563	0.490	0.360	0.250	0.160	0.090	0.063	0.040	0.010
	1	0.180	0.320	0.375	0.420	0.480	0.500	0.480	0.420	0.375	0.320	0.180
	2	0.010	0.040	0.063	0.090	0.160	0.250	0.360	0.490	0.563	0.640	0.810
3	0	0.729	0.512	0.422	0.343	0.216	0.125	0.064	0.027	0.016	0.008	0.001
	1	0.243	0.384	0.422	0.441	0.432	0.375	0.288	0.189	0.141	0.096	0.027
	2	0.027	0.096	0.141	0.189	0.288	0.375	0.432	0.441	0.422	0.384	0.243
	3	0.001	0.008	0.016	0.027	0.064	0.125	0.216	0.343	0.422	0.512	0.729
4	0	0.656	0.410	0.316	0.240	0.130	0.063	0.026	0.008	0.004	0.002	0.000
	1	0.292	0.410	0.422	0.412	0.346	0.250	0.154	0.076	0.047	0.026	0.004
	2	0.049	0.154	0.211	0.265	0.346	0.375	0.346	0.265	0.211	0.154	0.049
	3	0.004	0.026	0.047	0.076	0.154	0.250	0.346	0.412	0.422	0.410	0.292
	4	0.000	0.002	0.004	0.008	0.026	0.063	0.130	0.240	0.316	0.410	0.656
5	0	0.590	0.328	0.237	0.168	0.078	0.031	0.010	0.002	0.001	0.000	0.000
	1	0.328	0.410	0.396	0.360	0.259	0.156	0.077	0.028	0.015	0.006	0.000
	2	0.073	0.205	0.264	0.309	0.346	0.312	0.230	0.132	0.088	0.051	0.008
	3	0.008	0.051	0.088	0.132	0.230	0.312	0.346	0.309	0.264	0.205	0.073
	4	0.000	0.006	0.015	0.028	0.077	0.156	0.259	0.360	0.396	0.410	0.328
	5	0.000	0.000	0.001	0.002	0.010	0.031	0.078	0.168	0.237	0.328	0.590
6	0	0.531	0.262	0.178	0.118	0.047	0.016	0.004	0.001	0.000	0.000	0.000
	1	0.354	0.393	0.356	0.303	0.187	0.094	0.037	0.010	0.004	0.002	0.000
	2	0.098	0.246	0.297	0.324	0.311	0.234	0.138	0.060	0.033	0.015	0.001
	3	0.015	0.082	0.132	0.185	0.276	0.313	0.276	0.185	0.132	0.082	0.015
	4	0.001	0.015	0.033	0.060	0.138	0.234	0.311	0.324	0.297	0.246	0.098
	5	0.000	0.002	0.004	0.010	0.037	0.094	0.187	0.303	0.356	0.393	0.354
	6	0.000	0.000	0.000	0.001	0.004	0.016	0.047	0.118	0.178	0.262	0.531
7	0	0.478	0.210	0.133	0.082	0.028	0.008	0.002	0.000	0.000	0.000	0.000
	1	0.372	0.367	0.311	0.247	0.131	0.055	0.017	0.004	0.001	0.000	0.000
	2	0.124	0.275	0.311	0.318	0.261	0.164	0.077	0.025	0.012	0.004	0.000
	3	0.023	0.115	0.173	0.227	0.290	0.273	0.194	0.097	0.058	0.029	0.003
	4	0.003	0.029	0.058	0.097	0.194	0.273	0.290	0.227	0.173	0.115	0.023
	5	0.000	0.004	0.012	0.025	0.077	0.164	0.261	0.318	0.311	0.275	0.124
	6	0.000	0.000	0.001	0.004	0.017	0.055	0.131	0.247	0.311	0.367	0.372
	7	0.000	0.000	0.000	0.000	0.002	0.008	0.028	0.082	0.133	0.210	0.478

TABLE I (cont.)

Binomial probabilities:

$$\binom{n}{x} p^x (1-p)^{n-x}$$

| | | | | | | | p | | | | | |
n	x	0.1	0.2	0.25	0.3	0.4	0.5	0.6	0.7	0.75	0.8	0.9
8	0	0.430	0.168	0.100	0.058	0.017	0.004	0.001	0.000	0.000	0.000	0.000
	1	0.383	0.336	0.267	0.198	0.090	0.031	0.008	0.001	0.000	0.000	0.000
	2	0.149	0.294	0.311	0.296	0.209	0.109	0.041	0.010	0.004	0.001	0.000
	3	0.033	0.147	0.208	0.254	0.279	0.219	0.124	0.047	0.023	0.009	0.000
	4	0.005	0.046	0.087	0.136	0.232	0.273	0.232	0.136	0.087	0.046	0.005
	5	0.000	0.009	0.023	0.047	0.124	0.219	0.279	0.254	0.208	0.147	0.033
	6	0.000	0.001	0.004	0.010	0.041	0.109	0.209	0.296	0.311	0.294	0.149
	7	0.000	0.000	0.000	0.001	0.008	0.031	0.090	0.198	0.267	0.336	0.383
	8	0.000	0.000	0.000	0.000	0.001	0.004	0.017	0.058	0.100	0.168	0.430
9	0	0.387	0.134	0.075	0.040	0.010	0.002	0.000	0.000	0.000	0.000	0.000
	1	0.387	0.302	0.225	0.156	0.060	0.018	0.004	0.000	0.000	0.000	0.000
	2	0.172	0.302	0.300	0.267	0.161	0.070	0.021	0.004	0.001	0.000	0.000
	3	0.045	0.176	0.234	0.267	0.251	0.164	0.074	0.021	0.009	0.003	0.000
	4	0.007	0.066	0.117	0.172	0.251	0.246	0.167	0.074	0.039	0.017	0.001
	5	0.001	0.017	0.039	0.074	0.167	0.246	0.251	0.172	0.117	0.066	0.007
	6	0.000	0.003	0.009	0.021	0.074	0.164	0.251	0.267	0.234	0.176	0.045
	7	0.000	0.000	0.001	0.004	0.021	0.070	0.161	0.267	0.300	0.302	0.172
	8	0.000	0.000	0.000	0.000	0.004	0.018	0.060	0.156	0.225	0.302	0.387
	9	0.000	0.000	0.000	0.000	0.000	0.002	0.010	0.040	0.075	0.134	0.387
10	0	0.349	0.107	0.056	0.028	0.006	0.001	0.000	0.000	0.000	0.000	0.000
	1	0.387	0.268	0.188	0.121	0.040	0.010	0.002	0.000	0.000	0.000	0.000
	2	0.194	0.302	0.282	0.233	0.121	0.044	0.011	0.001	0.000	0.000	0.000
	3	0.057	0.201	0.250	0.267	0.215	0.117	0.042	0.009	0.003	0.001	0.000
	4	0.011	0.088	0.146	0.200	0.251	0.205	0.111	0.037	0.016	0.006	0.000
	5	0.001	0.026	0.058	0.103	0.201	0.246	0.201	0.103	0.058	0.026	0.001
	6	0.000	0.006	0.016	0.037	0.111	0.205	0.251	0.200	0.146	0.088	0.011
	7	0.000	0.001	0.003	0.009	0.042	0.117	0.215	0.267	0.250	0.201	0.057
	8	0.000	0.000	0.000	0.001	0.011	0.044	0.121	0.233	0.282	0.302	0.194
	9	0.000	0.000	0.000	0.000	0.002	0.010	0.040	0.121	0.188	0.268	0.387
	10	0.000	0.000	0.000	0.000	0.000	0.001	0.006	0.028	0.056	0.107	0.349
11	0	0.314	0.086	0.042	0.020	0.004	0.000	0.000	0.000	0.000	0.000	0.000
	1	0.384	0.236	0.155	0.093	0.027	0.005	0.001	0.000	0.000	0.000	0.000
	2	0.213	0.295	0.258	0.200	0.089	0.027	0.005	0.001	0.000	0.000	0.000
	3	0.071	0.221	0.258	0.257	0.177	0.081	0.023	0.004	0.001	0.000	0.000
	4	0.016	0.111	0.172	0.220	0.236	0.161	0.070	0.017	0.006	0.002	0.000
	5	0.002	0.039	0.080	0.132	0.221	0.226	0.147	0.057	0.027	0.010	0.000
	6	0.000	0.010	0.027	0.057	0.147	0.226	0.221	0.132	0.080	0.039	0.002
	7	0.000	0.002	0.006	0.017	0.070	0.161	0.236	0.220	0.172	0.111	0.016
	8	0.000	0.000	0.001	0.004	0.023	0.081	0.177	0.257	0.258	0.221	0.071
	9	0.000	0.000	0.000	0.001	0.005	0.027	0.089	0.200	0.258	0.295	0.213
	10	0.000	0.000	0.000	0.000	0.001	0.005	0.027	0.093	0.155	0.236	0.384
	11	0.000	0.000	0.000	0.000	0.000	0.000	0.004	0.020	0.042	0.086	0.314

(continued)

TABLE I (cont.)

Binomial probabilities:

$$\binom{n}{x} p^x (1-p)^{n-x}$$

							p					
n	x	0.1	0.2	0.25	0.3	0.4	0.5	0.6	0.7	0.75	0.8	0.9
12	0	0.282	0.069	0.032	0.014	0.002	0.000	0.000	0.000	0.000	0.000	0.000
	1	0.377	0.206	0.127	0.071	0.017	0.003	0.000	0.000	0.000	0.000	0.000
	2	0.230	0.283	0.232	0.168	0.064	0.016	0.002	0.000	0.000	0.000	0.000
	3	0.085	0.236	0.258	0.240	0.142	0.054	0.012	0.001	0.000	0.000	0.000
	4	0.021	0.133	0.194	0.231	0.213	0.121	0.042	0.008	0.002	0.001	0.000
	5	0.004	0.053	0.103	0.158	0.227	0.193	0.101	0.029	0.011	0.003	0.000
	6	0.000	0.016	0.040	0.079	0.177	0.226	0.177	0.079	0.040	0.016	0.000
	7	0.000	0.003	0.011	0.029	0.101	0.193	0.227	0.158	0.103	0.053	0.004
	8	0.000	0.001	0.002	0.008	0.042	0.121	0.213	0.231	0.194	0.133	0.021
	9	0.000	0.000	0.000	0.001	0.012	0.054	0.142	0.240	0.258	0.236	0.085
	10	0.000	0.000	0.000	0.000	0.002	0.016	0.064	0.168	0.232	0.283	0.230
	11	0.000	0.000	0.000	0.000	0.000	0.003	0.017	0.071	0.127	0.206	0.377
	12	0.000	0.000	0.000	0.000	0.000	0.000	0.002	0.014	0.032	0.069	0.282
13	0	0.254	0.055	0.024	0.010	0.001	0.000	0.000	0.000	0.000	0.000	0.000
	1	0.367	0.179	0.103	0.054	0.011	0.002	0.000	0.000	0.000	0.000	0.000
	2	0.245	0.268	0.206	0.139	0.045	0.010	0.001	0.000	0.000	0.000	0.000
	3	0.100	0.246	0.252	0.218	0.111	0.035	0.006	0.001	0.000	0.000	0.000
	4	0.028	0.154	0.210	0.234	0.184	0.087	0.024	0.003	0.001	0.000	0.000
	5	0.006	0.069	0.126	0.180	0.221	0.157	0.066	0.014	0.005	0.001	0.000
	6	0.001	0.023	0.056	0.103	0.197	0.209	0.131	0.044	0.019	0.006	0.000
	7	0.000	0.006	0.019	0.044	0.131	0.209	0.197	0.103	0.056	0.023	0.001
	8	0.000	0.001	0.005	0.014	0.066	0.157	0.221	0.180	0.126	0.069	0.006
	9	0.000	0.000	0.001	0.003	0.024	0.087	0.184	0.234	0.210	0.154	0.028
	10	0.000	0.000	0.000	0.001	0.006	0.035	0.111	0.218	0.252	0.246	0.100
	11	0.000	0.000	0.000	0.000	0.001	0.010	0.045	0.139	0.206	0.268	0.245
	12	0.000	0.000	0.000	0.000	0.000	0.002	0.011	0.054	0.103	0.179	0.367
	13	0.000	0.000	0.000	0.000	0.000	0.000	0.001	0.010	0.024	0.055	0.254
14	0	0.229	0.044	0.018	0.007	0.001	0.000	0.000	0.000	0.000	0.000	0.000
	1	0.356	0.154	0.083	0.041	0.007	0.001	0.000	0.000	0.000	0.000	0.000
	2	0.257	0.250	0.180	0.113	0.032	0.006	0.001	0.000	0.000	0.000	0.000
	3	0.114	0.250	0.240	0.194	0.085	0.022	0.003	0.000	0.000	0.000	0.000
	4	0.035	0.172	0.220	0.229	0.155	0.061	0.014	0.001	0.000	0.000	0.000
	5	0.008	0.086	0.147	0.196	0.207	0.122	0.041	0.007	0.002	0.000	0.000
	6	0.001	0.032	0.073	0.126	0.207	0.183	0.092	0.023	0.008	0.002	0.000
	7	0.000	0.009	0.028	0.062	0.157	0.209	0.157	0.062	0.028	0.009	0.000
	8	0.000	0.002	0.008	0.023	0.092	0.183	0.207	0.126	0.073	0.032	0.001
	9	0.000	0.000	0.002	0.007	0.041	0.122	0.207	0.196	0.147	0.086	0.008
	10	0.000	0.000	0.000	0.001	0.014	0.061	0.155	0.229	0.220	0.172	0.035
	11	0.000	0.000	0.000	0.000	0.003	0.022	0.085	0.194	0.240	0.250	0.114
	12	0.000	0.000	0.000	0.000	0.001	0.006	0.032	0.113	0.180	0.250	0.257
	13	0.000	0.000	0.000	0.000	0.000	0.001	0.007	0.041	0.083	0.154	0.356
	14	0.000	0.000	0.000	0.000	0.000	0.000	0.001	0.007	0.018	0.044	0.229

TABLE I (cont.)

Binomial probabilities:

$$\binom{n}{x} p^{x}(1-p)^{n-x}$$

n	x	0.1	0.2	0.25	0.3	0.4	0.5	0.6	0.7	0.75	0.8	0.9
15	0	0.206	0.035	0.013	0.005	0.000	0.000	0.000	0.000	0.000	0.000	0.000
	1	0.343	0.132	0.067	0.031	0.005	0.000	0.000	0.000	0.000	0.000	0.000
	2	0.267	0.231	0.156	0.092	0.022	0.003	0.000	0.000	0.000	0.000	0.000
	3	0.129	0.250	0.225	0.170	0.063	0.014	0.002	0.000	0.000	0.000	0.000
	4	0.043	0.188	0.225	0.219	0.127	0.042	0.007	0.001	0.000	0.000	0.000
	5	0.010	0.103	0.165	0.206	0.186	0.092	0.024	0.003	0.001	0.000	0.000
	6	0.002	0.043	0.092	0.147	0.207	0.153	0.061	0.012	0.003	0.001	0.000
	7	0.000	0.014	0.039	0.081	0.177	0.196	0.118	0.035	0.013	0.003	0.000
	8	0.000	0.003	0.013	0.035	0.118	0.196	0.177	0.081	0.039	0.014	0.000
	9	0.000	0.001	0.003	0.012	0.061	0.153	0.207	0.147	0.092	0.043	0.002
	10	0.000	0.000	0.001	0.003	0.024	0.092	0.186	0.206	0.165	0.103	0.010
	11	0.000	0.000	0.000	0.001	0.007	0.042	0.127	0.219	0.225	0.188	0.043
	12	0.000	0.000	0.000	0.000	0.002	0.014	0.063	0.170	0.225	0.250	0.129
	13	0.000	0.000	0.000	0.000	0.000	0.003	0.022	0.092	0.156	0.231	0.267
	14	0.000	0.000	0.000	0.000	0.000	0.000	0.005	0.031	0.067	0.132	0.343
	15	0.000	0.000	0.000	0.000	0.000	0.000	0.000	0.005	0.013	0.035	0.206
20	0	0.122	0.012	0.003	0.001	0.000	0.000	0.000	0.000	0.000	0.000	0.000
	1	0.270	0.058	0.021	0.007	0.000	0.000	0.000	0.000	0.000	0.000	0.000
	2	0.285	0.137	0.067	0.028	0.003	0.000	0.000	0.000	0.000	0.000	0.000
	3	0.190	0.205	0.134	0.072	0.012	0.001	0.000	0.000	0.000	0.000	0.000
	4	0.090	0.218	0.190	0.130	0.035	0.005	0.000	0.000	0.000	0.000	0.000
	5	0.032	0.175	0.202	0.179	0.075	0.015	0.001	0.000	0.000	0.000	0.000
	6	0.009	0.109	0.169	0.192	0.124	0.037	0.005	0.000	0.000	0.000	0.000
	7	0.002	0.055	0.112	0.164	0.166	0.074	0.015	0.001	0.000	0.000	0.000
	8	0.000	0.022	0.061	0.114	0.180	0.120	0.035	0.004	0.001	0.000	0.000
	9	0.000	0.007	0.027	0.065	0.160	0.160	0.071	0.012	0.003	0.000	0.000
	10	0.000	0.002	0.010	0.031	0.117	0.176	0.117	0.031	0.010	0.002	0.000
	11	0.000	0.000	0.003	0.012	0.071	0.160	0.160	0.065	0.027	0.007	0.000
	12	0.000	0.000	0.001	0.004	0.035	0.120	0.180	0.114	0.061	0.022	0.000
	13	0.000	0.000	0.000	0.001	0.015	0.074	0.166	0.164	0.112	0.055	0.002
	14	0.000	0.000	0.000	0.000	0.005	0.037	0.124	0.192	0.169	0.109	0.009
	15	0.000	0.000	0.000	0.000	0.001	0.015	0.075	0.179	0.202	0.175	0.032
	16	0.000	0.000	0.000	0.000	0.000	0.005	0.035	0.130	0.190	0.218	0.090
	17	0.000	0.000	0.000	0.000	0.000	0.001	0.012	0.072	0.134	0.205	0.190
	18	0.000	0.000	0.000	0.000	0.000	0.000	0.003	0.028	0.067	0.137	0.285
	19	0.000	0.000	0.000	0.000	0.000	0.000	0.000	0.007	0.021	0.058	0.270
	20	0.000	0.000	0.000	0.000	0.000	0.000	0.000	0.001	0.003	0.012	0.122

TABLE II
Areas under the
standard normal curve

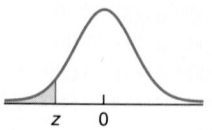

z	0.09	0.08	0.07	0.06	0.05	0.04	0.03	0.02	0.01	0.00
					Second decimal place in z					
−3.9										0.0000†
−3.8	0.0001	0.0001	0.0001	0.0001	0.0001	0.0001	0.0001	0.0001	0.0001	0.0001
−3.7	0.0001	0.0001	0.0001	0.0001	0.0001	0.0001	0.0001	0.0001	0.0001	0.0001
−3.6	0.0001	0.0001	0.0001	0.0001	0.0001	0.0001	0.0001	0.0001	0.0002	0.0002
−3.5	0.0002	0.0002	0.0002	0.0002	0.0002	0.0002	0.0002	0.0002	0.0002	0.0002
−3.4	0.0002	0.0003	0.0003	0.0003	0.0003	0.0003	0.0003	0.0003	0.0003	0.0003
−3.3	0.0003	0.0004	0.0004	0.0004	0.0004	0.0004	0.0004	0.0005	0.0005	0.0005
−3.2	0.0005	0.0005	0.0005	0.0006	0.0006	0.0006	0.0006	0.0006	0.0007	0.0007
−3.1	0.0007	0.0007	0.0008	0.0008	0.0008	0.0008	0.0009	0.0009	0.0009	0.0010
−3.0	0.0010	0.0010	0.0011	0.0011	0.0011	0.0012	0.0012	0.0013	0.0013	0.0013
−2.9	0.0014	0.0014	0.0015	0.0015	0.0016	0.0016	0.0017	0.0018	0.0018	0.0019
−2.8	0.0019	0.0020	0.0021	0.0021	0.0022	0.0023	0.0023	0.0024	0.0025	0.0026
−2.7	0.0026	0.0027	0.0028	0.0029	0.0030	0.0031	0.0032	0.0033	0.0034	0.0035
−2.6	0.0036	0.0037	0.0038	0.0039	0.0040	0.0041	0.0043	0.0044	0.0045	0.0047
−2.5	0.0048	0.0049	0.0051	0.0052	0.0054	0.0055	0.0057	0.0059	0.0060	0.0062
−2.4	0.0064	0.0066	0.0068	0.0069	0.0071	0.0073	0.0075	0.0078	0.0080	0.0082
−2.3	0.0084	0.0087	0.0089	0.0091	0.0094	0.0096	0.0099	0.0102	0.0104	0.0107
−2.2	0.0110	0.0113	0.0116	0.0119	0.0122	0.0125	0.0129	0.0132	0.0136	0.0139
−2.1	0.0143	0.0146	0.0150	0.0154	0.0158	0.0162	0.0166	0.0170	0.0174	0.0179
−2.0	0.0183	0.0188	0.0192	0.0197	0.0202	0.0207	0.0212	0.0217	0.0222	0.0228
−1.9	0.0233	0.0239	0.0244	0.0250	0.0256	0.0262	0.0268	0.0274	0.0281	0.0287
−1.8	0.0294	0.0301	0.0307	0.0314	0.0322	0.0329	0.0336	0.0344	0.0351	0.0359
−1.7	0.0367	0.0375	0.0384	0.0392	0.0401	0.0409	0.0418	0.0427	0.0436	0.0446
−1.6	0.0455	0.0465	0.0475	0.0485	0.0495	0.0505	0.0516	0.0526	0.0537	0.0548
−1.5	0.0559	0.0571	0.0582	0.0594	0.0606	0.0618	0.0630	0.0643	0.0655	0.0668
−1.4	0.0681	0.0694	0.0708	0.0721	0.0735	0.0749	0.0764	0.0778	0.0793	0.0808
−1.3	0.0823	0.0838	0.0853	0.0869	0.0885	0.0901	0.0918	0.0934	0.0951	0.0968
−1.2	0.0985	0.1003	0.1020	0.1038	0.1056	0.1075	0.1093	0.1112	0.1131	0.1151
−1.1	0.1170	0.1190	0.1210	0.1230	0.1251	0.1271	0.1292	0.1314	0.1335	0.1357
−1.0	0.1379	0.1401	0.1423	0.1446	0.1469	0.1492	0.1515	0.1539	0.1562	0.1587
−0.9	0.1611	0.1635	0.1660	0.1685	0.1711	0.1736	0.1762	0.1788	0.1814	0.1841
−0.8	0.1867	0.1894	0.1922	0.1949	0.1977	0.2005	0.2033	0.2061	0.2090	0.2119
−0.7	0.2148	0.2177	0.2206	0.2236	0.2266	0.2296	0.2327	0.2358	0.2389	0.2420
−0.6	0.2451	0.2483	0.2514	0.2546	0.2578	0.2611	0.2643	0.2676	0.2709	0.2743
−0.5	0.2776	0.2810	0.2843	0.2877	0.2912	0.2946	0.2981	0.3015	0.3050	0.3085
−0.4	0.3121	0.3156	0.3192	0.3228	0.3264	0.3300	0.3336	0.3372	0.3409	0.3446
−0.3	0.3483	0.3520	0.3557	0.3594	0.3632	0.3669	0.3707	0.3745	0.3783	0.3821
−0.2	0.3859	0.3897	0.3936	0.3974	0.4013	0.4052	0.4090	0.4129	0.4168	0.4207
−0.1	0.4247	0.4286	0.4325	0.4364	0.4404	0.4443	0.4483	0.4522	0.4562	0.4602
−0.0	0.4641	0.4681	0.4721	0.4761	0.4801	0.4840	0.4880	0.4920	0.4960	0.5000

† For $z \leq -3.90$, the areas are 0.0000 to four decimal places.

TABLE II (cont.)

Areas under the standard normal curve

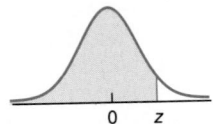

z	0.00	0.01	0.02	0.03	0.04	0.05	0.06	0.07	0.08	0.09
					Second decimal place in z					
0.0	0.5000	0.5040	0.5080	0.5120	0.5160	0.5199	0.5239	0.5279	0.5319	0.5359
0.1	0.5398	0.5438	0.5478	0.5517	0.5557	0.5596	0.5636	0.5675	0.5714	0.5753
0.2	0.5793	0.5832	0.5871	0.5910	0.5948	0.5987	0.6026	0.6064	0.6103	0.6141
0.3	0.6179	0.6217	0.6255	0.6293	0.6331	0.6368	0.6406	0.6443	0.6480	0.6517
0.4	0.6554	0.6591	0.6628	0.6664	0.6700	0.6736	0.6772	0.6808	0.6844	0.6879
0.5	0.6915	0.6950	0.6985	0.7019	0.7054	0.7088	0.7123	0.7157	0.7190	0.7224
0.6	0.7257	0.7291	0.7324	0.7357	0.7389	0.7422	0.7454	0.7486	0.7517	0.7549
0.7	0.7580	0.7611	0.7642	0.7673	0.7704	0.7734	0.7764	0.7794	0.7823	0.7852
0.8	0.7881	0.7910	0.7939	0.7967	0.7995	0.8023	0.8051	0.8078	0.8106	0.8133
0.9	0.8159	0.8186	0.8212	0.8238	0.8264	0.8289	0.8315	0.8340	0.8365	0.8389
1.0	0.8413	0.8438	0.8461	0.8485	0.8508	0.8531	0.8554	0.8577	0.8599	0.8621
1.1	0.8643	0.8665	0.8686	0.8708	0.8729	0.8749	0.8770	0.8790	0.8810	0.8830
1.2	0.8849	0.8869	0.8888	0.8907	0.8925	0.8944	0.8962	0.8980	0.8997	0.9015
1.3	0.9032	0.9049	0.9066	0.9082	0.9099	0.9115	0.9131	0.9147	0.9162	0.9177
1.4	0.9192	0.9207	0.9222	0.9236	0.9251	0.9265	0.9279	0.9292	0.9306	0.9319
1.5	0.9332	0.9345	0.9357	0.9370	0.9382	0.9394	0.9406	0.9418	0.9429	0.9441
1.6	0.9452	0.9463	0.9474	0.9484	0.9495	0.9505	0.9515	0.9525	0.9535	0.9545
1.7	0.9554	0.9564	0.9573	0.9582	0.9591	0.9599	0.9608	0.9616	0.9625	0.9633
1.8	0.9641	0.9649	0.9656	0.9664	0.9671	0.9678	0.9686	0.9693	0.9699	0.9706
1.9	0.9713	0.9719	0.9726	0.9732	0.9738	0.9744	0.9750	0.9756	0.9761	0.9767
2.0	0.9772	0.9778	0.9783	0.9788	0.9793	0.9798	0.9803	0.9808	0.9812	0.9817
2.1	0.9821	0.9826	0.9830	0.9834	0.9838	0.9842	0.9846	0.9850	0.9854	0.9857
2.2	0.9861	0.9864	0.9868	0.9871	0.9875	0.9878	0.9881	0.9884	0.9887	0.9890
2.3	0.9893	0.9896	0.9898	0.9901	0.9904	0.9906	0.9909	0.9911	0.9913	0.9916
2.4	0.9918	0.9920	0.9922	0.9925	0.9927	0.9929	0.9931	0.9932	0.9934	0.9936
2.5	0.9938	0.9940	0.9941	0.9943	0.9945	0.9946	0.9948	0.9949	0.9951	0.9952
2.6	0.9953	0.9955	0.9956	0.9957	0.9959	0.9960	0.9961	0.9962	0.9963	0.9964
2.7	0.9965	0.9966	0.9967	0.9968	0.9969	0.9970	0.9971	0.9972	0.9973	0.9974
2.8	0.9974	0.9975	0.9976	0.9977	0.9977	0.9978	0.9979	0.9979	0.9980	0.9981
2.9	0.9981	0.9982	0.9982	0.9983	0.9984	0.9984	0.9985	0.9985	0.9986	0.9986
3.0	0.9987	0.9987	0.9987	0.9988	0.9988	0.9989	0.9989	0.9989	0.9990	0.9990
3.1	0.9990	0.9991	0.9991	0.9991	0.9992	0.9992	0.9992	0.9992	0.9993	0.9993
3.2	0.9993	0.9993	0.9994	0.9994	0.9994	0.9994	0.9994	0.9995	0.9995	0.9995
3.3	0.9995	0.9995	0.9995	0.9996	0.9996	0.9996	0.9996	0.9996	0.9996	0.9997
3.4	0.9997	0.9997	0.9997	0.9997	0.9997	0.9997	0.9997	0.9997	0.9997	0.9998
3.5	0.9998	0.9998	0.9998	0.9998	0.9998	0.9998	0.9998	0.9998	0.9998	0.9998
3.6	0.9998	0.9998	0.9999	0.9999	0.9999	0.9999	0.9999	0.9999	0.9999	0.9999
3.7	0.9999	0.9999	0.9999	0.9999	0.9999	0.9999	0.9999	0.9999	0.9999	0.9999
3.8	0.9999	0.9999	0.9999	0.9999	0.9999	0.9999	0.9999	0.9999	0.9999	0.9999
3.9	1.0000[†]									

[†] For $z \geq 3.90$, the areas are 1.0000 to four decimal places.

TABLE III
Normal scores

Ordered position	n								
	5	6	7	8	9	10	11	12	13
1	−1.18	−1.28	−1.36	−1.43	−1.50	−1.55	−1.59	−1.64	−1.68
2	−0.50	−0.64	−0.76	−0.85	−0.93	−1.00	−1.06	−1.11	−1.16
3	0.00	−0.20	−0.35	−0.47	−0.57	−0.65	−0.73	−0.79	−0.85
4	0.50	0.20	0.00	−0.15	−0.27	−0.37	−0.46	−0.53	−0.60
5	1.18	0.64	0.35	0.15	0.00	−0.12	−0.22	−0.31	−0.39
6		1.28	0.76	0.47	0.27	0.12	0.00	−0.10	−0.19
7			1.36	0.85	0.57	0.37	0.22	0.10	0.00
8				1.43	0.93	0.65	0.46	0.31	0.19
9					1.50	1.00	0.73	0.53	0.39
10						1.55	1.06	0.79	0.60
11							1.59	1.11	0.85
12								1.64	1.16
13									1.68

TABLE III   (cont.)
Normal scores

Ordered position	n								
	14	15	16	17	18	19	20	21	22
1	−1.71	−1.74	−1.77	−1.80	−1.82	−1.85	−1.87	−1.89	−1.91
2	−1.20	−1.24	−1.28	−1.32	−1.35	−1.38	−1.40	−1.43	−1.45
3	−0.90	−0.94	−0.99	−1.03	−1.06	−1.10	−1.13	−1.16	−1.18
4	−0.66	−0.71	−0.76	−0.80	−0.84	−0.88	−0.92	−0.95	−0.98
5	−0.45	−0.51	−0.57	−0.62	−0.66	−0.70	−0.74	−0.78	−0.81
6	−0.27	−0.33	−0.39	−0.45	−0.50	−0.54	−0.59	−0.63	−0.66
7	−0.09	−0.16	−0.23	−0.29	−0.35	−0.40	−0.45	−0.49	−0.53
8	0.09	0.00	−0.08	−0.15	−0.21	−0.26	−0.31	−0.36	−0.40
9	0.27	0.16	0.08	0.00	−0.07	−0.13	−0.19	−0.24	−0.28
10	0.45	0.33	0.23	0.15	0.07	0.00	−0.06	−0.12	−0.17
11	0.66	0.51	0.39	0.29	0.21	0.13	0.06	0.00	−0.06
12	0.90	0.71	0.57	0.45	0.35	0.26	0.19	0.12	0.06
13	1.20	0.94	0.76	0.62	0.50	0.40	0.31	0.24	0.17
14	1.71	1.24	0.99	0.80	0.66	0.54	0.45	0.36	0.28
15		1.74	1.28	1.03	0.84	0.70	0.59	0.49	0.40
16			1.77	1.32	1.06	0.88	0.74	0.63	0.53
17				1.80	1.35	1.10	0.92	0.78	0.66
18					1.82	1.38	1.13	0.95	0.81
19						1.85	1.40	1.16	0.98
20							1.87	1.43	1.18
21								1.89	1.45
22									1.91

TABLE III (cont.)
Normal scores

Ordered position	n							
	23	24	25	26	27	28	29	30
1	−1.93	−1.95	−1.97	−1.98	−2.00	−2.01	−2.03	−2.04
2	−1.48	−1.50	−1.52	−1.54	−1.56	−1.58	−1.59	−1.61
3	−1.21	−1.24	−1.26	−1.28	−1.30	−1.32	−1.34	−1.36
4	−1.01	−1.04	−1.06	−1.09	−1.11	−1.13	−1.15	−1.17
5	−0.84	−0.87	−0.90	−0.93	−0.95	−0.98	−1.00	−1.02
6	−0.70	−0.73	−0.76	−0.79	−0.82	−0.84	−0.87	−0.89
7	−0.57	−0.60	−0.63	−0.66	−0.69	−0.72	−0.75	−0.77
8	−0.44	−0.48	−0.52	−0.55	−0.58	−0.61	−0.64	−0.67
9	−0.33	−0.37	−0.41	−0.44	−0.48	−0.51	−0.54	−0.57
10	−0.22	−0.26	−0.30	−0.34	−0.38	−0.41	−0.44	−0.47
11	−0.11	−0.15	−0.20	−0.24	−0.28	−0.31	−0.35	−0.38
12	0.00	−0.05	−0.10	−0.14	−0.18	−0.22	−0.26	−0.29
13	0.11	0.05	0.00	−0.05	−0.09	−0.13	−0.17	−0.21
14	0.22	0.15	0.10	0.05	0.00	−0.04	−0.09	−0.12
15	0.33	0.26	0.20	0.14	0.09	0.04	0.00	−0.04
16	0.44	0.37	0.30	0.24	0.18	0.13	0.09	0.04
17	0.57	0.48	0.41	0.34	0.28	0.22	0.17	0.12
18	0.70	0.60	0.52	0.44	0.38	0.31	0.26	0.21
19	0.84	0.73	0.63	0.55	0.48	0.41	0.35	0.29
20	1.01	0.87	0.76	0.66	0.58	0.51	0.44	0.38
21	1.21	1.04	0.90	0.79	0.69	0.61	0.54	0.47
22	1.48	1.24	1.06	0.93	0.82	0.72	0.64	0.57
23	1.93	1.50	1.26	1.09	0.95	0.84	0.75	0.67
24		1.95	1.52	1.28	1.11	0.98	0.87	0.77
25			1.97	1.54	1.30	1.13	1.00	0.89
26				1.98	1.56	1.32	1.15	1.02
27					2.00	1.58	1.34	1.17
28						2.01	1.59	1.36
29							2.03	1.61
30								2.04

TABLE IV
Values of t_α

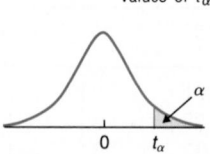

df	$t_{0.10}$	$t_{0.05}$	$t_{0.025}$	$t_{0.01}$	$t_{0.005}$	df
1	3.078	6.314	12.706	31.821	63.657	1
2	1.886	2.920	4.303	6.965	9.925	2
3	1.638	2.353	3.182	4.541	5.841	3
4	1.533	2.132	2.776	3.747	4.604	4
5	1.476	2.015	2.571	3.365	4.032	5
6	1.440	1.943	2.447	3.143	3.707	6
7	1.415	1.895	2.365	2.998	3.499	7
8	1.397	1.860	2.306	2.896	3.355	8
9	1.383	1.833	2.262	2.821	3.250	9
10	1.372	1.812	2.228	2.764	3.169	10
11	1.363	1.796	2.201	2.718	3.106	11
12	1.356	1.782	2.179	2.681	3.055	12
13	1.350	1.771	2.160	2.650	3.012	13
14	1.345	1.761	2.145	2.624	2.977	14
15	1.341	1.753	2.131	2.602	2.947	15
16	1.337	1.746	2.120	2.583	2.921	16
17	1.333	1.740	2.110	2.567	2.898	17
18	1.330	1.734	2.101	2.552	2.878	18
19	1.328	1.729	2.093	2.539	2.861	19
20	1.325	1.725	2.086	2.528	2.845	20
21	1.323	1.721	2.080	2.518	2.831	21
22	1.321	1.717	2.074	2.508	2.819	22
23	1.319	1.714	2.069	2.500	2.807	23
24	1.318	1.711	2.064	2.492	2.797	24
25	1.316	1.708	2.060	2.485	2.787	25
26	1.315	1.706	2.056	2.479	2.779	26
27	1.314	1.703	2.052	2.473	2.771	27
28	1.313	1.701	2.048	2.467	2.763	28
29	1.311	1.699	2.045	2.462	2.756	29
∞	1.282	1.645	1.960	2.326	2.576	∞

TABLE V
Values of χ_α^2

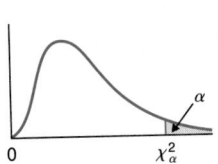

df	$\chi_{0.995}^2$	$\chi_{0.99}^2$	$\chi_{0.975}^2$	$\chi_{0.95}^2$	$\chi_{0.05}^2$	$\chi_{0.025}^2$	$\chi_{0.01}^2$	$\chi_{0.005}^2$	df
1	0.000	0.000	0.001	0.004	3.841	5.024	6.635	7.879	1
2	0.010	0.020	0.051	0.103	5.991	7.378	9.210	10.597	2
3	0.072	0.115	0.216	0.352	7.815	9.348	11.345	12.838	3
4	0.207	0.297	0.484	0.711	9.488	11.143	13.277	14.860	4
5	0.412	0.554	0.831	1.145	11.070	12.832	15.086	16.750	5
6	0.676	0.872	1.237	1.635	12.592	14.449	16.812	18.548	6
7	0.989	1.239	1.690	2.167	14.067	16.013	18.475	20.278	7
8	1.344	1.646	2.180	2.733	15.507	17.535	20.090	21.955	8
9	1.735	2.088	2.700	3.325	16.919	19.023	21.666	23.589	9
10	2.156	2.558	3.247	3.940	18.307	20.483	23.209	25.188	10
11	2.603	3.053	3.816	4.575	19.675	21.920	24.725	26.757	11
12	3.074	3.571	4.404	5.226	21.026	23.337	26.217	28.300	12
13	3.565	4.107	5.009	5.892	22.362	24.736	27.688	29.819	13
14	4.075	4.660	5.629	6.571	23.685	26.119	29.141	31.319	14
15	4.601	5.229	6.262	7.261	24.996	27.488	30.578	32.801	15
16	5.142	5.812	6.908	7.962	26.296	28.845	32.000	34.267	16
17	5.697	6.408	7.564	8.672	27.587	30.191	33.409	35.718	17
18	6.265	7.015	8.231	9.390	28.869	31.526	34.805	37.156	18
19	6.844	7.633	8.907	10.117	30.144	32.852	36.191	38.582	19
20	7.434	8.260	9.591	10.851	31.410	34.170	37.566	39.997	20
21	8.034	8.897	10.283	11.591	32.671	35.479	38.932	41.401	21
22	8.643	9.542	10.982	12.338	33.924	36.781	40.289	42.796	22
23	9.260	10.196	11.689	13.091	35.172	38.076	41.638	44.181	23
24	9.886	10.856	12.401	13.848	36.415	39.364	42.980	45.558	24
25	10.520	11.524	13.120	14.611	37.652	40.646	44.314	46.928	25
26	11.160	12.198	13.844	15.379	38.885	41.923	45.642	48.290	26
27	11.808	12.879	14.573	16.151	40.113	43.194	46.963	49.645	27
28	12.461	13.565	15.308	16.928	41.337	44.461	48.278	50.993	28
29	13.121	14.256	16.047	17.708	42.557	45.722	49.588	52.336	29
30	13.787	14.953	16.791	18.493	43.773	46.979	50.892	53.672	30

TABLE VI
Critical values for a
Wilcoxon signed-rank test

Sample size	Significance level, α		Critical value	
n	One-tailed	Two-tailed	W_l	W_r
3	0.125	0.250	0	6
4	0.063	0.125	0	10
	0.125	0.250	1	9
5	0.031	0.063	0	15
	0.063	0.125	1	14
	0.094	0.188	2	13
6	0.016	0.031	0	21
	0.031	0.063	1	20
	0.047	0.094	2	19
	0.109	0.219	4	17
7	0.008	0.016	0	28
	0.023	0.047	2	26
	0.055	0.109	4	24
	0.109	0.219	6	22
8	0.004	0.008	0	36
	0.012	0.023	2	34
	0.027	0.055	4	32
	0.055	0.109	6	30
	0.098	0.195	8	28
9	0.006	0.012	2	43
	0.010	0.020	3	42
	0.027	0.055	6	39
	0.049	0.098	8	37
	0.102	0.203	11	34
10	0.005	0.010	3	52
	0.010	0.020	5	50
	0.024	0.049	8	47
	0.053	0.105	11	44
	0.097	0.193	14	41
11	0.005	0.010	5	61
	0.009	0.019	7	59
	0.027	0.054	11	55
	0.051	0.102	14	52
	0.103	0.206	18	48
12	0.005	0.009	7	71
	0.010	0.021	10	68
	0.026	0.052	14	64
	0.046	0.092	17	61
	0.102	0.204	22	56

TABLE VI (cont.)
Critical values for a
Wilcoxon signed-rank test

| Sample size | Significance level, α | | Critical value | |
n	One-tailed	Two-tailed	W_l	W_r
13	0.005	0.010	10	81
	0.011	0.021	13	78
	0.024	0.048	17	74
	0.047	0.094	21	70
	0.095	0.191	26	65
14	0.005	0.011	13	92
	0.010	0.020	16	89
	0.025	0.049	21	84
	0.052	0.104	26	79
	0.097	0.194	31	74
15	0.005	0.010	16	104
	0.011	0.022	20	100
	0.024	0.048	25	95
	0.047	0.095	30	90
	0.104	0.208	37	83
16	0.005	0.009	19	117
	0.011	0.021	24	112
	0.025	0.051	30	106
	0.052	0.105	36	100
	0.096	0.193	42	94
17	0.005	0.009	23	130
	0.010	0.020	28	125
	0.025	0.051	35	118
	0.049	0.098	41	112
	0.103	0.207	49	104
18	0.005	0.010	28	143
	0.010	0.021	33	138
	0.024	0.048	40	131
	0.049	0.099	47	124
	0.098	0.196	55	116
19	0.005	0.009	32	158
	0.010	0.020	38	152
	0.025	0.049	46	144
	0.052	0.104	54	136
	0.098	0.196	62	128
20	0.005	0.009	37	173
	0.010	0.019	43	167
	0.024	0.048	52	158
	0.049	0.097	60	150
	0.101	0.202	70	140

TABLE VII

Critical values for a one-tailed Mann–Whitney test with $\alpha = 0.025$ or a two-tailed Mann–Whitney test with $\alpha = 0.05$

n_2 \ n_1	3		4		5		6		7		8		9		10	
	M_l	M_r	M_l	M_r	M_l	M_r	M_l	M_r	M_l	M_r	M_l	M_r	M_l	M_r	M_l	M_r
3	–	–														
4	6	18	11	25												
5	6	21	12	28	18	37										
6	7	23	12	32	19	41	26	52								
7	7	26	13	35	20	45	28	56	37	68						
8	8	28	14	38	21	49	29	61	39	73	49	87				
9	8	31	15	41	22	53	31	65	41	78	51	93	63	108		
10	9	33	16	44	24	56	32	70	43	83	54	98	66	114	79	131

TABLE VIII

Critical values for a one-tailed Mann–Whitney test with $\alpha = 0.05$ or a two-tailed Mann–Whitney test with $\alpha = 0.10$

n_2 \ n_1	3		4		5		6		7		8		9		10	
	M_l	M_r	M_l	M_r	M_l	M_r	M_l	M_r	M_l	M_r	M_l	M_r	M_l	M_r	M_l	M_r
3	6	15														
4	7	17	12	24												
5	7	20	13	27	19	36										
6	8	22	14	30	20	40	28	50								
7	9	24	15	33	22	43	30	54	39	66						
8	9	27	16	36	24	46	32	58	41	71	52	84				
9	10	29	17	39	25	50	33	63	43	76	54	90	66	105		
10	11	31	18	42	26	54	35	67	46	80	57	95	69	111	83	127

TABLE IX

Random numbers

Line number	Column number									
	00–09		*10–19*		*20–29*		*30–39*		*40–49*	
00	15544	80712	97742	21500	97081	42451	50623	56071	28882	28739
01	01011	21285	04729	39986	73150	31548	30168	76189	56996	19210
02	47435	53308	40718	29050	74858	64517	93573	51058	68501	42723
03	91312	75137	86274	59834	69844	19853	06917	17413	44474	86530
04	12775	08768	80791	16298	22934	09630	98862	39746	64623	32768
05	31466	43761	94872	92230	52367	13205	38634	55882	77518	36252
06	09300	43847	40881	51243	97810	18903	53914	31688	06220	40422
07	73582	13810	57784	72454	68997	72229	30340	08844	53924	89630
08	11092	81392	58189	22697	41063	09451	09789	00637	06450	85990
09	93322	98567	00116	35605	66790	52965	62877	21740	56476	49296
10	80134	12484	67089	08674	70753	90959	45842	59844	45214	36505
11	97888	31797	95037	84400	76041	96668	75920	68482	56855	97417
12	92612	27082	59459	69380	98654	20407	88151	56263	27126	63797
13	72744	45586	43279	44218	83638	05422	00995	70217	78925	39097
14	96256	70653	45285	26293	78305	80252	03625	40159	68760	84716
15	07851	47452	66742	83331	54701	06573	98169	37499	67756	68301
16	25594	41552	96475	56151	02089	33748	65289	89956	89559	33687
17	65358	15155	59374	80940	03411	94656	69440	47156	77115	99463
18	09402	31008	53424	21928	02198	61201	02457	87214	59750	51330
19	97424	90765	01634	37328	41243	33564	17884	94747	93650	77668

TABLE X

Values of $F_{0.01}$

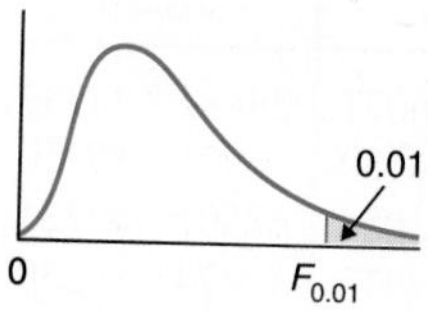

df for denominator	df for numerator: 1	2	3	4	5	6	7	8	9
1	4052	4999.5	5403	5625	5764	5859	5928	5981	6022
2	98.50	99.00	99.17	99.25	99.30	99.33	99.36	99.37	99.39
3	34.12	30.82	29.46	28.71	28.24	27.91	27.67	27.49	27.35
4	21.20	18.00	16.69	15.98	15.52	15.21	14.98	14.80	14.66
5	16.26	13.27	12.06	11.39	10.97	10.67	10.46	10.29	10.16
6	13.75	10.92	9.78	9.15	8.75	8.47	8.26	8.10	7.98
7	12.25	9.55	8.45	7.85	7.46	7.19	6.99	6.84	6.72
8	11.26	8.65	7.59	7.01	6.63	6.37	6.18	6.03	5.91
9	10.56	8.02	6.99	6.42	6.06	5.80	5.61	5.47	5.35
10	10.04	7.56	6.55	5.99	5.64	5.39	5.20	5.06	4.94
11	9.65	7.21	6.22	5.67	5.32	5.07	4.89	4.74	4.63
12	9.33	6.93	5.95	5.41	5.06	4.82	4.64	4.50	4.39
13	9.07	6.70	5.74	5.21	4.86	4.62	4.44	4.30	4.19
14	8.86	6.51	5.56	5.04	4.69	4.46	4.28	4.14	4.03
15	8.68	6.36	5.42	4.89	4.56	4.32	4.14	4.00	3.89
16	8.53	6.23	5.29	4.77	4.44	4.20	4.03	3.89	3.78
17	8.40	6.11	5.18	4.67	4.34	4.10	3.93	3.79	3.68
18	8.29	6.01	5.09	4.58	4.25	4.01	3.84	3.71	3.60
19	8.18	5.93	5.01	4.50	4.17	3.94	3.77	3.63	3.52
20	8.10	5.85	4.94	4.43	4.10	3.87	3.70	3.56	3.46
21	8.02	5.78	4.87	4.37	4.04	3.81	3.64	3.51	3.40
22	7.95	5.72	4.82	4.31	3.99	3.76	3.59	3.45	3.35
23	7.88	5.66	4.76	4.26	3.94	3.71	3.54	3.41	3.30
24	7.82	5.61	4.72	4.22	3.90	3.67	3.50	3.36	3.26
25	7.77	5.57	4.68	4.18	3.85	3.63	3.46	3.32	3.22
26	7.72	5.53	4.64	4.14	3.82	3.59	3.42	3.29	3.18
27	7.68	5.49	4.60	4.11	3.78	3.56	3.39	3.26	3.15
28	7.64	5.45	4.57	4.07	3.75	3.53	3.36	3.23	3.12
29	7.60	5.42	4.54	4.04	3.73	3.50	3.33	3.20	3.09
30	7.56	5.39	4.51	4.02	3.70	3.47	3.30	3.17	3.07
40	7.31	5.18	4.31	3.83	3.51	3.29	3.12	2.99	2.89
60	7.08	4.98	4.13	3.65	3.34	3.12	2.95	2.82	2.72
120	6.85	4.79	3.95	3.48	3.17	2.96	2.79	2.66	2.56
∞	6.63	4.61	3.78	3.32	3.02	2.80	2.64	2.51	2.41

Adapted from *Handbook of Statistical Tables* by D.B. Owen, Reading, MA: Addison-Wesley, 1962. Courtesy of the Atomic Energy Commission.

TABLE X (cont.)
Values of $F_{0.01}$

| df for numerator | | | | | | | | | | df for denominator |
10	12	15	20	24	30	40	60	120	∞	
6056	6106	6157	6209	6235	6261	6287	6313	6339	6366	1
99.40	99.42	99.43	99.45	99.46	99.47	99.47	99.48	99.49	99.50	2
27.23	27.05	26.87	26.69	26.60	26.50	26.41	26.32	26.22	26.13	3
14.55	14.37	14.20	14.02	13.93	13.84	13.75	13.65	13.56	13.46	4
10.05	9.89	9.72	9.55	9.47	9.38	9.29	9.20	9.11	9.02	5
7.87	7.72	7.56	7.40	7.31	7.23	7.14	7.06	6.97	6.88	6
6.62	6.47	6.31	6.16	6.07	5.99	5.91	5.82	5.74	5.65	7
5.81	5.67	5.52	5.36	5.28	5.20	5.12	5.03	4.95	4.86	8
5.26	5.11	4.96	4.81	4.73	4.65	4.57	4.48	4.40	4.31	9
4.85	4.71	4.56	4.41	4.33	4.25	4.17	4.08	4.00	3.91	10
4.54	4.40	4.25	4.10	4.02	3.94	3.86	3.78	3.69	3.60	11
4.30	4.16	4.01	3.86	3.78	3.70	3.62	3.54	3.45	3.36	12
4.10	3.96	3.82	3.66	3.59	3.51	3.43	3.34	3.25	3.17	13
3.94	3.80	3.66	3.51	3.43	3.35	3.27	3.18	3.09	3.00	14
3.80	3.67	3.52	3.37	3.29	3.21	3.13	3.05	2.96	2.87	15
3.69	3.55	3.41	3.26	3.18	3.10	3.02	2.93	2.84	2.75	16
3.59	3.46	3.31	3.16	3.08	3.00	2.92	2.83	2.75	2.65	17
3.51	3.37	3.23	3.08	3.00	2.92	2.84	2.75	2.66	2.57	18
3.43	3.30	3.15	3.00	2.92	2.84	2.76	2.67	2.58	2.49	19
3.37	3.23	3.09	2.94	2.86	2.78	2.69	2.61	2.52	2.42	20
3.31	3.17	3.03	2.88	2.80	2.72	2.64	2.55	2.46	2.36	21
3.26	3.12	2.98	2.83	2.75	2.67	2.58	2.50	2.40	2.31	22
3.21	3.07	2.93	2.78	2.70	2.62	2.54	2.45	2.35	2.26	23
3.17	3.03	2.89	2.74	2.66	2.58	2.49	2.40	2.31	2.21	24
3.13	2.99	2.85	2.70	2.62	2.54	2.45	2.36	2.27	2.17	25
3.09	2.96	2.81	2.66	2.58	2.50	2.42	2.33	2.23	2.13	26
3.06	2.93	2.78	2.63	2.55	2.47	2.38	2.29	2.20	2.10	27
3.03	2.90	2.75	2.60	2.52	2.44	2.35	2.26	2.17	2.06	28
3.00	2.87	2.73	2.57	2.49	2.41	2.33	2.23	2.14	2.03	29
2.98	2.84	2.70	2.55	2.47	2.39	2.30	2.21	2.11	2.01	30
2.80	2.66	2.52	2.37	2.29	2.20	2.11	2.02	1.92	1.80	40
2.63	2.50	2.35	2.20	2.12	2.03	1.94	1.84	1.73	1.60	60
2.47	2.34	2.19	2.03	1.95	1.86	1.76	1.66	1.53	1.38	120
2.32	2.18	2.04	1.88	1.79	1.70	1.59	1.47	1.32	1.00	∞

TABLE XI

Values of $F_{0.05}$

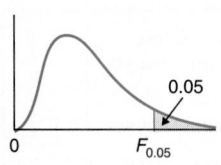

		df for numerator							
	1	*2*	*3*	*4*	*5*	*6*	*7*	*8*	*9*
1	161.4	199.5	215.7	224.6	230.2	234.0	236.8	238.9	240.5
2	18.51	19.00	19.16	19.25	19.30	19.33	19.35	19.37	19.38
3	10.13	9.55	9.28	9.12	9.01	8.94	8.89	8.85	8.81
4	7.71	6.94	6.59	6.39	6.26	6.16	6.09	6.04	6.00
5	6.61	5.79	5.41	5.19	5.05	4.95	4.88	4.82	4.77
6	5.99	5.14	4.76	4.53	4.39	4.28	4.21	4.15	4.10
7	5.59	4.74	4.35	4.12	3.97	3.87	3.79	3.73	3.68
8	5.32	4.46	4.07	3.84	3.69	3.58	3.50	3.44	3.39
9	5.12	4.26	3.86	3.63	3.48	3.37	3.29	3.23	3.18
10	4.96	4.10	3.71	3.48	3.33	3.22	3.14	3.07	3.02
11	4.84	3.98	3.59	3.36	3.20	3.09	3.01	2.95	2.90
12	4.75	3.89	3.49	3.26	3.11	3.00	2.91	2.85	2.80
13	4.67	3.81	3.41	3.18	3.03	2.92	2.83	2.77	2.71
14	4.60	3.74	3.34	3.11	2.96	2.85	2.76	2.70	2.65
15	4.54	3.68	3.29	3.06	2.90	2.79	2.71	2.64	2.59
16	4.49	3.63	3.24	3.01	2.85	2.74	2.66	2.59	2.54
17	4.45	3.59	3.20	2.96	2.81	2.70	2.61	2.55	2.49
18	4.41	3.55	3.16	2.93	2.77	2.66	2.58	2.51	2.46
19	4.38	3.52	3.13	2.90	2.74	2.63	2.54	2.48	2.42
20	4.35	3.49	3.10	2.87	2.71	2.60	2.51	2.45	2.39
21	4.32	3.47	3.07	2.84	2.68	2.57	2.49	2.42	2.37
22	4.30	3.44	3.05	2.82	2.66	2.55	2.46	2.40	2.34
23	4.28	3.42	3.03	2.80	2.64	2.53	2.44	2.37	2.32
24	4.26	3.40	3.01	2.78	2.62	2.51	2.42	2.36	2.30
25	4.24	3.39	2.99	2.76	2.60	2.49	2.40	2.34	2.28
26	4.23	3.37	2.98	2.74	2.59	2.47	2.39	2.32	2.27
27	4.21	3.35	2.96	2.73	2.57	2.46	2.37	2.31	2.25
28	4.20	3.34	2.95	2.71	2.56	2.45	2.36	2.29	2.24
29	4.18	3.33	2.93	2.70	2.55	2.43	2.35	2.28	2.22
30	4.17	3.32	2.92	2.69	2.53	2.42	2.33	2.27	2.21
40	4.08	3.23	2.84	2.61	2.45	2.34	2.25	2.18	2.12
60	4.00	3.15	2.76	2.53	2.37	2.25	2.17	2.10	2.04
120	3.92	3.07	2.68	2.45	2.29	2.17	2.09	2.02	1.96
∞	3.84	3.00	2.60	2.37	2.21	2.10	2.01	1.94	1.88

(df for denominator — left column)

Adapted from *Handbook of Statistical Tables* by D.B. Owen, Reading, MA: Addison-Wesley, 1962. Courtesy of the Atomic Energy Commission.

TABLE XI (cont.)

Values of $F_{0.05}$

				df for numerator						
10	12	15	20	24	30	40	60	120	∞	
241.9	243.9	245.9	248.0	249.1	250.1	251.1	252.2	253.3	254.3	1
19.40	19.41	19.43	19.45	19.45	19.46	19.47	19.48	19.49	19.50	2
8.79	8.74	8.70	8.66	8.64	8.62	8.59	8.57	8.55	8.53	3
5.96	5.91	5.86	5.80	5.77	5.75	5.72	5.69	5.66	5.63	4
4.74	4.68	4.62	4.56	4.53	4.50	4.46	4.43	4.40	4.36	5
4.06	4.00	3.94	3.87	3.84	3.81	3.77	3.74	3.70	3.67	6
3.64	3.57	3.51	3.44	3.41	3.38	3.34	3.30	3.27	3.23	7
3.35	3.28	3.22	3.15	3.12	3.08	3.04	3.01	2.97	2.93	8
3.14	3.07	3.01	2.94	2.90	2.86	2.83	2.79	2.75	2.71	9
2.98	2.91	2.85	2.77	2.74	2.70	2.66	2.62	2.58	2.54	10
2.85	2.79	2.72	2.65	2.61	2.57	2.53	2.49	2.45	2.40	11
2.75	2.69	2.62	2.54	2.51	2.47	2.43	2.38	2.34	2.30	12
2.67	2.60	2.53	2.46	2.42	2.38	2.34	2.30	2.25	2.21	13
2.60	2.53	2.46	2.39	2.35	2.31	2.27	2.22	2.18	2.13	14
2.54	2.48	2.40	2.33	2.29	2.25	2.20	2.16	2.11	2.07	15
2.49	2.42	2.35	2.28	2.24	2.19	2.15	2.11	2.06	2.01	16
2.45	2.38	2.31	2.23	2.19	2.15	2.10	2.06	2.01	1.96	17
2.41	2.34	2.27	2.19	2.15	2.11	2.06	2.02	1.97	1.92	18
2.38	2.31	2.23	2.16	2.11	2.07	2.03	1.98	1.93	1.88	19
2.35	2.28	2.20	2.12	2.08	2.04	1.99	1.95	1.90	1.84	20
2.32	2.25	2.18	2.10	2.05	2.01	1.96	1.92	1.87	1.81	21
2.30	2.23	2.15	2.07	2.03	1.98	1.94	1.89	1.84	1.78	22
2.27	2.20	2.13	2.05	2.01	1.96	1.91	1.86	1.81	1.76	23
2.25	2.18	2.11	2.03	1.98	1.94	1.89	1.84	1.79	1.73	24
2.24	2.16	2.09	2.01	1.96	1.92	1.87	1.82	1.77	1.71	25
2.22	2.15	2.07	1.99	1.95	1.90	1.85	1.80	1.75	1.69	26
2.20	2.13	2.06	1.97	1.93	1.88	1.84	1.79	1.73	1.67	27
2.19	2.12	2.04	1.96	1.91	1.87	1.82	1.77	1.71	1.65	28
2.18	2.10	2.03	1.94	1.90	1.85	1.81	1.75	1.70	1.64	29
2.16	2.09	2.01	1.93	1.89	1.84	1.79	1.74	1.68	1.62	30
2.08	2.00	1.92	1.84	1.79	1.74	1.69	1.64	1.58	1.51	40
1.99	1.92	1.84	1.75	1.70	1.65	1.59	1.53	1.47	1.39	60
1.91	1.83	1.75	1.66	1.61	1.55	1.50	1.43	1.35	1.25	120
1.83	1.75	1.67	1.57	1.52	1.46	1.39	1.32	1.22	1.00	∞

df for denominator

TABLE XII

Values of $q_{0.01}$

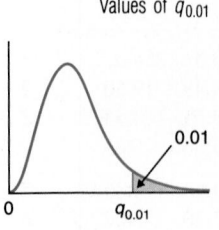

0.01

0          $q_{0.01}$

ν	2	3	4	5	6	7	8	9	10	ν
1	90.0	135	164	186	202	216	227	237	246	1
2	14.0	19.0	22.3	24.7	26.6	28.2	29.5	30.7	31.7	2
3	8.26	10.6	12.2	13.3	14.2	15.0	15.6	16.2	16.7	3
4	6.51	8.12	9.17	9.96	10.6	11.1	11.5	11.9	12.3	4
5	5.70	6.97	7.80	8.42	8.91	9.32	9.67	9.97	10.2	5
6	5.24	6.33	7.03	7.56	7.97	8.32	8.61	8.87	9.10	6
7	4.95	5.92	6.54	7.01	7.37	7.68	7.94	8.17	8.37	7
8	4.74	5.63	6.20	6.63	6.96	7.24	7.47	7.68	7.87	8
9	4.60	5.43	5.96	6.35	6.66	6.91	7.13	7.32	7.49	9
10	4.48	5.27	5.77	6.14	6.43	6.67	6.87	7.05	7.21	10
11	4.39	5.14	5.62	5.97	6.25	6.48	6.67	6.84	6.99	11
12	4.32	5.04	5.50	5.84	6.10	6.32	6.51	6.67	6.81	12
13	4.26	4.96	5.40	5.73	5.98	6.19	6.37	6.53	6.67	13
14	4.21	4.89	5.32	5.63	5.88	6.08	6.26	6.41	6.54	14
15	4.17	4.83	5.25	5.56	5.80	5.99	6.16	6.31	6.44	15
16	4.13	4.78	5.19	5.49	5.72	5.92	6.08	6.22	6.35	16
17	4.10	4.74	5.14	5.43	5.66	5.85	6.01	6.15	6.27	17
18	4.07	4.70	5.09	5.38	5.60	5.79	5.94	6.08	6.20	18
19	4.05	4.67	5.05	5.33	5.55	5.73	5.89	6.02	6.14	19
20	4.02	4.64	5.02	5.29	5.51	5.69	5.84	5.97	6.09	20
24	3.96	4.54	4.91	5.17	5.37	5.54	5.69	5.81	5.92	24
30	3.89	4.45	4.80	5.05	5.24	5.40	5.54	5.65	5.76	30
40	3.82	4.37	4.70	4.93	5.11	5.27	5.39	5.50	5.60	40
60	3.76	4.28	4.60	4.82	4.99	5.13	5.25	5.36	5.45	60
120	3.70	4.20	4.50	4.71	4.87	5.01	5.12	5.21	5.30	120
∞	3.64	4.12	4.40	4.60	4.76	4.88	4.99	5.08	5.16	∞

κ

TABLE XIII

Values of $q_{0.05}$

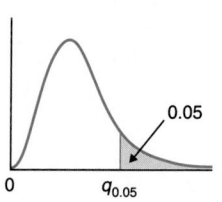

ν	κ									ν
	2	3	4	5	6	7	8	9	10	
1	18.0	27.0	32.8	37.1	40.4	43.1	45.4	47.4	49.1	1
2	6.08	8.33	9.80	10.9	11.7	12.4	13.0	13.5	14.0	2
3	4.50	5.91	6.82	7.50	8.04	8.48	8.85	9.18	9.46	3
4	3.93	5.04	5.76	6.29	6.71	7.05	7.35	7.60	7.83	4
5	3.64	4.60	5.22	5.67	6.03	6.33	6.58	6.80	6.99	5
6	3.46	4.34	4.90	5.30	5.63	5.90	6.12	6.32	6.49	6
7	3.34	4.16	4.68	5.06	5.36	5.61	5.82	6.00	6.16	7
8	3.26	4.04	4.53	4.89	5.17	5.40	5.60	5.77	5.92	8
9	3.20	3.95	4.41	4.76	5.02	5.24	5.43	5.59	5.74	9
10	3.15	3.88	4.33	4.65	4.91	5.12	5.30	5.46	5.60	10
11	3.11	3.82	4.26	4.57	4.82	5.03	5.20	5.35	5.49	11
12	3.08	3.77	4.20	4.51	4.75	4.95	5.12	5.27	5.39	12
13	3.06	3.73	4.15	4.45	4.69	4.88	5.05	5.19	5.32	13
14	3.03	3.70	4.11	4.41	4.64	4.83	4.99	5.13	5.25	14
15	3.01	3.67	4.08	4.37	4.59	4.78	4.94	5.08	5.20	15
16	3.00	3.65	4.05	4.33	4.56	4.74	4.90	5.03	5.15	16
17	2.98	3.63	4.02	4.30	4.52	4.70	4.86	4.99	5.11	17
18	2.97	3.61	4.00	4.28	4.49	4.67	4.82	4.96	5.07	18
19	2.96	3.59	3.98	4.25	4.47	4.65	4.79	4.92	5.04	19
20	2.95	3.58	3.96	4.23	4.45	4.62	4.77	4.90	5.01	20
24	2.92	3.53	3.90	4.17	4.37	4.54	4.68	4.81	4.92	24
30	2.89	3.49	3.85	4.10	4.30	4.46	4.60	4.72	4.82	30
40	2.86	3.44	3.79	4.04	4.23	4.39	4.52	4.63	4.73	40
60	2.83	3.40	3.74	3.98	4.16	4.31	4.44	4.55	4.65	60
120	2.80	3.36	3.68	3.92	4.10	4.24	4.36	4.47	4.56	120
∞	2.77	3.31	3.63	3.86	4.03	4.17	4.29	4.39	4.47	∞

		α	
n	0.10	0.05	0.01
5	0.903	0.880	0.832
6	0.911	0.889	0.841
7	0.918	0.897	0.852
8	0.924	0.905	0.862
9	0.930	0.911	0.871
10	0.935	0.917	0.879
11	0.938	0.923	0.887
12	0.942	0.927	0.894
13	0.945	0.931	0.900
14	0.948	0.935	0.905
15	0.951	0.938	0.910
16	0.953	0.941	0.914
17	0.955	0.944	0.918
18	0.957	0.946	0.922
19	0.959	0.949	0.925
20	0.960	0.951	0.928
21	0.962	0.952	0.931
22	0.963	0.954	0.933
23	0.964	0.956	0.936
24	0.966	0.957	0.938
25	0.967	0.958	0.940
26	0.968	0.960	0.942
27	0.969	0.961	0.944
28	0.969	0.962	0.945
29	0.970	0.963	0.947
30	0.971	0.964	0.949
40	0.977	0.972	0.958
50	0.981	0.976	0.966
60	0.983	0.980	0.971
70	0.985	0.982	0.975
80	0.987	0.984	0.978
90	0.988	0.986	0.980
100	0.989	0.987	0.982
200	0.994	0.993	0.990
300	0.996	0.995	0.993
400	0.997	0.996	0.995
500	0.998	0.997	0.996
1000	0.999	0.998	0.998

Focus Database

. . . . .

The following is a printout of a database obtained by randomly selecting 500 Arizona State University sophomores. The report was created by the *Focus Student Database* on December 8, 1991. Seven variables are considered for each student: sex, high-school GPA, SAT math score, cumulative GPA, SAT verbal score, age, and total hours. We will employ this sample data in the "Using the Focus Database" exercises appearing at the end of each chapter. *Note:* 1 = female, 2 = male.

OBS	SEX	HS GPA	SAT MATH	CUM GPA	SAT VERB	AGE	HOURS	OBS	SEX	HS GPA	SAT MATH	CUM GPA	SAT VERB	AGE	HOURS
1	1	3.86	640	3.45	580	19	35.0	26	1	3.86	560	3.88	590	19	32.0
2	1	3.43	530	3.20	330	19	47.0	27	2	2.56	490	2.68	420	19	28.0
3	1	2.78	380	3.00	480	18	37.0	28	1	3.35	580	2.88	430	20	32.0
4	2	2.46	510	2.77	430	20	52.0	29	2	2.88	510	2.04	370	21	47.0
5	1	2.47	480	3.48	470	19	27.0	30	1	3.42	640	2.76	410	19	25.0
6	1	2.84	460	2.35	430	20	29.0	31	1	2.92	500	1.75	450	19	33.0
7	2	3.73	510	3.57	440	19	34.7	32	2	3.01	520	2.40	360	20	52.0
8	1	3.11	410	2.30	410	20	54.0	33	1	3.03	370	2.54	500	19	25.0
9	2	2.49	650	3.00	480	21	50.0	34	2	3.35	610	4.00	530	19	32.0
10	1	2.88	590	2.76	580	19	25.0	35	2	2.50	600	2.58	510	19	31.0
11	1	3.10	480	2.38	490	20	51.0	36	2	2.77	610	2.62	510	20	52.0
12	2	3.08	660	2.51	520	19	43.0	37	1	2.89	420	2.40	360	19	26.0
13	1	2.84	520	2.16	320	20	31.0	38	1	2.65	430	3.25	520	20	49.0
14	2	2.53	510	2.81	550	19	32.0	39	2	2.50	410	2.48	320	21	43.0
15	1	3.51	590	3.30	470	19	31.0	40	1	3.18	550	2.48	510	19	25.0
16	1	3.13	520	2.02	400	20	52.0	41	1	3.93	630	3.90	650	20	48.0
17	1	2.92	540	2.89	540	18	28.0	42	1	2.82	590	2.70	510	19	30.0
18	1	3.00	390	2.80	400	19	36.0	43	2	2.90	400	2.86	300	19	25.0
19	1	2.90	570	2.13	470	19	37.0	44	2	2.36	480	2.00	410	21	35.0
20	2	3.29	570	3.12	450	19	33.0	45	2	2.76	470	3.04	430	19	28.0
21	1	3.00	410	2.32	310	19	25.0	46	2	3.93	680	3.43	430	19	28.0
22	2	2.67	550	2.41	440	20	51.0	47	2	3.30	590	2.69	560	20	32.0
23	1	3.63	500	3.12	510	19	36.7	48	1	2.61	350	2.33	590	20	46.0
24	1	3.11	600	2.42	550	20	50.0	49	1	3.70	650	3.21	660	19	53.0
25	1	3.77	580	2.42	550	21	46.0	50	1	3.28	440	3.20	480	19	30.0

OBS	SEX	HS GPA	SAT MATH	CUM GPA	SAT VERB	AGE	HOURS	OBS	SEX	HS GPA	SAT MATH	CUM GPA	SAT VERB	AGE	HOURS
51	2	1.72	620	2.15	580	19	30.0	97	1	3.60	550	3.36	470	19	29.0
52	1	2.90	400	2.24	480	19	25.0	98	1	0.00	570	2.93	530	19	29.0
53	2	3.46	610	2.52	420	19	25.0	99	1	2.88	660	3.00	530	19	36.0
54	1	3.86	470	3.62	510	19	26.0	100	1	2.58	480	2.88	500	19	25.0
55	1	3.07	540	2.62	440	19	26.0	101	1	3.37	640	4.00	250	20	28.0
56	2	3.61	540	3.48	520	20	25.0	102	2	3.01	550	2.30	590	21	44.0
57	1	3.10	480	3.26	380	19	46.0	103	2	3.76	720	4.00	670	19	36.0
58	1	2.95	370	3.20	360	21	30.0	104	2	3.50	380	2.64	430	20	52.0
59	2	3.00	480	2.89	440	18	28.0	105	2	2.60	540	2.80	440	21	28.0
60	1	2.00	370	4.00	300	20	50.0	106	2	2.93	410	2.00	290	19	25.0
61	2	2.68	500	2.18	410	20	46.0	107	2	2.97	480	2.88	560	21	46.0
62	1	3.66	570	3.52	470	19	31.0	108	1	3.03	430	2.78	550	19	32.0
63	1	3.90	570	3.61	520	18	31.0	109	1	3.03	540	2.56	470	19	27.0
64	1	3.00	550	2.60	430	20	45.0	110	1	2.25	320	0.29	290	21	54.0
65	2	3.81	530	3.16	530	19	26.0	111	1	3.42	500	3.00	570	20	52.0
66	1	3.42	420	2.88	440	19	29.0	112	1	2.64	480	2.45	420	19	37.0
67	2	2.69	310	2.00	340	20	37.0	113	2	3.54	570	3.63	540	19	36.0
68	2	3.24	600	2.81	630	19	32.0	114	2	3.05	500	2.08	360	19	26.0
69	1	3.74	590	3.31	630	20	26.0	115	1	2.69	290	2.22	450	20	54.0
70	2	2.95	520	3.00	440	19	28.0	116	1	2.88	460	1.96	420	20	52.0
71	1	3.04	610	2.96	460	19	34.0	117	2	2.68	500	2.38	570	19	34.0
72	2	1.81	560	1.75	450	20	38.0	118	1	3.24	550	2.59	470	19	27.0
73	1	3.55	340	2.84	360	20	25.0	119	1	3.85	630	3.64	650	19	34.0
74	1	2.24	610	2.28	500	20	25.0	120	2	2.99	460	2.42	540	19	26.0
75	2	3.28	580	0.82	460	22	30.0	121	2	1.93	470	3.15	450	21	41.0
76	2	2.93	490	1.75	520	20	27.0	122	1	2.66	370	2.09	330	20	34.0
77	1	2.57	600	2.96	510	19	26.0	123	2	3.53	480	2.10	470	20	46.0
78	1	3.61	520	2.58	560	20	54.0	124	2	3.29	490	2.48	440	21	31.0
79	1	3.71	470	2.19	360	21	38.0	125	1	3.16	630	2.52	550	19	36.0
80	2	3.07	540	2.69	540	20	41.0	126	1	3.34	510	2.67	390	19	32.0
81	2	3.78	650	3.00	510	20	31.0	127	1	2.98	650	2.96	610	18	27.0
82	1	2.72	540	2.64	490	19	28.0	128	1	2.86	480	1.72	560	19	25.0
83	2	2.05	700	3.53	390	19	30.0	129	2	2.55	530	2.00	500	20	32.0
84	2	3.03	460	3.59	310	19	37.0	130	2	2.43	480	2.63	450	21	50.0
85	1	3.30	650	2.65	390	19	37.0	131	2	0.00	540	3.24	540	26	39.0
86	1	3.40	420	2.78	530	20	39.0	132	1	3.46	670	3.29	600	19	38.0
87	2	0.00	520	1.83	330	21	46.0	133	2	2.21	430	2.17	430	22	50.0
88	2	2.51	630	2.81	500	19	30.0	134	2	2.52	380	2.67	360	21	35.0
89	2	2.62	460	2.31	480	20	54.0	135	2	2.41	420	2.39	470	20	49.0
90	1	2.63	460	2.36	450	20	28.0	136	2	2.15	380	1.73	410	20	31.0
91	1	0.00	590	1.80	560	22	49.0	137	1	2.66	550	2.60	420	19	30.0
92	2	2.80	570	2.67	460	19	27.0	138	1	3.14	590	3.04	490	19	27.0
93	1	3.20	400	2.60	450	22	40.0	139	1	2.68	500	2.24	390	19	29.0
94	1	3.45	470	1.43	420	20	25.0	140	1	3.64	540	2.90	480	19	31.0
95	1	3.45	440	3.06	330	19	34.0	141	2	2.45	480	3.43	360	21	42.0
96	2	2.95	450	2.39	380	19	31.0	142	1	3.02	350	2.18	330	20	44.4

OBS	SEX	HS GPA	SAT MATH	CUM GPA	SAT VERB	AGE	HOURS	OBS	SEX	HS GPA	SAT MATH	CUM GPA	SAT VERB	AGE	HOURS
143	2	2.72	590	2.49	440	20	53.0	189	1	3.22	440	2.83	430	20	49.0
144	2	2.76	480	2.33	450	19	27.0	190	2	2.52	490	2.21	470	19	29.0
145	2	3.17	520	3.29	520	19	32.0	191	1	3.72	600	3.35	460	19	32.0
146	2	3.29	480	3.31	420	19	26.0	192	2	3.33	450	3.14	320	19	34.0
147	2	1.96	430	2.32	340	19	25.0	193	1	3.29	450	3.06	400	20	42.0
148	2	2.81	460	3.18	660	27	39.0	194	2	2.91	620	2.82	480	20	50.0
149	2	3.01	530	2.27	480	20	30.0	195	2	2.60	400	2.12	470	19	25.0
150	1	2.67	560	2.38	490	19	33.0	196	1	3.82	480	3.50	440	19	26.0
151	1	2.83	390	2.45	490	20	49.0	197	2	2.58	320	1.77	340	21	52.0
152	1	2.71	310	1.84	260	19	26.0	198	2	3.54	490	2.34	410	20	50.0
153	2	2.82	550	2.56	480	19	29.0	199	2	3.29	520	3.18	560	22	27.0
154	2	3.32	590	2.48	510	21	54.0	200	2	2.72	420	2.24	460	21	49.0
155	1	3.31	500	2.90	450	19	26.0	201	1	2.50	480	1.17	440	20	46.0
156	2	3.19	520	2.43	590	20	39.0	202	2	3.60	710	3.07	430	19	38.0
157	1	2.58	500	3.00	620	19	39.0	203	2	1.97	560	2.15	480	21	51.0
158	2	2.83	430	2.33	360	20	30.0	204	2	2.65	710	2.46	410	20	46.0
159	2	2.34	430	2.50	490	20	28.0	205	1	2.87	560	3.57	510	19	30.0
160	2	3.54	430	1.96	440	21	27.0	206	2	3.18	510	3.74	530	20	50.0
161	1	3.71	570	3.28	500	19	25.0	207	1	0.00	340	0.69	410	22	33.0
162	1	3.41	520	3.61	550	19	33.0	208	1	3.29	420	2.36	360	19	32.0
163	2	2.39	510	1.94	400	20	38.0	209	2	2.66	420	2.38	400	19	28.0
164	1	2.71	380	2.81	420	19	26.0	210	1	3.90	580	3.00	450	19	25.0
165	1	2.78	610	2.59	680	19	27.0	211	2	2.24	600	2.39	380	20	48.0
166	2	3.00	480	3.14	370	19	37.0	212	1	3.94	540	3.64	510	19	25.0
167	1	3.03	510	3.00	540	19	28.0	213	1	2.70	520	2.67	380	19	27.0
168	1	2.05	480	1.79	380	20	43.0	214	1	2.92	450	2.03	480	19	28.0
169	2	1.89	560	1.66	450	19	26.0	215	1	3.26	370	2.77	270	19	26.0
170	2	2.21	690	2.46	460	19	28.0	216	1	3.20	570	2.83	590	19	29.0
171	2	2.51	370	1.84	310	20	41.0	217	2	2.58	350	2.90	440	19	34.0
172	1	2.67	420	2.86	430	19	25.0	218	2	2.51	500	2.17	420	20	51.0
173	1	3.13	400	2.68	410	21	50.0	219	2	3.35	410	1.95	330	21	40.0
174	2	2.89	390	3.40	410	19	40.0	220	1	2.91	520	2.62	380	20	50.0
175	2	3.43	600	3.60	460	19	30.0	221	1	2.88	380	1.90	460	19	27.0
176	2	0.00	540	3.31	550	18	32.0	222	2	0.00	490	1.87	510	25	34.0
177	2	3.64	640	3.91	660	19	36.0	223	1	2.49	570	2.41	560	20	52.0
178	2	3.46	660	2.82	580	19	32.0	224	1	2.98	330	2.41	370	19	25.0
179	1	3.60	460	2.14	510	21	33.0	225	2	2.82	450	3.43	390	19	28.0
180	1	2.69	560	3.38	350	20	27.0	226	1	3.50	580	2.85	420	19	27.0
181	1	2.88	450	3.42	480	20	47.0	227	2	3.67	720	3.55	560	20	50.0
182	1	3.51	510	3.35	530	19	29.0	228	2	2.71	540	3.64	340	19	31.0
183	2	2.52	390	2.26	360	21	43.0	229	2	2.78	710	3.15	470	19	27.0
184	1	3.44	610	3.36	500	19	33.0	230	2	3.40	430	2.42	470	20	48.0
185	1	2.06	320	2.20	490	19	25.0	231	2	3.90	640	3.50	600	20	37.7
186	2	3.97	660	3.47	510	18	44.0	232	2	3.87	640	3.30	440	19	34.0
187	1	2.15	440	1.68	440	19	34.0	233	2	0.00	730	4.00	450	20	45.0
188	1	3.70	600	3.41	460	19	32.0	234	2	2.88	550	3.10	450	19	29.0

OBS	SEX	HS GPA	SAT MATH	CUM GPA	SAT VERB	AGE	HOURS	OBS	SEX	HS GPA	SAT MATH	CUM GPA	SAT VERB	AGE	HOURS
235	2	3.08	500	3.15	440	20	52.0	281	2	3.25	570	2.54	560	19	52.0
236	1	3.07	370	2.50	320	19	27.0	282	2	2.67	540	1.84	520	21	54.0
237	2	2.60	390	2.31	340	20	49.0	283	2	2.08	680	2.44	480	19	27.0
238	2	3.38	530	2.57	460	19	28.0	284	2	2.47	520	2.50	320	20	29.0
239	2	3.96	570	4.00	460	19	45.0	285	1	3.35	470	2.98	590	25	49.0
240	1	3.26	490	2.35	420	19	26.0	286	2	3.80	590	3.80	660	19	36.0
241	2	3.11	580	2.50	390	19	30.0	287	2	2.52	430	2.48	440	19	25.0
242	1	2.82	560	2.25	530	20	45.0	288	2	3.00	550	3.17	660	19	36.0
243	2	3.35	550	3.02	390	21	53.0	289	2	2.87	500	2.07	350	20	51.0
244	2	2.67	420	1.82	380	20	25.0	290	2	3.00	590	3.42	600	19	40.0
245	2	2.65	530	2.56	460	20	25.0	291	1	3.68	490	3.11	430	19	30.0
246	1	2.88	390	2.43	400	22	46.0	292	2	2.99	580	3.05	470	20	41.0
247	2	3.58	610	3.73	610	20	51.0	293	2	2.82	570	2.18	430	20	48.0
248	2	0.00	520	2.50	600	26	44.4	294	1	2.74	520	2.63	460	19	27.0
249	1	3.95	610	3.90	590	18	38.0	295	2	3.28	490	3.08	350	19	30.0
250	2	2.49	610	2.13	380	21	40.0	296	2	3.50	430	2.79	450	19	28.0
251	2	2.74	560	2.86	450	19	28.0	297	2	3.94	700	3.87	640	19	31.0
252	1	3.22	460	3.19	400	18	30.0	298	1	3.27	510	2.08	520	20	45.0
253	1	2.83	490	2.76	440	20	25.0	299	2	4.00	750	4.00	540	19	40.0
254	1	3.32	490	2.80	360	19	26.0	300	2	2.62	490	2.64	430	19	28.0
255	1	3.73	520	3.14	510	19	28.0	301	2	2.99	680	3.92	500	23	42.4
256	2	2.23	420	1.67	450	20	38.0	302	2	3.11	500	1.96	460	20	40.0
257	1	2.80	350	2.76	400	19	37.0	303	2	2.85	530	2.76	420	19	25.0
258	1	3.37	440	2.11	490	19	25.0	304	2	2.70	690	2.27	460	19	26.0
259	1	2.84	440	3.26	400	19	27.0	305	1	2.98	490	2.66	390	19	29.0
260	1	3.72	530	3.21	450	19	33.0	306	2	2.92	430	1.95	280	21	47.0
261	1	2.88	530	3.53	530	21	47.0	307	1	3.63	460	3.32	460	19	31.0
262	1	3.44	450	3.14	420	19	36.0	308	2	2.96	590	2.38	470	20	42.0
263	1	2.99	430	2.02	430	20	51.0	309	1	2.14	410	2.51	440	20	53.0
264	1	0.00	410	1.75	350	20	53.5	310	1	3.50	700	1.81	580	19	34.0
265	1	2.46	570	3.58	510	19	39.0	311	2	1.83	520	2.35	500	20	50.0
266	2	2.00	500	2.79	340	21	52.0	312	2	3.28	660	2.33	610	20	53.0
267	1	3.10	490	2.62	410	21	39.0	313	2	2.65	500	3.59	480	23	43.0
268	1	3.78	630	3.19	480	19	32.0	314	2	2.35	560	2.48	450	20	51.0
269	1	3.53	510	2.38	350	21	51.0	315	1	3.79	500	3.36	420	19	34.0
270	2	3.70	610	3.18	530	19	28.0	316	2	3.00	400	2.84	370	19	31.0
271	2	3.19	520	3.79	380	19	37.0	317	1	2.75	350	2.71	330	20	52.0
272	2	2.33	310	1.93	360	21	49.0	318	1	2.75	320	3.40	370	22	47.0
273	2	3.71	620	3.71	450	19	35.0	319	1	3.10	530	2.19	420	20	54.0
274	2	3.75	640	2.38	370	21	29.0	320	2	3.52	450	2.78	390	20	27.0
275	2	1.81	630	2.50	470	21	30.0	321	2	3.71	580	3.60	430	19	25.0
276	2	3.72	520	3.88	500	19	51.0	322	2	3.50	680	2.60	470	19	29.0
277	2	3.62	740	2.00	540	19	30.0	323	1	2.95	410	2.21	440	21	43.0
278	2	3.37	600	3.00	470	19	31.0	324	1	2.81	490	2.12	430	19	25.0
279	2	3.46	640	3.77	630	20	52.0	325	1	3.20	480	2.94	360	20	34.0
280	1	2.35	350	3.13	480	19	28.0	326	1	2.97	580	2.90	430	19	30.0

OBS	SEX	HS GPA	SAT MATH	CUM GPA	SAT VERB	AGE	HOURS	OBS	SEX	HS GPA	SAT MATH	CUM GPA	SAT VERB	AGE	HOURS
327	2	2.44	530	2.07	400	19	29.0	373	2	2.25	580	2.56	400	20	54.0
328	2	3.38	600	2.76	530	19	33.0	374	2	2.81	590	2.76	420	19	31.0
329	1	3.72	740	3.87	560	19	39.0	375	1	3.71	520	3.38	420	18	35.0
330	2	2.98	520	2.32	380	21	47.0	376	1	3.02	410	1.25	470	20	49.0
331	1	3.02	350	1.90	380	19	32.0	377	2	3.63	660	2.35	440	20	31.0
332	1	3.36	520	2.54	310	19	37.0	378	1	2.61	400	2.08	410	19	25.0
333	2	2.34	590	2.96	490	20	27.0	379	2	1.98	420	2.32	400	21	49.0
334	1	2.85	400	2.15	330	20	39.4	380	1	3.41	570	2.47	390	20	45.0
335	2	2.41	390	1.46	400	21	29.0	381	1	2.46	390	2.80	540	19	25.0
336	2	2.69	330	2.00	320	20	32.5	382	2	2.64	630	2.29	520	20	53.0
337	2	3.63	460	2.57	350	19	28.0	383	2	3.53	530	2.76	360	20	53.0
338	1	3.18	410	1.96	380	19	28.0	384	2	2.77	490	2.11	410	20	52.0
339	2	2.84	640	2.52	400	19	29.0	385	1	2.50	670	2.58	590	19	31.0
340	2	2.66	460	2.28	310	20	50.0	386	2	3.44	700	2.86	520	19	43.0
341	2	2.98	430	2.23	470	21	45.0	387	1	3.60	480	3.47	440	20	53.0
342	2	2.85	520	2.96	560	20	53.0	388	1	2.82	430	1.98	430	20	54.0
343	1	3.33	590	1.70	610	19	45.0	389	1	2.98	520	3.22	420	19	44.1
344	2	3.60	520	2.70	460	19	33.0	390	2	1.82	540	0.94	450	22	39.0
345	2	2.49	450	2.30	500	19	27.0	391	2	2.32	570	2.50	360	21	49.0
346	2	2.26	650	2.36	440	19	25.0	392	2	3.27	650	2.74	520	19	40.0
347	1	2.93	470	1.37	400	20	49.0	393	2	2.89	570	1.91	370	20	33.0
348	1	3.38	510	2.72	530	19	29.0	394	2	2.89	550	2.29	410	19	31.0
349	2	1.86	590	2.10	510	19	37.0	395	1	2.91	490	2.27	450	21	40.0
350	2	3.81	700	3.54	500	19	32.0	396	1	3.38	380	2.42	420	20	54.0
351	2	2.09	270	2.73	530	21	52.0	397	2	3.07	350	2.65	500	19	26.0
352	1	3.55	350	2.61	340	19	34.0	398	2	3.52	600	2.89	570	20	53.0
353	1	3.15	520	2.73	460	19	35.0	399	1	3.20	410	2.83	360	20	30.0
354	2	3.95	700	0.00	650	18	27.0	400	2	3.38	510	3.63	450	19	27.0
355	1	3.40	650	2.10	390	20	34.0	401	1	3.33	540	3.00	460	21	45.0
356	1	2.16	410	2.00	510	21	39.0	402	2	2.66	400	2.72	300	19	27.0
357	2	2.68	470	2.28	500	20	47.0	403	2	2.28	450	2.11	350	20	38.0
358	2	2.87	460	2.62	360	19	26.0	404	1	3.34	420	2.41	390	19	29.0
359	2	3.58	520	2.37	480	20	43.0	405	1	3.84	520	2.62	410	20	29.0
360	2	2.47	590	1.91	390	20	37.0	406	2	2.69	530	2.73	480	21	50.0
361	1	3.78	490	3.88	510	19	39.0	407	2	3.26	580	2.50	450	19	36.0
362	2	2.75	540	2.14	430	19	33.0	408	2	2.55	460	2.88	550	20	41.0
363	2	2.38	490	2.40	380	20	41.0	409	1	3.46	490	2.96	440	19	31.0
364	1	2.93	610	2.57	450	19	33.0	410	1	2.54	510	1.58	390	19	26.0
365	1	3.65	680	3.67	600	19	28.0	411	2	3.18	520	2.00	540	19	28.0
366	1	3.49	590	3.07	660	24	42.0	412	1	3.29	610	3.50	460	19	30.0
367	2	3.13	430	3.26	540	19	27.0	413	2	2.60	550	2.72	440	19	25.0
368	2	3.31	640	2.52	500	20	41.0	414	2	2.92	460	2.84	440	19	25.0
369	2	2.36	540	2.00	400	29	35.0	415	2	2.75	470	2.47	440	20	51.0
370	2	3.30	600	1.97	430	19	25.0	416	2	3.51	640	3.62	650	20	29.0
371	2	2.68	470	2.00	470	20	34.0	417	1	3.31	460	2.19	500	20	31.0
372	1	2.56	360	1.00	280	21	51.0	418	2	2.80	440	2.03	360	19	27.0

OBS	SEX	HS GPA	SAT MATH	CUM GPA	SAT VERB	AGE	HOURS	OBS	SEX	HS GPA	SAT MATH	CUM GPA	SAT VERB	AGE	HOURS
419	1	3.15	470	3.02	450	20	49.0	460	1	3.06	600	2.78	440	20	48.0
420	1	2.87	400	2.23	250	20	52.0	461	2	3.47	690	2.34	610	19	41.0
421	2	3.83	650	3.32	650	20	40.0	462	1	3.77	630	3.12	550	19	32.0
422	2	3.28	600	3.12	490	20	52.0	463	2	3.83	650	3.86	570	19	35.0
423	2	3.19	550	1.94	620	19	29.0	464	1	3.64	480	3.08	570	19	50.0
424	2	2.85	570	3.25	420	23	29.0	465	2	2.00	570	2.08	490	22	47.0
425	2	3.42	380	3.04	420	20	53.0	466	2	3.35	420	2.62	350	20	51.0
426	2	3.75	630	3.69	490	19	46.0	467	2	2.62	560	2.62	380	21	36.0
427	2	3.32	560	3.75	410	20	53.0	468	1	3.26	410	2.36	390	19	25.0
428	1	2.49	430	1.88	310	20	38.0	469	2	3.71	660	3.42	550	19	31.0
429	1	1.67	390	2.89	450	20	52.0	470	2	2.23	550	2.79	520	24	44.0
430	1	3.50	380	2.71	320	19	44.0	471	2	3.10	610	2.46	530	20	48.0
431	2	3.28	550	2.28	440	19	43.0	472	2	2.64	510	2.04	420	20	28.0
432	1	3.62	580	3.60	460	20	30.0	473	2	2.95	530	2.32	390	19	25.0
433	2	3.09	600	2.13	400	20	27.0	474	1	2.57	550	2.83	510	19	27.0
434	1	2.58	450	1.48	420	18	25.0	475	2	3.26	600	1.92	490	21	25.0
435	2	3.96	590	3.11	340	19	27.0	476	2	3.65	610	3.41	440	20	51.0
436	1	2.33	340	2.50	400	20	49.0	477	1	2.63	340	2.12	430	19	33.0
437	1	3.18	470	2.97	500	19	32.0	478	1	2.73	690	2.52	510	19	27.0
438	1	4.00	460	3.42	410	19	32.0	479	1	3.35	410	2.48	490	19	31.0
439	1	2.97	420	2.53	450	20	51.0	480	1	2.86	520	3.15	440	19	33.0
440	1	3.87	450	3.04	540	19	27.0	481	1	2.85	470	2.80	440	19	30.0
441	2	2.48	260	3.00	200	19	25.0	482	1	3.02	460	2.68	500	19	28.0
442	1	3.00	450	3.00	400	19	39.0	483	1	3.54	660	1.92	590	21	47.0
443	1	3.50	540	2.50	490	20	47.0	484	2	3.43	420	2.70	410	19	30.0
444	1	2.34	520	2.86	500	25	54.0	485	2	3.39	540	2.08	420	19	26.0
445	2	2.77	540	2.29	460	19	31.0	486	2	2.82	720	2.00	460	20	45.0
446	2	2.53	500	2.92	410	20	36.0	487	2	3.60	560	3.00	480	19	30.0
447	2	3.21	460	2.36	360	20	53.0	488	1	3.60	600	3.50	450	19	26.0
448	2	2.54	480	2.19	450	19	27.0	489	2	2.85	540	1.83	390	21	45.0
449	2	3.36	570	2.00	440	19	28.0	490	2	2.73	440	1.68	390	21	52.4
450	2	2.39	590	1.90	480	20	51.0	491	1	1.75	430	1.82	450	20	52.0
451	2	2.49	530	2.72	440	19	25.0	492	1	2.95	500	2.71	370	19	28.0
452	2	3.05	540	2.33	410	21	46.0	493	1	3.89	560	3.63	520	19	30.0
453	1	3.58	440	3.33	400	19	33.0	494	2	2.84	570	2.21	460	20	42.0
454	2	2.87	520	2.00	530	19	25.0	495	2	2.50	510	2.00	440	25	40.0
455	2	3.11	700	2.59	340	20	49.0	496	2	3.24	630	2.23	640	20	50.0
456	1	3.23	400	2.24	370	20	54.0	497	1	3.70	600	3.04	600	19	38.0
457	2	2.75	510	2.30	440	19	27.0	498	1	2.81	490	2.29	520	20	32.0
458	2	2.73	560	2.54	410	20	46.0	499	2	2.95	530	2.81	440	19	32.0
459	1	3.81	560	3.57	550	19	28.0	500	2	3.29	610	2.42	420	20	34.0

Answers to
Selected Exercises

・　・　・　・　・

Note: Most of the numerical answers here were obtained using a computer. If you solve a problem by hand and do some intermediate rounding, your answer may differ somewhat.

CHAPTER 1

Exercises 1.1

1.1 See Definition 1.2 on page 5.

Exercises 1.2

1.3 inferential

1.5 inferential

1.7 descriptive

1.9 descriptive

1.11
a. inferential   b. descriptive   c. descriptive
d. inferential

1.13 descriptive, inferential

Exercises 1.6

1.19 Dentists form a high-income group whose incomes are not representative of the incomes of Seattle residents in general.

1.21

a.

Officials selected	Sample obtained
G, L, S	70, 40, 37
G, L, A	70, 40, 55
G, L, T	70, 40, 50
G, S, A	70, 37, 55
G, S, T	70, 37, 50
G, A, T	70, 55, 50
L, S, A	40, 37, 55
L, S, T	40, 37, 50
L, A, T	40, 55, 50
S, A, T	37, 55, 50

b. $\frac{1}{10}$, $\frac{1}{10}$, $\frac{1}{10}$

1.23

a.

Sample	Sample
AR, FS, BD, BI	AR, BD, JS, JC
AR, FS, BD, JS	AR, BI, JS, JC
AR, FS, BD, JC	FS, BD, BI, JS
AR, FS, BI, JS	FS, BD, BI, JC
AR, FS, BI, JC	FS, BD, JS, JC
AR, FS, JS, JC	FS, BI, JS, JC
AR, BD, BI, JS	BD, BI, JS, JC
AR, BD, BI, JC	

b. Write the initials of the six representatives on separate pieces of paper, place the six slips of paper into a box, and then, while blindfolded, pick four of the slips of paper. Or, number the representatives 1–6, and use a table of random numbers or a random-number generator to select four different numbers between 1 and 6.

c. $\frac{1}{15}$, $\frac{1}{15}$

1.25   Answers will vary.

1.27   Answers will vary.

Exercises 1.7

1.33

a. Answers will vary, but here is the procedure: (1) Divide the population size, 685, by the sample size, 25, and round down to the nearest whole number; this gives 27. (2) Use a table of random numbers (or a similar device) to select a number between 1 and 27; call it k. (3) List every 27th number, starting with k, until 25 numbers are obtained; thus the first number on the required list of 25 numbers is k, the second is $k + 27$, the third is $k + 54$, and so forth. (For example, if $k = 6$, then the numbers on the list are 6, 33, 60, ...)

b. systematic random sampling

c. Yes, unless there is some kind of cyclical pattern in the listing of the employees.

1.35   Yes, since there is no cyclical pattern in the listing. In fact, because the listing is by sales, one could argue that, in this case, systematic random sampling is preferable to simple random sampling.

1.37

a. Number the suites from 1 to 48, use a table of random numbers to randomly select 3 of the 48 suites, and take as the sample the 24 dormitory residents living in the 3 suites obtained.

b. Probably not, since friends often have similar opinions.

c. Proportional allocation dictates that the number of freshmen, sophomores, juniors, and seniors selected be, respectively, 8, 7, 6, and 3. Thus a stratified sample of 24 dormitory residents can be obtained as follows: Number the freshman dormitory residents from 1 to 128 and use a table of random numbers to randomly select 8 of the 128 freshman dormitory residents; number the sophomore dormitory residents from 1 to 112 and use a table of random numbers to randomly select 7 of the 112 sophomore dormitory residents; and so forth.

Exercises 1.8

1.39

a. An *observational study* is a study in which researchers simply observe.

b. A *designed experiment* is a study in which researchers design and control.

1.41   Observational studies can reveal only association; designed experiments can establish causation.

1.43   Here is one of several methods that could be used: Number the women from 1 to 4753; use a table of random numbers or a random-number generator to obtain 2376 different numbers between 1 and 4753; the 2376 women with those numbers are in one group, the remaining 2377 women are in the other group.

1.45   designed experiment

1.47   observational study

1.49   Double-blinding guards against bias, both in the evaluations and in the responses. In the Salk-vaccine experiment, double-blinding prevented a doctor's evaluation from being influenced by knowing which treatment (vaccine or placebo) a patient received; it also prevented a patient's response to the treatment from being influenced by knowing which treatment he or she received.

Review Test for Chapter 1

1. Answers will vary.

2. In conducting an inferential study, information will be obtained from a sample of the population. Generally, that information must be organized and summarized prior to applying inferential methods. Thus almost any inferential study involves aspects of descriptive statistics.

3. descriptive

4. inferential

5. inferential

6. descriptive

7. inferential

10. No, because parents of students at Yale tend to have higher incomes than parents of college students in general.

11. only (b)

12. a. Number the registered voters from 1 to 7246, use Table IX to obtain 50 different numbers between 1 and 7246, and take as the sample the 50 registered voters who are numbered with the numbers obtained.

b.

6293	7075	4096	1946	5945
3331	6679	7132	3335	6562
6151	4106	0189	6417	5109
0940	6899	3094	0102	2930
1928	5236	0529	5669	0353
4124	2293	3909	4865	0538
0219	6984	1966	5203	3098
0341	1424	4204	2200	5306
0208	0315	5802	0788	1793
5470	4198	1065	6875	4830

14. a. Answers will vary, but here is the procedure: (1) Divide the population size, 7246, by the sample size, 50, and round down to the nearest whole number; this gives 144. (2) Use a table of random numbers (or a similar device) to select a number between 1 and 144; call it k. (3) List every 144th number, starting with k, until 50 numbers are obtained; thus the first number on the required list of 50 numbers is k, the second is $k + 144$, the third is $k + 288$, and so forth. (For example, if $k = 86$, then the numbers on the list are 86, 230, 374, ...)

   b. Yes, unless for some reason there is a cyclical pattern in the listing of the registered voters.

15. a. Proportional allocation dictates that 10 full professors, 16 associate professors, 12 assistant professors, and 2 instructors be selected.

   b. The procedure is as follows: Number the full professors from 1 to 205, and use Table IX to randomly select 10 of the 205 full professors; number the associate professors from 1 to 328, and use Table IX to randomly select 16 of the 328 associate professors; and so on.

16. a. poverty status and IQ

   b. observational study

CHAPTER 2

Exercises 2.1

2.1 It may aid in choosing the correct statistical method.

2.3 quantitative, continuous; annual U.S. tobacco production in millions of pounds

2.5 quantitative, discrete; number of employees in millions

2.7

a. quantitative, continuous; land area in thousands of square miles

b. quantitative, discrete; land-area rank and population rank

c. quantitative, discrete; population in millions

d. qualitative; continent of birth

2.9

a. quantitative, discrete; rank by amount of deposits

b. quantitative, continuous (or discrete); amount of deposits in millions of dollars

2.11 qualitative

Exercises 2.2

2.15 No; data must be quantitative for class limits and class marks to make sense.

2.17

a. The frequency of a class is the number of data values in the class, whereas the relative frequency of a class is the ratio of the class frequency to the total number of pieces of data.

b. The relative frequency of a class is the percentage of the class expressed as a decimal.

2.19

Content (mL)	Frequency	Relative frequency	Class mark
910–929	1	0.033	919.5
930–949	1	0.033	939.5
950–969	3	0.100	959.5
970–989	9	0.300	979.5
990–1009	7	0.233	999.5
1010–1029	6	0.200	1019.5
1030–1049	2	0.067	1039.5
1050–1069	1	0.033	1059.5
	30	0.999	

2.21

Consumption (mil. BTU)	Frequency	Relative frequency	Class mark
40–49	1	0.02	44.5
50–59	7	0.14	54.5
60–69	7	0.14	64.5
70–79	3	0.06	74.5
80–89	6	0.12	84.5
90–99	10	0.20	94.5
100–109	5	0.10	104.5
110–119	4	0.08	114.5
120–129	2	0.04	124.5
130–139	3	0.06	134.5
140–149	0	0.00	144.5
150–159	2	0.04	154.5
	50	1.00	

2.23

Number of cars sold	Frequency	Relative frequency
0	7	0.135
1	15	0.288
2	12	0.231
3	9	0.173
4	5	0.096
5	3	0.058
6	1	0.019
	52	1.000

2.25

Number of days missed	Frequency	Relative frequency
0	4	0.050
1	2	0.025
2	14	0.175
3	10	0.125
4	16	0.200
5	18	0.225
6	10	0.125
7	6	0.075
	80	1.000

2.27

Starting salary ($thousands)	Frequency	Relative frequency	Class mark
16–under 17	3	0.086	16.5
17–under 18	3	0.086	17.5
18–under 19	5	0.143	18.5
19–under 20	9	0.257	19.5
20–under 21	9	0.257	20.5
21–under 22	4	0.114	21.5
22–under 23	1	0.029	22.5
23–under 24	1	0.029	23.5
	35	1.001	

2.29

Sales (millions)	Frequency	Relative frequency	Class mark
0–under 1	1	0.042	0.5
1–under 2	3	0.125	1.5
2–under 3	10	0.417	2.5
3–under 4	4	0.167	3.5
4–under 5	1	0.042	4.5
5–under 6	1	0.042	5.5
6–under 7	2	0.083	6.5
7–under 8	1	0.042	7.5
8–under 9	1	0.042	8.5
	24	1.002	

2.31

Champion	Frequency	Relative frequency
Oklahoma	2	0.067
Oklahoma State	6	0.200
Iowa State	7	0.233
Michigan State	1	0.033
Iowa	13	0.433
Arizona State	1	0.033
	30	0.999

Exercises 2.3

2.35 A frequency histogram displays the class frequencies on the vertical axis, whereas a relative-frequency histogram displays the class relative frequencies on the vertical axis.

2.37

a.

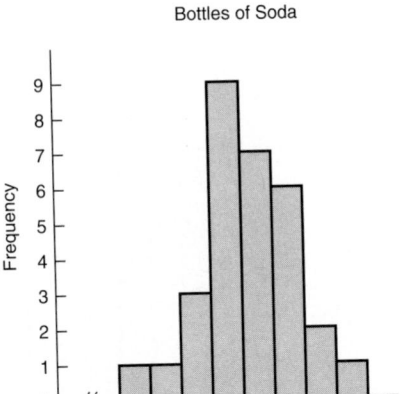

b.

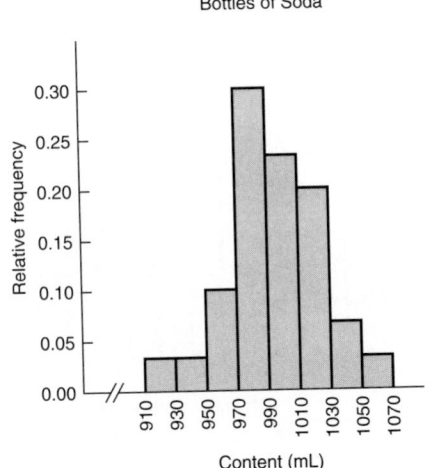

2.39

a.

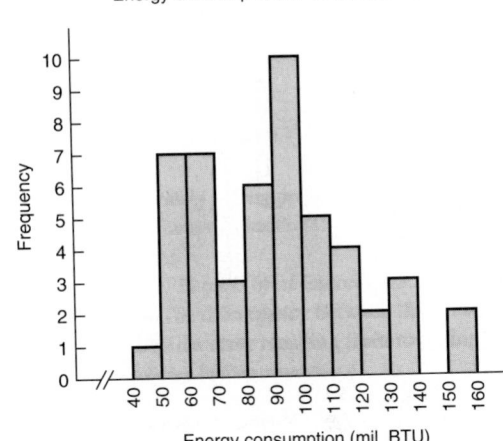

b.

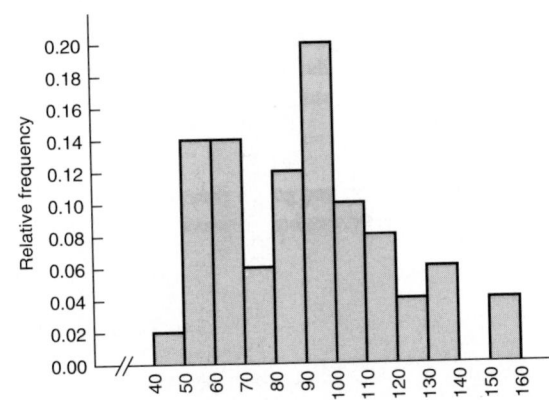

2.41

a.

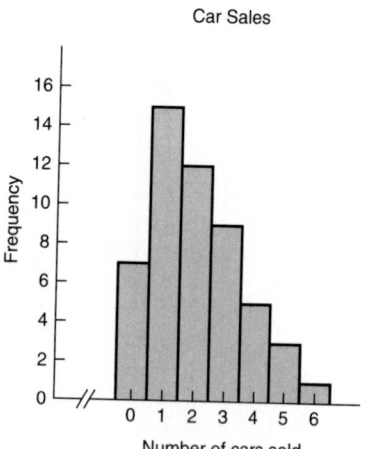

b.

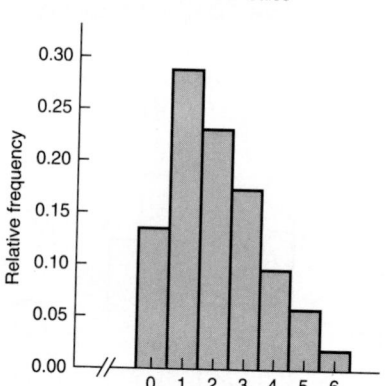

2.43

a.

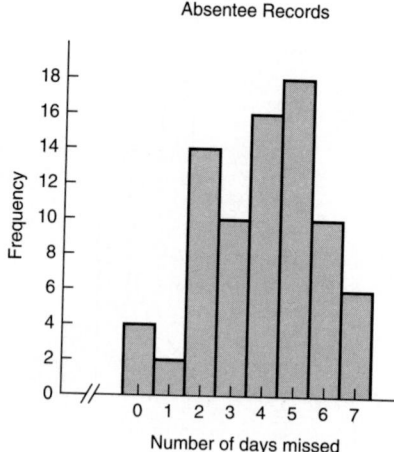

b.

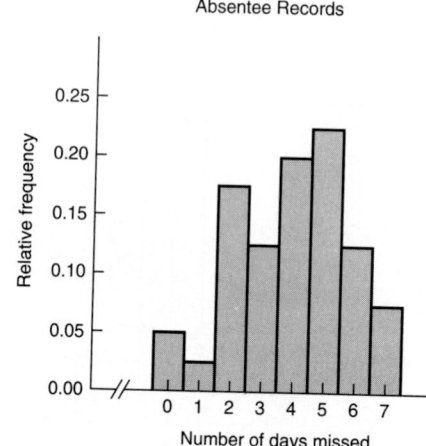

2.45

a.

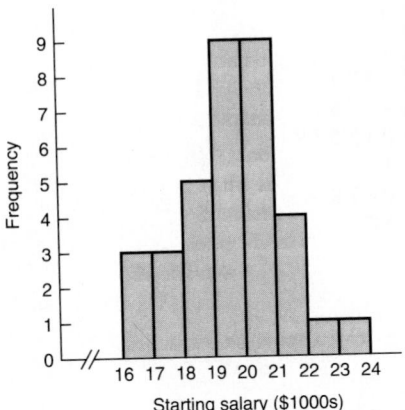

Salaries for Liberal-Arts Graduates

b.

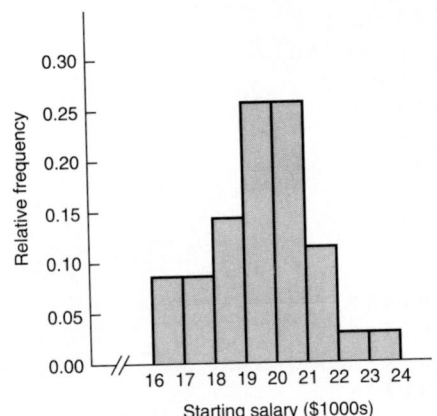

Salaries for Liberal-Arts Graduates

2.47

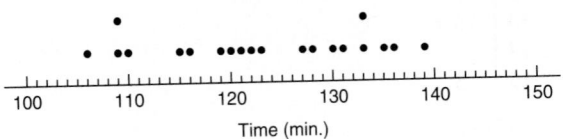

Drying Times

2.49

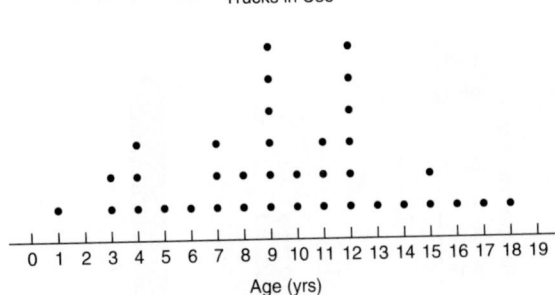

Trucks in Use

2.51

a.

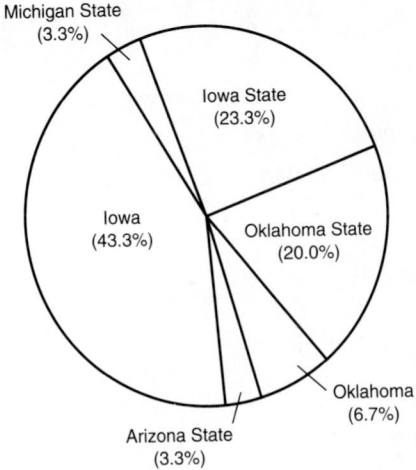

1963–1992 NCAA Wrestling Championships

b.

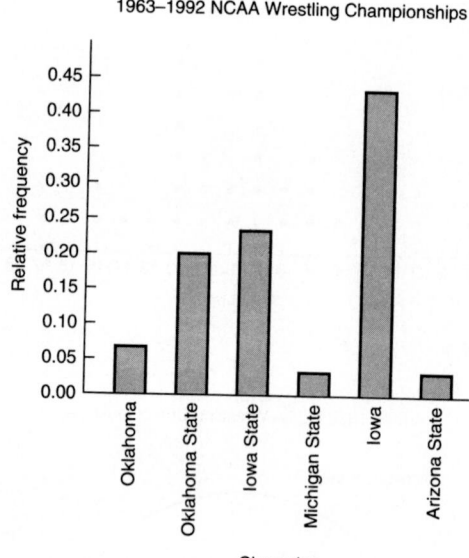

1963–1992 NCAA Wrestling Championships

2.53   a. 20%   b. 25%   c. 7

2.59   a. 50   b. 94.5   c. 10   d. 5

2.63   a. 37   b. 7

Exercises 2.4

2.71

a.

91	4
92	
93	
94	6
95	9 7
96	4
97	7 5 4 7
98	6 9 4 8 7
99	0 6 1 9 5 7
100	1
101	4 8 0 7
102	5 8
103	0 1
104	
105	
106	0

b.

91	4
92	
93	
94	6
95	7 9
96	4
97	4 5 7 7
98	4 6 7 8 9
99	0 1 5 6 7 9
100	1
101	0 4 7 8
102	5 8
103	0 1
104	
105	
106	0

2.73

a.

4	5
5	8 4 5 1 0 5 5
6	4 7 9 6 0 6 2
7	7 5 8
8	6 7 1 0 3 3
9	7 6 3 4 9 7 6 0 1 7
10	1 0 9 4 2
11	1 3 1 3
12	9 5
13	0 9 6
14	
15	5 1

b.

4	5
5	0 1 4 5 5 5 8
6	0 2 4 6 6 7 9
7	5 7 8
8	0 1 3 3 6 7
9	0 1 3 4 6 6 7 7 7 9
10	0 1 2 4 9
11	1 1 3 3
12	5 9
13	0 6 9
14	
15	1 5

2.75

a.
```
2 | 6 7 3
3 | 9 9 1 5 3 9 3 1 9
4 | 3 8 8 2 2 8 7 4 7 5 3 4 9 5 0 1 4
5 | 5 7 5 6 2 8 1 5
6 | 2 8 1 5 3 6 0 9
7 | 3 0 1 4
8 | 2
```

b.
```
2 | 3
2 | 6 7
3 | 1 3 3 1
3 | 9 9 5 9 9
4 | 3 2 2 4 3 4 0 1 4
4 | 8 8 8 7 7 5 9 5
5 | 2 1
5 | 5 7 5 6 8 5
6 | 2 1 3 0
6 | 8 5 6 9
7 | 3 0 1 4
7 |
8 | 2
8 |
```

2.77

a.
```
2 |
2 | 9 9
3 | 3 1 3 2 3 4 3
3 | 6 9 5 9 5 7 9 9 5 7 8
4 | 0 0 1 2 0 4 4 0 3 2 1 2 4 1 1 4 4
4 | 8 8 5 9 5 6 5 5 7 9 8 5 9 5 9 9 7
5 | 1 1 3 1 2 1
5 | 8 7 5 7 9 5 7
6 |
6 | 9
7 | 0 3
7 |
```

b.
```
2 |
2 |
2 |
2 |
2 | 9 9
3 | 1
3 | 3 3 2 3 3
3 | 5 5 4 5
3 | 6 7 7
3 | 9 9 9 9 8
4 | 0 0 1 0 0 1 1 1
4 | 2 3 2 2
4 | 5 5 5 4 4 5 5 4 5 4 4
4 | 6 7 7
4 | 8 8 9 9 8 9 9 9
5 | 1 1 1 1
5 | 3 2
5 | 5 5
5 | 7 7 7
5 | 8 9
6 |
6 |
6 |
6 |
6 | 9
7 | 0
7 | 3
7 |
7 |
7 |
```

2.81

a. 50   b. two   c. 7   d. 14   e. 83
f. 53, 54, 55, 55, 56, 56, 56, 56, 57, 58

Exercises 2.5

2.83 a. right skewed   b. right skewed

2.85 a. bell-shaped   b. symmetric

2.87 a. left skewed   b. left skewed
Note: The answers *bell-shaped* for part (a) and *symmetric* for part (b) are also acceptable.

2.89 a. bell-shaped   b. symmetric

2.91 a. right skewed   b. right skewed

Exercises 2.6

2.95

c. They give the misleading impression that the district average is much greater relative to the national average than it actually is.

2.97

a. It is a truncated graph.

b.

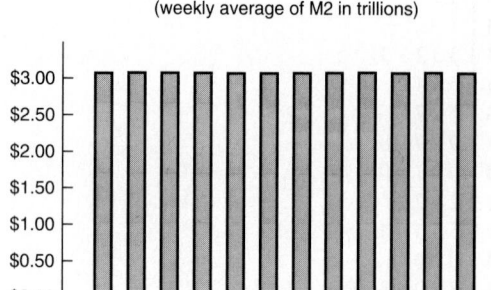

c.

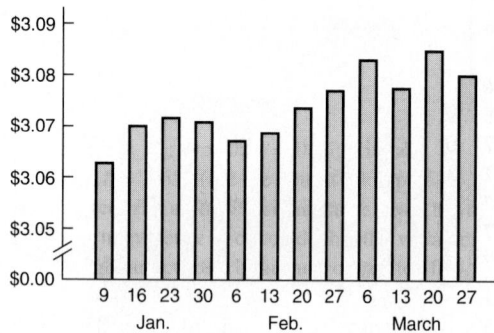

Review Test for Chapter 2

1. a. discrete quantitative
   b. continuous quantitative
   c. qualitative

2. a.

Age at inauguration	Frequency	Relative frequency	Class mark
40–44	2	0.048	42
45–49	6	0.143	47
50–54	12	0.286	52
55–59	12	0.286	57
60–64	7	0.167	62
65–69	3	0.071	67
	42	1.001	

b.

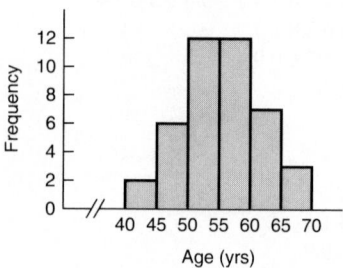

3.

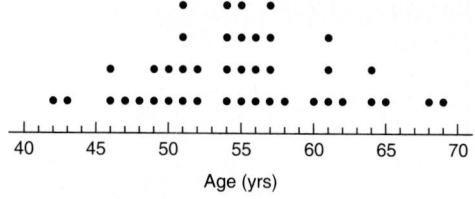

4. a. **4** | 2 3 6 6 7 8 9 9
   5 | 0 0 1 1 1 1 2 2 4 4 4 4 5 5 5 5 6 6 6 7 7 7 7 8
   6 | 0 1 1 1 2 4 4 5 8 9

   b. **4** | 2 3
   4 | 6 6 7 8 9 9
   5 | 0 0 1 1 1 1 2 2 4 4 4 4
   5 | 5 5 5 5 6 6 6 7 7 7 7 8
   6 | 0 1 1 1 2 4 4
   6 | 5 8 9

   c. The second one (i.e., the one with two lines per stem).

6. a. 22   b. 27   c. 3   d. TIMES

7. a. 2   b. 2   c. 4   d. 48 minutes
   e. 30, 31, 31, 37

8. a.

Number busy	Frequency	Relative frequency
0	1	0.04
1	2	0.08
2	2	0.08
3	4	0.16
4	5	0.20
5	7	0.28
6	4	0.16
	25	1.00

b.

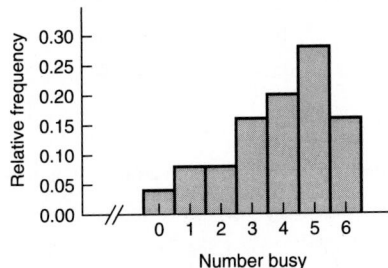

Busy Tellers

9. a.

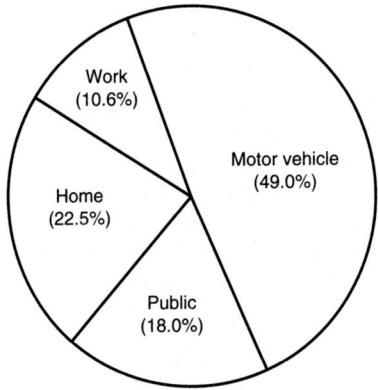

Accidental Deaths by Type, 1988*

*Data from National Safety Council

b.

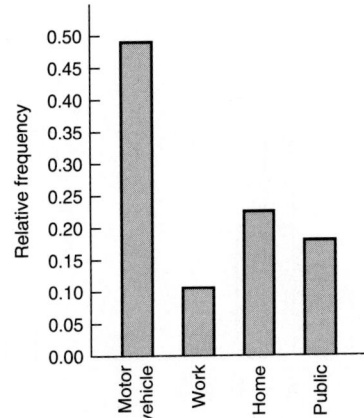

Accidental Deaths by Type, 1988*

*Data from National Safety Council

10. a.

High	Freq.	Relative frequency	Class mark
400–under 600	3	0.083	500
600–under 800	5	0.139	700
800–under 1000	13	0.361	900
1000–under 1200	6	0.167	1100
1200–under 1400	2	0.056	1300
1400–under 1600	1	0.028	1500
1600–under 1800	0	0.000	1700
1800–under 2000	1	0.028	1900
2000–under 2200	1	0.028	2100
2200–under 2400	0	0.000	2300
2400–under 2600	0	0.000	2500
2600–under 2800	2	0.056	2700
2800–under 3000	1	0.028	2900
3000–under 3200	1	0.028	3100
	36	1.002	

b.

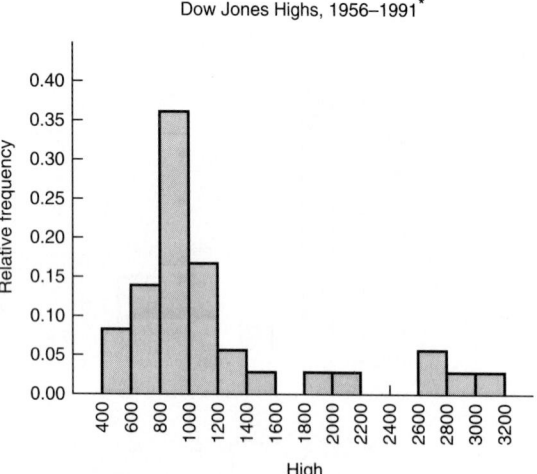

Dow Jones Highs, 1956–1991[*]

*Data from *The World Almanac. 1993*

11. a. bell-shaped   b. left skewed

12. Answers will vary, but here is one possibility:

13. b. The percentage of women in the labor force for 1985 is about two and two-thirds times that for 1970.
   c. About one and one-half.
   d. Because it is a truncated graph.
   e. Start the graph at 0 instead of 30.

CHAPTER 3

Exercises 3.1

3.1   To indicate where the center or most typical value of a data set lies.

3.3   mean = 33.5 years; median = 34.5 years; modes = 24, 28, 37 years

3.5   mean = 40.79¢/lb; median = 41.80¢/lb; no mode

3.7   mean = 306.0 mL; median = 306.5 mL; modes = 300, 311 mL

3.9   mean = 193.0 thousand; median = 79.0 thousand; no mode

3.11   The *median,* because unlike the mean it is not strongly affected by the relatively few homes that have an extremely large or small square footage.

3.13
a. Iowa
b. No, neither the mean nor the median can be used as a measure of central tendency for qualitative data.

3.15   a. 13.9   b. 4   c. median

Exercises 3.2

3.27   a. 46   b. 4   c. 11.5
3.29   a. 331,143 mi   b. 8   c. 41,392.9 mi
3.31   a. $31,061   b. 18   c. $1725.6
3.33   a. $97.5   b. 60,091, 0, 3053.5
3.35   a. 105.1   b. 111,153, 0, 692.9

Exercises 3.3

3.41   To indicate the amount of variation in a data set.
3.43   a. 20 yrs   b. 7.2 yrs   c. 7.2 yrs
3.45   a. 10.6¢/lb   b. 3.38¢/lb   c. 3.38¢/lb
3.47   a. 31 mL   b. 8.7 mL   c. 8.7 mL
3.49
a. 501 thousand   b. 205.5 thousand
c. 205.5 thousand
3.51
a. Brand A: $\bar{x} = 9.80$; Brand B: $\bar{x} = 9.80$
b. Brand A: median = 9.7; Brand B: median = 9.7
c. In the amount of variation.
d. The data set for Brand A.
e. Brand A: $s = 0.50$; Brand B: $s = 1.62$
f. Yes, because the data for Brand A, which has less variation than the data for Brand B, also has a smaller standard deviation.

3.53   a. 16.1   b. 16.1

Exercises 3.4

3.63
a. Data Set 4
b. Data Set 3: $\bar{x} = 83$, $s = 7.8$
   Data Set 4: $\bar{x} = 83$, $s = 22.3$
c. See Figs. A.1 and A.2.
e. Yes. In fact, all of the data do.

FIGURE A.1   Data Set 3. $\bar{x} = 83$, $s = 7.8$

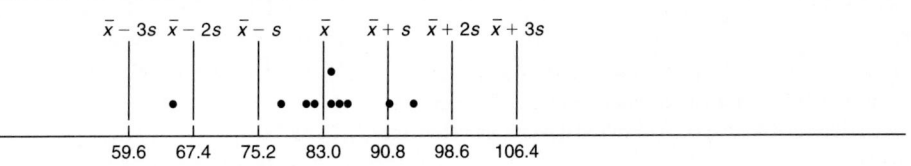

FIGURE A.2   Data Set 4. $\bar{x} = 83$, $s = 22.3$

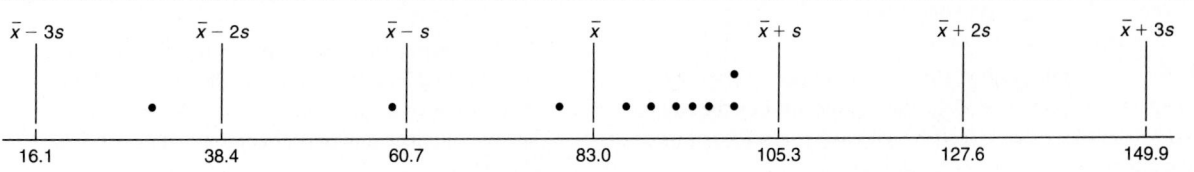

3.65

a. At least 36% of the data lie within 1.25 standard deviations to either side of its mean.

b. At least 92% of the data lie within 3.5 standard deviations to either side of its mean.

c. At least 96% of the data lie within 5 standard deviations to either side of its mean.

3.67

a. At least 75%; at least 89%.

b. 90%; 100%   c. 90%; 100%

d. Chebychev's rule does not necessarily give precise estimates for the percentage of data that lies within a specified number of standard deviations to either side of the mean.

3.69

a.

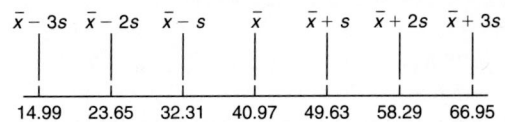

b. At least 225 of the 300 lodging establishments have room rates between $23.65 and $58.29.

c. At least 267 of the 300 lodging establishments have room rates between $14.99 and $66.95.

3.71

a.

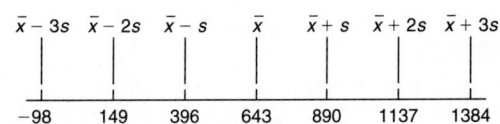

b. At least 188 of the 250 households have vegetable gardens that are between 149 and 1137 sq ft.

c. At least 223 of the 250 households have vegetable gardens that are at most 1384 sq ft.

3.73

a.

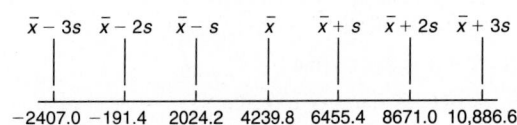

b. At least 23 of the 30 taxable returns have an income tax of at most $8671.0.

c. At least 27 of the 30 taxable returns have an income tax of at most $10,886.6.

3.75

a. 3.01, −2.42, −0.46. The three room rates are, respectively, 3.01 standard deviations above, 2.42 standard deviations below, and 0.46 standard deviations below the mean.

b. A room rate of $67 is greater than at least 89.0% ($= 1 - 1/3.01^2$) of the room rates in the sample; one of $20 is less than at least 82.9% ($= 1 - 1/2.42^2$) of the room rates in the sample; and one of $37 is near the mean and below it.

3.77

a. 4, −1.75, −0.10. The three vegetable-garden sizes are, respectively, 4 standard deviations above, 1.75 standard deviations below, and 0.10 standard deviations below the mean.

b. A vegetable garden whose size is 1631 sq ft is larger than at least 93.8% of the gardens in the sample; one whose size is 211 sq ft is smaller than at least 67.3% of the gardens in the sample; and one whose size is 618 sq ft is near the mean size and below it.

3.79   Yes. A sale price of $95,500 has a z-score of −2.82 relative to the sample of 25 sale prices obtained by the realtor and hence is less than at least 87.4% ($= 1 − 1/2.82^2$) of those sale prices.

3.81   The verbal, because the student's verbal score is 2.5 standard deviations above the mean and the student's math score is 2 standard deviations above the mean.

3.83   a. 1, 4, 7

Exercises 3.5

3.89   a. $\bar{x} = 3.2$ people   b. $s = 1.3$ people

3.91   a. $\bar{x} = 21.6$ yrs   b. $s = 3.8$ yrs

3.93

a. $\bar{x} = 39.5$ bu/acre, $s = 1.9$ bu/acre

b.

Yield	36	37	38	39	40	41	42	43
Freq.	1	3	5	8	5	3	1	4

c. $\bar{x} = 39.5$ bu/acre, $s = 1.9$ bu/acre

d. They are identical.

3.95 $\bar{x} \approx 105.0$, $s \approx 16.4$

3.97 $\bar{x} \approx 62.8$ min, $s \approx 10.7$ min

3.99

a. $\bar{x} = 89.7$ mil BTU, $s = 27.3$ mil BTU

b. $\bar{x} \approx 89.9$ mil BTU, $s \approx 27.3$ mil BTU

3.101   Because the class mark of each class provides only a typical value for the data values in the class and not the data values themselves.

Exercises 3.6

3.105 $Q_1 = 68.5$, $Q_2 = 83.0$, $Q_3 = 89.5$

3.107 $Q_1 = 8.0$, $Q_2 = 9.2$, $Q_3 = 11.2$

3.109 $Q_1 = 1168$, $Q_2 = 1247$, $Q_3 = 1393$

3.111 $Q_1 = 4$, $Q_2 = 7$, $Q_3 = 12$

3.113

a. IQR $= 21$ b. 34, 68.5, 83.0, 89.5, 100

c. 34 is a possible outlier.

d. Fig. A.3(a) shows a boxplot; Fig. A.3(b) shows a modified boxplot.

3.115

a. IQR $= 8$ b. 1, 4, 7, 12, 55

c. 55 is a probable outlier.

d. Fig. A.4(a) shows a boxplot; Fig. A.4(b) shows a modified boxplot.

3.119

a. The least variation occurs in the second quarter of the data, the next least in the third quarter, the next least in the fourth quarter, and the greatest in the first quarter. The five-number summary is approximately 0, 56, 62, 72, and 90.

b. Yes. Probable outlier(s) (approximately) at 0; possible outlier(s) (approximately) at 8; possible outlier(s) (approximately) at 20. Assuming no recording errors, the probable outlier may be a vegetarian or someone who does not eat beef for some other reason; the possible outliers may be people on low-beef diets.

c. 0, 55, 62, 72.5, and 89

3.121   Generally speaking, the accounting graduates have higher starting salaries than the liberal-arts graduates. In fact, more than half of the accounting graduates have higher starting salaries than the highest starting salary of the liberal-arts graduates, and the lowest starting salary of the accounting graduates exceeds more than half of the starting salaries of the liberal-arts graduates. Also, the variation in starting salaries of the accounting graduates is somewhat smaller than that of the liberal-arts graduates.

Exercises 3.7

3.125   To describe the entire population.

3.127

a. $\bar{x} = 75.0$ in.   b. $s = 6.2$ in.   c. $\mu = 75.0$ in.

d. $\sigma = 5.6$ in.

e. $\bar{x}$ and μ are computed in the same way: Sum the data and then divide by the total number of pieces of data.

f. s and σ are computed differently: In the defining formula for s, we divide by one less than the total number of pieces of data, whereas in the defining formula for σ, we divide by the total number of pieces of data.

3.129   a. $\mu = \$46.3$ b. $\sigma = \$17.8$

3.131   a. $\mu = \$101.1$ million   b. $\sigma = \$109.9$ million

3.133   It is a parameter, being a descriptive measure for a population.

FIGURE A.3   (a) Boxplot and (b) modified boxplot for Exercise 3.113(d)

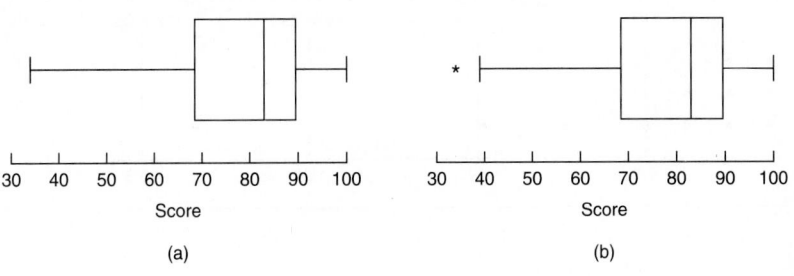

FIGURE A.4   (a) Boxplot and (b) modified boxplot for Exercise 3.115(d)

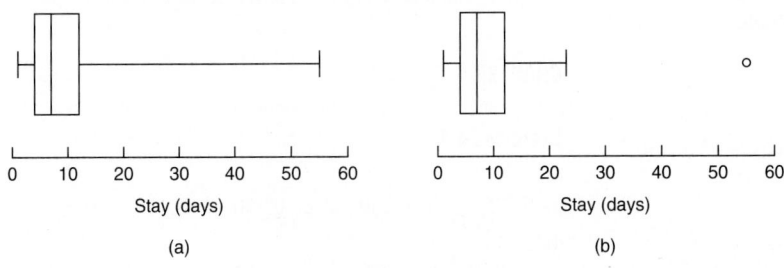

3.135   a. $\mu = 2$ b. $\sigma = 1.57$

3.137

a. $\mu \approx 12.9$ yrs   b. $\sigma \approx 6.3$ yrs

c. Because the class mark of each class provides only a typical value for the data values in the class and not the data values themselves.

3.139

a. 313.0, 366.2   b. 299.7, 379.5

c. -3.50, 2.14, -0.12

d. A gestation period of 293 days is shorter than at least 91.8% ($= 1 - 1/3.50^2$) of all gestation periods for the Morgan mare; one of 368 days is longer than at least 78.2% ($= 1 - 1/2.14^2$) of all gestation periods for the Morgan mare; and one of 338 days is near the mean gestation period for the Morgan mare and below it.

3.141

a. $z = -3.13$

b. Yes. A gas mileage of 21.4 mpg is less than at least 89.8% ($= 1 - 1/3.13^2$) of the gas mileages for all cars of this model.

Review Test for Chapter 3

1. a. 11.4 kg; 16.1 kg   b. 11.5 kg; 17.0 kg

c. 8, 12 kg; 12, 16, 19, 23 kg

2. The median, because it is resistant to outliers and other extreme values.

3. The mode; neither the mean nor the median can be used as a measure of central tendency for qualitative data.

4. a. $\bar{x} = 4.0$ min   b. Range $= 14$ min

   c. $s = 4.1$ min

5. a. See Fig. A.5 at the top of the next page.

   b. 75%   c. 91.7%

   d. Its generality; it holds for any data set.

6. a.

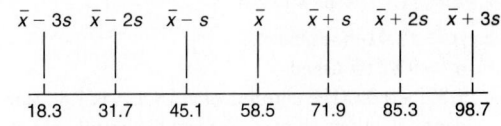

b. 31.7, 85.3   c. 89

7. a. 0.04, -2.05, 1.53. The three specified ages are, respectively, 0.04 standard deviations above, 2.05 standard deviations below, and 1.53 standard deviations above the mean.

FIGURE A.5   Graph for Problem 5(a)

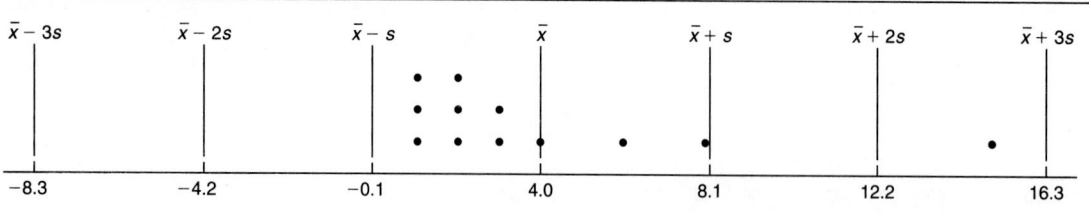

b. An age of 59 years is near the mean and above it; an age of 31 years is less than at least 76.2% $(= 1 - 1/2.05^2)$ of the ages in the sample; and an age of 79 years is greater than at least 57.3% $(= 1 - 1/1.53^2)$ of the ages in the sample.

8. a. $Q_1 = 48.0$, $Q_2 = 59.5$, $Q_3 = 68.5$
   b. 20.5
   c. 31, 48, 59.5, 68.5, 79
   d. Inner fences: 17.25, 99.25. Outer fences: -13.5, 130
   e. no potential outliers
   f.

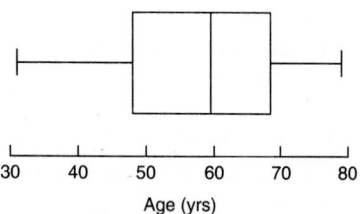

Age (yrs)

11. a. The least variation occurs in the third quarter, the next least in the fourth quarter, the next least in the second quarter, and the greatest in the first quarter. The approximate five-number summary is 31, 48, 60, 69, and 79.
    b. no
    c. The exact five-number summary is 31, 48, 59.5, 68.5, and 79.

12. a. $\bar{x} = 71.7$ b. $s = 2.5$

13. a. $\mu = 18.41$ thousand
    b. $\sigma = 9.82$ thousand
    c. 1.82, -0.05. The enrollment at UCLA is 1.82 standard deviations above the mean; the enrollment at UCSD is 0.05 standard deviations below the mean.

14. a. $\mu \approx \$86.0$ thousand
    b. $\sigma \approx \$18.5$ thousand
    c. Because the class mark of each class provides only a typical value for the data values in the class and not the data values themselves.

15. a. A sample mean. It is the mean price per gallon for the sample of 10,000 service stations.
    b. $\bar{x}$
    c. A statistic; it is a descriptive measure for a sample.

CHAPTER 4

Exercises 4.1

4.1
a. 0.125   b. 0.250   c. 0.750   d. 0   e. 1

4.3
a. 0.291   b. 0.288   c. 0.652   d. 0.192

4.5
a. 0.139   b. 0.500   c. 0.222   d. 0.111

4.7   0.020

4.9   a. 0.724   b. 0.912   c. 0.189

4.11   (b) and (d), since the probability of an event must always be between 0 and 1, inclusive.

4.13   The event in part (e) is certain; the event in part (d) is impossible.

Exercises 4.2

4.19

$A =$ ⚁ ⚃ ⚅

$B =$ ⚃ ⚄ ⚅

$C =$ ⚀ ⚀

$D =$ ⚁

4.21

$A = \{HHTT, HTHT, HTTH, THHT, THTH, TTHH\}$

$B = \{TTHH, TTHT, TTTH, TTTT\}$

$C = \{HHHH, HHHT, HHTH, HHTT, HTHH,$
    $HTHT, HTTH, HTTT\}$

$D = \{HHHH, TTTT\}$

4.23

a. $(not\ A) =$ ⚀ ⚂ ⚄

The event the die comes up odd.

b. $(A\ \&\ B) =$ ⚃ ⚅

The event the die comes up 4 or 6.

c. $(B\ or\ C) =$ ⚀ ⚁ ⚃ ⚄ ⚅

The event the die does not come up 3.

4.25

a. $(not\ B) = \{HHHH, HHHT, HHTH, HHTT,$
    $HTHH, HTHT, HTTH, HTTT,$
    $THHH, THHT, THTH, THTT\}$
The event that at least one of the first two tosses is heads.

b. $(A\ \&\ B) = \{TTHH\}$
The event that the first two tosses are tails and the last two are heads.

c. $(C\ or\ D) = \{HHHH, HHHT, HHTH, HHTT, HTHH$
    $HTHT, HTTH, HTTT, TTTT\}$
The event that the first toss is a head or all four tosses are tails.

4.27

a. $(not\ A)$ is the event the employee selected missed at least 4 days. Fifty employees missed at least 4 days.

b. $(A\ \&\ B)$ is the event the employee selected missed between 1 and 3 days, inclusive. Twenty-six employees missed between 1 and 3 days, inclusive.

c. $(C\ or\ D)$ is the event the employee selected missed at least 4 days. Fifty employees missed at least 4 days. [*Note:* From part (a), we see that $(not\ A) = (C\ or\ D)$.]

4.29

a. $(not\ C)$ is the event that the person selected is 45 years old or older. There are 28,116 thousand such people.

b. $(not\ B)$ is the event that the person selected is either under 20 or over 54. There are 18,799 thousand such people.

c. $(B\ \&\ C)$ is the event that the person selected is between 20 and 44, inclusive. There are 63,774 thousand such people.

d. $(A\ or\ D)$ is the event that the person selected is either under 20 or over 54. There are 18,799 thousand such people. [*Note:* From part (b), we see that $(not\ B) = (A\ or\ D)$.]

4.31

a. No   b. Yes   c. No

d. Yes, events $B, C,$ and D. No.

4.33

a. mutually exclusive   b. not mutually exclusive

c. mutually exclusive   d. not mutually exclusive

e. not mutually exclusive

Exercises 4.3

4.37 0.2; $P(E) = 0.2$

4.39

a. 0.69   b. $S = (A\ or\ B\ or\ C)$

c. 0, 0.23, 0.46   d. 0.69

4.41

a. 0.792   b. 0.064   c. 0.343

d. 79.2% of the businesses in the United States had receipts under $100,000; 6.4% had receipts of at least $500,000; and 34.3% had receipts between $25,000 and $499,999.

4.43 a. 1   b. 0.93
Note: You should have used the complementation rule to obtain these two probabilities.

4.45 a. 0.967   b. 0.3
Note: You should have used the complementation rule to obtain these two probabilities.

4.47

a. 0.167, 0.056, 0.028, 0.056, 0.028, 0.139, 0.167

b. 0.223   c. 0.112   d. 0.278, 0.278

4.49

a. 0.526, 0.070, 0.042

b. 0.554; 55.4% of U.S. adults are either female or divorced (or both).

c. 0.474

4.51 0.267

Exercises 4.4

4.55

a. 8   b. 3274   c. 863   d. 1471   e. 502

4.57

a. 12   b. 108,674.4 thousand

c. 44,670.4 thousand   d. 10,564.6 thousand

e. 686.0 thousand

4.59

a. The missing entries in the second, third, and fourth rows are, respectively, 157, 247, and 30.

b. 15   c. 690 thousand   d. 313 thousand

e. 128 thousand   f. 1026 thousand

g. 135 thousand

4.61

a. The institution selected is private; the institution selected is in the South; the institution selected is a public school in the West.

b. 0.551; 0.316; 0.096. 55.1% of institutions of higher education are private; 31.6% are in the South; and 9.6% are public schools in the West.

c.

Type

Region		Public T_1	Private T_2	$P(R_i)$
Northeast	R_1	0.081	0.170	0.251
Midwest	R_2	0.110	0.154	0.264
South	R_3	0.163	0.153	0.316
West	R_4	0.096	0.074	0.170
$P(T_j)$		0.449	0.551	1.000

4.63

a. The person selected is employed; the person selected has completed between 13 and 15 years of school; the person selected is employed and has completed between 13 and 15 years of school.

b. 0.903; 0.180; 0.168   c. (i) 0.915   (ii) 0.915

d.

Employment status

Years of school completed		Employed E_1	Unemployed E_2	$P(S_i)$
Less than 8	S_1	0.033	0.006	0.038
8	S_2	0.030	0.005	0.035
9–11	S_3	0.117	0.026	0.143
12	S_4	0.369	0.042	0.411
13–15	S_5	0.168	0.012	0.180
16 or more	S_6	0.186	0.006	0.192
$P(E_j)$		0.903	0.097	1.000

4.65

a. (i) A_3 (ii) T_2 (iii) $(T_1 \& A_5)$

b. (i) 0.241   (ii) 0.288   (iii) 0.015

c.

Tenure of operator

Acreage		Full owner T_1	Part owner T_2	Tenant T_3	Total
Under 50	A_1	21.5	3.0	3.4	27.9
50–179	A_2	22.7	6.3	3.8	32.9
180–499	A_3	10.6	10.0	3.5	24.1
500–999	A_4	2.3	5.2	1.2	8.7
1000+	A_5	1.5	4.3	0.7	6.5
Total		58.6	28.8	12.6	100.0

Exercises 4.5

4.69

a. 0.077   b. 0.333   c. 0.077   d. 0

e. 0.231   f. 1   g. 0.231   h. 0.167

4.71
a. 0.125 b. 0.132 c. 0.342
d. 12.5% of the employees at Cudahey Masonry missed exactly 3 days of work; 13.2% of those who missed at least 1 day of work missed exactly 3 days; and 34.2% of those who missed at least 1 day of work missed at most 3 days.

4.73
a. 0.251 b. 0.308 c. 0.676
d. 25.1% of all institutions of higher education are in the Northeast; 30.8% of all private institutions of higher education are in the Northeast; 67.6% of all institutions of higher education in the Northeast are private schools.

4.75
a. 0.126 b. 0.035 c. 0.278 d. 0.278

4.77
a. 0.187 b. 0.082 c. 0.439 d. 0.208
e. 18.7% of the members of the 102nd Congress are senators; 8.2% of the members of the 102nd Congress are Republican senators; 43.9% of the senators in the 102nd Congress are Republicans; and 20.8% of the Republicans in the 102nd Congress are senators.

4.79 33.0%

Exercises 4.6

4.83 0.008; 0.8% of all families are farm families making at least $25,000 per year.

4.85 a. 0.1 b. 0.111 c. 0.011 d. 0.222

4.87
a. $\frac{21}{50} \cdot \frac{27}{49} = 0.231$ b. $\frac{21}{50} \cdot \frac{20}{49} = 0.171$
c. Let $R1$, $D1$, and $I1$ denote, respectively, the events that the first governor selected is a Republican, Democrat, and Independent; and let $R2$, $D2$, and $I2$ denote, respectively, the events that the second governor selected is a Republican, Democrat, and Independent. See Fig. A.6.
d. 0.459

FIGURE A.6 Tree diagram for Exercise 4.87(c)

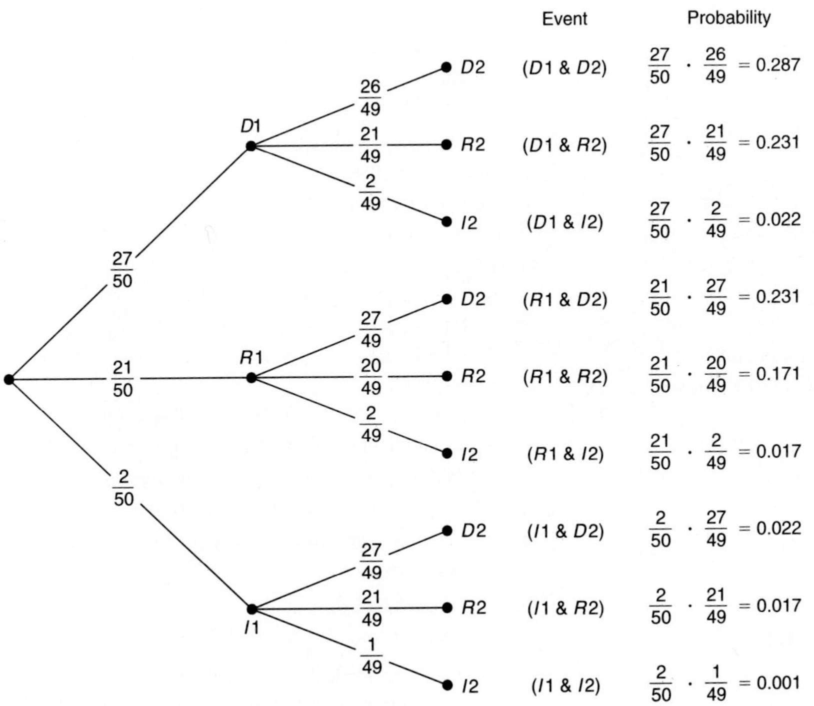
Event
Probability
27/50
21/50
2/50
D1
R1
I1
26/49
21/49
2/49
27/49
20/49
2/49
27/49
21/49
1/49
D2 (D1 & D2) 27/50 · 26/49 = 0.287
R2 (D1 & R2) 27/50 · 21/49 = 0.231
I2 (D1 & I2) 27/50 · 2/49 = 0.022
D2 (R1 & D2) 21/50 · 27/49 = 0.231
R2 (R1 & R2) 21/50 · 20/49 = 0.171
I2 (R1 & I2) 21/50 · 2/49 = 0.017
D2 (I1 & D2) 2/50 · 27/49 = 0.022
R2 (I1 & R2) 2/50 · 21/49 = 0.017
I2 (I1 & I2) 2/50 · 1/49 = 0.001

4.89
a. 0.151   b. 0.050
c. No, because $P(C_1 \mid S_2) \neq P(C_1)$.
d. No, because $P(S_1) = 0.580$ and $P(S_1 \mid C_2) = 0.458$, so that $P(S_1 \mid C_2) \neq P(S_1)$.

4.91
a. 0.5, 0.5, 0.375   b. 0.5
c. Yes, because $P(B \mid A) = P(B)$.   d. 0.25
e. No, because $P(C \mid A) \neq P(C)$.

4.93
a. 0.605, 0.187, 0.105
b. Events P_1 and C_2 are not independent because $P(P_1 \,\&\, C_2) \neq P(P_1) \cdot P(C_2)$ $(0.105 \neq 0.605 \cdot 0.187)$.

4.95   a. 0.006   b. 0.005

4.97   a. 0.083   b. 0.5

4.99
a. 0.928   b. 0.072
c. There was a 7.2% chance that at least one "criticality 1" item would fail; in the long run, at least one "criticality 1" item will fail in 7.2 out of every 100 such missions.

4.101
a. 0.105   b. 0.105   c. 0.032   d. 0.420

Exercises 4.7

4.111   a. 48.4%   b. 49.5%

4.113   a. 1.2%   b. 22.9%

4.115
a. 0.287   b. 0.332   c. 0.371
d. 28.7% of the people participating in the survey said that drug abuse is the nation's top problem; 33.2% were teenagers; and 37.1% of those who said that drug abuse is the nation's top problem were teenagers.

4.117
a. 93.4% of those having the disease will test positive; 96.8% of those not having the disease will test negative.
b. 0.055
c. Only 5.5% of those testing positive actually have the disease.

4.119   a. 84%   b. 4.8%

Exercises 4.8

4.123
a. 900   b. 9,000,000   c. 8,100,000,000

4.125   1,021,440

4.127
a. 24   b. 32,760   c. 30   d. 1   e. 40,320

4.129   657,720

4.131   a. 3,628,800   b. $\dfrac{1}{3,628,800}$

4.133
a. 4   b. 1365   c. 15   d. 1   e. 1

4.135
a. 75,287,520   b. 67,800,320   c. 0.901

4.137   8,145,060

4.139   a. 0.243   b. 0.972   c. 0.271

4.141   a. 0.0000002   b. 0.002   c. 0.971

Review Test for Chapter 4

1. a. 0.161   b. 0.376
   c. 0.161, 0.243, 0.187, 0.135, 0.096, 0.145, 0.032

2. a. (not J) is the event that the return selected shows an adjusted gross income of at least $100,000. There are 2873 thousand such returns.
   b. ($H \,\&\, I$) is the event that the return selected shows an adjusted gross income of between $20,000 and $49,999. There are 37,323 thousand such returns.
   c. (H or K) is the event that the return selected shows an adjusted gross income of at least $20,000. There are 53,156 thousand such returns.
   d. ($H \,\&\, K$) is the event that the return selected shows an adjusted gross income of between $50,000 and $99,999. There are 12,960 thousand such returns.

3. a. not mutually exclusive
   b. mutually exclusive
   c. mutually exclusive
   d. not mutually exclusive

4. a. 0.564, 0.822, 0.968, 0.178
   b. $H = (C$ or D or E or $F)$
      $I = (A$ or B or C or D or $E)$
      $J = (A$ or B or C or D or E or $F)$
      $K = (F$ or $G)$
   c. 0.563, 0.822, 0.967, 0.177 (*Note:* The slight discrepancies between the first, third, and fourth probabilities here and in part (a) are due to rounding errors.)

5. a. 0.032, 0.419, 0.596, 0.145   b. 0.968
   c. 0.597 (*Note:* The discrepancy with part (a) is due to rounding errors.)

6. a. 6   b. 13,615 thousand
   c. 48,778 thousand   d. 2562 thousand

7. a. (i) The student selected is in college; (ii) the student selected attends a public school; (iii) the student selected attends a public college.

b. 0.216; 0.866; 0.171. 21.6% of students attend college, 86.6% attend public schools, and 17.1% attend public colleges.

c.

Type

	Public T_1	Private T_2	$P(L_i)$
Elementary L_1	0.478	0.064	0.542
High school L_2	0.217	0.025	0.242
College L_3	0.171	0.045	0.216
$P(T_j)$	0.866	0.134	1.000

(Level)

d. (i) 0.911   (ii) 0.911

8. a. 0.197. 19.7% of students attending public schools are in college.
   b. 0.197

9. a. 0.134, 0.103
   b. No, because $P(T_2 \mid L_2) \neq P(T_2)$. We see that 10.3% of high-school students attend private schools, whereas 13.4% of all students attend private schools.
   c. No, because both events can occur if the student selected is any one of the 1400 thousand students who attend a private high school.
   d. $P(L_1) = 0.542$, $P(L_1 \mid T_1) = 0.553$. Because $P(L_1 \mid T_1) \neq P(L_1)$, the event a student is in elementary school is not independent of the event a student attends public school.

10. a. 0.023   b. 0.309   d. 0.451

11. a. 0.004   b. 0.012   c. 0.047

12. a. they cannot occur simultaneously.
    b. $P(B \mid A) = P(B)$ [or $P(A \mid B) = P(A)$].
    c. $P(A \text{ or } B) = P(A) + P(B) - P(A \& B)$.
    d. $P(A \& B) = P(A) \cdot P(B)$.

13. a. No, since $P(A \& B) \neq 0$.
    b. Yes, since $P(A \& B) = P(A) \cdot P(B)$.

14. a. 0.45   b. 0.507   c. 0.685   d. 0.608
    e. 45% of the women surveyed answered "no" to the question; 50.7% of the people surveyed answered "no" to the question; 68.5% of the people surveyed were women; 60.8% of the people who answered "no" to the question were women.
    f. The probabilities in parts (b) and (c) are prior; those in parts (a) and (d) are posterior.

15. a. 66   b. 1320   c. 28; 336

16. a. 635,013,559,600
    b. 0.213   c. 0.00045   d. 0.032   e. 0.013

CHAPTER 5

Exercises 5.1

5.1
a. 1, 2, 3, 4, 5, 6, 7   b. $\{x = 5\}$
c. 0.073. 7.3% of U.S. households consist of exactly five people.
d. 0.175
e.

x	1	2	3	4	5	6	7
$P(x)$	0.232	0.317	0.175	0.154	0.073	0.030	0.019

f.

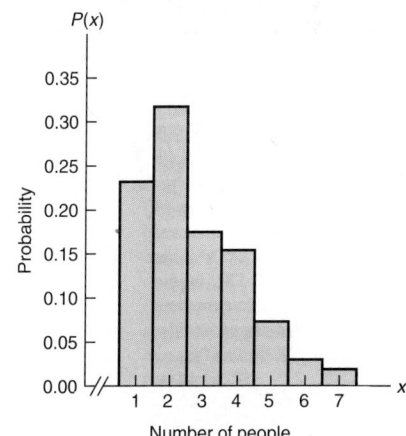

5.3
a. 2, 3, 4, 5, 6, 7, 8, 9, 10, 11, 12
b. $\{y = 7\}$ c. $\frac{1}{6}$ d. $\frac{1}{18}$
e.

y	2	3	4	5	6	7	8	9	10	11	12
$P(y)$	$\frac{1}{36}$	$\frac{1}{18}$	$\frac{1}{12}$	$\frac{1}{9}$	$\frac{5}{36}$	$\frac{1}{6}$	$\frac{5}{36}$	$\frac{1}{9}$	$\frac{1}{12}$	$\frac{1}{18}$	$\frac{1}{36}$

f.

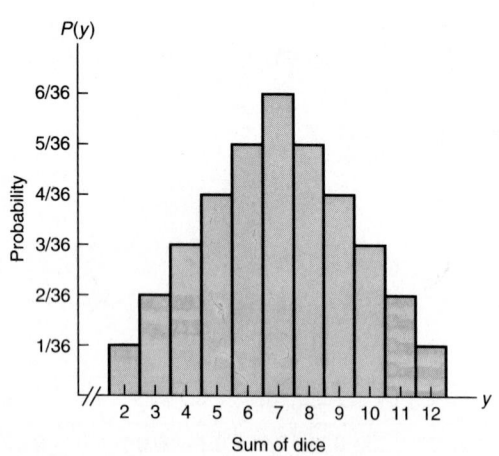

5.5
a. $\{x = 4\}$ b. $\{x \geq 2\}$ c. $\{x < 5\}$
d. $\{2 \leq x < 5\}$ e. 0.21   f. 0.92
g. 0.52   h. 0.44

5.7
a. $\{y = 2\}$ b. $\{y \leq 4\}$ c. $\{y > 1\}$
d. $\{2 \leq y \leq 4\}$ e. 0.134   f. 0.923
g. 0.415   h. 0.338   i. 0.338

5.9   0.975

Exercises 5.2

5.15   a. 2.7   b. 1.5   c. 1.5

5.17   a. 7   b. 2.4   c. 2.4

5.19   a. 4.1   b. 1.6   c. 1.6

5.23
a. $P(x = 1) = \frac{18}{38} = 0.474$, $P(x = -1) = \frac{20}{38} = 0.526$
b. −0.052   c. 5.2¢   d. $5.20, $52   e. No

5.25
a. $\mu_w = 0.25$, $\sigma_w = 0.536$ b. 0.25   c. 62.5

Exercises 5.3

5.29   5040; 40,320; 362,880

5.31   a. 10   b. 35   c. 120   d. 792

5.33   a. 10   b. 1   c. 1   d. 126

5.35
a. $p = 0.487$

b.

Outcome	Probability
sss	$(0.487)(0.487)(0.487) = 0.116$
ssf	$(0.487)(0.487)(0.513) = 0.122$
sfs	$(0.487)(0.513)(0.487) = 0.122$
sff	$(0.487)(0.513)(0.513) = 0.128$
fss	$(0.513)(0.487)(0.487) = 0.122$
fsf	$(0.513)(0.487)(0.513) = 0.128$
ffs	$(0.513)(0.513)(0.487) = 0.128$
fff	$(0.513)(0.513)(0.513) = 0.135$

d. *ssf, sfs, fss*
e. 0.122. Because each probability is obtained by multiplying two success probabilities of 0.487 and one failure probability of 0.513.

5.37
a. $p = 0.2$

b.

Outcome	Probability
ssss	$(0.2)(0.2)(0.2)(0.2) = 0.0016$
sssf	$(0.2)(0.2)(0.2)(0.8) = 0.0064$
ssfs	$(0.2)(0.2)(0.8)(0.2) = 0.0064$
ssff	$(0.2)(0.2)(0.8)(0.8) = 0.0256$
sfss	$(0.2)(0.8)(0.2)(0.2) = 0.0064$
sfsf	$(0.2)(0.8)(0.2)(0.8) = 0.0256$
sffs	$(0.2)(0.8)(0.8)(0.2) = 0.0256$
sfff	$(0.2)(0.8)(0.8)(0.8) = 0.1024$
fsss	$(0.8)(0.2)(0.2)(0.2) = 0.0064$
fssf	$(0.8)(0.2)(0.2)(0.8) = 0.0256$
fsfs	$(0.8)(0.2)(0.8)(0.2) = 0.0256$
fsff	$(0.8)(0.2)(0.8)(0.8) = 0.1024$
ffss	$(0.8)(0.8)(0.2)(0.2) = 0.0256$
ffsf	$(0.8)(0.8)(0.2)(0.8) = 0.1024$
fffs	$(0.8)(0.8)(0.8)(0.2) = 0.1024$
ffff	$(0.8)(0.8)(0.8)(0.8) = 0.4096$

d. *sssf, ssfs, sfss, fsss*
e. 0.0064. Because each probability is obtained by multiplying three success probabilities of 0.2 and one failure probability of 0.8.

5.39
a. $p = 0.232$

b.

Outcome	Probability
ssss	$(0.232)(0.232)(0.232)(0.232) = 0.003$
sssf	$(0.232)(0.232)(0.232)(0.768) = 0.010$
ssfs	$(0.232)(0.232)(0.768)(0.232) = 0.010$
ssff	$(0.232)(0.232)(0.768)(0.768) = 0.032$
sfss	$(0.232)(0.768)(0.232)(0.232) = 0.010$
sfsf	$(0.232)(0.768)(0.232)(0.768) = 0.032$
sffs	$(0.232)(0.768)(0.768)(0.232) = 0.032$
sfff	$(0.232)(0.768)(0.768)(0.768) = 0.105$
fsss	$(0.768)(0.232)(0.232)(0.232) = 0.010$
fssf	$(0.768)(0.232)(0.232)(0.768) = 0.032$
fsfs	$(0.768)(0.232)(0.768)(0.232) = 0.032$
fsff	$(0.768)(0.232)(0.768)(0.768) = 0.105$
ffss	$(0.768)(0.768)(0.232)(0.232) = 0.032$
ffsf	$(0.768)(0.768)(0.232)(0.768) = 0.105$
fffs	$(0.768)(0.768)(0.768)(0.232) = 0.105$
ffff	$(0.768)(0.768)(0.768)(0.768) = 0.348$

c. ssff, sfsf, sffs, fssf, fsfs, ffss
d. 0.032. Because each probability is obtained by multiplying two success probabilities of 0.232 and two failure probabilities of 0.768.

Exercises 5.4

5.45
a. $4 \cdot 0.0064 = 0.0256$
b. $\binom{4}{3}(0.2)^3(0.8)^1 = 0.0256$

5.47
a. 0.384   b. 0.519   c. 0.865   d. 0.749

e.

x	0	1	2	3
P(x)	0.135	0.384	0.365	0.116

f. Technically it is right skewed (see the italicized material in the first paragraph on page 304), but visually it is essentially symmetric.

5.49
a. 0.169   b. 0.957   c. 0.609   d. 0.609

e.

x	0	1	2	3	4	5	6
P(x)	0.000	0.000	0.005	0.038	0.169	0.398	0.391

f. left skewed

5.51
a. 0.177   b. 0.302   c. 0.996
d. 0.501   e. 0.125

5.53
a. 0.302   b. 0.738   c. 1
d. 0.846   e. 0.739

5.55
a. 0.267   b. 0.816   c. 0.451   d. 0.739

5.57   a. $p > 0.5$ b. $p = 0.5$

5.61
a. 7   b. 0.34   c. 0.2610   d. 0.7411
e. 0.9508   f. 0.1837

5.63
a. 0.3874   b. 0.6513   c. 0.7361
d. 0.2639   e. 0.6385

Exercises 5.5

5.71
a. $\mu_x = \Sigma x P(x) = 0.8$
$\sigma_x = \sqrt{\Sigma x^2 P(x) - \mu_x^2} = \sqrt{1.28 - (0.8)^2} = 0.8.$
Note: It is only a coincidence that $\mu_x = \sigma_x$.
b. $\mu_x = np = 4 \cdot 0.2 = 0.8$
$\sigma_x = \sqrt{np(1 - p)} = \sqrt{4 \cdot 0.2 \cdot 0.8} = 0.8.$
c. Much less work is required when the special formulas are used.

5.73 $\mu_x = 5.1, \sigma_x = 0.9$

5.75 $\mu_x = 4.4, \sigma_x = 1.7$

5.77 $\mu_x = 12.5, \sigma_x = 3.1$

Exercises 5.6

5.81
a. 0.224   b. 0.647   c. 0.950
d. 3   e. 1.7

5.83   a. 0.195   b. 0.102   c. 0.704

5.85   a. 0.311   b. 0.757   c. 0.507

5.87
a. $\mu_x = 3.87$ particles. On the average, 3.87 particles will reach the screen during an 8-minute interval.
b. $\sigma_x = 1.97$ particles.

5.89
a.

Number of calls x	Probability P(x)
0	0.183
1	0.311
2	0.264
3	0.150
4	0.064
5	0.022
6	0.006
7	0.001
8	0.000

5.91 a. 0.928   b. 0.072

5.93 a. 0.271   b. 0.677   c. 0.594

5.97 a. 2.1   b. 0.1890   c. 0.5999

Review Test for Chapter 5

1. a. 1, 2, 3, 4   b. $\{x = 3\}$
   c. 0.253. 25.3% of the undergraduates at this university are juniors.
   d. 0.211
   e.

x	1	2	3	4
$P(x)$	0.195	0.211	0.253	0.341

3. a. $\{y = 4\}$ b. $\{y \geq 4\}$
   c. $\{2 \leq y \leq 4\}$ d. $\{y \geq 1\}$
   e. 0.174   f. 0.322   g. 0.646   h. 0.948

4. a. 2.8   b. 2.8   c. 1.5   d. 1.5

6. $\mu; \sigma$

7. 1, 6, 24, 5040

8. a. 56   b. 56   c. 1
   d. 45   e. 91,390   f. 1

9. a. $p = 0.493$
   b.

Outcome	Probability
sss	$(0.493)(0.493)(0.493) = 0.120$
ssf	$(0.493)(0.493)(0.507) = 0.123$
sfs	$(0.493)(0.507)(0.493) = 0.123$
sff	$(0.493)(0.507)(0.507) = 0.127$
fss	$(0.507)(0.493)(0.493) = 0.123$
fsf	$(0.507)(0.493)(0.507) = 0.127$
ffs	$(0.507)(0.507)(0.493) = 0.127$
fff	$(0.507)(0.507)(0.507) = 0.130$

   d. *ssf, sfs, fss*
   e. 0.123. Each probability is obtained by multiplying two success probabilities of 0.493 and one failure probability of 0.507.

10. a. 0.410   b. 0.590   c. 0.819
    d.

x	0	1	2	3	4
$P(x)$	0.002	0.026	0.154	0.410	0.410

    e. left skewed

11. a. 0.410   b. 0.592   c. 0.820
    d. See answer to Problem 10(d).

13. a. $p = 0.5$ b. $p < 0.5$

14. a. 5   b. 0.65   c. 0.0488
    d. 0.5663   e. 0.0541   f. 0.9947

15. a. 8   b. 0.57   c. 0.4762
    d. 0.7765   e. 0.2527

16. a. 0.2613   b. 0.8413   c. 0.1586   d. 0.8225

17. $\mu_x = 3.2, \sigma_x = 0.8$

18. Because the trials are not independent and the success probability varies from trial to trial. Because the sample size is small relative to the size of the population.

19. a. 0.075   b. 0.494   c. 0.994
    d.

x	$P(x)$	x	$P(x)$
0	0.006	8	0.073
1	0.029	9	0.042
2	0.075	10	0.022
3	0.129	11	0.010
4	0.168	12	0.005
5	0.175	13	0.002
6	0.151	14	0.001
7	0.113	15	0.000

    f. Right skewed. Yes, all Poisson distributions are right skewed.

20. a. $\mu_x = 5.2$; on the average, the CPA firm receives 5.2 calls per hour.
    b. $\sigma_x = 2.3$

22. a. 2.4   b. 0.2613   c. 0.6916

23. a. 2.4   b. 0.2613   c. 0.6916

CHAPTER 6

Exercises 6.1

6.1 Because the total area under the standard normal curve equals 1 and the standard normal curve is symmetric about 0.

6.3
a. 0.9875   b. 0.0594   c. 0.5
d. 0.0000 (to four decimal places)

6.5
a. 0.8577   b. 0.2743   c. 0.5
d. 0.0000 (to four decimal places)

6.7 a. 0.9105   b. 0.0440   c. 0.2121   d. 0.1357

6.9 a. 0.0645   b. 0.7975

6.11 a. 0.7994   b. 0.8990   c. 0.0500   d. 0.0198

6.15 -1.96

6.17 0.67

6.19 -1.645

6.21 0.44

6.23   a. 1.88   b. 2.575

6.25   ±1.645

6.31   The four missing entries are 1.645, 1.96, 2.33, and 2.575.

Exercises 6.2

6.33   The one with parameters $\mu = 1$ and $\sigma = 2$; it has the larger σ-parameter.

6.35   True, because the parameter μ affects only where the normal curve is centered. The shape of the normal curve is determined by the parameter σ.

6.37

a.

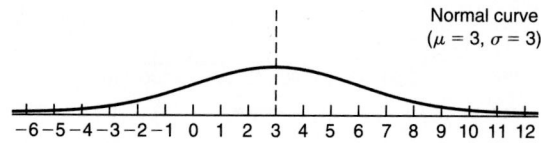

b.

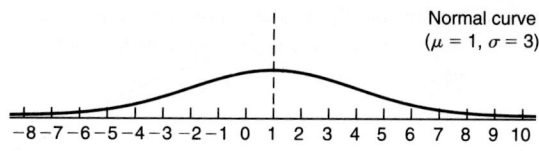

c.

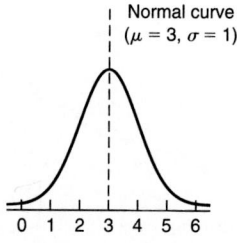

6.39   a. 0.6554   b. 0.1587   c. 0.5403

6.41   a. 0.9591   b. 0.1359   c. 0.0668

6.43   a. 0.9104   b. 0.1056

6.45   a. 0.9115   b. 0.2417

6.47   a. 0.2033   b. 0.0493

6.49   a. 77.29   b. 71.44   c. 70.08, 77.92

6.51   a. 337.50   b. 332.50

6.53   36.14, 40.90, 45.66

Exercises 6.3

6.65
a. Exact percentage = 11.70% (0.1170)
   Normal curve area = 12.23% (0.1223)
b. Exact percentage = 39.83% (0.3983)
   Normal curve area = 39.14% (0.3914)

6.67   a. 69.43%   b. 97.26%

6.69   a. 13.61%   b. 3.67%

6.71   a. 2   b. −1.75   c. 0

6.73
a. 68.26%   b. 95.44%   c. 99.74%
d. 86.64%

6.75
a. $114.80, $149.20   b. $97.60, $166.40
c. $80.40, $183.60   d. See Fig. A.7.

FIGURE A.7   Weekly food cost

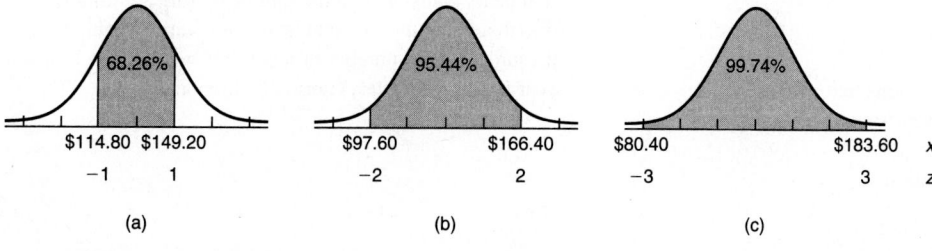

6.77   a. 95   b. 89.9

6.79   a. 2.575   b. 1.28

6.81

a. $Q_1 = \$521.01$, $Q_2 = \$586.00$, $Q_3 = \$650.99$

b. $P_{15} = \$485.12$ c. $P_{98} = \$784.85$

6.83

a. $Q_1 = \$120.48$, $Q_2 = \$132.00$, $Q_3 = \$143.52$

b. $D_3 = \$123.06$ c. $P_{85} = \$149.89$

Exercises 6.4

6.89

a. 0.2483   b. 0.2838

c. For a randomly selected secondary school teacher, the probability is 0.2483 that he or she makes less than \$28,000 per year; the probability is 0.2838 that he or she makes between \$35,000 and \$44,000 per year.

6.91

a. 0.9633   b. 0.9926

c. 96.33% of the batteries last longer than 20 hours. 99.26% of the batteries last between 15 and 45 hours.

6.93

a. 0.0594   b. 0.2699

c. For a randomly selected finisher, the probability is 0.0594 that his or her time exceeds 75 minutes; the probability is 0.2699 that his or her time is either less than 50 minutes or greater than 70 minutes.

6.95   For a randomly selected bolt, the probability is 0.6826 that it will have a diameter between 9.9 and 10.1 mm; the probability is 0.9544 that it will have a diameter between 9.8 and 10.2 mm; and the probability is 0.9974 that it will have a diameter between 9.7 and 10.3 mm.

6.97   For a randomly selected flashlight battery of this brand, the probability is 0.6826 that it will last between 24.4 and 35.6 hours; the probability is 0.9544 that it will last between 18.8 and 41.2 hours; and the probability is 0.9974 that it will last between 13.2 and 46.8 hours.

6.99

a. $z = (x - 16.3)/17.9$ b. 0.22, 2.70, −0.68

c. No

6.101

a. $z = (x - 61)/9$

b. The number of standard deviations that a randomly selected finisher's time is away from the mean of 61 minutes.

c. The standard normal distribution.

6.103

a. 1.96   b. 2.575

c. $P(\mu_x - 1.96\sigma_x < x < \mu_x + 1.96\sigma_x) = 0.95$

    $P(\mu_x - 2.575\sigma_x < x < \mu_x + 2.575\sigma_x) = 0.99$

Exercises 6.5

6.111

a.

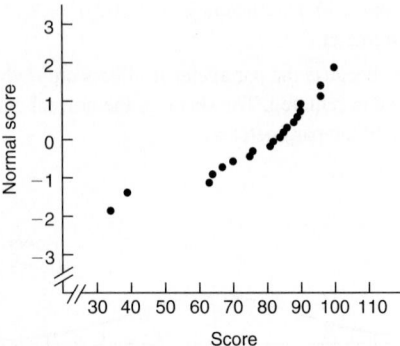

b. 34 and 39 are outliers.

c. The sample does not appear to be from a normally distributed population; that is, the final-exam scores in this introductory statistics class appear not to be normally distributed.

6.113

a.

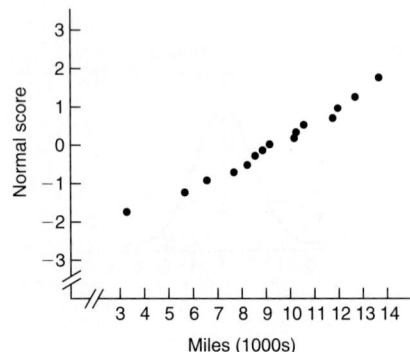

b. No outliers.

c. It appears plausible that the sample is from a normally distributed population; that is, it is not unreasonable to presume that the number of miles cars were driven last year is (approximately) normally distributed.

6.115

a.

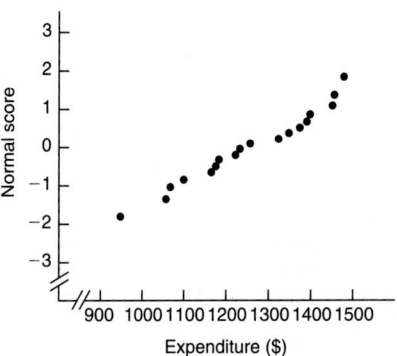

Expenditure ($)

b. No outliers.

c. It appears plausible that the sample is from a normally distributed population; that is, it is not unreasonable to presume that the year's energy expenditures for all U.S. households using electricity as their primary energy source were (approximately) normally distributed.

6.117

a.

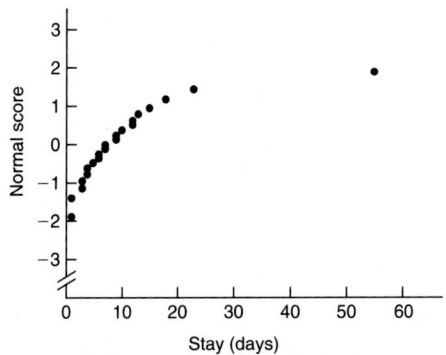

Stay (days)

b. Some would consider 55 an outlier.

c. The sample does not appear to be from a normally distributed population; that is, the length of stay by patients in short-term hospitals appears not to be normally distributed.

6.119 normal

Exercises 6.6

6.125 It is not practical to use the binomial probability formula when the number of trials is very large.

6.127 a. 0.4512, 0.8907   b. 0.4544, 0.8858

6.129 The normal curve with parameters $\mu = 15$ and $\sigma = 2.74$.

6.131 a. 0.0398   b. 0.2835   c. 0.0099

6.133 a. 0.0263   b. 0.8242   c. 0.9991

6.135 a. 0.0233   b. 0.1599   c. 0.0516

Review Test for Chapter 6

1. It is often appropriate to use the normal distribution as the distribution of a population or random variable; and the normal distribution is frequently employed in inferential statistics.

2. a. 0.0013   b. 0.2709   c. 0.1305
   d. 0.9803   e. 0.0668   f. 0.8426

4. a. −0.52   b. 1.28
   c. 1.96, 1.645, 2.33, 2.575   d. ±2.575

6. a.

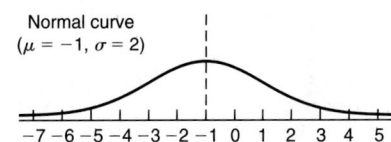

b.

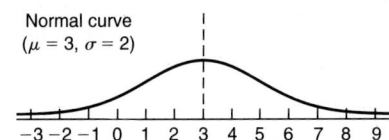

c.

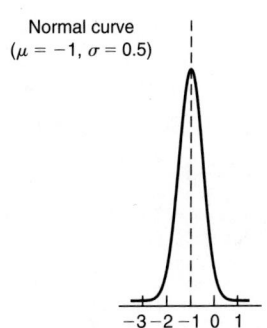

7. a. the second curve
   b. the first and second curves
   c. the first and third curves
   d. the third curve   e. the fourth curve

8. a. 0.1125   b. 0.9671   c. 0.0465

10. a. 3.90   b. −5.11   c. 1.48
    d. −5.11, 3.11

12. a. 82.76%   b. 89.44%   c. 0.62%

13. a. $Q_1 = 433$, $Q_2 = 500$, $Q_3 = 567$. Thus 25% of GRE scores are below 433, 25% are between 433 and 500, 25% are between 500 and 567, and 25% are above 567.

    b. $P_{99} = 733$. Thus 99% of GRE scores are below 733 and 1% are above 733.

14. a. 1.45   b. −1.80   c. 0

15. a. 400, 600   b. 300, 700   c. 200, 800

16. a. 99.4   b. 1.645

17. a. $\mu_x = 339.6$, $\sigma_x = 13.3$
    b. 0.1650   c. 0.0110
    d. For a randomly selected Morgan mare, the probability is 0.1650 that the gestation period will be between 320 and 330 days; the probability is 0.0110 that the gestation period will exceed 370 days.

19. a. 326.3, 352.9   b. 313.0, 366.2
    c. 299.7, 379.5

20. a. $z = (x − 339.6)/13.3$
    b. The number of standard deviations that the gestation period of a randomly selected Morgan mare is away from the mean.
    c. −1.10, 2.14, 0
    d. The standard normal distribution.
    e. No. Yes.

21. 2.33

22. a.

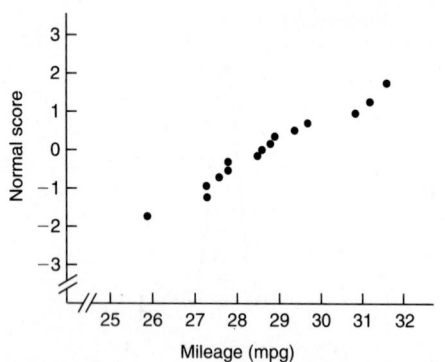

    b. No outliers.
    c. It appears plausible that the sample is from a normally distributed population; that is, it is not unreasonable to presume that the gas mileages of this particular model are (approximately) normally distributed.

24. a. 0.0076   b. 0.9505   c. 0.9988

CHAPTER 7

Exercises 7.1

7.1

a. $\mu = 68$ (i.e., $68,000)

b.

Sample	$\bar{x}$
76, 58	67.0
76, 64	70.0
76, 82	79.0
76, 60	68.0
58, 64	61.0
58, 82	70.0
58, 60	59.0
64, 82	73.0
64, 60	62.0
82, 60	71.0

c. See Fig. A.8.

d.

$\bar{x}$	59	61	62	67	68	70	71	73	79
$P(\bar{x})$	0.1	0.1	0.1	0.1	0.1	0.2	0.1	0.1	0.1

e. 0.1

f. 0.5. If we take a random sample of two salaries, there is a 50% chance that the mean of the sample selected will be within 4 (i.e., $4000) of the population mean.

7.3

b.

Sample	$\bar{x}$
76, 58, 64, 82	70.0
76, 58, 64, 60	64.5
76, 58, 82, 60	69.0
76, 64, 82, 60	70.5
58, 64, 82, 60	66.0

c. See Fig. A.9.

d.

$\bar{x}$	64.5	66.0	69.0	70.0	70.5
$P(\bar{x})$	0.2	0.2	0.2	0.2	0.2

e. 0

f. 1. If we take a random sample of four salaries, there is a 100% chance (i.e., it is certain) that the mean of the sample will be within 4 (i.e., $4000) of the population mean.

FIGURE A.8   Dotplot for Exercise 7.1(c)

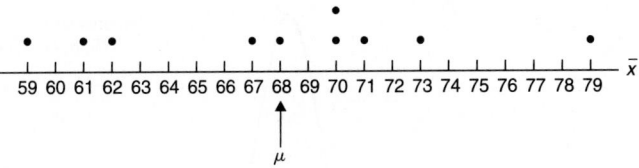

FIGURE A.9   Dotplot for Exercise 7.3(c)

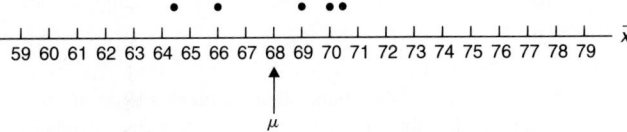

7.5

a. $\mu = 15$ cm

b.

Sample	$\bar{x}$
19, 14	16.5
19, 15	17.0
19, 9	14.0
19, 16	17.5
19, 17	18.0
14, 15	14.5
14, 9	11.5
14, 16	15.0
14, 17	15.5
15, 9	12.0
15, 16	15.5
15, 17	16.0
9, 16	12.5
9, 17	13.0
16, 17	16.5

d.

$\bar{x}$	$P(\bar{x})$
11.5	$\frac{1}{15}$
12.0	$\frac{1}{15}$
12.5	$\frac{1}{15}$
13.0	$\frac{1}{15}$
14.0	$\frac{1}{15}$
14.5	$\frac{1}{15}$
15.0	$\frac{1}{15}$
15.5	$\frac{2}{15}$
16.0	$\frac{1}{15}$
16.5	$\frac{2}{15}$
17.0	$\frac{1}{15}$
17.5	$\frac{1}{15}$
18.0	$\frac{1}{15}$

e. $\frac{1}{15} = 0.067$

f. $\frac{6}{15} = 0.4$. If we take a random sample of two bullfrogs, there is a 40% chance that their mean length will be within 1 cm of the population mean length.

7.7

b.

Sample	$\bar{x}$
19, 14, 15, 9	14.25
19, 14, 15, 16	16.00
19, 14, 15, 17	16.25
19, 14, 9, 16	14.50
19, 14, 9, 17	14.75
19, 14, 16, 17	16.50
19, 15, 9, 16	14.75
19, 15, 9, 17	15.00
19, 15, 16, 17	16.75
19, 9, 16, 17	15.25
14, 15, 9, 16	13.50
14, 15, 9, 17	13.75
14, 15, 16, 17	15.50
14, 9, 16, 17	14.00
15, 9, 16, 17	14.25

d.

$\bar{x}$	$P(\bar{x})$
13.50	$\frac{1}{15}$
13.75	$\frac{1}{15}$
14.00	$\frac{1}{15}$
14.25	$\frac{2}{15}$
14.50	$\frac{1}{15}$
14.75	$\frac{2}{15}$
15.00	$\frac{1}{15}$
15.25	$\frac{1}{15}$
15.50	$\frac{1}{15}$
16.00	$\frac{1}{15}$
16.25	$\frac{1}{15}$
16.50	$\frac{1}{15}$
16.75	$\frac{1}{15}$

e. $\frac{1}{15} = 0.067$

f. $\frac{10}{15} = 0.667$. If we take a random sample of four bullfrogs, there is a 66.7% chance that their mean length will be within 1 cm of the population mean length.

7.9

b.

Sample	$\bar{x}$
19, 14, 15, 9, 16, 17	15.0

d.

$\bar{x}$	$P(\bar{x})$
15.0	1

e. 1

f. 1. If we take a random sample of six bullfrogs, there is a 100% chance (i.e., it is certain) that their mean length will be within 1 cm of the population mean length.

The only possible sample here is identical to the population.

7.11 The larger the sample size, the smaller the sampling error tends to be in estimating a population mean, μ, by a sample mean, $\bar{x}$.

Exercises 7.2

7.15

a. $\mu = 68$ b. $\mu_{\bar{x}} = \Sigma \bar{x} P(\bar{x}) = 68$
c. $\mu_{\bar{x}} = \mu = 68$

7.17 b. $\mu_{\bar{x}} = \Sigma \bar{x} P(\bar{x}) = 68$ c. $\mu_{\bar{x}} = \mu = 68$

7.19

a. $\mu_{\bar{x}} = 6.9, \sigma_{\bar{x}} = 0.50$ b. $\mu_{\bar{x}} = 6.9, \sigma_{\bar{x}} = 0.19$

7.21

a. $\mu_{\bar{x}} = 7.4, \sigma_{\bar{x}} = 0.37$ b. $\mu_{\bar{x}} = 7.4, \sigma_{\bar{x}} = 0.18$

7.23 Increase the size of the sample.

Exercises 7.3

7.33

a. Approximately normally distributed with mean $\mu_{\bar{x}} = 100$ and standard deviation $\sigma_{\bar{x}} = 4$.
b. None
c. No, since the distribution of the population is not specified, we need a sample size of at least 30 to apply Key Fact 7.6.

7.35

a. Normal with parameters μ and $\sigma/\sqrt{n}$.
b. No, since the population being sampled is normally distributed.
c. $\mu_{\bar{x}} = \mu$ and $\sigma_{\bar{x}} = \sigma/\sqrt{n}$

7.37

a.

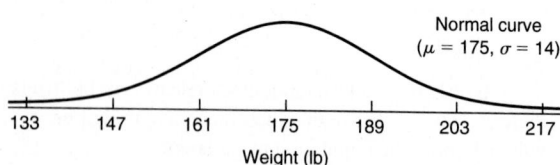

Normal curve
$(\mu = 175, \sigma = 14)$

Weight (lb)

b. Normal with $\mu_{\bar{x}} = 175$ and $\sigma_{\bar{x}} = 9.90$.

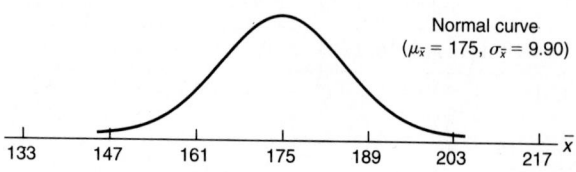

Normal curve
$(\mu_{\bar{x}} = 175, \sigma_{\bar{x}} = 9.90)$

c. Normal with $\mu_{\bar{x}} = 175$ and $\sigma_{\bar{x}} = 4.67$.

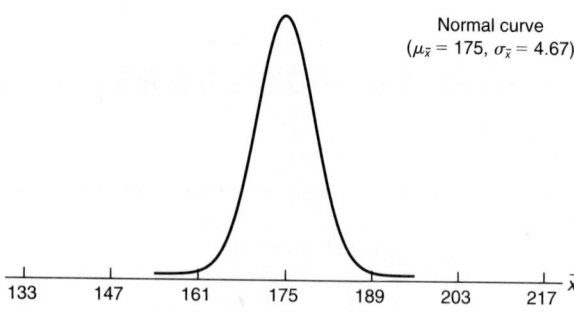

Normal curve
$(\mu_{\bar{x}} = 175, \sigma_{\bar{x}} = 4.67)$

7.39

a. 0.7154
b. There is a 71.54% chance that the mean weight of the nine males obtained will be within 5 lb of the population mean weight of 175 lb.
c. 71.54%   d. 0.95

7.41

a. 0.4972
b. No, because the sample size is at least 30.
c.

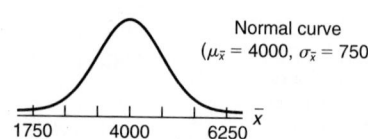

Normal curve
$(\mu_{\bar{x}} = 4000, \sigma_{\bar{x}} = 750)$

d. 0.9652
e. Because the population standard deviation is so large.

7.43 0.8064

7.45

a. $z = (\bar{x} - 100)/5.33$
b. standard normal distribution
c. Yes, because the sample size is less than 30.
d. 0.8990

7.47

a. 0.2514   b. 0.0174
c. No. There is about a 25% chance that a randomly selected bag of water-softener salt will weigh 39 lb or less, if the company's claim is correct.
d. Yes, because if the company's claim is correct, there is less than a 2% chance that 10 randomly selected bags of water-softener salt will have a mean weight of 39 lb or less.

Review Test for Chapter 7

1. a. The error resulting f͘
      tax, $\bar{x}$, of the 125,0͘
      timate of the me͘
      returns.
   b. $88
   c. No, not nec͘
      ple size fr͘
      likelihoc͘
   d. Increa͘

2. a. 18
   b. Th͘
      2͘
   c.

3. a. $\mu_͘$
   b. $\mu_{\bar{x}} = $͘
   c. Smaller, bec͘   e $\sigma_{\bar{x}}$ decreases
      with increasing sa͘

4. a. False   b. Not possibl͘   c. True

5. a. False   b. True   c. True

6. a.

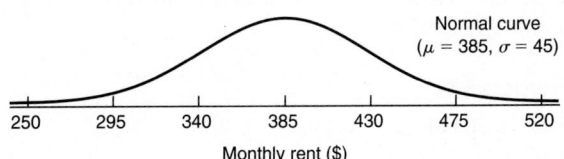

Normal curve
($\mu = 385$, $\sigma = 45$)

250   295   340   385   430   475   520

Monthly rent ($)

b. Normal with $\mu_{\bar{x}} = 385$ and $\sigma_{\bar{x}} = 25.98$.

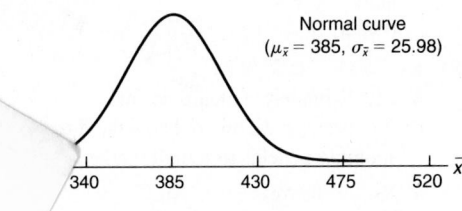

Normal curve
($\mu_{\bar{x}} = 385$, $\sigma_{\bar{x}} = 25.98$)

340   385   430   475   520 $\bar{x}$

with $\mu_{\bar{x}} = 385$ and $\sigma_{\bar{x}} = 15$.

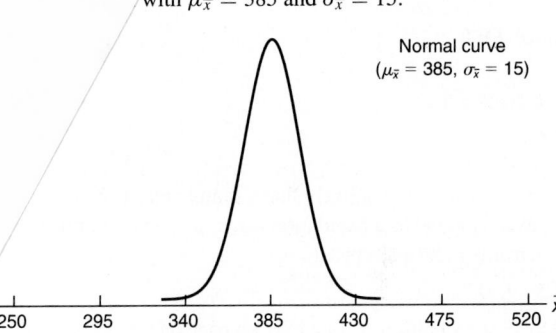

Normal curve
($\mu_{\bar{x}} = 385$, $\sigma_{\bar{x}} = 15$)

250   295   340   385   430   475   520 $\bar{x}$

7. a. 0.2960
   b. There is a 29.6% chance that the mean monthly
      rent, $\bar{x}$, of the three studio apartments obtained will
      be within $10 of the population mean monthly rent
      of $385.
   c. 29.6%   d. 0.9452

8. a. For a normally distributed population, the random
      variable $\bar{x}$ is normally distributed, regardless of the
      sample size. Also, we know that $\mu_{\bar{x}} = \mu$. Conse-
      quently, since the normal curve for a normally dis-
      tributed population or random variable is centered at
      its μ-parameter, all three curves are centered at the
      same place.
   b. Curve B. Since $\sigma_{\bar{x}} = \sigma/\sqrt{n}$, the larger the sample
      size, the smaller the value of $\sigma_{\bar{x}}$ and hence the smaller
      the spread of the normal curve for $\bar{x}$. Thus, Curve B,
      which has the smaller spread, corresponds to the
      larger sample size.
   c. Because $\sigma_{\bar{x}} = \sigma/\sqrt{n}$ and the spread of a normal
      curve is determined by $\sigma_{\bar{x}}$. Thus different sample
      sizes result in normal curves with different spreads.
   d. Curve B. The smaller the value of $\sigma_{\bar{x}}$, the smaller the
      sampling error tends to be.

9. a. 0.6212
   b. No, because the sample size is large and therefore
      $\bar{x}$ is approximately normally distributed, regardless

$\frac{1}{15}$

͘ 18.

of the distribution of the population of life-insurance amounts. Yes.

c. 0.9946

10. a. $z = (\bar{x} - 603)/8.32$

b. approximately standard normal   c. 0.0198

d. No, because we do not know the distribution of monthly benefits to retired workers.

11. a. No   b. Yes   c. No

CHAPTER 8

Exercises 8.1

8.1

a. 6.87 lb

b. No, because it is unlikely that a sample mean, $\bar{x}$, will be exactly equal to a population mean, μ; some sampling error is to be anticipated.

8.3 $70.7

8.5   See Definitions 8.1 and 8.2 on page 441.

8.7

a. 6.22 to 7.51 lb

b. We can be 95.44% confident that the mean weight, μ, of all newborns is somewhere between 6.22 and 7.51 lb.

c. It may or may not, but we can be 95.44% confident that it does.

8.9

a. $65.9 to $75.4

b. We can be 95.44% confident that the mean price of all science books is somewhere between $65.9 and $75.4.

Exercises 8.2

8.15

a. Confidence level $= 0.90$; $\alpha = 0.10$.

b. Confidence level $= 0.99$; $\alpha = 0.01$.

8.17

a. 617.3 to 668.7 sq ft

b. We can be 90% confident that the mean size, μ, of household vegetable gardens in the United States is somewhere between 617.3 and 668.7 sq ft.

8.19

a. $10.85 to $13.17

b. We can be 95% confident that the mean, μ, of the hourly earnings of all people employed in the aircraft industry is somewhere between $10.85 and $13.17.

8.21

a. 133.6 to 140.2 lb

b. We can be 90% confident that the mean weight of all U.S. women 5 feet 4 inches tall and in the age group 18–24 years is somewhere between 133.6 and 140.2 lb.

8.23 $9988.7 to $10,859.3

8.25

a. 131.7 to 142.1 lb

b. It is longer because the confidence level is greater.

c.

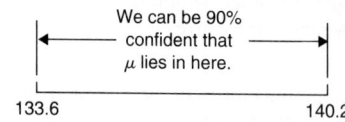

We can be 90% confident that μ lies in here.

133.6   140.2

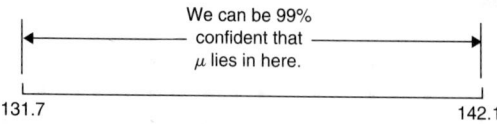

We can be 99% confident that μ lies in here.

131.7   142.1

8.27

a. 3.1 to 3.3 people

b. We can be 95% confident that the mean size, μ, of all U.S. families is somewhere between 3.1 and 3.3 people.

8.33

a. 20.8 lb   b. 21.47 lb   c. 80   d. 625.16 lb

e. 621.33 to 628.99 lb

Exercises 8.3

8.35   About $0.49

8.37

a. $1.16

b. We can be 95% confident that the margin of error in estimating μ by $\bar{x}$ is $1.16.

c. 163   d. $12.37 to $13.37

8.39

a. 3.3 lb   b. 271   c. Because σ is unknown.

d. 132.2 to 136.2 lb

Exercises 8.4

8.43   a. 1.440   b. 2.447   c. 3.143

8.45

a. 1.323   b. 2.518   c. -2.080 d. ± 1.721

8.47

a. $1614.98 to $1836.24

b. Yes, because the confidence interval does not contain, and is to the right of, the 1984 figure of $852.31.

8.49

a. 58.9 to 64.1 bu

9.37 Critical values: $\pm z_{0.05} = \pm 1.645$.

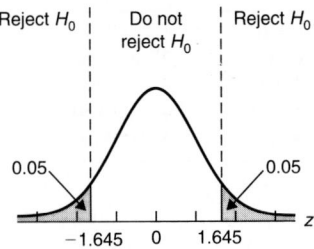

9.39 Critical value: $-z_{0.05} = -1.645$.

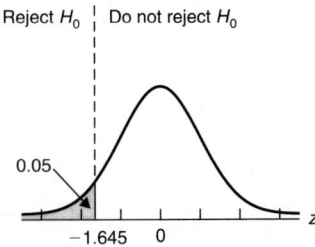

9.41 $H_0: \mu = \$280$, $H_a: \mu > \$280$; $\alpha = 0.05$; critical value $= 1.645$; $z = 0.26$; do not reject H_0; at the 5% significance level, the data do not provide sufficient evidence to conclude that last year's mean amount, μ, spent for maintenance and repairs has increased over the 1983 mean of $280.

9.43 $H_0: \mu = 18$ mg, $H_a: \mu < 18$ mg; $\alpha = 0.01$; critical value $= -2.33$; $z = -7.23$; reject H_0; at the 1% significance level, the data provide sufficient evidence to conclude that adult females under the age of 51 are, on the average, getting less than the RDA of 18 mg of iron.

9.45

a. $H_0: \mu = 50$ lb, $H_a: \mu \neq 50$ lb; $\alpha = 0.05$; critical values $= \pm 1.96$; $z = 1.13$; do not reject H_0; at the 5% significance level, the data do not provide sufficient evidence to conclude that the mean weight, μ, of all "50-lb" bags of this dog food differs from the advertised weight of 50 lb.

b. $H_0: \mu = 50$ lb, $H_a: \mu \neq 50$ lb; $\alpha = 0.05$; critical values $= \pm 1.96$; $z = 2.17$; reject H_0; at the 5% significance level, the data provide sufficient evidence to conclude that the mean weight, μ, of all "50-lb" bags of this dog food differs from the advertised weight of 50 lb.

9.47 $H_0: \mu = 26$ mpg, $H_a: \mu < 26$ mpg; $\alpha = 0.05$; critical value $= -1.645$; $z = -2.65$; reject H_0; at the 5% significance level, the data support the consumer group's conjecture.

9.49 For large samples, the sample standard deviation, s, is likely to be a good approximation of the population standard deviation, σ.

Exercises 9.4

9.53

a. If $\bar{x} \geq 348.9$, reject H_0; otherwise, do not reject H_0.

b. 0.05

c.

μ	β	μ	β
305	0.8531	405	0.0901
330	0.6736	430	0.0262
355	0.4404	455	0.0057
380	0.2296	480	0.0009

d.

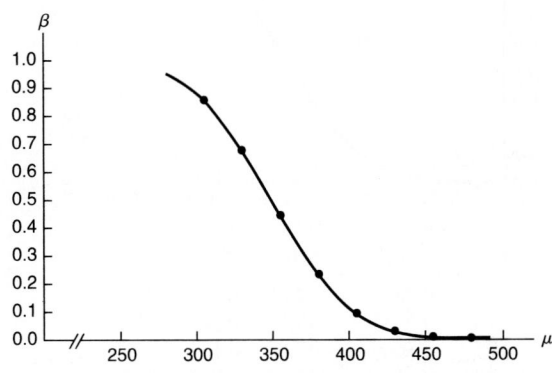

e.

μ	Power	μ	Power
305	0.1469	405	0.9099
330	0.3264	430	0.9738
355	0.5596	455	0.9943
380	0.7704	480	0.9991

f.

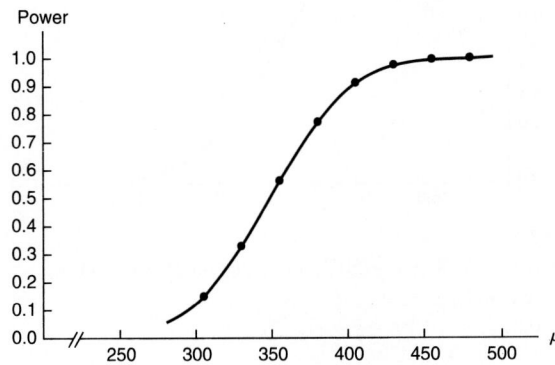

9.55

a. If $\bar{x} \leq 17.0$, reject H_0; otherwise, do not reject H_0.

b. 0.01

c.

μ	β	μ	β
16.00	0.0125	17.00	0.5000
16.25	0.0465	17.25	0.7123
16.50	0.1314	17.50	0.8686
16.75	0.2877	17.75	0.9535

d.

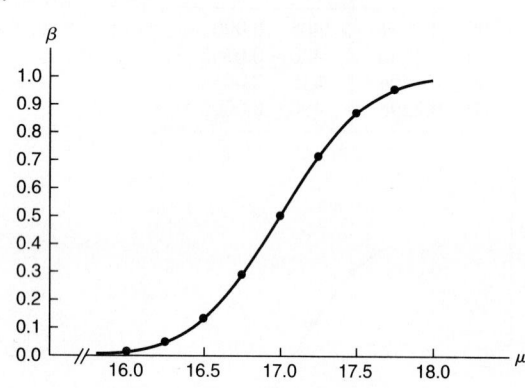

e.

μ	Power	μ	Power
16.00	0.9875	17.00	0.5000
16.25	0.9535	17.25	0.2877
16.50	0.8686	17.50	0.1314
16.75	0.7123	17.75	0.0465

f.

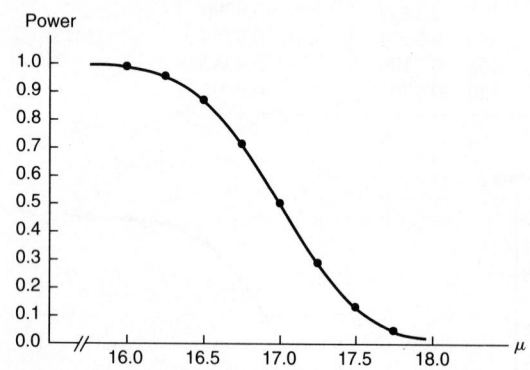

9.57

a. If $\bar{x} \leq 49.8$ or $\bar{x} \geq 50.2$, reject H_0; otherwise, do not reject H_0.

b. 0.05

c.

μ	β	μ	β
49.5	0.0006	50.1	0.8593
49.6	0.0150	50.2	0.5000
49.7	0.1401	50.3	0.1401
49.8	0.5000	50.4	0.0150
49.9	0.8593	50.5	0.0006

d.

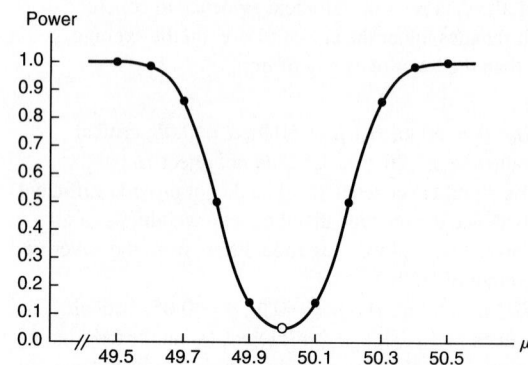

e.

μ	Power	μ	Power
49.5	0.9994	50.1	0.1407
49.6	0.9850	50.2	0.5000
49.7	0.8599	50.3	0.8599
49.8	0.5000	50.4	0.9850
49.9	0.1407	50.5	0.9994

f.

9.59

a. If $\bar{x} \geq 328.7$, reject H_0; otherwise, do not reject H_0.

b. 0.05

c.

μ	β	μ	β
305	0.7881	405	0.0049
330	0.4840	430	0.0003
355	0.1867	455	0.0000
380	0.0418	480	0.0000

d.

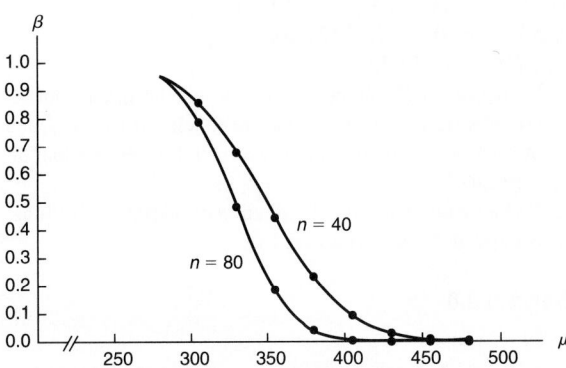

The principle being illustrated is that increasing the sample size for a hypothesis test without changing the significance level decreases the Type II error probabilities.

9.61 Procedure 2, because there is less chance of making a Type II error.

Exercises 9.5

9.65 a. 0.0212   b. 0.6217

9.67 a. 0.2296   b. 0.8770

9.69 a. 0.0970   c. 0.6030

9.71 H_0: $\mu = \$280$, H_a: $\mu > \$280$; $\alpha = 0.05$; $z = 0.26$; $P = 0.3974$; do not reject H_0; at the 5% significance level, the data do not provide sufficient evidence to conclude that last year's mean amount, μ, spent for maintenance and repairs has increased over the 1983 mean of $280; the data provide virtually no evidence against the null hypothesis.

9.73 H_0: $\mu = 18$ mg, H_a: $\mu < 18$ mg; $\alpha = 0.01$; $z = -7.23$; $P = 0.0000$ (to four decimal places); reject H_0; at the 1% significance level, the data provide sufficient evidence to conclude that adult females under the age of 51 are, on the average, getting less than the RDA of 18 mg of iron; the data provide very strong evidence against the null hypothesis.

9.75
a. H_0: $\mu = 50$ lb, H_a: $\mu \neq 50$ lb; $\alpha = 0.05$; $z = 1.13$; $P = 0.2584$; do not reject H_0; at the 5% significance level, the data do not provide sufficient evidence to conclude that the mean weight of all "50-lb" bags of this dog

food differs from the advertised weight of 50 lb; the data provide weak evidence at best against the null hypothesis.
b. H_0: $\mu = 50$ lb, H_a: $\mu \neq 50$ lb; $\alpha = 0.05$; $z = 2.17$; $P = 0.0300$; reject H_0; at the 5% significance level, the data provide sufficient evidence to conclude that the mean weight of all "50-lb" bags of this dog food differs from the advertised weight of 50 lb; the data provide strong evidence against the null hypothesis.

9.77 H_0: $\mu = 26$ mpg, H_a: $\mu < 26$ mpg; $\alpha = 0.05$; $z = -2.65$; $P = 0.0040$; reject H_0; at the 5% significance level, the data support the consumer group's conjecture; the data provide very strong evidence against the null hypothesis.

9.79
a. H_0: $\mu = 64$ lb, H_a: $\mu < 64$ lb; $\alpha = 0.05$; $z = -1.73$; $P = 0.0418$; reject H_0; at the 5% significance level, the data provide sufficient evidence to conclude that last year's mean beef consumption is less than the 1990 mean of 64 lb.
b. H_0: $\mu = 64$ lb, H_a: $\mu < 64$ lb; $\alpha = 0.05$; $z = 0.05$; $P = 0.5199$; do not reject H_0; at the 5% significance level, the data do not provide sufficient evidence to conclude that last year's mean beef consumption is less than the 1990 mean of 64 lb.
c. The null hypothesis can be rejected at the 5% level of significance when the potential outliers are retained but not when they are removed. Furthermore we see by referring to Table 9.12 on page 530 that the evidence against the null hypothesis is strong when the potential outliers are retained but essentially nonexistent when the potential outliers are removed.
d. Not advisable.

9.85
a. H_0: $\mu = 6.5$ days, H_a: $\mu < 6.5$ days
b. 7.70 days   c. 7.011 days   d. 40
e. 6.250 days   f. -0.21 g. 0.42   h. 0.42
i. At the 5% significance level, the data do not provide sufficient evidence to conclude that the mean hospital stay, μ, for this year will be less than the 1989 mean of 6.5 days.

Exercises 9.6

9.91 H_0: $\mu = 120$ min, H_a: $\mu > 120$ min; $\alpha = 0.05$; critical value $= 1.729$; $t = 1.386$; do not reject H_0; at the 5% significance level, the data do not provide sufficient evidence to conclude that the mean drying time, μ, is actually greater than the manufacturer's claim of 120 min. For the P-value approach, note that $0.05 < P < 0.10$.

9.93 H_0: $\mu = 36$ mo, H_a: $\mu < 36$ mo; $\alpha = 0.01$; critical value $= -2.821$; $t = -4.471$; reject H_0; at the 1% significance level, the data provide sufficient evidence to conclude that the mean life, μ, of the supplier's batteries is less than the claimed 36 months. For the *P*-value approach, note that $P < 0.005$.

9.95 H_0: $\mu = 38.5$¢/lb, H_a: $\mu \neq 38.5$¢/lb; $\alpha = 0.05$; critical values $= \pm 2.145$; $t = 2.624$; reject H_0; at the 5% significance level, the data provide sufficient evidence to conclude that the mean retail price, μ, for oranges now is different from the 1983 mean of 38.5 cents per pound. For the *P*-value approach, note that $P = 0.02$.

9.99

a. H_0: $\mu = \$550$, H_a: $\mu < \$550$
b. $\$68.624$ c. 15   d. $\$496.667$
e. -3.01 f. 0.0047   g. 0.0047
h. At the 5% significance level, the data provide sufficient evidence to conclude that, on the average, the new bumper will reduce repair costs resulting from front-end collisions at low speeds.

Exercises 9.7

9.109   Because the *D*-value for such a data value equals 0 and so we cannot attach a sign to the rank of $|D|$.

9.111

a. Wilcoxon signed-rank test   b. *t*-test
c. neither

9.113 H_0: $\eta = 33.1$ yrs, H_a: $\eta > 33.1$ yrs; $\alpha = 0.053$; critical value $= 44$; $W = 32$; do not reject H_0; at the 5.3% significance level, the data do not provide sufficient evidence to conclude that the median age of today's U.S. residents has increased over the 1991 median of 33.1 years.

9.115 H_0: $\mu = 12$ min, H_a: $\mu < 12$ min; $\alpha = 0.052$; critical value $= 26$; $W = 20.5$; reject H_0; at the 5.2% significance level, the data provide sufficient evidence to conclude that the new antacid tablet works faster.

9.117

a. H_0: $\mu = 38.5$¢/lb, H_a: $\mu \neq 38.5$¢/lb; $\alpha = 0.048$; critical values $= 25, 95$; $W = 99.5$; reject H_0; at the 4.8% significance level, the data provide sufficient evidence to conclude that the mean retail price for oranges now is different from the 1983 mean of 38.5 cents per pound.
b. Because a normally distributed population is symmetric.

9.119

a. H_0: $\mu = 310$ mL, H_a: $\mu < 310$ mL; $\alpha = 0.05$; critical value $= -1.753$; $t = -1.845$; reject H_0; at the 5% significance level, the data provide sufficient evidence to conclude that the mean content, μ, is less than the advertised content of 310 mL.
b. H_0: $\mu = 310$ mL, H_a: $\mu < 310$ mL; $\alpha = 0.052$; critical value $= 36$; $W = 36.5$; do not reject H_0; at the 5.2% sig-

nificance level, the data do not provide sufficient evidence to conclude that the mean content, μ, is less than the advertised content of 310 mL.
c. Since the population is normally distributed, the *t*-test is more powerful than the Wilcoxon signed-rank test; that is, the *t*-test is more likely to detect a false null hypothesis.

9.123

a. H_0: $\eta = 7.1$ yrs, H_a: $\eta < 7.1$ yrs
b. 50   c. 0   d. 7.531 yrs
e. 0.716   f. 0.716
g. Do not reject H_0; at the 5% significance level, the data do not provide sufficient evidence to conclude that last year's median marriage duration is less than the 1988 median of 7.1 years.
h. Probably not, since the distribution of marriage durations is most likely not symmetric.

Exercises 9.8

9.129   *t*-test

9.131   *z*-test

9.133   Wilcoxon signed-rank test

9.135   Consult a statistician.

Review Test for Chapter 9

1. Let μ denote last year's mean cheese consumption by Americans.
   a. H_0: $\mu = 24.7$ lb   b. H_a: $\mu > 24.7$ lb
   c. right-tailed

2. a. $z \geq 1.28$ b. $z < 1.28$ c. $z = 1.28$
   d. $\alpha = 0.10$
   e.

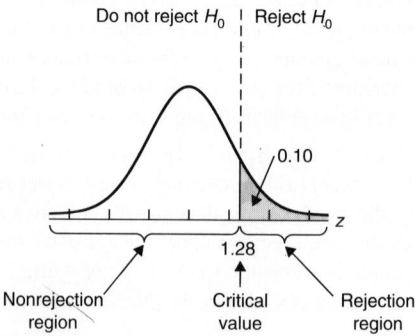

   f. right-tailed

3. a. A Type I error would occur if in fact $\mu = 24.7$ lb but the results of the sampling lead to the conclusion that $\mu > 24.7$ lb.

b. A Type II error would occur if in fact $\mu > 24.7$ lb but the results of the sampling fail to lead to that conclusion.

c. A correct decision would occur if in fact $\mu = 24.7$ lb and the results of the sampling do not lead to the rejection of that fact; or if in fact $\mu > 24.7$ lb and the results of the sampling lead to that conclusion.

d. Type I error   e. correct decision

4. a. If $\bar{x} \geq 26.2$, reject H_0; otherwise, do not reject H_0.

b. 0.10

c.

μ	β	μ	β
25.0	0.8485	28.0	0.0618
25.5	0.7257	28.5	0.0244
26.0	0.5675	29.0	0.0082
26.5	0.3974	29.5	0.0023
27.0	0.2451	30.0	0.0006
27.5	0.1335		

d.

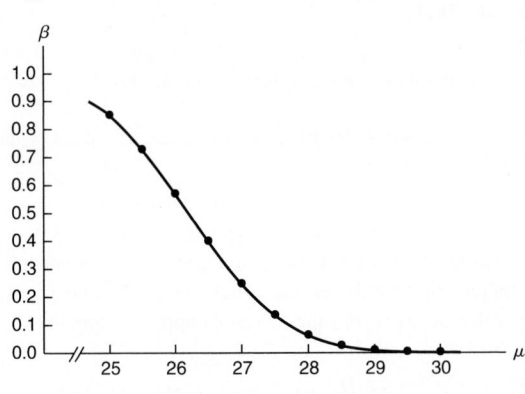

e.

μ	Power	μ	Power
25.0	0.1515	28.0	0.9382
25.5	0.2743	28.5	0.9756
26.0	0.4325	29.0	0.9918
26.5	0.6026	29.5	0.9977
27.0	0.7549	30.0	0.9994
27.5	0.8665		

f.

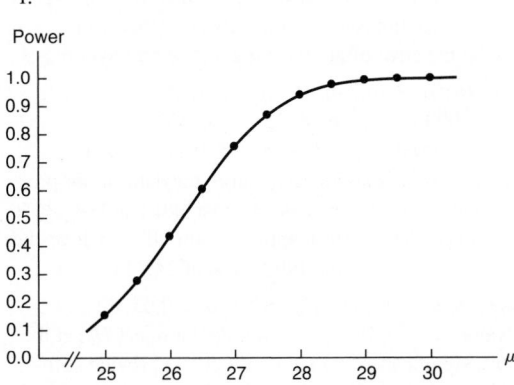

g. If $\bar{x} \geq 25.6$, reject H_0; otherwise, do not reject H_0.

h. 0.10

i.

μ	β	μ	β
25.0	0.8078	28.0	0.0003
25.5	0.5557	28.5	0.0000
26.0	0.2810	29.0	0.0000
26.5	0.0968	29.5	0.0000
27.0	0.0212	30.0	0.0000
27.5	0.0030		

j.

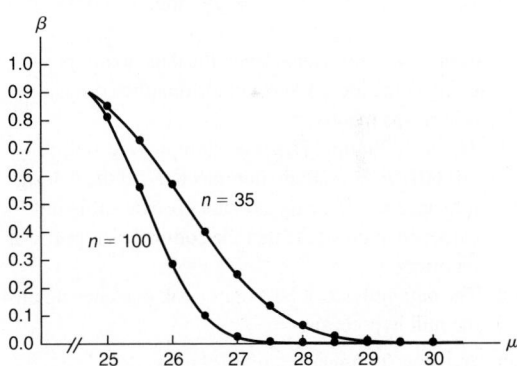

k. The principle being illustrated is that increasing the sample size for a hypothesis test without changing the significance level decreases the Type II error probabilities.

5. a. H_0: $\mu = 24.7$ lb, H_a: $\mu > 24.7$ lb; $\alpha = 0.10$; critical value $= 1.28$; $z = 0.94$; do not reject H_0; at the 10% significance level, the data do not provide sufficient evidence to conclude that last year's mean cheese consumption, μ, for all Americans has increased over the 1990 mean of 24.7 lb.

b. A Type II error, because given that the null hypothesis was not rejected, the only error that could be made is the error of not rejecting a false null hypothesis.

7. a. H_0: $\mu = 24.7$ lb, H_a: $\mu > 24.7$ lb
   b. 6.90 lb   c. 6.480 lb   d. 35
   e. 25.800 lb   f. 0.94   g. 0.17   h. 0.17
   i. At the 10% significance level, the data do not provide sufficient evidence to conclude that last year's mean cheese consumption, μ, for all Americans has increased over the 1990 mean of 24.7 lb.

8. H_0: $\mu = \$278$, H_a: $\mu < \$278$; $\alpha = 0.05$; critical value $= -1.645$; $z = -1.43$; do not reject H_0; at the 5% significance level, the data do not provide sufficient evidence to conclude that the mean value lost due to purse snatching has decreased from the 1990 mean of $278.

9. a. $P = 0.0764$
   b. H_0: $\mu = \$278$, H_a: $\mu < \$278$; $\alpha = 0.05$; $z = -1.43$; $P = 0.0764$; do not reject H_0; at the 5% significance level, the data do not provide sufficient evidence to conclude that the mean value lost due to purse snatching has decreased from the 1990 mean of $278.
   c. The data provide moderate evidence against the null hypothesis.

10. a. H_0: $\mu = 29$ mpg, H_a: $\mu \neq 29$ mpg; $\alpha = 0.05$; critical values $= \pm 2.145$; $t = -0.607$; do not reject H_0; at the 5% significance level, the data do not provide sufficient evidence to conclude that the company's report was incorrect.
    b. H_0: $\mu = 29$ mpg, H_a: $\mu \neq 29$ mpg; $\alpha = 0.05$; $t = -0.607$; $P > 0.20$; do not reject H_0; at the 5% significance level, the data do not provide sufficient evidence to conclude that the company's report was incorrect.
    c. The data provide at best only weak evidence against the null hypothesis.

12. a. H_0: $\mu = 29$ mpg, H_a: $\mu \neq 29$ mpg
    b. 1.595 mpg   c. 15   d. 28.753 mpg
    e. -0.60 f. 0.56   g. 0.56
    h. At the 5% significance level, the data do not provide sufficient evidence to conclude that the company's report was incorrect.

13. a. H_0: $\mu = 29$ mpg, H_a: $\mu \neq 29$ mpg; $\alpha = 0.048$; critical values $= 25, 95$; $W = 48.5$; do not reject H_0; at the 4.8% significance level, the data do not provide sufficient evidence to conclude that the company's report was incorrect.
    b. The distribution is symmetric.

c. Because a normally distributed population is symmetric.

15. a. H_0: $\eta = 29$ mpg, H_a: $\eta \neq 29$ mpg
    b. 15   c. 0   d. 48.5   e. 0.532   f. 0.532
    g. At the 5% significance level, the data do not provide sufficient evidence to conclude that the company's report was incorrect.
    h. 28.65 mpg

16. The t-test, because that hypothesis-testing procedure is designed specifically for a normally distributed population; the t-test is more powerful than the Wilcoxon signed-rank test for normal populations.

17. Wilcoxon signed-rank test

18. z-test

CHAPTER 10

Exercises 10.1

10.1

a. μ_1, σ_1, μ_2, and σ_2 are parameters; $\bar{x}_1, s_1, \bar{x}_2$, and s_2 are statistics.

b. μ_1, σ_1, μ_2, and σ_2 are fixed numbers; $\bar{x}_1, s_1, \bar{x}_2$, and s_2 are random variables.

10.3 H_0: $\mu_1 = \mu_2$, H_a: $\mu_1 > \mu_2$; $\alpha = 0.10$; critical value $= 1.28$; $z = 1.19$; do not reject H_0; at the 10% significance level, the data do not provide sufficient evidence to conclude that, on the average, males stay in the hospital longer than females. For the P-value approach, note that $P = 0.1170$.

10.5 H_0: $\mu_1 = \mu_2$, H_a: $\mu_1 \neq \mu_2$; $\alpha = 0.05$; critical values $= \pm 1.96$; $z = 6.32$; reject H_0; at the 5% significance level, the data provide sufficient evidence to conclude that last year's mean annual fuel expenditure for households using natural gas is different from that for households using only electricity. For the P-value approach, note that $P = 0.0000$ (to four decimal places).

10.7

a. H_0: $\mu_1 = \mu_2$, H_a: $\mu_1 < \mu_2$; $\alpha = 0.05$; critical value $= -1.645$; $z = -1.52$; do not reject H_0; at the 5% significance level, the data do not provide sufficient evidence to conclude that the training program increases sales. For the P-value approach, note that $P = 0.0643$.

b. H_0: $\mu_1 = \mu_2$, H_a: $\mu_1 < \mu_2$; $\alpha = 0.10$; critical value $= -1.28$; $z = -1.52$; reject H_0; at the 10% significance level, the data provide sufficient evidence to conclude that the training program increases sales. For the P-value approach, note that $P = 0.0643$.

10.9

a. −0.2 to 4.1 days

b. We can be 80% confident that the difference, $\mu_1 - \mu_2$, between the mean lengths of stay in short-term hospitals by males and females is somewhere between −0.2 and 4.1 days.

10.11

a. $175.25 to $332.75

b. We can be 95% confident that the difference, $\mu_1 - \mu_2$, between last year's mean fuel expenditures for households using natural gas and those using only electricity is somewhere between $175.25 and $332.75.

10.13

a. −$66.73 to $2.73   b. −$59.03 to −$4.97

Exercises 10.2

10.19

a. H_0: $\mu_1 = \mu_2$, H_a: $\mu_1 \neq \mu_2$; $\alpha = 0.05$; critical values = ± 2.101; $t = -3.657$; reject H_0; at the 5% significance level, the data provide sufficient evidence to conclude that there is a difference between the mean lasting times for the two brands of paint. For the P-value approach, note that $P < 0.01$.

b. designed experiment

10.21 H_0: $\mu_1 = \mu_2$, H_a: $\mu_1 < \mu_2$; $\alpha = 0.05$; critical value = −1.699; $t = -0.587$; do not reject H_0; at the 5% significance level, the data do not provide sufficient evidence to conclude that mine workers earn less, on the average, than construction workers. For the P-value approach, note that $P > 0.10$.

10.23 H_0: $\mu_1 = \mu_2$, H_a: $\mu_1 > \mu_2$; $\alpha = 0.05$; critical value = 1.714; $t = 1.186$; do not reject H_0; at the 5% significance level, the data do not provide sufficient evidence to conclude that males in the age group 25–34 years are, on the average, taller than those who were in that same age group 20 years ago. For the P-value approach, note that $P > 0.10$.

10.25

a. −2.93 to −0.79 months

b. We can be 95% confident that the difference, $\mu_1 - \mu_2$, between the mean lasting times of Brand A and Brand B paints is somewhere between −2.93 and −0.79 months.

10.27

a. −$1.91 to $0.93

b. We can be 90% confident that the difference, $\mu_1 - \mu_2$, between the mean hourly earnings of nonsupervisory mine and construction workers is somewhere between −$1.91 and $0.93.

10.29

a. −0.72 to 3.94 in.

b. We can be 90% confident that the difference between the mean height of males in the age group 25–34 years and the mean height of males in the age group 45–54 years is somewhere between −0.72 and 3.94 inches.

10.33

a. H_0: $\mu_1 = \mu_2$, H_a: $\mu_1 < \mu_2$

b. $s_1 = 55.3$ hr, $s_2 = 44.6$ hr

c. $n_1 = 20$, $n_2 = 10$

d. $\bar{x}_1 = 1023.2$ hr, $\bar{x}_2 = 1093.0$ hr

e. −3.46   f. 0.0009   g. 0.0009

h. At the 1% significance level, the data provide sufficient evidence to conclude that, on the average, the new bulb will outlast the bulb currently produced.

i. −126 to −14 hr   j. 52.1 hr

Exercises 10.3

10.39 H_0: $\mu_1 = \mu_2$, H_a: $\mu_1 \neq \mu_2$; $\alpha = 0.05$; critical value = ± 2.056; $t = -0.917$; do not reject H_0; at the 5% significance level, the data do not provide sufficient evidence to conclude that there is a difference between the mean GPAs of sophomores and juniors at the university. For the P-value approach, note that $P > 0.20$.

10.41

a. H_0: $\mu_1 = \mu_2$, H_a: $\mu_1 < \mu_2$; $\alpha = 0.01$; critical value = −2.473; $t = -3.210$; reject H_0; at the 1% significance level, the data provide sufficient evidence to conclude that Mirror-sheen has a longer effectiveness time, on the average, than Sureglow. For the P-value approach, note that $P < 0.005$.

b. Yes. In fact, since $P < 0.005$, the data provide very strong evidence against the null hypothesis and hence in favor of the alternative hypothesis that Mirror-sheen outlasts Sureglow on the average.

c. designed experiment

10.43 H_0: $\mu_1 = \mu_2$, H_a: $\mu_1 < \mu_2$; $\alpha = 0.05$; critical value = −1.645 (roughly); $t = -3.951$; reject H_0; at the 5% significance level, the data provide sufficient evidence to conclude that Idaho has a smaller mean potato yield than Nevada. For the P-value approach, note that $P < 0.005$.

10.45

a. −0.45 to 0.17

b. We can be 95% confident that the difference, $\mu_1 - \mu_2$, between the mean GPAs of sophomores and juniors at the university is somewhere between −0.45 and 0.17.

10.47

a. −7.2 to −0.9 days

b. We can be 98% confident that the difference, $\mu_1 - \mu_2$, between the mean effectiveness times of Sureglow and Mirror-sheen is somewhere between -7.2 and -0.9 days.

10.49

a. -47.9 to -19.7 cwt

b. We can be 90% confident that the difference, $\mu_1 - \mu_2$, between the mean yields per acre of potatoes for Idaho and Nevada is somewhere between -47.9 and -19.7 cwt.

10.53

a. $H_0: \mu_1 = \mu_2, H_a: \mu_1 > \mu_2$

b. $s_1 = 4.62$ yrs, $s_2 = 7.25$ yrs

c. $n_1 = 11, n_2 = 12$

d. $\bar{x}_1 = 35.69$ yrs, $\bar{x}_2 = 32.81$ yrs

e. 1.14   f. 0.13   g. 0.13

h. At the 5% significance level, the data do not provide sufficient evidence to conclude that the mean age at the time of first divorce for males is greater than that for females.

i. -1.5 to 7.2 yrs

j. The age distributions under consideration here are probably not normal. But, remember, the nonpooled-t procedures are robust to moderate deviations of the normality assumption. Therefore, whether it is appropriate to apply TwoSample here depends on how much the distributions deviate from normality.

Exercises 10.4

10.57 $H_0: \mu_1 = \mu_2, H_a: \mu_1 < \mu_2; \alpha = 0.05$; critical value $= 33; M = 33$; reject H_0; at the 5% significance level, the data provide sufficient evidence to conclude that students with fewer than 2 years of high-school algebra have a lower mean semester average in this teacher's chemistry courses than do students with 2 or more years of high-school algebra.

10.59 $H_0: \eta_1 = \eta_2, H_a: \eta_1 > \eta_2; \alpha = 0.05$; critical value $= 54; M = 47$; do not reject H_0; at the 5% significance level, the data do not provide sufficient evidence to conclude that the median number of volumes held by public colleges and universities is less than that held by private colleges and universities.

10.61

a. $H_0: \mu_1 = \mu_2, H_a: \mu_1 \neq \mu_2; \alpha = 0.05$; critical values $= 79, 131; M = 65.5$; reject H_0; at the 5% significance level, the data provide sufficient evidence to conclude that there is a difference between the mean lasting times for the two brands of paint.

b. Because two normal distributions with equal standard deviations have the same shape. The pooled t-test, because for normal distributions it is more powerful than the Mann–Whitney test.

10.63

a. nonpooled t-test   b. Mann–Whitney test

c. two-sample z-test

10.67

a. $H_0: \eta_1 = \eta_2, H_a: \eta_1 < \eta_2$

b. 1033.5 hr, 1105.0 hr   c. 20, 10

d. 242.5 (Recall that Minitab uses W instead of M.)

e. 0.0016   f. 0.0016

g. Reject H_0. At the 1% significance level, the data provide sufficient evidence to conclude that the new bulb will outlast the current bulb, on the average.

h. -109.0 to -25.0 hr

Exercises 10.5

10.73   paired samples and normal differences (i.e., the samples must be paired and the population of all possible paired differences must be approximately normally distributed)

10.75 $H_0: \mu_1 = \mu_2, H_a: \mu_1 > \mu_2; \alpha = 0.05$; critical value $= 1.761; t = 2.696$; reject H_0; at the 5% significance level, the data provide sufficient evidence to conclude that the running program will, on the average, reduce heart rates. For the P-value approach, note that $0.005 < P < 0.01$.

10.77 $H_0: \mu_1 = \mu_2, H_a: \mu_1 < \mu_2; \alpha = 0.05$; critical value $= -1.761; t = -1.420$; do not reject H_0; at the 5% significance level, the data do not provide sufficient evidence to conclude that nonsupervisory mine workers earn a smaller average hourly wage than nonsupervisory construction workers. For the P-value approach, note that $0.05 < P < 0.10$.

10.79 $H_0: \mu_1 = \mu_2, H_a: \mu_1 > \mu_2; \alpha = 0.05$; critical value $= 1.833; t = 2.213$; reject H_0; at the 5% significance level, the data provide sufficient evidence to conclude that the mean age of married men is greater than the mean age of married women. For the P-value approach, note that $0.025 < P < 0.05$.

10.81

a. 0.8 to 3.9

b. We can be 90% confident that the difference, $\mu_1 - \mu_2$, between the mean heart rates of all people before and after the running program is somewhere between 0.8 and 3.9.

10.83

a. $-\$1.02$ to $\$0.11$

b. We can be 90% confident that the difference, $\mu_1 - \mu_2$, between the mean hourly earnings of nonsupervisory mine and construction workers is somewhere between $-\$1.02$ and $\$0.11$.

10.85

a. 0.6 to 6.0 yrs

b. We can be 90% confident that the difference, $\mu_1 - \mu_2$, between the mean age of married men and the mean age of married women is somewhere between 0.6 and 6.0 years.

10.89

a. $H_0: \mu_1 = \mu_2$, $H_a: \mu_1 > \mu_2$; $\alpha = 0.047$; critical value = 70; $W = 78.5$; reject H_0; at the 4.7% significance level, the data provide sufficient evidence to conclude that the running program will, on the average, reduce heart rates.

b. Both tests reject the null hypothesis.

Exercises 10.6

10.99 Because the paired-sample tests are carried out by applying the corresponding one-sample tests to the sample of paired differences.

10.101 pooled t-test

10.103 Mann–Whitney test

10.105 paired t-test

10.107 nonpooled t-test

10.109 two-sample z-test

10.111 Consult a statistician.

Review Test for Chapter 10

1. $H_0: \mu_1 = \mu_2$, $H_a: \mu_1 > \mu_2$; $\alpha = 0.01$; critical value = 2.33; $z = 7.65$; reject H_0; at the 1% significance level, the data provide sufficient evidence to conclude that the mean resale price for homes in Phoenix, Arizona, exceeds that for homes in Flint, Michigan. For the P-value approach, note that $P = 0.0000$ to four decimal places.

2. $23,586 to $44,258. We can be 98% confident that the difference, $\mu_1 - \mu_2$, between the mean resale prices of homes in Phoenix, Arizona, and Flint, Michigan, is somewhere between $23,586 and $44,258.

3. $H_0: \mu_1 = \mu_2$, $H_a: \mu_1 \neq \mu_2$; $\alpha = 0.10$; critical values = ± 1.833; $t = -1.766$; do not reject H_0; at the 10% significance level, the data do not provide sufficient evidence to conclude that there is a difference in mean results for the two speed-reading programs. For the P-value approach, note that $0.10 < P < 0.20$.

4. −81.5 to 1.5 words per minute. We can be 90% confident that the difference, $\mu_1 - \mu_2$, between the mean reading speed of people using Program 1 and the mean reading speed of people using Program 2 is somewhere between −81.5 and 1.5 words per minute.

6. $H_0: \mu_1 = \mu_2$, $H_a: \mu_1 \neq \mu_2$; $\alpha = 0.05$; critical values = ± 1.96 (approximately); $t = 1.234$; do not reject H_0; at the 5% significance level, the data do not provide sufficient evidence to conclude that there is a difference between the mean IQs of male and female students at the university. For the P-value approach, note that $P > 0.20$.

7. −1.8 to 7.9. We can be 95% confident that the difference, $\mu_1 - \mu_2$, between the mean IQs of female and male students at the university is somewhere between −1.8 and 7.9.

9. a. $H_0: \mu_1 = \mu_2$, $H_a: \mu_1 \neq \mu_2$

   b. $s_1 = 7.61$, $s_2 = 8.02$

   c. $\bar{x}_1 = 118.45$, $\bar{x}_2 = 115.40$

   d. 1.23   e. 0.22   f. 0.22

   g. At the 5% significance level, the data do not provide sufficient evidence to conclude that there is a difference between the mean IQs of male and female students at the university.

   h. −2.0 to 8.1. *(Note:* This differs from the 95% confidence interval obtained in Problem 7 because Minitab uses the exact value for $t_{0.025}$ instead of the approximate value, 1.96, used in Problem 7.)

   i. 7.82

10. $H_0: \mu_1 = \mu_2$, $H_a: \mu_1 < \mu_2$; $\alpha = 0.05$; critical value = −1.721; $t = -2.740$; reject H_0; at the 5% significance level, the data provide sufficient evidence to conclude that last year the average German consumed less fish than the average Russian. For the P-value approach, note that $0.005 < P < 0.01$.

11. −7.6 to −1.7 kg. We can be 90% confident that the difference, $\mu_1 - \mu_2$, between last year's mean fish consumptions by Germans and Russians is somewhere between −7.6 and −1.7 kg.

13. a. $H_0: \mu_1 = \mu_2$, $H_a: \mu_1 < \mu_2$

    b. $s_1 = 2.84$ kg, $s_2 = 5.61$ kg

    c. $n_1 = 10$, $n_2 = 15$

    d. $\bar{x}_1 = 11.40$ kg, $\bar{x}_2 = 16.07$ kg

    e. −2.74   f. 0.0062   g. 0.0062

    h. At the 5% significance level, the data provide sufficient evidence to conclude that last year the average German consumed less fish than the average Russian.

    i. −7.60 kg to −1.7 kg

14. $H_0: \mu_1 = \mu_2$, $H_a: \mu_1 \neq \mu_2$; $\alpha = 0.05$; critical values = 54, 98; $M = 92$; do not reject H_0; at the 5% significance level, the data do not provide sufficient evidence to conclude that last year the mean number of miles driven per

truck differed from the mean number of miles driven per car.

16. a. H_0: $\eta_1 = \eta_2$, H_a: $\eta_1 \neq \eta_2$
    b. 13.90 thousand miles, 10.90 thousand miles
    c. $n_1 = 8, n_2 = 10$
    d. 92.0   e. 0.1684   f. 0.1684
    g. At the 5% significance level, the data do not provide sufficient evidence to conclude that last year the median (or mean) number of miles driven per truck differed from the median (or mean) number of miles driven per car.
    h. -2.10 to 18.31 thousand miles

17. H_0: $\mu_1 = \mu_2$, H_a: $\mu_1 > \mu_2$; $\alpha = 0.047$; critical value = 90; $W = 84$; do not reject H_0; at the 4.7% significance level, the data do not provide sufficient evidence to conclude that the mean monthly rent for renter-occupied housing units in the South exceeds that for those in the Midwest.

CHAPTER 11

Exercises 11.1

11.1 A population proportion, p, is a parameter since it is a descriptive measure for a population. A sample proportion, $\hat{p}$, is a statistic since it is a descriptive measure for a sample.

11.3
a. 0.626 to 0.674
b. We can be 95% confident that the percentage of Americans who drink beer, wine, or hard liquor, at least occasionally, is somewhere between 62.6% and 67.4%.

11.5
a. 78.4% to 83.6%
b. We can be 99% confident that the percentage of adult Americans who are in favor of "right to die" laws is somewhere between 78.4% and 83.6%.

11.7
a. 0.732 to 0.828
b. We can be 90% confident that the percentage of all real-estate experts who believe next year will be a good time to buy real estate is somewhere between 73.2% and 82.8%.

11.9 Not very well! I applied Procedure 11.1 without checking the assumption for its use; namely, that the number of successes, x, and the number of failures, $n - x$, are both at least 5. Since the number of successes here is only $0.008 \cdot 500 = 4$, I should not have used Procedure 11.1.

11.11 31.6% to 36.4%

11.13
a. 0.024   b. 2401   c. 0.611 to 0.649
d. 0.019, which is less than 0.02
e. 2305; 0.610 to 0.650; 0.02
f. By employing the guess for $\hat{p}$ in part (e), we reduced the required sample size by 96. Moreover, only $\frac{1}{10}$ of 1% of precision was lost—the margin of error rose from 0.019 to 0.02.

11.15
a. plus or minus 2.6 percentage points
b. 16,577   c. 0.817 to 0.833 (81.7% to 83.3%)
d. 0.008 (0.8%), which is less than 0.01 (1%)
e. 12,433; 0.816 to 0.834; 0.009 (0.9%)
f. By employing the guess for $\hat{p}$ in part (e), we reduced the required sample size by 4144. Moreover, only $\frac{1}{10}$ of 1% of precision was lost—the margin of error rose from 0.008 (0.8%) to 0.009 (0.9%).

11.17 Yes, because the two confidence intervals, 74% to 84% and 69% to 79%, overlap.

Exercises 11.2

11.25 H_0: $p = 0.5$, H_a: $p > 0.5$; $\alpha = 0.05$; critical value = 1.645; $z = 1.09$; do not reject H_0; at the 5% significance level, the data do not provide sufficient evidence to conclude that a majority of Arizona families who celebrate Christmas wait until Christmas Day to open their presents. For the P-value approach, note that $P = 0.1379$.

11.27 H_0: $p = 0.74$, H_a: $p > 0.74$; $\alpha = 0.05$; critical value = 1.645; $z = 6.29$; reject H_0; at the 5% significance level, the data provide sufficient evidence to conclude that the percentage of major U.S. firms that drug-tested in 1993 exceeds the 1992 figure of 74%. For the P-value approach, note that $P = 0.0000$ (to four decimal places).

11.29 H_0: $p = 0.169$, H_a: $p > 0.169$; $\alpha = 0.10$; critical value = 1.28; $z = 0.92$; do not reject H_0; at the 10% significance level, the data do not provide sufficient evidence to conclude that the proportion, p, of female physicians is higher now than in 1990. For the P-value approach, note that $P = 0.1788$.

11.31 H_0: $p = 0.474$, H_a: $p \neq 0.474$; $\alpha = 0.10$; critical values = ± 1.645; $z = -0.25$; do not reject H_0; at the 10% significance level, the data do not provide sufficient evidence to conclude that the ball is not landing on red the correct percentage of the time for a balanced wheel. For the P-value approach, note that $P = 0.8026$.

11.33 H_0: $p = 0.107$, H_a: $p < 0.107$; $\alpha = 0.05$; critical value = -1.645; $z = -0.77$; do not reject H_0; at the 5% significance level, the study does not provide sufficient evidence to conclude that in 1990, the proportion of northeastern families earning incomes below the poverty level was less than the national proportion of 0.107. For the P-value approach, note that $P = 0.2206$.

Exercises 11.3

11.37

a. p_1 and p_2 are parameters and the other quantities are statistics.

b. p_1 and p_2 are fixed numbers and the other quantities are random variables.

11.39

a. H_0: $p_1 = p_2$, H_a: $p_1 < p_2$; $\alpha = 0.01$; critical value $= -2.33$; $z = -2.61$; reject H_0; at the 1% significance level, the data provide sufficient evidence to conclude that women who take folic acid are at lesser risk of having children with major birth defects. For the P-value approach, note that $P = 0.0045$.

b. designed experiment

c. Yes, because for a designed experiment, it is reasonable to interpret statistical significance as a causal relationship.

11.41 H_0: $p_1 = p_2$, H_a: $p_1 > p_2$; $\alpha = 0.05$; critical value $= 1.645$; $z = 4.62$; reject H_0; at the 5% significance level, we can conclude that the percentage of employed workers who have registered to vote exceeds the percentage of unemployed workers who have registered to vote. For the P-value approach, note that $P = 0.0000$ (to four decimal places).

11.43 H_0: $p_1 = p_2$, H_a: $p_1 < p_2$; $\alpha = 0.10$; critical value $= -1.28$; $z = -0.46$; do not reject H_0; at the 10% significance level, the data do not provide sufficient evidence to conclude that the percentage of U.S. dentists practicing in the Northeast is smaller than the percentage of U.S. physicians practicing in the Northeast. For the P-value approach, note that $P = 0.3228$.

11.45

a. -0.019 to -0.001

b. We can be 98% confident that the rate of major birth defects for babies born to women who have not taken folic acid is somewhere between 1 per 1000 and 19 per 1000 higher than for babies born to women who have taken folic acid.

11.47

a. 0.102 to 0.212

b. We can be 90% confident that the difference, $p_1 - p_2$, between the proportions of employed and unemployed workers who have registered to vote is somewhere between 0.102 and 0.212.

11.49

a. -0.063 to 0.029

b. We can be 80% confident that the difference between the proportion of U.S. dentists who practice in the Northeast and the proportion of U.S. physicians who practice in the Northeast is somewhere between -0.063 and 0.029.

Review Test for Chapter 11

1. a. Approximately normally distributed with mean $\mu_{\hat{p}} = p$ and standard deviation $\sigma_{\hat{p}} = \sqrt{p(1-p)/n}$.

   b. Probabilities for $\hat{p}$ are approximately equal to areas under the normal curve with parameters p and $\sqrt{p(1-p)/n}$.

3. a. 0.400 to 0.461

   b. We can be 95% confident that the percentage of all Americans who thought news organizations were criticizing the Clinton administration unfairly was somewhere between 40.0% and 46.1%.

4. a. 0.031   b. 2401   c. 0.396 to 0.436

   d. 0.02, which is the same as that specified in part (b)

   e. 2377; 0.396 to 0.436; 0.02

   f. By employing the guess for $\hat{p}$ in part (e), we reduced the required sample size by 24 with (virtually) no sacrifice in precision.

5. 13% to 21%

6. a. H_0: $p = 0.25$, H_a: $p < 0.25$; $\alpha = 0.05$; critical value $= -1.645$; $z = -2.30$; reject H_0; at the 5% significance level, the data provide sufficient evidence to conclude that less than one in four Americans believe that juries "almost always" convict the guilty and free the innocent.

   b. H_0: $p = 0.25$, H_a: $p < 0.25$; $\alpha = 0.05$; $z = -2.30$; $P = 0.0107$; reject H_0; at the 5% significance level, the data provide sufficient evidence to conclude that less than one in four Americans believe that juries "almost always" convict the guilty and free the innocent.

   c. The data provide strong evidence against the null hypothesis.

7. a. observational study

   b. Being observational, the study established only an association between height and breast cancer; no causal relationship can be inferred, although there may be one.

8. a. H_0: $p_1 = p_2$, H_a: $p_1 < p_2$; $\alpha = 0.01$; critical value $= -2.33$; $z = -4.17$; reject H_0; at the 1% significance level, the data provide sufficient evidence to conclude that the percentage of Maricopa County

residents who thought Arizona's economy would improve over the next 2 years was less during the time of the 1992 poll than during the time of the 1993 poll.

b. $H_0: p_1 = p_2$, $H_a: p_1 < p_2$; $\alpha = 0.01$; $z = -4.17$; $P = 0.000$ (to four decimal places); reject H_0; at the 1% significance level, the data provide sufficient evidence to conclude that the percentage of Maricopa County residents who thought Arizona's economy would improve over the next 2 years was less during the time of the 1992 poll than during the time of the 1993 poll.

c. The data provide very strong evidence against the null hypothesis.

9. a. -0.187 to -0.053

b. We can be 98% confident that the difference between the percentages of Maricopa County residents who thought Arizona's economy would improve over the next 2 years during the time of the 1992 poll and during the time of the 1993 poll is somewhere between -18.7% and -5.3%.

10. a. 0.067; we can be 98% confident that the error in estimating the difference between the two population proportions, $p_1 - p_2$, by the difference between the two sample proportions, -0.12, is at most 0.067.

b. 0.067   c. 3017   d. -0.158 to -0.098

e. 0.03, which is the same as that specified in part (c)

CHAPTER 12

Exercises 12.1

12.1 a. 32.852   b. 10.117

12.3 a. 18.307   b. 3.247

12.5 a. 1.646   b. 15.507

12.7 a. 0.831, 12.832   b. 13.844, 41.923

Exercises 12.2

12.9 Because the hypothesis test is carried out by determining how well the observed frequencies fit the expected frequencies.

12.11 H_0: The current distribution of the U.S. resident population by region is the same as the 1983 distribution. H_a: The current distribution of the U.S. resident population by region is different from the 1983 distribution. Assumptions 1 and 2 are satisfied since all expected frequencies are at least 5. $\alpha = 0.05$; critical value $= 7.815$; $\chi^2 = 1.827$; do not reject H_0; at the 5% significance level, the data do not

provide sufficient evidence to conclude that the current distribution of the U.S. resident population by region is different from the 1983 distribution. For the P-value approach, note that $P > 0.05$.

12.13 H_0: The distribution of the number of years of school completed by Tennessee residents is the same as the national distribution. H_a: The distribution of the number of years of school completed by Tennessee residents is different from the national distribution. Assumptions 1 and 2 are satisfied since all expected frequencies are at least 5. $\alpha = 0.01$; critical value $= 13.277$; $\chi^2 = 20.037$; reject H_0; at the 1% significance level, the data provide sufficient evidence to conclude that the distribution of the number of years of school completed by Tennessee residents is different from the national distribution. For the P-value approach, note that $P < 0.005$.

12.15 H_0: The die is not loaded. H_a: The die is loaded. Assumptions 1 and 2 are satisfied since all expected frequencies are at least 5. $\alpha = 0.05$; critical value $= 11.070$; $\chi^2 = 2.480$; do not reject H_0; at the 5% significance level, the data do not provide sufficient evidence to conclude that the die is loaded. For the P-value approach, note that $P > 0.05$.

12.17 H_0: The distribution of types of violent crime in New Jersey is the same as that for the nation as a whole. H_a: The distribution of types of violent crime in New Jersey is different from that for the nation as a whole. Assumptions 1 and 2 are satisfied since all expected frequencies are at least 5. $\alpha = 0.01$; critical value $= 11.345$; $\chi^2 = 18.003$; reject H_0; at the 1% significance level, the data provide sufficient evidence to conclude that the distribution of types of violent crime in New Jersey is different from that for the nation as a whole. For the P-value approach, note that $P < 0.005$.

Exercises 12.3

12.21 Dividing each row in Table 12.5 on page 700 by its row total, we get the base-of-practice distributions within each specialty, as seen in Table A.1. Because the rows of Table A.1 are not identical, base-of-practice distributions within the specialty categories are not the same; so specialty and base of practice are statistically dependent. Figure A.10 shows a segmented bar graph.

TABLE A.1   Base-of-practice distributions within the specialty categories for Exercise 12.21

	Base of practice			
	Office B_1	Hospital B_2	Other B_3	Total
General surgery S_1	0.635	0.322	0.044	1.000
Obstetrics/Gynecology S_2	0.754	0.210	0.036	1.000
Orthopedics S_3	0.741	0.236	0.023	1.000
Ophthalmology S_4	0.793	0.173	0.033	1.000
Total	0.714	0.250	0.036	1.000

Specialty

FIGURE A.10   Segmented bar graph for Exercise 12.21

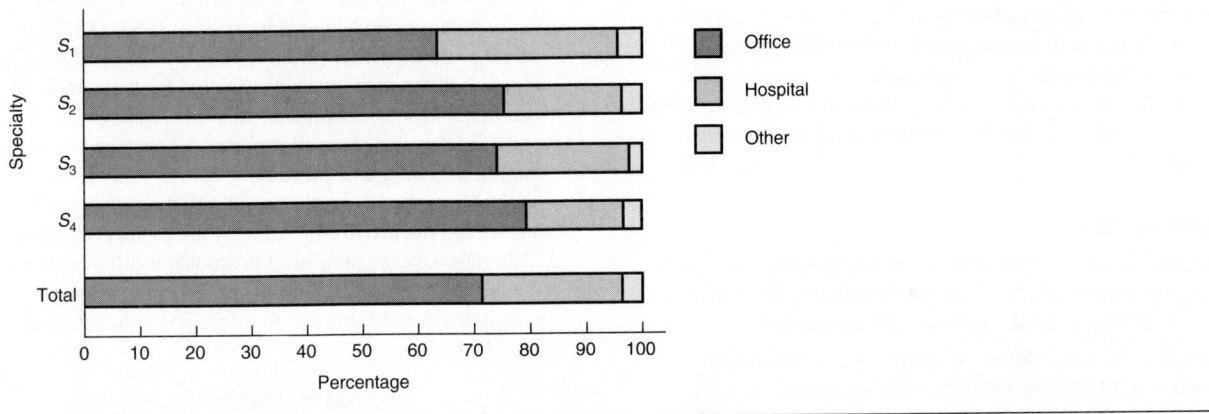

12.23 H_0: Annual income and educational level are statistically independent. H_a: Annual income and educational level are statistically dependent. Assumptions 1 and 2 are satisfied since all expected frequencies are at least 5. $\alpha = 0.01$; critical value $= 16.812$; $\chi^2 = 77.059$; reject H_0; at the 1% significance level, the data provide sufficient evidence to conclude that annual income and educational level are statistically dependent. For the P-value approach, note that $P < 0.005$.

12.25 H_0: Sex and occupation type for employed workers are nonassociated. H_a: Sex and occupation type for employed workers are associated. Assumptions 1 and 2 are

satisfied since all expected frequencies are at least 1 and only 12.5% (1/8) of the expected frequencies are less than 5. $\alpha = 0.01$; critical value $= 11.345$; $\chi^2 = 12.454$; reject H_0; at the 1% significance level, the data provide sufficient evidence to conclude that there is an association between the characteristics sex and occupation type for employed workers. For the P-value approach, note that $0.005 < P < 0.01$.

12.27

b. H_0: Type of violent crime and age of the person arrested are nonassociated. H_a: Type of violent crime and age of the person arrested are associated. Assumptions 1 and 2 are satisfied since all expected frequencies are at least 1 and only 16.7% (2/12) of the expected frequencies are less than 5. $\alpha = 0.01$; critical value $= 16.812$; $\chi^2 = 31.241$; reject H_0; at the 1% significance level, the data provide sufficient evidence to conclude that an association exists between the characteristics type of violent crime committed and age of the person arrested. For the P-value approach, note that $P < 0.005$.

12.29 H_0: Net worth and marital status for top wealthholders are statistically independent. H_a: Net worth and marital status for top wealthholders are statistically dependent. Assumption 2 is violated since 25% (3/12) of the expected frequencies are less than 5.

12.33
a. 34   b. 164   c. 9   d. 13.21
e. 1.343   f. 6
g. H_0: Grade and study time for intermediate algebra students at ASU are nonassociated. H_a: Grade and study time for intermediate algebra students at ASU are associated.
h. 10.574
i. Critical value $= 12.592$ (from Table V) and $\chi^2 = 10.574$ [from part (h)]; do not reject H_0; at the 5% significance level, the data do not provide sufficient evidence to conclude that an association exists between the characteristics grade and study time for intermediate algebra students at ASU.

Exercises 12.4

12.39 Because the procedures are based on the assumption that the population being sampled is normally distributed and are nonrobust to deviations from that assumption.

12.41 H_0: $\sigma = 100$, H_a: $\sigma \neq 100$; $\alpha = 0.05$; critical values $= 12.401, 39.364$; $\chi^2 = 17.545$; do not reject H_0; at the 5% significance level, the data do not provide sufficient evidence to conclude that the standard deviation of last year's verbal scores is different from the 1941 standard deviation of 100. For the P-value approach, note that $P > 0.10$.

12.43 H_0: $\sigma = 1$ sec, H_a: $\sigma > 1$ sec; $\alpha = 0.01$; critical value $= 36.191$; $\chi^2 = 39.222$; reject H_0; at the 1% significance level, the data provide sufficient evidence to conclude that the standard deviation, σ, of the weekly errors exceeds the 1-second claim made by the manufacturer. For the P-value approach, note that $P < 0.005$.

12.45 H_0: $\sigma = 0.2$ fl oz, H_a: $\sigma < 0.2$ fl oz; $\alpha = 0.05$; critical value $= 6.571$; $\chi^2 = 8.301$; do not reject H_0; at the 5% significance level, the data do not provide sufficient evidence to conclude that the standard deviation, σ, of the

amounts of coffee being dispensed is less than 0.2 fl oz. For the P-value approach, note that $P > 0.05$.

12.47 66.8 to 118.9. We can be 95% confident that the standard deviation of last year's verbal SAT scores is somewhere between 66.8 and 118.9.

12.49 1.04 to 2.27 sec. We can be 98% confident that the standard deviation, σ, of the weekly errors of the watches manufactured is somewhere between 1.04 and 2.27 seconds.

12.51 0.12 to 0.22 fl oz. We can be 90% confident that the standard deviation, σ, of the amounts of coffee being dispensed is somewhere between 0.12 and 0.22 fluid ounce.

12.53 To ensure that there will not be a large variation in the amount of coffee dispensed.

Review Test for Chapter 12

1. a. 6.408   b. 33.409   c. 27.587
   d. 8.672   e. 7.564, 30.191

2. H_0: The actual preference distribution for the number of bedrooms is the same as the distribution for the number of bedrooms in homes completed in 1987. H_a: The actual preference distribution for the number of bedrooms is different from the distribution for the number of bedrooms in homes completed in 1987. Assumptions 1 and 2 are satisfied since all expected frequencies are at least 5. $\alpha = 0.05$; critical value $= 5.991$; $\chi^2 = 29.821$; reject H_0; at the 5% significance level, the data provide sufficient evidence to conclude that the actual preference distribution for the number of bedrooms is different from the distribution for the number of bedrooms in homes completed in 1987. For the P-value approach, note that $P < 0.005$.

3. a.

Region of residence

Age	NE	MW	South	West	Total
Under 5	0.073	0.075	0.075	0.083	0.076
5–17	0.167	0.188	0.184	0.187	0.182
18–44	0.429	0.420	0.429	0.446	0.431
45–64	0.194	0.186	0.186	0.174	0.185
65 & over	0.137	0.131	0.125	0.109	0.126
Total	1.000	1.000	1.000	1.000	1.000

b. Yes, because the age distributions within the four regions of residence are not identical.
c. See Fig. A.11.

FIGURE A.11   Segmented bar graph for Problem 3(c)

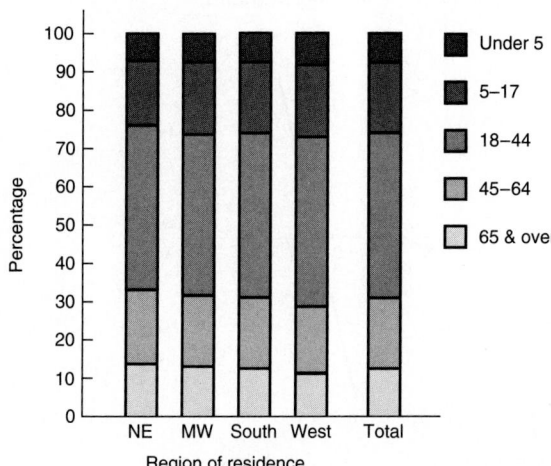

d. False; they are not identical because, as we know from part (b), an association exists between age and region of residence.

4. H_0: Response and educational level are statistically independent. H_a: Response and educational level are statistically dependent. Assumptions 1 and 2 are satisfied since all expected frequencies are at least 5. $\alpha = 0.01$; critical value $= 16.812$; $\chi^2 = 77.837$; reject H_0; at the 1% significance level, the data provide sufficient evidence to conclude that response and educational level are statistically dependent. For the P-value approach, note that $P < 0.005$.

6. a. 205   b. 238   c. 26   d. 31.93
   e. 1.102   f. 6   g. 77.837
   h. Critical value $= 16.812$ (from Table V) and $\chi^2 = 77.837$ [from part (g)]; reject H_0; at the 1% significance level, the data provide sufficient evidence to conclude that response and educational level are statistically dependent.

7. a. H_0: $\sigma = 16$, H_a: $\sigma \neq 16$; $\alpha = 0.10$; critical values $= 13.848, 36.415$; $\chi^2 = 21.111$; do not reject H_0; at the 10% significance level, the data do not provide sufficient evidence to conclude that the standard deviation of all IQs is not equal to 16. For the P-value approach, note that $P > 0.10$.
   b. Quite crucial, since the test is nonrobust to deviations from the normality assumption.

8. 12.2 to 19.8

CHAPTER 13

Exercises 13.1

13.1
a. $y = 41.88 + 0.33x$ b. $b_0 = 41.88$, $b_1 = 0.33$

c.

x	50	100	250
y	58.38	74.88	124.38

d.

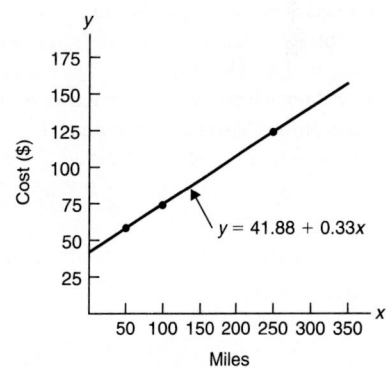

e. About $90; exact cost is $91.38.

13.3
a. $b_0 = 32$, $b_1 = 1.8$ b. $-40, 32, 68, 212$

c.

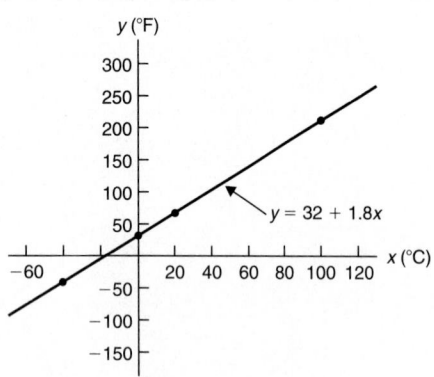

d. About 80° F; exact temperature is 82.4° F.

13.5

a. $b_0 = 41.88$, $b_1 = 0.33$

b. The y-intercept, $b_0 = 41.88$, gives the y-value at which the straight line, $y = 41.88 + 0.33x$, intersects the y-axis. The slope, $b_1 = 0.33$, indicates that the y-value increases by 0.33 units for every increase in x of 1 unit.

c. The y-intercept, $b_0 = 41.88$, is the cost (in dollars) for driving the car 0 miles. The slope, $b_1 = 0.33$, represents the fact that the cost per mile is $0.33; it is the amount the total cost increases for each additional mile driven.

13.7

a. $b_0 = 32$, $b_1 = 1.8$

b. The y-intercept, $b_0 = 32$, gives the y-value at which the straight line, $y = 32 + 1.8x$, intersects the y-axis. The slope, $b_1 = 1.8$, indicates that the y-value increases by 1.8 units for every increase in x of 1 unit.

c. The y-intercept, $b_0 = 32$, is the Fahrenheit temperature corresponding to 0° C. The slope, $b_1 = 1.8$, represents the fact that the Fahrenheit temperature increases by 1.8° for every increase of the Celsius temperature of 1°.

13.9

a. $b_0 = 3$, $b_1 = 4$ b. slopes upward

c.

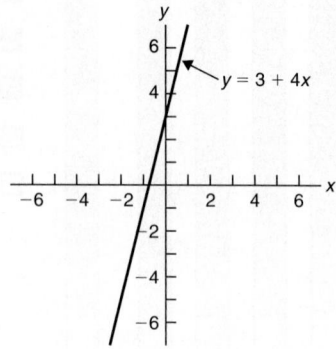

13.11

a. $b_0 = 6$, $b_1 = -7$ b. slopes downward

c.

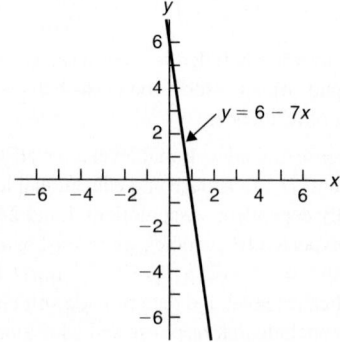

13.13   a. $b_0 = -2$, $b_1 = 0.5$ b. slopes upward

13.15   a. $b_0 = 2$, $b_1 = 0$ b. horizontal

13.17   a. $b_0 = 0$, $b_1 = 1.5$ b. slopes upward

13.19

a. slopes upward   b. $y = 5 + 2x$

c.

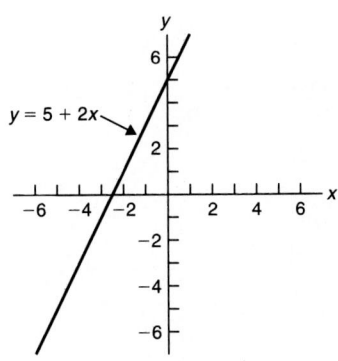

$y = 5 + 2x$

13.21
a. slopes downward   b. $y = -2 - 3x$
c.

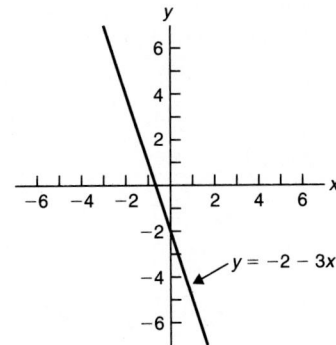

$y = -2 - 3x$

13.23   a. slopes downward   b. $y = -0.5x$
13.25   a. horizontal   b. $y = 3$

13.29
a.

Line A: $y = 3 - 0.6x$

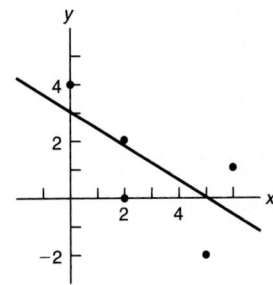

Line B: $y = 4 - x$

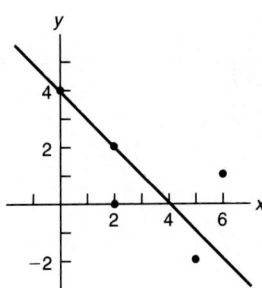

b. Line A: $y = 3 - 0.6x$

x	y	$\hat{y}$	e	e^2
0	4	3.0	1.0	1.00
2	2	1.8	0.2	0.04
2	0	1.8	−1.8	3.24
5	−2	0.0	−2.0	4.00
6	1	−0.6	1.6	2.56
				10.84

Line B: $y = 4 - x$

x	y	$\hat{y}$	e	e^2
0	4	4	0	0
2	2	2	0	0
2	0	2	−2	4
5	−2	−1	−1	1
6	1	−2	3	9
				14

c. Line A

13.31

a. $\hat{y} = 2.875 - 0.625x$

b.

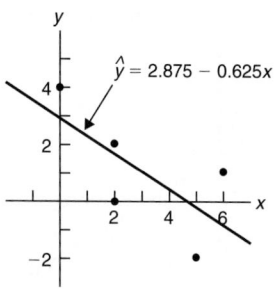

13.33

a. $\hat{y} = -174.49 + 4.84x$

b.

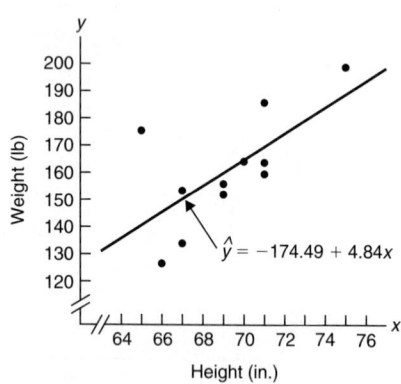

c. Weight tends to increase as height increases.

d. The weights of 18–24-year-old males increase an estimated 4.84 pounds for each increase in height of 1 inch.

e. 149.54 lb; 178.56 lb (see the note at the top of page A-31)

f. The predictor variable is height; the response variable is weight.

g. (65, 175) might be considered an outlier; (75, 198) is a potential influential observation.

13.35

a. $\hat{y} = 222.27 - 1.14x$

b.

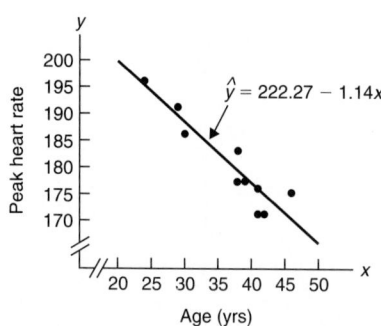

c. Peak heart rate tends to decrease as age increases.

d. The peak heart rate an individual can reach during intensive exercise decreases by an estimated 1.14 for each increase in age of 1 year.

e. 190.34 (see the note at the top of page A-31)

f. The predictor variable is age; the response variable is peak heart rate.

g. none

13.37

a. $\hat{y} = 12.86 + 1.21x$

b.

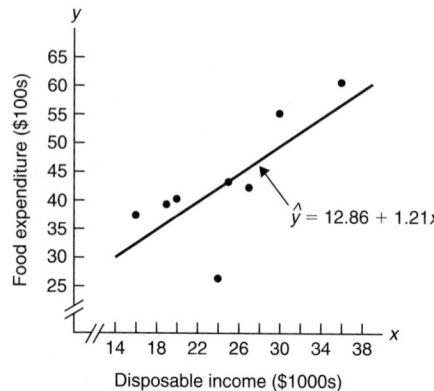

c. Annual food expenditure tends to increase as disposable income increases.

d. Annual food expenditure increases an estimated $121 (1.21 hundred dollars) for each increase in disposable income of $1000.

e. $4321 (see the note at the top of page A-31)

f. The predictor variable is disposable income; the response variable is annual food expenditure.

g. (24, 26) appears to be an outlier; no potential influential observations.

13.39   Only the second one.

13.41

a. It is acceptable to use the regression equation to predict the weight of an 18–24-year-old male who is 68 inches tall since that height lies within the range of the heights in the sample data. It is not acceptable (and would be extrapolation) to use the regression equation to predict the weight of an 18–24-year-old male who is 60 inches tall since that height lies outside the range of the heights in the sample data.

b. Heights between 65 and 75 inches, inclusive.

13.45

a. $\hat{y} = 2152 + 60.8x$ b. 3185.6 g

c. (24, 2800) is a potential outlier; no potential influential observations.

Exercises 13.3

13.53

a. The coefficient of determination, r^2.

b. See Key Fact 13.5 on page 766.

13.55

a. $SST = 20$, $SSR = 9.375$, $SSE = 10.625$

b. $20 = 9.375 + 10.625$ c. $r^2 = 0.469$

d. 46.9%   e. 46.9%   f. moderately useful

13.57

a. $SST = 4352.91$, $SSR = 1909.46$, $SSE = 2443.45$

b. 0.439

c. 43.9%; 43.9% of the variation in the weight data is explained by height.

d. moderately useful

13.59

a. $SST = 642.10$, $SSR = 553.60$, $SSE = 88.50$

b. 0.862

c. 86.2%; 86.2% of the variation in the peak-heart-rate data is explained by age.

d. extremely useful

13.61

a. $SST = 783.50$, $SSR = 429.95$, $SSE = 353.55$

b. 0.549

c. 54.9%; 54.9% of the variation in the data on food expenditure is explained by disposable income.

d. moderately useful

13.65

a. 0.372

b. $SSR = 2,505,745$, $SSE = 4,234,255$, $SST = 6,740,000$

c. 37.2%

Exercises 13.4

13.73

a. $r = -0.685$

b. Suggest a moderately strong negative linear correlation.

c. Data points are clustered moderately closely about the regression line.

d. $r^2 = 0.469$

13.75

a. $r = 0.662$

b. Suggests a moderately strong positive linear correlation.

c. Data points are clustered moderately closely about the regression line.

d. $r^2 = 0.438$ *(Note:* In part (b) of Exercise 13.57, we found that $r^2 = 0.439$. The discrepancy between these two values of r^2 is due to the error resulting from rounding r to three decimal places before squaring.)

13.77

a. $r = -0.929$

b. Suggest an extremely strong negative linear correlation.

c. Data points are clustered extremely closely about the regression line.

d. $r^2 = 0.863$ *(Note:* In part (b) of Exercise 13.59, we found that $r^2 = 0.862$. The discrepancy between these two values of r^2 is due to the error resulting from rounding r to three decimal places before squaring.)

13.79

a. $r = 0.741$

b. Suggests a moderately strong positive linear correlation.

c. Data points are clustered moderately closely about the regression line.

d. $r^2 = 0.549$

13.81

a. $r = 0$

b. No, only that there is no *linear* relationship between the variables.

d. No, because the data points are not scattered about a straight line.

e. For each data point (x, y), we have $y = x^2$.

Exercises 13.5

13.93

a. $\hat{y} = 266 + 6.73x_1 + 3.26x_2 + 4.51x_3$

b. $367.4 million

c. $R^2 = 0.911$; 91.1% of the variation in the sales data is explained by television, magazine, and radio advertising expenditures.

13.95

a. $\hat{y} = 52.4 + 0.210x_1 + 0.000220x_2 - 0.740x_3$

b. 57.2%
c. $R^2 = 0.166$
d. Not very useful

Review Test for Chapter 13

1. a. $y = 72 - 12x$ b. $b_0 = 72, b_1 = -12$
   c. The line slopes downward since $b_1 < 0$.
   d. $4800; $1200
   e.

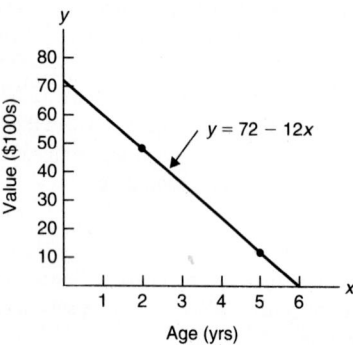

   f. About $2500; exact value is $2400.

2. a.

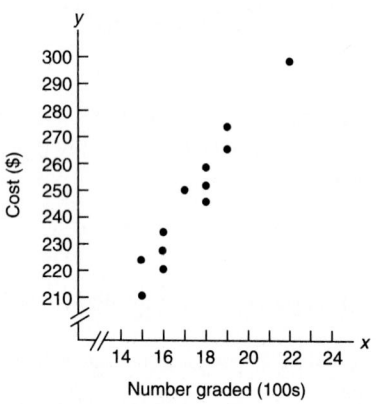

   b. Yes, because the data points appear to be scattered about a straight line.

c. $\hat{y} = 35.80 + 12.08x$

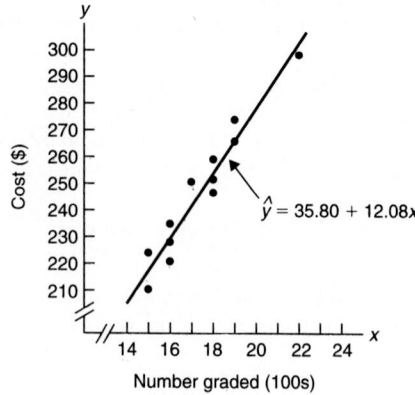

d. Cost of grading tends to increase as the number of papers graded increases.
e. Paper-grading costs increase an estimated $12.08 for each additional 100 papers graded.
f. $229.13 (see the note at the top of page A-31)
g. No outliers; (22, 298) is a potential influential observation.

3. a. $SST = 6984.25, SSR = 6558.31, SSE = 425.94$
   b. $r^2 = 0.939$ c. 93.9%   d. 93.9%
   e. extremely useful

5. a. $\hat{y} = 35.8 + 12.1x$ b. $r^2 = 0.939$
   c. $SSR = 6558.3, SSE = 425.9, SST = 6984.3$
   d. No potential outliers; (22, 298) is a potential influential observation.

6. a. $r = 0.969$
   b. Suggests an extremely strong positive linear correlation.
   c. Data points are clustered extremely closely about the regression line.
   d. $r^2 = (0.969)^2 = 0.939$

8. a. $\hat{y} = -40.9 + 0.772x_1 + 3.11x_2$
   b. $33,564
   c. $R^2 = 0.766$; 76.6% of the variation in the annual-income data is explained by age and number of years of school completed.

CHAPTER 14

Exercises 14.1

14.1 See Key Fact 14.1 on page 794.

14.3 There are constants, β_0, β_1, and σ, such that for each height, x, the weights of 18–24-year-old males of that height are normally distributed with mean $\beta_0 + \beta_1 x$ and standard deviation σ.

14.5 There are constants, β_0, β_1, and σ, such that for each age, x, the peak heart rates that can be reached during intensive exercise by individuals of that age are normally distributed with mean $\beta_0 + \beta_1 x$ and standard deviation σ.

14.7 There are constants, β_0, β_1, and σ, such that for each disposable-income level, x, the annual food expenditures made by middle-income families (father, mother, two children) at that level are normally distributed with mean $\beta_0 + \beta_1 x$ and standard deviation σ.

14.9

a. $s_e = 16.48$ lb; very roughly speaking we can say that, on the average, the predicted weight of an 18–24-year-old male in the sample differs from the observed weight by about 16.48 lb.

b. Presuming that the variables height (x) and weight (y) for 18–24-year-old males satisfy Assumptions 1–3 for regression inferences, the standard error of the estimate, $s_e = 16.48$ lb, provides an estimate for the common population standard deviation, σ, of weights for all 18–24-year-old males of any particular height.

c. See Fig. A.12.

d. It appears reasonable.

FIGURE A.12   (a) Residual plot and (b) normal probability plot of residuals for Exercise 14.9

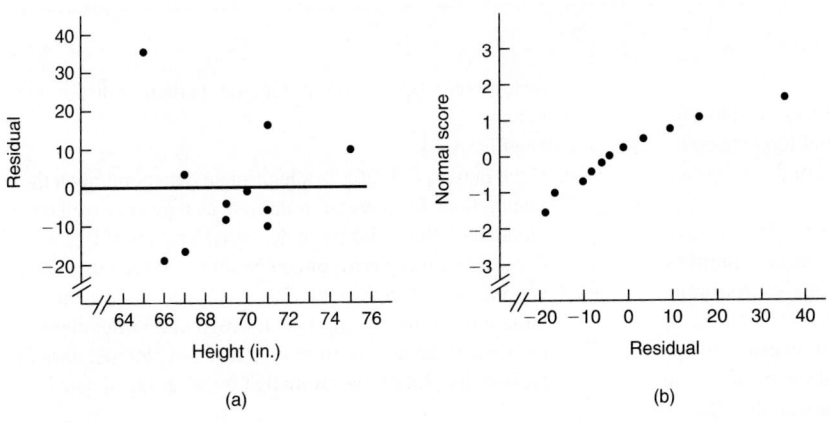

(a)   (b)

14.11

a. $s_e = 3.33$; very roughly speaking we can say that, on the average, the predicted peak heart rate of a person in the sample differs from the observed peak heart rate by about 3.33.

b. Presuming that the variables age (x) and peak heart rate (y) for individuals satisfy Assumptions 1–3 for regression inferences, the standard error of the estimate, $s_e = 3.33$, provides an estimate for the common population standard deviation, σ, of peak heart rates for all individuals of any particular age.

c. See Fig. A.13 at the top of the next page.

d. It appears reasonable, although the residual plot in Fig. A.13(a) casts some doubt on the assumption of equal standard deviations (Assumption 2).

FIGURE A.13   (a) Residual plot and (b) normal probability plot of residuals for Exercise 14.11

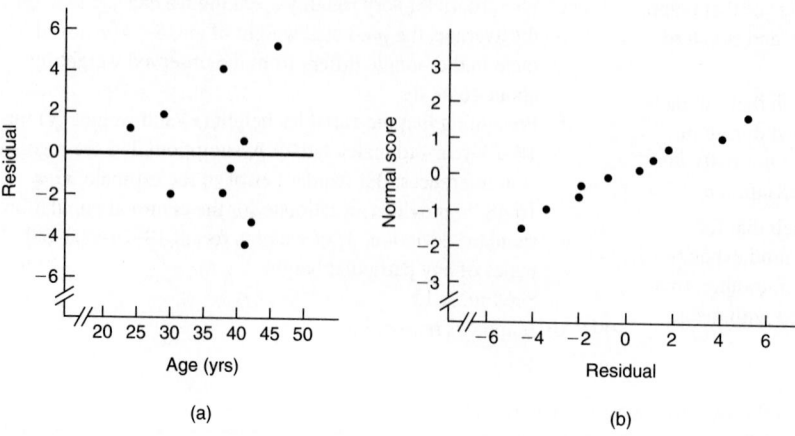

(a)   (b)

14.13

a. s_e = $7.68 hundred; very roughly speaking we can say that, on the average, the predicted annual food expenditure of a family in the sample differs from the observed annual food expenditure by about $768.

b. Presuming that the variables disposable income (x) and annual food expenditure (y) for middle-income families with a father, mother, and two children satisfy Assumptions 1–3 for regression inferences, the standard error of the estimate, s_e = 7.68 ($768), provides an estimate for the common population standard deviation, σ, of annual food expenditures for all middle-income families (fa-

ther, mother, two children) with any particular disposable income.

c. See Fig. A.14.

d. If the outlier, (24, 26), is a legitimate data point, then the assumptions for regression inferences may very well be violated by the variables under consideration; if the outlier is a recording error or can be removed for some other valid reason, then the resulting data reveal no obvious violations of the assumptions for regression inferences (as we can see by constructing a residual plot and normal probability plot of the residuals for the abridged data).

FIGURE A.14   (a) Residual plot and (b) normal probability plot of residuals for Exercise 14.13

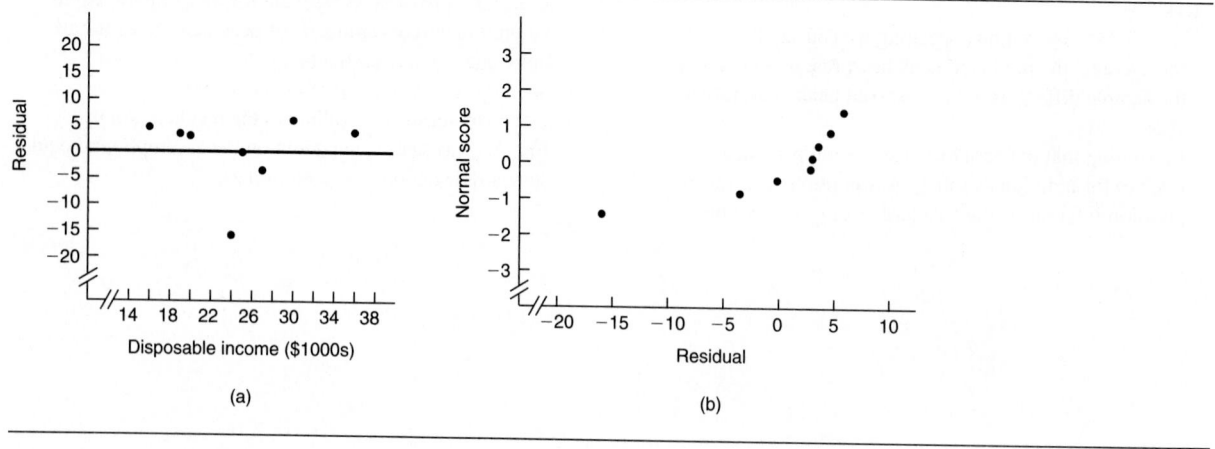

(a)   (b)

14.15 Part (a) is a tough call, but the assumption of linearity (Assumption 1) may be violated, as may the assumption of equal standard deviations (Assumption 2); in part (b) it appears that the assumption of equal standard deviations is violated; in part (d) it appears that the normality assumption (Assumption 3) is violated.

14.19
a. $s_e = 382.1$ g
b. No; the residual plot falls roughly in a horizontal band centered and symmetric about the x-axis, and the normal probability plot of the residuals is roughly linear.

Exercises 14.2

14.21 $H_0: \beta_1 = 0$, $H_a: \beta_1 \neq 0$; $\alpha = 0.10$; critical values $= \pm 1.833$; $t = 2.652$; reject H_0; at the 10% significance level, the data provide sufficient evidence to conclude that the slope of the population regression line is not 0 and hence that height is useful as a predictor of weight for 18–24-year-old males. For the P-value approach, note that $0.02 < P < 0.05$.

14.23 $H_0: \beta_1 = 0$, $H_a: \beta_1 \neq 0$; $\alpha = 0.05$; critical values $= \pm 2.306$; $t = -7.074$; reject H_0; at the 5% significance level, the data provide sufficient evidence to conclude that age is useful as a predictor of peak heart rate. For the P-value approach, note that $P < 0.01$.

14.25 $H_0: \beta_1 = 0$, $H_a: \beta_1 \neq 0$; $\alpha = 0.01$; critical values $= \pm 3.707$; $t = 2.701$; do not reject H_0; at the 1% significance level, the data do not provide sufficient evidence to conclude that disposable income is useful as a predictor of annual food expenditure for middle-income families with a father, mother, and two children. For the P-value approach, note that $0.02 < P < 0.05$.

14.27
a. 1.49 to 8.18 lb
b. We can be 90% confident that for 18–24-year-old males, the increase in mean weight per 1-inch increase in height is somewhere between 1.49 and 8.18 lb.

14.29
a. -1.51 to -0.77
b. We can be 95% confident that the drop in mean peak heart rate per 1-year increase in age is between 0.77 and 1.51.

14.31
a. -0.45 to 2.88
b. We can be 99% confident that for middle-income families with a father, mother, and two children, the change in mean annual food expenditure per $1000 increase in family disposable income is somewhere between $-$$45 and $288.

14.35
a. 60.82   b. 14.68   c. 4.14
d. 0.000 (to three decimal places)
e. At the 5% significance level, the data provide sufficient evidence to conclude that estriol level is useful for predicting birth weights.
f. 30.80 to 90.84. We can be 95% confident that the increase in mean birth weight per increase of estriol level by 1 mg/24 hr is somewhere between 30.80 and 90.84 g.
g. The assumptions for regression inferences, Assumptions 1–3 on page 794. By performing a residual analysis as indicated in Key Fact 14.2 on page 800.

Exercises 14.3

14.39
a. 164.05 lb
b. 154.54 to 173.56 lb. We can be 90% confident that the mean weight of all 18–24-year-old males who are 70 inches tall is between 154.54 and 173.56 lb.
c. 164.05 lb
d. 132.38 to 195.71 lb. We can be 90% certain that the weight of a randomly selected 18–24-year-old male who is 70 inches tall will be between 132.38 and 195.71 lb.
e. See Fig. A.15 at the top of the next page.
f. The error in the estimate of the mean weight of all 18–24-year-old males who are 70 inches tall is due only to the fact that the population regression line is being estimated by a sample regression line, whereas the error in the prediction of the weight of a randomly selected 18–24-year-old male who is 70 inches tall is due to that fact plus the variation in weights of such males.

14.41
a. 176.65
b. 173.95 to 179.35. We can be 95% confident that the mean peak heart rate of all 40-year-olds is somewhere between 173.95 and 179.35.
c. 176.65
d. 168.52 to 184.78. We can be 95% certain that the peak heart rate of a randomly selected 40-year-old will be somewhere between 168.52 and 184.78.

14.43
a. 33.13 to 53.29. We can be 99% confident that the mean annual food expenditure of all middle-income families consisting of a father, mother, and two children that have a disposable income of $25,000 is somewhere between $3313 and $5329.

FIGURE A.15   90% confidence and prediction intervals for Exercise 14.39(e)

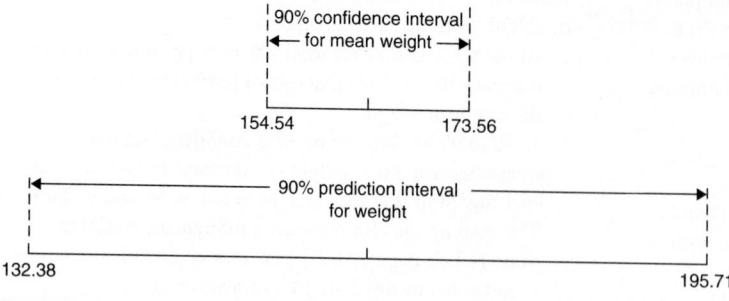

b. 13.02 to 73.39. We can be 99% certain that the annual food expenditure of a randomly selected middle-income family consisting of a father, mother, and two children that has a disposable income of $25,000 will be somewhere between $1302 and $7339.

14.47
a. 3064.6 g   b. 2909.1 to 3220.1 g
c. 2267.6 to 3861.6 g

Exercises 14.4

14.49 $H_0: \rho = 0, H_a: \rho > 0; \alpha = 0.05$; critical value $= 1.833$; $t = 2.652$ (see the note at the top of page A-31); reject H_0; at the 5% significance level, the data provide sufficient evidence to conclude that the variables height and weight are positively linearly correlated for 18–24-year-old males. For the P-value approach, note that $0.01 < P < 0.025$.

14.51 $H_0: \rho = 0, H_a: \rho < 0; \alpha = 0.025$; critical value $= -2.306$; $t = -7.074$ (see the note at the top of page A-31); reject H_0; at the 2.5% significance level, the data provide sufficient evidence to conclude that age and peak heart rate are negatively linearly correlated. For the P-value approach, note that $P < 0.005$.

14.53 $H_0: \rho = 0, H_a: \rho \neq 0; \alpha = 0.01$; critical values $= \pm 3.707$; $t = 2.701$ (see the note at the top of page A-31); do not reject H_0; at the 1% significance level, the data do not provide sufficient evidence to conclude that family disposable income and annual food expenditure are linearly correlated for middle-income families with a father, mother, and two children. For the P-value approach, note that $0.02 < P < 0.05$.

14.55   No. Yes. The population linear correlation coefficient is a fixed number, being a parameter. The sample linear correlation coefficient is a random variable since its value depends on chance, namely, on the sample obtained.

Exercises 14.5

14.57   It would mean there are constants, $\beta_0, \beta_1, \beta_2, \beta_3$, and σ, such that for each television advertising expenditure, x_1, magazine advertising expenditure, x_2, and radio advertising expenditure, x_3, the total sales is normally distributed with mean $\beta_0 + \beta_1 x_1 + \beta_2 x_2 + \beta_3 x_3$ and standard deviation σ.

14.59
a. $\hat{y} = 266 + 6.73x_1 + 3.26x_2 + 4.51x_3$, where x_1, x_2, and x_3 denote, respectively, television, magazine, and radio advertising expenditures (in millions of dollars) and $\hat{y}$ denotes predicted sales (in millions of dollars). Based on the sample data, this equation is the best estimate of the unknown population regression equation, $y = \beta_0 + \beta_1 x_1 + \beta_2 x_2 + \beta_3 x_3$.
b. $s_e = 4.418$. Based on the sample data, the best estimate for the common population standard deviation, σ, of sales, for any particular expenditures on television, magazine, and radio advertising, is $4.418 million.
c. Yes, because p $= 0.002$ in the line labeled Regression, $\alpha = 0.05$, and $0.002 \leq 0.05$.
d. Yes, because p $= 0.002$ in the line labeled TV, $\alpha = 0.05$, and $0.002 \leq 0.05$. At the 5% significance level, the data provide sufficient evidence to conclude that, in conjunction with magazine and radio advertising expenditures, television advertising expenditure is useful as a predictor of sales.
e. No, because p $= 0.269$ in the line labeled RADIO, $\alpha = 0.05$, and $0.269 > 0.05$. At the 5% significance level, the data do not provide sufficient evidence to conclude that, in conjunction with television and magazine advertising

expenditures, radio advertising expenditure is useful as a predictor of sales.

f. $367.58 million

g. $361.24 million to $373.92 million

h. $367.58 million

i. $355.05 million to $380.12 million

14.61

a. No, because $p = 0.061$ in the line labeled `Regression`, $\alpha = 0.05$, and $0.061 > 0.05$.

b. No, because $p = 0.071$ in the line labeled `TOP10%HS`, $\alpha = 0.05$, and $0.071 > 0.05$. At the 5% significance level, the data do not provide sufficient evidence to conclude that, in conjunction with total spending per student and student-to-faculty ratio, percentage of freshmen in the top 10% of their high-school class is useful for predicting graduation rate.

c. No for both variables. (P-values are 0.620 and 0.328, respectively.)

d. In view of the P-values observed in parts (b) and (c), it may be worthwhile to perform a regression analysis with percentage of freshmen in the top 10% of their high-school class as the only predictor variable. (*Note:* For this latter regression, the test for utility is significant at the 5% level; $P = 0.024$.)

e. 48.73% to 65.55%

f. 28.89% to 85.40%

g. The Assumptions 1–3 for multiple regression inferences, given in Key Fact 14.10 on page 834.

Exercises 14.6

14.69 If the population from which the sample was obtained is normally distributed, then the normal probability plot should be roughly linear, which means that the correlation between the sample data and its normal scores should be close to 1. Because the correlation can be at most 1, evidence against the null hypothesis of normality is provided if the correlation is "too small." Thus a correlation test for normality is always left-tailed.

14.71 H_0: Final-exam scores in the introductory statistics class are normally distributed. H_a: Final-exam scores in the introductory statistics class are not normally distributed. $\alpha = 0.05$; critical value = 0.951; $R_p = 0.939$; reject H_0; at the 5% significance level, the data provide sufficient evidence to conclude that the final-exam scores in the introductory statistics class are not normally distributed. For the P-value approach, note that $0.01 < P < 0.05$.

14.73 H_0: The number of miles cars were driven last year is normally distributed. H_a: The number of miles cars were driven last year is not normally distributed. $\alpha = 0.10$; critical value = 0.951; $R_p = 0.989$; do not reject H_0; at the 10% significance level, the data do not provide sufficient evidence to conclude that the number of miles cars were driven last year is not normally distributed. For the P-value approach, note that $P > 0.10$.

14.75 H_0: For the year in question, energy expenditures for households using electricity as their primary energy source are normally distributed. H_a: For the year in question, energy expenditures for households using electricity as their primary energy source are not normally distributed. $\alpha = 0.01$; critical value = 0.922; $R_p = 0.984$; do not reject H_0; at the 1% significance level, the data do not provide sufficient evidence to conclude that for the year in question, energy expenditures for households using electricity as their primary energy source are not normally distributed. For the P-value approach, note that $P > 0.10$.

Review Test for Chapter 14

1. There are constants, β_0, β_1, and σ, such that for each number of papers graded, x, the costs for grading are normally distributed with mean $\beta_0 + \beta_1 x$ and standard deviation σ.

2. a. $\hat{y} = 35.80 + 12.08x$

   b. $s_e = \$6.53$; very roughly speaking we can say that, on the average, the predicted cost for grading a number of papers in the sample differs from the observed cost by about $6.53.

   c. Presuming that the assumptions for regression inferences are met, the standard error of the estimate, $s_e = \$6.53$, provides an estimate for the common population standard deviation, σ, of all costs for grading any particular number of papers.

3. Referring to Fig. A.16 at the top of the following page, we find no obvious violations of the assumptions for regression inferences for the variables number of papers graded and cost.

4. a. H_0: $\beta_1 = 0$, H_a: $\beta_1 \neq 0$; $\alpha = 0.05$; critical values = ± 2.228; $t = 12.409$; reject H_0; at the 5% significance level, the data provide sufficient evidence to conclude that the number of papers graded is useful as a predictor of cost. For the P-value approach, note that $P < 0.01$.

   b. 9.91 to 14.25. We can be 95% confident that the increase in mean cost per increase in the number of papers graded by 100 is somewhere between $9.91 and $14.25.

FIGURE A.16   (a) Residual plot and (b) normal probability plot of residuals for Problem 3

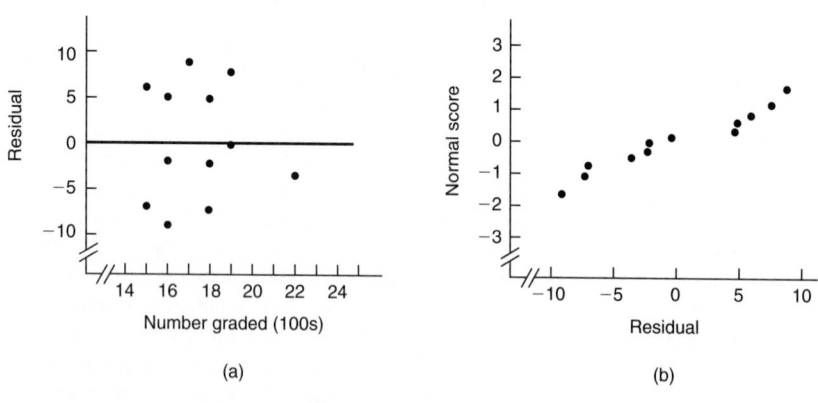

(a)   (b)

5. a. $229.13 (see the note at the top of page A-31)
    b. $223.93 to $234.33. We can be 95% confident that
       the mean cost of grading 1600 papers is somewhere
       between $223.93 and $234.33.
    c. $229.13
    d. $213.69 to $244.58. We can be 95% certain that
       the cost of grading 1600 papers will be somewhere
       between $213.69 and $244.58.
    e. The error in the estimate of the mean cost for grading
       1600 papers is due only to the fact that the population
       regression line is being estimated by a sample regres-
       sion line, whereas the error in the prediction of the
       cost for grading 1600 papers is due to that fact plus
       the variation in cost for grading 1600 papers.

9. a. $\hat{y} = 35.8 + 12.1x$ b. 6.526
    c. 12.0835   d. 0.9738   e. 12.41
    f. 0.000 (to three decimal places)
    g. At the 5% significance level, the data provide suffi-
       cient evidence to conclude that the number of papers
       graded is useful for predicting cost.
    h. 9.91 to 14.25. We can be 95% confident that the in-
       crease in mean cost per increase in the number of
       papers graded by 100 is somewhere between $9.91
       and $14.25.
    i. $229.13   j. $223.93 to $234.34
    k. $229.13   l. $213.68 to $244.58

10. H_0: $\rho = 0$, H_a: $\rho > 0$; $\alpha = 0.025$; critical value =
    2.228; $t = 12.409$ (see the note at the top of page A-31);
    reject H_0; at the 2.5% significance level, the data pro-
    vide sufficient evidence to conclude that the variables
    number of papers graded and cost are positively lin-

early correlated. For the P-value approach, note that
$P < 0.005$.

11. a. $\hat{y} = -40.9 + 0.772x_1 + 3.11x_2$, where x_1 and x_2 de-
       note, respectively, age and number of years of school
       completed, and $\hat{y}$ denotes predicted annual income
       (in thousands of dollars). Based on the sample data,
       this equation is the best estimate of the unknown
       population regression equation, $y = \beta_0 + \beta_1 x_1 + \beta_2 x_2$.
    b. $s_e = 7.886$. Based on the sample data, the best es-
       timate for the common population standard devia-
       tion, σ, of annual incomes, for any particular age and
       number of years of school completed, is $7886.
    c. Yes, because p = 0.000 in the line labeled Regres-
       sion, $\alpha = 0.05$, and $0.000 \leq 0.05$.
    d. Yes, because p = 0.000 in the line labeled AGE, $\alpha =$
       0.05, and $0.000 \leq 0.05$. At the 5% significance level,
       the data provide sufficient evidence to conclude that,
       in conjunction with the number of years of school
       completed, age is useful as a predictor of annual in-
       come for males (between the ages of 25 and 50 with
       at least a ninth-grade education).
    e. Yes, because p = 0.000 in the line labeled EDUC,
       $\alpha = 0.05$, and $0.000 \leq 0.05$. At the 5% significance
       level, the data provide sufficient evidence to conclude
       that, in conjunction with age, the number of years of
       school completed is useful as a predictor of annual
       income for males (between the ages of 25 and 50 with
       at least a ninth-grade education).

12. a. $33,528   b. $31,342 to $35,714
    c. $33,528   d. $17,653 to $49,403

13. No. All three residual plots fall roughly in a horizon-
    tal band centered and symmetric about the horizontal

axis, and the normal probability plot of the residuals is roughly linear.

15. H_0: Gas mileages for this model car are normally distributed. H_a: Gas mileages for this model car are not normally distributed. $\alpha = 0.05$; critical value $= 0.938$; $R_p = 0.981$; do not reject H_0; at the 5% significance level, the data do not provide sufficient evidence to conclude that the gas mileages for this model car are not normally distributed. For the P-value approach, note that $P > 0.10$.

CHAPTER 15

Exercises 15.1

15.1 a. 12 b. 7

15.3 a. 1.79 b. 2.29

15.5 a. 2.88 b. 2.10

15.7 a. 2.25 b. 3.21

Exercises 15.2

15.11 The pooled-t procedure of Section 10.2.

15.13 The procedure for comparing the means involves analyzing the variation in the data.

15.15
a. the treatment mean square, $MSTR$
b. the error mean square, MSE
c. the F-statistic, $F = MSTR/MSE$

15.17 It signifies that the ANOVA compares the means of populations that are classified in one way.

15.19 No, because the variation among the sample means is not large relative to the variation within the samples.

Exercises 15.3

15.25 $SST = SSTR + SSE$. The total variation among all the sample data can be partitioned into a component representing variation among the sample means and a component representing variation within the samples.

15.27

Source	df	SS	MS	F
Treatment	2	2.8	1.4	1.56
Error	12	10.8	0.9	
Total	14	13.6		

15.29

Source	df	SS	MS	F
Treatment	3	132,508.2	44,169.40	7.37
Error	16	95,877.6	5,992.35	
Total	19	228,385.8		

15.31 The missing entries in the first row are 43.304 and 3.08; in the second, 12 and 7.033; and in the third, 127.704.

15.33 H_0: $\mu_1 = \mu_2 = \mu_3 = \mu_4$, H_a: Not all the means are equal. $\alpha = 0.05$; critical value $= 3.24$; $SST = 560.2$, $SSTR = 68.2$, $SSE = 492.0$; $F = 0.74$; do not reject H_0; at the 5% significance level, the data do not provide sufficient evidence to conclude that there is a difference in mean lifetimes among the four brands of batteries. For the P-value approach, note that $P > 0.05$.

15.35 H_0: $\mu_1 = \mu_2 = \mu_3 = \mu_4 = \mu_5$, H_a: Not all the means are equal. $\alpha = 0.05$; critical value $= 2.82$; $SST = 369,461.407$, $SSTR = 326,631.207$, $SSE = 42,830.200$; $F = 41.94$; reject H_0; at the 5% significance level, the data provide sufficient evidence to conclude that a difference exists in mean weekly earnings among nonsupervisory workers in the five industries. For the P-value approach, note that $P < 0.01$.

15.37 H_0: $\mu_1 = \mu_2 = \mu_3 = \mu_4 = \mu_5$, H_a: Not all the means are equal. $\alpha = 0.01$; critical value $= 3.65$; $SST = 1487.739$, $SSTR = 412.909$, $SSE = 1074.830$; $F = 5.76$; reject H_0; at the 1% significance level, the data provide sufficient evidence to conclude that a difference exists in mean time served by prisoners among the five offense groups. For the P-value approach, note that $P < 0.01$.

15.41
a. 6545.7, 7236.0, 13,781.7 b. 3272.9, 47.0
c. 69.65
d. H_0: $\mu_1 = \mu_2 = \mu_3$, H_a: Not all the means are equal.
e. 0.000
f. At the 1% significance level, the data provide sufficient evidence to conclude that a difference exists in mean annual incomes among women in the three categories of educational attainment.
g. $n_1 = 40, \bar{x}_1 = \$11,099, s_1 = \$6720$
   $n_2 = 62, \bar{x}_2 = \$16,587, s_2 = \$7324$
   $n_3 = 55, \bar{x}_3 = \$27,190, s_3 = \$6388$
h. approximately \$15,000 to \$18,600

Exercises 15.4

15.47 Because the family confidence level is the confidence we have that all the confidence intervals contain the differences between the corresponding population means, whereas the individual confidence level is the confidence we have that any particular confidence interval contains the difference between the corresponding population means.

15.49 Family confidence level $= 0.95$; $q_{0.05} = 4.05$; simultaneous 95% CIs are as follows.

Means difference	Confidence interval
$\mu_1 - \mu_2$	-7.24 to 12.84
$\mu_1 - \mu_3$	-6.24 to 13.84
$\mu_1 - \mu_4$	-5.04 to 15.04
$\mu_2 - \mu_3$	-9.04 to 11.04
$\mu_2 - \mu_4$	-7.84 to 12.24
$\mu_3 - \mu_4$	-8.84 to 11.24

The above table shows that no two populations means can be declared different. This is summarized in the following diagram.

Brand D	Brand C	Brand B	Brand A
(4)	(3)	(2)	(1)
28.6	29.8	30.8	33.6

15.51 Family confidence level $= 0.95$; $q_{0.05} = 4.20$; simultaneous 95% CIs are as follows.

Means difference	Confidence interval
$\mu_1 - \mu_2$	50.8 to 209.4
$\mu_1 - \mu_3$	246.8 to 398.2
$\mu_1 - \mu_4$	13.9 to 183.1
$\mu_1 - \mu_5$	86.3 to 237.7
$\mu_2 - \mu_3$	113.1 to 271.7
$\mu_2 - \mu_4$	-119.5 to 56.3
$\mu_2 - \mu_5$	-47.4 to 111.2
$\mu_3 - \mu_4$	-308.6 to -139.4
$\mu_3 - \mu_5$	-236.2 to -84.8
$\mu_4 - \mu_5$	-21.1 to 148.1

The above table shows that we can declare the following pairs of means different: μ_1 and μ_2, μ_1 and μ_3, μ_1 and μ_4, μ_1 and μ_5, μ_2 and μ_3, μ_3 and μ_4, μ_3 and μ_5; all other pairs of means are not declared different. This is summarized in the following diagram.

Retail trade	Services	Wholesale trade	Finance, Insurance, Real estate	Transp. and Pub. util.
(3)	(5)	(2)	(4)	(1)
178.0	338.5	370.4	402.0	500.5

Interpreting this diagram, we conclude with 95% confidence that the mean weekly earnings of transportation/public-utility workers exceeds those of the other four industries; the mean weekly earnings of retail-trade workers is less than those of the other four industries; no other means can be declared different.

15.53 Family confidence level $= 0.99$; $q_{0.01} = 4.82$; simultaneous 99% CIs are as follows.

Means difference	Confidence interval
$\mu_1 - \mu_2$	-9.01 to 1.21
$\mu_1 - \mu_3$	-9.29 to 1.89
$\mu_1 - \mu_4$	-6.99 to 4.79
$\mu_1 - \mu_5$	-2.73 to 8.73
$\mu_2 - \mu_3$	-5.24 to 5.64
$\mu_2 - \mu_4$	-2.95 to 8.55
$\mu_2 - \mu_5$	1.32 to 12.48
$\mu_3 - \mu_4$	-3.58 to 8.78
$\mu_3 - \mu_5$	0.68 to 12.72
$\mu_4 - \mu_5$	-2.20 to 10.40

The above table shows that we can declare the following pairs of means different: μ_2 and μ_5, μ_3 and μ_5; all other pairs of means are not declared different. This is summarized in the following diagram.

Fraud	C-feiting	Forgery	Firearms	Drug laws
(5)	(1)	(4)	(3)	(2)
11.5	14.5	15.6	18.2	18.4

Interpreting this diagram, we conclude with 99% confidence that for prisoners released from federal institutions for the first time, the mean time served for firearms and drug-law offenses exceeds that for fraud offenses; no other means can be declared different.

15.57

a. 0.99; we can be 99% confident that all the confidence intervals contain the differences between the corresponding population means.

b. 0.99641; we can be 99.641% confident that any particular confidence interval contains the difference between the corresponding population means.

c. 4.18   d. $-\$14{,}356$ to $-\$6{,}851$

e. $\$11{,}881$ to $\$20{,}301$ f. all pairs

g. With 99% confidence, we can state that the mean annual incomes of women in the three educational-attainment groups are all different—the mean for women whose educational attainment is at the elementary level is smallest, the mean for women whose educational attainment is at the secondary level is next smallest, and the mean for women whose educational attainment is at the college level is largest.

Exercises 15.5

15.61 Independent samples, same-shape populations, and all sample sizes are 5 or greater.

15.63 $H_0: \mu_1 = \mu_2 = \mu_3$, H_a: Not all the means are equal. $\alpha = 0.01$; critical value $= 9.210$; $H = 17.305$; reject H_0; at the 1% significance level, the data provide sufficient evidence to conclude that there is a difference in mean consumption of low-fat milk for the years 1970, 1980, and 1990. For the P-value approach, note that $P < 0.005$.

15.65 $H_0: \eta_1 = \eta_2 = \eta_3 = \eta_4$, H_a: Not all the medians are equal. $\alpha = 0.05$; critical value $= 7.815$; $H = 4.212$; do not reject H_0; at the 5% significance level, the data do not provide sufficient evidence to conclude that a difference exists among the median asking rents in the four U.S. regions. For the P-value approach, note that $P > 0.05$.

15.67

a. $H_0: \mu_1 = \mu_2 = \mu_3 = \mu_4 = \mu_5$, H_a: Not all the means are equal. $\alpha = 0.05$; critical value $= 9.488$; $H = 22.126$; reject H_0; at the 5% significance level, the data provide sufficient evidence to conclude that a difference exists in mean weekly earnings among nonsupervisory workers in the five industries. For the P-value approach, note that $P < 0.005$.

b. Because normal populations having equal standard deviations have the same shape. It is better to use the one-way ANOVA test because when the assumptions for that test are met it is more powerful than the Kruskal–Wallis test.

15.71

a. 75.08

b. $H_0: \mu_1 = \mu_2 = \mu_3$, H_a: Not all the means are equal.

c. 0.000

d. At the 1% significance level, the data provide sufficient evidence to conclude that a difference exists in mean annual incomes among women in the three categories of educational attainment.

e. $n_1 = 40$, $M_1 = \$11,560$, $\overline{R}_1 = 41.3$
   $n_2 = 62$, $M_2 = \$17,600$, $\overline{R}_2 = 67.5$
   $n_3 = 55$, $M_3 = \$27,650$, $\overline{R}_3 = 119.5$
   where M_j and $\overline{R}_j$ denote, respectively, the median and the mean rank of the jth sample.

f. Both tests easily reject the null hypothesis (the P-value is 0 to three decimal places for both tests).

Exercises 15.6

15.75

a. 4   b. 2   c. 8

d.

| | Factor B | |
	Level 1	Level 2
Level 1	Population 1	Population 2
Level 2	Population 3	Population 4
Level 3	Population 5	Population 6
Level 4	Population 7	Population 8

(Factor A labels the rows)

15.77

a. *MSA*   b. *MSB*   c. *MSAB*   d. *MSE*

15.79

a. The missing entries by row are: 2, 0.37; 2, 3021.033; 4, 12.518, 1.28; 264.789, 9.807; 35.

b. Each factor has three levels.   c. 4

15.81

a. H_0: The two factors do not interact.
   H_a: The two factors interact.

b. H_0: There is no main effect due to Factor A.
   H_a: There is a main effect due to Factor A.

c. H_0: There is no main effect due to Factor B.
   H_a: There is a main effect due to Factor B.

15.83 Because the two tests for main effects must be interpreted much more carefully when the test for interaction is statistically significant than when it is not.

15.85 For each part we have conjectured on the results of the tests for interaction, Factor A main effects, and Factor B main effects, respectively. A "no" indicates "do not reject H_0"; a "yes" indicates "reject H_0."

a. no, yes, yes   b. yes, yes, no

c. no, no, no   d. yes, no, yes

15.87 For each part, (i), (ii), and (iii) refer, respectively, to the tests for interaction, Factor A main effects, and Factor B main effects.

a. (i) The two factors do not interact. (ii) The mean for Level 1 of Factor A exceeds the mean for Level 2 of Factor A, and the same is true at each level of Factor B. (iii) The mean for Level 1 of Factor B exceeds the mean for Level 2 of Factor B, and the same is true at each level of Factor A.

b. (i) The two factors interact. (ii) The mean for Level 1 of Factor *A* exceeds the mean for Level 2 of Factor *A*; this is also true at Level 1 of Factor *B*, although at Level 2 of Factor *B* the means are equal. (iii) The mean for Level 1 of Factor *B* equals the mean for Level 2 of Factor *B*; but at Level 1 of Factor *A* the mean for Level 1 of Factor *B* exceeds the mean for Level 2 of Factor *B*, whereas at Level 2 of Factor *A* the mean for Level 1 of Factor *B* is smaller than the mean for Level 2 of Factor *B*.

c. (i) The two factors do not interact. (ii) The mean for Level 1 of Factor *A* equals the mean for Level 2 of Factor *A*, and the same is true at each level of Factor *B*. (iii) The mean for Level 1 of Factor *B* equals the mean for Level 2 of Factor *B*, and the same is true at each level of Factor *A*.

d. (i) The two factors interact. (ii) The mean for Level 1 of Factor *A* equals the mean for Level 2 of Factor *A*; but at Level 1 of Factor *B* the mean for Level 1 of Factor *A* exceeds the mean for Level 2 of Factor *A*, whereas at Level 2 of Factor *B* the mean for Level 1 of Factor *A* is smaller than the mean for Level 2 of Factor *A*. (iii) The

mean for Level 1 of Factor *B* exceeds the mean for Level 2 of Factor *B*; this is also true at Level 1 of Factor *A*, but at Level 2 of Factor *A* the means are equal.

Exercises 15.7

15.91 $\alpha = 0.05$. A table of sample means is displayed in Table A.2; a plot of cell means is as follows.

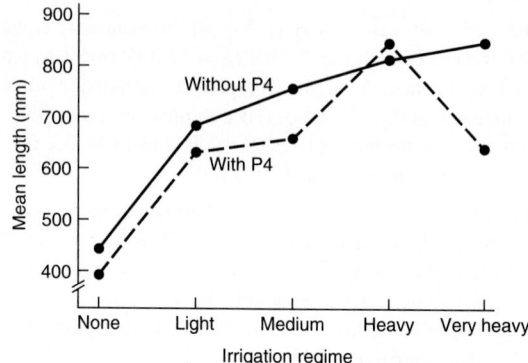

TABLE A.2   Table of sample means for Exercise 15.91

		Irrigation regime					
		None	Light	Medium	Heavy	Very heavy	All
Polymer	Without P4	442.50	684.25	757.00	813.50	846.75	708.80
	With P4	392.50	631.25	659.00	845.50	639.50	633.55
	All	417.50	657.75	708.00	829.50	743.13	671.17

Test for interaction: H_0: Polymer and irrigation regime do not interact in their effect on mean total length of cuttings. H_a: Polymer and irrigation regime interact in their effect on mean total length of cuttings. $P = 0.276$; do not reject H_0.

Test for Factor A main effects: H_0: Polymer does not affect mean total length of cuttings. H_a: Polymer affects mean total length of cuttings. $P = 0.033$; reject H_0.

Test for Factor B main effects: H_0: Irrigation regime does not affect mean total length of cuttings. H_a: Irrigation regime affects mean total length of cuttings. $P = 0.000$; reject H_0.

15.93 $\alpha = 0.05$. A table of sample means is given in Printout 15.16 on page 928; a plot of cell means is as follows.

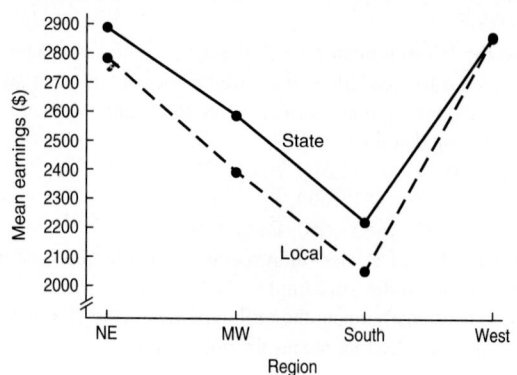

Test for interaction: H_0: Government and region do not interact in their effect on mean earnings. H_a: Government and region interact in their effect on mean earnings. $P = 0.969$; do not reject H_0.

Test for Factor A main effects: H_0: Government does not affect mean earnings. H_a: Government affects mean earnings. $P = 0.430$; do not reject H_0.

Test for Factor B main effects: H_0: Region does not affect mean earnings. H_a: Region affects mean earnings. $P = 0.004$; reject H_0.

15.95 $\alpha = 0.05$. A table of sample means is given in Printout 15.18 on page 930; a plot of cell means is as follows.

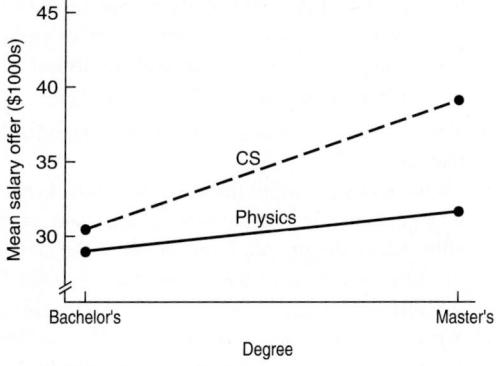

Test for interaction: H_0: Major and degree do not interact in their effect on mean salary offer. H_a: Major and degree interact in their effect on mean salary offer. $P = 0.028$; reject H_0.

Review Test for Chapter 15

1. a. 24   b. 5   c. 4.53   d. 9.47   e. 4.53

2. a. *MSTR* (or *SSTR*)   b. *MSE* (or *SSE*)

3. a. The total sum of squares, *SST*, represents the total variation among all the sample data; the treatment sum of squares, *SSTR*, represents the variation among the sample means; and the error sum of squares, *SSE*, represents the variation within the samples.

   b. $SST = SSTR + SSE$; the one-way ANOVA identity shows that the total variation among all the sample data can be partitioned into a component representing variation among the sample means and a component representing variation within the samples.

4. a. $\bar{x}_1 = 635.80$, $s_1^2 = 8{,}630.20$
   $\bar{x}_2 = 506.83$, $s_2^2 = 15{,}890.97$
   $\bar{x}_3 = 393.17$, $s_3^2 = 19{,}312.97$

   b. $MSTR = 80{,}300.708$, $MSE = 15{,}038.605$ *(Note: These were obtained using the nonrounded values of the means and variances.)*

   c. The variation among the sample means.

   d. The variation within the samples.

5. a. The three assumptions for one-way ANOVA, given in Key Fact 15.2 on page 868: independent samples, normal populations, and equal standard deviations. Assumption 1 on independent samples is essential to the one-way ANOVA procedure. Assumption 2 on normality is not too critical as long as the populations are not too far from being normally distributed. Assumption 3 on equal standard deviations is also not that important provided the sample sizes are roughly equal.

   c. The ratio of the largest to the smallest sample standard deviation is roughly 1.5, which is less than 2.

   e. Yes. We are told that the samples are independent; so Assumption 1 is met. Normal probability plots of the three samples and of the residuals reveal no outliers and are roughly linear. Furthermore, the rule of 2 is satisfied, and a plot of the residuals against the sample means falls roughly in a horizontal band centered and symmetric about the horizontal axis. So we can consider Assumptions 2 and 3 met.

6. H_0: $\mu_1 = \mu_2 = \mu_3$, H_a: Not all the means are equal. $\alpha = 0.05$; critical value $= 3.74$; $SST = 371{,}141.882$, $SSTR = 160{,}601.416$, $SSE = 210{,}540.467$; $F = 5.34$; reject H_0; at the 5% significance level, the data provide sufficient evidence to conclude that a difference in mean losses exists among the three types of robberies. For the P-value approach, note that $0.01 < P < 0.05$.

8. a. 160,601, 210,540, 371,142

   b. 80,301, 15,039

   c. 5.34   d. 0.019

   e. Reject H_0; at the 5% significance level, the data provide sufficient evidence to conclude that a difference in mean losses exists among the three types of robberies.

   f. $n_1 = 5$, $\bar{x}_1 = 635.8$, $s_1 = 92.9$
   $n_2 = 6$, $\bar{x}_2 = 506.8$, $s_2 = 126.1$
   $n_3 = 6$, $\bar{x}_3 = 393.2$, $s_3 = 139.0$

   g. approximately \$285 to \$495

9. a. Using multiple comparison we can obtain simultaneous confidence intervals for the differences between all pairs of means; this, in turn, permits us to ascertain the relationship among all the means.

b. The family confidence level is the confidence we have that all the confidence intervals contain the differences between the corresponding population means, whereas the individual confidence level is the confidence we have that any particular confidence interval contains the difference between the corresponding population means.

c. Family confidence level, since multiple-comparison techniques deal simultaneously with all the confidence intervals.

d. studentized range distribution (or q-distribution)

10. Family confidence level $= 0.95$; $q_{0.05} = 3.70$; simultaneous 95% CIs are as follows.

Means difference	Confidence interval
$\mu_1 - \mu_2$	-65.3 to 323.2
$\mu_1 - \mu_3$	48.4 to 436.9
$\mu_2 - \mu_3$	-71.6 to 298.9

The above table shows that we can declare as different only μ_1 and μ_3. This is summarized in the following diagram.

Convenience store (3)	Gas station (2)	Highway (1)
393.17	506.83	635.80

Interpreting this diagram, we conclude with 95% confidence that the mean loss due to convenience-store robberies is less than that due to highway robberies; no other means can be declared different.

12. a. 0.95   b. 0.9797   c. 3.70
d. $-\$72$ to $\$299$ e. $-\$323$ to $\$65$
f. only μ_1 and μ_3
g. With 95% confidence we can state that the mean loss due to convenience-store robberies is less than that due to highway robberies; no other means can be declared different.

13. a. H_0: $\mu_1 = \mu_2 = \mu_3$, H_a: Not all the means are equal. $\alpha = 0.05$; critical value $= 5.991$; $H = 6.819$; reject H_0; at the 5% significance level, the data provide sufficient evidence to conclude that a difference in mean losses exists among the three types of robberies. For the P-value approach, note that $0.025 < P < 0.05$.

b. Because normal populations having equal standard deviations have the same shape. It is better to use the one-way ANOVA test because when the assumptions for that test are met it is more powerful than the Kruskal–Wallis test.

15. a. 6.82 (6.83 when adjusted for ties)
b. H_0: $\mu_1 = \mu_2 = \mu_3$, H_a: Not all the means are equal.
c. 0.033
d. At the 5% significance level, the data provide sufficient evidence to conclude that a difference in mean losses exists among the three types of robberies.
e. $n_1 = 5$, $M_1 = \$652.0$, $\overline{R}_1 = 13.4$
$n_2 = 6$, $M_2 = \$522.5$, $\overline{R}_2 = 8.9$
$n_3 = 6$, $M_3 = \$407.0$, $\overline{R}_3 = 5.4$
where M_j and $\overline{R}_j$ denote, respectively, the median and the mean rank of the jth sample.
f. Both tests reject the null hypothesis, but the one-way ANOVA test provides stronger evidence against the null hypothesis than the Kruskal–Wallis test (P-values are 0.019 and 0.033, respectively).

16. a. The *levels of Factor B* are the various categories of that factor.
b. *No interaction* signifies that the two factors do not interact in their effect on the mean of the response variable; that is, for any two levels of one of the factors, the difference between the means of the response variable is the same for all levels of the other factor.
c. We say there is *a main effect due to Factor A* if the means of the levels for Factor A are not identical.

17.

Source	df	SS	MS = SS/df	F-statistic
Vehicle	2	1045.698	522.849	171.87
Year	3	5.757	1.919	0.63
Interaction	6	2.914	0.486	0.16
Error	48	146.023	3.042	
Total	59	1200.392		

18. $\alpha = 0.05$. A table of sample means is given in Printout 15.22 on page 937; a plot of cell means is as follows.

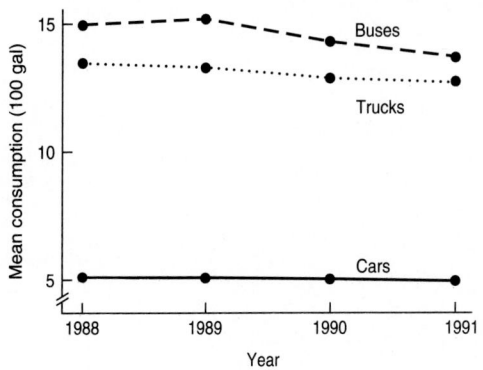

Test for interaction: H_0: Vehicle and year do not interact in their effect on mean fuel consumption. H_a: Vehicle and year interact in their effect on mean fuel consumption. $P = 0.986$; do not reject H_0.

Test for Factor A main effects: H_0: Vehicle does not affect mean fuel consumption. H_a: Vehicle affects mean fuel consumption. $P = 0.000$; reject H_0.

Test for Factor B main effects: H_0: Year does not affect mean fuel consumption. H_a: Year affects mean fuel consumption. $P = 0.599$; do not reject H_0.

Index

....